GEOSYSTEMS

AN INTRODUCTION TO PHYSICAL GEOGRAPHY

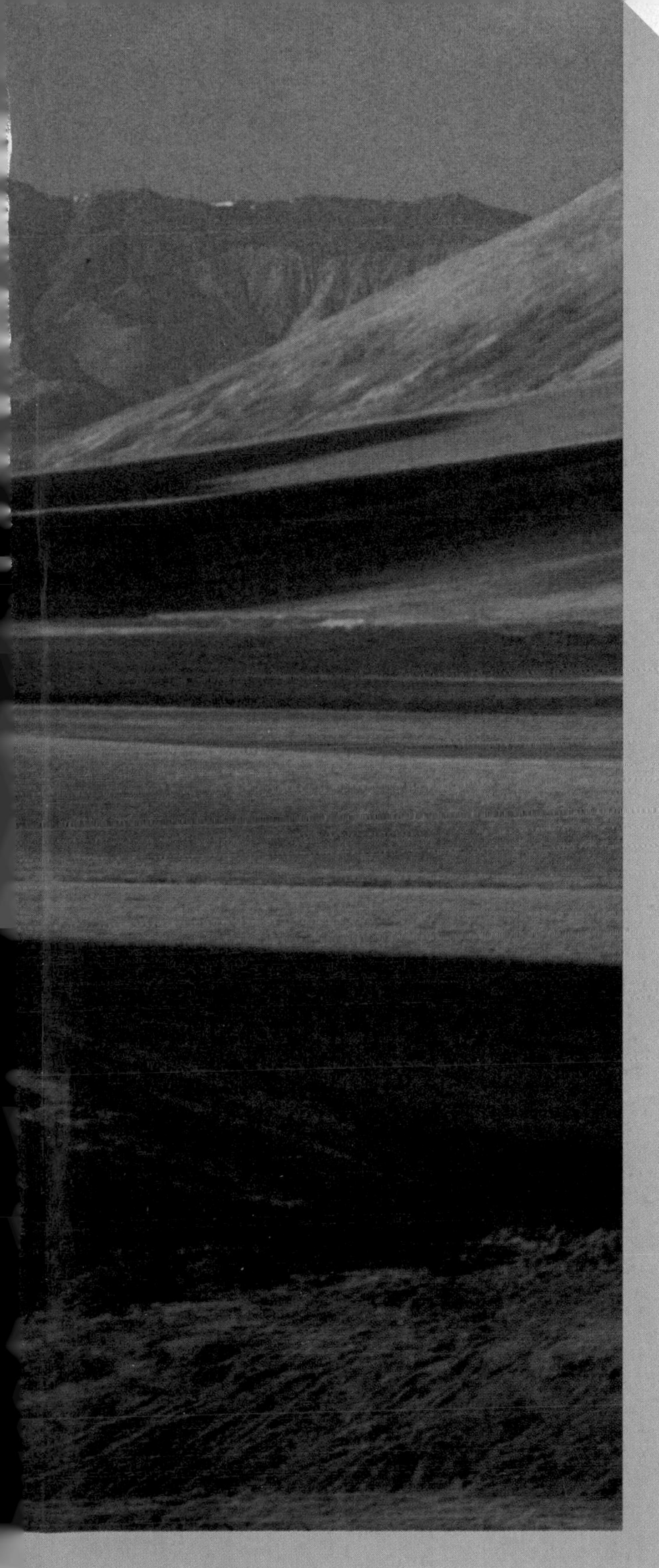

GEOSYSTEMS

AN INTRODUCTION TO PHYSICAL GEOGRAPHY

SECOND EDITION

Robert W. Christopherson

MACMILLAN COLLEGE PUBLISHING COMPANY
Englewood Cliffs, New Jersey 07632

To all the students and teachers of Earth,
our home planet,
and its sustainable future.

Front Cover Photo: Mitre Peak (1692 m, 5550 ft), Milford Sound (45.5° S 168° E), Fjordland National Park, South Island, New Zealand, by Chris Arnesen.
Back Cover Photo: Full Earth photo by Apollo 17 astronauts, December 1972, from NASA.
Frontispiece: Mount Chamberlain through Sunset Pass (70° N 145° W), Arctic National Wildlife Refuge, North Slope, Alaska, by Scott T. Smith.

Editor: Paul F. Corey
Production Supervisor: Elisabeth Belfer
Production Assistant: Bobbé Christopherson
Production Manager: Nicholas Sklitsis
Art Director: Patricia Smythe
Page Layout: Robert Freese
Cover Designer: Leslie Baker
Photo Editor: Chris Migdol
Photo Researcher: Yvonne Gerin
Illustrations: Maryland CartoGraphics Inc., Precision Graphics, and Tasa Graphic Arts Inc.

Macmillan College Publishing Company
Englewood Cliffs, New Jersey 07632

Macmillan College Publishing Company is part of the Paramount Communications Group of Companies.

LIBRARY OF CONGRESS CATALOGING-IN-PUBLICATION DATA
Christopherson, Robert W.
Geosystems: an introduction to physical geography / Robert W. Christopherson. — 2nd ed.
p. cm.
Includes bibliographical references and index.
ISBN 0-02-322451-7
1. Physical geography. I. Title.
GB54.5.C48 1994 93-34288
910'.02—dc20 CIP

PRINTING 2 3 4 5 6 7 8 YEAR 4 5 6 7 8 9 0 1 2 3

PREFACE

Welcome to the second edition of *Geosystems*. During the span of the first edition significant environmental events and changes transpired, including the Earth Summit that was held in Rio de Janeiro in 1992. This largest-ever gathering of nations acted on an agenda relevant to the environment. The 1993 floods in America's Midwest, another record tornado year, powerful earthquakes in Japan and Guam, drought in the southeastern states, dramatic and variable weather patterns across Canada, rising levels of the Great Lakes, further thinning of the stratospheric ozone layer, and continuing losses of biodiversity, all serve as reminders of Earth's systems. A dramatic survey of current physical geography concerns introduces Chapter 1. This edition is thoroughly updated and revised to reflect the dynamic science of physical geography.

The goal of physical geography is to explain the spatial dimension of Earth's natural systems—its energy, air, water, weather, climates, landforms, soils, plants, and animals. Earth is a place of great physical and cultural diversity, yet people generally know little of it. We must work to reverse this trend. An informed citizenry requires relevant education about the life-sustaining environment that infuses life. A goal of *Geosystems* is to serve this need.

THE SECOND EDITION

From across the United States and Canada we received many positive reviews and letters about the first edition. Readers and adopters complimented our systems organization, subject coverage, readability, the functional beauty of the photographs and art, the text-related glossary, and the relevancy of content, among other items. The second edition improves on these established qualities.

To make this text both informative and enjoyable, a clear writing style is further enhanced by a careful rewrite and particular attention to the sequence of outline headings. Several sections are rearranged to improve the flow of topics; as examples: seasons now appear in Chapter 2; periglacial processes are in Chapter 17.

Coverage of essential physical geography and inclusion of nonquantitative analyses of Earth's systems results in a text appropriate for both major and nonmajor science students, regardless of your science background. This sensitivity to your needs makes *Geosystems* an accessible science text. One adopter noted, "It is scientific without being pedantic, stuffy, or boring . . . It carries students through the complexities of the physical environment."

Systems Organization. Our second edition maintains the careful organization of the first through a logical order and flow of topics. *Geosystems* is structured in four parts, each containing related chapters according to the flow of individual systems, or consistent with time and the flow of events. Chapter 1 presents the essentials of physical geography as a foundation, including a discussion of geography, systems analysis, latitude, longitude, time, and the science of mapmaking (cartography). With these essentials learned, each of the parts then can be covered, either in their presented order or in any convenient sequence.

Part One exemplifies the systems organization of the text, beginning with the origin of the Solar System and the Sun. Solar energy passes across space to Earth's atmosphere varying in intensity by latitude and seasonal variations of Sun angle and daylength (Chapter 2). Energy is traced through the atmosphere to Earth's surface (Chapter 3). Earth's atmospheric and surface energy balances are generated by this cascade of solar energy (Chapter 4), producing patterns of world temperatures (Chapter 5), and general and local atmospheric and oceanic circulations (Chapter 6). Thus, the sciences of astronomy, geodesy, physics, and other atmospheric sciences are dealt with in Chapters 2 through 6.

Part Two presents hydrology, meteorology and weather, oceanography, and climate, in a flowing sequence through Chapters 7 to 10. Physical geography is linked to Earth sciences, an influence seen in Chapters 11 through 17 of *Part Three* where we discuss the physical planet and related processes that affect the crust. Earth's surface is a place of an enormous ongoing struggle between the processes that build, warp, and fault the landscape and those that wear it down through the action of rivers, wind, waves, and ice.

Finally, *Part Four* brings the content of the first three parts together in biogeography, including soils, plants, animals, and Earth's major terrestrial biomes (Chapters 18 through 20). The text culminates with Chapter 21, "The Human Denominator," a unique capstone chapter that summarizes human-environment interactions pertinent to physical geography. This chapter is sure to stimulate further thought and discussion, dealing as it does with the most profound issue of our time, Earth stewardship. The historic 1992 Earth Summit—the United Nations Conference on Environment and Development held in Rio de Janeiro—is detailed in a new focus study in this chapter.

Important Learning Aids and Features. The second edition contains many features to assist you as a student of physical geography:

- Each chapter includes a heading outline, review questions, key terms list, updated suggested readings, and rewritten and expanded summaries.
- Key chapters (climate, soils, biomes) present large, integrative tables to help you synthesize content.
- The Glossary provides basic definitions for 700 key terms and concepts that are printed in **boldface** where they are defined within the text; entries include chapter location. Other important terms and ideas in the text are in *italics* for emphasis.
- A *Student Study Guide* is available to provide additional learning tools, objectives, glossary exercises, examples, and self-tests.
- The Macmillan Geodisc. This exciting new videodisc features 1500 still images, over 50 minutes of motion video, and multiple full-color animations. A separate instructor's manual facilitates classroom use.
- The text and all figures use metric/English measurement equivalencies appropriate to this transition period in the United States and for science courses in general. A complete set of measurement conversions is presented in an easy-to-use arrangement inside the back cover.
- Appendix A contains weather, water balance, and other data for 40 cities worldwide, arranged by climate. Appendix B provides address listings for important geographic and environmental organizations, agencies, and general reference works.
- Twenty-one focus study essays, several completely revised and updated for this edition, provide additional explanation of key topics as diverse as the scientific method, the strato-

spheric ozone predicament, the Mount Saint Helens eruption, the status of the Colorado River, fire ecology, and biosphere reserves.

- Our widely praised cartography program is updated to reflect the rapid pace of change in political boundaries. All the maps in this edition reflect the breakup of the former Soviet state into separate republics.
- Our planet is consistently treated in *Geosystems* as a proper noun—in this text Earth is capitalized. This is consistent with the treatment of other planets and is an effective cue to think of Earth as our significant home planet.

MAJOR CHANGES IN THE SECOND EDITION

The second edition features major additions and revisions to photographs, maps, art, tables, and coverage of Canada. The many new figures added to this edition are fully integrated within text discussions.

- *Geosystems* features 224 color photographs (110 new to this edition); 79 remote sensing images (47 new); 172 maps (including 35 new or rerendered maps); 288 art drawings (125 new or rerendered); and 52 tables; all selected and prepared to enhance your learning experience.
- Expanded coverage of Canadian physical geography includes text, figures, and maps of periglacial landscapes and Canadian soils. Canadian data on a variety of subjects are portrayed specifically on 20 maps in combination with the United States.
- The global implications of the June 1991 Mount Pinatubo eruption are woven throughout (Chapters 1, 3, 4, 5, 10, 12, and 21), not just in the section on volcanoes. In this way *Geosystems* analyzes the worldwide impact of this event, synthesizing varied physical factors into a complete picture.
- Updated treatment of global climate change and related potential effects are woven through six chapters of the text. Sections are developed from current research including the Intergovernmental Panel on Climate Change (IPCC) 1990 reports and the 1992 IPCC supplementary update.
- A sample of new remote sensing in this edition includes: computer images produced from the Earth Radiation Budget experiments (ERBE), AVHRR images of Hurricane Andrew from *NOAA 10* and *11, Nimbus-7* measurements of the sulfuric acid produced by the Mount Pinatubo eruption, images taken by the *Galileo* spacecraft during its fly-by of Earth, X-ray and visible light images of the Sun from the *Yohkoh* satellite, the latest *Nimbus-7* images of stratospheric ozone depletion, and several comparative sets of *Landsat* images to show environmental changes and human impacts over time (acid rain damage to forests, the Kuwait oil-well fires, and equatorial rain forest losses).
- Updated discussions include: tornado data through 1992; an accurate discussion of standard time and UTC and the new *NIST-7* primary clock; new imagery of the Pacific Warm Pool; the December 1992 nor'easter storm that struck the Eastern seabord portrayed in weather map, satellite image, and discussion; river flooding and floodplain management; worldwide sea level rise; and the 1992 California Landers earthquakes.

ACKNOWLEDGMENTS

I begin by thanking my supportive family, Mom and Dad, my sister and brothers Lynne, Randy, and Marty, and our children Keri, Matt, Reneé, and Steve, and the next generation—Chavon, Bryce, and our newest Payton. Despite the missed time together that projects such as this entail, they were never far from my thoughts. This text is a teaching text, so past mentors are remembered for their commitment to learning and to students.

Thanks to the authors and scientists who published research, articles, and books that I used in refining this revision. My gratitude goes to all the students over these past 24 years at American River College for defining the importance of Earth's future, for their questions and enthusiasm—*to them this text is dedicated.*

I owe many thanks to the following individuals who consulted with Macmillan Publishing for the first and second editions, reviewed portions of the manuscript in several development stages, and made contributions and suggestions; several completed comprehensive manuscript reviews. These most helpful geographers and their affiliations are:

Ted J. Alsop, Utah State University

Ward Barrett, University of Minnesota

David R. Butler, University of North Carolina

Ian A. Campbell, University of Alberta–Edmonton

Armando M. da Silva, Towson State University

Dirk H. de Boer, University of Saskatchewan

Mario P. Delisio, Boise State University

Joseph R. Desloges, University of Toronto

Lee R. Dexter, North Arizona University

Don W. Duckson, Jr., Frostburg State University

Christopher H. Exline, University of Nevada–Reno

Michael M. Folsom, Eastern Washington University

Glen Fredlund, University of Wisconsin–Milwaukee

David E. Greenland, University of Oregon

John W. Hall, Louisiana State University–Shreveport

David A. Howarth, University of Louisville

Patricia G. Humbertson, Youngstown State University

David W. Icenogle, Auburn University

Philip L. Jackson, Oregon State University

J. Peter Johnson, Jr., Carleton University

Guy King, California State University–Chico

Peter W. Knightes, Central Texas College

Richard Kurzhals, Grand Rapids Junior College

Robert D. Larson, Southwest Texas State University

Joyce Lundberg, Carleton University

W. Andrew Marcus, Montana State University

Elliot G. McIntire, California State University–Northridge

Norman Meek, California State University–San Bernardino

Lawrence C. Nkemdirim, University of Calgary

John E. Oliver, Indiana State University

Bradley M. Opdyke, University of Michigan

Robin J. Rapai, University of North Dakota

Philip D. Renner, American River College

William C. Rense, Shippensburg University

Wolf Roder, University of Cincinnati

Dorothy Sack, University of Wisconsin–Madison

Glenn R. Sebastian, University of South Alabama

Daniel A. Selwa, U.S.C. Coastal Carolina College

Thomas W. Small, Frostburg State University

Daniel J. Smith, University of Saskatchewan

Stephen J. Stadler, Oklahoma State University

Susanna T. Y. Tong, University of Cincinnati

David Weide, University of Nevada–Las Vegas

Brenton M. Yarnal, Pennsylvania State University

My thanks to the staff of Macmillan Publishing Company. Their dedication to the craft of modern textbook publishing is in evidence throughout this text. A special thanks to executive editor Paul Corey for his abiding friendship and expert stewardship of our project. My deep gratitude to Elisabeth Belfer, my production supervisor, who imparted her great experience and wisdom to every facet of this book. Also, thanks go to Gary Carlson and Madalyn Stone; Chris Migdol for assisting with our extensive photo research for this edition, and his assistant Yvonne Gerin; the skill of art director Pat Smythe; and to developmental editor Fred Schroyer for his creative ability and hours of dialogue.

Finally, to the continuing partnership of my special collaborator, Bobbé Christopherson, I express extraordinary gratitude for her tireless work as my production assistant. She worked many hours researching and cataloging photographs, updating tables, obtaining permissions, proofing art, and reading through drafts, galleys, and page proofs for this second edition. Her unique sensitivity to

Earth and love for the smallest of living things are an ever-present wonder and inspiration.

Physical geography teaches us about the intricate supporting web that is Earth's environment. Dramatic changes that demand our attention and understanding are occurring in many human-Earth relationships as we approach the new millennium. All things considered, this is an important time to be enrolled in a relevant geography course! The best to you in your studies.

Robert W. Christopherson

Folsom, California

BRIEF CONTENTS

CONTENTS

PART ONE
The Energy-Atmosphere System

4 Atmosphere and Surface Energy Balances 95

5 World Temperatures 119

6 Atmospheric and Oceanic Circulations 145

PART TWO
The Water, Weather, and Climate System

PART THREE
The Earth-Atmosphere Interface

16 Coastal Processes and Landforms 477

17 Glacial and Periglacial Processes and Landforms 507

PART FOUR
Soils, Ecosystems, and Biomes

Geosystems

An Introduction to Physical Geography

An astronomer uses a simple device to sight the Sun's angle above the horizon and determine latitude. [*From Jacques de Vault,* Cosmographia, *1583.*]

1

ESSENTIALS OF GEOGRAPHY

Welcome to physical geography! Consider the following events and conditions:

- A hurricane devastates southern Florida; five days later its remnants drench Ontario and Quebec. Within a few weeks similar storms strike Kauai and Guam. Rare tornadoes hit New Brunswick in November 1992. Weather patterns, both expected and unusual, constantly occur.
- Lake Eyre in central Australia fills to unusual depths; Lake Chad in western Africa continues to recede; and the Great Salt Lake rises to record levels—all in the last seven years. Tributaries that supply Mono Lake and the Aral Sea are diverted for irrigation—and levels decline. Summer 1992 rainfall in the Great Lakes basin is 145% of normal.
- Lush rain forests straddle the equator, stark deserts bake in the subtropics, and cool moist climates dominate northwestern coastlines. Most natural ecosystems bear the imprint of civilization: only a dozen white rhinos remain in the northern portion of their African habitat, record numbers of species face extinction in the tropics, agricultural soils lose productivity, and forests decline due to acid rain.
- The Appalachians and the Atlas mountains, in North America and Africa respectively, once joined, drift apart as vast plates of continental crust continue to migrate. Rocks found in northwestern Canada are 3.96 billion years old, predating all others on our 4.6 billion-year-old planet. Earthquakes strike Ontario, Canada, southern California, and Turkey. Volcanoes erupt in Chile, Japan, Alaska, and the Philippines.
- Mount Pinatubo explodes in the Philippines: colorful afterglows dominate twilight worldwide, global temperatures drop slightly, and resulting sulfuric acid mists further contribute to stratospheric ozone losses.
- Dynamic patterns of energy, air, water, weather, climates, landforms, soils, plants, and animals vary from place to place and through the seasons, and evolve over time.

Where are these places? Why are they there? Why do physical conditions differ from equator to midlatitudes, between deserts and polar regions? How does solar energy contribute to the distribution of trees, soils, climates, and lifestyles? To the patterns of wind, weather, and ocean currents? How do aspects of nature affect human populations? What impact do humans have on natural systems? Let us explore these questions, and others, through geography's unique perspective—welcome to physical geography!

A knowledge of physical geography is important to answer the where, the why, and the how questions about Earth's physical systems and their interaction with living things. Thus, our study of geosystems—Earth systems—begins with a look at the science of physical geography and the geographic tools we will need.

Physical geographers use systems analysis to study the environment. Therefore, we introduce Earth's interrelated natural and human systems. We then consider location, a key theme of geographic inquiry—the latitude, longitude, and time coordinates that delimit Earth's surface. The pursuit of longitude and a world time system provide us with interesting insights into geography. Finally, we examine maps as critical tools that geographers create to portray physical and cultural information.

THE SCIENCE OF GEOGRAPHY

Geography (from *geo*, "Earth," and *graphein*, "to write") is the science that studies the interdependence among geographic areas, natural systems, society, and cultural activities *over space*. The term *spatial* refers to the nature and character of physical space; to measurements, relations, locations, and the distribution of things. Human beings are spatial actors, both affecting and being affected by Earth. We influence vast areas because of our technology, mobility, and numbers. For example, think of your own route to the classroom or library today, your knowledge of street patterns, traffic trouble spots, parking spaces, or bike rack locations used to minimize walking distances.

The Association of American Geographers

(AAG) and the National Council for Geographic Education (NCGE) have categorized the discipline into five key themes for modern geographic education: **location**, **place**, **movement**, **region,** and **human-Earth relationships** (Figure 1-1). *Geosystems* draws on traditions within each theme.

FIGURE 1-1
Fundamental themes in geography defined. *Location*: geographic freeway sign; *Place*: Uluru (Ayers) Rock, Northern Territory, Australia; *Movement:* global and deep-space communications through modern satellites; *Regions:* intensive use of land for rice paddies in China; and *Human-Earth relationships:* thatched houses on stilts along the Amazon River, Brazil. [Regions photo by Frank Balthis; Place photo by Michael Dunning/TSW; Human-Earth relationships photo by Gael Summer-Hebdon; Movement and Location photos by author.]

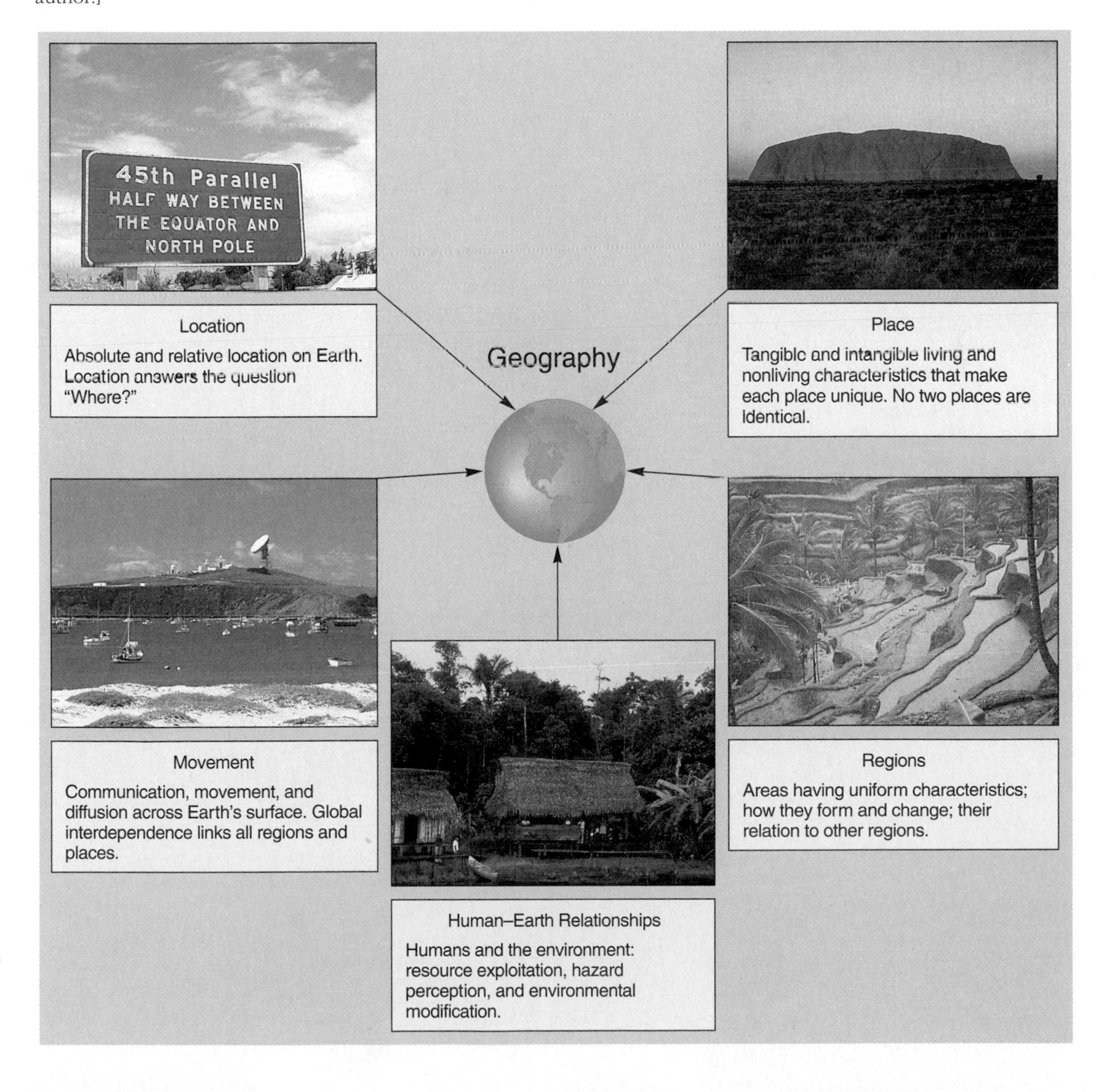

As a discipline that synthesizes knowledge from many fields, geography integrates spatial elements to form a coherent picture of Earth. Geography is governed by a method—a spatial approach —rather than by a specific body of knowledge. Various words denote this geographic context of **spatial analysis**: space, territory, zone, pattern, distribution, place, location, region, sphere, province, and distance. Geographers analyze the differences and similarities between places and locations. Because few physical phenomena are purely spatial, **process** is central to geographic synthesis—that is, analyzing a set of actions or mechanisms that operate in some special order. As an example, think of the numerous processes involved in the study of Earth's vast water system.

Spatial analysis, then, is a process-oriented integrative approach central to geographic inquiry. **Physical geography**, therefore, centers on the spatial analysis of all the physical elements and processes that make up the environment: energy, air, water, weather, climate, landforms, soils, animals, plants, and Earth itself. We add to this the oldest theme in geographic tradition, that of human activity, impact, and the shared human-Earth relationship.

The Geographic Continuum

Geography is eclectic, integrating a wide range of subject matter from diverse fields; virtually any subject can be examined geographically. Figure 1-2 shows a continuous distribution—a continuum—along which the content of geography is arranged. Disciplines in the physical and life sciences are listed on the continuum's left; those in the cultural/human sciences are listed on the right. Specialties within geography that draw from these disciplines are listed within the appropriate spheres at either end of the continuum.

This continuum reflects a basic duality within geography—physical geography versus human/cultural geography. This duality is paralleled in society by the tendency of those in developed countries to distance themselves from their supportive life-sustaining environment and to think of themselves as exempt from the physical functions of Earth. In contrast, many people in Third World countries live closer to nature and are acutely aware of its importance in their daily lives. Regardless of our philosophies toward Earth, we all depend on Earth's systems to provide oxygen, water, nutrients, and other resources.

Our modern world requires that we shift our study of geographic processes, and perhaps our philosophies, toward the center of the continuum to attain a more *holistic*, or complete, perspective. Modern environmental problems call for geographers to examine relations among Earth, society, and cultures throughout space and time. Critical environmental concerns that are integral to physical geography include:

FIGURE 1-2
Distribution of geographic content along a continuum. *Geosystems'* focus is physical geography, but with an integration of the human component into the study of physical geography. Synthesis of Earth topics and human topics is suggested by movement toward the middle of the continuum.

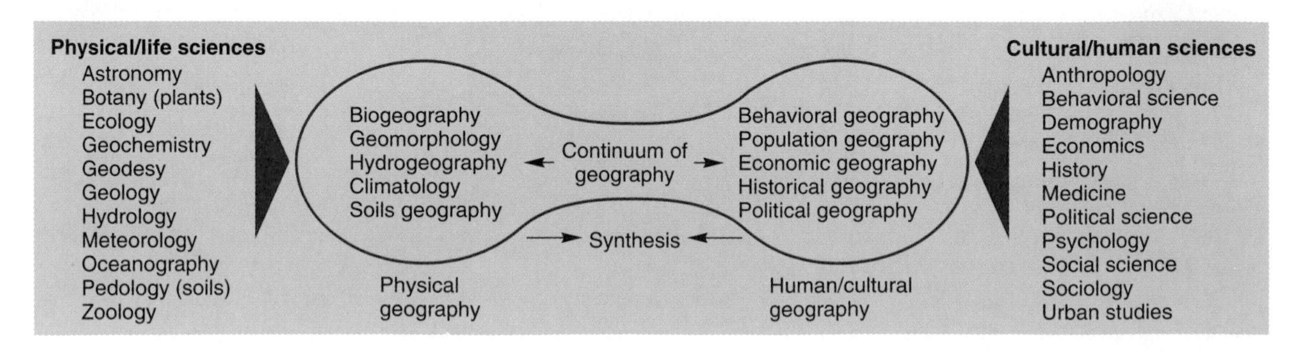

1. Global ozone depletion in the upper atmosphere that allows increasing amounts of ultraviolet radiation to reach Earth's surface.
2. Possible global warming through human-caused increases of carbon dioxide in the atmosphere.
3. Worsening air pollution, particularly in metropolitan areas.
4. Identification of natural hazards that threaten society, such as hurricanes, earthquakes, droughts, and floods.
5. Deliberate destruction of Earth's forests.
6. Increasing losses of plant and animal diversity as habitats disappear.
7. Accounting for natural resources as environmental assets on national economic balance sheets.

Furthermore, geography is in a unique position to synthesize the environmental, spatial, and human aspects of catastrophic events. Examples are the 1986 Chernobyl nuclear power station accident in Ukraine; the 1989 *Exxon Valdez* oil tanker catastrophe in Prince William Sound, Alaska; the 1991 eruption of Mount Pinatubo in the Philippines; the environmental impact of the 1991 Persian Gulf War; or weather disasters such as hurricanes Andrew and Iniki and Typhoon Omar in 1992.

The *United Nations Conference on Environment and Development*, held at Rio de Janeiro in 1992, brought tens of thousands of delegates from over 160 countries and 1000 nongovernmental organizations to the largest world summit ever conducted. On the agenda were global climate change, biodiversity, sustainable forestry, and an "Earth Charter"—themes at the heart of geographic studies (see Focus Study 21-1).

The spatial patterns of change we face today are unique, for we are taxing Earth's systems in new ways. Some past civilizations adapted to crises, whereas others failed. Perhaps this ability to adapt is the key. If so, understanding of our relationship to Earth's physical geography is of great importance to human survival, for the innumerable physical processes in the environment operate as interacting life-support systems.

EARTH SYSTEMS CONCEPTS

You no doubt are familiar with the word *system*: "check the cooling system"; "how does the grading system work?"; "there is a weather system approaching." Systems surround us. Beginning with studies of energy and temperature in systems (thermodynamics) in the 19th century, and developed by engineering disciplines during World War II, *systems analysis* has moved to the forefront as a method for understanding operational behavior in many disciplines. Geographers use systems methodology as an analytical tool at varying levels of complexity.

Systems Theory

Simply stated, a **system** is any ordered, interrelated set of things, and their attributes, linked by flows of energy and matter, as distinct from their surrounding environment. The elements within a system may be arranged in a series or interwoven with one another. A system can comprise any number of subsystems. Within Earth's systems both matter and energy are stored and retrieved, and energy is transformed from one type to another.

A natural system generally is not self-contained: inputs of energy and matter flow into the system, and outputs flow from the system. Such a system is called an **open system**. Earth is an open system *in terms of energy*, as are most natural systems. A system that is shut off from the surrounding environment so that it is self-contained is a **closed system**. Earth is essentially a closed system *in terms of physical matter and resources*. Figure 1-3 illustrates open and closed systems.

Figure 1-4 illustrates a simple open-flow system, using plant photosynthesis as an example. Photosynthesis is the growth process by which plants transform inputs of solar energy, water, nutrients, and carbon dioxide into stored chemical energy (carbohydrates), releasing an output of oxygen as a by-product. Plants derive energy for their operations from a reverse of photosynthesis, a process called respiration. In respiration, the plant consumes inputs of chemical energy (carbohydrates) and releases outputs of carbon dioxide, water, and heat into the environment. Thus, a plant acts as an

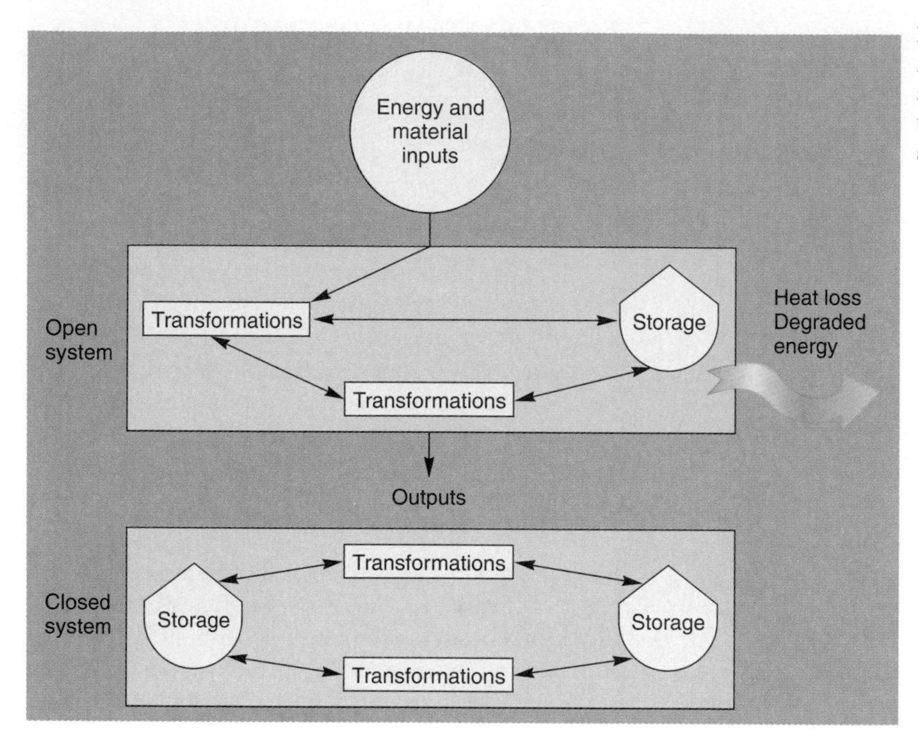

FIGURE 1-3
An open system and a closed system. In either type, inputs are transformed and stored as the system operates.

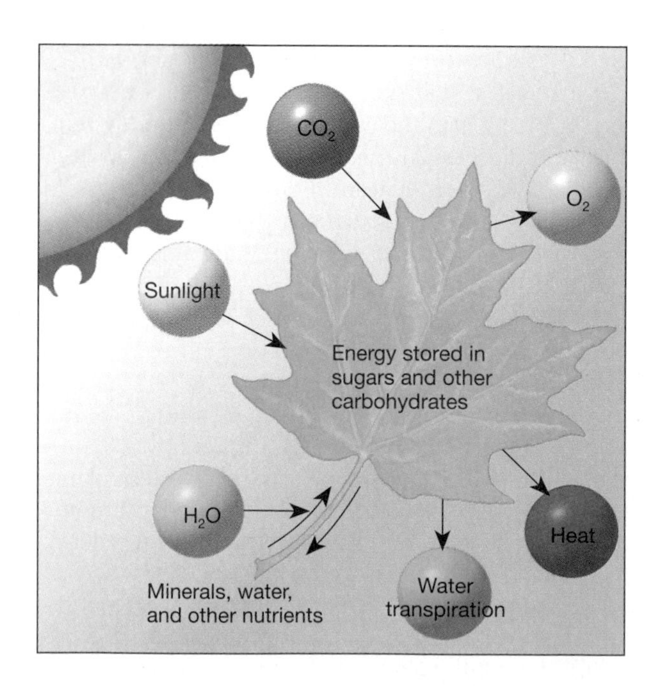

FIGURE 1-4
Simple flow system of a plant leaf. Plants take light, carbon dioxide (CO_2), water (H_2O), and nutrients as system inputs to produce outputs of oxygen (O_2) and carbohydrates (sugars) through the photosynthesis process. Plant respiration is simply a reverse of this process.

open system. (Photosynthesis and respiration processes are discussed in Chapter 19.)

As a system operates, outputs that can influence continuing system operations are generated. This "information" is returned to various points in the system via pathways called **feedback loops**. Feedback can guide further system operations. If the "information" amplifies or encourages response in the system, it is called **positive feedback**. If the "information" slows or discourages response in the system, it is called **negative feedback**. Such negative feedback causes self-regulation in a natural system, maintaining the system within its performance range, and in a dynamic steady state. In the leaf-flow system (Figure 1-4), any increase or decrease in daylength (photoperiod), carbon dioxide, or water availability will produce feedback information that effects specific responses in plant operations.

In Earth's systems, negative feedback is far more common than positive feedback. Unchecked positive feedback in a system can create a runaway ("snowballing") condition until a critical limit is reached, leading to instability, disruption, or death.

Most systems maintain structure and character over time. An energy and material system that remains balanced over time is at **equilibrium**. When the rates of inputs and outputs in the system are equal and the amounts of energy and matter in storage within the system are constant (or more realistically, as they fluctuate around a stable average), the system is in **steady-state equilibrium**.

However, a system fluctuating around an average value may demonstrate a trend over time and represent a changing *dynamic equilibrium* state through time. Such is the status of long-term climatic changes and the present global warming. A landscape, such as a hillside, adjusts after a landslide to a new equilibrium among slope, materials, and energy, operates as a system that demonstrates a dynamic equilibrium over time. Figure 1-5 illustrates these two states.

A **model** is a simplified version of a system and represents an idealized part of the real world. Models are designed with varying degrees of abstraction. Physical geographers construct simple system models to demonstrate complex associations in the environment, such as the hydrologic cycle that models Earth's entire water system. The simplicity of a model makes a system easier to comprehend. It also allows predictions, but such predictions are only as good as the assumptions and accuracy built into the model. Imposing a model too rigidly on a natural setting can lead to misinterpretation, so it is best to view a model for

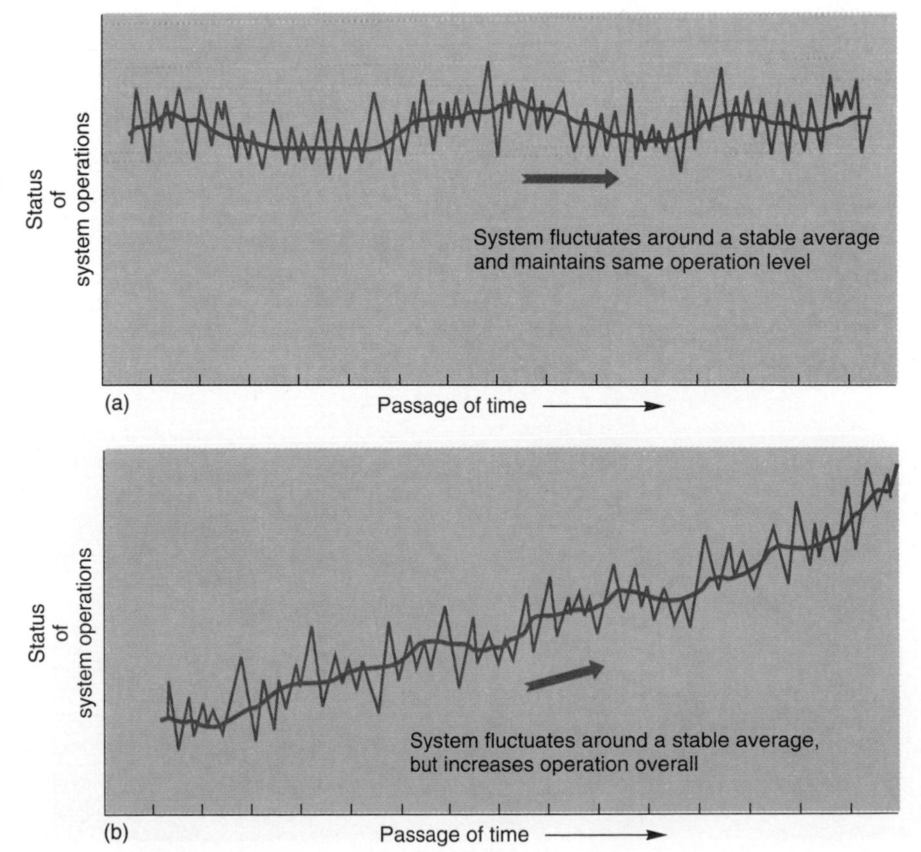

FIGURE 1-5
Steady-state equilibrium in two systems. Rather than a static equilibrium over time (a), some systems exhibit an increasing (or decreasing) operational trend (b). A system exhibiting such trends and adjustments is often referred to as dynamic.

what it is—a simplification. We discuss many system models in this text, including the hydrologic cycle, water balance, surface energy budgets, earthquakes and faulting, glacier mass budgets, soil profiles, and various ecosystems. Computer-based models are in use to study most natural systems—from climate change to Earth's interior.

Earth as a System

Because it receives energy from an outside source—the Sun—and radiates energy into space, Earth operates as an open system. In terms of matter, however, Earth is essentially a closed system: closed to any large amounts of incoming matter.

Earth's Energy Equilibrium: An Open System. Most Earth systems are dynamic because of the tremendous infusion of radiant energy from thermonuclear reactions deep within the Sun. This energy penetrates the outermost edge of Earth's atmosphere and cascades through the terrestrial systems, transforming along the way into various other forms of energy, such as kinetic energy, mechanical energy, or potential energy. Eventually, Earth radiates this energy back to the cold vacuum of space in an amount essentially equal to that which entered the system.

Researchers are examining the dynamics of Earth's energy equilibrium in more detail than ever before to distinguish natural changes from those produced by human intervention. Tremendous breakthroughs in understanding and prediction should occur as general circulation models of Earth's energy-atmosphere-water system become even more accurate. By the late 1990s, at least four polar-orbiting satellites—part of an Earth Observation System, or EOS—are expected to introduce a new era in the monitoring of Earth's open energy system.

Earth's Physical Matter: A Closed System. In terms of physical matter—air, water, and material resources—Earth is nearly a closed system. The only exceptions to this are occasional meteors, cosmic and meteoric dust, and outgassing of water from deep within Earth's crust. Since the initial formation of the planet, no significant quantities of matter have entered the system. Just as importantly, no significant quantities have left the system, either. This is it! Earth's physical materials are finite. No matter how numerous and daring the technological reorganizations of matter become, the physical base is fixed.

As Earth's people understand this fact of limited resources, a serious effort to recycle and to make efficient use of energy and materials can begin. Society simply loses track of its resources in what is essentially a once-through resource stream from virgin materials to the landfill (Figure 1-6). Recycling costs for some materials may be prohibitive. However, about half the contents of an average landfill are extractable economically—glass, newspaper, aluminum, and some other metals. The fact that Earth is a closed material system makes such recycling efforts inevitable.

FIGURE 1-6
Typical urban landfill that essentially is an "urban ore mine." [Photo by Tom and Pat Leeson/DRK Photo.]

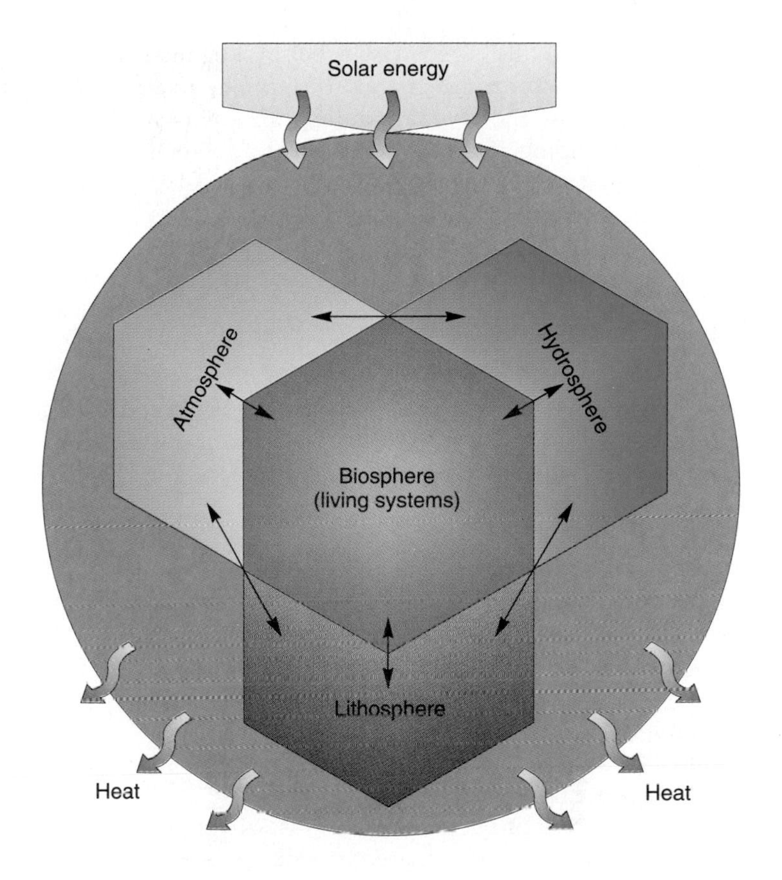

FIGURE 1-7
Earth's four spheres. Each sphere is a model for the four "parts" used to organize *Geosystems*.
Part 1–atmosphere
Part 2–hydrosphere
Part 3–lithosphere
Part 4–biosphere

Earth's Four Spheres

Earth's surface is where four immense open systems interface and interact. Figure 1-7 shows three **abiotic** (nonliving) systems overlapping to form the realm of the **biotic** (living) system. The abiotic spheres are the **atmosphere**, **hydrosphere**, and **lithosphere**. The biotic sphere is called the **biosphere**. Because these four models are not independent units in nature, their boundaries must be understood as transition zones rather than sharp delimitations.

Atmosphere. The atmosphere is a thin, gaseous veil surrounding Earth, held to the planet by the force of gravity. Formed by gases arising from within Earth's crust and interior, and the exhalations of all life over time, the lower atmosphere is unique in the Solar System. It is a combination of nitrogen, oxygen, argon, carbon dioxide, water vapor, and small amounts of trace gases.

Hydrosphere. Earth's waters exist in the atmosphere, on the surface, and in the crust near the surface, in liquid, solid, and gaseous forms. Water occurs in two forms, fresh and saline (salty), and exhibits important heat properties as well as playing its extraordinary role as a solvent. Among the planets in the Solar System, only Earth possesses water in any quantity.

Lithosphere. Earth's crust and a portion of the upper mantle directly below the crust form the lithosphere. The crust is quite brittle compared to the layers beneath it, which are in motion in response to an uneven distribution of heat and pressure. (In a broad sense, the term lithosphere sometimes refers to the entire solid planet.)

Biosphere. The intricate, interconnected web that links all organisms with their physical environment is the biosphere. Sometimes called the **ecosphere**, the biosphere is the area in which physical and chemical factors form the context of life. The biosphere exists in the overlap among the abiotic spheres, extending from the sea floor to about 8 km (5 mi) into the atmosphere. Life is sustainable within these natural limits. In turn, life processes have powerfully shaped the other three spheres through various interactive processes. The biosphere has evolved, reorganized itself at times, faced extinction, gained new vitality, and managed to flourish overall. Earth's biosphere is the only one known in the Solar System; thus, life as we know it is unique to Earth.

A SPHERICAL PLANET

Information about Earth's shape and size is fundamental to physical geography. The roundness or sphericity of Earth is not as modern a concept as many believe. For instance, the Greek mathematician and philosopher Pythagoras (ca. 580–500 B.C.) determined through observation that Earth is round. We don't know what Pythagoras observed to conclude that Earth is a sphere, but we can speculate. He might have noticed ships sailing beyond the horizon and apparently sinking below the surface, only to arrive back at port with dry decks. Perhaps he noticed Earth's curved shadow cast on the lunar surface during an eclipse of the Moon. He might have deduced that the Sun and Moon are not just disks but are spherical, and that Earth must be a sphere as well.

Earth's sphericity was generally accepted by the educated populace as early as the first century A.D. Christopher Columbus, for example, knew he was sailing around a sphere in 1492; that is why he expected to reach the East Indies.

In 1687 Sir Isaac Newton postulated that the round Earth, along with the other planets, could not be perfectly spherical. Until that time the spherical-perfection model was a basic assumption of **geodesy**, the science that attempts to determine Earth's shape and size by surveys and mathematical calculations. Newton reasoned that the more rapid rotational speed at the equator—the equator being farthest from the central axis of the planet and therefore moving faster—would produce an equatorial bulge in response to a greater centrifugal force pulling Earth's surface outward. He was convinced that Earth is slightly misshapen into what he termed an *oblate* (flattened) *spheroid*, with the oblateness occurring at the poles.

Earth's equatorial bulge and its polar oblateness are universally accepted and confirmed by satellite observations. The modern era of Earth measurement is one of tremendous precision and is called the "geoidal epoch" because Earth is considered a **geoid**, meaning literally that "the shape of Earth is Earth-shaped." Imagine Earth's geoid as a mean sea-level surface extended beneath the continents.

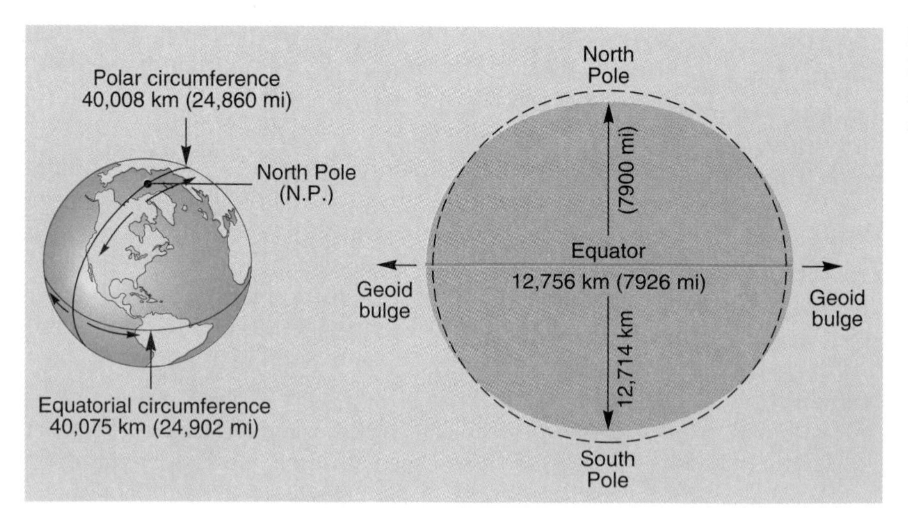

FIGURE 1-8
Earth's diameter and circumference. The dashed line is a perfect circle for reference.

Heights on land and depths in the oceans are measured from this hypothetical surface. Think of the geoid surface as a balance between the gravitational attraction of Earth's mass and the centrifugal pull caused by Earth's rotation.

Figure 1-8 gives Earth's polar and equatorial circumferences and diameters. Earth's circumference was first measured over 2200 years ago by the Greek geographer, astronomer, and librarian Eratosthenes (ca. 276–195 B.C.). The ingenious reasoning by which he arrived at his calculation is presented in Focus Study 1-1.

LOCATION AND TIME ON EARTH

To know specifically where something is located on Earth's surface, a coordinated grid system is needed, one that is agreed to by all peoples. The terms latitude and longitude were in use on maps as early as the first century A.D., with the concepts themselves dating back to Eratosthenes and others.

Ptolemy (ca. A.D. 90–168) contributed greatly to modern maps, and many of his terms and configurations are still used today. On his map of the known world (Figure 1-9), Ptolemy described 8000 places according to their north-south and east-west coordinates. He also placed north at the top and east at the right, orienting maps in a manner familiar to us.

Ptolemy divided the circle into 360 *degrees* (360°), with each degree comprising 60 *minutes* (60′), and each minute including 60 *seconds* (60″), in a manner adapted from the Babylonians. He located places using these degrees, minutes, and seconds, although the precise linear value of a degree of latitude and a degree of longitude on Earth's surface remained unresolved for the next 17 centuries. Ptolemy's values for longitude were in error because he accepted a shorter estimate for Earth's circumference. We'll now examine each of these grid coordinate elements.

Latitude

Latitude is an angular distance north or south of the equator, measured from the center of Earth (Figure 1-10a). On a map or globe, the lines designating these angles of latitude run east and west, parallel to the equator. Because Earth's equator divides the distance between the North Pole and the South Pole exactly in half, it is assigned the value

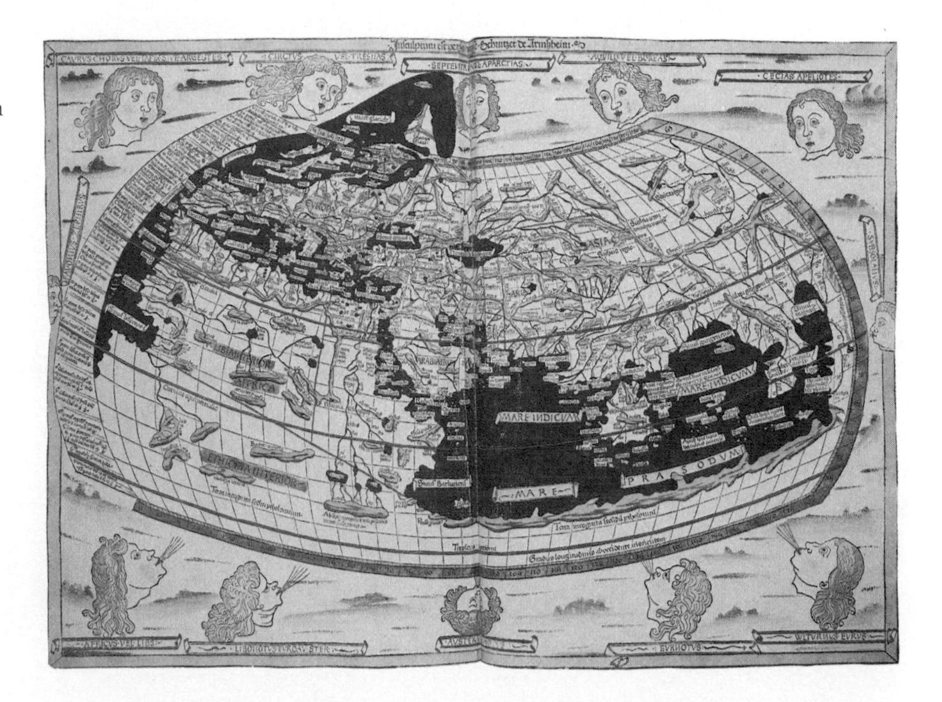

FIGURE 1-9
Ptolemy published an early version of this map of the known world in A.D. 140. Shown is a Renaissance interpretation. [© National Maritime Museum, Greenwich, England.]

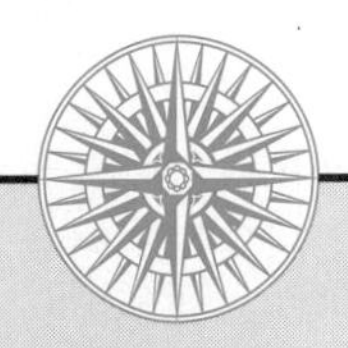

Focus Study 1-1 Measuring Earth in 247 B.C.

Eratosthenes, foremost among early geographers, served as the librarian of Alexandria in Egypt during the third century B.C. He was in a position of scientific leadership, for Alexandria's library was the greatest in the ancient world. Among his achievements was calculation of Earth's circumference to a high level of accuracy, quite a feat for 247 B.C.

Travelers told Eratosthenes that on June 21 they had seen the Sun's rays shine directly to the bottom of a well at Syene, the location of present-day Aswan, Egypt (Figure 1). Eratosthenes knew from his own observations that the Sun's rays never were directly overhead at Alexandria, even at noon on June 21, the longest day of the year and the day on which the Sun is at its most northward position in the sky. Instead, objects always cast a daytime shadow in Alexandria. Using the considerable geometric knowledge of the era, Eratosthenes conducted an experiment.

In Alexandria at noon on June 21, Eratosthenes carefully measured the angle of a shadow cast by an obelisk, a perpendicular column used for telling time by the Sun. Knowing the height of the obelisk and measuring the length of the shadow from its base, he solved the triangle for the angle of the Sun's rays, which he determined to be 7.2°. However, at Syene on the same day, the angle of the Sun's rays was 0° from a perpendicular—that is, the Sun was directly overhead. Geometric principles told Eratosthenes that the distance on the ground between Alexandria and Syene formed an arc of Earth's circumference equal to the angle of the Sun's rays at Alexandria. Since 7.2° is roughly 1/50 of the 360° in Earth's total circumference, the distance between Alexandria and Syene must represent approximately 1/50 of Earth's total circumference.

Camel caravans took 50 days to trek from Syene to Alexandria, covering about 100 stadia a day—5000 stadia one way. Eratosthenes determined the surface distance between the two cities using this estimate. He then multiplied 5000 stadia by 50 to determine that Earth's circumference is about 250,000 stadia. Assuming that a stadion, a Greek unit of measure, equals approximately 185 m (607 ft), Eratosthenes's calculations convert to roughly 46,250 km (28,738 mi), which is remarkably close to the correct answer of 40,075 km (24,902 mi) for Earth's equatorial circumference.

Eratosthenes's work teaches the value of observing carefully and integrating all observations with previous learning. Measuring Earth's circumference required application of his knowledge of Earth-Sun relationships, geometry, and geography to his keen observations to prove his suspicions about Earth's size.

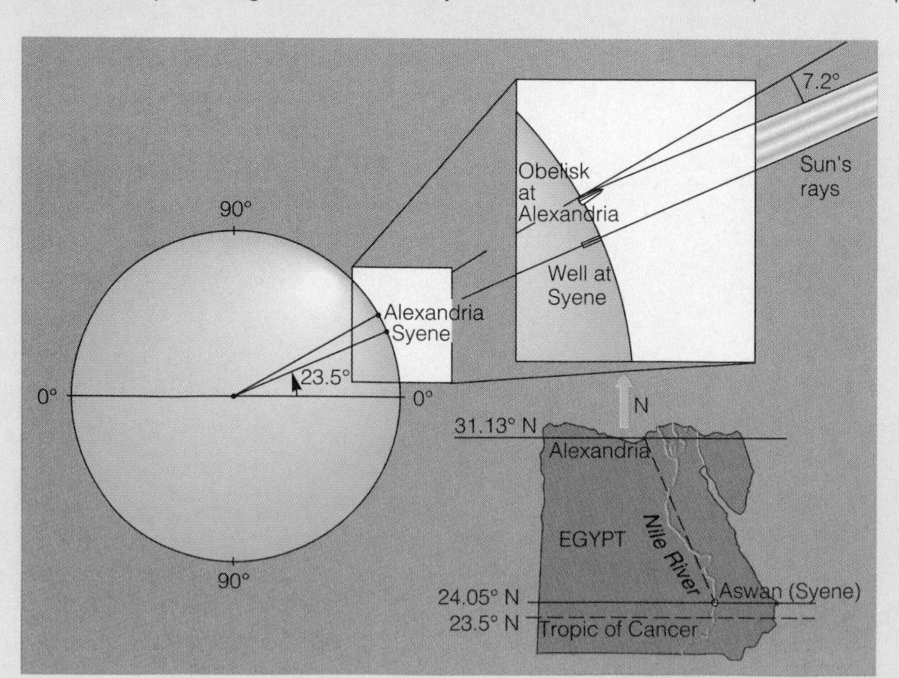

FIGURE 1
Eratosthenes's determination of Earth's circumference. This remarkably accurate measurement was based on a precise geometry and approximate camel-caravan speed.

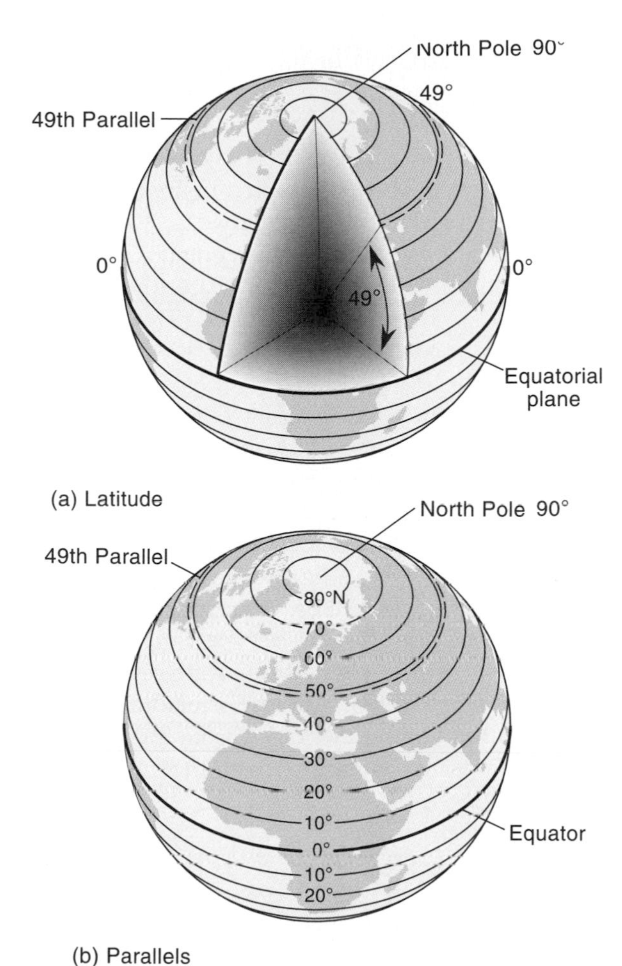

FIGURE 1-10
Angles of latitude measured north or south of the equator (a) determine parallels (b).

of 0° latitude. The North Pole is 90° north latitude, and the South Pole is 90° south latitude.

A line connecting all points along the same latitudinal angle is called a **parallel**. Thus, in Figure 1-10b, latitude is the name of the angle (49° north latitude), parallel names the line (49th parallel), and both indicate distance north of the equator. In the figure, an angle of 49° north latitude is measured and, by connecting all points at this latitude, the 49th parallel is designated. The 49th parallel is a significant one in the Western Hemisphere, for it forms the boundary between Canada and the United States from Minnesota to the Pacific.

Latitude is readily determined by using the Sun or the stars as points of reference, a method dating to ancient times. During daylight hours the angle of the Sun above the horizon indicates the observer's latitude, if adjustment is made for the season and time of day. Because Polaris, the North Star, is almost directly overhead at the North Pole, persons anywhere in the Northern Hemisphere can determine their latitude simply by sighting Polaris and measuring its angular distance above the local horizon (Figure 1-11). The angle of elevation of

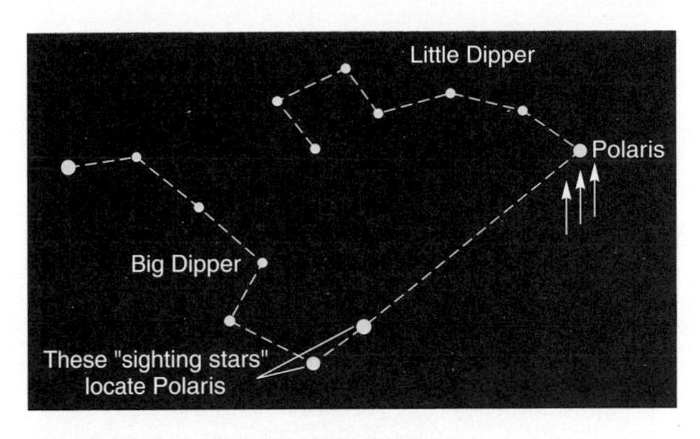

FIGURE 1-11 (above and below)
Determining latitude by using Polaris (the North Star). You can locate Polaris anywhere in the Northern Hemisphere using the "sighting stars" in the Big Dipper. All lines of sight to Polaris appear parallel when viewed from different points, as simulated in this figure.

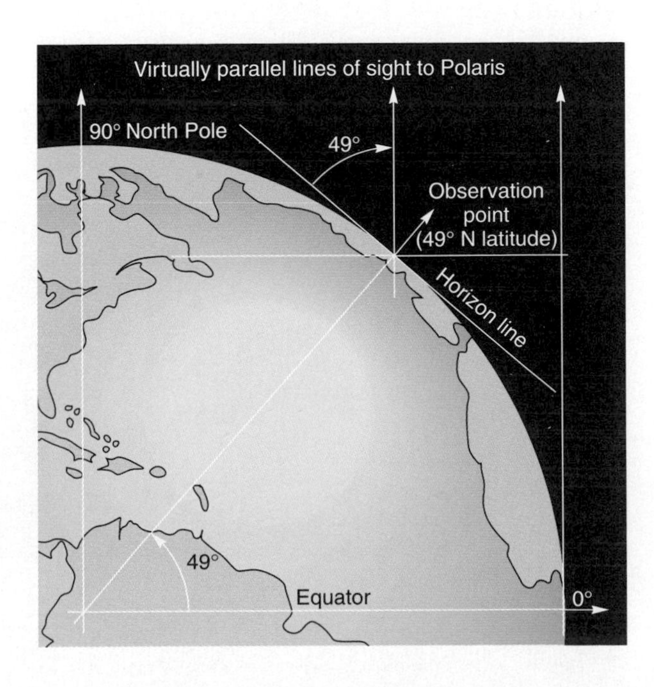

Polaris above the horizon equals the latitude of the observation point. In the Southern Hemisphere, Polaris cannot be seen (it is below the horizon), so such measurements are accomplished by sighting on the Southern Cross (Crux Australis) constellation.

Latitudinal Geographic Zones. Natural environments differ in operation and appearance from the equator to the poles. These differences basically are generated by the amount of solar energy received, which varies by latitude and season of the year. As a convenience, geographers identify *latitudinal geographic zones* as regions with fairly consistent qualities. Figure 1-12 portrays these zones, their locations, and their names: **equatorial**, **tropical**, **subtropical**, **midlatiude**, **subarctic** or **subantarctic**, and **arctic** or **antarctic**.

These generalized latitudinal zones are useful concepts, but are not rigid delineations. The *Tropic of Cancer* (23.5° north parallel) and the *Tropic of Capricorn* (23.5° south parallel), discussed further in Chapter 2, are the most extreme north and south parallels that experience perpendicular (directly overhead) rays of the Sun at local noon at some time during the year. The Arctic Circle (66.5° north parallel) and the Antarctic Circle (66.5° south parallel) are the parallels farthest from the poles that experience 24 uninterrupted hours of night, or day, at some time during the year. "Lower latitudes" are those nearer the equator, whereas "higher latitudes" refer to those nearer the poles.

Longitude

Longitude is an angular distance east or west of a point on Earth's surface, measured from the center of Earth (Figure 1-13a). On a map or globe, the lines designating these angles of longitude run north and south at right angles (90°) to the equator and to all parallels. A line connecting all points along the same longitude is a **meridian**. Thus, longitude is the name of the angle, meridian names the line, and both indicate distance east or west of an arbitrary **prime meridian** (Figure 1-13b). Earth's prime meridian through Greenwich, England, was not generally agreed to by most nations until 1884.

FIGURE 1-12
Latitudinal geographic zones. These generalized zones phase into one another over broad transitional areas.

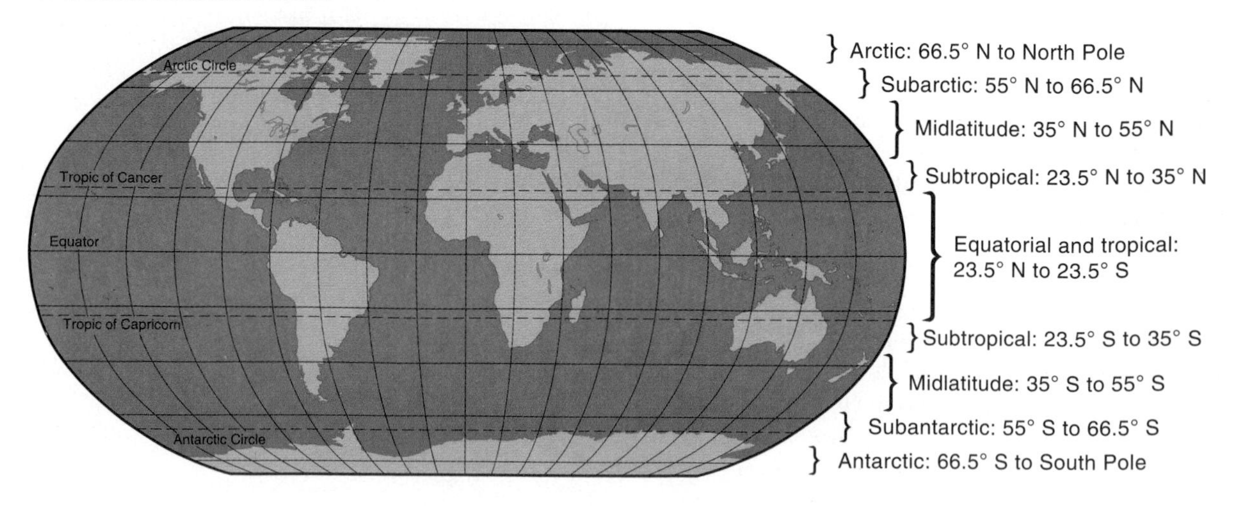

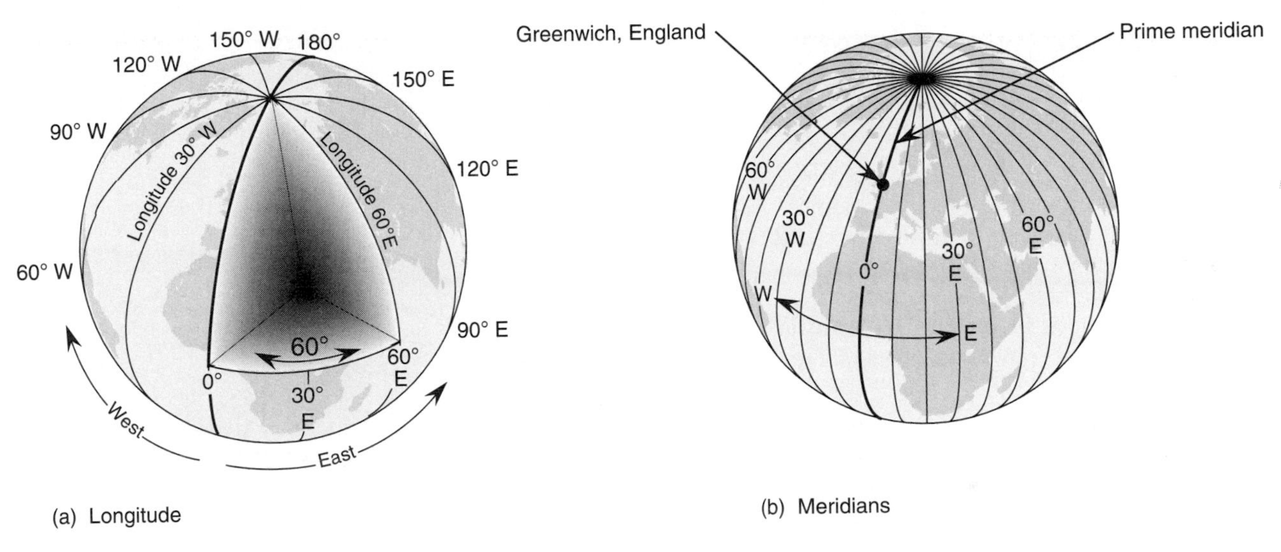

FIGURE 1-13
(a) Longitude is measured in degrees east or west of a 0° starting line. (b) Angles of longitude measured from this prime meridian, which is the line drawn from the North Pole through the observatory in Greenwich, England, to the South Pole, determine other meridians.

Because meridians of longitude converge toward the poles, the actual distance on the ground spanned by a degree of longitude is greatest at the equator (where meridians separate to their widest distance apart) and diminishes to zero at the poles (where they converge). The distance covered by a degree of latitude, however, varies only slightly, owing to deviations in Earth's shape. Table 1-1 compares latitude and longitude degree length. It also shows the similarity of ground distance for a degree of latitude and a degree of longitude at the equator.

We have noted that latitude is easily determined by sighting the North Star or the Southern Cross, but a method of accurately determining longitude at sea remained a major difficulty in navigation until the mid-1700s. In addition, a specific determination of longitude was needed before world standard time and time zones could be established. The interesting geographic quest for longitude is the topic of Focus Study 1-2.

TABLE 1-1
Physical Distances Represented by Degrees of Latitude and Longitude.

Latitudinal Location	Latitude Degree Length km (mi)	Longitude Degree Length km (mi)
90° (poles)	117.70 (69.41)	0 (0)
60°	111.42 (69.23)	55.80 (34.67)
50°	111.23 (69.12)	71.70 (44.55)
40°	111.04 (69.00)	85.40 (53.07)
30°	110.86 (68.89)	96.49 (59.96)
0° (equator)	110.58 (68.71)	111.32 (69.17)

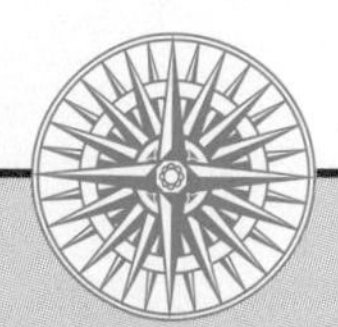

FOCUS STUDY 1-2 The Search for Longitude

Unlike latitude, longitude cannot readily be determined from celestial bodies owing to Earth's rotation. Even with precise observations of the Moon and a sophisticated use of mathematics, results are significantly inaccurate. This inability to know longitude at sea persisted far into the eighteenth century. Amerigo Vespucci, the explorer for whom America is named, made this note in his diary, as quoted by Daniel J. Boorstin in *The Discoverers*:

> 5 September 1499—As to longitude, I declare that I found so much difficulty in determining it that I was put to great pains to ascertain the east-west distance I had covered. (p. 247)

In his historical novel *Shogun*, author James Clavell expresses a similar perspective through his pilot Blackthorn:

> Find how to fix longitude and you're the richest man in the world. . . . the Queen, God bless her, 'll give you ten thousand pound and dukedom for answer to the riddle. . . . Out of sight of land you're always lost, lad. (p. 10)*

In the early 1600s Galileo explained that longitude could be measured by using two clocks. Any point on Earth takes 24 hours to travel around the full 360° of one rotation (one day). If you divide 360° by 24 hours, you find that any point on Earth travels through 15° of longitude every hour. Thus, if there were a way to measure time accurately at sea, a comparison of two clocks could give a value for longitude. One clock could be reset at local noon each day, as determined by the highest Sun position in the sky (solar zenith). The other clock would not be reset; it would indicate the time back at home port (Figure 1).

If the shipboard clock reads local noon and the clock set for home port reads 3:00 P.M., ship time is three hours earlier than home time. Therefore, calculating three hours at 15° per hour puts the ship at 45° west longitude from home port. Unfortunately, the pendulum clock invented by Christian Huygens in 1656 did not work on the rolling deck of a ship at sea!

In 1707 the British lost four ships and 2000 men in a sea tragedy that was blamed specifically on the longitude problem. Parliament passed an act in 1714—"Publik Reward . . . to Discover the Longitude at Sea"—and authorized a prize of £20,000 sterling (equal to over $1 million today) to the first successful inventor. The Board of Longitude was established to judge any devices submitted. John Harrison, a self-taught country clockmaker, began work on the problem in 1728 and finally produced his marine chronometer, known as Number 4, in 1760. The clock was tested during a nine-week trip to Jamaica in November 1761. When taken ashore for testing against Jamaica's land-based longitude, Harrison's Number 4 was only five seconds slow, an error that translates to only 1.25′, or 2.3 km (1.4 mi), well within Parliament's standard. Following many delays, Harrison finally received most of the prize money in his last years of life. From that time it was possible to determine longitude accurately

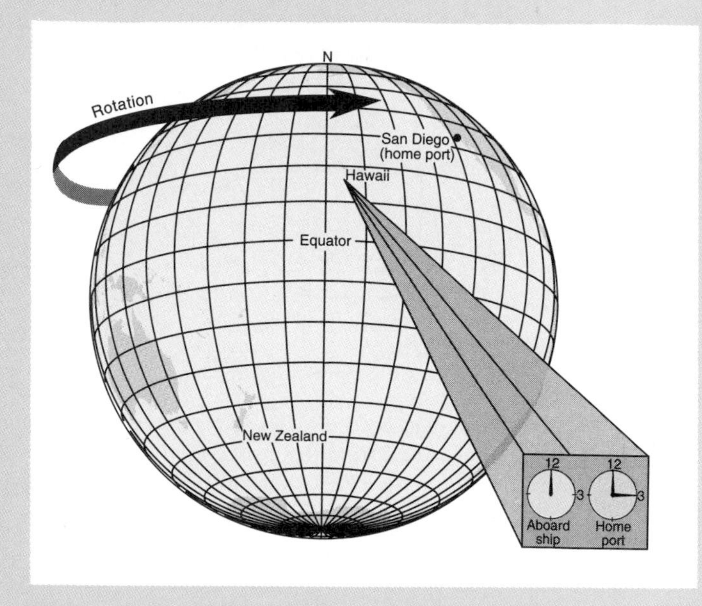

FIGURE 1
Using two clocks to determine longitude. For convenience Pacific Time and Hawaiian Standard Time are set 2 hours different despite the 45° longitudinal distance.

*From *Shogun* by James Clavell, copyright © 1975 by James Clavell, reprinted by permission of Delacorte Press, a division of Bantam, Doubleday, Dell Publishing Group, Inc.

on land and sea, as long as everyone agreed upon a prime meridian to use as the reference for time comparisons. Figure 2 shows the Royal Observatory, Greenwich, England, established by international treaty in 1884 as Earth's prime meridian.

In this modern era of atomic clocks and satellites in mathematically precise orbits, we have far greater accuracy available for the determination of longitude on Earth's surface. Measurements of Earth's surface from satellites are accurate within millimeters, and surface laser surveys are precise within micrometers.

FIGURE 2
Courtyard of the old Royal Observatory, Greenwich, which is still used as Earth's prime meridian—0° longitude. [© the National Maritime Museum, Greenwich, England.]

Prime Meridian and Standard Time

Today we take for granted international standard time zones and an agreed-upon prime meridian. If you live in Oklahoma City and it is 3:00 P.M. you know that it is 4:00 P.M. in Baltimore, 2:00 P.M. in Salt Lake City, and 1:00 P.M. in Seattle and Los Angeles. You also probably realize that it is 9:00 P.M. in London and midnight in Riyadh, Saudi Arabia. (The designation A.M. is for *ante meridiem*, "before noon," whereas P.M. is for *post meridiem*, meaning "after noon.") Coordination of international trade, airline schedules, business activities, and daily living depend on a common time system. Today's orderly management of time zones is only a century old. Until 1884, the world lacked an international agreement on a standard prime meridian.

Because Earth revolves 360° every 24 hours, or 15° per hour (360° ÷ 24 = 15°), a time zone of one hour is established for each 15° of longitude. Most nations in the early 1800s used their own national prime meridians for land maps, creating confusion in global mapping and clock time, as exemplified in Table 1-2. As a result, no standard time existed between or even within countries. The importance of Great Britain as a world power at the time is evident in that Greenwich, an observatory near London, was used on many nations' maritime charts as the prime meridian.

Setting time was not so great a problem in small European countries, most of which are less than 15° wide. In North America, which covers more than 90° of longitude (the equivalent of six 15° time zones), the problem was serious. In 1870, travelers going from Maine to San Francisco made 22 adjustments to their watches to stay consistent with local railroad time. Sir Sanford Fleming led the fight in Canada for standard time and an international agreement upon a prime meridian. His

TABLE 1-2
Sample of Prime Meridians Used in the 1800s

Country	Prime Meridians on: Land Maps	Marine Charts
Austria	Ferro*	Greenwich
Belgium	Brussels	Greenwich
Brazil	Rio de Janeiro	Rio de Janeiro
France	Paris	Ferro
Italy	Rome	Greenwich
Portugal	Lisbon	Lisbon
Spain	Madrid	Cádiz
United States	Washington	Greenwich

*Ferro (Spanish: Hierro) Island is the westernmost of the Canary Islands in the eastern Atlantic Ocean. Ferro was the most westerly place known to early geographers. Ptolemy chose it as an initial prime meridian, ca. A.D. 150. France used the Ferro prime meridian until 1911.

struggle led the United States and Canada to adopt a standard time in 1883.

In 1884 the International Meridian Conference was held in Washington, DC, attended by 27 nations. After lengthy debate, most participating nations chose the Royal Observatory at Greenwich as the place for the prime meridian of 0° longitude. (The Royal Observatory is pictured in Focus Study 1-2; note the line in the courtyard indicating the division of Eastern and Western hemispheres and the open roof used for sighting on the Sun.) The Observatory was highly respected, and more than 70% of the world's merchant ships already used the London prime meridian.

Thus, a world standard was established—**Greenwich Mean Time (GMT)**—and international time was set (Figure 1-14). Each time zone theoretically covers 7.5° on either side of a controlling meridian and represents one hour. From the map of global

FIGURE 1-14
Modern international standard time zones. Numbers indicate how many hours each zone is earlier (minus sign) or later (plus sign) than Greenwich Mean Time. The United States has five time zones: Canada is divided into six zones.

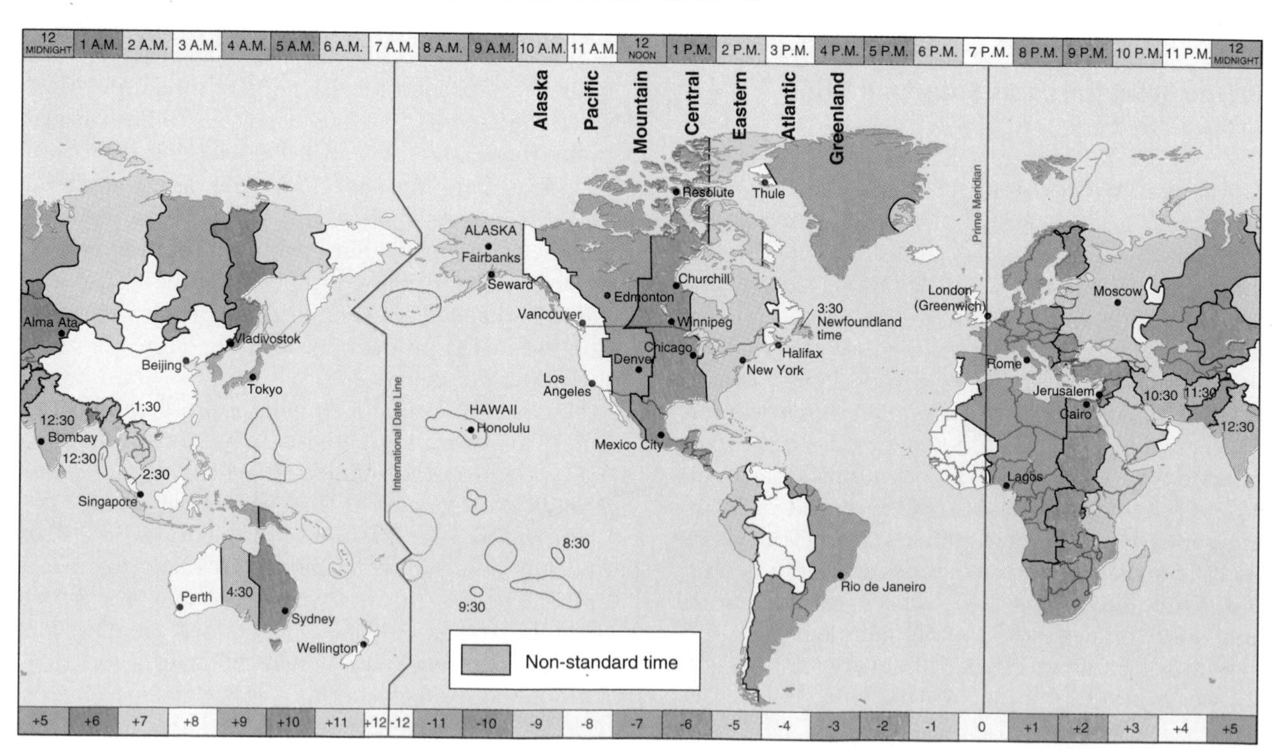

time zones, can you determine the present time in Moscow? London? Halifax? Chicago? Winnipeg? Denver? Los Angeles? Fairbanks? Honolulu? Tokyo? Singapore?

International Date Line. An important corollary of the prime meridian is the 180° meridian on the opposite side of the planet. This meridian is called the **International Date Line** and marks the place where each day officially begins (12:01 A.M.) and sweeps westward across Earth. This westward movement of time is created by the planet's turning eastward on its axis. At the International Date Line, the west side of the line is always one day ahead of the east side. No matter what time of day it is when the line is crossed, the calendar changes a day (Figure 1-15).

Locating the date line in the sparsely populated Pacific Ocean minimizes most local confusion. However, the consternation of early explorers before the date-line concept was understood is interesting. For example, Magellan's crew returned from the first circumnavigation of Earth in 1522, confident from their ship's log that the day of their arrival was a Wednesday. They were shocked when informed by insistent local residents that it was actually a Thursday!

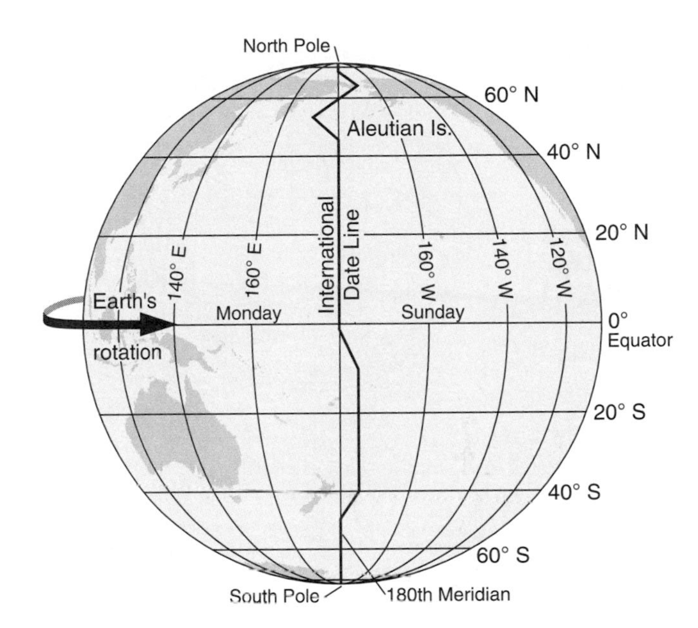

FIGURE 1-15
International Date Line location, the 180th meridian (see also Figure 1-14).

Coordinated Universal Time. Greenwich Mean Time was supplanted by a universal time system in 1928, and this system was expanded in 1964 when **Coordinated Universal Time (UTC)** was instituted. Today, UTC is the reference for official time in all countries. Although the prime meridian still runs through Greenwich, UTC is based on average time calculations collected in Paris and broadcast worldwide.

UTC and the official length of the second are measured today by the very regular vibrations of cesium atoms in five *primary standard clocks.* Three are operated by the Time Standards group of the Institute for Measurement Standards, National Research Council, Ottawa, and the other two are operated by the Physikalisch-Technische Bundesanstalt (PTB) in Germany. These five are to be joined by the new *NIST-7* primary clock being built and tested by Time and Frequency Services of the National Institute for Standards and Technology in Boulder, Colorado.

Added to these primary clocks, 200 *normal standard clocks* of commercial accuracy, operated in 70 nations, report to the Bureau International de l'Heure (BIH) in Paris, the international headquarters where computers bring together all measurements for "keeping the time." This level of precision has permitted scientists to add a leap second each year since January 1, 1972 to maintain UTC accuracy in coordination with Earth's slightly irregular rate of rotation.

Daylight Saving Time. Time is set ahead one hour in the spring and set back one hour in the fall in the Northern Hemisphere—a practice known as **daylight saving time**. The idea to extend daylight in the evening was first proposed by Benjamin Franklin, although not until World War I did Great Britain, Australia, Germany, Canada, and the United States adopt the practice.

Clocks in the United States were left advanced an hour from 1942 to 1945 during World War II, producing an added benefit of energy savings.

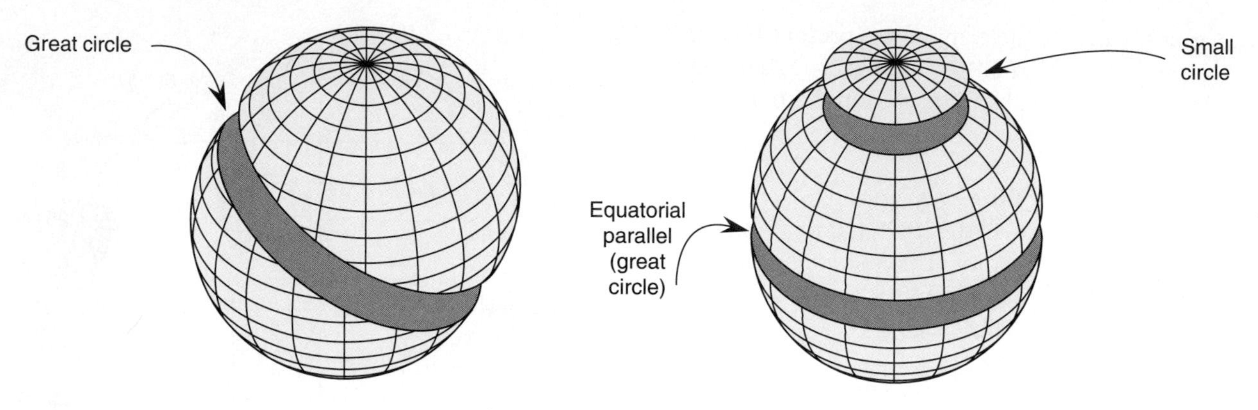

FIGURE 1-16
Great circles and small circles on Earth.

Again for energy conservation, a year-round daylight saving time was applied from January 1974 to October 1975.

Daylight saving time was increased in length in the United States and Canada in 1986. Time is set ahead on the first Sunday in April and set back on the last Sunday in October—only Hawaii, Arizona, portions of Indiana, and Saskatchewan exempt themselves. In Europe the last Sundays in March and September generally are used.

Great Circles and Small Circles

Great circles and small circles are important concepts that help summarize latitude and longitude (Figure 1-16). A **great circle** is any circle of Earth's circumference whose center coincides with the center of Earth. An infinite number of great circles can be drawn on Earth. Every meridian is one-half of a great circle that passes through the poles. On flat maps, airline and shipping routes appear to arch their way across oceans and landmasses. These are great circle routes, the shortest distance between two points on Earth. Only one parallel is a great circle—the equatorial parallel. All other parallels diminish in length toward the poles and, along with any other non-great circle that one might draw, constitute **small circles**—circles whose centers do not coincide with Earth's center.

MAPS AND MAP PROJECTIONS

The earliest known graphic map presentations date to 2300 B.C., when the Babylonians used clay tablets to record information about the region of the Tigris and Euphrates rivers (the area of modern-day Iraq). Much later, when sailing vessels dominated the seas, a pilot's *rutter*—a descriptive diary of locations, places, coastlines, and collected maps—became a critical reference. *Portolan* charts, describing harbor locations and coastal features and showing compass directions, also were prepared from these many navigational experiences. Today, the making of maps and charts is a specialized science as well as an art, blending aspects of geography, engineering, mathematics, graphics, computer science, and artistic specialties. It is similar in ways to architecture, in which aesthetics and utility are combined to produce an end product.

A map is a generalized view of an area, usually some portion of Earth's surface, as seen from above and greatly reduced in size. The part of geography that embodies mapmaking is called **cartography**. Maps are critical tools with which geographers depict spatial information and analyze spatial relationships. We all use maps at some time to visualize our location and our relationship to other places, or maybe to plan a trip, or to coordinate commercial and economic activities. Have you found yourself looking at a map, planning real and imagined adventures to far-distant places? Maps are wonderful tools! Learning a few basics about maps is essential to our study of physical geography.

The Scale of Maps

Architects and mapmakers have something in common: they both create scale models. They both reduce real things and places to the more convenient scale of a model, diagram, or map. An architect renders a blueprint of a building to guide the contractors, selecting a scale so that one centimeter (or inch) on the drawing represents so many meters (or feet) on the proposed building. Often, the drawing is 1/50 to 1/100 of real size.

The cartographer does the same thing in preparing a map. The ratio of the image on a map to the real world is called **scale**; it relates a unit on the map to a similar unit on the ground. A 1:1 scale would mean that a centimeter on the map represents a centimeter on the ground (although this certainly would be an impractical map scale, for the map would be as large as the area being mapped!). A more appropriate scale for a local map is 1:24,000, in which 1 unit on the map represents 24,000 identical units on the ground.

Map scales may be presented in several ways: as a written scale, a representative fraction, or a graphic scale (Figure 1-17). A *written scale* simply states the ratio—for example, "one centimeter to one kilometer" or "one inch to one mile." A *representative fraction* (RF, or fractional scale) can be expressed with either a : or a /, as in 1:125,000 or 1/125,000. No actual units of measurement are mentioned because any unit is applicable as long as both parts of the fraction are in the same unit: 1 cm to 125,000 cm, 1 in. to 125,000 in., or even 1 arm length to 125,000 arm lengths, and so on.

A *graphic scale*, or bar scale, is expressed as a bar graph with units noted to allow measurement of distances on the map. An important advantage of a graphic scale is that it changes to match the map during enlargement or reduction. In contrast, written and fractional scales become incorrect with enlargement or reduction: you can shrink a map from 1:24,000 to 1:63,360, but the scale will still say "1 in. = 2000 ft," instead of the new correct scale of 1 in. = 1 mi.

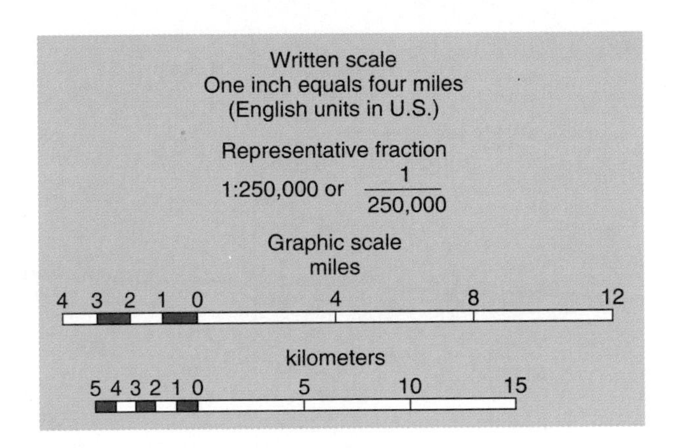

FIGURE 1-17
Three common expressions of map scale.

Scales are called "small," "medium," and "large," depending on the ratio described. Thus, in relative terms, a scale of 1:24,000 is a large scale, whereas a scale of 1:50,000,000 is a small scale. The greater the size of the denominator in a fractional scale (or of the number on the right in a ratio scale), the smaller the scale and the more abstract the map must be in relation to what is being mapped. Examples of selected representative fractions and written scales are listed in Table 1-3 for small-, medium-, and large-scale maps. In Chapter 12, Figure 12-13 presents a small-scale map of a portion of Pennsylvania at a 1:3,500,000 scale. Enlarged from this in the figure is a *Landsat* image and topographic map at a medium scale of 1:250,000. Note the increased detail at the larger scale.

If a world globe is 61 cm (24 in.) in diameter and we know Earth has an equatorial diameter of 12,756 km (7926 mi), then the scale of the globe is the ratio of 61 cm to 12,756 km. So, we divide Earth's actual diameter by the globe's diameter and determine that 1 cm of the globe's diameter equals about 20,900,000 cm of Earth's diameter. Thus, the representative fraction for the globe is expressed in centimeters as 1:20,900,000. This representative

TABLE 1-3
Sample Representative Fractions and Written Scales for Small-, Medium-, and Large-scale Maps

Units	Scale Size	Representative Fraction	Written Scale
English	Small	1:3,168,000	1 in. = 50 mi
		1:2,500,000	1 in. = 40 mi
		1:1,000,000	1 in. = 16 mi
		1:500,000	1 in. = 8 mi
		1:250,000	1 in. = 4 mi
	Medium	1:125,000	1 in. = 2 mi
		1:63,360 (or 1:62,500)	1 in. = 1 mi
		1:31,680	1 in. = 0.5 mi
		1:30,000	1 in. = 2500 ft
	Large	1:24,000	1 in. = 2000 ft
Units		**Representative Fraction**	**Written Scale**
Metric		1:1,100,000	1 cm = 10.0 km
		1:50,000	1 cm = 0.50 km
		1:25,000	1 cm = 0.25 km
		1:20,000	1 cm = 0.20 km

fraction can now be expressed in *any* unit of measure, metric or English, as long as both numbers are in the same units. Thus, this 61 cm globe is a small-scale representation of Earth with little local detail. If there is a globe available in your library or classroom, check to see the scale at which it was drawn. See if you can find examples of written, representative, and graphic scales on wall maps, highway maps, and in atlases.

Map Projections

A globe is not always a helpful representation of Earth. Travelers, for example, need more detailed information than a globe can provide, and large globes don't fit well in cars and aircraft. Consequently, to provide localized detail, cartographers prepare large-scale flat maps, which are two-dimensional representations (scale models) of our three-dimensional Earth. This reduction of the spherical Earth to a flat surface in some orderly and systematic realignment of the latitude and longitude grid is called a **map projection.**

Because a globe is the only true representation of distance, direction, area, shape, and proximity, the preparation of a flat version means that decisions must be made as to the type and amount of distortion that users will accept. To understand this problem, consider these important properties of a globe: On a globe, parallels always are parallel to each other, always are evenly spaced along meridians, and always decrease in length toward the poles. On a globe, meridians converge at both poles and are evenly spaced along any individual parallel. On a globe, the distance between meridians decreases toward poles, with the spacing between meridians at the 60th parallel equal to one half the equatorial spacing. And on a globe, parallels and meridians always cross each other at right angles.

The problem is that all these qualities cannot be reproduced on a flat surface. Simply taking a globe apart and laying it flat on a table illustrates the problem faced by cartographers in constructing a flat map (Figure 1-18). You can see the empty spaces that open up between the sections, or gores, of the globe. Thus, no flat map projection of Earth can ever have all the features of a globe. Flat maps always possess some degree of distortion—much less for large-scale maps representing a few kilometers; much more for small-scale maps covering individual countries, continents, or the entire world.

Properties of Projections. *The best projection is always determined by its intended use.* The major

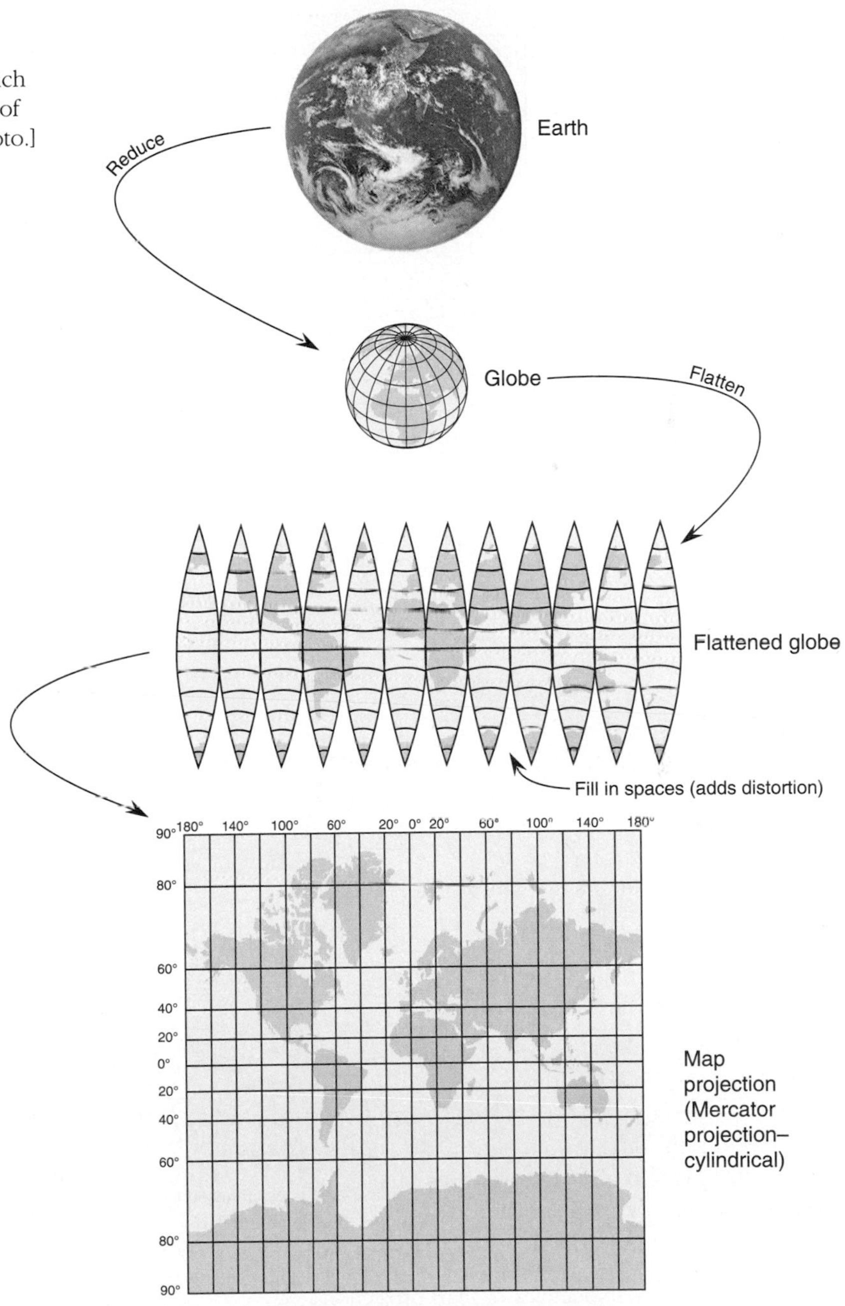

FIGURE 1-18
Conversion of the globe to a flat map projection requires decisions about which properties to preserve and the amount of distortion that is acceptable. [NASA photo.]

decisions in selecting a map projection involve the properties of **equal area** (equivalence) and **true shape** (conformality). If a cartographer selects equal area as the desired trait, as for a map showing the distribution of world climates, then shape must be sacrificed by *stretching* and *shearing* (allowing parallels and meridians to cross at other than right angles). On an equal-area map, a coin covers the same amount of surface area, no matter where you place it on the map.

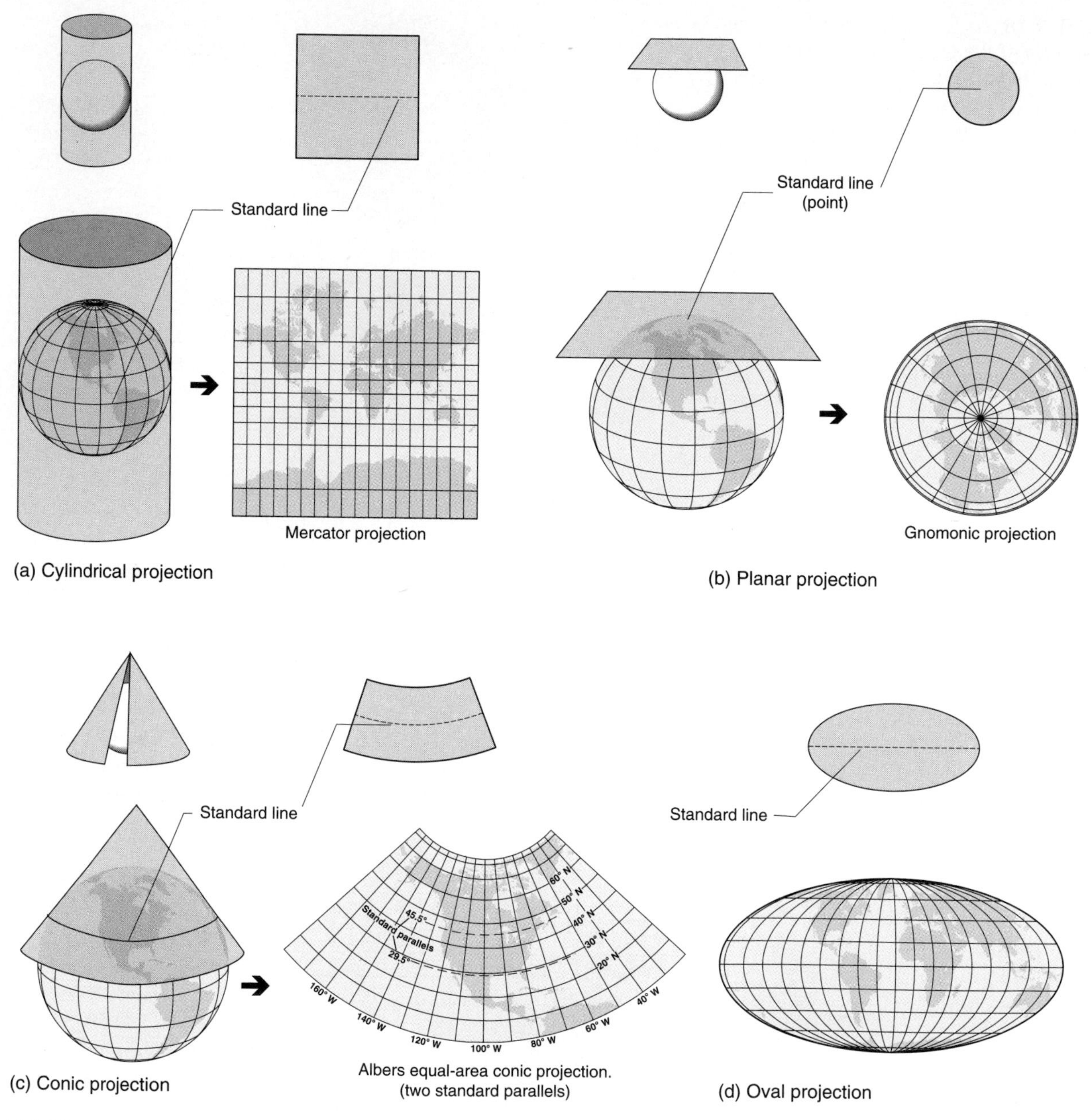

FIGURE 1-19
General classes and perspectives of map projections.

If, on the other hand, a cartographer selects the property of true shape, as for a map used for navigational purposes, then equal area must be sacrificed, and the scale will actually change from one region of the map to another. Two additional properties related to map projections are true direction (azimuth) and true distance (equidistance).

The Nature and Classes of Projections. Despite modern cartographic technology utilizing mathematical constructions and computer-assisted graphics, the word "projection" is still used. In times past, actual perspective methods were used to project the globe onto a geometric surface. A physical perspective approach uses a globe constructed of wire parallels and meridians. A light source strategically placed at varying distances then casts a pattern of latitude and longitude lines from the globe onto various geometric surfaces, such as a cylinder, plane, or, cone.

Figure 1-19 illustrates the derivation of the general classes of map projections and the perspectives from which they are generated. The classes shown include the cylindrical, planar (or azimuthal), and conic. Another class of projections that cannot be derived from this physical-perspective approach is the nonperspective oval-shaped. Still others are derived from purely mathematical calculations.

With all projections, the contact line or point between the wire globe and the projection surface—called a *standard line*—is *the only place where all globe properties are preserved*; thus, a *standard parallel* or *standard meridian* is a standard line true to scale along its entire length without any distortion. Areas away from this critical tangent line or point become increasingly distorted. Consequently, this area of optimum spatial properties should be centered on the region of immediate interest so that greatest accuracy is preserved there.

The commonly used Mercator projection (from Gerardus Mercator, A.D. 1569) is a cylindrical projection (Figure 1-19a). Figure 1-14 utilizes a Mercator projection for the presentation of standard time zones. The Mercator is a true-shape projection, with meridians appearing as equally spaced straight lines and parallels appearing as straight lines that come closer together near the equator. The poles are infinitely stretched, with the 84th north parallel and 84th south parallel fixed at the same length as that of the equator. Locally, the shape is accurate and recognizable for navigation; however, the scale varies with latitude.

The advantage of the Mercator projection is that lines of constant direction, called **rhumb lines**, are straight and thus facilitate the task of plotting compass directions between two points (see Figure 1-20). Thus, the Mercator projection is useful in navigation and has been the standard for nautical charts prepared by the National Ocean Service (formerly U.S. Coast and Geodetic Survey) since 1910.

Unfortunately, Mercator classroom maps present false notions of the size (area) of midlatitude and poleward landmasses. A dramatic example on the cylindrical projection in Figure 1-19a is Greenland, which looks bigger than all of South America. In reality, Greenland is only one-eighth the size of South America and is actually 20% smaller than Argentina alone!

The gnomonic or planar projection in Figure 1-19b is generated with a light source at the center of a globe projecting onto a plane tangent to the globe's surface. The resulting severe distortion prevents showing a full hemisphere on one projection. However, a valuable feature is derived: all great-circle routes, which are the shortest distances between two points on Earth's surface, are projected as straight lines. The great-circle routes plotted on a gnomonic projection then can be transferred to a true-direction projection, such as the Mercator, for determination of precise compass headings (Figure 1-20).

Maps Used in This Text. Two projections generally are in use in this text. One is an interrupted world map designed in 1925 by Dr. J. Paul Goode of the University of Chicago and called Goode's homolosine equal-area projection (Figure 1-21). Goode's projection is a combination of two oval projections (*homolo*graphic and *sin*usoidal projections).

To improve the rendering of landmass shapes, two equal-area projections are cut and pasted together. One type of projection is used between 40° N and S latitudes (a *sinusoidal projection*—central meridian is a straight line, with all other meridians spaced as sinusoidal curves and with parallels evenly spaced). The other type of projection is used from 40° N to the North Pole and from 40° S to the South Pole (a *Mollweide*, or *homolographic, projection*—central meridian is a straight line, with all other meridians spaced as elliptical arcs and with parallels unequally spaced: farther apart at the equator, closer together poleward). Areal size relationships are preserved, so the map is excellent for mapping spatial distributions when interruptions do not pose a problem. The world

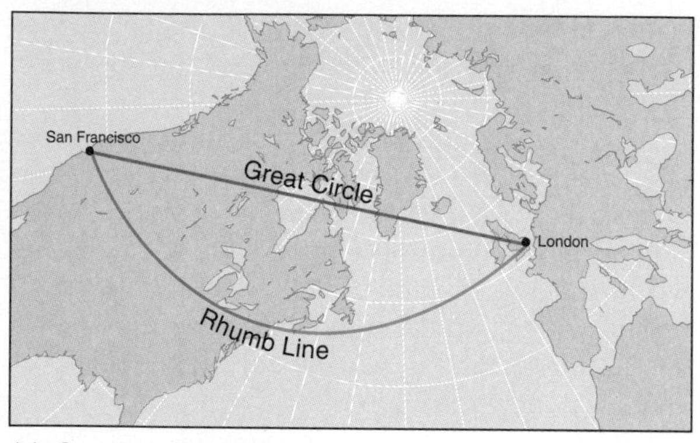

(a) Gnomonic Projection

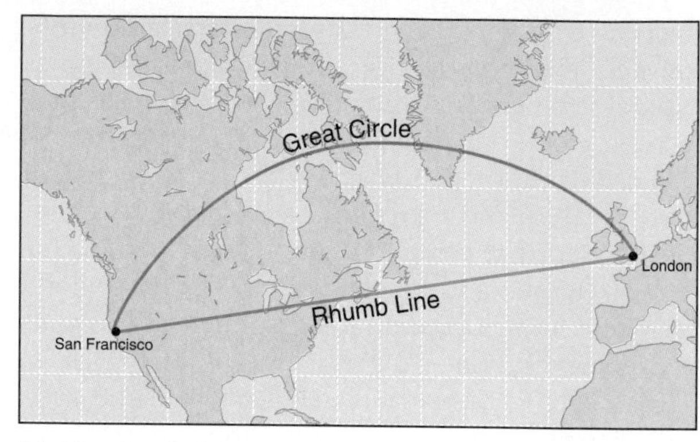

(b) Mercator Projection

FIGURE 1-20
Comparison of rhumb lines and great-circle routes from San Francisco to London on Mercator and gnomonic projections. All great-circle routes are the shortest distances between any two points on Earth.

climate map in Chapter 10, the soils map in Chapter 18, and the terrestrial ecosystems map in Chapter 20 use a Goode's homolosine projection.

Another projection used in this text is the Robinson projection, designed by Arthur Robinson in 1963 (Figure 1-22). This projection is neither equal area nor true shape, but is a compromise between both considerations. The North and South Poles appear as lines slightly more than half the length of the equator, thus higher latitudes are exaggerated less than on other oval and cylindrical projections. Examples in this text are the latitudinal geographic zones map (Figure 1-12), the world temperature maps in Chapter 5 (Figures 5-8, 5-10 and 5-11), the world pressure maps in Chapter 6 (Figure 6-9), and the maps of lithospheric plates of crust (Figure 11-17) and volcanoes and earthquakes (Figure 11-19). The National Geographic Society adopted this projection for their primary world map in 1988.

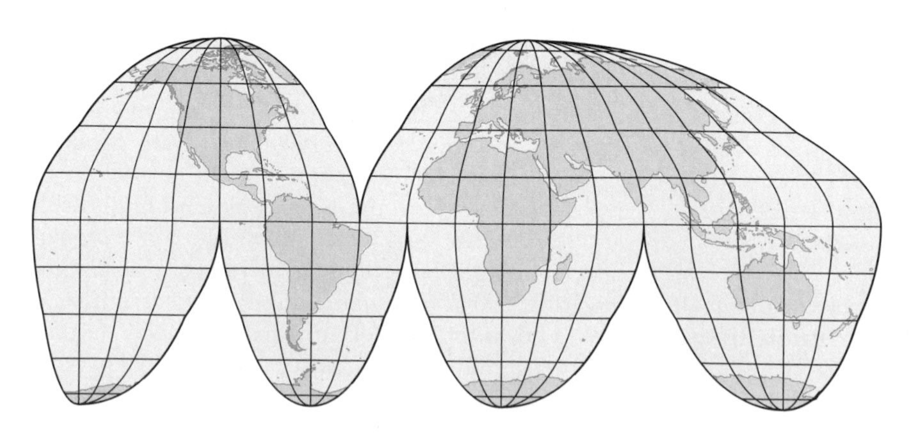

FIGURE 1-21
Goode's homolosine equal-area projection. [Copyright by the University of Chicago. Used by permission of the University of Chicago Press.]

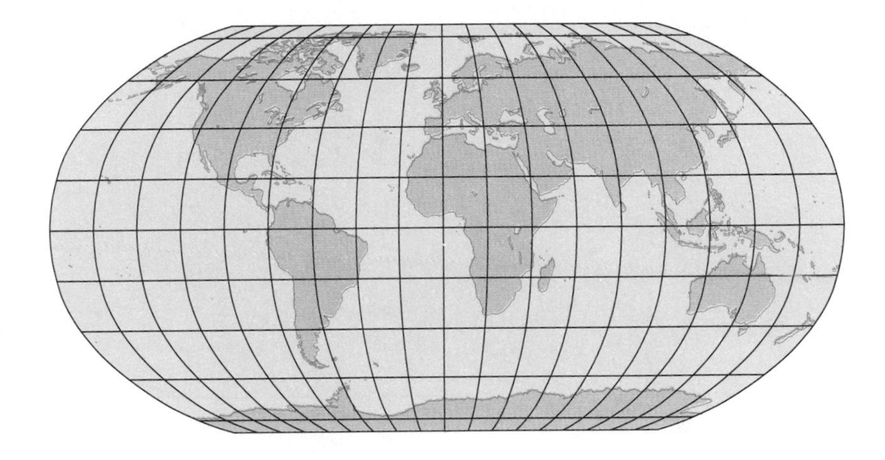

FIGURE 1-22
Robinson projection. Developed by Arthur H. Robinson, 1963.

Mapping and Topographic Maps

The Public Lands Survey System (1785) delineated locations in the United States. Although the Bureau of Land Management was responsible for subdividing public lands and keeping records of all related surveys, actual preparation and recording of this information fell to the U.S. Geological Survey (USGS), a branch of the Department of the Interior.

Detailed mapping of a variety of features is carried on continuously by the USGS. The USGS published the *National Atlas of the USA* in 1970. In Canada, national mapping is conducted by the Energy, Mines, and Resources Department (EMR). The EMR prepares base maps, thematic maps, aeronautical charts, federal topographic maps, and the *National Atlas of Canada*, now in its fifth edition (1986).

The USGS depicts the information on quadrangle maps, which are rectangular maps bounded by parallels and meridians rather than by political boundaries. From the conic class of map projections, the Albers equal-area projection is used as a base for these quadrangle maps (Figure 1-19c). Two standard parallels (where the cone intersects the globe's surface) are used to improve the accuracy in conformality and scale for the conterminous United States (the 48 states with shared boundaries)—29.5° N and 45.5° N latitudes. The standard parallels are shifted for conic projections of Alaska (55° N and 65° N) and Hawaii (8° N and 18° N).

In mapping, a basic **planimetric map** is first prepared, showing the horizontal position of boundaries, land-use aspects, and political, economic, and social features. A highway map is a common example of a planimetric map. Then, a vertical scale of physical features is added to portray the terrain. The most popular and widely used maps of this type are the detailed **topographic maps** prepared by the USGS. Topographic maps portray physical relief through the use of elevation **contour lines**. A contour line connects all points at the same elevation above or below a stated reference level, as you can see from the contour lines in Figure 1-23b. This reference level—usually mean sea level—is called the *vertical datum*. The *contour interval* is the difference in elevation between two adjacent contour lines (20 ft or 6.1 m in Figure 1-23).

The topographic map in Figure 1-23 shows a hypothetical landscape that demonstrates how contour lines and intervals depict slope and relief. Slope is indicated by the pattern of lines and the space between them. The steeper a slope or cliff, the closer together the contour lines appear—in Figure 1-23b, note the narrowly spaced contours representing the cliffs. A more gradual slope is portrayed by a wider spacing of these contour lines, as you can see from the widely spaced lines on the beach. Actual USGS topographic maps appear in several chapters of this text because they are so useful in depicting features of the physical landscape. As examples, see Figure 15-15 portray-

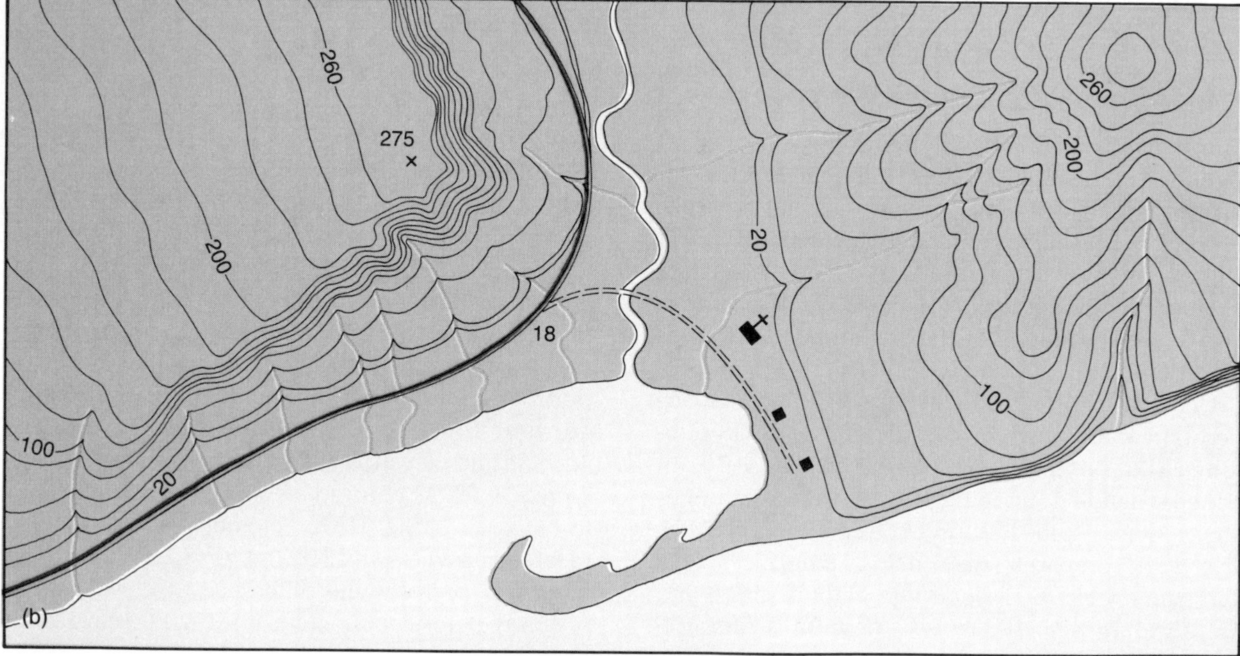

FIGURE 1-23
(a) Perspective view of a hypothetical landscape; (b) depiction of that landscape on a topographic map (contours in feet). [(a) Reprinted with the permission of Macmillan Publishing Company, from *The Earth: An Introduction to Physical Geology,* 4th ed., by Edward J. Tarbuck and Frederick K. Lutgens. Copyright © 1993 by Macmillan Publishing Company.]

ing the Cedar Creek alluvial fan, or Figure 17-2 showing glaciers in Alaska.

Because a single topographic map of the United States at 1:24,000 scale would be over 200 m wide (more than 600 ft), some system had to be devised for dividing the country up into maps of manageable size. Thus, a quadrangle system based on latitude and longitude coordinates is used for map classification and layout (Figure 1-24). Because meridians converge toward the poles, the width of

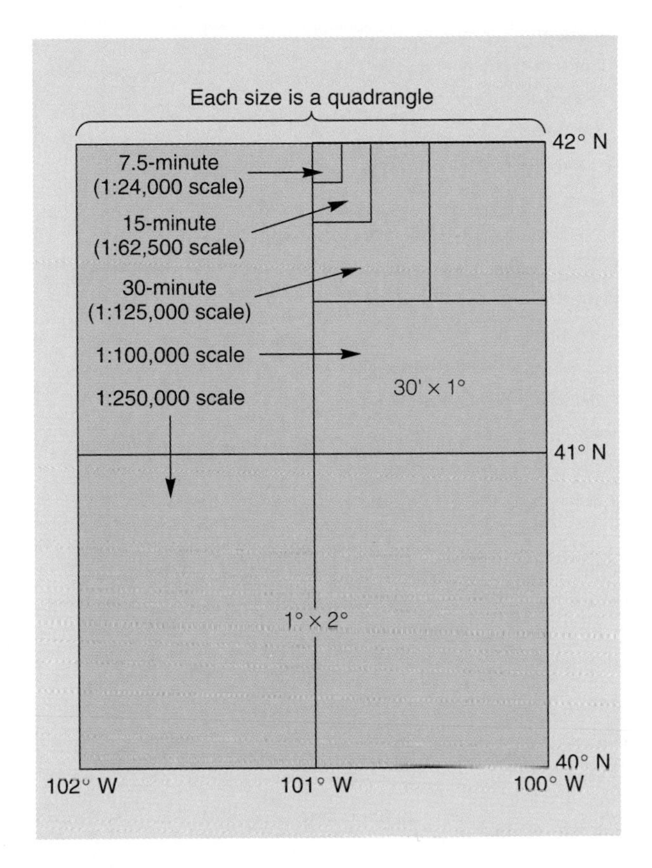

FIGURE 1-24
Quadrangle system of maps used by the USGS.

quadrangles also narrows noticeably toward the poles.

Each map quadrangle is referred to by its angular dimensions (Figure 1-24). Thus, a map that is one-half a degree (30′) on each side is a 30-minute quadrangle, and a map one-fourth of a degree (15′) on each side is a 15-minute quadrangle (the USGS standard from 1910 to 1950). A map that is one-eighth of a degree (7.5′) on each side is a 7.5-minute quadrangle, the most widely produced of all USGS topographic maps, and the standard since 1950.

The USGS National Mapping Program recently completed coverage of the entire country (except Alaska) on 7.5-minute maps. The standard 7.5-minute quadrangle is scaled at 1:24,000 (1 in. = 2000 ft, a large scale). It takes 53,838 separate 7.5-minute quadrangles to cover the lower 48 states, Hawaii, and U.S. territories. Alaska is covered with a series of more general 15-minute topographic maps. In Canada, the entire country is mapped at a scale of 1:250,000 (1.0 cm to 2.5 km), and about half the country also is mapped at 1:50,000. In the United States, the eventual changeover to the metric system someday will necessitate revision of the units used on all maps, with the 7.5-minute series eventually changing to a scale of 1:25,000, although this program was halted in 1991.

The principal symbols used on USGS topographic maps are presented in Figure 1-25. Colors are standard on all USGS topographic maps: black for human constructions, blue for water features, brown for relief features and contours, pink for urbanized areas, and green for woodlands, orchards, brush, and the like. The margins of a topographic map contain a wealth of information about the concept and content of the map: quadrangle name, names of adjoining quads, quad series and type, relative positioning on the latitude-longitude and other grid coordinate systems, title, legend, magnetic declination and compass information, datum plane, symbols used for roads and trails, the dates and history of the survey of that particular quad, and more. Many outdoor-product stores and state geological surveys sell topographic maps to assist people in planning their activities. These maps also may be purchased directly from the USGS or EMR.

SUMMARY—Essentials of Geography

Geography is a science of method, a special way of analyzing phenomena. Geography's method is spatial analysis, used to study the interdependence among geographic areas, natural systems, society, and cultural activities over space. The science is eclectic, integrating a wide range of subject matter. Geography synthesizes disciplines from the physical/life sciences with the cultural/human sciences to attain a holistic view of Earth. Geographic education recognizes five major themes in geography: location, place, human-Earth relationships (including environmental concerns), movement, and regions.

Physical geography applies spatial analysis to all the physical elements and processes that make up

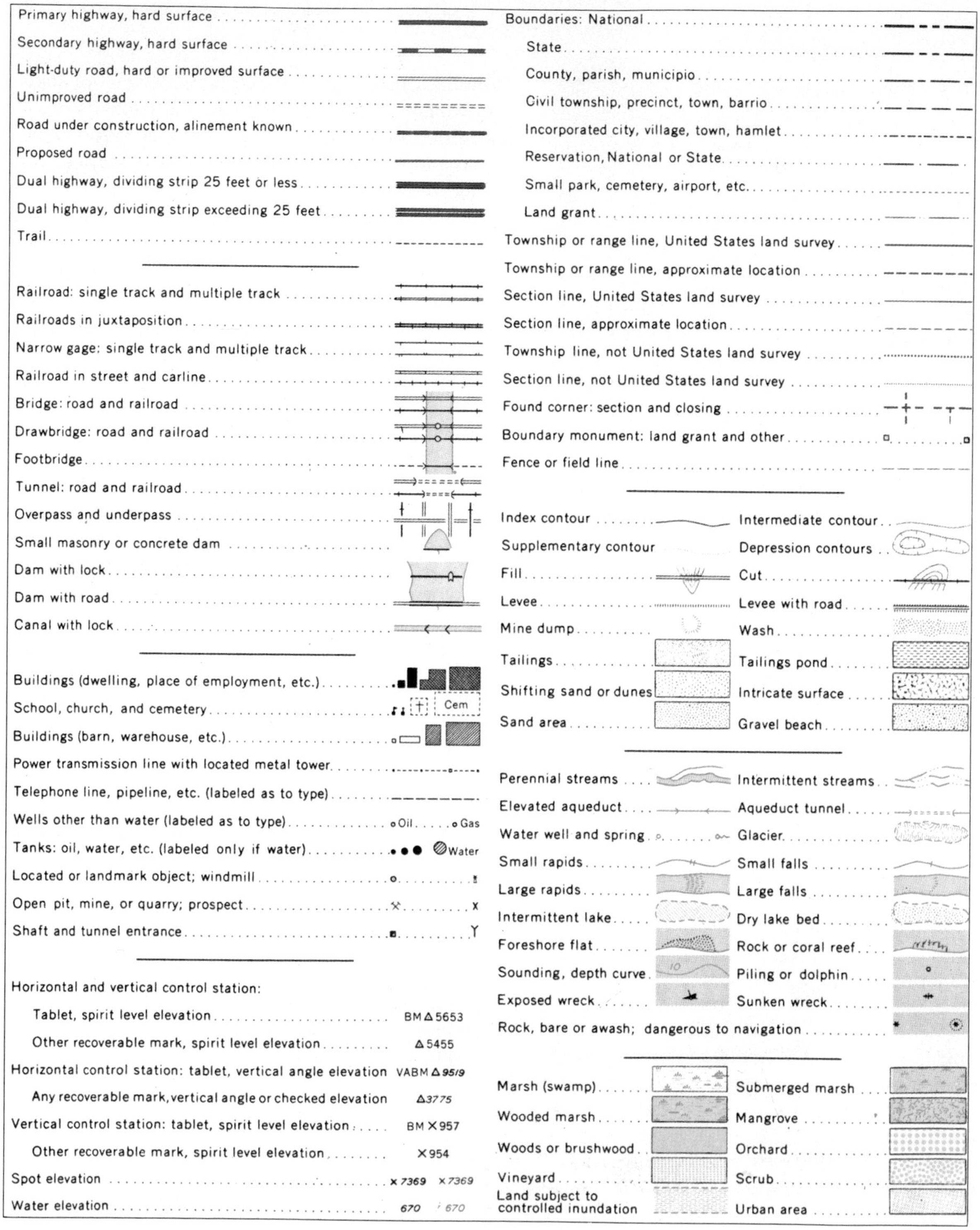

Primary highway, hard surface
Secondary highway, hard surface
Light-duty road, hard or improved surface
Unimproved road
Road under construction, alinement known
Proposed road
Dual highway, dividing strip 25 feet or less
Dual highway, dividing strip exceeding 25 feet
Trail

Railroad: single track and multiple track
Railroads in juxtaposition
Narrow gage: single track and multiple track
Railroad in street and carline
Bridge: road and railroad
Drawbridge: road and railroad
Footbridge
Tunnel: road and railroad
Overpass and underpass
Small masonry or concrete dam
Dam with lock
Dam with road
Canal with lock

Buildings (dwelling, place of employment, etc.)
School, church, and cemetery — Cem
Buildings (barn, warehouse, etc.)
Power transmission line with located metal tower
Telephone line, pipeline, etc. (labeled as to type)
Wells other than water (labeled as to type) — Oil ..... Gas
Tanks: oil, water, etc. (labeled only if water) — Water
Located or landmark object; windmill
Open pit, mine, or quarry; prospect
Shaft and tunnel entrance

Horizontal and vertical control station:

Tablet, spirit level elevation — BM Δ 5653
Other recoverable mark, spirit level elevation — Δ 5455
Horizontal control station: tablet, vertical angle elevation — VABM Δ *9519*
Any recoverable mark, vertical angle or checked elevation — Δ *3775*
Vertical control station: tablet, spirit level elevation — BM × 957
Other recoverable mark, spirit level elevation — × 954
Spot elevation — × *7369* × *7369*
Water elevation — *670* *670*

Boundaries: National
State
County, parish, municipio
Civil township, precinct, town, barrio
Incorporated city, village, town, hamlet
Reservation, National or State
Small park, cemetery, airport, etc.
Land grant
Township or range line, United States land survey
Township or range line, approximate location
Section line, United States land survey
Section line, approximate location
Township line, not United States land survey
Section line, not United States land survey
Found corner: section and closing
Boundary monument: land grant and other
Fence or field line

Index contour | Intermediate contour
Supplementary contour | Depression contours
Fill | Cut
Levee | Levee with road
Mine dump | Wash
Tailings | Tailings pond
Shifting sand or dunes | Intricate surface
Sand area | Gravel beach

Perennial streams | Intermittent streams
Elevated aqueduct | Aqueduct tunnel
Water well and spring | Glacier
Small rapids | Small falls
Large rapids | Large falls
Intermittent lake | Dry lake bed
Foreshore flat | Rock or coral reef
Sounding, depth curve — *10* | Piling or dolphin
Exposed wreck | Sunken wreck
Rock, bare or awash; dangerous to navigation

Marsh (swamp) | Submerged marsh
Wooded marsh | Mangrove
Woods or brushwood | Orchard
Vineyard | Scrub
Land subject to controlled inundation | Urban area

FIGURE 1-25
Standardized topographic map symbols used on USGS maps. English units still prevail, although some newer USGS maps are metric. [From USGS, *Topographic Maps*, 1969.]

the environment: energy, air, water, weather, climate, landforms, soils, animals, plants, and Earth itself. Understanding the complex relationships of these elements is important to human survival because Earth's physical systems and human society are so intertwined.

Systems methodology is used by geographers as an analytical tool. A system is any ordered, related set of things and their attributes, as distinct from their surrounding environment. Geographers often construct models of systems to better understand them. Earth is an open system for energy, receiving energy from the Sun, but essentially a closed system for matter and physical resources. Four immense open systems powerfully interact at Earth's surface: three nonliving abiotic systems (atmosphere, hydrosphere, and lithosphere) and a living biotic system (biosphere).

Earth bulges slightly through the equator and is oblate at the poles. Geographers call our misshapen spheroid a geoid. Absolute location on Earth is described with a specific reference grid of parallels of latitude and meridians of longitude. A historic breakthrough in navigation and time keeping occurred with the establishment of an international prime meridian (0° through Greenwich, England) and the invention of precise chronometers that enabled accurate measurement of longitude. Today, Coordinated Universal Time (UTC) is the worldwide standard and the basis for international time zones.

Maps are used by geographers for the spatial portrayal of Earth's physical systems. Cartographers create maps for specific purposes, selecting the best compromise of projection and scale for each application. Compromise is always necessary because Earth's round surface cannot be exactly replicated on a flat map. Equal area, true shape, true direction, and true distance are all considerations in selecting a projection. The U.S. Geological Survey publishes maps using various projections and scales for all parts of the nation, as does the Energy, Mines, and Resources Department in Canada. Best known are topographic maps, which use contour lines to portray elevation and which show detailed physical and cultural features.

The science of physical geography is in a unique position to synthesize the spatial environmental and human aspects of our increasingly complex relationship with our home planet—Earth.

KEY TERMS

abiotic
antarctic geographic zone
arctic geographic zone
atmosphere
biosphere
biotic
cartography
closed system
contour lines
Coordinated Universal Time (UTC)
daylight saving time
ecosphere
equal area
equatorial geographic zone
equilibrium
feedback loops
geodesy
geography
geoid
great circle
Greenwich Mean Time (GMT)
human-Earth relationships
hydrosphere
International Date Line
latitude
lithosphere
location
longitude
map projection
meridian
midlatitude geographic zone
model

movement
negative feedback
open system
parallel
physical geography
place
planimetric map
positive feedback
prime meridian
process
region
rhumb lines
scale
small circles
spatial analysis
steady-state equilibrium,
subantarctic geographic zone
subarctic geographic zone
subtropical geographic zone
system
topographic maps
tropical geographic zone
true shape

REVIEW QUESTIONS

1. What is geography? Based on information in this chapter, define physical geography and give a specific instance of the geographic approach.
2. Make a list of decisions you made today or items you've read recently that involve geographic concepts.
3. Assess your geographic literacy by examining available atlases and maps. What types of maps have you used—political? physical? topographic? Do you know what projections they employed? Do you know the names and locations of the four oceans, seven continents, and individual countries?
4. Suggest a representative example for each of the five geographic themes and use that theme in a sentence.
5. Define systems theory as an organizational strategy. What are open systems and closed systems, negative feedback, and a system at a steady-state equilibrium condition? What type of system is a human body? A lake? A wheat plant?
6. Describe Earth as a system in terms of energy and of matter.
7. What are the three abiotic spheres that comprise Earth's environment? Relate these to the biosphere.
8. Describe Earth's shape and size with a diagram.
9. What are the latitude and longitude coordinates (in degrees, minutes, and seconds) of your present location? Where can you find this information?
10. Define latitude and parallel. Define longitude and meridian.
11. Draw a diagram showing Eratosthenes's method of determining Earth's circumference. Use your diagram to explain his calculations.
12. Identify the various latitudinal geographic zones. In which zone do you live?
13. What do clocks have to do with longitude? Explain this relationship. How is standard time determined on Earth?
14. What and where is the prime meridian? How was it selected? Describe the meridian opposite the prime meridian.
15. Define a great circle and a small circle. In terms of these concepts, describe the equator, other parallels, and meridians.
16. In what way is cartography an integrative discipline?
17. What is map scale? How is it expressed on a map?

18. Identify each of the following ratios as large scale, medium scale, or small scale: 1:3,168,000, 1:24,000, 1:250,000.
19. Describe the differences between the characteristics of a globe and a flat map.
20. What type of map projection is used in Figure 1-12? Figure 1-14?

SUGGESTED READINGS

Boorstin, Daniel J. *The Discoverers.* New York: Random House, 1983.

Committee on Map Projections, American Cartographic Association. *Choosing a World Map* and *Which Map Is Best.* Falls Church, VA: American Congress on Surveying and Mapping, 1988.

Dunn, Margery G., and Donald J. Crump. *Exploring Your World—The Adventure of Geography.* Washington, DC: National Geographic Society, 1989.

Gaile, Gary L., and Cort J. Willmott, eds. *Geography in America.* Columbus, OH: Macmillan Publishing Company, 1989.

Greenwood, David. *Mapping.* Chicago: The University of Chicago Press, 1964.

Howse, Derek. *Greenwich Time and the Discovery of the Longitude.* London: Oxford University Press, 1980.

Making, Joel, and Laura Bergen, eds. *The Map Catalog—Every Kind of Map and Chart on Earth.* New York: Vintage Books, 1986.

Miller, Victor C., and Mary E. Westerback. *Interpretation of Topographic Maps.* Columbus, OH: Macmillan Publishing Company, 1989.

Odum, Howard T. *Systems Ecology—An Introduction.* New York: John Wiley & Sons, 1983.

Robinson, Arthur H., J. Morrison, and P. Muehrcke. *Elements of Cartography,* 5th ed. New York: John Wiley & Sons, 1985.

Snyder, John P. *Map Projections—A Working Manual.* U.S. Geological Survey Professional Paper 1395. Washington, DC: U.S. Government Printing Office, 1987.

Thompson, Morris M. *Maps for America.* 3d ed. U.S. Geological Survey. Washington, DC: U.S. Government Printing Office, 1987.

PART 1

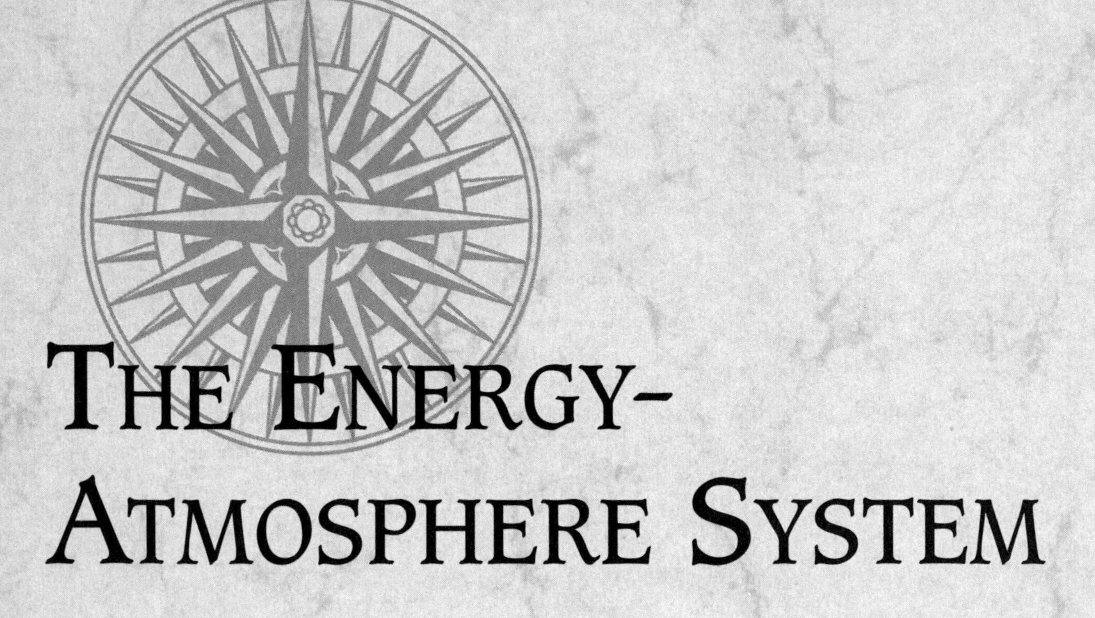

THE ENERGY-ATMOSPHERE SYSTEM

Alaskan sunset. [*Photo by Galen Rowell.*]

Our planet and our lives are powered by radiant energy from the star that is closest to Earth—the Sun. For more than 4.6 billion years, solar energy has traveled across interplanetary space to Earth, where a small portion of the solar output is intercepted. Because of Earth's curvature, the energy at the top of the atmosphere is unevenly distributed, creating imbalances from the equator to each pole. This unevenness of energy receipt empowers circulations in the atmosphere and on the surface below. The pulse of seasonal change varies the distribution of energy during the year.

Earth's atmosphere acts as an efficient filter, absorbing most harmful radiation, charged particles, and space debris so that they do not reach Earth's surface. Surface energy balances are established, giving rise to global patterns of temperature, winds, and ocean currents. Each of us depends on many systems that are set into motion by energy from the Sun. These systems are the subjects of Part One.

Aurora borealis in Alaska. [*Photo by Johnny Johnson/Allstock.*]

2

SOLAR ENERGY TO EARTH

Our immediate home is North America, a major continent on planet Earth, third planet from a typical yellow star in a solar system. That star, our Sun, is only one of billions in the Milky Way Galaxy, which is one of millions of galaxies in the universe. This chapter examines the flow of energy from the Sun to Earth's atmosphere.

The uneven receipt of solar energy at the top of the atmosphere sets into motion the winds, weather systems, and ocean currents below. These Earth-Sun relationships and the nature of Earth's tilt and rotation produce daily, weekly, and yearly seasonal pulses of energy.

The modern technological age of computer-based information systems and remote-sensing capabilities adds a new and exciting dimension to physical geography research. Chapter 2 concludes with a brief overview of the rapidly expanding fields of remote sensing and geographic information systems.

SOLAR SYSTEM: STRUCTURE AND FORMATION

Our Sun is unique to us, and yet commonplace in our galaxy. It is only average in temperature, size, and color when compared to other stars, yet it is the ultimate energy source for most life processes in our biosphere.

Solar System Location

Our Sun is located on a remote, trailing edge of the **Milky Way Galaxy**, a flattened, disk-shaped mass estimated to contain up to 400 billion stars (Figure 2-1). Our Solar System is embedded more than halfway out from the center, in one of the Milky Way's spiral gravitational arms, called the Orion Arm. From our Earth-bound perspective in the Milky Way, the galaxy appears to stretch across the night sky like a narrow band of hazy light. On a clear night the unaided eye can see only a few thousand of these billions of stars.

Dimensions. The **speed of light** is 299,792 km per second (186,282 mi/sec), which is more than 9 trillion km (9,454,000,000,000 km or 5,875,000,000,000 mi). This tremendous distance that light travels in a year is known as a *light-year* and is used as a unit of measurement for the vast universe. For spatial comparison, our Moon is an average distance of 384,400 km (238,866 mi) from Earth, or about 1.28 seconds in terms of light speed. Our entire Solar System is approximately 11 hours in diameter, measured by light speed. In contrast, the Milky Way is about 100,000 light-years from side to side, and the known universe that is observable from Earth stretches approximately 12 billion light-years in all directions.

Earth's Orbit. Earth's orbit around the Sun is presently elliptical—a closed, oval-shaped path (Figure 2-2). Earth's average distance from the Sun is approximately 150 million km (93 million mi), which means that light reaches Earth from the Sun in an average of 8 minutes and 20 seconds. A plane including all points of Earth's orbit is termed the **plane of the ecliptic**, and Earth's axis remains fixed relative to this plane as Earth revolves around the Sun.

Earth is at **perihelion** (its closest position to the Sun) on January 3 at 147,255,000 km (91,500,000 mi) and at **aphelion** (its farthest position from the Sun) on July 4 at 152,083,000 km (94,500,000 mi). This seasonal difference in distance from the Sun causes a variation of 3.4% in the solar energy intercepted by Earth.

The structure of Earth's orbit is not a constant but instead exhibits changes over long periods. As shown in Focus Study 17-1, Figure 2, Earth's distance from the Sun varies more than 17.7 million km (11 million mi) during a 100,000-year cycle, placing it closer or farther at different periods in the cycle. This variation is thought to be one of several factors that create Earth's cyclical pattern of glaciations and interglacial periods, which are

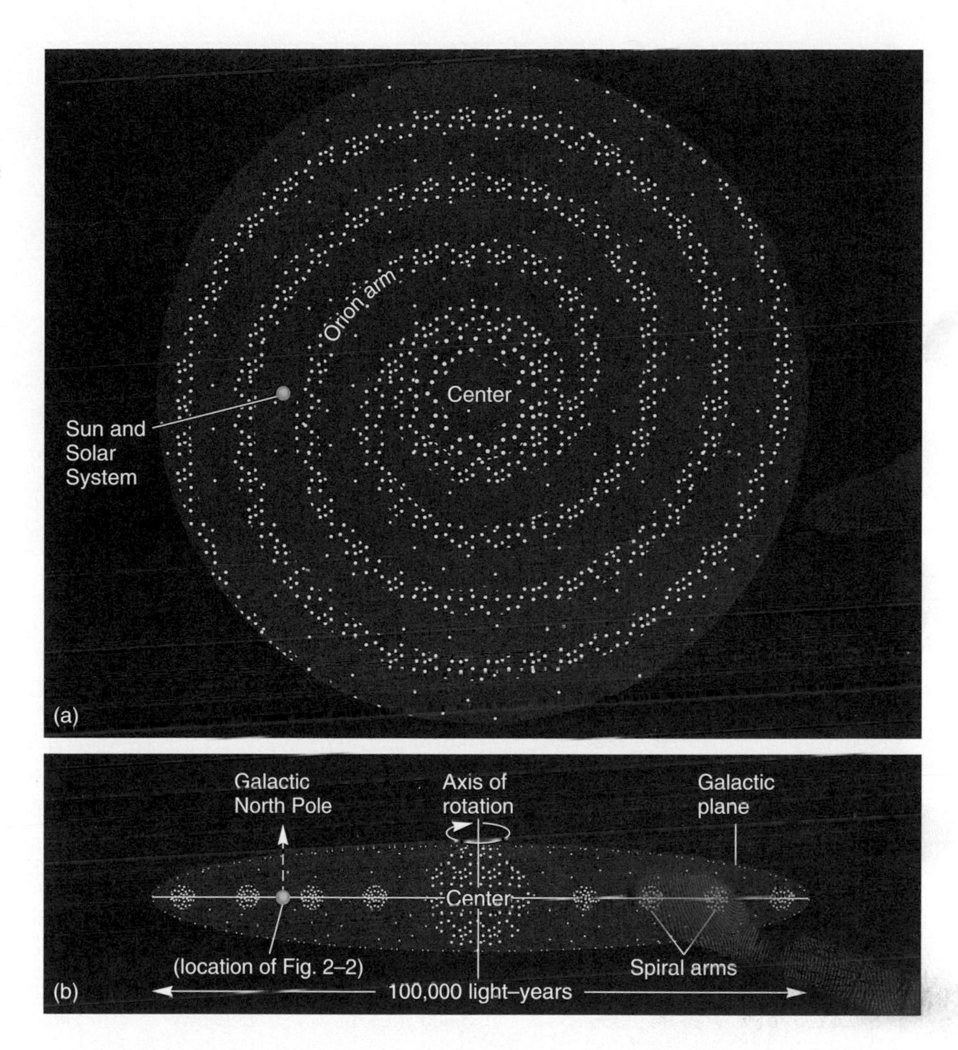

FIGURE 2-1
Milky Way Galaxy viewed from above (a) and cross-section side view (b). Our Solar System is some 30,000 light-years from the center of the galaxy.

FIGURE 2-2
Structure of Earth's elliptical orbit.

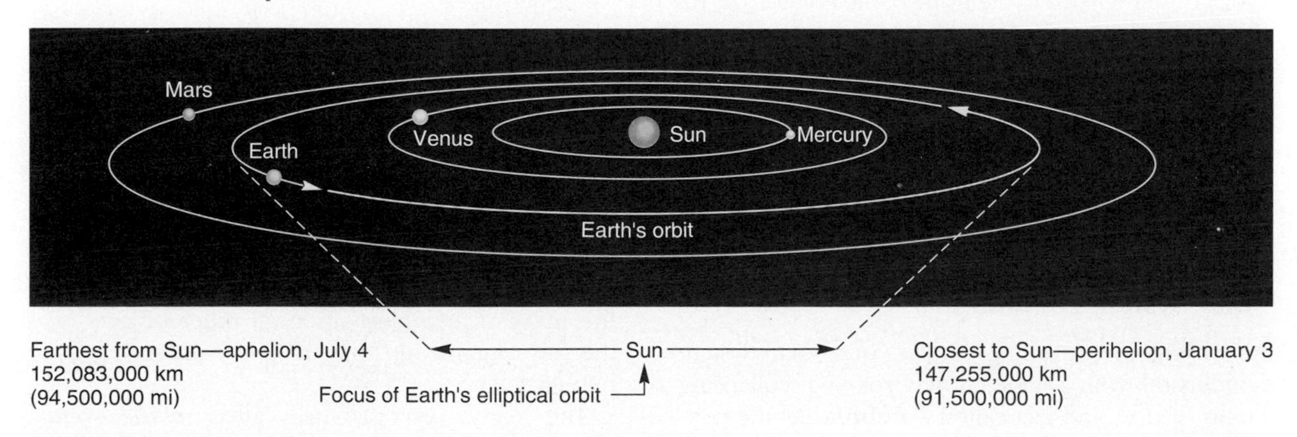

	Average Distance from Sun, millions of km (mi)	Diameter, km (mi)	Tilt of Axis, in degrees	Revolution, in Earth years	Rotation, in hr, min, or Earth days	Natural Satellites
Mercury	58 (36)	4,878 (3,032)	0 (?)	0.24	59 d	0
Venus	108 (67)	12,100 (7,520)	3	0.62	243 d	0
Earth	150 (93)	12,756 (7,926)	23.5	1.00	1 d	1
Mars	228 (142)	6,796 (4,225)	25	1.88	1 d	2
Jupiter	778 (483)	142,800 (88,730)	3.1	11.86	9 h 50 m	16+
Saturn	1472 (913)	120,660 (74,975)	26.7	29.46	10 h 39 m	20+
Uranus	2870 (1783)	51,400 (31,940)	82	84.01	17 h 18 m	5
Neptune	4486 (2787)	50,950 (31,660)	28.8	164.97	≅ 18 h	2
Pluto	5900 (3666)	3,500 (2,170)	unknown	248.40	6.4 d	1

FIGURE 2-3
A comparison of planets in our Solar System. All of the planets except Pluto have orbits closely aligned to the plane of the ecliptic. The size of each planet appears approximately to scale.

colder and warmer times respectively. Figure 2-3 compares Earth's orbital distance and other information with that for the other planets in the Solar System.

Solar System Formation

According to prevailing theory, our Solar System condensed from a large, slowly rotating, collapsing cloud of dust and gas called a **nebula**. As the nebular cloud organized and flattened into a disk shape, the early proto-Sun grew in mass at the center, drawing more matter to it. Small accretion (accumulation) eddies swirled at varying distances from the center of the solar nebula; these were the protoplanets. **Gravity**, the natural force exerted by the mass of an object upon all other objects, was the key organizing force in this condensing solar nebula.

The early protoplanets, called *planetesimals,*

were orbiting at approximately the same distances from the Sun as the planets are today. The beginnings of the Sun and its Solar System are estimated to have occurred more than 4.6 billion years ago, and the Sun finally achieved its present brightness only after a long period of development. This same process—suns condensing from nebular clouds with planetesimals forming in orbits around their central masses (the **planetesimal hypothesis**, or dust-cloud hypothesis)—is actually being observed in other parts of the galaxy. The planets appear to accrete from dust, gases, and icy comets that are drawn into collision and coalescence.

The development of such hypotheses and theories is an exercise of the **scientific method**, a methodology important to physical geography research. Focus Study 2-1 explains this essential process of science.

Earth's Development

The development of Earth is closely related to the growth of the Sun, and the long-term future of Earth likewise is tied closely to the Sun's life cycle. The evolution of Earth's atmosphere and surface, the formation of free oxygen gas, and the development of the biosphere all represent complex interactions that were in operation from the beginning of Earth's environment. Processes set into motion as the solar nebula condensed and the planetesimals congealed proceeded at rates imperceptible on a scale of human events.

Earth's Past Atmospheres. A principal component of Earth's history is the evolution of its modern atmosphere. This evolution occurred in four broad stages that blended, each into the next. The constituents of Earth's **primordial atmosphere** were derived from the original solar nebula. This atmosphere and the second stage—**evolutionary atmosphere**—are thought to have persisted for relatively short periods. The third and fourth stages—the **living atmosphere** and the **modern atmosphere**—have existed over much greater time spans. Table 2-1 summarizes present scientific views regarding the time, duration, composition, and dominant features of all four atmospheres. The times and durations listed in the table are estimates that represent long, gradual periods of transition.

A long process of chemical evolution was necessary as a forerunner to biological evolution, beneath the shield of newly collected surface waters. In still pools at least 10 m (33 ft) deep, water provided protection from the high radiation levels that existed in the environment at that time (both solar and terrestrial in origin). What a fascinating ecology was involved in the development of the atmosphere and hydrosphere, with organisms modifying their environment and in turn being modified by it! Evidence suggests that primitive cells called purple sulfur bacteria were functioning 3.6 billion years ago, producing organic materials from inorganic elements in the nonoxygen environment of the time. That process is known as **chemosynthesis**, in which an organism uses chemical energy to synthesize organic compounds. Similar chemosynthetic activity is observed today in certain areas of the cold, dark environment on the ocean floor. Vents of hot, mineral-rich water sustain bacterial activity that provide the basis of simple food chains through chemosynthesis.

Approximately 3.3 billion years ago, the first organism to perform *photosynthesis* evolved in Earth's shallow waters. Photosynthesis releases oxygen into the air, so this marked the beginning of the third distinct atmosphere, the living atmosphere. Cyanobacteria (blue-green algae) began this production of oxygen, but oxygen did not reach levels comparable to today until about 500 million years ago. The modern atmosphere defined here is the setting for Chapters 3 through 10 of this text.

SOLAR ENERGY: FROM SUN TO EARTH

The Sun captured about 99.9% of the matter (dust and gas) from the original nebula. The remaining 0.1% gave rise to all the planets, their satellites, asteroids, comets, and debris. Clearly, the dominant object in our region of space is the Sun. In the entire Solar System, it is the only object having the enormous mass needed to create the internal conditions of temperature and pressure required to produce significant energy.

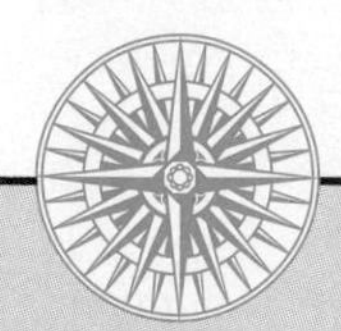

Focus Study 2-1 The Scientific Method

The term *scientific method* may have an aura of complexity that it should not. The scientific method is simply the application of common sense in an organized and objective manner. A scientist conducts observations, makes generalizations, formulates hypotheses, performs tests and experiments, and develops theories that such formulations demand. Sir Isaac Newton (1642–1727) developed this method of discovering the patterns of nature, although the term scientific method was applied later.

The scientific method begins with a specific question to be answered by scientific inquiry. Scientists who study the physical environment turn to nature for clues that they can observe and measure empirically. Then, observations and data are analyzed to identify coherent patterns that may be present. If patterns are discovered, the researcher may formulate a *hypothesis*—a formal generalization—deduced from a specific group of observations (e.g., planetesimal hypothesis, nuclear-winter hypothesis, moisture-benefits-from-hurricanes hypothesis). Further observations are related back to the general principles established by the hypothesis. The fact that a hypothesis can be made about things not yet observed produces a useful flexibility within the scientific method. Finally, if data gathered are found to support the hypothesis, and if predictions made according to it prove accurate, the hypothesis may be elevated to the status of a theory.

A *theory* is constructed on the basis of several hypotheses that have been extensively tested. It represents a truly broad general principle (e.g., theory of relativity, theory of evolution, atomic theory, Big Bang theory, or the theory of plate tectonics discussed in Chapter 11). A theory is a powerful device with which to understand both the orderliness and chaos in nature. Predictions based on a theory can be made about things not yet known, the effects of which can be tested and verified or disproved through tangible evidence. The ultimate value is the continued observation, testing, understanding, and pursuit of knowledge that the theory stimulates.

Important to consider is that pure science does not make value judgments. Instead, pure science provides people and their institutions with information on which to base their own value judgments. These social and political judgments about the applications of science are increasingly critical as Earth's natural systems respond to the impact of modern civilization.

Solar Operation and Fusion

The solar mass produces tremendous pressure and high temperatures deep in its dense interior. Under these conditions, pairs of hydrogen nuclei, the lightest of all the natural elements, are forced to fuse together. This process of forcibly joining positively charged hydrogen nuclei is called **fusion**. In the fusion reaction, hydrogen nuclei form helium, the second-lightest element in nature, and liberate enormous quantities of energy. During each second of operation the Sun consumes 657 million tons of hydrogen, converting it into 652.5 million tons of helium. The difference of 4.5 million tons is the quantity that is converted directly to energy—literally, disappearing solar mass.

A sunny day can seem so peaceful, certainly belying the violence proceeding on the Sun. Before the lunar voyages of the late 1960s, there was serious scientific concern that humans traveling above the protective layers of the atmosphere would be killed by solar radiation. Of course, this proved not to be the case. The Sun's principal outputs consist of the solar wind and portions of the electromagnetic spectrum of radiant energy. Let us trace each of these emissions across space to Earth.

TABLE 2-1
Summary of Earth's Past Atmospheres

Name	Approximate Duration (billions of years ago)	Composition (slow transitionary phases)	Probable Dominant Features
Primordial atmosphere	4.6–4.0	Water (H_2O), hydrogen cyanide (HCN), ammonia (NH_3), methane (CH_4), sulfur, iodine, bromine, chlorine	Character derived from the nebula Lighter gases of hydrogen and helium escaping to space. An unstable, hot surface with no liquid water collecting
Evolutionary atmosphere	4.0–3.3	At 4.0 billion years ago: H_2O, CO_2, N_2, sulfurous fumes, hydrocarbons, little or no free O_2	Terrigenic origins (outgassing from Earth) Surface water accumulation Earth thought to have been shrouded in clouds Anaerobic (nonoxygen) environment Chemosynthetic bacteria (at 3.6 billion years ago)
Living atmosphere	3.3–0.6	At 3.0 billion years ago: CO_2, H_2O vapor, N_2, $<1\%$ O_2	Continued outgassing First photosynthesis in cyanobacteria (at 3.3 billion years ago) Heavy global rains, ocean basins filling Slow evolution of gaseous constituents toward the modern atmosphere; increasing O_2 levels
Modern atmosphere	0.6–present	Today: 78% N_2, 21% O_2, 0.9% argon, 0.036% CO_2, trace gases	Gradual development to the modern atmosphere Abundance of life Moderately fluctuating climate Beginning of anthropogenic atmosphere (human impact)

Solar Wind

The Sun constantly emits clouds of ionized (electrically charged) gases. Streams of these charged particles (principally electrons and protons) travel in all directions from the Sun's surface at much less than the speed of light—at about 50 million km (31 million mi) a day—taking approximately 3 days to reach Earth. The term **solar wind** was first applied to this phenomenon in 1958. Solar wind may extend from the Sun as much as 50 times the distance from the Sun to Earth, a distance beyond the planets of our Solar System.

Sunspots and Solar Activity. The most conspicuous features on the Sun are large magnetic disturbances called **sunspots**, which are caused by magnetic storms that indicate unusual solar activi-

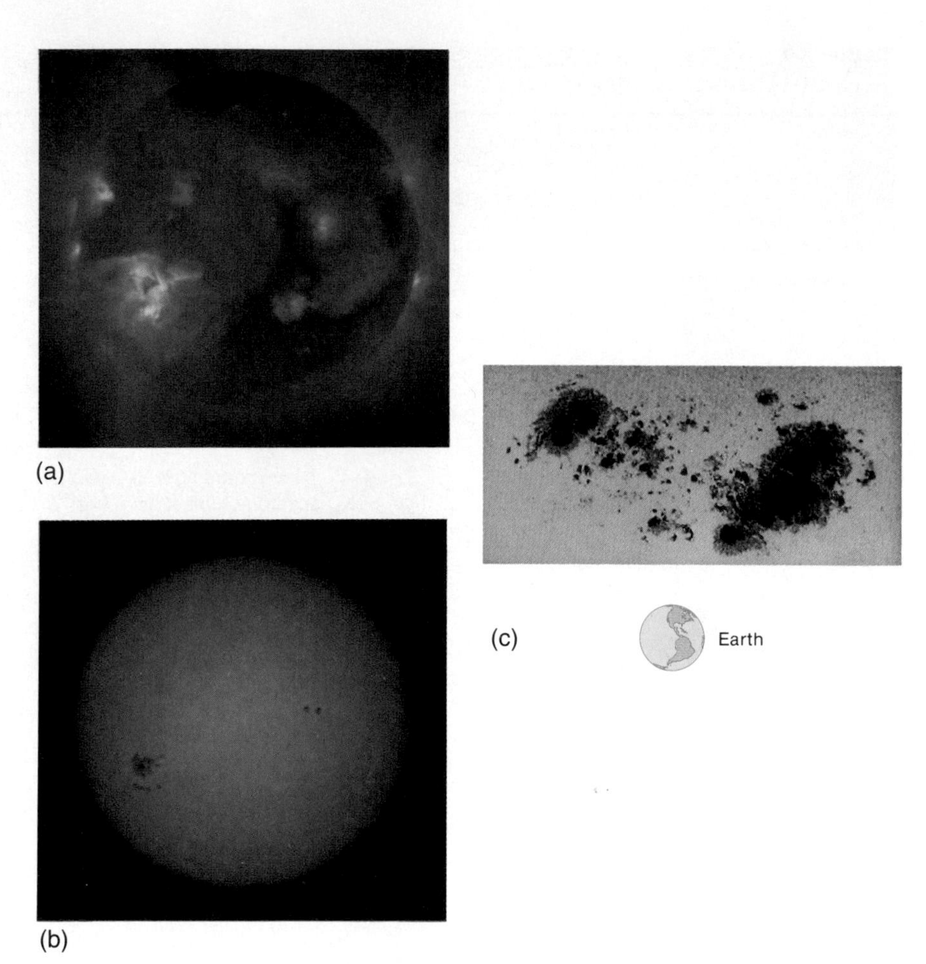

FIGURE 2-4
The Sun in X-ray (a) and visible light (b) wavelengths on April 21, 1992, as imaged by the *Yohkoh* satellite (launched August 30, 1991). Sunspots appear as visible dark patches across the Sun, and also as areas of intense X-ray activity. Earth is shown for scale (c). *Yohkoh* is a joint effort of the National Astronomical Observatory of Japan, the University of Tokyo, Lockheed Palo Alto Research Laboratory, and NASA. [Images courtesy of Dr. Kieth T. Strong of Lockheed and Dr. Yutaka Uchida of the *Yohkoh* Science Committee and the University of Tokyo.]

ty. Individual sunspots may range in diameter from 10,000 to 50,000 km (6200 to 31,100 mi), with some growing as large as 160,000 km (100,000 mi), which is more than 12 times Earth's diameter (Figure 2-4). Flares, prominences, and outbreaks of charged material are associated with these surface disturbances; this material surges into space as the solar wind. Although sunspots have been described and recorded in some detail for almost 400 years, a full explanation of their occurrence is still evolving.

Sunspot Cycles. The solar wind grows stronger during periods of increased sunspot activity and weaker during times of lowered activity. A regular cycle exists for sunspot occurrences, averaging 11 years from maximum to maximum; however, the cycle may vary from 7 to 17 years. In recent cycles, a solar minimum occurred in 1976 and a solar maximum took place during 1979, with over 100 sunspots visible. Another minimum was reached in 1986, and an extremely active solar maximum followed in 1990 with over 200 sunspots visible at some time during the year. In fact, the 1990–1991 maximum was the most intense ever observed. This happened 11 years after the previous maximum, maintaining the average.

Earth's Magnetosphere

Earth's outer defense against the solar wind is the **magnetosphere,** which is a magnetic field surrounding Earth, generated by dynamolike motions within our planet. The magnetosphere affects the solar wind, deflecting its path (Figure 2-5). When it was first discovered and named in 1958, scientists believed the magnetosphere to be doughnut-shaped and symmetrical. Instead, sensitive instru-

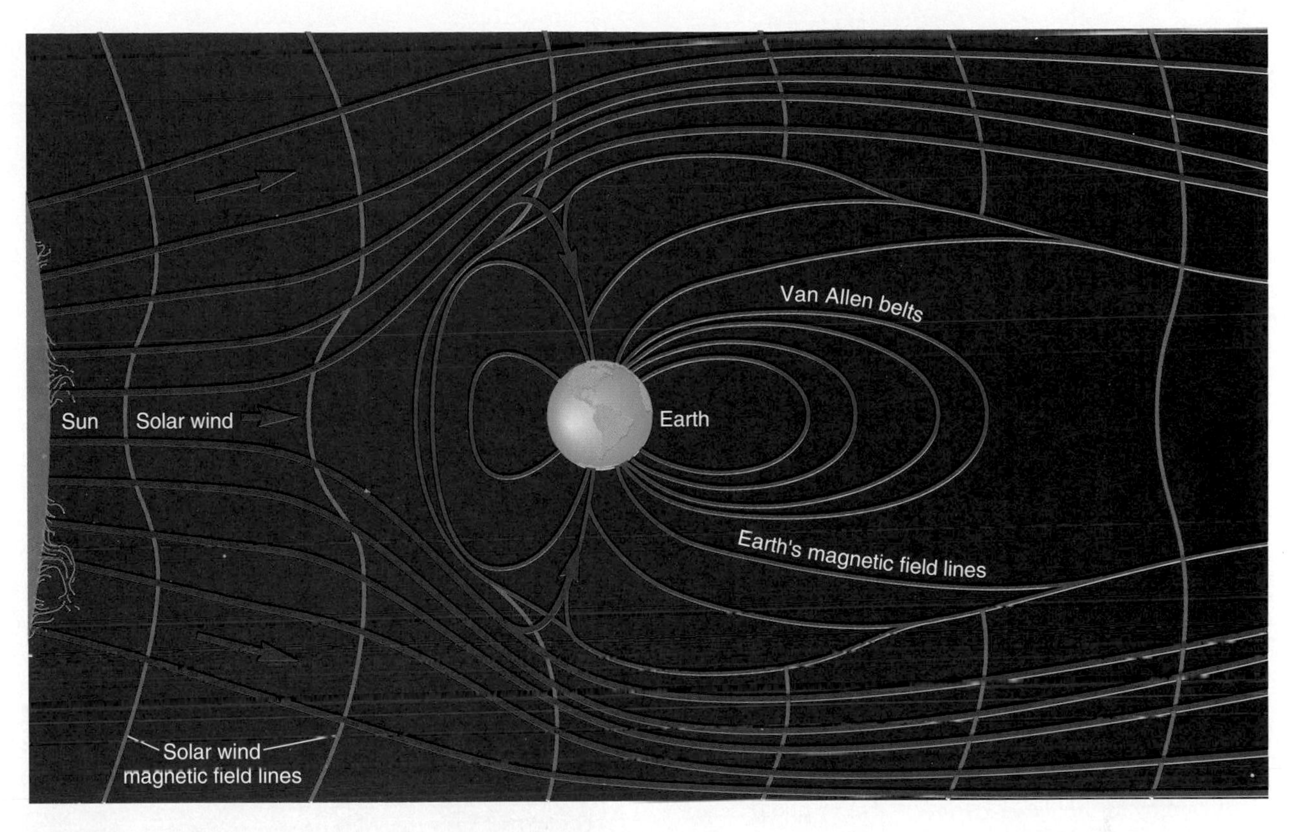

FIGURE 2-5
Earth's magnetosphere and solar wind interaction. Note how the streams of charged particles are swept toward the poles by Earth's magnetic field.

ments aboard space probes found it to be teardrop-shaped.

As the solar wind approaches Earth, the streams of charged particles are deflected by the magnetosphere and course along the magnetic-field lines. As shown in Figure 2-5, the extreme northern and southern polar regions of the upper atmosphere are the points of entry for the solar wind stream. However, the solar wind does not completely penetrate the atmosphere and therefore does not reach Earth's protected surface. Thus, research on this phenomenon must be conducted in space. The *Apollo XI* astronauts deployed a solar wind experiment on the lunar surface on July 20, 1969 (Figure 2-6). A piece of foil was exposed to the solar wind while the astronauts worked on the Moon. When returned to Earth, the exposed foil was examined for particle impacts that confirmed the presence and character of the solar wind.

Solar Wind Effects

The interaction of the solar wind and the upper layers of Earth's atmosphere produces some remarkable phenomena relevant to physical geography: the auroras that occur toward both poles, disruption of certain radio broadcasts and some satellite transmissions, overloads on Earth-based electrical systems, and possible effects on weather patterns.

Auroras. The solar wind creates the auroral phenomenon in the upper atmosphere, 80–480 km (50–300 mi) above Earth's surface. In this portion

FIGURE 2-6
Without a protective atmosphere, the lunar surface allows direct solar wind and electromagnetic radiation to reach the surface. Here a solar wind experiment is being deployed by an *Apollo XI* astronaut. [NASA photo.]

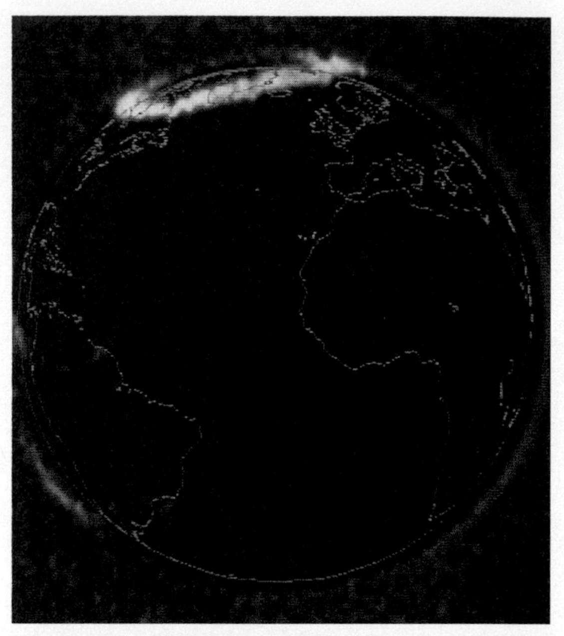

(a)

(b)

FIGURE 2-7
Satellite image of auroral halo over Earth's North Pole (a). Surface view of the aurora borealis in the night sky over Alaska (b). [(a) *Dynamics Explorer 1* satellite image courtesy of L. A. Frank and J. D. Craven, The University of Iowa; (b) aurora photo by Mark Kelly/Allstock.]

of the atmosphere, certain atoms and molecules are ionized, or electrically charged, by interactions with the solar wind. This absorbed energy then is radiated as light energy of varying colors, depending on which atoms and molecules are stimulated. These lighting effects are the **auroras**: *aurora borealis* (northern lights) and *aurora australis* (southern lights).

The auroras generally are visible poleward of 65° latitude when the solar wind is active, as shown by the dramatic polar-orbiting satellite image in Figure 2-7a—the auroral halo crowning Earth. Ground-based observations are equally dramatic, as folded sheets of green, yellow, blue, and red light undulate across the skies of high latitudes as shown in Figure 2-7b. However, at times of intense solar activity, these glowing colored lights in the night sky can be seen far into the middle latitudes. In March 1989, following an enormous solar outbreak, the aurora borealis was visible as far south as Key West, Florida, and Kingston, Jamaica, below 20° N latitude.

Weather Effects. Another effect of the solar wind in the atmosphere involves possible influence on weather. Much scientific inquiry is in progress regarding the relationship between the Sun and Earth's weather and climate cycles: Why do wetter periods in some areas of the midlatitudes tend to coincide with every other solar maximum? Why do droughts often occur near the time of every other solar minimum? For example, sunspot cycles during the 250 years from 1740 to 1990 coincide with periods of wetness and drought, as estimated by an analysis of tree growth rings for that period throughout the western United States. These variations in weather tend to occur within two or three years after the solar maximum or minimum.

Although the correlation is interesting—and controversial, and still undergoing research—the specific causative link in the atmosphere remains to be discovered. Speculation as to the weather connection is centered on heating and wind changes in the upper atmosphere that may be attributable to a conversion of energy from the solar wind. Other explanations also have been proposed.

A remarkable failure in current planning worldwide is the lack of attention given these evident cyclical patterns of drought and wetness, regardless of their cause. If such patterns were prepared for, it would be possible to accommodate their consequences. For instance, cyclical drought could be offset through implementation of widespread water conservation and efficiency practices. Wet spells might require strengthening of levees along river channels and more careful reservoir management. As knowledge of the solar wind/weather phenomenon increases, the ability to forecast the consequences of solar magnetic storms for Earth systems may become feasible.

Electromagnetic Spectrum of Radiant Energy

The key solar input to life is electromagnetic energy. Solar radiation occupies a portion of the **electromagnetic spectrum** of radiant energy. This radiant energy travels at the speed of light, transmitting energy to Earth. The total spectrum of this radiant energy is made up of wavelengths, a **wavelength** being the distance between corresponding points on any two successive waves (Figure 2-8). The number of waves passing a fixed point in one second is the *frequency.* The Sun emits radiant energy composed of 8% ultraviolet, X-ray, and gamma ray wavelengths; 47% visible light wavelengths; and 45% infrared wavelengths. A portion of the electromagnetic spectrum is illustrated in Figure 2-9. Examine the spectrum and note the wavelengths at which various phenomena occur.

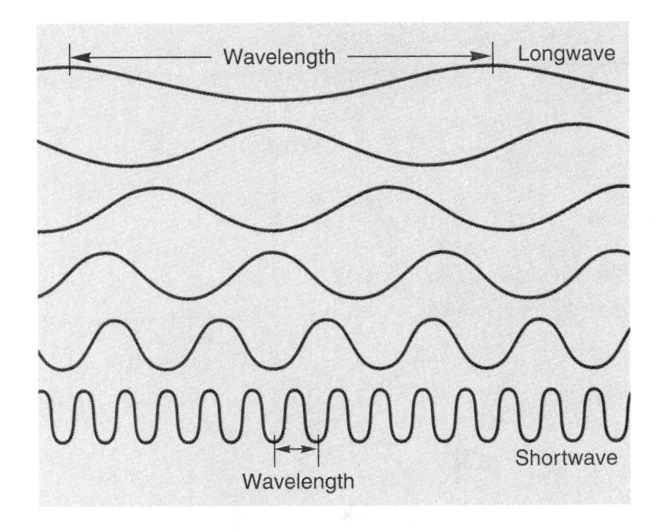

FIGURE 2-8
Wavelength and frequency are used to describe the electromagnetic spectrum. These are two ways of describing the same phenomenon—electromagnetic wave motion. Shorter wavelengths are higher in frequency, whereas longer wavelengths are lower in frequency.

An important physical law applicable to all bodies is that they radiate energy in wavelengths related to their individual surface temperatures: the hotter the body, the shorter the wavelengths emitted. Figure 2-10 shows that the hot Sun radiates energy primarily in shorter wavelengths, concentrated around 0.4–0.5 micrometer (µm or micron). The Sun's surface temperature is about 6000°C (11,000°F), and its emission curve, as shown in the figure, is similar to that predicted for an idealized 6000°C surface or *black body radiator.* The Sun

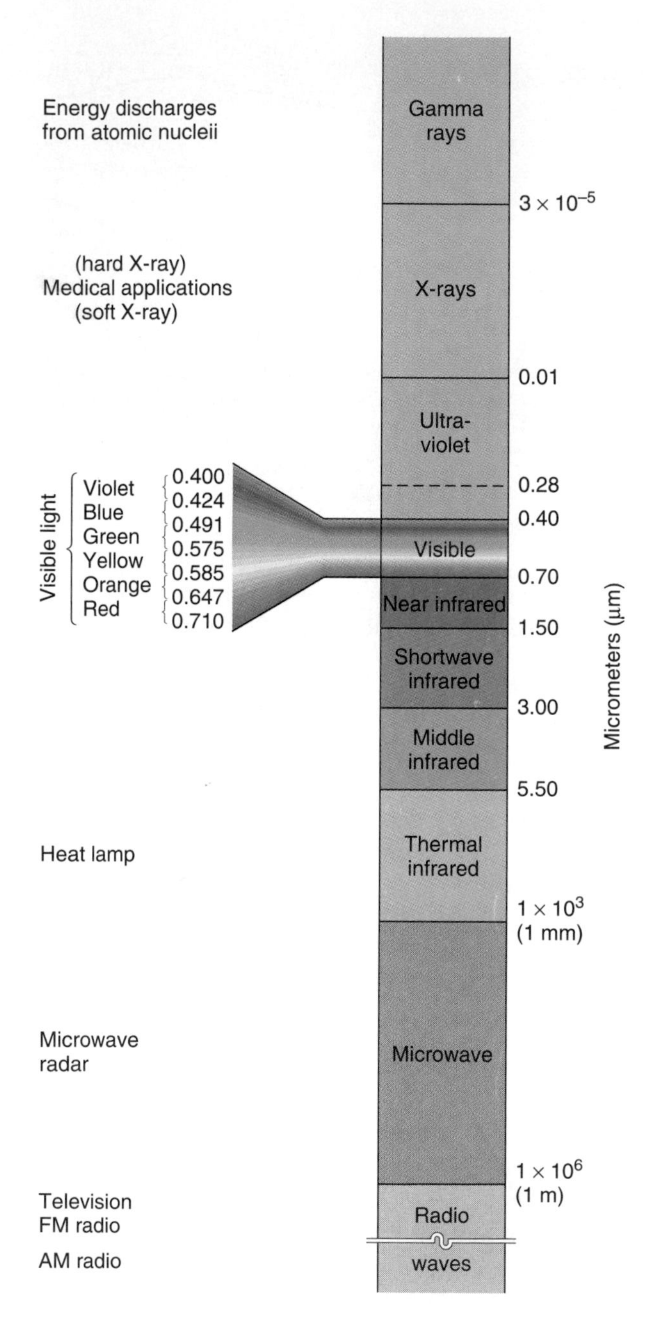

FIGURE 2-9
A portion of the electromagnetic spectrum of radiant energy.

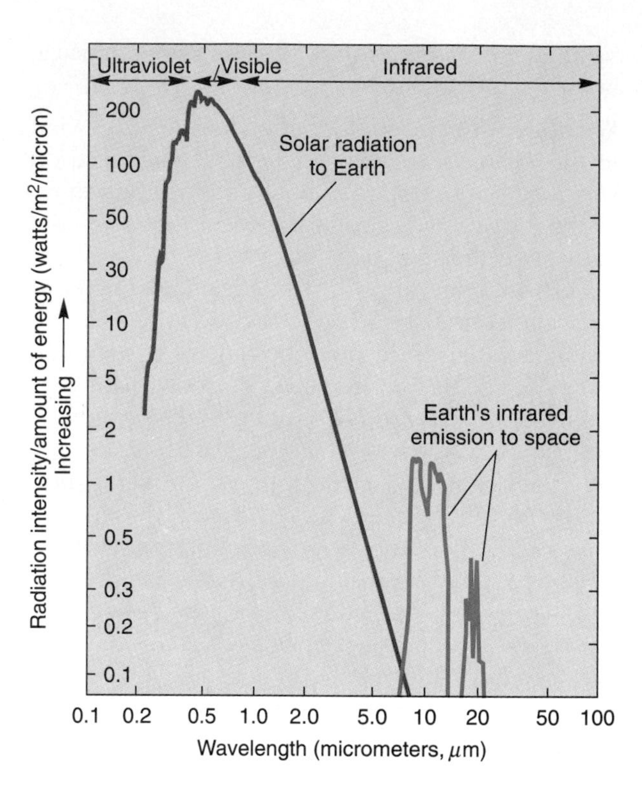

FIGURE 2-10
Solar and terrestrial energy distribution by wavelength.

emits a much greater amount of energy per unit area of its surface than does a similar area of Earth's environment.

Earth, on the other hand, demonstrates that the cooler the radiating body, the longer are the wavelengths emitted. Figure 2-10 shows that the radiation emitted by Earth occurs in longer wavelengths, centered around 10 micrometers and entirely within the infrared portion of the spectrum. To summarize, the solar spectrum is *shortwave radiation* because it peaks in the visible wavelengths, and Earth's spectrum is *longwave radiation* because its outgoing energy is concentrated in infrared wavelengths. In Chapter 4, we will see that Earth, clouds, sky, ground, and all things that are terrestrial are cool-body radiators.

ENERGY AT THE TOP OF THE ATMOSPHERE

The region at the top of the atmosphere, approximately 480 km (300 mi) above Earth's surface, is termed the **thermopause**. It is the outer boundary of Earth's energy system and provides a useful point at which to assess the arriving solar radiation before it is diminished by passage through the atmosphere.

Intercepted Energy

Earth's distance from the Sun results in Earth's interception of only one two-billionth of the Sun's total energy output. Nevertheless, this tiny fraction of the Sun's overall output represents an enormous amount of energy input into Earth's systems. Intercepted solar radiation is called **insolation**, a term specifically applied to radiation arriving at Earth's surface and atmosphere. Insolation at the top of the atmosphere is expressed as the solar constant.

Solar Constant. Knowing the amount of insolation intercepted by Earth is important to climatologists and other scientists. The **solar constant** is the average value of insolation received at the thermopause (on a plane surface perpendicular to the Sun's rays) when Earth is at its average distance from the Sun. That value of the solar constant is 1372 watts per square meter (W/m^2).* As we follow insolation through the atmosphere to Earth's surface (Chapters 3 and 4) we see the value of the solar constant reduced by half or more through reflection and absorption of shortwave radiation.

The constancy of this value over time is important, for small variations of even 0.5 or 1.0% could prove dramatic for Earth's energy system. Paleontologists, who deal with the life of past geologic periods, estimate that solar energy levels have varied slightly, perhaps less than 10%, over the past several billion years. The Solar Maximum Mission (1980–1989), dubbed Solar Max, was a satellite launched to measure total solar output. Solar Max found average variations of ±0.04% in the solar constant, with the largest changes ranging up to 0.3%, essentially in correlation with sunspot activity.

*A *watt* is equal to one joule (a unit of energy) per second and is the standard unit of power in the SI-metric system. (See the inside back cover of this text for more information on measurement conversions.) In nonmetric calorie heat units, the solar constant is expressed as approximately 2 calories per cm^2 per minute, or 2 *langleys* per minute (a langley being 1 cal per cm^2). A *calorie* is the amount of energy required to raise the temperature of one gram of water (at 15°C) one degree Celsius and is equal to 4.184 joules.

The *Nimbus-7* satellite was launched in late 1978 and placed at an altitude of 955 km (595 mi). All instruments on board are still operating, including the Earth Radiation Budget package (ERB). The ERB determined that solar irradiance, or luminous brightness of solar radiation, is directly correlated to the sunspot cycle discussed earlier. During the solar maximums of 1979 and 1991, the solar constant exceeded 1374 W/m^2; the 1986 minimum produced a constant of 1371 W/m^2.

Uneven Distribution of Insolation. Earth's curved surface presents a continually varied angle to the incoming parallel rays of insolation (Figure 2-11). The only point receiving insolation perpendicular to the surface (from directly overhead) is

FIGURE 2-11
Solar insolation angles determine the concentration of energy receipts by latitude. Lower latitudes receive more concentrated energy. Higher latitudes receive slanting rays and more diffuse energy.

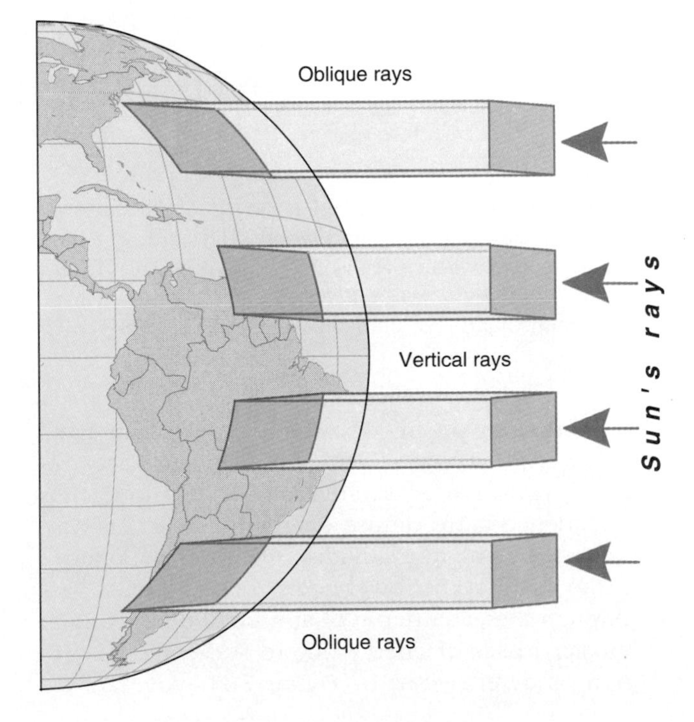

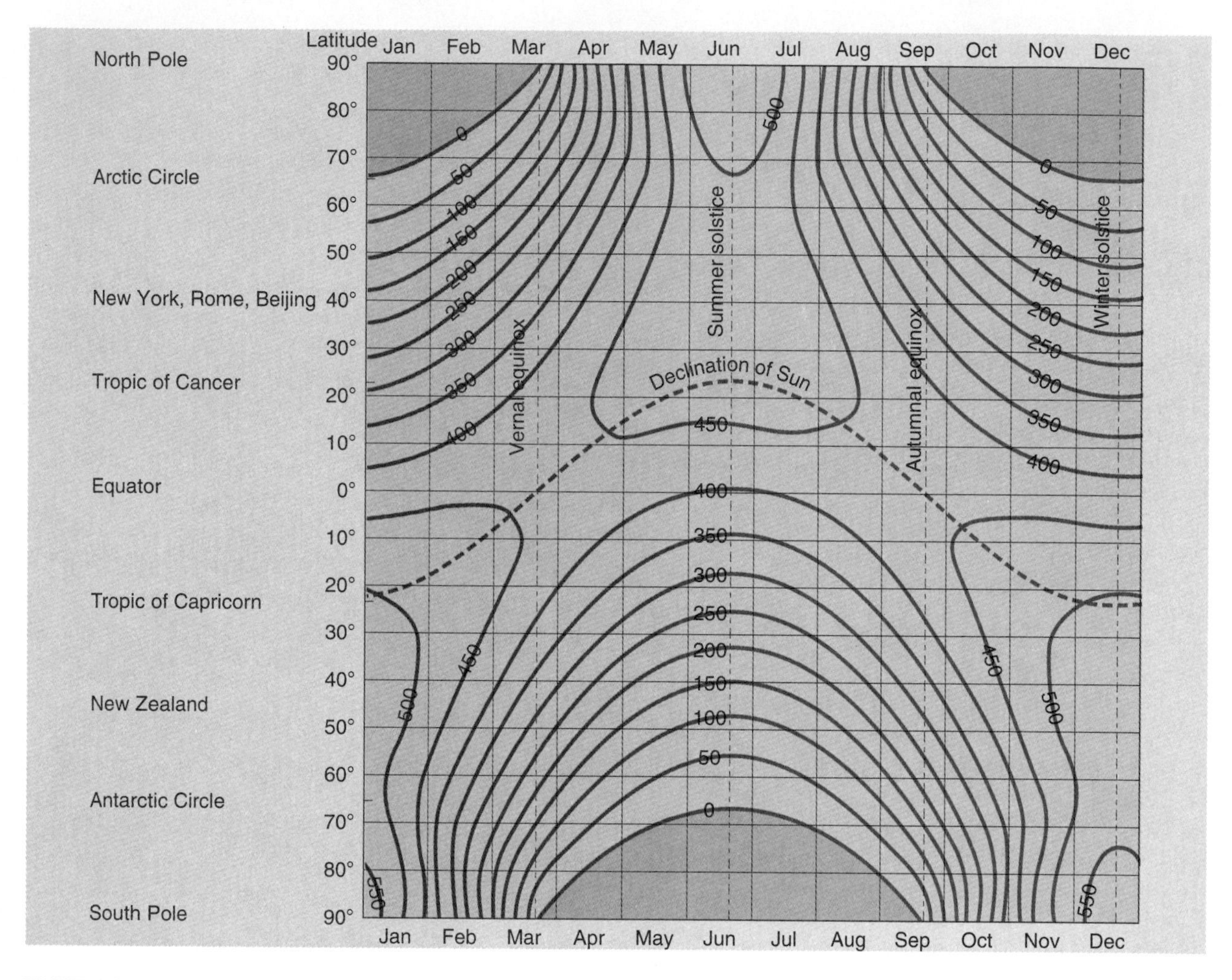

FIGURE 2-12
Total daily insolation received at the top of the atmosphere, charted in watts per square meter by latitude and month. (1 watt/m^2 = 2.064 cal/cm^2/day.) [Reproduced by permission of the Smithsonian Institution Press from *Smithsonian Miscellaneous Collections: Smithsonian Meteorological Tables*, vol. 114, 6th Edition. Robert List, ed. Smithsonian Institution, Washington, DC, 1984, p. 419, Table 134.]

the **subsolar point**, which is intensely illuminated and occurs at lower latitudes. All other places receive insolation at an angle—obliquely—and thus experience more diffuse energy; this effect is pronounced at higher latitudes. Of lesser importance is the fact that the lower-angle solar rays must pass through a greater depth of atmosphere, resulting in greater losses of energy due to scattering, absorption, and reflection. In Figure 2-11, you can see that at lower latitudes the smaller the area covered by the same amount of insolation, the more concentrated the energy receipt. During a year's time, the thermopause above the equatorial region receives 2.5 times more insolation than the thermopause above the poles.

The latitudinal variation in the angle of solar rays results in an uneven global distribution of insolation. Figure 2-12 illustrates the daily variation

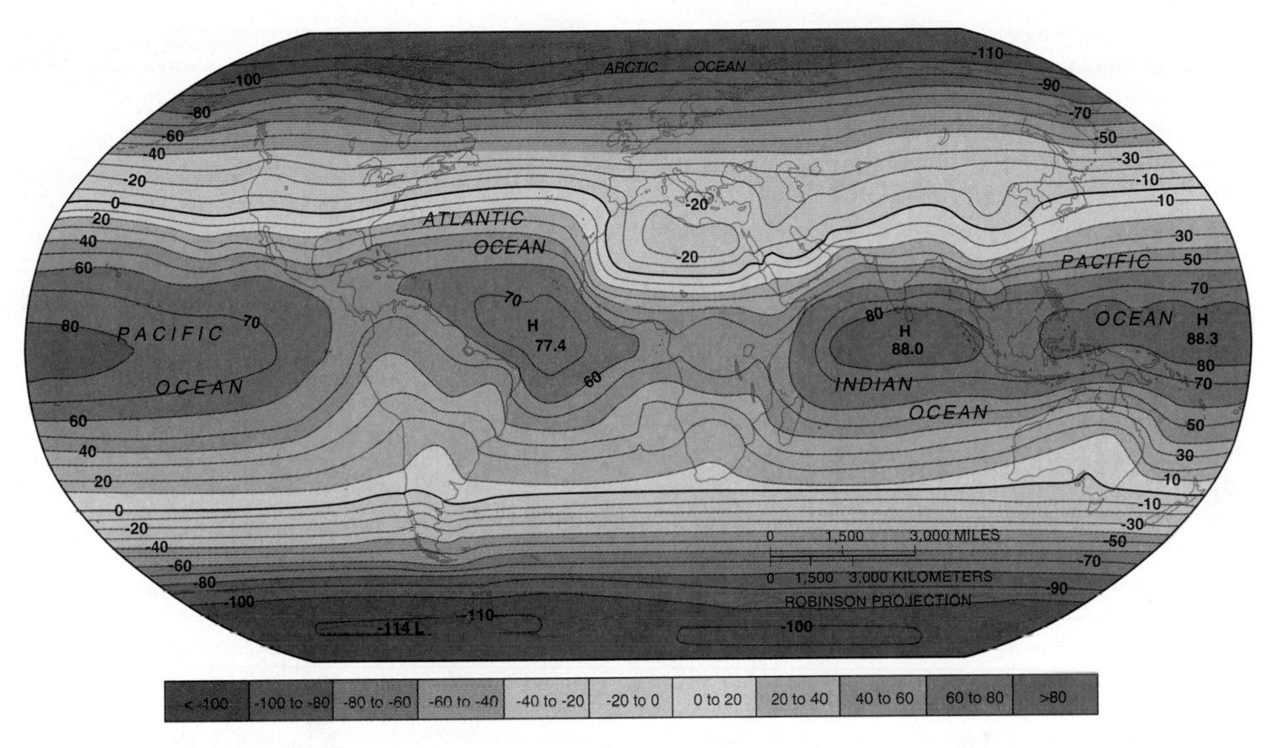

FIGURE 2-13
Diurnally averaged net radiation fluxes for the 9-year period (1979 to 1987) measured at the top of the atmosphere by the Earth Radiation Budget (ERB) instrument aboard the *Nimbus-7* satellite. Units are given in W/m^2. [Map courtesy of Dr. H. Lee Kyle, Goddard Space Flight Center, NASA.]

in energy at the top of the atmosphere for various latitudes in W/m^2 per day. The chart shows a decrease in insolation from the equatorial regions northward and southward toward the poles. However, in June, the North Pole receives more than 500 W/m^2 per day, which is more than is received at 40° N latitude or at the equator. Such high values result from the 24-hour duration of exposure, as compared with only 15 hours of daylight at 40° N latitude and 12 hours at the equator. Even with the summer noon Sun low in the sky at the poles, a day twice as long as at the equator yields only about 100 W/m^2 difference. In December, the pattern reverses. Note that the top of the atmosphere at the South Pole receives even more insolation than the North Pole does in June (over 550 W/m^2). This is a function of Earth's closer proximity to the Sun at perihelion.

Along the equator, two maximum periods of approximately 430 W/m^2 occur at the equinoxes when the subsolar point is at the equator. Find your latitude and follow across the months to determine the seasonal variation of insolation where you live.

The Earth Radiation Budget (ERB) instrument aboard *Nimbus-7* measures shortwave and longwave *fluxes*, or flows of energy per unit area per unit of time, at the top of the atmosphere. A pattern emerges from daily average radiation measurements made between 1979 and 1987. *Net radiation* is the balance between shortwave and longwave radiation fluxes and is shown on the map in Figure 2-13. First note the latitudinal energy imbalance in net radiation: positive values in lower latitudes and negative values toward the poles. In middle and high latitudes, approximately poleward of 36° north and south latitudes, net radiation is negative. Earth's climate system loses more heat to space than it gains from the Sun as measured at the top of the atmosphere for these higher latitudes. In the lower atmosphere, these polar deficits are offset by flows of heat from tropical surpluses, as we shall see in Chapters 4 and 6. The largest net radiation values are above the tropical oceans along a narrow equatorial zone, averaging 80 W/m^2. Net radiation minimums are lowest over Antarctica.

Of interest is the -20 W/m^2 area over the Sahara desert region, where usually clear skies—which permit large longwave radiation losses from Earth's surface—and reflective surfaces work together to reduce net radiation values at the thermopause. Clouds in the lower atmosphere also affect radiation patterns at the top of the atmosphere by reflecting greater amounts of shortwave energy to space.

This overall imbalance of insolation and heating from equator to the poles is the basis for major circulations within the lower atmosphere and in the ocean. The atmosphere is like a giant heat engine, driven by differences in insolation and heating from place to place. As you go about your daily activities, use the dynamic natural systems as reminders of the constant flow of solar energy.

THE SEASONS

The periodic rhythm of warmth and cold, rain and drought, dawn and daylight, twilight and night have fascinated humans for centuries. In fact, many ancient societies demonstrated a greater awareness of seasonal change than modern peoples and formally commemorated natural energy rhythms with festivals and monuments. At Stonehenge, rocks weighing 25 metric tons (28 tons) were hauled 480 km (300 mi) and placed in patterns that evidently mark seasonal changes—specifically, the sunrise on or about June 21 at the summer solstice (Figure 2-14). Other seasonal events and eclipses of the Sun and the Moon apparently are predicted by this 3500-year-old calendar monument. Such ancient seasonal monuments and markings occur worldwide, including thousands of sites in North America.

FIGURE 2-14
Stonehenge, Salisbury Plain in Wiltshire, England. [Photo by Robert Llewellyn.]

Seasonality

Seasonality refers to both the seasonal variation of the Sun's rays above the horizon and changing daylengths during the year. Seasonal variations are a response to changes in the Sun's **altitude,** or the angular difference between the horizon and the Sun. The Sun is directly overhead (at zenith or 90° altitude) only at the subsolar point, with all other surface points receiving a lower angle of more diffuse insolation. The Sun's **declination** is the angular distance from the equator to the latitude of the subsolar point. Declination annually migrates through 47° of latitude between the *Tropic of Cancer* at 23.5° N and the *Tropic of Capricorn* at 23.5° S latitude.

Seasonality also means a changing duration of exposure, or **daylength**, which varies during the year depending on latitude. The equator always receives equal hours of day and night, whereas people living along 40° N or S latitude experience about 6 hours' difference in daylight hours between winter and summer. Those at 50° N or S latitude experience almost 8 hours of annual daylength variation. At the polar extremes, the range extends from a six-month period of no insolation to a six-month period of continuous 24-hour days (note the illumination of the North Pole [June] and South Pole [December] in Figure 2-18). Can you determine in what month the Earth photo on the back cover was taken?

TABLE 2-2
Five Reasons for Seasons

Factor	Description
Revolution	Orbit around the Sun; requires 365.24 days to complete at 107,280 kmph (66,660 mph)
Rotation	Earth turning on its axis; takes approximately 24 hours to complete at 1675 kmph at equator (1041 mph)
Tilt	Axis is aligned at a 23.5° angle from a perpendicular to the plane of the ecliptic (the plane of Earth's orbit)
Axis	Remains in a fixed alignment of axial parallelism, with Polaris directly overhead at the North Pole throughout the year
Sphericity	Appears as an oblate spheroid to the Sun's parallel rays; the geoid

Contributing Physical Factors

Variations in the Sun's altitude, declination, and daylength patterns at the surface are created by several contributing physical factors that operate together: Earth's revolution and rotation, its tilt and fixed-axis orientation, and its sphericity (Table 2-2). Of course, the essential ingredient is having a single source of illumination—the Sun.

Revolution. The structure of Earth's orbit and **revolution** about the Sun is shown in Figure 2-2. At an average distance from the Sun of 150 million km (93 million mi), Earth completes its annual orbit in 365.2422 days at speeds averaging 107,280 kmph (66,660 mph) in a counterclockwise direction when viewed from above Earth's North Pole (Figure 2-15). This calculation is based on a *tropical year*, measured from March to March (or the elapsed time between two crossings of the equator by the Sun). Earth's revolution determines the length of the year and therefore the duration of the seasons. The Earth-to-Sun distance is not a contributing seasonal factor even though it varies some 4.8 million km (3 million mi) during the year.

Rotation. Earth's **rotation**, or turning, is a complex motion that averages 24 hours in duration. Rotation determines daylength, produces the apparent deflection of winds and ocean currents, and produces the twice-daily action of the tides in relation to the gravitational pull of the Sun and the Moon. Earth's **axis** is an imaginary line extending through the planet from the geographic North Pole to the South Pole. When viewed from above the North Pole, Earth rotates counterclockwise around this axis; viewed from above the equator it moves west to east, or eastward. This west-to-east rotation creates the Sun's apparent daily journey from east to west, even though the Sun remains in a fixed position in the center of the Solar System. The Moon revolves around Earth and rotates counterclockwise on its axis.

Although points at all latitudes take the same 24 hours to complete one rotation, the *linear velocity* of rotation at any point on Earth's surface varies with latitude. The equator is 40,075 km (24,902 mi) in length; therefore, rotational velocity at the equator must be approximately 1675 kmph (1041 mph) to cover that distance in one day. At 60° latitude, a parallel is only half the length of the equator, or 20,038 km (12,451 mi) long, so the rotational ve-

FIGURE 2-15
Earth's revolution about the Sun and rotation on its axis, the Moon's rotation on its axis and revolution about Earth, as viewed from above Earth's North Pole.

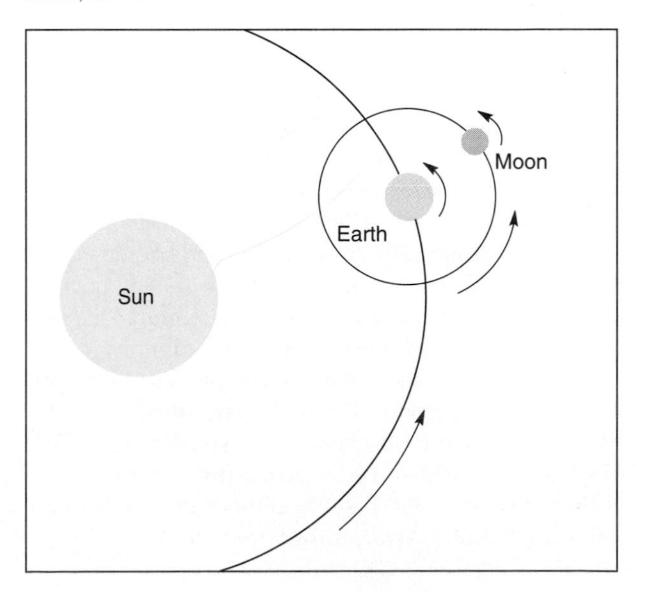

TABLE 2-3
Rotational Velocities at Selected Latitudes

Latitude	kmph (mph)	Cities at Approximate Latitudes
0°	1675 (1041)	Quito, Ecuador; Pontianak, Indonesia
30°	1452 (902)	Porto Alegre, Brazil; New Orleans
40°	1284 (798)	Valdivia, Chile; Columbus, Ohio; Beijing
50°	1078 (670)	Chibougamau, Quebec; Kiev, Ukraine
60°	838 (521)	Seward, Alaska; Oslo, Norway; Saint Petersburg, Russia
90°	0 (0)	(North Pole)

locity there is 838 kmph (521 mph). This variation in rotational velocity establishes the Coriolis force, discussed in Chapter 6. Table 2-3 lists the speed of rotation for several selected latitudes.

Earth's rotation produces the diurnal pattern of day and night. The dividing line between day and night is called the **circle of illumination**, as illustrated in Figure 2-18. Because this day-night dividing circle of illumination is a great circle that intersects the equator, which is another great circle, daylength at the equator is always evenly divided—12 hours of day and 12 hours of night. (Any two great circles on a sphere bisect one another.) Except for two days a year, on the equinoxes, all other latitudes experience uneven daylength through the seasons.

A true day varies slightly from 24 hours; but by international agreement a day is considered to be exactly 24 hours (86,400 seconds) in length. This average, called *mean solar time,* eliminates predictable rotational and orbital (revolution) variations that cause the solar day to change in length throughout the year. However, an *apparent solar day* is based on observed successive passages of the Sun over a given meridian. Any difference between this observed solar time and mean solar time is called the *equation of time.*

On successive days, if the Sun arrives overhead at a meridian *after* 12:00 noon local standard time (that is, taking an interval longer than the 24 hours of a mean solar day), the equation of time is *negative.* The Sun is described as "slow," much like an airliner arriving later than scheduled. If, on the other hand, the Sun arrives overhead *before* 12:00 noon local time on successive days (that is, taking an interval shorter than 24 hours), the equation of time is *positive.* The Sun then is described as "fast," like an airliner arriving ahead of schedule. The equation of time can range up to 16 minutes slow and 14 minutes fast at different times during the year.

The **analemma** is useful for demonstrating the positive and negative equations of time, the periods of fast Sun times (during October and November) and slow Sun times (during February and March). It is also a convenient device to track the passage of the Sun's path and declination throughout the year (Figure 2-16). Can you identify the days of fast and slow Sun on the analemma in the figure? Can you determine the Sun's declination on your birthday?

Tilt of Earth's Axis. To understand Earth's tilt, imagine Earth's elliptical orbit about the Sun as a plane, with half of the Sun and Earth above the plane and half below. (It may help to envision two spheres floating in water, where the water's surface forms a plane.) This level surface is termed the *plane of the ecliptic,* and Earth's orbital plane is slightly inclined from the Sun's equatorial plane. Now, imagine a perpendicular line passing through the plane. Earth's axis is tilted 23.5° from the perpendicular to this plane, forming a 66.5° angle from the plane itself (Figure 2-17).

The axis through Earth's two poles points just slightly off Polaris, which is appropriately called the North Star (see Figure 1-11). Hypothetically, if Earth was tilted on its side with its axis parallel to the plane of the ecliptic, we would experience a maximum variation in seasons worldwide. On the other hand, if Earth's axis was perpendicular to the plane of its orbit—that is, with no tilt—we would experience no seasonal changes, just a perpetual spring/fall season, and all latitudes would experience 12-hour days and nights.

Axial Parallelism. Throughout our annual journey around the Sun, Earth's axis maintains the same alignment relative to the plane of the ecliptic

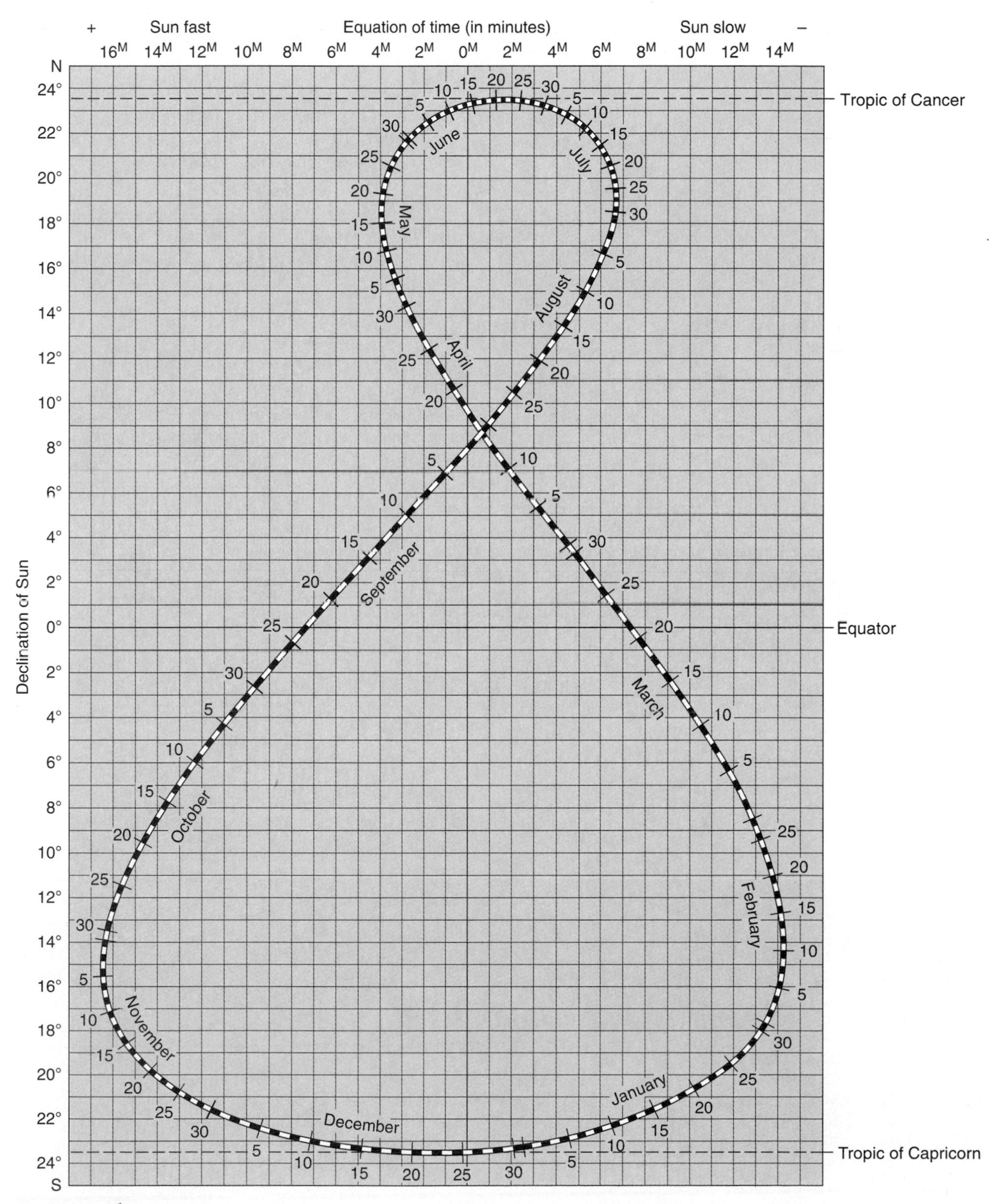

FIGURE 2-16
The analemma.

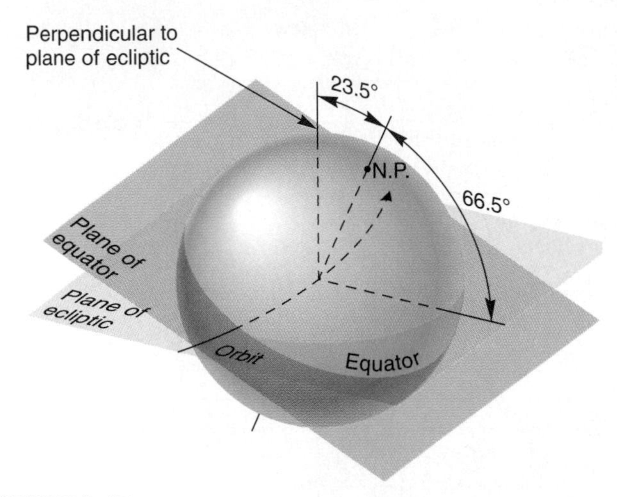

FIGURE 2-17
Earth's tilt and the plane of the ecliptic.

and to Polaris and the other stars. In other words, if we compared the axis in different months, it would always appear parallel to itself, a condition known as **axial parallelism**.

Annual March of the Seasons

The combined effect of all these physical factors is the annual march of the seasons on Earth. Daylength is the most evident way of sensing changes in season at latitudes away from the equator. **Sunrise** is that moment when the disk of the Sun first appears above the horizon, whereas **sunset** is that moment when it totally disappears. Table 2-4 lists the times of sunrise and sunset and the duration of daylength for various latitudes and seasons in the Northern Hemisphere.

The extremes of daylength occur in December and June. The times around December 21 and June 21 are termed the solstices. They mark the times of the year when the Sun's declination is at one of the two tropics, its farthest northerly or southerly position. ("Tropic" is from the Latin *tropicus,* meaning a turn or change.)

Table 2-5 presents the key seasonal anniversary dates, their names, and the Sun's declination. Compare the dates and terminology in this table to the analemma presented in Figure 2-16. During the year, places on Earth outside of the equatorial region experience a continuous but gradual shift in daylength, a few minutes each day, and the Sun's altitude increases or decreases a portion of a degree daily for any particular location. You may have noticed that these daily variations become more pronounced in spring and autumn, when the declination is changing at a faster rate.

Figure 2-18 demonstrates the annual march of the seasons and illustrates Earth's relationship to the Sun during the year. At the **winter solstice** (literally, "winter Sun stance"), or **December solstice,** on December 21 or 22, the circle of illumination excludes the North Pole region from sunlight and includes the South Pole region. The subsolar point is at 23.5° S latitude, a parallel known

TABLE 2-4
Daylength (Sunrise and Sunset) Times at Selected Latitudes (Northern Hemisphere)

	Winter Solstice (December Solstice) December 21–22			Vernal Equinox (March Equinox) March 20–21			Summer Solstice (June Solstice) June 20–21			Autumnal Equinox (September Equinox) September 22–23		
	A.M.	P.M.	Daylength	A.M.	P.M.	Daylength	A.M.	P.M.	Daylength	A.M.	P.M.	Daylength
0°	6:00	6:00	12:00	6:00	6:00	12:00	6:00	6:00	12:00	6:00	6:00	12:00
30°	6:58	5:02	10:04	6:00	6:00	12:00	5:02	6:58	13:56	6:00	6:00	12:00
40°	7:26	4:34	9:00	6:00	6:00	12:00	4:34	7:26	15:00	6:00	6:00	12:00
50°	8:05	3:55	8:00	6:00	6:00	12:00	3:55	8:05	16:00	6:00	6:00	12:00
60°	9:15	2:45	5:30	6:00	6:00	12:00	2:45	9:15	18:30	6:00	6:00	12:00
90°	No sunlight			Rising Sun			Continuous sunlight			Setting Sun		

TABLE 2-5
Annual March of the Seasons

Approximate Date	Northern Hemisphere Name	Location of the Subsolar Point
December 21–22	Winter solstice (December solstice)	23.5° S latitude (Tropic of Capricorn)
March 20–21	Vernal equinox (March equinox)	0° (equator)
June 20–21	Summer solstice (June solstice)	23.5° N latitude (Tropic of Cancer)
September 22–23	Autumnal equinox (September equinox)	0° (equator)

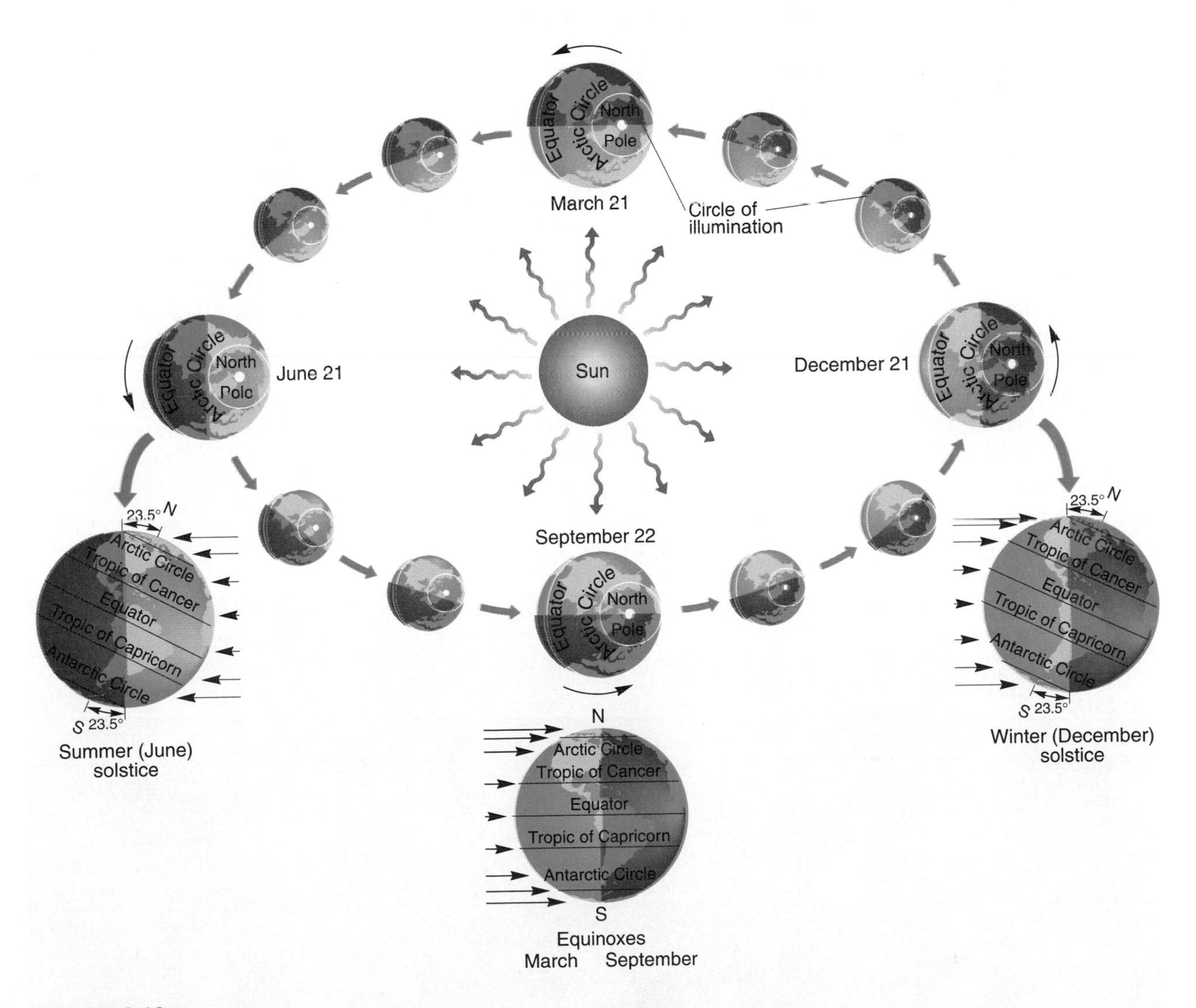

FIGURE 2-18
Annual march of the seasons as Earth revolves about the Sun. Shading indicates the changing position of the circle of illumination in the Northern Hemisphere.

as the **Tropic of Capricorn,** at the moment of the solstice. The Northern Hemisphere is tilted away from these more direct rays of sunlight, thereby creating a lower angle for the incoming solar rays and thus a more diffuse pattern of insolation.

From 66.5° N latitude to 90° N (the North Pole), the Sun remains below the horizon the entire day. This latitude (66.5° N) marks the *Arctic Circle*, the southernmost parallel (in the Northern Hemisphere) that experiences a 24-hour period of darkness. During the following three months, daylength and solar angles gradually change as Earth completes one-fourth of its orbit.

The **vernal equinox**, or **March equinox,** occurs on March 20 or 21. At that time, the circle of illumination passes through both poles so that all locations on Earth experience a 12-hour day and a 12-hour night. Those living at 40° N latitude have gained three hours of daylight since the December solstice. At the North Pole, the Sun rises for the first time since the previous September; at the South Pole the Sun is setting. Figure 2-12 clearly illustrates these moments of beginning and ending insolation at the poles.

From March, the seasons move on to June 20 or 21 and the **summer solstice**, or **June solstice.** The subsolar point is at 23.5° N latitude, the **Tropic of Cancer**. Because the circle of illumination now includes the North Polar region, everything north of the Arctic Circle receives 24 hours of daylight. (Figure 2-19 is a multiple-image photo of the midnight Sun as seen north of the Arctic Circle.) In contrast, the region from the *Antarctic Circle* to the South Pole (66.5–90° S latitude) is in darkness the entire day.

Although the orientation of Earth's axis has remained fixed relative to the heavens and parallel to its previous position, the Northern Hemisphere in June is tilted toward the Sun. Consequently, that hemisphere receives higher Sun angles and experiences longer days—and therefore more insolation—than the Southern Hemisphere. Those living at 40° N latitude now receive more than 15 hours of sunlight, which is 6 hours more than in December.

September 22 or 23 is the time of the **autumnal equinox**, or **September equinox,** when Earth's orientation is such that the circle of illumination again passes through both poles so that all parts of the globe receive a 12-hour day and a 12-hour night. The subsolar point has returned to the equator, with days growing shorter to the north and

FIGURE 2-19
Midnight Sun north of the Arctic Circle. [Photos © Wiancko-Braasch/Allstock.]

FIGURE 2-20
Fall foliage and seasonal change in a mixed forest create a scene of colorful beauty, Humber River valley, Newfoundland. [Photo by John Eastcott/Yva Momatiak/DRK Photo.]

longer to the south. Researchers stationed at the South Pole see the disk of the Sun just rising, ending their six months of night. In the Northern Hemisphere, fall arrives, a time of many colorful changes in the landscape (Figure 2-20).

Seasonal Observations. In the Northern Hemisphere midlatitudes, the point of sunrise migrates from the southeast in December to the northeast in June, whereas sunset moves from the southwest to the northwest during those same times. The Sun's altitude at local noon at 40° N latitude migrates from a 26° angle above the horizon at the winter (December) solstice to a 73° angle above the horizon at the summer (June) solstice—a range of 47°.

Dawn and Twilight Concepts. The diffused light that occurs before sunrise and after sunset represents useful work time for humans. Light is scattered by the molecules of atmospheric gases and reflected by dust and moisture so that the atmosphere is illuminated. Such effects may be enhanced by the presence of pollution aerosols and other suspended particles, such as those from volcanic eruptions or forest fires. Dawn and twilight are roughly equal in extent on any given day at any particular latitude. The duration of these Sun effects is a function of latitude; that is, the angle of the Sun's path relative to the horizon determines the thickness of the atmosphere through which the Sun's rays must pass and thus the length of dawn and twilight.

At the equator, where the Sun's rays are nearly perpendicular to the horizon throughout the year, dawn and twilight are limited to 30–45 minutes each. These times increase to 1–2 hours at 40° latitude, and at 60° latitude these times range upward from 2.5 hours, with little true night in summer. The poles experience about 7 weeks of dawn and 7 weeks of twilight, leaving only 2.5 months of near darkness during the six months when the Sun is completely below the horizon.

Technology enhances the observation and analysis of solar energy from the Sun to Earth, the spatial patterns of warmth and cold, humid and arid lands, vegetation, snow and ice, the seasonal variation of atmospheric and oceanic circulation, and the impact of human activities. Science is developing new techniques to probe the home planet through remote sensing and geographic information systems.

REMOTE SENSING

In this era of observations from orbit outside the atmosphere and from aircraft within it, scientists are obtaining a wide array of remotely sensed data. Remote sensing is an important tool for the spatial analysis of Earth's environment. When we observe the environment with our eyes, we are sensing the shape, size, and color of objects from a distance, utilizing the visible-wavelength portion of the electromagnetic spectrum. Similarly, the film in a camera senses the wavelengths for which it was designed (visible light or infrared) and is exposed by the energy that is reflected and emitted from a scene.

Our eyes and cameras both represent familiar means of **remote sensing** information about a distant subject, without physical contact. Aerial photographs have been used for years to improve the accuracy of surface maps at less expense and with greater ease than is permitted by actual on-site surveys. Such accurate measurements derived from photographs is the realm of **photogrammetry**, an important application of remote sensing.

Various surface materials absorb and reflect insolation in different and characteristic ways. These differences can be remotely sensed by their reflected wavelengths: ultraviolet (0.3–0.4 µm), visible light (0.4–0.7 µm), reflected infrared (0.7–3.0 µm), thermal infrared (3.0–5.0 µm and 8.0–14.0 µm), and microwave radars (0.3–300 cm) (see Figures 2-9 and 2-10).

Many of the remotely sensed images used in this text were initially recorded in digital form for later processing, enhancement, and generation. Satellites do not take conventional film photographs. Rather, they record *images* that are transmitted to Earth-based receivers in a manner similar to television broadcasts. A scene is scanned and broken down into *pixels* (*pic*ture *el*ements) identified by coordinates known as *lines* (horizontal rows) and *samples* (vertical columns). A grid of 6000 lines and 7000 samples produces 30–120 m of ground resolution depending on lenses used and camera altitude above the ground. You can see that the pixel count for such an image runs well into the millions. Digital data are processed in many ways to enhance their utility: false and simulated natural color, enhanced contrast, signal filtering, and different levels of sampling and resolution. Two types of remote-sensing systems are used: active and passive.

Active systems direct a beam of energy at a surface and analyze the reflected energy. Radar (*Ra*dio *D*etection *a*nd *R*anging) is an example of an active remote sensor. It transmits radiation of relatively long wavelengths (1 to 10 m) in short bursts to the subject terrain, penetrating clouds and darkness. Reflected radiation, known as backscatter, is received by the sensing instrument and analyzed. NASA sent imaging radar systems into orbit on the *Seasat* satellite (1978) and on two Space Shuttles in 1981 and 1984 (SIR-A and SIR-B). Oceanography, landforms and geology, and biogeography were all subjects of study. The computer image of wind and sea-surface patterns over the Pacific used in Chapter 6 (Figure 6-4) was developed from 150,000 radar-derived measurements made on a single day. Side-Looking Airborne Radar (SLAR) also is such an active system. Its radar energy produces high-resolution images of surfaces it scans. Chapter 8 presents an analysis of Hurricane Gilbert using a SLAR image (Figure 8-21).

Passive remote sensing systems record radiant energy transmitted from a surface, as in the visible-light photograph of Earth on the back cover of this text. The United States *Landsat* series of five satellites, launched between 1972 and 1984, provided a variety of visible and infrared data, as shown in images of the Appalachians in Chapter 12 (Figure 12-13), river deltas in Chapter 14 (Figures 14-25 and 14-26), and the Malaspina and Hubbard glaciers in Alaska featured in Chapter 17 (Figures 17-3 and 17-8). All *Landsats* carried a Multispectral Scanner (MSS) device that sampled in four spectral bands of visible and near-infrared wavelengths. The two *Landsats* (4 and 5) that remain operational are equipped with the Thematic Mapper (TM) as a principal sensor. The TM is designed to provide high resolution from seven spectral bands of visible and infrared wavelengths.

The National Oceanic and Atmospheric Administration (NOAA) series of polar-orbiting satellites carry the Advanced Very High Resolution Radiometer (AVHRR) sensors (*NOAA 10* and *11* are currently operating). AVHRR is sensitive in visible, near-infrared, and thermal-infrared wavelengths in five spectral bands. The incredible images of Hurricane Andrew in Chapter 8 (Figure 8-22) are examples, among others in the text, produced by an AVHRR system. AVHRR is primarily used for sensing day or night clouds, snow and ice; monitoring forest fires, clouds, and surface temperature; and for determining natural and planted vegetation patterns and various plant indices. In Chapter 19, an AVHRR image portrays clear-cutting in the Pacific northwest (Figure 19-19). These examples of resource analysis were impossible to do at such a scale just a few years ago.

The highest resolution commercial system, the French satellite called *SPOT* (Systeme Probatoire d' Observation de la Terre), can resolve from 10 to 20 m, depending on which of its sensors is used. The dramatic image in Chapter 15 of sand dunes in the Rub al Khali Erg, Saudi Arabia (Figure 15-6), and in Chapter 17 of floating ice in the Weddell Sea, Antarctica (Figure 17-31), are examples of *SPOT* images. Diverse applications for such remote sensing include agricultural and crop monitoring, terrain analysis, documentation of change in an ecosystem over time, river studies to reduce flood hazards, analysis of logging practices, identification of development and construction impacts, and surveillance.

The Geostationary Operational Environmental Satellite, known as *GOES-7* (Central), provides the daily infrared and visible images of weather in the Western Hemisphere that you have seen on television. This satellite is positioned at 98° W during the spring and early-summer tornado season; then moved eastward to monitor late-summer and fall hurricanes. Geostationary satellites stay in semipermanent positions because they keep pace with Earth's rotational speed at their altitude of 35,400 km (21,995 mi). Three *GOES* images are used with the weather maps for April 1–3, 1988, that appear in Chapter 8 (Figure 8-15). Other specific weather satellites include the *GMS* weather satellite run by the Japan Weather Association and covering the Far East, and *METEOSAT* for Europe and Africa operated by the European Space Agency.

Geographic Information System (GIS)

Remote sensing is an important acquisition tool that generates large volumes of spatial data. Storage, processing, and retrieval in useful ways are important if there is to be any benefit from these scientific inputs. The value of remote sensing and the data derived rests on an ability to establish new information handling–systems. Computers have allowed an integration of geographic information from direct surveys and remote sensing in complex ways never before possible. A **geographic information system (GIS)** is a computer-based data-processing tool or methodology for gathering, manipulating, and analyzing geographic information. An example is shown in Figure 2-21.

The method involves applications of almost unlimited variety. A common terrestrial reference system, such as the latitude and longitude coordinates provided on a map, is an essential beginning component for any GIS. Such maps can be converted into digital data of areas, points, and lines, depending on the task. Remotely sensed imagery and data are then transferred to the coordinate system on the reference map.

A GIS is capable of analyzing patterns and relationships within one data plane, such as the floodplain or soil layer in the Figure 2-21. A GIS also can produce an *overlay analysis* where two or more data planes interact. With locational aspects of the base map held constant, various assumptions, comparisons, and policies can be tested. The resulting synthesis, a *composite overlay*, follows specific points or areas through the complex of overlay planes for analysis.

Prior to the advent of computers, an environmental-impact analysis required someone to gather various data and painstakingly hand-produce overlays of information to determine positive and negative aspects of a project or event. Today, this layered information can be handled by a computer-driven GIS, which assesses the complex interconnections among different components. In this way, subtle changes in one element of a landscape may be identified as having a powerful impact elsewhere.

GIS applications are useful for the analysis of environmental events, both natural and human-caused. For example, the USGS completed a GIS to help analyze the spatial impact of the 1989 *Exxon Valdez* oil spill in Alaska. Scientists at NASA's Goddard Space Flight Center recently completed a three-year comprehensive GIS of Brazil in an effort to better understand land-use patterns—specifically, loss of the rain forest and agricultural practices.

One of the most extensive and longest-operating systems is the Canada Geographic Information Sys-

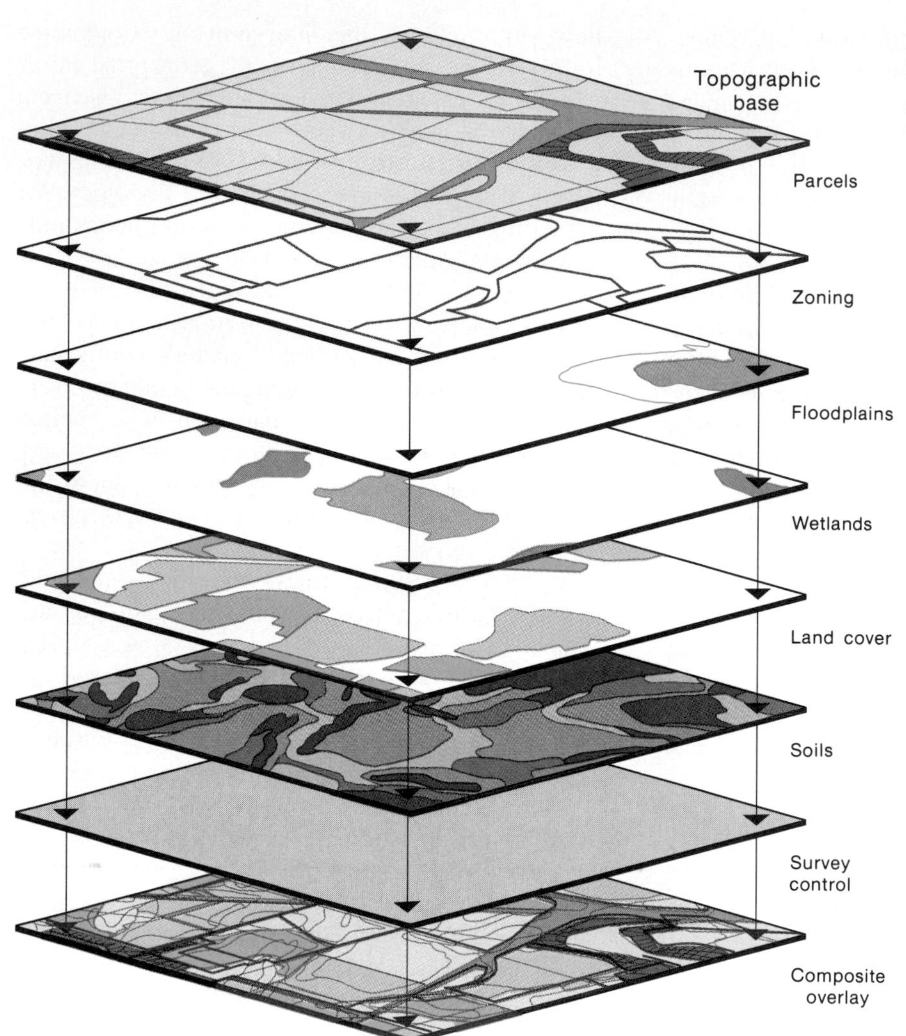

FIGURE 2-21
Layered spatial data in a geographic information system (GIS) format. [After USGS.]

tem (CGIS). Environmental data about natural features, resources, and land use were taken from maps, aerial photographs, and orbital sources and reduced to map segments. These were divided into polygons and entered into the CGIS. The development of this system has progressed hand in hand with the ongoing Canada land-inventory project.

Simulations and construction of hypothetical models are other important capabilities of a GIS. This is evident from the general circulation models used to analyze the atmosphere-ocean interaction, which is so important to forecasting future weather and climate patterns. An important planning and analytical element will be an improved interface between GIS and remote-sensing capabilities.

The entire GIS methodology represents expanding career opportunities in many fields and an important step in better understanding Earth's systems. GIS-degree programs are now available at many colleges and universities. A National Center for Geographic Information Analysis is established at the University of California-Santa Barbara.

SUMMARY—Solar Energy to Earth

The Solar System, planets, and Earth began to condense from clouds of dust, gas, debris, and icy comets approximately 4.6 billion years ago. Over the span of these years Earth's atmosphere evolved in 4 stages—primordial, evolutionary, living, and modern. Solar energy in the form of charged particles of solar wind and the electromagnetic spectrum of radiant energy travel outward in all directions from the Sun. Fusion processes in the Sun's interior generate these incredible quantities of energy.

Solar wind is deflected by Earth's magnetosphere, producing various effects in the upper atmosphere, including spectacular auroras that surge across the skies at higher latitudes. The solar wind–atmosphere interaction disrupts certain radio transmissions and appears to influence Earth's weather. Electromagnetic radiation from the Sun passes through Earth's magnetosphere to the top of the atmosphere—the thermopause. The average amount of incoming solar radiation, or insolation, is measured as 1372 W/m^2 (2.0 calories/cm^2/min; 2 langleys/min)—a value called the solar constant.

The dynamic flows in the atmosphere and ocean, the patterns of temperature and weather, and the vibrant biosphere result from the uneven distribution of insolation. The angle of the Sun's rays is greater and the rays are more direct and thus more concentrated at lower latitudes; the angle is more oblique and the rays more diffuse at higher latitudes. Insolation received at any specific place varies from day to day. Seasonality is the change in daylength and varying altitude of the Sun above the horizon. Earth's distinct seasons are produced by interactions of revolution, rotation, sphericity, axial tilt, and alignment of the axis throughout the year.

The operation of Earth's systems is being disclosed through orbital and aerial remote sensing. Satellites do not take photographs; rather, they record images that are transmitted to Earth-based receivers. Satellite images are recorded in digital form for later processing, enhancement, and generation. This flood of data has lead to the development of geographic information systems (GIS). Computers have allowed an integration of geographic information from direct surveys and remote sensing in complex ways never before possible. GIS methodology is an important step in better understanding Earth's complex systems and is a vital career opportunity for geographers.

KEY TERMS

altitude
analemma
aphelion
auroras
autumnal (September) equinox
axial parallelism
axis
chemosynthesis
circle of illumination
daylength
declination
electromagnetic spectrum
evolutionary atmosphere
fusion
geographic information system (GIS)
gravity
insolation
living atmosphere
magnetosphere
Milky Way Galaxy
modern atmosphere
nebula
perihelion
photogrammetry
plane of the ecliptic
planetesimal hypothesis
primordial atmosphere
remote sensing
revolution
rotation
scientific method
solar constant
solar wind
speed of light
subsolar point
summer (June) solstice

sunrise
sunset
sunspots
thermopause
Tropic of Cancer
Tropic of Capricorn
vernal (March) equinox
wavelength
winter (December) solstice

REVIEW QUESTIONS

1. Describe the Sun's status among stars in the Milky Way Galaxy. Describe the Sun's location, size, and relationship to its planets.
2. If you have seen the Milky Way, briefly describe it. Use specifics from the text in your description.
3. Briefly describe Earth's origin as part of the Solar System.
4. How far is Earth from the Sun in terms of light speed? In terms of kilometers and miles? Relate this distance to the shape of Earth's orbit during the year.
5. How many distinct atmospheres have there been? How does Earth's present atmosphere compare to past atmospheres?
6. Within which of Earth's atmospheres did photosynthesis begin?
7. Define the scientific method and give an example of its application.
8. How does the Sun produce such tremendous quantities of energy? Write out a simple fusion reaction, using the quantities (in tons) of hydrogen involved.
9. What is the sunspot cycle? At what stage in the cycle were we in 1990?
10. Describe Earth's magnetosphere and its effects on the solar wind and the electromagnetic spectrum.
11. Summarize the presently known and hypothesized effects of the solar wind relative to Earth's environment.
12. Describe the segments of the electromagnetic spectrum, from shortest to longest wavelength.
13. What is the solar constant? Why is it important to know?
14. Select 40° or 50° north latitude on Figure 2-12, and describe the amount of energy in W/m^2 per day characteristic of each month throughout the year.
15. If Earth were flat and oriented perpendicular to the incoming solar radiation, what would be the latitudinal distribution of solar energy at the top of the atmosphere?
16. In what possible way does the Stonehenge monument relate to seasons?
17. The concept of seasonality refers to what specific factors? How do the variables of seasonality change with latitude? 0°? 40°? 90°?
18. Differentiate between the Sun's altitude and its declination.
19. For the latitude at which you live, how does daylength vary during the year? How does the Sun's altitude vary? Does your local newspaper publish a weather calendar containing such information?
20. List the five physical factors that operate together to produce seasons.
21. Describe revolution and rotation, and differentiate between them.
22. Define Earth's present tilt relative to its orbit about the Sun.
23. What is an analemma? How does it relate to the Sun's declination and the equation of time during the year?
24. Describe seasonal conditions at each of the four key seasonal anniversary dates during the year. What are the solstices and equinoxes?

SUGGESTED READINGS

Aronoff, S. *Geographic Information Systems: A Management Perspective.* Ottawa: WDL Publications, 1989.

Avery, Thomas E. and Graydon Berlin. *Fundamentals of Remote Sensing and Airphoto Interpretation,* 5th ed. New York: Macmillan Publishing Company, 1992.

Briggs, Geoffrey, and Frederic Taylor. *The Cambridge Photographic Atlas of the Planets.* Cambridge: Cambridge University Press, 1982.

Colwell, Robert N., ed. *Manual of Remote Sensing,* 2d ed. Vol. 1, *Theory, Instruments, and Techniques*; Vol. 2, *Interpretation and Applications.* Falls Church, VA: American Society of Photogrammetry, 1983.

Eddy, John A. *A New Sun—The Results from Skylab.* U.S. National Aeronautics and Space Administration. Washington, DC: U.S. Government Printing Office, 1979.

Editors of Time-Life Books. *Stars.* Voyage Through the Universe Series. Alexandria, VA: Time-Life Books, 1988.

Frazier, Kendrick, and the editors of Time-Life Books. *Solar System.* Time-Life Planet Earth Series. Alexandria, VA: Time-Life Books, 1985.

Gallant, Roy A. *National Geographic Picture Atlas of Our Universe.* Washington, DC: National Geographic Society, 1986.

Herman, John R., and Richard Goldberg. *Sun, Weather, and Climate.* U.S. National Aeronautics and Space Administration. Washington, DC: U.S. Government Printing Office, 1978.

Meinel, Aden, and Marjorie Meinel. *Sunsets, Twilights, and Evening Skies.* Cambridge: Cambridge University Press, 1983.

Rabenhorst, Thomas D., and Paul D. McDermott. *Applied Cartography—Introduction to Remote Sensing.* Columbus, OH: Macmillan Publishing Company, 1989.

Royal Astronomical Society of Canada. *The Observer's Handbook.* Toronto: University of Toronto Press, published annually.

"The Solar System," *Scientific American* 233, no. 3 (September 1975): entire issue.

Sullivan, Walter. "Curtains of Light, Horsemen of Night." *Audubon* 89, no. 1 (January 1987): 40–51.

United States Naval Observatory. *The Air Almanac.* Washington, DC: U.S. Government Printing Office, published annually.

New Zealand sunset. [*Photo by John Eastcott/Yva Momatiak/DRK Photo.*]

3

EARTH'S MODERN ATMOSPHERE

Earth's atmosphere is a unique reservoir of gases, the product of nearly 5 billion years of development. We all participate in the atmosphere with each breath we take. This chapter examines the atmosphere's structure, function, and composition. Our consideration of the modern atmosphere also must include the spatial aspects of human-induced problems that affect it, such as air pollution, the stratospheric ozone predicament, and the blight of acid deposition. We consider these critical topics carefully, for we are participating in producing the atmosphere of the future.

ATMOSPHERIC STRUCTURE AND FUNCTION

The modern atmosphere probably is the fourth general atmosphere in Earth's history. (The modern atmosphere and the other three—the primordial, evolutionary, and living atmospheres—were discussed in Chapter 2.) This modern atmosphere is a gaseous mixture of ancient origin, the sum of all the exhalations and inhalations of life on Earth throughout time. The principal substance of this atmosphere is air, the medium of life as well as a major industrial and chemical raw material. *Air* is a simple additive mixture of gases that is naturally odorless, colorless, tasteless, and formless, blended so thoroughly that it behaves as if it were a single gas.

In his book *The Lives of a Cell,* physician and self-styled "biology-watcher" Lewis Thomas compared the atmosphere of Earth to an enormous cell membrane. The membrane around a cell regulates the interactions between the cell's delicate inner workings and the potentially disruptive outer environment. Each cell membrane is very selective as to what it will and will not allow to pass through. The modern atmosphere acts as Earth's protective membrane, as Thomas describes so vividly (Figure 3-1).

The atmospheres of the other three inner planets bear no resemblance to Earth's atmosphere. Mars has a cold, thin atmosphere of carbon dioxide, equivalent in pressure to Earth's atmosphere 32 km (20 mi) above the surface. Venus also has an atmosphere dominated by carbon dioxide and is shrouded in clouds of sulfuric acid. Surface pressures on Venus are about 90 times those on Earth, and surface temperatures average 500°C (932°F). Mercury has no appreciable atmosphere and, in fact, appears similar to the Moon, pockmarked by innumerable craters.

Earth's modern atmosphere is arranged in a series of imperfectly shaped concentric spheres, all bound to the planet by gravity. As critical as the atmosphere is to us, it represents only a thin-skinned envelope amounting to less than one-millionth of Earth's total mass. Let's begin our study of the atmosphere by analyzing the pressure it exerts.

Air Pressure

The gases that comprise air create pressure through their motion, size, and number. This pressure is exerted on all surfaces in contact with the air. The weight of the atmosphere, or **air pressure,** crushes in on all of us; fortunately, that same pressure also exists inside us, pushing outward. The atmosphere exerts an average force of approximately 1 kg/cm^2 (14.7 lb/in.^{2}) at sea level. Under the acceleration of gravity, air is compressed and therefore denser near Earth's surface (Figure 3-2). It rapidly thins with increased altitude, a decrease that is measurable because air exerts its weight as pressure. Consequently, over half the total mass of the atmosphere is compressed below 5500 m (18,000 ft), 75% is compressed below 10,700 m (35,105 ft), and 90% is below 16,000 m (52,500 ft). All but 0.1% is accounted for at an altitude of 50 km (31 mi), as shown in the pressure profile (orange line) in Figure 3-5.

In A.D. 1643, a pupil of Galileo's named Evangelista Torricelli was working on a mine drainage problem. This led him to discover a method for measuring air pressure (Figure 3-3a). He knew that pumps in the mine were able to draw water upward only about 10 m (33 ft), but did not know

FIGURE 3-1
Earthrise over the lunar surface. [NASA photo.] Quotation from "The World's Biggest Membrane" from *The Lives of a Cell* by Lewis Thomas. Copyright © 1973 by the Massachusetts Medical Society. Originally published in the *New England Journal of Medicine*. Reprinted by permission of Viking Penguin, a division of Penguin Books USA, Inc.]

why. Careful observation led him to discover that this limitation was not caused by the pumps but by the atmosphere itself. Torricelli noted that the water level in the vertical pipe fluctuated from day to day. He surmised correctly that air pressure varied over time and with changing weather conditions. A pump of the type shown works by creating a vacuum above the water, which allows the pressure of the atmosphere that is pushing down on water in the mine to force water up the pipe.

To simulate the problem at the mine, Torricelli devised an instrument at the suggestion of Galileo. Instead of water, he used a much denser fluid, mercury (Hg), so that he could use a glass laboratory tube only 1 m high. He sealed the glass tube at one end, filled it with mercury, and then inverted it into a dish of mercury (Figure 3-3b). Torricelli established that the average height of the column of mercury in the tube was 760 mm (29.92 in.). He concluded that the column of mercury was counterbalanced by the mass of surrounding air exerting an equivalent pressure on the mercury in the

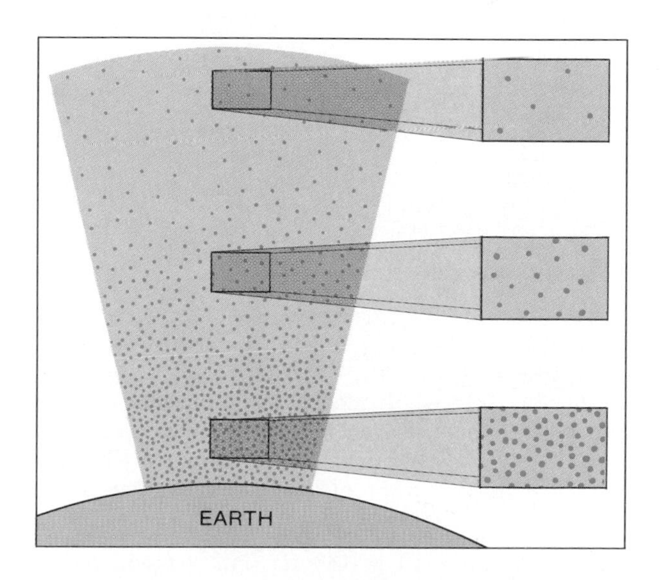

FIGURE 3-2
The atmosphere, denser near Earth's surface, rapidly decreases in density with altitude. This difference in density is easily measured because air exerts its weight as pressure.

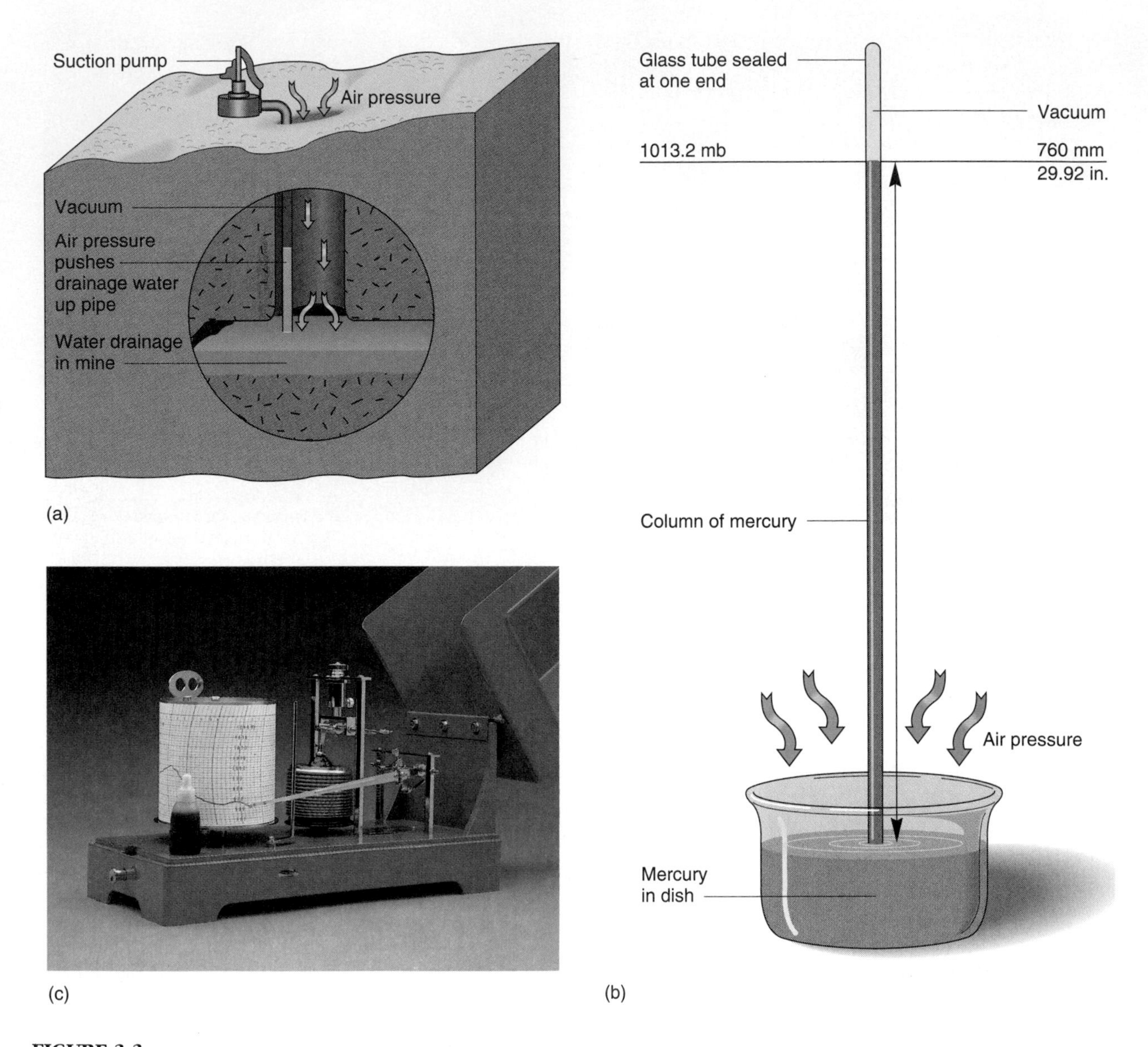

FIGURE 3-3
Evangelista Torricelli developed the barometer to measure air pressure as a by-product of trying to solve a mine drainage problem (a). Instruments used to measure atmospheric pressure are (b) a mercury barometer and (c) an aneroid barometer. [(c) Courtesy of Qualimetrics, Inc., Sacramento, CA.]

dish. Thus, the 10 m (33 ft) limit to which the mine pumps could draw water was controlled by the mass of the atmosphere. Using similar instruments today, normal sea level pressure is expressed as 1013.2 millibars of mercury (a way of expressing force per square meter of surface area), or 29.92 in. of mercury, or 101.32 kPa (kilopascal; 1 kilopascal = 10 millibars).

Any instrument that measures air pressure is called a barometer. Torricelli developed a **mercury barometer**. A more compact design that works without a meter-long tube of mercury is

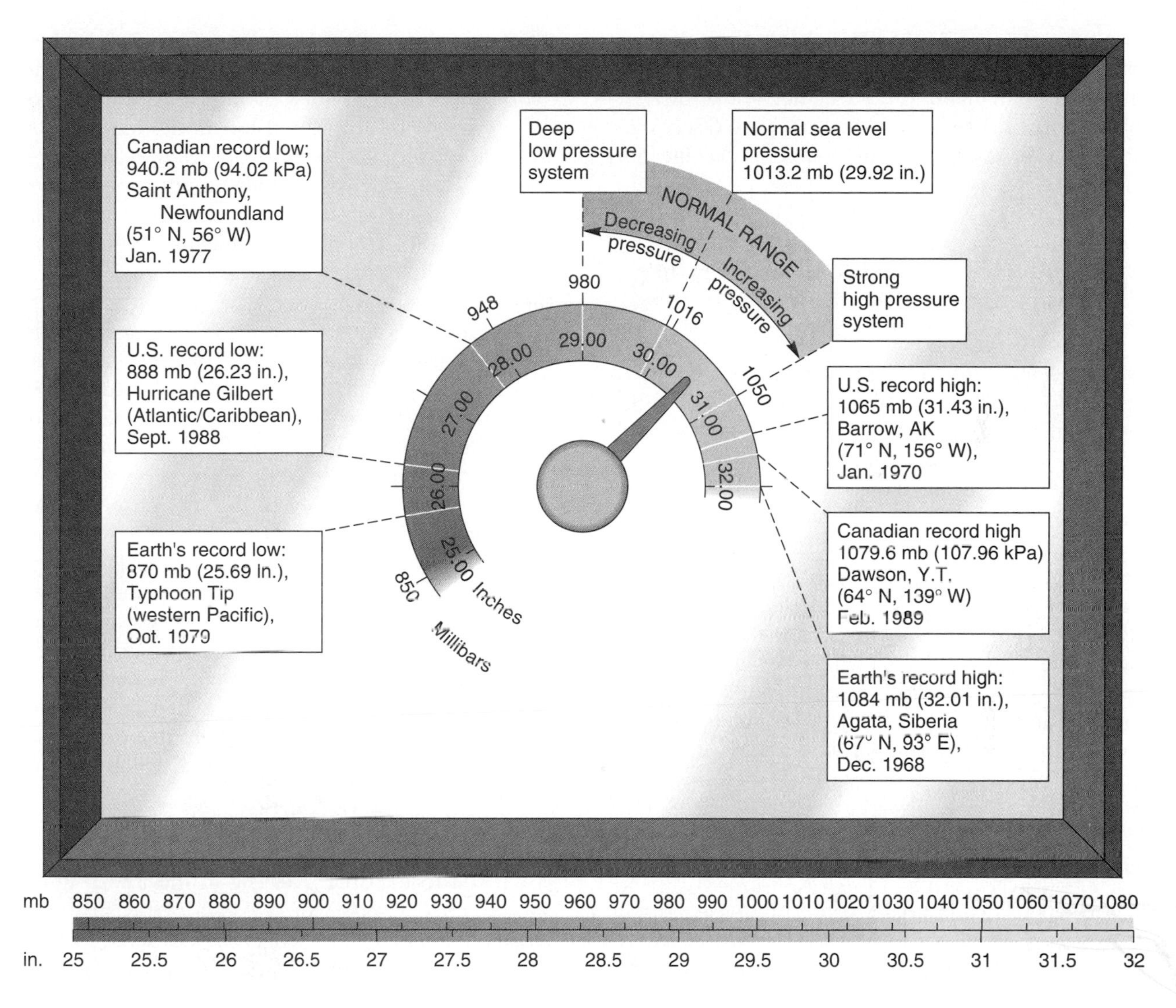

FIGURE 3-4
Scales for expressing barometric air pressure in millibars and inches, with average air pressure values and recorded pressure extremes. Canadian values include kilopascal equivalents (10 mb = 1 kPa).

called an **aneroid barometer** (Figure 3-3c). Inside this type of barometer, a small chamber is partially emptied of air, sealed, and connected to a mechanism that is sensitive to changes in the chamber. As air pressure varies, the mechanism responds to the difference and moves the needle on the dial. Altimeters used in aircraft are a type of aneroid barometer that is able to accurately measure altitude because air pressure diminishes with elevation above sea level.

Figure 3-4 illustrates comparative scales in millibars and inches used to express air pressure and its relative force. Strong high pressure to deep low pressure can range from 1050 to 980 mb. The figure also indicates the extreme highest and lowest pressures ever recorded in the United States,

Canada, and on Earth. The concept of air pressure, its measurement, and its application to atmospheric circulation and weather are further explained in Chapters 6 and 8. Let's now discuss the atmosphere, using several different classification criteria.

Atmospheric Composition, Temperature, and Function Classifications

Important to the following discussion is Figure 3-5, which charts essential aspects of the entire atmosphere. We simplify our discussion of the atmosphere through the use of three criteria: composition, temperature, and function. These criteria are noted along the left side of Figure 3-5. On the basis of chemical *composition*, the atmosphere is divided into two broad regions, the heterosphere and the homosphere. Based on *temperature*, the atmosphere is divided into four distinct zones, the thermosphere, mesosphere, stratosphere, and troposphere. Finally, two specific zones are identified on the basis of *function* for their role in removing most of the harmful wavelengths of solar radiation: the ionosphere and the ozonosphere (ozone layer).

Heterosphere

Based on the criterion of composition, the upper atmosphere ranging from 80 to over 10,000 km in altitude (50–6000 mi) is called the **heterosphere**, as shown in Figure 3-5. As the prefix *hetero-* implies, this region is not uniform (not evenly mixed). Instead, distinct layers of molecules and atoms (mainly nitrogen and oxygen) are distributed according to their atomic weight. The heterosphere has a distinctive layered composition, thermal properties (thermosphere), and an important function (ionosphere).

Although individual hydrogen and helium atoms are weakly bound by gravity as far as 32,000 km (19,900 mi) from Earth, for all practical purposes the top of our atmosphere is at roughly 480 km (300 mi)—the same as the boundary we used in Chapter 2 for determining the solar constant. What lies beyond that altitude is the vacuum of space, commonly called the **exosphere**.

Thermosphere. Based on the criterion of temperature, the atmosphere also can be divided into four temperature regions. Dominating the heterosphere from 80 km (50 mi) up to 480 km (50–300 mi) is the **thermosphere** (from *thermo-*, meaning "heat"). The upper part of the thermosphere is a transition zone into space, termed the thermopause (the suffix *-pause* means "to cause to change"). High temperatures are generated in the thermosphere by the absorption of shortwave solar radiation. During periods of a less active Sun (fewer sunspots and coronal bursts), the thermopause may be as low as 250 km altitude (155 mi). An active Sun will cause the upper atmosphere to swell to an altitude of 550 km (340 mi).

The temperature profile in Figure 3-5 shows that temperatures rise sharply in the thermosphere, up to 1200°C (2200°F) and higher. However, we must use different concepts of "temperature" and "heat" to understand this effect. The intense radiation in this portion of the atmosphere excites individual molecules (nitrogen and oxygen) and atoms (oxygen) to high levels of vibration. This **kinetic energy,** the energy of motion, is the vibrational energy stated as "temperature." However, the density of the molecules is so low that little actual heat is produced.

In the exosphere, molecules are so scarce and molecular collisions so infrequent that temperature is difficult to measure. Heating in the lower atmosphere near Earth's surface differs because the greater number of molecules in the denser atmosphere transmit their kinetic energy as **sensible heat,** meaning that we can sense it. (Density, temperature, and heat capacity determine the sensible heat of a substance.)

Ionosphere. The third criterion for atmospheric classification identifies two functional layers, so-called because both function to filter harmful wavelengths of solar radiation, protecting Earth's surface from bombardment in any significant quantity. The upper functional layer, the **ionosphere,** extends throughout the thermosphere. It is composed of charged particles produced by the absorption of cosmic rays, gamma rays, X-rays, and shorter wavelengths of ultraviolet radiation, all of which change the atoms to positively charged ions, giving the ionosphere its name. Radiation

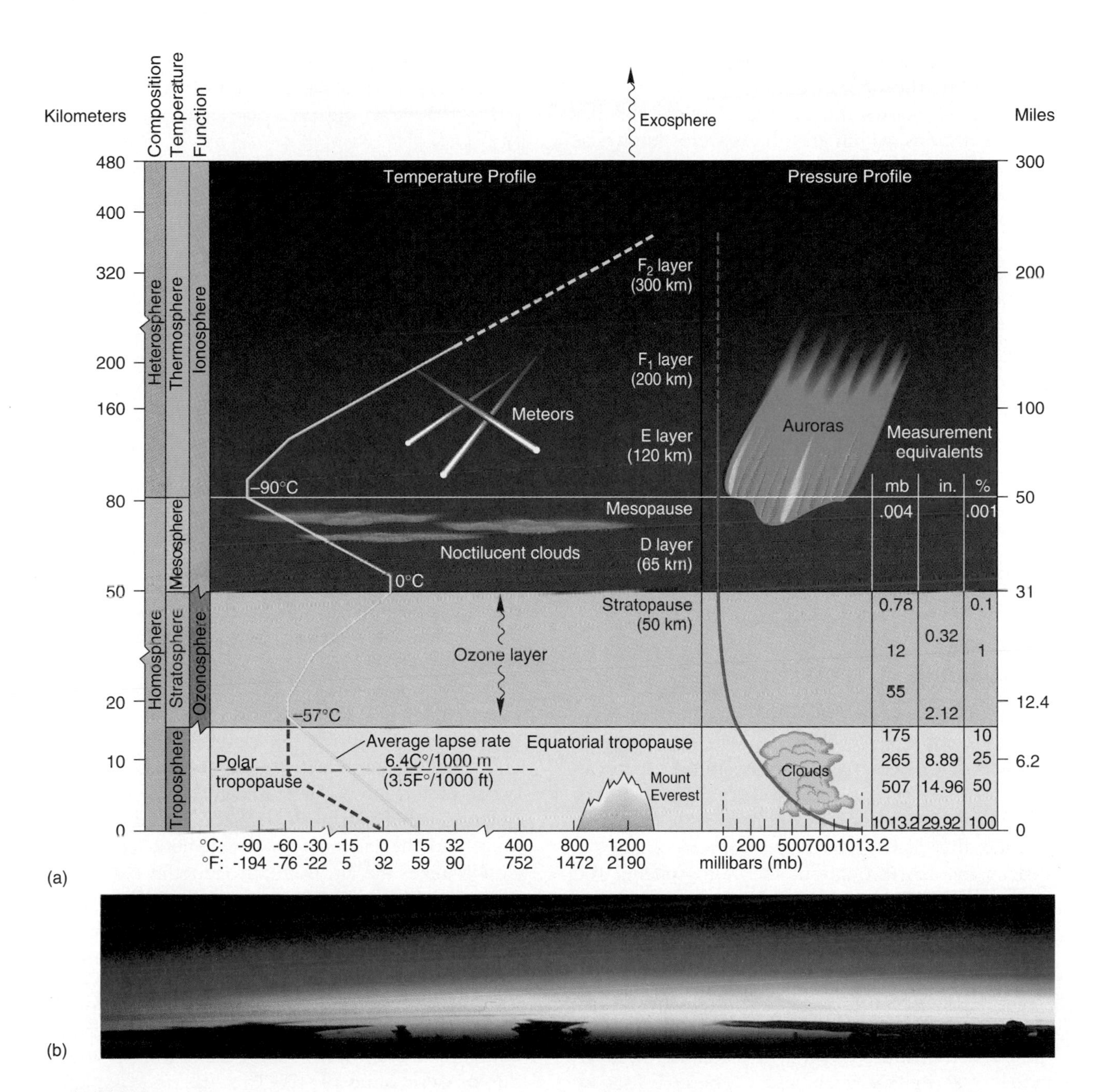

FIGURE 3-5
Modern atmosphere integrated chart (a) showing composition (heterosphere and homosphere), temperature (thermosphere, mesosphere, stratosphere, and troposphere), and function (ionosphere and ozonosphere). The ranges in altitude for the three criteria are along the left side of the chart. The pressure profile of the atmosphere is the orange line. The chart covers from Earth's surface to the thermopause at 480 km. Space Shuttle astronauts captured a dramatic sunset (b) through various atmospheric layers across the edge of our planet–called Earth's limb. A silhouetted cumulonimbus thunderhead cloud is seen rising to the tropopause. [Space Shuttle photo from NASA.]

bombards the ionosphere constantly, producing a constant flux (flow) of electrons.

Figure 3-5 shows the average daytime altitudes of four regions within the ionosphere, known as the D, E, F_1, and F_2 layers, which are important to broadcast communications. At night the D layer virtually disappears, and the F_1 and F_2 layers usually merge and reflect certain wavelengths of AM radio, returning them at predictable angles to receivers hundreds of kilometers from the transmitters. However, during the day, the D and E regions in the ionosphere actively absorb arriving radio signals, preventing such distant reception. Unaffected are FM or TV broadcast wavelengths, which pass through to space, thus necessitating use of communication satellites for retransmission of these signals to the surface. During times of intense solar wind and auroral activity, the ionosphere becomes so active that radio communications can be disrupted. The glowing auroral lights discussed in Chapter 2 occur principally within the ionosphere.

Homosphere

Below the heterosphere is the second compositional region of the atmosphere, the **homosphere.** It extends from Earth's surface up to an altitude of 80 km (50 mi). Even though the atmosphere rapidly decreases in density with increasing altitude, the blend (relative proportion) of gases is nearly uniform throughout the homosphere. The only exceptions are the concentration of ozone (O_3) from 19 to 50 km (12–31 mi), and the variations in water vapor and pollutants in the lowest portion of the atmosphere near Earth's surface.

The stable mixture of gases throughout the homosphere has evolved slowly. The present proportion, which includes oxygen, was attained approximately 600 million years ago. Table 3-1 lists by volume the stable ingredients that comprise the homosphere.

The homosphere is a vast reservoir of relatively inert nitrogen, principally originating from volcanic sources. Nitrogen is a key element of life, yet we exhale all the nitrogen that we inhale. This apparent contradiction is explained because the input of nitrogen to our body systems comes instead through compounds in food. In the soil, nitrogen is bound into compounds by nitrogen-fixing bacteria and is returned to the atmosphere by denitrifying bacteria that remove nitrogen from organic materials.

TABLE 3-1
Stable Components of the Modern Homosphere

Gas (Symbol)	% by Volume	Parts per Million (ppm)
Nitrogen (N_2)	78.084	780,840
Oxygen (O_2)	20.946	209,460
Argon (Ar)	0.934	9,340
Carbon dioxide (CO_2)	0.036	360
Neon (Ne)	0.001818	18
Helium (He)	0.000525	5
Methane (CH_4)	0.00014	1.4
Krypton (Kr)	0.00010	1.0
Ozone (O_3)	variable	
Nitrous oxide (N_2O)	trace	
Hydrogen (H)	trace	
Xenon (Xe)	trace	

Oxygen, a by-product of photosynthesis, also is essential for life processes. Along with other elements, oxygen forms compounds that compose about half of Earth's crust. Slight spatial variations occur in the atmospheric oxygen percentage because of variations in photosynthetic rates with latitude and the lag time as atmospheric circulation slowly mixes the air. Both nitrogen and oxygen reserves in the atmosphere are so extensive that, at present, they far exceed human capabilities to disrupt or deplete them.

Argon is a residue from the radioactive decay of an isotope (form) of potassium called potassium-40 (symbolized ^{40}K). Slow accumulation over millions of years accounts for all the argon present in the modern atmosphere. This gas is completely inert (and thus is called a "noble" gas) and therefore is unusable in life processes. Because industry has found uses for inert argon (in light bulbs, welding, and some lasers), it is extracted from the atmosphere, along with nitrogen and oxygen, for commercial and industrial uses (Figure 3-6).

Although **carbon dioxide** in the atmosphere has gradually increased over the past 200 years,

FIGURE 3-6
Air is a major industrial and chemical raw material that is extracted from the atmosphere by air mining companies. [Photo by author.]

it is basically a stable atmospheric component and thus is included in Table 3-1. It is a natural by-product of life processes, and the implications of its current increase are critical to society and the future. The role of carbon dioxide in the gradual warming of Earth is discussed in Chapters 5 and 10.

Based on temperature criteria, we can subdivide the homosphere into three layers: the mesosphere, the stratosphere, and the one closest to Earth's surface, the troposphere.

Mesosphere. The **mesosphere** is the area from 50 up to 80 km (30–50 mi). The mesopause is the coldest portion of the atmosphere, averaging −90°C (−130°F), although that temperature may vary by ±25–30C° (±45–54F°). Very low pressures ranging from 0.1 mb to 0.001 mb exist in the mesosphere. The lower limit for the auroral glow is the upper mesosphere. In addition, the mesosphere sometimes receives cosmic or meteoric dust, which acts as nuclei for forming fine ice crystals. At high latitudes an observer may see these bands of crystals glow in rare and unusual night clouds called **noctilucent clouds**.

The *Upper Atmosphere Research Satellite (UARS),* launched and operational during the fall 1991, is detecting large, continent-sized windstorms in the mesosphere. The very rarefied air is moving in vast waves at speeds in excess of 320 kmph (200 mph). The importance of this new discovery is being considered.

Stratosphere and Ozonosphere. The **stratosphere** extends from 20 to 50 km (12.4–31 mi) above Earth's surface. Temperatures increase throughout the stratosphere, from −57°C (−70°F) at 18 to 20 km (tropopause) warming to 0°C (32°F) at 50 km (stratopause). Meanwhile, pressures decrease from 175 mb at the tropopause down to only 0.78 mb at the stratopause. The heat source throughout the stratosphere is the other functional layer, called the **ozonosphere**, or **ozone layer**. The ozone layer is composed of ozone, highly reactive oxygen molecules made up of three oxygen atoms (O_3) instead of the usual two atoms (O_2) that make up oxygen gas. Ozone absorbs wavelengths of ultraviolet light (0.1–0.3 micrometer) and subsequently reradiates that energy at longer wavelengths, as infrared energy. Through this process, most harmful ultraviolet radiation is effectively "filtered" from the incoming solar radiation, safeguarding life at Earth's surface and heating the stratosphere.

The ozone layer, presumed to be relatively stable over the past several hundred million years (allowing, of course, for daily and seasonal fluctuations), is now in a state of continuous change. Focus Study 3-1 presents a background analysis of the developing crisis in this critical section of our atmosphere. The spatial implications for humanity of losses in stratospheric ozone make this an important topic for applied studies in physical geography and atmospheric sciences.

Troposphere. The home of the biosphere is the fourth temperature region of the atmosphere nearest Earth's surface, the **troposphere.** Approximately 90% of the total mass of the atmosphere and the bulk of all water vapor, clouds, weather, and air pollution are contained within the troposphere. The *tropopause* is defined by an average temperature of –57°C (–70°F), but its exact elevation varies with the season, latitude, and surface temperatures and pressures. Near the equator, because of intense heating from below, the tropopause occurs at 18 km (11 mi); in the middle latitudes, it occurs at 13 km (8 mi); and at the North and South poles it is only 8.0 km (5.0 mi) or less above Earth's surface.

Figure 3–7 illustrates the temperature profile within the troposphere during the daytime. As the graph shows, temperatures decrease with increasing altitude at an average of 6.4C° per km (3.5F° per 1000 ft), a rate known as the **normal lapse rate.** The *actual* lapse rate at any particular time and place under local weather conditions, called the **environmental lapse rate,** may vary greatly from the normal lapse rate. This variance in temperature in the lower troposphere is important to our discussion of weather processes (Chapter 8).

The marked warming in the stratosphere with increasing altitude causes the tropopause to act like a lid, essentially preventing whatever is in the cooler air below from mixing into the stratosphere. However, the tropopause is disrupted above the midlatitudes wherever jet streams occur (Chapter 6), and the resulting vertical turbulence does allow some interchange between the troposphere and the stratosphere. Also, hurricanes occasionally inject moisture above this inverted temperature layer, and powerful volcanic eruptions may push ash and sulfuric acid mists into the stratosphere, as did the Mount Pinatubo eruptions that began in 1991.

Within the troposphere, both natural and human-caused variable gases, particles, and other chemicals are part of the atmosphere. The spatial aspects of these variable pollution components are a significant applied topic in physical geography.

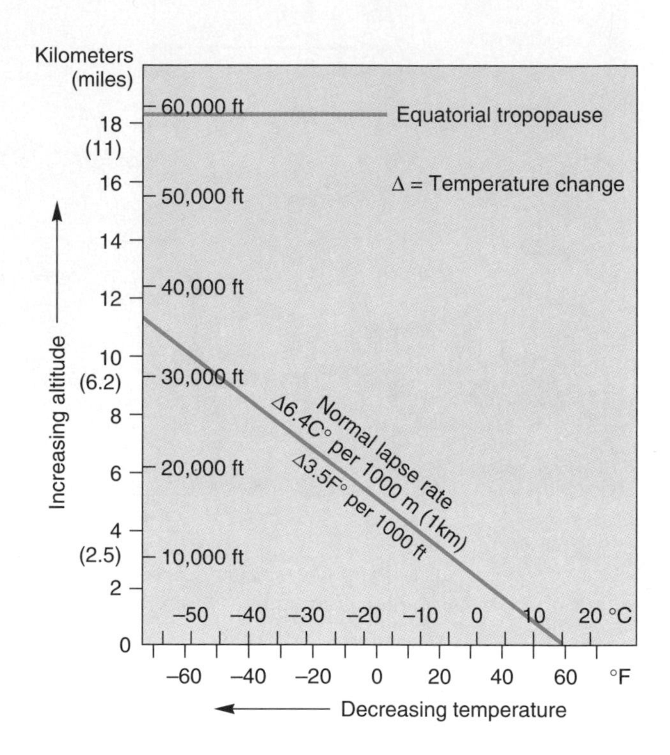

FIGURE 3-7
The temperature profile of the troposphere. During daytime it decreases at a rate known as the normal lapse rate.

VARIABLE ATMOSPHERIC COMPONENTS

Air pollution is not a new problem. Romans complained over 2000 years ago about the foul air of their cities. The stench of open sewers, fires, kilns, and smelters filled the air. In human experience, cities are always the place where the environment's natural ability to process and recycle waste is most taxed. English diarist John Evelyn recorded in 1684 that the air in London was so filled with smoke that ". . . one could hardly see across the streets, and this filling the lungs with its gross particles exceedingly obstructed the breast so as one could hardly breathe."* Air pollution is closely

*W. Wise, *Killer Smog: The World's Worst Air Pollution Disaster,* Chicago: Rand McNally, 1968.

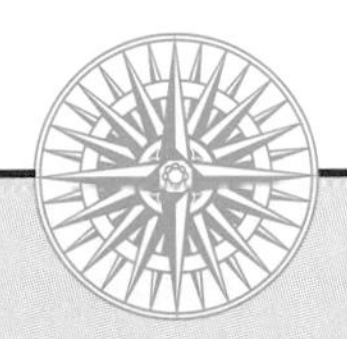

Focus Study 3-1 The Stratospheric Ozone Predicament

Why is the ozone layer so important? As noted in the text, it reduces ultraviolet radiation levels from the Sun to relatively safe levels at Earth's surface. Even a 1% decrease in stratospheric ozone increases surface exposures to ultraviolet light by 1–3%, leading to a 2–5% increase in skin cancers. Most affected are light-skinned persons who live at higher elevations and those who work principally outdoors. However, eventually everyone is affected.

Skin cancers and related precursory ailments are steadily increasing in North America, now totaling more than 700,000 cases a year, including over 10,000 malignant melanoma deaths annually. Skin cancer occurrences are increasing at an alarming 10% per year. Damage to eye tissues and reduction in the effectiveness of certain body immune systems are considered other possible effects of reduced stratospheric ozone. Also, potential damage to crop yields and certain aquatic life forms are probable consequences of ozone loss.

An oceanographer at Texas A&M, in addition to scientists at Scripps Institution of Oceanography, completed studies in 1989 that proved the vulnerability of phytoplankton to such severe increases in ultraviolet light in the oceans and seas of polar regions. In 1990, scientists at the University of California at Santa Barbara, San Diego, and San Francisco campuses, and at Moss Landing Marine Labs, measured a 6–12% reduction in primary production by phytoplankton in Antarctic waters due to increased ultraviolet radiation reaching the surface. The scientists concluded:

> **We find that the O_3-induced loss to a natural community of phytoplankton in the marginal ice zone is measurable, and the subsequent ecological consequence . . . of this early spring loss [August to October in the Southern Hemisphere] remains to be determined.***

This does not bode well for essential food chains in affected oceans.

Ozone Monitoring

Stratospheric ozone has a long history of measurement from the ground, balloons, aircraft, and orbiting satellites. The Dobson instrument, a spectrophotometer, for ground-based measurement of ozone layer thickness, was designed in the 1920s. Measurements have been made daily at a station in Switzerland since 1926. Balloon-launched ozonosondes measure ozone from the ground to approximately 30 km altitude (18.6 mi). A Total Ozone Mapping Spectrometer (TOMS) went into operation in November 1978 aboard the *Nimbus*-7 satellite. The new *Upper Atmosphere Research Satellite (UARS)* is now actively monitoring stratospheric ozone losses and chlorine monoxide (ClO) concentrations in refined detail.

A 30-station ozone monitoring network has functioned since 1957, principally in the Northern Hemisphere, although worldwide coverage is expanding. The entire network is coordinated with a "world standard" set each summer at Mauna Loa, Hawaii. Monthly reports are published by the World Ozone Data Center in Toronto.

A Fragile Veil

The stratospheric ozone layer is rarefied. A sample from the densest part of the ozone layer—at an altitude of 29 km (18 mi)—contains only 1 part ozone per 4 million parts of air. Yet this layer has maintained a steady-state equilibrium for at least 500 million years. Since World War II, human activities have begun to alter this long-standing system into a dynamic condition that exhibits an overall declining ozone concentration. The possible depletion of the ozone layer by human activity was first suggested during the summer of 1974 by professors F. Sherwood Rowland and Mario J. Molina of the University of California at Irvine. Their article appeared in the British journal *Nature* and triggered the great ozone debate.*

Rowland and Molina proposed that chlorine atoms in manufactured chemicals were depleting Earth's ozone layer. Because measurement and confirmation are difficult on a scale as vast as the atmosphere, the industries producing these products successfully delayed action for 15 years, claiming that hard evidence was lacking to substantiate the ozone-depletion theory. Now, after much scientific evidence and verification of actual depletion of stratospheric ozone by chlorine atoms, even industry admits the problem is real.

---

*R. C. Smith, B. B., Prézelin, K. S., Baker, et al., "Ozone Depletion: Ultraviolet Radiation and Phytoplankton Biology in Antarctic Waters," *Science* 255 (12 February 1992): 952–58.

*F. Sherwood Rowland and Mario J. Molina, *Nature* 249 (1974): 810.

Chlorofluorocarbon compounds, or CFCs, are large manufactured molecules containing chlorine, fluorine, and carbon; they are still marketed under many trade names. CFCs are inert, meaning that they do not readily react with anything, and they possess remarkable heat properties useful for heat-transfer systems. These two traits make them valuable as aerosol propellants and refrigerants. By 1991 almost 18 million metric tons (19.8 million tons) of CFCs had been sold worldwide and subsequently released into the atmosphere.

Rowland and Molina hypothesized that the inert CFC molecules remained intact in the atmosphere, eventually migrating upward, working their way into the stratosphere. Being stable, CFCs do not dissolve in water and do not break down biologically. However, in the stratosphere, the increased ultraviolet light dissociates (splits) the CFC molecules, releasing chlorine (Cl) atoms and forming chlorine oxide (ClO) molecules. This chlorine then acts as a catalyst, stimulating a complex reaction that results in a reduction of ozone. (Chlorine and fluorine are halogens, so this is called halogen-catalyzed chemistry.) A single chlorine atom destroys tens of thousands of ozone molecules. Because each chlorine atom remains at ozone-layer altitudes for 40–100 years, the possible long-term consequences of these reactions pose an increasing threat through the next century. The rarefied ozone layer that protects us is a fragile veil indeed.

How Much Damage?

Rowland and Molina initially postulated losses of ozone that ranged from 7 to 13% over the next 100 years. Four different reports from the National Academy of Sciences (NAS) postulated varying estimates of depletion, ranging from 16.5% to only 2–4% (1979 and 1984 reports, respectively). With little action taking place in response to these measurements, Dr. Rowland stated his frustration, still applicable today:

> What's the use of having developed a science well enough to make predictions, if in the end all we are willing to do is stand around and wait for them to come true. . . . Unfortunately, this means that if there is a disaster in the making in the stratosphere, we are probably not going to avoid it.*

By 1991 ozone losses above the midlatitudes were running 6–8% per decade for the period December through March, much faster than scientists previously realized. Significant losses in stratospheric ozone now are being measured in both hemispheres at middle and high latitudes.

*Roger B. Barry, "The Annals of Chemistry," *The New Yorker* (9 June 1986): 83.

An International Response

Canada, Sweden, Norway, and the state of Oregon instituted bans on CFC propellants in aerosols between 1976 and 1979. The U.S. federal ban commenced in the fall of 1978. However, even with the U.S. ban, more than half of the country's production was not outlawed, including CFCs used as blowing agents in making polyurethane foam and as a refrigerant in commercial, residential, and automotive air conditioning systems. Sales of CFCs initially dropped after the aerosol ban, but started upward again by 1981 following an executive order by President Reagan that permitted the export of banned products. Spurred by this action, CFC sales continued to rise until a peak was hit in 1987 when 1.2 million metric tons (1.32 million tons) were sold. Fortunately, sales have started to decline as many nations implement measures to reduce demand, and as industry begins production of safer alternative chemicals. Production of CFCs in 1991 was down to 682,000 metric tons (750,000 tons).

In 1985, an initial meeting of 31 nations produced agreement on a short-term freeze of CFC production and a general consensus for a long-term phaseout. This led to the Montreal Protocol, signed in 1987 and revised in 1990. By 1992 some 92 nations, consuming and producing 90% of all CFCs, had signed the Protocol. Under the auspices of the United Nations Environment Program, the nations initially agreed to freeze production at 1986 levels beginning in 1990. The 12 nations of the European Community have committed to total elimination of CFCs, a commitment likely to be followed by the rest of the world. The revised treaty currently calls for an accelerated complete ban of all ultraviolet-sensitive halogens by the year 2000. With new support from the United States, a fund is being set up to assist developing countries with the phaseout. Developing countries have a decade grace period to assist them in meeting the phaseout goal.

Ozone Losses Over the Poles

British scientists at their Halley Base in Antarctica have recorded steady losses of stratospheric ozone from 1970 to the present, with huge temporary decreases

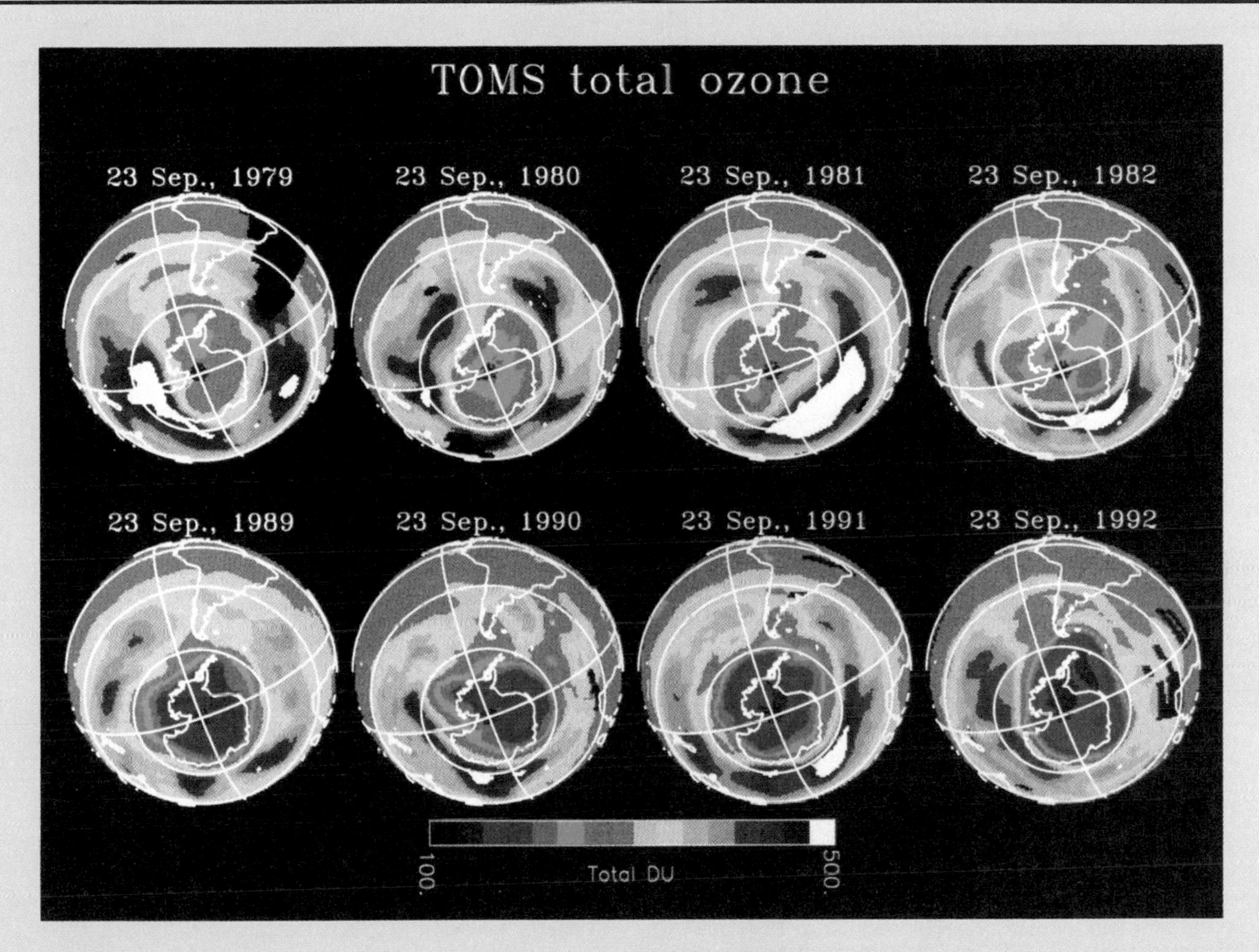

FIGURE 1
Nimbus-7 TOMS (Total Ozone Mapping Spectrometer) measurements for each September 23 from 1979 to 1982 and 1989 to 1992. The color scale represents ozone concentrations in Dobson units with blues and purples for amounts of less than 200. (One Dobson unit equals 2.69×10^{16} molecules of O_3 per cm^2.) Measurements dropped as low as 105 Dobson units in 1992 (1987 decreased to 120, 1991 was down to 110). The 1992 hole covered a record 23 million km^2 (9 million mi^2), or 15% larger than in 1991. This is an area larger than Canada, the United States, and Mexico combined. The ozone hole reached the southern tip of South America. [Images courtesy of Goddard Space Flight Center, NASA.]

occurring every Antarctic spring. In each succeeding year of measurement, the "hole" in the ozone has widened and deepened. NOAA satellites have recorded this alarming situation (Figure 1). Chlorine freed in the Northern Hemisphere midlatitudes evidently is redistributed by atmospheric winds, concentrating over Antarctica.

In Antarctica each year, stratospheric ozone concentrations approach total depletion between altitudes of 12 and 24 km (7.5–14.9 mi) over a surface area about twice the size of the Antarctic continent. The 1989–1992 holes exceeded the record depletion that occurred in 1987. The depletion deepens to a low in late September and October. Then, in November, the polar vortex dissipates, and bundles of ozone-depleted air migrate to lower latitudes. Thus is protective ozone depleted over southern portions of South America and Africa, and Australia and New Zealand.

An important catalyst that releases chlorine for ozone-depleting reactions is *polar stratospheric clouds* (PSCs). PSCs form over the poles during the long, cold winter months as tight circulation patterns form over the Antarctic continent—the *polar vortex.* Droplets in these clouds trigger the breakdown of otherwise inert molecular chlorine (Cl_2), freeing chlorine (Cl) to react with ozone (O_3), forming chlorine oxide (ClO) and oxygen molecules (O_2). The Cl is then freed by sunlight and continues to break apart more O_3. Figure 2 graphs the anticorrelation between ClO and O_3 over the Antarctic continent within the polar vortex. Such evidence confirmed the CFC-related catalyzed chemistry that is devastating the ozonosphere.

In the Arctic, the polar vortex mechanism is variable because atmospheric conditions are more changeable than over the South Pole. Extremely cold, persistent temperatures are needed if an ozone hole is to form in the Arctic similar in severity to the one over Antarctica. Arctic ozone depletion exceeded 35% in 1989 and again in 1990, 1991, and 1992. With continuing increases of CFCs in the stratosphere, scientists are closely watching the ozone layer each Northern Hemisphere spring season.

These holes in the polar ozone layer and general depletion overall provide an early warning to civilization about the spatial implications of the ozone predicament. Antarctic explorers and scientists in 1990 experienced what was described as "standing under an ultraviolet lamp 24 hours a day." Any exposed skin quickly burned and swelled, and protective clothing that was not carefully selected deteriorated rapidly. Hopefully, taking action now will reduce the hazard for life in the next century in more populated lower latitudes.

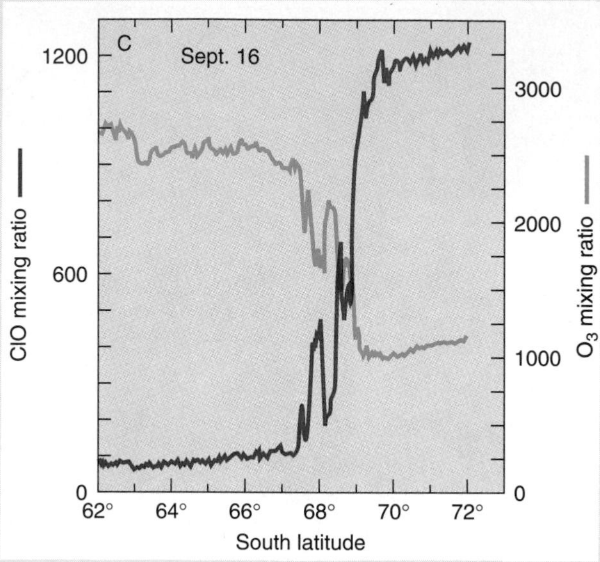

FIGURE 2
The anticorrelation between ClO and O_3 over the Antarctic continent poleward of 68° S latitude. The data were collected from flights during September 1987 at stratospheric altitudes. [Data from NASA.]

related to our production and consumption of energy, and as human occupation of a region increases in duration and complexity, the environmental impact on the troposphere also increases.

Natural Sources

Natural sources produce a greater quantity of nitrogen oxides, carbon monoxide, and carbon dioxide than do human-made sources. Table 3-2 lists some of these natural sources and the substances they contribute to the air. However, any attempt to diminish the impact of human-made air pollution through a comparison with natural sources is irrelevant, for we have coevolved with and adapted to the *natural* ingredients in the air, and we have *not* evolved in relation to the comparatively recent concentrations of anthropogenic (human-caused) contaminants in our metropolitan areas.

A dramatic natural source of pollution was produced by the 1991 eruption of Mount Pinatubo in the Philippines (15° N 120° E) that injected between 15 and 20 million tons of sulfur dioxide (SO_2) into the stratosphere. The spread of these emissions is shown in a sequence of satellite images presented in Chapter 6. This event gives scientists an opportunity to study naturally produced sulfuric acid mists in the atmosphere. Scientists

TABLE 3-2
Sources of Natural Variable Gases and Materials

Sources	Contribution
Volcanoes	Sulfur oxides, particulates
Forest fires	Carbon monoxide and dioxide, nitrogen oxides, particulates
Plants	Hydrocarbons, pollens
Decaying plants	Methane, hydrogen sulfides
Soil	Dust and viruses
Ocean	Salt spray and particulates

theorize that the smallest sulfuric acid droplets act as catalysts to free chlorine from CFC compounds and subsequently form chlorine monoxide—the destroyer of stratospheric ozone. Sulfuric acid aerosols mimic the role played by polar stratospheric clouds discussed in Focus Study 3-1. Evidently, the Antarctic ozone hole of 1992 was slightly worsened at lower altitudes (13–16 km, or 8–10 mi) by the debris from Pinatubo. Human-caused disruption produces total ozone destruction at higher altitudes, usually above 16 km.

Natural Factors That Affect Air Pollution

Augmenting the problems resulting from both natural and human-made atmospheric contaminants are several important natural factors. Among these are wind, local and regional landscape characteristics, and temperature inversions in the troposphere.

Winds gather and move pollutants from one area to another, sometimes reducing the concentration of pollution in one location while increasing it in another. Such air movements make the atmosphere's condition an international issue. Indeed, the movement of air pollution from the United States to Canada is the subject of much complaint and negotiation between these two governments. In Europe, the cross-boundary drift of pollution is a major issue because of the close proximity of nations and has led in part to Europe's unification.

Wind can produce dramatic episodes of dust movement (Figure 3-8). (Dust is defined as particles less than 62 microns, or 0.0025 in.). In 1977, a *GOES* weather satellite followed a 1 million km^2 (390,000 mi^2) dust cloud from Colorado and New

FIGURE 3-8
Natural variable pollution. (a) A dust storm in central Nevada. (b) Wind-blown alkali dust rising from high interior-drainage basins in the Andes Mountains of Chile and Argentina and blowing far over the Atlantic Ocean. [(a) Photo by author. (b) Space Shuttle photo from NASA.]

(a)

(b)

Mexico eastward. Former soil was lofted 5000 m (3.1 mi) in altitude. Chemical analysis is frequently employed by scientists to track such dust to its source area: dust from Africa contributes to the soils of South America and Texas dust ends up in Europe. An example of a point-source dust storm is shown in Figure 3-8b. Silt and alkaline dust are picked up by high-altitude winds in the Andes of South America and eventually carried out over the Atlantic.

Local and regional landscapes are another important factor in air pollution. Surrounding mountains and hills can form barriers to air movement or can direct pollutants from one area to another. Some of the worst incidents have resulted when local landscapes have trapped and caused a concentration of air pollution.

Temperature Inversion. Vertical temperature distribution in the troposphere also can contribute to conditions of pollution. A **temperature inversion** occurs when the normal decrease of temperature with increasing altitude (normal lapse rate) reverses at any point from ground level up to several thousand meters. Figure 3-9 compares a normal temperature profile with that of a temperature inversion. The normal profile illustrated in part (a) permits lifting air to ventilate surface pollution, but the warm air inversion shown in part (b) prevents the vertical mixing of pollutants with other atmospheric gases. Thus, instead of being carried away, pollutants are trapped under the inversion layer.

Inversions most often result from certain weather conditions, such as when the air near the ground is radiatively cooled on clear nights, or from topographic situations that can produce cold air drainage into valleys. In addition, the air above snow-covered surfaces or beneath subsiding air in a high-pressure system may cause a temperature inversion. As an example, in winter months in midwestern and eastern North America, high-pressure areas created by subsiding cold air masses produce inversion conditions that trap air pollution. In the western United States, summer subtropical high-pressure systems also cause inversions that produce air stagnation.

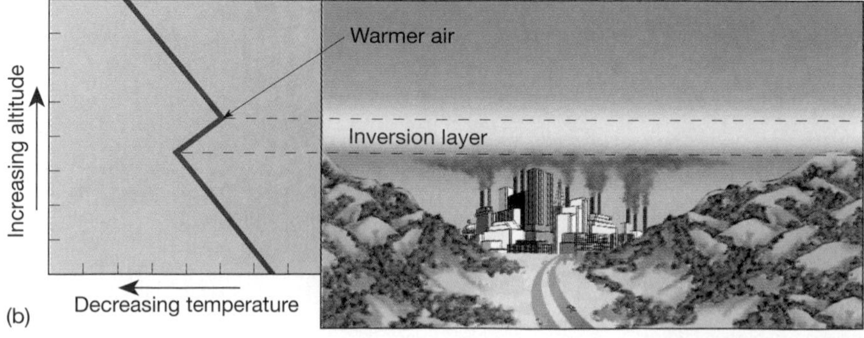

FIGURE 3-9
A comparison of a normal temperature profile (a) with a temperature inversion in the lower atmosphere (b).

TABLE 3-3
Anthropogenic Gases and Materials in the Lower Atmosphere

Name	Symbol	Source
Carbon Monoxide	CO	Incomplete combustion of fuels
Nitrogen oxides	NO_x (NO, NO_2)	High temperature/pressure combustion
Hydrocarbons	HC	Incomplete combustion of fuels
Ozone	O_3	Photochemical reactions
Peroxyacetyl nitrates	PAN	Photochemical reactions
Sulfur oxides	SO_x (SO_2, SO_3)	Combustion of sulfur-containing fuels
Particulates	—	Dust, dirt, soot, salt, metals, organics
Exotics	—	Fission products, other toxics
Carbon dioxide	CO_2	Complete combustion
Water vapor	H_2O vapor	Combustion processes, steam
Methane	CH_4	Organic processes

Anthropogenic Pollution

Anthropogenic, or human-caused, air pollution remains most prevalent in urbanized regions. Table 3-3 lists the names, symbols, and principal sources of variable anthropogenic components in the air. Seven of the components in Table 3-3 (carbon monoxide, nitrogen oxides, hydrocarbons, ozone, peroxyacetyl nitrates, sulfur oxides, and particulates) result from combustion of fossil fuels in transportation (specifically automobiles) and at stationary sources such as power plants and factories. Overall, automobiles contribute about 60% of American and 40% of Canadian human-caused air pollution. Environment Canada, an agency similar to the U.S. Environmental Protection Agency, reports that automobile emissions contribute 45% of the carbon monoxide, 20% of the nitrogen oxides, and 24% of the hydrocarbons in Canada. Figure 3-10 identifies major human-caused pollutants and their proportional sources in the United States. These pollutants are described in the following sections.

The last three gases shown in Table 3-3 are discussed elsewhere in this text: water vapor is examined with weather and the hydrologic cycle (Chapters 7 and 8), carbon dioxide with temperature and climate (Chapters 5 and 10), and methane with greenhouse gases (Chapter 5).

Carbon Monoxide. Carbon monoxide (CO) is an odorless, colorless, and tasteless combination of one atom of carbon and one of oxygen. CO is produced by incomplete combustion of fuels or other carbon-containing substances. A log decaying in the woods produces CO, as does a forest fire or other organic decomposition. Natural sources produce up to 90% of existing carbon monoxide, whereas anthropogenic sources, principally transportation, produce the other 10%. In the tropics of south-central Africa, an interesting source of CO is widespread burning of biomass (living material) during the dry season from August to October. This CO spreads throughout the Southern Hemisphere—Africa, Australia, Antarctica, and the south Atlantic. Generally, though, the carbon monoxide from human sources is concentrated in urban areas, where it directly impacts human health.

The toxicity of CO is well known; one sometimes hears of a death from carbon monoxide poisoning in an enclosed space with a running automobile engine or in a tent with an unvented heater. The toxicity of CO is due to its affinity for blood hemoglobin, which is the oxygen-carrying pigment in red blood cells. In the presence of CO, the oxygen is displaced and the blood becomes deoxygenated.

CO pollution did not become a problem until many fossil fuel–burning cars, trucks, and buses became concentrated in urban areas. Normal CO levels range from 10 to 30 parts per million (ppm) in metropolitan areas, but freeways and parking garages can reach 50 ppm CO, with an additional 30 ppm possible under a temperature inversion. These common levels in cities are dangerous because of the resultant reduction of oxygen in the

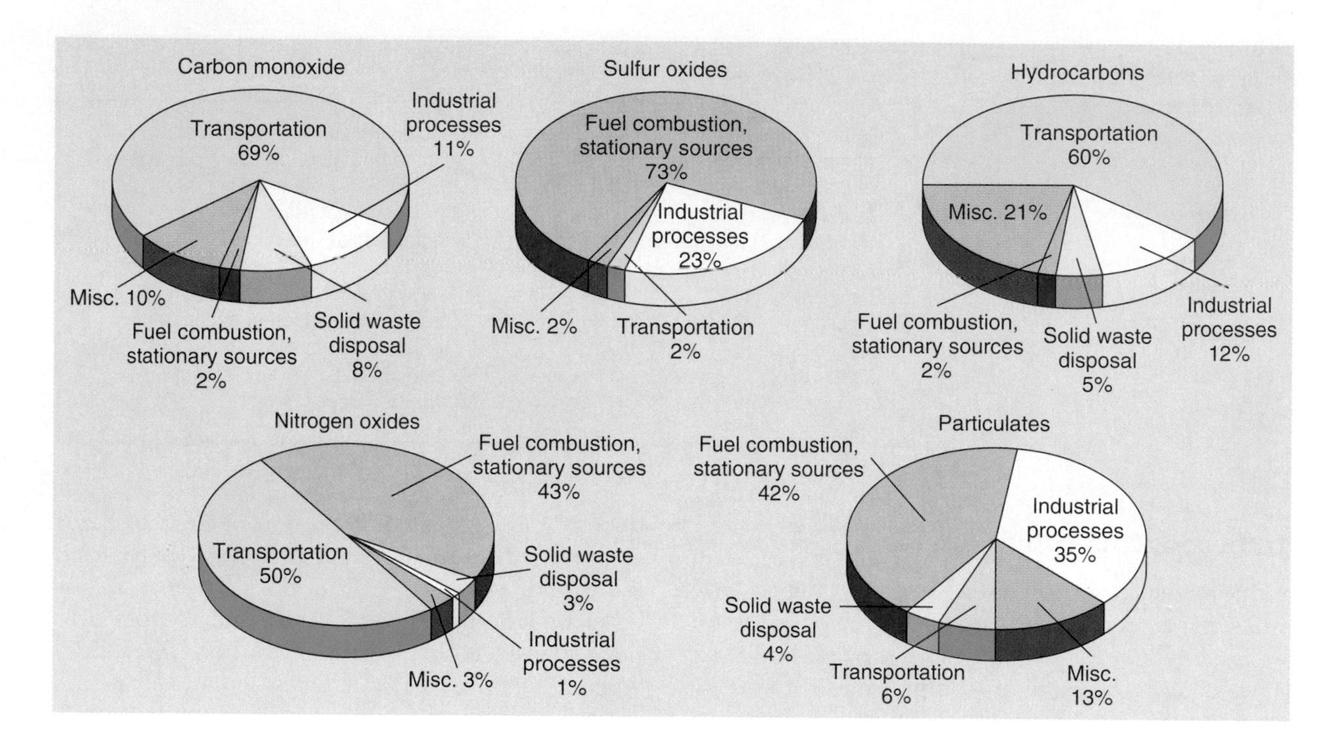

FIGURE 3-10
Major types of human-caused air pollution and their sources in the United States.

blood and the health symptoms experienced, including headaches, some vision and judgment loss, and impaired performance on simple tests.

Photochemical Smog Reactions. Photochemical smog is another type of pollution that was not generally experienced in the past but developed with the advent of the automobile. Today it is the major component of anthropogenic air pollution. **Photochemical smog** results from the interaction of sunlight and the products of automobile exhaust. Although the term "smog"—a combination of the words "smoke" and "fog"—is a misnomer, it is generally used to describe this phenomenon. Smog is responsible for the hazy sky and reduced sunlight in many of our cities.

Mexico City is notorious for poor air quality as its 20 million inhabitants work, commute, and live in the world's second largest metropolitan region. Conditions are worsened by frequent subtropical high-pressure systems that act as effective air traps over the Valley of Mexico. New laws were enacted in 1990 to reduce these unhealthy photochemical smog conditions: more public rapid transportation, controls on automobiles, limitations on factory operations, and more pollution-absorbing park space and trees. The contrast between a rare, clear day and frequent polluted days is dramatic (Figure 3-11).

The connection between automobile exhaust and smog was not determined until 1953, well after society had established its dependence upon individualized transportation. Despite this discovery, widespread mass transit has declined, the railroads have dwindled, and the polluting individual automobile remains America's preferred transportation mode.

Please examine Figure 3-12 as you read this paragraph. The increased temperatures in modern automobile engines cause reactions that produce **nitrogen dioxide** (NO_2). This nitrogen dioxide derived from automobiles, and to a lesser extent from power plants, is highly reactive with ultraviolet light. The reaction liberates atomic oxygen (O)

(a)

(b)

FIGURE 3-11
A rare day of clear air (a) in contrast to a more typical day of photochemical smog pollution (b) in the skies over Mexico City. [Photos by Larry Reider/PIX/Sipa Press.]

and a nitric oxide (NO) molecule from the NO_2 (Figure 3-12). The free oxygen atom combines with an oxygen molecule (O_2) to form the oxidant ozone (O_3); this same gas that is so beneficial in the stratosphere is an air-pollution hazard at Earth's surface. In addition, the nitric oxide (NO) molecule reacts with hydrocarbons (HC) to produce a whole family of chemicals generally called **peroxyacetyl nitrates (PAN)**. PAN produces no known health effect in humans but is particularly damaging to plants, including both agriculture and forests. Damage in California is estimated to exceed 1 billion dollars a year in the farming sector.

Nitrogen dioxide (NO_2) is a reddish-brown gas that damages and inflames human respiratory systems, destroys lung tissue, and damages plants. Concentrations of 3 ppm are dangerous enough to require alerting parents to keep children indoors; a level of 5 ppm is extremely serious. Worldwide, the problem with nitrogen dioxide production is its concentration in metropolitan regions. North American urban areas may have from 10 to 100 times higher nitrogen dioxide concentrations than nonurban areas. Nitrogen dioxide interacts with water vapor to form nitric acid (HNO_3), a contributor to acid deposition by precipitation, the subject of Focus Study 3-2. Emission levels of nitrogen oxides characteristic of the 1980s and wet deposition of nitrogen oxides during 1988 are shown in Figure 3-13.

Ozone Pollution. Ozone is highly reactive and results in various negative consequences. Paint oxidizes, painted surfaces peel, elastic and rubber dry and crack, paper dries and yellows, and plants are damaged at levels of only 0.01–0.09 ppm. Extensive studies have found that exposure to 0.1 ppm for a few hours a day for several days to several weeks causes significant crop damage and reduces yields as much as 50% for a wide variety of agricultural crops. At 0.3 ppm eye, nose, and throat irritation is noticeable. As concentration increases to 1.0–3.0 ppm, extreme fatigue and poor coordination can appear in humans.

The economics of ozone damage to plants was studied by the U.S. National Crop Loss Assessment Network (NCLAN) and reported through the U.S. Environmental Protection Agency (EPA). Similar

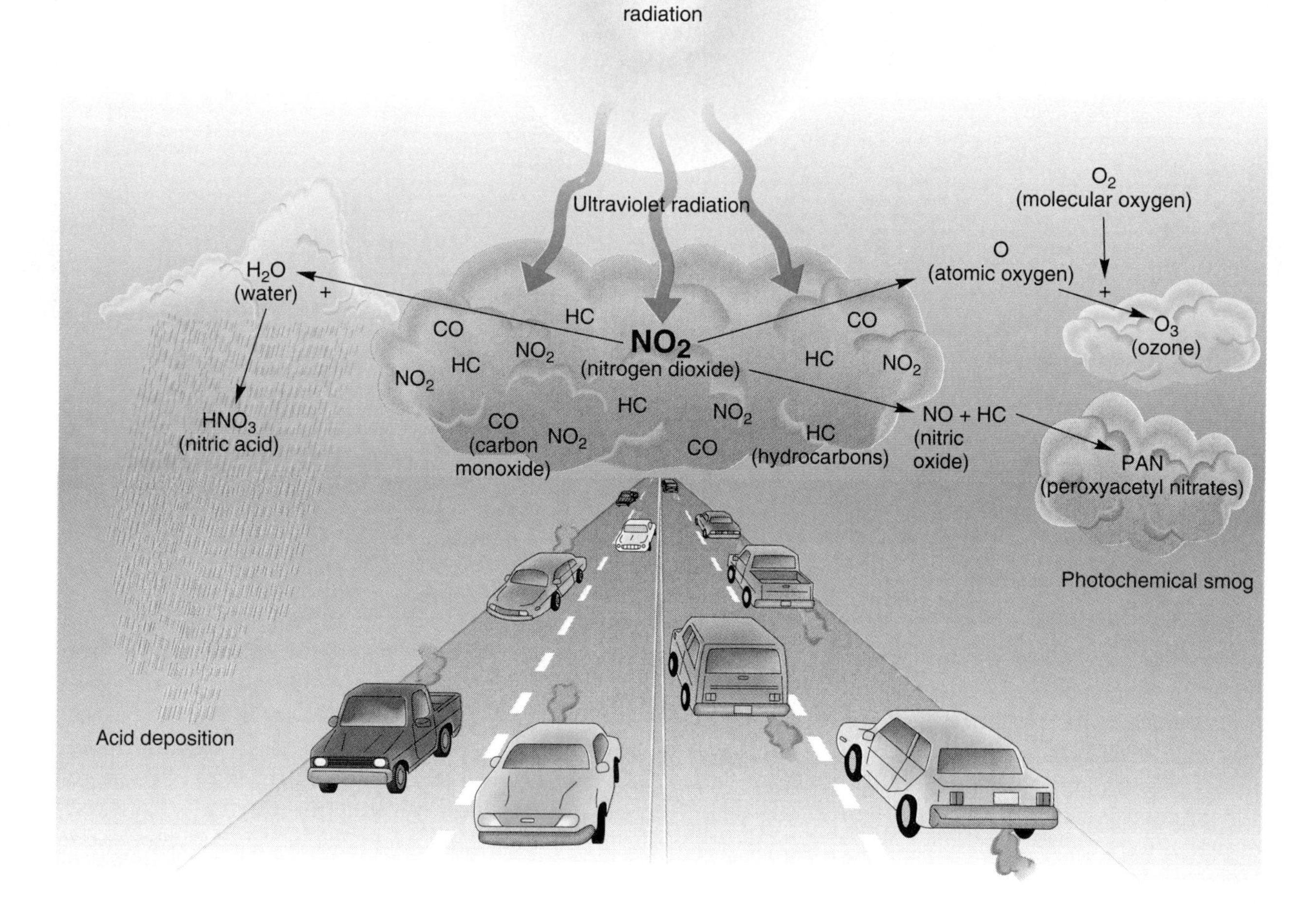

FIGURE 3-12
Photochemical reactions are produced through the interaction of automobile exhaust (NO_2, HC, CO) and ultraviolet radiation in sunlight. The resulting photochemical pollution includes ozone and PAN.

studies have examined the damage done by photochemical smog to human health and property, effect on mortality rate, hospital usage, and indirect health consequences. An 11-year study completed at UCLA found significant deterioration of lung function after chronic exposure to oxidants and air pollution.

The benefits of implementing air-pollution-abatement strategies far outweigh the implementation cost of known solutions. The South Coast Air Quality Management District, which oversees the Los Angeles Basin, estimated in 1991 that full implementation of their clean air plan would cost about $6.1 billion and generate at least $9.4 billion

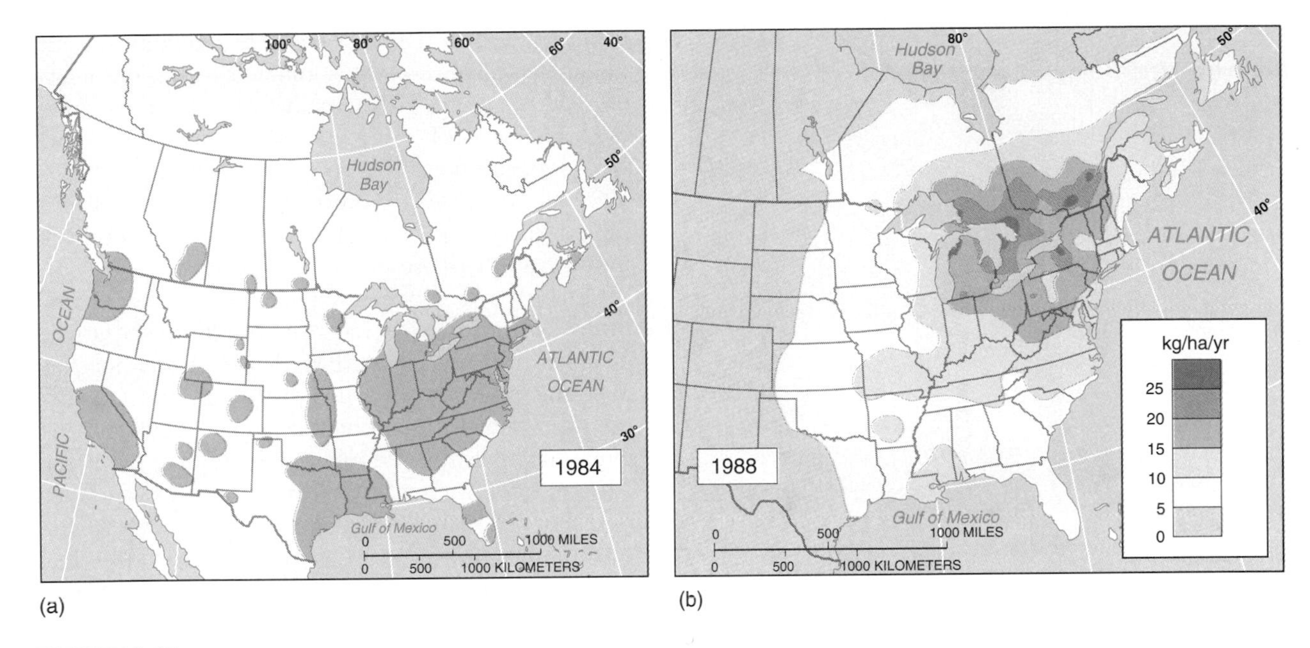

FIGURE 3-13
Spatial portrayal of (a) distribution of emissions of nitrogen oxides in 1984, and (b) annual wet deposition of nitrogen oxides (principally NO_3) on the landscape for the year 1988 (in kilograms per hectare per year). [Data courtesy of Dr. Chul-Un Ro, National Atmospheric Chemistry Data Base (NAtChem), Air Quality Measurements and Analysis Research Division, Environment Canada.]

in benefits from ozone reduction alone. Thus, taking action actually is a cheaper strategy than allowing the situation to degrade further.

Industrial Smog and Sulfur Oxides. Over the past 300 years, coal has slowly replaced wood as the basic fuel used by society. The industrial revolution required high-grade energy to run machines and to facilitate the conversion from animate energy (energy from living sources, such as animals powering farm equipment) to inanimate energy (energy from nonliving sources, such as coal, steam, and water). The air pollution associated with coal-burning industries is known as **industrial smog** (Figure 3-14). The term smog was coined by a London physician at the turn of this century to describe the combination of fog and smoke containing sulfur, an impurity in fossil fuels.

Sulfur dioxide (SO_2), which forms during combustion, is colorless but can be detected by its pungent odor. At low concentrations of from 0.1 to 1 ppm, the air is dangerous to society; at 0.3 ppm the air takes on a definite taste. Human respiratory systems can become irritated, especially in individuals with asthma, chronic bronchitis, and emphysema.

Once in the atmosphere, SO_2 reacts with oxygen to form sulfur trioxide (SO_3), which is highly reactive and, in the presence of water or water vapor, forms sulfuric acid (H_2SO_4). Sulfuric acid can occur in even moderately polluted air, and its synthesis does not require high temperatures. In this manner, a sulfur dioxide–laden atmosphere is dangerous to health, corrodes metals, and deteriorates stone building materials at accelerated rates. Sulfuric acid deposition has increased in severity since it was analyzed and described in the 1970s and is a growing blight in the biosphere. Focus Study 3-2 discusses this vital atmospheric issue.

In the United States, electric utilities and steel manufacturing are the main sources of sulfur dioxide. And, because of the prevailing movement of air masses, they are the main sources of sulfur dioxide in Canada, too. This conclusion comes

FIGURE 3-14
Industrial smog. Pollution generated by industry differs from that produced by transportation. Industrial pollution has high concentrations of sulfur oxides, particulates, and carbon dioxide. [Photo by author.]

from detailed studies that show as much as 70% of Canadian sulfur dioxide is initiated within the United States. Figure 3-15 portrays representative production levels of sulfur dioxide in the 1980s and wet deposition of sulfur oxides during 1988. In 1988 alone, 20,000 industrial facilities poured 1.1 billion kilograms (2.4 billion pounds) of chemicals into the atmosphere in the United States.

Certainly, society cannot simply halt two centuries of industrialization; the resulting economic chaos would be devastating. However, neither can it permit pollution production to continue unabated, for catastrophic environmental feedback inevitably will result. People have been on Earth for such a brief fragment of Earth's history, and yet our activities can be a major influence on the future of the atmosphere. We are now contributing significantly to the creation of the **anthropogenic atmosphere**, a tentative label for Earth's next (fifth) atmosphere. The urban air we breathe today may be just a preview.

FIGURE 3-15
Spatial portrayal of (a) sulfur dioxide emissions in 1984, and (b) wet deposition of sulfur oxides (principally SO_4) on the landscape for the year 1988 (in kilograms per hectare per year). In (a), the dots locate sources producing 100–500 metric kilotons (110,000–550,000 tons) per year; the squares locate sources producing 500 metric kilotons (500,000 tons) or more per year. [Data courtesy of Dr. Chul-Un Ro, National Atmospheric Chemistry Data Base (NAtChem), Air Quality Measurements and Analysis Research Division, Environment Canada.]

(a)

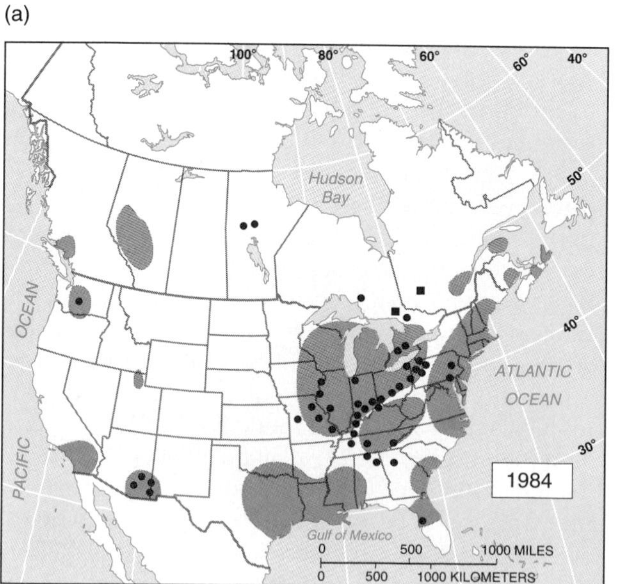

(b)

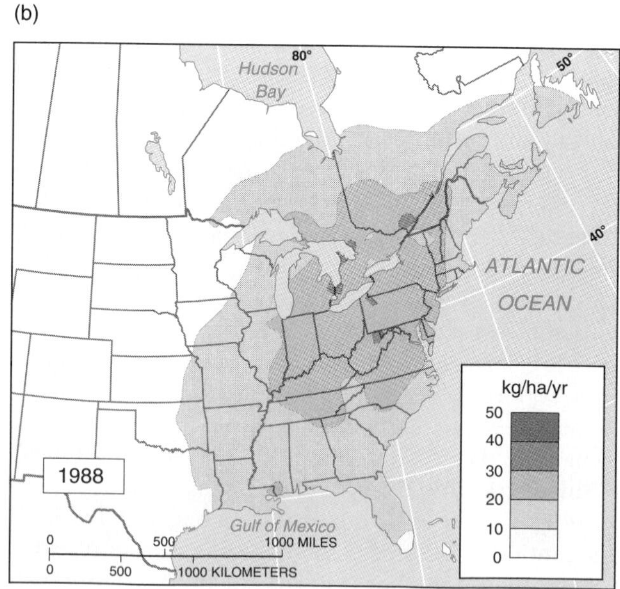

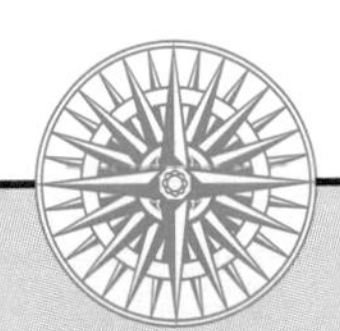

FOCUS STUDY 3-2 Acid Deposition: A Blight on the Landscape

Acid deposition is a major environmental problem in some areas of the United States, Canada, Europe, and Asia. Such deposition occurs in a dry form as dust or aerosols or wet in the form of what is called acid rain or snow. In addition, winds can carry the acid-producing chemicals many kilometers from their sources before they settle on the landscape, entering streams and lakes as runoff and groundwater flows.

Acid deposition is causally linked to declining fish populations and fish kills in the northeastern United States, southeastern Canada, Sweden, and Norway; to widespread forest damage in these same places and Germany; and to damage to buildings, sculptures, and historical artifacts. Corrective action has been delayed because of the complexity of the problem, which the U.S. General Accounting Office calls a "combination of meteorological, chemical, and biological phenomena."

The acidity of precipitation is measured on the pH scale, which expresses the relative abundance of free hydrogen ions ($H+$) in a solution. Free hydrogen ions in a solution are what make an acid corrosive, for they easily combine with other ions. The pH scale is logarithmic: each whole number represents a 10-fold change. A pH of 7.0 is neutral (neither acidic nor basic), values below 7.0 are increasingly acidic, and values above 7.0 are increasingly basic or alkaline. (A pH scale for soil acidity and alkalinity is graphically portrayed in Chapter 18.)

Natural precipitation dissolves carbon dioxide from the atmosphere, a process that releases hydrogen ions and produces an average pH reading of 5.65. The normal range for precipitation is 5.3–6.0. Thus, normal precipitation is always slightly acidic.

Certain anthropogenic gases are converted in the atmosphere into acids that are removed by wet and dry deposition processes. Nitrogen and sulfur oxides (NO_x and SO_x) released in the combustion of fossil fuels can produce nitric acid (HNO_3) and sulfuric acid (H_2SO_4) in the atmosphere. Figures 3-13 and 3-15 depict the patterns of NO_x and SO_x emissions and wet deposition in the United States and Canada. Precipitation as acidic as pH 2.0 has fallen in the eastern United States, Scandinavia, and Europe. By comparison, vinegar and lemon juice register slightly less than 3.0. Aquatic plant and animal life perishes when lakes drop below pH 4.8.

More than 50,000 lakes and some 96,500 km (60,000 mi) of streams in the United States and Canada are at a pH level below normal (i.e., below 5.3), with several hundred lakes incapable of supporting any aquatic life. Acid deposition causes the release of aluminum and magnesium from clay minerals in the soil, and both of these elements negatively impact fish and plant communities. Also, relatively harmless mercury deposits in lake-bottom sediments are converted by acidified lake waters into highly toxic methylmercury, which also is deadly to aquatic life. Local health advisories in two provinces and 22 states are regularly issued to warn those who fish of the methylmercury problem. Mercury atoms rapidly bond with carbon and move through biological systems as an *organometallic compound.*

Damage to forests results from the rearrangement of soil nutrients, the death of soil microorganisms, and an aluminum-induced calcium deficiency that is currently under investigation. Regional-scale decline in forest cover is significant. Europe is experiencing a more advanced impact on their forests than is evidenced elsewhere, principally due to the burning of coal and density of industrial activity. In Germany, up to 50% of the forests are dead or damaged; in Switzerland 30% are afflicted. The percentage of forest loss is higher in the nations of eastern Europe. Earth Satellite Corporation analyzed a small area of Poland that is heavily industrialized. The region produces and burns about 98% of Poland's coal. The two *Landsat* images allow you to do a GIS-type comparison of changes between 1981 and 1989 (Figure 1). Brighter colors on the left depict a healthier vegetation cover. The predominately blue colors on the right show the decline in forests and vegetation. Over 50% of Poland's forests are dead or dying. A high correlation exists between these devastated areas and acid rain–producing industrial activity.

In the United States, trees at higher elevations in the Appalachians are being injured by acid-laden cloud cover. In New England, some stands of spruce are as much as 75% affected, as evidenced through analysis of tree growth rings, which become narrower in adverse growing years. Another possible indicator of forest damage is the reduction by almost half of the annual production of U.S. and Canadian maple sugar.

The National Academy of Sciences (NAS) and the National Research Council have identified causal relationships between fossil fuel combustion and acidic deposition, citing evidence that is "overwhelming." In-

1981

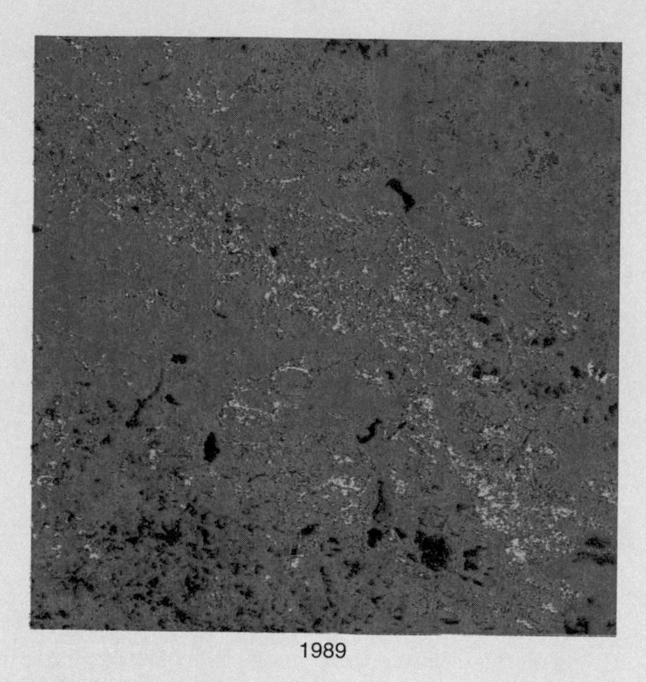

1989

FIGURE 1
The harm done to forests and crops by acid deposition is well established, especially in Europe. A heavily industrialized area of Poland was scanned by a *Landsat* satellite in 1981 and again in 1989 for a GIS-type comparison of the forest cover. False coloration helps in the analysis: brighter colors on the 1981 image (left) depict healthier vegetation, whereas the predominately blue patterns on the 1989 image denote over 50% of the forests in decline and death. [Image courtesy of Earth Satellite Corporation, Rockville, MD.]

spite of this consensus, politics and special interests have halted preventive measures. Government estimates place damage in the United States, Canada, and in Europe at over $50 billion annually. Because wind and weather patterns are international, efforts at reducing acidic deposition also must be international in scope. The NAS stated that a reduction of 50% in combustion emissions would reduce acid deposition by 50%.

At best, acid deposition is an unfolding international political issue of global spatial significance for which science is providing strong incentives for action. *Climatic Perspectives,* a publication from Environment Canada, now regularly reports acid rain data for various sites, including pH measurement and "air path to site." Not until March 1991 did the United States finally agree to a landmark agreement with Canada to lower sulfur dioxide and nitrogen oxide emissions—a first.

SUMMARY—Earth's Modern Atmosphere

Each breath we take is a reminder of Earth's physical and natural history, for the atmosphere is a product of the planet and the inhalations and exhalations of all life that has existed. Our modern atmosphere is a gaseous mixture of ancient origins so evenly mixed it behaves as if it was a single gas, naturally odorless, colorless, tasteless, and formless. The principal substance of this atmosphere is air—the medium of life. The weight of the atmosphere, or air pressure, crushes in on all of us and averages around 1 kg/cm^2; we measure it with a barometer.

Our atmosphere is unique among the planets. Earth is the home of the only biosphere in the Solar System. The atmosphere is defined by various criteria. Using *composition*, we can divide the atmosphere into the heterosphere and the homosphere. With *temperature*, we can divide it into the thermosphere, mesosphere, stratosphere, and troposphere. Within the troposphere occur the biosphere and almost all weather, water vapor, and the bulk of pollution. Using *function* as a criterion, we can distinguish two regions: the ionosphere and the ozonosphere, both of which absorb life-threatening radiation and convert it into heat energy. The overall reduction of stratospheric ozone during the past several decades is actively being measured and analyzed by scientists. This ozone loss represents an unhealthy condition for society and many natural systems.

We coevolved with natural "pollution" and thus are adapted to it. However, we are not adapted to cope with our own anthropogenic pollution that constitutes a major health threat, particularly where people concentrate in cities. Photochemical and industrial smogs do not respect political boundaries and are carried by winds far from their point of origin to other nations. The distributions of human-produced carbon monoxide, nitrogen dioxide, ozone, PAN, and sulfur dioxide over North America, Europe, and Asia are related to transportation and energy production. So, energy conservation and efficiency, and cleaning of emissions are essential strategies for abating air pollution. Earth's next atmosphere most likely will be anthropogenic, or human-influenced in character. The air we breathe in our cities just might be a preview of the future.

KEY TERMS

air pressure
aneroid barometer
anthropogenic atmosphere
carbon dioxide
carbon monoxide
chlorofluorocarbon compounds (CFCs)
environmental lapse rate
exosphere
heterosphere
homosphere
industrial smog
ionosphere
kinetic energy
mercury barometer
mesosphere
nitrogen dioxide
noctilucent clouds
normal lapse rate
ozone layer
ozonosphere
peroxyacetyl nitrates (PAN)
photochemical smog
sensible heat
stratosphere
sulfur dioxide
temperature inversion
thermosphere
troposphere

REVIEW QUESTIONS

1. What is air? Where did the components in Earth's present atmosphere originate?
2. In view of the analogy by Lewis Thomas, characterize the various functions the atmosphere performs.
3. How does air exert pressure? Describe the basic instrument used to measure air pressure. Compare the two different types of instruments discussed.
4. What is normal sea level pressure in mm? mb? in.? kPa?

5. What three distinct criteria are employed in dividing the atmosphere for study?
6. Describe the overall temperature profile of the atmosphere.
7. Name the four most prevalent stable gases in the homosphere. Where did each originate? Is the prevalence of any of these changing at this time?
8. Why is stratospheric ozone so important? Summarize the ozone predicament and the present findings and treaties.
9. Evaluate Rowland and Molina's use of the scientific method in investigating stratospheric ozone depletion.
10. What is the difference between industrial smog and photochemical smog?
11. Describe the relationship between automobiles and the production of ozone and PAN in city air. What are the principal negative impacts of these gases?
12. How are sulfur impurities in fossil fuels related to the formation of acid in the atmosphere and acid deposition on the land?
13. Why are anthropogenic gases more significant to human health than those produced from natural sources?
14. In what ways does a temperature inversion worsen an air pollution episode?
15. In your opinion, what are the solutions to the growing problems in the anthropogenic atmosphere?

SUGGESTED READINGS

Albritton, D. L., R. Monastersky, J. A. Eddy, and others. "Our Ozone Shield," *Report to the Nation, Our Changing Planet*. Boulder, Colorado: University for Atmospheric Research, Fall 1992.

Anderson, J. G., D. W. Toohey, and W. H. Brune. "Free Radicals Within the Antarctic Vortex: The Role of CFCs in Antarctic Ozone Loss," *Science* 251 (4 January 1991): 39–46.

Barry, Roger B., and Richard J. Chorley. *Atmosphere, Weather, and Climate*, 5th ed. London: Methuen, 1987.

Brodeur, Paul. "Annals of Chemistry—In the Face of Doubt," *The New Yorker* (9 June 1986): 70–87.

Brune, W. H., J. G. Anderson, D. W. Toohey, et al. "The Potential for Ozone Depletion in the Arctic Polar Stratosphere," *Science* 252 (31 May 1991): 1260–66.

The Conservation Foundation. *State of the Environment—A View Toward the Nineties*. Washington, DC: The Conservation Foundation and the Charles Stewart Mott Foundation, 1987.

Ember, Lois R., Patricia L. Layman, Will Lepkowski, and Pamela S. Zurer. "The Changing Atmosphere—Implications for Mankind," *Chemical and Engineering News* (24 November 1986): entire issue.

Federal/Provincial Research and Monitoring Coordinating Committee. *The 1990 Canadian Long-Range Transport of Air Pollutants and Acid Deposition Assessment Report,* Part 3: *Atmospheric Sciences*. Ottawa: Environment Canada, 1990.

Hall, Jane V., A. M. Winer, M. T. Kleinman, et al. "Valuing the Health Benefits of Clean Air," *Science 255* (14 February 1992): 812–17.

Miller, G. Tyler, Jr. *Living in the Environment,* 5th ed. Belmont, CA: Wadsworth, 1988.

Miller, John M. "Acid Rain," *Weatherwise* 37, no. 5 (October 1984): 232–49.

Office of Technology Assessment. *Acid Rain and Transported Pollutants—Implications for Public Policy*. Washington, DC: U.S. Government Printing Office, 1984.

Office of Technology Assessment. *Catching Our Breath: Next Steps for Reducing Urban Ozone*. OTA-0-412. Washington, DC: U.S. Government Printing Office, July 1989.

Pope, C. Arden, Joal Schwartz, and Michael Ransom. "Daily Mortality and PM_{10} Pollution in Utah Valley," *Archives of Environmental Health 47*, no. 3 (May/June 1992): 211–17.

Stolarski, Richard, R. Bojkov, et al. "Measured Trends in Stratospheric Ozone," *Science 256* (17 April 1992): 342–49.

Stole, Richard S. "The Antarctic Ozone Hole," *Scientific American* 258, no. 1 (January 1988): 30–36.

Wayne, Richard P. *Chemistry of Atmospheres*. London: Oxford University Press, 1985.

Sandstone fins and billowing clouds in Arches National Park, Utah. [*Photo by author.*]

4

Atmosphere and Surface Energy Balances

Earth's biosphere pulses with flows of energy. Think for a moment of the annual pace of your own life and activities—all reflect atmosphere and surface energy patterns. The shifting seasonal rhythms of these energy patterns are discussed in Chapter 2. This chapter examines the cascade of energy through the atmosphere to Earth's surface, as insolation is absorbed and redirected along various pathways in the troposphere.

The input of insolation and outputs of reflected light and emitted heat from our atmosphere and surface environment, together, determine the net radiation available to perform work. We examine surface energy budgets, analyzing how net radiation is expended at different locales. The cities in which we live alter surface energy characteristics, so the climates of our urban areas differ measurably from those of surrounding rural areas. A focus study concludes the chapter with an examination of solar energy as a renewable resource of great potential to society.

ENERGY BALANCE OF THE TROPOSPHERE

Earth's atmosphere and surface are heated by solar energy, which is unevenly distributed by latitude and which fluctuates seasonally. Our budget of atmospheric energy comprises shortwave radiation inputs (ultraviolet light, visible light, and near-infrared wavelengths) and longwave radiation outputs (thermal infrared, or heat). **Transmission** refers to the passage of shortwave and longwave energy, through either the atmosphere or water. The atmosphere and surface eventually radiate heat energy back to space, and this energy, together with reflected energy, equals the initial solar input. Thus a balance of energy input and output exists in our atmosphere.

Energy in the Atmosphere: Some Basics

When viewing a photograph of Earth taken from space, the pattern of surface response to incoming insolation is clearly visible (see the back cover of this text). Solar energy is intercepted by land and water surfaces, clouds, and atmospheric gases and dust. The flows of energy are evident in weather patterns, oceanic flows, and the distribution of vegetation. Specific energy patterns differ for deserts, oceans, mountain tops, rain forests, and ice-covered landscapes. In addition, clear or cloudy weather might mean a 75% difference in the amount of energy reaching the surface.

Figure 4-5 is a detailed flow diagram of the energy balance between Earth and the atmosphere. Please refer to it as you read this discussion.

Insolation Input. Insolation is the basic energy input driving the Earth-atmosphere system. The world map in Figure 4-1 shows the distribution of average annual solar energy at Earth's surface. It includes total radiation, both direct and diffuse (or downward scattered). The energy units are watts per square meter (100 W/m^2 = 75 $kcal/cm^2/year$).

Several patterns are notable on the map. Insolation decreases poleward from about 25° latitude in both the Northern and Southern hemispheres. Average annual values of 180–200 W/m^2 occur throughout the equatorial and tropical latitudes. In general, greater insolation of 240–280 W/m^2 occurs in low-latitude deserts worldwide because of frequently cloudless skies. Note the energy pattern in the cloudless subtropical deserts in both hemispheres (for example, the Sonoran, Saharan, Arabian, Gobi, Atacama, Namib, Kalahari, and Australian deserts).

Albedo and Reflection. A portion of arriving energy bounces directly back into space without being converted into heat or performing any work. This returned energy is called **reflection,** a term that applies to both visible and ultraviolet light. The reflective quality of a surface is its **albedo.** Albedo expresses the relationship of reflected energy to insolation as a percentage. Albedo is an extremely important control over the amount of insolation that is available to heat a surface. In the visible wavelengths, darker colors have lower albedos, and lighter colors have higher albedos. On water surfaces, the angle of the solar rays also affects albedo values; lower angles produce a greater reflection than do higher angles (Figure 4-2).

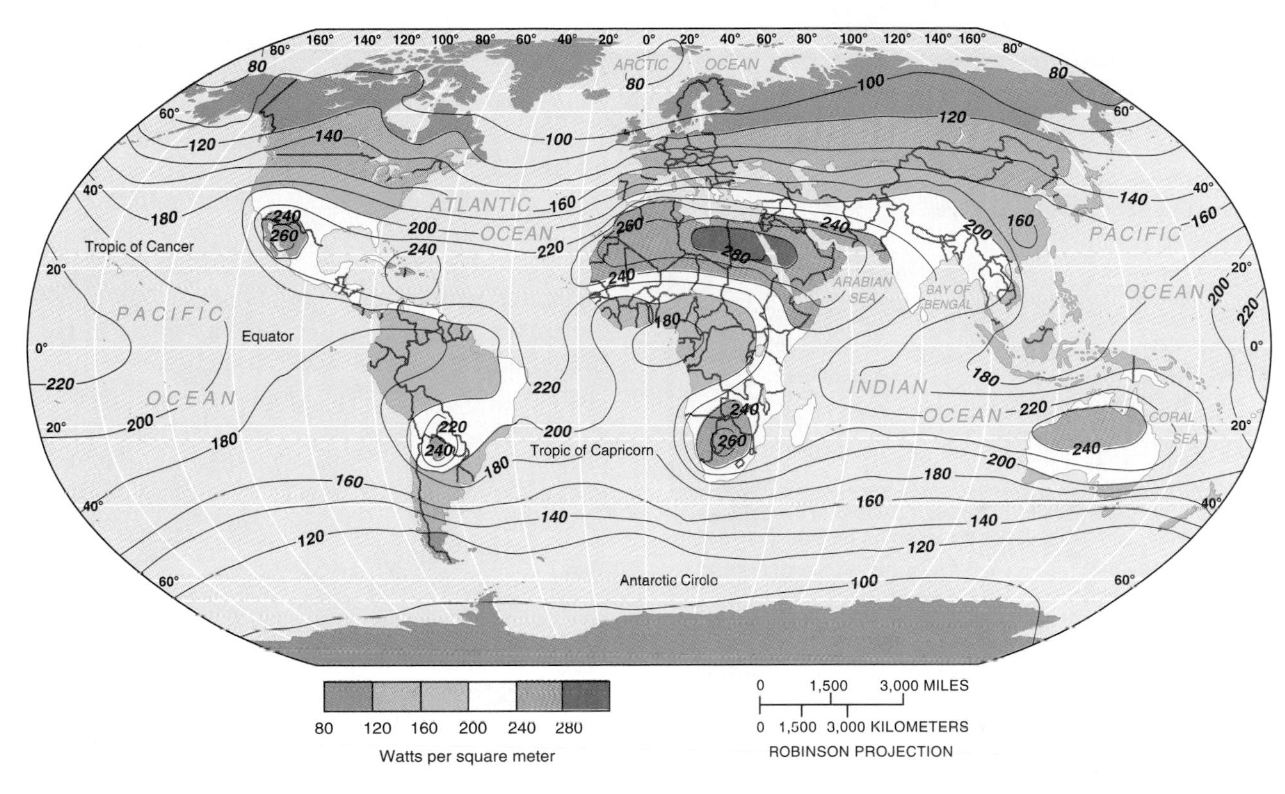

FIGURE 4-1
Average annual solar radiation receipt on a horizontal surface at ground level in watts per square meter (100 W/m^2 = 75 $kcal/cm^2/yr$). [After M. I. Budyko, 1958, and *Atlas Teplovogo Balansa,* Moscow, 1963.]

FIGURE 4-2
An astronaut view of sun glints off the Mozambique Channel near the east coast of Africa. The southern tip of the island Madagascar is visible to the right. Albedo values increase with lower sun angles and calmer seas. [Space Shuttle photo from NASA.]

TABLE 4-1
Selected Albedo Values of Various Surfaces

Type of Surface	Albedo (% Reflected)
Fresh snow	80–95
Snow, polluted, several days old	40–70
Sea ice	30–40
Forests	10–20
Crops, general	10–25
Dry, light, sandy soils	25–45
Dark soils	5–15
Concrete, dry	17–27
Asphalt, black top	5–10
Earth (average)	31+
Moon	6–8

SOURCE: M. I. Budyko, 1958, *The Heat Balance of the Earth's Surface,* Washington, D.C: U.S. Department of Commerce, p. 36.

In addition, smooth surfaces increase albedo, whereas rougher surfaces reduce it. Table 4-1 presents albedo values for various surfaces.

Specific locations experience highly variable conditions during the year in response to changes in cloud and ground cover. Earth Radiation Budget (ERB) sensors aboard the *Nimbus-7* satellite measure average albedos of 19–38% between the tropics to as high as 80% in the polar regions. Earth and its atmosphere reflect 31% of all insolation when averaged over a year. Figure 4-5 shows that Earth's average albedo is a combination of 21% reflected by clouds, 3% reflected by the ground (combined continental and oceanic surfaces), and 7% reflected and scattered by the atmosphere. By comparison, a full Moon, which is bright enough to read by under clear skies, has only a 6–8% albedo value. Thus, with earthlight being four times greater in albedo than moonlight, and with Earth four times larger than the Moon, it is no surprise that astronauts report how startling our planet looks from space.

Figure 4-3 portrays total albedos for July 1985 (a) and January 1986 (b) as measured by the Earth Radiation Budget Experiments (ERBE) aboard several satellites. These patterns are typical of most years. As compared to July, note the higher January albedos poleward of 40° N caused by snow and ice covering the ground. Tropical forests are characteristically lower in albedo (15%), whereas generally cloudless deserts have higher albedos (35%). The southward-shifting cloud cover over equatorial Africa is quite apparent on the January map.

A major uncertainty in the tropospheric energy budget, and therefore in refining climatic models, is the role of clouds. Clouds reflect *insolation,* thus cooling Earth's surface. Yet, clouds act as *insulation,* thus trapping longwave radiation and raising minimum temperatures. Important is not merely the percentage of cloud cover but the cloud type, height, and thickness (water content and density). High-altitude, ice-crystal clouds have albedos of about 50%, whereas thick, lower cloud cover reflects about 90%. Understanding the nature of global cloud cover is crucial in refining computer models of predicted global warming. A section on clouds is in Chapter 7.

Other mechanisms affect atmospheric albedo and therefore temperature patterns. The eruption of Mount Pinatubo in the Philippines, beginning explosively during June 1991, illustrates how Earth's internal processes can affect the atmosphere. Approximately 15–20 megatons of SO_2 was injected into the stratosphere; winds rapidly spread this aerosol worldwide. As a result, atmospheric albedo increased and produced a worldwide average temperature decrease of 0.5C° (0.9F°).

Scattering (Diffuse Reflection). Insolation encounters an increasing density of atmospheric gases as it travels down to the surface. These molecules redirect the insolation, changing the direction of light's movement without altering its wavelengths. This phenomenon is known as **scattering** and represents 7% of Earth's albedo (Figure 4-5). Dust particles, ice, cloud droplets, and water vapor produce further scattering. A sky filled with smog and haze appears almost white because the larger particles associated with air pollution act to scatter all wavelengths of the visible light.

Why is Earth's sky blue? And why are sunsets and sunrises often red? These simple questions have an interesting explanation, based upon a principle known as Rayleigh scattering (named for English physicist Lord Rayleigh, who stated the principle in 1881). This principle relates wavelength to the size of molecules or particles that cause the scattering. The general rule is: the shorter the wavelength, the greater the scattering, and

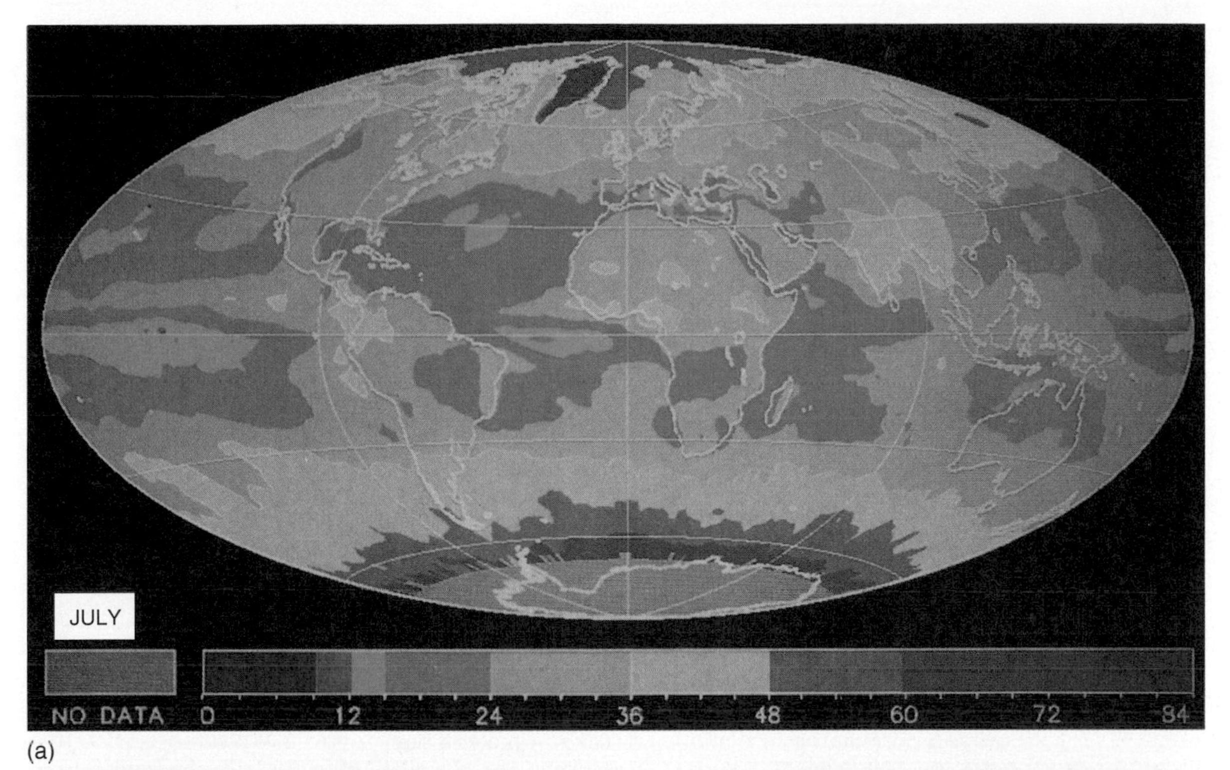

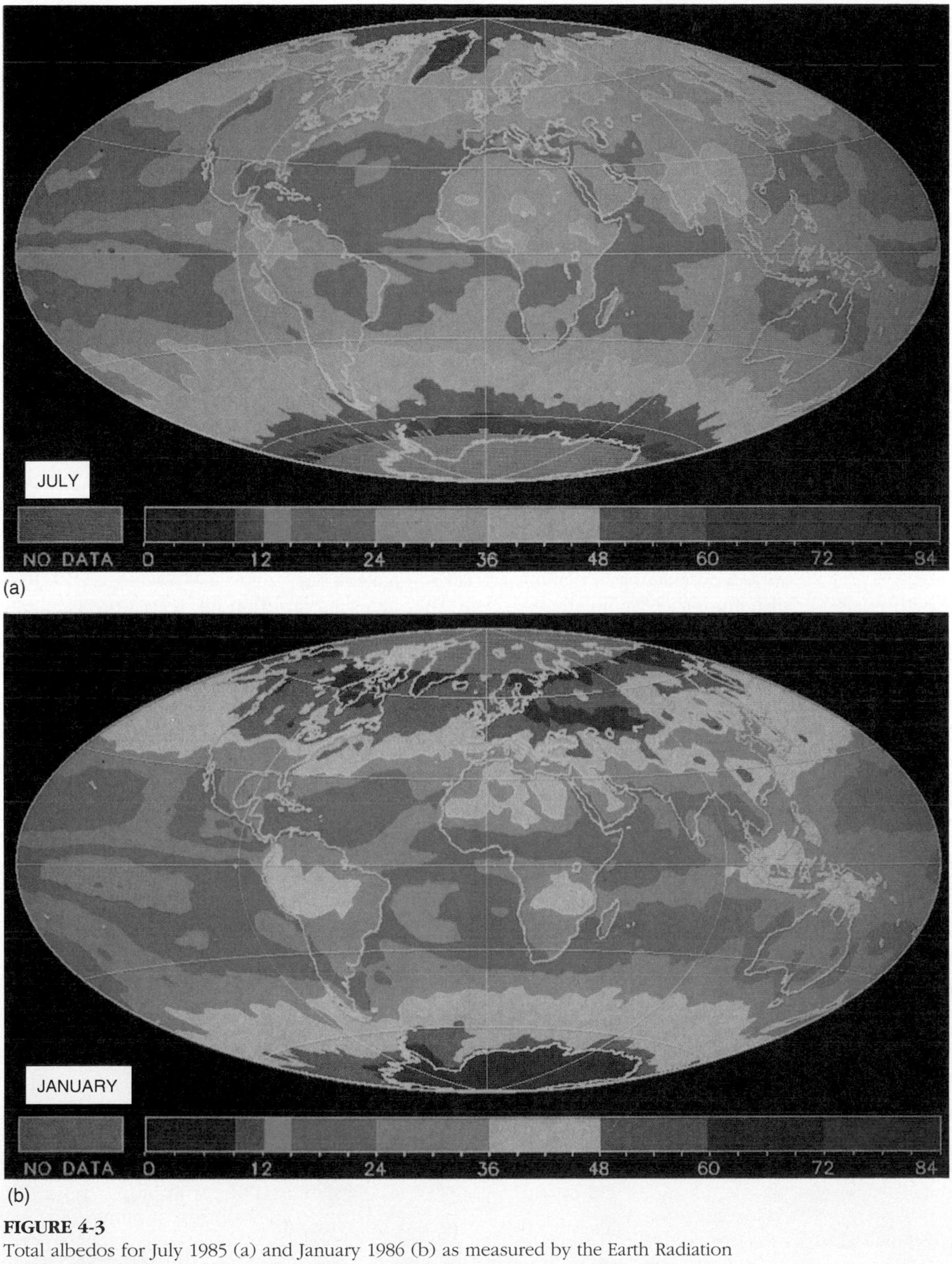

(a)

(b)

FIGURE 4-3
Total albedos for July 1985 (a) and January 1986 (b) as measured by the Earth Radiation Budget Experiments (ERBE) aboard satellites *Nimbus-7, NOAA-9,* and *ERBS*. Each color represents a 12% interval in albedo values. A modified elliptical equal-area projection is used. [Courtesy of Radiation Services Branch, Langley Research Center, NASA.]

the longer the wavelength, the less the scattering. Shorter wavelengths of light are scattered by small gas molecules in the air. Thus, the shorter wavelengths of visible light—the blues and violets—are scattered the most and dominate the lower atmosphere. And because there are more blue than violet wavelengths in sunlight, a blue sky prevails.

The angle of the Sun's rays determines the depth of atmosphere they must pass through to reach the surface. Direct rays experience less scattering and absorption than do low, oblique-angle rays that must travel farther through the atmosphere. As for red sunsets and sunrises, insolation from the low-altitude Sun must pass through an increasing mass of atmosphere. The effective scattering of shorter wavelengths leaves the residual oranges and reds to reach the observer (see Part 1 opening photograph).

Figure 4-5 shows that some incoming insolation is diffused by clouds and atmosphere and is transmitted to Earth as **diffuse radiation,** the downward component of scattered light. This light is multidirectional and thus casts shadowless light on the ground.

Refraction. When insolation enters the atmosphere, it passes from one medium to another (from virtually empty space to atmospheric gas) and is subject to a bending action called **refraction.** In the same way, a crystal or prism refracts light passing through it, bending different wavelengths to different degrees, separating the light into its component colors to display the spectrum. A rainbow is created when visible light passes through myriad raindrops and is refracted and reflected toward the observer at a precise angle (Figure 4-4). Another example of refraction is a *mirage,* an image that appears near the horizon where light waves are refracted by layers of air of differing temperatures (and resulting differing densities) on a hot day.

An interesting function of refraction is that it adds approximately eight minutes of daylight for

FIGURE 4-4
Double rainbow near Poco Butte, Arizona. Raindrops refract and reflect light to produce the primary rainbow. The secondary rainbow is caused by the same process only with two internal reflections within the raindrops. The color order in the secondary rainbow is reversed from that of the primary rainbow. [Photo by Dick Camby/DRK Photo.]

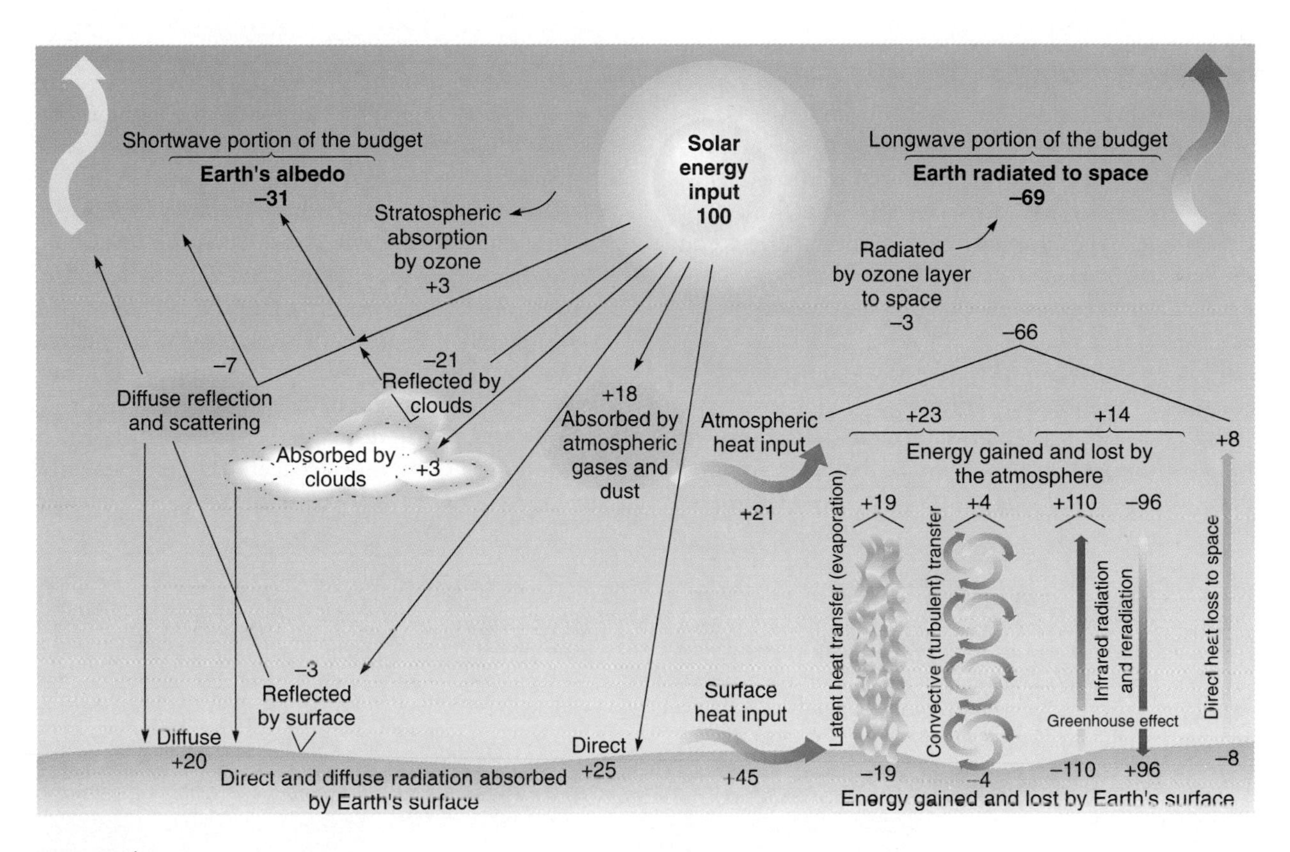

FIGURE 4-5
Earth-atmosphere energy balance. An input of 100 units of solar energy cascades through the lower atmosphere. Transformation and storage of this energy occur within the system. Eventually Earth radiates absorbed energy in infrared wavelengths that, when added to Earth's average albedo, approximates a total energy output to space of 100 units.

us. The Sun's image is refracted in its passage from space through the atmosphere, and so, at sunrise, we see the Sun about four minutes before it actually peeks over the horizon. Similarly, the Sun actually sets at sunset, but its image is refracted from over the horizon for about four minutes afterward. To this day, modern science cannot predict the exact time of sunrise or sunset within these four minutes because the degree of refraction continually varies with temperature, moisture, and pollutants.

Absorption. Absorption is the *assimilation* of radiation and its *conversion* from one form to another. Insolation (both direct and diffuse) that is not part of the 31% reflected from Earth's surfaces is absorbed and converted into infrared radiation (heat) or is used by plants in photosynthesis. The temperature of the absorbing surface is raised in the process, causing that surface to radiate more total energy at shorter wavelengths. In addition to absorption by land and water surfaces (about 45%), absorption also occurs in atmospheric gases, dust, clouds, and stratospheric ozone (about 24%). Figure 4-5 summarizes the pathways of insolation.

Earth Reradiation and the Greenhouse Effect. Previously we characterized Earth as a cool-body radiator, emitting energy in infrared wavelengths from its surface and atmosphere back toward space. However, some of this infrared radiation is absorbed by carbon dioxide, water vapor, methane, CFCs, and other gases in the lower atmosphere and is then reradiated to Earth, thus delaying heat loss to space. This counterradiation process is an important factor in warming the lower atmosphere. The approximate similarity between this process and the way a greenhouse operates gives the process its name—the **greenhouse effect.**

In a greenhouse, the glass is transparent to shortwave insolation, allowing light to pass through to the soil, plants, and materials inside. The absorbed energy is then radiated as infrared energy back toward the glass, but the glass effectively traps both the infrared wavelengths and the warmed air inside the greenhouse. Thus, the glass acts as a one-way filter, allowing the light in but not allowing the heat out. The same process also can be observed in a dwelling that has large areas of windows or in a car parked in direct sunlight.

Opening a roof vent allows the air inside the greenhouse to mix with the outside environment, thereby removing heat by moving air *physically* from one place to another. Rolling down your car window accomplishes the same thing. Such movement is called **convection** if it involves a strong vertical motion and **advection** if a lateral (horizontal) motion is dominant. Heat in a greenhouse also can be conducted by exposed surfaces. **Conduc-**

FIGURE 4-6
Longwave radiation leaving the Earth-atmosphere system (net outgoing longwave) in average monthly values for April 1985. Data are in watts per square meter. Latitudinal distribution of longwave radiation is apparent, with cooler regions toward the poles and warmer regions in the tropics. [Courtesy of Radiation Services Branch, Langley Research Center, NASA.]

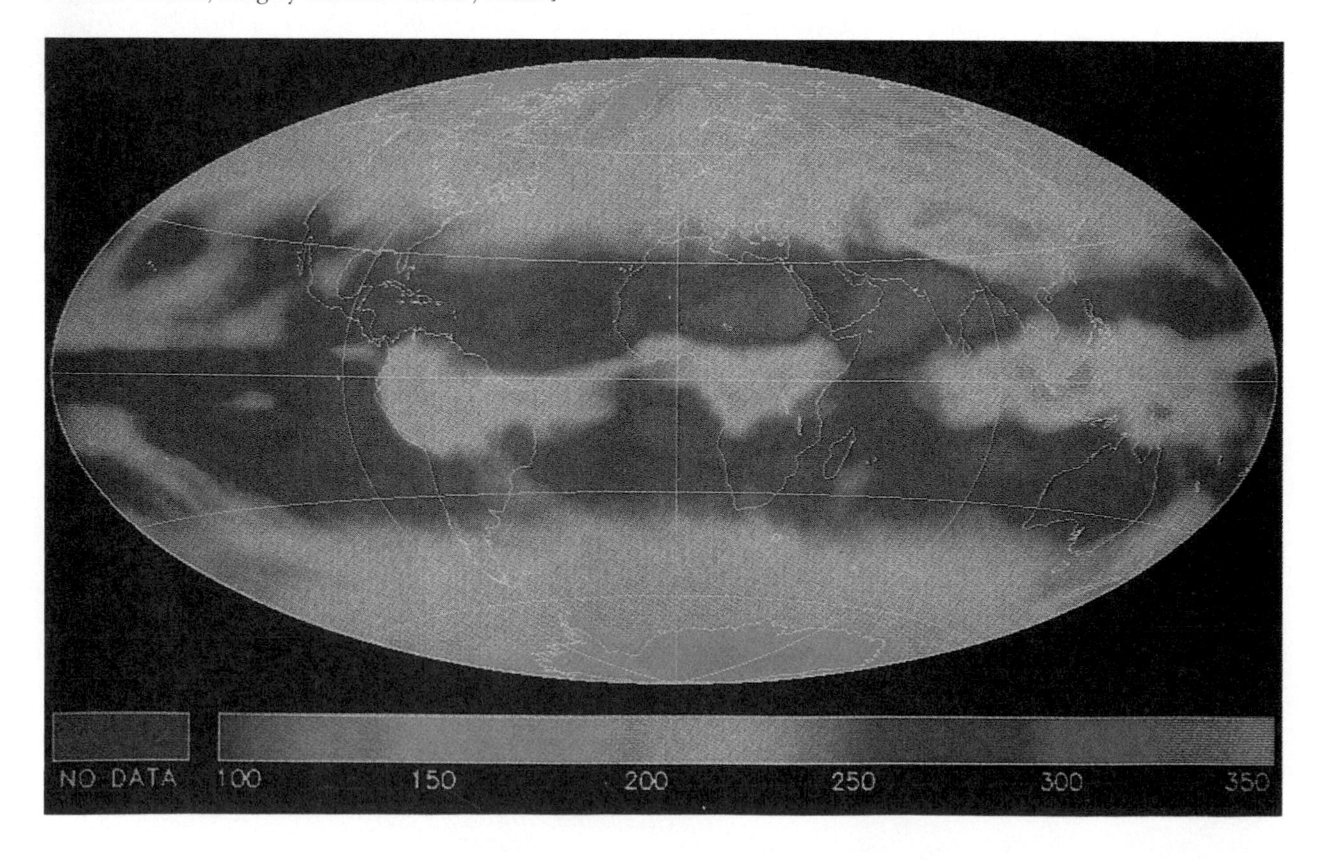

tion is the molecule-to-molecule transfer of heat (or sound or electricity) as it diffuses through a substance. As molecules warm, their vibration increases, causing collisions that produce motion in neighboring molecules and thus transferring heat energy from warmer to cooler materials.

In the atmosphere, the greenhouse analogy is not fully applicable because infrared radiation is not trapped as it is in a greenhouse. Rather, its passage to space is delayed as the heat is radiated and reradiated back and forth between Earth's surface and certain gases and particulates in the atmosphere. Today's increasing carbon dioxide concentration is causing more infrared radiation absorption in the lower atmosphere, thus forcing a warming trend and disruption of the Earth-atmosphere energy system, according to many scientists.

Earth-Atmosphere Radiation Balance

The Earth-atmosphere energy system naturally balances in a steady-state equilibrium. But if Earth's surface and its atmosphere are considered separately, neither exhibits a balanced radiation budget. The average annual distribution of energy demonstrates an overall positive radiation balance (energy surplus) for Earth's surface and an overall negative radiation balance (energy deficit) for the atmosphere. (Together, these two balance each other.) The natural energy balance occurs through both nonradiative and radiative energy transfers. Nonradiative transfers include the latent heat of evaporation (heat absorbed by water vapor when water evaporates), convection, and conduction. Radiative transfer is by infrared radiation between the surface and the atmosphere.

A representative pattern of longwave radiation from the Earth-atmosphere system (net outgoing longwave) is shown in Figure 4-6. Data gathered from ERBE instruments aboard several satellites show that variations in longwave radiation are generally latitudinal (zonal). An exception to latitudinal distribution, with lower values at lower latitudes, is the frequently cloud-covered region of the tropics (Amazon, equatorial Africa, and Indonesia). Major desert areas have greater longwave radiation emissions owing to the presence of little cloud cover and greater radiative heat loss.

Figure 4-5 summarizes the Earth-atmosphere radiation balance. It brings together all the elements discussed to this point in the chapter by following 100% of energy through the system. To summarize: of 100% of solar energy arriving, Earth's average albedo reflects 31% into space; absorption by atmospheric clouds, dust, and gases involves another 21% (which, when added to the 3% absorbed by stratospheric ozone, totals the atmospheric heat input). This leaves about 45% of the incoming insolation to actually reach Earth's surface as direct and diffuse radiation. Earth eventually radiates 69% back into space as heat.

Global Net Radiation. Figure 4-7 summarizes the radiation balance for all shortwave and longwave energy by latitude. In the equatorial zone, energy surpluses dominate, for in those areas more energy is received than is lost. The solar angle is high, with consistent daylength. However, deficits exist in the polar regions, where more energy is lost than gained. At the poles, the Sun is extremely low in the sky, surfaces are light and reflective, and for six months during the year no insolation is received. *Nimbus-7* measurement of net radiation for the Earth-atmosphere system shows a balance at around 36° latitude. Input of insolation minus output of longwave radiation and shortwave reflection establishes this pattern of surpluses and deficits.

This imbalance of net radiation from the equator to the poles drives vast global circulation of both energy and mass. The meridional (north-south) transfer agents are global wind circulation, ocean currents, dynamic weather systems, and related phenomena. Tropical cyclones (hurricanes and typhoons) represent dramatic examples of such energy and mass transfers. Forming near the equator, they mature and migrate to higher latitudes, carrying with them surplus heat energy, water, and water vapor. Having examined the Earth-atmosphere radiation balance, let us now focus on energy characteristics at Earth's surface.

EARTH'S SURFACE ENERGY BALANCE

With only minor exceptions, solar energy is the principal heat source at Earth's surface. The direct and diffuse radiation and infrared radiation arriving

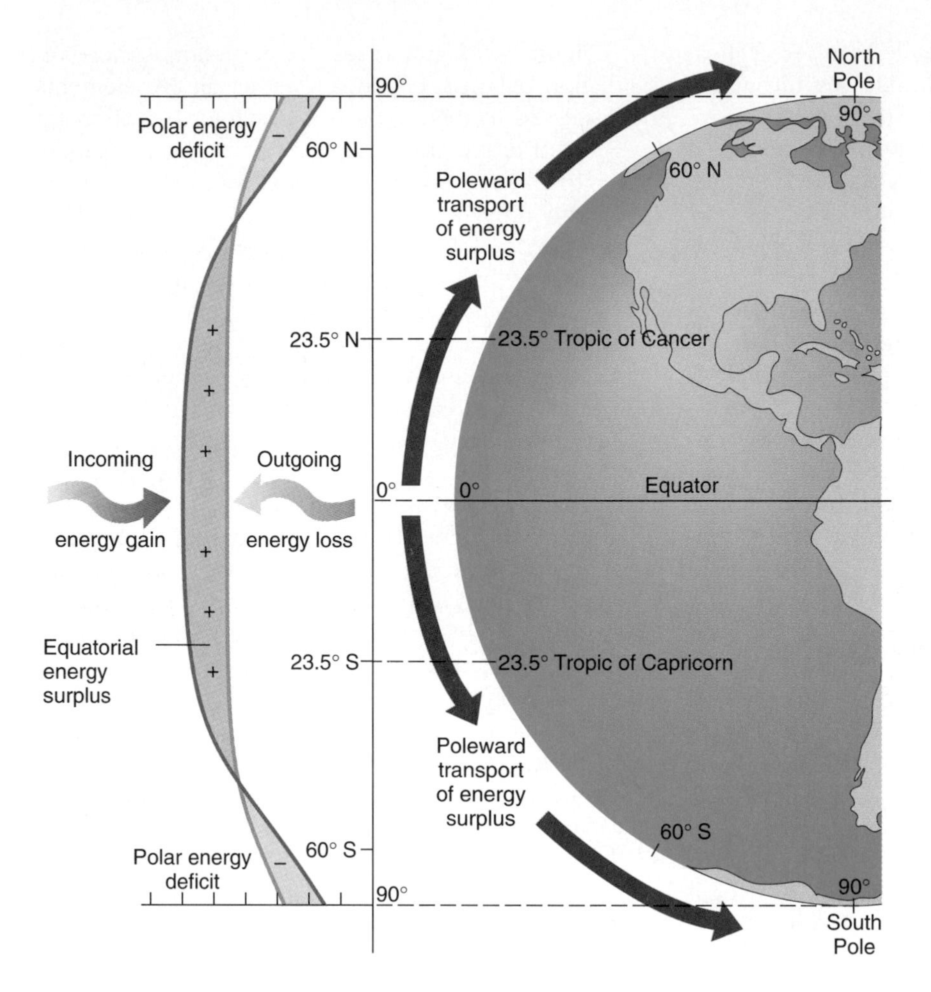

FIGURE 4-7
Earth's energy surpluses and deficits by latitude.

at the ground surface are shown in Figure 4-5. These daily radiation patterns at the surface are of great interest to geographers.

Daily Radiation Curves

The fluctuating daily pattern of incoming shortwave energy and outgoing longwave energy at the surface, and resultant air temperature, are shown in Figure 4-8. This graph represents conditions for bare soil on a cloudless day in the middle latitudes. Incoming energy arrives throughout the illuminated part of the day, beginning at sunrise, peaking at noon, and ending at sunset. The shape and height of this insolation curve varies with season and latitude. The highest trend for such a curve occurs at the time of the summer solstice (around June 21 in the Northern Hemisphere and December 21 in the Southern Hemisphere). The temperature plot also responds to seasons and variations in input. Within a 24-hour day, temperature peaks at around 3:00–4:00 P.M. and dips to its lowest point right at or slightly after sunrise.

The relationship between the insolation curve and the temperature curve on the graph is interesting—they do not align. As long as the incoming energy exceeds the outgoing energy, temperature continues to increase during the day, not peaking until the incoming energy begins to diminish in the afternoon as the Sun loses altitude. The warmest time of day occurs not at the moment of maximum insolation but at that moment when a maximum of insolation is absorbed. Thus, this temperature lag places the warmest time of day

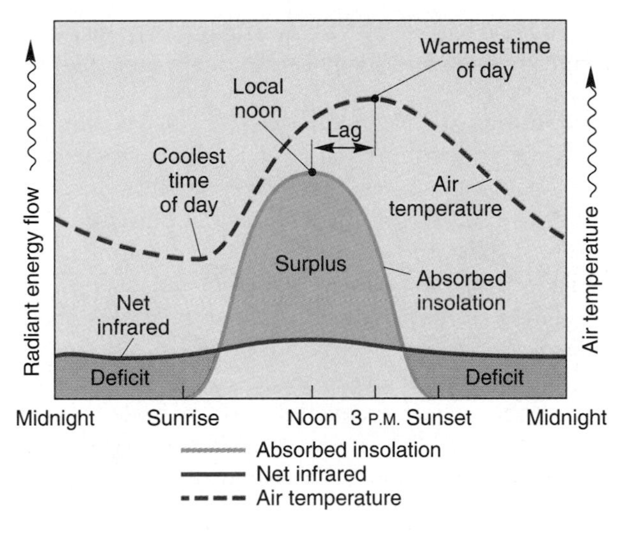

FIGURE 4-8
Daily radiation curves demonstrate the lag between local noon and the warmest time of day. The gradual heating of the air by convective transfer from the ground produces a lag of several hours following maximum insolation input.

three to four hours after solar noon as absorbed heat is supplied to the atmosphere from the ground. Then, as the insolation input decreases toward sunset, the amount of heat lost exceeds the input, and temperatures begin to drop until the surface has radiated away the maximum amount of energy, just at dawn.

The annual pattern of insolation and temperature exhibits a similar lag. For the Northern Hemisphere, January is usually the coldest month, occurring after the winter solstice, the shortest day in December. Similarly, the warmest months of July and August occur after the summer solstice, the longest day in June.

Simplified Surface Energy Balance

Earth's surface is quite energetic, causing varied climatic patterns. The climate at or near Earth's surface is generally termed the *boundary layer climate*. **Microclimatology** is the study of this portion of the atmosphere. The following discussion will be more meaningful if you keep in mind an actual surface—perhaps a park, a front yard, or a place on campus.

The surface receives light and heat, and it reflects light and radiates heat according to this basic scheme:

$$+SW\downarrow \; - \; SW\uparrow \; + \; LW\downarrow \; - \; LW\uparrow \; = \; \text{NET R}$$

(Insolation, light) (Albedo value of surface) (Heat) (Heat) — (Infrared radiation) (Net Radiation)

Figure 4-9 shows how net radiation is derived for a midlatitude location on a typical summer day. The components are direct and diffuse insolation, reflected energy (surface albedo), and infrared radiation arriving at and leaving from the surface. **Net radiation (NET R),** or the balance of all radiation at Earth's surface—shortwave (SW) and longwave (LW)—is derived from these radiation inputs and outputs.

Energy moving toward the surface is regarded as positive (a gain), and energy moving away from the surface is considered negative (a loss). The amount of insolation to the surface ($+SW\downarrow$) varies with season, cloudiness, and latitude. The albedo

FIGURE 4-9
Input and output components of a surface energy balance for a typical soil column.

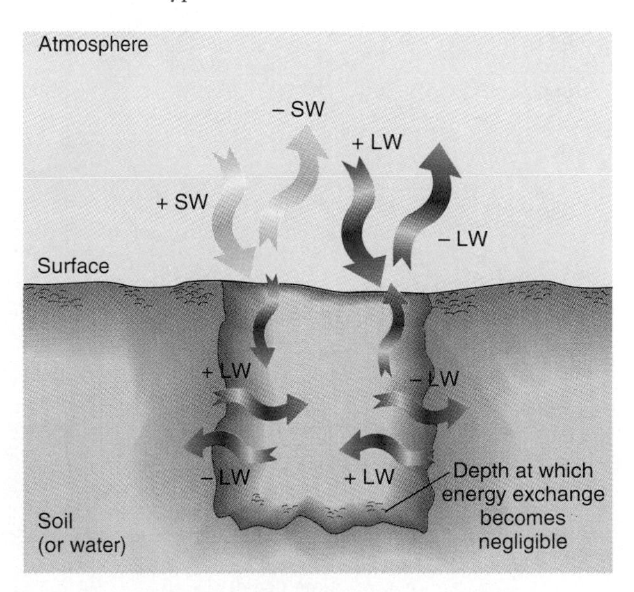

dictates the amount of insolation reflected and therefore not absorbed at the surface (–SW↑). Darker and rougher surfaces reflect less energy, absorb more energy, and therefore produce more net radiation. Lighter and smoother surfaces reflect more shortwave energy, leading to less net radiation. Snow-covered surfaces reflect most of the light reaching them, whereas urban areas tend to be darker in color and therefore convert more insolation to infrared radiation (review Table 4-1). At night the net radiation value is negative because the SW component ceases at sunset and the surface continues to lose radiative heat energy to the atmosphere. The surface rarely reaches a zero net radiation value—a perfect balance—at any one moment, but over time Earth's surface generally balances incoming and outgoing energies.

Along the surface, as illustrated in Figure 4-10, are the components of a surface energy balance. The column of soil continues to a depth at which energy exchange with surrounding materials becomes negligible, usually less than a meter.

FIGURE 4-10
Radiation components and the derivation of net radiation on a typical summer day for a midlatitude location (Matador in southern Saskatchewan, about 51° N, on 30 July 1971). [Adapted by permission from T. R. Oke, *Boundary Layer Climates*. New York: Methuen & Co., 1978, p. 21.]

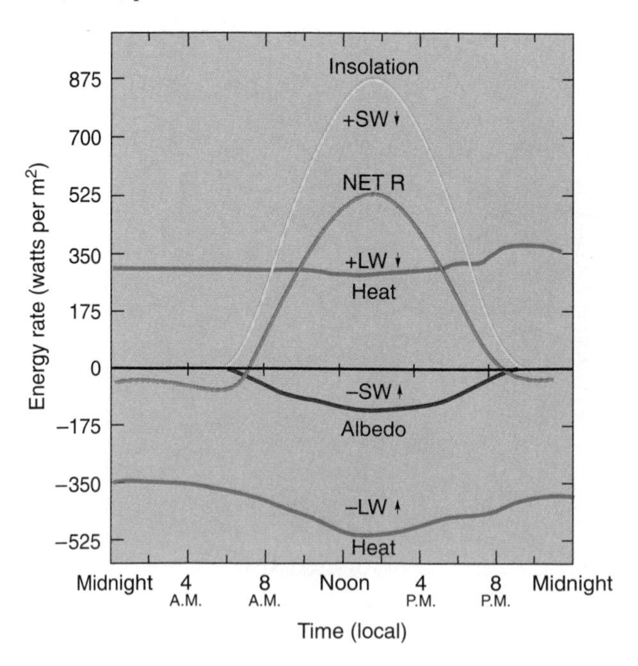

Net Radiation. Net radiation (NET R) is the net all-wave radiation available at Earth's surface; it is the final outcome of the entire radiation-balance process discussed in this chapter. The map in Figure 4-11 plots the distribution of mean annual net radiation in watts per square meter at ground level. The abrupt change in radiation balance from ocean to land surfaces is quite evident. Note that all values are positive; negative values probably occur only over ice-covered surfaces poleward of 70° latitude in both hemispheres. The highest net radiation occurs north of the equator in the Arabian Sea at 185 W/m^2 per year. Aside from the obvious interruption caused by landmasses, the pattern of values appears generally zonal, or parallel to the equator.

Net radiation leaves a nonvegetated surface through three components:

- *H,* or turbulent sensible heat transfer. H is the quantity of sensible heat energy mechanically transferred from air to surface, and surface to air, through convection and conduction as turbulent heat. This quantity depends on surface and boundary-layer temperatures and on the intensity of convective motion in the atmosphere. About 18% of Earth's entire NET R is reradiated as such sensible heat.
- *LE,* or latent heat of evaporation. LE refers to heat energy that becomes stored in water vapor as water evaporates. Large quantities of latent heat are absorbed into water vapor during its change of state from liquid to gas. Conversely, this heat is released in its change of state back to a liquid (Chapter 7). Because the evaporation of water is the principal method of transferring and dissipating heat surpluses vertically into the atmosphere, latent heat is the *dominant expenditure* of Earth's entire NET R. Latent heat links Earth's energy and water (hydrologic) systems and for most landscapes is the key component in surface energy budgets.
- *G,* or ground heating and cooling. Energy that flows into and out of the ground surface by conduction is symbolized by G.

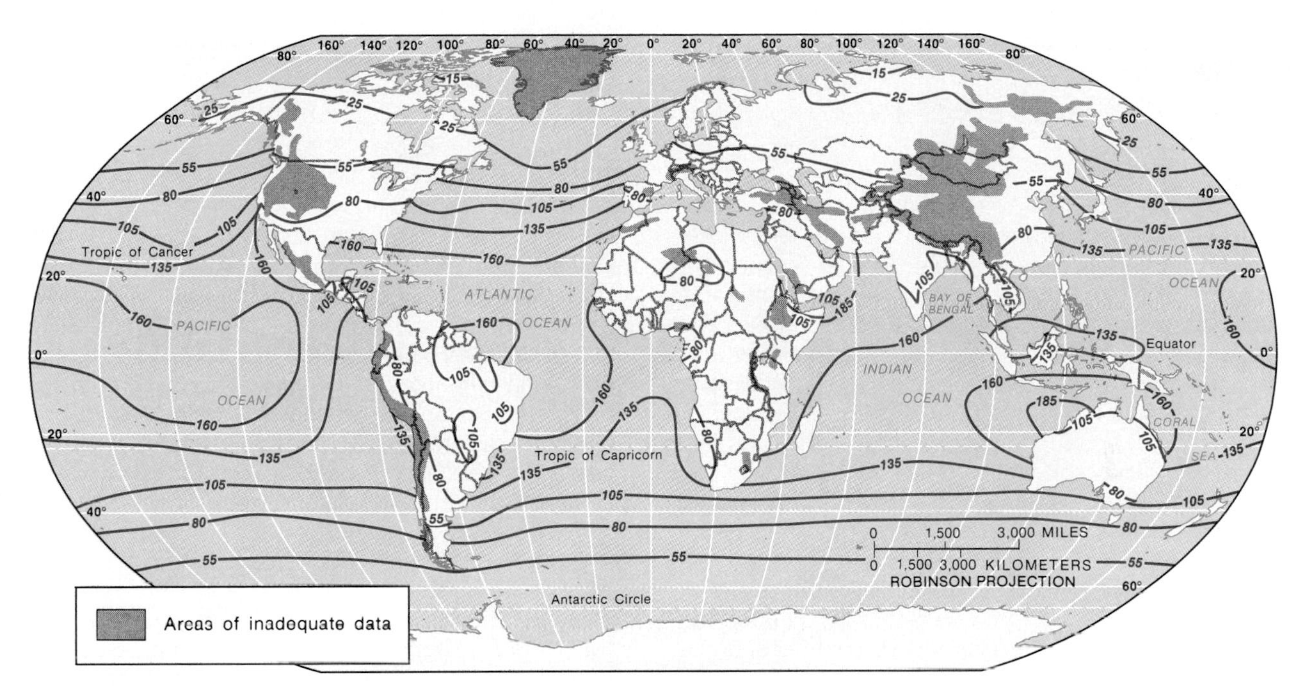

FIGURE 4-11
Distribution of global mean annual net radiation at ground level in watts per square meter (100 W/m^2 = 75 kcal/cm^2/year). [After M. I. Budyko, 1958, and *Atlas Teplovogo Balansa,* Moscow, 1963.]

As you might expect, the highest annual values on land for LE (latent heat of evaporation) occur between 10° N and 10° S latitude and decrease with distance toward each of the poles. The values for H (sensible heat) are distributed differently, being highest in the subtropics where nearly waterless and cloudless expanses of subtropical deserts stretch almost vegetation-free across vast regions. Over the oceans, the highest LE values are at those very same subtropical latitudes where hot, dry air comes into contact with warm ocean water.

The overall G value is zero because the stored energy from spring and summer is equaled by losses in fall and winter. The input for G is absorption of insolation and infrared radiation from above and by horizontal transfer from the surrounding environment. The output for G is heat lost by radiation to the atmosphere, sensible heat transfer when the air is cooler than the ground, energy absorbed for evaporation, and by horizontal transfer.

Moist surfaces covered with vegetation modify ground heating. Net radiation available at the surface is decreased because incoming shortwave energy is intercepted by a forest canopy of leaves. The amount of light reflected and absorbed depends upon tree height and the density of the leaves. Photosynthesis by plants absorbs approximately 8% of net radiation, converting it into biochemical energy. The variation in vegetation (biomass) over Earth reflects the energy expended for photosynthesis, as discussed in Chapters 19 and 20. Latitudes with abundant insolation and water are the most photosynthetically productive.

Another factor in ground heating is energy absorbed at the surface to melt snow or ice. In snow- or ice-covered landscapes, almost all available energy is absorbed as sensible and latent heat used in the melting and warming process.

Understanding the net radiation balance is essential to solar energy technologies that concentrate shortwave and longwave energy for use. Solar energy offers great potential worldwide and will see expanded use in the near future. Focus Study 4-1 briefly reviews solar energy strategies.

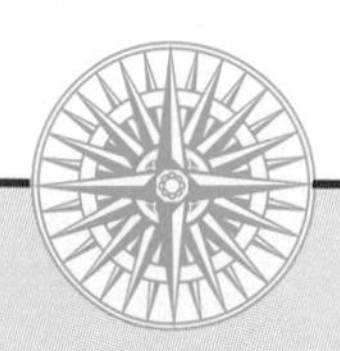

FOCUS STUDY 4-1 Solar Energy Collection and Concentration

Insolation not only warms Earth's surface; it provides the closest thing to an inexhaustible supply of energy for humanity. Yet, it is essentially untapped. Sunlight is direct, pervasive, and renewable and has been collected for centuries through various technologies. Turn-of-the-century photographs of residences in Southern California and Florida show solar water heaters on many rooftops. Early twentieth-century ads in newspapers and merchandise catalogs featured the Climax solar water heater (1905) and the Day and Night water heater (1909). However, low-priced natural gas and oil displaced many of these early applications.

By the early 1970s, the depletion of fossil fuels, their rising prices, and our growing dependence upon foreign sources spurred a reintroduction of tried-and-true solar technologies and energy-efficiency strategies. Cities such as Davis, California, passed energy-conservation and solar-utilization ordinances to guide construction and to encourage solar-energy installations. The country seemed to be rediscovering old, proven techniques. Unfortunately, during the early 1980s, much of this progress was halted by political decisions and severe budget cuts made by President Reagan's administration. A new energy plan in 1992 promises to renew the push for solar applications.

Any surface that receives light from the Sun is a solar collector. In fact, an average building in the United States receives 6 to 10 times more energy from the Sun than is required to heat the inside! But the diffuse nature of solar energy received at the surface requires that it be collected, transformed, and stored to be useful. Space heating is the simplest application. Windows that are carefully designed and placed allow sunlight to shine into the building, where it is absorbed and converted into sensible heat. This is an everyday application of the greenhouse effect.

A *passive solar system* captures heat energy and stores it in a "thermal mass," such as water-filled tanks, adobe, tile, or concrete. Three key features of passive solar design are (1) large areas of glass facing south (or facing north in the Southern Hemisphere) (2) a thermal storage medium, and (3) shades or screens to prevent light entry in the warm summer months. At night, shades or other devices can cover glass areas to prevent heat loss from the structure. An *active solar system* involves heating water or air in a collector and then pumping it through a plumbing system to a tank where it can provide hot water for direct use or for space heating.

Solar energy systems are capable of generating low-grade heat of an appropriate scale for approximately half the present domestic applications in the United States (space heating and water heating). In marginal climates, solar assisted water and space heating is feasible as a backup; even in New England and the Northern Plains states, solar-efficient collection systems have proven effective.

Reflective surfaces, such as focusing mirrors or parabolic troughs, can be used to attain very high temperatures to heat water or other fluids. Kramer Junction, California, about 225 km (140 mi) northeast of Los Angeles, is the location of the world's largest operating solar electric-generating facility, with 194 megawatts capacity. Long troughs of computer-guided curved mirrors concentrate sunlight to create temperatures of 390°C (735°F) in vacuum-sealed tubes filled with synthetic oil. This is then used to heat water, which produces steam that rotates turbines to generate cost-effective electricity. Thirty years of contracts to buy this electrical power already are signed, and utility companies are seeking the construction of more installed capacity (Figure 1).

Electricity is also being produced by photovoltaic cells, a technology that has been used in spacecraft since 1958. When sunlight shines upon a semiconductor material in these cells, it stimulates a flow of electrons (an electrical current). The efficiency of these cells has gradually improved and they now are cost-competitive. A drawback of both solar heating and solar electric systems is that periods of cloudiness and night inhibit operations. Research is going forward to enhance energy storage techniques, such as hydrogen fuel production (using energy to extract hydrogen from water for later use in producing more energy) and improved battery technology.

Solar energy is a wise choice for the future because it is directly available to the consumer, it is based upon renewable sources of an appropriate scale for end use needs, and most solar strategies are labor intensive (rather than capital intensive). Solar seems preferable to further development of our decreasing fossil fuel re-

FIGURE 1
Luz International solar-thermal energy installation in southern California. [Photos courtesy of Luz International Limited, Los Angeles.]

serves, oil imports, and troubled nuclear power. Large centralized power production is less desirable in that it is indirect, nonrenewable, high-grade (high temperature and cost) energy that is capital intensive. Whether the alternative path of solar energy is taken depends principally upon political decisions, for many of the technological innovations are presently ready for installation, requiring only leadership from the government.

Sample Stations. An energy balance equation can be structured for any time frame. This section examines daily energy balances at two locations, El Mirage in California and Pitt Meadows in British Columbia.

El Mirage, at 35° N, is a hot desert location characterized by bare, dry soil with sparse vegetation (Figure 4-12a and b). The day selected is in summer, a clear day, with a light wind in the late afternoon. The NET R value is lower than might be expected, considering the Sun's position close to zenith and the absence of clouds. But the income of energy at this site is countered by surfaces of higher albedo value than a forest or cropland and by hot soil surfaces radiating heat back to the atmosphere throughout the afternoon.

El Mirage has little or no energy expenditure for LE. With little water and sparse vegetation, most of the available radiant energy is dissipated as turbulent sensible heat (H), warming air and soil to high temperatures. Over a 24-hour period, H is 90% of NET R, the remaining 10% being G. The G component is greater in the morning, when winds are light and turbulent transfers are reduced. In the afternoon, heated air rises off the hot ground, and convective heat expenditures are accelerated as winds increase.

Compare (a) and (c) of Figure 4-12 and note that the NET R at the El Mirage desert location is quite similar to that of the Pitt Meadows, British Columbia, site. However, Pitt Meadows is midlatitude (49° N), vegetated, and moist, and its energy expenditures differ greatly from those at El Mirage. The Pitt Meadows landscape is able to retain much more of its energy because of a lower albedo (less reflection), the presence of more water and plants, and lower surface temperatures than those of El Mirage. The energy balance data for Pitt Meadows are plotted for a cloudless summer day. The higher LE values are attributed to the moist environment of rye grass and irrigated mixed-orchard ground cover for the sample area, contributing to the more moderate sensible heat (H) levels during the day.

Urban-Induced Energy Balances

Most of you live in or near a city. It is well established that urban microclimates generally differ from those of nearby nonurban areas. In fact, the surface energy characteristics of urban areas possess unique properties similar to energy balance traits of desert locations. Because more than 60% of the world's population will be living in cities by the year 2000, the study of **urban heat islands** and other specific environmental alterations related to cities is an important topic for physical geographers.

At least six factors contribute to urban microclimates:

1. Urban surfaces typically are metal, glass, asphalt, concrete, or stone. These city surfaces conduct up to three times more heat than wet sandy soil. The heat storage capacity of these materials also exceeds that of most natural surfaces. During the day and evening, higher temperatures exist above these surfaces. At night such surfaces rapidly radiate this stored heat to the atmosphere, producing minimum temperatures some 5–8C° (9–14F°) warmer, especially on calm, clear nights. As a result, both maximum and minimum temperatures are higher than in nonurban areas, although the effect of heat-island characteristics is more profound during nighttime cooling.
2. Urban surfaces behave differently from natural surfaces in energy balance. Albedo values are lower, leading to a higher net radiation value. However, urban surfaces expend more of that energy as sensible heat than nonurban areas; more than 70% of the net radiation in an urban setting is spent in this way. Figure 4-13 illustrates a generalized cross section of a typical urban heat island, showing increasing temperatures toward the center of the city.
3. Urban surfaces generally are sealed (built on and paved) so that water cannot reach the soil. Central business districts are on the average 50% sealed, whereas suburbs are about 20% sealed. Because urban precipitation cannot penetrate the soil, water runoff increases. Urban areas respond much as a desert landscape: a storm may cause a flash flood over the hard, sparsely vegetated surfaces, only to be followed by a return to dry conditions a

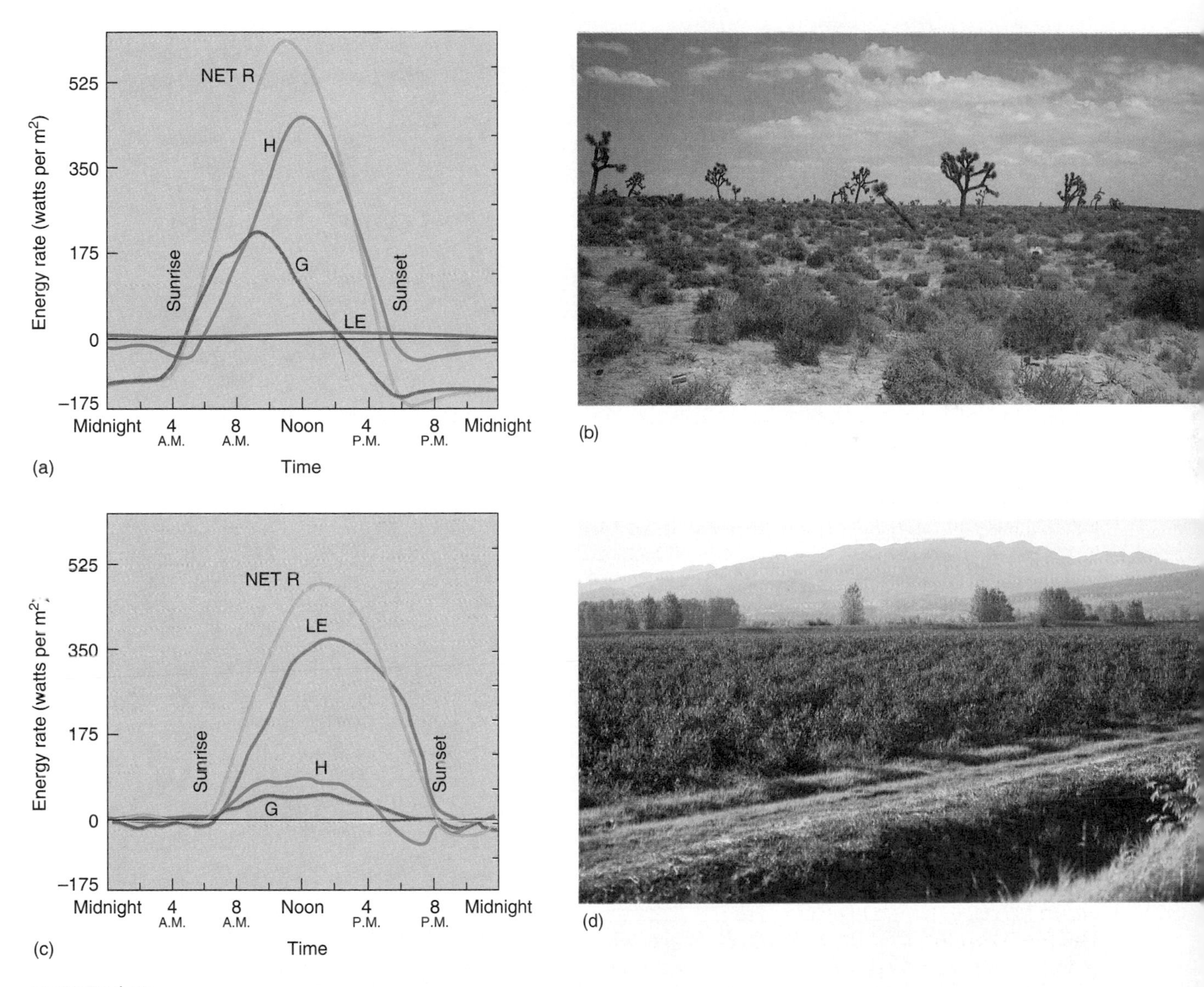

FIGURE 4-12
Daily net radiation expenditure for (a) El Mirage, California (east of Los Angeles, about 35° N). The typical desert landscape near El Mirage (b).
Daily net radiation expenditure for (c) Pitt Meadows in southern British Columbia (about 49° N). Irrigated blueberry orchards are characteristic of the agricultural activity at Pitt Meadows (d). H = turbulent sensible heat transfer; LE = latent heat of evaporation; G = ground heating and cooling. [(a) Adapted by permission from William D. Sellers, *Physical Climatology,* fig. 33, copyright © 1965 by The University of Chicago. All rights reserved. (b) Adapted by permission from T. R. Oke, *Boundary Layer Climates*. New York: Methuen & Co., 1978, p. 23. Photos by author.]

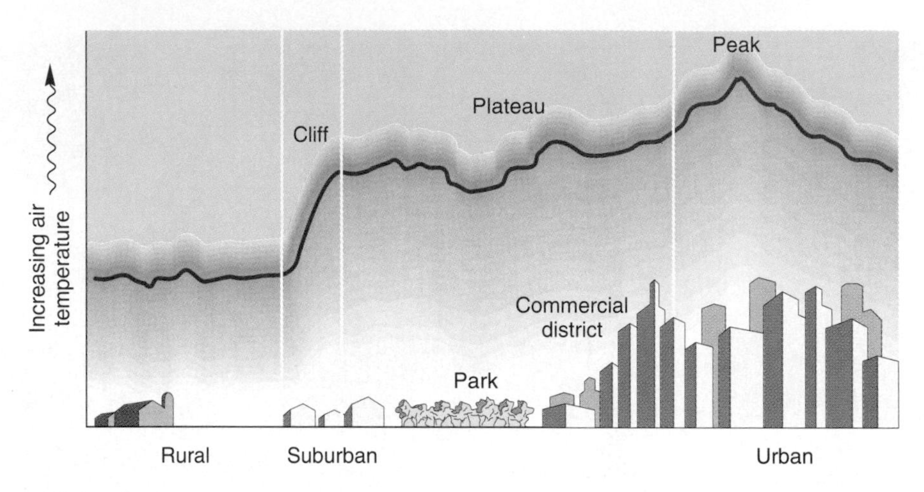

FIGURE 4-13
Generalized cross section of a typical urban heat island. The trend of the temperature gradient is described by the terms "cliff" at the edge of the city, where temperatures steeply rise, "plateau" over the suburban built-up area, and "peak" where temperature is highest at the urban core. [Reprinted by permission from T. R. Oke, *Boundary Layer Climates*. New York: Methuen & Co., 1978, p. 254.]

few hours later. Little of the net radiation is expended for evaporation on such a surface.

4. The irregular geometric shapes of a modern city affect radiation patterns and winds. Incoming insolation is caught in mazelike reflection and radiation "canyons," which tend to trap energy for conduction into surface materials, thus increasing temperatures (Figure 4-14). In natural settings, insolation is more readily reflected, stored in plants, or converted into latent heat through evaporation. Buildings tend to interrupt wind flows,

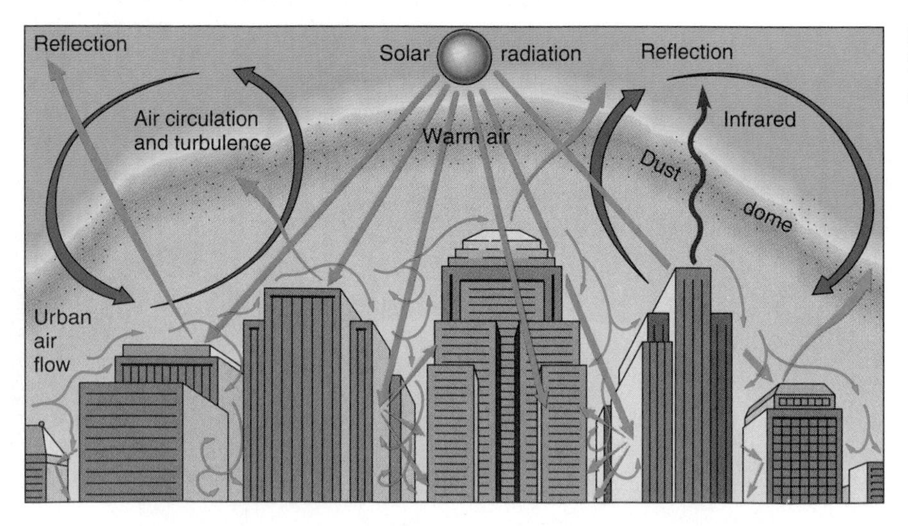

FIGURE 4-14
Insolation, wind movements, and dust dome in city environments.

FIGURE 4-15
Urbanized areas, such as Chicago, Illinois, possess unique physical attributes different from surrounding natural areas. Studies in the 1970s established the effects of the Chicago metropolitan region on weather downwind. [Photo by author.]

thereby diminishing heat loss through advective (horizontal) movement. On the average, winds have 25% less velocity in cities than in rural areas, although the buildings do create local turbulence and funneling effects. Wind flows can significantly reduce heat-island effects by increasing turbulent heat loss. Thus, maximum heat-island effects in the most built-up areas of a city occur on calm, clear days and nights.

5. Human activity alters the heat characteristics of cities. In New York City, for example, during the summer months, production of electricity and use of fossil fuels release an amount of energy equivalent to 25–50% of the arriving insolation. In the winter, urban-generated heat is on the average 250% greater than arriving insolation! In fact, for many northern cities in North America, winter heating requirements are actually reduced by this enhanced community heat-island effect.
6. The amount of air pollution, including gases and aerosols, is greater in urban areas than in comparable natural settings. Pollution actually increases reflection (albedo) in the atmosphere above the city, thus reducing insolation reaching the ground. The pollution blanket also absorbs infrared radiation, radiating the heat downward. Every major city produces its own **dust dome** of airborne pollution, which can be blown from the city in elongated plumes, depending on wind direction and speed (Figure 4-14). The increased particulates present in pollution act as condensation nuclei for water vapor. Convection created by the heat island lifts the air and particulates, producing increased cloud formation and the potential for increased precipitation. Although these precipitation effects of the urban heat island are difficult to isolate and prove, research suggests that urban-stimulated precipitation occurs downwind from cities (Figure 4-15).

TABLE 4-2
Average Differences in Climatic Elements Between Urban and Rural Environments

Element	Urban Compared to Rural Environs
Contaminants	
Condensation nuclei	10 times more
Particulates	10 times more
Gaseous admixtures	5–25 times more
Radiation	
Total on horizontal surface	0–20% less
Ultraviolet, winter	30% less
Ultraviolet, summer	5% less
Sunshine duration	5–15% less
Cloudiness	
Clouds	5–10% more
Fog, winter	100% more
Fog, summer	30% more
Precipitation	
Amounts	5–15% more
Days with <5 mm (0.2 in.)	10% more
Snowfall, inner city	5–10% less
Snowfall, downwind (lee) of city	10% more
Thunderstorms	10–15% more
Temperature	
Annual mean	0.5–3.0C° (0.9–5.4F°) more
Winter minima (average)	1.0–2C° (1.8–3.6F°) more
Summer maxima	1.0–3C° (1.8–3.0F°) more
Heating degree days	10% less
Relative humidity	
Annual mean	6% less
Winter	2% less
Summer	8% less
Wind speed	
Annual mean	20–30% less
Extreme gusts	10–20% less
Calm	5–20% more

SOURCE: Helmut E. Landsberg, 1981, *The Urban Climate,* Vol. 28. International Geophysics Series, p. 258. Reprinted by permission from Academic Press.

Table 4-2 summarizes various climatic factors and the average changes brought about by urbanization. Specific temperature differences of 6C° (11F°) are common between urban and nonurban areas, with maximum differences of 11C° (20F°) possible on calm, clear nights.

SUMMARY—Atmosphere and Surface Energy Balances

Earth's biosphere is powered by radiant energy from the Sun that cascades through complex circuits to the surface. This energy links Earth's hydrosphere, atmosphere, and lithosphere together in an intricate system that supports the biosphere—all life. Satellite measurements are a valuable contribution to understanding our planet's energy system.

Physical geographers use spatial models that effectively budget energy inputs, pathways, and outputs in the troposphere, on the ground, and within the ocean. Approximately 31% of insolation reflects back to space without performing any work, as Earth's average albedo. Insolation is partially absorbed by atmospheric clouds, gases, and dust, with the remaining energy eventually reaching land and water surfaces. This shortwave energy is transformed by Earth and emitted as longwave radiation back to space. The input of insolation and outputs of reflected light and emitted heat from our atmosphere and surface environments determine net radiation values available to perform work.

Surface energy balances are used to summarize the energy patterns for various locations. Surface energy measurements are used as an analytical tool of microclimatology, an important component of physical geography. A natural Earth-atmosphere energy balance occurs through a combination of nonradiative energy transfers from the surface as latent heat of evaporation, convection, and conduction, and radiative transfers of infrared radiation. Surface energy balances produce a net radiation (NET R) value. NET R is expended as sensible heat, latent heat, ground heating, and conversion of heat energy to biochemical energy through photosynthesis.

A growing percentage of Earth's human inhabitants live in cities. These cities possess a unique set of altered microclimatic effects, distinct from natural landscapes: increased conduction, lower albedos, higher NET R values, increased runoff, complex radiation and reflection patterns, anthropogenic heating, and the gases, dusts, and aerosols that are part of urban pollution.

Sunlight is direct, pervasive, and renewable and is a viable energy resource for society through known technologies. With development, solar energy applications can provide an inexhaustible, nonpolluting source of heat. Yet we find sunlight an essentially untapped resource.

KEY TERMS

absorption
advection
albedo
conduction
convection
diffuse radiation
dust dome
greenhouse effect
microclimatology
net radiation (NET R)
reflection
refraction
scattering
transmission
urban heat islands

REVIEW QUESTIONS

1. Diagram a simple energy balance for the troposphere. Label each shortwave and longwave component and the directional aspects of related flows.
2. List several types of surfaces and their albedos. Explain the differences. What determines the reflectivity of a surface?
3. Using Figure 4-3, explain the seasonal differences in albedo values for each hemisphere. Be specific, using the table of albedos (Table 4-1) when appropriate.
4. Why is the lower atmosphere blue? What would you expect the sky color to be at 50 km (30 mi) altitude? Why?
5. What role do clouds play in the Earth-atmosphere radiation balance? Is cloud type important?
6. Define conduction, convection, absorption, transmission, and diffuse radiation.

7. Define refraction. How does this relate to daylength? To a rainbow? To the beautiful colors of a sunset?
8. What are the similarities and differences between an actual greenhouse and the gaseous atmospheric greenhouse? Why is Earth's greenhouse changing?
9. Why is there a temperature lag between the highest Sun altitude and the warmest time of day? Relate your answer to the insolation and temperature curves.
10. Generalize the pattern of global net radiation. How might this pattern drive the atmospheric weather machine?
11. In terms of surface energy balance, explain the term *net radiation*.
12. What is the key role played by latent heat in surface energy budgets?
13. What are the expenditure pathways for surface net radiation? What kind of work is accomplished?
14. Compare the daily surface energy balances of El Mirage and Pitt Meadows. Explain the differences.
15. What is the basis for the urban heat-island concept? Describe the climatic effects attributable to urbanization as compared to nonurban environments.
16. Which of the items in Table 4-2 have you yourself experienced? Explain.

SUGGESTED READINGS

Ackerman, B., S. A. Changnon, Jr., et al. *Summary of METROMEX*. Vol. 2, *Causes of Precipitation Anomalies*. State of Illinois, Department of Registration and Education Bulletin 63. Urbana: Illinois State Water Survey, 1978.

Ardanuy, P. E., H. Lee Kyle, and D. Hoyt. "Global Relationships Among the Earth's Radiation Budget, Cloudiness, Volcanic Aerosols, and Surface Temperature," *Journal of Climate 5*, no. 10 (October 1992): 1120–39.

Budyko, M. I. *The Heat Balance of the Earth's Surface*. Translated by Nina A. Stepanova. Washington, DC: U.S. Department of Commerce, U.S. Weather Bureau, 1958.

Budyko, M. I. *Climate and Life*. Edited by David H. Miller. International Geophysics Series, vol. 18. New York: Academic Press, 1974.

Gates, David M. "The Flow of Energy in the Biosphere," *Scientific American* 224, no. 3 (September 1971): 88–100.

Harrison, E. F., David R. Brooks, P. Minnis, et. al. "First Estimates of the Diurnal Variation of Longwave Radiation from the Multiple-Satellite Earth Radiation Budget Experiment (ERBE)," *Bulletin of the American Meteorological Society* 69, no. 10 (October 1988): 1144–51.

Harrison, E. F., P. Minnis, B. R. Barkstrom, et. al. "Seasonal Variation of Cloud Radiative Forcing Derived from the Earth Radiation Budget Experiment," *Journal of Geophysical Research* 95, no. D11 (20 October 1990): 18,687–703.

Hobbs, John E. *Applied Climatology–A Study of Atmospheric Resources*. Kent, England: Dawson Westview Press, 1980.

Hubbard, H. M. "The Real Cost of Energy," *Scientific American* 264, no. 4 (April 1991): 36–42.

Kyle, H. Lee, R. R. Hucek, and B. J. Vallette. *Atlas of the Earth's Radiation Budget as Measured by Nimbus-7: May 1979 to May 1989*. NASA Reference Publication 1263, May 1991. Greenbelt, MD: Space Data and Computing Division, Goddard Space Flight Center, National Aeronautics and Space Administration.

Landsberg, Helmut E. *The Urban Climate.* International Geophysics Series, vol. 28. New York: Academic Press, 1981.

Meinel, Aden, and Marjorie Meinel. *Sunsets, Twilights, and Evening Skies.* Cambridge, England: Cambridge University Press, 1983.

Miller, David H. *Energy at the Surface of the Earth—An Introduction to the Energetics of Ecosystems.* International Geophysics Series, vol. 27. New York: Academic Press, 1981.

Oke, T. R. *Boundary Layer Climates,* 2d ed. London: Methuen, 1987.

Ramanathan, V., B. R. Barkstrom, and E. F. Harrison. "Climate and the Earth's Radiation Budget," *Physics Today* 42, no. 5 (May 1989): 22–32.

Sellers, William D. *Physical Climatology.* Chicago: University of Chicago Press, 1965.

Spring breakup on the Turnagain Arm inlet, Alaska. [*Photo by Galen Rowell.*]

5

WORLD TEMPERATURES

What temperature is it now (both indoors and outdoors) as you read these words? How is it measured and what does the value mean? How is today's air temperature controlling your plans for the day? Our bodies sense temperature and subjectively judge comfort, reacting to changing temperatures with predictable responses. Air temperature has a remarkable influence upon our lives, both at the microlevel and at the macrolevel.

A variety of temperature regimes worldwide affect cultures, decision making, and resources consumed. Global temperature patterns presently are changing in a warming trend that is affecting us all and is the subject of much scientific, geographic, and political interest. International cooperation and research are progressing to eliminate uncertainty in forecasts for the next century.

Of course, Earth's temperatures have not always been as they are today. In fact, humans are too recent as inhabitants of Earth to have experienced the planet's "normal" climate—those climate patterns characteristic of Earth's first 3 billion years. (Chapter 17 examines past Earth temperatures and some of the temperature controls that influence such long-term patterns.) Focus Study 5-1 introduces some essential temperature concepts to begin our study of world temperatures.

PRINCIPAL TEMPERATURE CONTROLS

Temperature, a measure of sensible heat energy present in matter indicates the average kinetic energy of individual molecules within the atmosphere and other media. Principal controls and influences upon temperature patterns include latitude, altitude, cloud cover, land-water heating differences, ocean currents and sea-surface temperatures, and general surface conditions. Let us begin by examining each of the principal temperature controls.

Latitude

Insolation received is the single most important influence on temperature variations. The distribution of intercepted solar energy at the top of the atmosphere exhibits an overall imbalance among equatorial, midlatitude, and polar locations. As discussed previously (and shown in Figures 2-11 and 2-12), insolation intensity decreases as surface distance increases from the subsolar point, a point that migrates between the Tropic of Cancer and Tropic of Capricorn—that is, between 23.5° N and 23.5° S—during the year. In addition, daylength and Sun angle change throughout the year, increasing in seasonal effect with increasing latitude. From equator to poles, Earth ranges from continually warm, to seasonally variable, to continually cold.

Altitude

Within the troposphere, temperatures decrease with increasing altitude above Earth's surface (recall that the normal lapse rate of temperature change with altitude is 6.4C°/1000 m or 3.5F°/1000 ft—see Figure 3-7). Thus, worldwide, mountainous areas experience lower temperatures than do regions nearer sea level, even at similar latitudes. The density of the atmosphere also diminishes with increasing altitude, as discussed in Chapter 3. In fact, air pressure at an elevation of 5500 m (18,000 ft) is about half of that at sea level. As the atmosphere thins, its ability to absorb and radiate heat is reduced.

The consequences are that average air temperatures at higher elevations are lower, nighttime cooling increases, and the temperature range between day and night and between areas of sunlight and shadow also is greater. Temperatures may decrease noticeably in the shadows and shortly after sunset. Surfaces both gain heat rapidly and lose heat rapidly to the thinner atmosphere. Also, at higher elevations, the insolation received is more intense because of the reduced mass of atmospheric gases. As a result of this intensity, the ultraviolet energy component makes sunburn a distinct hazard.

The snowline seen in mountain areas indicates where winter snowfall exceeds the amount of snow lost through summer melting and evaporation. The snowline's location is a function both of elevation and latitude, which means that glaciers

Focus Study 5-1 Temperature Concepts, Terms, and Measurements

The amount of heat energy present in any substance can be measured and expressed as its temperature. Temperature actually is a reference to the speed at which atoms and molecules that make up a substance are moving. Changes in temperature are caused by the absorption or emission (gain or loss) of energy. The temperature at which all motion in a substance stops is called 0° absolute temperature. Its equivalent in different temperature-measuring schemes is –273° Celsius (C), –459.4° Fahrenheit (F), or 0 Kelvin (K).

The Celsius scale (formerly called centigrade) is named after Swedish astronomer Anders Celsius, 1701–1744. It divides the difference between the freezing and boiling temperatures of water at sea level into 100 degrees on a decimal scale.

The Fahrenheit scale places the freezing point of water at 32°F (0°C, 273 K) and boiling point of water at 212°F (100°C, 373 K). The United States remains the only major country still using the Fahrenheit scale, named after the German physicist who invented the alcohol and mercury thermometers and who developed this thermometric scale in the early 1700s.

The Kelvin scale, named after British physicist Lord Kelvin (born William Thomson, 1824–1907) who proposed the scale in 1848, is used in science because temperature readings are proportional to the actual kinetic energy in a material.

Most countries use the Celsius scale to express temperature. This text presents Celsius (with Fahrenheit equivalents in parentheses) throughout in an effort to bridge this transitional era in the United States. The continuing pressure of the scientific community and corporate interests, not to mention the rest of the world, makes adoption of the Celsius and the SI (International System of Units) system inevitable for the United States. Figure 1 compares these two scales. Formulas for converting between Celsius, SI, and English units are on the inside back cover of this text.

Outdoor temperature usually is measured with a mercury thermometer or alcohol thermometer in a sealed glass tube. Alcohol is preferred in cold climates because it freezes at –112°C (–170°F), whereas mercury freezes at –39°C (–38.2°F). The principle of these thermometers is simple: when these fluids are heated, they expand; upon cooling, they contract. A thermometer stores fluid in a small reservoir at one end and is marked with calibrations to measure the expansion or contraction of the fluid, which reflects the temperature of the thermometer's environment.

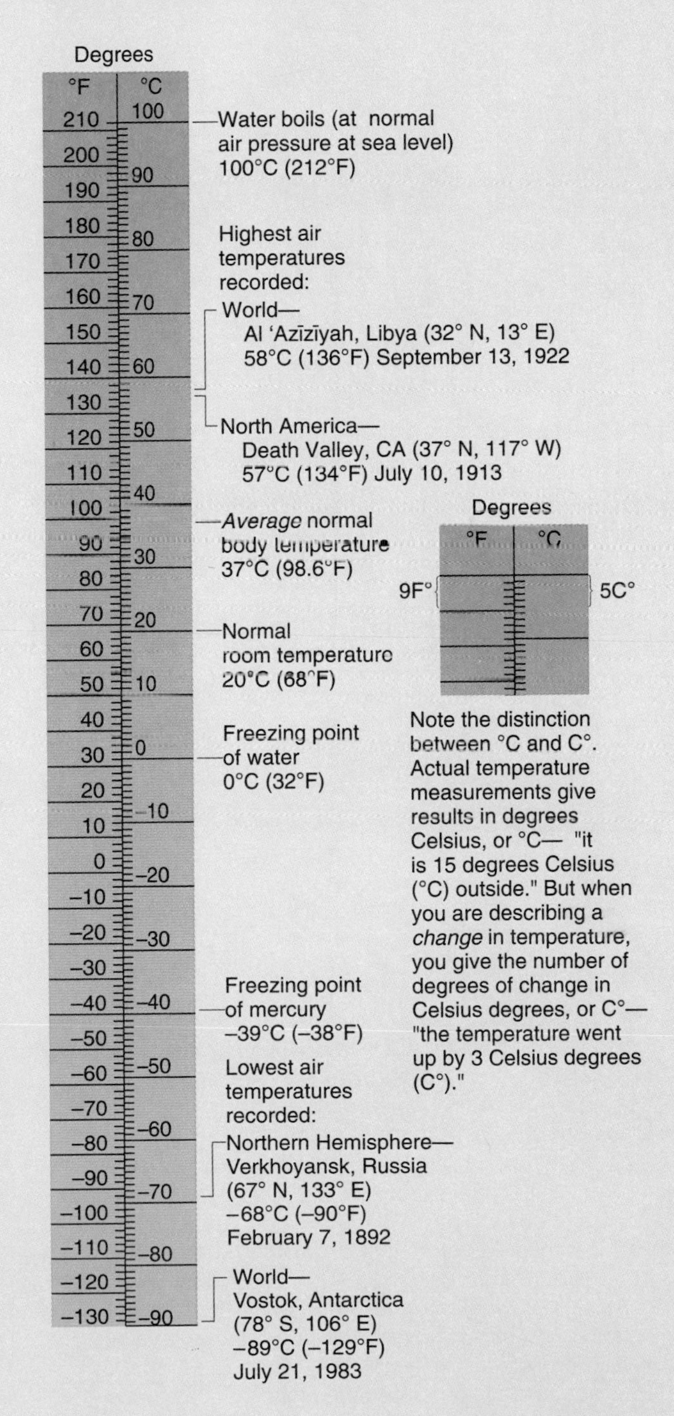

FIGURE 1
Scales for expressing temperature in degrees Celsius and Fahrenheit.

FIGURE 2
Minimum-maximum thermometer. [Photo by Bobbé Christopherson.]

FIGURE 3
Standard weather instrument shelter. [Photo by author.]

Figure 2 shows a mercury minimum-maximum thermometer, designed to record the day's highest and lowest temperatures and preserve the readings until it is reset. Resetting is accomplished by moving the markers with a magnet. Another type, the recording thermometer, creates an inked record on a turning drum; it is usually set for one full rotation every 24 hours or every 7 days.

Thermometers for standardized official readings are deployed inside white (for high albedo) louvered (for ventilation) weather stations (Figure 3), designed to avoid excessive heating of the instruments inside. They are placed at least 1.2–1.8 m (4–6 ft) above a surface, usually on turf; in the United States 1.2 m (4 ft) is standard. Official temperatures are measured in the shade to prevent the effect of direct insolation. Temperature readings are taken daily, sometimes hourly, at more than 10,000 weather stations worldwide. Some stations with recording equipment also report the duration of temperatures, rates of rise or fall, and variation over time throughout the day and night.

Daily minimum-maximum readings are averaged to get the *daily mean temperature*. The *monthly mean temperature* is the total of daily mean temperatures for the month, divided by the number of days in the month. An *annual temperature* range expresses the difference between the lowest and highest monthly mean temperatures for a given year. If you install a thermometer for outdoor temperature reading, be sure to avoid direct sunlight on the instrument and place it in an area of good ventilation.

can exist even at equatorial latitudes if the elevation is great enough. In equatorial mountains, the snowline occurs at approximately 5000 m (16,400 ft) because of the altitude, and permanent ice fields and glaciers exist on equatorial mountain summits in the Andes and East Africa. With increasing latitude toward the poles, snowlines gradually descend in elevation from 2700 m (8850 ft) in the midlatitudes to lower than 900 m (2953 ft) in southern Greenland.

Two cities in Bolivia illustrate the interaction of the two temperature controls, latitude and altitude. Figure 5-1 displays temperature data for the cities of Concepción and La Paz, which are near the same latitude (about 16° S). Note the elevation, average annual temperature, and precipitation for each location. The hot, humid climate of Concepción at its much lower elevation stands in marked contrast to the cool, dry climate of highland La Paz. People living around La Paz actually grow wheat, barley, and potatoes—crops characteristically grown in the cooler midlatitudes—despite the fact that La Paz is 4103 m (13,461 ft) above sea level. (For comparison, the summit of Pikes Peak in Colorado is at 4301 m or 14,111 ft.) The combination of elevation and equatorial location guarantee La Paz nearly constant daylength and moderate temperatures, averaging about 9°C (48°F) for every month. Such moderate temperature and moisture conditions lead to the formation of more fertile soils than those found in the warmer, wetter climate of Concepción.

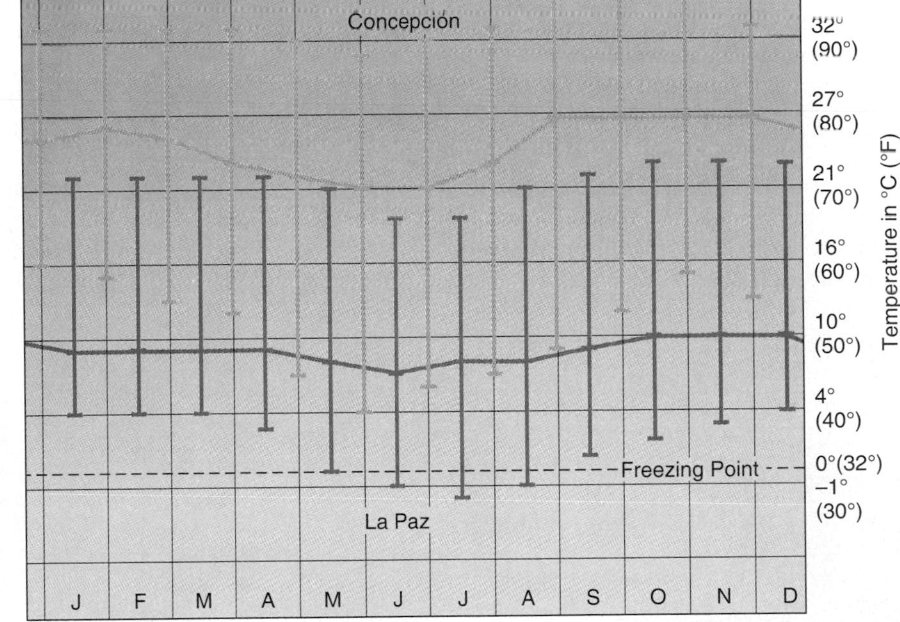

FIGURE 5-1
Comparison of temperature patterns in La Paz and Concepción, Bolivia.

	Station	
	Concepción, Bolivia	**La Paz, Bolivia**
Latitude/longitude	16°15′ S, 62°03′ W	16°30′ S, 68°10′ W
Elevation	490 m (1607.6 ft)	4103 m (13,461 ft)
Avg. annual temperature	24°C (75.2°F)	9°C (48.2°F)
Ann. temperature range	5C° (9F°)	3C° (5.4F°)
Annual precipitation	121.2 cm (47.7 in.)	55.5 cm (21.9 in.)
Population	768,000	993,000

Cloud Cover

The type, height, and density of cloud cover are important to temperature patterns. Orbital surveys reveal that approximately 50% of Earth is cloud-covered at any one time. Clouds are the most variable factor influencing Earth's radiation budget, and they are the subject of much investigation and effort to improve computer models of atmospheric behavior. The International Satellite Cloud Climatology Project, part of the World Climate Research Program, is presently in the midst of such research. As part of the Earth Radiation Budget Experiment mentioned in Chapters 2 and 4, cloud effects on longwave, shortwave, and net radiation patterns are assessed.

Clouds are moderating influences on temperature, producing lower daily maximums and higher nighttime minimums. Acting as insulation, clouds hold heat energy below them at night, preventing more rapid radiative losses, whereas during the day, clouds reflect insolation as a result of their high albedo values. The moisture in clouds both absorbs and liberates large amounts of heat energy, yet another factor in moderating temperatures at the surface. Relative to seasonal changes and the distribution of temperatures from equator to poles, clouds reduce latitudinal and seasonal temperature differences. Overall, the Earth-atmosphere energy system responds with slightly lower temperatures as a result of cloud cover.

Land-Water Heating Differences

The irregular arrangement of landmasses and water bodies on Earth contributes to the overall pattern of temperature. The physical nature of the substances themselves—rock and soil vs. water—is the reason for these **land-water heating differences.** More moderate temperature patterns are associated with water bodies, compared to more extreme temperatures inland. Figure 5-2 visually summarizes the following discussion.

Evaporation. More of the energy arriving at the ocean's surface is expended for evaporation than is expended over a comparable area of land. An estimated 84% of all evaporation on Earth is from the oceans. When water evaporates and thus changes to water vapor, heat energy is absorbed in the process and is stored in the water vapor. This stored heat energy is called *latent heat* and is discussed fully in Chapter 7. You can experience this evaporative heat loss (cooling) by wetting the back of your hand and then blowing on the moist skin. Sensible heat energy is drawn from your skin to supply some of the energy for the water evaporation, and you feel the cooling. Similarly, as surface water evaporates, substantial energy is absorbed, resulting in a lowering of nearby air temperatures. The land, being dry, is not substantially moderated by evaporative cooling.

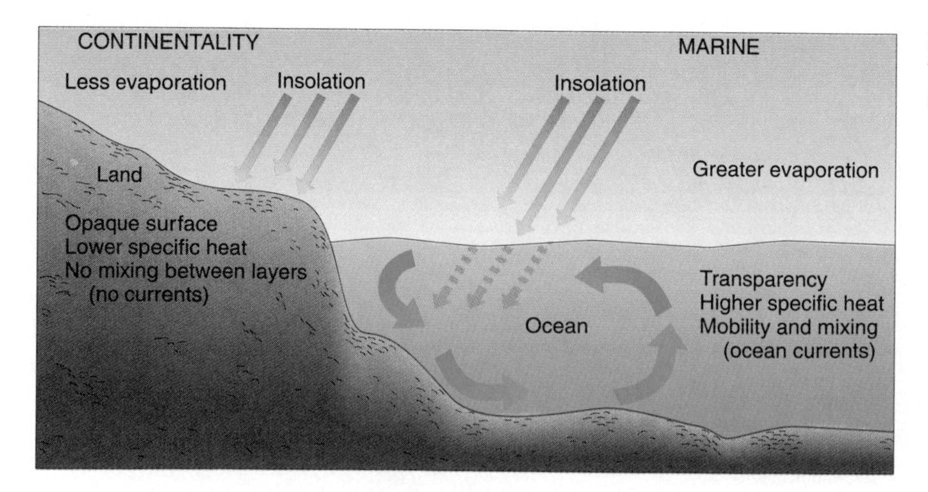

FIGURE 5-2
The differential heating of land and water.

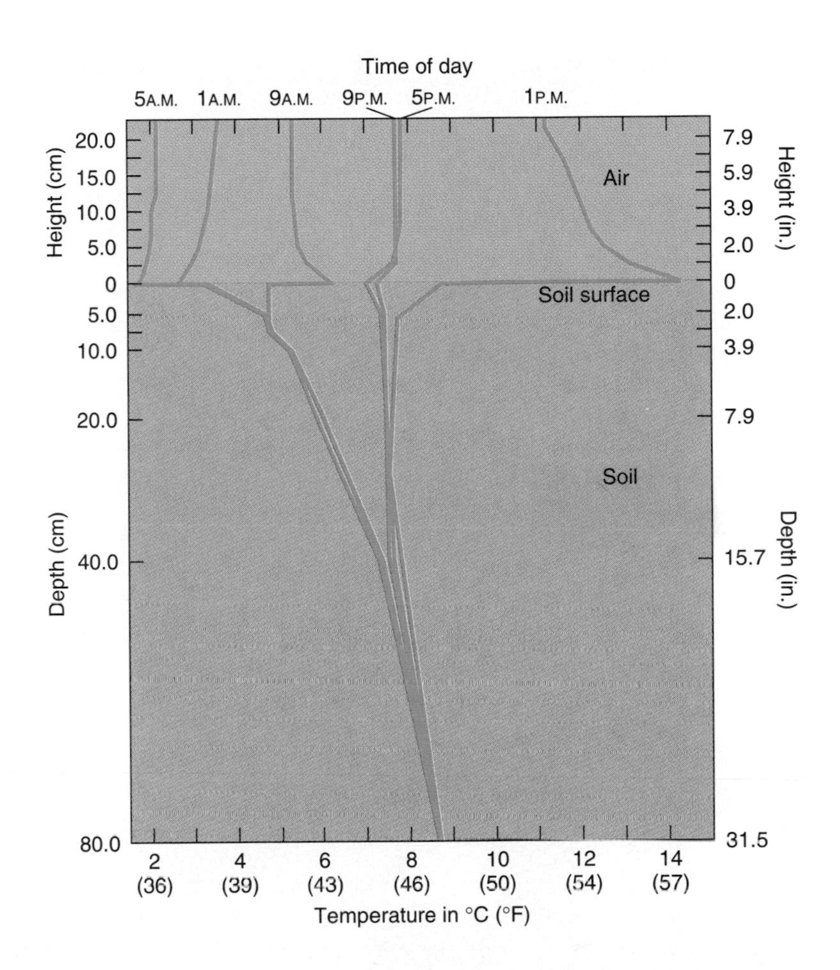

FIGURE 5-3
Profile of air and soil temperatures in Seabrook, New Jersey. [From John R. Mather, *Climatology: Fundamentals and Applications*. New York: McGraw-Hill 1974, p. 36. Adapted by permission.]

Transmissibility. The transmission of light obviously differs between soil and water; solid ground is opaque, water is transparent. Consequently, light striking a soil surface does not penetrate but is absorbed, heating the ground surface. That heat is accumulated during times of exposure and is rapidly lost at night or in shadows. Figure 5-3 shows the profile of diurnal temperatures for a column of soil and the atmosphere above it at a midlatitude location. You can see that maximum and minimum temperatures generally are experienced right at ground level. Below the surface, even at shallow depths, temperatures remain relatively static throughout the day. This same situation often exists at a beach, where surface sand may be painfully hot to your feet but the sand a few centimeters below the surface feels cooler and offers relief.

In contrast, when light reaches a body of water it penetrates the surface because of water's **transparency,** transmitting light to an average depth of 60 m (200 ft) in the ocean. This illuminated zone is known as the *photic layer* and has been recorded in some ocean waters to depths of 300 m (1000 ft). This characteristic of water results in the distribution of available heat energy over a much greater depth and volume, forming a larger heat reservoir than that made up of the surface layers of the land.

Specific Heat. When equal volumes of water and land are compared, water requires far more heat to raise its temperature than does land. In other words, water can hold more heat than can soil or rock, and therefore water is said to have a higher **specific heat.** A given volume of water represents a more substantial heat reservoir than an equal volume of land, so that changing the temperature of the oceanic heat reservoir is a slower process than changing the temperature of land.

Movement. Land is a rigid, solid material, whereas water is a fluid and is capable of movement. Differing temperatures and currents result in a mixing of cooler and warmer waters, and that mixing spreads the available heat over an even greater volume than if the water were still. Surface water and deeper waters mix, redistributing heat energy. Both ocean and land surfaces radiate heat at night, but land loses its heat energy more rapidly than does the moving mass of the oceanic heat reservoir.

Ocean Currents and Sea-Surface Temperatures. Although ocean currents are discussed in greater detail in the next chapter, the influence of currents on temperature and weather requires mention here. In an air mass, water vapor content is affected by ocean temperatures, for warm water tends to energize overlying air through high evaporation rates and transfers of latent heat.

As a specific example, the **Gulf Stream** (described in Chapter 6) moves northward off the east coast of North America, carrying warm water far into the North Atlantic (Figure 5-4). As a result, the southern third of Iceland experiences much milder temperatures than would be expected for a latitude of 65° N, just below the Arctic Circle (66.5°). In Reykjavik, on the southwestern coast of Iceland, monthly temperatures average above freezing during all months of the year. The Gulf Stream affects Scandinavia and northwestern Europe in the same manner.

In the western Pacific Ocean, the *Kuroshio* or Japan Current, similar to the Gulf Stream, functions much the same in its warming effect on Japan, the Aleutians, and the northwestern margin of North America. In contrast, along midlatitude west coasts,

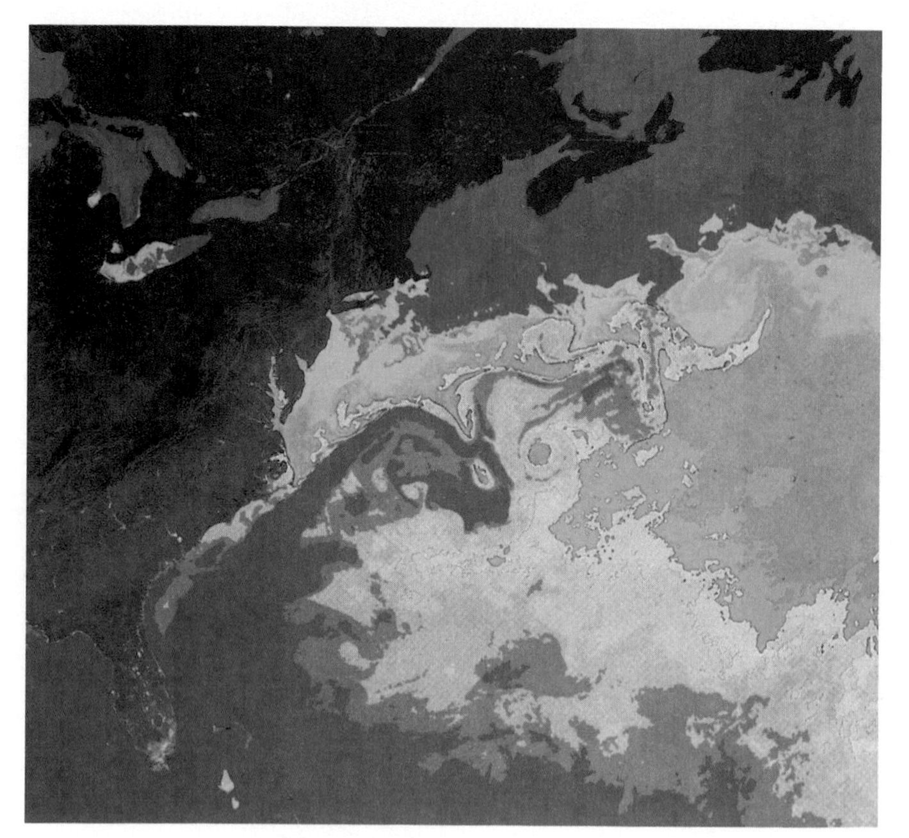

FIGURE 5-4
Satellite image of the warm Gulf Stream as it flows northward along the U.S. east coast. Instruments sensitive to infrared wavelengths produced this remote-sensing image. Approximately 11.4 million km^2 (4.4 million mi^2) are covered in the image. Temperature differences are distinguished by computer-enhanced coloration: reds/oranges = 25–29°C (76–84°F), yellows/greens = 17–24°C (63–75°F); blues = 10–16°C (50–61°F); and purples = 2–9°C (36–48°F). [Imagery by NASA and NOAA.]

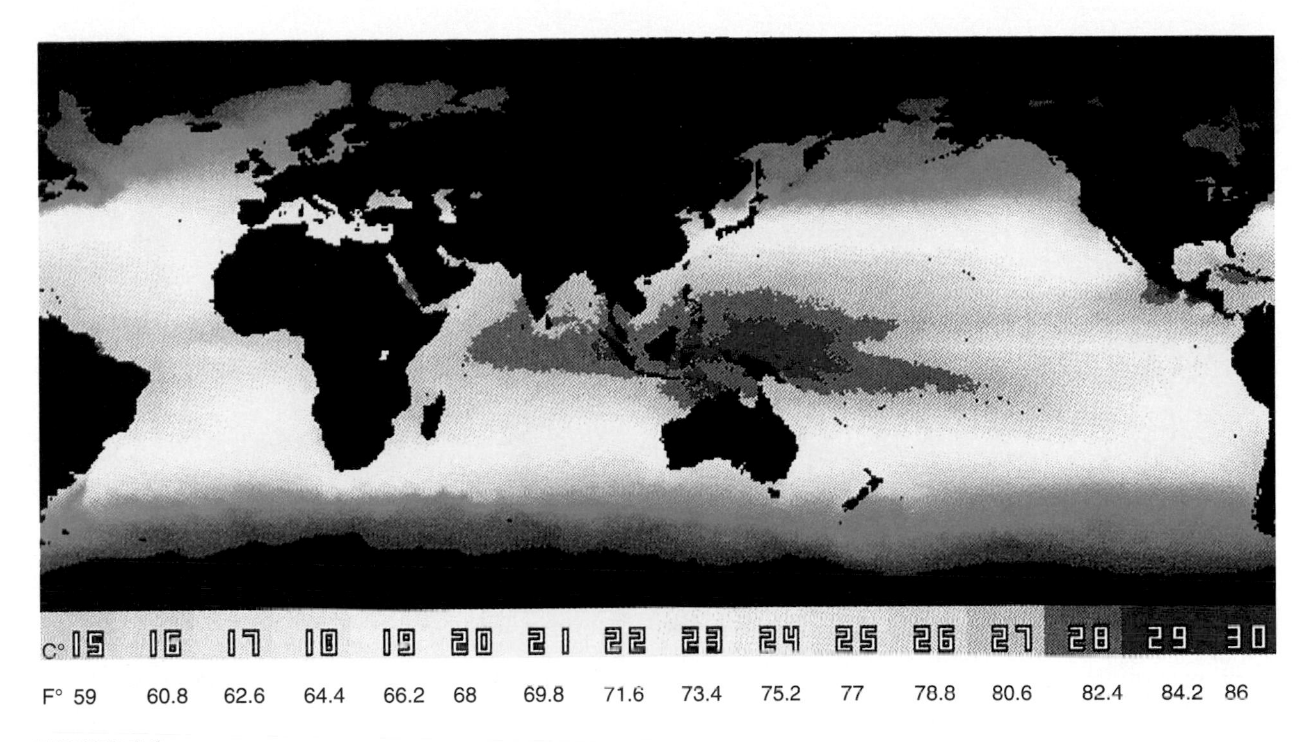

FIGURE 5-5
Average annual sea-surface temperatures for the period 1982–1991 as measured by several satellites. The Western Pacific Warm Pool, warmest area of all oceans, is well defined. These remote sensed data correlate closely with actual measurements of the ocean's surface temperature. [Satellite image processed by Dr. Xiao-Hai Yan, Center for Remote Sensing, University of Delaware. Reprinted by permission.]

cool ocean currents influence air temperatures. When conditions in these regions are warm and moist, fog frequently forms in the chilled air over the cooler currents.

Remote sensing is providing sea-surface temperature (SST) data that correlate well with actual sea-surface measurements. This correlation permits a thorough global assessment of SSTs in programs such as Tropical Ocean Global Atmosphere (TOGA) and Coupled Ocean-Atmosphere Response Experiment (COARE).

TIROS-N and *NOAA* satellites provided scientists at the University of Delaware with a 10-year record of SSTs. Mean annual SSTs increased from 1982 to 1991, with some fluctuations after 1987. Figure 5-5 displays these satellite data. The red and orange area in the southwestern Pacific Ocean with temperatures above 28°C (82.4°F), occupying a region larger than the United States, is called the *Western Pacific Warm Pool*. These are the highest average ocean temperatures in the world and appear to be important in the early stages of El Niño–Southern Oscillation development (see the focus study in Chapter 10). Sea-surface temperatures are affected by these recurring El Niño events and by volcanic eruptions such as El Chichón in 1982 and Mount Pinatubo in 1991. Understanding the linkage between ocean and atmosphere systems is important in establishing accurate general circulation models for forecasting.

Summary of Marine vs. Continental Conditions. Figure 5-2 summarizes the operation of all these land-water temperature controls: evaporation, transmissibility, specific heat, movement, and ocean currents and sea-surface temperatures. The

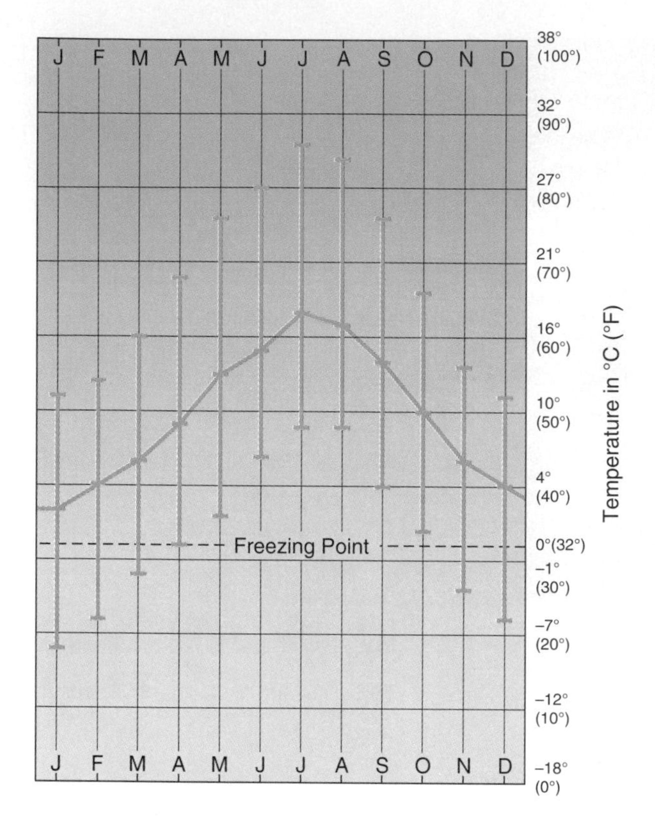

Station: Vancouver, British Columbia
Lat/long: 49°11'N, 123°10'W
Avg. Ann. Temp.: 10°C (50°F)
Total Ann. Precip.: 104.8 cm (41.3 in.)
Elevation: sea level
Population: 431,000
Ann. Temp. Range: 16C° (28.8F°)

Station: Winnipeg, Manitoba
Lat/long: 49°54'N, 97°14'W
Avg. Ann. Temp.: 2°C (35.6°F)
Total Ann. Precip.: 51.7 cm (20.3 in.)
Elevation: 248 m (813.6 ft)
Population: 561,000
Ann. Temp. Range: 38C° (68.4F°)

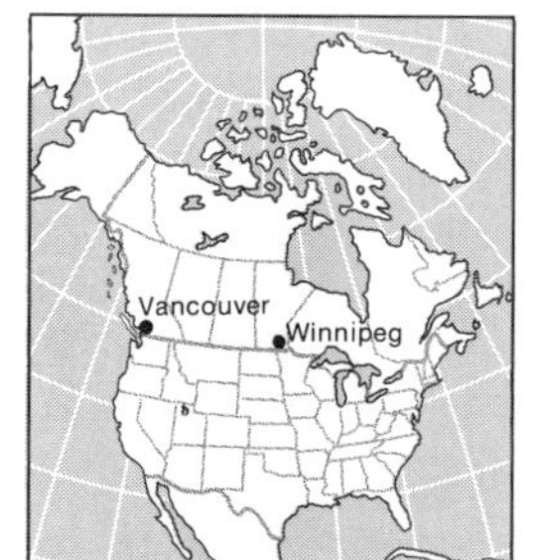

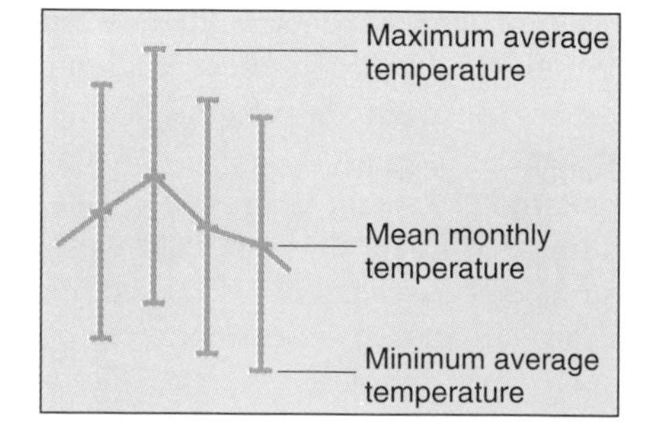

FIGURE 5-6
Comparison of temperatures in coastal Vancouver, British Columbia, and continental Winnipeg, Manitoba.

term **marine,** or maritime, is used to describe locations that exhibit the moderating influences of the ocean, usually along coastlines or on islands. **Continentality** refers to the condition of areas that are less affected by the sea and therefore have a greater range between maximum and minimum temperatures diurnally and yearly.

The Canadian cities of Vancouver, British Columbia, and Winnipeg, Manitoba, exemplify these marine and continental conditions (Figure 5-6). Both cities are at approximately 49° N latitude. Respectively, they are at sea level and 248 m (814 ft) elevation. However, Vancouver has a more moderate pattern of average maximum and minimum temperatures than Winnipeg. Vancouver's annual range of 16.0C° (28.8F°) is far below the 38.0C° (68.4F°) temperature range in Winnipeg. In fact, Winnipeg's continental temperature pattern is more extreme in every aspect than that of maritime Vancouver.

A similar comparison of San Francisco and Wichita, Kansas, is presented in Figure 5-7. Both cities are at approximately 37° 40' N latitude and are at 5 m (16.4 ft) and 403 m (1321 ft) in elevation, respectively. Summer fog plays a role in delaying until September the warmest summer month in San Francisco. In 90 years of weather records there, summer maximum temperatures have risen

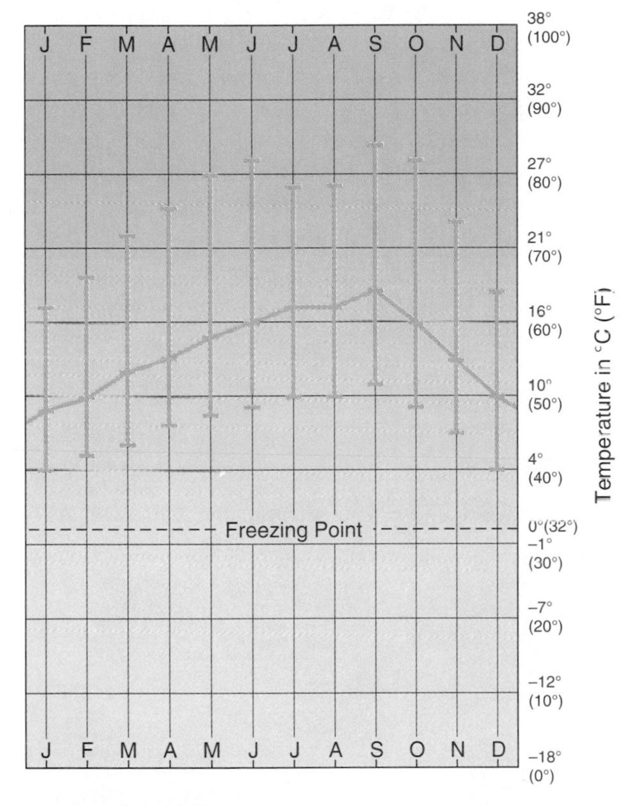

Station: San Francisco, California
Lat/long: 37°37'N, 122°23'W
Avg. Ann. Temp.: 14°C (57.2°F)
Total Ann. Precip.: 47.5 cm (18.7 in.)
Elevation:5 m (16.4 ft)
Population: 750,000
Ann. Temp. Range: 9C° (16.2F°)

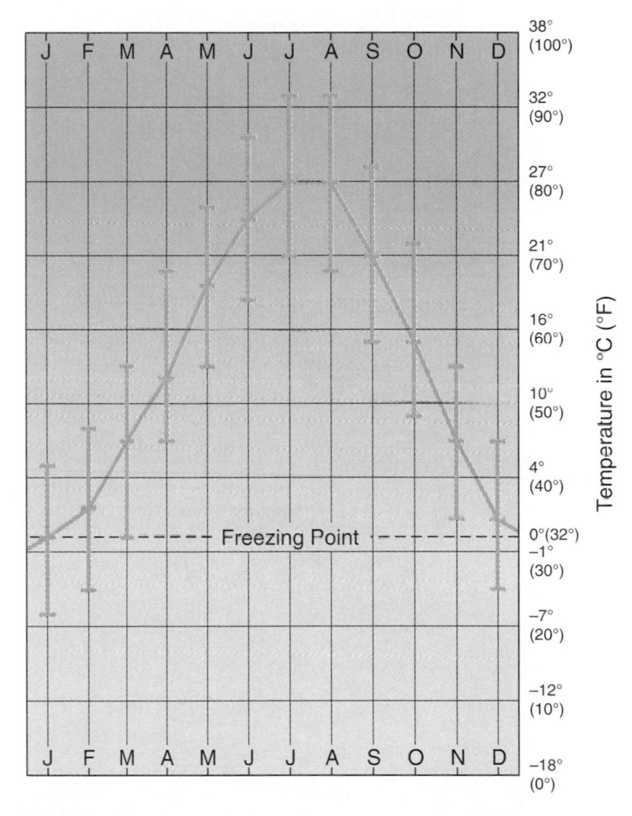

Station: Wichita, Kansas
Lat/long: 37°39'N, 97°25'W
Avg. Ann. Temp.: 13.7°C (56.6°F)
Total Ann. Precip.: 72.2 cm (28.4 in.)
Elevation:402.6 m (1321ft)
Population: 280,000
Ann. Temp. Range: 27C° (48.6F°)

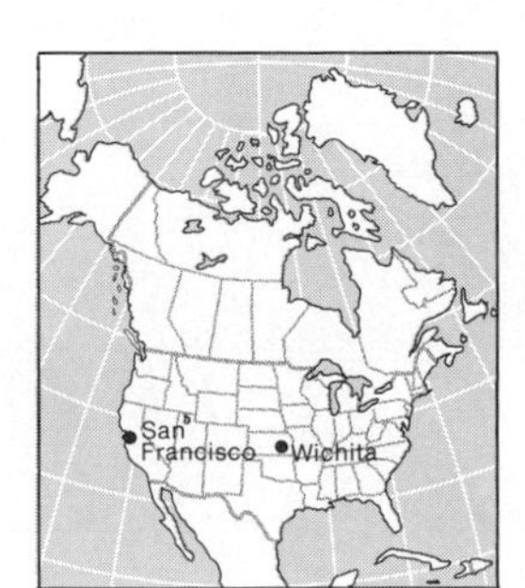

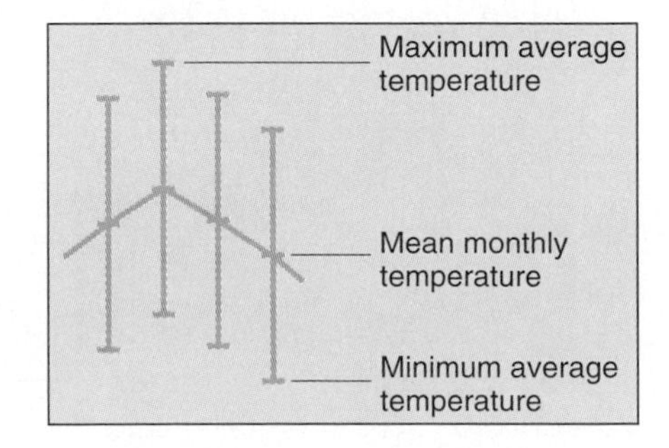

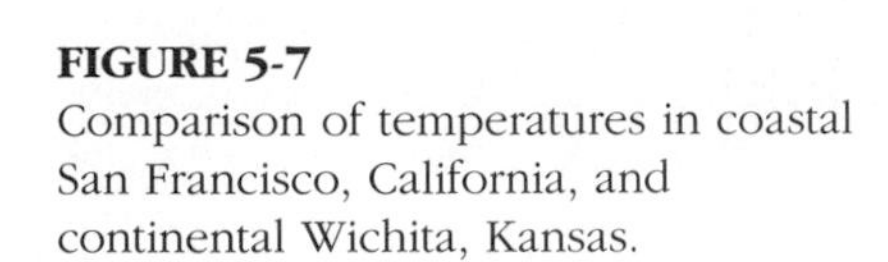

FIGURE 5-7
Comparison of temperatures in coastal San Francisco, California, and continental Wichita, Kansas.

above 32.2°C (90°F) an average of only once per year and have dropped below freezing an average of once per year. In contrast, Wichita is susceptible to freezes from late October to mid-April, with diurnal variations slightly enhanced by its elevation. Wichita has temperatures of 32.2°C (90°F) or higher at least 65 days each year. West of Wichita, winters increase in severity with increasing distance from the moderating influences of invading air masses from the Gulf of Mexico.

EARTH'S TEMPERATURE PATTERNS

Both the pattern and variety of temperatures on Earth result from the global energy system. The interaction of the principal temperature control factors is portrayed in the temperature maps in this section, which show worldwide mean monthly air temperatures for January (Figure 5-8) and July (Figure 5-10). These maps, along with Figure 5-11, which shows January–July ranges, also are useful in identifying areas that experience the greatest annual extremes. We use maps for January and July instead of for the solstice months (December and June) because, as explained in Chapter 4, a lag occurs between insolation received and maximum or minimum temperatures experienced.

The lines on temperature maps are known as isotherms. An **isotherm** connects points of equal temperature and portrays the temperature pattern, just as a contour line on a topographic map portrays points of equal elevation. Geographers are concerned with spatial analysis of temperatures, and isotherms facilitate this analysis.

FIGURE 5-8
Worldwide mean temperatures for January. Temperatures are in Celsius (converted to Fahrenheit by means of the scale) and are equated to sea level to compensate for the effects of landforms.

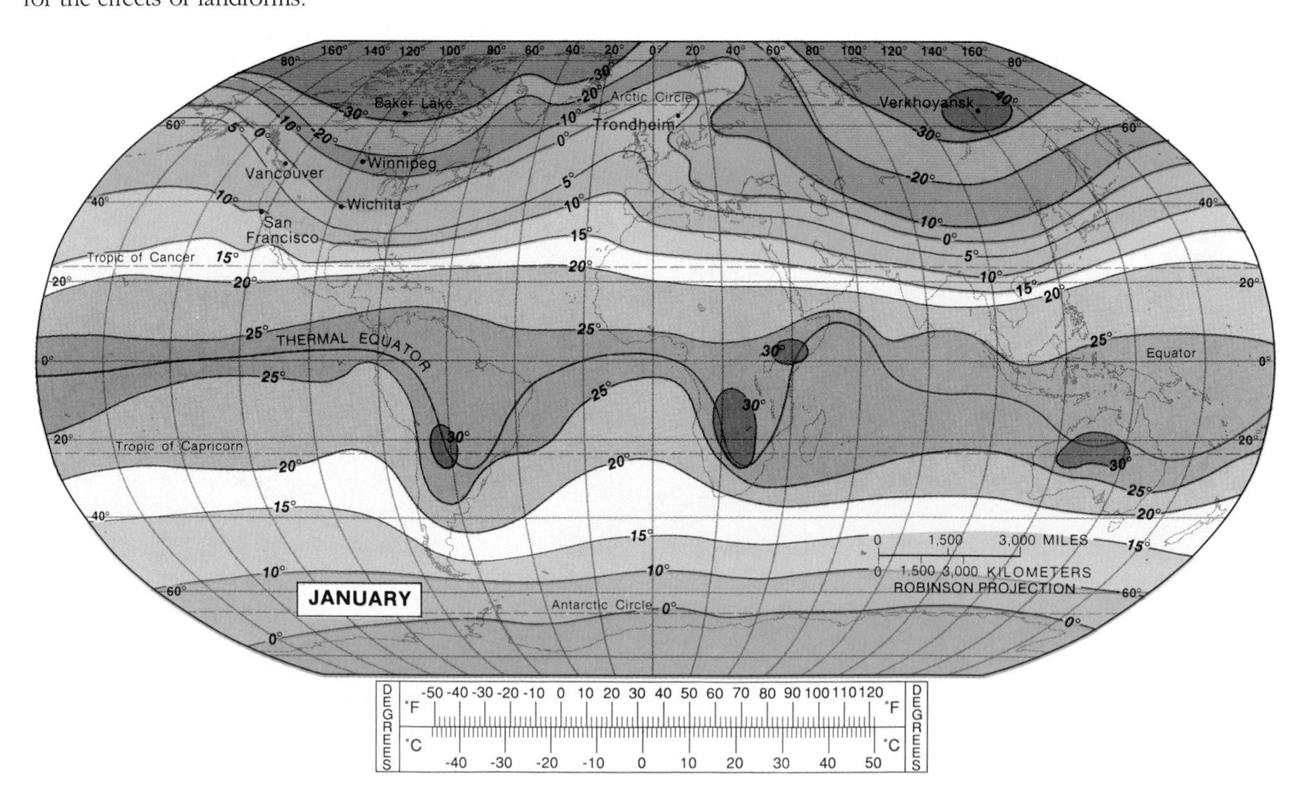

With each map, begin by finding your own city or town and noting the temperatures indicated by the isotherms (the small scale of these maps will permit only a general determination). Record the information from these maps in your notebook. As you work through the different maps throughout this text, note atmospheric pressure and winds, precipitation, climate, landforms, soil orders, vegetation, and terrestrial biomes. By the end of the course you will have recorded a complete profile of your specific locale.

January Temperature Map

Figure 5-8 is a map of mean temperatures for January. In the Southern Hemisphere, the higher Sun altitude means longer days and summer weather conditions, whereas in the Northern Hemisphere, January marks winter's shortened daylength and lower Sun angles. Isotherms generally trend east-west (zonal), parallel to the equator, marking the general decrease in insolation and net radiation with distance from the equator. The **thermal equator,** a line connecting all points of highest mean temperature (the red line on the map), trends southward into the interior of South America and Africa, indicating higher temperatures over landmasses. In the Northern Hemisphere, isotherms trend equatorward as cold air chills the continental interiors. The oceans, on the other hand, appear more moderate, with warmer conditions extending farther north than over land at comparable latitudes.

The coldest area on the map is in northeastern Siberia in Russia. The cold experienced there relates to consistent clear, dry, calm air, small insolation input, and an inland location far from any moderating maritime effects. Verkhoyansk and Oymyakon, Russia, each have experienced a minimum temperature of −68°C (−90°F) and a daily average of −50.5°C (−58.9°F) for January (Figure 5-9). Verkhoyansk experiences at least seven months of temperatures below freezing, including at least four months below −34°C (−30°F)! People do live and work in Verkhoyansk, which has a population of 1400; it has been continuously occupied since 1638 and is today a minor mining district. In contrast, these same towns experience maximum temperatures of +37°C (+98°F) in July.

Trondheim, Norway, is near the latitude of Verkhoyansk and at a similar elevation. However, Trondheim's coastal location moderates its annual temperature regime (Figure 5-9): January minimum and maximum temperatures range between −17 and +8°C (+1.4 and +46°F), and the minimum/maximum range for July is from +5 to +27°C (+41° to +81°F). The most extreme minimum and maximum temperatures ever recorded in Trondheim are −30 and +35°C (−22 and +95°F)—quite a difference from the extremes at Verkhoyansk.

July Temperature Map

Average July temperatures are presented in Figure 5-10. The longer days of summer and higher Sun altitude now are in the Northern Hemisphere. Winter dominates the Southern Hemisphere, although it is milder there because large continental landmasses are absent and oceans and seas dominate. The thermal equator shifts northward with the high summer Sun and reaches the Persian Gulf–Pakistan–Iran area. The Persian Gulf is the site of the highest recorded sea-surface temperature of 36°C (96°F), difficult to imagine for a sea or ocean body.

July is a time of 24 hour long nights in Antarctica. The lowest natural temperature reported on Earth occurred on July 21, 1983, at the Russian research base at Vostok, Antarctica (78° 27' S, elevation 3420 m or 11,220 ft): a frigid −89.2°C (−128.6°F). For comparison, such a temperature is 11C° (19.8F°) colder than dry ice (solid carbon dioxide)! During July in the Northern Hemisphere, isotherms trend poleward over land, as higher temperatures dominate continental interiors. July temperatures in Verkhoyansk average more than 13°C (56°F), which represents a 63C° (113F°) seasonal variation between winter and summer averages. The Verkhoyansk region of Siberia is probably the greatest example of continentality on Earth.

The hottest places on Earth occur in Northern Hemisphere deserts during July. These deserts are areas of clear skies and strong surface heating, with virtually no surface water and few plants. Lo-

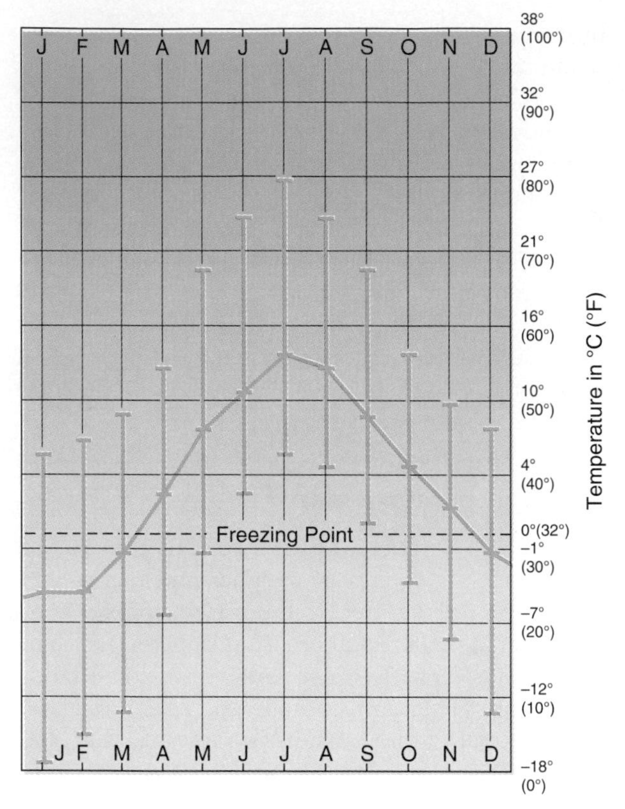

Station: Trondheim, Norway
Lat/long: 63°25'N, 10°27'E
Avg. Ann. Temp.: 5°C (41°F)
Total Ann. Precip.: 85.7 cm (33.7 in.)
Elevation: 115 m (377.3 ft)
Population: 135,000
Ann. Temp. Range: 17C° (30.6F°)

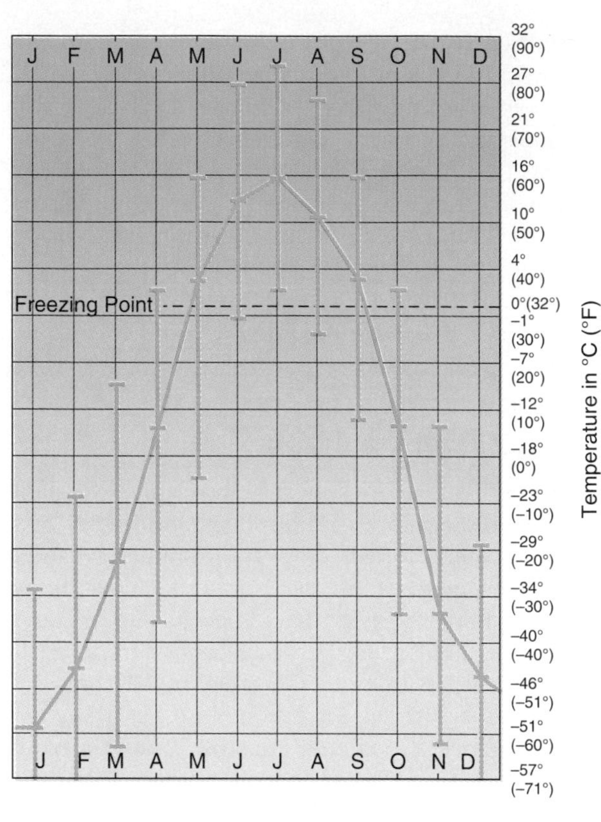

Station: Verkhoyansk, Russia
Lat/long: 67°33′N, 135°23′E
Avg. Ann. Temp.: −15°C (5°F)
Total Ann. Precip.: 15.5 cm (6.1 in.)
Elevation: 137 m (449.5 ft)
Population: 1400
Ann. Temp. Range: 63C° (113.4F°)

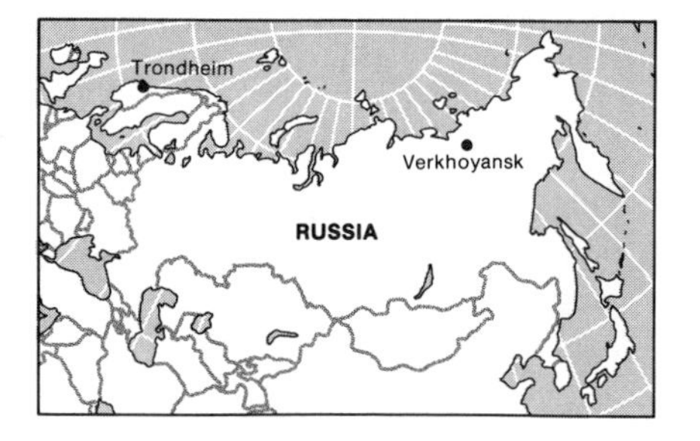

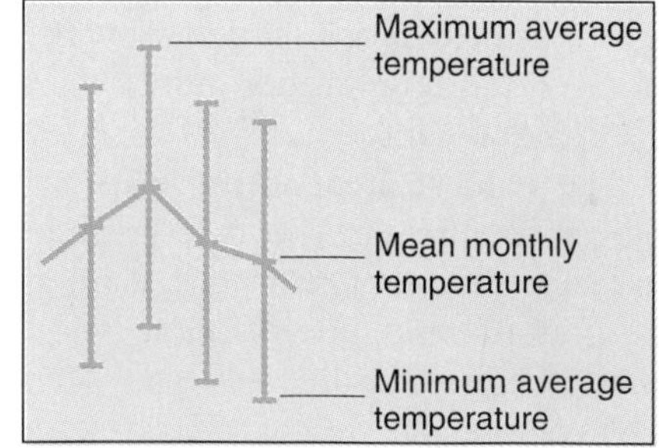

FIGURE 5-9
Comparison of temperatures in coastal Trondheim, Norway, and continental Verkhoyansk, Russia.

cations such as portions of the Sonoran Desert of North America and the Sahara of Africa are prime examples. Africa has recorded shade temperatures in excess of 58°C (136°F), a record set on September 13, 1922 at Al 'Azīzīyah, Libya (32° 32' N; 112 m or 367 ft elevation). The highest maximum and annual average temperatures in North America occurred in Death Valley, California, where the Greenland Ranch Station (37° N; −54.3 m or −178 ft below sea level) reached 57°C (134°F) in 1913. Such arid and hot lands are discussed further in Chapter 15.

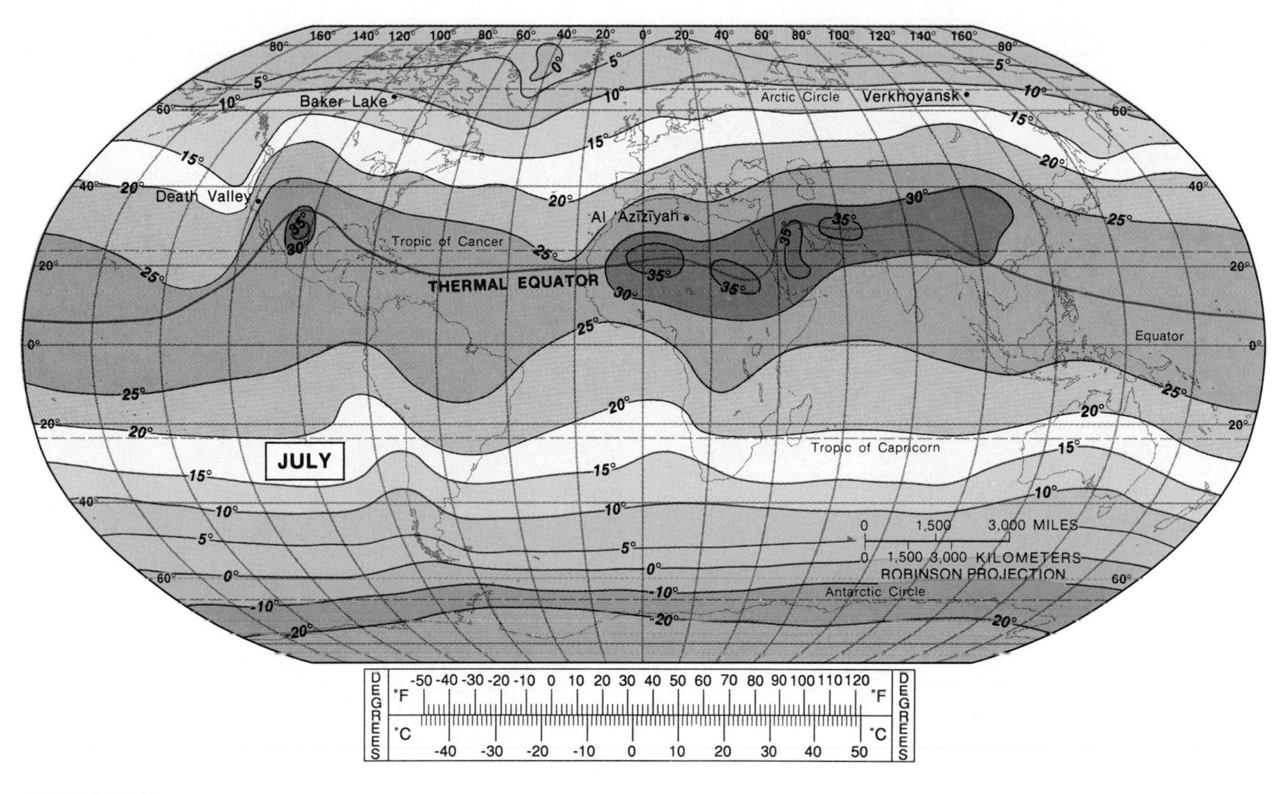

FIGURE 5-10
Worldwide mean sea level temperatures for July. Temperatures are in Celsius (converted to Fahrenheit by means of the scale) and are equated to sea level to compensate for the effects of landforms.

Annual Range of Temperatures

The pattern of marine vs. continental influence emerges dramatically when the range in averages is mapped for a full year (Figure 5-11). As you might expect, the largest temperature ranges occur in subpolar locations in North America and Asia, where average ranges of 64C° (115F°) are recorded. The Southern Hemisphere, on the other hand, produces little seasonal variation in mean temperatures, although the landmasses present there do produce some increase in temperature range.

For example, in January (see Figure 5-8) Australia is dominated by isotherms of 20–30°C (68°–86°F), whereas in July (see Figure 5-10) Australia experiences isotherms of 10–20°C (50°–68°F). Generally speaking, Southern Hemisphere patterns are more marine, whereas Northern Hemisphere patterns are more continental. The Northern Hemisphere, with grteater land area overall, registers a slightly higher average surface temperature than does the Southern Hemisphere: 15°C (59°F) as opposed to 12.7°C (55°F).

A logical question to ask about these patterns of world temperatures is what effect they have on the human body. What would you experience if you went to some of these locations?

Air Temperature and the Human Body

We humans are capable of sensing small changes in temperature in the environment around us. Our

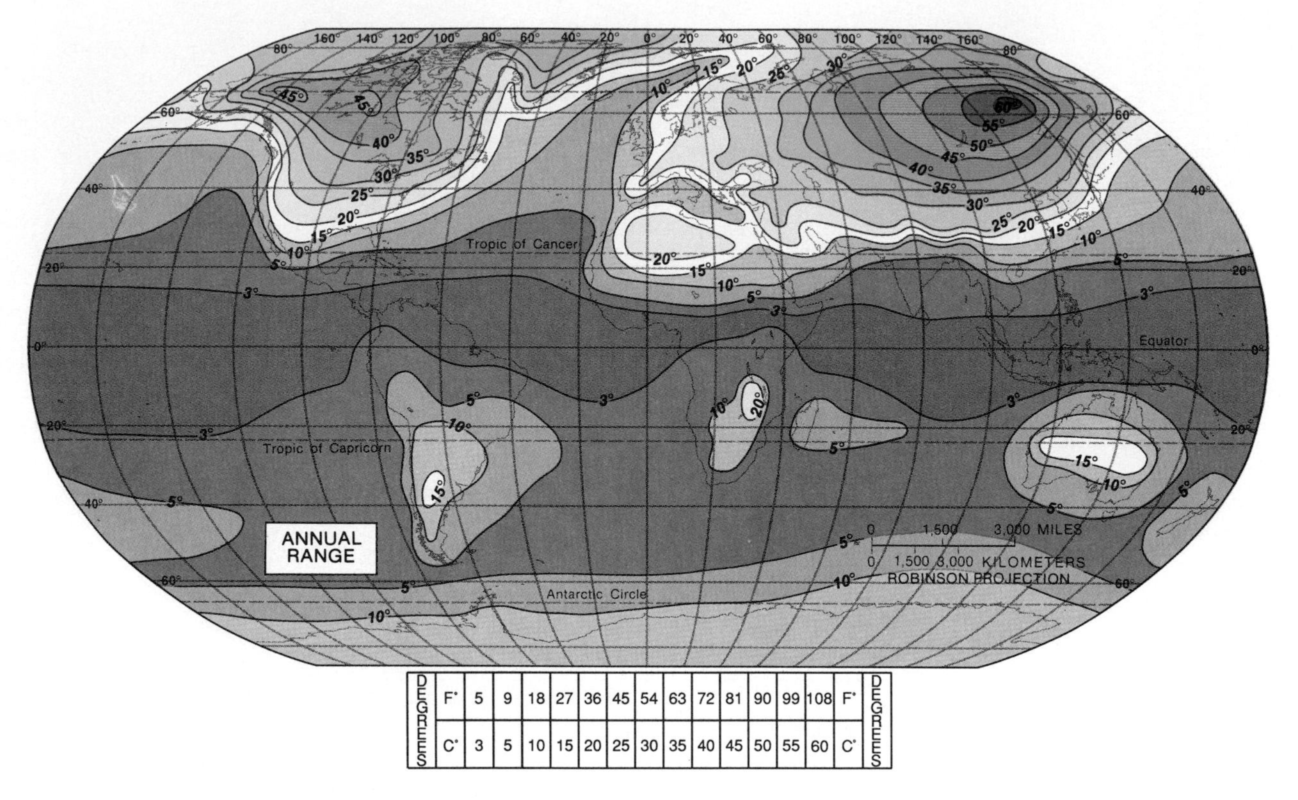

FIGURE 5-11
Annual range of global temperatures in Celsius (Fahrenheit) degrees.

perception of temperature is described by the terms **apparent temperature** or sensible temperature. This perception of temperature varies among individuals and cultures. Through complex mechanisms, our bodies maintain an average internal temperature ranging within a degree of 37°C (98.6°F), slightly lower in the morning or in cold weather and slightly higher at emotional times or during exercise and work.

The water vapor content of air, the wind speed, and air temperature taken together affect each individual's sense of comfort. High temperatures, high humidity, and low winds produce the most heat discomfort, whereas low humidity and strong winds enhance cooling rates. Although modern heating and cooling systems can reduce the impact of uncomfortable temperatures indoors, the danger to human life from excessive heat or cold persists outdoors. When changes occur in the surrounding air, the human body reacts in various ways to maintain its core temperature and to protect the brain at all cost. Table 5-1 summarizes the human body's response to stress induced by low and high temperatures.

The wind chill index is important to those who experience winters with freezing temperatures (Figure 5-12). The **wind chill factor** indicates the enhanced rate at which body heat is lost to the air. As wind speeds increase, heat loss from the skin increases. For example, if the air temperature is −1°C (30°F) and the wind is blowing at 32 kmph (20 mph), skin temperatures will be −16°C (4°F). The lower wind chill values present a serious freezing hazard to exposed flesh.

A measured index of the human body's reaction to air temperature and water vapor is called the *heat index (HI)*. The water vapor in air is expressed as relative humidity, a concept presented in Chapter 7. For now, we simply can assume that the amount of water vapor in the air affects the

TABLE 5-1
The Human Body's Response to Temperature Stress

At Low Temperatures	At High Temperatures
Temperature Regulation Methods	
Heat-gaining mechanism	Heat-dissipation mechanism
Constriction of surface blood vessels	Dilation of surface blood vessels
Concentration of blood	Dilution of blood
Flexing to reduce surface exposure	Extending to increase exposure
Increased muscle tone	Decreased muscle tone
Decrease in sweating	Sweating
Inclination to increase activity	Inclination to decrease activity
Shivering	
Increased cell metabolism	
Consequent Disturbances	
Increased urine volume	Decreased urine volume
Danger of inadequate blood supply to exposed parts; frostbite	Reduced blood supply to brain; dizziness, nausea, fainting
Discomfort leading to neuroses	Discomfort leading to neuroses
Increased appetite	Decreased appetite
	Mobilization of tissue fluid
	Thirst and dehydration
	Reduced chloride balance; heat cramps
Failure of Regulation	
Falling body temperature	Rising body temperature
Drowsiness	Impaired heat-regulating center
Cessation of heartbeat and respiration	Failure of nervous regulation; cessation of breathing

SOURCE: *Climatology*—Arid Zone Research X. © UNESCO 1958. Reproduced by permission of UNESCO.

FIGURE 5-12
Wind chill factor for various temperatures and wind velocities. A wind chill of −20°C results in a loss of 1420 W/m^2; −40°C, a loss of 1955 W/m^2; −60°C, a loss of 2490 W/m^2.

Actual Air Temperatures in °C (°F)

Temperature Effects on Exposed Flesh

Wind speed, kmph (mph)	10° (50°)	4° (40°)	−1° (30°)	−7° (20°)	−12° (10°)	−18° (0°)	−23° (−10°)	−29° (−20°)	−34° (−30°)	−40° (−40°)	−46° (−50°)	−51° (−60°)
Calm	10° (50°)	4° (40°)	−1° (30°)	−7° (20°)	−12° (10°)	−18° (0°)	−23° (−10°)	−29° (−20°)	−34° (−30°)	−40° (−40°)	−46° (−50°)	−51° (−60°)
8 (5)	9° (48°)	3° (37°)	−2° (28°)	−9° (16°)	−14° (6°)	−21° (−5°)	−26° (−15°)	−32° (−26°)	−38° (−36°)	−44° (−47°)	−49° (−57°)	−56° (−68°)
16 (10)	4° (40°)	−2° (28°)	−9° (16°)	−16° (4°)	−23° (−9°)	−29° (−21°)	−36° (−33°)	−43° (−46°)	−50° (−58°)	−57° (−70°)	−64° (−83°)	−71° (−95°)
24 (15)	2° (36°)	−6° (22°)	−13° (9°)	−21° (−5°)	−28° (−18°)	−38° (−36°)	−43° (−45°)	−50° (−58°)	−58° (−72°)	−65° (−85°)	−73° (−99°)	−74° (−102°)
32 (20)	0° (32°)	−8° (18°)	−16° (4°)	−23° (−10°)	−32° (−25°)	−39° (−39°)	−47° (−53°)	−55° (−67°)	−63° (−82°)	−71° (−96°)	−79° (−110°)	−87° (−124°)
40 (25)	−1° (30°)	−9° (16°)	−18° (0°)	−26° (−15°)	−34° (−29°)	−42° (−44°)	−51° (−59°)	−59° (−74°)	−64° (−83°)	−76° (−104°)	−81° (−113°)	−92° (−133°)
48 (30)	−2° (28°)	−11° (13°)	−19° (−2°)	−28° (−18°)	−36° (−33°)	−44° (−48°)	−53° (−63°)	−62° (−79°)	−70° (−94°)	−78° (−109°)	−87° (−125°)	−96° (−140°)
56 (35)	−3° (27°)	−12° (11°)	−20° (−4°)	−29° (−20°)	−37° (−35°)	−45° (−49°)	−53° (−64°)	−63° (−82°)	−72° (−98°)	−84° (−119°)	−89° (−129°)	−98° (−145°)
64 (40)	−3° (26°)	−12° (10°)	−21° (−6°)	−29° (−21°)	−38° (−37°)	−47° (−53°)	−56° (−69°)	−65° (−85°)	−74° (−102°)	−82° (−116°)	−91° (−132°)	−100° (−148°)

Low danger with proper clothing — Exposed flesh in danger of freezing

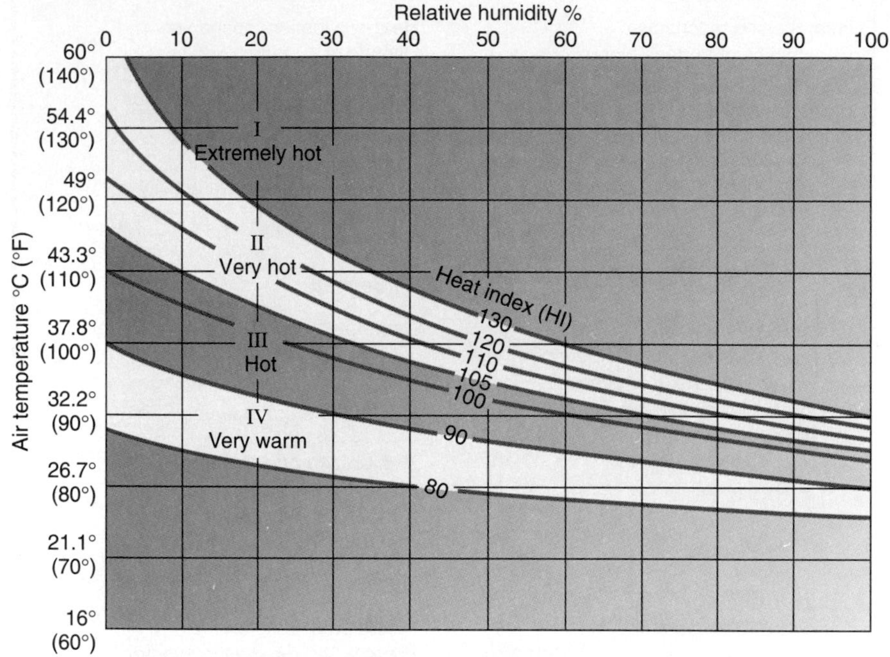

Category	Heat index Apparent temperature	General effect of heat index on people in higher risk groups
I	54°C (130°F) or higher	Heat/sunstroke highly likely with continued exposure
II	41–54°C (105–130°F)	Sunstroke, heat cramps, or heat exhaustion likely and heatstroke possible with prolonged exposure and/or physical activity
III	32–41°C (90–105°F)	Sunstroke, heat cramps, and heat exhaustion possible with prolonged exposure and/or physical activity
IV	27–32°C (80–90°F)	Fatigue possible with prolonged exposure and/or physical activity

FIGURE 5-13
Heat index graph for various temperatures and relative humidity levels. [Courtesy of National Weather Service.]

evaporation rate of perspiration on the skin because the more water vapor in the air (the higher the humidity), the less water from perspiration the air can absorb through evaporation. The heat index indicates how the air *feels to an average person*–in other words, its apparent temperature.

Figure 5-13 is an abbreviated version of the heat index used by the National Weather Service (NWS) and now included in its daily weather summaries during appropriate months. The inset describes the effects of heat-index categories on higher-risk groups, because a combination of high temperature and humidity can severely reduce the body's natural ability to regulate internal temperature. There is a distinct possibility that future humans may experience greater temperature-related challenges due to complex changes now underway in the lower atmosphere.

FUTURE TEMPERATURE TRENDS

Significant climatic change has occurred on Earth in the past and most certainly will occur in the future. There is nothing society can do about long-term influences that cycle Earth through swings from ice ages to warmer periods. However, our global society must address possible short term changes that are influencing global temperatures within the life span of present generations.

A cooperative global network of nations, under the auspices of the United Nations Environment Programme (UNEP), participates in the World Weather Watch System and the Global Change Program for the gathering of temperature and other weather and climatic information. Evidence of this cooperation is the study completed by the Intergovernmental Panel on Climate Change (IPCC) in 1990 and 1992.

Computers make possible the processing of global data with which to assess trends in average temperatures and sea level. In Chapter 10 the broader issues of climate change and computer-based general circulation models are discussed. What follows is an overview of current scientific efforts to determine future short-term temperature trends.

Global Warming

Human activities are enhancing the greenhouse effect described in Chapter 4. Average air temperatures worldwide are increasing. Several world conferences on global warming and climate change now are held each year. Representatives at the Villach Conference, held as an initial assessment by the United Nations Environment Programme in 1985 at Villach, Austria, offered some direction:

> **The rate and degree of future warming [of Earth's climates] could be profoundly affected by governmental policies on energy conservation, use of fossil fuels, and the emission of some greenhouse gases. . . . understanding of the greenhouse question is sufficiently developed that scientists and policy makers should begin an active collaboration to explore the effectiveness of alternative policies and adjustments.***

Five years later these findings were mirrored in a pledge agreement to reduce carbon dioxide emissions, signed in November 1990. This historic agreement was reached by 130 nations, or about two-thirds of Earth's governments. This set the stage for the historic United Nations Conference on Environment and Development (UNCED)—the Earth Summit—held in Rio de Janeiro in 1992. The *U.N. Framework Convention on Climate Change* was signed by over 160 nations and is a legally binding agreement to evaluate and address global warming. The success or failure of this agreement will be assessed in the coming years; action taken by the United States government is critical to progress. We are in a unique position to alter, perhaps reverse, Earth's probable anthropogenic warming trend by controlling the production of certain radiatively active gases. But first, we must understand the problem and its roots.

Carbon Dioxide and Global Warming. Radiatively active gases are atmospheric gases, such as carbon dioxide, methane, CFCs, and water vapor, that absorb and radiate infrared wavelengths. Carbon dioxide is the principal radiatively active gas causing Earth's natural greenhouse effect. CO_2 is transparent to light but opaque to the infrared wavelengths radiated by Earth, and thus its presence delays heat loss to space. While detained, that heat energy is absorbed and reradiated over and over in the lower atmosphere, causing average temperatures to increase. The Industrial Revolution, beginning in the mid-1700s, initiated a tremendous surge in the burning of fossil fuels. This, coupled with the destruction and inadequate replacement rate of harvested forests, is causing atmospheric CO_2 levels to increase 7.3–9.1 billion metric tons (8–10 billion tons) per year (Figure 5-14). CO_2 alone is thought to be responsible for almost 60% of the global warming trend.

Table 5-2 shows the increasing percentage of CO_2 in the lower atmosphere from past to present

TABLE 5-2
Lower Atmosphere Concentration of CO_2

Date	CO_2 Concentration (%)	Parts per Million
1825	0.021	210
1888	0.028	280
1970	0.032	320
1985	0.035	350
1995 (estimate)	0.037	370
2020 (estimate)	0.055	550
2050 (estimate)	0.060	600

*United Nations Environment Programme, *An Assessment of the Role of Carbon Dioxide and of Other Greenhouse Gases in Climate Variations and Associated Impacts,* Geneva: World Meteorological Organization, 1985.

FIGURE 5-14
Industrial landscapes contribute to the increasing output of CO_2 into the atmosphere. [Photo by author.]

and gives estimates for the future. CO_2 is presently increasing in concentration at the rate of 0.4% per year. This increase appears sufficient to override any natural climatic tendencies toward a cooling trend, as well as to produce possible unwanted global warming in the next century.

Figure 5-15 shows past and projected sources of excessive (non-natural) carbon dioxide by country or region, clearly identifying developing nations as the sector with the greatest probable growth in fossil-fuel consumption and new CO_2 production. However, national and corporate policies could alter this forecast by actively steering developing countries toward more appropriately scaled alternative energy sources (low temperature, renewable, labor intensive).

Especially important in the United States are policies to eliminate energy waste and inefficiency, both in terms of reducing carbon dioxide production and in setting a positive example. Yet relatively simple solutions are not being promoted, and throughout the 1980s the United States was uncooperative in worldwide efforts to halt this human-

FIGURE 5-15
Countries and regions of origin (a) for excessive CO_2 in 1980 and (b) forecast for 2025.

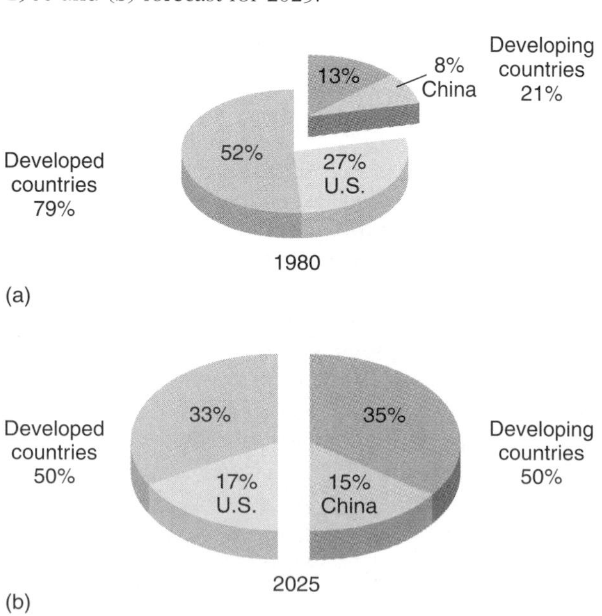

induced increase in CO_2. The United States was not a signatory to the November 1990 agreement mentioned, despite the fact that this country alone produces 27% of the excess CO_2 annually!

The Office of Technology Assessment and the National Academies of Science and Engineering, in separate assessments, concluded that the United States could hold to 1990 levels by the year 2015 at little or no additional cost. (See OTA's "Changing by Degrees: Steps to Reduce Greenhouse Gases," February 5, 1991, and "Policy Implications of Greenhouse Warming," by the Committee on Science, Engineering, and Public Policy of NAS, 1991.) The OTA report concluded that

> Some reductions may even be at a net savings if the proper policies are implemented. . . . The efficiency of practically every end use of energy can be improved relatively inexpensively.

Initially, the goal at the 1992 UNCED was to set specific timetables and limits for cutting emissions of greenhouse gases. The United States objected to any specific targets for controlling CO_2 emissions throughout the presummit sessions, citing economic uncertainties and unknown costs. The European Community and a majority of nations favored a goal to stabilize emissions at 1990 levels by the year 2000. The United States, however, failed to pledge specific emission cutbacks. U.S. emissions instead are predicted to rise 15% over 1990 levels by the year 2000.

Methane. Another radiatively active gas contributing to the overall greenhouse effect is **methane** (CH_4), which also is increasing in concentration at more than 1% per year. Air bubbles in ice show that concentrations of methane in the past—500–27,000 years ago—were approximately 0.7 ppm, whereas current atmospheric concentrations are 1.6 ppm. Methane is generated by organic processes, such as digestion and rotting in the absence of oxygen (anaerobic processes). About 50% of the excess methane being produced comes from bacterial action in the intestinal tracts of livestock and from underwater bacteria in rice paddies. Burning of vegetation causes another 20% of the excess, and bacterial action inside the digestion systems of increasing termite populations also is a significant source. Methane is now believed responsible for at least 12% of the total atmospheric warming, complementing the warming caused by the buildup of CO_2 and equaling about one-half the contribution of CFCs.

Chlorofluorocarbons (CFCs). CFCs are thought to contribute about 25% of the global warming. CFCs absorb infrared in wavelengths missed by carbon dioxide and water vapor in the lower troposphere. As radiatively active gases, CFCs enhance the greenhouse effect, and also play a negative role in stratospheric ozone depletion.

Present Indications and the Future. Global mean temperatures between 1980 and 1992 registered eight of the warmest years in the history of instrumental measurement, including the four warmest years in order: 1990, 1991, 1988, and 1983. A global average temperature of 15.4°C (59.8°F) was reached in 1990. The eruption of Mount Pinatubo lowered temperatures in 1991; otherwise, that year would have placed first.

Comparing annual temperatures and 5-year mean temperatures gives a sense of overall trends. Figure 5-16 shows observed temperatures from 1880 through 1992. The overall pattern is clear—we are entering a period of record warmth. The 1990 Scientific Assessment of Climate Change, an international working group of 170 scientists from 30 nations, reached a consensus that the warming will be socially significant, and that warming rates will accelerate in the near future. This agreement among scientists was echoed by the Intergovernmental Panel on Climate Changes in their three 1990 reports and 1992 update, discussed in Chapter 10.

The consequences of this potential warming are complex. During the decade of the 1980s, average ocean surface temperatures rose at an annual rate of 0.11C° (0.2F°). Of course, any change in ocean temperature has a profound effect on weather and, indirectly, on agriculture and soil moisture. Specific biological effects presently thought to be a result of these warmer waters are discussed in Chapter 16. Even the warming of the ocean itself will contribute about 25% of sea-level rise due to thermal expansion of the water.

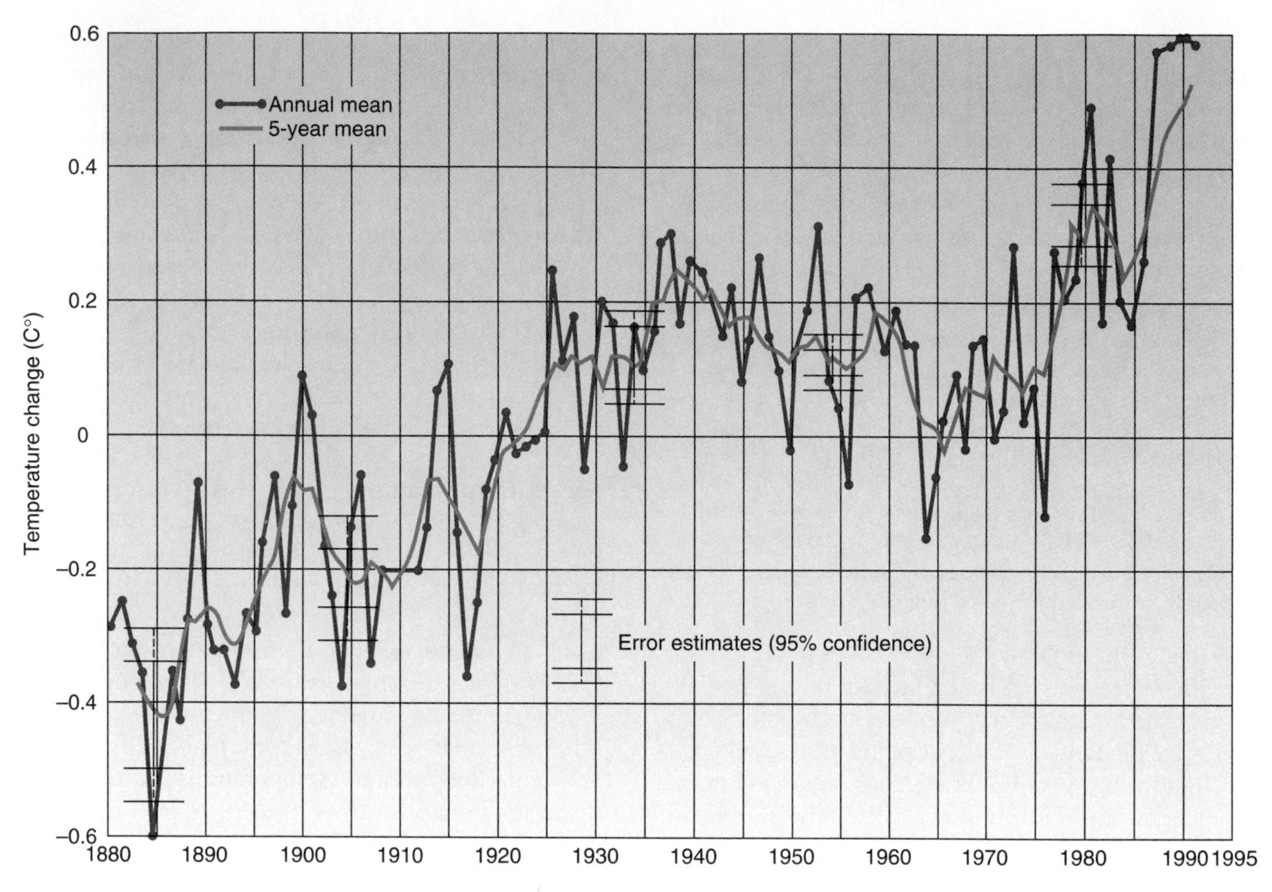

FIGURE 5-16
Global temperature trends from 1880 to 1992. The 0 baseline represents the 1950–1980 global average temperature. [Courtesy of Dr. James E. Hansen, Goddard Institute, NASA.]

The Scripps Institution of Oceanography, in La Jolla, California, has kept ocean temperature records since 1916. From 1950 to the present, they registered a 0.8C° (1.44F°) warming of the upper 100 m (330) ft). Significant increases are being recorded to depths of over 300 m (985 ft). Five of the ten warmest years on record were during the 1980s. Such concentrated increases are cause for concern.

Perhaps the most pervasive effect to monitor is high-latitude and polar warming, which could lead to a rapid escalation of ice melting. In the past century alone, sea-level rise is estimated at 10–20 cm (4–8 in.). Additional meltwater, especially from grounded ice masses (those that rest entirely on land) in Antarctica and Greenland, could raise worldwide sea level in the future.

Global Cooling

In the 1980s and early 1990s, several volcanic eruptions also affected global temperatures. The 1982 El Chichón eruption in south-central Mexico sent a cloud of sulfuric acid and ash 25 km (15.5 mi) into the atmosphere. Within 21 days the cloud had circled the globe, increasing the atmospheric albedo and lowering temperatures by 0.5C° (0.9F°). Some of the finer particles remained in the stratosphere through 1985.

The largest eruption so far in this century began during June 1991. Mount Pinatubo in the Philippines caused significant global temperature effects by injecting a blanket of aerosols into the atmosphere. This increased atmospheric albedo, which decreased insolation reaching the surface. The volcanic aerosols increased shortwave reflectance by 4.3 W/m^2, whereas longwave absorption increased only 1.8 W/m^2, producing a net radiation reduction of 2.5 W/m^2. This change in net radiation could lower average temperatures in the Northern Hemisphere from 0.5C° to 1C° (0.9F° to 1.8F°), and about half that decrease in the Southern Hemisphere, over several years following the eruption.

Such volcanic effects are well documented in history. In addition, worldwide air pollution has increased the turbidity, or murkiness, of the atmosphere (Chapter 3). Perhaps a combination of increasing air pollution and several volcanic eruptions explains why the warming trend during this century was interrupted slightly between 1940 and 1970, only to resume after 1970 (see Figure 5-16).

Another potential cooling effect was postulated by Paul Crutzen (a Dutch atmospheric scientist with the Max Planck Institute of Germany) and John Birks (a chemist at the University of Colorado) in "The Atmosphere After a Nuclear War: Twilight at Noon." Their study, published in the Royal Swedish Academy of Sciences publication *Ambio* in June 1982, launched an avalanche of scientific analyses, assessments, and general confirmations (see Suggested Readings for additional references on this subject).

The **nuclear winter hypothesis,** as it has come to be known, now encompasses a whole range of ecological, biological, and climatic impacts associated with the detonation of a relatively small number of nuclear warheads within the biosphere. Resulting urban fires and fire storms would produce such a pall of insolation-obscuring smoke that surface heating would be drastically reduced. Temperatures could drop 25C° (45F°) and perhaps more in the midlatitudes. Scientists are anxiously studying the localized temperature effects caused by the oil-well fires in Kuwait and Iraq, which were started in early 1991 (see Figure 21-5 for a satellite image of these fires). This potential global impact of modern warfare affects literally all aspects of Earth systems and therefore is appropriate to our study of physical geography.

SUMMARY—
World Temperatures

Probably the first thing you notice when you step outside is the air temperature. It affects your daily decisions on clothing and activities. Worldwide, varied temperature regimes affect individuals and cultures. The principal factors thatoperate to produce the pattern of world temperatures include latitude, altitude, cloud cover, and the numerous physical properties that create differences in heating and cooling rates for land and water.

The pattern of present Earth temperatures is best portrayed on isothermal maps, which permit geographic analysis. The annual range of mean temperatures between January and July demonstrates the interaction of the principal temperature controls. The large continental regions experience a greater annual range in temperature, or continentality, as compared with areas nearer a coastline, where conditions are more marine.

As humans, we need to be aware of our sensitivity to temperature differences and the physiological responses produced in our bodies by exposure to temperature extremes. The heat index and wind chill charts are two useful indicators of apparent temperature.

Human activities apparently are affecting short-term temperature trends by producing increases in certain radiatively active gases. An increase in carbon dioxide concentration over the past two centuries is probably enhancing the natural atmospheric greenhouse effect. Eight of the warmest years in the history of instrumental measurement were recorded between 1980 and 1992. The pattern of world temperatures produced by specific natural and unnatural factors knows no political boundaries and is truly an international issue. Further international cooperation seems mandatory, based on existing evidence of this escalating condition. The largest-ever international meeting of nations occurred in Rio de Janeiro in 1992. At this United Nations Conference on Environment and Development (UNCED), a Convention on Climate Change was agreed to by more than 160 nations. The spatial implications of human-caused climate change are of key interest to physical geographers.

KEY TERMS

apparent temperature
continentality
Gulf Stream
isotherm
land-water heating differences
marine
methane
nuclear winter hypothesis
radiatively active gases
specific heat
temperature
thermal equator
transparency
wind chill factor

REVIEW QUESTIONS

1. Distinguish between sensible heat and sensible temperature.
2. What does air temperature indicate about energy in the atmosphere?
3. Explain the effect of altitude upon air temperature.
4. What noticeable effect does air density have on the absorption and radiation of heat? What role does altitude play in that process?
5. How is it possible to grow moderate-climate crops such as wheat, barley, and potatoes at an elevation of 4103 m (13,461 ft) near La Paz, Bolivia, so near the equator?
6. Describe the effect of cloud cover with regard to Earth's temperature patterns.
7. List the physical aspects of land and water that produce their different responses to heating. What is the specific effect of transmissibility?
8. Describe the pattern of sea-surface temperatures as determined by satellite remote sensing. Where is the warmest ocean region on Earth?
9. Differentiate between marine and continental temperatures. Give geographical examples of each from the text.
10. What is the thermal equator? Describe the location of the thermal equator in January and in July.
11. Explain the extreme temperature range experienced in north-central Siberia between January and July.
12. Where are the hottest places on Earth? Are they near the equator or elsewhere? Explain. Where is the coldest place on Earth?
13. From the maps in Figures 5-8, 5-10, and 5-11, determine the average temperature values and annual range of temperatures for your location.
14. Identify the different responses of the human body to low-temperature and high-temperature stress.
15. Describe the interaction between air temperature and wind speed and their affect on skin chilling. Select several temperatures and wind speeds from Figure 5-12 and determine the wind chill factor.
16. Define the heat index. What characteristic patterns are experienced throughout the year where you live, relative to the graph in Figure 5-13?
17. Explain the factors that are creating the global warming trend. What are the indications that such a pattern is occurring?
18. What is the record of the United States on reducing greenhouse gas emissions? Have target dates been set for emission cutback goals? Describe the findings of the Office of Technology Assessment in their 1991 report.
19. Briefly, what strategies might be implemented to reduce the buildup of radiatively active gases?
20. What could cause a global cooling trend? What are some real and potential cooling factors that might affect global temperatures?

SUGGESTED READINGS

Budyko, M. I. *The Earth's Climate: Past and Future.* International Geophysics Series, vol. 29. New York: Academic Press, 1982.

Graedel, Thomas E., and Paul J. Crutzen. "The Changing Atmosphere," *Scientific American* 261, no. 3 (September 1989): 58–68.

Hansen, James E., and others. "Prediction of Near-Term Climate Evolution: What Can We Tell the Decision-Makers Now?" *Proceedings of the First North American Conference on Preparing for Climate Changes.* Washington. DC: Government Institutes, 1987.

Hansen, James E., and Sergej Lebedeff. "Global Trends of Measured Surface Air Temperature," *Journal of Geophysical Research 92,* no. D11 (20 November 1987): 13,345–72.

Houghton, Richard A., and George M. Woodwell. "Global Climate Change," *Scientific American* 260, no. 4 (April 1989): 36–44.

Kerr, Richard A. "New Greenhouse Report Puts Down Dissenters," Science 249 (3 August 1990): 481.

Mather, John R. Climatology—*Fundamentals and Applications.* New York: McGraw-Hill, 1974.

Nordhaus, W. D. "An Optimal Transition Path for Controlling Greenhouse Gases," *Science* 258, no. 5086 (20 November 1992): 1315–19.

Riordan, Pauline, and Paul G. Bourget. *World Weather Extremes.* Fort Belvoir, VA: U.S. Army Corp of Engineers, 1985.

Rudloff, Willy. *World Climates.* Stuttgart, Germany: Wissenschaftliche Verlagsgesellschaft, 1981.

Schneider, Stephen H. "The Changing Climate," *Scientific American* 261, no. 3 (September 1989): 70–79.

Turco, R. P., et al. "Nuclear Winter: Global Consequences of Multiple Nuclear Explosions," *Science* 222, no. 4630 (23 December 1983): 1283–92.

Turco, R. P., et al. "Climate and Smoke: An Appraisal of Nuclear Winter," *Science* 247, no. 4939 (12 January 1990): 166–76.

Wernstedt, Frederick L. *World Climatic Data.* Lemont, PA: Climatic Data Press, 1972.

Willmott, Cort J., John R. Mather, and Clinton M. Rowe. *Average Monthly and Annual Surface Air Temperature and Precipitation Data for the World.* Part 1, "The Eastern Hemisphere," and Part 2, "The Western Hemisphere." Elmer, NJ: C. W. Thornthwaite Associates and the University of Delaware, 1981.

Yan, Xiao-Hai, Chung-Ru Ho, Quanan Zheng, and Vic Klemas. "Temperature and Size Variabilities of the Western Pacific Warm Pool," *Science* 258, no. 5088 (4 December 1992): 1643–45.

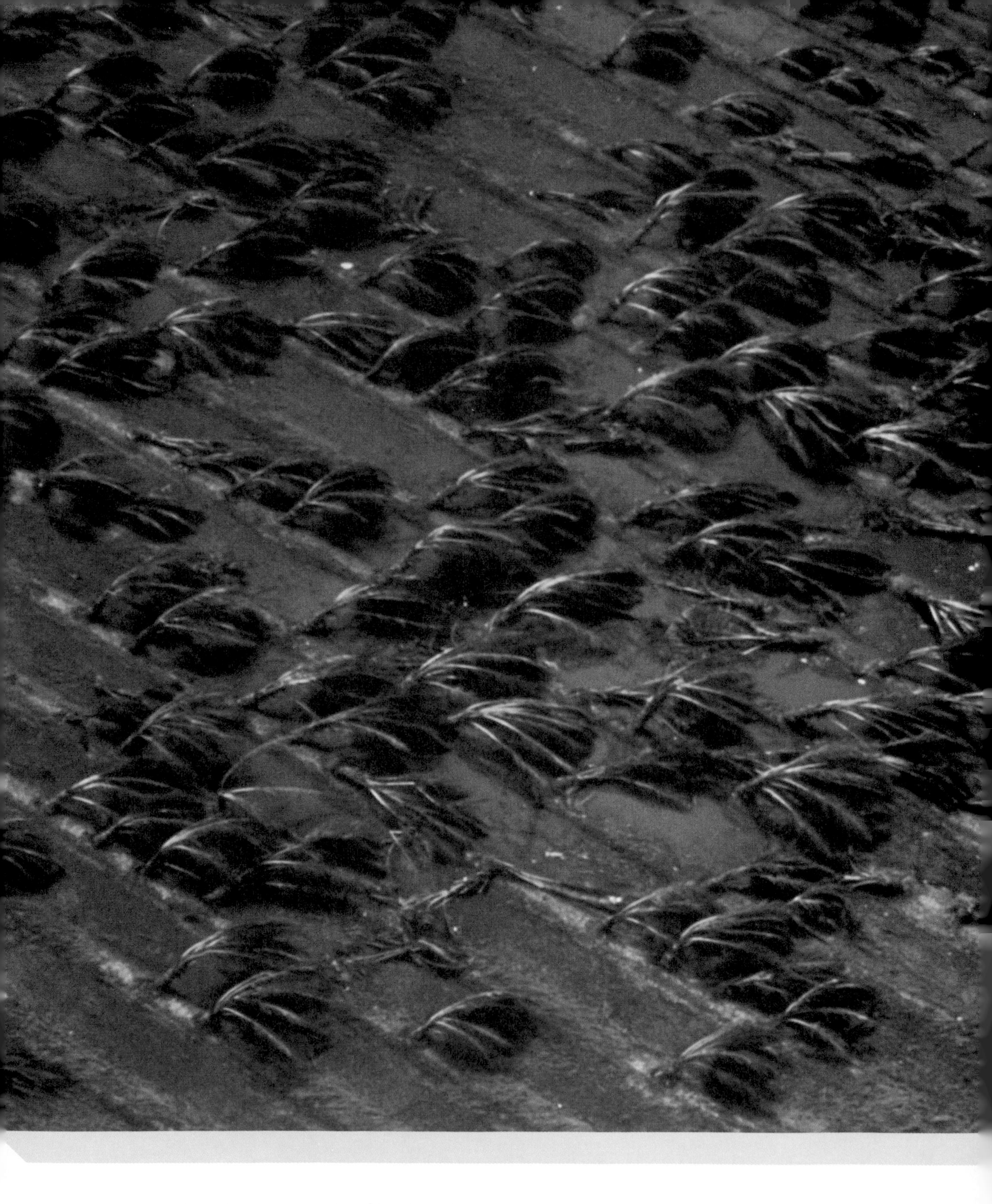

The powerful winds of Hurricane Andrew blew down thousands of trees in southern Florida. [*Photo by John Lopinot/Palm Beach Post/Sygma.*]

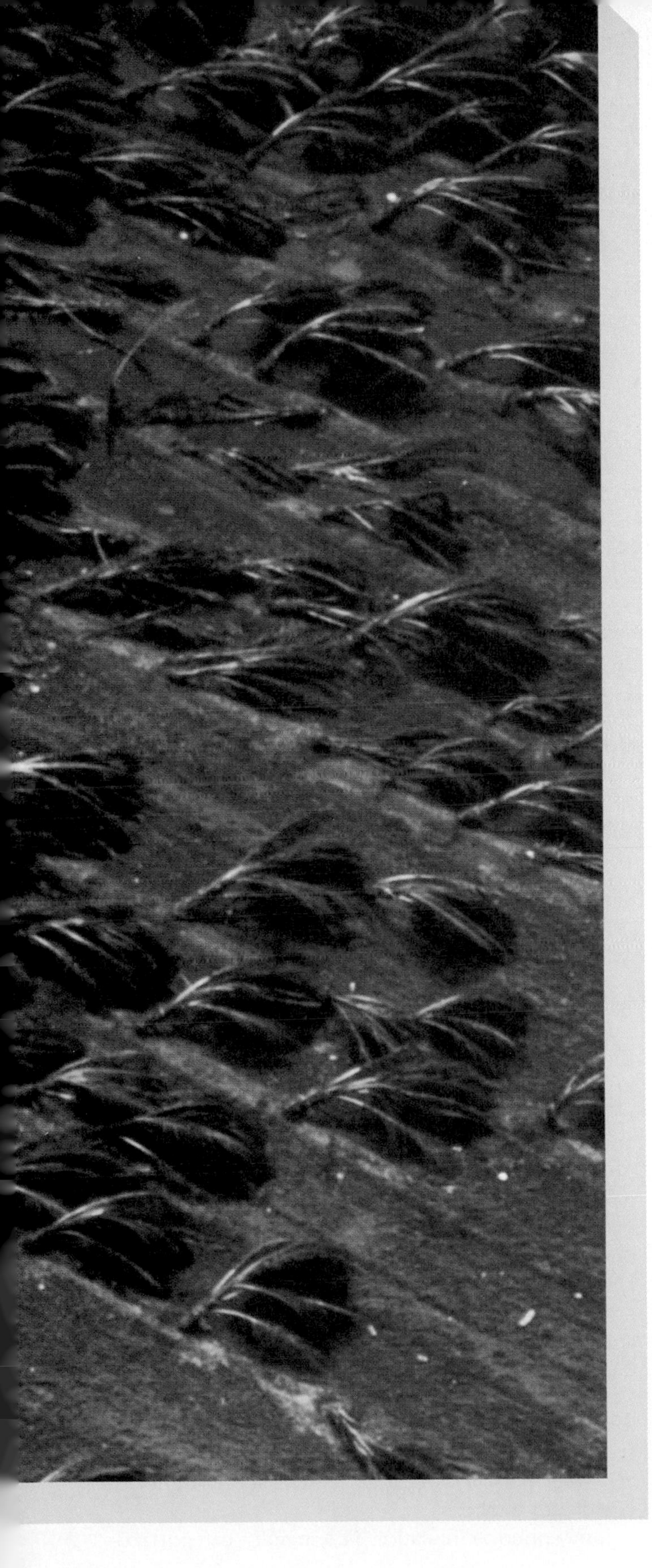

6

ATMOSPHERIC AND OCEANIC CIRCULATIONS

Early in April 1815, on an island named Sumbawa in present-day Indonesia, the volcano Tambora erupted violently, releasing an estimated 150 km^3 (36 mi^3) of material, 25 times the volume produced by the 1980 Mount Saint Helens eruption in Washington State. Some material from Tambora—the aerosols and acid mists—were carried worldwide by global atmospheric circulation, creating a stratospheric dust veil. The result was both a higher atmospheric albedo and absorption of energy by the particulate materials injected into the stratosphere. Remarkable optical and meteorological phenomena resulting from the spreading dust were noted for months and years after the eruption.

Scientists in 1815 lacked the remote-sensing capability of satellite technology, and had no way of knowing the global impact of Tambora's eruption. Today, technology permits a depth of analysis unknown in the past—satellites now track the atmospheric effects from dust storms, forest fires, industrial haze, warfare aftereffects, and the dispersal of volcanic explosions, among many things.

After 635 years of dormancy, Mount Pinatubo in the Philippines erupted on June 15, 1991, an event of tremendous atmospheric impact. This volcanic explosion lofted 15–20 million tons of ash, dust, and SO_2 into the atmosphere. The sulfur compound quickly formed H_2SO_4 (sulfuric acid) aerosols as they were lifted into the stratosphere, concentrating at 16–25 km (10–15.5 mi) altitude. The increase in atmospheric albedo (about 1.3%) produced by these aerosols is analogous to the aerosol optical thickness (AOT) of the atmosphere. This gave scientists an estimate of the overall aerosol load generated by Mount Pinatubo and a unique insight into the dynamics of atmospheric circulation.

The AVHRR (Advanced Very High Resolution Radiometer) instrument aboard *NOAA-11* monitored the reflected solar radiation from these aerosols as they were swept around Earth by global winds. Figure 6-1 shows combined images made at two-week intervals that clearly track this handiwork of our atmospheric circulation. The earliest image, near the time of the eruption, shows winds blowing dust storms from Africa, smoke from the Kuwaiti oil well fires of the Persian Gulf War, and some haze off the east coast of the United States. Also visible is the Pinatubo aerosol layer beginning to emerge north of Indonesia. Atmospheric winds spread the debris worldwide in just 21 days, dusting skies in a band from 15° S to 25° N in width. By the last image in the sequence, some 60 days after the eruption, the aerosol cloud spanned from 20° S to 30° N and covered about 42% of the globe. Most of the world's population experienced spectacular sunrises and sunsets and a lowering of average temperatures during the two years that followed.

In this chapter we examine the dynamic circulation of Earth's atmosphere that carried Tambora's debris and Mount Pinatubo's acid aerosols worldwide, and that carries the common ingredients oxygen, carbon dioxide, and water vapor around the globe. We also consider Earth's wind-driven oceanic currents.

WIND ESSENTIALS

Earth's atmospheric circulation is an important transfer mechanism for both energy and mass. In the process, the imbalance between equatorial energy surpluses and polar energy deficits is partly resolved, Earth's weather patterns are generated, and ocean currents are produced. Human-caused pollution also is spread worldwide by this circulation, far from its points of origin—certainly an important reason why the United States, the former Soviet Union, and Great Britain signed the 1963 Limited Test Ban Treaty. That treaty banned aboveground testing of nuclear weapons because atmospheric circulation was spreading radioactive contamination worldwide. This is an illustration of how the atmosphere socializes humanity more than any other natural or cultural factor. Our atmosphere makes all the world a spatially linked society—one person's or nation's exhalation is another's breath intake.

Atmospheric circulation is generally categorized

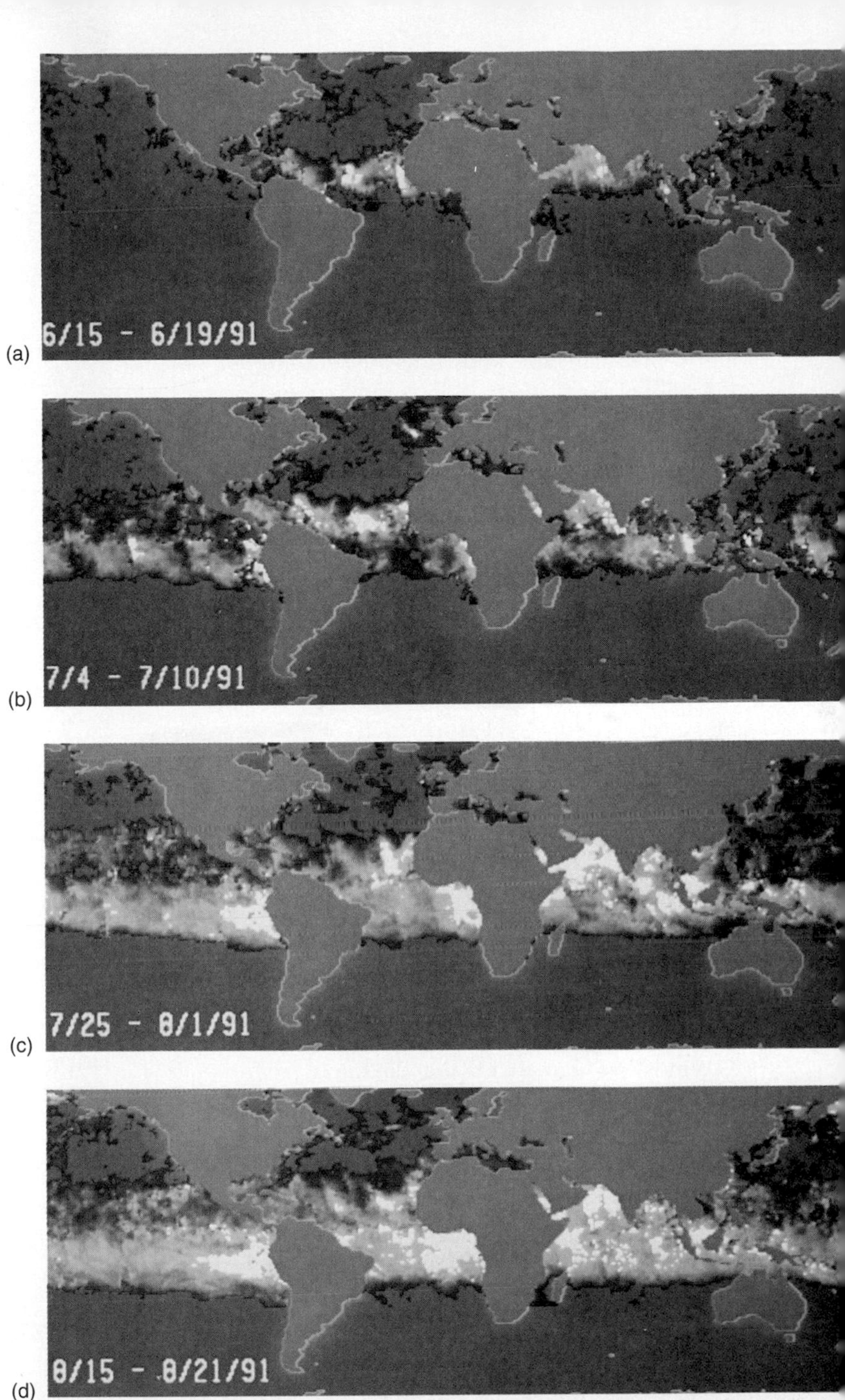

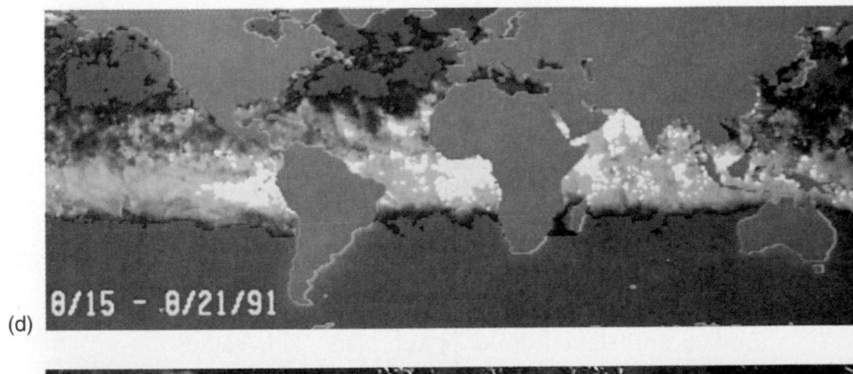

FIGURE 6-1
The eruption of Mount Pinatubo, June 15, 1991, produced aerosols that were swept around the globe many times by the general atmospheric circulation. False-color denotes the atmosphere's aerosol optical thickness (AOT): dull-yellow areas indicate lowest values for AOT; bright yellow indicates medium values, and white shades indicate highest AOT values—that is, the highest accumulations of volcanic aerosols. Also note on the last composite image (8/15–21/91) wind-blown smoke from forest fires in Siberia. [(a–d) AVHRR satellite images of aerosols courtesy of Dr. Larry L. Stowe, National Environmental Satellite, Data, and Information Service, National Oceanic and Atmospheric Administration. Used by permission. (e) AVHRR satellite eruption image courtesy of EROS Data Center.]

at three levels: *primary* (general) *circulation, secondary circulation* of migratory high-pressure and low-pressure systems, and *tertiary circulation* that includes local winds and temporal weather patterns. Winds that move principally north or south along meridians are known as *meridional flows* or meridional circulations. Winds moving east or west along parallels of latitude are called *zonal flows,* or zonal circulations.

Wind: Description and Measurement

Simply stated, **wind** is the horizontal motion of air across Earth's surface. It is produced by differences in air pressure from one location to another and is influenced by several variables. Wind's two principal properties are speed and direction, and instruments are used to measure each. Wind speed is measured with an **anemometer** and is expressed in kilometers per hour (kmph), miles per hour (mph), meters per second (m/sec), or knots. (A *knot* is a nautical mile per hour, covering 1 minute of Earth's arc in an hour, equivalent to 1.85 kmph or 1.15 mph.) Wind direction is determined with a **wind vane;** the standard measurement is taken 10 m (33 ft) above the ground to avoid, as much as possible, local effects of topography upon wind direction (Figure 6-2).

Winds are named for the direction *from which they originate.* For example, a wind from the west is a *westerly* wind (it blows eastward); a wind out of the south is a *southerly* wind (it blows northward). Figure 6-3 illustrates a simple wind compass, naming the 16 principal wind directions.

A descriptive scale useful in visually estimating winds is the traditional **Beaufort wind scale.** Originally established in 1806 by Admiral Beaufort of the British Navy for use at sea, the scale was expanded to include wind speeds by G. C. Simpson in 1926 and standardized by the National Weather Service (Weather Bureau) in 1955. The scale presented in Table 6-1 is modernized and adapted from these earlier versions. It still is referenced on ocean charts and is presented here with descriptions of visual wind effects on land and sea. This observational scale makes possible the estimation of wind speed without instruments, although most modern ships use sophisticated equipment to perform such measurements.

FIGURE 6-2
Instruments used to measure wind direction (wind vane) and wind speed (anemometer). [Photo by author.]

FIGURE 6-3
A wind compass.

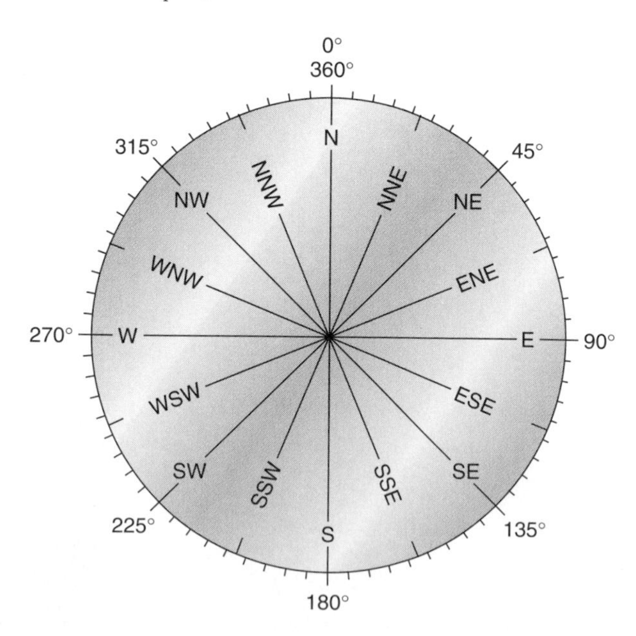

TABLE 6-1
Beaufort Wind Scale

Wind Speed			Beaufort Number	Wind Description	Observed Effects at Sea	Observed Effects on Land
kmph	mph	knots				
<1	<1	<1	0	Calm	Glassy calm, like a mirror	Calm, no movement of leaves
1–5	1–3	1–3	1	Light air	Small ripples, wavelet scales, no foam on crests	Slight leaf movement, smoke drifts, wind vanes still
6–11	4–7	4–6	2	Light breeze	Small wavelets; glassy look to crests, which do not break	Leaves rustling; wind felt, wind vanes moving
12–19	8–12	7–10	3	Gentle breeze	Large wavelets, dispersed whitecaps as crests break	Leaves and twigs in motion, small flags and banners extended
20–29	13–18	11–16	4	Moderate breeze	Small, longer waves; numerous whitecaps	Small branches moving; raising dust, paper litter,and dry leaves
30–38	19–24	17–21	5	Fresh breeze	Moderate, pronounced waves; many whitecaps; some spray	Small trees and branches swaying, wavelets forming on inland waterways
30 40	25–31	22–27	6	Strong breeze	Large waves, white foam crests everywhere; some spray	Large branches swaying, overhead wires whistling, difficult to control an umbrella
50–61	32–38	28–33	7	Moderate (near) gale	Sea mounding up, foam and sea spray blown in streaks in the direction of the wind	Entire trees moving, difficult to walk into wind
62–74	39–46	34–40	8	Fresh gale (or gale)	Moderately high waves of greater length, breaking crests forming seaspray, well-marked foam streaks	Small branches breaking, difficult to walk, moving automobiles drifting and veering
75–87	47–54	41–47	9	Strong gale	High waves, wave crests tumbling and the sea beginning to roll, visibility reduced by blowing spray	Roof shingles blown away, slight damage to structures, broken branches littering the ground
88–101	55–63	48–55	10	Whole gale (or storm)	Very high waves and heavy, rolling seas; white appearance to foam-covered sea; overhanging waves; visibility reduced	Uprooted and broken trees, structural damage, considerable destruction, seldom occurring
102–116	64–73	56–63	11	Storm (or violent storm)	White foam covering a breaking sea of exceptionally high waves, small and medium-sized ships lost from view in wave troughs, wave crests frothy	Widespread damage to structures and trees, a rare occurrence
>117	>74	>64	12–17	Hurricane	Driving foam and spray filling the air, white sea, visibility poor to nonexistent	Severe to catastrophic damage, devastation to affected society

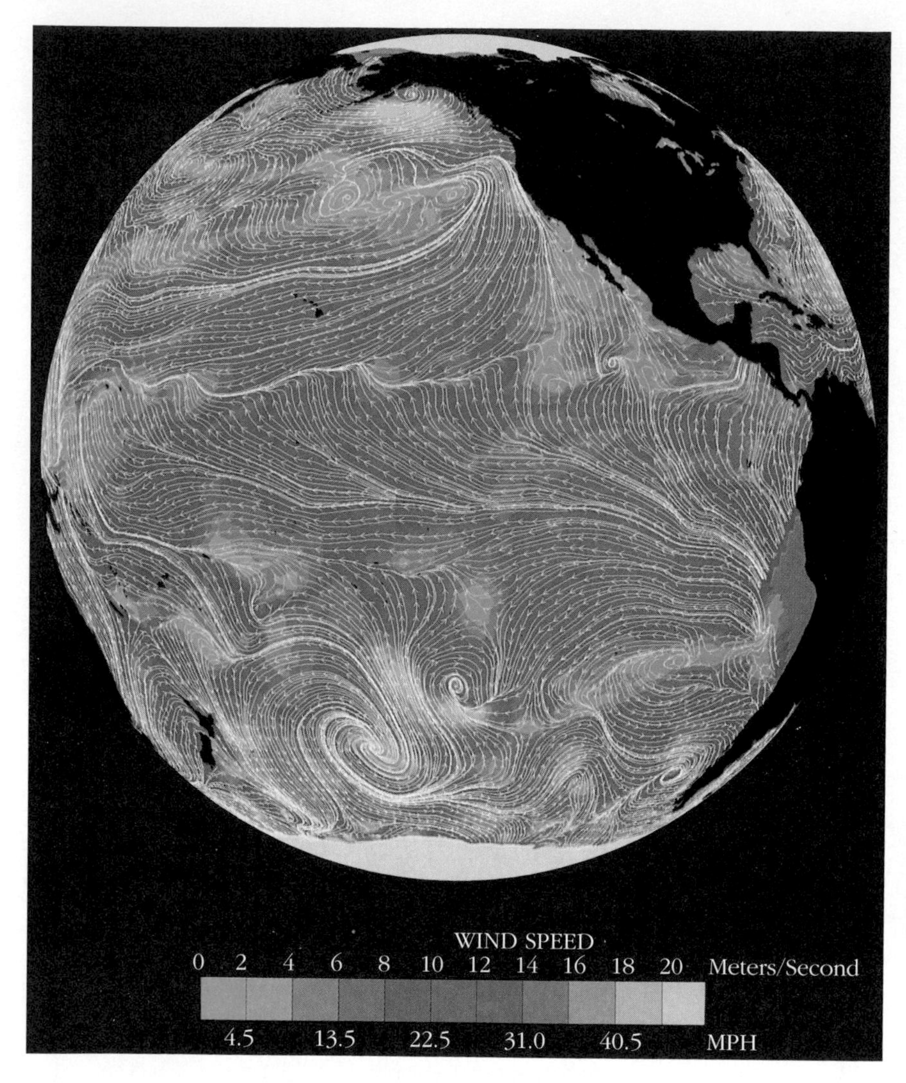

FIGURE 6-4
Surface wind fields measured by radar scatterometer aboard the *Seasat* satellite on September 14, 1978. Scientists analyzed 150,000 measurements to produce this image. Colors correlate to wind speeds, and the white arrows note wind direction. [Image courtesy of Dr. Peter Woiceshyn, Jet Propulsion Laboratory, Pasadena, California. Satellite image below from Laboratory of Planetary Studies, Cornell University. Used by permission.]

Global Winds

The primary circulation of winds across Earth has fascinated travelers, sailors, and scientists for centuries, although only in the modern era is a true picture emerging of the pattern and cause of global winds. With the assistance of scientific breakthroughs in space-based observations and earthbound computer technology, models of total atmospheric circulation are becoming more refined.

A NASA *Seasat* satellite image was painstakingly assembled by scientists at the Jet Propulsion Laboratory and the University of California–Los Angeles from 150,000 measurements made during a single day (Figure 6-4). Wind drives waves on the ocean surface that combine to form patterns of currents. *Seasat* used radar to measure the motion and direction of waves from their backscatter to the satellite, producing this portrait of surface wind fields over the Pacific Ocean. The patterns you see are the response to specific forces at work in the atmosphere, and these forces are our next topic. As we progress through this chapter, you may want to refer to this *Seasat* image to identify the winds, eddies, and vortexes it portrays.

DRIVING FORCES WITHIN THE ATMOSPHERE

Several forces determine both speed and direction of winds:

- Earth's *gravitational force* on the atmosphere is practically uniform, equally compressing the atmosphere near the ground worldwide, with decreasing density occurring with increasing altitude.
- The **pressure gradient force** drives air from areas of higher barometric pressure to areas of lower barometric pressure.
- The **Coriolis force,** a deflective force, appears to deflect wind from a straight path in relation to Earth's rotating surface—to the right in the Northern Hemisphere and to the left in the Southern Hemisphere.
- The **friction force** drags on the wind as it moves across surfaces; it decreases with height above the surface.

All four of these forces operate on moving air and water at Earth's surface and affect the circulation patterns of global winds. The following sections describe the actions of the pressure gradient, Coriolis, and friction forces.

Pressure Gradient Force

High- and low-pressure areas that exist in the atmosphere principally result from unequal heating at Earth's surface and from certain dynamic forces in the atmosphere. A pressure gradient is the difference in atmospheric pressure between areas of higher pressure and lower pressure. Air pressure is measured with a barometer, as described in Chapter 3. Lines plotted on a weather map connecting points of equal pressure are called **isobars;** the distance between isobars indicates the degree of pressure difference. Isobars on a weather map facilitate a spatial analysis of pressure patterns.

Just as closer contour lines on a topographic map mark a steeper slope, so do closer isobars denote steepness in the pressure gradient. In Figure 6-5, note the spacing of the isobars (purple lines). A steep gradient causes faster air movement from a high-pressure area to a low-pressure area. Isobars spaced at greater distances from one another mark a more gradual pressure gradient, one that creates a slower air flow. Along a horizontal surface, the pressure gradient force acts at right angles to the isobars (Figure 6-5).

Figure 6-6 illustrates the combined effect of the forces that direct the wind. The pressure gradient force acting alone is shown in Figure 6-6a from two perspectives. As air descends in the high-pressure area, a field of subsiding, or sinking, air develops. Air moves out of the high-pressure area in a flow described as diverging. High-pressure areas feature *descending, diverging* air flows. On the other hand, air moving into a low-pressure area does so with a converging flow. Thus, low-pressure areas feature *converging, ascending* air flows.

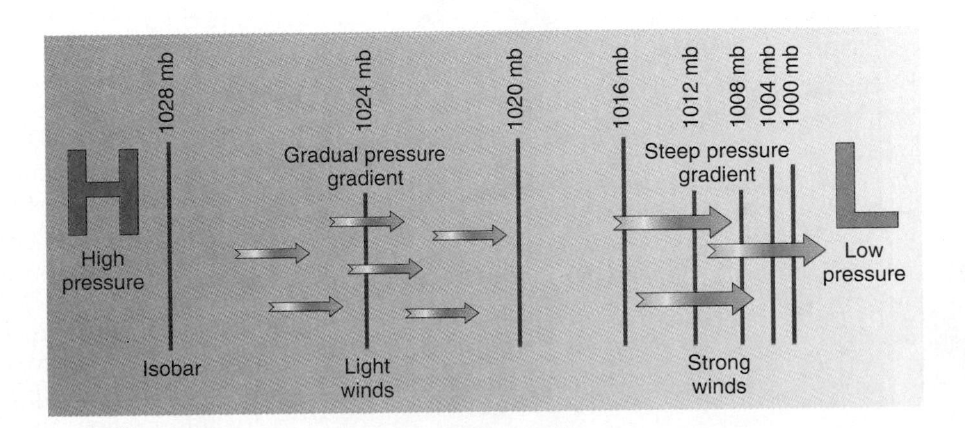

FIGURE 6-5
Pressure gradient.

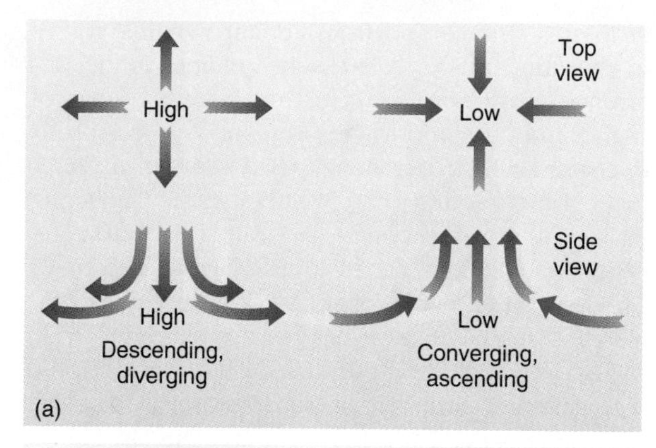

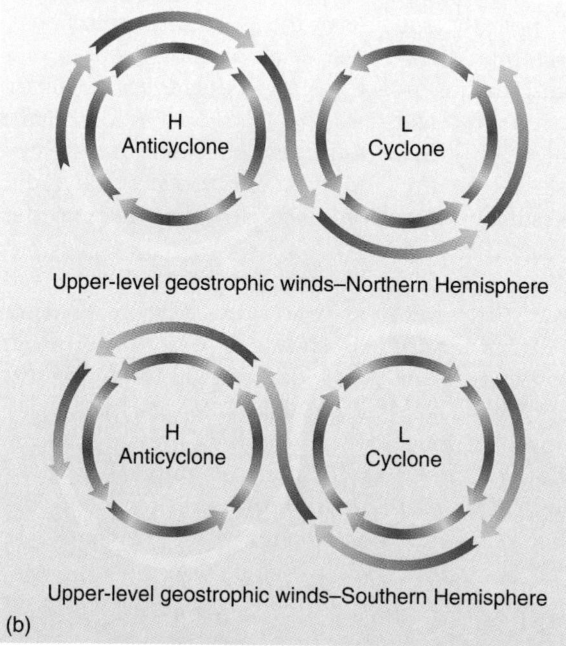

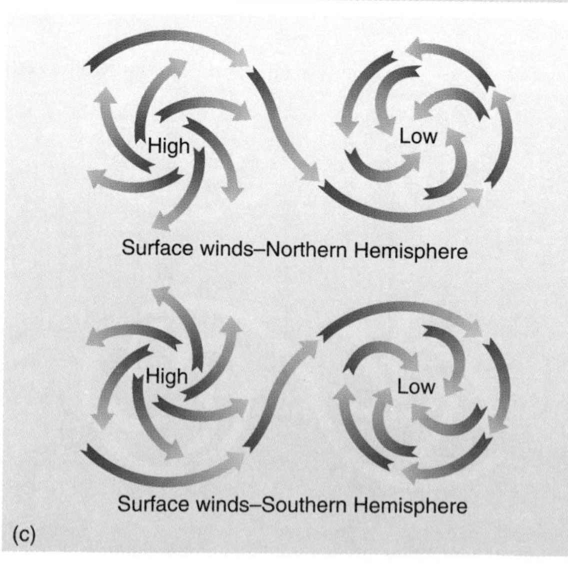

FIGURE 6-6
Three physical forces that integrate to produce wind patterns at the surface and aloft: (a) the pressure gradient force, (b) the Coriolis force, and (c) the friction force. Note the reverse circulation pattern in the Southern Hemisphere. (The gravitational force is assumed.)

Coriolis Force

You might expect surface winds to move in a straight line from areas of higher pressure to areas of lower pressure, but they do not. The reason for this is the Coriolis force. The *Coriolis force* deflects from a straight path any object that flies or flows across Earth's surface, be it the wind, a plane, or ocean currents. The deflection produced by the Coriolis force is caused by Earth's rotation.

Earth's rotational speed varies with latitude, increasing from 0 kmph at the poles to 1675 kmph (1041 mph) at the equator (see Table 2-3). Because Earth rotates eastward, objects that move in an absolute straight line over a distance (such as winds and ocean currents) appear to curve to the right in the Northern Hemisphere and to the left in the Southern Hemisphere. The effect of the Coriolis force increases as the speed of the moving object increases; thus, the higher the wind speed, the greater its apparent deflection.

Viewed from an object passing over Earth's surface, the surface rotates beneath the object. But viewed from the surface, the object appears to curve off-course. The object does not actually deviate from a straight path, but it appears to because of the rotational movement of Earth's surface beneath it. The deflection occurs regardless of the direction in which the object is moving. The Coriolis force is zero along the equator, increasing to half the maximum deflection at 30° N and S latitudes, and to maximum deflection flowing away from the poles (Figure 6-7a).

As an example of the effect of this force, a pilot leaves the North Pole and flies due south toward Quito, Ecuador, traveling a direct route (Figure 6-7b). If Earth were not rotating, the aircraft would simply travel along a meridian of longitude and arrive at Quito. However, Earth is rotating beneath the aircraft's flight path. If the pilot does not allow for this rotation, the plane will reach the equator

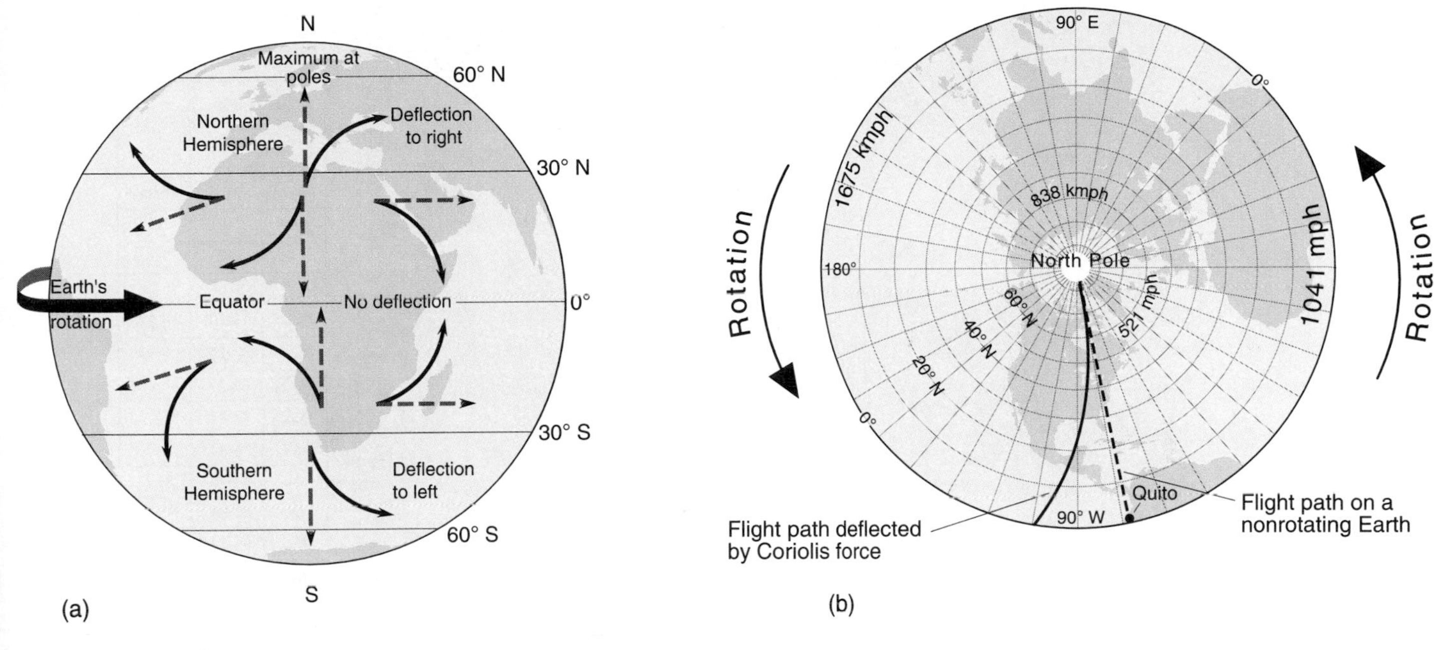

FIGURE 6-7
Distribution of the Coriolis force on Earth: (a) apparent deflection to the right of a straight line in the Northern Hemisphere; apparent deflection to the left in the Southern Hemisphere. (b) Coriolis deflection of a flight path between the North Pole and Quito, Ecuador. Note the latitudinal variations in speed of rotation.

over the ocean far to the west of the intended destination. Pilots must correct for this Coriolis deflection in their navigation calculations.

How does the Coriolis force affect wind? As air rises from the surface through the lowest levels of the atmosphere, leaving the drag of surface friction behind, its speed increases, thus increasing the Coriolis force, spiraling the winds to the right in the Northern Hemisphere or to the left in the Southern Hemisphere. In the upper troposphere the Coriolis force just balances the pressure gradient force. Consequently, the winds between high-pressure and low-pressure areas aloft flow parallel to the isobars.

Figure 6-6(b) illustrates the combined effect of the pressure gradient force and the Coriolis force on the atmosphere, producing **geostrophic winds.** Geostrophic winds are characteristic of upper tropospheric circulation. The air does not flow directly from high to low, but around the pressure areas instead, remaining parallel to the isobars and producing the characteristic pattern shown on the upper-air weather chart in Figure 6-8.

The label "force" is appropriate here because the physicist Sir Isaac Newton stated that, if something is accelerating over a space, a force is in operation. This apparent force is named for Gaspard Coriolis, a French mathematician and researcher of applied mechanics, who described the phenomenon in 1831.

Friction Force

The effect of surface friction extends to a height of about 500 m (1640 ft) and varies with surface texture, wind speed, time of day and year, and atmospheric conditions. Generally, rougher surfaces produce more friction. Figure 6-6c adds the effect of friction to the Coriolis and pressure gradient forces on wind movements. Because surface friction decreases wind speed, it reduces the effect of the Coriolis force and causes winds to move across

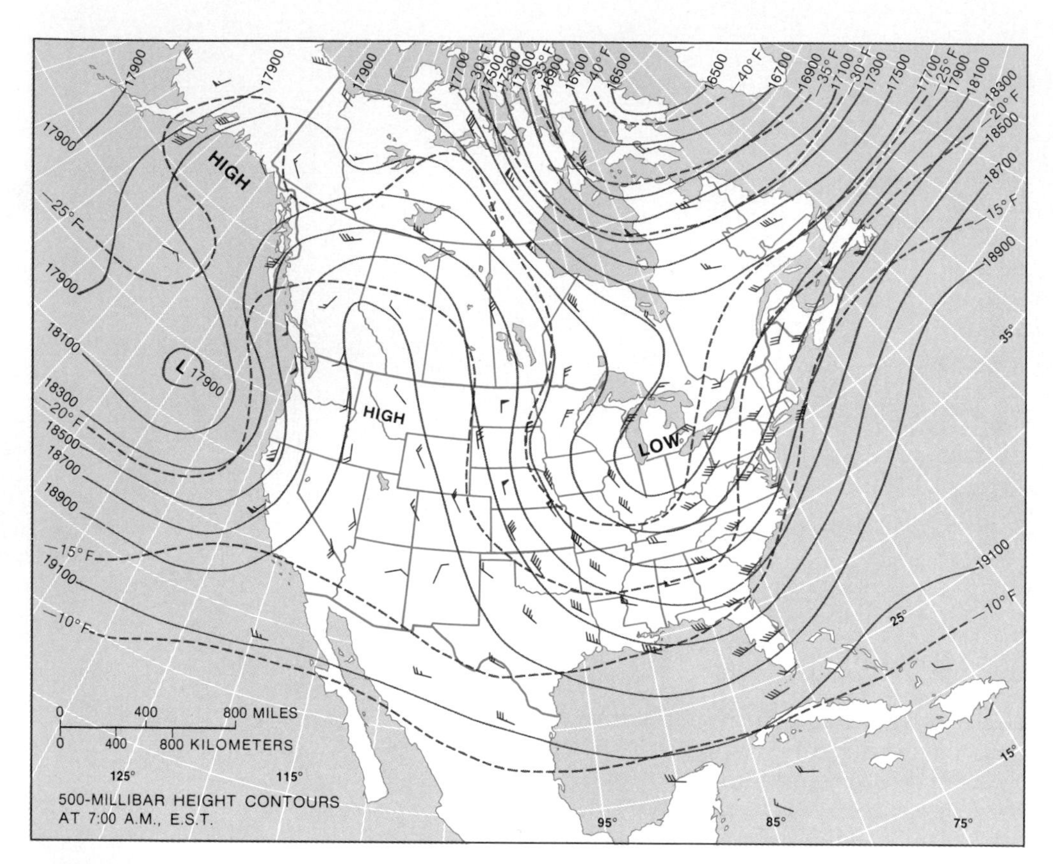

wind speed symbol	miles (statute) per hour	knots
◎	Calm	Calm
	1-2	1-2
	3-8	3-7
	9-14	8-12
	15-20	13-17
	21-25	18-22
	26-31	23-27
	32-37	28-32
	38-43	33-37
	44-49	38-42
	50-54	43-47
	55-60	48-52
	61-66	53-57
	67-71	58-62
	72-77	63-67
	78-83	68-72
	84-89	73-77
	119-123	103-107

FIGURE 6-8
Isobaric chart for April 25, 1977, 7:00 A.M. EST. Contours show height at which 500 mb pressure occurs (in feet). The pattern of contours reveals geostrophic wind patterns in the troposphere at approximately 5500 m (18,000 ft) altitude. Note the "ridge" of high pressure over the Intermountain West and the "trough" of low pressure over the Great Lakes region. [Courtesy of National Weather Service.]

isobars at an angle, spiraling out from a high-pressure area clockwise to form an **anticyclone** and spiraling into a low-pressure area counterclockwise to form a **cyclone.** In the Southern Hemisphere these circulation patterns are reversed, flowing out of high-pressure cells counterclockwise and into low-pressure cells clockwise.

ATMOSPHERIC PATTERNS OF MOTION

With these forces and motions in mind, we are ready to understand the *Seasat* image of winds (see Figure 6-4) and build a general model of total atmospheric circulation. If Earth did not rotate, then the warmer, less dense air of the equatorial regions would rise, creating low pressure at the surface, and the colder and denser air of the poles would sink, creating high pressure at the surface. The net effect would be a meridional flow of winds established by this simple pressure gradient. But in reality, because Earth does rotate, a much more complex system exists for the transfer of energy and mass from equatorial surpluses to polar deficits—one with waves, streams, and eddies on a planetary scale. And instead of this hypothetical meridional flow, the flow is predominantly zonal (latitudinal) at the surface and aloft: westerly (eastward) in the middle and high latitudes and easterly (westward) in the low latitudes of both hemispheres.

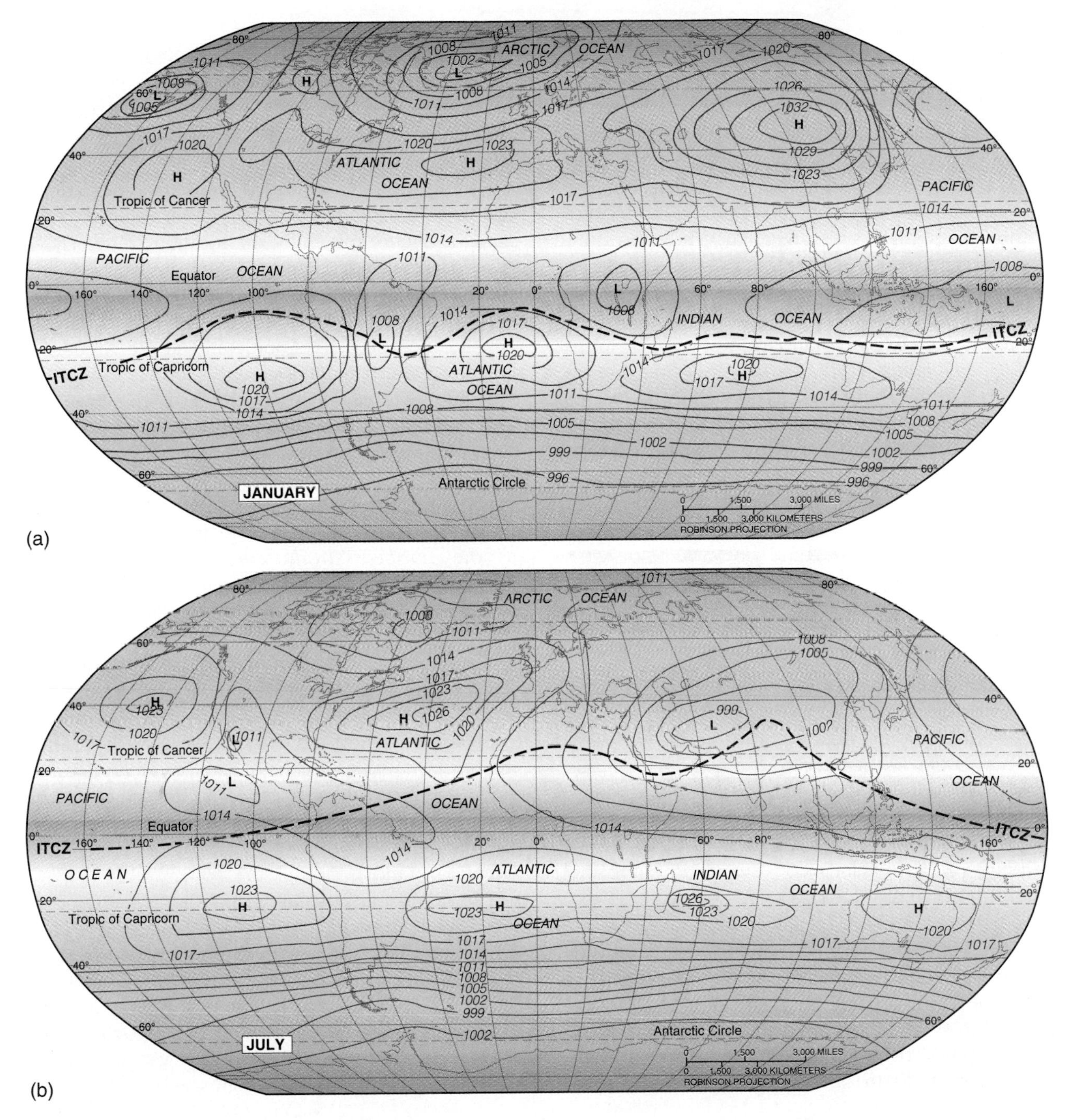

FIGURE 6-9
Average surface barometric pressure (millibars) for (a) January and (b) July. Dashed line marks the intertropical convergence zone (ITCZ).

Primary High-Pressure and Low-Pressure Areas

The following discussion of Earth's pressure and wind patterns refers often to Figure 6-9, which shows January and July isobaric maps of average surface barometric pressure. These maps indicate prevailing surface winds, suggested by the isobars. The primary high- and low-pressure areas on Earth appear on these isobaric maps as interrupted cells or uneven belts of similar pressure that stretch across the face of the planet. Between these areas flow the primary winds, which have been noted in adventure and myth throughout human experience. Secondary highs and lows, from a few hundred to a few thousand kilometers in diameter and hundreds to thousands of meters high, are formed within the primary pressure areas. These pressure systems seasonally migrate to produce changing weather and climate patterns in the regions over which they pass.

Four identifiable pressure areas cover the Northern Hemisphere; a similar set is found in the Southern Hemisphere. In each hemisphere, two of the pressure areas are specifically stimulated by thermal (temperature) factors: the **equatorial low-pressure trough** and the weak **polar high-pressure cells,** both north and south. The other two areas are formed by dynamic factors: the **subtropical high-pressure cells** and **subpolar low-pressure cells.** Table 6-2 summarizes the characteristics of these pressure areas, and we now examine each.

Equatorial Low-Pressure Trough. The equatorial low-pressure trough is an elongated, narrow band of low pressure that nearly girdles Earth, following an undulating linear axis. Constant high Sun altitude and consistent daylength make large amounts of energy available in this region throughout the year. The warming creates lighter, less-dense, ascending air, with winds converging all along the extent of the trough. This converging air is extremely moist and full of latent heat energy. Vertical cloud columns frequently reach the tropopause, and precipitation is heavy throughout this zone. The combination of heating and convergence forces air aloft and forms the **intertropical convergence zone (ITCZ).** Figure 6-10 is a satellite image of Earth showing the equatorial low-pressure trough. The ITCZ is identified by bands of clouds associated with the convergence of winds along the equator.

During summer, a marked wet season accompanies the shifting ITCZ over various regions. The maps in Figure 6-9 show the location of the ITCZ (dashed line) in January and July. In January, the zone crosses northern Australia and dips southward in eastern Africa and South America, whereas by July the zone shifts northward in the Americas and as far north as Pakistan and southern Asia.

Figure 6-11 shows two views of the **Hadley cell,** named for the 18th-century English scientist who described the trade winds. Such a cell is generated by this equatorial low-pressure system of converging, ascending air. These winds converging on the equatorial low-pressure trough are known as the *northeast trade winds* (in the Northern Hemisphere) and *southeast trade winds* (in the Southern Hemisphere), or generally as **trade winds.** The trade winds pick up large quantities of moisture as they return through the Hadley circulation cell for another cycle of uplift and condensation. Within the ITCZ, winds are calm or mildly variable because of the even pressure gradient and the strong vertical ascent of air. These equatorial calms are the **doldrums,** a name formed from an older English word meaning foolish, because of the difficulty sailing ships encountered when attempting to move through this zone.

The rising air from the equatorial low-pressure area spirals upward into a geostrophic flow to the north and south. These upper-air winds turn eastward, flowing from west to east, beginning at about 20° N and S, forming descending anticyclonic flows above the subtropics. In each hemisphere, this circulation pattern appears most vertically symmetrical near the equinoxes.

TABLE 6-2
Four Hemispheric Pressure Areas

Name	Cause	Location	Air Temperature/ Moisture
Polar high-pressure cells	Thermal	90° N, 90° S	Cold/dry
Subpolar low-pressure cells	Dynamic	60° N, 60° S	Cool/wet
Subtropical high-pressure cells	Dynamic	20° to 35° N and S	Hot/dry
Equatorial low-pressure trough	Thermal	10° N to 10° S	Warm/wet

FIGURE 6-10
An interrupted band of clouds denotes the intertropical convergence zone (ITCZ), flanked on either side by several regions dominated by subtropical high-pressure systems. This natural color image was taken by the Solid State Imaging instrument (violet, green, and red filters) aboard the *Galileo* spacecraft during its December 8, 1990, flyby of Earth on its voyage to Jupiter. [Image courtesy of Dr. W. Reid Thompson, Laboratory of Planetary Studies, Cornell University. Used by permission.]

Subtropical High-Pressure Cells. Between 20° and 35° latitude in both hemispheres, a broad high-pressure zone of hot, dry air is evident across the globe (Figure 6-9). It is represented by the clear, frequently cloudless skies of the satellite imagery in Figure 6-10. The dynamic cause of these subtropical anticyclones is too complex to detail here, but generally they form as air in the Hadley cell descends in these latitudes, constantly supported by a dynamic mechanism in the upper-air circulation. Air above the subtropics is mechanically pushed downward and is heated as it is compressed on its descent to the surface, as illustrated in Figure 6-11. Warmer air has a greater capacity to hold water vapor than does cooler air, making this descending warm air relatively dry. The air is also dry because heavy precipitation along the equatorial circulation removed moisture.

Surface air diverging from the subtropical high-pressure cells generates Earth's principal surface winds: the westerlies and the trade winds. The **westerlies** are the dominant surface winds from the subtropics to high latitudes. They diminish somewhat in summer and are stronger in winter.

If you examine Figure 6-9, you will find several high-pressure areas. In the Northern Hemisphere, the Atlantic subtropical high-pressure cell is called the **Bermuda high** (in the western Atlantic) or the **Azores high** (when it migrates to the eastern Atlantic). The area under this subtropical high features clear, warm waters and large quantities of *Sargassum* (a seaweed), which gives the area its

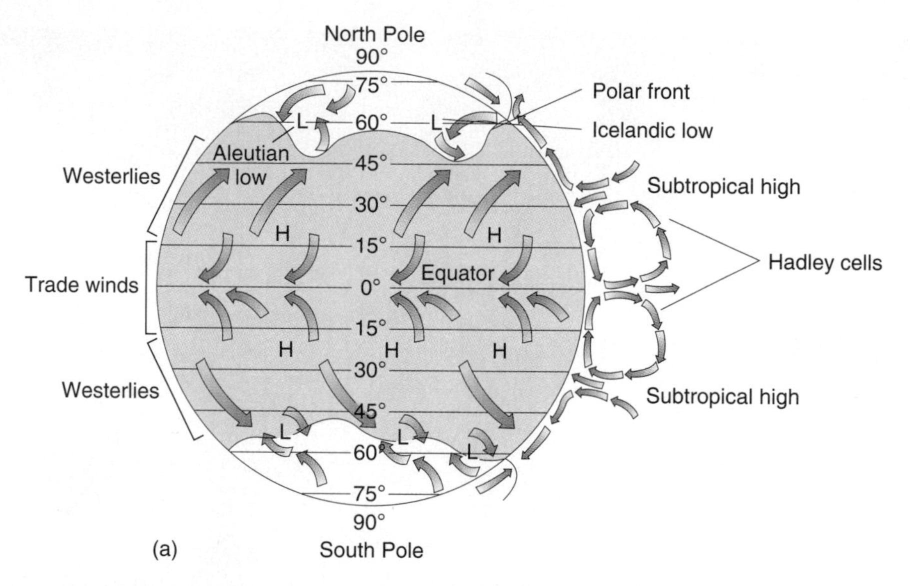

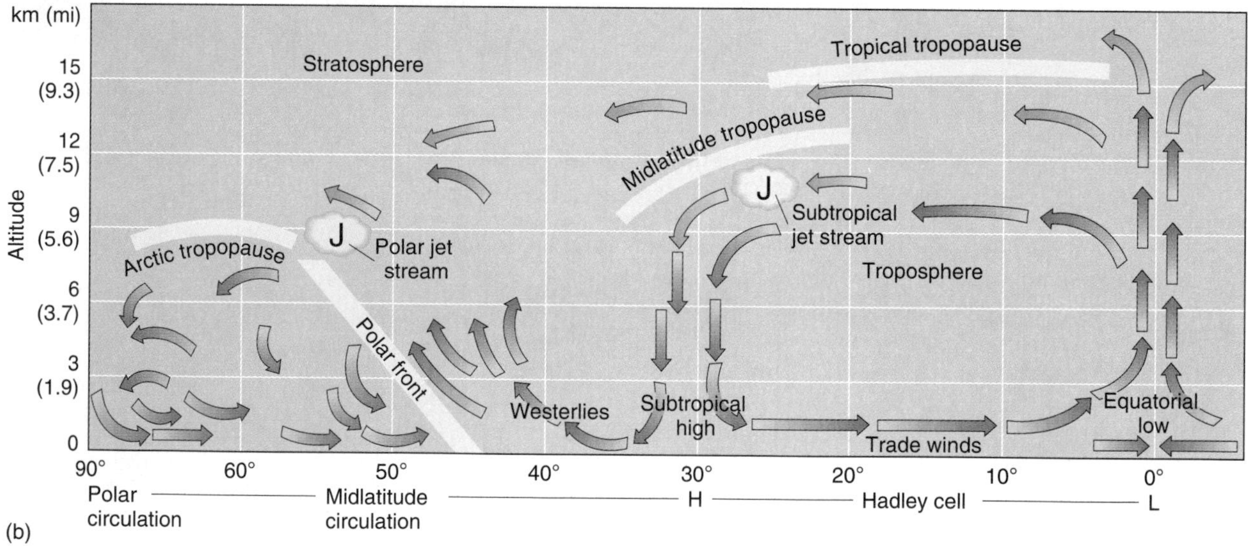

FIGURE 6-11
Two views of the general atmospheric circulation showing Hadley cells, subtropical highs, polar front, the subpolar low-pressure cells, and approximate locations of the subtropical and polar jet streams: (a) general circulation schematic; (b) equator-to-pole cross-section for the Northern Hemisphere. [Cross-section adapted from Joseph E. Van Riper, *Man's Physical World,* p. 211. Copyright 1971 by McGraw-Hill. Adapted by permission.]

name—the Sargasso Sea. The **Pacific high,** or Hawaiian high, dominates the Pacific in July, retreating southward in January. In the Southern Hemisphere, three large high-pressure centers dominate the Pacific, Atlantic, and Indian oceans and tend to move along parallels in shifting zonal positions.

The equatorward boundaries of the subtropical

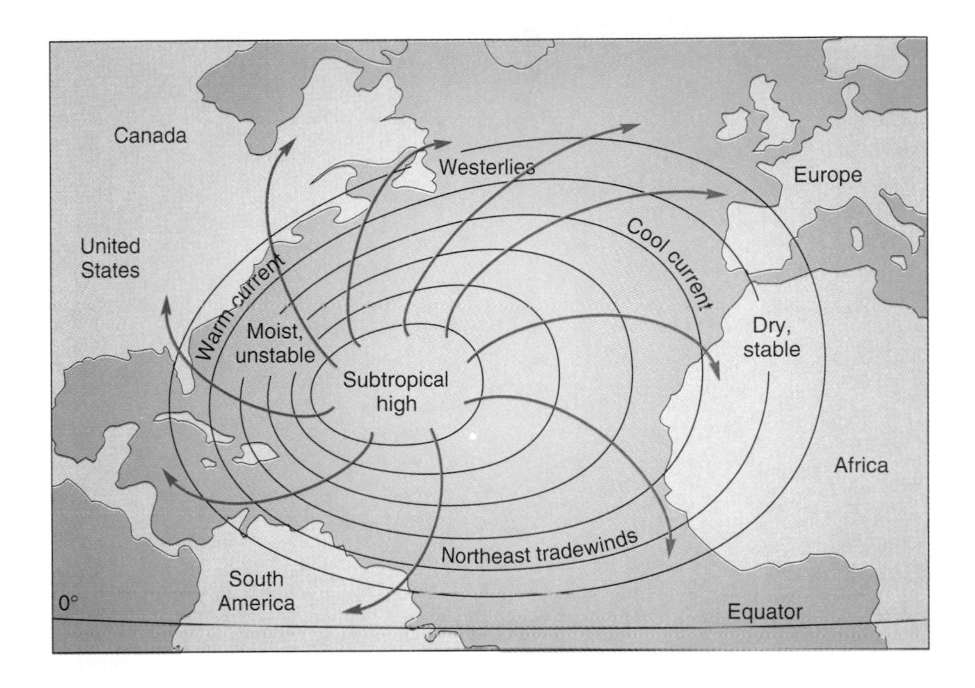

FIGURE 6-12
Characteristic circulation and climate conditions related to Atlantic subtropical high-pressure anticyclone in the Northern Hemisphere.

highs are generally stable and cloud-free, but the poleward boundaries are quite variable. The entire high-pressure system migrates with the summer high Sun, fluctuating about 5–10° in latitude. The eastern sides of these anticyclonic systems are drier and more stable, and feature cooler ocean currents than do the western sides. Thus, subtropical and midlatitude west coasts are influenced by the eastern side of these systems and therefore experience dry-summer conditions (Figure 6-12). In fact, Earth's major deserts generally occur within the subtropical belt and extend to the west coast of each continent except Antarctica.

The western sides of subtropical high-pressure cells tend to be moist and unstable, as characterized in Figure 6-12. This causes warm, moist conditions in Hawaii, Japan, southeastern China, and the southeastern United States.

Because the cores of the subtropical belts are near 25° N and S latitudes, these areas sometimes are known as the calms of Cancer and the calms of Capricorn. The name **horse latitudes** also is used for these zones of windless, hot, dry air, so deadly in the era of sailing ships. This label is popularly attributed to becalmed sailing crews of past centuries who destroyed the horses on board, not wanting to share food or water with the livestock. The term's true origin may never be known; the *Oxford English Dictionary* calls it "uncertain."

Subpolar Low-Pressure Cells. The January map in Figure 6-9a shows two low-pressure cyclonic cells over the oceans around 60° N latitude. The North Pacific **Aleutian low** is near the Aleutian Islands; the North Atlantic **Icelandic low** is near Iceland. Both cells are dominant in winter and weaken significantly or disappear altogether in the summer with the strengthening of high pressure systems in the subtropics. The area of contrast between cold and warm air masses forms a contact zone known as the **polar front,** which encircles Earth at this latitude and is focused in these low-pressure areas.

Figure 6-11 illustrates this confrontation between warmer, moist air from the westerlies and colder, drier air from the polar and Arctic regions. The warmer air is displaced upward by the heavier cold air, forcing cooling and condensation in the lifted air. Low-pressure cyclonic storms migrate out of the Aleutian and Icelandic frontal areas and may produce precipitation in North America and Europe, respectively. Northwestern sections of North America and Europe generally are cool and moist as a result of the passage of these cyclonic systems

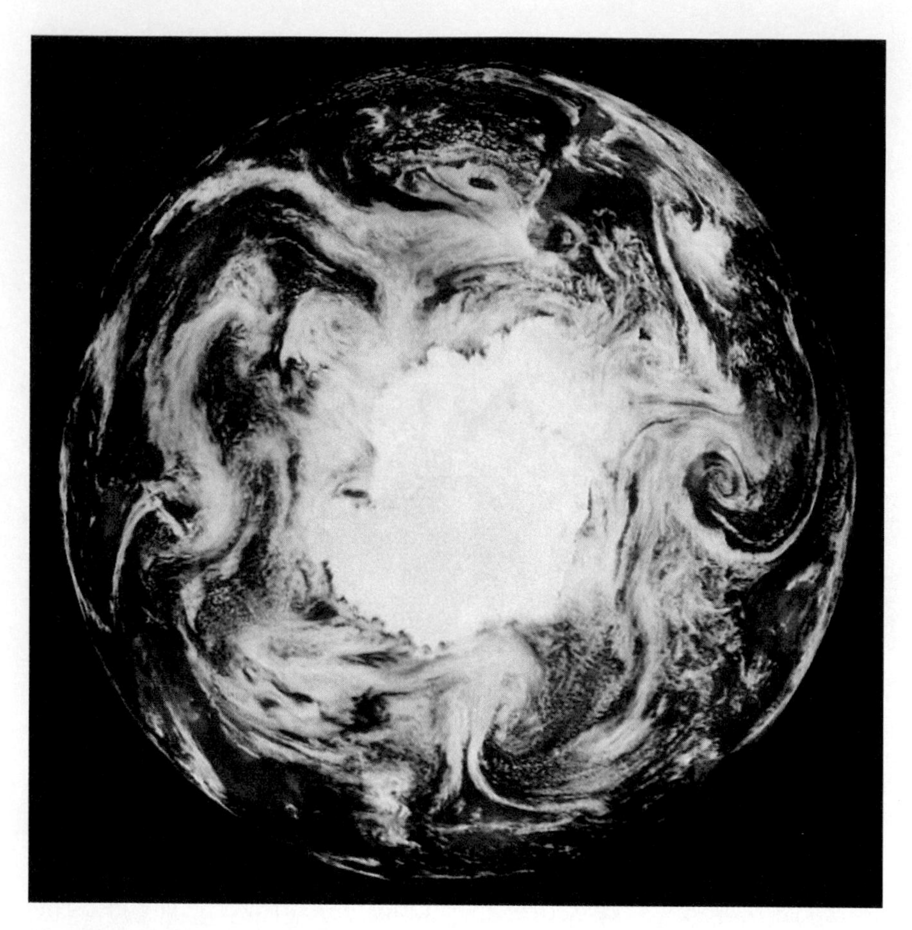

FIGURE 6-13
Centered on Antarctica, this satellite image shows a circumpolar series of subpolar low-pressure cyclones in the Southern Hemisphere. Antarctica is fully illuminated by a midsummer Sun. This natural color image was taken by the Solid State Imaging instrument (violet, green, and red filters) aboard the *Galileo* spacecraft during its December 8, 1990, flyby of Earth on its voyage to Jupiter. [Image courtesy of Dr. W. Reid Thompson, Laboratory of Planetary Studies, Cornell University. Used by permission.]

onshore—consider the weather in Washington State and Ireland.

In the Southern Hemisphere, a noncontinuous belt of subpolar cyclonic pressure systems surrounds Antarctica. The spiraling cloud patterns produced by these cyclonic systems are visible in Figure 6-13. Severe cyclonic storms can cross Antarctica, producing strong winds and new snowfall.

Polar High-Pressure Cells. The weakness of the polar high-pressure cells seems to argue against their inclusion among pressure areas on Earth. The polar atmospheric mass is small, receiving little energy to put it into motion. The variable cold, dry winds moving away from the polar region are anticyclonic, descending and diverging clockwise in the Northern Hemisphere (counterclockwise in the Southern Hemisphere) and forming weak, variable winds called **polar easterlies.** Of the two polar regions, the **Antarctic high** predominates in terms of both strength and persistence. In the Arctic, a polar high-pressure cell is less pronounced, and

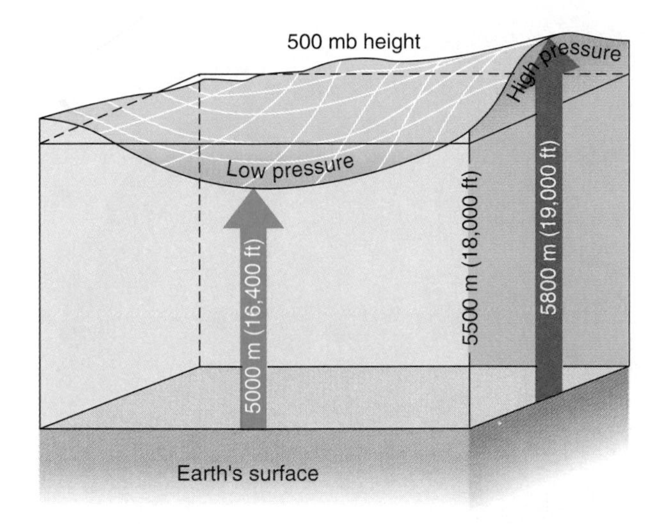

FIGURE 6-14
Representation of the 500-mb isobaric surface is used for analyzing pressure patterns at this height above the surface. [From *Meteorology: The Atmosphere in Action* by Joe R. Eagleman, © 1980 by Litton Educational Publishing, Inc. Reprinted by permission of Wadsworth, Inc.]

when it does form, it tends to locate over the colder northern continental areas in winter (Canadian and Siberian highs) rather than directly over the relatively warmer Arctic Ocean.

Upper Atmospheric Circulation

Middle and upper tropospheric circulation is an important component of the atmosphere's general circulation. As described previously, these upper atmosphere winds tend to blow west to east from the subtropics to the poles. The isobars that mark the paths of upper-air winds are based upon elevation above sea level; these form an **isobaric surface** along which all points have the same pressure (Figure 6-14). Just as sea level is a reference datum for evaluating air pressure at the surface, this isobaric surface is a reference datum used to evaluate air-pressure characteristics at altitude.

Closer spacing of the isobars indicates greater winds; wider spacing indicates lesser winds. In this isobaric pressure surface, altitude variations from the reference datum are called *ridges* for high pressure (with isobars on the map bending poleward) and *troughs* for low pressure (with isobars on the map bending equatorward). Looking at Figure 6-8, can you identify such ridges and troughs on the isobaric chart of geostrophic winds?

Within the westerly flow of geostrophic winds are great waving undulations called **Rossby waves,** named for meteorologist C. G. Rossby, who first described them mathematically. The polar front is the contact between colder air to the north and warmer air to the south (Figure 6-15). The Rossby waves bring tongues of cold air southward, with warmer tropical air moving northward. The development of Rossby waves in the upper-air circulation is shown in the figure. As these disturbances mature, distinct cyclonic circulations form, with warmer air and colder air mixing along distinct fronts. The development of cyclonic storm systems at the surface is supported by these wave-and-eddy formations and other upper-air flows. These Rossby waves develop along the flow axis of a jet stream.

Jet Streams. The most prominent movement in these upper-level westerly wind flows is the **jet stream,** an irregular, concentrated band of wind occurring at several different locations. Rather flattened in vertical cross section, the jet streams normally are 160–480 km (100–300 mi) wide by 900–2150 m (3000–7000 ft) thick, with core speeds that can exceed 300 kmph (190 mph).

Airline schedules reflect the presence of these upper-level westerly winds, for they allot shorter flight times from west to east and longer flight times from east to west. Also important to both military and civilian aircraft is the effect the jet streams have on fuel consumption and the existence of air turbulence. Jet streams in each hemisphere tend to weaken during the hemisphere's summer and strengthen during winter as the streams shift closer to the equator.

The **polar jet stream** is located at the tropopause along the polar front, at altitudes between 7600 and 10,700 m (24,900–35,100 ft), meandering between 30° and 70° N latitude. The polar jet stream can migrate as far south as Texas, steering colder air masses into North America and influencing surface storm paths traveling eastward. In

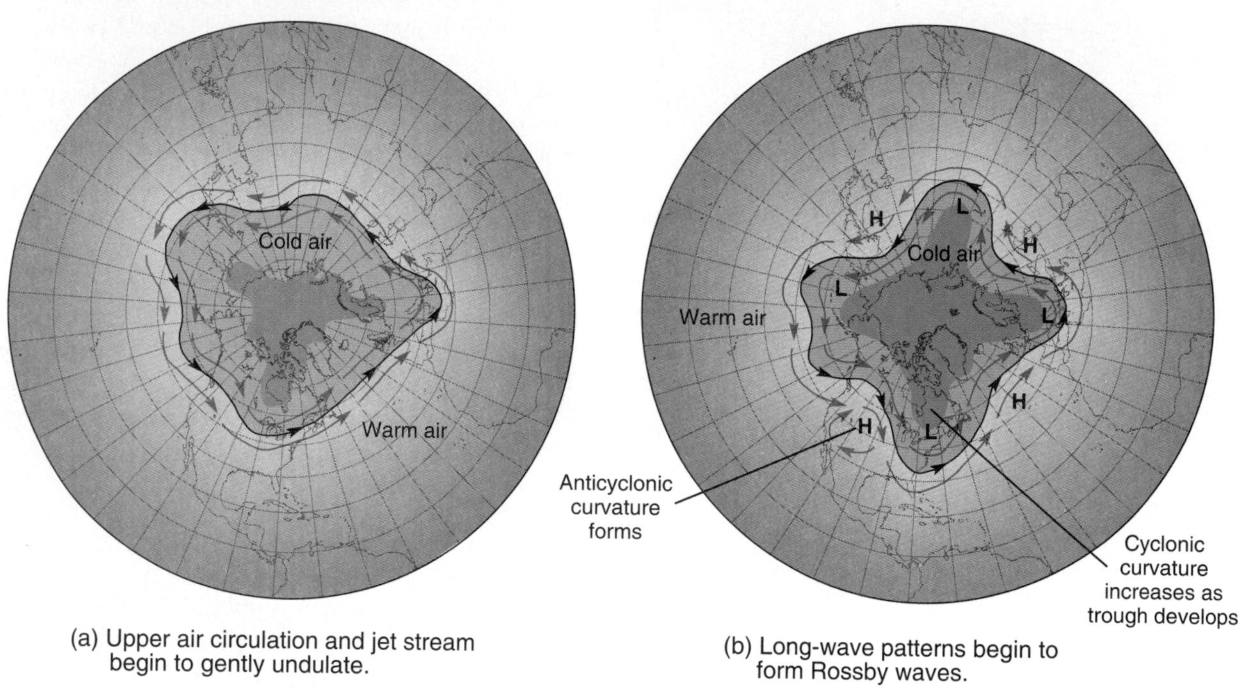

(a) Upper air circulation and jet stream begin to gently undulate.

(b) Long-wave patterns begin to form Rossby waves.

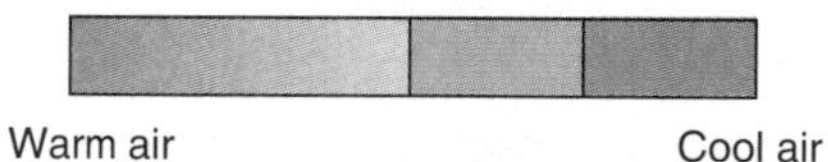

Polar stereographic projections azimuthal and conformal

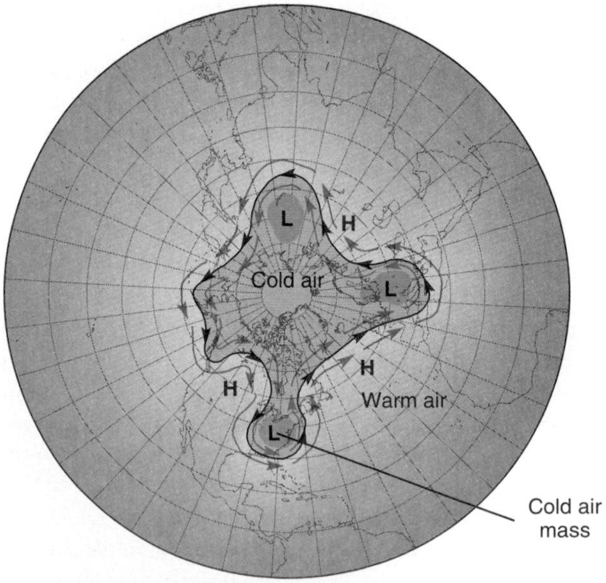

L
H
Cold air
L
H
H
Warm air
L
Cold air mass

(c) Strong development of waves produce cells of cold and warm air—high-pressure ridges and low-pressure troughs.

FIGURE 6-15
Development of longwaves in the upper-air circulation first described by C. G. Rossby in 1938 and detailed by J. Namias in 1952. [Adapted from J. Namias, National Oceanic and Atmospheric Administration.]

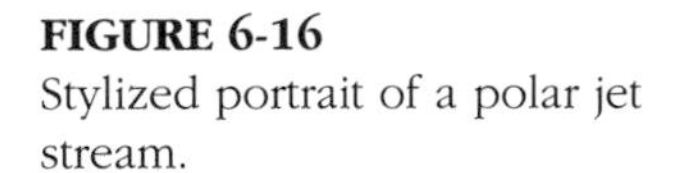

FIGURE 6-16
Stylized portrait of a polar jet stream.

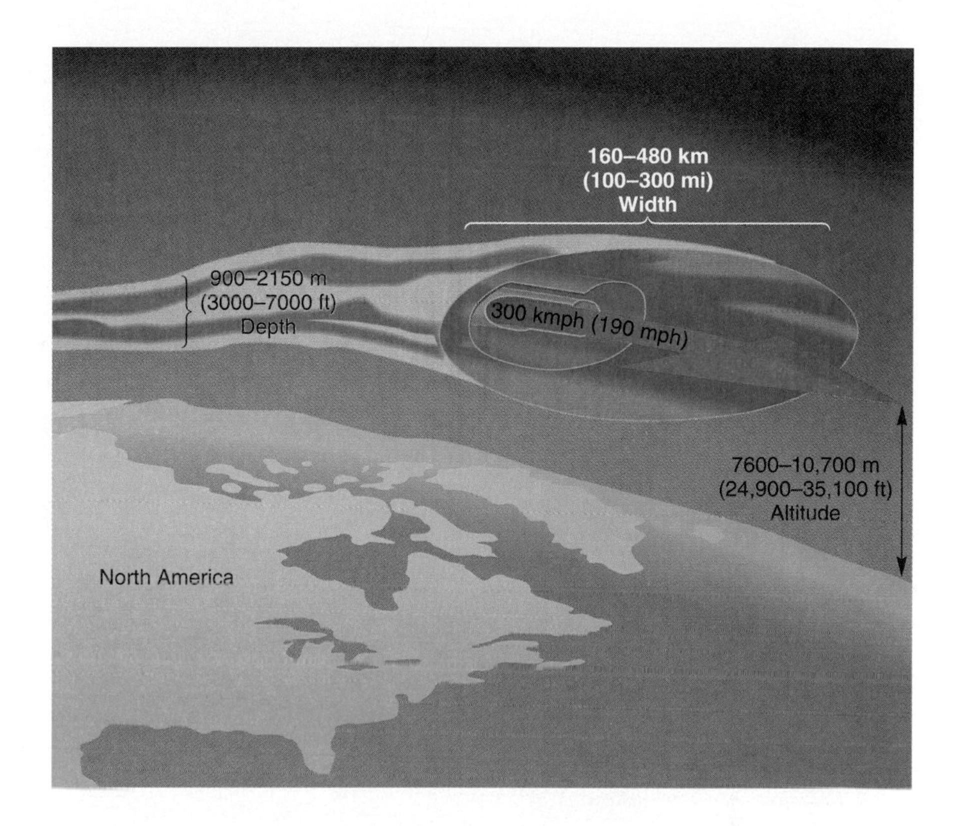

the summer, the polar jet stream exerts less influence on storms by staying far poleward. Figure 6-16 shows a stylized view of a polar jet stream flow.

In subtropical latitudes, near the boundary between tropical and midlatitude air, another jet stream flows near the tropopause. This **subtropical jet stream** ranges from 9100 to 13,700 m (29,850–45,000 ft) in altitude, and although it is generally weaker, it can reach greater speeds than those of the polar jet stream. The subtropical jet stream meanders from 20° to 50° latitude and may occur over North America simultaneously with the polar jet stream. The cross section of the troposphere in Figure 6-11b depicts the general location of these two jet streams.

Global Circulation Model

The general circulation of the atmosphere is schematically presented in Figure 6-17, which summarizes the chapter to this point. You can use this idealized model to locate and identify each of Earth's principal pressure and wind systems we have discussed. Atmospheric circulation is drawn in dimensional perspective, with the vertical dimension greatly exaggerated to better exhibit the anatomy of the circulation. Surface wind patterns are seen through the cutaway view.

Local Winds

Several winds that form in response to local terrain deserve mention in our discussion of atmospheric circulation. These local effects can, of course, be overwhelmed by weather systems passing through an area.

Land-sea breezes occur on most coastlines (Figure 6-18). The different heating characteristics of land and water surfaces create these winds. During the day, land heats faster and becomes warmer than the water offshore. Because warm air is less dense, it rises and triggers an onshore flow of cooler marine air, usually stronger in the afternoon. At night, inland areas cool (radiate heat)

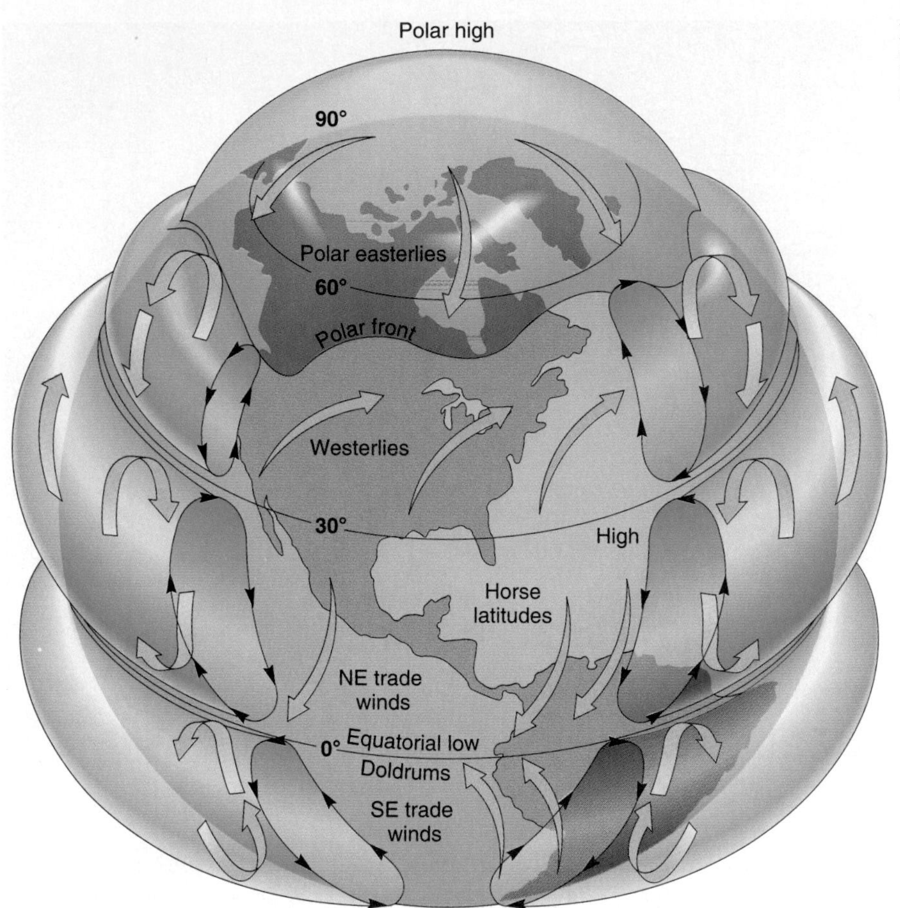

FIGURE 6-17
Idealized global circulation model.

faster than offshore waters. As a result, the cooler air over the land subsides and flows offshore over the warmer water, where the air is lifted. This night pattern reverses the process that developed during the day.

Well inland from the Pacific Ocean—160 km (100 mi)—Sacramento, California, demonstrates the sea-breeze effect. The city is at 39° N and 5 m (17 ft) elevation. The average July maximum and minimum temperatures are 34°C and 14°C (93°F and 57°F). The evening cooling by the natural flow of marine air establishes a monthly mean of only 24°C (75°F), despite high daytime temperatures. In some similar locations (for example, Hyderabad, Pakistan), large deflectors are placed over roof openings to direct onshore sea-to-land breezes into dwellings to take advantage of the natural cooling.

Mountain-valley breezes are created in a somewhat similar exchange. Mountain air cools rapidly at night, and valley air heats rapidly during the day (Figure 6-19). Thus, warm air rises upslope during the day, particularly in the afternoon; at night, cooler air subsides downslope into the valleys. Other downslope winds are discussed in Chapter 8.

Katabatic winds, or gravity drainage winds, are significant on a larger regional scale than mountain-valley breezes, under certain conditions. These winds usually are stronger than mountain-valley winds. An elevated plateau or highland is essential, where layers of air at the surface cool, become denser, and flow downslope. The fero-

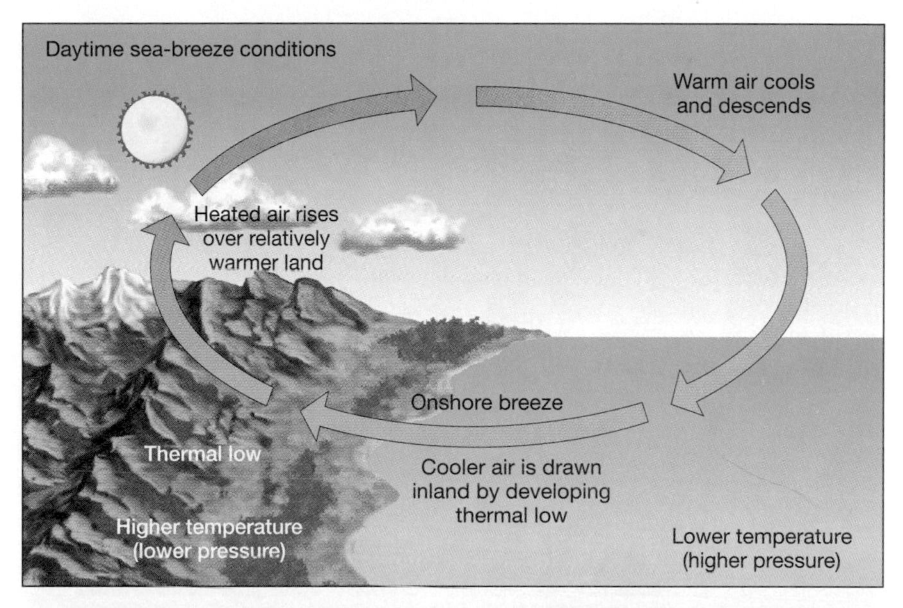

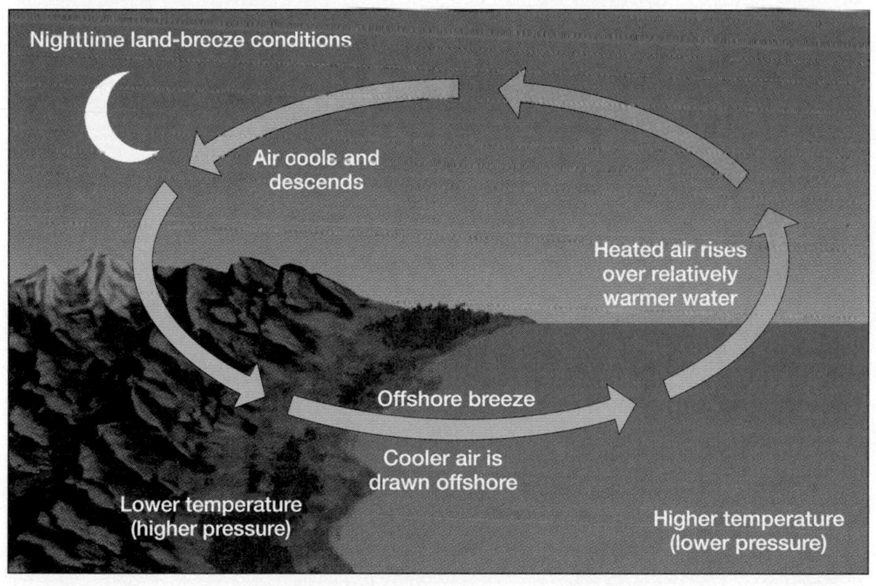

FIGURE 6-18
Land-sea breezes characteristic of day and night.

cious winds that can blow off the ice sheets of Antarctica and Greenland are katabatic in nature. Worldwide, a variety of terrains produce such winds and bear many local names. The *mistral* of the Rhône Valley in southern France can cause frost damage to vineyards as the cold north winds move over the region to the Gulf of Lions and the Mediterranean Sea. The frequently stronger *bora,* driven by the cold air of winter high-pressure systems inland, flows across the Adriatic Coast to the west and south. In Alaska such winds are called the *taku*.

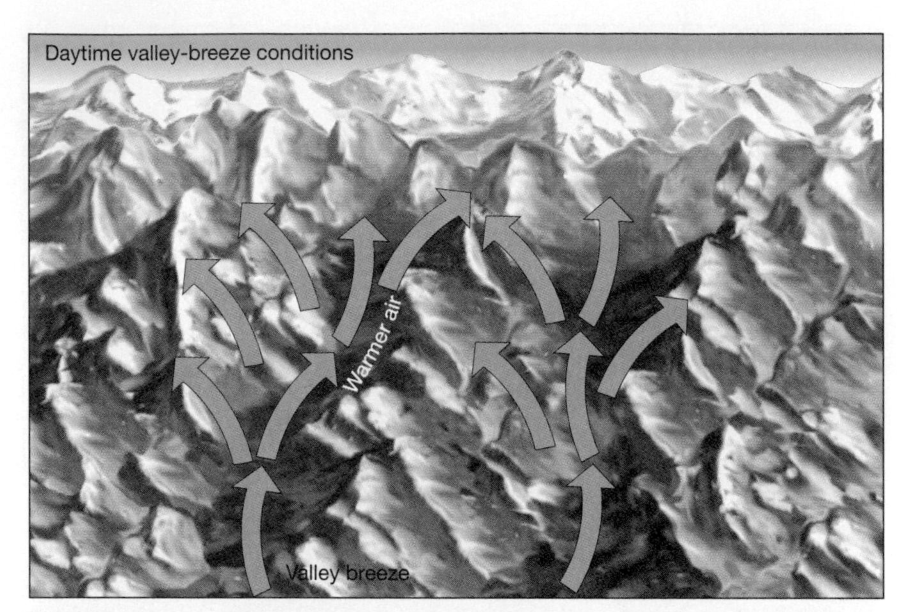

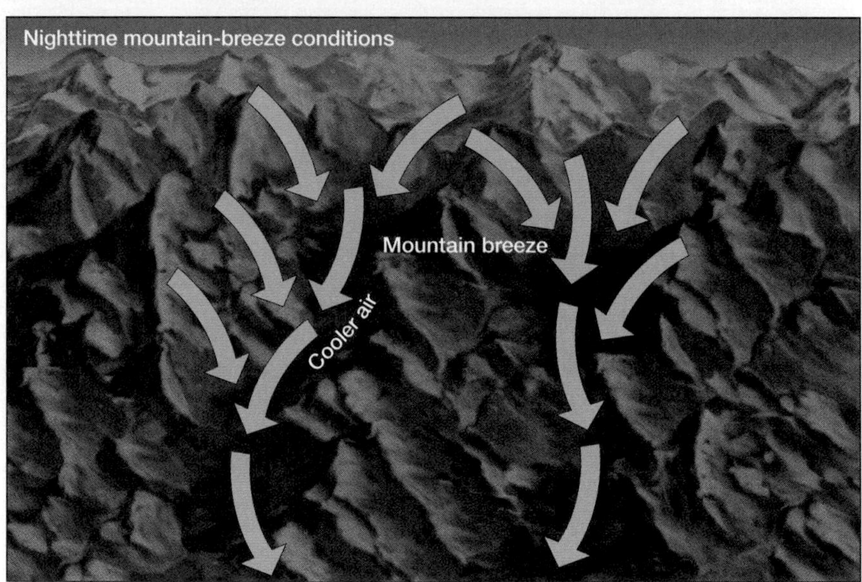

FIGURE 6-19
Pattern of mountain-valley breezes during day and night.

Santa Ana winds are another local wind type, generated by high pressure over the Great Basin of the western United States. A strong, dry wind is produced that flows out across the desert to southern California coastal areas. The air is heated by compression as it flows from higher to lower elevations and with increasing speed it moves through constricting valleys to the southwest. These winds irritate the population with their dust, dryness, and heat.

Locally, wind represents a source of renewable energy. Focus Study 6-1 briefly explores the potential for development of wind resources.

Monsoonal Winds. Regional wind systems that seasonally change direction are important in some

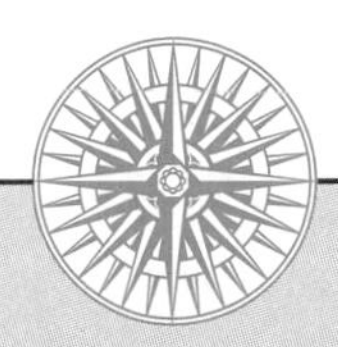

Focus Study 6-1 Wind Power: An Energy Resource

The principles may be old, but the technology is modern and the prospects favorable. More than half of the world's population relies on renewable energy, derived from resources that are not depleted in the span of a human lifetime. That energy in developing countries—for cooking, heating, and pumping—comes principally from small hydroelectric plants, wind-power systems, and wood resources. In developed nations, energy sources are focused more upon nonrenewable fuels, although some electricity is generated at wind farms, where groups of wind machines are massed, and at solar installations.

More than 50,000 wind turbines have been installed worldwide since 1974, with 13,000 of them in California alone (Figure 1). Denmark and California lead the world in this technology, with Denmark moving forward under its own official National Wind Strategy. Both California and Denmark have a goal to provide 10% of their electrical generating capacity with wind energy technology by A.D. 2000. Sweden is implementing plans to dismantle nuclear power plants and no doubt will increase use of its appreciable wind resources.

Power generation from wind is site specific, meaning that the localized conditions that produce adequate winds are confined to certain areas. In places where wind is reliable less than 25–30% of the time, only small-scale uses are economically feasible. Wind resources are greatest along coastlines influenced by trade winds and westerly winds; where mountain passes constrict the flow of air and interior valleys develop thermal low-pressure areas, thus drawing air across the landscape; or where localized winds such as katabatic and monsoonal flows occur. In addition, the mountains of North America, northwestern Europe, southwestern Australia, and the Arctic coastlands of Russia all offer favorable sites.

FIGURE 1
Wind farm in the Tehachapi Mountains of southern California. [Photo by Steve Mulligan.]

Problems with further deployment of wind farms in the United States arose because of political decisions in the 1980s. All alternative energy programs, including those involving wind, were cut or canceled, and tax incentives eliminated, even though tax incentives and research for nonrenewable fossil fuels and nuclear power were increased.

One factor that may accelerate wind applications to replace oil-fired electrical generation at appropriate sites is an estimate of remaining domestic oil reserves made by the U.S. Geological Survey in 1989. With 51 billion barrels of recoverable oil left, including expected undiscovered reserves, and with the rate of domestic consumption at over 5.4 billion barrels per year, the United States has only a 16-year supply left. Imports at the 50% or greater level will stretch this reserve, but with unknown economic and military consequences. And the domestic natural gas supply estimate is not much better, with a 35-year supply left until depletion, at present consumption rates.

These economic realities no doubt will override any further delaying actions, especially in regions where peak winds are in concert with peak electrical demand for air cooling, space heating, or agricultural water pumping. Long-term and marginal costs also favor wind-energy deployment, for it does not produce carbon dioxide, radioactivity, sulfur dioxide, toxic wastes, or ash, nor does it require mining or the importation of foreign fuels—all of which represent significant cost with nonrenewable fuels. A Wind Energy Association study determined that the wind resource in 12 midcontinent states exceeds by 300% the total electrical consumption in the United States in 1987. The utility of wind power will be enhanced by improvements in energy storage. In addition, wind farm sites can have multiple uses—for example, for pasture and farming.

History is rich with the accomplishments of wind-powered sails. Experiments are going forward today to see whether the wind can still drive ships across the sea as in days past—an obvious answer awaits! In an experiment more than 10 years ago, the Japanese oil tanker *Shin Aitoku Maru* installed two computer-guided sails of 300 m^2 (3228 ft^2) each. Fuel consumption was reduced by 50% in early tests. If we consider the world's cargo fleet of more than 25,000 vessels, consuming 4–5 million barrels of oil a day, a savings of just 10–15% would be significant.

Cash-poor, energy-poor developing countries already have a decentralized demand for energy, so for them large, centralized, capital-intensive electrical generation seems inappropriate. The key issue appears to be whether societies will develop nonrenewable-centralized or renewable-decentralized energy resources. And, as for sails on ships: wind is a renewable resource, its use waiting to be realized as in the sailing days of old.

areas. Examples of intense, seasonally shifting wind systems occur in the tropics over Southeast Asia, Indonesia, India, northern Australia, and equatorial Africa. These winds involve an annual cycle of returning precipitation with the summer Sun and are named after the Arabic word for season, *mausim,* or **monsoon.** (Specific monsoonal weather, associated climate types, and vegetation regions are discussed in Chapters 8, 10, and 20.)

The monsoons of southern and eastern Asia (Figure 6-20) are driven by the location and size of the Asian landmass and its proximity to the Indian Ocean. Also important to the generation of monsoonal flows are wind and pressure patterns in the upper-air circulation. The extreme temperature range from summer to winter over the Asian landmass is due to its continentality, reflecting its isolation from the modifying effects of the ocean. This continental landmass is dominated by an intense high-pressure anticyclone in winter (Figures 6-9a and 6-20), whereas the central area of the Indian Ocean is dominated by the equatorial low-pressure trough.

Resultant cold, dry winds blow from the Asian interior over the Himalayas, downslope, and across India, producing average temperatures between 15 and 20°C (60–68°F) at lower elevations. These dry winds desiccate (dehydrate) the landscape and then give way to hot weather from March through May. When the monsoonal rains arrive from June to September, they are welcome relief from the dust, heat, and parched land of Asia's springtime.

During the June-September wet period, the subsolar point shifts northward to the Tropic of Cancer, near the mouths of the Indus and Ganges rivers. The intertropical convergence zone is shifted northward over southern Asia and the Asian continental interior develops a thermal low pressure, associated with high average temperatures. Meanwhile, the Indian Ocean, with surface temperature of 30°C (86°F), is under the influence of subtropical high pressure. As a result, hot, dry subtropical air sweeps over the warm ocean, producing extremely high evaporation rates (Figure 6-20b).

By the time this mass of air reaches India and the convergence zone, it is laden with moisture in thunderous, dark clouds. The warmth of the land lends additional lifting to the incoming air, as do the Himalayas, forcing the air mass to higher altitudes. These conditions produce the wet monsoon of India, where world-record rainfalls have occurred: both the second-highest average rainfall (1143 cm or 450 in.) and the highest single-year rainfall (2647 cm or 1042 in.) were measured at Cherrapunji, India. This city's record rainfall is discussed further in Chapter 8.

The annual monsoon is an integral part of Indian culture and life, as writer Khushwant Singh describes:

> To know India and her peoples, one has to know the monsoon. . . . It has to be a personal experience because nothing short of living through it can fully convey all it means. . . . What the four seasons of the year mean to the European, the one season of the monsoon means to the Indian. It is preceded by desolation; it brings with it the hopes of spring; it has the fullness of summer and the fulfillment of autumn all in one. . . . much of India's art, music, and literature is concerned with the monsoon. . . . First it

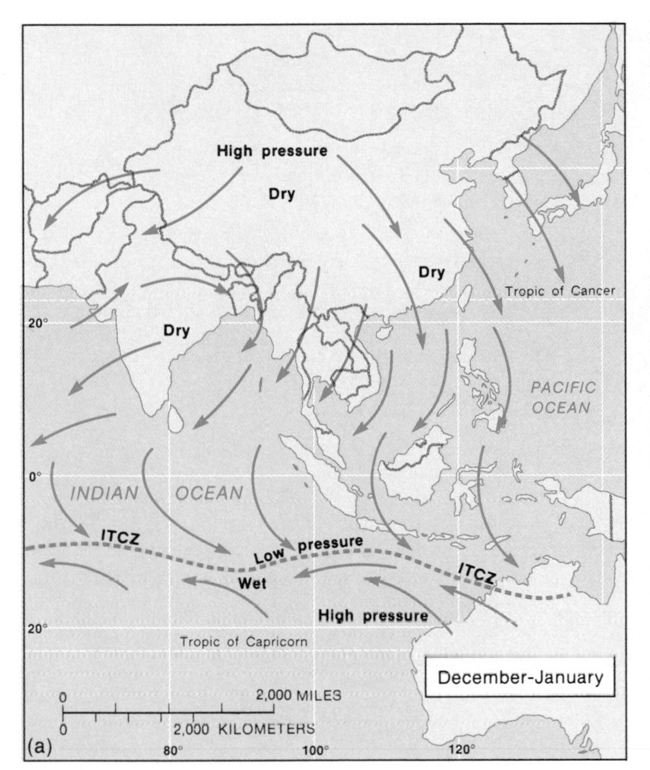

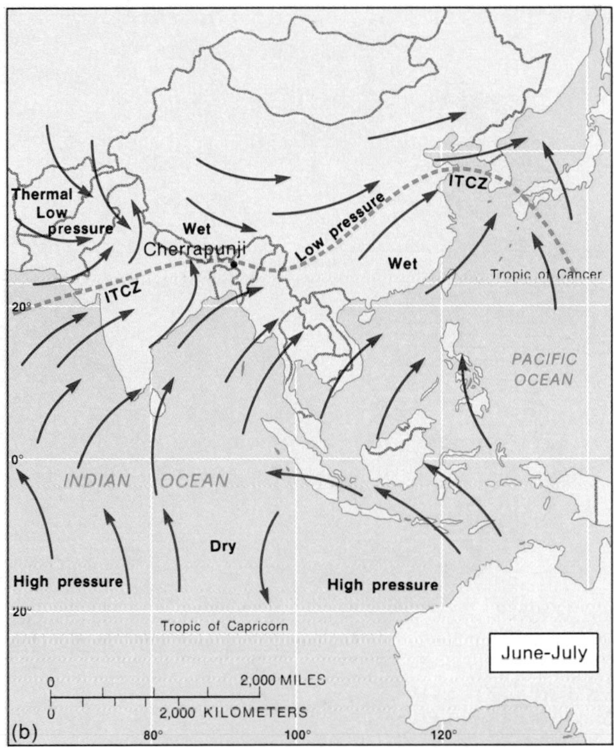

FIGURE 6-20
Asian monsoon patterns: (a) winter pressure and wind patterns over Asia; (b) summer pressure and wind patterns over Asia and the location of the ITCZ. [Adapted from Joseph E. Van Riper, *Man's Physical World*, p. 215. Copyright 1971 by McGraw Hill. Adapted by permission.]

falls in fat drops; the earth rises to meet them. She laps them up thirstily and is filled with fragrance. It brings the odor of the earth and of green vegetation to the nostrils.*

OCEANIC CURRENTS

Earth's atmospheric and oceanic circulations are interrelated. The driving force for ocean currents is the frictional drag of the winds. Also important in shaping these currents is the interplay of the Coriolis force, density differences associated with temperature and salinity, the configuration of the continents and ocean floor, and astronomical forces (the tides). You may be surprised to learn that the ocean surface is not level, for it features small height differences in response to these currents, atmospheric pressure differences, slight variations in gravity, and the presence of waves.

The general patterns of major ocean currents are shown in Figure 6-21. Because ocean currents flow over distance and through time, they come under the influence of the deflective Coriolis force. However, their pattern of deflection is not as tightly circular as that of the atmosphere. If we compare this map with that of Earth's pressure and wind systems (Figure 6-9), we find the ocean currents driven by the circulation around subtropical high-pressure cells in both hemispheres. These circulation systems are known as **gyres** and generally appear offset toward the western side of each ocean

*Khushwant Singh, *I Hear the Nightingale*. New York: Grove Press, 1959, pp. 101–102. Reprinted by permission.

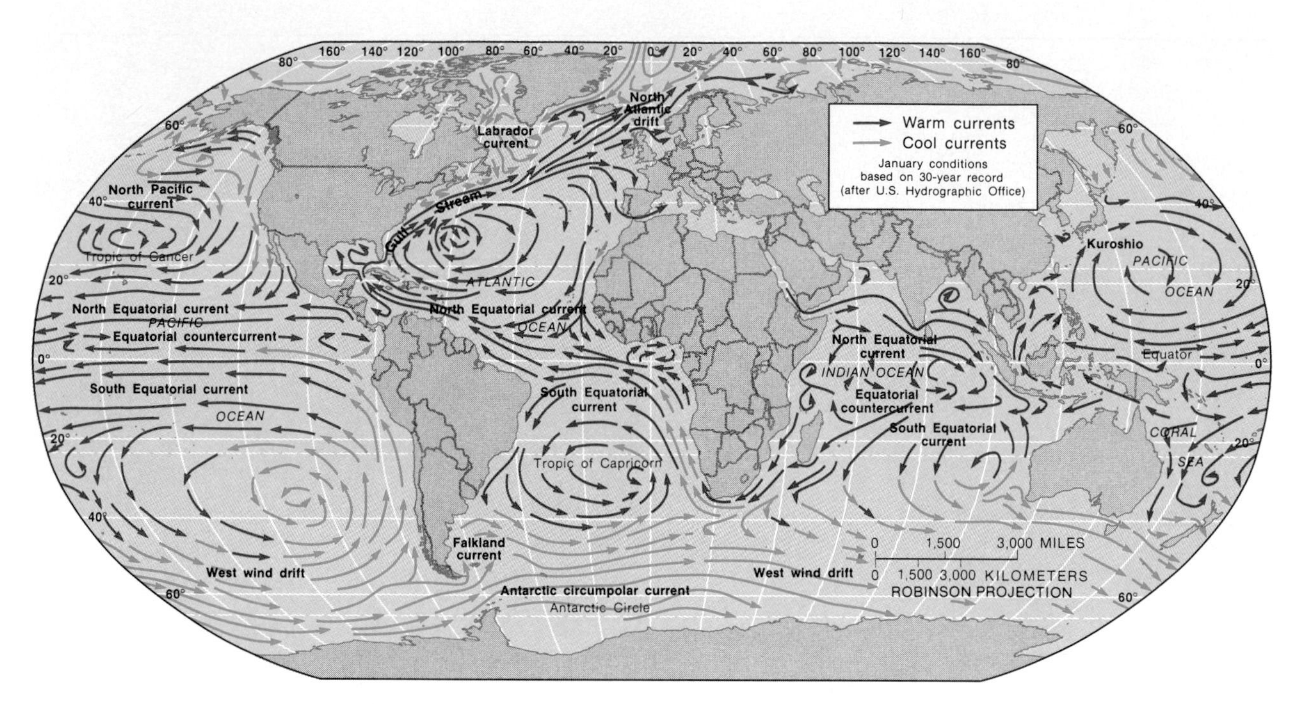

FIGURE 6-21
Major ocean currents.

basin. To understand these gyres and their western margins, which feature slightly stronger currents, we must understand a process known as **western intensification.**

Along the full extent of ocean areas adjoining the equator, the trade winds drive the oceans westward in a concentrated channel. These currents are kept near the equator by a Coriolis force influence that weakens near the equator. As the surface current approaches the western margins of the oceans, the water actually piles up an average of 15 cm (6 in.). From this western edge, ocean water then spills northward and southward in strong currents, flowing in tight channels along the western edges of the ocean basins (eastern shorelines of continents). Additionally, a strong countercurrent is generated that flows through the full extent of the Pacific, Atlantic, and Indian oceans. This occasionally strong, somewhat sporadic eastward **equatorial countercurrent** may be alongside or just beneath the surface current, at depths of 100 m (300 ft).

In the Northern Hemisphere, the Gulf Stream and the Kuroshio move forcefully northward as a result of western intensification, with their speed and depth increased by the constriction of the area they occupy. The warm, deep-blue water of the ribbon-like Gulf Stream (Figure 5-4) usually is 50–80 km (30–50 mi) wide and 1.5–2.0 km (0.9–1.2 mi) deep, moving at 3–10 kmph (1.8–6.2 mph). In 24 hours, ocean water can move 70–240 km (40–150 mi) in the Gulf Stream, although a complete circuit around an entire gyre may take a year.

Deep Currents

Where surface water is swept away from a coast, either by surface divergence (induced by the Coriolis force) or by offshore winds, an **upwelling current** occurs. This cool water generally is nutrient-rich and rises from great depths to replace the vacating water. Such cold upwelling currents occur off the Pacific coasts of North and South America and the subtropical and midlatitude west coast of Africa. In other portions of the sea where there is an accumulation of water—as at the western end of an equatorial current or along the margins of Antarctica—excess water is thrust downward in a **downwelling current.** Important mixing currents along the

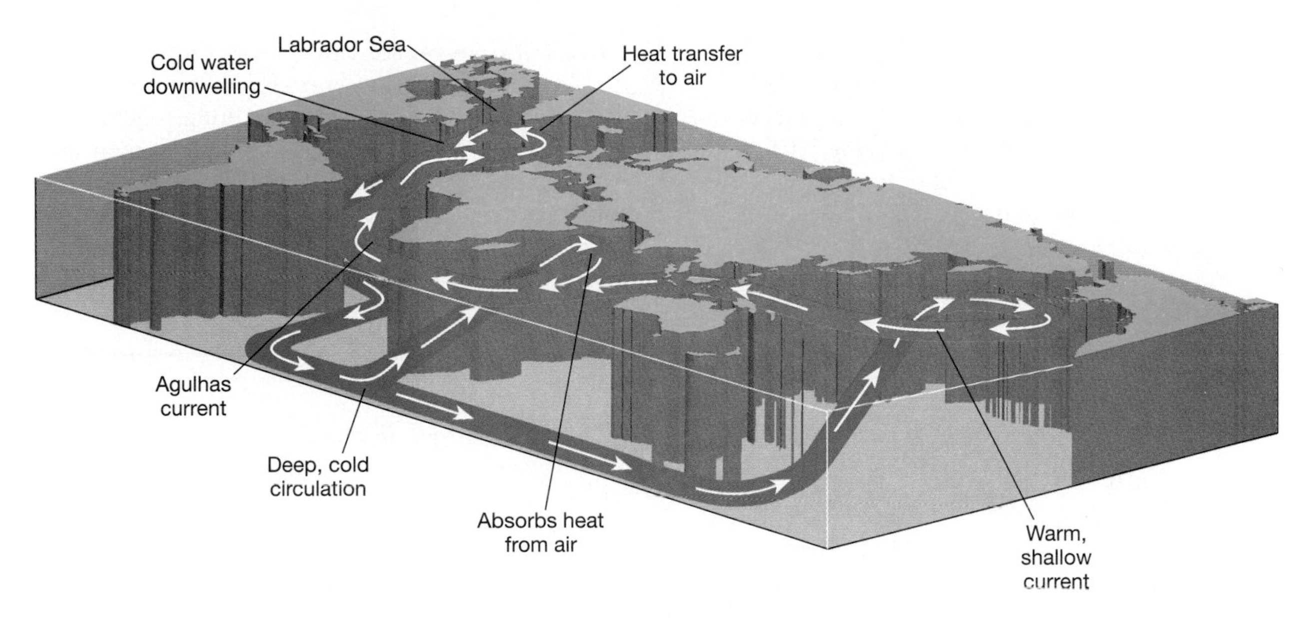

FIGURE 6-22
Centuries-long deep circulation in the oceans is being deciphered by scientists. This global circulation mimics a vast conveyor belt of water drawing heat from some regions and transporting it for release in others.

ocean floor are generated from such downwelling zones and travel the full extent of the ocean basins, carrying heat energy and salinity.

Imagine a continuous channel of water beginning with cold water downwelling in the North Atlantic, flowing deep and strong to upwellings in the Indian Ocean and North Pacific (Figure 6-22). Here it warms and then is carried in surface currents back to the North Atlantic. A complete circuit of the current may require 1000 years from downwelling in the Labrador Sea off Greenland to its reemergence in the southern Indian Ocean and return. Even deeper Antarctic bottom water flows northward in the Atlantic Basin beneath these currents.

Scientists at the Institute of Oceanographic Sciences, Wormley, England, have found that the returning surface warm currents from the Indian Ocean flow into the Atlantic—as the aggressive Agulhas current that rounds South Africa—bringing large quantities of heat into Atlantic waters and into the Gulf Stream current (Figure 5-4). Their work is beautifully portrayed in *The FRAM Atlas of the Southern Ocean* (see Suggested Readings).

SUMMARY—Atmospheric and Oceanic Circulations

Volcanic eruptions such as Tambora in 1815 and Mount Pinatubo in 1992 dramatically demonstrate the power of global winds to disperse aerosols worldwide in a matter of weeks. Atmospheric circulation facilitates important transfers of energy and mass on Earth, thus maintaining Earth's natural balances. Earth's atmospheric and oceanic circulations represent a vast heat engine powered by the Sun.

Wind is the horizontal movement of air across Earth's surface, directed and driven by pressure differences, the Coriolis force, and frictional drag. The pattern of high and low pressures on Earth in generalized belts in each hemisphere produces the distribution of specific wind systems. A generalized model of the overall atmospheric circulation system is helpful in visualizing wind circulation both at the surface and in the upper troposphere.

Constructing an accurate multidimensional computer model of the linked atmospheric and oceanic circulation remains an elusive and actively pursued goal for science. Ocean currents are primarily caused by the frictional drag of wind and occur worldwide at varying intensities, temperatures, and speeds. Radar imaging from satellites detects the pattern of ocean currents by analyzing backscatter off the ocean waves. This technology provides images of atmospheric and oceanic circulation never before possible.

Wind itself is an important energy source and is a renewable resource with potential for increasing development and application in the future. Earth's dynamic atmosphere and ocean bring together all the elements of Part 1 in this text. More than any other spatial factor, natural or cultural, the atmosphere brings together humanity—one person's or nation's exhalation is another's breath intake.

KEY TERMS

Aleutian low
anemometer
Antarctic high
anticyclone
Azores high
Beaufort wind scale
Bermuda high
Coriolis force
cyclone
doldrums
downwelling current
equatorial countercurrent
equatorial low-pressure trough
friction force
geostrophic winds
gyres
Hadley cell
horse latitudes
Icelandic low
intertropical convergence zone (ITCZ)
isobaric surface
isobars
jet stream
katabatic winds
monsoon
Pacific high
polar easterlies
polar front
polar high-pressure cells
polar jet stream
pressure gradient force
Rossby waves
subpolar low-pressure cells
subtropical high-pressure cells
subtropical jet stream
trade winds
upwelling current
westerlies
western intensification
wind
wind vane

REVIEW QUESTIONS

1. What is a possible explanation for the beautiful sunrises and sunsets during the summer of 1816 in New England? Relate your answer to global circulation.
2. Explain the statement, "the atmosphere socializes humanity, making all the world a spatially linked society." Illustrate your answer with some examples.
3. Define wind. How is it measured? How is its direction determined?
4. Distinguish among primary, secondary, and tertiary classifications of global atmospheric circulation.
5. What is the purpose of the Beaufort wind scale? Characterize winds given Beaufort numbers of 4, 8, and 12, giving both water and land effects.
6. Describe the horizontal and vertical air motions in a high-pressure cell and in a low-pressure cell.
7. Describe the effect of the Coriolis force. How does it apparently deflect atmospheric and oceanic circulations? Explain.

8. What are geostrophic winds, and where are they encountered in the atmosphere?
9. Construct a simple diagram of Earth's general circulation, including the four principal pressure belts or zones and the three principal wind systems.
10. How does the intertropical convergence zone (ITCZ) relate to the equatorial low-pressure trough? How might it appear on a satellite image?
11. What is one source of migratory low-pressure cyclonic storms in North America? In Europe?
12. Relate the jet-stream phenomenon to general upper-air circulation. How does the presence of this circulation relate to airline schedules from New York to San Francisco and the return trip to New York?
13. People living along coastlines generally experience variations in day-night winds. Explain the factors that produce these changing wind patterns.
14. Describe the seasonal pressure patterns that produce the Asian monsoonal wind and precipitation patterns.
15. What is the relationship between global atmospheric circulation and ocean currents? Relate oceanic gyres to patterns of subtropical high pressure.
16. Define the western intensification. How is it related to the Gulf Stream and Kuroshio currents?
17. Where on Earth are upwelling currents experienced?
18. What is meant by deep-ocean circulation? At what rates do these currents flow? How might this relate to the Gulf Stream in the western Atlantic Ocean?
19. What is the present status of wind-energy technologies? Describe deployment delays. How do these delays relate to the nature of the technology?

SUGGESTED READINGS

Brasseur, Guy, and Claire Granier. "Mount Pinatubo Aerosols, Chlorofluorocarbons, and Ozone Depletion." *Science* 257 (August 28, 1992): 1239–42.

Eagleman, Joe R. *Meteorology—The Atmosphere in Action*. New York: D. Van Nostrand, 1980.

Institute of Oceanographic Sciences, Deacon Laboratory. *The FRAM (Fine Resolution Antarctic Model) Atlas of the Southern Ocean*. Wormley, Surrey, England: Natural Environment Research Council, 1991.

Marr, John C., ed. *The Kuroshio—A Symposium on the Japan Current*. Honolulu: East-West Center Press, 1970.

McDonald, James E. "The Coriolis Effect." *Scientific American* 186 (May 1952): 72–76.

Neiburger, Morris, James G. Edinger, and William D. Bonner. *Understanding Our Atmospheric Environment*. San Francisco: W. H. Freeman, 1973.

Starr, Victor P. "The General Circulation of the Atmosphere." *Scientific American* 195 (December 1956): 40–45.

Stowe, L.L., R. M. Carey, and P. P. Pellegrino. "Monitoring the Mt. Pinatubo Aerosol Layer with *NOAA -11* AVHRR Data," *Geophysical Research Letters* 19, no. 2 (January 24, 1992): 159–62.

Thurman, Harold V. *Introductory Oceanography*. 5th ed. Columbus, OH: Macmillan Publishng Co., 1988.

Vesilind, Priit J. "Monsoon—Life Breath of Half the World," *National Geographic* 166, no. 6 (December 1984): 712–47.

Whipple, A. B. C., and the editors of Time-life Books. *Restless Oceans*. Alexandria, VA: Time-Life Books, 1983.

Wiebe, Peter H. "Rings of the Gulf Stream." *Scientific American* 246 (March 1982): 60–70.

PART 2

THE WATER, WEATHER, AND CLIMATE SYSTEM

Winter sunset, Yosemite National Park. [*Photo by Galen Rowell.*]

Earth is the water planet. Its surface waters are unique in the solar system. Answers to why water exists here in such quantity, what qualities and properties it possesses, and how it circulates over Earth are the topics that begin this section. The dynamics of daily weather phenomena follow: the effects of moisture and energy in the atmosphere, the interpretation of cloud forms, conditions of stability or instability, the interaction of air masses, and the occurrence of violent weather. The specifics of the hydrologic cycle are explained through the water-balance concept, which is useful in understanding water-resource relationships, whether global, regional, or local. Important water resources include rivers, lakes, groundwater, and oceans. The spatial implications over time of this water-weather system lead to the final topic in Part 2—world climate patterns.

Burney Falls, McArthur-Burney Falls State Park, California. [*Photo by author.*]

7

WATER AND ATMOSPHERIC MOISTURE

Walden is blue at one time and green at another, even from the same point of view. Lying between the earth and the heavens, it partakes of the color of both. . . . A lake is the landscape's most beautiful and expressive feature. It is earth's eye; looking into which the beholder measures the depth of his own nature. . . . Sky water. It needs no fence. Nations come and go without defiling it. It is a mirror which no stone can crack . . . Nature continually repairs . . . a mirror in which all impurity presented to it sinks, swept and dusted by the sun's hazy brush. A field of water . . . is continually receiving new life and motion from above. It is intermediate in its nature between land and sky.*

Thus did Thoreau speak of the water so dear to him—Walden Pond in Massachusetts, along whose shore he lived.

Water is the essential medium of our daily lives and a principal compound in nature. It covers 71% of Earth (by area) and in the solar system occurs in such significant quantities only on our planet. Water is naturally colorless, odorless, and tasteless and weighs 1 gram per cm^3 or 1 kg per liter (62.3 lb/ft^3 or 8.337 lb/gal).

Water constitutes nearly 70% of our bodies by weight and is the major ingredient in plants, animals, and our food. A human being can survive 50 to 60 days without food, but only 2 or 3 days without water. The water we use must be adequate in quantity as well as quality for its many tasks—everything from personal hygiene to vast national water projects. Water indeed occupies that place between land and sky, mediating energy and shaping both the lithosphere and atmosphere, which Thoreau revealed in his thoughtful observations.

This chapter examines water and the dynamics of atmospheric moisture in particular. The clouds that form as air becomes water-saturated are more than whimsical, beautiful configurations; they are important indicators of overall atmospheric conditions.

*Reprinted by permission of Merrill, an imprint of Macmillan Publishing Company, from *Walden* by Henry David Thoreau, pp. 192, 202, 204. Copyright ©1969 by Merrill Publishing. Originally published 1854.

BEGINNINGS

Earth's hydrosphere contains about 1,359,208,000 km^3 (326,074,000 mi^3) of water. According to the best evidence, water on Earth was formed within the planet, reaching Earth's surface in an ongoing process called **outgassing,** by which water and water vapor emerge from layers deep within and below the crust—25 km (15.5 mi) or more below Earth's surface (Figure 7-1). Much of this water originated from icy comets—a portion of the planetesimals that accreted to form the planet. Various geophysical factors explain the timing of this outgassing over the past 4 billion years. In the early atmosphere, massive quantities of outgassed water vapor condensed and fell in torrents, only to vaporize again because of high temperatures at Earth's surface. For water to remain on Earth's surface, land temperatures had to drop below the boiling point of 100°C (212°F), something that occurred about 3.8 billion years ago.

The lowest places across the face of Earth then began to fill with water: first ponds, then lakes and seas, and eventually ocean-sized bodies of water. (A sea is generally smaller than an ocean and is near a landmass; sometimes the term refers to a large, inland, salty body of water.) Massive flows of water washed over the landscape, carrying both dissolved and undissolved elements to these early seas and oceans.

Quantity Equilibrium

Today, water is the most common compound on the surface of Earth, having achieved the present volume of 1.36 billion km^3 (326 million mi^3) approximately 2 billion years ago. This quantity has remained relatively constant, even though water is continuously being lost from the system, escaping to space or breaking down and forming new compounds with other elements. Lost water is replaced by pristine water not previously at the surface, water that emerges from within Earth. The net re-

FIGURE 7-1
Outgassing of water from Earth's crust in Chilean Andes. [Photo by Stephen Cunha.]

sult of these inputs and outputs to water quantity is a steady-state equilibrium in Earth's hydrosphere.

Despite this overall net balance in quantity, worldwide changes in sea level do occur, called **eustasy**—specifically related to changes in volume of water in the oceans. Some of these changes are explained by **glacio-eustatic** factors (see Chapter 17). Glacio-eustatic factors are based on the amount of water stored on Earth as ice. As more water is bound up in glaciers (in mountains worldwide) and in ice sheets (Greenland and Antarctica), sea level lowers, whereas a warmer era reduces the quantity of water stored as ice and thus raises sea level. Some 18,000 years ago, during the most recent ice-age pulse, sea level was more than 100 m (330 ft) lower than today; 40,000 years ago it was 150 m (490 ft) lower.

Over the past 100 years, mean sea level has steadily risen and is still rising worldwide at this time. Apparent changes in sea level also are related to actual physical changes in landmasses and isostatic change, such as continental uplift or subsidence (*isostasy* is discussed in Chapter 11).

Distribution of Earth's Water Today. Most of Earth's continental land is in the Northern Hemisphere, whereas the Southern Hemisphere is dominated by water. The present location of all of Earth's water, in terms of fresh vs. saline, surface vs. underground, and frozen vs. liquid, is illustrated in Figure 7-2. Data are first divided between

FIGURE 7-2
Ocean and freshwater distribution on Earth.

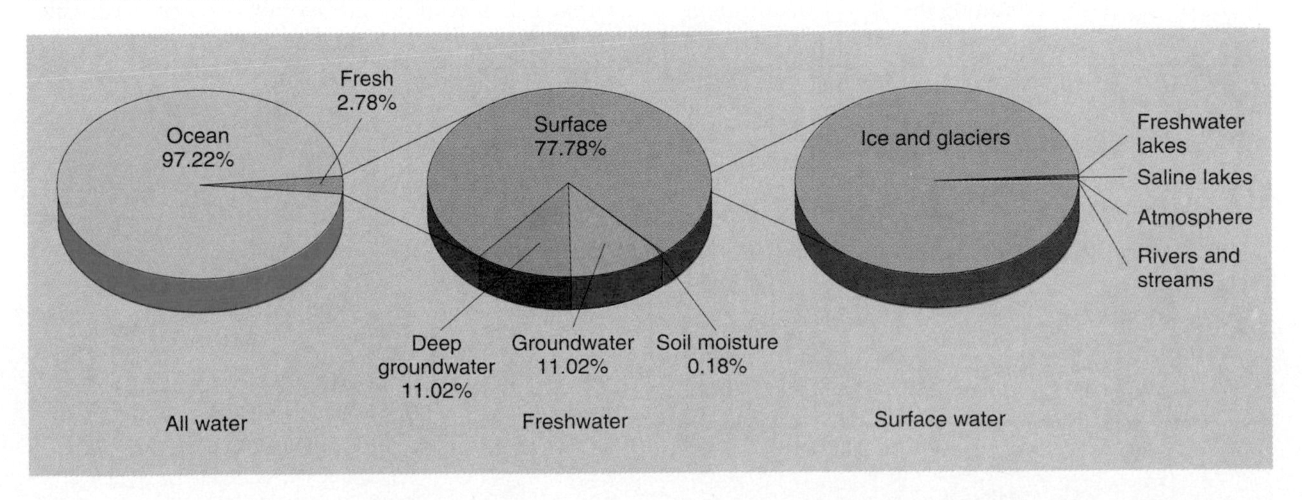

TABLE 7-1
Distribution of Fresh Water

Location	Amount in km³ (mi³)	Percent of Fresh Water	Percent of Total Water
Surface Water			
Ice sheets and glaciers	29,180,000 (7,000,000)	77.14	2.146
Freshwater lakes	125,000 (30,000)	0.33	0.009
Saline lakes and inland seas	104,000 (25,000)	0.28	0.008
Atmosphere	13,000 (3,100)	0.03	0.001
Rivers and streams	1,250 (300)	0.003	0.0001
Total surface water	29,423,250 (7,058,400)	77.78	2.164
Subsurface Water			
Groundwater—surface to 762 m (2500 ft) depth	4,170,000 (1,000,000)	11.02	0.306
Groundwater—762–3962 m (2500–13,000 ft) depth	4,170,000 (1,000,000)	11.02	0.306
Soil moisture storage	67,000 (16,000)	0.18	0.005
Total subsurface water	8,407,000 (2,016,000)	22.22	0.617
TOTAL (rounded)	37,800,000 (9,070,000)	100.00%	2.78%

ocean water and fresh water, with oceans containing 97.22% of all water, or about 1.321 billion km³ (317 million mi³). The remaining 2.78% is classified as fresh, or nonoceanic, water.

Table 7-1 divides this fresh water into two categories—surface and subsurface water. The greatest single repository of surface fresh water is ice. Ice sheets and glaciers account for 77.14% of all fresh water on Earth. Add to this the subsurface groundwater, and we have accounted for 99.36% of all fresh water. The remaining fresh water, although very familiar to us, and present in seemingly huge amounts in lakes, rivers, and streams, actually represents but a small quantity, less than 1%. All the freshwater lakes in the world total only 125,000 km³ (30,000 mi³), with 80% of this in just 40 of the largest lakes, and about 50% contained in just 7 lakes (Table 7-2).

The greatest volume of water resides in 25-million-year-old Lake Baykal in Siberian Russia. This lake contains almost as much water as all the Great Lakes combined. Overall, 70% of lake water

TABLE 7-2
Major Freshwater Lakes

Lake	Volume in km³ (mi³)	Surface Area in km² (mi²)	Depth in m (ft)
Baykal (Russia)	22,000 (5280)	31,500 (12,160)	1620 (5315)
Tanganyika (Africa)	18,750 (4500)	39,900 (15,405)	1470 (4923)
Superior (U.S./Canada)	12,500 (3000)	83,290 (32,150)	397 (1301)
Michigan (U.S.)	4920 (1180)	58,030 (22,400)	281 (922)
Huron (U.S./Canada)	3545 (850)	60,620 (23,400)	229 (751)
Ontario (U.S./Canada)	1640 (395)	19,570 (7550)	237 (777)
Erie (U.S./Canada)	485 (115)	25,670 (9910)	64 (210)

is in North America, Africa, and Asia, with about a fourth of lake water worldwide in innumerable small lakes. Over 3 million lakes exist in Alaska alone, and Canada has over 750 km^2 of lake surface.

Saline lakes and inland seas, which are neither fresh nor connected to the ocean, contain 104,000 km^3 (25,000 mi^3) of water. Such lakes as the Great Salt Lake, Mono Lake, and the Caspian Sea exist as remnants of past wetter climatic eras and are usually in regions of interior river drainage.

Think of the weather occurring in the atmosphere worldwide at this moment, and the many flowing rivers and streams, which combined amount to only 14,250 km^3 (3400 mi^3), or only 0.033% of fresh water, or 0.0011% of all water! Yet, this small amount is very dynamic. A water molecule traveling along atmospheric and surface-water paths moves through the entire hydrologic cycle in less than two weeks, whereas a water molecule located in deep ocean circulation, groundwater, or a glacier moves slowly, taking thousands of years to migrate through the hydrologic system.

Quality (Composition) Equilibrium

Water often is called the "universal solvent," dissolving at least 57 of the elements found in nature. In fact, most natural elements and the compounds they form are found in the seas as dissolved solids, or solutes. Thus, seawater is a solution, and the concentration of dissolved solids is called **salinity.** Long ago, during the evolutionary and living atmosphere periods, heavy precipitation produced runoff from the lands that delivered enormous quantities of dissolved solids to the ocean. Later, about a billion years ago and after much accumulation, the salinity of the oceans achieved a steady-state equilibrium. Any *influx* of material from physical, chemical, and biological processes was countered by an *efflux,* or removal, in the form of sediments, sulfide and silicate compounds, clays, and animal shells (calcium carbonate).

Salinity. The oceans remain a remarkably homogeneous mixture, and the ratio of individual salts does not change, despite minor fluctuations in overall salinity. In 1874 the British ship *HMS Challenger* sailed around the world, taking surface and depth measurements and collecting samples of seawater. Analyses of those samples demonstrated the uniform composition of seawater. Ocean chemistry is a result of complex exchanges among seawater, the atmosphere, influx of minerals, bottom sediments, and living organisms. In addition, significant amounts of mineral-rich water flow into the ocean through *hydrothermal vents* in the ocean floor ("black smokers"). The uniformity of seawater results from complementary chemical reactions and continuous mixing.

Seven elements comprise more than 99% of the dissolved solids in sea water: chlorine (Cl), sodium (Na), magnesium (Mg), sulfur (S), calcium (Ca), potassium (K), and bromine (Br). Seawater also contains dissolved gases (such as carbon dioxide, nitrogen, and oxygen), solid and dissolved organic matter, and a multitude of trace elements. Only sodium chloride (common table salt), magnesium, and a little bromine are commercially extracted in any significant amount. Future mining of minerals from the sea floor is technically feasible, although it remains uneconomical.

There are several ways to express salinity in seawater. Here are examples, using the worldwide average value:

- 3.5% (% = parts per hundred)
- 35,000 ppm (parts per million)
- 35,000 mg per liter
- 35 g/kg
- 35‰ (‰ = parts per thousand), the most common notation

Salinity worldwide normally varies between 34‰ and 37‰; variations are attributable to atmospheric conditions above the water and to the quantity of freshwater inflows. In equatorial water, precipitation is high throughout the year, diluting salinity values to slightly lower than average (34.5‰). In subtropical oceans—where evaporation rates are high due to the influence of hot, dry subtropical high-pressure cells—salinity is more concentrated, increasing to 36.5‰. Figure 7-3 plots evaporation/precipitation and salinity by latitude to illustrate this slight spatial variability.

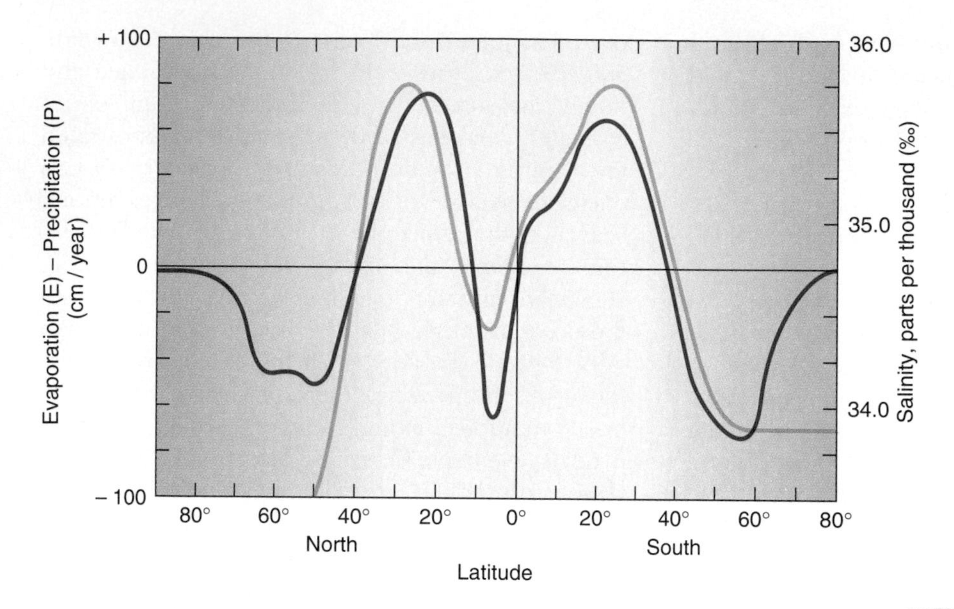

FIGURE 7-3
The variation in salinity by latitude. Salinity (green line) is principally a function of climatic conditions, specifically the difference between evaporation and precipitation (E − P), shown by the purple line. [From G. Wüst, 1936, pp. 347–59.]

The term **brine** is applied to water that exceeds the average of 35‰ salinity, whereas **brackish** applies to water that is less than 35‰. Generally, oceans are lower in salinity near landmasses because of river discharges and runoff. Extreme examples include the Baltic Sea (north of Poland and Germany) and the Gulf of Bothnia (between Sweden and Finland), which average 10‰ or less salinity because of heavy freshwater runoff and low evaporation rates. On the other hand, the Sargasso Sea, within the North Atlantic subtropical gyre, averages 38‰. The Persian Gulf has a salinity of 40‰ as a result of high evaporation rates in an almost-enclosed basin. Deep pockets near the floor of the Red Sea register a very salty 225‰.

GLOBAL OCEANS AND SEAS

The ocean is one of Earth's last great scientific frontiers and is of great interest to oceanographers and geographers. From a geographical point of view, ocean and land surfaces are not evenly distributed. In fact, from certain perspectives, Earth appears to have a distinct oceanic hemisphere and a distinct land hemisphere (Figure 7-4). The present distribution of the world's oceans and major seas is illustrated in Figure 7-5. Specific surface areas, volumes, and depths are given in Table 7-3. (Names and locations of major ocean currents were shown in Figure 6-21.)

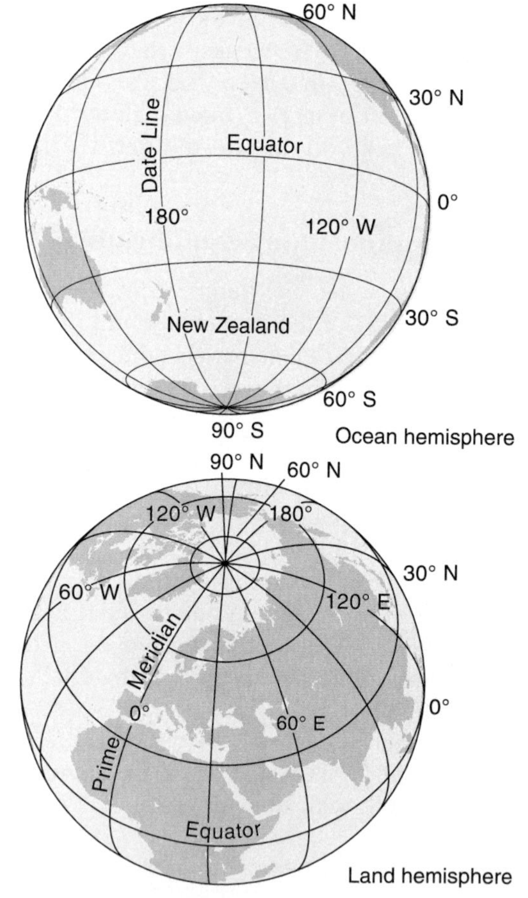

FIGURE 7-4
Two perspectives that divide Earth's surface into (a) an ocean hemisphere and (b) a land hemisphere.

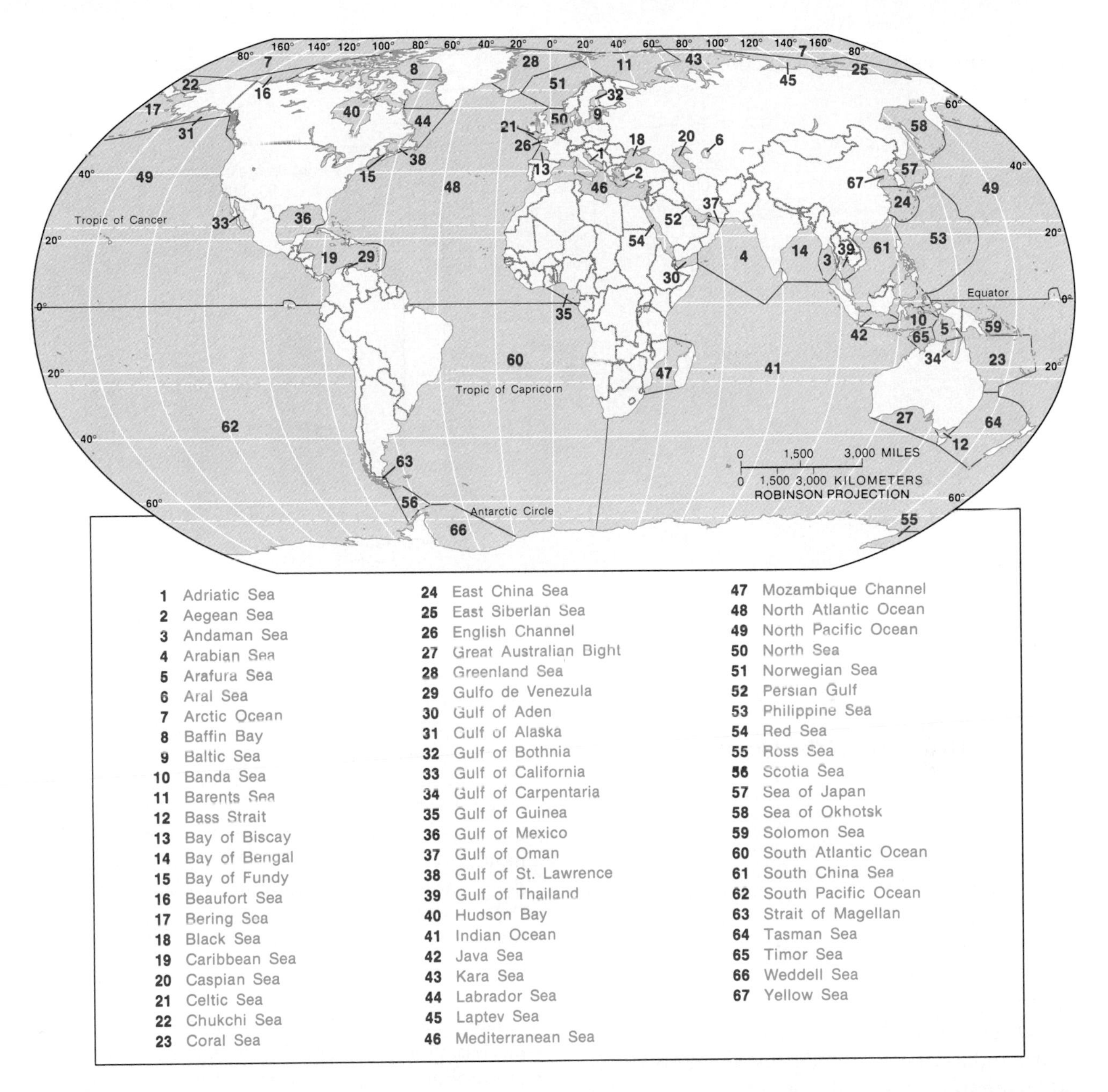

FIGURE 7-5
The names of principal oceans and seas of the world.

TABLE 7-3
Area, Volume, and Depth of Earth's Oceans

Ocean	% of Earth's Ocean Area	*Area in km^2 (mi^2)	*Volume in km^3 (mi^3)	Mean Depth of Main Basin in m (ft)	Deepest Point in m (ft)	
Pacific	48%	179,670 (69,370)	724,330 (173,700)	4280 (14,040)	Mariana Trench	11,033 (36,198)
Atlantic	28%	106,450 (41,100)	355,280 (85,200)	3930 (12,890)	Puerto Rico Trench	8605 (28,224)
Indian	20%	74,930 (28,930)	292,310 (70,100)	3960 (12,900)	Java Trench	7125 (23,376)
Arctic	4%	14,090 (5440)	17,100 (4100)	1205 (3950)	Eurasian Basin	5450 (17,876)

*Data in thousands (,000); includes all marginal seas.

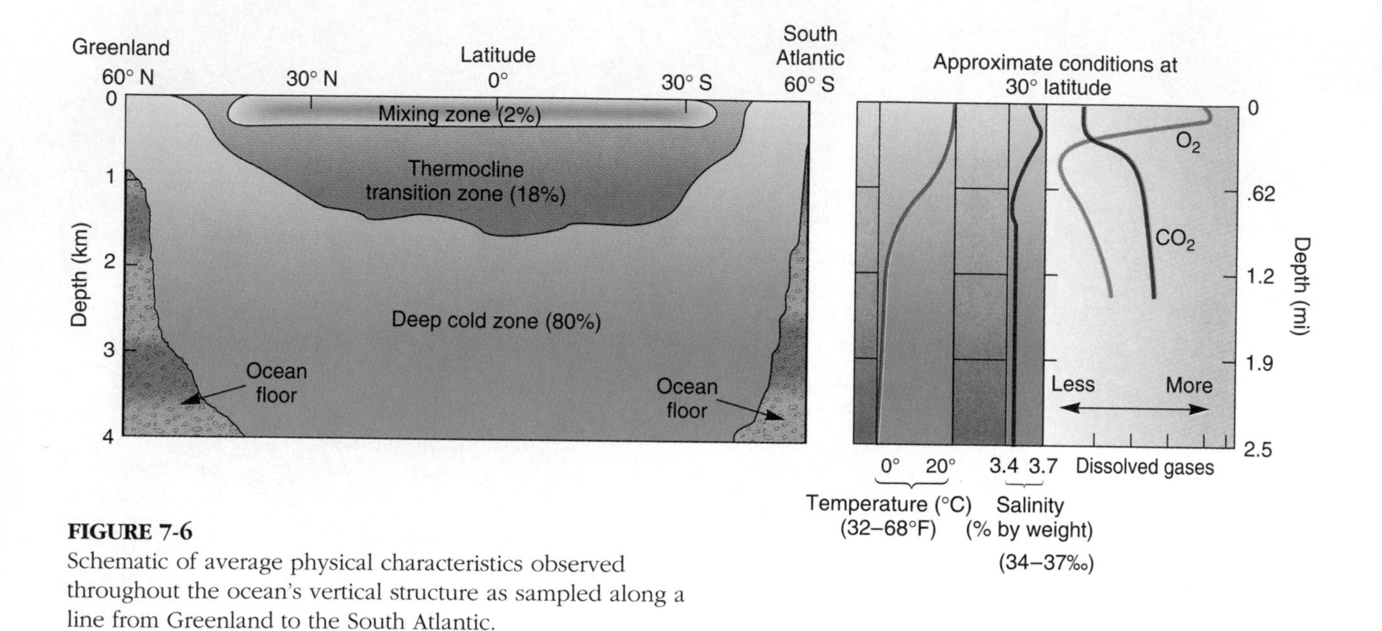

FIGURE 7-6
Schematic of average physical characteristics observed throughout the ocean's vertical structure as sampled along a line from Greenland to the South Atlantic.

Physical Structure of the Ocean

The basic physical structure of the ocean is layered (Figure 7-6). The graph illustrates average temperature, salinity, and dissolved carbon dioxide and oxygen levels with increasing depth. The ocean's surface layer is warmed by the Sun and is wind-driven. Variations in water temperature and solutes are blended rapidly in a **mixing zone** that represents only 2% of the oceanic mass. Below this is the **thermocline transition zone,** a region of strong temperature gradient that lacks the motion of the surface. Friction dampens the effect of surface currents, with colder water temperatures at the lower margin tending to inhibit any convective movements.

Starting at a depth of 1–1.5 km (0.6–0.9 mi) and going down to the bottom, temperature and salinity values are quite uniform (Figure 7-6, graph). Temperatures in this **deep cold zone** are near 0°C (32°F). Water in the deep cold zone does not freeze, however, because of its salinity and intense pressures at those depths; seawater freezes at about −2°C (28.4°F). The coldest water is at the bottom, except near the poles, where the coldest water may be near or at the surface.

This physical structure exposes the inaccuracy of the supposition that pollution pumped into the sea is mixed throughout its mass (the solution-to-pollution-is-dilution rationale). In reality, only the surface layers actively mix. Deep currents are slow-moving masses. It takes thousands of years for deep currents to flow among the North Atlantic, Indian, and North Pacific oceans, or for the colder, saltier flows that sink near Antarctica to flow northward along the ocean floor past the equator (see Figure 6-22).

The United States enacted the Ocean Dumping Act Ban (1988) to immediately halt industrial waste disposal at sea. The dumping of municipal sewage sludge was to cease completely by the end of 1991. New York City, at the cost of large fines, continued to dump through 1992. There are 110 dumping sites in operation, principally for dredged (excavated) materials. A constant political pressure exists to open the oceans for dumping of most municipal, industrial, and radioactive wastes.

UNIQUE PROPERTIES OF WATER

Earth's distance from the Sun places it within a most remarkable temperate zone, compared to the other planets. This temperate location allows all states of water—ice, liquid, and vapor—to occur naturally on Earth. Water is composed of two atoms of hydrogen and one of oxygen, which readily bond. The resulting water molecule exhibits a unique stability, is a versatile solvent, and possesses extraordinary heat characteristics. As the most common compound on the surface of Earth, water exhibits the most uncommon of properties.

Once hydrogen and oxygen atoms combine, they are quite difficult to separate, thereby producing a water molecule that remains stable in Earth's environment. The nature of the hydrogen-oxygen bond results in the hydrogen side having a positive charge and the oxygen side a negative charge (see inset in Figure 7-7). This polarity of the water molecule explains why water "acts wet" and dissolves so many other molecules and elements. Because of this solvency ability, pure water is rare in nature.

Water molecules are attracted to each other because of their polarity; the positive (hydrogen) side of one water molecule is attracted to the negative (oxygen) side of another. (Note this bonding pattern in the molecular illustrations in Figure 7-7b and c.) This bonding between water molecules is called *hydrogen bonding*. The effects of hydrogen bonding in water are observable in everyday life. Hydrogen bonding creates the *surface tension* that allows you to float a steel needle on the surface of water, even though steel is denser than water. This surface tension also allows you to slightly overfill a glass with water, so that the water surface actually is above the rim of the glass, retained by millions of hydrogen bonds.

Hydrogen bonding also is the cause of capillarity, which you observe when you use a paper towel. The towel draws water upward through its fibers because each molecule is pulling on its neighbor. In chemistry laboratory classes, students observe the curved meniscus that forms in a cylinder or a test tube because hydrogen bonding allows the water to slightly "climb" the glass walls. Capillary action is an important component of soil moisture processes, discussed in Chapters 9 and 18. It is important to note that, without hydrogen bonding, water would be a gas at ambient surface temperatures.

Heat Properties

For water to change from one state to another, heat energy must be added to it or released from it. The amount of heat energy must be sufficient to affect the hydrogen bonds between the molecules. Figure 7-7 presents the three states of water and the terms used to describe each **phase change.** The term **sublimation** refers to the direct change of water vapor to ice or ice to water vapor; sometimes the term *deposition* is used if water vapor attaches itself directly to an ice crystal. The deposition of water vapor to ice may form *frost* on surfaces. *Melting* and *freezing* describe the phase change between solid and liquid. The terms *condensation* and *evaporation,* or *vaporization,* at boiling temperature, apply to the change between liquid and vapor. The heat exchanged in the phase changes of water provides over 30% of the energy powering the general circulation of the atmosphere.

Ice, the Solid Phase. As water cools, it behaves like most compounds and contracts in volume, reaching its greatest density at 4°C (39°F). But below that temperature, water behaves very differently from other compounds, and begins to expand as more hydrogen bonds form among the slower-moving molecules, creating the hexagonal structures shown in Figure 7-7c. This expansion continues to a temperature of –29°C (–20°F), with up to a 9% increase in volume possible. As shown in Figure 7-8a, the rigid internal structure of ice dictates the six-sided appearance of all ice crystals, which can loosely combine to form snowflakes. This six-sided preference applies to ice crystals of all shapes: plates, columns, needles, and dendrites (branching or treelike forms)—a unique interaction of randomness and the determinism of physical principles.

The expansion in volume that accompanies the freezing process results in a decrease in density. Specifically, ice has 0.91 times the density of water, and so it floats. Without this change in density, much of Earth's fresh water would be bound in masses of ice on the ocean floor. Instead, we have floating icebergs, with approximately 1/11 (9%) of their mass exposed and 10/11 (91%) hidden beneath the ocean's surface (Figure 7-8b).

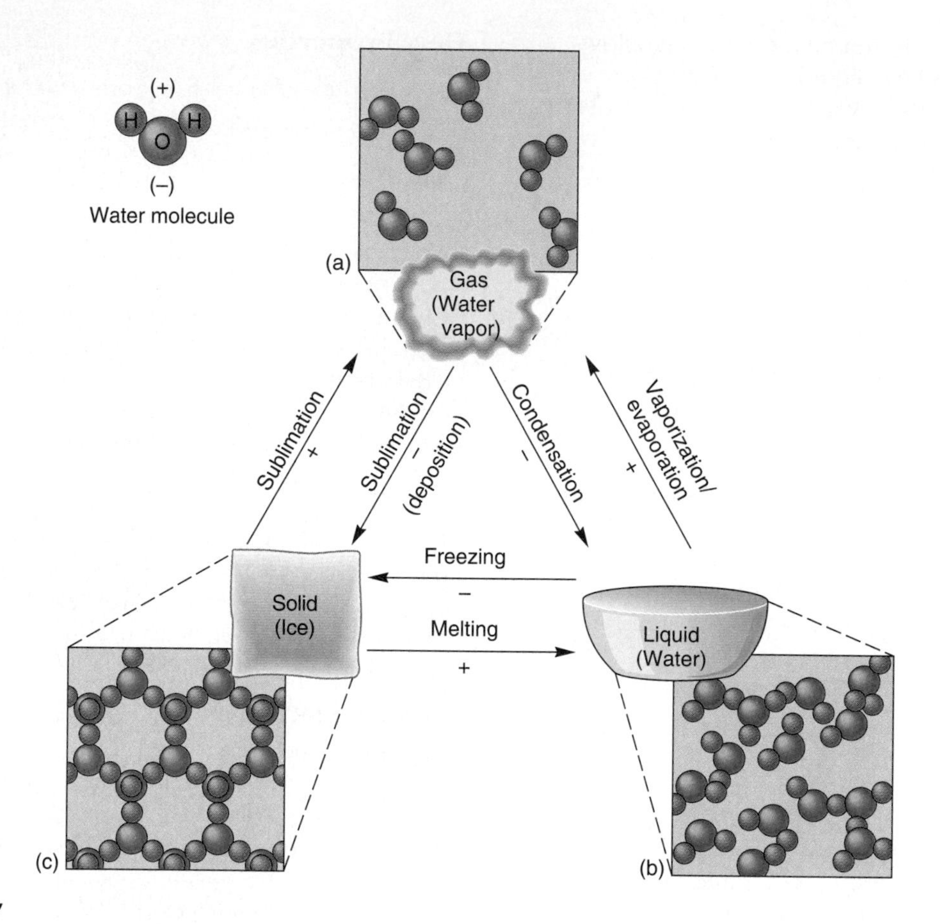

FIGURE 7-7
The three physical states of water: (a) water vapor, (b) liquid water, and (c) ice. Note the molecular arrangement in each state and the terms that describe the changes from one phase to another. Also note how the polarity of water molecules bonds them to one another, loosely in the liquid state and firmly in the solid state.

Winter damage to highways is often the result of precisely this phase change from water to ice. Rainwater seeps into roadway cracks and then expands as it freezes, thus breaking up the pavement. Historically, this physical property of water was put to useful work in quarrying rock for building materials. Holes were drilled and filled with water before winter, so that when cold weather arrived the water would freeze and expand, cracking the rock into manageable shapes.

The freezing action of ice as an important physical weathering process is discussed in Chapter 13. Approximately 30% of Earth's surface is affected by the freeze-thaw action of surface and subsurface water, producing a variety of processes and landforms. Such periglacial landscapes are discussed in Chapter 17.

Water, the Liquid Phase. As a liquid, water assumes the shape of its container and is a noncompressible fluid, quite different from solid, rigid ice. For ice to melt, heat energy must increase the motion of the water molecules to break some of the hydrogen bonds (Figure 7-7b). Despite the fact

(a)

(b)

(c)

FIGURE 7-8
(a) Ice crystal patterns are dictated by the internal structure between water molecules. This structure also explains the lower density of ice and why ice floats. (b) Icebergs floating in Tracy Arm, Tongass National Forest Wilderness Area, Alaska. (c) Ice crystals forming on these branches exhibit a remarkable struggle between physical principles of randomness and order. [(a) Photos by W. A. Bentley and W. J. Humphreys, 1961, *Snow Crystals,* New York: Dover Publications, pp. 91, 117, 137, and 183; (b) photo by Tom Bean; (c) photo by author.]

that there is no change in sensible temperature between ice at 0°C and water at 0°C, 80 calories are required for the phase change of 1 gram of ice to 1 gram of water (Figure 7-9, upper left). This heat, called **latent heat,** is stored within the water and is liberated whenever phase reverses and a gram of water freezes. These 80 calories are the *latent heat of fusion,* or the *latent heat of melting* and of *freezing.*

To raise the temperature of 1 gram of water from freezing at 0°C (32°F) to boiling at 100°C (212°F), we must add 100 additional calories, gaining an increase of 1C° (1.8F°) for each calorie added.

Water Vapor, the Gas Phase. Water vapor is an invisible and compressible gas in which each molecule moves independently (Figure 7-7a). To accomplish the phase change from liquid to vapor at boiling temperature, under normal sea-level pressure, 540 calories must be added to 1 gram of boiling water to achieve a phase change to water vapor (Figure 7-9). Those calories are the **latent heat of vaporization.** When water vapor condenses to a liquid, each gram gives up its hidden 540 calories as the **latent heat of condensation.**

To summarize, taking 1 gram of ice at 0°C and changing it to water vapor at 100°C—changing it from a solid, to a liquid, to a gas—*absorbs* 720 calories (80 cal + 100 cal + 540 cal). Or, reversing the process, changing 1 gram of water vapor at 100°C to ice at 0°C *liberates* 720 calories.

Heat Properties of Water in Nature. In a lake, stream, or in soil water, at 20°C (68°F), every gram of water that breaks away from the surface through evaporation must absorb from the environment approximately 585 calories as the *latent heat of evaporation* (Figure 7-9). This is slightly more energy than would be required if the water were boiling (540 cal). You can feel this absorption of latent heat as evaporative cooling on your skin when it is wet. This latent heat exchange is the dominant cooling process in Earth's energy budget.

The process reverses when air that contains water vapor is cooled. The vapor eventually condenses back into the liquid state, forming moisture droplets and thus liberating 585 calories as the *latent heat of condensation* for every gram of water.

The *latent heat of sublimation* absorbs 680 calories as a gram of ice transforms into vapor. A comparable amount of energy is released as vapor transforms back to ice.

When you realize that a small, puffy, fair-weather cumulus cloud holds 500–1000 metric tons (550–1100 tons) of moisture droplets, you can appreciate the tremendous latent energy available for release in the atmosphere! Weather events such as Hurricane Camille (1969), Hurricane Gilbert (1988), Hurricane Hugo (1989), and Hurricane Andrew (1992) involve a staggering amount of energy.

Government meteorologists estimate that the moisture in Camille weighed 27 trillion metric tons (30 trillion tons) at its maximum power and mass. The latent heat liberated each day from those hurricane clouds approximately equaled the total energy consumption in the United States for six months! These unique heat properties and latent heat exchanges in water, empowered by solar energy, drive weather systems worldwide.

THE HYDROLOGIC CYCLE

Vast currents of water, water vapor, ice, and energy are flowing about us continuously in an elaborate, open global plumbing system. Together they form the **hydrologic cycle,** or hydrologic system, which has operated for billions of years, from the lower atmosphere down to several kilometers beneath Earth's surface, circulating and transforming water throughout Earth's various spheres.

Modern study of the hydrologic cycle involves theory, computer modeling, and both direct and remote sensing observations. A better understanding of the hydrologic cycle is central to understanding global climate change. Progress has proven difficult owing to the inherent chaos of such a nonlinear system as the hydrologic cycle.

A Hydrologic Cycle Model

A simplified model of this complex system is useful to our study of the hydrologic cycle. The ocean provides a starting point, where more than 97% of

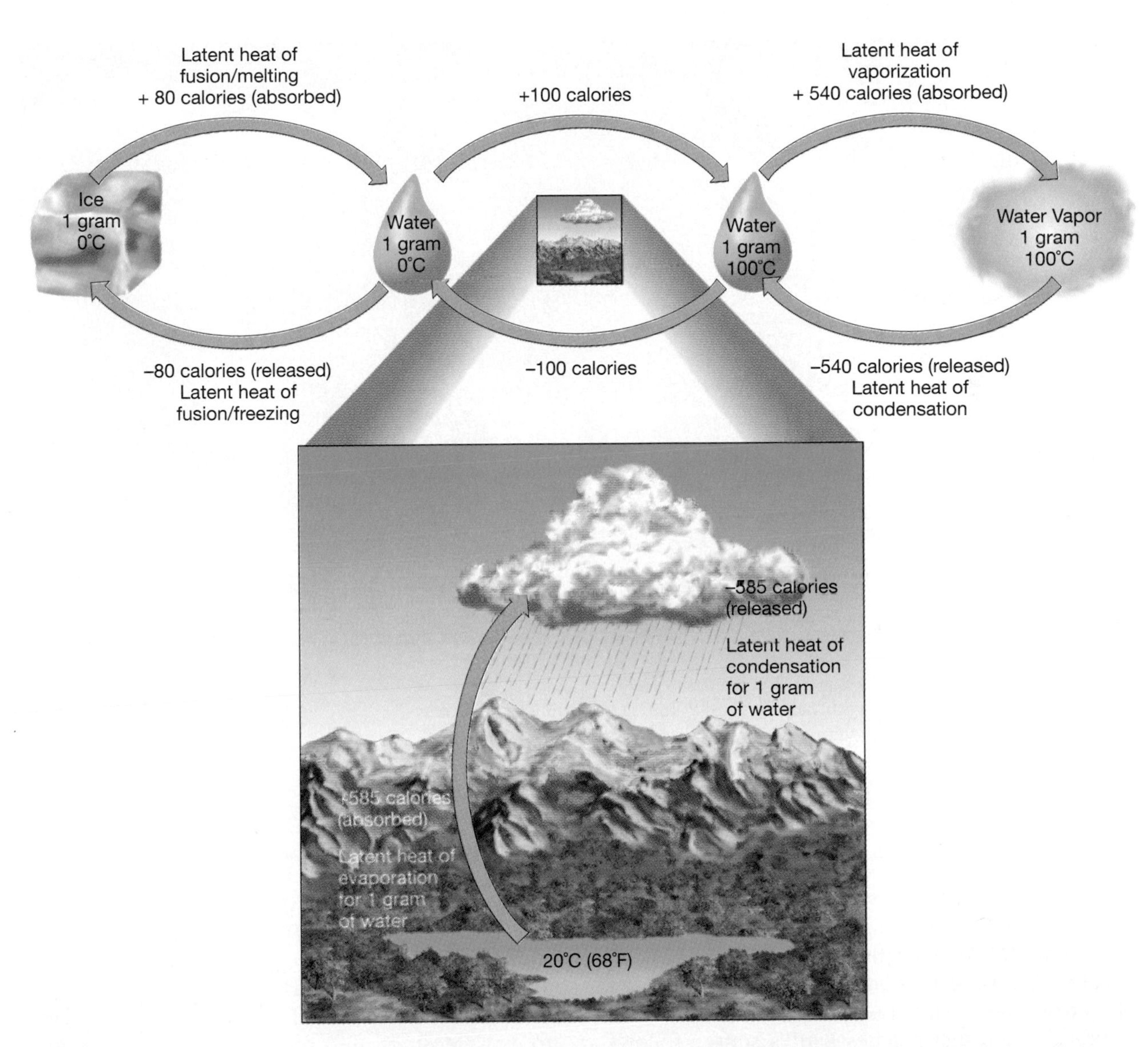

FIGURE 7-9
Significant latent heat energy is involved in the phase changes of water. The inset landscape illustrates average conditions in the environment.

all water is located and most evaporation and precipitation occur. If we assume that the mean annual global evaporation equals 100%, we can trace 86% to the ocean. The other 14% comes from the land, including water moving from the soil into plant roots and passing through their leaves (a process called transpiration, described further in Chapter 9). In the following discussion, refer to Figure 7-10 to follow each flow path in the cycle.

Of the ocean's evaporated 86%, 66% combines with 12% advected from the land to produce the 78% of precipitation that falls back into the ocean.

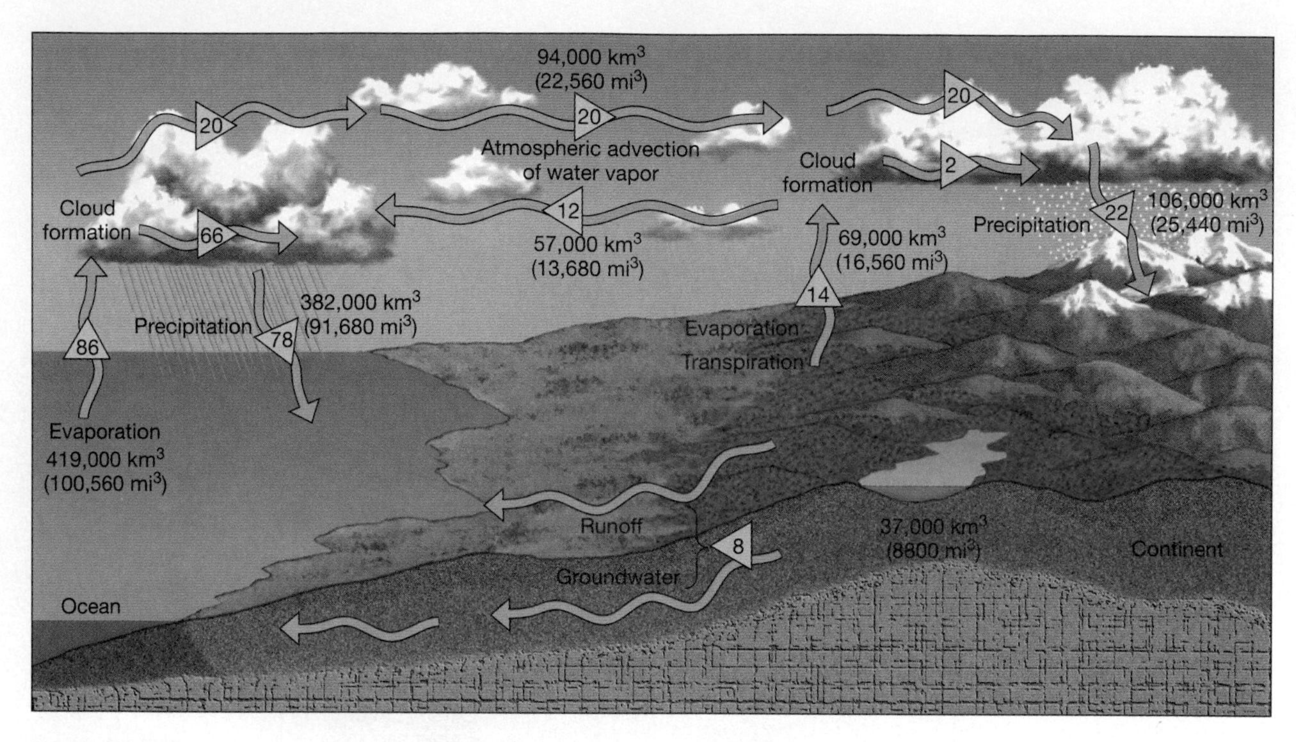

FIGURE 7-10
The hydrologic cycle model with percentages, volumes, and directional arrows denoting flow paths. Global average values are listed as percentages in the directional triangles; note that all evaporation (86% + 14% = 100%) equals all precipitation (78% + 22% = 100%).

The remaining 20% of moisture evaporated from the ocean, plus 2% of land-derived moisture, produces the 22% of precipitation that falls over land. Clearly, the bulk of continental precipitation derives from the oceanic portion of the cycle.

Precipitation that reaches Earth's surface follows a variety of pathways. The process of precipitation striking vegetation or other ground cover is called **interception.** Precipitation that falls directly to the ground, coupled with that which drips onto the ground from vegetation, constitutes *throughfall.* Intercepted water that drains across plant leaves and down plant stems is termed *stem flow* and can represent an important moisture route to the surface. Water reaches the subsurface through **infiltration,** or penetration of the soil surface. It then permeates soil or rock through vertical movement called **percolation.**

The atmospheric advection of water vapor from sea to land and land to sea in Figure 7-10 appears to be unbalanced: 20% moving inland but only 12% moving out to sea. However, this exchange is balanced by 8% as runoff that also flows from land to sea. Most of this runoff—about 95%—comes from surface waters washing across land as overland flow and streamflow, leaving only 5% of the runoff attributable to slow-moving subsurface groundwater. These percentages indicate that the small amount of water in rivers and streams is very dynamic, whereas the large quantity of subsurface water is sluggish in comparison and represents only a small portion of total runoff.

The residence time for water in any portion of the hydrologic cycle determines its relative importance in affecting Earth's climates. The short time spent by water in transit through the atmosphere

(10-day average) is reflected in temporary fluctuations in regional climatic patterns. Long residence times, such as the 3000–10,000 years in deep ocean circulations, groundwater aquifers, and glacial ice, act to moderate temperatures and climates. These slower parts of the cycle work as a system memory; heat is stored and released to buffer change.

To observe and describe the cycle and its related energy budgets, scientists established the Global Energy and Water Cycle Experiment (GEWEX) in 1988. This is part of the World Climate Research Program and represents a central focus in climate-change studies. Now that you are acquainted with the hydrologic cycle and surface water, let's examine atmospheric moisture.

HUMIDITY

The water vapor content of air is termed **humidity.** The capacity of air to hold water vapor is primarily a function of temperature—the temperature of both the air and the water vapor, which are usually the same. Warmer air has a greater capacity for water vapor, whereas cooler air has a lesser capacity. We are all aware of humidity in the air, for its relationship to air temperature determines our sense of comfort. North Americans spend billions of dollars a year to adjust humidity, either with air conditioning (extracting water vapor and cooling) or with air humidifying (adding water vapor). To determine the energy available for weather activities, it is essential to know the water vapor content of air.

Relative Humidity

Next to air temperature and barometric pressure, the most common piece of information in local weather broadcasts is **relative humidity.** Relative humidity is not a direct measurement of water vapor. Rather, it is a ratio (expressed as a percentage) of the amount of water vapor that is actually in the air (content), compared to the maximum water vapor the air could hold at a given temperature (capacity). If the air is relatively dry in comparison to its capacity, the percentage is lower; if the air is relatively moist, the percentage is higher.

$$\text{relative humidity} = \frac{\text{actual water vapor content of the air}}{\text{max. water vapor capacity of the air}} \times 100$$

Relative humidity varies because of evaporation, condensation, or temperature changes, all of which affect both the moisture content and the capacity of the air to hold water vapor. Relative humidity is an expression of an ongoing process between air and moist surfaces, for condensation and evaporation operate continuously—water molecules move back and forth between air and water. Air and water vapor temperatures are the principal regulators of these processes and thus of relative humidity.

Relative humidity indicates the nearness of air to a saturated condition and indicates when active condensation begins. Air is said to be **saturated,** or full, if it is holding all the water vapor that it can hold at a given temperature; under such conditions the net transfer of water molecules between surface and air achieves a saturation equilibrium. Saturated air has a relative humidity of 100%. Saturation indicates that any further addition of water vapor (change in content) or any decrease in temperature (change in capacity) will result in active condensation.

The temperature at which a given mass of air becomes saturated is termed the **dew-point temperature.** In other words, *air is saturated when the dew-point temperature and the air temperature are the same.* A cold drink in a glass provides a common example of these conditions: the water droplets that form on the outside of the glass condense from the air because the air layer next to the glass is chilled to below its dew-point temperature and thus is saturated.

During a typical day, air temperature and relative humidity relate inversely (Figure 7-11a). Relative humidity is highest at dawn, when air temperature is lower and the capacity of the air to hold water vapor is less. Relative humidity is lowest in the late afternoon, when higher air temperatures increase the capacity of the air to hold water vapor. The actual water vapor content in the air may have remained the same throughout the day, but because the temperature varied, relative humidity percentages changed from morning until afternoon.

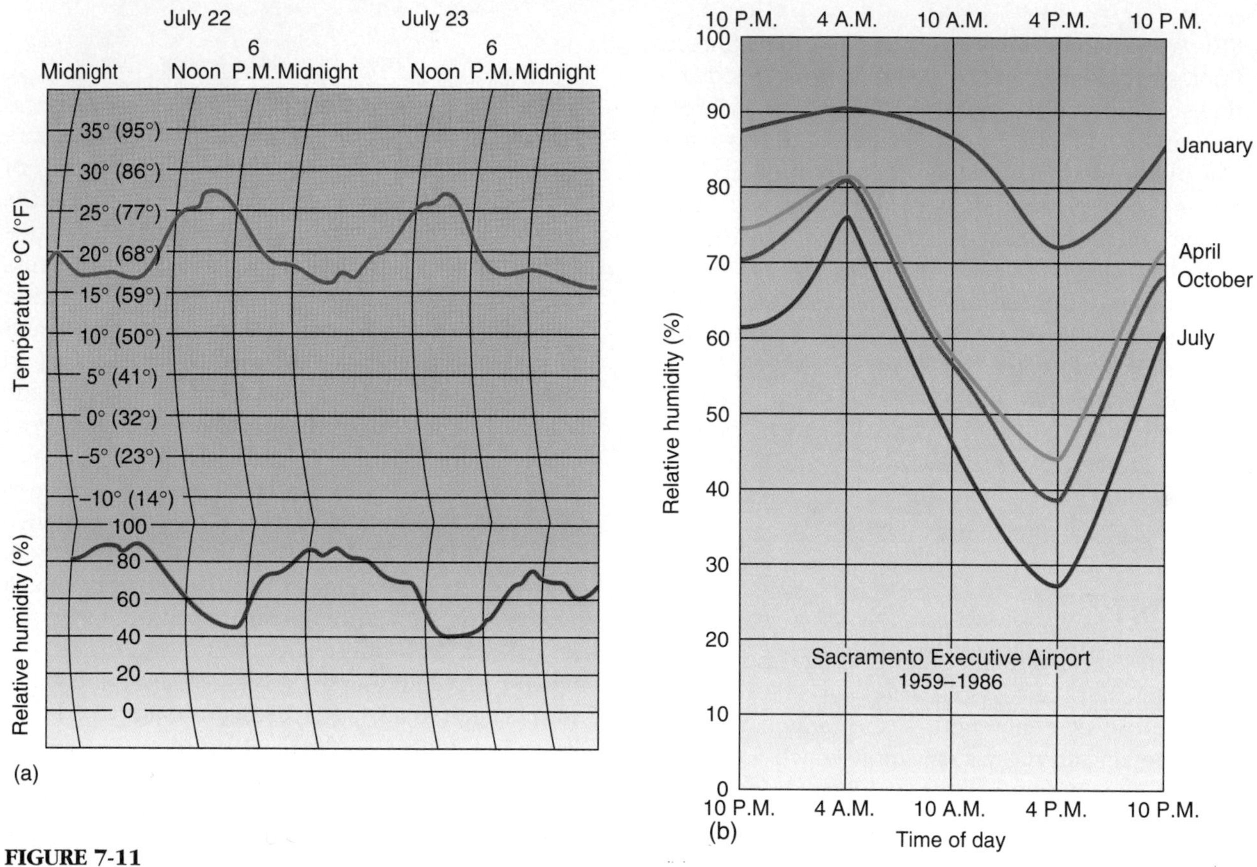

FIGURE 7-11
(a) Typical daily variations in relative humidity and temperature and (b) seasonal daily variations in relative humidity for Sacramento, California.

Think of the capacities of warm air and cooler air to hold water vapor as being like sponges of varying sizes: warmer air has in it a larger sponge; cooler air a smaller sponge. A cooler body of air can be saturated—filled to capacity—by an amount of water vapor that will only partially fill the capacity of the warmer air.

Weather records for Sacramento, California, show the seasonal variation in relative humidity by time of day, confirming the relationship of temperature and relative humidity (Figure 7-11b). Beyond the expected similarity in overall daily pattern for the four months shown, January readings are higher than July readings because air temperatures are lower overall in winter. Similar relative humidity records at most weather stations demonstrate the same relationship among season, temperature, and relative humidity.

Expressions of Relative Humidity

Three ways are used to express humidity and relative humidity, each with its own utility and application: vapor pressure, specific humidity, and absolute humidity.

Vapor Pressure. One way of describing humidity is related to air pressure. As free water molecules evaporate from a surface into the atmosphere, they become water vapor, one of the gases in air. That portion of total air pressure that is made up of water vapor molecules is termed **vapor pressure** and is expressed in millibars (mb). Water vapor molecules continue to evaporate from a moist surface, slowly diffusing into the air, until the increasing vapor pressure in the air causes some molecules to return to the surface. Saturation is reached when the movement of water molecules between surface and air is in equilibrium. The maximum capacity of the air at a given temperature is termed the *saturation vapor pressure* and indicates the maximum pressure that water vapor molecules can exert. Any increase or decrease in temperature causes the saturation vapor pressure to change. Figure 7-12 graphs the saturation vapor pressure at varying air temperatures.

FIGURE 7-12
Saturation vapor pressure of air at various temperatures. Inset compares saturation vapor pressure over water surfaces and over ice surfaces at subfreezing temperatures.

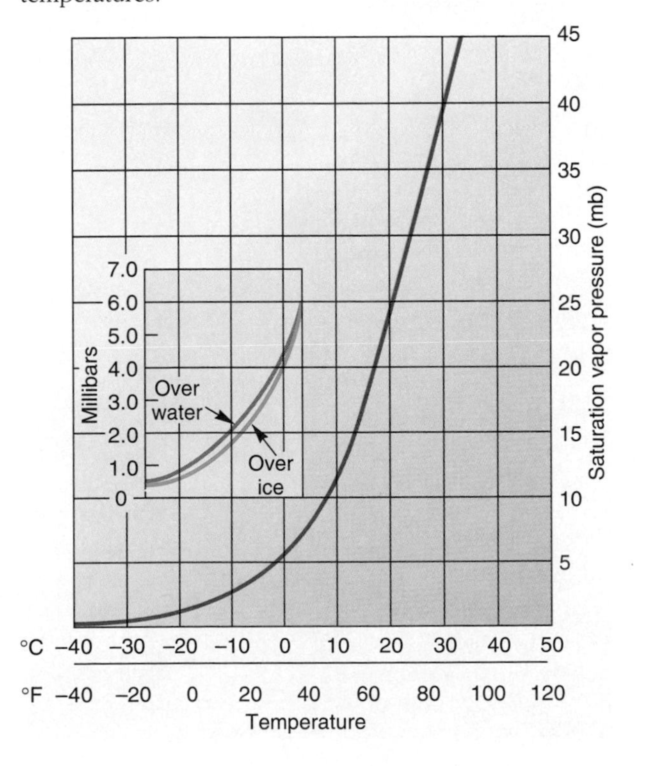

As the graph shows, air at 20°C (68°F) has a saturation vapor pressure of 24 mb; that is, the air is saturated if the water vapor portion of the air pressure is at 24 mb. Thus, if the water vapor content actually present is exerting a vapor pressure of only 12 mb in 20°C air, the relative humidity is 50% (12 mb ÷ 24 mb = 0.50 × 100 = 50%). The graph illustrates that, for every temperature increase of 10C° (18F°), the vapor pressure capacity of air nearly doubles.

The inset in Figure 7-12 compares saturation vapor pressure over water and over ice surfaces at subfreezing temperatures. You can see that saturation vapor pressure is greater above a water surface than over an ice surface—that is, it takes more water vapor molecules to saturate air above water than it does above ice. This fact is important to condensation processes and rain droplet formation, both of which are discussed in the section on clouds later in this chapter.

Specific Humidity. A humidity measure that remains constant as temperature and pressure change is useful. **Specific humidity** refers to the mass of water vapor (in grams) per mass of air (in kilograms) at any specified temperature. Because it is measured in mass, specific humidity is not affected by changes in temperature or pressure such as occur when an air parcel rises to higher elevations, and is therefore a more valuable term in weather considerations. Specific humidity stays constant despite volume changes.

The maximum mass of water vapor that a kilogram of air can hold at any specified temperature is termed the *maximum specific humidity* and is plotted in Figure 7-13. The graph shows that a kilogram of air could hold a maximum specific humidity of 47 g of water vapor at 40°C (104°F), 15 g at 20°C (68°F), and 4 g at 0°C (32°F). Therefore, if a kilogram of air at 40°C has a specific humidity of 12 g, its relative humidity is 25.5% (12 g ÷ 47 g = 0.255 × 100 = 25.5%). Specific humidity is useful in describing the moisture content of large air masses that are interacting in a weather system.

Absolute Humidity. Absolute humidity is the actual amount of water vapor in the air, expressed

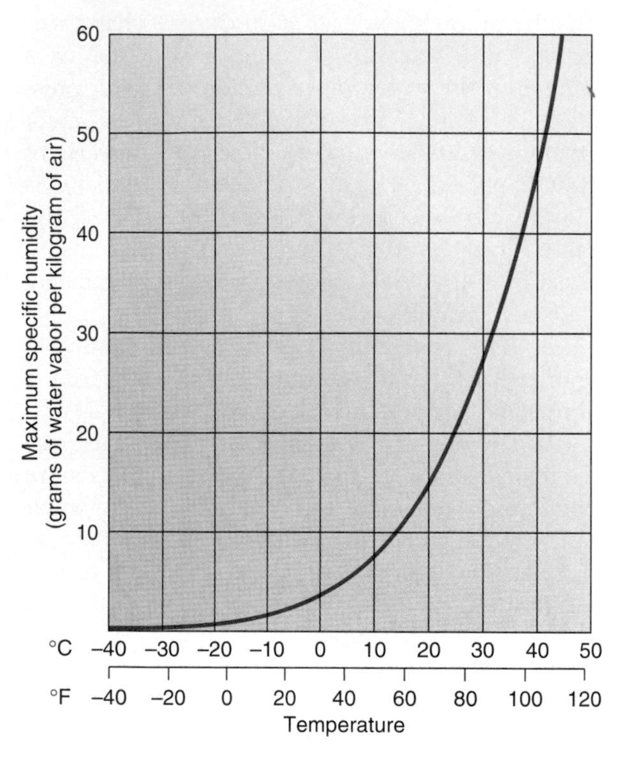

FIGURE 7-13
Maximum specific humidity for a mass of air at various temperatures.

as the weight of water vapor (grams) per unit volume (cubic meter) of air—that is, an expression of weight per volume. The maximum weight of water vapor that a cubic meter of air can hold at a specified temperature is the *saturation absolute humidity*. Air at 30°C (86°F) has the capacity to hold 30 g of water vapor per cubic meter; at 20°C (68°F), 17 g; and at 10°C (50°F), 10 g.

A distinct limitation of using absolute humidity is that volume changes with pressure. Absolute humidity declines as air rises (expands), even though no water vapor has been removed; it increases as air falls (compresses), even though no new vapor has been added. Absolute humidity, therefore, has limited utility in weather analysis, but it is used for engineering and air-conditioning calculations.

Instruments for Measurement. Relative humidity is measured with various instruments. The **hair hygrometer** uses the principle that human hair changes as much as 4% in length between 0 and 100% relative humidity. The instrument connects a standardized bundle of human hair through a mechanism to a gauge and a graph to indicate relative humidity.

Another instrument used to measure relative humidity is a **sling psychrometer.** Figure 7-14 shows this device, which has two thermometers mounted side-by-side on a metal holder. One is called the dry-bulb thermometer; it simply records the ambient (surrounding) air temperature. The other thermometer is called the wet-bulb thermometer; it is set lower in the holder and its bulb

FIGURE 7-14
Sling psychrometer used for measurement of relative humidity. [Photo by Bobbé Christopherson.]

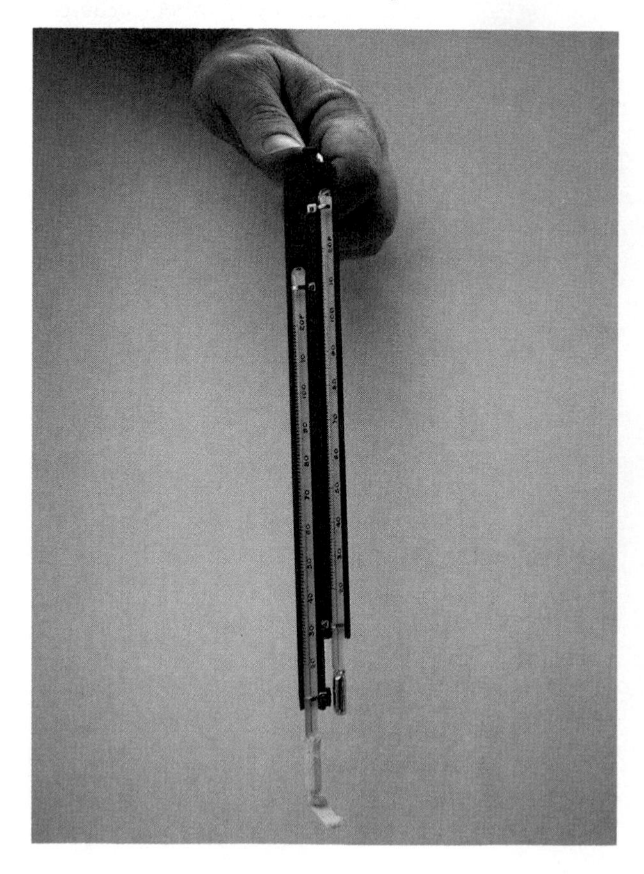

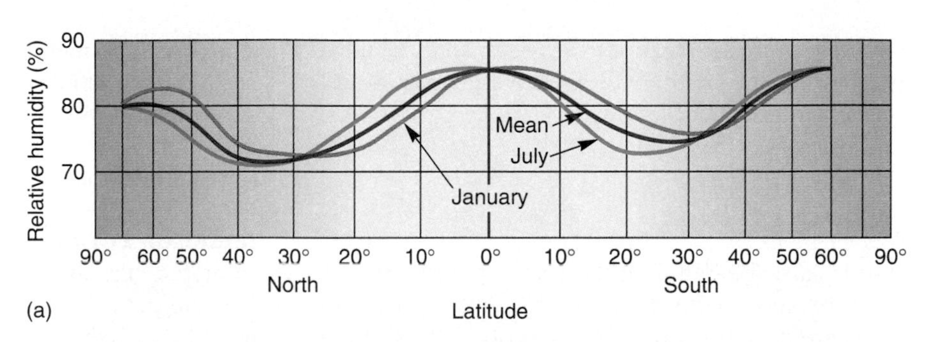

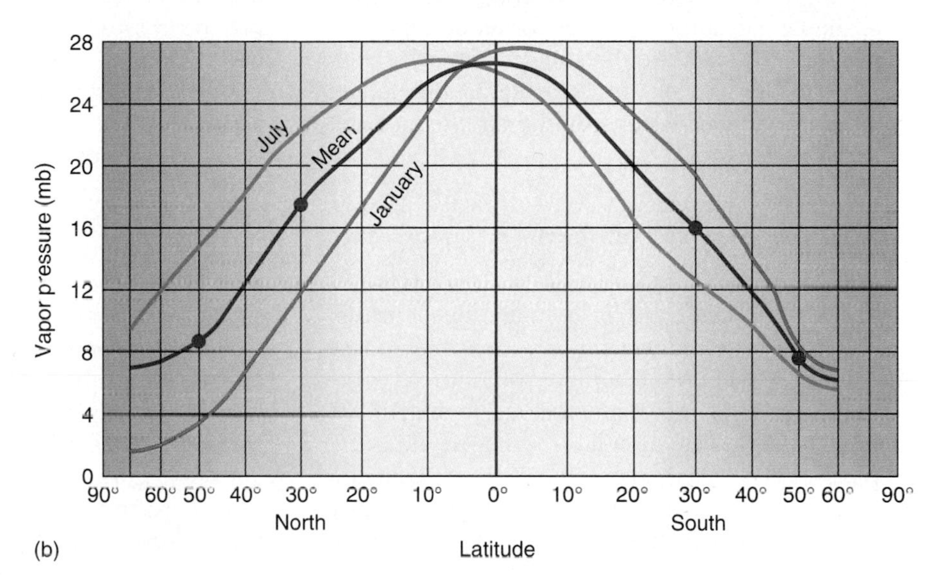

FIGURE 7-15
(a) Approximate mean latitudinal distribution of relative humidity and (b) distribution of vapor pressure by latitude. [After Haurwitz and Austin, 1944, *Climatology*. New York: McGraw-Hill, p. 45. Adapted by permission.]

is covered by a cloth wick, which is moistened with distilled water. The psychrometer is then spun by its handle. (Other versions mechanically spin the thermometers or place them where a fan forces air over them.)

The rate at which water evaporates from the wick depends on the relative saturation of the surrounding air. If the air is dry, water evaporates quickly, absorbing the latent heat of evaporation from the wet-bulb thermometer, causing the temperature to drop. In an area of high humidity, much less water evaporates from the wick. After spinning or ventilating the psychrometer for a minute or two, the temperature on each bulb is noted and compared on a relative humidity (psychrometric) chart, from which relative humidity can be determined.

Global Distribution of Relative Humidity

Relative humidity varies across Earth's surface according to temperature and moisture availability. Figure 7-15a plots the approximate mean distribution of relative humidity and Figure 7-15b plots the mean distribution of vapor pressure by latitude for January and July. The graphs illustrate the relative dryness of the subtropical regions and the relative wetness of equatorial areas and higher latitudes.

In terms of vapor pressure, the water vapor actually in the air above subtropical deserts at 30° latitude is about double the amount at higher latitudes such as 50°, yet the relative humidity in the subtropical regions is lower because the warmer air has a greater capacity to hold water vapor. Thus, the relatively dry air over the generally cloudless Sahara (warmer air) actually contains more water vapor than the relatively moist air in the midlatitudes (cooler air).

CLOUDS AND FOG

Clouds are beautiful indicators of atmospheric moisture and weather conditions. Although cloud types are too numerous to describe all of them here, the classification schemes for cloud types are quite simple. Clouds are the subject of much scientific inquiry as part of the Earth Radiation Budget Experiment, especially regarding their effect on net radiation patterns.

Cloud Formation Processes

Clouds are not initially composed of raindrops. Instead they are made up of a multitude of **moisture droplets,** each individually invisible to the human eye without magnification. An average raindrop, at 2000 μm diameter (0.2 cm or 0.078 in.), is made up of a million or more moisture droplets, each approximately 20 μm diameter (0.002 cm or 0.0008 in.). After more than a century of debate, science has determined how these droplets form during condensation and what processes cause them to coalesce into raindrops.

Under unstable conditions (discussed at the beginning of the next chapter), a parcel of air may rise to where it becomes saturated—i.e., the air cools to the dew-point temperature and 100% relative humidity (although under certain conditions condensation may occur at slightly less or more than 100% relative humidity). Further cooling of the air parcel due to lifting produces active condensation of water vapor to water. This condensation requires microscopic particles called **condensation nuclei,** which always are present in the atmosphere.

In continental air masses, which average 10 billion nuclei per cubic meter, condensation nuclei are typically derived from ordinary dust, volcanic and forest-fire soot and ash, and particles from fuel combustion. Given the air composition over cities, great concentrations of such nuclei are available. In maritime air masses, which average 1 billion nuclei per cubic meter, the nuclei are formed from a high concentration of sea salts derived from ocean sprays. Salts are particularly attracted to moisture and thus are called **hygroscopic nuclei.** The lower atmosphere never lacks such nuclei.

Raindrop Formation. Given the preconditions of saturated air, availability of condensation nuclei, and the presence of cooling mechanisms in the atmosphere, active condensation occurs. Two principal processes account for the majority of the world's raindrops. The **collision-coalescence process** predominates in clouds that form at above-freezing temperatures, principally in the warm clouds of the tropics. Initially, simple condensation takes place on small nuclei, some of which are larger and produce larger water droplets. As those larger droplets respond to gravity and fall through a cloud, they combine with smaller droplets, gradually coalescing until their weight is beyond the ability of air circulation in the cloud to hold them aloft. Towering tropical clouds keep a growing raindrop aloft longer, thus producing more collisions and more drops of greater diameter (Figure 7-16a).

The existence of ice crystals and supercooled water droplets also is a significant mechanism for raindrop production. Water in bulk freezes at 0°C (32°F), but fine, pure (distilled) water droplets don't freeze until the temperature drops to –40°C (–40°F). Consequently, **supercooled droplets** may be found in clouds at temperatures from –15 to –35°C (+5 to –31°F).

Figure 7-12 showed that the saturation vapor pressure near an ice surface is lower than that near a water surface; therefore, supercooled water droplets will evaporate more rapidly near ice crystals, which then absorb the vapor (the vapor sublimates on the ice crystal). This **ice-crystal process** proposed by meteorologist Alfred Wegener—noted in Chapter 11 for his theory of the drifting continents—was elaborated on by Swedish scientist Tor Bergeron. The ice crystals feed on the supercooled cloud droplets, grow in size, and eventually fall as snow or rain (Figure 7-16b). Precipitation in middle and high latitudes begins as ice and snow high in the clouds, then melts and gathers moisture as it falls through the warmer portions of the cloud.

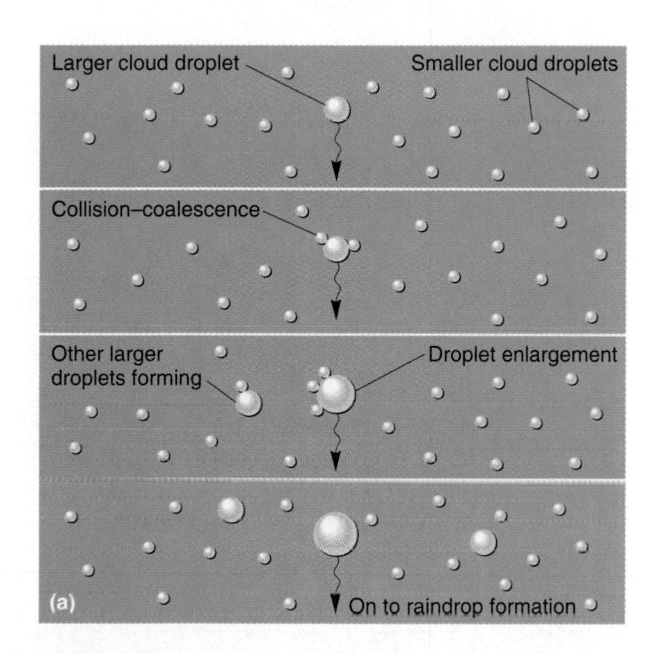

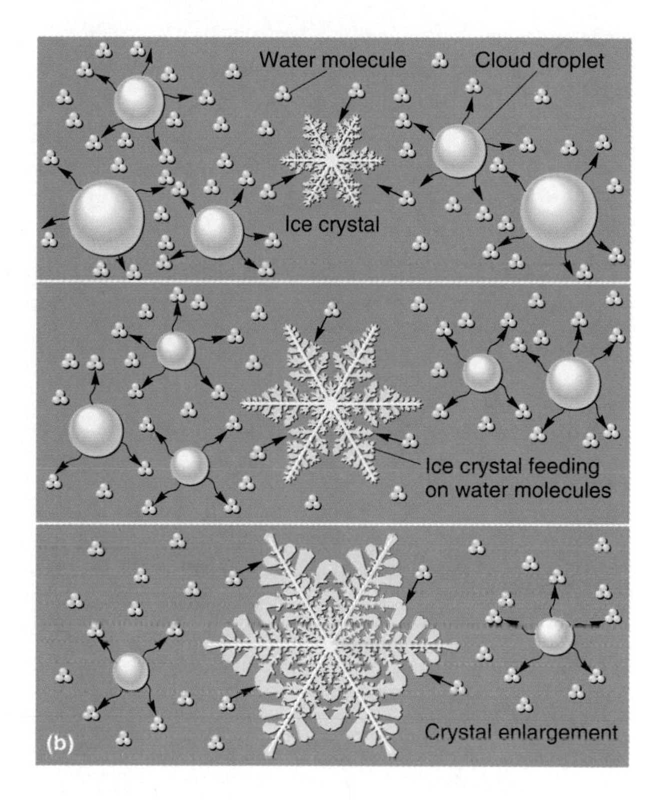

FIGURE 7-16
Principal processes for raindrop and snowflake formation: (a) the collision-coalescence process; (b) the ice-crystal process. [From Frederick K. Lutgens and Edward J. Tarbuck, *The Atmosphere: An Introduction to Meteorology,* 3rd ed., copyright © 1986, p. 127. Reprinted by permission of Prentice Hall, Inc., Englewood Cliffs, NJ.]

Cloud Types and Identification

A cloud is simply an aggregation, or grouping, of moisture droplets and ice crystals that are suspended in air and are great enough in volume and density to be visible to the human eye. In 1803 an English biologist named Luke Howard established a classification system for clouds and coined Latin names for them that we still use today.

Clouds usually are classified by altitude and by shape. They come in three basic forms—flat, puffy, and wispy—which occur in four primary altitude classes and ten basic types. Clouds that are developed horizontally—flat and layered—are called *stratiform* clouds. Those that are developed vertically—puffy and globular—are termed *cumuliform*. Wispy clouds usually are quite high, composed of ice crystals, and are labeled *cirroform*. These three basic forms occur in four altitudinal classes: low, middle, high, and those that are vertically developed through these altitude classes. Table 7-4 presents the basic cloud classes and types. Figure 7-17 illustrates the general appearance of each type, and Figure 7-18 includes representative photographs.

Cloud Classes. Low clouds, ranging from the surface up to 2000 m (6500 ft) in the middle latitudes, are simply called **stratus** or **cumulus** (Latin for "layer" and "heap," respectively). Stratus clouds ap-

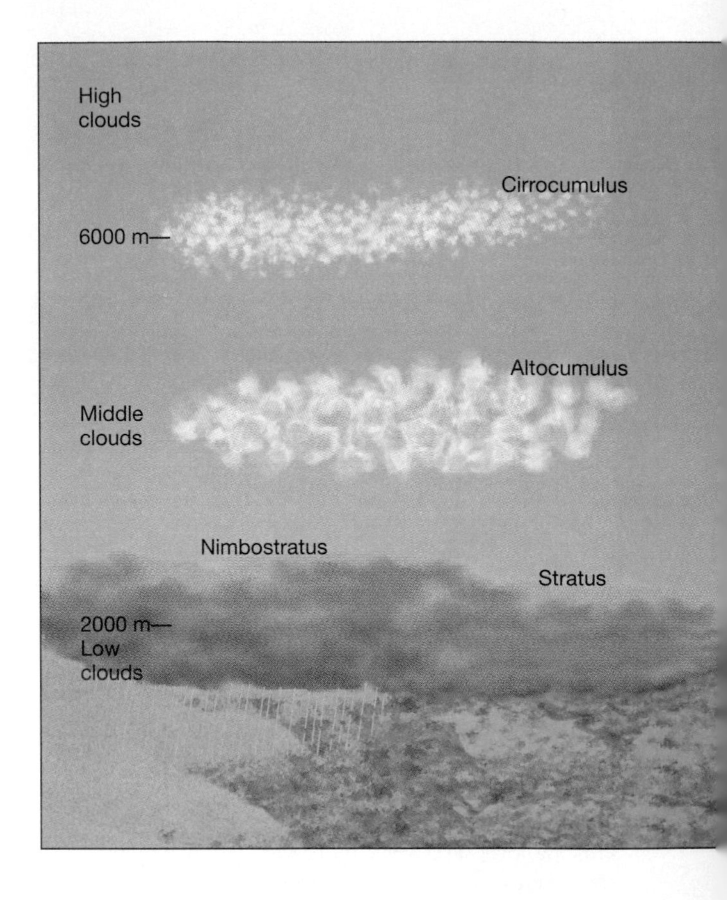

FIGURE 7-17
Principal cloud types, classified by form and altitude: low, middle, high, and vertically developed.

FIGURE 7-18
Principal cloud types: (a) stratus, (b) nimbostratus, (c) altostratus, (d) altocumulus, (e) cirrus, (f) cirrostratus, (g) cumulus, (h) cumulonimbus. [Photos by author.]

(a)

(b)

(e)

(f)

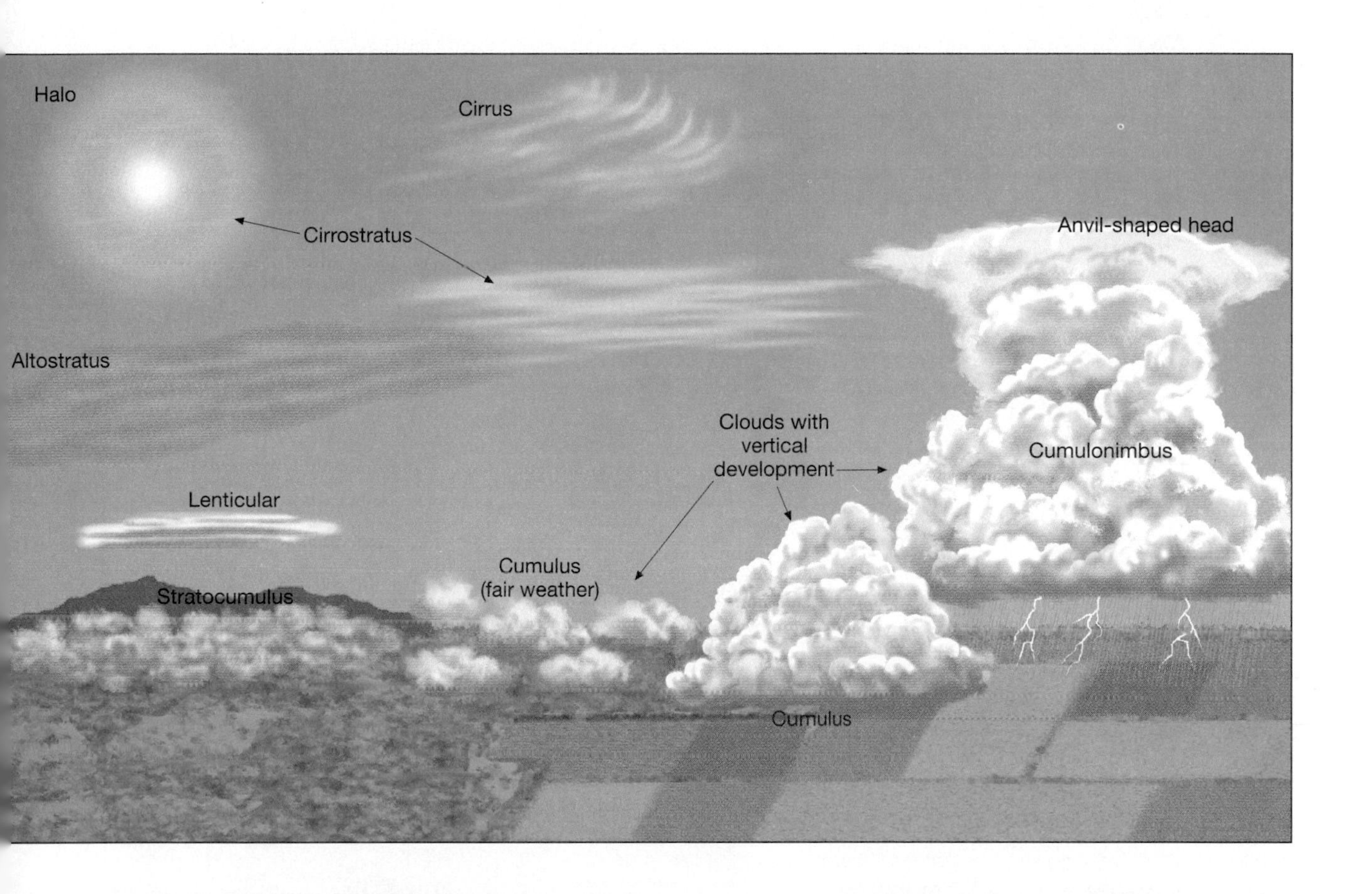
Halo
Cirrus
Cirrostratus
Anvil-shaped head
Altostratus
Clouds with vertical development
Cumulonimbus
Lenticular
Cumulus (fair weather)
Stratocumulus
Cumulus

(c)
(d)
(g)
(h)

TABLE 7-4
Cloud Classes and Types

Class	Altitude/Composition at Midlatitudes	Type	Symbol	Description
Low clouds (C_L)	Up to 2000 m (6500 ft) Water	Stratus (St)		Uniform, featureless, gray, like high fog
		Stratocumulus (Sc)		Soft gray globular masses in lines, groups, or waves, heavy rolls, irregular overcast patterns
		Nimbostratus (Ns)		Gray, dark, low, with drizzling rain
Middle clouds (C_M)	2000–6000 m (6500–20,000 ft) Ice and Water	Altostratus (As)		Thin to thick, no halos, Sun's outline just visible, gray day
		Altocumulus (Ac)		Patches of cotton balls, dappled, arranged in lines or groups, rippling waves, the lenticular clouds associated with mountains
High clouds (C_H)	6000–13,000 m (20,000–43,000 ft). Ice	Cirrus (Ci)		Mare's tails, wispy, feathery, hairlike,delicate fibers, streaks, or plumes
		Cirrostratus (Cs)		Veil of fused sheets of ice crystals, milky, with Sun and Moon halos
		Cirrocumulus (Cc)		Dappled, "mackerel sky," small white flakes, tufts, in lines or groups, sometimes in ripples
Vertically developed clouds	Near surface to 13,000 m (43,000 ft) Water below, ice above	Cumulus (Cu)		Sharply outlined, puffy, billowy, flat-based, swelling tops, fair weather
		Cumulonimbus (Cb)		Dense, heavy, massive, dark thunderstorms, hard showers, explosive top, great vertical development, towering, cirrus-topped plume blown into anvil-shaped head

pear dull, gray, and featureless. When they yield precipitation, they are called **nimbostratus** (*nimbo-* denotes precipitation or heavy rain), and their showers typically fall as drizzling rain (Figure 7-18b). Cumulus clouds appear bright and puffy, like cotton balls. When they do not cover the sky, they float by in infinitely varied shapes. Vertically developed clouds are listed in a separate class in Table 7-4 because further vertical development can produce clouds that extend beyond low altitudes into middle and high altitudes.

Sometimes near the end of the day, lumpy, grayish, low-level clouds called **stratocumulus** may fill the sky in patches. Near sunset, these spreading puffy stratiform remnants may catch and filter the Sun's rays, sometimes indicating clearing weather.

Stratus and cumulus middle-level clouds are denoted by the prefix *alto-*. They are made of water droplets and, when cold enough, can be mixed with ice crystals. *Altocumulus* clouds, in particular, represent a broad category that occurs in many different styles.

Clouds occurring above 6000 m (20,000 ft) are composed principally of ice crystals in thin concentrations. These wispy filaments, usually white except when colored by sunrise or sunset, are termed **cirrus** clouds (Latin for "curl of hair"), sometimes dubbed mare's-tails. Cirrus clouds look as if an artist has taken a brush and placed delicate feathery strokes high in the sky. Often, cirrus clouds are associated with an oncoming storm, especially if they thicken and lower in elevation. The prefix *cirro-* (cirrostratus, cirrocumulus) is used for other high clouds.

A cumulus cloud can develop into a towering giant called **cumulonimbus** (again, *-nimbus* denotes precipitation). Such clouds generally are referred to as thunderheads because of their shape and their associated lightning and thunder (Figure 7-19). When the rising cloud column enters a re-

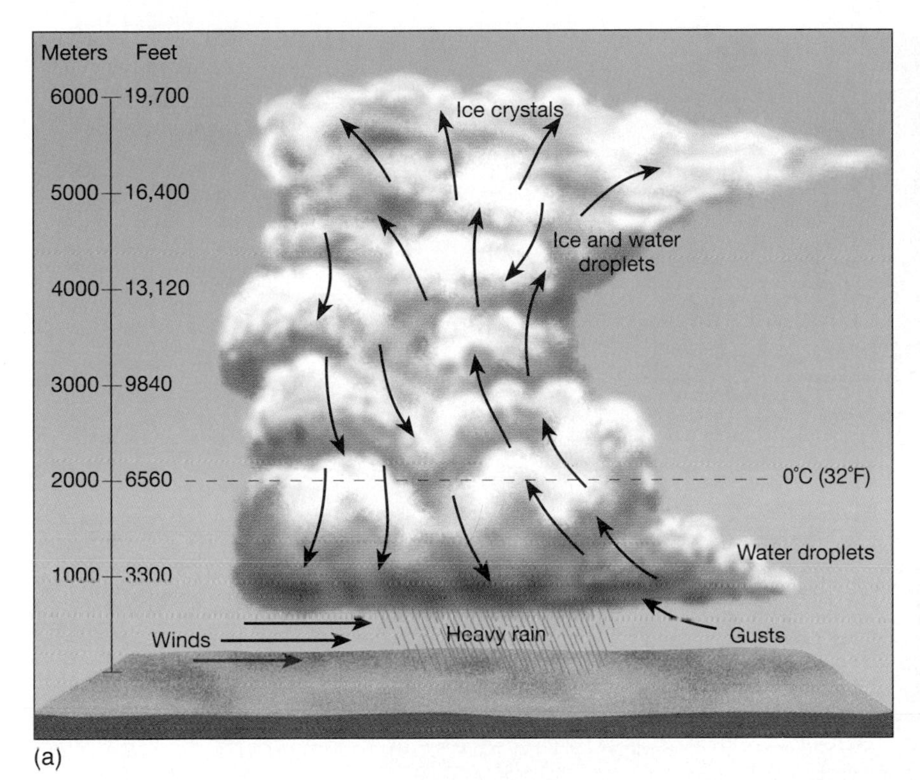

FIGURE 7-19
(a) Structure and form of a cumulonimbus cloud. Violent updrafts and downdrafts mark the circulation within the cloud. Blustery wind gusts occur along the ground. (b) Shuttle astronauts capture a dramatic cumulonimbus thunderhead as it moves over Galveston Bay, Texas. [Space Shuttle photo From NASA.]

gion of temperatures well below freezing, ice crystals form and consume available moisture droplets. It is possible to witness this transformation from a water-droplet cloud to an ice-crystal cloud, a process called **glaciation,** because the sharply delineated form of the cloud top softens in appearance. High-altitude winds may then shear the top of the cloud into the characteristic anvil shape of the mature thunderhead.

Fog

A cloud in contact with the ground is commonly referred to as **fog.** The presence of fog tells us that the air temperature and the dew-point temperature at ground level are nearly identical, producing saturated conditions. Generally, fog is capped by an inversion layer, with as much as 30C° (50F°) difference in air temperature between the ground under the fog and the clear, sunny skies above.

By international agreement, fog is officially described as a cloud layer on the ground, with visibility restricted to less than 1 km (3300 ft). Almost all fog is warm—that is, its moisture droplets are above freezing. Supercooled fog, which occurs when the moisture droplets are below freezing, is special because it can be dispersed through artificial seeding with ice crystals or other crystals that mimic ice, following the principles of the ice-crystal formation process described earlier.

FIGURE 7-20
San Francisco's Golden Gate Bridge shrouded by an advection fog. [Photo by author.]

Two principal forms of fog related to cooling are advection fog and radiation fog. As the name implies, **advection fog** forms when air in one place migrates to another place where saturated conditions exist. When warm, moist air overlays cooler ocean currents, lake surfaces, or snow masses, the layer of air directly above the surface is chilled to the dew point. Off all subtropical west coasts in the world, summer fog forms in this manner (Figure 7-20). Organisms have adapted to this special condition, as they adapt to all Earthly phenomena. For example, sand beetles that have adapted to summer advection fog live in the Namib Desert in extreme southwestern Africa. In an otherwise dry environment, they harvest water from the fog by holding up their wings so condensation runs down to their mouths.

In the Atacama Desert of Chile and Peru, residents stretch large nets on frames in the air to intercept the fog; moisture condenses on the netting and drips into barrels. Chungungo, Chile, receives water from fog harvesting in a pilot program developed by Canadian and Chilean interests. Large sheets of plastic mesh along a ridge of the El Tofo mountains harvest water from advection fog. The captured water flows through pipes to the fishing village below. At least 22 countries, distributed on each continent, experience conditions suitable for this water resource technology.

Another type of fog forms when cold air flows over the warm water of a lake, ocean surface, or even a swimming pool. An **evaporation fog,** or steam fog, may form as the water molecules evaporate from the water surface into the cold overly-

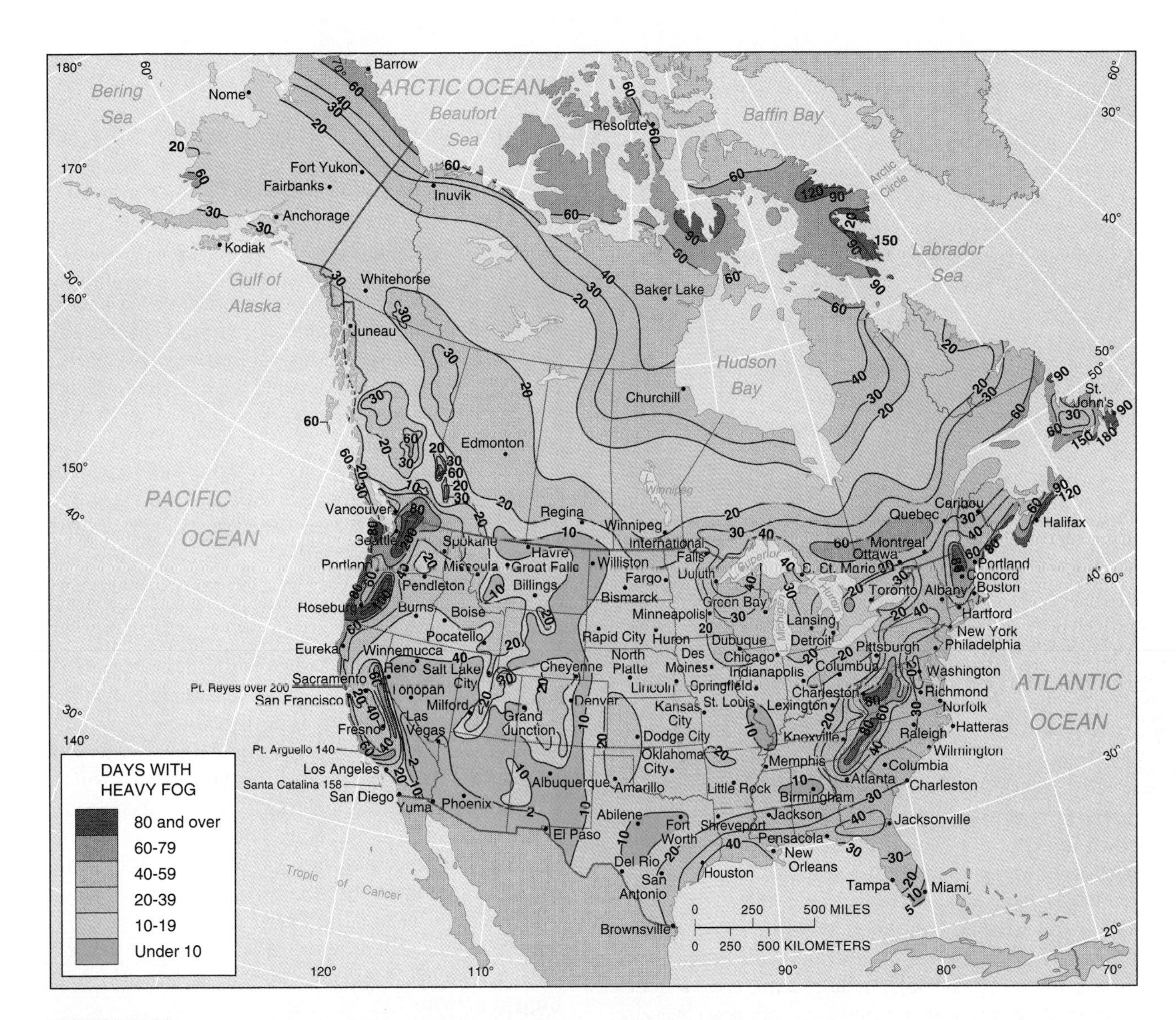

FIGURE 7-21
Mean annual number of days with heavy fog in the United States and Canada. Officially fog is declared if visibility is less than 1 km (3300 ft). The foggiest spot in the United States is the mouth of the Columbia River where it enters the Pacific Ocean at Cape Disappointment, Washington. One of the foggiest places in the world is Newfoundland's Avalon Peninsula, specifically Argentia and Belle Isle, which regularly exceed 200 days. [Data courtesy of National Weather Service; Map Series 3, *Climate Atlas of Canada,* Atmospheric Environment Service; and *The Climates of Canada,* compiled by David Philips, Senior Climatologist, Atmospheric Environment Service, Environment Canada, 1990.]

ing air, effectively humidifying the air. When visible at sea, the term *sea smoke* is applied to this shipping hazard.

A type of advection fog involving the movement of air forms when moist air is forced to higher elevations along a hill or mountain. This upslope lifting leads to cooling by expansion as the air rises. The resulting **upslope fog** forms a stratus cloud at the level of saturation. Along the Appalachians and the eastern slopes of the Rockies, such fog is common in winter and spring. Another fog associated with topography is called **valley fog.** Because cool air is denser, it settles in low-lying areas, producing a fog in the chilled, saturated layer near the ground.

Radiation fog forms when radiative cooling of a surface chills the air layer directly above that surface to the dew-point temperature, creating saturated conditions and fog. This fog occurs especially on clear nights over moist ground; it does not occur over water, because water does not cool appreciably overnight. Slight movements of air deliver even more moisture to the cooled area for more fog formation of greater depth.

Fog poses a hazard to various human activities, including driving automobiles, flying aircraft, navigating ships, and for pedestrians or cyclists negotiating traffic. In addition, fog droplets can combine with atmospheric pollutants to form airborne acid pollution and even a "killer fog" in extreme cases, with stagnant air leading to unhealthy concentrations and possible fatalities. The prevalence of fog throughout the United States and Canada is shown in Figure 7-21. The spatial implications of fog occurrence on a local scale should represent a key component in any location analysis for a proposed airport or harbor facility.

SUMMARY—Water and Atmospheric Moisture

The next time it rains where you live, pause and reflect on the journey each of those water molecules has made in terms of distance and time. Water molecules came from within Earth over billions of years, in a process called outgassing, and began endless cycling through the hydrologic system of evaporation-condensation-precipitation. Water is the most common compound on the surface of Earth, and, besides its solvency, it possesses great stability and unusual heat characteristics. The present salinity of the sea is a result of water's great dissolving ability.

Part of Earth's uniqueness is that water naturally exists in all three states—solid, liquid, and gas. The vast hydrologic cycle is a valuable model used for the study of Earth's waters. The observation and description of the hydrologic cycle and related energy budgets are a central focus in climate-change studies.

In many ways water vapor is the most important gas in the atmosphere. The transfer of massive amounts of energy, powering the atmospheric and oceanic circulations and the weather, is accomplished through the absorption and release of latent heat. The measurement of water vapor in the air is expressed as humidity and relative humidity. Saturated air and the presence of condensation nuclei create conditions that produce condensation and eventual cloud formation, which occurs in a variety of classes and types. Fog is a cloud that occurs at the surface and can prove hazardous to human activities.

The inherent power of a cloud is a constant reminder that clouds constitute a major heat exchange system in the environment. Clouds are the subject of much scientific inquiry related to their effect on net radiation patterns.

KEY TERMS

absolute humidity
advection fog
brackish
brine
cirrus
collision-coalescence process
condensation nuclei
cumulonimbus
cumulus
deep cold zone
dew-point temperature
eustasy
evaporation fog
fog
glaciation
glacio-eustatic
hair hygrometer
humidity
hydrologic cycle
hygroscopic nuclei
ice-crystal process
infiltration
interception
latent heat
latent heat of condensation
latent heat of vaporization
mixing zone
moisture droplets
nimbostratus
outgassing
percolation
phase change
radiation fog
relative humidity
salinity
saturated
sling psychrometer
specific humidity
stratocumulus
stratus
sublimation
supercooled droplets
thermocline transition zone
upslope fog
valley fog
vapor pressure

REVIEW QUESTIONS

1. Approximately where and when did Earth's water originate?
2. If the quantity of water on Earth has been quite constant in volume for at least 2 billion years, how can sea level have fluctuated? Explain.
3. Describe the locations of Earth's water, both oceanic and fresh. What is the largest repository of fresh water at this time? In what ways is this significant to modern society?
4. Evaluate the salinity of seawater: its composition, amount, and distribution.
5. Analyze the latitudinal distribution of salinity shown in Figure 7-3. Why is salinity lower along the equator and higher in the subtropics?
6. What are the three general zones relative to physical structure within the ocean? Characterize each in terms of temperature, salinity, oxygen, and carbon dioxide.
7. Describe the three states of matter as they apply to ice, water, and water vapor.
8. What happens to the physical structure of water as it cools below 4°C (39°F)? What are some of the visible indications of these physical changes?
9. What is latent heat? How is it involved in the phase changes of water? What amounts of energy are involved?

10. Explain the simplified model of the complex flows of water on Earth—the hydrologic cycle.
11. What are the possible routes that a raindrop may take on its way to and into the soil surface?
12. Where does the bulk of evaporation and precipitation on Earth take place?
13. Define relative humidity. What does the concept represent? What is meant by saturation and dew-point temperature?
14. Using several different measures of humidity in the air, derive relative humidity values (vapor pressure/saturation vapor pressure; specific humidity/maximum specific humidity).
15. How do the two instruments described in this chapter measure relative humidity?
16. Evaluate the latitudinal distribution of relative humidity as shown in Figure 7-15.
17. How does the latitudinal distribution of relative humidity compare with the actual water vapor content of air?
18. Specifically, what is a cloud? Describe the droplets that form a cloud.
19. Explain the condensation process: what are the necessary requirements? What two principal processes are discussed in this chapter?
20. What are the basic forms of clouds? Using Table 7-4, describe how the basic cloud forms vary with altitude.
21. What type of cloud is fog? List and define the principal types of fog.
22. Why might you describe Earth as the water planet? Explain.

SUGGESTED READINGS

Allen, Oliver E., and the editors of Time-Life Books. *Atmosphere.* Planet Earth Series. Alexandria, VA: Time-Life Books, 1983.

Buswell, Arthur M., and Worth H. Rodebush. "Water," *Scientific American* 194 (April 1956): 76–80.

Chahine, Moustafa T. "The Hydrologic Cycle and Its Influence on Climate," *Nature* 359 (October 1, 1992): 373–80.

Hidore, John J., and John E. Oliver. *Climatology—An Atmospheric Science.* New York: Macmillan Publishing Company, 1993.

Knight, Charles, and Nancy Knight. "Snow Crystals," *Scientific American* 228 (January 1973): 100–107.

Leopold, Luna B. *Water—A Primer.* San Francisco: W. H. Freeman, 1974.

Leopold, Luna B., Kenneth S. Davis, and the editors of Time-Life Books. *Water.* Life Science Library. New York: Time-Life Books, 1966.

Miller, Albert, Jack C. Thompson, Richard E. Peterson, and Donald R. Haragan. *Elements of Meteorology.* Columbus, OH: Macmillan Publishing Company, 1983.

Moran, Joseph M., and Michael D. Morgan. *Meteorology—The Atmosphere and the Science of Weather,* 4th ed. New York: Macmillan Publishing Company, 1994.

Neiburger, Morris, James G. Edinger, and William D. Bonner. *Understanding Our Atmospheric Environment.* San Francisco: W. H. Freeman, 1973.

Pringle, Laurence, and the editors of Time-Life Books. *Rivers and Lakes.* Planet Earth Series. Alexandria, VA: Time-Life Books, 1985.

Rind, D., C. Rosenweig, and R. Goldberg. "Modelling the Hydrologic Cycle in Assessments of Climate Change," *Nature* 253 (July 9, 1992): 119–22.

Runnels, L. K. "Ice," *Scientific American* 215 (December 1966): 118–24.

Scorer, Richard, and Arjen Verkaik. *Spacious Skies*. London: David & Charles, 1989.

Scorer, Richard, and Harry Wexler. *Cloud Studies in Colour*. London: Pergamon Press, 1967.

van der Leeden, Frits, Fred L. Troise, and D. K. Todd. *The Water Encyclopedia,* 2nd ed. Chelsea, MI: Lewis Publishers, Inc., 1990.

Whipple, A. B. C., and the editors of Time-Life Books. *Restless Oceans*. Alexandria, VA: Time-Life Books, 1983.

Lightning over the Grand Canyon. [*Photo by Dick Dietrich.*]

8

WEATHER

Every day a vast drama plays out on Earth's stage, involving the stability or instability of interacting air mass actors, and featuring powerful dialogue and beautiful special effects in the lower atmosphere. Large air masses come into conflict, moving and shifting to dominate different regions in response to the controls of insolation, latitude, patterns of atmospheric temperature and pressure, the arrangement of land, water, and mountains, and migrating cyclonic systems that cross the middle latitudes. Tropical cyclones travel northward and southward from equatorial latitudes, hitting susceptible coastal lowlands as hurricanes and typhoons. This was dramatically demonstrated in 1992 by Hurricane Andrew in Florida and Louisiana, Hurricane Iniki in Hawaii, and Typhoon Omar in Guam.

Temperature, air pressure, relative humidity, wind speed and direction, daylength, and Sun angle are important measurable elements that contribute to the weather. **Weather** is the short-term condition of the atmosphere, as compared to *climate,* which reflects long-term atmospheric conditions and extremes. We tune to a local station for the day's weather report from the National Weather Service (in the United States) or Atmospheric Environment Service (in Canada) to see the current satellite images and to hear tomorrow's forecast.

Meteorology is the scientific study of the atmosphere (*meteor* means "heavenly" or "of the atmosphere"). Embodied within this science is a study of the atmosphere's physical characteristics and motions, related chemical, physical, and geological processes, the complex linkages of atmospheric systems, and weather forecasting. Computers permit the handling of enormous amounts of data on water vapor, clouds, precipitation, and radiation for accurate forecasting of near-term weather. New developments in supercomputers and ground- and space-based observation systems are making the 1990s a time of dramatic developments. The spatial implications of these atmospheric sciences and their relationship to human activities strongly link meteorology to physical geography.

We begin our study of weather with a discussion of atmospheric stability and instability. We will follow huge air masses across North America, observe powerful lifting mechanisms in the atmosphere, examine cyclonic systems, and conclude with a portrait of the violent and dramatic weather that occurs in the atmosphere.

ATMOSPHERIC STABILITY

Stability refers to the tendency of a parcel of air either to remain as it is or to change its initial position by ascending or descending. An air parcel is termed *stable* if it resists displacement upward or, when disturbed, it tends to return to its starting place. On the other hand, an air parcel is considered *unstable* when it continues to rise until it reaches an altitude where the surrounding air has a density similar to its own. This difference between an air parcel and the surrounding environment produces a buoyancy that contributes to further lifting. Determining the degree of stability or instability involves measuring simple temperature relationships between the air parcel and the surrounding air. Such temperature measurements are made daily with balloon soundings (instrument packages called radiosondes) at thousands of weather stations.

Adiabatic Processes

The normal lapse rate, as introduced in Chapter 3, is the average decrease in temperature with increasing altitude, a value of 6.4C° per 1000 m (3.5F° per 1000 ft). This rate of temperature change is for still, calm air, but it can differ greatly under varying weather conditions, and so the actual lapse rate at a particular place and time is labeled the *environmental lapse rate.* In contrast, an ascending parcel of air tends to cool by expansion, responding to the reduced pressure at higher altitudes. Descending air tends to heat by compression.

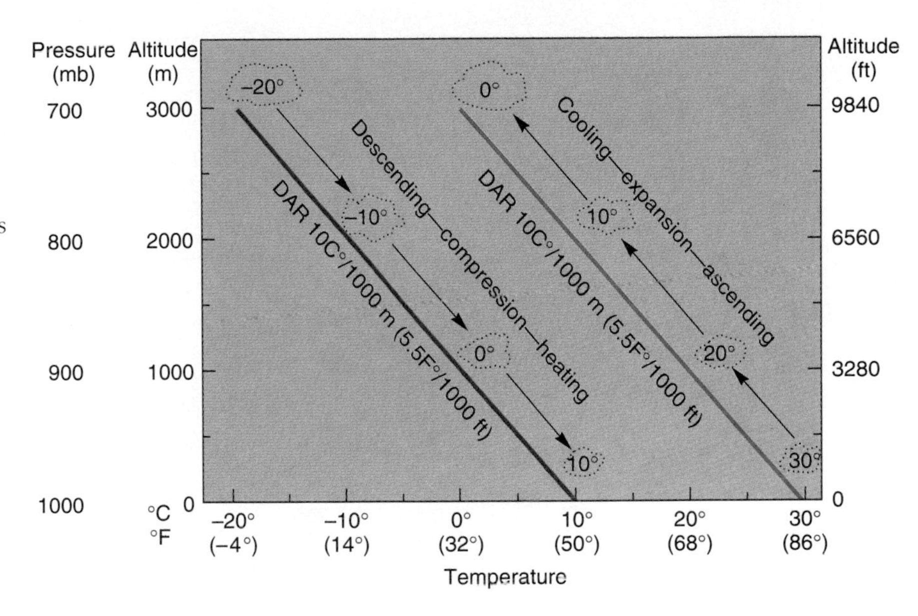

FIGURE 8-1
Vertically moving air parcels that are not saturated expand when they rise and are compressed when they descend. An ascending parcel is cooled by expansion; a descending parcel is heated by compression. Temperature change in both expansion and compression occurs at the dry adiabatic rate (DAR)—10C° per 1000 m or 1C° per 100 m (5.5F° per 1000 ft)—because the parcels are not saturated.

Both ascending and descending temperature changes are assumed to occur without any heat exchange between the surrounding environment and the vertically moving parcel of air. The warming and cooling rates for a parcel of expanding or compressing air are termed **adiabatic.** *Diabatic* means occurring *with* an exchange of heat; *adiabatic* means occurring *without* a loss or gain of heat.

Adiabatic rates are measured with one of two dynamic rates, depending on moisture conditions in the parcel: *dry* adiabatic rate (DAR) and *moist* adaibatic rate (MAR).

Dry Adiabatic Rate (DAR). The **dry adiabatic rate** is the rate at which "dry" air cools by expansion (if ascending) or heats by compression (if descending). "Dry" air is less than saturated, with a relative humidity less than 100%. The DAR is 10C° per 1000 m (5.5F° per 1000 ft), as illustrated in Figure 8-1. For example, consider an unsaturated parcel of air at the surface, whose temperature measures 27°C (80°F). It is lifted, expands, and cools at the DAR as it rises from the ground to 2500 m (approximately 8000 ft). What happens to the temperature of the parcel? Its temperature at that altitude is 2°C (36°F), cooling by expansion as it is lifted.

$$10\text{C}^\circ \text{ per } 1000 \text{ m} \times 2500 \text{ m} = 25\text{C}^\circ \text{ of total cooling}$$
$$(5.5\text{F}^\circ \text{ per } 1000 \text{ ft} \times 8000 \text{ ft} = 44\text{F}^\circ)$$

Subtracting the 25C° of cooling from the starting temperature of 27°C gives the temperature at 2500 m of 2°C.

Moist Adiabatic Rate (MAR). The **moist adiabatic rate** is the average rate at which ascending air that is moist (saturated) cools by expansion. The *average* MAR is 6C° per 1000 m (3.3F° per 1000 ft), or roughly 4C° (2F°) less than the DAR. However, the MAR varies with moisture content and temperature, and can range from 4C° to 10C° per 1000 m (2F° to 6F° per 1000 ft). The reason is that, in a saturated air parcel, latent heat of condensation is liberated as sensible heat, which reduces the adiabatic rate of cooling. The release of latent heat may vary, which affects the MAR. The MAR is much lower than the DAR in warm air, whereas the two rates are more similar in cold air.

Stable and Unstable Atmospheric Conditions

The relationship among the dry adiabatic rate (DAR), moist adiabatic rate (MAR), and the environmental (actual) lapse rate is complex and determines the stability of the atmosphere over an area.

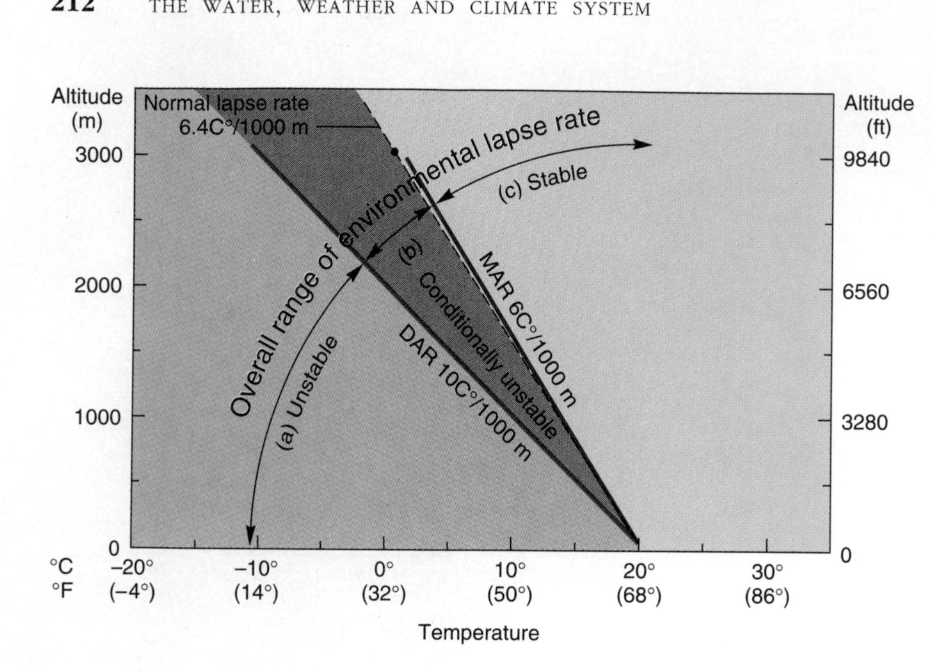

FIGURE 8-2
The relationship between dry and moist adiabatic rates and environmental lapse rates produces three conditions: (a) unstable atmosphere—environmental lapse rate exceeds the DAR; (b) conditionally unstable atmosphere—environmental lapse rate is between the DAR and MAR; (c) stable atmosphere—environmental lapse rate is less than DAR and MAR.

You can see the range of possible relationships in Figure 8-2.

Let's apply this concept to a specific situation. Figure 8-3 illustrates three temperature relationships in the atmosphere that lead to three different conditions: unstable, conditionally unstable, and stable. For the sake of illustration, all three examples begin with a parcel of air at the surface at 25°C (77°F). In each of the three examples, examine the temperature relationships between the parcel of air and the surrounding environment. Assume that a lifting mechanism is present (we will examine lifting mechanisms shortly).

Under the unstable conditions in Figure 8-3a, the air parcel continues to rise through the atmosphere because it is warmer than the surrounding environment, which is cooler by an environmental lapse rate of 12C° per 1000 m (6.6F° per 1000 ft). By 1000 m (3280 ft), the parcel has adiabatically cooled by expansion to 15°C (59°F), but the surrounding air is at 13°C (55.4°F). So, the temperature in the parcel is 2C° (3.6F°) warmer than the surrounding air.

On the other hand, the stable conditions in Figure 8-3c are caused by an environmental lapse rate of 5C° per 1000 m (3F° per 1000 ft), which is less than both the DAR and the MAR, thus forcing the parcel of air to settle back to its original position because it is cooler (denser) than the surrounding environment.

In Figure 8-3b, the environmental lapse rate is between the DAR and the MAR at 7C° per 1000 m. Under these conditions, a parcel resists upward movement if it is less than saturated. However, if it is saturated and cools at the MAR, it eventually becomes unstable and continues to rise.

With these stability relationships in mind, let's examine specific air masses—each with their own specific humidity, temperature, and stability characteristics.

AIR MASSES

Each area of Earth's surface imparts its varying characteristics to the air it touches. Such a distinctive body of air is called an **air mass,** and it initially reflects the characteristics of its *source region*. The longer an air mass remains stationary over a region, the more definite its physical attributes become. Within each air mass there is a homogeneity of temperature and humidity that sometimes extends through the lower half of the troposphere.

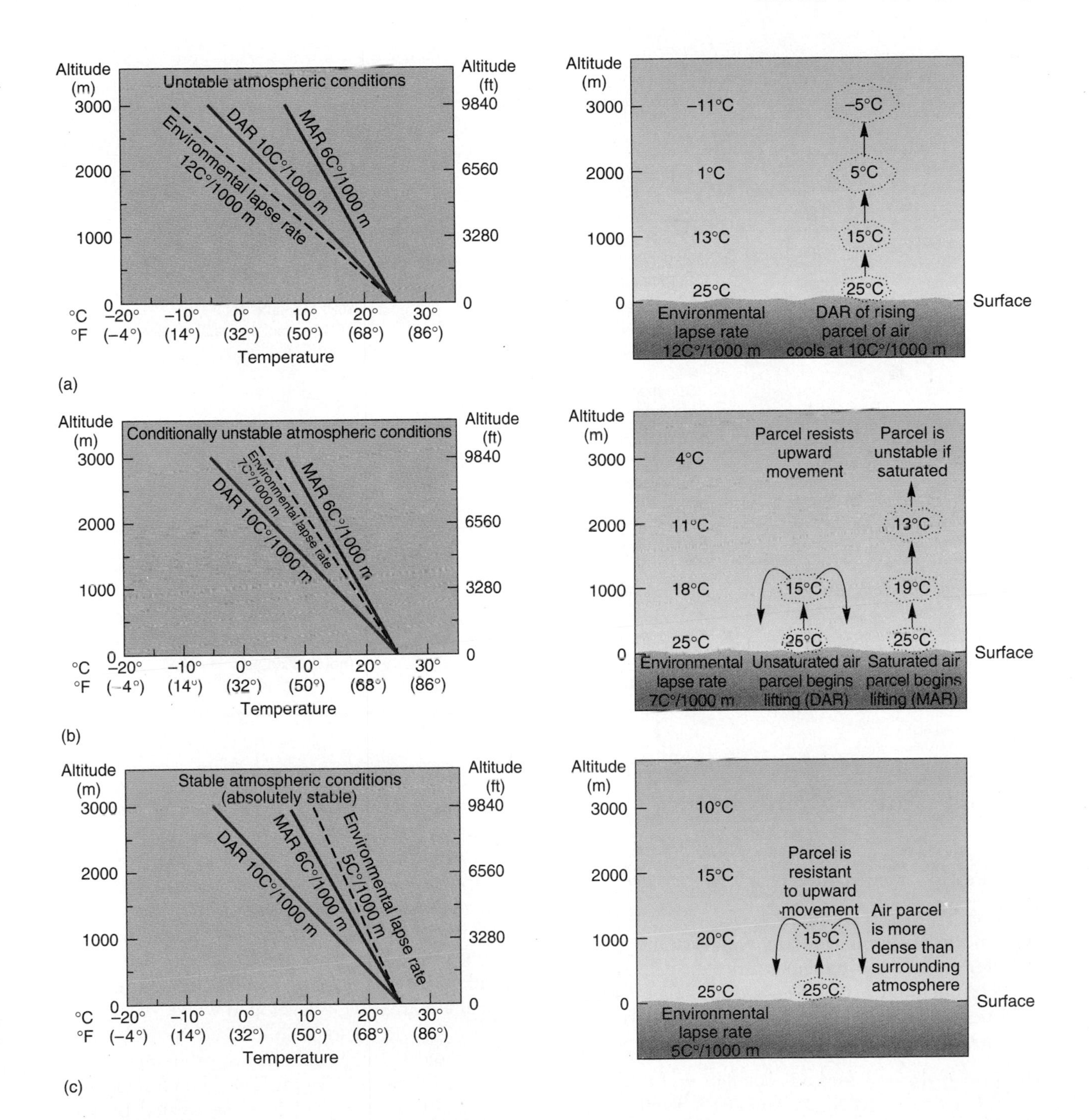

FIGURE 8-3
Specific examples of (a) unstable, (b) conditionally unstable, and (c) stable conditions in the lower atmosphere.

TABLE 8-1
Principal Air Masses

Latitudinal Source Region	Moisture Source Region: c: Continental	Moisture Source Region: m: Marine
A: Arctic, 60–90° N	cA: Continental Arctic *Winter:* very cold, very dry, stable (avg. SH = 0.1 g/kg) *Summer:* cold, dry	(not applicable)
P: Polar, 40–60° N or S	cP: Continental polar *Winter:* (N.H. only) cold, dry, stable, with high pressure (avg. SH = 1.4 g/kg) *Summer:* warm to cool, dry, moderately stable	mP: Maritime polar Cool, moist, unstable all year (avg. SH = 4.4 g/kg)
T: Tropical 15–35° N or S	cT: Continental tropical Hot, very dry, stable aloft, unstable at surface, turbulent in summer (avg. SH = 10 g/kg	mT: Maritime tropical mT Gulf/Atlantic *Winter:* warm, wet, unstable *Summer:* warm, wet, very unstable (avg. SH = 17 g/kg) mT Pacific Weak, conditionally unstable
E: Equatorial, 15° N to 15° S	(not applicable)	mE: Maritime equatorial Unstable, very warm, very wet, winterless (avg. SH = 19 g/kg)
AA: Antarctic, 60–90° S	cAA: Continental Antarctic Very cold, very dry, very stable	(not applicable)

SH = specific humidity, g/kg

Air Masses Affecting North America

Air masses generally are classified according to the moisture and temperature characteristics of their source regions:

1. **Moisture—designated m for maritime (wetter) and c for continental (drier).**
2. **Temperature (latitude)—designated A (arctic), P (polar), T (tropical), E (equatorial), and AA (antarctic).**

The principal air masses are detailed in Table 8-1, and those that affect North America in winter and summer are mapped in Figure 8-4.

Continental polar (cP) air masses form only in the Northern Hemisphere and are most developed in winter. The Southern Hemisphere lacks the necessary continental masses (continentality) at high latitudes to produce continental polar characteristics.

Marine polar (mP) air masses in the Northern Hemisphere are northwest and northeast of the North American continent, and within them cool, moist, unstable conditions prevail throughout the year. The Aleutian and Icelandic subpolar low-pressure cells reside within these mP air masses, especially in their well-developed winter state.

North America also is influenced by two maritime tropical (mT) air masses—the *mT Gulf/Atlantic* and the *mT Pacific*. The humidity experienced in the East and Midwest is created by the mT Gulf/Atlantic air mass, which is particularly unstable and active from late spring to early fall. In contrast, the mT Pacific is stable to conditionally unstable and generally much lower in moisture content and available energy. As a result, the west-

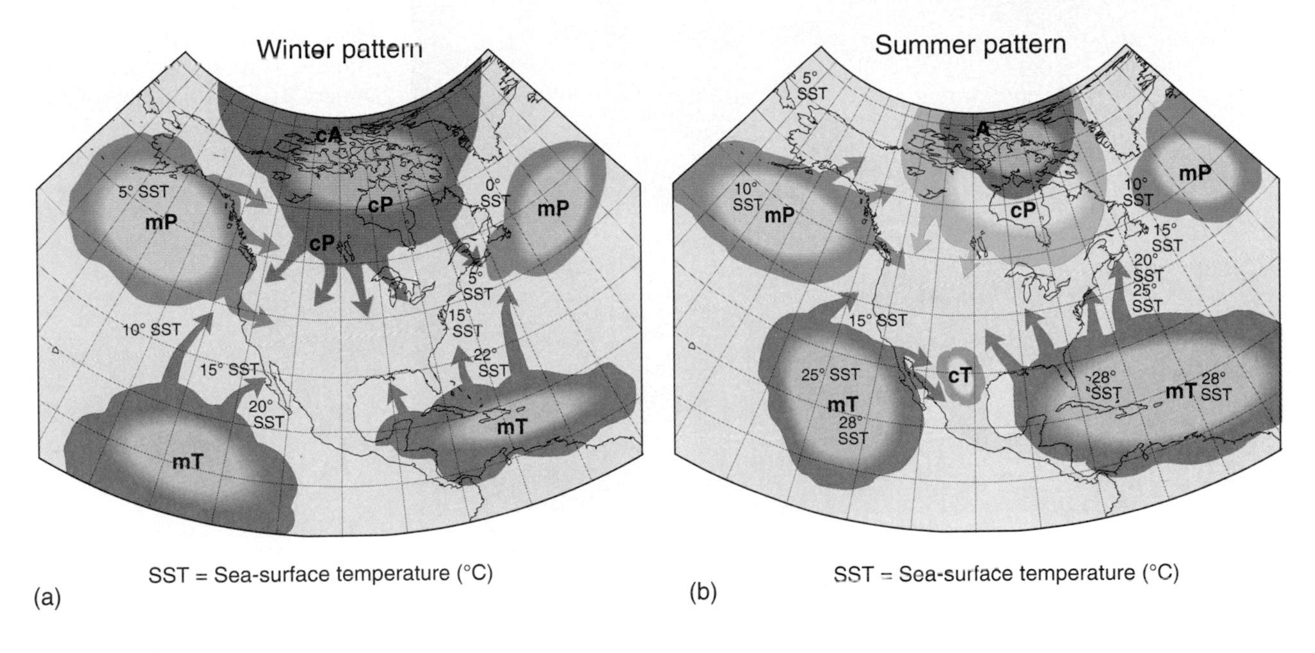

FIGURE 8-4
Air masses that influence North America and their characteristics in both (a) winter and (b) summer patterns. Temperatures shown are sea-surface temperatures (SSTs).

ern United States, influenced by this weaker Pacific air mass, receives lower average precipitation than the rest of the country.

Please refer to Figure 6-12 and the discussion of subtropical high-pressure cells off the coast of North America—the moist, unstable conditions on the western edge (east coast), and the drier, stable conditions on the eastern edge (west coast). These conditions, coupled respectively with warmer and cooler ocean currents, produce the characteristics of each source region.

Air Mass Modification

As air masses migrate from their source regions, their temperature and moisture characteristics are modified. For example, an mT Gulf/Atlantic air mass may carry humidity to Chicago and on to Winnipeg, but gradually will lose its initial characteristics with each day's passage northward. Similarly, temperatures below freezing occasionally reach into southern Texas and Florida, influenced by an invading winter cP air mass from the north. However, that air mass will have warmed from the –50°C (–58°F) of its source region in central Canada. In winter, as a cP air mass moves southward over warmer land, it is modified, especially warming after it crosses the snowline. If it passes over the warmer Great Lakes, it will absorb heat and moisture and produce heavy *lake-effect* snowfall downwind as far east as New York, Pennsylvania, New England, and the Maritime Provinces of Canada. These lake-effect snow conditions are most pronounced in fall and early winter.

On the opposite side of the globe, winter monsoon winds bring cold, dry air from the interior of Siberia over the surrounding lands and oceans. The air mass is modified as it passes southward and eastward. Figure 8-5 shows a satellite image of such an air mass as it moves out over the Sea of Japan and the China Sea. Over warmer water the moisture content of the mass is increased, and its clouds form elongated bands. The result is a cool, moist cP air mass that is humidified, warmed, and destabilized. It has thus become a **secondary air mass** to the initial cA or cP mass that left Siberia.

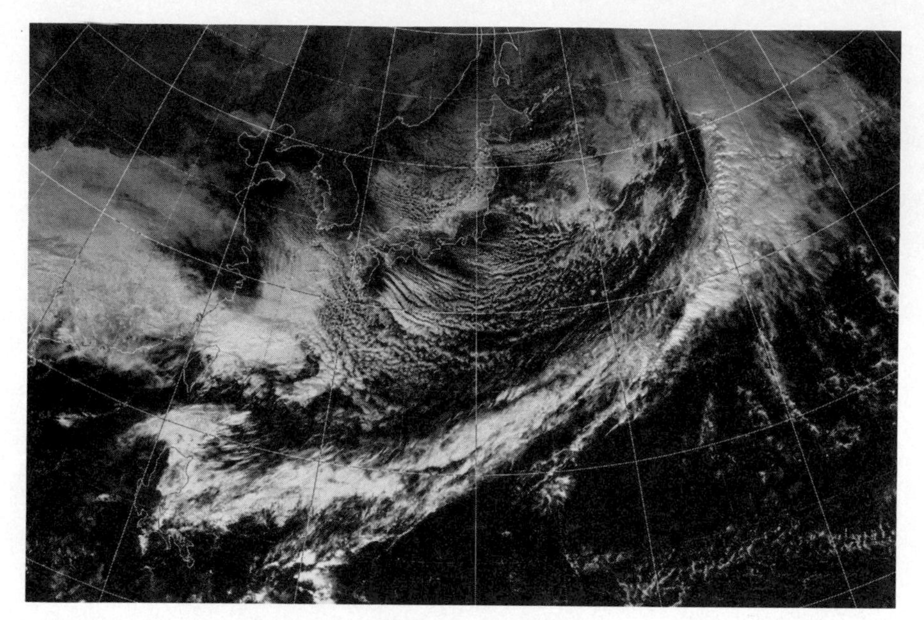

FIGURE 8-5
Air mass modification over the Sea of Japan and the North Pacific, as evidenced by cloud patterns. [GMS satellite image for January 24, 1988, courtesy of the Japan Meteorological Association.]

ATMOSPHERIC LIFTING MECHANISMS

If air masses are to cool adiabatically (by expansion), and if they are to reach the dew-point temperature and saturate, condense, form clouds, and perhaps precipitate, then they must be lifted. Three principal lifting mechanisms operate in the atmosphere: convectional lifting (stimulated by local surface heating), orographic lifting (when air is forced over a barrier such as a mountain range), and frontal lifting (along the leading edges of contrasting air masses). Descriptions of all three mechanisms follow.

Convectional Lifting

When an air mass passes from a maritime source region to a warmer continental region, heating from the warmer land causes lifting in the air mass. If conditions are unstable, initial lifting is sustained and clouds develop. Figure 8-6 illustrates convectional action stimulated by local heating, with unstable conditions present in the atmosphere. The rising parcel of air continues its ascent because it is warmer than the surrounding environment.

The MAR is used above the *lifting condensation level,* visible at the elevation of the cloud base

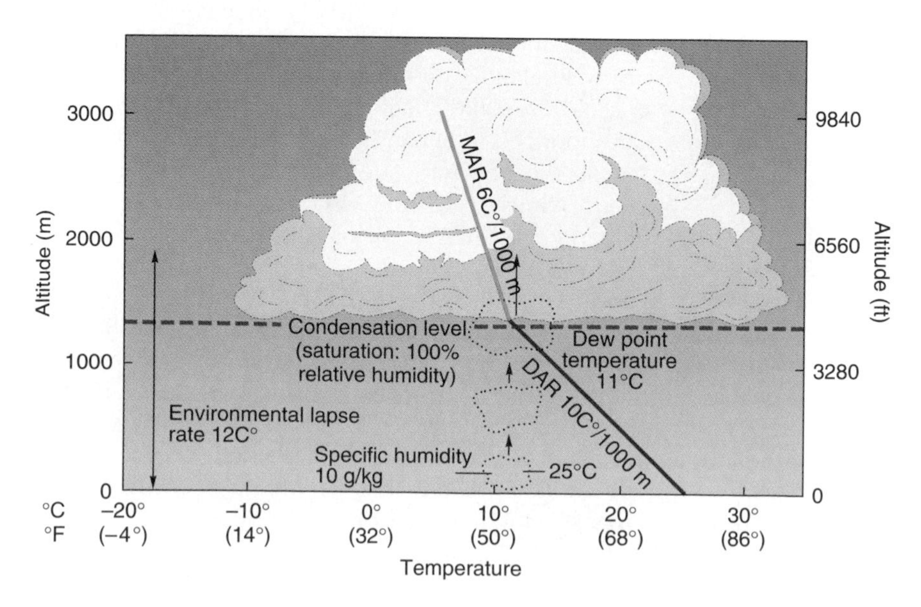

FIGURE 8-6
Local heating and convection under unstable atmospheric conditions. The air parcel has a specific humidity of 10 g/kg and thus an initial dew-point temperature of 11°C (51.8°F). The environmental lapse rate is 12C° per 1000 m (7F° per 1000 ft) so that the parcel is warmer and continues to rise. At 1400 m (4590 ft) the parcel reaches the dew point and becomes saturated.

FIGURE 8-7
Convectional activity over Florida. Cumulus clouds outline the land, with several cells developing into cumulonimbus thunderheads. [Project Gemini photo from NASA.]

where the rising air mass becomes saturated. Buoyancy is added at this level through the release of latent heat. Continued lifting produces cooling of the air temperature and the dew-point temperature at the MAR.

As an example of local heating and convectional lifting, Figure 8-7 depicts a day on which the landmass of Florida was warmer than the surrounding Gulf of Mexico and Atlantic Ocean. Because the land gradually heats throughout the day, convectional showers tend to form in the afternoon and early evening. The land becomes covered by clouds formed by this process. Towering cumulonimbus clouds are an expected summertime feature in the regions of North America affected by mT Gulf/Atlantic air masses, and to a lesser extent by the weaker mT Pacific air mass. Convectional precipitation also dominates along the intertropical convergence zone (ITCZ), over tropical islands, and anywhere that moist, unstable air is heated from below, or where the inflowing trade winds produce a dynamic convergence.

Orographic Lifting

The physical presence of a mountain acts as a topographic barrier to migrating air masses. **Orographic lifting** occurs when air is lifted upslope as it is pushed against a mountain and cools adiabatically. Stable air that is forced upward in this manner may produce stratiform clouds, whereas unstable or conditionally unstable air usually forms a line of cumulus and cumulonimbus clouds. An orographic barrier enhances convectional activity and causes additional lifting during the passage of weather fronts and cyclonic systems, thereby extracting more moisture from passing air masses. Figures 8-8 and 8-9 illustrate the operation of orographic lifting.

The wetter intercepting slope is termed the **windward slope,** as opposed to the drier far-side slope, known as the **leeward slope.** Moisture is condensed from the lifting air mass on the windward side of the mountain; on the leeward side the descending air mass is heated by compression, causing evaporation of any remaining water in the air. Thus, air can begin its ascent up a mountain warm and moist but finish its descent on the leeward slope hot and dry. In North America, **chinook winds** (called föhn or foehn winds in Europe) are the warm, downslope air flows characteristic of the leeward side of mountains. Such winds can bring a 20C° (36F°) jump in temperature and greatly reduced relative humidity.

The term **rain shadow** is applied to dry regions leeward of mountains. East of the Cascade Range,

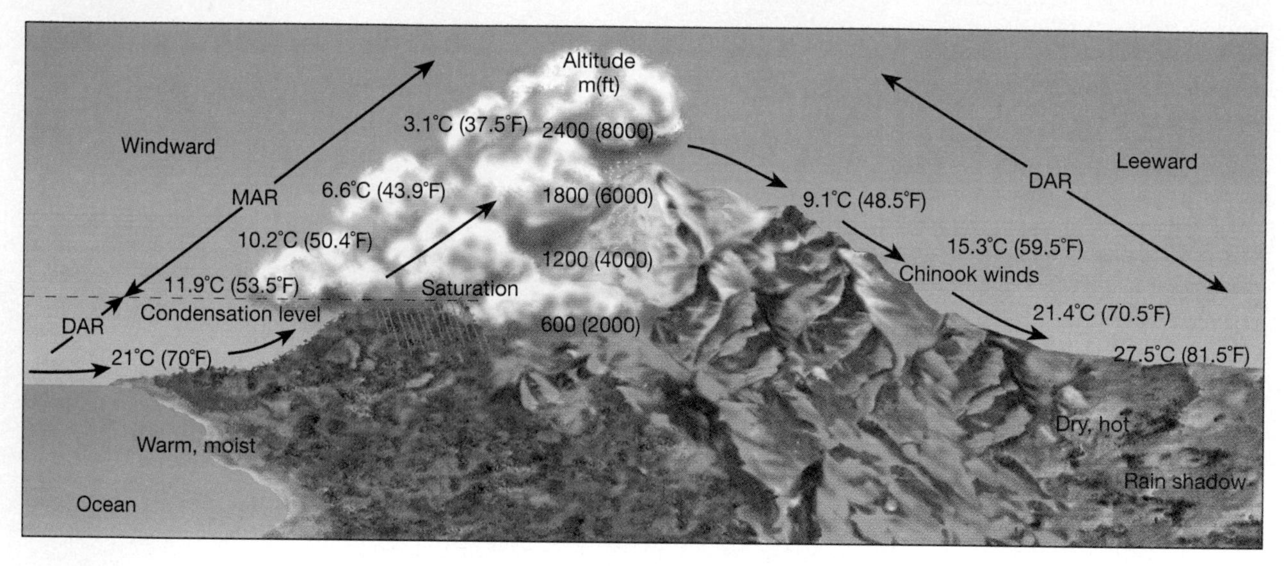

FIGURE 8-8
Orographic barrier and precipitation patterns—unstable conditions assumed. Warm, moist air is forced upward against a mountain range, producing adiabatic cooling and eventual saturation. On the leeward slopes, compressional heating as the air travels downward leads to hot, relatively dry air in the rain shadow of the mountain.

FIGURE 8-9
Orographic lifting along the Pacific Coast. [Photo by author.]

Sierra Nevada, and Rocky Mountains, such rain-shadow patterns predominate. In fact, the precipitation pattern of windward and leeward slopes persists worldwide, as confirmed by the precipitation maps for North America (Figure 9-2) and the world (Figure 10-2).

Orographic Precipitation World Records. Orographic precipitation is limited in areal extent because it can occur only where a topographic barrier exists, thus making it the least dominant lifting mechanism in terms of total worldwide precipitation. Nonetheless, because mountains are fixed features on the landscape, they are the most consistent precipitation-inducing mechanism. Thus, the greatest average annual precipitation and maximum annual precipitation on Earth occur on the windward slopes of mountains that intercept tropical trade winds.

The world's greatest average annual precipitation occurs in the United States on the windward slopes of Mount Waialeale, on the island of Kauai, Hawaii, which rises 1569 m (5147 ft) above sea level. It averages 1234 cm (486 in. or 40.5 ft) a year (records for 1941–present). In contrast, the rain-shadow side of Kauai receives only 50 cm (20 in.) of rain annually. If no islands existed at this location, this portion of the Pacific Ocean would receive only an average 63.5 cm (25 in.) of precipitation a year. Kauai is pictured in the opening photo for Chapter 10.

Another place receiving world-record precipitation is Cherrapunji, India, 1313 m (4309 ft) above sea level at 25° N latitude, in the Assam Hills south of the Himalayas. It is located just north of Bangladesh, where tropical cyclones, torrential rains, and flooding in 1970 and 1991 killed hundreds of thousands of people. Because of the summer monsoons that pour in from the Indian Ocean and the Bay of Bengal, Cherrapunji has received 930 cm (366 in. or 30.5 ft) of rainfall in one month and a total of 2647 cm (1042 in. or 86.8 ft) in one year. Not surprisingly, Cherrapunji is the all-time precipitation record holder for a single year and for every other time interval from 15 days to 2 years. The average annual precipitation there is 1143 cm (450 in.), placing it second only to Mount Waialeale.

Frontal Lifting

The leading edge of an advancing air mass is called its **front.** That term was applied by Vilhelm Bjerknes (1862–1951) and a team of meteorologists working in Norway during World War I, because it seemed to them that migrating air-mass "armies" were doing battle along their fronts. A front is a place of atmospheric discontinuity, of considerable difference in temperature, pressure, humidity, wind direction and speed, and cloud development. The leading edge of a cold air mass is a **cold front,** whereas the leading edge of a warm air mass is a **warm front.**

Cold Front. On weather maps, such as those illustrated in Figure 8-15, a cold front is identified by a line marked with triangular spikes pointing in the direction of frontal movement along an advancing cP air mass in winter. The steep face of the cold air mass suggests its ground-hugging nature, caused by its density and unified physical character (Figure 8-10).

Warmer, moist air in advance of the cold front is lifted upward abruptly and is subjected to the same adiabatic rates and factors of stability/instability that pertain to all lifting air parcels. A day or two ahead of the cold front's passage, high cirrus clouds appear, telling observers that a lifting mechanism is on the way. The cold front's advance is marked by a wind shift, temperature drop, and lowering barometric pressure. Air pressure reaches a local low as the line of most intense lifting passes, usually just ahead of the front itself. Clouds may build up along the cold front into characteristic cumulonimbus types and may appear as an advancing wall of clouds. Precipitation usually is hard, containing large droplets, and can be accompanied by hail, lightning, and thunder. The aftermath usually brings northerly winds (in the Northern Hemisphere, southerly winds in the Southern Hemisphere), lower temperatures, increasing air pressure in the cooler, denser air, and broken cloud cover.

The particular shape and size of the North American landmass and its latitudinal position represent conditions where cP and mT air masses are best developed and have the most direct access to

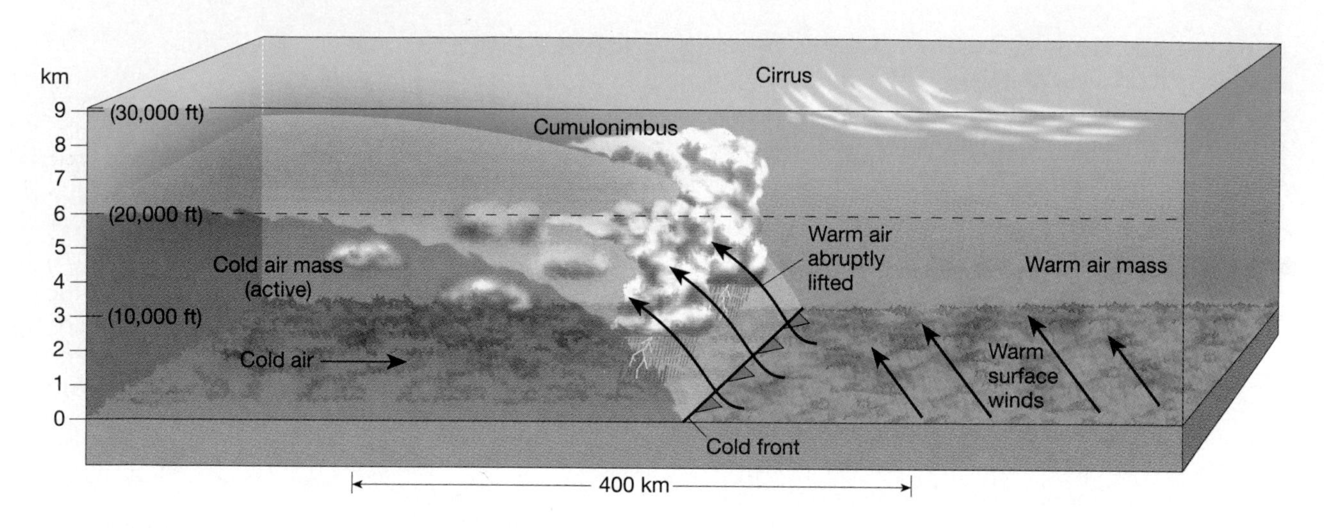

FIGURE 8-10
A typical cold front. Warm moist air is forced to lift abruptly by denser advancing cold air. As the air lifts, it cools by expansion at the DAR, cooling to the dew-point temperature as it rises to a level of active condensation and cloud formation. Cumulonimbus clouds may produce large raindrops, heavy showers, lightning and thunder, and hail.

each other. The resulting contrast can lead to dramatic weather, particularly in lat spring, with sizable temperature differences from one side of a cold front to the other. A fast-advancing cold front can cause violent lifting and create a zone slightly ahead of the front called a **squall line.** Along a squall line such as the one in the Gulf of Mexico shown in Figure 8-11, wind patterns are turbulent and wildly changing and precipitation is intense. The well-defined wall of clouds rises abruptly to 16,800 m (55,110 ft), with new thunderstorms forming along the front. Tornadoes also may develop along such a squall line.

FIGURE 8-11
Squall line in the Gulf of Mexico with clouds rising to 16,800 m (55,120 ft). The passage of such a frontal system over land often produces strong winds and possibly tornadoes. [Space Shuttle photo from NASA.]

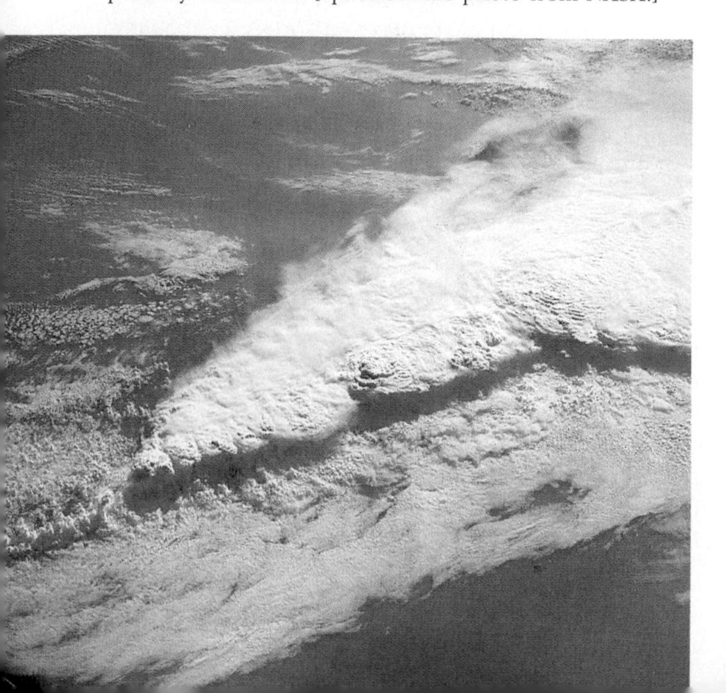

Warm Front. A warm front is denoted on weather maps by a line marked with semicircles facing the direction of frontal movement (Figure 8-15). The leading edge of an advancing warm air mass is unable to displace cooler, passive air, which is more dense. Instead, the warm air tends to push the cooler, underlying air into a characteristic wedge shape, with the warmer air sliding up over the cooler air. Thus, in the cooler-air region, a temperature inversion is present, sometimes causing poor air drainage and stagnation.

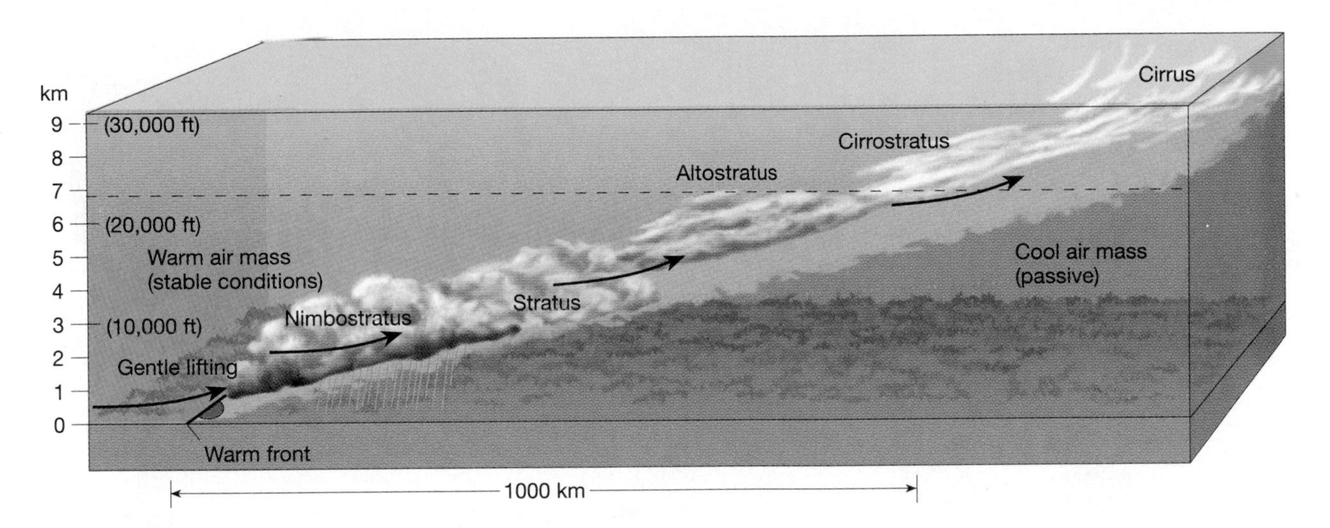

FIGURE 8-12
A typical warm front. Note the sequence of cloud development as the warm front approaches. Warm air slides upward over a wedge of cooler, passive air near the ground. Gentle lifting of the warm, moist air produces stratus and nimbostratus clouds and drizzly rain showers in contrast to the more dramatic precipitation associated with the passage of a cold front.

Figure 8-12 illustrates a typical warm front in which mT air is gently lifted, leading to stratiform cloud development and characteristic nimbostratus clouds and drizzly precipitation. Associated with the warm front are the high cirrus and cirrostratus clouds that announce the advancing frontal system. Closer to the front, the clouds lower and thicken to altostratus, and then to stratus within several hundred kilometers of the front.

MIDLATITUDE CYCLONIC SYSTEMS

The conflict between contrasting air masses along fronts leads to conditions appropriate for the development of a midlatitude **wave cyclone,** a migrating center of low pressure, with converging, ascending air, spiraling inward counterclockwise in the Northern Hemisphere (or converging clockwise in the Southern Hemisphere). This cyclonic motion is generated by the pressure gradient force, the Coriolis force, and surface friction, discussed in Chapter 6.

Wave cyclones form a dominant type of weather pattern in the middle and higher latitudes of both the Northern and Southern hemispheres and act as a catalyst for air-mass conflict. Such a *midlatitude cyclone,* or extratropical cyclone, can be born along the polar front, particularly in the region of the Icelandic and Aleutian subpolar low-pressure cells in the Northern Hemisphere (see Figures 6-15 and 6-17). Development and strengthening of a midlatitude wave cyclone is known as *cyclogenesis.* In addition to the polar front, certain other areas are associated with wave cyclone development and intensification: the eastern slope of the Rockies and other north-south mountain barriers, the Gulf Coast, and the east coasts of North America and Asia. Bjerknes, mentioned earlier, first proposed the polar front theory of cyclonic development.

Figure 8-13 shows the birth, maturity, and death of a typical midlatitude cyclone. On the average, a midlatitude cyclone takes 3–10 days to progress through these stages and cross the North American continent.

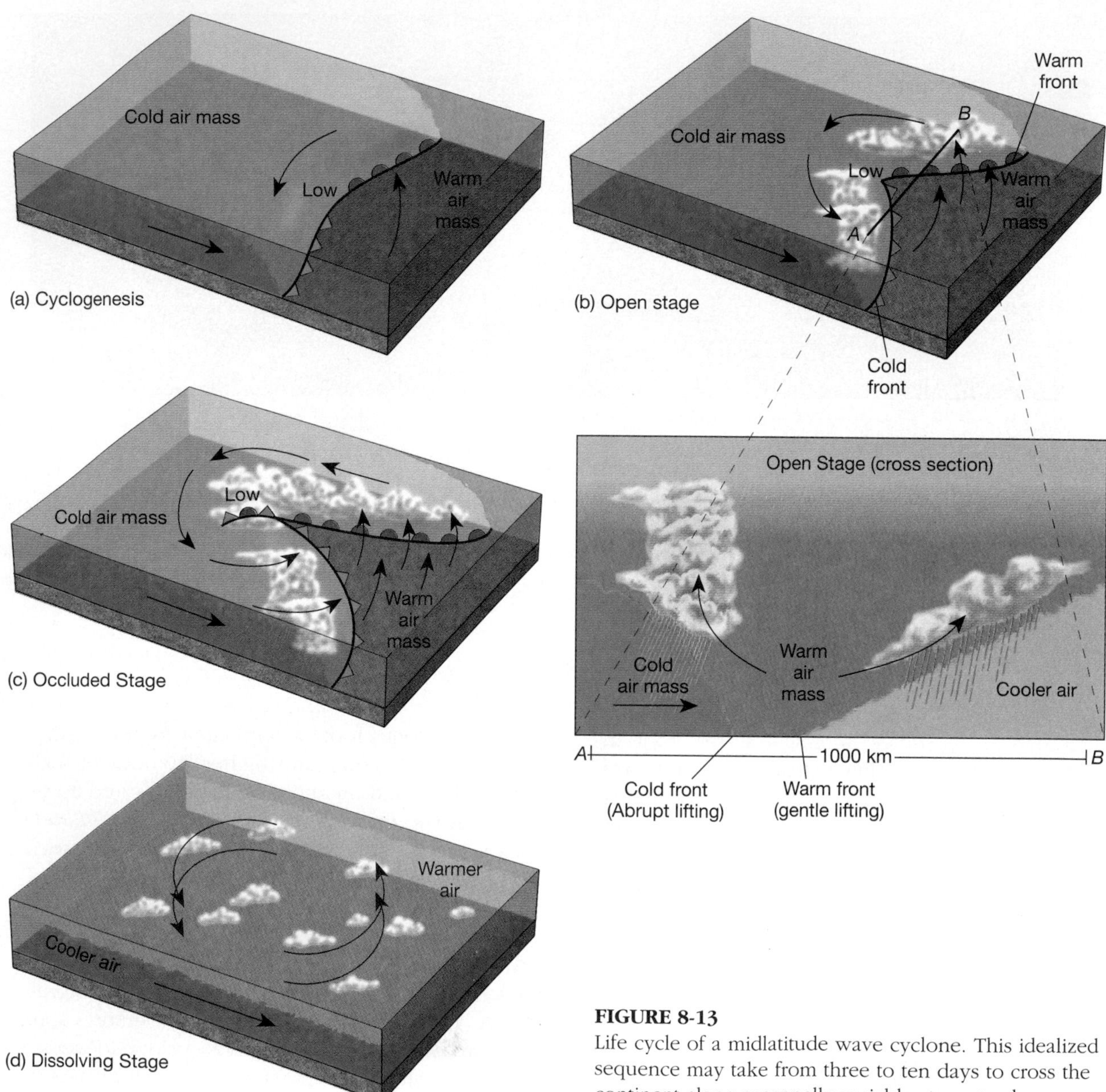

FIGURE 8-13
Life cycle of a midlatitude wave cyclone. This idealized sequence may take from three to ten days to cross the continent along seasonally variable storm tracks.

Cyclogenesis: The Birth of a Cyclone

Cyclogenesis is the atmospheric process in which low-pressure systems originate and strengthen (Figure 8-13a). Along the polar front, cold and warm air masses converge and conflict. The polar front forms a discontinuity of temperature, moisture, and winds that establishes potentially unstable conditions. However, for a wave cyclone to form along the polar front, a point of *convergence* at the surface must be matched by a compensating area of *divergence* aloft. Even a slight disturbance along the polar front, perhaps a small change in the path of the jet stream, can initiate the converging, asending motion of a surface low-pressure system.

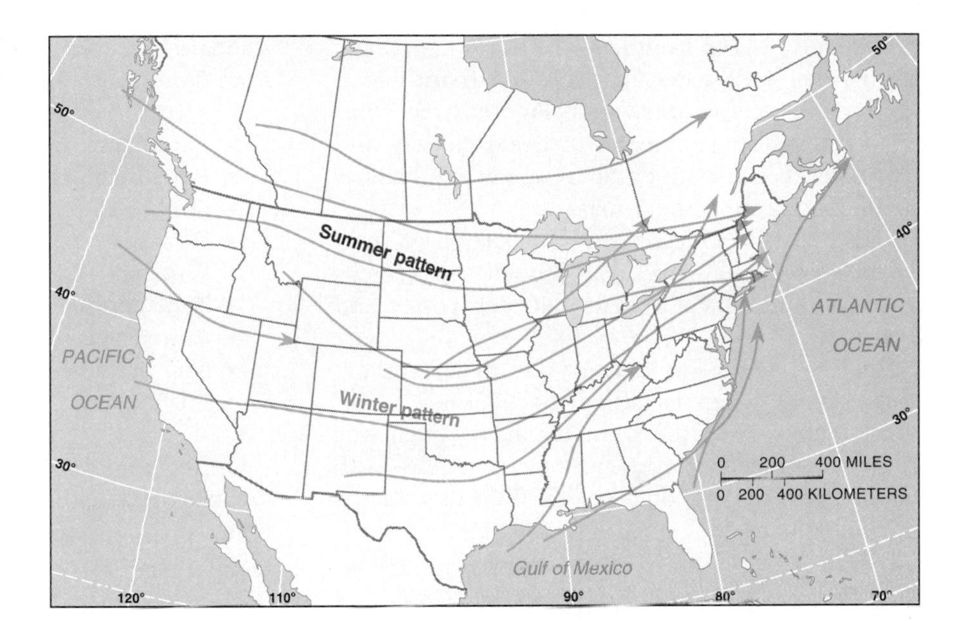

FIGURE 8-14
Cyclonic storm tracks for North America vary seasonally.

The growing circulation system then is vented into the streams of upper-level winds. To the east of the developing low-pressure center, warm air begins to move along an advancing front, while cold air advances to the west of the center (Figure 8-13b). As the midlatitude cyclone matures, the counterclockwise flow in the Northern Hemisphere draws the cold air mass from the north and west and the warm air mass from the south.

These cyclonic storms—1600 km (1000 mi) wide—and their attendant air masses move across the continent along storm tracks, which shift latitudinally with the Sun, crossing North America farther to the north in summer and farther to the south in winter (Figure 8-14). As the storm tracks begin to shift northward in the spring, cP and mT air masses are brought into their clearest conflict. This is the time of strongest frontal activity, with associated thunderstorms and tornadoes. Storm tracks follow the path of upper-air winds which direct storm systems across the continent.

Cyclonic circulation also frequently develops on the lee side of mountain ranges, as along the Rockies from Alberta to Colorado. As air moves downslope, the vertical axis of the air column extends, shrinking the system horizontally and intensifying wind speeds. As the air travels downslope, it is deflected in a cyclonic flow, thus developing new cyclonic systems or intensifying existing ones. By crossing the mountains, such systems gain access to the moisture- and energy-rich mT air mass from the Gulf of Mexico.

An interesting type of cyclonic polar-low circulation can form in a hurricane-like structure: a highly symmetrical, spiral-cloud pattern around a clear central eye, with intense cumulonimbus cloud development and strong winds near the core. These small-scale "arctic hurricanes" develop over open water where cold air flowing off pack ice passes over warm ocean currents. This disequilibrium between atmosphere and ocean in terms of vertical temperature and pressure profiles is key to storm formation. These conditions are quite different from those that lead to the formation of true hurricanes that we discuss shortly. Scientists are studying these arctic storms to reduce the danger they represent to shipping lanes and society.

Occluded Fronts and a Dissolving Cyclone

Because the cP air mass is more homogeneous in temperature and pressure than the mT air mass, the cold front leads a denser, more unified mass, and therefore moves faster than the warm front. Cold fronts can average 40 kmph (25 mph), whereas warm fronts average 16–24 kmph (10–15 mph). Thus, a cold front may overrun the cyclonic warm front, producing an **occluded front** within the overall cyclonic circulation Figure 8-13c). Precipitation may be moderate to heavy initially and then taper as the warmer air wedge is lifted higher by the advancing cold air mass.

Also note the designation of a stationary front on these weather maps. This frontal symbol tells you that there is a stalemate between cooler and warmer air masses.

The final dissolving stage of the midlatitude cyclone occurs when its lifting mechanism is completely cut off from the warm air mass, which was its source of energy and moisture (Figure 8-13d). Remnants of the cyclonic system then dissipate in the atmosphere.

Daily Weather Map and the Midlatitude Cyclone

The four-stage sequence in Figure 8-13 illustrates the evolution in an ideal midlatitude cyclone model, but the actual pattern of cyclonic passage over North America demonstrates a great variety of shape, form, and duration. However, we can apply this model along with our understanding of warm and cold fronts to the actual midlatitude cyclone shown in the weather-map sequence for April 1–3, 1988 (Figure 8-15).

Building a data base of wind, pressure, temperature, and moisture conditions is key to *numerical weather prediction* and the development of weather forecasting models. *Synoptic analysis* involves the characterization of these weather elements over a region within the same time frame, as shown on the three weather maps presented in Figure 8-15. The difficulty in achieving accuracy with such numerical models is that the atmosphere operates as a nonlinear system, tending toward chaotic behavior. Slight variations in input data produce widely divergent results only a few days later.

Preparing a weather report and forecast requires analysis of a daily weather map and satellite images that depict atmospheric conditions. Figure 8–15 presents weather maps and satellite images for a typical three-day period and explains the standard weather station symbol. Weather data from these sources include the following:

- Barometric pressure
- Pressure tendency
- Surface air temperature
- Dew-point temperature
- Wind speed and direction
- Type and movement of clouds
- Current weather
- State of the sky
- Precipitation since last observation

On the map for April 2 in Figure 8-15b, you can identify the temperatures reported by various stations, the patterns created, and the location of warm and cold fronts. The distribution of air pressure is defined by the use of *isobars,* the lines connecting points of equal pressure on the map. Although the low-pressure center identified is not intense, it is well defined over eastern Nebraska, with winds following in a counterclockwise, cyclonic path around the low pressure. The maps and satellite images also illustrate the evolving pattern of precipitation over the three-day period as the warm and cold fronts sweep eastward. The mT air mass, advancing northward from the Gulf/Atlantic source region, provides the moisture for afternoon and evening thundershowers. Can you identify and plot the storm track for the three days shown in the sequence?

VIOLENT WEATHER

Weather provides a continuous reminder of nature's power. The flow of energy across the latitudes can at times set into motion violent and potentially destructive conditions. Thunderstorms, tornadoes, and tropical cyclones are such forms of violent weather.

An example of violent weather occurred in the eastern United States and Canada during December 1992. A classic "nor'easter" struck as a powerful midlatitude cyclone that some called "the storm of the century" for the region. This type of storm produces some of the worst weather for the northeast U.S. and Maritime Provinces. Heavy rains and winds gusting to 145 kmph (90 mph) disrupted cities and destroyed coastal dwellings. Figure 8-16 presents a *GOES-7* image and synoptic weather map for the northeastern region for December 12, 1992. A similar storm that hit between March 11 and 14, 1888, appeared in weather maps published in the first issue of *National Geographic Magazine,* October 1888.

Thunderstorms

Tremendous energy is liberated by the condensation of large quantities of water vapor. This process is accompanied by violent updrafts and downdrafts as rising parcels of air pull surrounding air into the column and as the frictional drag of raindrops pulls air toward the ground. As a result, giant cumulonimbus clouds can create dramatic weather moments—squall lines of heavy precipitation, lightning, thunder, hail, blustery winds, and tornadoes. Thunderstorms may develop within an air mass, along a front (particularly a cold front), or where mountain slopes cause orographic lifting.

Thousands of thunderstorms occur on Earth at any given moment. Equatorial regions and the ITCZ experience many of them, as exemplified by the city of Kampala in Uganda, East Africa (north of Lake Victoria), which sits virtually on the equator and averages 242 days a year of thunderstorms—a record. Figure 8-17 shows the distribution of annual days with thunderstorms across the United States and Canada. You can see that, in North America, most thunderstorms occur in areas dominated by mT air masses.

Lightning and Thunder. An estimated 8 million lightning strikes occur each day on Earth (see the chapter opening photo). **Lightning** refers to flashes of light caused by enormous electrical discharges—tens to hundreds of millions of volts—which briefly superheat the air to temperatures of 15,000°–30,000°C (27,000°–54,000°F). The violent expansion of this abruptly heated air sends shock waves through the atmosphere—the sonic bangs known as **thunder.** The greater the distance a lightning stroke travels, the longer the thunder echoes as the waves of sound arrive. Lightning at great distance from the observer may not be accompanied by thunder and is called *heat lightning.* Lightning is created by a buildup of electrical energy between areas within a cumulonimbus cloud or between the cloud and the ground, with sufficient electrical potential to overcome the high electrical resistance of the atmosphere and leap from one surface to the other—a giant spark.

The beginning of a lightning discharge, called the *leader stroke,* goes from the cloud to the surface, following the route of least resistance, usually in a succession of forked steps. The path of charged ions remaining in the air behind this relatively slow-moving stroke then becomes the lower-resistance pathway for the main stroke, which surges in an enormous rush of energy along that path a fraction of a second later. This *return stroke,* traveling at a tenth the speed of light, produces most of the light seen as lightning. From 2 to 40 repetitions of this sequence occur in a second or two along the same path and produce *forked lightning.* A discharge of forked lightning from one portion of a cumulonimbus cloud to another, usually masked by the cloud and appearing as a broad flash of light, is termed *sheet lightning.*

Lightning poses a hazard to aircraft and to people, animals, trees, and structures. Individuals should acknowledge that certain precautions are mandatory when the threat of a discharge is near. When such lightning is imminent, agencies such as the National Weather Service issue severe storm warnings and caution people to remain indoors.

The United States and Canada operate a National Lightning Detection Network of over 150 sensors to detect electromagnetic fields and related lightning discharges. Atmospheric scientists at the State University of New York–Albany manage the research program.

Hail. Hail is common in the United States and Canada, although somewhat infrequent at any given location, hitting a specific location perhaps every one or two years in the highest frequency areas. Annual hail damage in the United States tops $750 million. **Hail** generally forms within a cumulonimbus cloud when a raindrop circulates repeatedly above and below the freezing level in the cloud, adding layers of ice until its weight no

FIGURE 8-15
Weather maps and *GOES*-7 satellite images for April 1, 2, and 3, 1988. For reference the original April 2, 1988, weather map appears in Appendix C. [National Weather Service and National Oceanic and Atmospheric Administration, National Climatic Center, Satellite Data Services Division.]

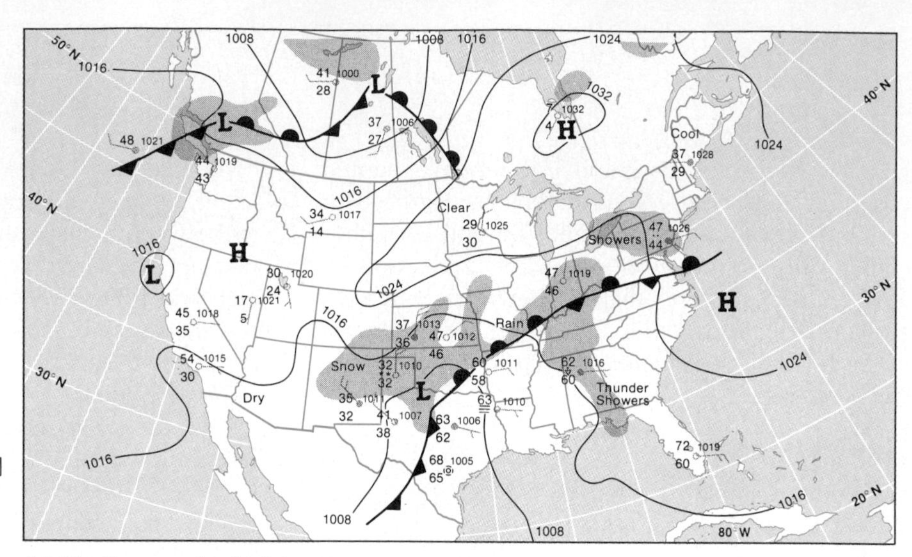

(a) Weather map for April 1, 1988

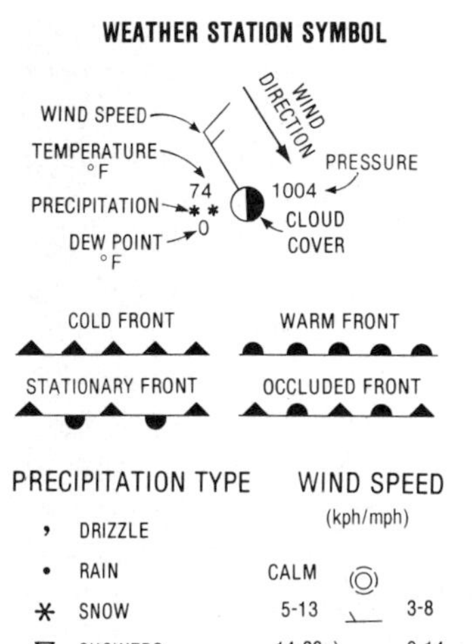

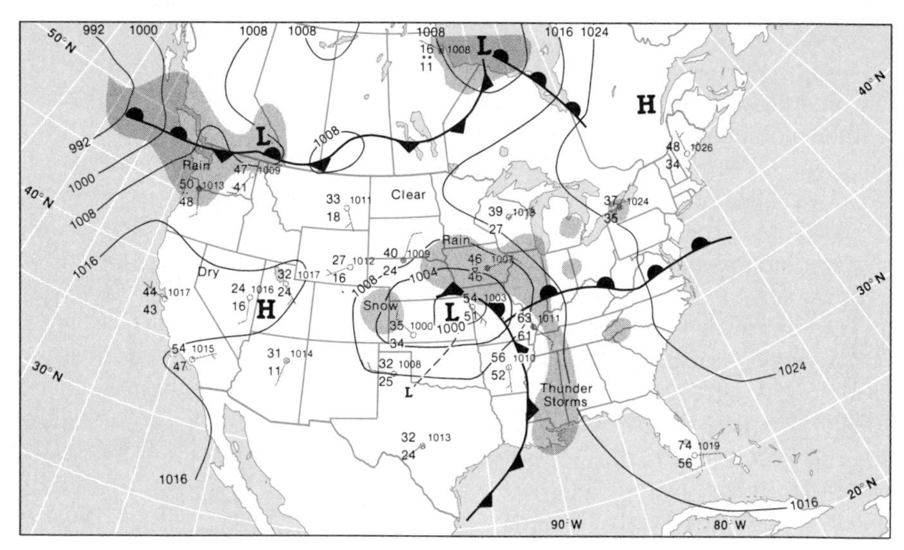

(b) Weather map for April 2, 1988

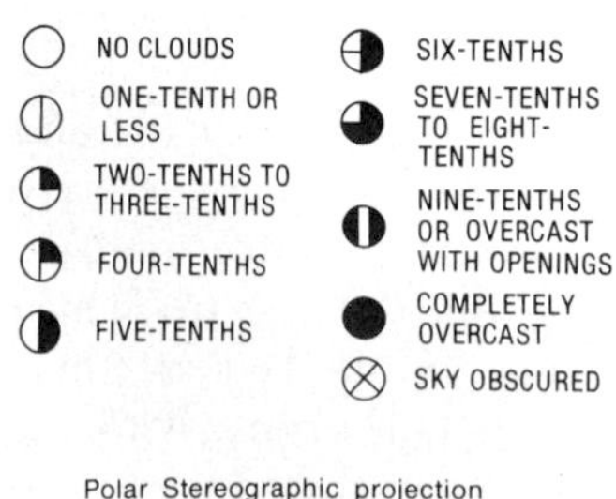

Polar Stereographic projection true at 60° N

0 400 800 km
60°
40°
20°
0 250 500 mi
Scale at different latitudes

SURFACE WEATHER MAP AND STATION WEATHER AT 7:00 A.M., E.S.T.

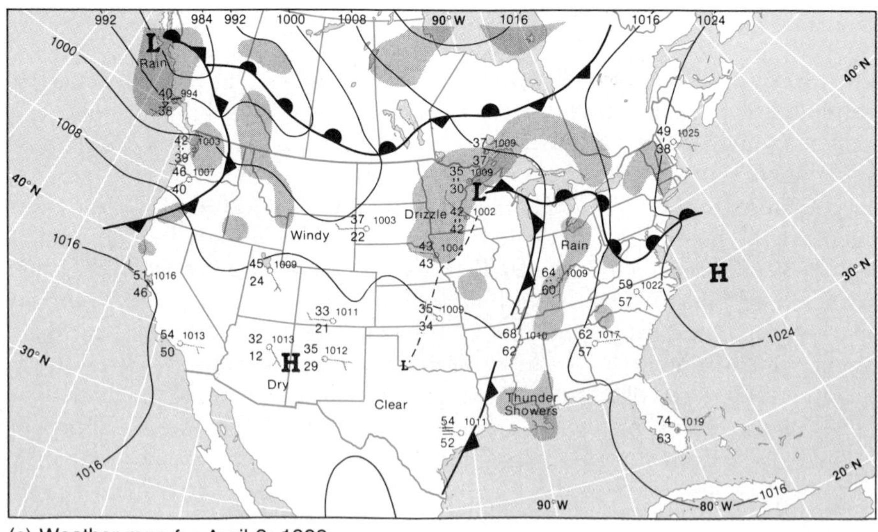

(c) Weather map for April 3, 1988

(a) Satellite image for April 1, 1988

(b) Satellite image for April 2, 1988

(c) Satellite image for April 3, 1988

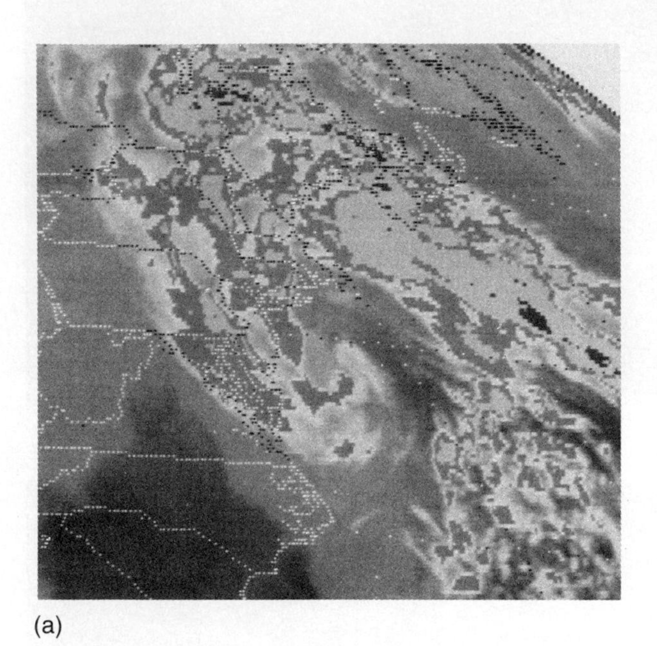

(a)

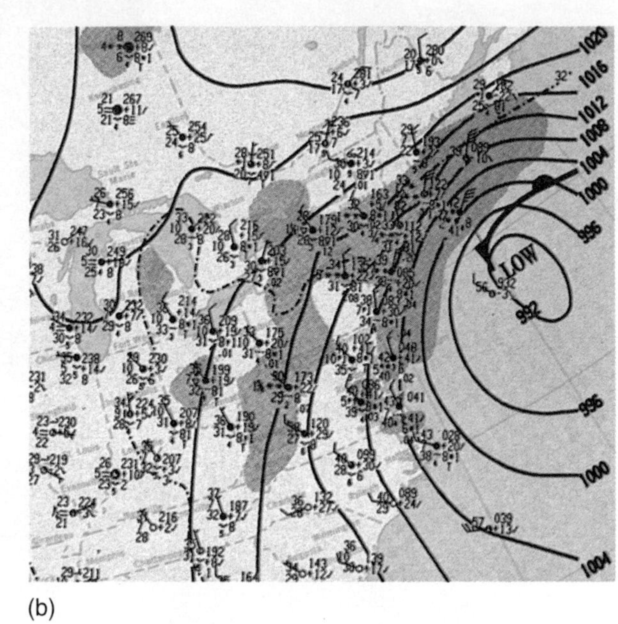

(b)

FIGURE 8-16
A "nor'easter" with hurricane-force winds struck the eastern seaboard on December 12, 1992. (a) This infrared *GOES-7* image was taken at 10:54 A.M. EST; note the cyclonic swirl in the cloud patterns. The synoptic weather map for the same date at 7:00 A.M. EST is presented in (b); note the pattern of isobars (4 mb intervals) and the position of the fronts. [(a) Image from National Climatic Data Center, Satellite Data Services Division, National Oceanic and Atmospheric Administration; (b) weather map from *Daily Weather Maps Weekly Series* for December 6–13, 1992, compiled by the Meteorological Operations Division and Climate Analysis Center, National Weather Service.]

longer can be supported by the circulation in the cloud. Pea-sized and marble-sized hail are common, although hail the size of golf balls and baseballs also is reported at least once or twice a year somewhere in North America. For larger stones to form, the frozen pellets must stay aloft for longer periods. The largest authenticated hailstone in the world landed in Kansas in 1970; it measured 14 cm (5.6 in.) in diameter and weighed 758 grams (1.67 pounds)! The pattern of hail occurrence across the United States and Canada is similar to that of thunderstorms shown in Figure 8-17, dropping to zero approximately by the 60th parallel.

Tornadoes

The updrafts associated with a cumulonimbus cloud appear on satellite images as pulsing bubbles of clouds. Because winds in the troposphere blow stronger above Earth's surface than they do at the surface due to friction, a body of air pushes forward faster at altitude than at the surface, thus creating a rotation in the air along a horizontal axis that is parallel to the ground. When that rotating air encounters the strong updrafts associated with frontal activity, the axis of rotation is shifted to a vertical alignment, perpendicular to the ground.

According to one hypothesis, this spinning, cy-

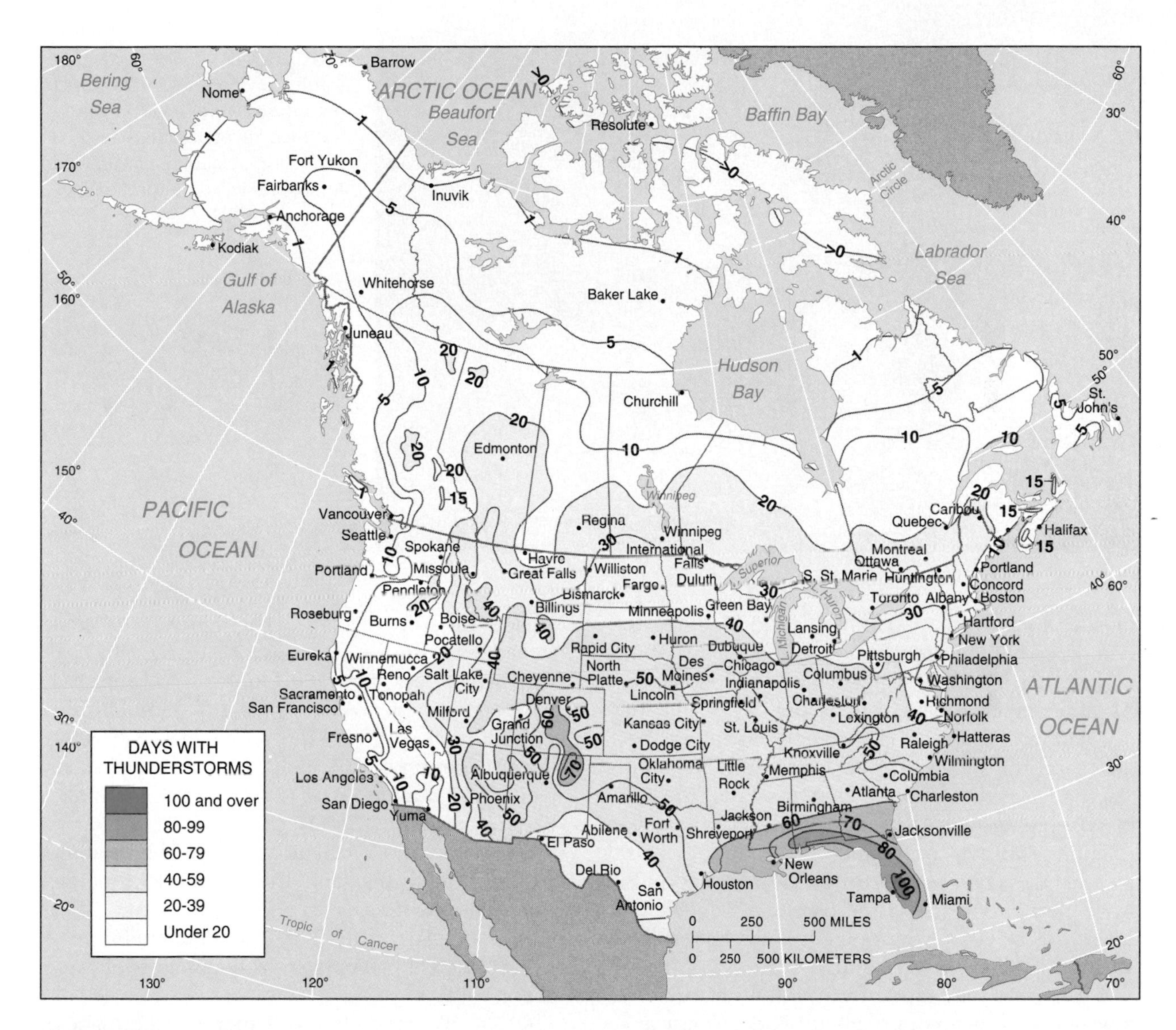

FIGURE 8-17

Mean annual number of days with thunderstorms. [Courtesy of National Weather Service and Map Series 3, *Climatic Atlas of Canada,* Atmospheric Environment Service.]

clonic, rising column of mid-troposphere-level air forms a **mesocyclone.** A mesocyclone can range up to 10 km (6 mi) in diameter and rotate over thousands of meters vertically within the parent cloud. As a mesocyclone extends vertically and contracts horizontally, wind speeds accelerate in an inward vortex (much as ice skaters accelerate while spinning by pulling their arms in closer to their bodies or, in miniature, as water accelerates as it spirals down a sink drain). A well-developed

FIGURE 8-18
A fully developed tornado locks onto the ground. [Photo from National Severe Storms Laboratory, National Oceanic and Atmospheric Administration.]

mesocyclone most certainly will produce heavy rain, large hail, blustery winds, and lightning; some mature mesocyclones will generate tornado activity. As more moisture-laden air is drawn up into the circulation of a mesocyclone, more energy is liberated, and the rotation of air becomes more rapid. The narrower the mesocyclone, the faster the spin of converging parcels of air being sucked into the rotation. The swirl of the mesocyclone itself is visible, as are dark gray **funnel clouds** that pulse from the bottom side of the parent cloud. The terror of this stage of development is the lowering of such a funnel to Earth (Figure 8-18).

A **tornado** can range from a few meters to a few hundred meters in diameter and can last anywhere from a few moments to tens of minutes. Pressures inside a tornado funnel usually are about 10% less than those in the surrounding air. This disparity is similar to the pressure difference between sea level and an altitude of 1000 m (3280 ft). The in-rushing convergence created by such a horizontal pressure gradient causes severe wind speeds that can exceed 485 kmph (300 mph), with at least a third of all tornadoes exceeding 182 kmph (113 mph).

Dr. Theodore Fujita, a noted meteorologist from the University of Chicago, designed a scale for classifying tornadoes according to wind speed and related property damage. The *Fujita Scale* ranks tornadoes from designations of *F0* and *F1* (weak, with winds less than 180 kmph or 112 mph); *F2* and *F3* (strong, with winds 181–332 kmph or 113–206 mph); and *F4* and *F5* (violent, with winds over 333 kmph or 207 mph). About 65% of all tornadoes are ranked F0 or F1, about 33% are ranked F2 or F3, and only 2% are classed as F4 or F5.

When tornado circulation occurs over water, a **waterspout** forms, and the sea or lake surface water is drawn up into the funnel some 3–5 m (10–16 ft). The rest of the waterspout funnel is made visible by the rapid condensation of water vapor.

Of the 50 states, 49 have experienced tornadoes, as have all the Canadian provinces and territories. May is the peak month. A small number of tornadoes are reported in other countries each year, but North America receives the greatest share because its latitudinal position and shape permit conflicting and contrasting air masses access to each other. In the United States, 26,458 tornadoes were recorded in the 33 years between 1959 and 1992, with 723 causing 2615 deaths. In addition, these tornadoes resulted in over 52,000 injuries and damages over $16 billion. The yearly average of $360 million in damage is rising at about $20 million a year. Figure 8-19 presents various tornado data for 1959–1991.

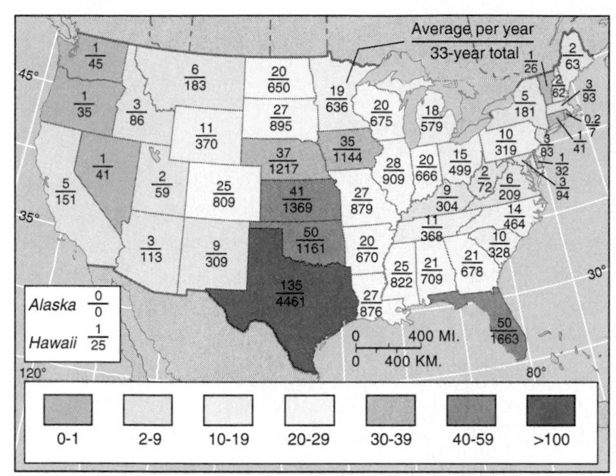

(a) Average tornadoes per year and 33-year totals (1959-1991)

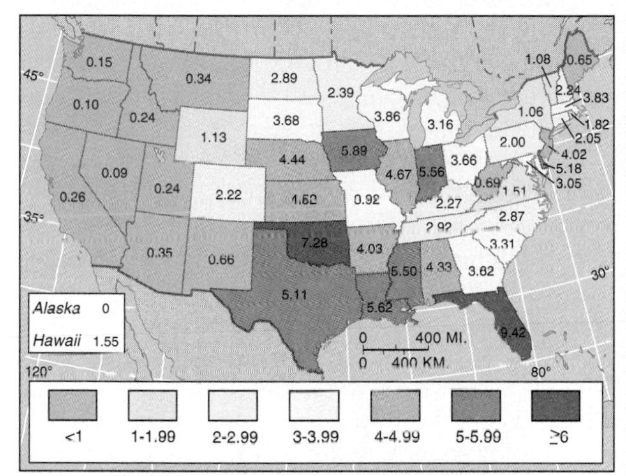

(b) Tornado frequency per 10,000 mi² (1959-1991)

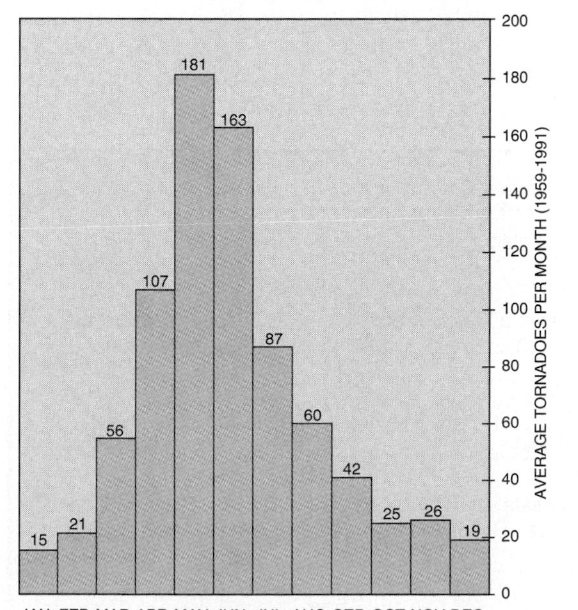

(c)

STATE	RANK BY 33-YR TOTAL	33 YEAR TOTAL	AVERAGE PER YEAR	FREQUENCY PER 10,000 MI² PER YR.	RANK BY FREQUENCY PER 10,000 MI²
TX	1	4461	135	5.11	8
FL	2	1663	50	9.42	1
OK	3	1661	50	7.28	2
KS	4	1369	41	4.52	10
NE	5	1217	37	4.44	11
IA	6	1144	35	5.89	3
IL	7	909	28	3.68	18
SD	8	895	27	3.92	15
MO	9	879	27	4.67	9
LA	10	876	27	5.50	6
MS	11	822	27	5.62	4
CO	12	809	25	2.22	30
AL	13	709	21	4.33	12
GA	14	678	21	3.62	20
WI	15	675	20	4.03	13
AR	16	670	20	3.86	16
IN	17	666	20	2.89	25
ND	18	650	20	5.56	5
MN	19	636	19	2.39	27
MI	20	579	18	3.16	22
OH	21	499	15	3.66	19
NC	22	464	14	2.87	26
WY	23	370	11	2.92	24
TN	24	368	11	1.13	36
SC	25	328	10	3.31	21
PA	26	319	10	2.00	32
NM	27	309	9	2.27	28
KY	28	304	8	0.66	40
VA	29	209	6	1.51	35
MT	30	183	6	1.06	38
NY	31	181	5	0.34	43
CA	32	151	5	0.26	44
AZ	33	113	3	0.35	42
MD	34	94	3	3.83	17
MA	35	93	3	3.05	23
ID	36	86	3	4.02	14
NJ	37	83	3	0.69	39
WV	38	72	2	0.24	45
ME	39	63	2	2.24	29
NH	40	62	2	0.65	41
UT	41	59	2	0.24	46
WA	42	45	1	0.15	47
CT	43	41	1	0.09	49
NV	44	41	1	2.05	31
OR	45	35	1	5.18	7
DE	46	32	1	0.10	48
VT	47	26	1	1.08	37
HI	48	25	1	1.55	34
RI	49	7	0.2	1.82	33
AK	50	1	0	0	50

(d) 33 Years of tornadoes in the United States (1959-1991)

FIGURE 8-19
(a) Average tornadoes per year during 1959–1991; (b) tornado frequency per 10,000 km^2 during 1959–1991; (c) average tornadoes per month, 1959–1991; (d) tornado rankings and frequency by state. During an average year, 820 tornadoes strike throughout the United States and Canada. [Data courtesy of John Halmstad, National Severe Storm Forecast Center, National Weather Service.]

Interestingly, 1992 set a record for the most tornadoes in a single year for the United States, reaching an all-time high of 1384. In June alone, 421 occurred. 1990 ranked second with 1133 tornadoes; 1991 ranked third with 1132. The annual average for tornado occurrences for the preceding ten years was only 820. The reason for this apparent marked increase in tornado frequency in North America is being researched, with explanations ranging from global climate change to better reporting by a larger population.

It is not enough to account statistically for past tornadoes; the need is for accurate forecasting to avoid loss of life and to promote safety measures in vulnerable areas. The National Severe Storms Forecast Center in Kansas City, Missouri, is the U. S. government's headquarters for such activities. Increasing use of Doppler radar is enabling scientists to detect the specific direction and flow of moisture droplets in mesocyclones, allowing a warning period of 30 minutes to 1 hour in areas at risk. A severe storm watch sometimes is possible a few hours in advance. But at other times, this most intense phenomenon of the atmosphere can strike with destructive suddenness and little warning. On April 27, 1991, a four-state area was devastated when over 70 tornadoes touched down from Texas to Nebraska, killing over 30 people and causing millions of dollars in property damage. Compare this to the 53 casualties that occurred during the entire 1990 season.

Tropical Cyclones

A powerful manifestation of Earth's energy and moisture systems is a tropical storm known as a **tropical cyclone,** which originates entirely within tropical air masses. The tropics extend from the Tropic of Cancer at 23.5° N to the Tropic of Capricorn at 23.5° S, encompassing the equatorial zone between 10° N and 10° S. Approximately 80 tropical cyclones occur annually worldwide. Table 8-2 identifies the varying stages of cyclonic development designated by wind speeds and notes some characteristic effects. When you hear meteorologists speak of a "category four" hurricane, they are using the Saffir-Simpson Hurricane Damage Potential Scale to estimate possible damage from hurricane-force winds. This scale ranks hurricanes and typhoons in five categories, which are summarized in Table 8-3.

The cyclonic systems forming in the tropics are quite different from midlatitude cyclones because the air of the tropics is essentially homogeneous, with no fronts or conflicting air masses. In addition, the warm air and warm seas ensure an abundant supply of water vapor and thus the necessary latent heat to fuel these storms. Tropical meteorologists now think that the cyclonic motion begins to form as disturbances associated with slow-moving easterly waves of low pressure in the trade wind belt of the tropics. It is along the eastern (leeward) side of these migrating troughs

TABLE 8-2
Tropical Cyclone Classification

Designation	Winds	Features
Tropical disturbance	Variable, low	Definite area of surface low pressure; patches of clouds
Tropical depression	Up to 30 knots (63 kmph, 39 mph)	Gale force, organizing circulation, light to moderate rain
Tropical storm	30 to 64 knots (63–117.5 kmph, 39–73 mph)	Closed isobars, definite circular organization, heavy rain, assigned a name
Hurricane (Atlantic and E. Pacific) Typhoon (W. Pacific) Cyclone (Indian Ocean, Australia) Baguios (Phillippines)	Greater than 65 knots (119 kmph, 74 mph)	Circular, closed isobars; heavy rain, storm surges, tornadoes in right-front quadrant

TABLE 8-3
Saffir-Simpson Hurricane Scale

Category	Wind Speed	Central Pressure
1	64–82 knots (74–95 mph)	>979 mb
2	83–95 knots (96–110 mph)	965–979 mb
3	96–113 knots (111–130 mph	945–964 mb
4	114–135 knots (131–155 mph	920–944 mb
5	>135 knots (>155 mph)	<920 mb

FIGURE 8-20
Worldwide pattern of the most intense tropical cyclones, typical tropical storm tracks, with principal months of occurrence and regional names. For the North Atlantic region from 1871 to 1986, 934 tropical cyclones (storms and hurricanes) developed. The peak day, given a 9-day moving average calculation, is September 10, with 60 cyclones for the 116-year period.

of low pressure, a place of convergence and rainfall, that tropical cyclones are formed. Surface air flow then spins in toward the low-pressure area, ascends, and flows outward aloft. This important divergence aloft acts as a chimney, pulling more moisture-laden air into the developing system. Tropical cyclones also may result from extratropical cyclonic flows extending equatorward as well as along the ITCZ.

Atmospheric scientists at Colorado State University have found an interesting correlation between the timing and severity of the western African rainy season and the intensity of hurricanes that make landfall along the East Coast of the United States. An analysis of weather records for the past 92 years disclosed a significant causative relationship between the intensity of Sahelian monsoon and Atlantic hurricane development.

The map in Figure 8-20 shows tropical cyclone formation areas and some of the characteristic storm tracks traveled, and indicates the range of months during which tropical cyclones are most

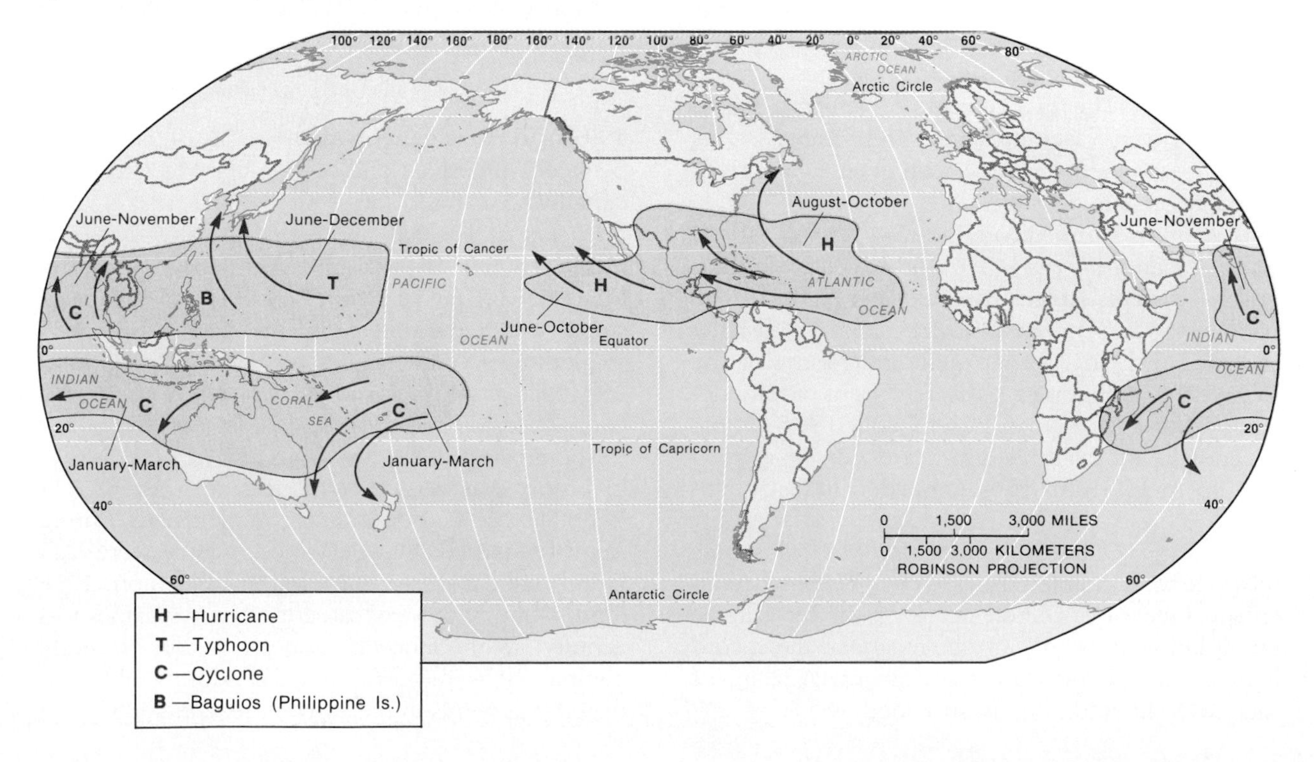

likely to appear. As the map shows, tropical cyclones tend to occur when the equatorial low-pressure trough is farthest from the equator, that is, during the months that follow the summer solstice in each hemisphere. About 10% of all tropical disturbances have the right ingredients to intensify into a full-fledged **hurricane** or **typhoon.**

Hurricanes and Typhoons. Tropical cyclones are potentially the most destructive storms experienced by humans, costing an average of several hundred million dollars in property damage and thousands of lives each year. The tropical cyclone that struck Bangladesh in 1970 killed an estimated 300,000 people and the one in 1991 claimed over 200,000. The Galveston, Texas, hurricane of 1900 killed 6000; Hurricane Audry (1957), 400; Hurricane Gilbert (1988), 318; Hurricane Camille (1969), 256; and Hurricane Agnes (1972), 117.

Fully organized tropical cyclones possess an intriguing physical appearance. They range in diameter from a compact 160 km (100 mi), to 960 km (600 mi), to some western Pacific typhoons that reach 1300–1600 km (800–1000 mi). Vertically, these storms dominate the full height of the troposphere. The inward-spiraling clouds form dense rain bands, with a central area designated the *eye,* around which a thunderstorm cloud called the *eyewall* swirls, producing the area of most intense precipitation. The eye remains an enigma, for in the midst of devastating winds and torrential rains the eye has quiet, warm air with even a glimpse of blue sky or stars possible.

The entire storm travels at 16–40 kmph (10–25 mph). When it moves ashore, or makes **landfall,** it pushes **storm surges** of seawater inland. The strongest winds of a tropical cyclone are usually recorded in its right-front quadrant (relative to the storm's directional path), where dozens of fully developed tornadoes may be embedded at the time of landfall. For example, Hurricane Camille in 1969 had up to 100 tornadoes embedded in that quadrant.

Hurricane Gilbert (September 1988) illustrates these general characteristics. For a Western Hemisphere hurricane, Gilbert achieved the record size (1600 km or 1000 mi in diameter) and the record low barometric pressure (888 mb or 26.22 in. of mercury). In addition, it sustained winds of 298 kmph (185 mph) with peak winds exceeding 320 kmph (200 mph). Figure 8-21 portrays Gilbert from overhead and in cross section. From these images you can see the vast counterclockwise circulation, the central eye, eyewall, rain bands, and overall hurricane appearance.

Typhoons form in the vast region of the Pacific, where the annual frequency of tropical cyclones is greatest. The magnitude and intensity of typhoons generally exceed those of hurricanes. In fact, the lowest sea-level pressure recorded on Earth was 870 mb (25.69 in.) in the center of Typhoon Tip, in October 1979, 837 km (520 mi) northwest of Guam (17° N 138° E). The Geostationary Meteorological Satellite (GMS), operated by the Japan Weather Association, provides images for tracking these storms.

Researchers now are scrambling to understand why tropical cyclones have been so strong during the 1988–93 seasons and what relation might exist between planetary warming trends in the atmosphere and ocean and these intense storms. During Hurricane Andrew, tentative global-weather forecasting models and more accurate limited-area dynamic models were in full operation. This produced accurate landfall predictions along the Florida and Gulf coasts. Let us briefly examine Hurricane Andrew.

1992 and Hurricane Andrew. The 19-day period between August 24 and September 11, 1992, was certainly one for displays of nature's power, and for personal, societal, and economic tragedy. Hurricane Andrew (Florida and Louisiana), Hurricane Iniki (Kauai, Hawaii), and Typhoon Omar (Guam) struck with record-breaking fury. Figure 8-22 presents three AVHRR images for August 24, 25, and 26, 1992, as Andrew moved westward across the Gulf of Mexico and onshore in Louisiana.

The greatest dollar loss from any natural disaster in history was caused by Hurricane Andrew as it swept across Florida and on to Louisiana during August 24–27, 1992. Sustained winds were 225 kmph, with gusts to 282+ kmph (140 mph, 175+ mph)—one of the few category five hurricanes this century (Saffir-Simpson scale). Studies recently completed by meteorologist Theodore Fujita estimated that winds in the eyewall reached 320 kmph

FIGURE 8-21
Hurricane Gilbert, September 13, 1988: (a) *GOES-7* satellite image; (b) side-view radar image (Side-looking Airborne Radar—SLAR) taken from an aircraft flying through the central portion of the storm. Rain bands of greater cloud density are false-colored in yellows and reds. The clear sky in the central eye is dramatically portrayed. [National Oceanic and Atmospheric Administration and the National Hurricane Center, Miami.]

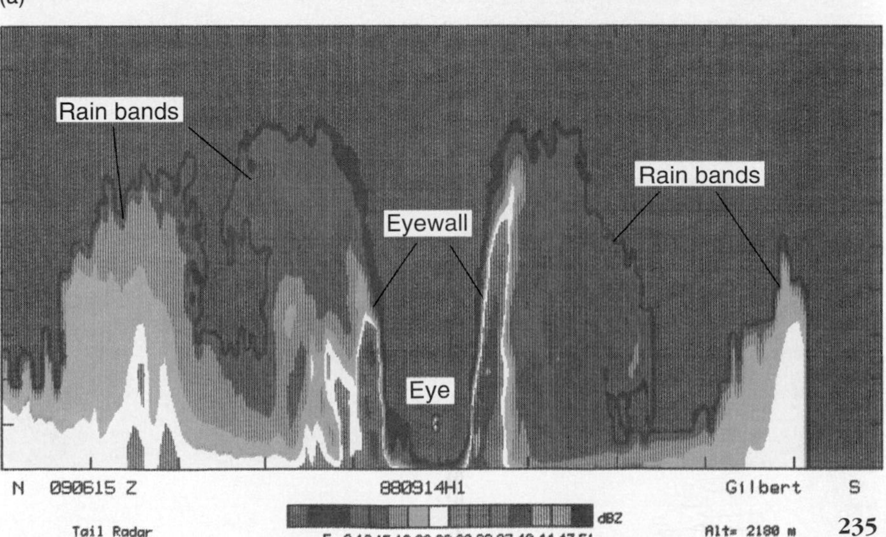

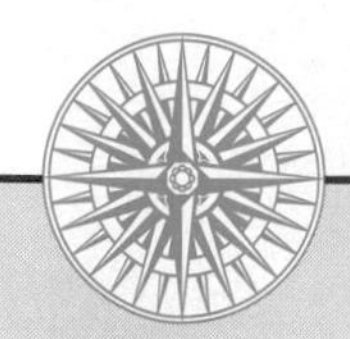

Focus Study 8-1 Hurricane Camille, 1969: Disaster and Benefit

One of the most devastating hurricanes to strike the United States coastline this century was Hurricane Camille. Ironically, it was significant not only for the disaster it brought (256 dead, $1.5 billion damage), but also for the moisture benefit it delivered and the drought it abated. Figure 1 presents two satellite images from August 17 and 18, 1969, as Camille made landfall on the Gulf Coast west of Biloxi, Mississippi, and then moved north. Hurricane-force winds diminished sharply shortly after landfall, leaving a vast rainstorm that traveled from the Gulf Coast through Mississippi, western Tennessee, Kentucky, and into central Virginia. Severe flooding occurred along the Gulf Coast near the point of landfall and in the James River basin of Virginia, where torrential rains produced record floods. However, beyond this flooding and immediate coastal damage, Camille's effect actually was beneficial, ending year-long drought conditions along major portions of its track.

According to the weather records of the past 40 years, about one-third of all hurricanes making landfall in the United States have created beneficial conditions inland in terms of the precipitation contributed to local water budgets. Figure 2 shows two maps of the precipitation attributable to Camille (moisture supply) and the moisture deficits abated (moisture shortages avoided) due to rains generated during Camille's passage. Dry soil conditions and low reservoirs were recharged by the rains. Camille's monetary benefits inland outweighed costs by 2 to 1, it is estimated. The tragic loss of life must not be forgotten, of course. But, in terms of overall effect, hurricanes should be viewed as normal and natural meteorological events that certainly have destructive potential but also contribute to the precipitation regimes of the southern and eastern United States.

A closing thought on our society's response to these powerful weather events: despite the increased

FIGURE 1
Hurricane Camille (a) made landfall on the night of August 17, 1969, and (b) continued inland on August 18. These two images were made exactly 24 hours apart. [Photos from National Oceanic and Atmospheric Administration.]

(a)

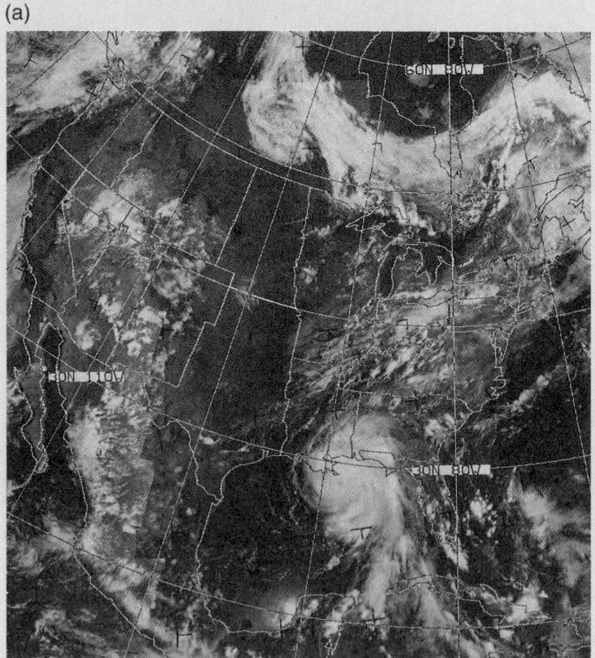

(b)

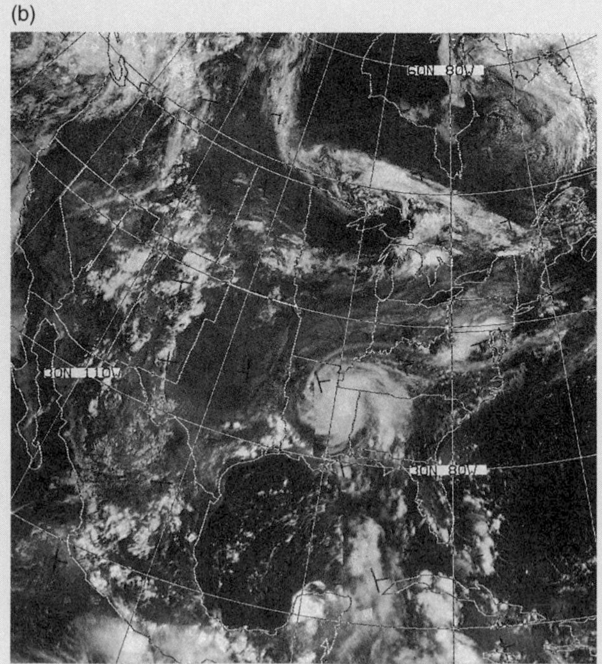

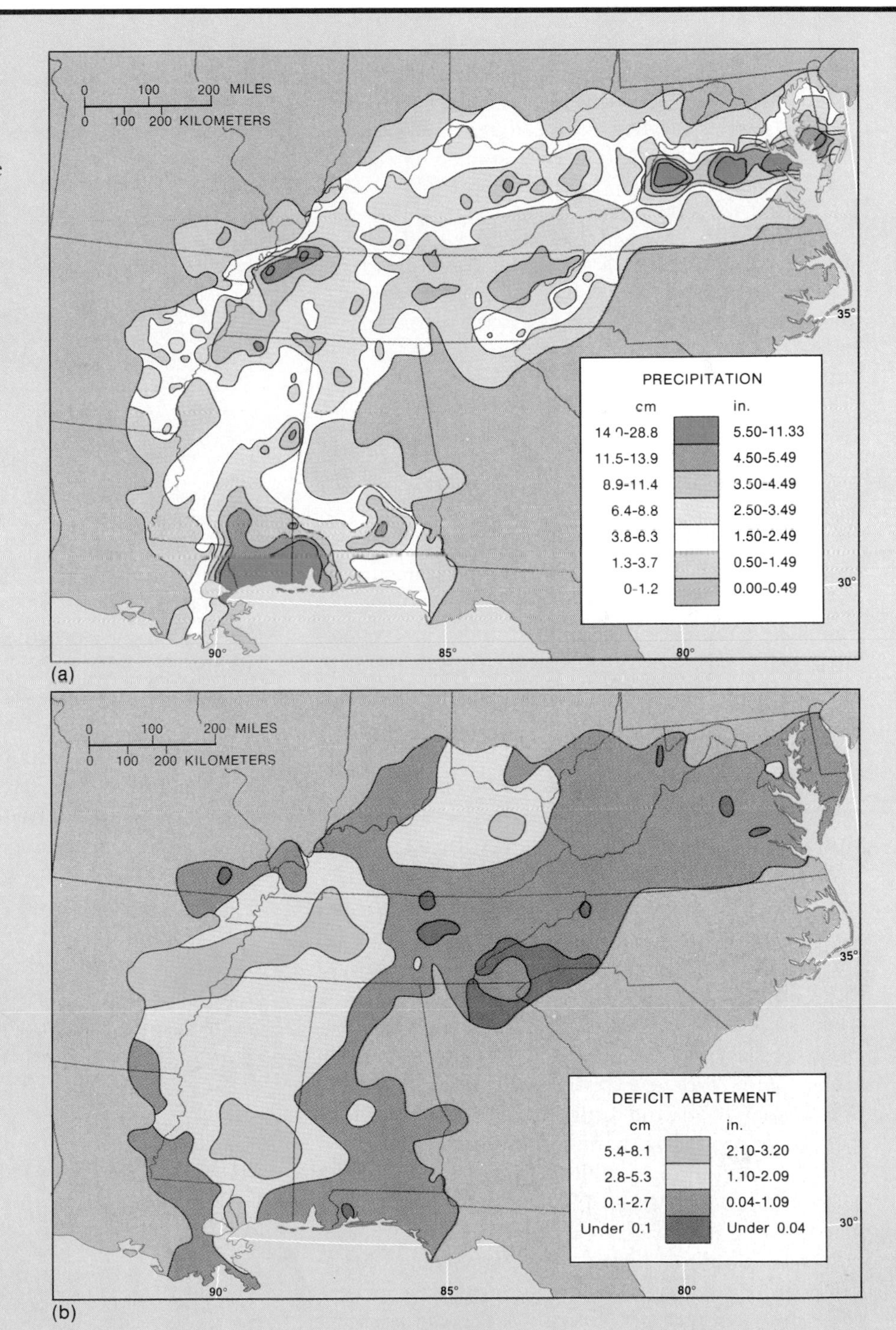

FIGURE 2
A water resource view of Hurricane Camille's impact on local water budgets: (a) precipitation attributable to the storm (moisture supply); (b) resulting deficit abatement (avoided moisture shortages) attributable to Camille.

damage from hurricanes, we see evidence of increased construction and development in vulnerable coastal lowlands. Continued development in vulnerable areas helps explain the estimated $20 billion in damage caused by Hurricane Andrew in 1992. Even after Camille's devastation, a building boom hit the region. Ironically, new buildings, apartments, and governmental offices opened right next to still-visible rubble and bare foundation pads. Unfortunately, careful hazard planning to guide the settlement of these high-risk areas has never been policy. Appropriate and effective hazard perception is rarely put into practice by public or private decision makers. Coastal lowlands, river floodplains, and earthquake fault zones are at risk because of poor perception and lack of planning.

The consequence is that our entire society bears the financial cost of poor perception, no matter where in the country it occurs, not just those victims who directly shoulder the physical, emotional, and economic hardship of the event. This recurrent, yet avoidable cycle—construction, devastation, reconstruction, devastation—was portrayed in *Fortune:*

> Before long the beachfront is expected to bristle with new motels, apartments, houses, condominiums, and office buildings. Gulf Coast businessmen, incurably optimistic, doubt there will ever be another hurricane like Camille, and even if there is, they vow, the Gulf Coast will rebuild bigger and better after that.*

**Fortune,* October 1969, p. 62.

(200 mph) in small vortices. Property damage exceeded $20 billion. For comparison, Hurricane Agnes in 1972 and Hurricane Camille in 1969 totaled $3.1 billion and $1.5 billion respectively. Hurricane Hugo in 1989 caused several hundred deaths and more than $4 billion in damages.

Despite these tragic losses and human suffering, hurricanes can produce some measurable benefits. Focus Study 8-1 presents a moisture analysis of Hurricane Camille that highlights benefits from such a storm.

The tragedy from Andrew is that the storm destroyed, or seriously damaged 70,000 homes and left 200,000 people homeless between Miami and the Florida Keys. By April 1993, 60,000 people were still homeless and reconstruction was progressing slowly. The storm is causing continued losses from reduced property assessments. Approximately 8% of the agricultural industry in Florida's Dade County (Miami region) was destroyed outright–exceeding $1 billion in lost sales. About 25% of Louisiana's sugar cane crop was lost. Many plants were killed by wind-driven saltwater that desiccated (dried) leaves that were not already stripped by the winds. See the opening photograph for Chapter 6 for a sense of this devastation.

The remnants of Hurricane Andrew continued northward, reaching Quebec and eastern Canada by August 28. Locally, heavy rains were produced as the remnants of Andrew interacted with weather fronts moving through the area. Many local records in Quebec were broken for 24-hour precipitation totals.*

The Everglades, the coral reefs north of Key Largo, some 10,000 acres of mangrove wetlands, and coastal southern pine forests all took significant hits from Andrew. Natural recovery will take years, but it does offer opportunities for study. Initial scientific assessments judge the Everglades to be quite resilient because the region's ecosystems evolved in relation to such hurricane occurrences. An important thing to remember is that storms damage human structures more than they damage natural systems. *Threats from urbanization, agriculture, and water diversion represent a greater ongoing crisis to the Everglades than these great storms.*

SUMMARY—Weather

Three record-breaking tornado years for the United States, powerful hurricanes and typhoons, occurrences of persistent drought, and dramatic storms brought weather to the headlines in the 1990s. Weather is the short-term condition of the atmo-

**Climate Perspectives,* vol. 14, no. 35, 24–30 August 1992. Ottawa: Environment Canada.

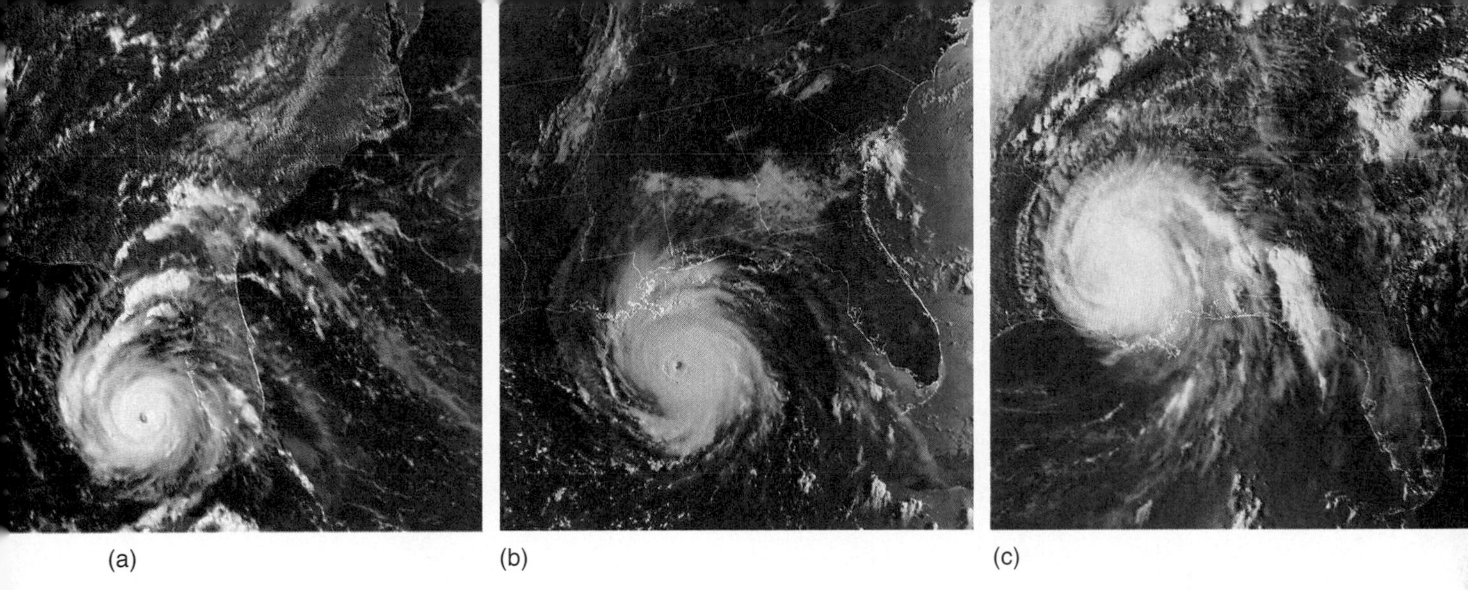
(a) (b) (c)

Figure 8-22
NOAA-10 and 11 AVHRR images for (a) August 24 (4:41 P.M. EDT), (b) 25 (9:28 A.M. EDT), and (c) 26 (4:17 P.M. EDT), 1992, as Hurricane Andrew moved westward across the Gulf of Mexico and onshore in Louisiana. Note the time of day and direction of sunlight as it highlights the features of the hurricane. [Images from National Oceanographic and Atmospheric Administration as supplied by the EROS Data Center.]

sphere; meteorology is the scientific study of the atmosphere. The spatial implications of these atmospheric sciences and their relationship to human activities strongly link meteorology to physical geography. Analyzing and understanding patterns of wind, pressure, temperature, and moisture conditions portrayed on daily weather maps is key to numerical weather forecasting.

The stability of the lower atmosphere determines whether active lifting of air parcels will occur. A simple comparison of the dry adiabatic rate (DAR) and moist adiabatic rate (MAR) in a vertically moving parcel of air with that of the environmental lapse rate in the surrounding air determines the atmosphere's stability. The interaction of air masses of differing temperature and moisture characteristics is the essence of weather phenomena. Unstable conditions create lifting air masses that cool, saturate and condense, and form clouds. These masses can be lifted by convection heating, topographic barriers, and frontal contrasts. Midlatitude wave cyclones organize air masses in conflicting patterns as they migrate across the continents and oceans from west to east.

The violent power of some weather phenomena is displayed in thunderstorms, lightning, tornadoes, hurricanes, and typhoons. Hurricane and typhoon damage in the United States exceeded $30 billion during 1992, with Hurricane Andrew alone accounting for $20 billion. As forecasting and the public's perception of weather-related hazards improves, loss of life and property could stabilize rather than continue to increase with each event.

KEY TERMS

- adiabatic
- air mass
- chinook winds
- cold front
- cyclogenesis
- dry adiabatic rate (DAR)
- front
- funnel clouds
- hail
- hurricane

landfall
leeward slope
lightning
mesocyclone
meteorology
moist adiabatic rate (MAR)
occluded front
orographic lifting
rain shadow
secondary air mass
squall line
stability
storm surges
thunder
tornado
tropical cyclone
typhoon
warm front
waterspout
wave cyclone
weather
windward slope

REVIEW QUESTIONS

1. Differentiate between stability and instability relative to a parcel of air rising vertically in the atmosphere.
2. How do the adiabatic rates of heating or cooling differ from the normal lapse rate and environmental lapse rate?
3. Why is there a difference between the dry adiabatic rate (DAR) and the moist adiabatic rate (MAR)?
4. How does a source region influence the type of air mass that forms over it? Give specific examples of each basic classification.
5. Of all the air masses, which are of greatest significance to the United States and Canada? What happens to them as they migrate to locations different from their source regions? Give an example of air-mass modification.
6. Explain why it is necessary for an air mass to rise if there is to be precipitation. What are the three main lifting mechanisms described in this chapter?
7. When an air mass passes across a mountain range, many things happen to it. Describe each aspect of a mountain crossing by a moist air mass. What is the pattern of precipitation that results?
8. Differentiate between a cold front and a warm front.
9. How does a wave cyclone act as a catalyst for conflict between air masses?
10. What is meant by cyclogenesis? Where does it occur and why?
11. Diagram a midlatitude cyclonic storm during its open stage. Label each of the components in your illustration, and add arrows to indicate wind patterns in the system.
12. What constitutes a thunderstorm? What type of cloud is involved? What type of air masses would you expect in an area of thunderstorms in North America?
13. Lightning and thunder are powerful phenomena in nature. Briefly describe how they develop.
14. Describe the formation process of a mesocyclone. How is this development associated with that of a tornado?
15. Evaluate the pattern of tornado activity in the United States from Figure 8-19. What generalizations can you make about the distribution and timing of tornadoes?
16. What are the different classifications for tropical cyclones? List the various names used worldwide for hurricanes.
17. What factors contributed to the incredible damage figures produced by Hurricane Andrew?

18. Was Hurricane Camille detrimental, or beneficial, or both? Explain your answer and give specific support for your assessment of the storm's impact.

SUGGESTED READINGS

Alper, Joe. "Everglades Rebound from Andrew," *Science* 257 (25 September 1992): 1852–54.

Businger, Steven. "Arctic Hurricanes," *American Scientist* 79 (January–February 1991): 18–33.

Ferguson, Edward, and Frederick Ostby. "Tornadoes of 1990," *Weatherwise* 44, no. 2 (April 1991): 19–28.

Fincher, Jack. "Summer's Light and Sound Show: A Deadly Delight," *Smithsonian* 19, no. 4 (July 1988): 110–21.

Gray, William M. "Strong Association Between West African and U.S. Landfall of Intense Hurricanes," *Science* 249 (14 September 1990): 1251–56.

Gray, William M., Paul W. Mielke, and Kenneth Berry. "Long-term Variations of Western Sahelian Monsoon Rainfall and Intense U.S. Landfalling Hurricanes," *Journal of Climate* 5, no. 12 (December 1992): 1528–34.

Grenier, Leo A., John Halmstad, and Preston Leftwich, Jr. *Severe Local Storm Warning Verification: 1989.* National Oceanic and Atmospheric Administration Technical Memorandum NWS NSSFC-27. Kansas City: Severe Storms Forecast Center, May 1990.

Hughes, Patrick. *American Weather Stories.* Environmental Data Service and National Oceanic and Atmospheric Administration. Washington, DC. U.S. Government Printing Office, 1976.

Lockhart, Gary. *The Weather Companion.* New York: John Wiley & Sons, 1988.

Miller, Peter. "Tracking Tornadoes!" *National Geographic* 171, no. 6 (June 1987): 690–715.

Neumann, Charles, Brian Jarvinen, Arthur Pike, and Joe D. Elms. *Tropical Cyclones of the North Atlantic Ocean, 1871–1986* (updated through 1989), Historical Climatology Series 6-2. Asheville, NC: National Climatic Data Center in cooperation with the National Hurricane Center, Coral Gables, Florida, July 1992.

Newcott, W. R. "Lightning, Nature's High-Voltage Spectacle," *National Geographic* 184, no. 1 (July 1993): 83:103.

Oliver, John E., and Rhodes W. Fairbridge, eds. *The Encyclopedia of Climatology.* New York: Van Nostrand Reinhold Co., 1987.

Roberts, Walter Orr. "We're Doing Something About the Weather," *National Geographic* 141, no. 4 (April 1972): 518–55.

Schaeffer, Vincent J., and John A. Day. *A Field Guide to the Atmosphere.* Peterson Field Guide Series. Boston: Houghton Mifflin, 1981.

Snow, John T. "The Tornado," *Scientific American* 250, no. 4 (April 1984): 86–96.

Thompson, Philip D., Robert O'Brien, and the editors of Time-Life Books. *Weather.* Life Science Library. New York: Time-Life Books, 1965.

Whipple, A. B. C., and the editors of Time-Life Books. *Storm.* Planet Earth Series. Alexandria, VA: Time-Life Books, 1982.

Williams, Earle R. "The Electrification of Thunderstorms," *Scientific American* 259, no. 5 (November 1988): 88–99.

Williams, Jack, and Jack Faidley. "Hurricane Andrew in Florida," *Weatherwise* 45, no. 6 (December 1992/January 1993): 7–17.

Desert farmlands on the Hopi Reservation, Arizona. [*Photo by author.*]

9

WATER BALANCE AND WATER RESOURCES

The availability of water is controlled by atmospheric processes and climate patterns. Because water is not always naturally available when and where it is needed, humans must rearrange water resources. The maintenance of a houseplant, the distribution of local water supplies, an irrigation program on a farm, the rearrangement of river flows—all involve aspects of the water balance and water-resource management.

We begin this chapter by examining the water balance, which is an accounting of the hydrologic cycle for a specific area with emphasis on plants and soil moisture. We discuss the nature of groundwater and look at several examples of this generally abused resource. Groundwater resources are closely tied to surface-water budgets. We also consider the water we withdraw and consume from available resources, in terms of both quantity and quality. For many countries, water availability looms as a key environmental issue in the near future, for others, it is already a problem.

THE WATER-BALANCE CONCEPT

A water balance can be established for any defined area of Earth's surface—a continent, nation, region, or field—by calculating the total precipitation input and the total water output. The water-balance is a portrait of the hydrologic cycle at a specific site or area for any period of time, including estimation of streamflow. Charles Warren Thornthwaite (1899–1963), a geographer pioneering in applied water-resource analysis, worked with others to develop a water-balance methodology. They related water-balance concepts to geographical problems, especially to accurately determining irrigation quantity and timing. Thornthwaite also developed methods for estimating evaporation and transpiration from a variety of ground covers and surfaces. He identified the relationship between a given supply of water and the local demand as an important climatic element. In fact, his initial application of the techniques discussed here was to develop a new climatic classification system.

The Water-Balance Equation

The water balance describes how the water supply is expended. To understand Thornthwaite's water-balance methodology and "accounting" or "bookkeeping" procedures, we must first understand the terms and concepts in the simple water-balance equation:

$$\text{PRECIP} = \underbrace{(\text{POTET} - \text{DEFIC})}_{\substack{\text{ACTET} \\ \text{Actual evapotranspiration}}} + \text{SURPL} \pm \Delta\,\text{STRGE}$$

$$\text{Supply} = (\text{Demand} - \text{Shortage}) + \text{Oversupply} \pm \text{Soil moisture utilization or recharge}$$

in which:

- PRECIP (precipitation) is rain, sleet, snow, and hail—*the moisture supply.*
- POTET (potential evapotranspiration) is the amount of moisture that would evaporate and transpire through plants if the moisture were available; the amount that would become output under optimum moisture conditions—*the moisture demand.*
- DEFIC (deficit) is the amount of unsatisfied POTET; the amount of demand that is not met either by PRECIP or by soil moisture storage—*the moisture shortage.*
- SURPL (surplus) is the amount of moisture that exceeds POTET, when soil moisture storage is at field capacity (full)—*the moisture oversupply.*
- ± ΔSTRGE (soil moisture storage change) is the use (decrease) or recharge (increase) of soil moisture, snow pack, or lake and surface storage or detention of water—*the moisture savings.*
- ACTET (actual evapotranspiration) is the actual amount of evaporation and transpiration that occurs, derived from POTET–DEFIC; thus, if all the demand is satisfied, POTET will equal ACTET—*the actual satisfied demand.*

Precipitation. The supply of moisture to Earth's surface is **precipitation** (PRECIP)—the rain, sleet,

FIGURE 9-1
A rain gauge installation. A standard rain gauge is cylindrical, 20.3 cm (8.0 in.) in diameter and 58 cm (23 in.) deep. Modern styles include the tipping-bucket and weighing-type gauges. The water is guided by a funnel into a measuring tube. The device is designed to minimize evaporation, which could lead to inaccurate readings. [Photo by author.]

snow, and hail that arrive at the surface. In some climates, additional deposits of water at Earth's surface occur in the form of dew and fog. Precipitation is measured with the **rain gauge,** which receives precipitation through an open top so that water can be measured by depth, weight, or volume (Figure 9-1).

Various conditions affect rain gauge performance. For instance, wind can cause an undercatch of precipitation because the drops are not falling vertically; a wind of 37 kmph (23 mph) produces an undercatch as high as 40%. Consequently, windshields are installed above the opening of a gauge to reduce error by catching raindrops that arrive at an angle to the opening in the gauge. In addition, when excessively heavy rains hit an area, various containers such as buckets or pans may be used to augment the rain gauge.

According to the World Meteorological Organization, more than 40,000 weather monitoring stations are operating worldwide, with over 100,000 places measuring precipitation. Precipitation patterns over the United States and Canada are shown in Figure 9-2. The greatest amounts are in the East, Southeast, Northwest, and western mountain areas. Less precipitation is found in the southwestern and western interiors and in the extreme north. A map displaying the pattern of world precipitation is included in the next chapter as Figure 10-2.

Potential Evapotranspiration. Evaporation is the net movement of free water molecules away from a wet surface into air that is less than saturated. **Transpiration** takes place in plants. It is the outward movement of water through small openings called stomata in the underside of leaves. Both evaporation and transpiration are water balance outputs and directly respond to climatic conditions of temperature and humidity. Additionally, transpiration rates are partially controlled by the plants themselves as they conserve or release water by controlling small cells around the stomata. Transpired quantities can be significant: on a hot day, a single tree may transpire hundreds of liters of water. The overall coincidence of evaporation and transpiration makes it convenient to combine them into one term—**evapotranspiration.**

The demand for water at Earth's surface is called **potential evapotranspiration** (POTET). It represents the amount of water that would evaporate and transpire under optimum moisture conditions (adequate precipitation and full soil moisture capacity). Filling a bowl with water and letting it evaporate illustrates this concept: when the bowl becomes dry, is there still an evaporation demand?

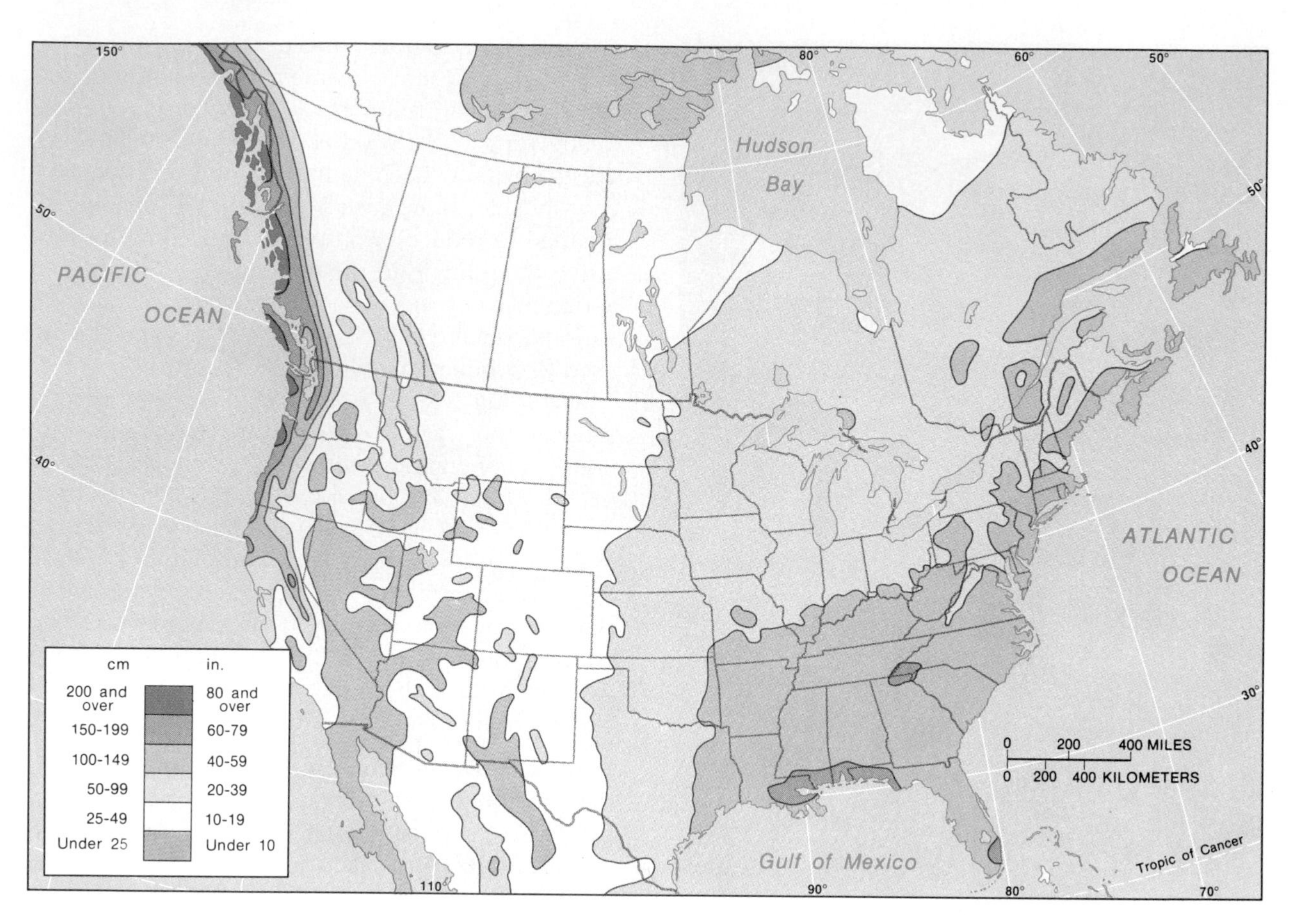

FIGURE 9-2
Annual precipitation (supply) in the United States and Canada. [Adapted from U.S. National Weather Service, U.S. Department of Agriculture, and Environment Canada.]

A demand of course is present, even when the bowl is dry. Similarly, the amount of water that would evaporate and/or transpire if water always were available is the POTET, or the amount that would evaporate from the bowl if it constantly were supplied with water.

Determining POTET. For the empirical measurement of POTET, the easiest method employs an **evaporation pan,** or evaporimeter (Figure 9-3). As evaporation occurs, water in measured amounts is automatically replaced in the pan. Mesh screens over the pan protect against overmeasurement created by wind, which accelarates evaporation.

A **lysimeter** is a relatively elaborate device for measuring POTET. It is a buried tank, open at the surface and approximately a cubic meter in size. Using an actual portion of a field, a lysimeter isolates soil, subsoil, and plants so that the moisture moving through the sampled area can be measured. Deep-rooted crops require a larger tank so that root growth is not inhibited. Conditions must be similar in the lysimeter and the surrounding field to maximize accuracy. A rain gauge next to the lysimeter measures the precipitation input. Some of the water remains as soil moisture, is incorporated into the vegetation cover, or is measured as gravitational water which drains from the

FIGURE 9-3
Standard evaporation pan.
[Photo by author.]

bottom of the lysimeter. The remainder is credited to evapotranspiration. Given natural conditions, the lysimeter measures actual evapotranspiration. If adequate water is supplied through irrigation and the soil moisture available to the plants is optimum, the rate of water loss will be equivalent to POTET.

A *weighing lysimeter* is more complex. It isolates a large portion of a field in an embedded tank, with the entire mass on a scale, sometimes constructed to allow access through an underground room where weight is recorded. The utility of lysimeters and even evaporation pans is greatly limited by their scarcity across North America. They do, however, provide a necessary data base with which to develop other methods of estimating POTET.

The Thornthwaite method of estimating POTET is based on empirical measurements from watersheds and lysimeters in the central and eastern United States. Thornthwaite found that POTET is best approximated using mean monthly air temperature and daylength, the latter a function of the measuring station's latitude. Both factors are easily determined, making it relatively simple to calculate POTET and the other water-balance components in the equation indirectly. Water-balance results are obtained with fair accuracy for most midlatitude locations.

Thornthwaite's method has limitations and is more useful in some climates than others. His method does not consider warm or cool air movements (wind) across a surface, sublimation effects and surface-water retention when temperatures are below freezing, areas of poor drainage, or areas of frozen soil (*permafrost,* discussed in Chapter 17). Nevertheless, we use the method here because of its great utility as a teaching tool and its overall ease of application in geographic studies to analyze water budgets over large areas with acceptable results.

Figure 9-4 presents POTET values derived by the Thornthwaite method for the United States and Canada. Higher values are found in the South, with highest readings in the Southwest. Lower POTET values are found at higher latitudes and elevations.

Please compare this POTET (demand) map with Figure 9-2, which is the PRECIP (supply) map for the same region. The relationship between PRECIP supplies and POTET demands determines the remaining components of the water-balance equation. Can you identify from the two maps regions where PRECIP is higher than POTET? Or where POTET is higher than PRECIP? Where you live, is the water demand usually met by the precipitation supply? Or, does your area experience a natural shortage?

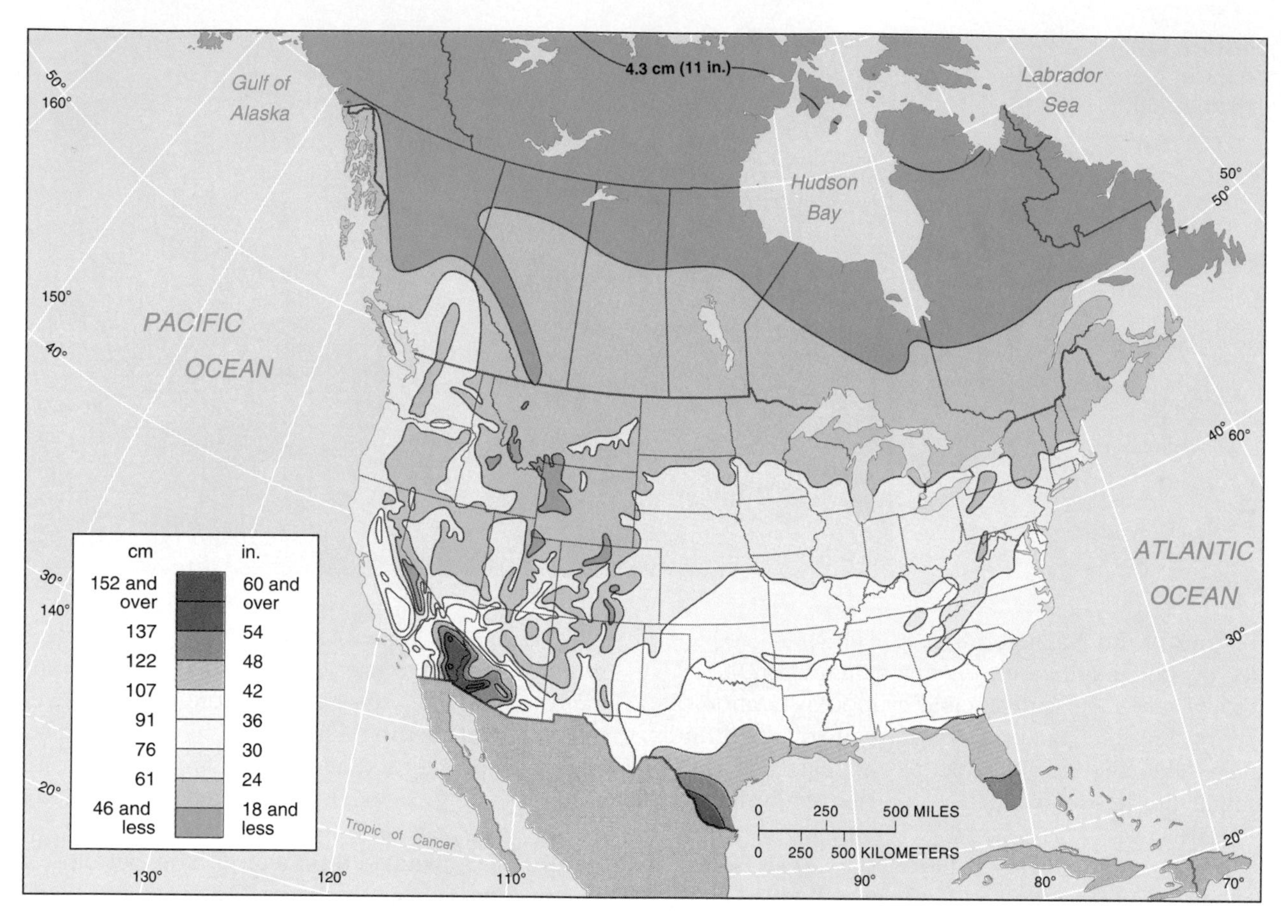

FIGURE 9-4
Potential evapotranspiration (demand) for the United States and Canada. [From Charles W. Thornthwaite, 1948, "An Approach Toward a Rational Classification of Climate," *Geographical Review* 38, p. 64. Adapted by permission from the American Geographical Society. Canadian data adapted from Marie Sanderson, "The Climates of Canada According to the New Thornthwaite Classification," *Scientific Agriculture* 28 (1948): 501–517.]

Deficit. POTET can be satisfied by PRECIP or by moisture derived from the soil, or it can be artificially met through irrigation. If these sources are inadequate, the site experiences a moisture shortage. This unsatisfied POTET is **deficit** (DEFIC). By subtracting DEFIC from POTET, we determine the **actual evapotranspiration,** or ACTET, that takes place. Under ideal conditions POTET and ACTET are about the same, so that plants do not experience a water shortage; prolonged deficits would lead to drought conditions.

Surplus. If POTET is satisfied and soil moisture is full, then additional water input becomes **surplus** (SURPL), or water oversupply. This excess water may sit on the surface in puddles, a situation known as **water detention,** or flow through the soil to groundwater storage. Surplus water that flows across the surface toward stream channels is termed **overland flow;** such flow combines with other precipitation and subsurface flows into river channels to make up the **total runoff** from the area. Because streamflow, or runoff, is generated

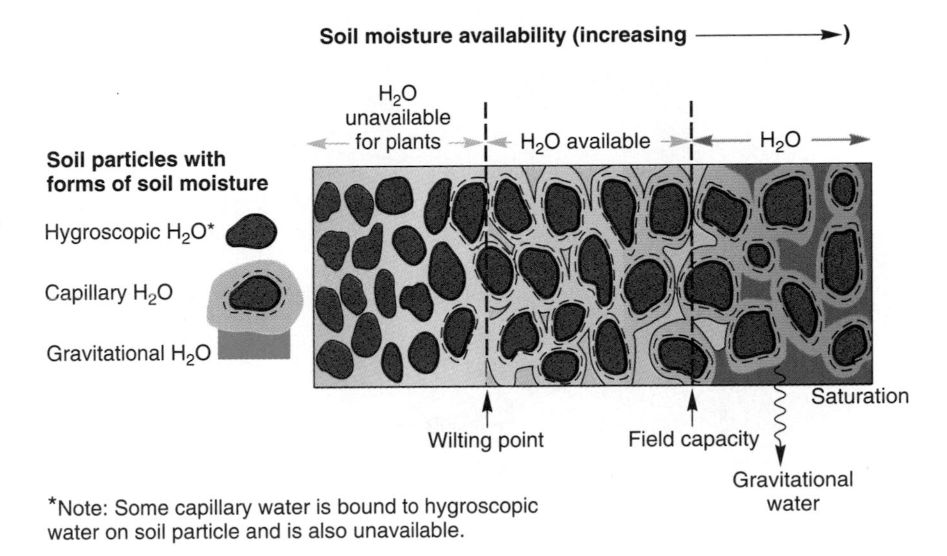

FIGURE 9-5
Classes of soil moisture: hygroscopic, capillary, available, and gravitational. [After Donald Steila, *The Geography of Soils*, © 1976, p. 45. Reprinted by permission of Prentice Hall, Inc., Englewood Cliffs, NJ.]

mostly from surplus water, the water-balance methodology can be thought of as a useful technique for indirectly estimating streamflow.

Soil Moisture Storage. Soil moisture storage is a "savings account" of water that can receive deposits and allow withdrawals as conditions change in the water balance. Soil moisture storage (ΔSTRGE) refers to the amount of water that is stored in the soil and is accessible to plant roots. (Soil moisture is discussed in more detail in Chapter 18.)

The moisture retained in the soil comprises two classes of water, only one of which is accessible to plants (Figure 9-5). The inaccessible class is **hygroscopic water,** the thin molecular layer that is tightly bound to each soil particle by the natural hydrogen bonding of water molecules. Such water exists even in the desert, and is not available to meet POTET demands. Soil is said to be at the **wilting point** when all that is left in the soil is unextractable water; the plants wilt and eventually die after a prolonged period of such moisture stress.

The soil moisture that is generally accessible to plant roots is **capillary water,** held in the soil by surface tension and hydrogen-bonding between the water and the soil. Almost all capillary water is **available water** in soil moisture storage and is removable for POTET demands through the action of plant roots and surface evaporation.

When soil becomes saturated after a precipitation event, water surpluses become **gravitational water** that percolates downward from the shallower capillary zone to the deeper groundwater zone. After water drains from the larger pore spaces, the available water remaining for plants is termed **field capacity,** or storage capacity. This water is held in the soil against the pull of gravity by hydrogen bonding. Field capacity is specific to each soil type and is an amount that can be determined by soil surveys.

Figure 9-6 is a simplified graph of soil water capacity, showing the relationship of soil texture to soil moisture content. Various plant types send roots to different depths and therefore are exposed to varying amounts of soil moisture. For example, shallow-rooted crops such as spinach, beans, and carrots send roots down about 65 cm (25 in.) in a silt loam, whereas deep-rooted crops such as alfalfa and shrubs exceed a depth of 125 cm (50 in.) in such a soil. A soil blend that maximizes available water is best for supplying plant water needs. Based on Figure 9-6, can you determine the soil texture with the greatest quantity of available water?

Plant roots work the soil for their water needs and exert a tensional force on soil moisture. As

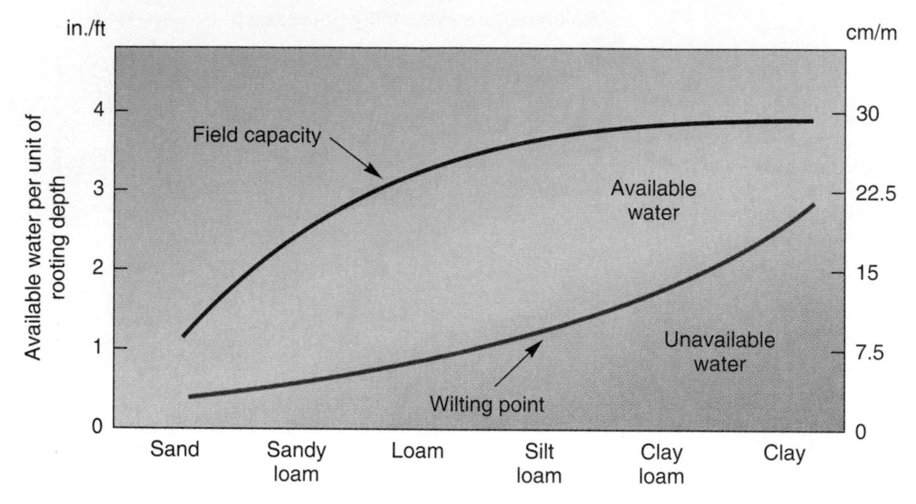

FIGURE 9-6
The relationship between soil moisture availability and soil texture. [After U.S. Department of Agriculture, *1955 Yearbook of Agriculture—Water,* p. 120.]

water evaporates from the leaves, a pressure gradient is established throughout the plant, thus "pulling" water into the plant roots. When soil moisture is at field capacity, plant roots are able to obtain water with less effort, and water is thus rapidly available to them. As the soil water is reduced by *soil moisture utilization,* the plants must exert greater effort to extract the same amount of moisture. As a result, even though a small amount of water may remain in the soil, plants may be unable to exert enough pressure to utilize it. The unsatisfied demand resulting from this situation is calculated as a deficit. Avoiding a deficit and reducing plant growth inefficiencies are the goals of a proper irrigation program, for the harder plants must work to get water, the less their yield and growth will be.

Whether naturally occurring or artificially applied, water infiltrates soil and replenishes available water content, a process known as *soil moisture recharge.* The texture and the structure of the soil and deeper groundwater zones dictate available pore spaces, or **porosity.** The flow of water through the soil, or **permeability,** determines the rate of soil moisture recharge. Permeability depends on particle sizes and the shape and packing of soil grains. It is expressed as a rate, generally lower for clays and higher for sands and gravels. (Chapter 18 presents more on soil texture, structure, and moisture characteristics.)

Thornthwaite assumed that the soil moisture recharge rate is 100% efficient as long as the soil is below field capacity and above a temperature of –1°C (30.2°F). Under real conditions we know that infiltration actually proceeds rapidly in the first minutes of precipitation, slowing as the upper layers of soil become saturated even though the soil below is still dry. Several agricultural practices, such as plowing and the addition of materials to loosen soil structure, can improve both soil permeability and the depth to which moisture can efficiently penetrate to recharge soil storage.

Sample Water Balances

Using all of these concepts, we now can apply the water-balance equation to a specific station. Table 9-1 presents the long-term average water supply and demand conditions for the city of Kingsport in the extreme northeast corner of Tennessee. Kingsport is at 36.6° N 82.5° W, with an elevation of 390 m (1280 ft), and has maintained weather records for more than half a century. (A complete set of data for Kingsport appears in Appendix A.) The monthly values in Table 9-1 assume that PRECIP and POTET are evenly distributed throughout each month, smoothing the actual daily and hourly conditions.

Let's first compare PRECIP with POTET by month to determine whether there is a net supply

TABLE 9-1
Water Balance for Kingsport, Tennessee

	Jan	Feb	Mar	Apr	May	Jun	Jul	Aug	Sep	Oct	Nov	Dec	Total
PRECIP	97	99	97	84	104	97	132	112	66	66	66	99	1119
	(3.8)	(3.9)	(3.8)	(3.3)	(4.1)	(3.8)	(5.2)	(4.4)	(2.6)	(2.6)	(2.6)	(3.9)	(44.1)
POTET	7	8	24	57	97	132	150	133	99	55	12	7	781
	(0.3)	(0.3)	(0.9)	(2.2)	(3.8)	(5.2)	(5.9)	(5.2)	(3.9)	(2.2)	(0.5)	(0.3)	(30.7)
PRECIP – POTET	+90	+91	+73	+27	+7	–35	–18	–21	–33	+11	+54	+92	—
	(+3.5)	(+3.6)	(+2.9)	(+1.1)	(+0.3)	(–1.4)	(–0.7)	(–0.8)	(–1.3)	(+0.4)	(+2.1)	(+3.6)	
STRGE	100	100	100	100	100	67	51	34	9	20	74	100	—
	(4.0)	(4.0)	(4.0)	(4.0)	(4.0)	(2.7)	(2.1)	(1.4)	(0.4)	(0.8)	(2.9)	(4.0)	
±ΔSTRGE	0	0	0	0	0	–33	–16	–17	–25	+11	+54	+26	±0
						(–1.3)	(–0.6)	(–0.7)	(–1.0)	(+0.4)	(+2.1)	(+1.1)	
ACTET	7	8	24	57	97	130	148	129	91	55	12	7	765
	(0.3)	(0.3)	(0.9)	(2.2)	(3.8)	(5.1)	(5.8)	(5.1)	(3.6)	(2.2)	(0.5)	(0.3)	(30.6)
DEFIC	0	0	0	0	0	2	2	4	8	0	0	0	16
						(0.1)	(0.1)	(0.2)	(0.3)				(0.6)
SURPL	90	91	73	27	7	0	0	0	0	0	0	66	354
	(3.5)	(3.6)	(2.9)	(1.1)	(0.3)							(2.6)	(14.0)

All quantities in millimeters (inches)

(+) or net demand (–) for water. (PRECIP, POTET, and their difference appear in color shading in Table 9-1.) We can see that Kingsport experiences a net supply from January through May and from October through December. However, the warm days and months from June through September result in net demands for water. Assuming a soil moisture storage capacity of 100 mm (4.0 in.), typical of shallow-rooted plants, those months of net demand for moisture are satisfied through soil moisture utilization.

Figure 9-7 graphs the water-balance components of these same monthly averages. Table 9-1 shows that soil moisture estimates are at field capacity through May, with the excess supply crediting surplus each month. In June the net demand is –35 mm (–1.4 in.), most of which is withdrawn from soil moisture storage, bringing soil moisture storage down to 67 mm (2.7 in.) at the end of June. The other 2 mm (0.1 in.) of net demand appears as DEFIC; that amount remains in the soil, reflecting

FIGURE 9-7
Annual average water balance components graphed for Kingsport, Tennessee.

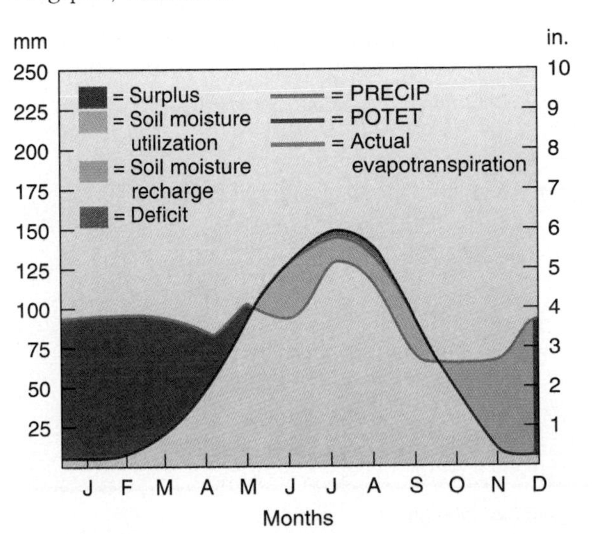

the increasing inefficiency of water removal by plants as soil moisture drops below field capacity. From June through September the ACTET (actual evapotranspiration) is progressively less than the POTET, and slight deficits continue to be recorded.

July, August, and September feature soil moisture utilization of –16 mm (–0.6 in.), –17 mm (–0.7 in.), and –25 mm (–1.0 in.), so that by the end of September shallow-rooted soil moisture is near the wilting point. During October, however, 11 mm of net supply is generated and is credited as soil moisture recharge. Recharge occurs again in November and December, bringing the soil moisture back up to field capacity by the end of the year, with a surplus of 66 mm (2.6 in.) estimated for December.

The Kingsport water-balance equations for the year and for the months of March and September are isolated in Table 9-2 to show the interaction of these various components. Check these three equations against Table 9-1 and perform the functions indicated to see whether we have accounted for all the PRECIP received.

Obviously, not all stations experience the surplus moisture patterns of this humid continental city and surrounding region; different climatic regimes experience different relationships among their water-balance components. Also, medium- and deep-rooted plants produce different results. Figure 9-8 presents water-balance graphs for several other cities in North America and the Caribbean Sea region. (Appendix A presents PRECIP, POTET, and temperature values, along with other information, for a wide variety of stations worldwide, grouped by climatic classification.)

Water Management: An Example

Because water is distributed unevenly over space and time, most water management projects attempt to rearrange water resources either geographically or across the calendar. Substantial human effort has been expended to override natural water-balance regimes and recurring drought. In North America, several large water-management projects are already in operation: the Bonneville Power Authority in Washington State, the Tennessee Valley Authority in the Southeast, the California Water Project, the Central Arizona Project, and the Churchill Falls and Nelson River Projects of Manitoba. Controversy and battles among conflicting interests invariably accompany water projects of this magnitude, especially because most ideal sites for multipurpose hydroelectric projects already are taken.

A particularly ambitious regional water project is the Snowy Mountains Hydroelectric Authority Scheme of Australia, where water-balance variations created the need to relocate water. The precipitation patterns in southeastern Australia are portrayed on the world precipitation map in Figure 10-2. In the Snowy Mountains, part of the Great Dividing Range in extreme southeastern Australia, precipitation ranges from 100 to 200 cm (40–80 in.) a year, whereas interior Australia receives under 50 cm (20 in.), and drops to less than 25 cm (10 in.) farther inland. POTET values are high throughout the Australian interior and lower in the higher elevations of the Snowy Mountains.

The Snowy Mountains Scheme was begun in the 1950s and completed over 20 years later. The plan

TABLE 9-2
Annual, March, and September Water Balances for Kingsport

	PRECIP	=	POTET	–	DEFIC	+	SURPL	±	Δ STRGE
Annual	1119	=	(781	–	16)	+	354	±	0
	(44.1)	=	(30.7	–	0.6)	+	(14.0)	±	0
March	97	=	(24	–	0)	+	73	±	0
	(3.8)	=	(0.9	–	0)	+	(2.9)	±	0
September	66	=	(99	–	8)	+	0	–	25
	(2.6)	=	(3.9	–	0.3)	+	0	–	(1.0)

All quantities in millimeters (inches).

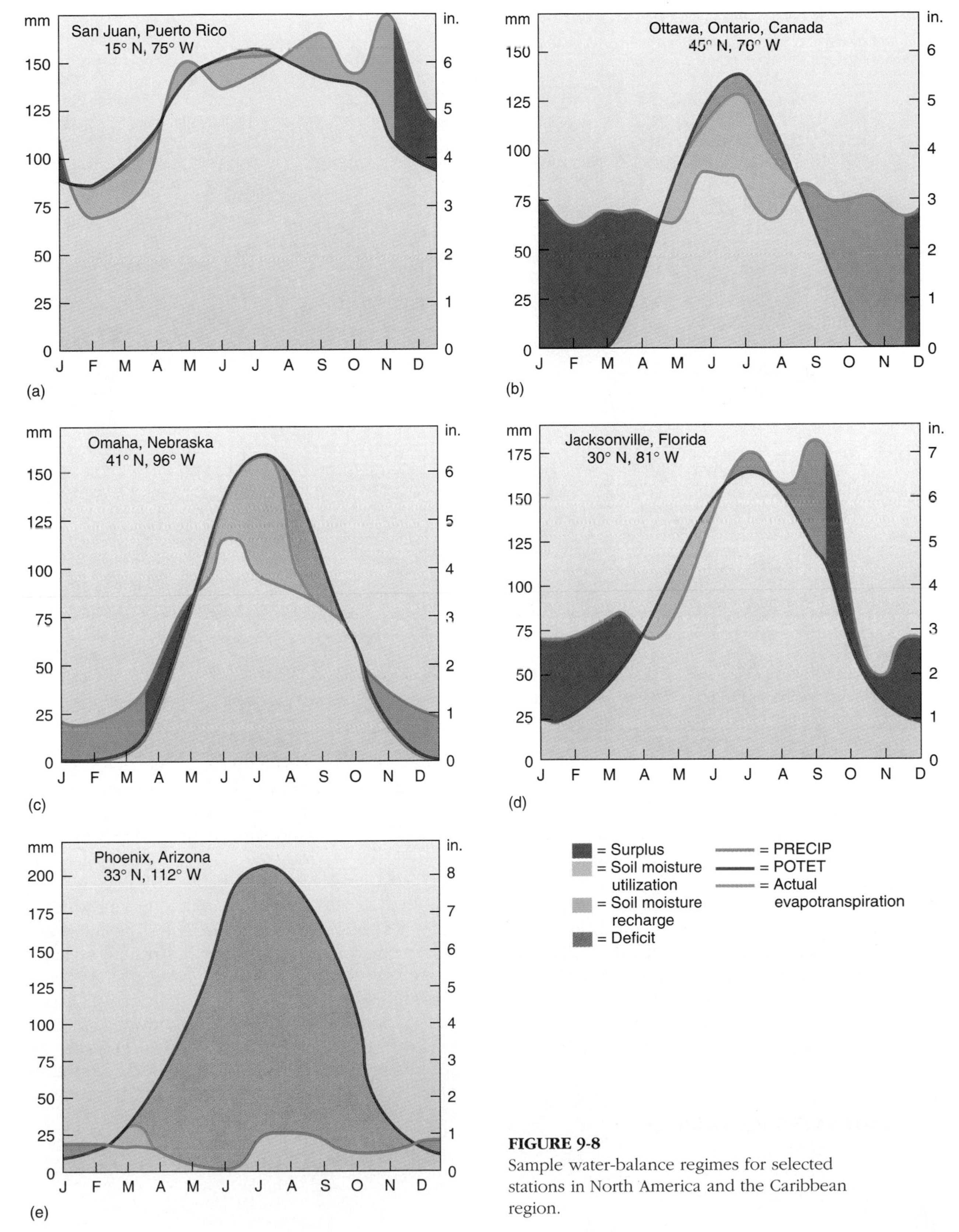

FIGURE 9-8
Sample water-balance regimes for selected stations in North America and the Caribbean region.

FIGURE 9-9
Snowy Mountains irrigation water makes possible the opening of new lands in Australia's eastern interior. [Photo by Snowy Mountains Hydroelectric Authority.]

was to take surplus water that flowed down the Snowy River eastward to the Tasman Sea and reverse the flow to support newly irrigated farmland in the interior of New South Wales and Victoria. Through some of the longest tunnels ever built, vast pumping systems, numerous reservoirs, and power plants, this plan was completed and even expanded to include many tributary streams and drainage systems high in the mountains. The westward-flowing rivers—the Murray, Tumut, and Murrumbidgee—have had their flows augmented, and as a result, new acreage is now in production in what was dry outback, formerly served only by wells drawing on meager groundwater (Figure 9-9).

GROUNDWATER RESOURCES

Groundwater is a part of the hydrologic cycle that lies beneath the surface and therefore is tied to surface supplies. Groundwater is the largest potential source of freshwater in the hydrologic cycle—larger than all surface lakes and streams combined. Between Earth's land surface and a depth of 4 km (13,100 ft) worldwide, some 8,340,000 km^3 (2,000,000 mi^3) of water resides. Despite this volume and its obvious importance, groundwater is subject to abuse. In many areas it is polluted and is consumed in quantities beyond natural recharge rates.

About 50% of the U.S. population derives a portion of its freshwater from groundwater sources. Between 1950 and 1990, the annual groundwater withdrawal increased 160%. In Canada, about 4% of all water withdrawals come from groundwater sources—4.2 million m^3/day, 1.5 billion m^3/year (148.3 million ft^3/day, 53 billion ft^3/year). Figure 9-10 shows potential groundwater resources in the United States and Canada.

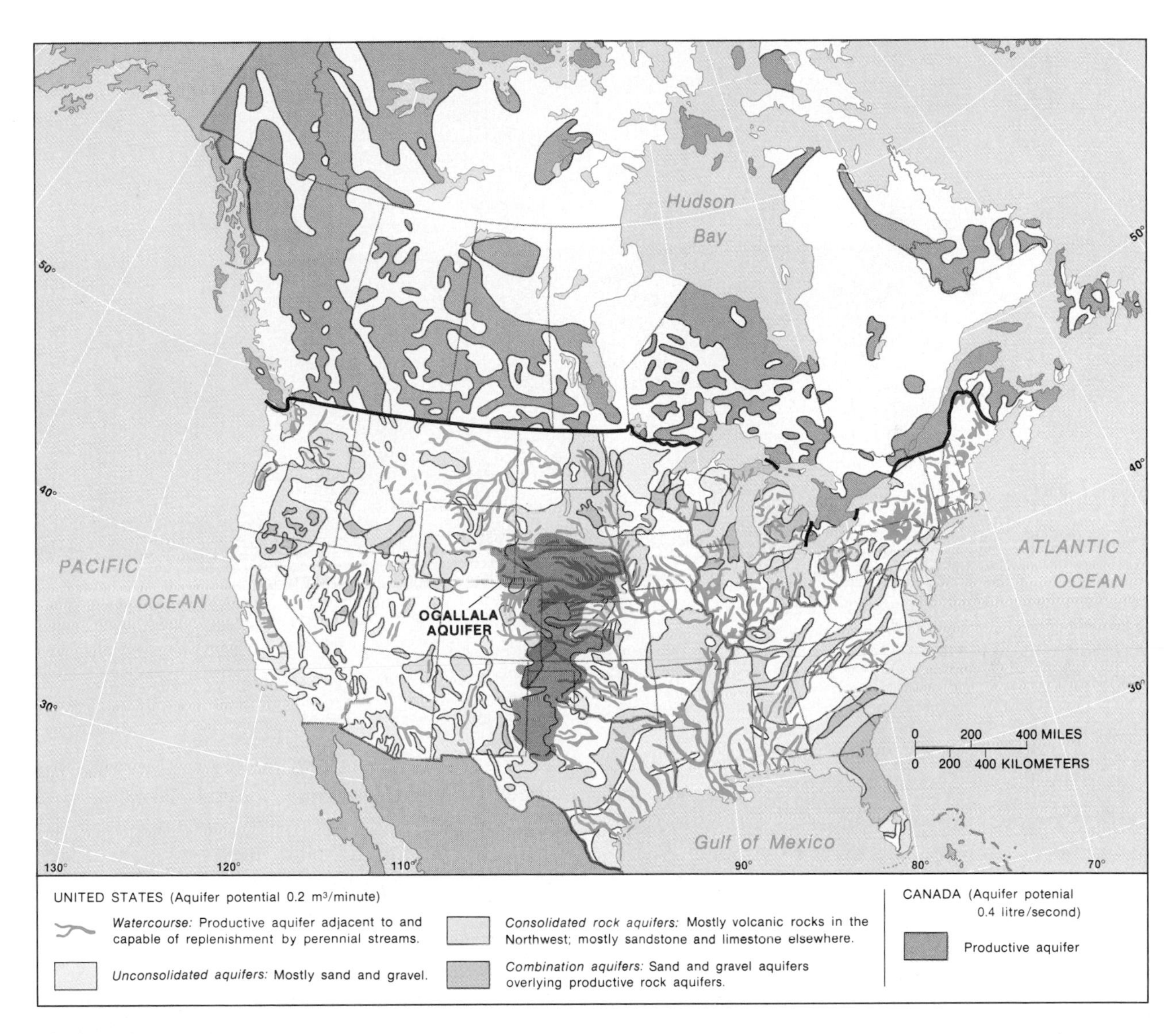

FIGURE 9-10
Groundwater resource potential for the United States and Canada. For the United States, we highlight areas underlain by productive aquifers capable of yielding fresh water to wells at 0.2 m^3/minute or more (for Canada, 0.4 liter/second). [Courtesy of Water Resources Council for the United States and the *Inquiry on Federal Water Policy* for Canada.]

Groundwater Description

Figure 9-11 illustrates the following discussion. Groundwater is fed by surplus water, which percolates downward from the zone of capillary water as gravitational water. This excess surface water moves through the **zone of aeration** where soil and rock are less than saturated. Eventually water reaches an area of collected subsurface water known as the **zone of saturation.** Like a sponge made of sand, gravel, and rock, this zone stores

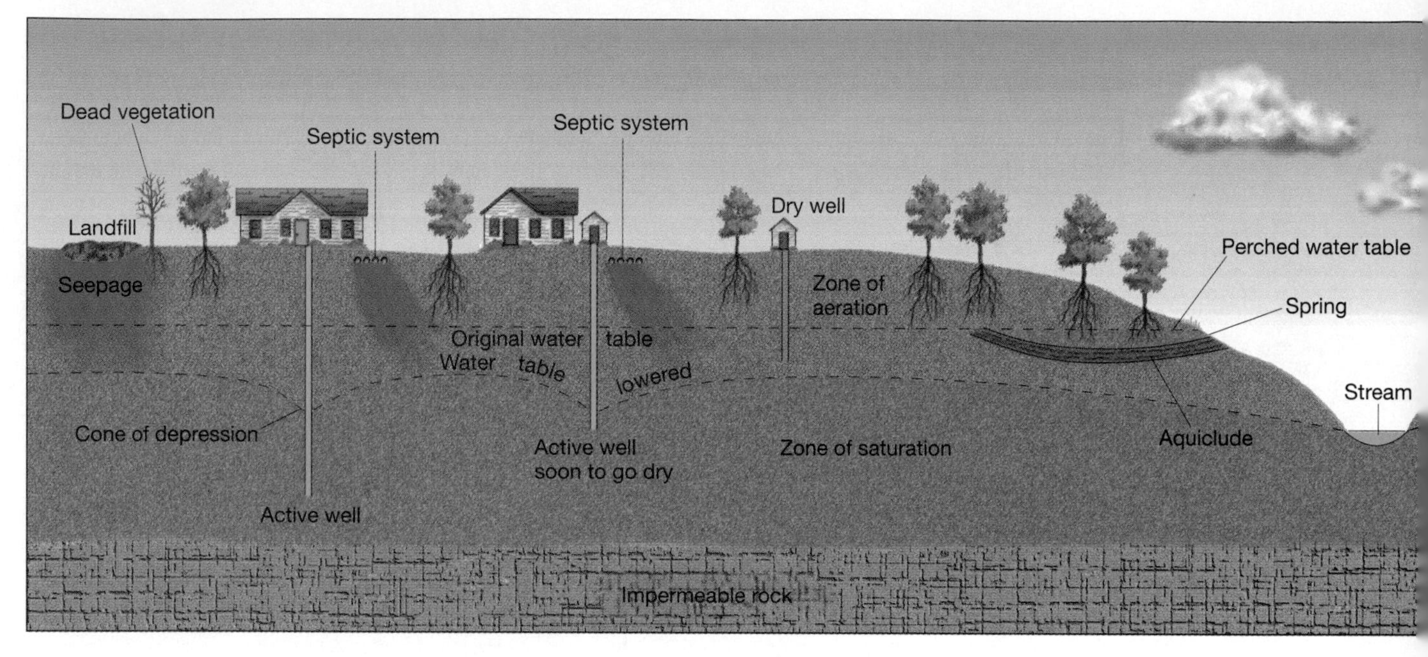

FIGURE 9-11
Subsurface groundwater characteristics, processes, water flows, and human interactions.

water within its structure, filling all available pores and voids. The *porosity* of this layer is dependent on the arrangement, size, and shape of individual component particles, the cement between them, how compacted they are, and the presence of noncompressed bits of rock. Subsurface structures are referred to as permeable or impermeable, depending on whether they permit or obstruct water flows.

An **aquifer** is a rock layer that is permeable to groundwater flow in significant amounts, whereas an **aquiclude** is a body of rock that does not conduct water in usable amounts. Technically, the zone of saturation is an **unconfined aquifer,** a water-bearing stratum that is not confined by an impermeable overburden. The upper limit of the water that collects in the zone of saturation is called the **water table;** it is the contact point between the zones of saturation and aeration.

Water movement is controlled by the slope of the water table, which generally follows the contours of the land surface (Figure 9-11). To optimize their potential flow, water wells must penetrate the water table. When the water table intersects the surface, it creates springs or feeds lakes or river beds (Figure 9-12). Ultimately, groundwater may enter stream channels and flow as surface water. During dry periods the water table can act to sustain river flows. Groundwater tends to move toward areas of lower pressure and elevation in the subsurface environment.

A confined aquifer differs from an unconfined aquifer in several important ways. A **confined aquifer** is bounded above and below by impermeable layers of rock or sediment, while an unconfined aquifer is not (Figure 9-11, right). In addition, the two aquifers differ in **aquifer recharge area,** which is the surface area where water enters an aquifer to recharge it. For an unconfined aquifer, that area generally extends above the entire aquifer, but in the case of a confined aquifer the recharge area is far more restricted, as you can see in the figure. Once these recharge areas are identified and mapped, the appropriate government body should zone them to prohibit pollution discharges, septic and sewage system installations, or

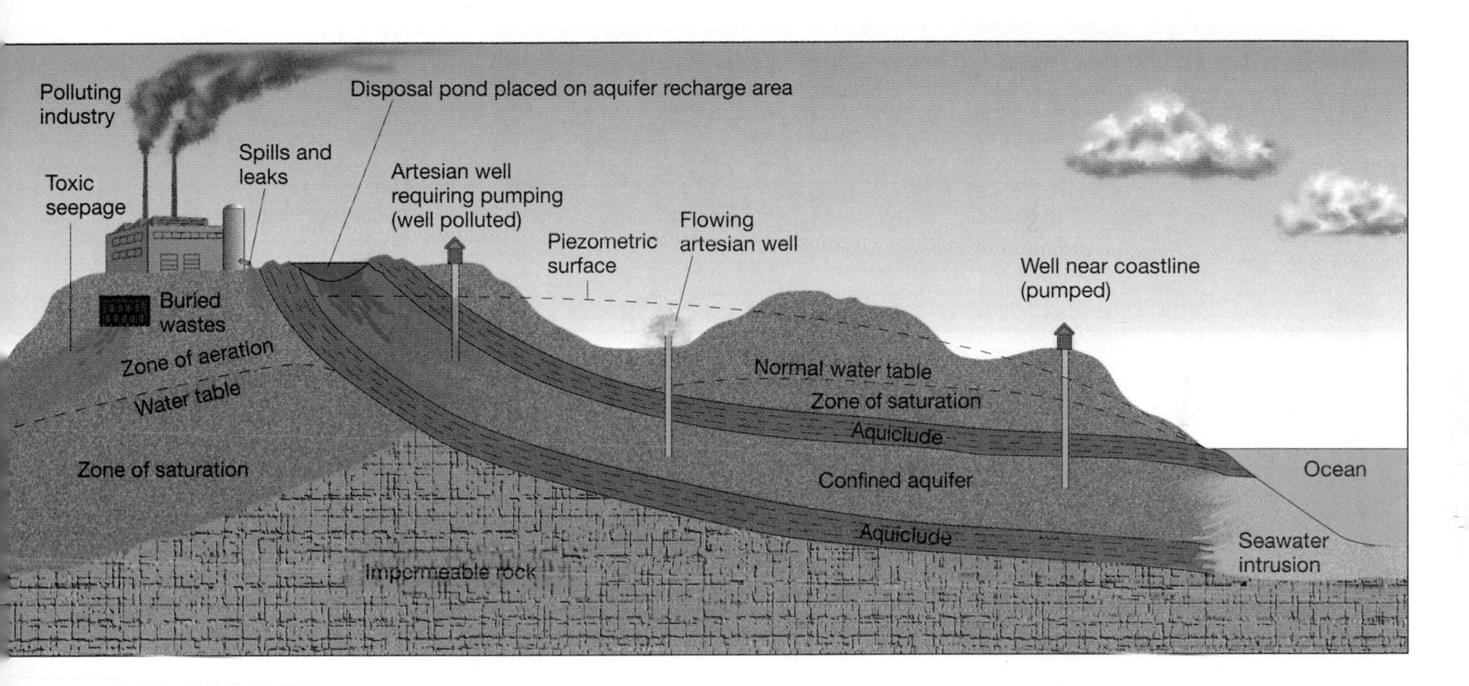

hazardous-material dumping. Of course, an informed population would insist on such action.

Aquifers also differ in pressure (Figure 9-11, right). A well drilled into an unconfined aquifer must be pumped to make the water level rise above the water table. In contrast, the water in a confined aquifer is under the pressure of its own weight, creating a pressure level called the **piezometric surface,** which actually can be above ground level. Under this condition, **artesian water,** or groundwater confined under pressure, may rise up in wells and even flow out at the surface if the head of the well is below the piezometric surface. In other wells, however, pressure may be inadequate, and the artesian water must be pumped the remaining distance to the surface.

Groundwater Utilization

Specific yield is the groundwater available for utilization at the surface. It is expressed as a percentage of the total volume of the saturated aquifer (when it can be determined). As water is pumped from a well, the adjacent water table within an unconfined aquifer will **draw down,** or become

FIGURE 9-12
An active spring, evidence of groundwater and a probable exposed aquifer. [Photo by author.]

FIGURE 9-13
Central-pivot irrigation in the Palouse region of east-central Washington is used in the production of alfalfa. Each irrigated circle covers a quarter section (66 hectares, 160 acres). Since 1954 irrigated lands in the region increased by 49% to over 615,000 ha (1.5 million acres). At the time of this photograph ash from the 1980 eruption of Mount Saint Helens covered surrounding wind-blown sandy soils. [Photo by Georg Gerster/Comstock. Used by permission. All rights reserved.]

lower, if the rate of pumping exceeds the horizontal flow of water in the aquifer. The resultant shape of the water table around the well is called a **cone of depression** (Figure 9-11, left).

Aquifers frequently are pumped beyond their flow and recharge capacities, resulting in **groundwater mining.** Large tracts experience chronic groundwater overdrafts in the Midwest, West, lower Mississippi Valley, Florida, and the intensely farmed Palouse region of eastern Washington State (Figure 9-13). In many places the water table or artesian water level has declined more than 12 m (40 ft). Groundwater mining is of special concern today in the Ogallala aquifer, which is the topic of Focus Study 9-1.

Another possible effect of water removal from an aquifer is that the stratum loses its internal support—the water in the pore spaces between the rock grains—and overlying rock may crush the aquifer, resulting in land subsidence and lower surface elevations. For example, within an 80 km (50 mi) radius of Houston, Texas, land has subsided more than 3 m (10 ft). In the Fresno area of the San Joaquin Valley in central California, after years of intensive irrigation, land levels have dropped almost 10 m (32 ft) because of water removal and soil compaction. Unfortunately, collapsed aquifers may not be rechargeable even if surplus gravitational water becomes available, because pore spaces may be permanently closed. Aquifers also are destroyed by strip mining for coal and lignite in such places as eastern Texas, northeastern Louisiana, Wyoming, Arizona, and other locales.

An additional problem arises when aquifers are overpumped near the ocean. In such a circumstance the groundwater interface between freshwater and seawater often migrates inland. As a result, wells may become contaminated with saltwater, and the aquifer may become useless as a freshwater source (Figure 9-11 illustrates this *seawater intrusion* on the far right margin). Aquifers also can be invaded by ancient water that was trapped in sediments at the time they were deposited (known as *connate water).* The contamination may be halted by pumping freshwater back into the aquifer, although the contaminated aquifer usually cannot be reclaimed.

This problem plagues the Persian Gulf states, which soon may run out of freshwater. Their vast groundwater resource is being overpumped to such an extent that saltwater is filling aquifers tens of kilometers inland. By the year 2000, water on

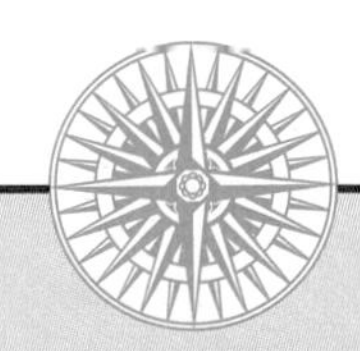

Focus Study 9-1 Ogallala Aquifer Overdraft

The largest known aquifer in the world lies below a six-state area extending from southern South Dakota and Nebraska to Texas, a region known as the American High Plains (Figure 1). The sand and gravel of the Ogallala aquifer (High Plains aquifer) were charged with meltwaters from retreating glaciers for several hundred thousand years. However, for the past 100 years, Ogallala groundwater has been heavily mined. The mining has been especially intense since World War II, when center-pivot irrigation devices were introduced (Figure 2). These large circular devices facilitate production of wheat, sorghums, cotton, corn, and about 40% of the grain fed to cattle in the United States.

The Ogallala aquifer is used to irrigate an area representing about one-fifth of all U.S. cropland: some 5.7 million hectares (14 million acres) are provided with water from 150,000 wells. In this process, water is being removed from the aquifer at 26 billion m^3 (21 million acre-feet) a year, an increase of more than 300% since 1950. In the past four decades the water table in the aquifer has dropped more than 30 m (100 ft), and throughout the 1980s it has averaged a 2 m (6 ft) drop per year because of extensive groundwater mining. U.S. Geological Survey scientists estimate that recharging the recoverable portions of the Ogallala aquifer (those portions that have not been crushed or subsided) would take at least 1000 years if mining stopped today!

The aquifer is most imperiled in Texas, where the saturated layer is shallower and some 75,000 wells are in operation. The declining water table in Floyd County, Texas, illustrates the overdraft problem (Figure 3). Between 1952 and 1982, pumping costs in Floyd County have increased 221% (adjusted for increases in the price index for crops), even though irrigated acreage dropped during that same time by almost 15%.

Obviously, billions of dollars of agricultural activity cannot be abruptly halted, but neither can profligate water mining continue. This issue raises tough questions involving the management of cropland, reliance on extensive irrigation, the demand to produce exports, and the need to assess the importance of certain crops that are in chronic oversupply as well as various crop subsidies. Present agricultural practices, if continued, will result in the loss of about half the total Ogallala aquifer resource, and a two-thirds loss in the Texas portion alone, by the year 2020.

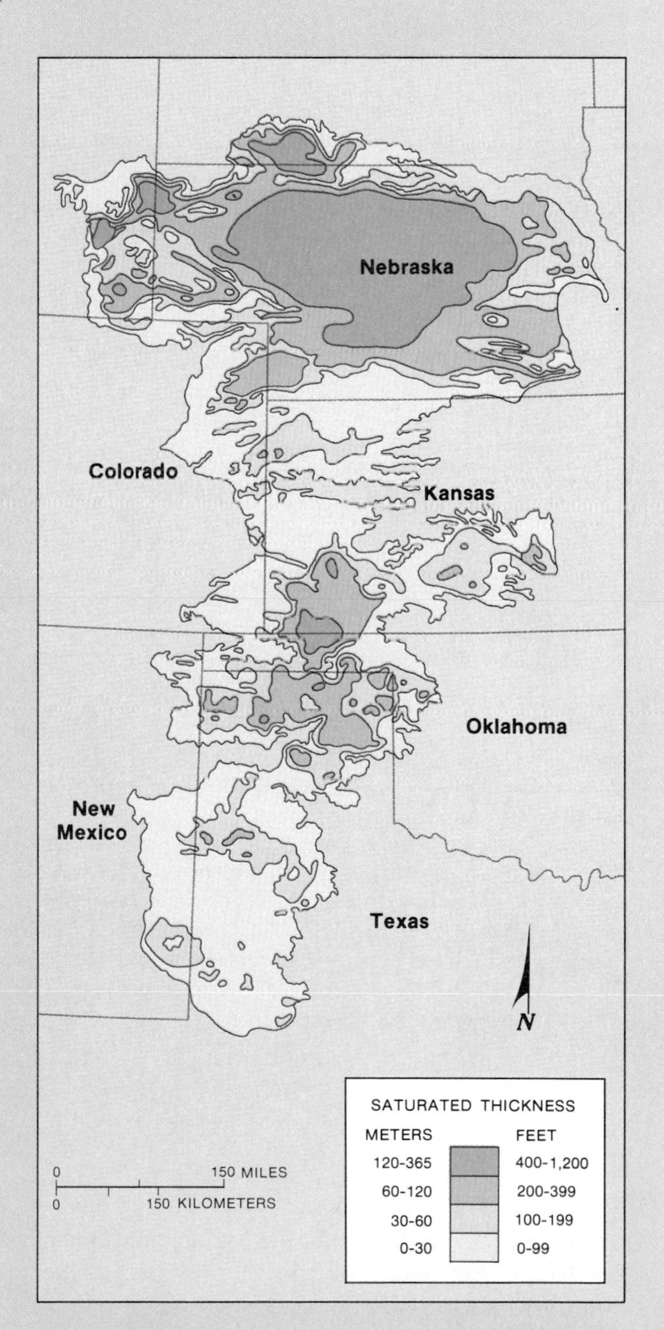

FIGURE 1
Map of the Ogallala acquifer. [After D. E. Kromm and S. E. White, "Interstate Groundwater Management Preference Differences: The Ogallala Region," *Journal of Geography* 86, no. 1 (Jan.-Feb. 1987): 5.]

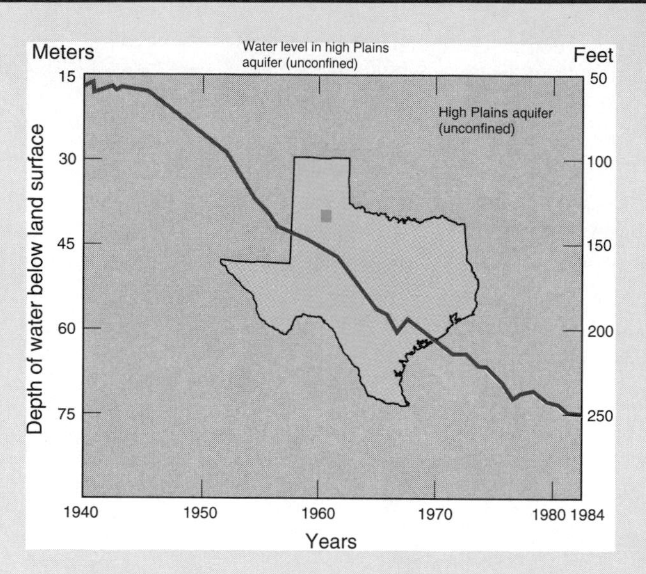

FIGURE 3
Water levels in Floyd County, Texas, wells. [After John E. Schefter, *Declining Ground-Water Levels and Pumping Costs: Floyd County, Texas,* U.S. Geological Survey Water Supply Paper No. 2275, Washington, DC: Government Printing Office, 1985, p. 114.]

FIGURE 2
Fields of corn are watered by myriad central-pivot irrigation systems in north-central Nebraska. Each irrigated circle covers a quarter section (66 hectares, 160 acres) in the U.S. Township and Range survey system. A growing season for corn requires from 10 to 20 revolutions of the sprinkler arm, depending on the weather, applying about 3 cm of water per revolution. The Ogallala aquifer is at depths of over 76 m (250 ft) in this part of Nebraska. [Photo by Georg Gerster/Comstock. Used by permission. All rights reserved.]

the Arabian Peninsula may be undrinkable. Imagine having the main issue in the Middle East become water, not oil!

Remedies for groundwater overutilization are neither easy nor cheap. In the Persian Gulf area additional freshwater is being obtained through desalinization plants along the coasts. In fact, approximately 60% of the world's 4000 desalination plants are presently operating in Saudi Arabia and other Persian Gulf states. A very costly water pipeline that will carry water overland from Turkey some 1500 km (930 mi, the distance from New York City to St. Louis) is being seriously considered to import 6 million m^3 (1.68 billion gallons) a day through two branching pipes—one toward Israel, the other through the Arabian Peninsula. Estimates predict the cost of water piped through the "Peace Pipeline" to be about the same as that of the desalinated water.

In addition, end-use patterns could be altered, for example, by upgrading traditional agricultural practices and adding greater efficiency to urban usage, to help reduce the demand for groundwater. Aquifers and rivers could be shared, but at present no negotiated accords exist for this purpose in the Middle East. The 1991 Persian Gulf War posed a grave threat to desalinization facilities in Saudi Arabia, and many in Iraq and Kuwait were damaged or destroyed.

Pollution of the Groundwater Resource

When surface water is polluted, groundwater also becomes contaminated because it is recharged from surface-water supplies. Groundwater migrates very slowly compared to surface water. Surface water flows rapidly and flushes pollution down stream, but sluggish groundwater, once contaminated, remains polluted virtually forever. Pollution can enter groundwater from industrial injection wells, septic tank outflows, seepage from hazardous-waste disposal sites, industrial toxic-waste dumps, residues of agricultural pesticides, herbicides, fertilizers, and residential and urban wastes in landfills. Thus, pollution can come either from a point source or from a large general area (a nonpoint source), and it can spread over a great distance, as illustrated in Figure 9-11.

Despite widespread publicity about groundwater pollution in the United States, Canada, and Europe, few new protective laws have been established. Laws such as the U.S. Safe Drinking Water Act (1974) and the U.S. Resource Conservation and Recovery Act (1976, 1984) either neglect or do not explicitly direct action for groundwater protection.

In 1988 the General Accounting Office (GAO) completed a nationwide survey of groundwater protection standards. It found that 38 states have nonspecific narrative standards (general standards that do not specify a numeric maximum concentration), which are difficult to characterize because of their lack of uniformity. Only 26 states have any numerical or empirical groundwater standards. In the face of government inaction, serious contamination of groundwater continues nationwide.

> **One characteristic is the practical irreversibility of groundwater pollution, causing the cost of clean-up to be prohibitively high. . . . It is a questionable ethical practice to impose the potential risks associated with groundwater contamination on future generations when steps can be taken today to prevent further contamination.***

DISTRIBUTION OF STREAMS

Streams always have been important to human beings. They represent only 1250 km^3 (300 mi^3) of water, the smallest volume of any of the freshwater categories we have discussed. Yet, streams are the portion of the hydrologic cycle on which we most depend, representing four-fifths of all the water we use. Whether streams are perennial (constantly flowing) or intermittent, the total runoff that moves through them comes from surplus surface-water runoff, subsurface throughflow, and groundwater (Figure 9-14). Because of their rapid renewal, as compared to lake water or groundwater, streams are the best single measure of available water supply.

A stream's flow rate is termed its **discharge,** usually expressed as the volume of water that passes a given point per unit of time. In volume of runoff, the Western Hemisphere exceeds the Eastern Hemisphere, primarily because the Amazon River (Figure 9-15) carries a far greater volume than any other, and because four of the world's highest-discharge river systems are in the Western Hemisphere (Mississippi-Missouri, Orinoco, St. Lawrence, and Mackenzie river systems). The Atlantic Ocean receives about 1.5 times more runoff than the Pacific Ocean. Table 9-3 lists the outflow locations, lengths, and discharge values for the world's largest rivers. (Chapter 14 examines more closely the hydrologic aspects of streams and their interaction with the landscape.)

---

*James Tripp, "Groundwater Protection Strategies," *Groundwater Pollution, Environmental and Legal Problems*. Washington, DC: American Association for the Advancement of Science 1984, p. 137. Reprinted by permission.

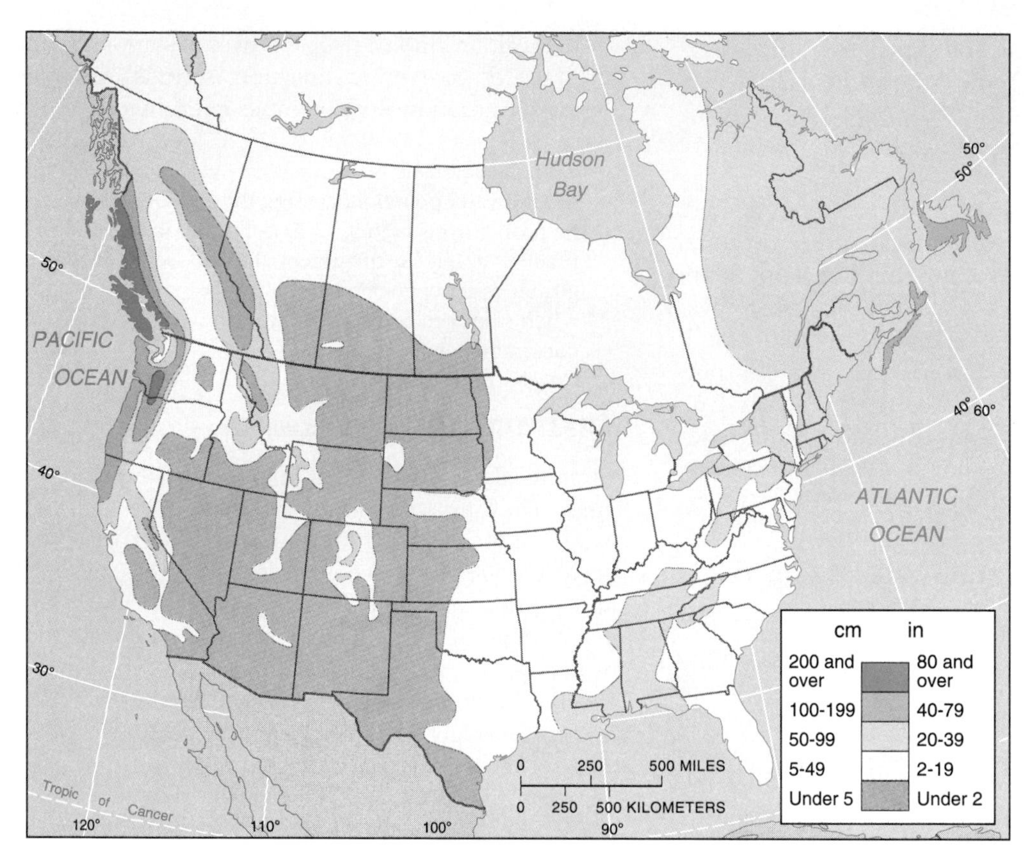

FIGURE 9-14
Distribution of runoff in the United States and Canada. [Data courtesy of Water Resources Council for the United States and *Currents of Change* by Environment Canada.]

FIGURE 9-15
The Amazon River discharges a fifth of all fresh water that enters the world's oceans through its 160 km (100 mi) wide mouth. Millions of tons of sediments are derived from a drainage basin as large as the Australian continent. [Space Shuttle photograph from NASA.]

TABLE 9-3
Largest Rivers on Earth Ranked by Discharge Volume

Rank by Volume	Average Discharge at Mouth in Thousands of cms (cfs)	River (with Tributaries)	Outflow/Location	Length km (mi)	Rank by length
1	212.5 (7500)	Amazon (Ucayali, Tambo, Ene, Apurimac)	Atlantic Ocean/Amapá-Pará, Brazil	6570 (4080)	2
2	79.3 (2800)	La Plata (Paraná)	Atlantic Ocean/Argentina	3945 (2450)	16
3	39.7 (1400)	Congo, also known as the Zaire (Lualaba)	Atlantic Ocean/Angola, Zaire	4630 (2880)	10
4	38.5 (1360)	Ganges (Brahmaputra)	Bay of Bengal/India	2898 (1800)	23
5	21.8 (770)	Yangtze (Chàng Chiang)	East China Sea/Kiangsu, China	5980 (3720)	4
6	17.4 (614)	Yenisey (Angara, Selenga or Selenge, Ider)	Yenisey Gulf of Kara Sea/Siberia	5870 (3650)	5
7	17.3 (611)	Mississippi (Missouri, Ohio, Tennessee, Jefferson, Beaverhead, Red Rock)	Gulf of Mexico/Louisiana	6020 (3740)	3
8	17.0 (600)	Orinoco	Atlantic Ocean/Venezuela	2737 (1700)	27
9	15.5 (547)	Lena	Laptev Sea/Siberia	4400 (2730)	11
10	14.2 (500)	St. Lawrence	Gulf of St. Lawrence/Canada and U.S.	3060 (1900)	21
17	7.9 (280)	Mackenzie (Slave, Peace, Finlay)	Beaufort Sea, Northwest Territories	4240 (2630)	12
33	2.83 (100)	Nile (Kagera, Ruvuvu, Luvironza)	Mediterranean Sea/Egypt	6690 (4160)	1

In the United States, the conventional unit used by hydrologists to measure streamflow is *cubic feet per second (cfs);* in Canada it is *cubic meters per second (cms).* In the eastern United States, or for large-scale assessments such as Figure 9-16, water managers speak in terms of *millions of gallons a day (MGD)* or *billions of gallons a day (BGD)* or *billions of liters per day.* In the western United States, where irrigated agriculture is so important, the measure frequently used is *acre-feet per year;* one acre-foot contains 325,872 gallons (43,560 ft^3; 1234 m^3; 1,233,429 liters).

Streamflows generally increase downstream because the area being drained increases. On the other hand, a stream that originates in a humid region and subsequently flows through an arid region usually decreases in discharge with distance because of high POTET rates. Such a stream is called an **exotic stream** and is exemplified by the Nile River (Table 9-3). In the United States the Colorado River flow also decreases with distance and, in fact, no longer reaches its mouth in the Gulf of California! The exotic Colorado River is depleted by upstream appropriation of water for agricultural and municipal uses, which is the subject of a Focus Study in Chapter 15.

In other regions, stream drainage that does not flow into the ocean, but instead outflows through

TABLE 9-4
Estimates of Available Global Water Supply

	Land Area in Thousands of km^2 (mi^2)	Mean Annual Discharge (Water Supply) in km^3 per yr (BGD)	Mean Annual Runoff in mm (in.)	Population in Millions for	
				1990	2000
Africa	30,600 (11,800)	4,220 (3,060)	139 (5.5)	645	877
Asia	44,600 (17,200)	13,200 (9,540)	296 (12.0)	3057	3544
China	9,560 (3,690)	2,880 (2,080)	300 (12.0)	1120	1256
India	3,290 (1,270)	1,590 (1,150)	485 (19.0)	832	962
Phillippines	300 (116)	390 (282)	1300 (51.0)	61	75
Commonwealth of Independent States (former USSR)	22,400 (8,650)	4,350 (3,150)	194 (7.6)	291	315
Australia-Oceania	8,420 (3,250)	1,960 (1,420)	245 (9.6)	27	30.4
Europe	9,770 (3,770)	3,150 (2,280)	323 (13.0)	500	513
North America (Canada, Mexico, U.S.)	22,100 (8,510)	5,960 (4,310)	286 (11.0)	364	397
United States (48 states)	7,820 (3,020)	1,620 (1,231)	182 (7.1)	—	—
United States (50 states)	9,360 (3,620)	2,340 (1,700)	250 (9.9)	250	268
South America	17,800 (6,880)	10,400 (7,510)	583 (23.0)	299	360
Global (excluding Antarctica)	134,000 (51,600)	38,900 (28,100)	290 (11.0)	5249	6127

evaporation or subsurface gravitational flow, terminates in areas of **internal drainage.** Portions of Asia, Africa, Australia, Mexico, and the western United States have regions with such internal drainage patterns.

Estimates of available water supplies for the continents and a few selected nations are presented in Table 9-4. Population figures are included to give an idea of the demand for the resource. Comparing the 182 mm (7.2 in.) mean annual runoff in the United States (48 states) to the 1300 mm (51.0 in.) in the Philippines indicates the runoff variability that occurs worldwide under varying climatic regimes.

OUR WATER SUPPLY

Water is critical for survival. Just to produce our food requires enormous quantities of water. For example, growing 77g (2.7 oz) of broccoli requires 42 L (11 gal) of water; producing 250 ml (8 oz) of milk consumes 182 L (48 gal) of water; 28 g (1 oz) of cheese, 212 L (56 gal); and 1 egg, 238 L (63 gal). Yet the quality and quantity of our supply often is taken for granted. The U.S. water supply is derived from surface and groundwater sources that are fed by an average daily precipitation of 4244 BGD (Figure 9-16). That sum is based on an average annual precipitation value of 76.2 cm (30 in.) divided evenly among the 48 contiguous states (excluding Alaska and Hawaii). The Canadian resource greatly exceeds that of the United States.

Daily Water Budget

The 4244 BGD that make up the U.S. water supply is not evenly distributed across the country or during the calendar year. Abundant water supplies in New England result in only about 1% of renewable available water being consumed each year, whereas in the dry Colorado River basin mentioned earlier, the discharge is completely consumed. In fact, by treaty and compact agreements, the Colorado actually is budgeted beyond its average discharge through political misunderstanding of the water resource.

Figure 9-16 shows that the daily water balance has two general outputs: 71% ACTET (nonirrigated

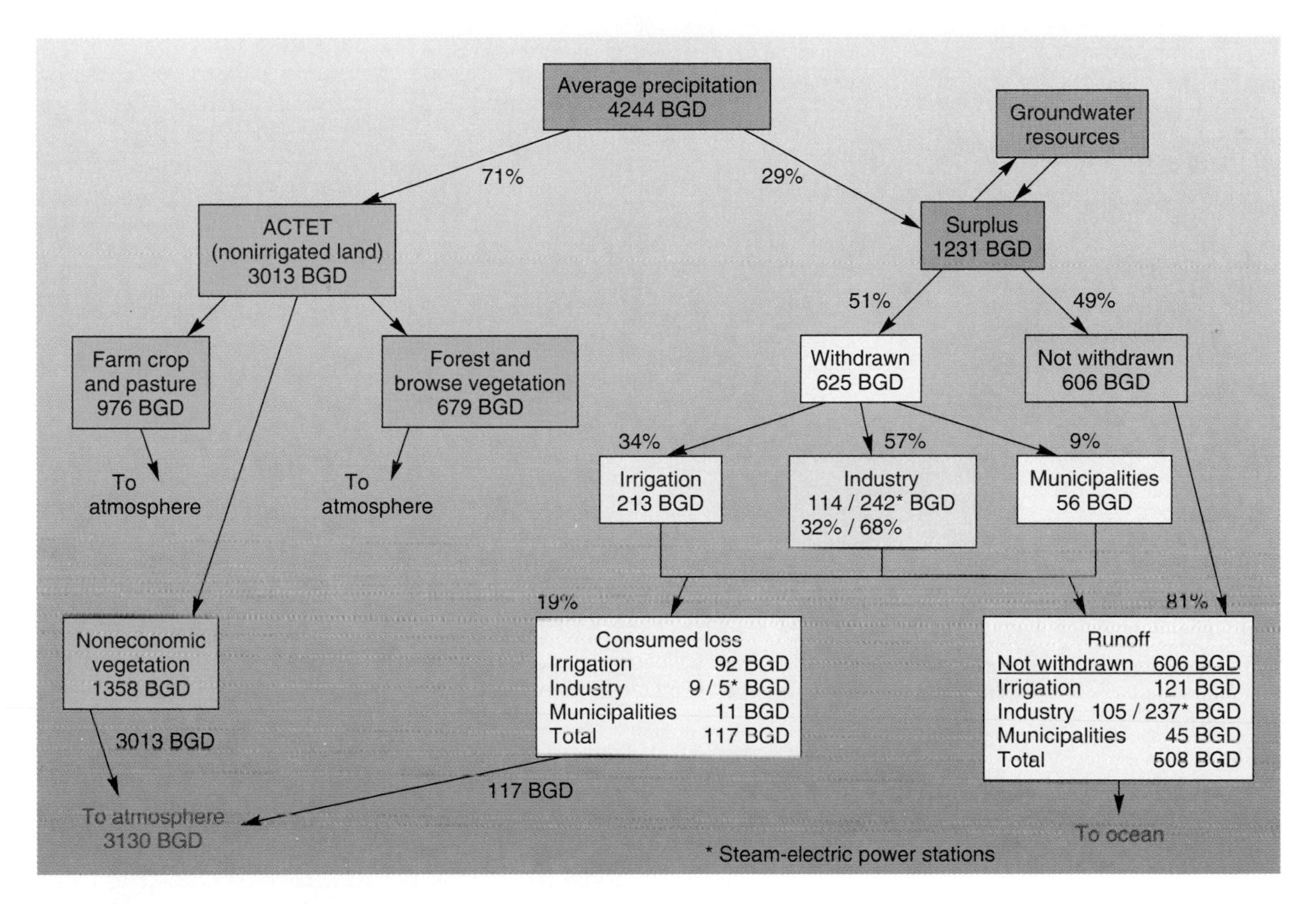

FIGURE 9-16
Daily water budget for the United States (lower 48 states) in billions of gallons a day (BGD). [Adapted and updated from Abel Wolman, 1962, *Water Resources: A Report to the Committee on Natural Resources of the National Academy of Sciences,* Publication 1000-B. Washington, DC: Natural Resources Council.]

land) and 29% surplus. About 71% (3013 BGD) of the daily supply passes through nonirrigated land—farm lands and pastures; forest and browse (young leaves, twigs, and shoots); and noneconomic vegetation—and eventually returns to the atmosphere to continue its journey through the hydrologic cycle. Only 29% (1231 BGD) becomes surplus runoff, available for withdrawal and consumption. (This same surplus value appears in Table 9-4 under mean annual discharge for the 48 states.) In keeping with our earlier discussion, groundwater is linked to surface surpluses and thus is not presented as an independent water resource.

Consumptive uses are those that remove water from the stream at some point, without returning it farther downstream. **Withdrawal** refers to water that is removed from the supply, used for various purposes, and then returned to the supply. Withdrawn water represents the opportunity to extend the resource through reuse and recycling. Often, however, withdrawn water is returned after being contaminated with chemical pollutants, waste (from industry, agriculture, and municipalities), or heat. Contaminated or not, because this water is returned to streams, it becomes a part of all water systems downstream. Thus, for example, the entire Missouri-Ohio-Mississippi River drainage is of im-

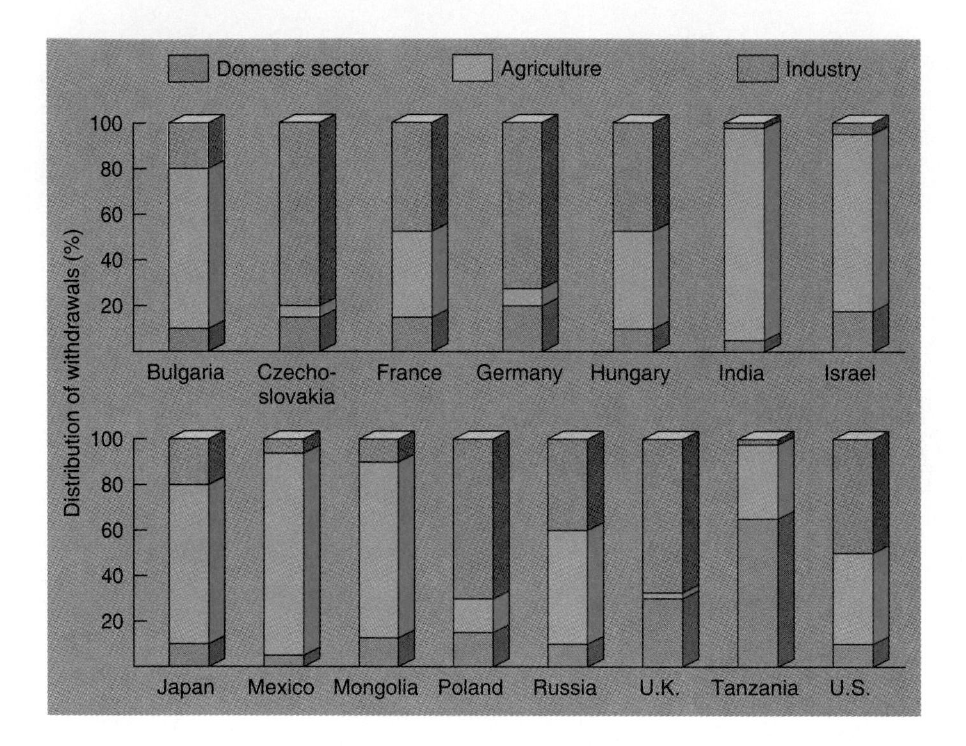

FIGURE 9-17
Water withdrawal by sector for selected countries. [After Gilbert R. White and others, 1982, *Resources and Needs: Assessment of the World Water Situation,* U.N. Water Conference, 1977.]

mediate concern to residents of New Orleans, the last city to withdraw municipal water from the polluted river.

The estimated U.S. withdrawal of water for the mid-1990s is 625 BGD, a sizable increase from the 450 BGD withdrawn in 1980. Figure 9-16 illustrates the three main flows of this withdrawn water in the United States: irrigation (34%), industry (57%), and municipalities (9%). In contrast, Canada uses only 7% of its withdrawn water for irrigation, and 84% for industry. Figure 9-17 compares several nations in their use of withdrawn water and graphically illustrates the differences between developed and developing countries.

In Figure 9-16 the 606 BGD that are not withdrawn flow as stream runoff, performing other important functions—transportation, waste removal and dilution, recreation, hydroelectric power generation, and provision of fish and wildlife habitats. Together with discharge water from irrigation, industry, and municipalities, this runoff returns to the ocean.

On an individual level, statistics suggest that urban dwellers directly use an average 680 liters (180 gallons) of water per person per day, whereas rural populations average only 265 liters (70 gallons) per person per day. However, each of us indirectly accounts for a much greater water demand because of our use of food and products produced with water. For the United States, dividing the total water withdrawal (625 BGD) by a population of 259 million people (estimate for 1995) yields a per capita direct and indirect use of 9080 liters (2400 gallons) of water per day! *Clearly, the average American lifestyle depends on enormous quantities of water and thus is vulnerable to shortfalls or quality problems.*

Future Considerations

When the seemingly large supply of precipitation is budgeted as in Figure 9-16, the limits of the water resource become readily apparent, leading to obvious questions about how to satisfy the growing demand for water. Water supplies per person decline as population increases, thus world population growth since 1970 has reduced per capita water supplies by a third. Also, pollution limits the water-resource base in both the United States and Canada, so that even before quantity constraints are felt, quality problems may limit the health and growth of a region. Of course, the overall situation in Canada is quite different than in the United States because of the large water supply and the lower overall demand generated by a smaller population of 27 million (1990 estimate).

Internationally, the world's water future appears even more complex. About one-fourth of the annual renewable water worldwide will be actively utilized by the year 2000. By that time, one-half billion people will be depending on the polluted Ganges River alone. Furthermore, of the 200 major river basins in the world (basins where rivers drain into an ocean, lake, or inland sea), 148 are shared by two nations, and 52 are shared by from three to ten nations. Clearly, water issues are international in scope, yet we continue toward a water crisis without a concept of a *world water economy* as a frame of reference.

SUMMARY—Water Balance and Water Resources

Sustainable management of water resources is of growing importance to society. Scientists employ a water-balance accounting approach to the hydrologic cycle with emphasis on plants and soil moisture. Using the water-balance approach, we can compare the supply of PRECIP to the natural demands of POTET for any area of land—a yard, farm, or region—or for a global perspective. The uneven geographical distribution and seasonal variation of water supply and demand often require intervention if civilization is to succeed in an area. Society interacts with surplus water supplies through the use of groundwater and surface runoff.

Groundwater replenishment is tied to surface surpluses because groundwater does not exist as an independent water source. In our present era, groundwater is being mined beyond natural recharge rates in many areas. Pollution of groundwater is being reported to a greater extent than ever before. Once contaminated, an aquifer remains permanently ruined, making this an issue of national significance because of society's dependence on groundwater.

The streams that drain the landscape often constitute our immediate water supply. Americans in the 48 contiguous states withdraw approximately one-half of the available surplus runoff for irrigation, industry, and municipal uses. Water-resource planning on a regional and global scale, using water-balance principles, is an important consideration for the future if societies everywhere are to have enough water of adequate quality.

KEY TERMS

actual evapotranspiration
aquiclude
aquifer
aquifer recharge area
artesian water
available water
capillary water
cone of depression
confined aquifer
consumptive uses
deficit
discharge
draw down
evaporation
evaporation pan
evapotranspiration
exotic stream
field capacity
gravitational water
groundwater mining
hygroscopic water
internal drainage
lysimeter
overland flow
permeability
piezometric surface
porosity
potential evapotranspiration

precipitation
rain gauge
soil moisture storage
specific yield
surplus
total runoff
transpiration
unconfined aquifer
water detention
water table
wilting point
withdrawal
zone of aeration
zone of saturation

REVIEW QUESTIONS

1. What are the components that make up the water-balance concept? How does the concept relate to the hydrologic cycle model?
2. Briefly describe the work and contribution of C. W. Thornthwaite in water-resource analysis.
3. Write out the water-balance equation, and work through it for one or two months, using specific data from Table 9-1. Try this "bookkeeping" method using data from Appendix A.
4. Explain how to derive actual evapotranspiration in the water-balance equation.
5. What is potential evapotranspiration? How do we go about measuring this potential rate? What method did Thornthwaite use to determine this value?
6. Explain the operation of soil moisture storage, soil moisture utilization, and soil moisture recharge. Include discussion of field capacity, capillary water, and wilting point concepts.
7. In the case of silt loam soil from Figure 9-6, roughly what is the available water capacity? How is this value derived?
8. What does the PRECIP-POTET bookkeeping procedure tell us about the individual months in the Kingsport, Tennessee, example?
9. Describe the different water balances in March and September in Kingsport.
10. In terms of water balance and water management, explain the logic behind the Snowy Mountains Scheme in southeastern Australia.
11. Are groundwater resources independent of surface supplies or interrelated with them? Explain your answer.
12. Make a simple sketch of the subsurface environment, labeling zones of aeration and saturation and the water table in an unconfined aquifer. Next, add a confined aquifer to the sketch.
13. At what point does groundwater utilization become groundwater mining? Use the Ogallala aquifer example to explain your answer.
14. What is the nature of groundwater pollution? Can contaminated groundwater be cleaned up easily? Explain.
15. Are there any differences in the freshwater inflow into the Atlantic Ocean and that into the Pacific Ocean? Why?
16. What are the five largest rivers on Earth in terms of discharge? Relate these to the weather patterns in each area. Can these flows be explained through an analysis of regional water budgets?
17. Describe the principal pathways involved in the water budget of the contiguous 48 states. What is the difference between consumptive use and withdrawal of water resources?
18. Characterize each of the sectors withdrawing water: irrigation, industry, and municipalities. What are the present usage trends in developed and developing nations?

SUGGESTED READINGS

Adler, Jerry, and William J. Cook, Stryker McGuire, Gerald C. Lubenow, Martin Kasindorf, Frank Maier, and Holly Morris. "Are We Running Out of Water? The Browning of America," *Newsweek* 23 (February 1981): 26–36.

Black, Peter E. *Thornthwaite's Mean Annual Water Balance*. Silviculture General Utility Library, Program GU-101. Syracuse, NY: State University College of Forestry, 1966.

Cairns, John, Jr., and Ruth Patrick, eds. *Managing Water Resources*. Environmental Regeneration Series. New York: Praeger, 1986.

Chelimsky, Eleanor. *The Nation's Water—Key Unanswered Questions About the Quality of Rivers and Streams*. Report to the Chairman, Subcommittee on Investigations and Oversight, Committee on Public Works and Transportation, House of Representatives, September 1986. Washington, DC: General Accounting Office.

Chelimsky, Eleanor. *Groundwater Standards—States Need More Information from EPA*. Report to the Chairman, Subcommittee on Hazardous Wastes and Toxic Substances, Committee on Environment and Public Works, U.S. Senate, March 1988. Washington, DC: General Accounting Office.

Chelimsky, Eleanor. "Fighting Groundwater Contamination." Testimony given before Subcommittee on Water Resources, Transportation, and Infrastructure, Committee on Environment and Public Works, U.S. Senate, May 1988. Washington, DC: General Accounting Office.

Chow, Ven Te, ed. *Handbook of Applied Hydrology—A Compendium of Water-Resources Technology*. New York: McGraw-Hill, 1964.

Cobb, Charles E., Jr. "The Great Lakes' Troubled Waters," *National Geographic* 172, no. 1 (July 1987): 2–31.

Mather, John R., ed. *The Measurement of Potential Evapotranspiration*. Publications in Climatology, vol. 1. Seabrook, NJ: Johns Hopkins University, Laboratory of Climatology, 1954.

Postel, Sandra. *Water: Rethinking Management in an Age of Scarcity*. Worldwatch Paper 62. Washington, DC: Worldwatch Institute, 1984.

Postel, Sandra. *Last Oasis: Facing Water Scarcity*. New York: W. W. Norton and Company, 1992.

Sidney, Hugh, Dan Goodgame, and David Brand. "The Big Dry," "Just Enough to Fight Over," and "Is the Earth Warming Up?" *Time* 132, no. 1 (July 4, 1988): 12–18.

Smith, Richard A., Richard B. Alexander, and M. Gordon Wolman. "Water Quality Trends in the Nation's Rivers," *Science* 235 (March 27, 1987): 1607–15.

Thornthwaite, Charles W., and John R. Mather. *Instructions and Tables for Computing Potential Evapotranspiration and the Water Balance*. Publications in Climatology, vol. 3. Centerton, NJ: Drexel Institute of Technology, Laboratory of Climatology, 1957.

Thornthwaite, Charles W., and John R. Mather. *The Water Balance*. Publications in Climatology, vol. 1. Centerton, NJ: Drexel Institute of Technology, Laboratory of Climatology, 1955.

Travis, Curtis, and Elizabeth L. Etnier, eds. *Groundwater Pollution—Environmental and Legal Problems*. American Association for the Advancement of Science Selected Symposia Series no. 95. Boulder, CO: Westview Press, 1984.

U.S. Geological Survey. *National Water Summary 1984*. Water-Supply Paper No. 2275. Washington, DC: Government Printing Office, 1985.

U.S. Geological Survey *National Water Summary 1985*. Water-Supply Paper No. 2300. Washington, DC: Government Printing Office, 1986.

Hawaiian landscape, Kauai, Hawaii. [*Photo by A. R. Christopherson.*]

10

GLOBAL CLIMATE SYSTEMS

Earth experiences an almost infinite variety of weather. Even the same location may go through periods of changing weather. This variability, when considered along with the average conditions at a place over time, constitutes climate. Climates are so diverse that no two places on Earth's surface experience exactly the same climatic conditions, although general similarities permit grouping and classification for regional studies.

In a traditional framework, early climatologists faced the challenge of identifying patterns as a basis for establishing climatic classifications. Currently at the forefront of scientific effort by climatologists is developing models that can simulate complex interactions and causal relationships of the atmosphere and hydrosphere. Vast linkages exist in the Earth-atmosphere system: for example, strong monsoonal rains in West Africa correlate with the development of intense Atlantic hurricanes; an El Niño in the Pacific is tied to drought-breaking rains in California, a drought in Australia, and floods in Louisiana; and many other examples.

A modern emphasis for climatologists is the complex change occurring in global patterns of climate. Prediction of such change and its implications for society is an essential goal of contemporary climatology.

CLIMATE SYSTEM COMPONENTS

Our study of climate synthesizes the energy-atmosphere phenomena and water-weather phenomena discussed in the first nine chapters of this text. The condition of the atmosphere at any given place and time is called *weather*. The consistent behavior of weather over time, including its variability, represents a region's **climate.** Therefore, climate includes not just the averages but the extremes experienced in a place, for a climate may have temperatures that average above freezing every month, yet still threaten severe frost problems for agricultural activities. Think of climate as dynamic rather than static, or fixed and inactive.

Climatology, the study of climate, involves analysis of the patterns in time or space created by various physical factors in the environment. One type of climatic analysis involves discerning areas of similar weather statistics and grouping these into **climatic regions** that contain characteristic weather patterns. Climate classifications represent an effort to formalize these patterns and determine their related implications to humans.

Weather components that combine to produce Earth's climates include insolation patterns, temperature, humidity, seasonal distribution and amounts of precipitation, atmospheric pressure and winds, types of weather disturbances, and cloud coverage and occurrence. Climates may be humid with distinct seasons, dry with consistent warmth, moist and cool—almost any combination is possible. There are places where it rains more than 20 cm (8 in.) each month and average monthly temperatures remain higher than 27°C (80°F) throughout the year. Other places may go without rain for 10 years at a time. Expected climatic patterns often are disrupted, as is the case with the recurring El Niño phenomenon, discussed in Focus Study 10-1.

Climates form an essential basis for *ecosystems,* the natural, self-regulating communities formed by plants and animals in their nonliving environment. On land, the basic climatic regions determine to a large extent the location of the world's major ecosystems. These regions, called *biomes,* include forest, grassland, savanna, tundra, and desert (Figure 10-1). Plant, soil, and animal communities are associated with these biomes. (In Chapters 19 and 20 we discuss global ecosystems and principal biomes in detail.)

Because climate is continuously cycling through periodic change and is never really stable, these ecosystems should be thought of as being in a constant state of adaptation and response. The present global climatic warming trend probably is producing such changes in plant and animal distributions at this time. Climatologists speculate that

FIGURE 10-1
A northern coniferous forest biome, indicative of a cool, moist climate at this elevation in the Canadian Rockies. [Photo by author.]

changes in climate and natural vegetation in the next 50 years could exceed the total of all changes since the peak of the last ice-age episodes, some 18,000 years ago. The possibility exists that the operative factor behind these climatic changes is human activity.

Radiation, Temperature, Pressure, and Air Masses

Before we look at climatic regions, let's briefly review key concepts from the first nine chapters. Uneven insolation over Earth's surface, varying with latitude, is the energy input for the climate system (Chapter 2 and Figures 2-12 and 2-13). Daylength and temperature patterns vary diurnally and seasonally. The principal controls of temperature are latitude, altitude, land-water heating differences, and the amount and duration of cloud cover. The pattern of world temperatures and annual temperature ranges are discussed in Chapter 5 and portrayed in Figures 5-8, 5-10, and 5-11.

Average temperatures and daylength are the basic factors that help us approximate POTET (potential evapotranspiration). These temperature variations are coupled with dynamic forces in the atmosphere to form Earth's patterns of atmospheric pressure and its global wind systems (see Figures 6-11 and 6-17). Important too are the locations and physical characteristics of air masses, those vast bodies of homogeneous air that form over oceanic and continental source regions.

Precipitation Input

The hydrologic cycle provides the basic means of transferring energy and mass through Earth's climate system. The moisture input to climate is precipitation in the form of rain, sleet, snow, and hail. Figure 9-2 portrays the distribution of mean annual precipitation in North America; Figure 10-2 shows the worldwide distribution of precipitation and identifies several patterns. As shown in Figure 10-2, precipitation decreases from western Europe inland toward the heart of Asia, whereas high precipitation values dominate southern Asia where the landscape receives moisture-laden monsoonal winds in the summer. Southern South America reflects a pattern of wet western (windward) slopes and dry eastern (leeward) slopes.

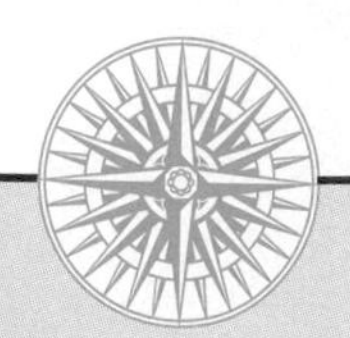

Focus Study 10-1 The El Niño Phenomenon

Climate is defined as the consistent behavior of weather over time. However, average weather conditions may include extremes that depart from what is considered normal. The El Niño-Southern Oscillation phenomenon in the Pacific Ocean is such a disruption of expected climate patterns.

Normally, as shown in Figure 6-21, the region off the west coast of South America is dominated by the northward-flowing Peru Current. These cold waters move toward the equator and join the westward movement of the south equatorial current. The Peru Current is part of the overall counterclockwise circulation that normally guides the winds and surface ocean currents around the subtropical high-pressure cell dominating the eastern subtropical Pacific in the Southern Hemisphere (visible on the world pressure maps in Figure 6-9). As a result, Guayaquil, Ecuador, normally receives 91.4 cm (36 in.) of precipitation per year, with high pressures dominant, whereas islands in the western Pacific and the maritime landmasses of the Indonesian archipelago receive over 254 cm (100 in.), with low pressures dominant.

Occasionally, and for as-yet unexplained reasons, pressure patterns alter and shift from their usual locations, thus affecting surface ocean currents and weather on both sides of the Pacific. Unusually high pressure develops in the western Pacific and lower pressure in the eastern Pacific. This regional change is an indication of large-scale ocean-atmosphere interactions. Trade winds normally moving from east to west weaken and can be replaced by an eastward (west-to-east) flow. Sea-surface temperatures of the central and eastern Pacific then rise above normal, sometimes becoming more than 8C° (14F°) warmer, replacing the normally cold, upwelling, nutrient-rich water along Peru's coastline. Such ocean-surface warming, the "warm pool," may extend to the International Date Line.

People fishing the waters off Peru have come to expect a slight warming of surface waters that lasts for a few weeks at irregular intervals. The name El Niño ("the boy child") is used by Peruvians because these episodes periodically occur around the traditional December celebration time of Christ's birth (although they have occurred as early as spring and summer).

The shifting of atmospheric pressure and wind patterns across the Pacific, known as the Southern Oscillation, both initiates and supports El Niño (Figure 1). Thus, the oceanic El Niño/Southern Oscillation (ENSO) designation is derived. National Oceanographic and Atmospheric Administration (NOAA) scientists speculate that ENSO events historically occurred in 1877–1878, 1884, 1891, 1899–1900, 1925, 1931, and 1941–1942, among other times. And, they are certain that such ENSO events occurred in 1953, 1957–1958, 1965, 1969–1970, 1972–1973, 1976–1977, 1986–1987, 1991–1992, and 1993, with the most intense episode of this century during 1982–1983. The expected interval for occurrence is 3–5 years, but may range from 2 to 12 years.

The anchovy catch off Peru dropped from 12 million tons in 1970 to a low of 0.5 million tons in 1983, following the warm water blockage of the cold, upwelling current by several ENSO events, principally in 1972–1973 and 1976–1977, culminating in the 1982–1983 episode. In addition, migratory birds did not arrive at expected places and times. Related effects occurred into the midlatitudes: droughts in South Africa, southern India, Australia, and the Philippines; strong hurricanes in the Pacific, including Tahiti and French Polynesia; and flooding in the southwestern United States and mountain states, Bolivia, Cuba, Ecuador, and Peru. The Colorado River flooding shown in the Focus Study in Chapter 15 was in large part attributable to the 1982–1983 El Niño.

The 1982–1983 El Niño, clearly noted on Figure 1c, strengthened by an extremely strong Southern Oscillation, led scientists to further clarification of complex global interconnections among pressure patterns in the Pacific, sea-surface temperatures, occurrences of drought in some places, excessive rainfall in others, and the disruption of fisheries and wildlife. Estimates place the overall damage at more than $8 billion worldwide for that one event. It appears that the climate of one location is related to climates elsewhere—although it should be no surprise that Earth operates as a vast integrated system. "It is fascinating that what happens in one area can affect the whole world. As to why this happens, that's the question of the century. Scientists are trying to make order out of chaos."*

---

*Alan Strong of NOAA's National Environmental Satellite, Data, and Information Service, quoted by Barry Siegel in "El Niño: Global Climatic Patterns Become Chaotic," *Los Angeles Times*, August 17, 1983.

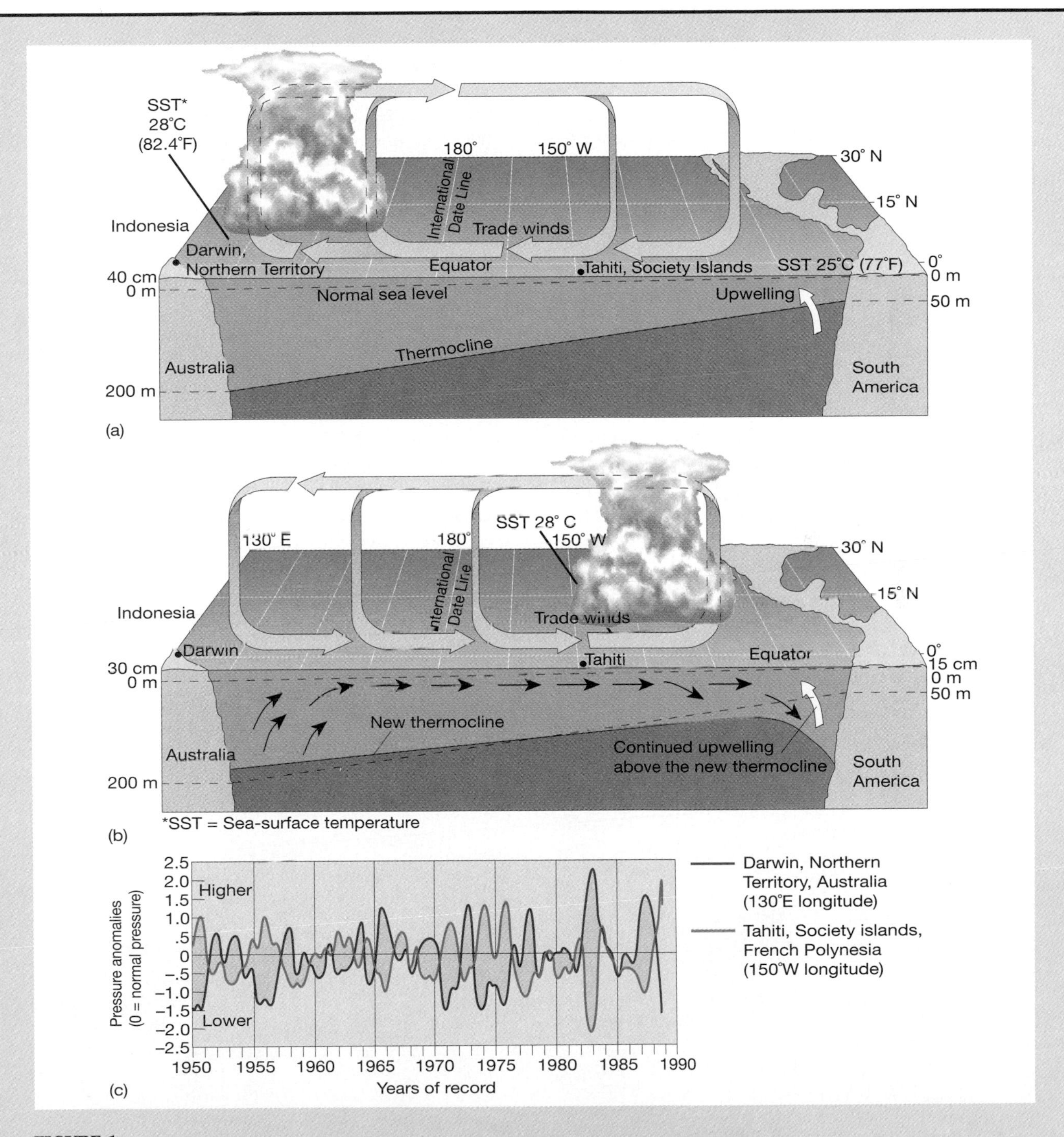

FIGURE 1
Normal (a) and reversed (b) pressure, wind, and current patterns across the Pacific Ocean. Air pressure over Darwin, Australia, and Tahiti is the driving force behind the Southern Oscillation (c). Scientists have linked these pressure and related wind changes to occurrences of El Niños. [After C. S. Ramage, "El Niño." Copyright © 1985 by Scientific American, Inc. (c) from K. E. Trenberth, "Signal Versus Noise in the Southern Oscillation," *Monthly Weather Review* 112 (1984): 326–32.

A six-year dry spell in the western United States was ended in early 1993 by record rain and snowfall produced by lingering effects from the 1992 El Niño. These unpredictable changes in expected climatic patterns present a challenge to science to decipher the Earth-atmosphere system.

Currently, remote-sensing capabilities of satellites and improved surface observations are permitting scientists to monitor sea-surface temperatures and to better predict the timing and intensity of each El Niño—as an example, see the sea-surface temperature image in Figure 5-5. Understanding will involve many principles of physical geography and will require a global-scale spatial perspective.

Most of Earth's deserts and regions of permanent drought are located in lands dominated by subtropical high-pressure cells, with bordering lands grading to grasslands and to forests as precipitation increases. The most consistently wet places on Earth straddle the equator in the Amazon region of South America, the Zaire region of Africa, and the Indonesian and Southeast Asian area, all of which are influenced by equatorial low pressure and the intertropical convergence zone (Figure 6-10).

Simply comparing the two principal climatic components—temperature and precipitation—reveals important relationships indicative of the distribution of climate types (Figure 10-3). Synoptic analysis of weather elements—that is, the characterization of weather over a region at a specific

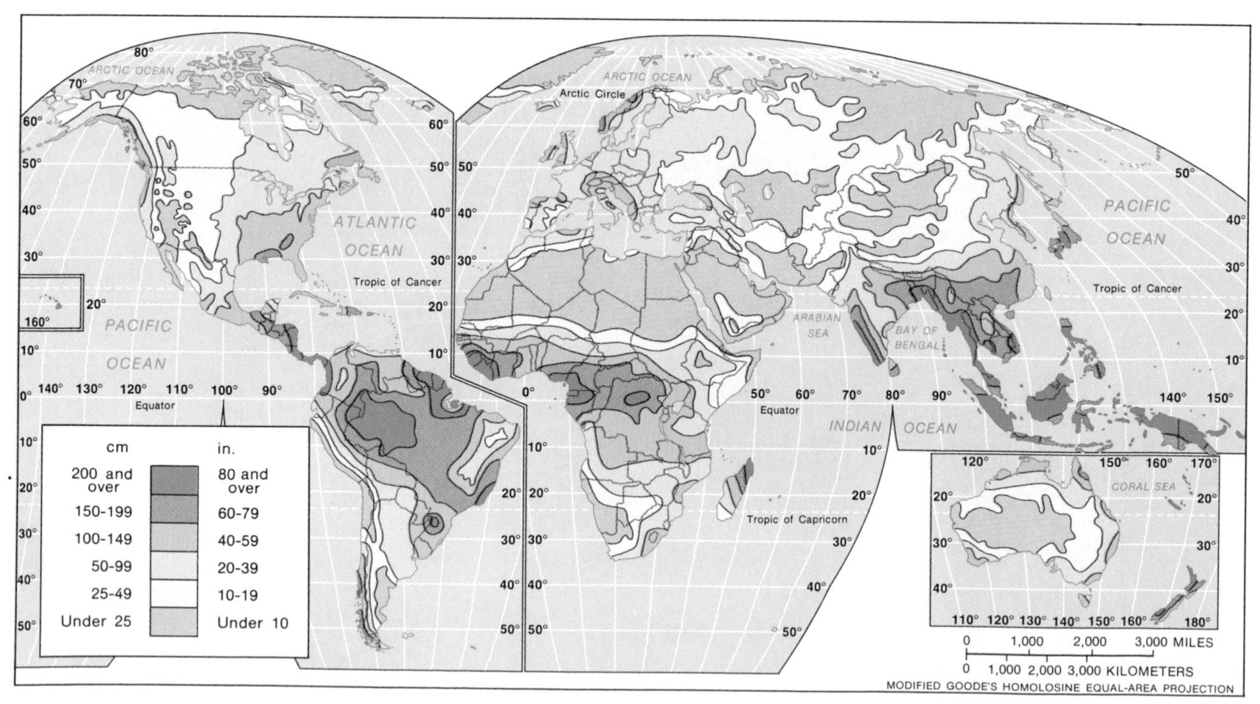

FIGURE 10-2
Worldwide average annual precipitation.

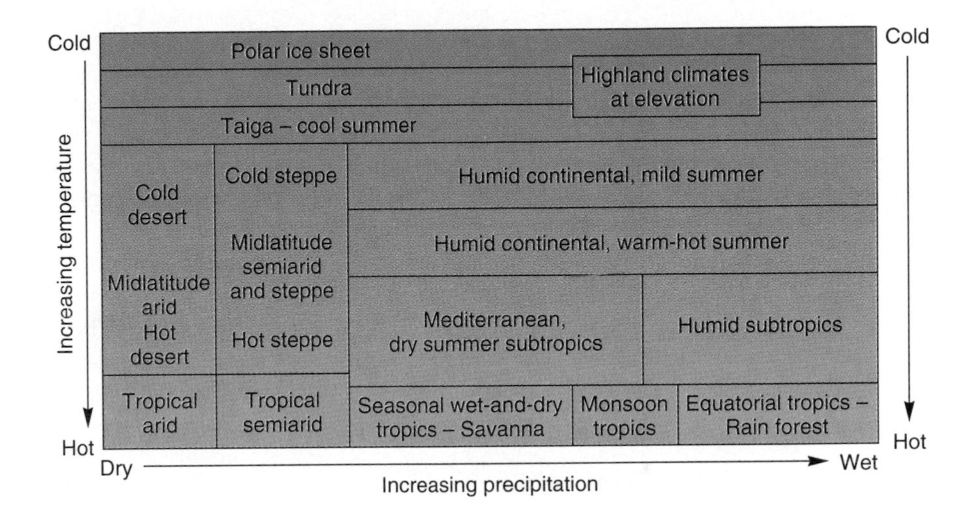

FIGURE 10-3
Temperature and precipitation schematic showing climatic relationships.

time—is like a snapshot of prevailing conditions. These simultaneous measurements and observations of weather elements are plotted on maps for analysis, important for the determination of climate types and the development of computer models to simulate the world's climates.

Developing Climate Models

Imagine the tremendous complexity of building a mathematical model of all these interactive components over different time frames and at various scales! The challenge is to predict future climatic patterns in what is essentially a nonlinear, chaotic natural system. The most complex climate models, known as **general circulation models (GCMs),** are essentially based on mathematical models that were originally established for forecasting weather. At best they are only generalizations of reality; the complex ocean-atmosphere interface and the role of clouds still elude climatologists. Adequate data availability is a problem that is slowly being resolved.

Climatic models can be only as accurate as the data input and assumptions built into them. Thermometers have been in use since the late 1600s, but a significant worldwide density of readings has been available only from around the turn of this century. Today, approximately 1000 stations worldwide balloon-launch automated instrument packages twice a day. These provide temperature, humidity, and pressure measurements at varying altitudes. New land-based methods employing lasers (amplified light) to obtain these readings are gradually increasing the data base.

To define manageable portions of Earth's climatic system for study, climatologists define statistical three-dimensional "boxes" that range from beneath the ocean to the tropopause in multiple layers. Computer analysis must deal not only with the climatic components within each box, but also with the interaction among boxes. Even on a modern high-speed computer, a program can run for months and then only give general indications of climatic patterns.

An important goal of the ongoing multinational Global Change Program is to improve computer hardware and software and the data base to create more accurate simulations and a better predictive capability of future climatic change. The development of coupled general circulation models (CGCMs) that link atmospheric and oceanic components is critical to improving climate predictions.

At present, four complex GCMs (three-dimensional models) are in use in the United States and four in other countries: CCM—National Center for Atmospheric Research; GISS—Goddard Institute for Space Studies (NASA); GFDL—Geophysical Fluid Dynamics Laboratory (Princeton); OSU—Oregon State University; CCC–Canadian Climate Center; MRI—Meteorological Research Institute, Japan; CSIRO—Commonwealth Scientific and Industrial

Research Organization, Australia; and UKMO—Meteorological Office, United Kingdom.

Climatic sensitivity to doubling of CO_2 levels in the atmosphere acts as a comparative benchmark among the operational GCMs. These GCMs do not predict specific temperatures but rather offer various scenarios of global warming. Interestingly, they tend to agree on future global-scale climate changes, but vary in specific changes at the regional level.

CLASSIFICATION OF CLIMATIC REGIONS

Consider this: climate cannot actually be observed and really does not exist at any particular moment. Climate is therefore a conceptual *statistical construction* from measured weather elements according to the assumptions of the classification scheme employed. The ancient Greeks simplified their view of world climates into three zones: the "torrid zone" referred to warmer areas south of the Mediterranean, the "frigid zone" was to the north, and the area where they lived was labeled the "temperate zone," which they considered the optimum climate.

Classification is the process of ordering or grouping data or phenomena in related classes. A classification based on causative factors—for example, the interaction of air masses—is called a **genetic classification.** An **empirical classification** is based on statistics or other data used to determine general categories. Climate classifications based on temperature and precipitation data are empirical classifications. Generalizations are important organizational tools in science and are especially useful for the spatial analysis of climatic regions.

Just as there is no single universal climate classification system, neither is there a single set of empirical or genetic elements to which everyone agrees. Designing a comprehensive climate classification system is a most difficult task because there is such a great interactive complexity among weather elements. Any classification should be viewed as developmental, because it is always open to change and improvement. Such an evolutionary view is healthiest scientifically, because a dynamic classification can provide the same stimulus as a theory in generating further exploration and research.

Many climate classifications have been proposed over the years, based upon a variety of assumptions. Some are genetic and explanatory, based upon net radiation, thermal regimes, or air-mass dominance over a region. Others are empirical and descriptive. One empirical classification system of particular interest was published by C. W. Thornthwaite in 1948. The essentials of Thornthwaite's system are the determination of moisture regions using aspects of the water-balance approach (discussed in Chapter 9) and vegetation types. Another empirical classification system, and the one we describe here, is the Köppen classification system.

The Köppen Classification System

The Köppen classification system, widely used for its ease of comprehension, was designed by Wladimir Köppen (1846–1940), a German climatologist and botanist. First published in stages, his classification began with an article on heat zones in 1884. By 1900 he was considering plant communities in his selection of some temperature criteria, using a world vegetation map prepared by French plant physiologist A. de Candolle in 1855. Letter symbols then were added to designate climate types. Later he reduced the role played by plants in setting boundaries and moved his system strictly toward climatological empiricism. The first wall map showing world climates, coauthored with his student Rudolph Geiger, was introduced in 1928 and soon was widely used. Never completely satisfied with his system, Köppen continued to refine it until his death.

Classification Criteria. The basis of any empirical classification system is the choice of criteria used to draw lines between categories. Köppen chose mean monthly temperatures, mean monthly precipitation, and total annual precipitation to devise his spatial categories and boundaries. But we must remember that boundaries really are gray areas; they are transition zones of gradual change. The trends and overall patterns of boundary lines are more important than their precise placement, especially with the small scales generally used for world maps. The modified Köppen-Geiger system does not consider winds, temperature extremes, precipitation intensity, amount of sunshine, cloud cover, or net radiation. Yet the system is important because its *correlations with the actual world are reasonable* and the *input data are standardized and readily available.* For our purposes of general understanding, a modified Köppen-Geiger classification is useful.

Köppen's Climatic Designations. The Köppen system uses capital letters (A, B, C, D, E, H) to designate climatic categories from the equator to the poles. Five of the climate classifications are based upon thermal criteria:

- *A* Tropical (equatorial regions)
- *C* Mesothermal (Mediterranean, humid subtropical, marine west coast regions)
- *D* Microthermal (humid continental, subarctic regions)
- *E* Polar (polar regions)
- *H* Highland (compared to lowlands at the same latitude, highlands have lower temperatures—recall the normal lapse rate—and more efficient precipitation due to lower POTET demand, justifying separation of highlands from surrounding climates)

Only one is based on moisture as well:

- *B* Dry, (deserts and steppes).

Additional lowercase letters are used to signify temperature and moisture conditions, as illustrated in Figure 10-4. Appendix A presents weather data for a selection of stations worldwide, grouped by Köppen climate type.

Figure 10-4 shows the criteria for each letter symbol within that climate type. For example, in an *Af tropical rain forest* climate, the *A* tells us that the average coolest month is above 18°C (64.4°F), and the *f* indicates that the weather station recording the measurement is constantly wet, with the driest month receiving at least 6 cm (2.4 in.) of precipitation. The *Af tropical rain forest* climate dominates along the equator and equatorial rain forest.

In a *Dfa* climate, the *D* means that the average warmest month is above 10°C (50°F), with at least one month falling below 0°C (32°F); the *f* says that at least 3 cm (1.2 in.) of precipitation falls during every month; and the *a* indicates a warmest summer month averaging above 22°C (71.6°F). Thus, a *Dfa* climate is a humid-continental, hot-summer climate in the microthermal category. Using Figure 10-4 and the weather data for your city or town, see if you can determine the Köppen classification for where you live.

Let's not get lost in the alphabet with this system, for it is meant to help you understand climate through a simplified set of criteria. Always say the descriptive name of the climate as well as its symbol; for example: *Csa Mediterranean summer dry, ET tundra, or BSk cold midlatitude steppe.*

Global Climate Patterns. A simplified Köppen-Geiger classification system is presented on the world map in Figure 10-5. In the following sections, **climographs** are presented for cities that exemplify particular climates. These climographs show monthly temperature and precipitation, location coordinates, annual temperature range, total annual precipitation, annual hours of sunshine (as an indication of cloudiness), the local population, and a location map.

Discussions of soils, vegetation, and major terrestrial biomes that fully integrate these global climate patterns are presented in Chapters 18, 19, and 20. Table 20-1 integrates all this information and will enhance your understanding of this chapter, so please refer to it as you read.

Tropical A Climates

Tropical *A* climates are the most extensive, occupying about 36% of Earth's surface, including both ocean and land areas (Figure 10-4). The *A* climate classification extends along all equatorial latitudes, straddling the tropics from about 20° N to 20° S and stretching as far north as the tip of Florida and south-central Mexico, central India, and Southeast Asia. The key temperature criterion for an *A* climate is that the coolest month must be warmer than 18°C (64.4°F), making these climates truly winterless. The consistent daylength and almost perpendicular Sun angle throughout the year generate this warmth.

THERMAL CLASSIFICATIONS

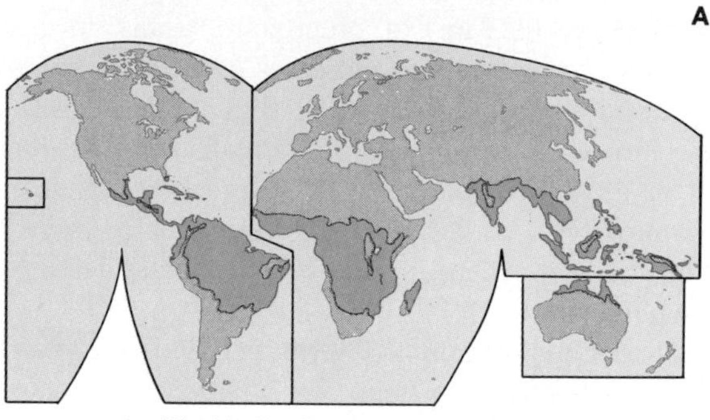

A World distribution of A climates.

A Tropical Climates

Consistently warm with all months averaging above 18°C (64.4°F); annual PRECIP exceeds POTET.

Af - Tropical rain forest:
f = All months receive PRECIP in excess of 6 cm (2.4 in.).

Am - Tropical monsoon:
m = A marked short dry season with 1 or more months receiving less than 6 cm (2.4 in.) PRECIP; an otherwise excessively wet rainy season. ITCZ 6-12 months dominant.

Aw - Tropical savanna:
w = Summer wet season, winter dry season; ITCZ dominant 6 months or less, winter water-balance deficits.

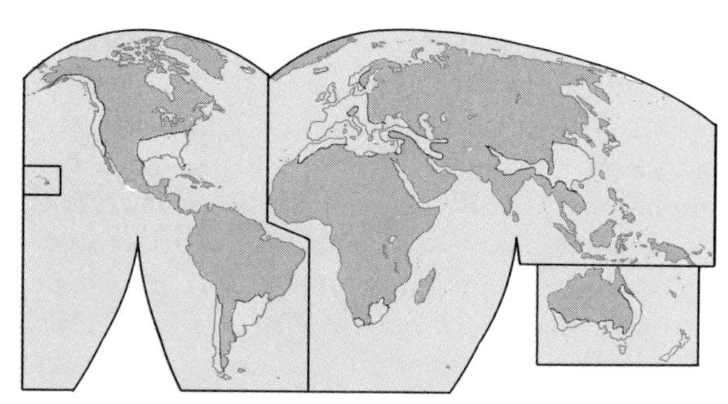

C World distribution of C climates.

C Mesothermal Climates

Warmest month above 10°C (50°F); coldest month above 0°C (32°F) but below 18°C (64.4°F); seasonal climates.

Cfa, Cwa - Humid subtropical:
a = Hot summer; warmest month above 22°C (71.6°F).
f = Year-round PRECIP.
w = Winter drought, summer wettest month 10 times more PRECIP than driest winter month.

Cfb, Cfc - Marine west coast:
Mild-to-cool summer.
f = Receives year-round PRECIP
b = Warmest month below 22°C (71.6°F) with 4 months above 10°C.
c = 1 to 3 months above 10°C.

Csa, Csb - Mediterranean summer dry:
s = Pronounced summer drought with 70% of PRECIP in winter.
a = Hot summer with warmest month above 22°C (71.6°F).
b = Mild summer; warmest month below 22°C.

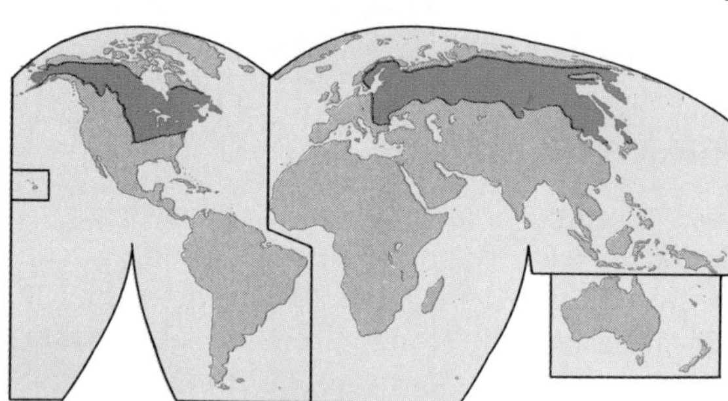

D World distribution of D climates.

D Microthermal Climates

Warmest month above 10°C (50°F); coldest month below 0°C (32°F); cool temperate-to-cold conditions; snow climates. In Southern Hemisphere, only in highland climates.

Dfa, Dwa - Humid continental:
a = Hot summer; warmest month above 22°C (71.6°F).
f = Year-round PRECIP.
w = Winter drought.

Dfb, Dwb - Humid continental:
b = Mild summer; warmest month below 22°C (71.6°F).
f = Year-round PRECIP.
w = Winter drought.

Dfc, Dwc, Dwd - Subarctic:
Cool summers, cold winters.
f = Year-round PRECIP.
w = Winter drought.
c = 1 to 4 months above 10°C.
d = Coldest month below -38°C (-36.4°F), in Siberia only.

FIGURE 10-4
Köppen-Geiger world climate classification system with locator maps for each of the principal climatic regions.

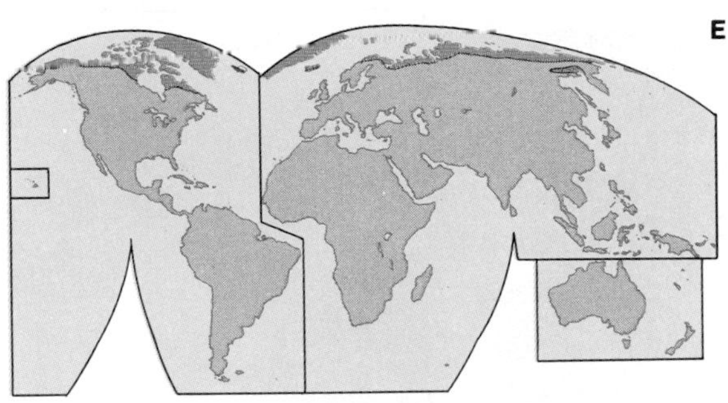

E World distribution of E climates.

E Polar Climates

Warmest month below 10°C (50°F); always cold; ice climates.

EF - Ice cap:
Warmest month below 0°C (32°F); PRECIP exceeds very small POTET demand; the polar regions.

ET - Tundra:
Warmest month between 0-10°C (32-50°F); PRECIP exceeds small POTET demand; snow cover 8-10 months.

EM - Polar marine:
All months above -7°C (20°F), warmest month above 0°C; annual temperature range <17°C (30°F).

MOISTURE CLASSIFICATION

B Dry Arid and Semiarid Climates

POTET exceeds PRECIP in all **B** climates. Subdivisions based on PRECIP timing and amount and mean annual temperature. Boundaries determined by formulas and graphs.

Earth's arid climates.

BWh - Hot low-latitude desert
BWk - Cold midlatitude desert

BW = PRECIP less than ½ POTET.
h = Mean annual temperature >18°C (64.4°F).
k = Mean annual temperature <18°C.

Earth's semiarid climates.

BSh - Hot low latitude steppe
BSk - Cold midlatitude steppe

BS = PRECIP MORE THAN ½ POTET but not equal to it.
h = Mean annual temperature >18°C.
k = Mean annual temperature <18°C.

B World distribution of B climates.

B - Climate determinations: formulas in metric units [in English units]

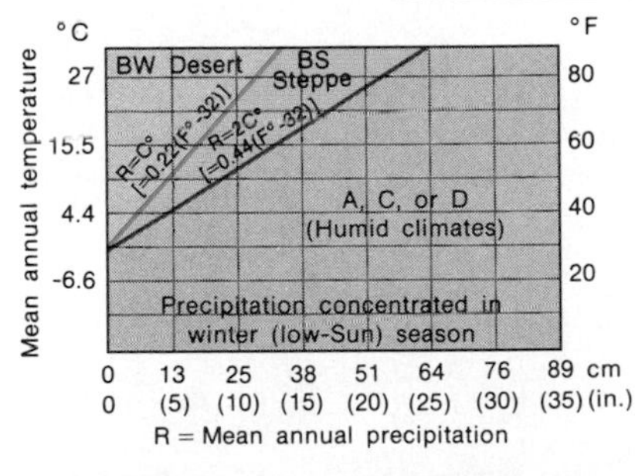

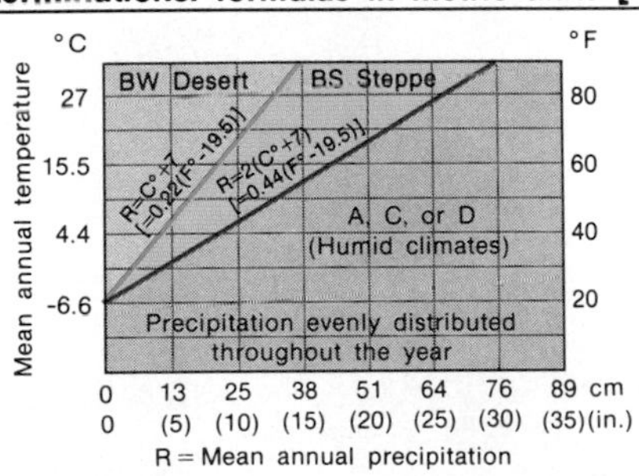

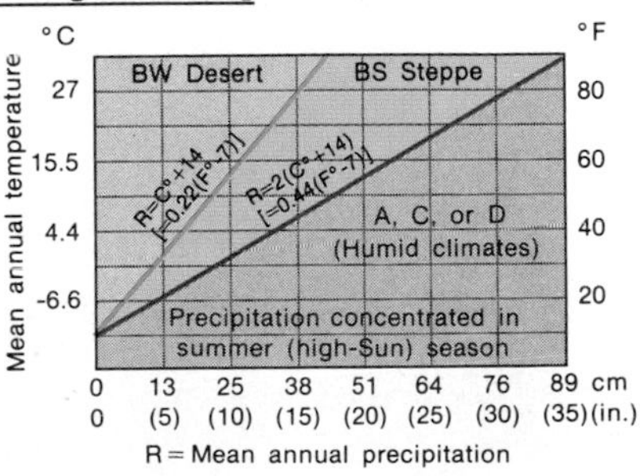

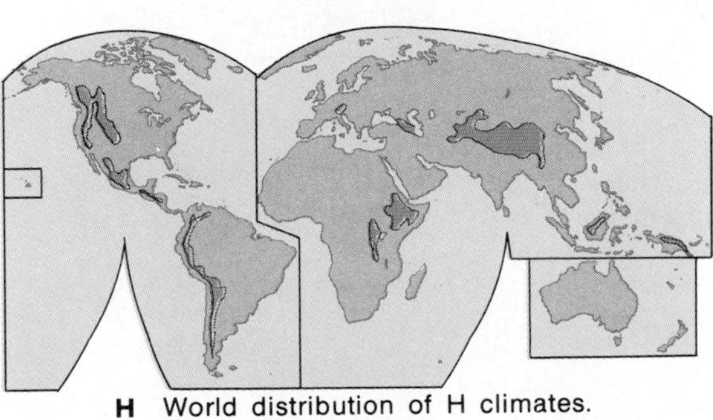

H World distribution of H climates.

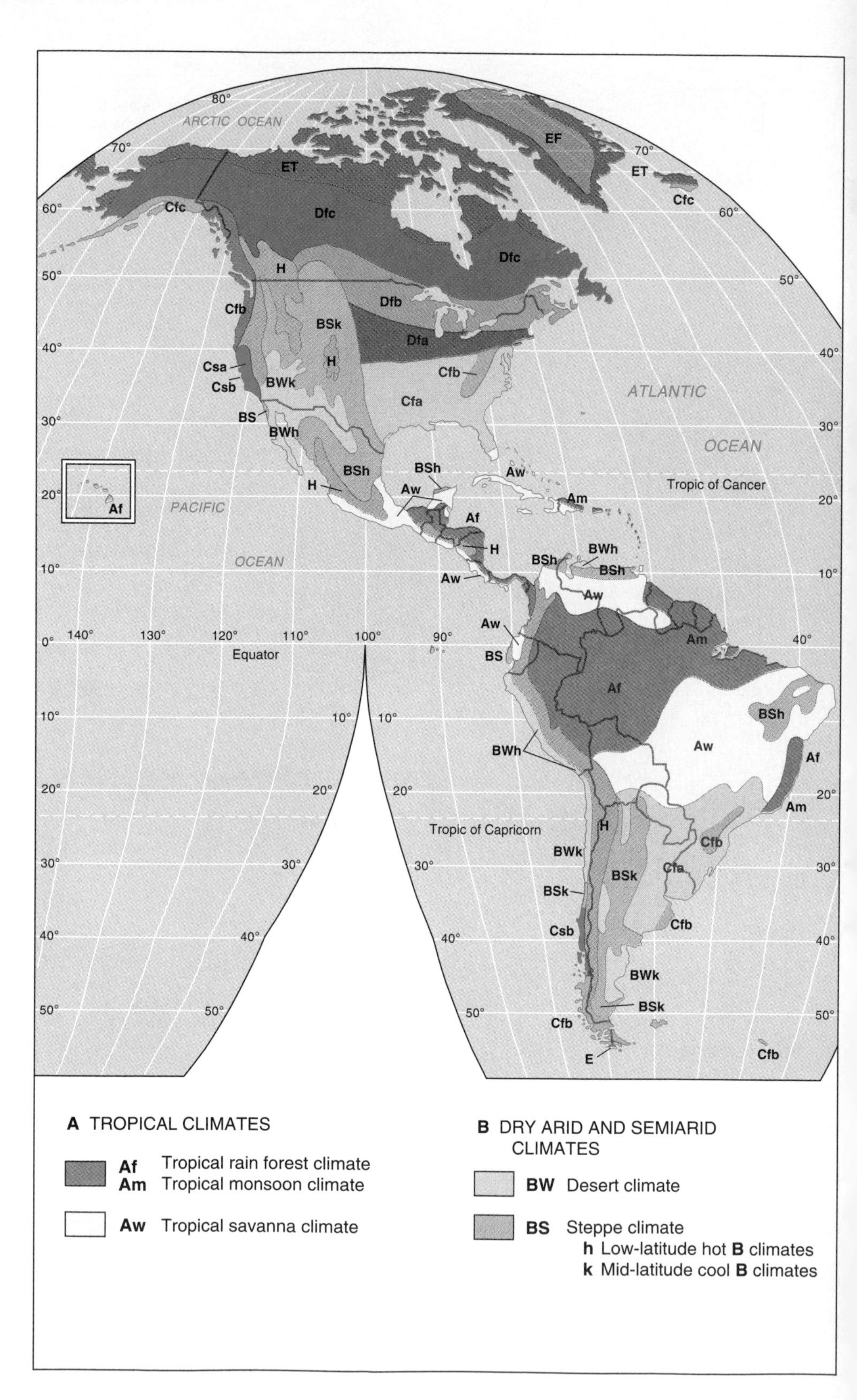

FIGURE 10-5
World climates according to the Köppen-Geiger classification system.

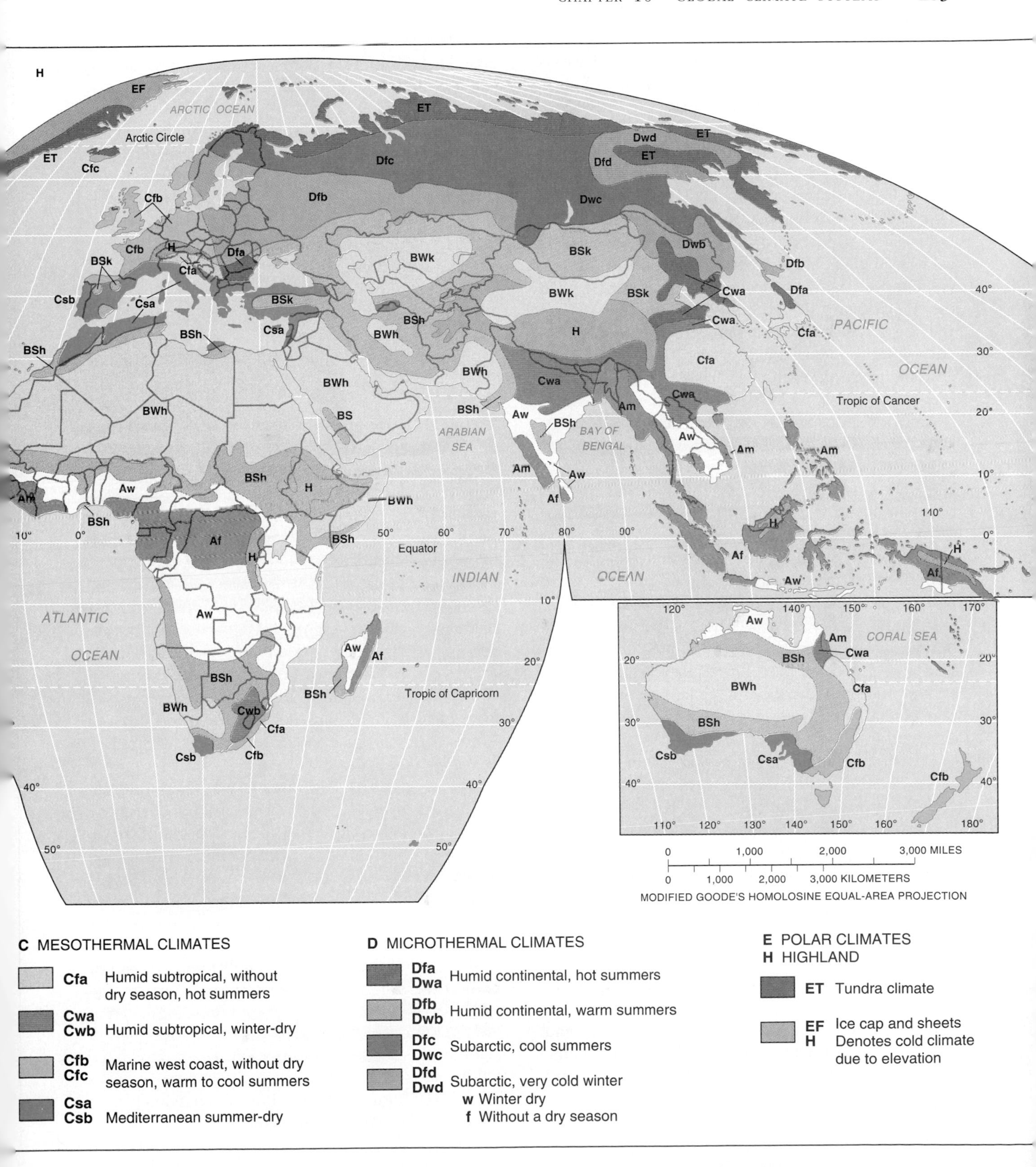
ARCTIC OCEAN
Arctic Circle
PACIFIC OCEAN
ARABIAN SEA
BAY OF BENGAL
INDIAN OCEAN
ATLANTIC OCEAN
CORAL SEA
Tropic of Cancer
Equator
Tropic of Capricorn
0 1,000 2,000 3,000 MILES
0 1,000 2,000 3,000 KILOMETERS
MODIFIED GOODE'S HOMOLOSINE EQUAL-AREA PROJECTION
C MESOTHERMAL CLIMATES
Cfa Humid subtropical, without dry season, hot summers
Cwa Cwb Humid subtropical, winter-dry
Cfb Cfc Marine west coast, without dry season, warm to cool summers
Csa Csb Mediterranean summer-dry
D MICROTHERMAL CLIMATES
Dfa Dwa Humid continental, hot summers
Dfb Dwb Humid continental, warm summers
Dfc Dwc Subarctic, cool summers
Dfd Dwd Subarctic, very cold winter
w Winter dry
f Without a dry season
E POLAR CLIMATES
H HIGHLAND
ET Tundra climate
EF Ice cap and sheets
H Denotes cold climate due to elevation

Tropical Rain Forest Climates (Af). Subdivisions of the *A* climates are based upon the distribution of precipitation during the year (Figure 10-5). Thus, in addition to consistent warmth, an *Af tropical rain forest* climate is constantly moist, with no month recording less than 6 cm (2.4 in.) of precipitation. Indeed, most stations in *Af tropical rain forest* climates receive in excess of 250 cm (100 in.) of rainfall a year. Not surprisingly, the water balances in these regions exhibit enormous water surpluses, creating the world's largest stream discharges in the Amazon and Congo (Zaire) rivers.

The source of this abundant precipitation is convectional thunderstorms triggered by local heating, tending to peak each day from midafternoon to late evening inland and earlier in the day where marine influence is strong. This precipitation pattern coincides with the location of the intertropical convergence zone (ITCZ) discussed in Chapter 6. The ITCZ shifts with the high summer Sun throughout the year but influences *Af tropical rain forest* regions during all 12 months.

The world climate map shows that the only interruption in the equatorial extent of *Af tropical rain forest* climates occurs in the highlands of the Andes and in East Africa, where higher elevations produce lower temperatures. Mount Kilimanjaro, at 5895 m (19,340 ft), is within 4° south of the equator but has permanent snow near its summit. Köppen placed such sites in the *H* (highland) climate designation.

The high rainfall of *Af tropical rain forest* climates sustains lush evergreen broadleaf tree growth, producing Earth's equatorial and tropical rain forests. Their leaf canopy is so dense that light does not easily reach the forest floor, leaving the ground surface in low light and sparse plant cover. The main areas of dense surface vegetation occur along river banks, where light shines through to the surface. A major ecological issue facing the world society today is the rampant deforestation of these rain forests, discussed in more detail in Chapter 20.

The high temperatures of this climate contribute to high bacterial action in the soil so that organic material is quickly lost, and certain minerals and nutrients are washed away by the heavy precipitation. The resulting soils are somewhat sterile and unable to support intensive agricultural activity without the addition of fertilizers.

Uaupés, Brazil (Figure 10-6), is characteristic of the *Af tropical rain forest* classification. The month with the lowest precipitation receives nearly 15 cm (6 in.), and the annual range of temperature is barely 2C° (3.6F°). In all *Af tropical rain forest* climates, the diurnal temperature range exceeds the annual minimum–maximum range; in fact, day-night temperatures can range more than 11C° (20F°), which is more than five times the annual range. Other characteristic *Af tropical rain forest* stations include Iquitos, Peru; Kiribati, Nigeria; Mbandaka, Zaire; Jakarta, Indonesia; and Singapore.

Tropical Monsoon Climates (Am). Tropical *A* climates that have a dry season—one or more months with rain of less than 6 cm (2.4 in.)—are candidates for this classification. The migration of the ITCZ affects these areas from 6 to 12 months of the year, with the dry season occurring when the convergence zone is not overhead. *Am tropical monsoon* climates lie principally along coastal areas within the *Af tropical rain forest* climatic realm and experience seasonal variation of winds and precipitation. Evergreen trees grade into thorn forests on the drier margins near the adjoining savanna climates. Typical *Am tropical monsoon* stations include Freetown, Sierra Leone (west coast of Africa); Columbo, Sri Lanka; and Cairns, Australia.

Tropical Savanna Climates (Aw). The world climate map in Figure 10-5 shows the pattern of *Aw tropical savanna* climates poleward of the *Af tropical rain forest* climates. *Aw tropical savanna* climates occur in Africa, South America, India, Southeast Asia, and the northern portion of Australia. The ITCZ dominates *Aw tropical savanna* climates for approximately six months or less of the year as it shifts with the high Sun. Summers are wetter than winters in *Aw tropical savanna* climates because of the influence of convectional rains that accompany the shifting ITCZ. The *w* designation signifies this winter-dry condition, when rains have left the area and high pressures dominate. As a result, POTET rates exceed PRECIP in winter, and deficits in the water balance result.

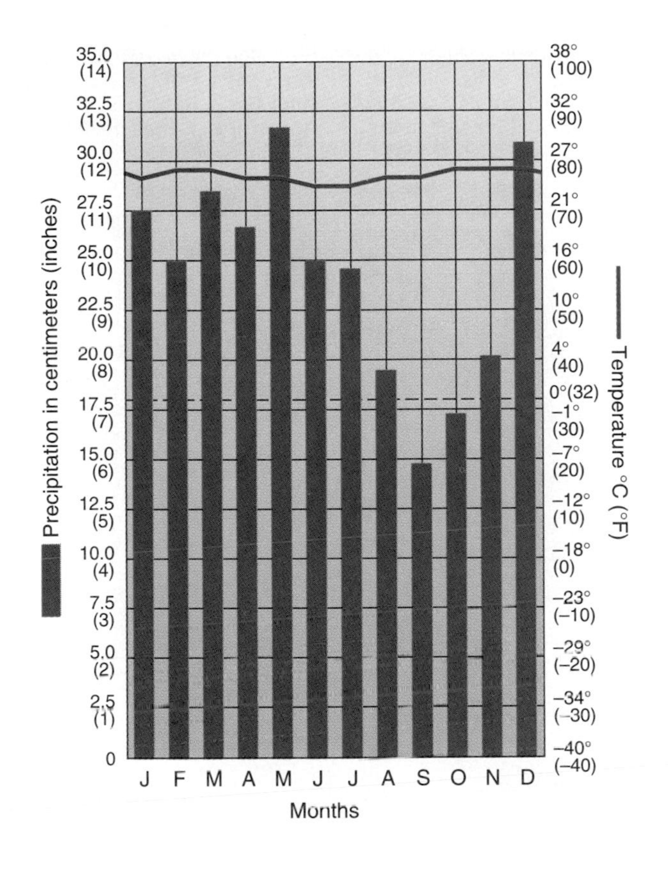

Station: Uaupés, Brazil **Af**
Lat/long: 0°08' S, 67°05' W
Avg. Ann. Temp.: 25°C (77°F)
Total Annual Precipitation: 291.7 cm (114.8 in.)
Elevation: 86 m (282.2 ft)
Population: 7500
Ann. Temp. Range: 2C° (3.6F°)
Ann. Hrs. of Sunshine: 2018

FIGURE 10-6
Climograph for Uaupés, Brazil (Af)

Temperatures also fluctuate more in *Aw tropical savanna* than in *Af tropical rain forest* climates. The *Aw* climate regime can feature two temperature maximums because the Sun's direct rays are overhead twice during the year. Tropical savanna vegetation is associated with *Aw* climates, characterized by scattered trees throughout dominant grasslands. The trees, of course, must be drought resistant to cope with the highly variable precipitation (Figure 10-7).

FIGURE 10-7
Characteristic tropical savanna landscape in Kenya with plants adapted to seasonally dry water budgets. [Photo by Stephen J. Krasemann/DRK Photo.]

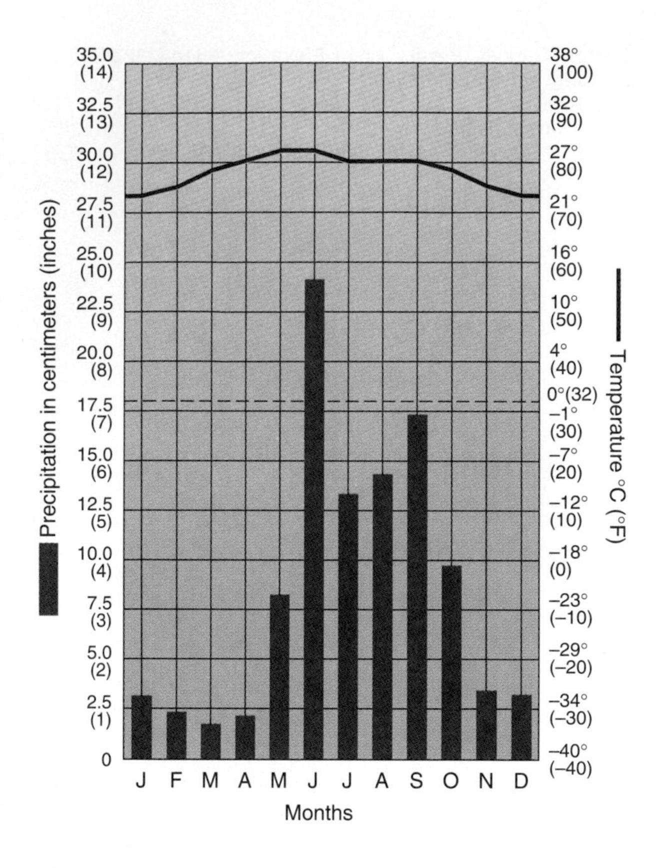

Station: Mérida, Mexico **Aw**
Lat/long: 20°59' N, 89°39' W
Avg. Ann. Temp.: 26°C (78.8°F)
Total Annual Precipitation: 93 cm (36.6 in.)
Elevation: 22 m (72.2 ft)
Population: 400,000
Ann. Temp. Range: 5C°(9F°)
Ann. Hrs. of Sunshine: 2379

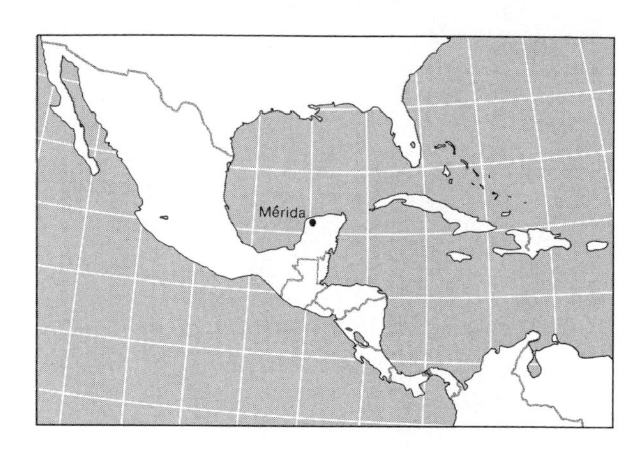

FIGURE 10-8
Climograph for Mérida, Mexico (Aw climate).

Mérida, Mexico, on the northern edge of the Yucatán Peninsula, is a characteristic *Aw tropical savanna* station (Figure 10-8). Temperatures in Mérida are consistent with the pattern of tropical *A* climates, although the comparative winter dryness offers a few months of lower humidity. The summer rain, associated with thunderstorms, is sometimes augmented by tropical cyclones. In September of 1988, Hurricane Gilbert presented an extreme example of this moisture source, devastating most of Mérida with rain and strong winds.

Mesothermal C Climates

The mesothermal *C* climates occupy the second-largest percentage of Earth's land-and-sea surface, totaling about 27% (Figure 10-4). However, they rank only fourth when land area alone is considered. Together, *A* and *C* climates dominate over half of Earth's oceans and about one-third of its land area. The largest percentage of the world's population—approximately 55%—resides in the *C* climates.

The word mesothermal ("middle temperature") suggests warm and temperate conditions, with the coldest month averaging below 18°C (64.4°F) but with all months averaging above 0°C (32°F). Originally, Köppen proposed that the coldest month be considered at the −3°C (26.6°F) isotherm as the boundary with colder climates, perhaps a more accurate criterion for Europe. However, for conditions in North America, the 0°C isotherm is more appropriate. The difference between the 0°C and −3°C isotherms covers an area about the width of the state of Ohio. A line denoting at least one month below freezing runs from New York City roughly along the Ohio River, trending westward until it meets the dry climates in the southeastern corner of Colorado. The line marking −3°C as the coldest month runs farther north along Lake Erie and the southern tip of Lake Michigan.

The mesothermal *C* climates, and nearby portions of the microthermal *D* climates, are regions of great weather variability, for these are the latitudes of greatest air-mass conflict. The *C* climatic region marks the beginning of true seasonality;

contrasts in temperature are evidenced by vegetation, soils, and human lifestyle adaptations. Subdivisions of the *C* classification are based on precipitation variability as given in Figure 10-4.

Humid Subtropical Hot Summer Climates (Cfa, Cwa). The humid subtropical hot summer climates, with a warmest month above 22°C (71.6°F), form two principal types. The *Cf* type is moist all year, with the *f* indicating that all months receive precipitation above 3 cm (1.2 in.); the *Cw* type has a pronounced winter-dry period, designated by a *w*.

The *Cfa humid subtropical hot summer* climate is influenced during the summer by the maritime tropical air masses that are generated over warm coastal waters off eastern coasts. The warm, moist, unstable air forms convectional showers over land. In fall, winter, and spring, maritime tropical and continental polar air masses interact, generating frontal activity and frequent midlatitude cyclonic storms. Overall, precipitation averages 100–200 cm (40–80 in.) per year.

Nagasaki, Japan (Figure 10-9), is a characteristic station with no month receiving less than 7.5 cm (3.0 in.) of precipitation. Unlike *Cfa humid subtropical hot summer* stations in the United States, however, Nagasaki's summer precipitation is quite a bit higher than its winter precipitation because of the effects of the Asian monsoon.

Cfa humid subtropical hot summer climates are

FIGURE 10-9
Climograph for Nagasaki, Japan (Cfa climate), and the landscape of the Nagasaki region. [Photo by Noboru Komine/Photo Researchers.]

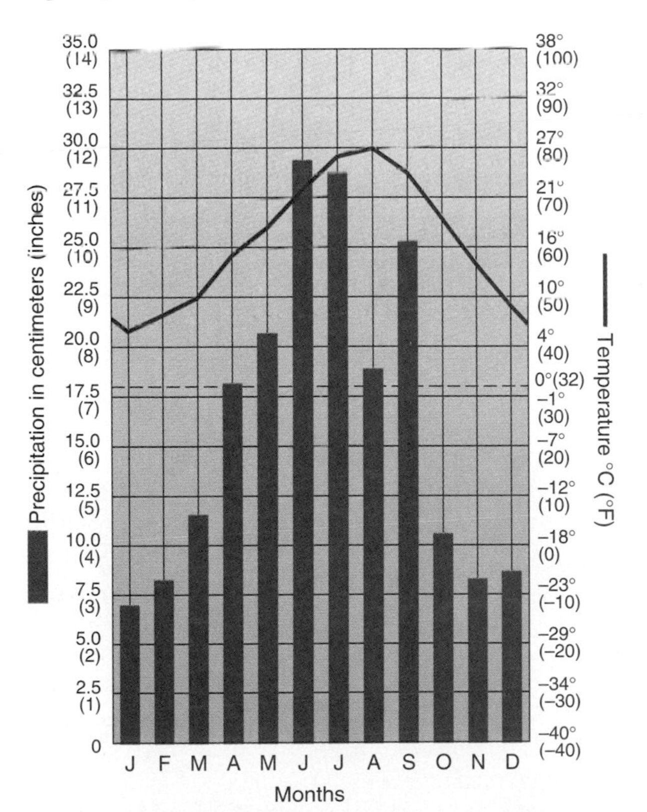

Station: Nagasaki, Japan **Cfa**
Lat/long: 32°44' N, 129°52' E
Avg. Ann. Temp.: 16°C (60.8°F)
Total Annual Precipitation: 195.7 cm (77 in.)
Elevation: 27 m (88.6 ft)
Population: 449,000
Ann. Temp. Range: 21C° (37.8F°)
Ann. Hrs. of Sunshine: 2131

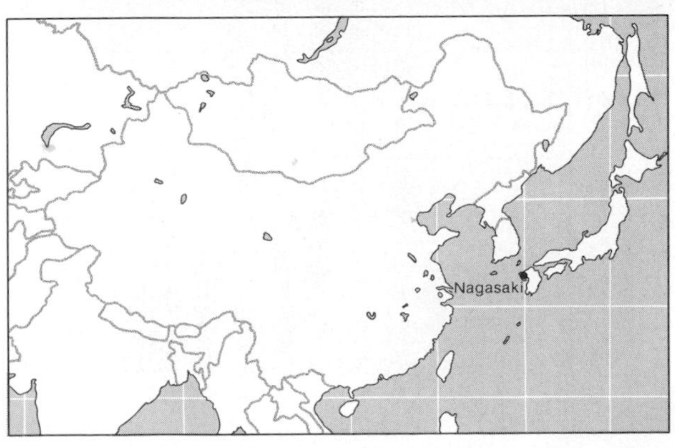

located in the eastern and east-central portions of the continents: areas of the United States, Argentina, and Brazil; southern European lowlands; eastern China; portions along the east coast of Australia; and a narrow strip along the far southeast African coast. Other *Cfa humid subtropical hot summer* climates include New Orleans; Nashville; Atlanta; Baltimore; Tokyo; Rosario, Argentina; and Sydney, Australia.

Cwa humid subtropical winter drought climates are related to the winter-dry, seasonal pulse of the monsoons and extend poleward from the *Aw tropical savanna* climates. Köppen identified the wettest *Cwa* summer month as receiving 10 times more precipitation than the driest winter month. Cherrapunji, India, mentioned in Chapter 8 as receiving the most precipitation in a single year, is an extreme example of this classification. In that location the contrast between the dry and wet monsoons is most severe; weather conditions range from dry winds in the winter to torrential rains and floods in the summer. Downstream from the Assam Hills, such heavy rains produced floods in Bangladesh in 1988 and 1991, among other years.

A representative station of this humid subtropical wet summer and dry winter regime is Ch'engtu, China. Figure 10-10 demonstrates the strong correlation between precipitation and the high Sun of summer. Other examples of *Cwa humid subtropical winter drought* climates include Allahabad, India; Lashio, Myanmar (Burma); Hong Kong; Tsinan, China; and Bandeirantes, Brazil.

The habitability of the *Cfa humid subtropical hot summer* and *Cwa humid subtropical winter drought* climates and their ability to sustain populations are borne out by the concentrations in north-central India, the bulk of China's 1.1 billion people, and the many who live in climatically similar portions of the United States.

FIGURE 10-10
Climograph for Ch'engtu, China (Cwa climate).

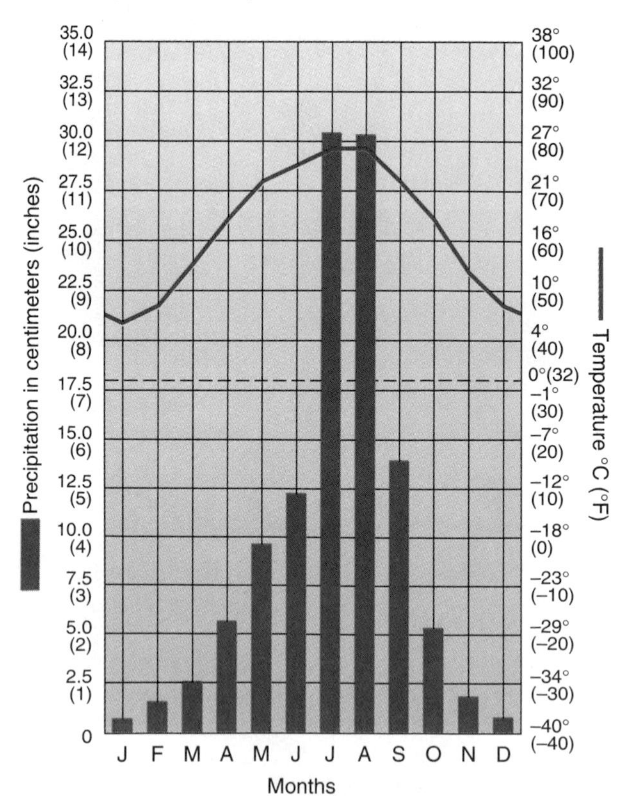

Station: Ch'engtu, China **Cwa**
Lat/long: 30°40' N, 104°04' E
Avg. Ann. Temp.: 17°C (62.6°F)
Total Annual Precipitation: 114.6 cm (45.1 in.)
Elevation: 498 m (1633.9 ft)
Population: 2,260,000
Ann. Temp. Range: 20C° (36F°)
Ann. Hrs. of Sunshine: 1058

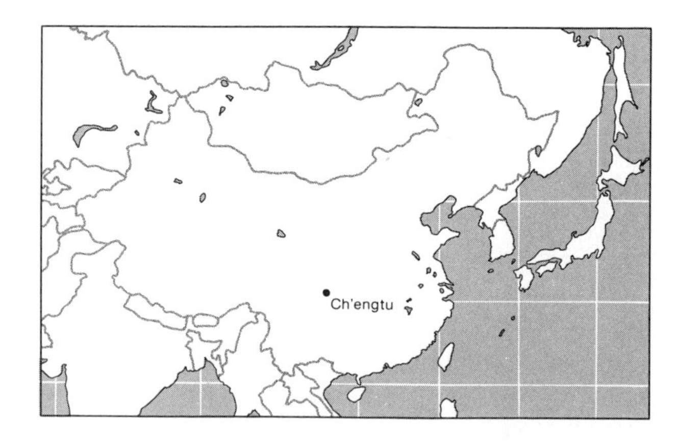

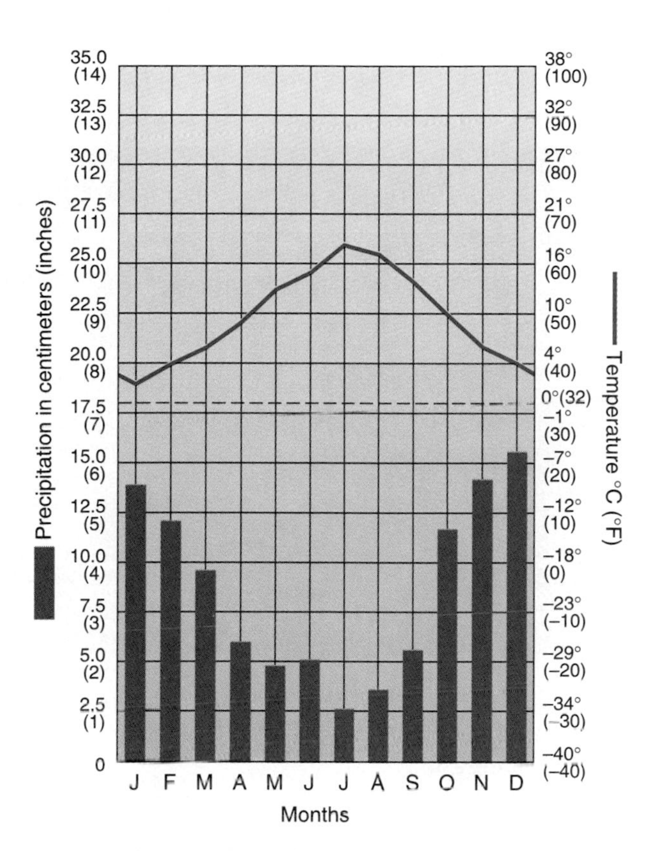

Station: Vancouver, British Columbia
Lat/long: 49°11' N, 123°10' W
Avg. Ann. Temp.: 10°C (50°F) **Cfb**
Total Annual Precipitation: 104.8 cm (41.3 in.)
Elevation: sea level
Population: 431,000
Ann. Temp. Range: 16C°(28.8F°)
Ann. Hrs. of Sunshine: 1723

FIGURE 10-11
Climograph of Vancouver, British Columbia. The natural vegetation of nearby Cape Scott, on Vancouver Island, is representative of the marine west coast climate (Cfb climate). [Photo by Michael Collier.]

Marine West Coast Climates (Cfb, Cfc). Certainly the dominant climates of Europe and other middle-to-high-latitude west coasts are the marine west coast climates, featuring mild winters and cool summers (Figure 10-5). The *b* indicates that the warmest summer month is below 22°C (71.6°F), with at least four months averaging above 10°C (50°F); the *c* indicates that only one to three months are above 10°C; again, the *f* denotes consistent moisture throughout the year. Chapter 5 discusses the moderating effects of a marine location on temperature patterns. Such climates demonstrate an unusual mildness for their latitude. These marine west coast climates extend all along the coastal margins of the Aleutians in the North Pacific, covering the southern third of Iceland in the North Atlantic, as well as Scandinavia and the British Isles. Unlike those in Europe, these climates in Canada, Alaska, Chile, and Australia are backed by mountains and remain restricted to coastal environs.

Representative marine west coast stations include Seattle; Vancouver, British Columbia; Valdivia, Chile; Hobart, Tasmania; Dublin, Ireland; Greenwich and London; Bordeaux, France; and Stuttgart, Germany. The cooler *Cfc* version of the marine west coast regime occurs at higher latitudes and elevations, such as Kodiak, Alaska; Reykjavik, Iceland; and Lerwick, Scotland. The climograph for Vancouver demonstrates the moderate temperature patterns and the annual temperature range for a *Cfb marine west coast* station (Figure 10-11).

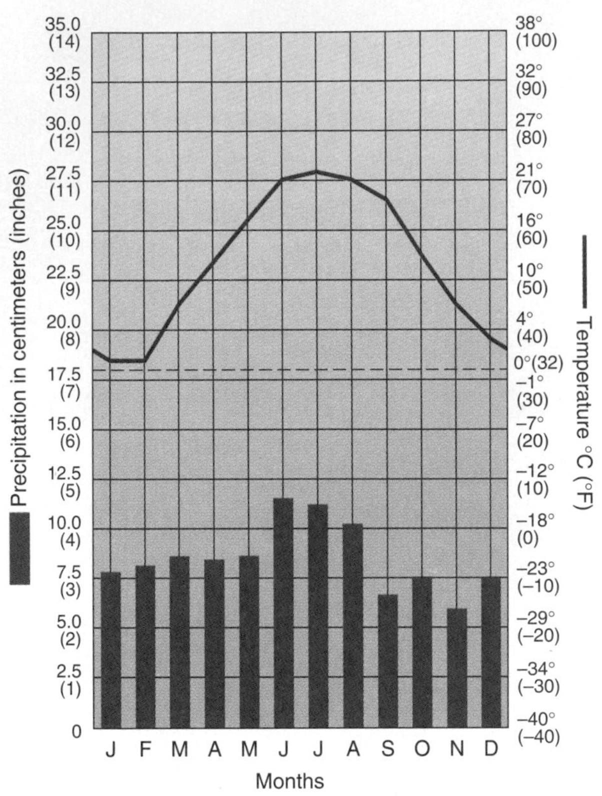

Station: Bluefield, West Virginia
Lat/long: 37°16' N, 81°13' W
Avg. Ann. Temp.: 12°C (53.6°F) **Cfb**
Total Annual Precipitation: 101.9 cm (40.1 in.)
Elevation: 780 m (2559 ft)
Population: 16,000
Ann. Temp. Range: 21C° (37.8F°)
Ann. Hrs. of Sunshine: not available

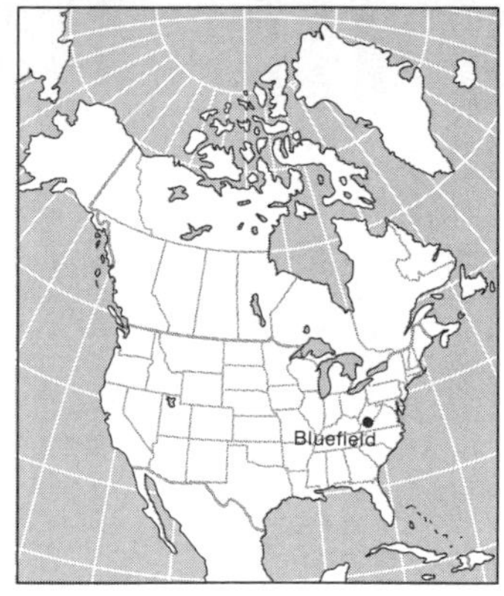

FIGURE 10-12
Climograph for Bluefield, West Virginia (Cfb climate) and the characteristic mixed forest appearance of the Appalachian highland region in the Dolly Sods Wilderness. [Photo by David Muench.]

The dominant air mass in *Cfb* and *Cfc marine west coast* climates is the maritime polar, which is cool, moist, and unstable. Weather systems forming along the polar front access these regions throughout the year, making weather quite unpredictable. Coastal fog, totaling a month or two of days, is a part of the moderating marine influence. Frosts are possible and tend to shorten the growing season.

An interesting anomaly relative to the marine west coast climate occurs in the eastern United States. In portions of the Appalachian highlands, increased elevation moderates summer temperatures in the *Cfa humid subtropical hot summer* classification, producing a *Cfb marine west coast* designation. The climograph for Bluefield, West Virginia (Figure 10-12), shows that its temperature and precipitation patterns are a marine west coast type, despite its location in the east. Vegetation similarities between the Appalachians and the Pacific Northwest are quite noticeable, enticing many emigrants who relocate from the East to settle in these climatically familiar environments in the West.

Mediterranean Dry Summer Climates (Csa, Csb). Across the planet during summer months, shifting cells of subtropical high pressure block moisture-bearing winds from adjacent regions. As an example, in summer the continental tropical air mass over the Sahara in Africa shifts northward over the Mediterranean region and blocks maritime air masses and cyclonic systems. This shifting of stable, warm-to-hot, dry air over an area in summer and away from these regions in the winter creates a pronounced dry summer and wet winter pattern. The designation *s* specifies that at least 70% of annual precipitation occurs during the winter months.

Cool offshore currents (the California Current, Canary Current, Peru Current, Benguela Current, and West Australian Current) produce stability in overlying maritime tropical air masses along west coasts, poleward of subtropical high pressure. The world climate map shows *Csa* and *Csb Mediterranean dry summer* climates along the western margins of North America, central Chile, and the southwestern tip of Africa, as well as across southern Australia and the Mediterranean basin—the climate's namesake region.

Figure 10-13 compares the climographs of Mediterranean dry summer cities Sevilla, Spain *(Csa)* and San Francisco *(Csb)*. Coastal maritime effects moderate San Francisco's summer so that the warmest month falls below 22°C (71.6°F). With greater refinement of the 22°C boundary, the *b* warm summer designation usually warms into an *a* hot summer designation no more than 24–32 km (15–20 mi) inland. Along most west coasts, the warm, moist air overlying cool ocean water produces frequent summer fog, discussed in Chapter 8. This type of fog is not associated with the *Csa* climates of the Mediterranean region because offshore water temperatures there are higher than water temperatures near *Csb* designations during the summer months. Instead, the Mediterranean regime features high humidity values.

The Mediterranean dry summer climate brings water balance deficits in summer. Winter precipitation recharges soil moisture, but water utilization usually exhausts soil moisture by late spring. Thus, local water balances must be augmented by irrigation for large-scale agriculture, although some subtropical fruits, nuts, and vegetables are uniquely suited to this warm-to-hot, dry summer condition. Natural vegetation is a hard-leafed, drought-resistant variety known as *sclerophyllous vegetation*; in the western United States it is locally known as *chaparral*. Chapter 20 discusses the other types and local names for this type of vegetation in other parts of the world.

A few representative *Csa* stations include Sacramento; Rome; Adelaide and Perth, Australia; Izmir, Turkey; and Tunis, Tunisia. Coastal *Csb* stations include Santa Monica, California; Portland, Oregon; Valparaíso, Chile; and Lisbon, Portugal.

Microthermal D Climates

Humid microthermal climates experience a long winter season with some summer warmth. Microthermal signifies conditions that are cool temperate to cold, with at least one month averaging below 0°C (32°F) and at least one month above 10°C (50°F). Approximately 21% of Earth's land surface is influenced by *D* climates, equaling about 7% of Earth's total surface. These climates occur poleward of *C* climates and experience severe ranges of temperature related to continentality and air-mass conflicts. Because the Southern Hemisphere lacks substantial landmasses, *D* climates do not develop there except in highland areas. (Check Figure 10-5 and you can see that no *D* climates exist in the Southern Hemisphere.) The tertiary letters *a, b,* and *c* (hot, warm, and cool, respectively) that are used with *D* climates carry the same meaning that they had with *C* climates. The *D* climates range from a *Dfa* in Chicago to the formidable extremes of a *Dwd* in the region of Verkhoyansk, Siberia.

Humid Continental Hot Summer Climates (Dfa, Dwa). The humid continental hot summer climates are distinguished from each other by the distribution of precipitation during the year. The dry winter associated with the vast Asian landmass, specifically Siberia, is exclusively assigned *w* because it is dominated by an extremely dry and frigid anticyclone in winter. The dry monsoons of southern and eastern Asia are produced in the winter months by this high-pressure system, with winds blowing outward to the Pacific and Indian oceans. The intruding deep cold of continental

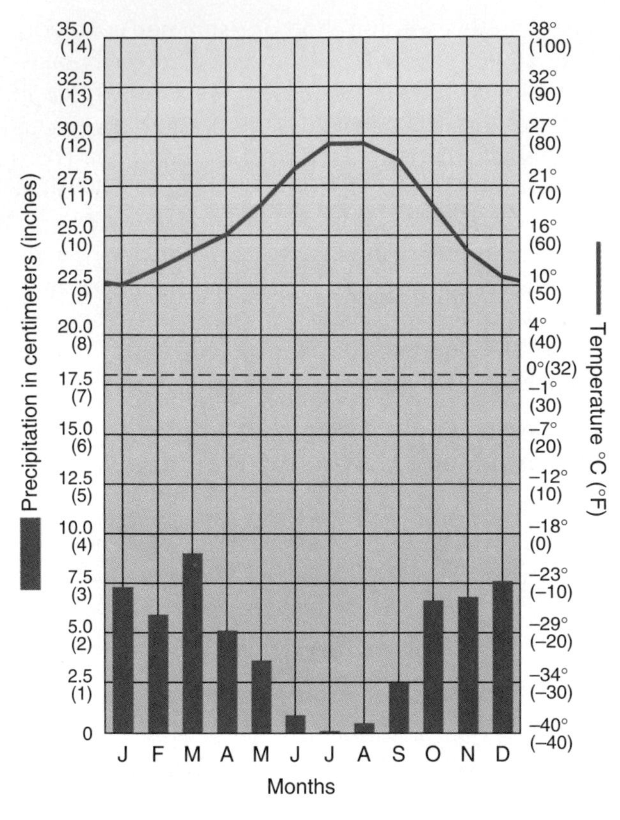

Station: Sevilla, Spain
Lat/long: 37°22' N, 6°00' W
Avg. Ann. Temp.: 18°C (64.4°F) **Csa**
Total Annual Precipitation: 55.9 cm (22 in.)
Elevation: 13 m (42.6 ft)
Population: 651,000
Ann. Temp. Range: 16C° (28.8F°)
Ann. Hrs. of Sunshine: 2862

(a)

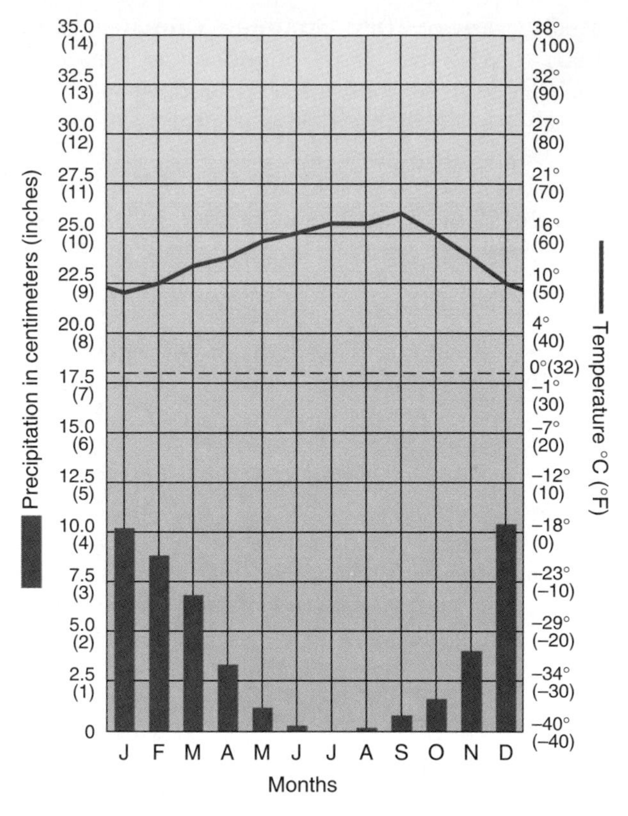

Station: San Francisco, California
Lat/long: 37°37' N, 122°23' W
Avg. Ann. Temp.: 14°C (57.2°F) **Csb**
Total Annual Precipitation: 47.5 cm (18.7 in.)
Elevation: 5 m (16.4 ft)
Population: 750,000
Ann. Temp. Range: 9C° (16.2F°)
Ann. Hrs. of Sunshine: 2975

(b)

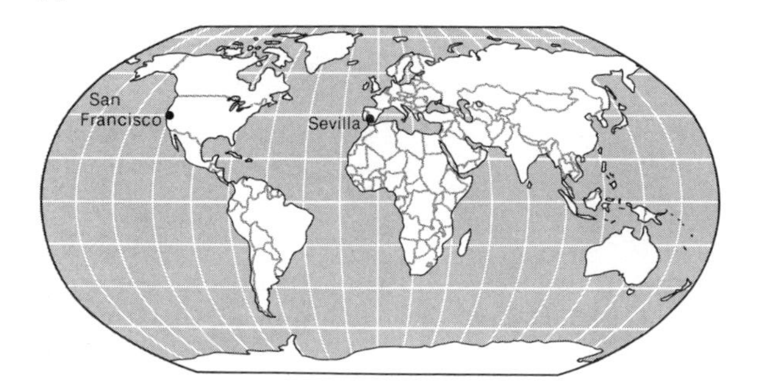

FIGURE 10-13
Climographs for (a) Sevilla, Spain (Csa climate), and (b) San Francisco (Csb climate).

polar air is a significant winter feature. Both *Dfa* and *Dwa humid continental hot summer* climates are influenced by maritime tropical air masses in the summer. In North America, frequent frontal activity is possible between conflicting maritime tropical and continental polar air masses, especially in winter.

The climographs for New York City and Dalian, China, illustrate these two *Da* climates (Figure 10-14). Other examples of *Dfa* climates include Chica-

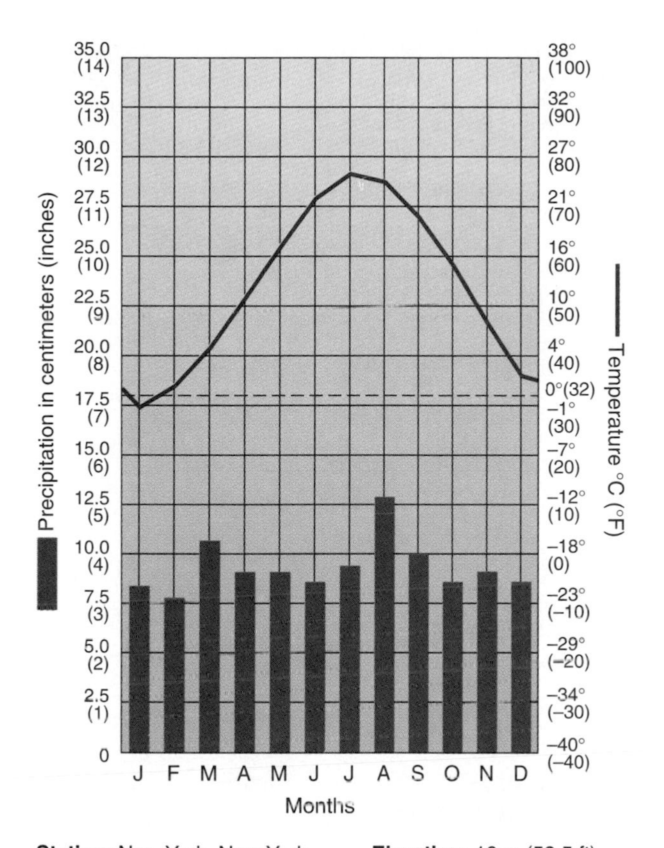

Station: New York, New York
Lat/long: 40°46' N, 74°01' W
Avg. Ann. Temp.: 13°C (55.4°F)
Dfa
Total Annual Precipitation: 112.3 cm (44.2 in.)
Elevation: 16 m (52.5 ft)
Population: 7,100,000
Ann. Temp. Range: 24C° (43.2F°)
Ann. Hrs. of Sunshine: 2564

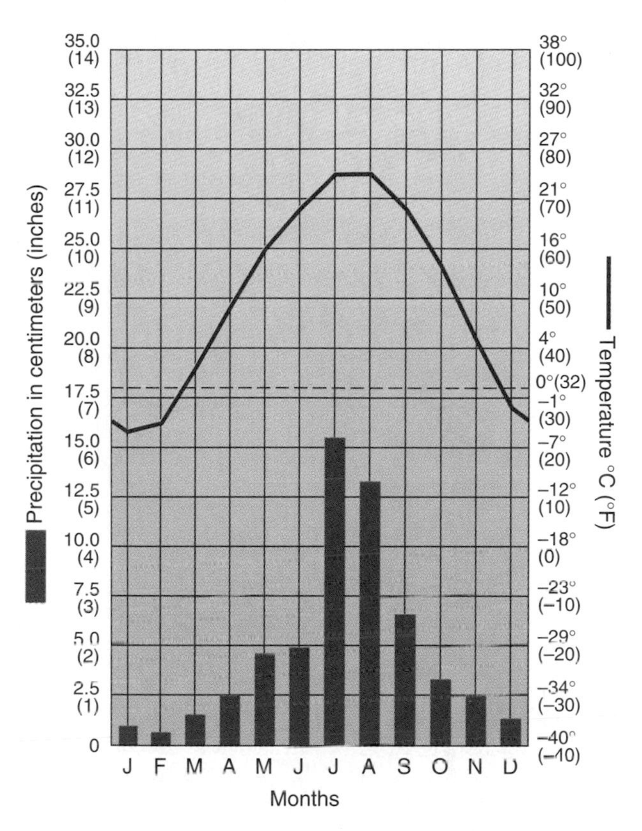

Station: Dalian, China
Lat/long: 38°54' N, 121°54' E
Avg. Ann. Temp.: 10°C (50°F)
Dwa
Total Annual Precipitation: 57.8 cm (22.8 in.)
Elevation: 96 m (314.9 ft)
Population: 4,270,000
Ann. Temp. Range: 29C° (52.2F°)
Ann. Hrs. of Sunshine: 2762

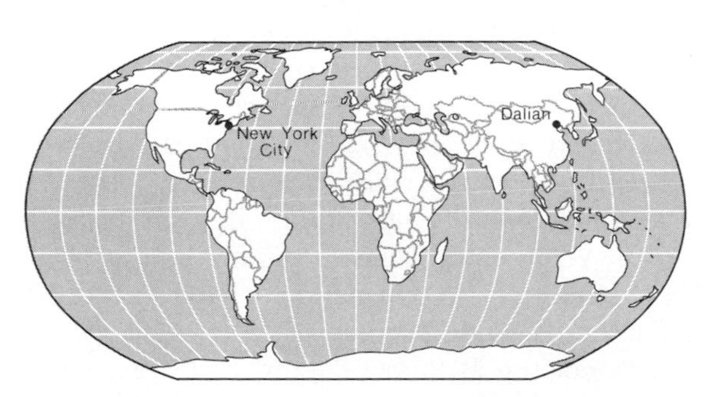

FIGURE 10-14
Climographs for (a) New York City (Dfa climate) and (b) Dalian, China (Dwa climate).

go; Omaha; Columbus, Ohio; Huron, South Dakota; and Varna, Bulgaria. *Dwa* climates are typified by Beijing and Tsingtao in China.

Originally, forests covered the *Dfa humid continental hot summer* climatic region of the United States as far west as the Indiana-Illinois border. Beyond that approximate line, the tall grass prairies extended westward to about the 98th meridian (98° W) and the approximate location of the 51 cm (20 in.) *isohyet* (line of equal precipitation), with the shortgrass prairies beyond to the west. The deep sod posed problems for the first emigrant

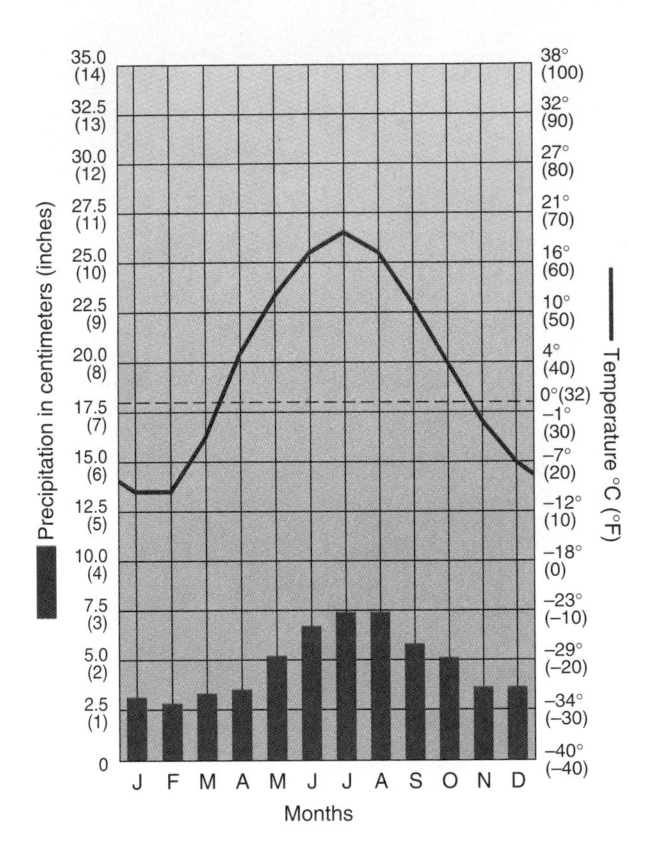

Station: Moscow, Russia **Dfb**
Lat/long: 55°45' N, 37°34' E
Avg. Ann. Temp.: 4°C (39.2°F)
Total Annual Precipitation: 57.5 cm (22.6 in.)
Elevation: 156 m (511.8 ft)
Population: 9,900,000
Ann. Temp. Range: 29C° (52.2F°)
Ann. Hrs. of Sunshine: 1597

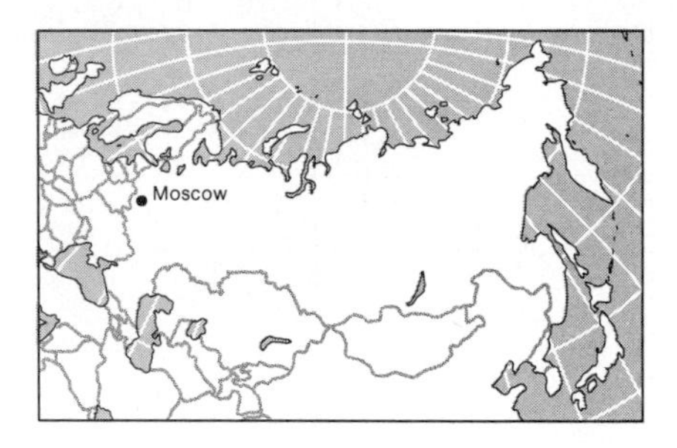

FIGURE 10-15
Climograph for Moscow, Russia (Dfb climate).

settlers, as did the climate. However, native grasses soon were replaced with domesticated wheat and barley, and various eastern inventions (barbed wire, the self-scouring steel plow, well-drilling techniques, railroads, etc.) helped open the region further. In the United States today, the *Dfa humid continental hot summer* climatic region is the location of corn, soybean, hog, and cattle production. The soybean was first domesticated in China and is now widely grown in this region, exceeded only by corn and wheat production.

Humid Continental Mild Summer Climates (Dfb, Dwb). Soils are thinner and less fertile in these cooler *D* climates, yet agricultural activity is important and includes dairy cattle, poultry, flax, sunflowers, sugar beets, wheat, and potatoes. Frost-free periods range from fewer than 90 days in the north to as many as 225 days in the south. Overall, precipitation is lower than it is in the hot summer regions to the south; however, heavier snowfall is notable and important to soil moisture recharge when it melts. Various snow-capturing strategies have been used, including fences and taller stubble left in fields to create snow drifts and thus more moisture retention on the ground.

The dry winter aspect of this cool summer climate occurs only in Asia, in a far-eastern area poleward of the winter-dry *C* climates. A representative *Dwb humid continental mild summer* climate is Vladivostok, Russia, usually one of only two ice-free ports in that nation. Characteristic *Dfb* stations are Duluth, Minnesota, and Moscow and Saint Petersburg in Russia. Figure 10-15 presents a climograph for Moscow, which is at 55° N, or about the same latitude as the southern shore of Hudson Bay in Canada.

Subarctic Climates (Dfc, Dwc, Dwd). As we move farther poleward, seasonality becomes greater. The growing season, though short, is more intense during long summer days, with at least one

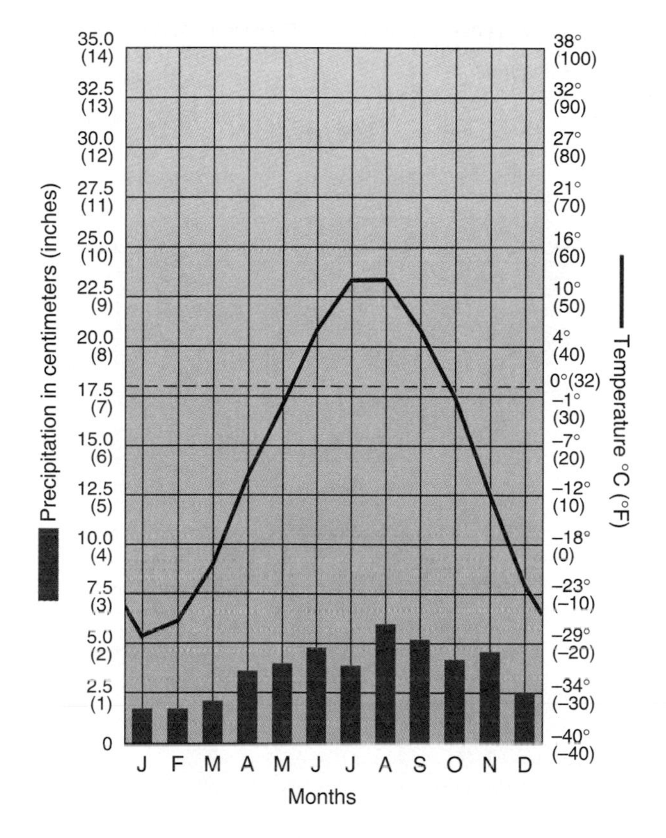

Station: Churchill, Manitoba **Dfc**
Lat/long: 58°45' N, 94°04' W
Avg. Ann. Temp.: −7°C (19.4°F)
Total Annual Precipitation: 44.3 cm (17.4 in.)
Elevation: 35 m (114.8 ft)
Population: 1400
Ann. Temp. Range: 40C° (72F°)
Ann. Hrs. of Sunshine: 1732

FIGURE 10-16
Climograph for Churchill, Manitoba (Dfc climate).

to four months averaging above 10°C (50°F). The world climate map illustrates the land covered by these three cold climates: vast stretches of Alaska, Canada, northern Scandinavia, and Russia. Discoveries of minerals and petroleum reserves have led to new interest in portions of these regions.

Those subarctic areas that receive 25 cm (10 in.) or more of precipitation per year on the northern continental margins are covered by the so-called "snow forest" of fir, spruce, larch, and birch—the boreal forests of Canada and the taiga of Russia. (*Taiga* is the more general term used in ecological studies.) These forests are a transition to the more northern open woodlands and to the tundra region of the far north. Köppen used the 10°C (50°F) temperature criterion for the warmest month in these climates because it is generally the lowest average temperature regime that can support a forest cover. Soils are thin in these lands once scoured by glaciers, and precipitation is low. However, POTET is also low, so soils are generally moist and either partially or totally frozen beneath the surface, a phenomenon known as *permafrost*.

The Churchill, Manitoba climograph (Figure 10-16) shows average monthly temperatures below freezing for seven months of the year, during which time light snow cover and frozen ground persist. Churchill is representative of the *Dfc subarctic* climate, with an annual temperature range of 40C° (72F°) and a low precipitation total of 44.3 cm (17.4 in.), spread fairly evenly over the year but reaching a maximum in late summer. High pressure dominates Churchill during its cold winter—this is the source area for the continental polar air mass. Other *Dfc subarctic* climates are found at Fairbanks and Eagle, Alaska; Dawson, Yukon; Fort Vermillion, Alberta; and Trondheim, Norway. The *Dwc* and *Dwd* subarctic climates occur only within Russia. Köppen selected the tertiary letter *d* for the intense cold of Siberia and north-central and eastern Asia; it designates a coldest month with an average temperature lower than −38°C (−36.4°F).

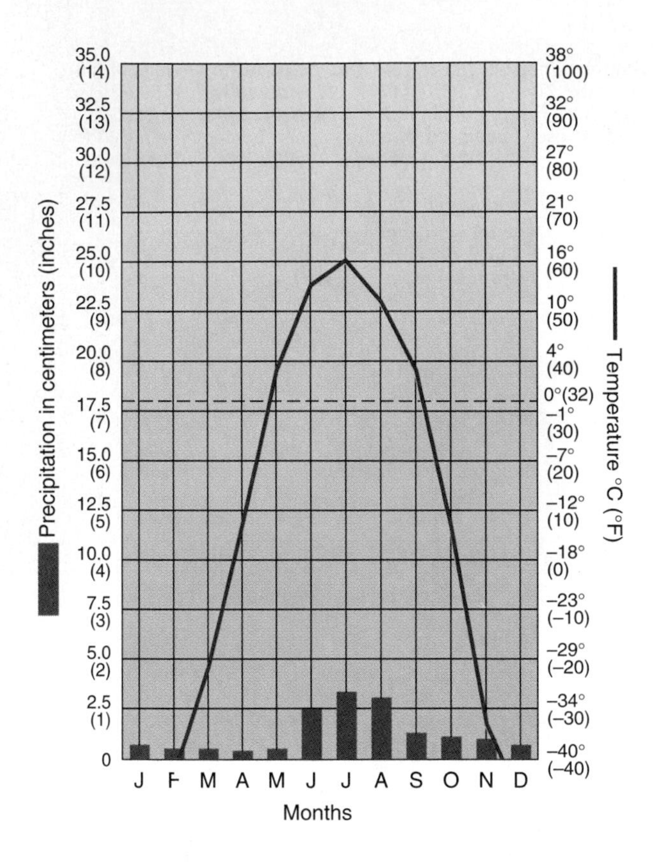

Station: Verkhoyansk, Russia **Dwd**
Lat/long: 67°33′ N, 135°23′ E
Avg. Ann. Temp.: −15°C (5°F)
Total Annual Precipitation: 15.5 cm (6.1 in.)
Elevation: 137 m (449.5 ft)
Population: 1400
Ann. Temp. Range: 63C° (113.4F°)
Ann. Hrs. of Sunshine: not available

FIGURE 10-17
Climograph for Verkhoyansk, Russia (Dwd climate).

A typical *Dwd subarctic* station is Verkhoyansk, Siberia (Figure 10-17). For four months of the year average temperatures fall below −34°C (−29.2°F). Verkhoyansk frequently reaches minimum winter temperatures that are lower than −68°C (−90°F). However, as pointed out in Chapter 5 (Figure 5-9), summer temperatures in the same area produce the world's greatest annual temperature range from winter to summer, a remarkable 63C° (113.4F°). Winter lifestyles feature brittle metals and plastics, triple-thick window panes, and temperatures that render straight antifreeze a solid.

Polar Climates

The polar climates—*ET tundra, EF ice cap,* and *EM polar marine*—cover about 19% of Earth's total surface, about 17% of its land area. These climates have no true summer similar to that of the lower latitudes. Poleward of the Arctic and Antarctic circles, daylength increases in summer until daylight becomes continuous, yet average monthly temperatures never rise above 10°C (50°F). Daylength, which in part determines the amount of insolation received, and low Sun altitude are the principal climatic factors in these frozen and barren regions.

In an *ET tundra* climate, the temperature of the warmest month is between 0°C and 10°C (32°F and 50°F), and PRECIP is in excess of a very small POTET demand. The land is under continuous snow cover for 8–10 months, but when the snow melts and spring arrives, numerous plants appear on the landscape. Characteristic tundra vegetation is composed of stunted sedges, mosses, flowering plants, and lichens. Much of the area experiences permafrost conditions. The tundra is also the summer home of mosquitoes of legend and black gnats. Like the *D* climates, *ET tundra* climates are strictly a Northern Hemisphere occurrence, except for elevated mountain locations in the Southern Hemisphere and a portion of the Antarctic Peninsula. A narrow band of *ET tundra* climate stretches along the northern extreme of Canada, Scandinavia, and Asia. As an example, at lower latitudes,

FIGURE 10-18
Ever-frozen Antarctic landscape in the daylight of midnight—Earth's largest repository of freshwater. [Photo by Wolfgang Kaehler.]

the summit of Mount Washington in New Hampshire (1914 m or 6280 ft) statistically qualifies as a highland *ET tundra* climate of small scale.

Most all of Antarctica falls within the *EF ice cap* climate, as does the North Pole, with all months averaging below freezing. Both regions are dominated by dry, frigid air masses, with vast expanses that never warm above freezing. The area of the North Pole is actually a sea covered by ice, whereas Antarctica is a substantial continental landmass covered by Earth's greatest ice sheet. For comparison, winter minimums at the South Pole (July) frequently drop below the temperature of frozen CO_2 or "dry ice" (−78°C; −109°F).

Antarctica is constantly snow covered but receives less than 8 cm (3 in.) of precipitation each year. Because of consistent winds, it is difficult to discern what is new snow and what is old snow being blown around. Antarctic ice is several kilometers thick and is the largest repository of freshwater on Earth. This ice represents a vast historical record of Earth's atmosphere, within which thousands of volcanic eruptions worldwide have deposited ash layers, and ancient combinations of atmospheric gases lie trapped in frozen bubbles (Figure 10-18). An analysis of an ice core taken from Antarctica is presented in Focus Study 17-1 in Chapter 17.

The *EM polar marine* climate was not proposed by Köppen, but by later researchers. *EM polar marine* stations are more moderate than other polar climates in winter, with no month below −7°C (20°F), yet they are not as warm as *ET tundra* climates. Because of marine influences, annual temperature ranges do not exceed 17C° (30F°). *EM polar marine* stations occur along the Bering Sea, the tip of Greenland, northern Iceland, Norway, and in the Southern Hemisphere, generally over oceans between 50° and 60° S. Precipitation, which frequently falls as sleet, is greater in these regions than in continental *E* climates.

Dry Arid and Semiarid Climates

The *B* climates are the only ones that Köppen classified according to the *amount and distribution of precipitation*. These are the world's arid and semiarid regions—deserts and steppes possessing unique plants, animals, and physical features. The mountains, rock strata, long vistas, and the resilient struggle for life are all magnified by the dryness. Because of the sparseness of vegetation, the landscape is bare and exposed; POTET exceeds PRECIP in all parts of *B* climates, creating varying permanent water deficits the sizes of which distinguish the different subdivisions within this climatic group. Vegetation is typically xerophytic; drought resistant, waxy, hard leafed, and adapted to aridity and low transpiration loss. Along water

FIGURE 10-19
Desert plants are particularly adapted to the harsh environment of the Mojave desert. [Photo by author.]

courses, plants called phreatophytes, or "water-well plants," penetrate to great depths for the water they need (Figure 10-19).

The *B* climates occupy more than 35% of Earth's land area, clearly the most extensive climate over land. The world climate map reveals the pattern of Earth's dry climates, which cover broad regions between 15° and 30° N and S. In these areas the subtropical high-pressure cells predominate, with subsiding, stable air and low relative humidity. Under generally cloudless skies, these subtropical deserts extend to western continental margins, where cool, stabilizing ocean currents operate offshore and summer advection fogs form. The Atacama Desert of Chile, the Namib Desert of Namibia, the Western Sahara of Morocco, and the Australian Desert each lie adjacent to a coastline. Extension of these dry regions into higher latitudes is associated with rain shadows, induced by orographic lifting over western mountains in North America and Argentina. Interior Asia, far distant from any moisture-bearing air masses, is also included within the *B* climates.

The major subdivisions are the *BW deserts,* where PRECIP is *less* than one-half of POTET, and the *BS steppes,* where PRECIP is *more* than one-half of POTET. In an effort to better describe the dry climates, Köppen developed simple formulas to determine the usefulness of rainfall, based on the season in which it falls—whether it falls principally in the summer with a dry winter, or in the winter with a dry summer, or whether the rainfall is evenly distributed. Winter rains are the most effective because they fall at a time of lower POTET.

Köppen considered the summer dry if the wettest winter month was three times wetter than the driest summer month. The reverse reflects the role of POTET: winter is considered dry if the wettest summer month receives 10 times the precipitation of the driest winter month. Small graphs on Figure 10-4 help demonstrate these determinations. A tertiary lowercase letter is defined by the average annual temperature: temperatures above 18°C (64.4°F) are signified by an *h* for hot; those below 18°C are represented by a *k* for cold.

Hot Low-Latitude Desert Climates (BWh). The *BWh hot low-latitude desert* climates are Earth's true tropical and subtropical deserts. They are concentrated on the western sides of continents, although Egypt, Somalia, and Saudi Arabia also fall within this classification. Rainfall, usually the result of local convectional showers during the summer season, is very unreliable in *BWh hot low-latitude desert* climates. Some regions receive near zero amounts, whereas others may receive up to 35 cm (14 in.). Specific annual and daily desert temperature regimes, including the highest record temperatures, are discussed in Chapter 5; surface energy budgets are covered in Chapter 4. Representative

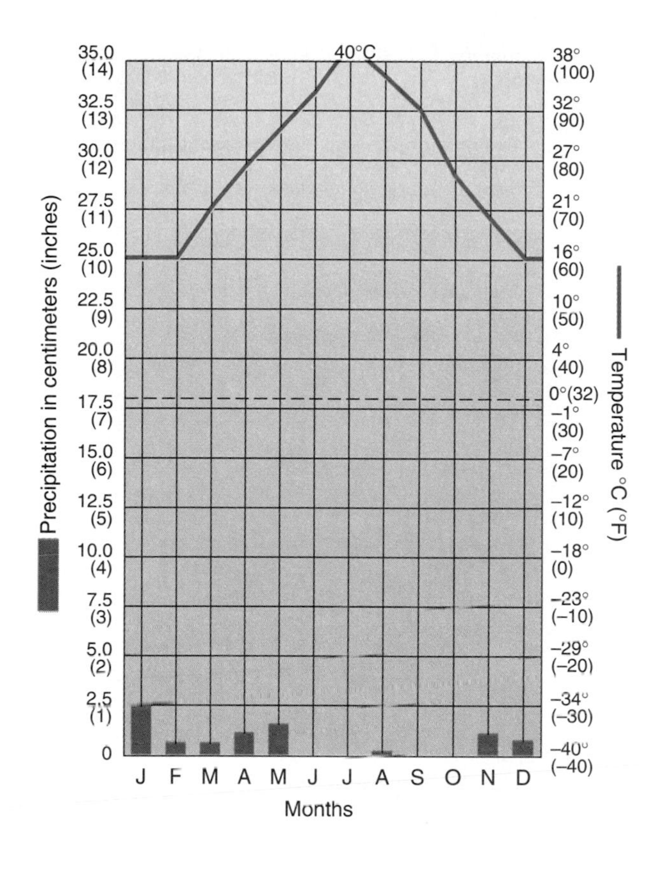

Station: Riyadh, Saudi Arabia **BWh** **Elevation:** 609 m (1998 ft)
Lat/long: 24°42' N, 46°43' E **Population:** 1,308,000
Avg. Ann. Temp.: 26°C (78.8°F) **Ann. Temp. Range:** 24C° (43.2F°)
Total Annual Precipitation: 8.2 cm (3.2 in.) **Ann. Hrs. of Sunshine:** not available

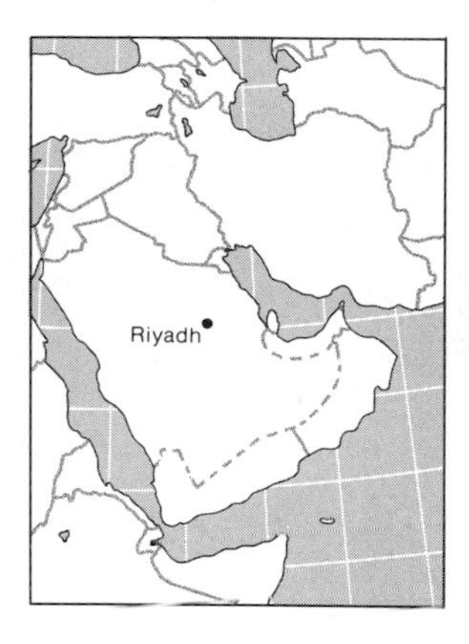

FIGURE 10-20
Climograph for Riyadh, Saudi Arabia (BWh climate).

BWh hot low-latitude desert stations include Yuma, Arizona; Aswan, Egypt; Lima, Peru; Bahrain Island in the Persian Gulf; and Riyadh, Saudi Arabia (Figure 10-20).

Cold Midlatitude Desert Climates (BWk). On the world climate map, this classification covers only a small area: the southern countries of the former USSR, the Gobi Desert, and Mongolia in Asia; the central third of Nevada and areas of the American Southwest, particularly at high elevations; and Patagonia in Argentina. Because of lower temperatures and lower POTET values, rainfall must be low for stations to meet the *BWk cold midlatitude desert* criteria; consequently, total average annual rainfall is only about 15 cm (6 in.). Albuquerque, New Mexico, with 20.7 cm (8.1 in.) of precipitation and an annual average temperature of 14°C (57.2°F) (which is 4C° or 7.2F° below the standard), is representative of this climate type (Figure 10-21). A comparison of the *BWk cold midlatitude desert* and *BWh hot low-latitude desert* stations shows an interesting similarity in the annual range of temperatures and a distinct difference in precipitation patterns.

Hot Low-Latitude Steppe Climates (BSh). These climates are generally located around the periphery of hot deserts, where shifting subtropical high-pressure cells create a distinct summer-dry and winter-wet precipitation pattern. This climate type is best identified around the periphery of the Sahara and in Iran, Afghanistan, and the southern portion of the former USSR. Along the southern margin of the Sahara lies the Sahel, a drought-tortured region of expanding desert conditions. Human populations in this region have suffered great hardship as desert conditions have gradually expanded over their countries. Average annual precipitation in *BSh hot low-latitude steppe* areas is usually below 60 cm (23.6 in.), less than the Köppen criterion for this classification of 86 cm (33.9

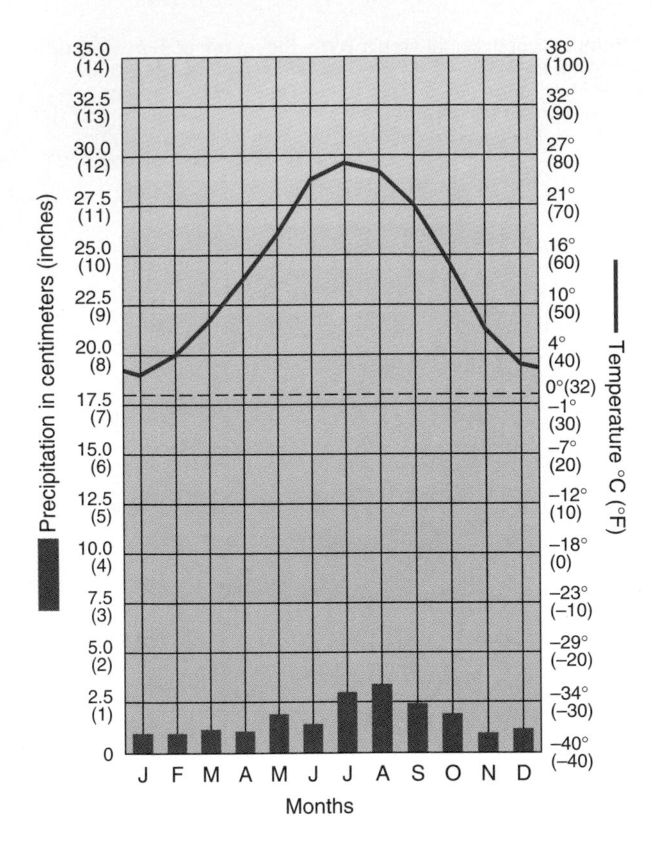

Station: Albuquerque, New Mexico **BWk**
Lat/long: 35°03' N, 106°37' W
Avg. Ann. Temp.: 14°C (57.2°F)
Total Annual Precipitation: 20.7 cm (8.1 in.)
Elevation: 1620 m (5315 ft)
Population: 370,000
Ann. Temp. Range: 24C° (43.2F°)
Ann. Hrs. of Sunshine: 3420

FIGURE 10-21
Climograph for Albuquerque, New Mexico (BWk climate).

in.). A few sample stations include Santiago, Chile; Kayes, Mali; Las Palmas, Canary Islands; Daly Waters, Australia; and Poona, India.

Cold Midlatitude Steppe Climates (BSk). The *BSk cold midlatitude steppe* climates occur poleward of about 30° latitude and the *BWk cold midlatitude desert* climates. Such midlatitude steppes are not generally found in the Southern Hemisphere. As with other *B* climate classifications, *BSk cold midlatitude steppe* climate rainfall variability is large and undependable, ranging from 20 to 40 cm (7.9–15.7 in.). Not all rainfall is convectional, for cyclonic systems penetrate the continents; however, most produce little precipitation in this climate region. A sample of *BSk cold midlatitude steppe* stations includes Denver; Salt Lake City; Cheyenne, Wyoming; Lethbridge, Alberta; Ulaanbaatar, Mongolia; and Semipalatinsk, Kazakhstan (Figure 10-22).

FUTURE CLIMATE PATTERNS

The greenhouse effect and the contributive role of radiatively active gases such as carbon dioxide, CFCs, and methane are discussed in Chapters 3 and 5. Earlier in this chapter we mentioned the complexity of developing and running global circulation models to predict the climatic implications of increases in these gases, but there is little scientific doubt that air temperatures are the highest they have been since air-temperature recordings were begun in earnest more than 100 years ago. In terms of **paleoclimatology**, the science that studies past climates (Chapter 17), Earth is within 1C° of equaling the highest average temperature of the past 125,000 years. Nine of the warmest years in this century were in the past decade (1991, 1990, 1988, and 1987 broke all records) and are thought to be attributable to a buildup of greenhouse gases. Climatologists Richard Houghton and

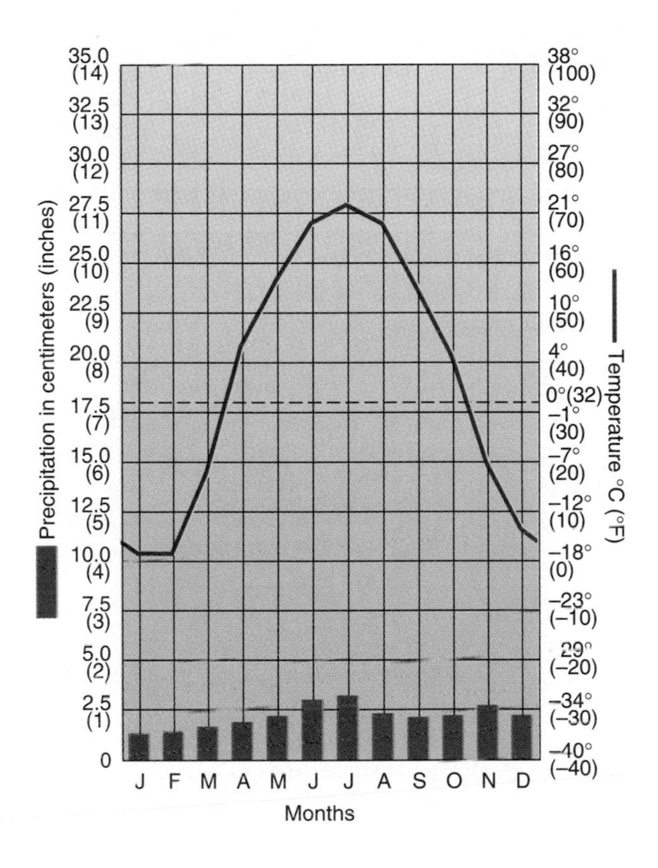

Station: Semipalatinsk, Kazakhstan **BSk**
Lat/long: 50°21' N, 80°15' E
Avg. Ann. Temp.: 3°C (37.4°F)
Total Annual Precipitation: 26.4 cm (10.4 in.)
Elevation: 206 m (675.9 ft)
Population: 330,000
Ann. Temp. Range: 39C° (70.2F°)
Ann. Hrs. of Sunshine: not available

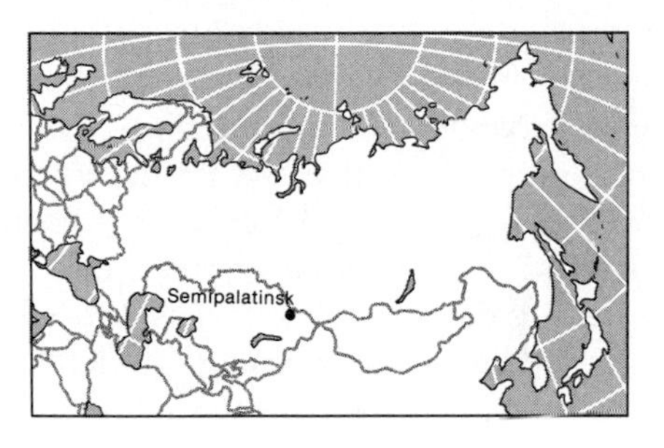

FIGURE 10-22
Climograph for Semipalatinsk, Kazakhstan (BSk climate).

George Woodwell describe the present climatic condition:

> The world is warming. Climatic zones are shifting. Glaciers are melting. Sea level is rising. These are not hypothetical events from a science fiction movie; these changes and others are already taking place, and we expect them to accelerate over the next years as the amounts of carbon dioxide, methane, and other trace gases accumulating in the atmosphere through human activities increase.*

Scientists are attempting to determine the difference between forced fluctuations (human-caused) and unforced fluctuations (natural) as a key to unraveling future climate trends. Because the gases generating any forced fluctuations are anthropogenic, various management strategies are possible to reduce their input. Several scenarios have been developed using computer-driven general circulation models (GCMs). These climate models produce results generally consistent with empirical evidence, although uncertainty remains as to actual overall global climate sensitivity, the role of oceans in heat absorption and transport, and the changeable role that clouds may or may not play.

Nevertheless, the confidence level is high for the current GCM projections of greenhouse warming substantiated by a data base of billions of observations of the atmosphere. The 1990 report of the Intergovernmental Panel on Climate Change (IPCC), a broad-based group of 380 scientists established in 1988 by the United Nations Environment Programme and representing the efforts of 63 coun-

*Richard Houghton and George Woodwell, "Global Climate Change," *Scientific American,* April 1989, p. 36.

tries and 18 nongovernmental organizations, concluded that there is ". . . virtual unanimity among greenhouse experts that a warming is on the way and that the consequences will be serious."

IPCC affirmed their major conclusions in a 1992 update: human-produced emissions of CO_2, CH_4, CFCs, and N_2O (nitrous oxide) are substantially increasing atmospheric concentrations of these gases. Also, they affirm that a warming trend is occurring. However, uncertainties as to magnitude, timing, and specific regional patterns are still acknowledged. A minority of scientists persist as critics of the greenhouse warming model, and although no member of the IPCC agrees with these critics, all points of view should be considered.

Scientists forecasted temperature changes by programming their GCMs with different assumptions and through consideration of different policy strategies. Let us assume three such *scenarios* from worst-possible case (Scenario A) to best-possible case (Scenario C). *Scenario A* assumes that activities continue with no real change in the behavior of society, producing a doubling of CO_2 by 2030 and the most climatic warming (assuming that no large volcanic eruptions inject into the atmosphere material that reflects or absorbs insolation). *Scenario B*, a middle case, represents certain modifications in human activity that lead to limits of CO_2 and other greenhouse gas emissions. This moderate set of assumptions merely delays the doubling of CO_2 to the middle of the century. An unrealistic, yet most desirable, *Scenario C* involves drastic reductions in greenhouse gas emissions that would produce significant delays in climatic warming. This scenario requires the implementation of more resource conservation and efficiency, renewable energy technology, and mass transportation.

There is a wide range of uncertainty in these scenario estimates, from 1.5 to 4.5C° (2.7–8.1F°). According to the updated IPCC assessment, an average warming on the low side of forecasts of 2.5C° (4.5F°) is predicted for the timeframe 1990 to 2029. If carbon dioxide concentrations reach 550 ppm by the year 2020, a temperature increase of 3C° (5.4F°) is forecast for the equatorial regions. This equatorial increase translates to a high-latitude increase of about 10C° (18F°), notable on the maps. (See the 8 pages of GCM forecast maps, after page 146, in the 1992 IPCC report listed under the Suggested Readings for this chapter.)

Figure 10-23 presents three computer-generated predictions for July of 1990, 2000, and 2029, using the assumptions of the moderate *Scenario B*. The map for 1990 proved accurate when measured against actual data for that year. During 1991–92 the GCM operated by James Hansen and his staff at the Goddard Institute for Space Studies (GISS) processed Mount Pinatubo eruption data. Year-end temperature records for 1992 confirmed the ability of this GCM to accurately predict the temporary cooling produced by this June 1991 event. As millions of tons of eruption debris clear from the atmosphere, the GISS predicts a return to the greenhouse warming trend of the past two decades. The maps for the years 2000 and 2029 still appear reasonable in their prediction of average global warming.

Consistent with other computer-generated studies, unambiguous warming appears over low-latitude oceans and interior areas of Asia, especially China, with the greatest warming assumed to occur over Arctic and Antarctic regions of sea ice. In the United States, the East shows greater warming than the West as a result of maritime effects, pressure patterns, and winds—northerly winds are expected to increase in the West, whereas southerly winds are expected to increase in the East, because of a strengthening of the related subtropical high-pressure cells.

These forecasts are valuable for the trends they show and the need for future research and appropriate public policy which they demand. Such climatic changes, if they occur, would have a major impact on the biosphere and challenging implications for society.

Consequences of Climatic Warming

The consequences of uncontrolled atmospheric warming are complex. Regional climate responses are expected as temperature, precipitation, soil moisture, and air-mass characteristics change. Although the ability to accurately forecast such regional changes is still evolving, some implications of warming are forecasted. For example, modern single-crop agriculture is more delicate and susceptible to changes in temperature, water demand

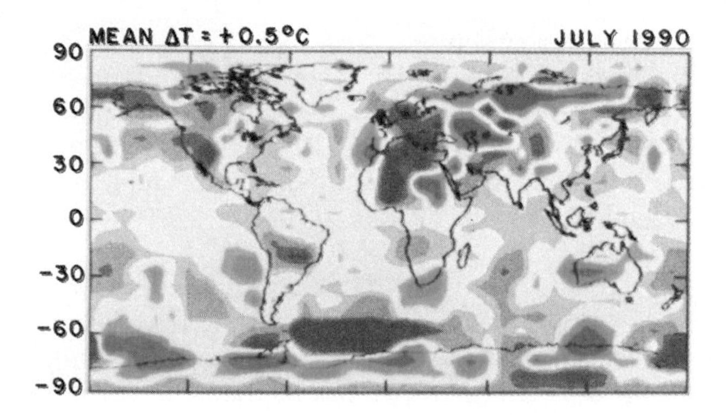

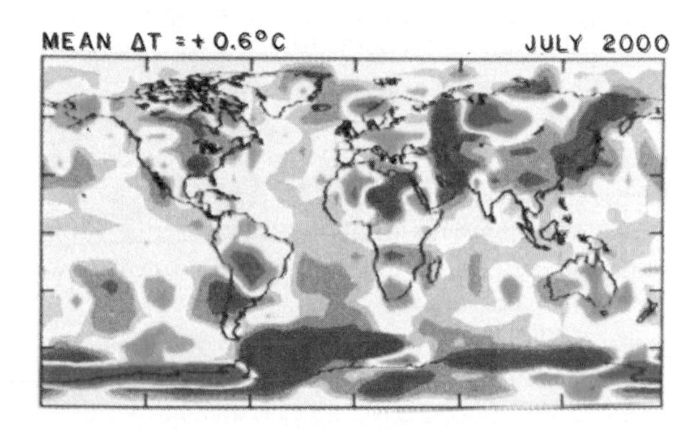

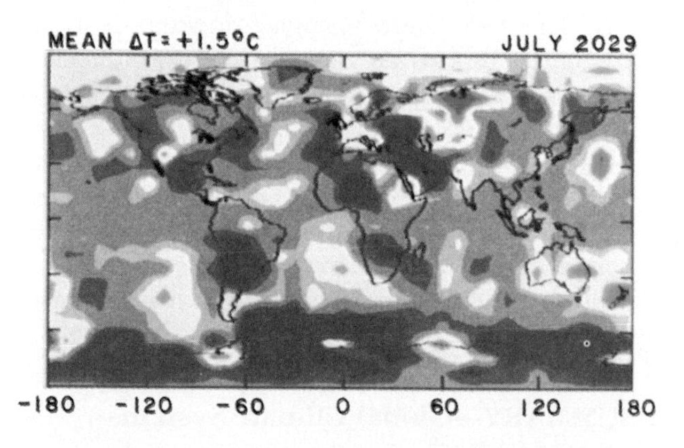

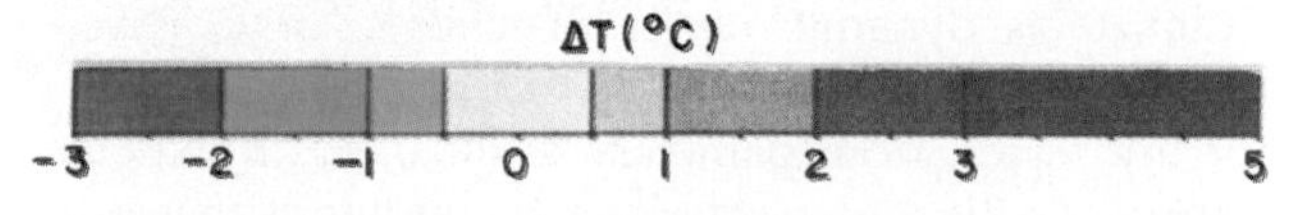

FIGURE 10-23
July temperature projections for 1990, 2000, and 2029 for *Scenario B.* [From J. Hansen et al., "Global Climate Changes as Forecast by Goddard Institute for Space Studies Three-Dimensional Model," *Journal of Geophysical Research* 43, no. 8 (August 1988): plate 6. Published by the American Geophysical Union.]

and irrigation needs, and soil chemistry than traditional multicrop agriculture.

Specifically, the southern and central grain-producing areas of North America are forecast to experience hot and dry weather by the middle of the next century as a result of higher temperatures. An increased probability of extreme heat waves is forecast for these U.S. grain regions. Available soil moisture is projected to be at least 10% less throughout the midlatitudes in the next 30 years.

Today, the United States and Canada sell wheat to China and the Commonwealth of Independent States (former USSR). If the foreign grain areas become wetter and gain a longer frost-free period, grain sales could shift in the other direction. Specifically, as U.S. production decreases, the longer growing seasons, increased temperatures and thermal energy patterns, and possible increased precipitation at higher latitudes will benefit Canadian cropping patterns as well. The possibility exists that billions of dollars of agricultural losses in one region could be countered by billions of dollars of gain in another.

Scientists have postulated possible adaptations, such as changing crop varieties to late-maturing, heat-resistant types, and adjustments in soil-management practices, such as fertilizer applications and irrigation schedules. Studies completed at the universities of Toronto and Wisconsin–Madison postulate, with an admitted degree of uncertainty, that greenhouse warming will increase the temperature of the Great Lakes, which in turn will be an advantage for various species of fish. On the other hand, a warming of small Canadian lakes might prove harmful to aquatic life.

Also, land-based animals will have to adapt to changing patterns of available forage, for animals (mammals in particular) are habituated and adapted to the climate zones in which they live. Increased temperatures already are causing wide-ranging thermal stress to embryos as they reach their thermal limits. Warming of ocean water is endangering corals worldwide, and they appear to be deteriorating at record levels. This developing condition is discussed further in Chapter 16.

Polar and High-Latitude Warming Effects. Perhaps the most pervasive climatic effect of increased warming would be the rapid escalation of ice melt. The additional water, especially from continental ice masses that are grounded, would raise sea level worldwide. Satellite remote-sensing technology allows monitoring of global sea-ice and continental-ice distributions.

Scientists currently are studying the ice sheets of Greenland and Antarctica for possible changes in the operation of the hydrologic cycle, including snowlines and the rate at which icebergs break off (calve) into the sea. The key area being watched is the **West Antarctic ice sheet**, where the Ross Ice Shelf holds back vast grounded ice masses. The flow from several visible ice streams (channels of ice) is accelerating from the ice sheet. During the late 1980s some of the largest icebergs ever noted broke away from Antarctica. In September 1991 Space Shuttle Discovery's astronauts observed one of these huge icebergs at 57° S latitude, where it had drifted since breaking off the Antarctic ice shelf in 1986! Several interpretations for these changes are possible.

The possible sea-level rise must be expressed as a range of values, based on a variety of assumptions that are under constant reassessment. Estimates of future rise in sea level range from a low of 30–110 cm (1–4 ft) to a high of 6 m (20 ft). These levels could be further augmented by melting of alpine and mountain glaciers, which have experienced about a 30% decrease in overall ice mass during this century. The Intergovernmental Panel on Climate Change in *Climate Change 1992* (p. 158) states:

> **there is conclusive evidence for a worldwide recession of mountain glaciers. . . . This is among the clearest and best evidence for a change in energy balance at the Earth's surface since the end of the last century.**

A quick survey of world coastlines shows that even moderate changes could bring disaster of unparalleled proportions. At stake are the river deltas, lowland coastal farming valleys, and low-lying mainland areas. In the worst possible scenario, a rise of 6 m (20 ft) would flood 20% of Florida; inundate the Mississippi floodplain as far inland as St. Louis; flood the Pampas of Argentina; drown Venice; submerge the Bahamas, The Netherlands, the Maldives; and more. All surviving coastal areas would contend with high water and high tides—but again this is with a worst possible scenario. Worldwide measurements confirm that sea level has been rising this century at 10–20 mm per decade.

Solutions. The position is held by some government leaders that society has adapted to many problems in the past; capital can be raised and technology harnessed to block the flood and guard the coastlines and deltas. Although such adaptive strategies may be needed, no substantive long-range planning is going forward. Why are we waiting? Why not take action now? Whatever the actual impact—from major financial and cultural irritation to outright catastrophe—solutions exist to reduce the severity of the hazard. In addition, for Americans and Canadians, individual conservation and a demand for energy efficiency remain cheap ways to stretch present energy resources, reduce imports, and cut carbon dioxide emissions.

The four IPCC volumes and participants at many world conferences have agreed that governmental policies can profoundly alter the greenhouse effect. The 1989 Cairo Compact and the 1990 and 1992 IPCC reports are steps in the right direction. One of the significant agreements that emerged from the 1992 United Nations Conference on Environment and Development (the "Earth Summit" detailed in Focus Study 21-1) was the *U.N. Framework Convention on Climate Change*—a legally binding agreement that represents a first-ever attempt to evaluate and address global warming on an international scale.

SUMMARY—Global Climate Systems

Climate is dynamic rather than static or inactive. Climate is a synthesis of weather phenomena at many scales, from planetary to local. Earth experiences a wide variety of climatic conditions that can be grouped by general similarities for studies of climatic regions. A modified empirical climate classification system developed by Köppen provides a basis for describing the various distinct environments on Earth: tropical *A*, mesothermal *C*, microthermal *D*, polar *E*, and arid *B* climates. Highland climates are assigned an *H* to distinguish them from the others and to denote the conditions created by elevation. The main climatic groups are divided into subgroups, based upon temperature and the seasonal timing of precipitation.

Various activities of present-day society are producing climatic changes, particularly a global warming trend. The 1980s and early 1990s were dominated by the highest average annual temperatures experienced since the advent of instrumental measurements. There is little scientific doubt that this warming is related to the greenhouse effect and that various control options need to be activated in a concerted global effort, although some uncertainties remain to be resolved.

These findings were further supported by a 1992 update report by the Intergovernmental Panel on Climate Change. The IPCC predicted surface-temperature response to a doubling of CO_2 with a range of increase between 1.5C° and 4.5C° (2.7–8.1F°) for the next century. Climatologists are constructing and refining general circulation models (GCMs) to better forecast future climate patterns. People and their political institutions then may use this information to form policies aimed at reducing unwanted climate change.

KEY TERMS

classification
climate
climatic regions
climatology
climographs
empirical classification
general circulation models (GCMs)
genetic classification
paleoclimatology
West Antarctic ice sheet

REVIEW QUESTIONS

1. Define climate and compare it to weather. What is climatology?
2. What created the climatic disruption associated with El Niño? What were some of the changes and effects that occurred worldwide?
3. Evaluate the relationships among a climatic region, ecosystem, and biome.
4. How do radiation receipts, temperature, air pressure inputs, and precipitation patterns interrelate to produce climate types?
5. What are the differences between a genetic and an empirical classification system?
6. Describe Köppen's approach to climatic classification. What was the basis of the system he devised?
7. List and discuss each of the principal climate designations. In which one of these general types do you live? Which classification is the only type associated with the distribution and amount of precipitation?
8. What is a climograph, and how is it used to display climatic information?
9. Which of the major climate types occupies the most land and ocean area on Earth?
10. Characterize the tropical *A* climates in terms of temperature, moisture, and location.
11. Using Africa's tropical climates as an example, characterize the climates produced by the seasonal shifting of the ITCZ with the high Sun.
12. Mesothermal *C* climates occupy the second largest portion of Earth's entire surface. Describe their temperature, moisture, and precipitation characteristics.

13. Explain the distribution of the *Cfa humid continental* and *Csa Mediterranean dry summer* climates at similar latitudes and the difference in precipitation patterns between the two types. Describe the difference in vegetation associated with these two climate types.
14. Which climates are characteristic of the Asian monsoon region? What do each of the letters in that Köppen classification indicate?
15. How can a marine west coast climate type (*Cfb*) occur in the Appalachian region of the eastern United States? Explain.
16. What role do offshore currents play in the distribution of the *Csb* climate designation? What type of fog is formed in these regions?
17. Discuss the climatic designation for the coldest places on Earth outside the poles. What do each of the letters in the Köppen classification indicate?
18. What is the largest single group of climates on Earth in terms of land area alone? Analyze the general pattern of this climate type.
19. In general terms, what are the differences among the four desert classifications? How are moisture and temperature distributions used to differentiate these subtypes?
20. Explain climate forecast scenarios. What is implied in *Scenario A* as compared to *Scenario C?*
21. Describe the potential climatic effects of global warming on polar and high-latitude regions. What are the implications of these climatic changes for persons living at lower latitudes? The effects on agricultural activity?

SUGGESTED READINGS

Griffiths, John F., and Dennis M. Driscoll. *Survey of Climatology.* Columbus, OH: Macmillan Publishing Company, 1982.

Haurwitz, B., and J. M. Austin. *Climatology,* 2d ed. Englewood Cliffs, NJ: Prentice Hall, 1966.

Houghton, J. T., G. J. Jenkins, and J. J. Ephraums, eds. *Climate Change 1992—The Supplementary Report to the IPCC Scientific Assessment.* World Meteorological Organization/United Nations Environment Programme. New York: Cambridge University Press, 1992.

Houghton, J. T., G. J. Jenkins, and J. J. Ephraums, eds. *Climate Change—The IPCC Scientific Assessment.* Working Group I, World Meteorological Organization/United Nations Environment Programme. Port Chester, NY: Cambridge University Press, 1991.

Intergovernmental Panel on Climate Change. *Climate Change—The IPCC Response Strategies.* Working Group III. World Meteorological Organization/United Nations Environment Programme. Covelo, CA: Island Press, 1991.

Lydolph, Paul E. *The Climate of the Earth.* Totowa, NJ: Rowman & Allanheld, 1985.

Matthews, Samuel W. "Under the Sun—Is Our World Warming?" *National Geographic* 178, no. 4 (October 1990): 66–99.

Mintzer, Irving M. *A Matter of Degrees: The Potential for Controlling the Greenhouse Effect.* Research Report No. 5. Washington, DC: World Resources Institute, 1987.

Oliver, John E., and Rhodes W. Fairbridge, eds. *The Encyclopedia of Climatology.* New York: Van Nostrand Reinhold Co., 1987.

Revkin, Andrew C. "Endless Summer: Living with the Greenhouse Effect," *Discover* 9, no. 10 (October 1988): 50–61.

Strahler, Arthur N., and Alan H. Strahler. *Modern Physical Geography,* 3d ed. New York: John Wiley & Sons, 1987.

Tegart, W. J. McG., G. W. Sheldon, and D. C. Griffiths. *Climate Change—The IPCC Impacts Assessment.* Working Group II. World Meteorological Organization/United Nations Environment Programme. Portland, OR: International Specialized Book Service/Australian Government, 1991.

Thomas, Robert. "Future Sea Level Rise and Its Early Detection by Satellite Remote Sensing." In James Titus, ed. *Effects of Changes in Stratospheric Ozone and Global Climate.* Vol. 4, *Sea Level Rise.* Washington, DC: Environmental Protection Agency, 1986.

Thornthwaite, C. W. "An Approach Toward a Rational Classification of Climate," *Geographical Review* 38 (1948): 55–94.

Tooley, Michael J., and Ian Shennan, eds. *Sea-Level Changes.* Oxford: Basil Blackwell & the Institute of British Geographers, 1987.

Topping, John C., Jr., ed. "Coping with Climate Change," *Proceedings of the Second North American Conference on Preparing for Climate Change.* Washington, DC: Climate Institute, 1989.

Trewartha, Glenn T., and Lyle H. Horn. *An Introduction to Climate,* 5th ed. New York: McGraw-Hill, 1980.

Wilcox, Arthur A. "Köppen After Fifty Years," *Annals of the Association of American Geographers* 58, no. 1 (March 1968): 12–28.

PART 3

THE EARTH-ATMOSPHERE INTERFACE

Mount Hayden, from north rim of the Grand Canyon. [*Photo by Dick Dietrich.*]

Earth is a dynamic planet whose surface is actively shaped by physical agents of change. Part 3 is organized around two broad systems of these agents—endogenic and exogenic. The **endogenic system** (Chapters 11 and 12) encompasses internal processes that produce flows of heat and material from deep below the crust, powered by radioactive decay—the solid realm of Earth. Earth's surface responds by moving, warping, and breaking, sometimes in dramatic episodes of earthquakes and volcanic eruptions.

At the same time, the **exogenic system** (Chapters 13–17) involves external processes that set air, water, and ice into motion, powered by solar energy—the fluid realm of Earth's environment. These media are sculpting agents that carve, shape, and reduce the landscape. One such process—weathering—breaks up and dissolves the crust, making materials available for erosion, transport, and deposition by rivers, winds, wave action along coastlines, and flowing glaciers. Thus, Earth's surface is the interface between two systems, one that builds the landscape and one that reduces it.

Little Skellig Rocks off the coast of Ireland. [*Photo by Tom Bean/DRK Photo.*]

11

THE DYNAMIC PLANET

The twentieth century is a time of great discovery about Earth's internal structure and dynamic crust, yet much remains undiscovered. This is a time of revolution in our understanding of how the present arrangement of continents and oceans evolved. One task of physical geography is to explain the spatial implications of all this new knowledge and its effect on Earth's surface and society.

Earth's interior is highly structured, with uneven heating generated by the radioactive decay of unstable elements. The results are irregular patterns of surface fractures, the occurrence of earthquakes and volcanic activity, and the formation of mountain ranges. The U.S. Geological Survey reports that, in an average year, continental margins and seafloors expand by 1.9 km^3 (0.46 mi^3). But, at the same time, 1.1 km^3 (0.26 mi^3) are consumed, resulting in a net addition of 0.8 km^3 (0.2 mi^3) to Earth's crust. This chapter describes Earth's endogenic system, for it is within Earth that we find the driving forces that affect the crust.

A new era of Earth-systems science is emerging, effectively combining various disciplines within the study of physical geography. Remember that geography is an approach—a spatial way of looking at things—the spatial science that examines place, location, human-Earth relationships, the uniqueness of regions, and movement (Figure 1-1). The geographic essence of geology, geophysics, paleontology, seismology, and geomorphology are all integrated by geographers to produce an overall picture of Earth's surface environment and of specific physical phenomena that affect human society. Improved remote-sensing capabilities and the development of computer-based geographic information systems are enhancing these efforts.

THE PACE OF CHANGE

The **geologic time scale** (Figure 11-1) reflects currently accepted names and the relative and absolute time intervals that encompass Earth's history (eons, eras, periods, and epochs). The *sequence* in this scale is based upon the relative positions of rock strata above or below each other. An important general principle is that of *superposition,* which states that rock and sediment always are arranged with the youngest beds "superposed" near the top of a rock formation and the oldest at the base—if they have not been disturbed. The *absolute ages* on the scale, determined by scientific methods such as dating by radioactive isotopes, are also used to refine the time-scale sequence. Some highlights of Earth's history are listed on the geologic time scale in the figure.

The guiding principle of Earth science is uniformitarianism, first proposed by James Hutton in his *Theory of the Earth* (1795) and later amplified by Charles Lyell in *Principles of Geology* (1830). **Uniformitarianism** assumes that *the same physical processes active in the environment today have been operating throughout geologic time.* The phrase "the present is the key to the past" is an expression coined to describe this principle. Uniformitarianism is a valuable generalization supported by unfolding evidence from modern exploration and from the landscape record of volcanic eruptions, earthquakes, and processes that shape the landscape.

In contrast, the notion of **catastrophism** attempts to fit the vastness of Earth's age and the complexity of its rocks into a shortened time span. Catastrophism holds that Earth is young and that mountains, canyons, and plains formed through catastrophic events that did not require eons of time. Because there is little physical evidence to support this idea, catastrophism is more appropriately considered a belief than a serious scientific hypothesis. Ancient landscapes, such as the Appalachians, represent a much broader time scale of events (Figure 11-2).

However, geologic time is punctuated by dramatic events, such as massive landslides, earthquakes, and volcanic episodes. Within the principle of uniformitarianism, these localized catastrophic events occur as small interruptions in the generally uniform processes that shape the slowly evolving landscape.

We now start our journey deep within the planet. A knowledge of Earth's internal structure and energy is key to understanding the surface.

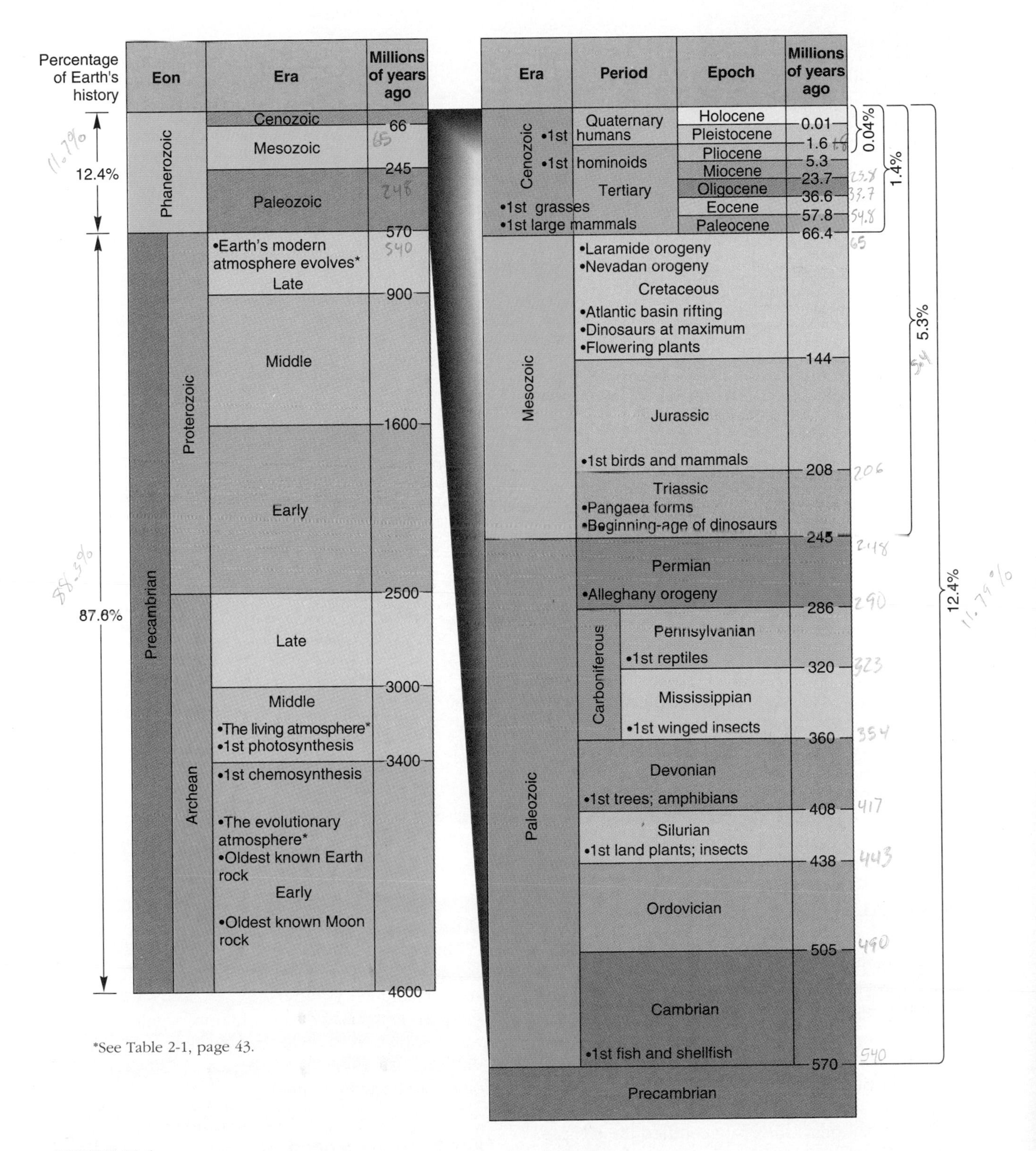

*See Table 2-1, page 43.

FIGURE 11-1
The geologic time scale was organized using relative-dating methods, with absolute dating determined through modern technological means. Note that over 87% of geologic time occurred during the Precambrian. [Data from *1983 Geologic Time Scale,* Geological Society of America.]

FIGURE 11-2
Ancient rock formations of the Appalachians, which began forming during the Alleghany Orogeny over 250 million years ago. [Photo by John B. Yates/ Rainbow.]

EARTH'S STRUCTURE AND INTERNAL ENERGY

Along with the other planets and the Sun, Earth is thought to have condensed and congealed from a nebula of dust, gas, and icy comets about 4.6 billion years ago (Chapter 2). The oldest surface rock yet discovered on Earth (known as the Acasta Gneiss) lies in northwestern Canada and dates back 3.96 billion years. Scientists from Saint Louis University, the Canadian Geological Survey, and Australian National University certified the sample's age in November 1989. Previously, rocks from Greenland held the record at 3.8 billion years old. These finds tell us something very significant: Earth was forming continental crust nearly 4 billion years ago.

Throughout Earth's formation process, heat was accumulating. As the protoplanet compacted into a smaller volume, energy was transformed into heat through compression. In addition, significant heat was trapped in Earth from the decay of unstable isotopes of uranium, thorium, potassium, and other radioactive elements.

Earth in Cross Section

As Earth solidified, heavier elements slowly gravitated toward the center, and lighter elements slow-

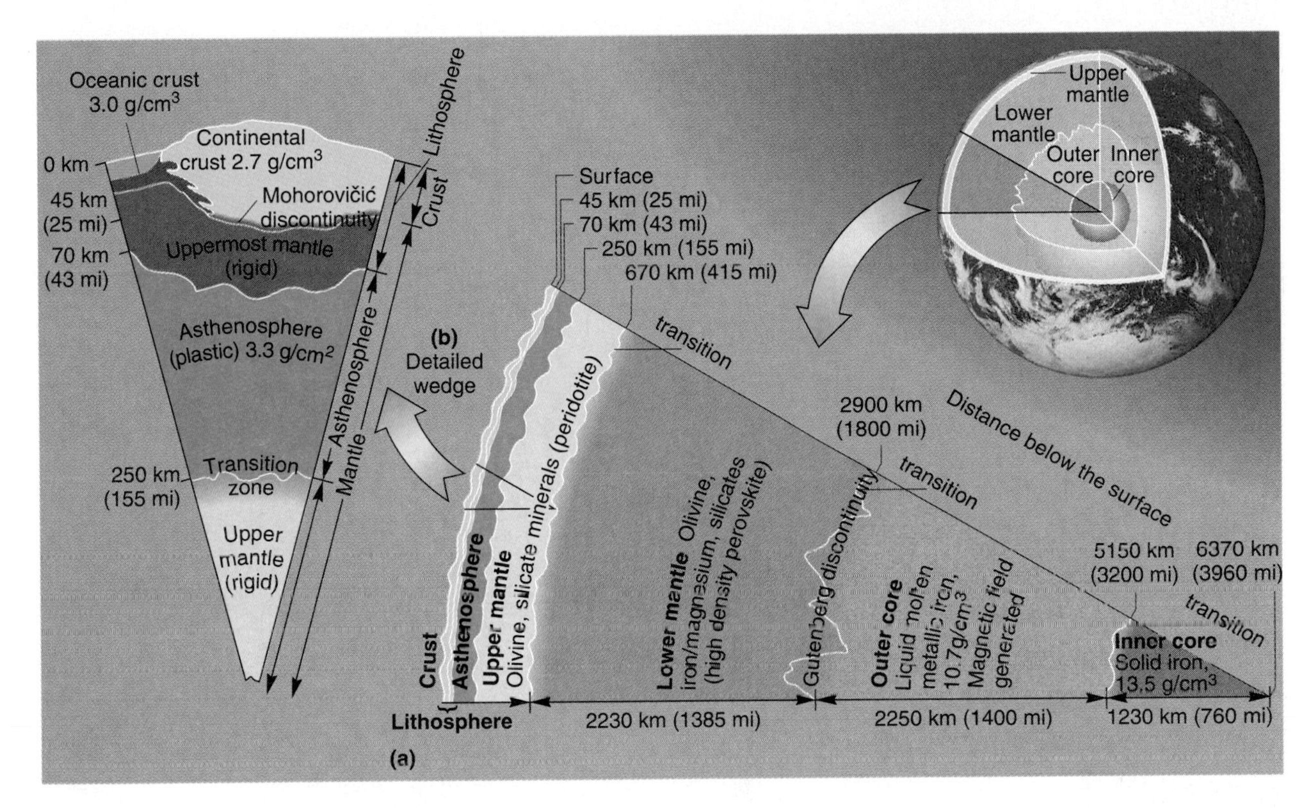

FIGURE 11-3
Earth's interior from the central core to the crust is illustrated in cross section (a). The structure of the lithosphere and its relation to the asthenosphere is detailed in the enlarged wedge (b).

ly welled upward to the surface, concentrating in the crust. Consequently, Earth's interior is arranged roughly in concentric layers, each one distinct either chemically or thermally, with heat radiating outward from the center by conduction and then by physical convection at levels nearer the surface.

Our knowledge of Earth's *internal differentiation* into these concentric layers has been acquired entirely through indirect evidence, because we are unable to drill more than a few kilometers into Earth. There are several physical properties of Earth materials that enable us to approximate the nature of the interior. For example, when an earthquake or underground nuclear test sends shock waves through the planet, the cooler areas, which generally are more rigid, transmit these **seismic waves** at a higher velocity than do the hotter areas. Density also has an effect on seismic velocities, and fluid or plastic zones simply do not transmit some types of seismic waves (they absorb them). Some seismic waves are reflected as densities change; others are refracted, or bent, as they travel through Earth. Thus, the distinctive ways in which seismic waves pass through Earth and the time they take to travel between two surface points help *seismologists* deduce the structure of Earth's interior (Figure 11-3). Important to this research are an increased knowledge of gravitational, magnetic, and earthquake data; the development of laboratory devices that can recreate temperatures and pressures similar to those inside the planet; and an expanded network of scientific instruments, improvements in their sensitivity, and the increasing speed of modern computers.

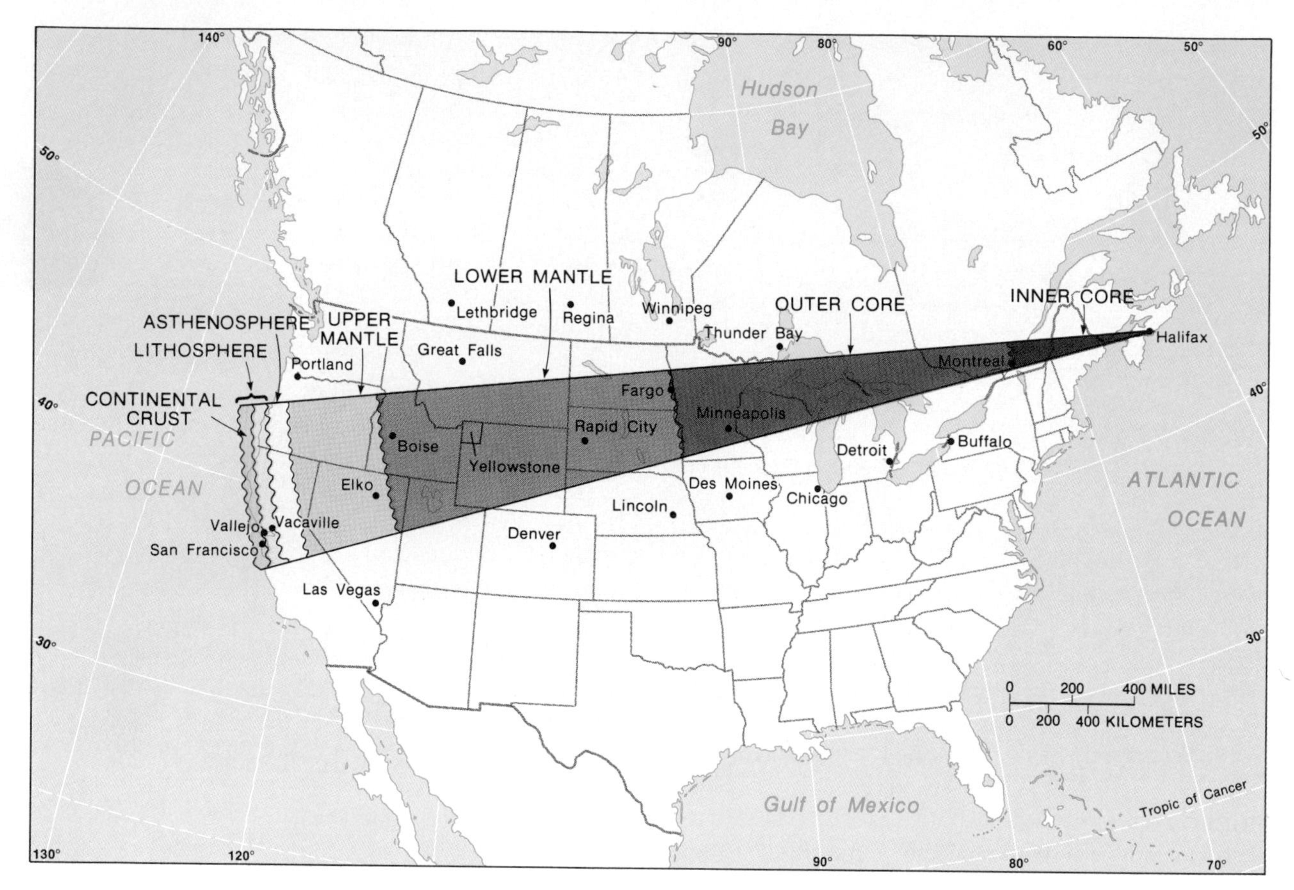

FIGURE 11-4
Surface map, using the distance from Halifax, Nova Scotia, to San Francisco to compare the distance from Earth's center to the surface. The continental crust thickness is the same as the distance between Vallejo (in the eastern portion of San Francisco Bay) and the city of San Francisco.

Figure 11-4 illustrates the dimensions of Earth's interior in comparison to surface distances in North America to give you a sense of size and scale.

Earth's Core. A third of Earth's entire mass, but only a sixth of its volume, lies in the core. The **core** is differentiated into two regions—inner core and outer core—divided by a transition zone of several hundred kilometers (Figure 11-3). The inner core is thought to be solid iron, with a density of 13.5 g per cm^3. (The metric relationship of grams per cubic centimeter is used to express such density. For comparison purposes, the density of water is 1.0 g/cm^3, and mercury is 13.0 g/cm^3.) The inner core remains solid even though it is at high temperature because of the tremendous pressures present. The iron in the core is not pure, but probably is combined with silicon, and possibly oxygen and sulfur. The outer core is molten, metallic iron with lighter densities averaging 10.7 g/cm^3. Estimated core temperatures range from 3000°C (5400°F) to as high as 6650°C (12,000°F).

Earth's Magnetism. The fluid outer core generates at least 90% of Earth's magnetic field and the magnetosphere that surrounds and protects Earth from the solar wind. A present hypothesis by scientists from Cambridge University details spiraling circulation patterns in the outer core region that are influenced by Earth's rotation; this circulation generates electric currents, which in turn induce the magnetic field. Think of flow rates of several kilometers per year, over a million times faster than in the overlying mantle.

An intriguing feature of Earth's magnetic field is that its polarity sometimes fades to zero and then returns to full strength with north and south magnetic poles reversed! In the process, the field does not blink on and off but instead oscillates slowly to low intensity and then slowly regains its full strength. This **magnetic reversal** has taken place nine times during the past 4 million years and hundreds of times over Earth's history. The average period of a magnetic reversal is 500,000 years; occurrences possibly vary from as short as several thousand years to as long as tens of millions of years.

There appears to be a trend in recent geologic time toward more frequent reversals with shorter intervals. When Earth is without polarity in its magnetic field, a random pattern of magnetism results. Transition periods last from 4000 to 8000 years. The effects of these low intensity episodes on life are still speculative, but without a magneto sphere, the surface environment is unprotected from cosmic radiation and solar wind. Given present rates of magnetic field decay, we are perhaps 2000 years away from the next reversal.

The reasons for these magnetic reversals are unknown; however, the spatial patterns they create at Earth's surface represent an important diagnostic tool in understanding the evolution of landmasses and the movements of the continents. When new iron-bearing rocks solidify from molten material at Earth's surface, the small magnetic particles in the rocks align according to the orientation of the magnetic poles at that time. This pattern is then locked in place as the rocks solidify. All across Earth, rocks bear this identical record of reversals as measurable magnetic stripes. In addition, they record the migration of the continents over time in relation to Earth's magnetic poles, which do not coincide with its geographic poles (axis). (The difference in location of the magnetic and geographic poles is shown in Figure 17-30 for both the north and south polar regions.) Later in this chapter we shall show the importance of this migration and these magnetic reversals.

Earth's Mantle. A transition zone of several hundred kilometers marks the top of the outer core and the beginning of the mantle. Scientists at the California Institute of Technology analyzed the behavior of more than 25,000 earthquakes and determined that this transition area is bumpy and uneven, with ragged peak-and-valleylike formations. Some of the motions in the mantle may be created by this rough texture at what is called the **Gutenberg discontinuity**. A *discontinuity* is a place where a change in physical properties occurs between two regions deep in Earth's interior.

The **mantle** represents about 80% of Earth's total volume, with average densities of 4.5 g/cm^3. The mantle is rich in oxides and silicates of iron and magnesium (FeO, MgO, and SiO_2) and is more dense and tightly packed in the lower region, grading to lesser densities in the upper region. Of the mantle's volume, 50% is in the lower mantle, which is composed of rock that probably has never been at Earth's surface. The upper mantle is separated from the lower mantle by a broad transition zone of several hundred kilometers centered about 670 km (415 mi) below the surface. The entire mantle experiences a gradual temperature increase with depth.

The upper mantle is divided into three fairly distinct layers. Outermost is a high-velocity zone just below the crust (approximately 45–70 km thick) where seismic waves are transmitted through a rigid, cooler layer of solidity and strength. This uppermost mantle, along with the crust, makes up the *lithosphere*.

From about 70 km down to 250 km is the **asthenosphere**, or plastic layer, from the Greek *asthenos*, meaning weak. It contains pockets of increased heat from radioactive decay and is susceptible to convective currents in these hotter (and therefore less dense) materials. The depths affected by these convection currents are the subject of much scientific speculation.

Because of this dynamic condition, the asthenosphere is the least rigid region of the mantle, with densities averaging 3.3 g/cm^3. About 10% of the asthenosphere is molten in asymmetrical patterns and hot spots. The resulting slow movement in this zone causes movement of the crust and creates tectonic activity—the folding, faulting, and general deformation of surface rocks. In return, the movement of the crust enhances the geometry of the mantle flows.

From 250 km to the transition region at a depth of 670 km resides the third layer, the rest of the upper mantle. There the rocks are solid again, a zone of greater density and high seismic velocity, because of increasing pressures.

The lower mantle is thought to be almost uniform in composition and distinct from the upper mantle, containing high-density minerals of iron, magnesium, and silicates, with some calcium and aluminum—known as silicate perovskite. However, much scientific speculation surrounds the composition of the lower mantle. Some scientists think the upper and lower mantle remain separate and distinct. Others think that the 670 km transition zone presents no barrier to mixing and that convection occurs throughout the mantle.

Lithosphere and Crust. The lithosphere includes the entire **crust**, which represents only 0.01% of Earth's mass, and the upper mantle down to a depth of about 70 km (43 mi), with an imprecise transition marking the lower boundary. The boundary between the crust and the high-velocity portion of the lithospheric upper mantle is another discontinuity called the **Mohorovičić discontinuity**, or *Moho* for short, named for the Yugoslavian seismologist who determined that seismic waves change at this depth, owing to sharp contrasts of materials and densities.

Figure 11-3b illustrates the relationship of the crust to the rest of the lithosphere and the asthenosphere below. Crustal areas below mountain masses extend farther downward, perhaps to 50–60 km (31–37 mi), whereas the crust beneath continental interiors averages about 30 km (19 mi) in thickness and oceanic crust averages only 5 km (3 mi). The composition and texture of continental and oceanic crust are quite different. Continental crust is basically **granite**; it is crystalline and high in silica, aluminum, potassium, calcium, sodium, and other constituents. (Sometimes continental crust is called *sial*, shorthand for silica and aluminum.) Oceanic crust is **basalt**; it is granular, and high in silica, magnesium, and iron. (Sometimes oceanic crust is called *sima*, shorthand for silica and magnesium.)

The principle of buoyancy (that something less dense, like wood, floats in something denser, like water) and the principle of balance were further developed in the 1800s into the important principle of **isostasy** to explain certain movements of Earth's crust. Think of Earth's outer crust as floating on the denser layers beneath, much as a boat floats on water. With a greater load (e.g., ice, sediment, mountains), the crust tends to ride lower in the asthenosphere. Without that load (e.g., when ice melts), the crust rides higher, in a recovery uplift known as an *isostatic rebound.* Thus, the entire crust is in a constant state of compensating adjustment, or isostasy, slowly rising and sinking in response to its own weight, and pushed and dragged about by currents in the asthenosphere (Figure 11-5).

After decades of effort, humans have yet to drill as deep as the Moho discontinuity at the base of the crust. Scientists want to sample mantle material directly. The longest-lasting deep-drilling program is on the northern Kola Peninsula in Russia, 250 km (155 mi) north of the Arctic Circle. Twenty years of high-technology drilling has produced a hole 12 km (7.5 mi or 39,400 ft) deep. Presently, crystalline rocks 1.4 billion years old at 180°C (356°F) are being penetrated by the bite of diamond-tipped drill bits. Oceanic crust is thinner than continental crust and is the object of several drilling attempts. The International Ocean Drilling Program (ODP), a cooperative effort directed by Texas A & M University, worked a 2 km hole in an oceanic rift near the Galápagos Islands from 1975 to 1992, with further drilling planned. The ocean at the drill site is 2.5 km (8200 ft) deep. An ocean floor of distinct layers of lava and deeper molten rock chambers were found.

Earth's crust is the outermost irregular shell that resides restlessly on a dynamic and diverse interior. Let us now examine the processes at work on this crust and the variety of rock types that compose the landscape.

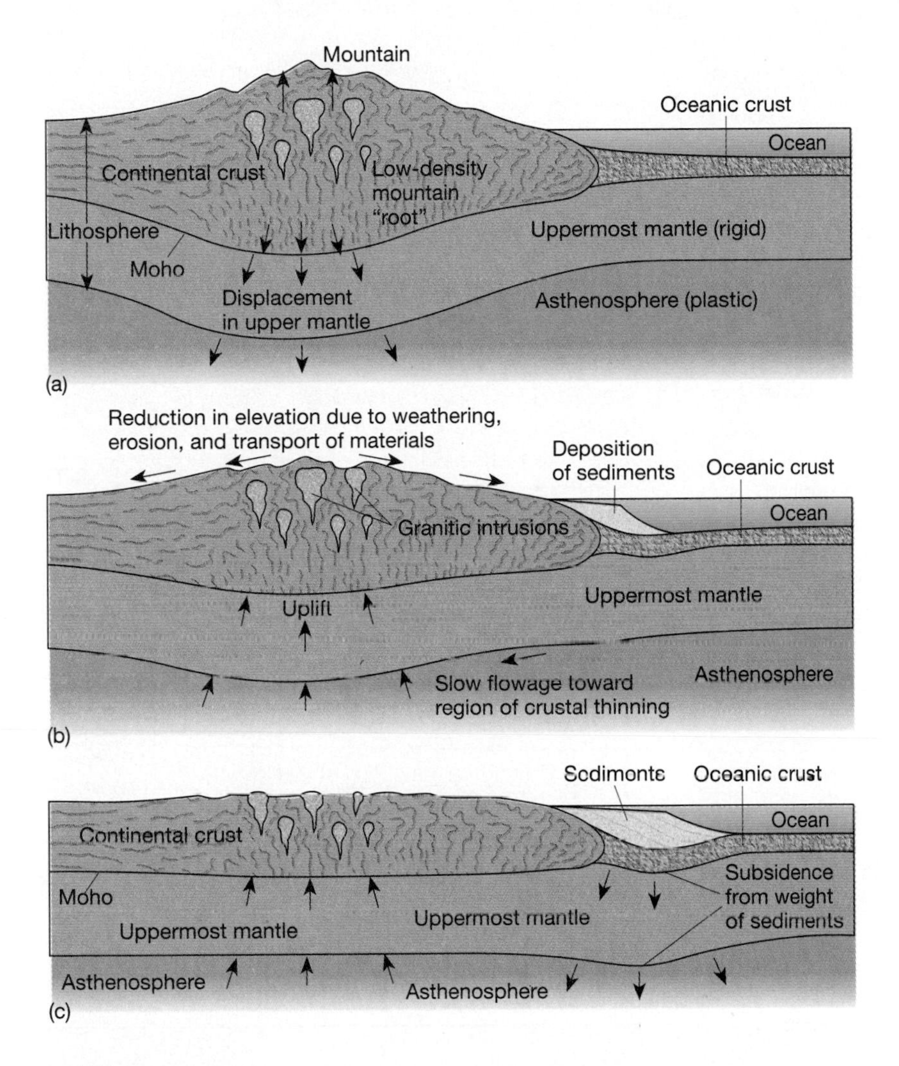

FIGURE 11-5
Isostatic adjustment of the crust. Earth's entire crust is in a constant state of compensating adjustment as suggested by these three sequential stages. In (a) the mountain mass slowly sinks, displacing mantle material. In (b), because of the loss of mass through erosion and transportation, the crust isostatically adjusts upward and sediments accumulate in the ocean. As the continent thins, the heavy sediment load offshore begins to deform the lithosphere beneath the ocean (c).

GEOLOGIC CYCLE

Earth's crust is in an ongoing state of change, being formed, deformed, moved, and broken down by physical, chemical, and biological processes. The endogenic (internal) system is at work building landforms, while the exogenic (external) system works to wear them down. This vast give-and-take at the Earth-atmosphere interface is called the **geologic cycle**. It is fueled by Earth's internal heat and by solar energy, influenced by the leveling force of Earth's gravity. Figure 11-6 illustrates the geologic cycle, combining many of the elements discussed in this text.

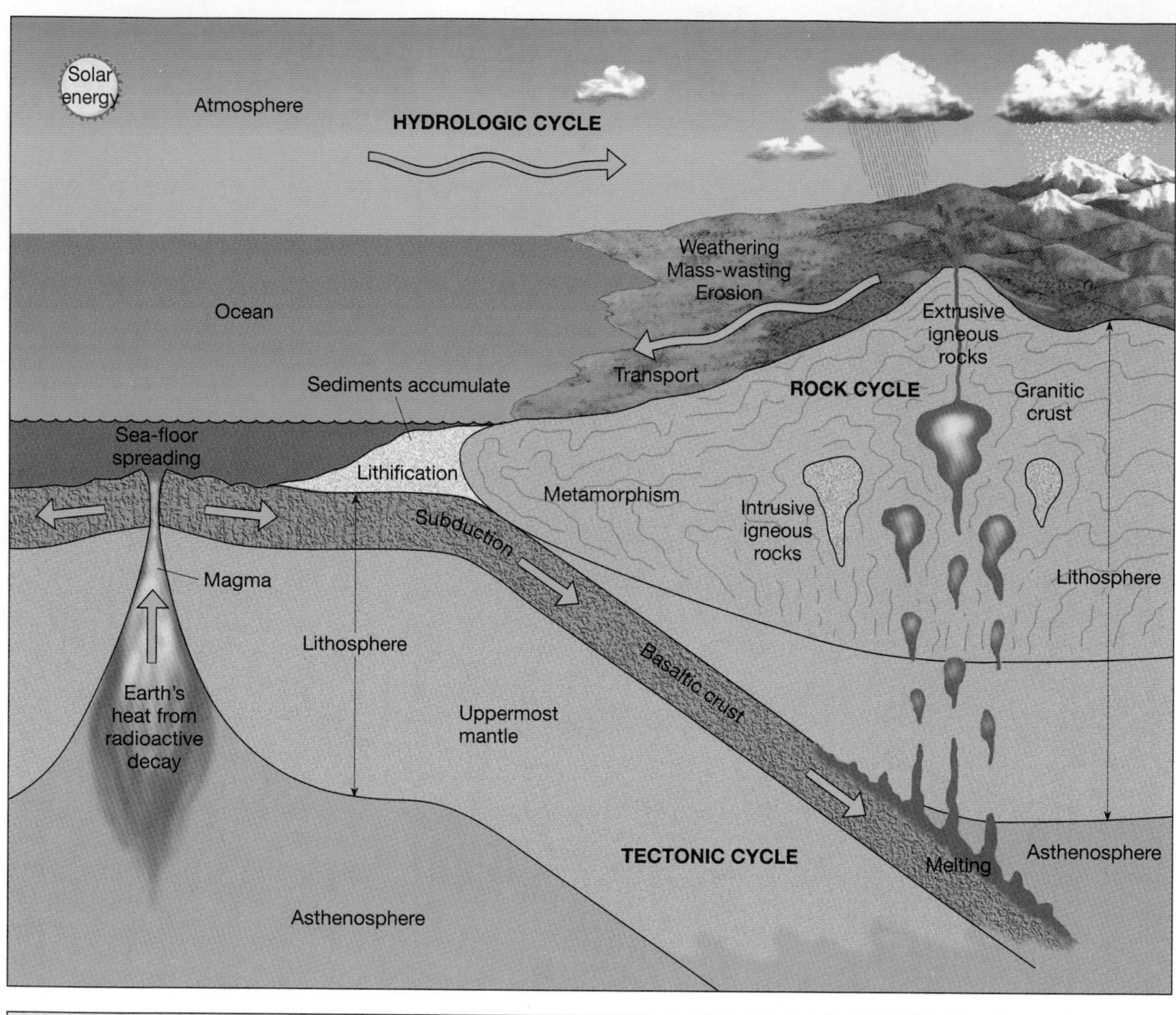

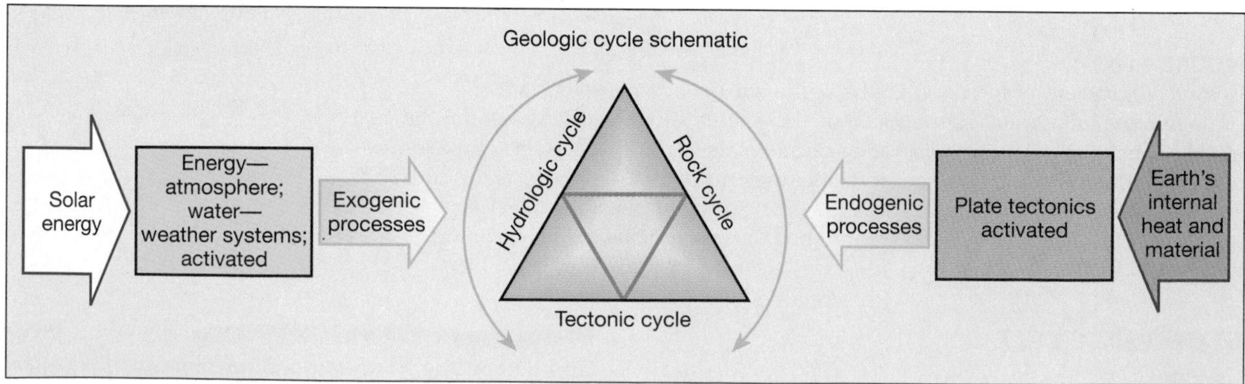

FIGURE 11-6
Geologic cycle, showing the interactive relationship of the rock cycle, tectonic cycle, and hydrologic cycle, as endogenic (internal) and exogenic (external) processes operate at or near Earth's surface.

As you can see, the geologic cycle itself is composed of three principal cycles—hydrologic, rock, and tectonic. The *hydrologic cycle*, detailed in Chapter 7, transports and deposits Earth's varied materials. The *rock cycle* produces the three basic rock types found in the crust. The *tectonic cycle* brings heat energy and new material to the surface, creating movement and deformation of the crust.

Rock Cycle

Of Earth's crust, 99% is composed of only eight natural elements, and only two of these—oxygen and silicon—account for 74% (Table 11-1). Oxygen is the most reactive gas in the lower atmosphere, readily combining with other elements. For this reason, the percentage of oxygen is higher in the crust than in the atmosphere. The relatively large percentages of lightweight elements such as silicon and aluminum in the crust are explained by the internal differentiation process in which less dense elements migrated toward the surface, as discussed earlier.

A **mineral** is an element or combination of elements that forms an inorganic natural compound. A mineral can be described with a specific symbol or formula and possesses specific qualities, including a crystalline structure. Silicon (Si) readily combines with other elements to produce the *silicate* mineral family, which includes quartz, feldspar, amphibole, and clay minerals, among others. Another important mineral family is the *carbonate* group, which features carbon in combination with oxygen and other elements such as calcium, magnesium, and potassium. Of the nearly 3000 minerals, only 20 are common, with just 10 of those making up 90% of the minerals in the crust.

TABLE 11-1
Common Elements in Earth's Crust

Element	Percent of Earth's Crust by Weight
Oxygen (O)	46.60
Silicon (Si)	27.72
Aluminum (Al)	8.13
Iron (Fe)	5.00
Calcium (Ca)	3.63
Sodium (Na)	2.83
Potassium (K)	2.70
Magnesium (Mg)	2.09
Total	98.70

A *rock* is an assemblage of minerals bound together or an aggregate of pieces of a single mineral. Literally thousands of rocks have been identified, all the result of three kinds of rock-forming processes: *igneous, sedimentary,* and *metamorphic.* Figure 11-7 illustrates the interrelationships among these three processes in the **rock cycle.** The next three sections examine each process.

Igneous Processes. Rocks that solidify and crystallize from a molten state are called **igneous rocks**. They form from **magma**, which is molten rock beneath the surface (hence the name *igneous*, which means "fire formed" in Latin). Magma is fluid, highly gaseous, and under tremendous pressure. It is either *intruded* into preexisting crustal rocks, known as *country rock*, or *extruded* onto the surface as **lava**. The cooling history of the rock—how fast it cooled, and how steadily the temperature dropped—determines its texture and degree of crystallization. These range from coarse-grained (slower cooling, with more time for larger crystals to form) to fine-grained or glassy (faster cooling). Most rocks in the crust are igneous. Figure 11-8 illustrates the variety of occurrences of igneous rocks, both on and beneath Earth's surface.

Intrusive igneous rock that cools slowly in the crust forms a **pluton**, a general term for any intrusive igneous rock body, regardless of size or shape, named after the Roman god of the underworld, Pluto. The largest pluton form is a **batholith** (Figure 11-8), defined as an irregular-shaped mass with a surface exposure greater than 100 km^2 (40 mi^2) that has invaded layers of crustal rocks. Batholiths form the mass of many large mountain ranges, for example, the Sierra Nevada batholith, the Idaho batholith, and the Coast Range

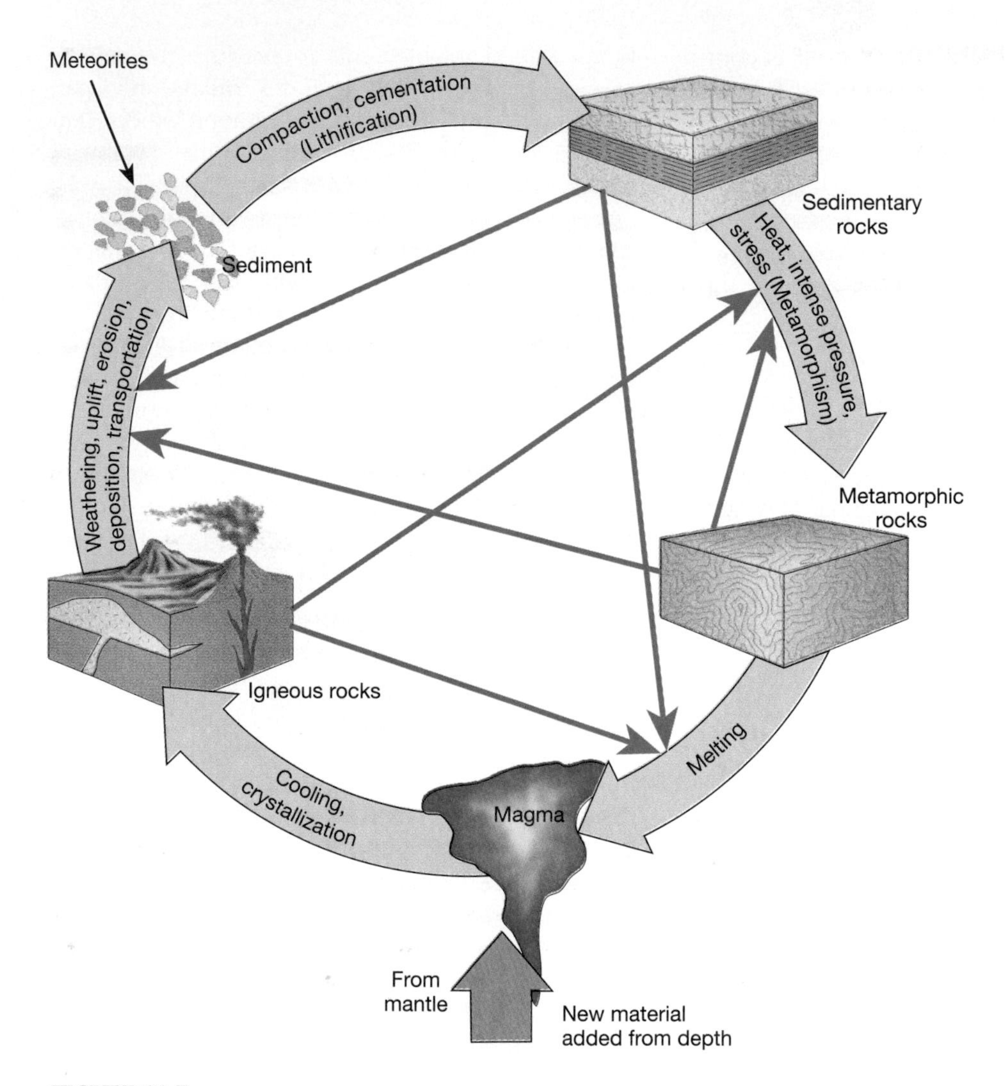

FIGURE 11-7
A rock-cycle schematic demonstrating the relationships among igneous, sedimentary, and metamorphic processes. [Adapted by permission from *Laboratory Manual in Physical Geology,* 3d ed., Richard M. Busch, editor, p. 48. Copyright © 1993 by Macmillan Publishing Company.]

batholith of Canada and extreme northern Washington State (Figure 11-9a).

Smaller plutons include the magma conduits of ancient volcanoes, which have cooled and hardened in the crust. Those that form parallel to layers of sedimentary rock are termed *sills;* those that cross layers of invaded country rock are called *dikes*. Magma also can bulge between rock strata and produce a lens-shaped body called a *laccolith*, a form of sill. In addition, magma conduits themselves may solidify in roughly cylindrical forms that stand starkly above the landscape when finally exposed by weathering and erosion. Ship Rock, New Mexico, is such a feature, as suggested in Figure 11-8. All of these intrusive forms can be exposed by weathering actions of air, water, and ice.

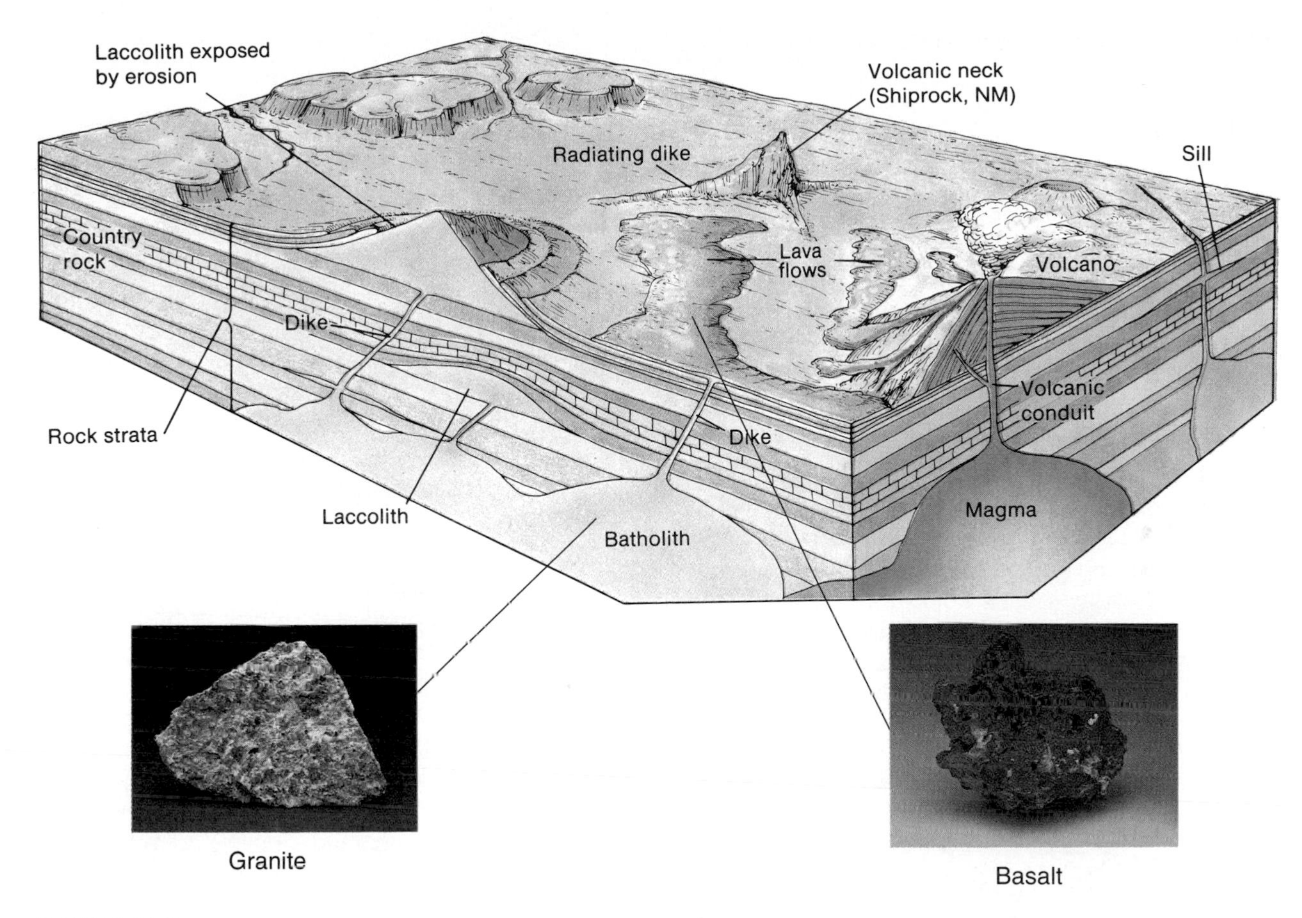

FIGURE 11-8
The variety of occurrences of igneous rocks, both intrusive (below the surface) and extrusive (on the surface). Inset photographs show samples of granite (intrusive) and basalt (extrusive). [Photos from *Laboratory Manual in Physical Geology,* 3rd ed., Richard M. Busch, editor, pp. 61, 63. Copyright © 1993 by Macmillan Publishing Company.]

FIGURE 11-9
(a) Exposed granites of the Sierra Nevada batholith; (b) basaltic lava flows on James Island of the Galápagos Islands, Ecuador. [(a) Photo by author; (b) photo by Galen Rowell.]

The categorization of igneous rocks is usually done by mineral composition and texture (Table 11-2). The two broad categories are:

1. ***Felsic*** **igneous rocks—derived both in composition and name from *fel*dspar and *silica* (SiO_2). Felsic minerals are generally high in silica, aluminum, potassium, and sodium, with low melting points. Rocks formed from felsic minerals generally are lighter in color and density than mafic mineral rocks.**
2. ***Mafic*** **igneous rocks—derived both in composition and name from *ma*gnesium and *ferric* (Latin for iron). Mafic minerals are low in silica and high in magnesium and iron, with high melting points. Rocks formed from mafic minerals are darker in color and of greater density than felsic mineral rocks.**

The table shows that the same magma which produces coarse-grained granite when slowly cooled beneath the surface also forms a fine-grained rhyolite as its rapidly cooled volcanic counterpart. If it cools rapidly, magma having a silica content comparable to granite and rhyolite may form the dark, smoky, glassy-textured rock called *obsidian* or volcanic glass. Another glassy rock called *pumice* forms when escaping gases bubble a frothy texture into the lava. Pumice is full of small openings, is light in weight, and low enough in density to float in water.

On the mafic side, basalt is the most common fine-grained extrusive igneous rock. It makes up the bulk of the ocean floor and appears in lava flows such as those on the Galápagos Islands (Figure 11-9b). Its intrusive counterpart, formed by slow cooling of the parent magma, is *gabbro.*

Sedimentary Processes. Most sedimentary rocks are derived from preexisting rocks, or from organic materials (such as bone and shell) that form limestone. The exogenic processes of weathering and erosion generate the material sediments needed to form these rocks. Bits and pieces of former rocks—principally quartz, feldspar, and clay minerals—are eroded and then mechanically transported (by water, ice, wind, and gravity) to other sites

TABLE 11–2
Igneous Rock Minerals

	Felsic Minerals (feldspars and silica)				Mafic Minerals (magnesium and iron)			Ultramafic Minerals (low silica)
General characteristics	Higher ◄— Silica content —► Lower Higher ◄— Resistance to weathering —► Lower ◄— Increased potassium and sodium Increased calcium, iron, magnesium —► Lower ◄— Melting temperatures —► Higher Lighter ◄— Coloration —► Darker							
Mineral families	Quartz SiO_2	Feldspars: Potassium feldspars K, Al, Si, (orthoclase)	Sodium feldspars Na, Al, Si, Ca (plagioclase) (aluminosilicates)	Calcium feldspars Ca, Al, Si	Mica K, Fe, Mg, Al, Si (biotite: black; muscovite: white)	Amphibole Fe, Mg, Al, Si (complex) (hornblende: black)	Pyroxene Fe, Mg, Si (dark)	Olivine Mg, Fe (dark green) (no quartz, no feldspars)
Coarse-grained texture (intrusive) slower cooling rate			Granite	Diorite			Gabbro	Peridotite
Fine-grained texture (extrusive) faster cooling rate			Rhyolite	Andesite Dacite (sodic feldspar) (Mount Saint Helens)			Basalt	
Other textures			Obsidian (glassy)	Pumice (vesicular)			Scoria (vesicular)	

FIGURE 11-10
Two types of sedimentary rock: (a) sandstone formation with sedimentary strata subjected to differential erosion; note weaker underlying siltstone. (b) Limestone formation, an important chemical sedimentary rock. [(a) Photo by author; (b) photo by Tom Scott, Florida Geological Society; insets © Macmillan Publishing Company.]

where they are deposited. In addition, some minerals are dissolved into solution and form sedimentary deposits by precipitating from those solutions; this is an important process in the oceanic environment.

Various sedimentation forms are created under different environmental regimes—deserts, glaciers, beaches, tropics, and so on. Characteristically, sedimentary rocks are laid down by wind, water, and ice in horizontally layered beds. These layered strata are important records of past ages, and **stratigraphy** is the study of their sequence (superposition), spacing, and spatial distribution for clues to the age and origin of the rocks. Figure 11-10a is a sandstone sedimentary rock in a desert landscape. Note the various layers in the formation and how differently they resist erosional processes.

The cementation, compaction, and hardening of sediments into **sedimentary rocks** is called **lithification**. Various cements fuse rock particles together; lime ($CaCO_3$, or calcium carbonate) is the most common, followed by iron oxides (Fe_2O_3), and silica (SiO_2). Particles also can be united by drying (dehydration), heating, or chemical reactions. The two primary sources of sedimentary rocks are the mechanically transported bits and pieces of former rock, called *clastic sediments*, and the dissolved minerals in solution, called *chemical sediments*. Clastic sedimentary rocks are derived from weathered and fragmented rocks that are further worn in transport. Table 11-3 lists the range of clast sizes—everything from boulders to microscopic clay particles—and the form they take as lithified rock.

Chemical sedimentary rocks are not formed from physical pieces of broken rock, but instead are dis-

TABLE 11-3
Clastic Sediment Sizes and Related Rock Type

Unconsolidated Sediment	Grain Size	Lithified Rock
Boulders, cobbles	>80 mm	Conglomerate (breccia, if pieces are angular)
Pebbles, gravel	>2 mm	
Coarse sand	0.5–2.0 mm	Sandstone
Medium-to-fine sand	0.05–0.5 mm	Sandstone
Silt	0.002–0.05 mm	Siltstone (mudstone)
Clay	<0.002 mm	Shale

SOURCE: U.S. Department of Agriculture, Soil Conservation Service.

solved, transported in solution, and chemically precipitated out of solution. The most common chemical sedimentary rock is **limestone**, which is lithified calcium carbonate ($CaCO_3$) as shown in Figure 11-10b. A similar form is dolomite, which is lithified calcium-magnesium carbonate, $CaMg(CO_3)_2$. About 90% of all limestone comes from organic sources—deposits of shell and bone that have been lithified. Once formed, these rocks are susceptible to being dissolved in solution and assume unique characteristic forms, as discussed in the weathering section of Chapter 13.

Another sedimentary rock of organic origin is coal. Coal is an important nonrenewable fossil-fuel resource and is the topic of Focus Study 11-1.

Chemical sediments also form from inorganic sources when water evaporates and leaves behind a residue of salts. These **evaporites** may exist as common salts, such as gypsum or sodium chloride (table salt), to name only two, and often appear as flat, layered deposits across a dry landscape. This process is dramatically demonstrated in the pair of photographs in Figure 11-11, taken in Death Valley National Monument one day and one month after a record 2.57 cm (1.01 in.) rainfall.

FIGURE 11-11
A Death Valley landscape (a) one day after a record rainfall event when the valley was covered by water of several square kilometers yet only a few centimeters deep. (b) One month later the same valley is coated with evaporites (borated salts). [Photos by author.]

(a)

(b)

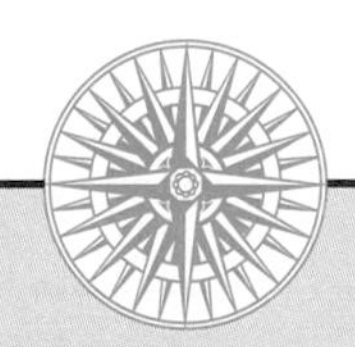

Focus Study 11-1 Coal An Important Sedimentary Rock

Coal is an important sedimentary rock that develops from living organisms; it is really a biochemical rock, very rich in carbon. Plants incorporate carbon into their tissues by taking in CO_2 during photosynthesis, and it is this process that fixes organic carbon into the rock cycle. During the Carboniferous period (360–286 million years ago in Figure 11-1) lush vegetation flourished and died in tropical swamps.In addition, photosynthesizing organisms flourished in tropical waters, as well as in bogs and forests in more temperate latitudes. As these carbon-fixing life forms perished, organic debris collected in water-soaked bogs. The water was important, for this low-oxygen environment was necessary to limit natural decomposition in the accumulating layers. Bacterial action freed various elements and thereby concentrated the carbon in the debris.

Coal forms in these bogs through progressive stages. First, pieces of former plants can be easily recognized in peat, a brownish, compacted fibrous organic sediment. Peat has been burned as a low-grade fuel for centuries, producing a low-heat energy output and a great deal of air pollution. Following more compaction and hardening, the soft, coal-like substance known as *lignite* forms. Further lithification leads to *bituminous coal,* the first true sedimentary rock in this sequence. Burial of coal deposits under sedimentary overburden results in the formation of coal seams, usually between beds of shale or sandstone. Finally, where conditions are right, metamorphism of bituminous coal results in *anthracite coal,* a metamorphic rock that is extremely hard and therefore difficult to mine.

Sedimentary and metamorphic coal deposits are found from the Arctic to Antarctica and on every continent, yet another confirmation of continental drift. About 30–35% of the world's reserves are in the United States and Canada (Figure 1), 25% in the Commonwealth of Independent States,15% in China, and lesser amounts in Australia, Great Britain, India, and Poland, among others. The largest consumer of coal in the United States is the electric power industry, which uses 70% of mined coal for the production of steam in power plant boilers. (The steam passes through turbines that generate the electricity.)

Estimates of how long these known reserves of coal will last vary and are based on assumptions regarding the annual rate of increase in the extraction and consumption of coal. There are at least 500 years of coal remaining in the United States and Canada if extraction and consumption rates stay about where they were in the 1980s (about 1 billion tons a year), with bituminous coal constituting 65% of the production. However, if these reserves are subjected to an annual production growth of 3–5%, the lifetime of the reserves will be reduced to 100 years or less. Coal is a valuable source of chemicals for many important uses, including medicines, so its depletion for the production of electricity— a high-grade source of energy used generally for low-grade energy applications—would be a tragedy for future generations.

Another concern related to coal is an impurity commonly found within it, sulfur. Sulfur emissions are a major air pollutant when released to the atmosphere through power plant and industrial smokestacks. In addition, coal is baked to produce coke, a purified carbon that is a basic raw material for steelmaking; In the coking process, large amounts of sulfur oxides are released into the atmosphere. The combination of sulfur oxides and water vapor forms acidic precipitation, which causes environmentally damaging acid deposition. The amount of sulfur in coal varies spatially, even over a few kilometers, and as a result of these environmental concerns low-sulfur coal is in great demand.

A further problem associated with coal is that, when burned, it produces more carbon dioxide per unit consumed than any other fossil fuel. The role of increased carbon dioxide in stimulating the greenhouse effect is detailed in Chapters 5 and 10. Strategies to reduce atmospheric carbon dioxide must include energy conservation and reduced coal use, which would in turn extend the lifetime of the resource—preserving it for more valuable uses and applications.

One environmental aspect of coal extraction is its impact on the landscape and water quality. Many coal seams lie deep beneath the surface, roughly 100 m (300 ft) or deeper, and are mined by drilling shafts and tunnels. These weaken the overlying rock strata, which crack, causing subsidence of the land surface and disrupting natural drainage. Other coal seams lie near the surface and are surface mined (strip mined) by removing the overburden, extracting the coal, and replacing the overburden, as prescribed by law. However, in many cases the disturbed area is not returned to

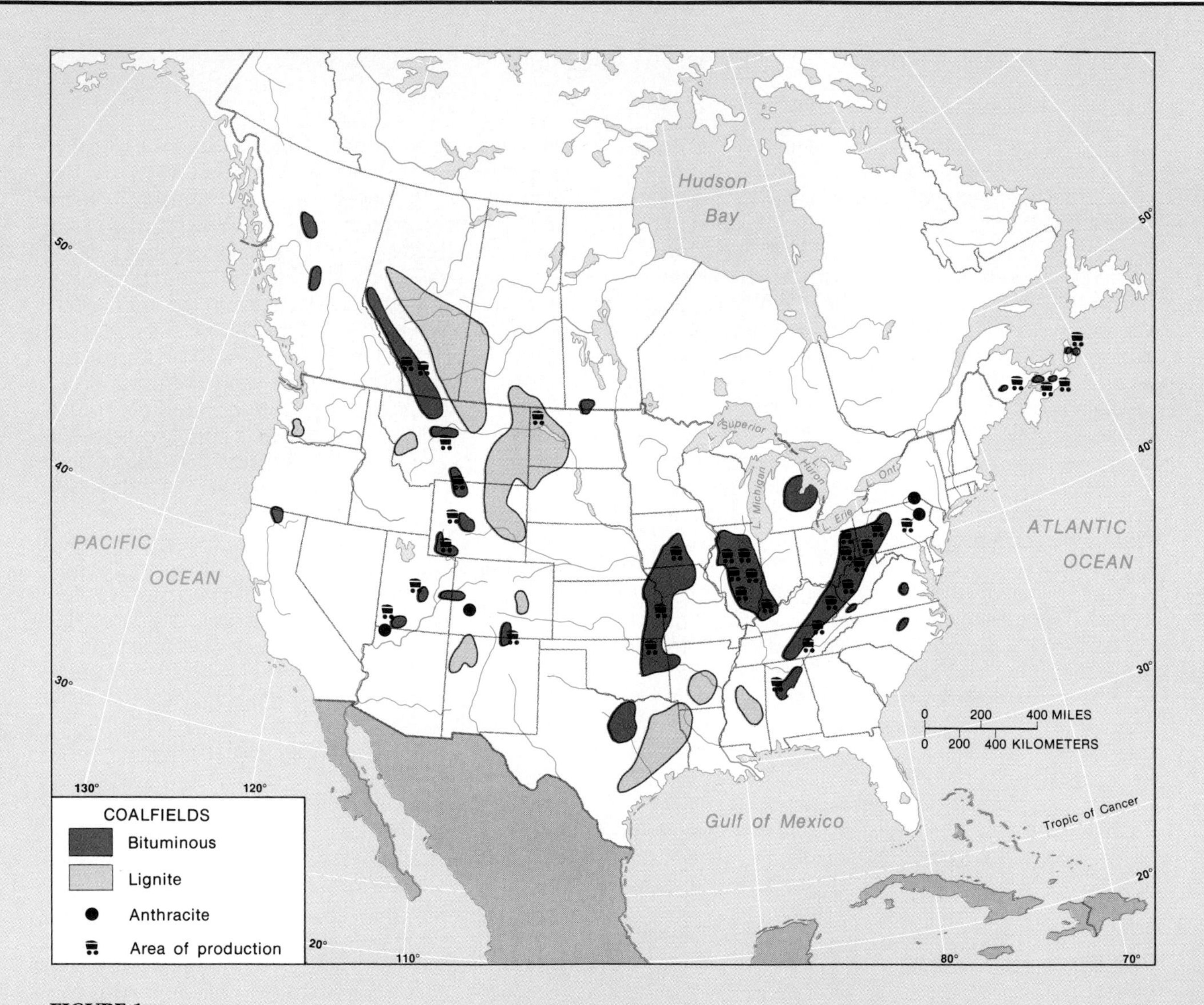

FIGURE 1
Types of coal deposits in North America. [Adapted from *Geography and Development,* 4th ed., by James S. Fisher, p. 121. Copyright © 1992 by Macmillan Publishing Company.]

its original topography, and the landscape remains marred by deposits of mining debris called "spoil banks." Another problem with both types of mining is that water and air combine with sulfur in the mine debris to form bright orange "acid mine drainage" that pollutes streams. Major regional controversies revolve around these mining scars and related reclamation issues (Figure 2).

FIGURE 2
Strip mine spoil banks in southern West Virginia. [Photo by author.]

Chemical deposition also occurs in the water of natural hot springs from chemical reactions between minerals and oxygen; this is a sedimentary process related to **hydrothermal activity.** A deposit of travertine, a form of calcium carbonate, at Mammoth Hot Springs in Yellowstone National Park is an example.

Another hydrothermal area occurs near the Nevada-California border in Mono Lake, an alkaline lake high in carbonates and sulfates that presently has an overall salinity of 95‰ (the ocean average is 35‰, or 35 parts per thousand). Carbonates in Mono Lake interact with calcium-rich hot springs to produce a rock called tufa. The tufa deposits grew tall in Mono Lake as long as the springs were beneath the surface of the lake. However, since the 1940s, tributary streams have been diverted out of the region, lowering the lake's surface more than 20 m (65 ft), reducing its surface area by more than half, exposing these chemical sedimentary deposits and alkaline shorelines, and damaging the lake's natural ecology and habitats for wildlife, particularly birds. Some tufa towers now rise over 10 m (30 ft) above the shoreline, as shown in the foreground of Figure 7-18g. Strong winds in the area blow across the newly formed alkali flats of evaporites, producing days of severe air pollution. Interestingly, tributary flows were restored by court order in April 1991, and the lake level is again rising.

Metamorphic Processes. Any rock, either igneous or sedimentary, may be transformed into a **metamorphic rock,** by going through profound physical and/or chemical changes under increased pressure and temperature. (The name metamorphic comes from the Greek, meaning to "change form.") Metamorphic rocks generally are more

FIGURE 11-12
An example of metamorphic rock at Black Balsam Knob, in the southern Appalachians. In zones of intense mountain building, as in the Appalachian Mountains, metamorphic rocks such as marble underlie the landscape. [Photo by George Wuerthner.]

compact than the original rock and therefore are harder and more resistant to weathering and erosion (Figure 11-12).

Several conditions can cause metamorphism, particularly when subsurface rock is subjected to high temperatures and high compressional stresses occurring over millions of years. Igneous rocks are compressed during collisions of portions of Earth's crust (described under "Plate Tectonics" later in this chapter). Or, igneous rocks may be sheared and stressed along earthquake fault zones. Sometimes they simply are crushed under great weight when a crustal area is thrust beneath other crust.

Another metamorphic condition occurs when sediments collect in broad depressions in Earth's crust and, because of their own weight, create enough pressure in the bottommost layers to transform the sediments into metamorphic rock. Also, molten magma rising within the crust may initiate metamorphic processes in adjacent rock, which is "cooked" by the intruding magma, a process called *contact metamorphism.*

Metamorphic rocks may be changed both physically and chemically from the original rocks. If the mineral structure demonstrates a particular alignment after metamorphism, the rock is **foliated** and some minerals may appear as wavy striations (streaks or lines) in the new rock. On the other hand, parent rock with a more homogeneous makeup may produce a *nonfoliated* rock. Table 11-4 lists some metamorphic rock types, their parent rocks, and resultant textures. The two inset photographs demonstrate foliated and nonfoliated textures.

The ancient roots of mountains are composed predominantly of metamorphic rocks. For example, at the bottom of the inner gorge of the Grand Canyon in Arizona, a Precambrian metamorphic rock called the Vishnu Schist is over 2 billion years in age. Despite the fact that this schist is harder than steel, the Colorado River has cut down into the uplifted Colorado Plateau, exposing and eroding these old rocks. Thus, the canyon itself is much younger than the schist. Within the overall 1.6 km (1.0 mi) deep canyon, this river-cut inner gorge represents the bottom 455 m (1500 ft).

You have seen how the rock-forming processes yield the igneous, sedimentary, and metamorphic materials of Earth's crust. Next we look at how vast **tectonic processes** press, push, and drag portions of the crust in large-scale movements, causing the continents to "drift." The facts that the continents drift and the young seafloor spreads from rifts are revolutionary discoveries in the *tectonic cycle* that came to light only in this century.

TABLE 11-4
Metamorphic Rocks

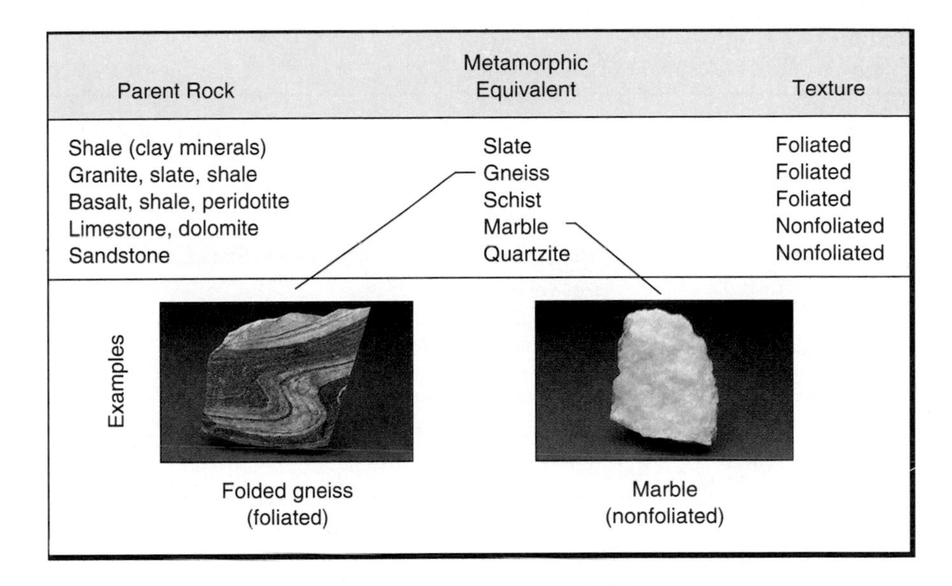

Parent Rock	Metamorphic Equivalent	Texture
Shale (clay minerals)	Slate	Foliated
Granite, slate, shale	Gneiss	Foliated
Basalt, shale, peridotite	Schist	Foliated
Limestone, dolomite	Marble	Nonfoliated
Sandstone	Quartzite	Nonfoliated

CONTINENTAL DRIFT

Have you ever looked at a world map and noticed that a few of the continental landmasses appear to have matching shapes like pieces of a jigsaw puzzle—particularly South America and Africa? The incredible reality is that the continental pieces indeed once were fitted together! Continental landmasses not only migrated to their present locations but continue to move at speeds up to 6 cm (2.4 in.) per year. We say that the continents are "adrift" because convection currents in the asthenosphere and upper mantle are dragging them around. The key point is that the arrangement of continents and oceans we see today is not permanent but is in a continuing state of change.

A continent such as North America is actually a collage of pieces and fragments of crust that have migrated to form the landscape we know. Through a historical chronology, let us trace the discoveries that produced this major revolution in science, called *plate tectonics*.

A Brief History

In 1620, the English philosopher Sir Francis Bacon noted gross similarities between the shapes of Africa and South America (although he did not suggest that they had drifted apart). Others wrote—unscientifically—about such apparent relationships, but it was not until much later that a valid explanation was proposed. In 1912, German geophysicist and meteorologist Alfred Wegener publicly presented in a lecture his idea that Earth's landmasses migrate. His book, *Origin of the Continents and Oceans,* appeared in 1915. Wegener today is regarded as the father of this concept, which he called **continental drift.** But scientists at the time, knowing little of Earth's interior structure and bound by an inertial mindset as to how continents and mountains were formed, were unreceptive of Wegener's revolutionary proposal and, in a nonscientific spirit, rejected it outright.

Wegener postulated that all landmasses were united in one supercontinent approximately 225 million years ago, during the Triassic period—see Figure 11-16b. (As you read this, it may be helpful to refer to the geologic time scale in Figure 11-1.) This one landmass he called **Pangaea,** meaning "all Earth." Although his initial model kept the landmasses together far too long and his idea about the driving mechanism was incorrect, Wegener's overall configuration of Pangaea was right. To come up with his Pangaea fit, he began studying the research of others, specifically the geologic record, the fossil record, and the climatic record for the continents. He concluded that South America and Africa were related in many complex ways. He also concluded that the large midlatitude coal deposits, which stretch from North America to Europe to China, and date to the Permian and

TABLE 11-5
Stages in the Development of a Revolution

Date	Significant Events
1912–1915	Continental drift proposed by Wegener
1915–1930	Great debate
1930–1950	Stalemate (a lost cause in the United States; continuing debate elsewhere)
1950–1960	Revival of interest Exploration of ocean floor Fossil magnetism and polar wandering
1960–1962	Sea-floor spreading
1963	Oceanic magnetic anomalies associated with sea-floor spreading; the magnetic "tape recorder" Hot spots
1963–1966	Polarity reversals of Earth's magnetic field; fossil magnetism and accurate rock dating for lavas on land and for deep-sea sediment cores
1966–1967	Revolution proclaimed by scientists
1967–1968	Spreading and drift incorporated in plate tectonics Earthquake synthesis
1968–1970	Deep-sea drilling by *Glomar Challenger*
1970	"Geopoetry becomes geofact," *Time* magazine (and other more scientific sources)
Currently	Recent geological features are being related to present plates and plate boundaries; ancient geological features are being analyzed to reconstruct plate history

SOURCE: Peter J. Wyllie, *The Way the Earth Works.* Copyright © 1976 by John Wiley & Sons, Inc.

Carboniferous periods (245–360 million years ago), existed because these regions once were more equatorial in location and therefore covered by lush vegetation that became coal.

As modern scientific capabilities built the case for continental drift, the 1950s and 1960s saw a revival of interest in Wegener's concepts, and finally confirmation. Table 11-5 highlights developments from the time of Wegener's proposal to the present. Each of the principal developments is discussed in the balance of this chapter. Aided by an avalanche of discoveries, the theory today is nearly universally accepted as an accurate model of the way Earth's surface evolves, and virtually all Earth scientists accept the fact that continental masses move about.

Sea-Floor Spreading

The key to establishing the fact of continental drift was a better understanding of the sea floor. The sea floor has a remarkable feature: an interconnected worldwide mountain chain (ridge) some 64,000 km (40,000 mi) in extent and averaging more than 1000 km (600 mi) in width. How did it get there? In the early 1960s, geophysicists Harry H. Hess and Robert S. Dietz proposed **sea-floor spreading** as the mechanism that builds this mountain chain and drives continental movement.

Hess said that these submarine mountain ranges, called the **mid-ocean ridges,** were the direct result of upwelling flows of magma from hot areas in the upper mantle and asthenosphere and perhaps deeper sources. When mantle convection brings magma up to the crust, the crust is fractured and the magma cools to form new sea floor, building the ridges and spreading laterally. Figure 11-13 suggests that the ocean floor is rifted and scarred along a mid-ocean ridge, with arrows indicating the direction in which the floor spreads.

As new crust is generated and the sea floor

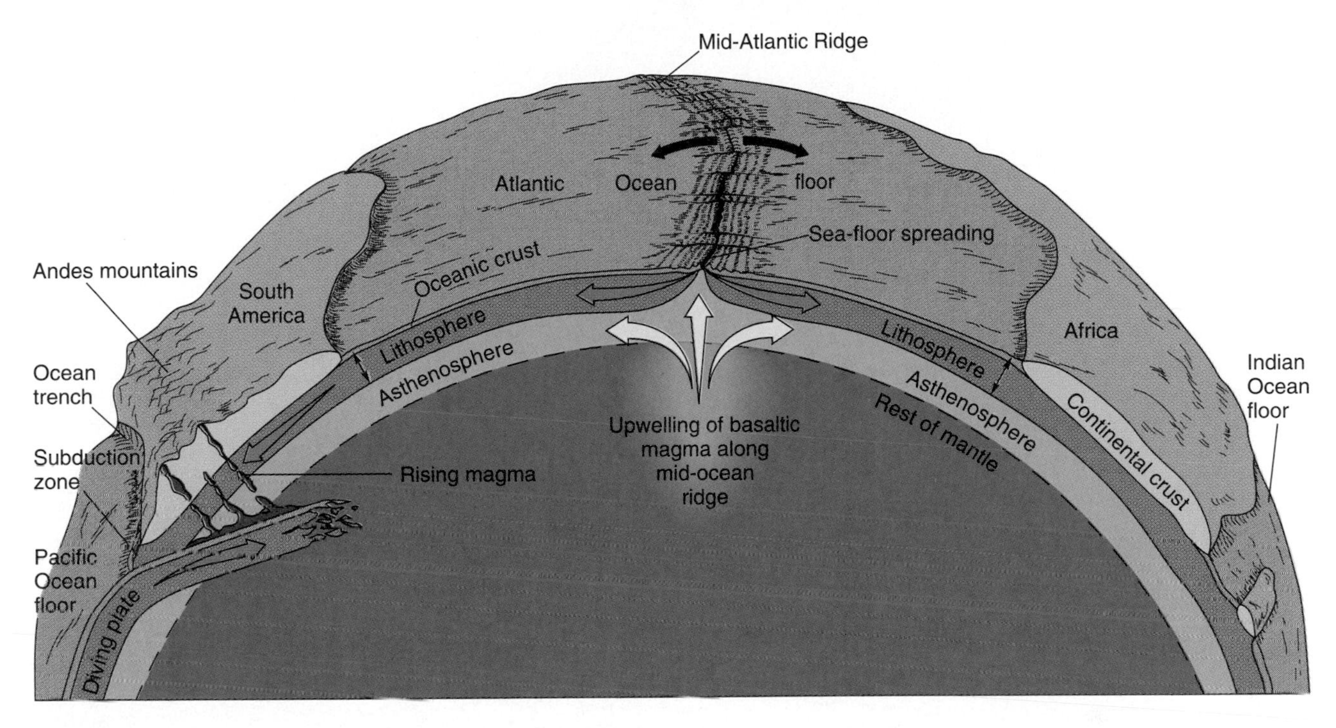

FIGURE 11-13
Sea-floor spreading, upwelling currents, subduction, and plate movements, shown in cross section. [After Peter J. Wyllie, *The Way The Earth Works*. Copyright © 1976, by John Wiley & Sons, Inc. Adapted by permission of John Wiley & Sons, Inc.]

spreads, the alignment of the magnetic field in force at the time dictates the alignment of magnetic particles in the cooling rock, creating a kind of magnetic tape recording in the sea floor. Each magnetic reversal and reorientation of Earth's polarity is recorded in the oceanic crust. Figure 11-14 illustrates the mirror images that develop on either side of the sea-floor rift as a result of these magnetic reversals recorded in the rock. If you measure off the same distance from the rift to either side, you find the same orientation of magnetic content in the rock. Thus, the periodic reversals of Earth's magnetic field have proven a valuable clue to understanding sea-floor spreading. In essence, scientists were provided with a global pattern of magnetic stripes to use in fitting together pieces of Earth's crust.

As the age of the sea floor was determined from these sea-floor recordings of Earth's magnetic-field reversals and other measurements, the complex harmony of the two concepts of continental drift and sea-floor spreading became clearer. The youngest crust anywhere on Earth is at the spreading centers of the mid-ocean ridges (Figure 11-15). With increasing distance from these centers, Earth's surface gets increasingly older. The oldest sea floor is in the western Pacific near Japan, an area called the Pigafetta Basin that dates to 170 million years ago (during the Jurassic period).

Overall, the sea floor is relatively young—nowhere does it exceed 200 million years in age, remarkable when you remember that Earth's age is 4.6 billion years. The reason is that oceanic crust is not permanent—it is being consumed by subduction beneath continental crust along Earth's deep oceanic trenches. The discovery that the sea floor is so young demolished earlier geologic thinking that the oldest rocks would be found there.

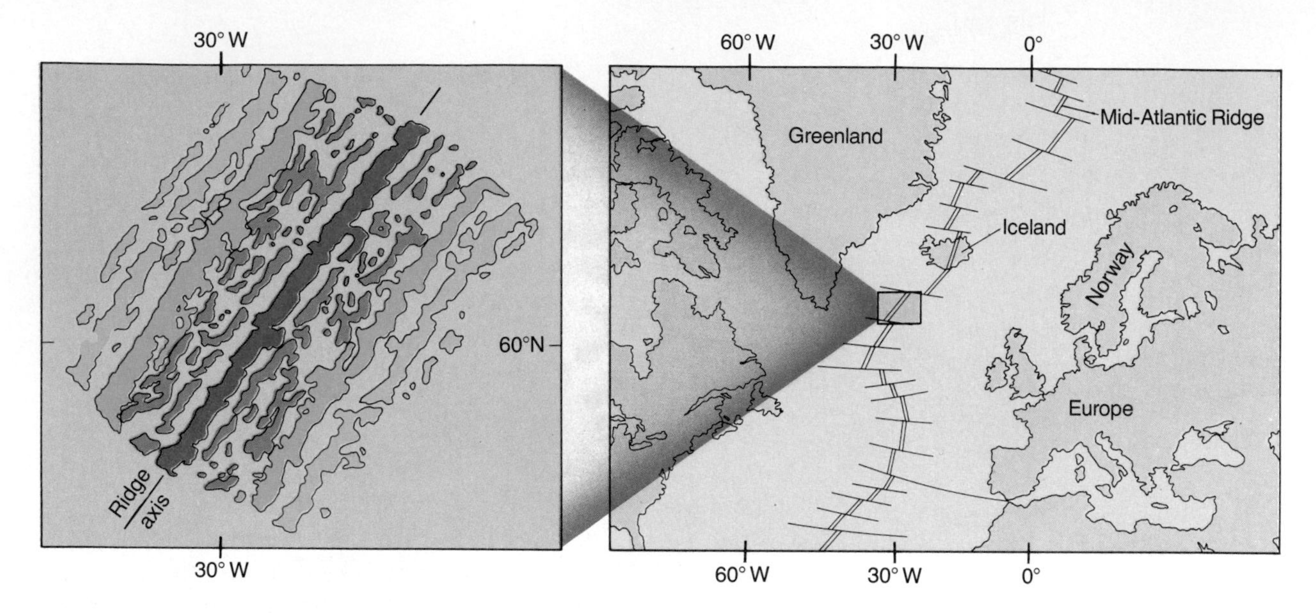

FIGURE 11-14
Magnetic reversals recorded in the sea floor south of Iceland along the Mid-Atlantic Ridge. [Magnetic reversals reprinted from *Deep-Sea Research* 13, J. R. Heirtzler, S. Le Pichon, and J. G. Baron. Copyright 1966, Pergamon Press, p. 247.]

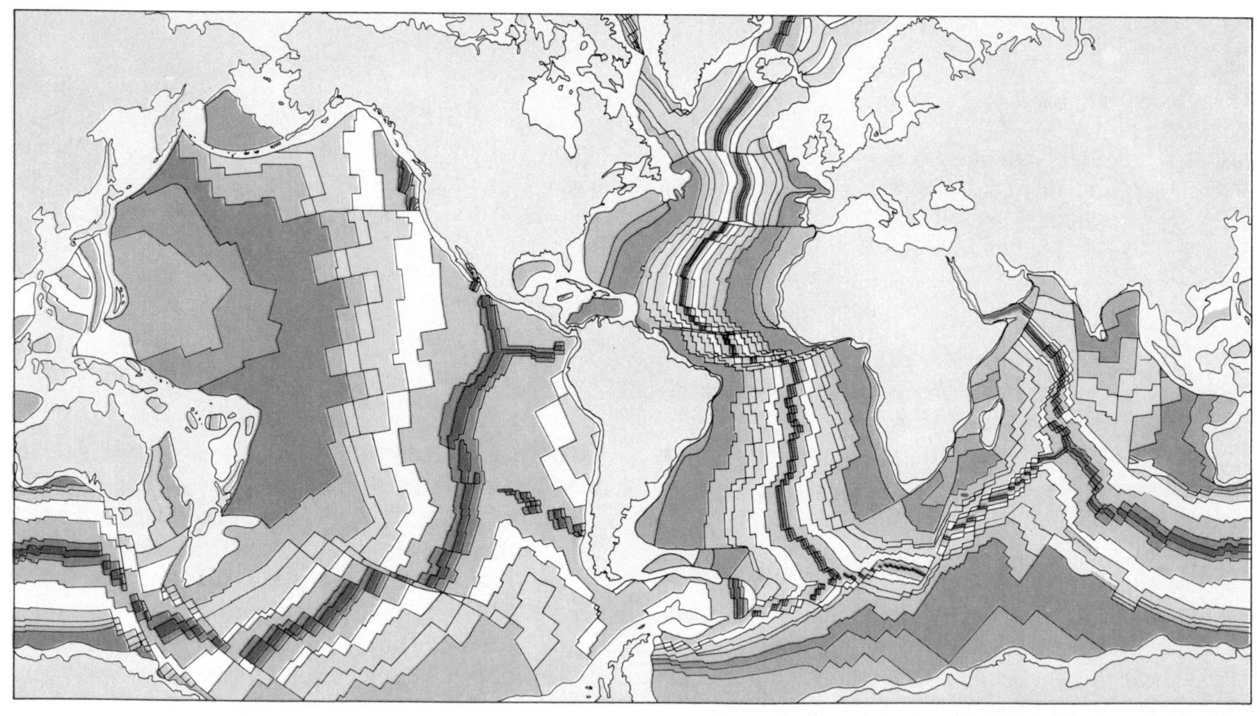

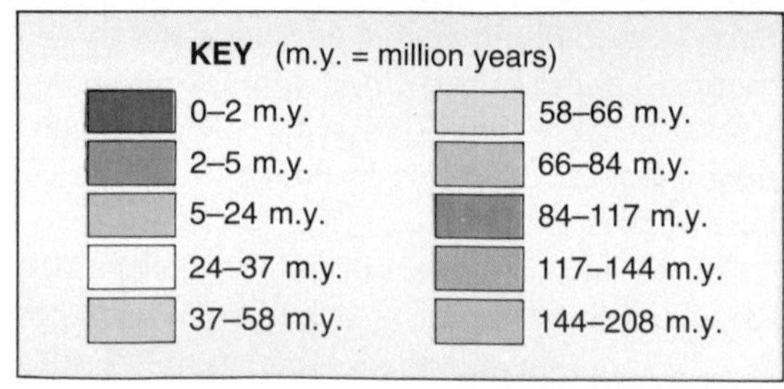

FIGURE 11-15
Relative age of the oceanic crust. [Adapted with the permission of W. H. Freeman and Company from *The Bedrock Geology of the World* by R. L. Larson et al. Copyright © 1985.]

Subduction

In contrast to the upwelling zones along the mid-ocean ridges are the areas of descending crust elsewhere. We must remember that the basaltic ocean crust has a density of 3.0 g/cm^3, whereas the continental crust averages a significantly lighter 2.7 g/cm^3. As a result, when continental crust and oceanic crust collide, the heavier ocean floor will dive beneath the lighter continent, thus forming a descending **subduction zone** (Figure 11-13). The world's **oceanic trenches** coincide with these subduction zones and are the deepest features on Earth's surface. (The deepest is the Mariana Trench near Guam, which descends below sea level to −11,033 m or −36,198 ft).

The subducted portion of crust is dragged down into the mantle, where it remelts, is recycled, and eventually migrates back toward the surface through deep fissures and cracks in crustal rock. Volcanic mountains like the Andes in South America and the Cascade Range from northern California to the Canadian border form inland from subduction zones, providing an important clue for the study of continental drift. The fact that spreading ridges and subduction zones are areas of earthquake and volcanic activity provides further evidence.

Sea-floor spreading, subduction, and mantle convection all combine to produce *plate tectonics,* which by 1968 had become the all-encompassing term for the concepts that began with Wegener's original continental drift proposal. Using current scientific findings, let us go back and reconstruct the past, and Pangaea.

The Formation and Breakup of Pangaea

The supercontinent of Pangaea and its subsequent breakup into today's continents represent only the last 225 million years of Earth's 4.6 billion years, or only the most recent 1/23 of Earth's existence. During the other 22/23 of geologic time, other things were happening. The landmasses as we know them are barely recognizable. Much scientific work is being done to establish even more ancient alignments of the primitive crust.

Pre-Pangaea. Figure 11-16a shows the pre-Pangaea arrangement of 465 million years ago (during the Middle Ordovician period). The outlines of South America, Africa, India, Australia, and southern Europe appear within the continent called Gondwana in the Southern Hemisphere. What is to become North America and Greenland is the inverted continent of Laurentia, which straddles the equator near the location of the present-day Fiji and Samoa island groups. The other land groups are identified.

Pangaea. Figure 11-16b shows an updated version of Wegener's Pangaea, 225–200 million years ago (Triassic-Jurassic periods). Areas of North America, North Africa, the Middle East, and Eurasia were near the equator and therefore were covered by plentiful vegetation. The landmasses focused near the South Pole and attached to the sides of Antarctica were covered by ice. Today, those portions of South America, southern Africa, India, Australia, and Antarctica bear the imprint of this shared glacial blanket.

The India plate (the slab of Earth's crust that included modern India) appears larger on the map than the same area today because the Tibetan Plateau formed the northern portion of the plate. We also can see that Africa shared a common connection with both North and South America. Today, the Appalachian Mountains in the eastern United States and the Atlas Mountains of northwestern Africa reflect this common ancestry; they are, in fact, portions of the same mountain range. At the time of Figure 11-16b, the Atlantic Ocean did not exist. The only oceans were Panthalassa ("all seas"), which would become the Pacific, and the Tethys Sea (partly enclosed by the African and the Eurasian plates), which formed the Mediterranean Sea, with trapped portions becoming the present-day Caspian Sea.

Pangaea Breaks Up. Figure 11-16c shows the movement of plates that had occurred by 135 million years ago (the beginning of the Cretaceous period). New sea floor that had formed is highlighted. The active spreading center had rifted North America away from landmasses to the east, shaping the coast of Labrador. India was further along in its journey, with a spreading center to the south, and a subduction zone to the north where the leading edge of the India plate was diving beneath Eurasia. At that time a rift also began in what

FIGURE 11-16
The formation and breakup of Pangaea and the types of motions occurring at plate boundaries. [(a) from R. K. Bambach, Scotese, and Ziegler, "Before Pangaea: The Geography of the Paleozoic World," *American Scientist* 68 (1980): 26–38, reprinted by permission; (b) through (e) from Robert S. Dietz and John C. Holden, *Journal of Geophysical Research* 75, no. 26 (10 September 1970): 4939–56, copyright by the American Geophysical Union.]

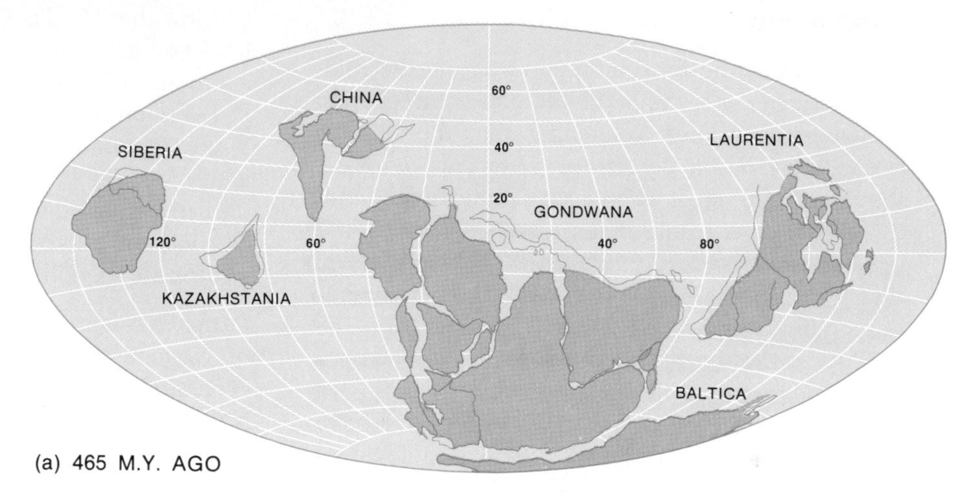

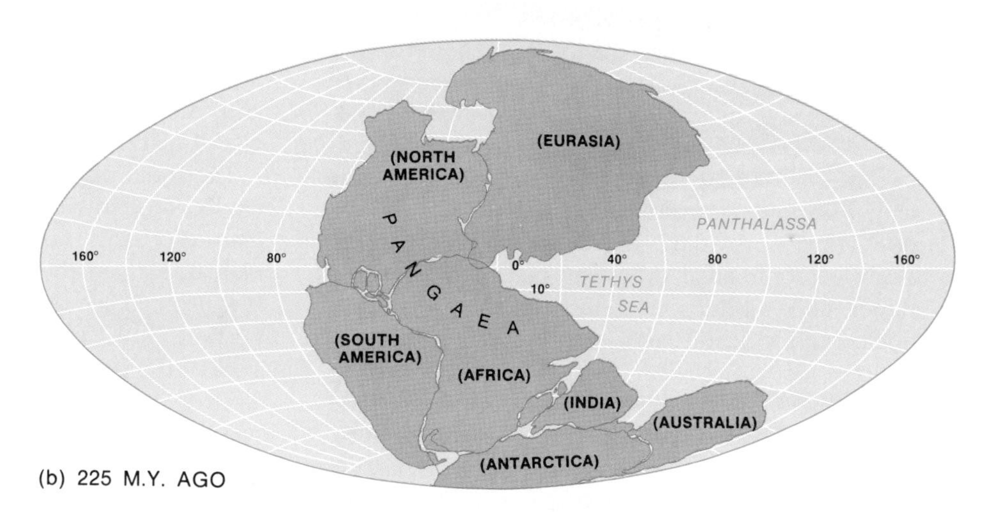

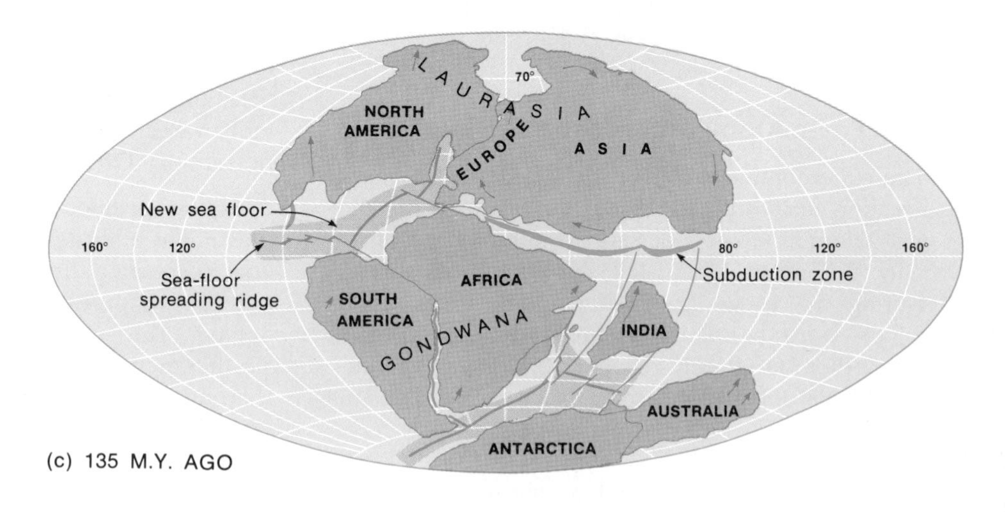

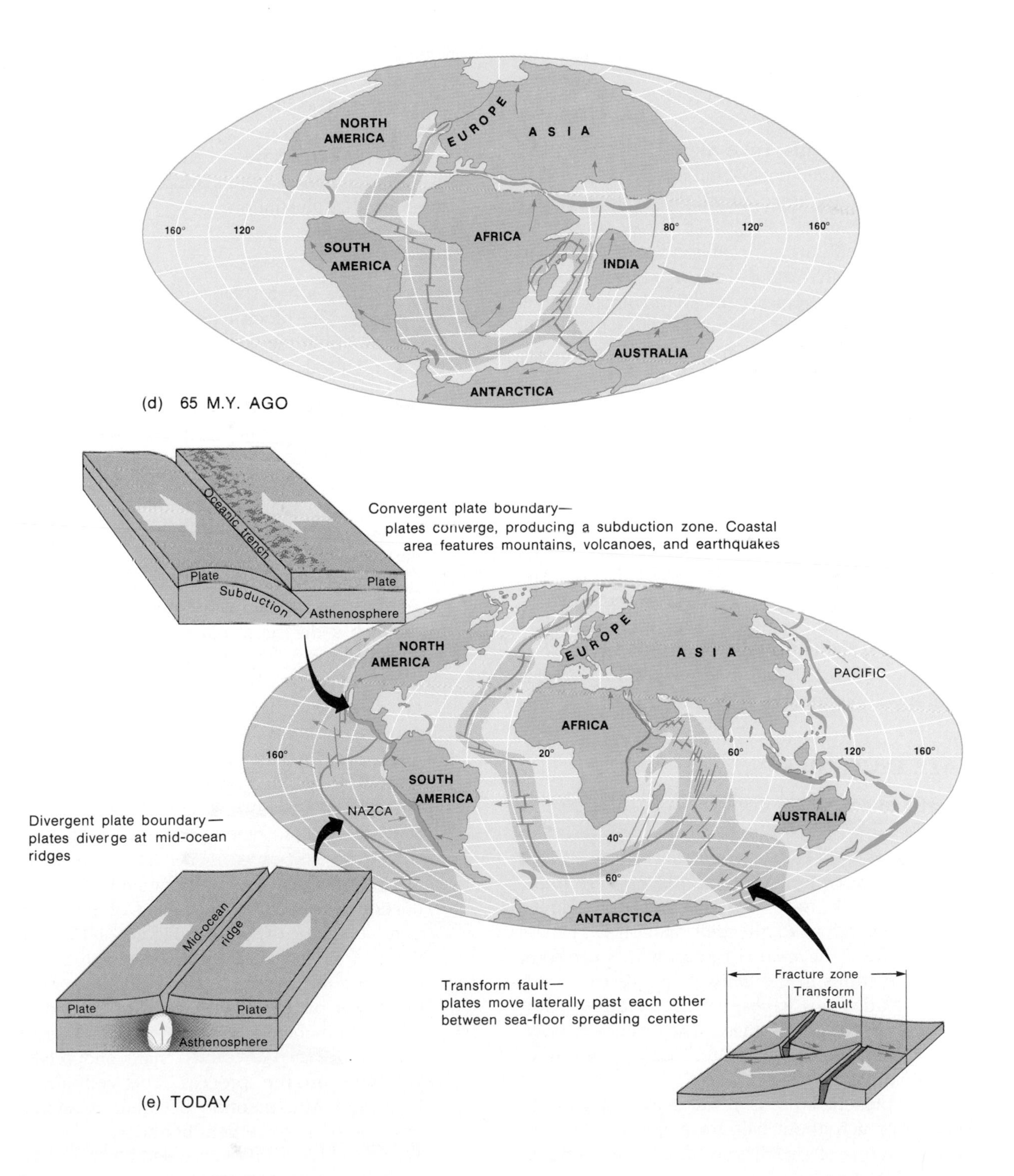
NORTH AMERICA
EUROPE
ASIA
160°
120°
80°
120°
160°
SOUTH AMERICA
AFRICA
INDIA
AUSTRALIA
ANTARCTICA
(d) 65 M.Y. AGO
Oceanic trench
Plate
Plate
Subduction
Asthenosphere
Convergent plate boundary—
plates converge, producing a subduction zone. Coastal
area features mountains, volcanoes, and earthquakes
NORTH AMERICA
EUROPE
ASIA
PACIFIC
AFRICA
160°
20°
60°
120°
160°
SOUTH AMERICA
NAZCA
AUSTRALIA
40°
60°
ANTARCTICA
Divergent plate boundary—
plates diverge at mid-ocean
ridges
Mid-ocean ridge
Plate
Plate
Asthenosphere
Transform fault—
plates move laterally past each other
between sea-floor spreading centers
Fracture zone
Transform fault
(e) TODAY

was to be the South Atlantic, and Antarctica and Australia were pulled into isolation from the other masses.

Modern Continents Take Shape. Figure 11-16d shows the arrangement 65 million years ago (beginning of the Tertiary period). Sea-floor spreading along the Mid-Atlantic Ridge had grown some 3000 km (almost 1900 mi) in 70 million years. Africa had moved northward about 10° in latitude, leaving Madagascar split from the mainland and opening up the Gulf of Aden. The rifting along what would be the Red Sea had begun. The India plate had moved three-fourths of the way to Asia, as Asia continued to rotate clockwise. Of all the major plates, India traveled the farthest. From this time until the present, according to Robert Dietz and John Holden, more than half of the ocean floor was renewed.

The Continents Today. Figure 11-16e shows modern geologic time (the late Cenozoic Era). The northern reaches of the India plate have underthrust the southern mass of Asia through subduction, forming the Himalayas in the upheaval created by the collision. Plate motions continue to this day and will proceed into the future.

PLATE TECTONICS

Plate tectonics is the inclusive conceptual model for Wegener's continental drift proposal, sea-floor spreading, and other related aspects of this twentieth-century revolution in Earth sciences. *Tectonic,* from the Greek *tektonikós* meaning building or construction, refers to the deformation of Earth's crust as a result of internal forces, which can form various structures in the lithosphere. Tectonic processes include folds, faults, warps, fractures, earthquakes, and volcanic activity. The system is powered by density and temperature differences in an unevenly mixed mantle.

Earth's present crust is divided into at least 14 plates, of which about half are major and half are minor, in terms of size. These broad plates (Figure 11-17) are composed of literally hundreds of smaller pieces and perhaps dozens of microplates that have migrated together. The arrows in the figure indicate the direction in which each plate is presently moving, and the length of the arrows suggests the rate of movement during the past 20 million years. The illustration of the sea floor that begins Chapter 12 also may be helpful in identifying the plate boundaries in the following discussion. Try to correlate this map of the tectonic plates with that illustration.

Plate Boundaries

The boundaries where plates meet clearly are dynamic places. The block diagram inserts in Figure 11-16e show the three general types of motion and interaction that occur along the boundary areas:

- *Divergent boundaries* are characteristic of sea floor spreading centers, where upwelling material from the mantle forms new sea floor, and crustal plates are spread apart. These are zones of tension. An example noted in the figure is the divergent boundary along the East Pacific rise, which gives birth to the Nazca plate and the Pacific plate. Figure 11-14 illustrates patterns of magnetic reversals on such a divergent plate boundary south of Iceland along the Mid-Atlantic Ridge. While most divergent boundaries occur at mid-ocean ridges, there are a few within continents themselves. An example is the Great Rift Valley of East Africa, where continental crust is being rifted apart.
- *Convergent boundaries* are characteristic of collision zones, where areas of continental and/or oceanic crust collide. These are zones of compression. Examples include the subduction zone off the west coast of South and Central America (noted in Figure 11-16e) and the area along the Japan and Aleutian trenches. Along the western edge of South America, the Nazca plate collides with and is subducted beneath the South American plate, creating the Andes Mountains chain and related volcanoes in the process. The collision of India and Asia mentioned earlier is another example of a convergent boundary.
- *Transform boundaries* occur where plates slide laterally past one another at right angles to a sea-floor spreading center, neither diverging nor converging, and usually with no volcanic eruptions. Examples include the right-angle

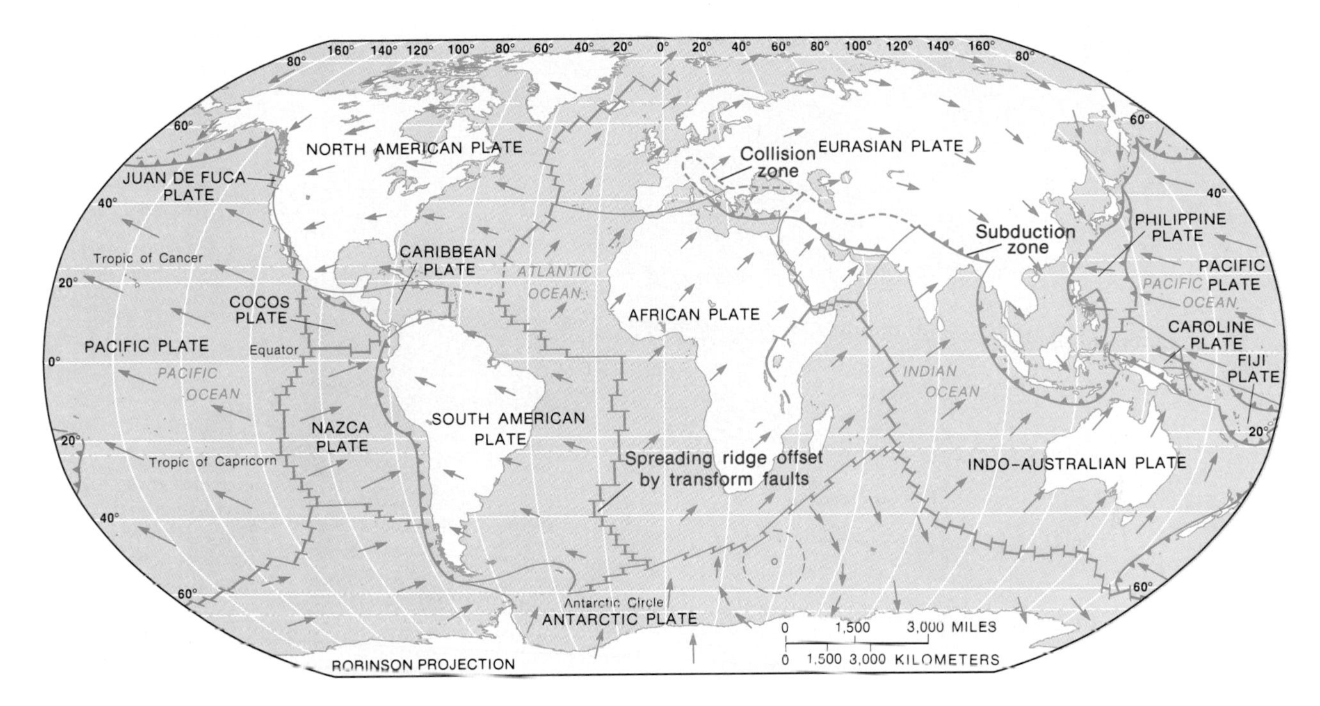

FIGURE 11-17
Earth's 14 major lithospheric plates and their motions. Each arrow represents 20 million years of movement, the longer arrows indicating that the Pacific and Nazca plates are moving more rapidly than the Atlantic plates. [Adapted from U.S. Geodynamics Committee and Sheldon Judson, Kenneth S. Deffeyes, and Robert B. Hargraves, *Physical Geology*, © 1976, p. 28. Reprinted by permission of Prentice Hall, Inc., Englewood Cliffs, NJ.]

fractures stretching across the Mid-Atlantic Ridge, the East Pacific Rise, and the southeast Indian Ridge (noted in Figure 11-16e).

Transform Faults. In 1965, another piece of the tectonic puzzle fell into place when University of Toronto geophysicist Tuzo Wilson first described the nature of transform boundaries and their relationship to earthquake activity. All the spreading centers on Earth's crust feature these lateral scars that stretch out on both sides of spreading centers (Figure 11-18). Some are a few hundred kilometers long; others, such as those along the East Pacific rise, stretch out 1000 km or more (over 600 mi). Across the entire ocean floor these boundaries are the location of **transform faults,** always parallel to the direction in which the plate is moving. How do they occur?

You can see in Figure 11-18 that mid-ocean ridges are not simple straight lines. When a mid-ocean rift begins, it opens at points of weakness in the crust. The fractures you see in the figure began as a series of offset breaks in the crust in each portion of the spreading center. As new material rises to the surface, building the mid-ocean ridges and spreading the plates, these offset areas slide past each other, in horizontal faulting motions. The resulting long *fracture zone,* which follows these breaks in Earth's crust, is active only along the fault section between ridges of spreading centers, as shown in Figure 11-18.

Along transform faults (the fault section between C and D in the figure) the motion is purely one of horizontal displacement—no new crust is formed or old crust subducted. Beyond the spreading centers, the two sides of the fracture zones are joined and inactive. In fact, the plate pieces on either side of the fracture zone are moving in the same direction, away from the spreading center (the fracture zones between A and B and between E and F in the figure). The name "transform" was assigned because of this apparent transformation in the direction of the fault movement. The famous San An-

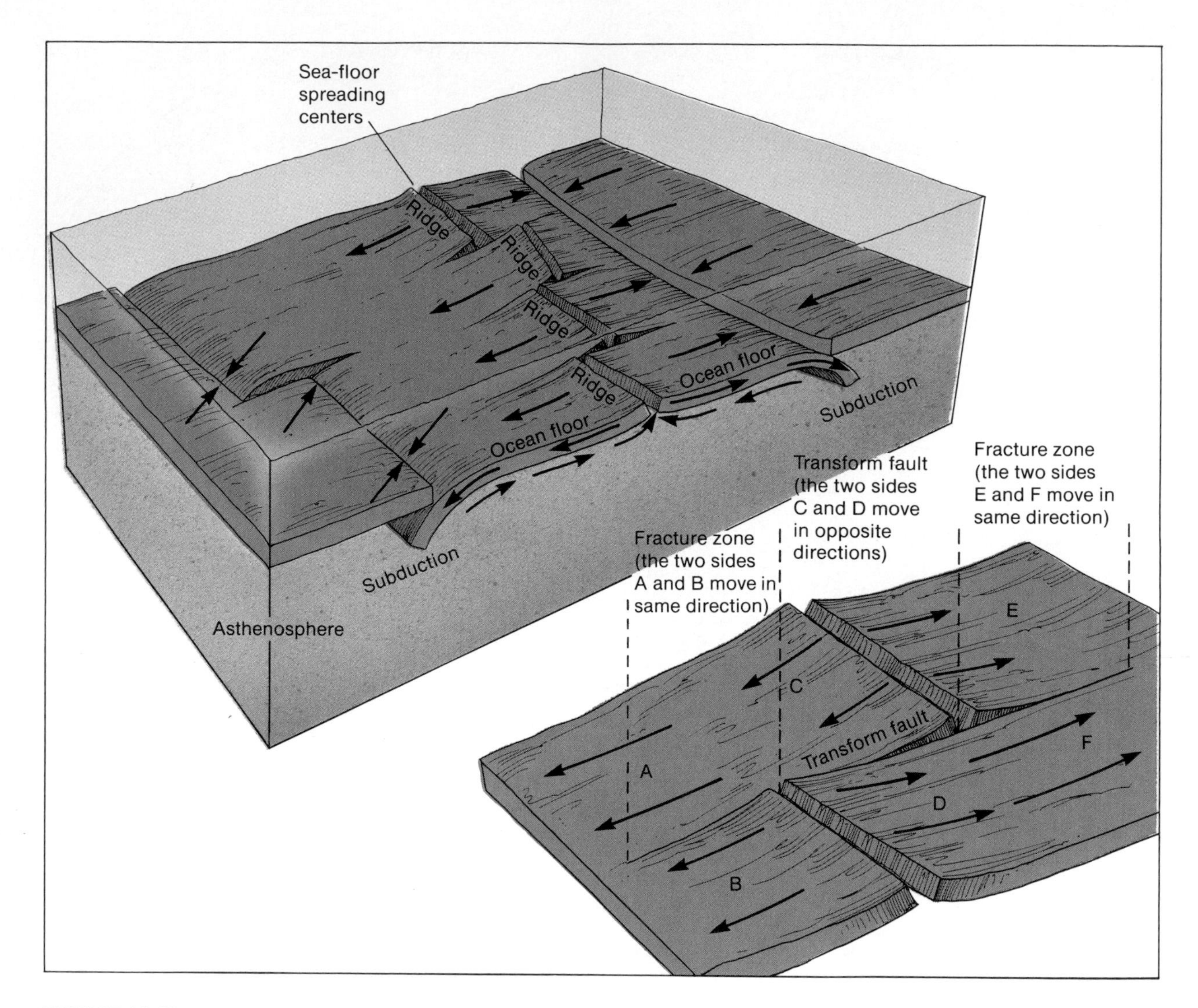

FIGURE 11-18
Transform faults. Appearing along fracture zones, a transform fault is only that section between spreading centers where adjacent plates move in opposite directions. [Adapted from B. Isacks, J. Oliver, and L. R. Sykes, *Journal of Geophysical Research* 73, (1968): 5855–99, copyright by the American Geophysical Union.]

dreas fault system in California, where continental crust has overridden a transform system, is related to this type of motion and is discussed in the next chapter.

Earthquake and Volcanic Activity. Plate boundaries are the primary location of Earth's earthquake and volcanic activity, and the correlation of these phenomena is an important aspect of plate tectonics. Earthquakes and volcanic activity are discussed in more detail in the next chapter, but their general relationship to the tectonic plates is important to point out here. Earthquake zones and volcanic sites are identified in Figure 11-19. The "ring of fire" surrounding the Pacific basin, named for the frequent incidence of volcanoes, is most evident. The subducting edge of the Pacific plate thrusts deep into the crust and upper mantle, producing molten material that makes its way back toward the surface, causing active volcanoes along the Pacific Rim. Such processes occur similarly at plate boundaries throughout the world.

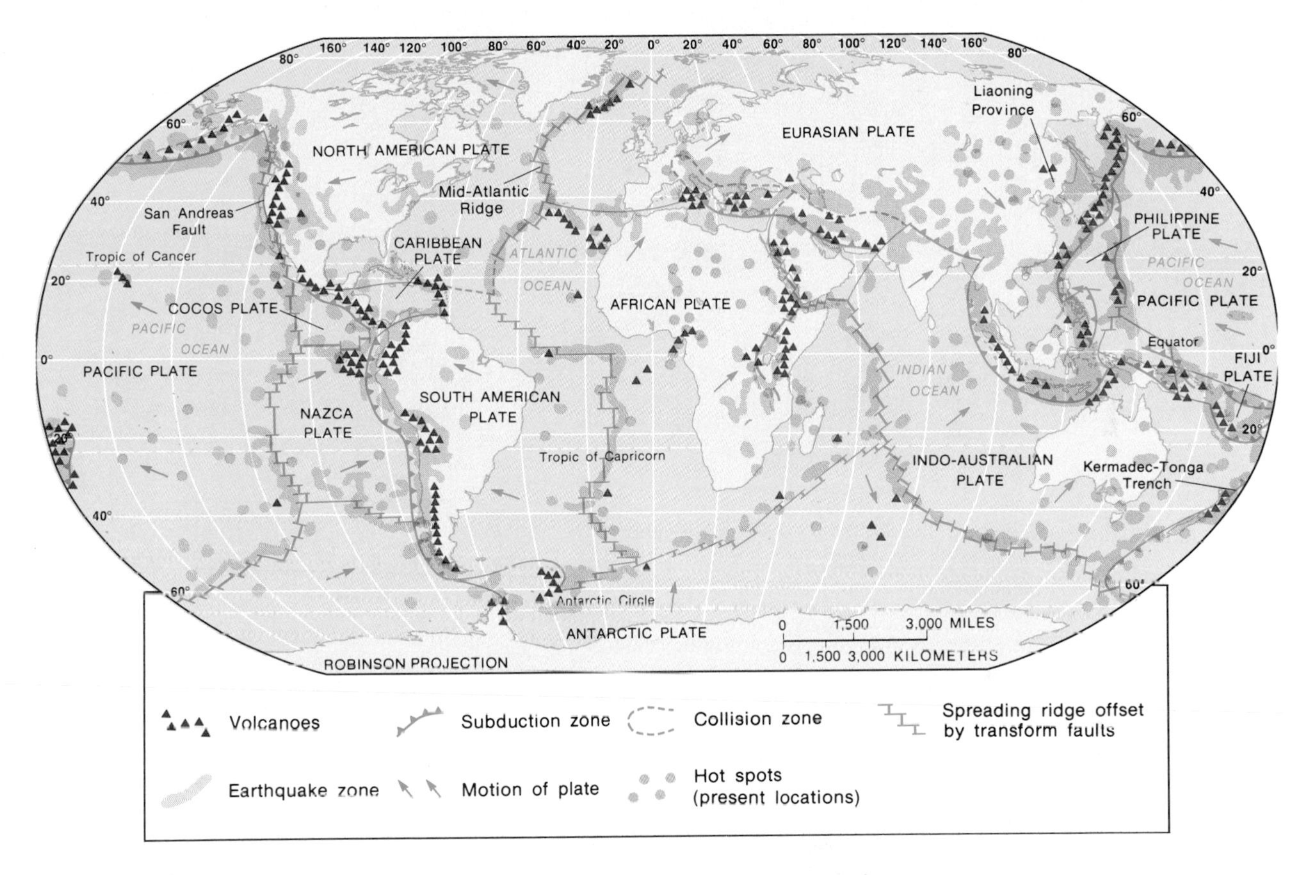

FIGURE 11-19
Earthquake and volcanic activity in relation to major tectonic plate boundaries, and principal hot spots. [Earthquake/volcano data from *Earthquakes* by Bruce A. Bolt. Copyright © 1978, 1988 W. H. Freeman and Company; reprinted with permission. Hot spot locations from "Hot Spots on the Earth's Surface," Kevin C. Burke and J. Tuzo Wilson, Copyright © 1976 by Scientific American, Inc. All rights reserved.]

Hot Spots

A dramatic aspect of plate tectonics is the estimated 50–100 **hot spots** across Earth's surface. These are individual sites of upwelling material from the mantle, noted on Figure 11-19. Hot spots occur beneath both oceanic and continental crust and appear to be deeply anchored in the mantle, tending to remain fixed relative to migrating plates. Unlike convectional currents in the asthenosphere that principally drive plate motions, scientists now think hot spots are initiated by upwelling plumes that are rooted below the 670 km transition zone. Thus, the area of a plate that is above a hot spot is locally heated for the brief geologic time it is there (a few hundred thousand or million years).

An example of an isolated hot spot is the one that has formed the Hawaiian-Emperor islands chain (Figure 11-20). The Pacific plate has moved across this hot, upward-erupting plume for almost 80 million years, with the resulting string of volcanic islands moving northwestward away from the hot spot. Thus, the age of each island or seamount in the chain increases northwestward from the island of Hawaii, as you can see from the ages marked in the figure. The oldest island in the Hawaiian part of the chain is Kauai, approximately

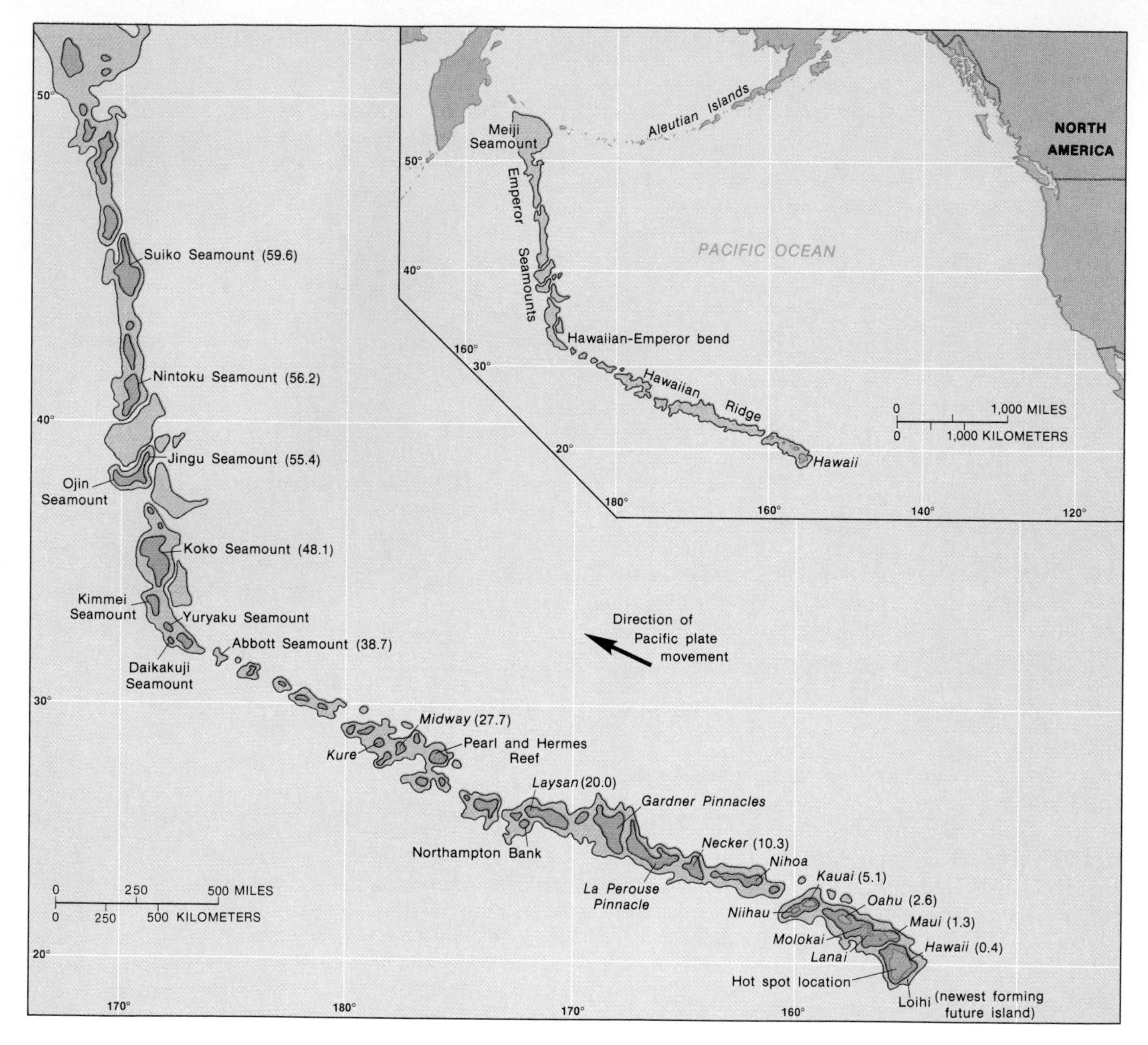

FIGURE 11-20
Hawaii and the linear volcanic chain of islands known as the Emperor Seamounts. Ages of islands and seamounts in the chain are shown in millions of years. [After David A. Clague, "Petrology and K–Ar (Potassium–Argon) Ages of Dredged Volcanic Rocks from the Western Hawaiian Ridge and the Southern Emperor Seamount Chain," *Geological Society of America Bulletin* 86 (1975): 991.]

5 million years old; it is weathered, eroded, and deeply etched with canyons and valleys.

The big island of Hawaii actually took less than 1 million years to build to its present stature. The youngest island in the chain is still a *seamount,* a submarine mountain that does not reach the surface. It rises 3350 m (11,000 ft) from the ocean floor but is still 975 m (3200 ft) beneath the ocean surface. Even though this new island will not experience the Sun for about 10,000 years, it is already named Loihi.

To the northwest of this active hot spot in

Hawaii, the island of Midway rises as a part of the same system. From there the Emperor Seamounts stretch northwestward until they reach about 40 million years of age. At that point, this linear island chain shifts direction northward, evidently reflecting a change in the movement of the Pacific plate at that earlier time. At the northernmost extreme, the seamounts that formed about 80 million years ago are now approaching the Aleutian Trench, where they eventually will be subducted beneath the Eurasian plate.

Most linear volcanic chains have a hot spot at one end, where mantle material works its way up through the passing plate. The track left by these eruptions matches the direction of the plate. Yellowstone National Park, in northwestern Wyoming, is above such a hot spot, and because the North American plate is moving westward, the hot spot track left behind is visible in formations through Idaho, Oregon, and Washington. If the plate continues drifting westward, the phenomenon of Yellowstone, including thermal springs, lava flows, and earthquake and volcanic activity, eventually will appear in the Dakotas.

Iceland is an example of an active hot spot sitting astride a mid-ocean ridge (Figure 11-14). It also is the best example of a segment of mid-ocean ridge rising above sea level. This hot spot has generated enough material to form Iceland and continues to cause eruptions from deep in the asthenosphere. As a result, Iceland is still growing in area and volume. The youngest rocks are near the center of Iceland, with rock age increasing toward the eastern and western coasts. Iceland and Hawaii are active examples of our changing world on a most dynamic planet.

SUMMARY — The Dynamic Planet

The Earth-atmosphere interface is where the endogenic (internal) system, powered by heat energy from within the planet, and the exogenic (external) system, powered by insolation and influenced by gravity, work together to produce Earth's diverse landscape. The geologic time scale is an effective device for organizing the vast span of geologic time and the sequence of events that produced our planet as we know it today. The nature of the endogenic system lies hidden beneath Earth's crust and is studied by scientists indirectly through analysis of seismic (earthquake) wave behavior within the planet. The geologic cycle is a model of these internal and external interactions. It comprises three other cycles: the hydrologic cycle, tectonic cycle, and rock cycle. The rock cycle describes the three principal rock-forming processes—igneous, sedimentary, and metamorphic.

A knowledge of Earth's interior is important in understanding its magnetic field and the dynamics of crustal motion: upwelling, sea-floor spreading, and subduction. Polarity reversals in Earth's magnetism are recorded in cooling magma and settling sediments that contain iron minerals. The pattern of changing magnetism frozen in rock helps scientists piece together the story of Earth's drifting crust. Earth's crust is broken into huge slabs or plates, each in motion in response to flowing currents in the mantle below the crust. The present configuration of the oceans and the continents bears the imprint of these vast tectonic forces. Likewise, occurrences of often damaging earthquakes and volcanoes correlate with plate boundaries. The theory of plate tectonics has brought a twentieth-century revolution in Earth sciences.

KEY TERMS

asthenosphere
basalt
batholith
catastrophism
continental drift
core
crust
endogenic system
evaporites
exogenic system
foliated
geologic cycle
geologic time scale
granite

Gutenberg discontinuity
hot spots
hydrothermal activity
igneous rocks
isostasy
lava
limestone
lithification
magma
magnetic reversal
mantle
metamorphic rock
mid-ocean ridges
mineral
Mohorovičić discontinuity
oceanic trenches
Pangaea
plate tectonics
pluton
rock cycle
sea-floor spreading
sedimentary rocks
seismic waves
stratigraphy
subduction zone
tectonic processes
transform faults
uniformitarianism

REVIEW QUESTIONS

1. To what extent is Earth's crust actively building at this time in its history?
2. Define the endogenic and the exogenic systems. Describe their energy sources.
3. How is geologic time organized? What is the basis for the time scale? What era, period, and epoch are we living in?
4. Contrast uniformitarianism and catastrophism.
5. Make a simple sketch of Earth's interior, label each layer, and list its physical characteristics, temperature, composition, and range of size on your drawing.
6. What is the present thinking on how Earth generates its magnetic field? Is this field constant, or does it change? Explain the implications of your answer.
7. Describe the asthenosphere. Why and under what circumstances is it mobile? What are the consequences of this movement?
8. What is a discontinuity? Describe the two principal discontinuities within Earth.
9. Define isostasy and isostatic rebound, and explain the crustal equilibrium concept.
10. Diagram the upper mantle and crust. Label the density of the layers and include generalizations as to composition.
11. Illustrate the geologic cycle and define each component: rock cycle, tectonic cycle, and hydrologic cycle.
12. What is a mineral? A mineral family? Name the most common minerals on Earth.
13. Describe igneous processes. What is the difference between intrusive and extrusive forms?
14. Characterize felsic and mafic minerals. Give examples of both coarse- and fine-grained textures.
15. From what sources and particle sizes are the materials for sedimentary processes derived? Briefly describe sedimentary processes and lithification.
16. What is metamorphism? Name some original parent rocks and their metamorphic equivalents.
17. Briefly review the history of continental drift, sea-floor spreading, and the all-inclusive plate tectonics theory. What was Alfred Wegener's role?

18. Define upwelling and describe related features associated with such action on the ocean floor. Define subduction and explain the process.
19. What was Pangaea? What happened to it during the past 225 million years?
20. Characterize the three types of plate boundaries and the probable conditions associated with each type.
21. Why is there a relationship between plate boundaries and volcanic and earthquake activity?

SUGGESTED READINGS

Alexander, Tom. "A Revolution Called Plate Tectonics Has Given Us a Whole New Earth," *Smithsonian Magazine* 5, no. 10 (January 1975): 30–40.

Alexander, Tom. "Plate Tectonics Has a Lot to Tell Us About the Present and Future Earth," *Smithsonian Magazine* 5, no. 11 (February 1975): 38–47.

Anderson, Don L., Toshiro Tanimoto, and Yu-shen Zhang. "Plate Tectonics and Hotspots: The Third Dimension," *Science* 256 (June 19, 1992): 1645–51.

Bloxham, Jeremy, and David Gubbins. "The Evolution of the Earth's Magnetic Field," *Scientific American* 261, no. 6 (December 1989): 68–75.

Bolt, Bruce A. *Inside the Earth—Evidence from Earthquakes*. San Francisco: W. H. Freeman, 1982.

Continents Adrift and Continents Aground—Readings from Scientific American. San Francisco: W. H. Freeman, 1976. "The Dynamic Earth"; "The Earth's Core"; "The Earth's Mantle"; "The Continental Crust." *Scientific American* 249, no. 3 (September 1983): entire issue.

Gore, Rick. "Our Restless Planet Earth," *National Geographic* 168 (August 1985): 142–81.

Hamblin, W. Kenneth. *The Earth's Dynamic Systems,* 6th ed. New York: Macmillan Publishing Company, 1992.

Hill, R. I., I. H. Campbell, G. F. Davies, and R. W. Griffiths. "Mantle Plumes and Continental Tectonics," *Science* 256 (April 10, 1992): 186–95.

Hurley, Patrick M. "The Confirmation of Continental Drift," *Scientific American* 218, no. 4 (April 1968): 52–64.

Miller, Russel, and the editors of Time-Life Books. *Continents in Collision*. Planet Earth Series. Alexandria, VA: Time-Life Books, 1983.

Overbye, Dennis. "The Shape of Tomorrow—How Earth Will Look in 250 Million Years," *Discover* 3, no. 11 (November 1982): 20–25.

Powell, Corey S. "Trends in Geophysics—Peering Inward," *Scientific American* 254, no. 6 (June 1991): 100–11.

Press, Frank, and Raymond Siever. *Earth*, 4th ed. New York: W. H. Freeman, 1986.

Tarbuck, Edward J., and Frederick K. Lutgens. *The Earth—An Introduction to Physical Geology,* 4th ed. Columbus, OH: Macmillan Publishing Company, 1993.

Uyeda, Seiya. *The New View of the Earth—Moving Continents and Moving Oceans*. Translated by Masako Ohnuki. San Francisco: W. H. Freeman, 1978.

Vink, Gregory E., W. Jason Morgan, and Peter R. Vogt. "The Earth's Hot Spots," *Scientific American* 252, no. 4 (April 1985): 50–57.

Wyllie, Peter J. *The Way the Earth Works: An Introduction to the New Global Geology and Its Revolutionary Development*. New York: John Wiley & Sons, 1976.

Floor of the Oceans, 1975, by Bruce C. Heezen and Marie Tharp. [Copyright © 1980 by Marie Tharp. Reproduced by permission of Marie Tharp, 1 Washington Ave., South Nyack, NY 10960.]

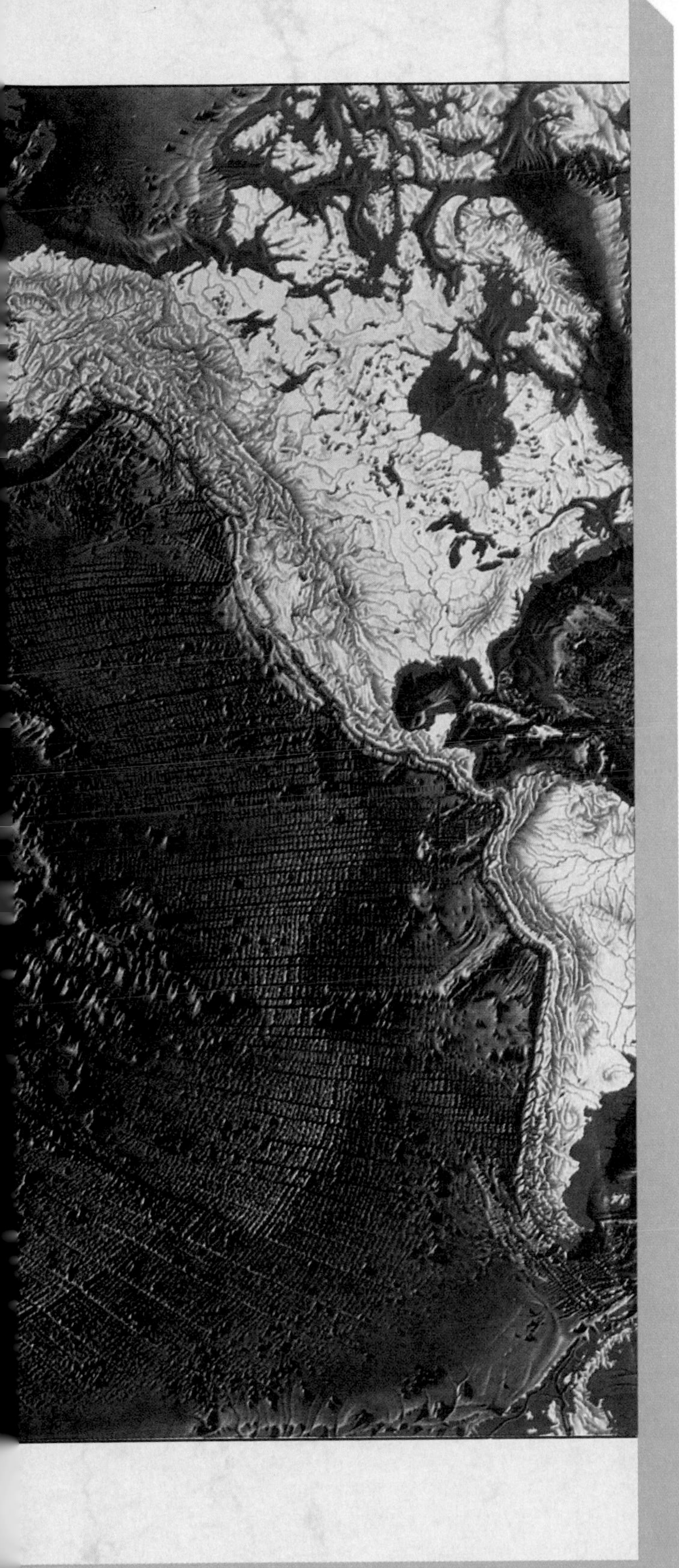

12

TECTONICS AND VOLCANISM

Tectonic forces were brought home to hundreds of millions of television viewers as they watched the 1989 baseball World Series from Candlestick Park near San Francisco. Less than half an hour before the call to "play ball," a powerful earthquake rocked the region and turned sportscasters into newscasters and sports fans into disaster witnesses. Such tectonic activity has repeatedly deformed, recycled, and reshaped Earth's crust during its 4.6-billion-year existence. The principal tectonic and volcanic zones lie along plate boundaries or in areas where Earth's crustal plates are affected by processes in the mantle. The arrangement of continents and oceans, the origin of mountain ranges, orders of relief, and the locations of earthquake and volcanic activity are all the result of these dynamic Earth processes.

THE OCEAN FLOOR

The illustration that opens this chapter is a striking representation of Earth with its blanket of water removed. The ocean floor is revealed to us through decades of direct and indirect observation. Careful examination of this portrayal provides a helpful review of concepts learned in the previous chapter, laying the foundation for this and subsequent chapters. The scarred ocean floor is clearly visible, its sea-floor spreading centers marked by over 64,000 km (40,000 mi) of oceanic ridges, its subduction zones indicated by deep oceanic trenches, and its transform faults stretching at angles between portions of oceanic ridges.

On the floor of the Indian Ocean you can locate the track along which the India plate traveled to its collision with the Eurasian plate. You also can spot the vast sediment patterns in the Indian Ocean, south of the Indus and Ganges rivers on either side of India. Sediments derived from the Himalayan Range blanket the floor of the Bay of Bengal (south of Bangladesh) to a depth of 20 km (12.4 mi). The principle of isostasy can be examined in central and west-central Greenland, where the weight of ice has pressed portions of the land far below sea level.

In the area of the Hawaiian Islands, the hot spot track is marked by a chain of islands and seamounts which you can follow along the Pacific plate from Hawaii to the Aleutians. You can find Iceland's position on the Mid-Atlantic Ridge. Subduction zones south and east of Alaska and Japan as well as along the western coast of South and Central America are quite visible. In addition, you can follow the East Pacific Rise northward as it trends beneath the west coast of the North American plate, disappearing under earthquake-prone California. The continents, submerged continental shelves and slopes, and the expanse of the sediment-covered abyssal plain are all identifiable on this illustration. (See if you can correlate this sea-floor illustration with the map of crustal plates and plate boundaries shown in Figures 11-17 and 11-19.) Geographers use a simple system to characterize Earth's continental and ocean floor features.

EARTH'S SURFACE RELIEF FEATURES

Earth's surface landscape is the place where we see its solid and fluid realms interact—the dynamic interplay of Earth's internal and external systems. This landscape is a composite of separate landforms composed of various rock structures evolving in the struggle between these systems.

Crustal Orders of Relief

Relief refers to vertical elevation differences in a local landscape. The undulating form that gives Earth's surface its character and general configuration is called **topography,** the feature portrayed so effectively on topographic maps. Earth's crustal landforms have played a vital role in human history: high mountain passes have both protected and isolated societies; ridges and valleys have dictated transportation routes; and vast plains have encouraged better methods of communication and travel.

Earth's topography has stimulated human invention and adaptation.

Understanding Earth's relief and topography is enhanced by new computer-based capabilities. Scientists put elevation data in digital form; the data then are available for computer processing, manipulation, and display. Analysis is aided by these *digital elevation models* (DEMs) of topography, area-altitude distributions, slope aspects, and local stream-drainage characteristics. The map in Figure 12-13a is a portion of a digitized shaded-relief map of the United States prepared by the United States Geological Survey in 1992. The map is made up of 12 million spot elevations, each less than a kilometer apart.

For convenience, we group the landscape's topography into three "orders of relief." These classify landscapes by scale, from enormous ocean basins and continents down to local hills and valleys.

First Order of Relief. The broadest category of landforms includes huge **continental platforms** and **ocean basins.** Continental platforms are the masses of crust that reside above or near sea level, including the undersea continental shelves along the coastlines. The ocean basins are entirely below sea level. Approximately 71% of Earth's surface is covered by water, with only 29% of its surface appearing as continents. The distribution of land and water in evidence today demonstrates a distinct water hemisphere and continental hemisphere, as shown in Chapter 7 (see Figure 7-4).

Second Order of Relief. Continental features that are classified in the second order of relief include continental masses, mountain masses, plains, and lowlands. A few examples are the Alps, Rocky Mountains (both Canadian and American), west Siberian lowland, and Tibetan Plateau. The great shields that form the heart of each continental mass are of this second order. In the ocean basins, second order of relief includes continental rises, slopes, abyssal plains, mid-ocean ridges, submarine canyons, and subduction trenches—all visible in the sea-floor illustration that opens this chapter.

Third Order of Relief. Individual mountains, cliffs, valleys, hills, and other landforms of smaller scale are included in the third order of relief. These features are identifiable as local landscapes.

Figure 12-1 is a *hypsographic curve,* a generalized curve showing the distribution of Earth's surface by area and elevation, or hypsometry (the Greek *hypsos* means height). Relative to Earth's size, the surface generally is of low relief–in other words, Earth's topography is gentle when compared to the planet's diameter. The average elevation of Earth's overall surface is under water: –2070 m (–6790 ft) below mean sea level. The average elevation just for exposed land is +875 m (+2870 ft). For the ocean depths, the average elevation is –3800 m (–12,470 ft). From this you can see that, on the average, the oceans are much deeper than continental regions are high. Overall, the underwater ocean basins, ocean floor, and submarine mountain ranges represent the largest areal percentage of crust.

Earth's Topographical Regions

These landform orders can be further generalized into six topographic regions—plains, high tablelands, hills and low tablelands, mountains, widely spaced mountains, and depressions (Figure 12-2). Each type of topography is delineated by an arbitrary elevation or descriptive limit that is in common use (see legend in the figure). Four of the continents possess extensive plains, which are identified as areas with local relief of less than 100 m (325 ft) and slope angles of 5° or less. Some plains have high elevations of over 600 m (2000 ft); in the United States the high plains achieve elevations above 1220 m (4000 ft). The Colorado Plateau and Antarctica are notable high tablelands, with elevations exceeding 1520 m (5000 ft). Africa is dominated by hills and low tablelands.

Mountain ranges are characterized by local relief exceeding 600 m (2000 ft) and appear in locations on each continent that probably are familiar to you. Earth's relief and topography are dynamic and undergoing constant change as a result of crustal processes.

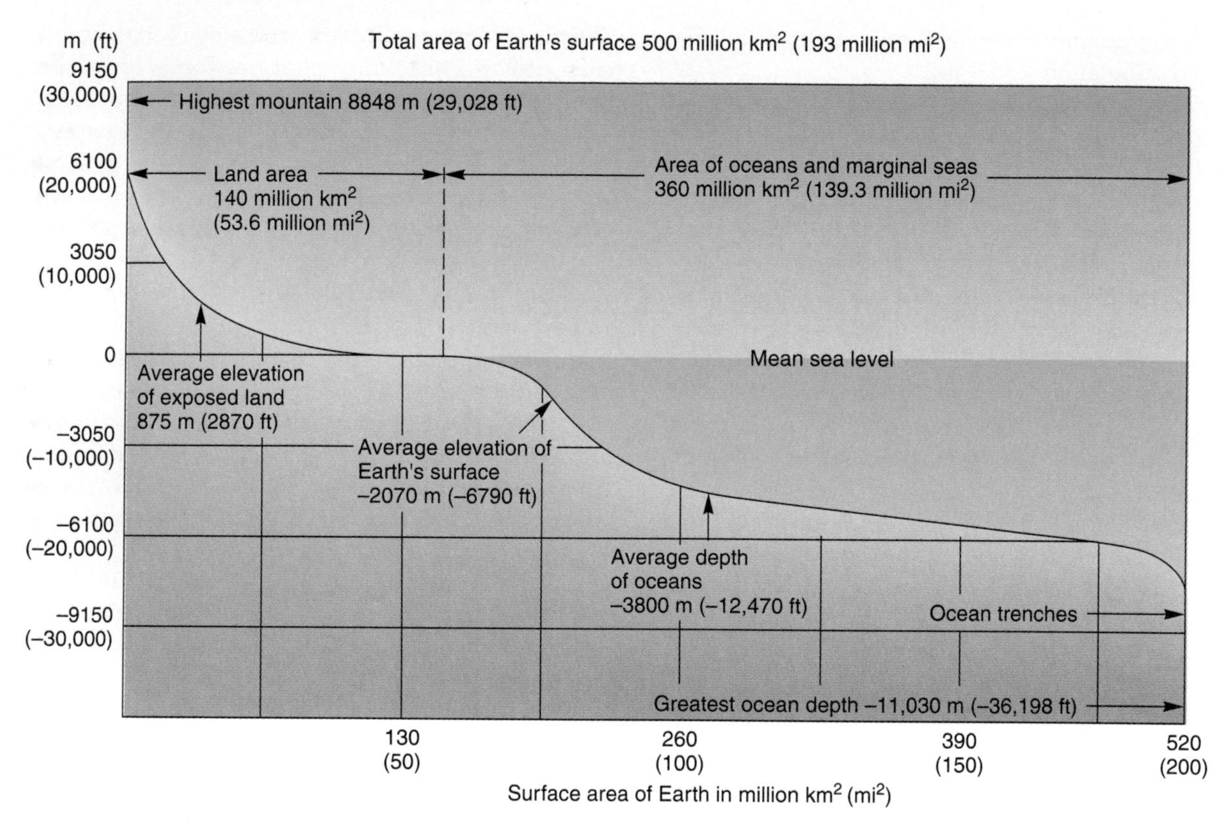

FIGURE 12-1
Hypsographic curve of Earth's surface, charting elevation as related to mean sea level. [After Joseph E. Van Riper, *Man's Physical World*, p. 389, copyright © 1971 by McGraw-Hill. Adapted by permission.]

CRUSTAL FORMATION PROCESSES

How did Earth's continental crust form? What gives rise to the three orders of relief just discussed? Ultimately, the answer is tectonic activity, driven by our planet's internal energy.

Tectonic activity is generally slow when viewed in human terms. Endogenic (internal) processes result in slow uplift and new landforms, with major mountain building occurring along plate boundaries. These uplifted crustal regions are quite varied, depending on composition, structure, and operative processes. Nonetheless, they generally can be thought of in three categories, all of which are discussed in this chapter: (1) residual mountains and shields, formed from inactive remnants of older rock—the ancient heart of continental crust; (2) tectonic mountains and landforms, produced by active folding and faulting movements that deform the crust; and (3) volcanic features, formed by the surface accumulation of molten rock from eruptions of subsurface materials. Thus, several distinct processes operate in concert to produce continental crust.

Continental Crust

Continental crust is more complex in both content and structure than the relatively homogeneous oceanic crust. All continents have a nucleus of crystalline rock on which the continent "grows"

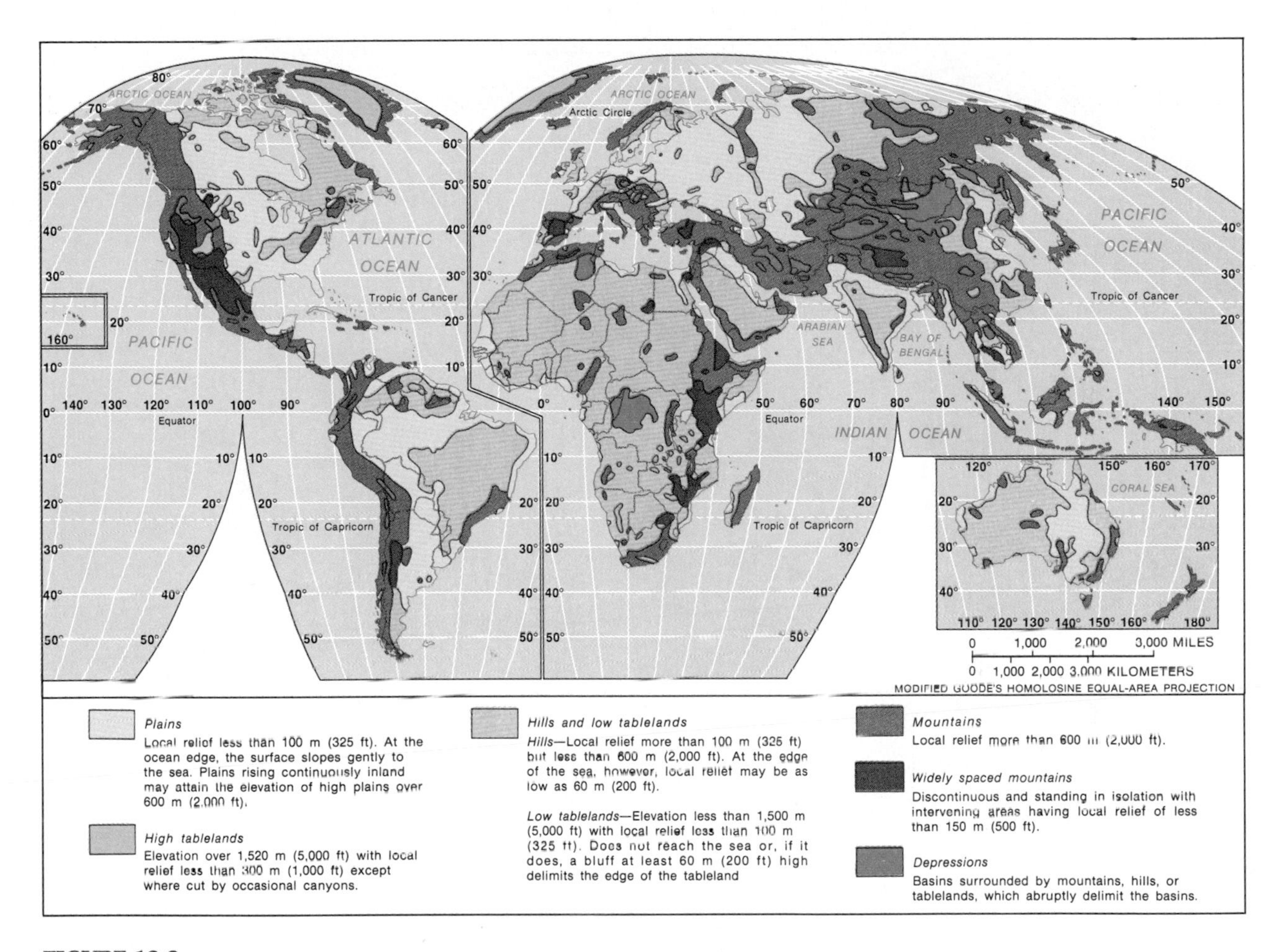

FIGURE 12-2
Earth's topographic regions. [After Richard E. Murphy, "Landforms of the World," Map Supplement No. 9, *Annals of the Association of American Geographers* 58, no. 1 (March 1968). Adapted by permission.]

with additions of other crust and sediments. The nucleus is the *craton,* or heartland region, of the continental crust. Cratons generally are low in elevation and relief and are old (most exceed two billion years in age but all are Precambrian, or older than 570 million years). A region where a craton is exposed at the surface is called a **continental shield.** Figure 12-3 shows the principal areas of exposed shields.

Portions of cratons are covered with layers of sedimentary rock that are quite stable over time. An example of such a stable *platform* is the region that stretches from the Rockies to the Appalachians and northward into central Canada.

Continental crust results from a complex system involving the sea-floor spreading of oceanic crust, its subduction, remelting, and subsequent production of rising magma. The magma that originates in the asthenosphere and wells up along the mid-ocean ridges is less than 50% silica and is rich in iron and magnesium. This material rises at the spreading centers, flows outward as ocean floor,

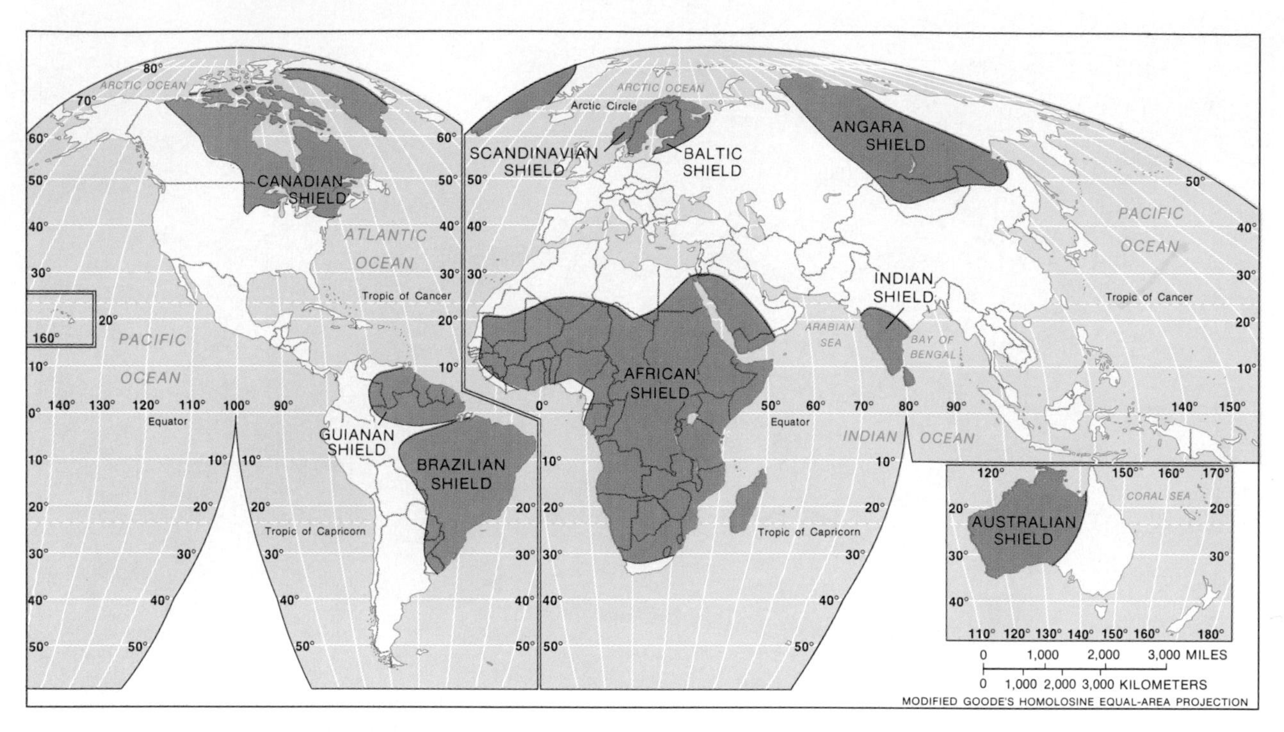

Figure 12-3
Major continental shields. [After Richard E. Murphy, "Landforms of the World," Map Supplement No. 9, *Annals of the Association of American Geographers* 58, no. 1 (March 1968). Adapted by permission.]

and collides with the continents, only to plunge back into the mantle, remelt, and rise to form more continent as intrusive igneous rock. The subducting oceanic plate works its way under a continental plate, taking with it sediment and trapped water, melting and incorporating various elements from the crust that are drawn into the mixture. As a result, the magma (generally called a "melt") that migrates upward from a subducted plate contains 50–75% silica, is high in aluminum, and has a high-viscosity (thick) texture. Bodies of such magma may reach the surface in explosive volcanic eruptions, or they may stop short and become subsurface intrusive bodies in the crust, cooling slowly to form crystalline plutons such as batholiths (see Figures 11-8 and 11-9a).

Terranes. A surprising discovery is that each of Earth's major plates actually is a collage of many crustal pieces acquired from a variety of sources. Accretion, or accumulation, has occurred as crustal fragments of ocean floor, curved chains (or arcs) of volcanic islands, and other pieces of continental crust have been swept aboard the edges of continental shields and platforms. These migrating crustal pieces, which have become attached to the plates, are called **terranes** (not to be confused with "terrain," which refers to the topography of a tract of land). Figure 12-4 shows an interesting example of how western North America has accreted various terranes.

Displaced terranes have histories different from those of the continents that capture them, are usually framed by fractures, and are distinguished in rock composition and structure from their new continental homes.

This process of continental construction is envisioned in James Michener's novel *Alaska*:

> **But the immediate task is to understand how this trivial ancestral [continental] nucleus could aggregate to itself the many additional segments of rocky land which would ultimately unite to comprise the Alaska**

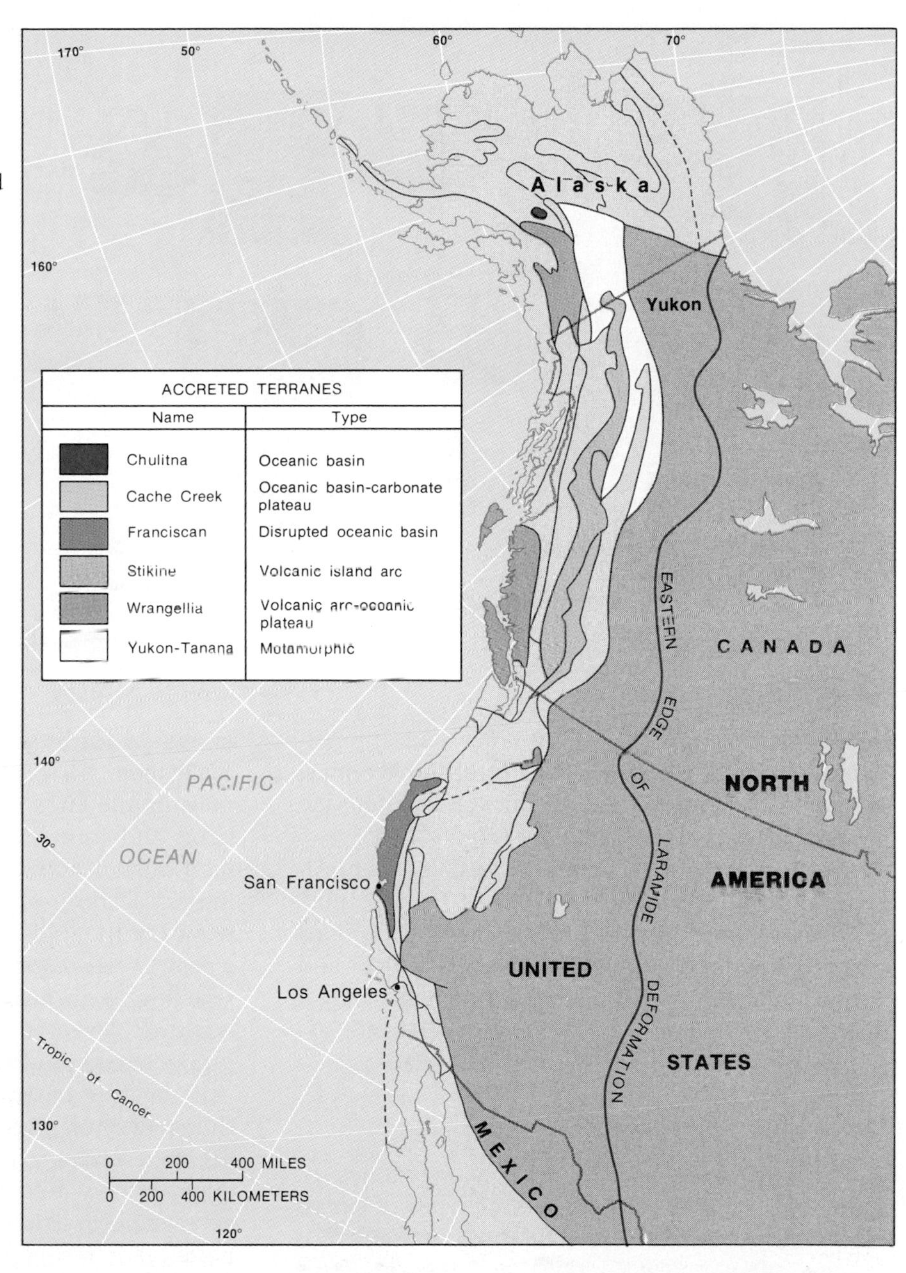

Figure 12-4
Wrangellia terranes, highlighted among the other terranes along the western margin of North America. [After "The Growth of Western North America" by David L. Jones, Allan Cox, Peter Coney, and Myrl Beck, illustration by Andrew Tomko. Copyright © 1982 by Scientific American, Inc. All rights reserved.]

we know. Like a spider waiting to grab any passing fly, the nucleus remained passive but did accept any passing terranes—those unified agglomerations of rock considerable in size and adventurous in motion—that wandered within reach . . . . And the great plates of Earth's crust never rest . . . .*

*James A. Michener, *Alaska* (New York: Random House, 1988), pp. 5 and 9. Reprinted by permission.

In the region surrounding the Pacific, accreted terranes are particularly prevalent. At least 25% of the growth of western North America can be attributed to the accretion of terranes since the early Jurassic period (190 million years ago). A good example is the Wrangell Mountains, which lie just east of Prince William Sound and the city of Valdez, Alaska. The **Wrangellia terranes**—a former volcanic island arc and associated marine sed-

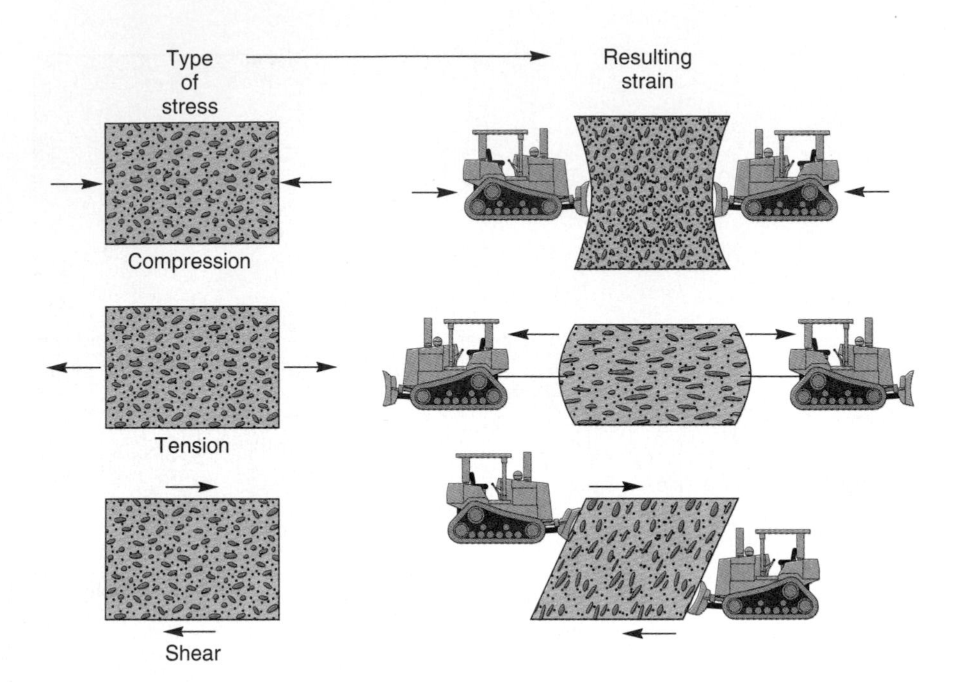

Figure 12-5
General types of stress and resulting strain.

iments from near the equator—migrated approximately 10,000 km (6200 mi) to form the Wrangell Mountains and three other distinct areas along the western margin of the continent (Figure 12-4).

The Appalachian Mountains, extending from Alabama to the Maritime Provinces of Canada, possess bits of land once attached to ancient portions of Europe, Africa, South America, Antarctica, and various oceanic islands. These discoveries, barely a decade old, demonstrate how continents are assembled.

Deformation Processes

Earth's crust bends and breaks along zones of strain. The strain is caused by three types of stress: *compression* (shortening or folding, and faulting), *tension* (stretching or faulting), and *shearing* (tangential or transform stress when two pieces slide past each other), as shown in Figure 12-5. The patterns created by these processes are evident in the landforms we see today.

Folding. When layered flat strata are subjected to compressional forces, the original structure is bent and deformed. Convergent plate boundaries worldwide create intensely deformed rock in a process known as **folding.** If we take pieces of thick cloth, stack them flat on a table, and then slowly push on opposite ends of the stack, the cloth layers will bend and form similar folds (Figure 12-6). If we then draw a line down the center axis of a resulting ridge and trough, we are able to see how the names of the folds are assigned. Along the ridge of a fold, layers *slope downward away from the axis,* resulting in an **anticline**. In the trough of a fold, however, layers *slope downward toward the axis,* producing a **syncline.**

If the axis of either type of fold is not "level" (horizontal, or parallel to Earth's surface), the layers then *plunge* (dip down) at an angle. Knowledge of how folds are angled to Earth's surface and where individual structures are located is important for resource recovery. Petroleum, for example, collects in the upper portions of folds in permeable rock layers such as sandstone.

Figure 12-6a further illustrates different kinds of folds that have been weathered and reduced by exogenic processes. Within a syncline, a residual ridge may form, caused by varying resistance of the different rock strata to exogenic processes (note the "synclinal ridge" near the right side of the figure). Compressional forces often push folds

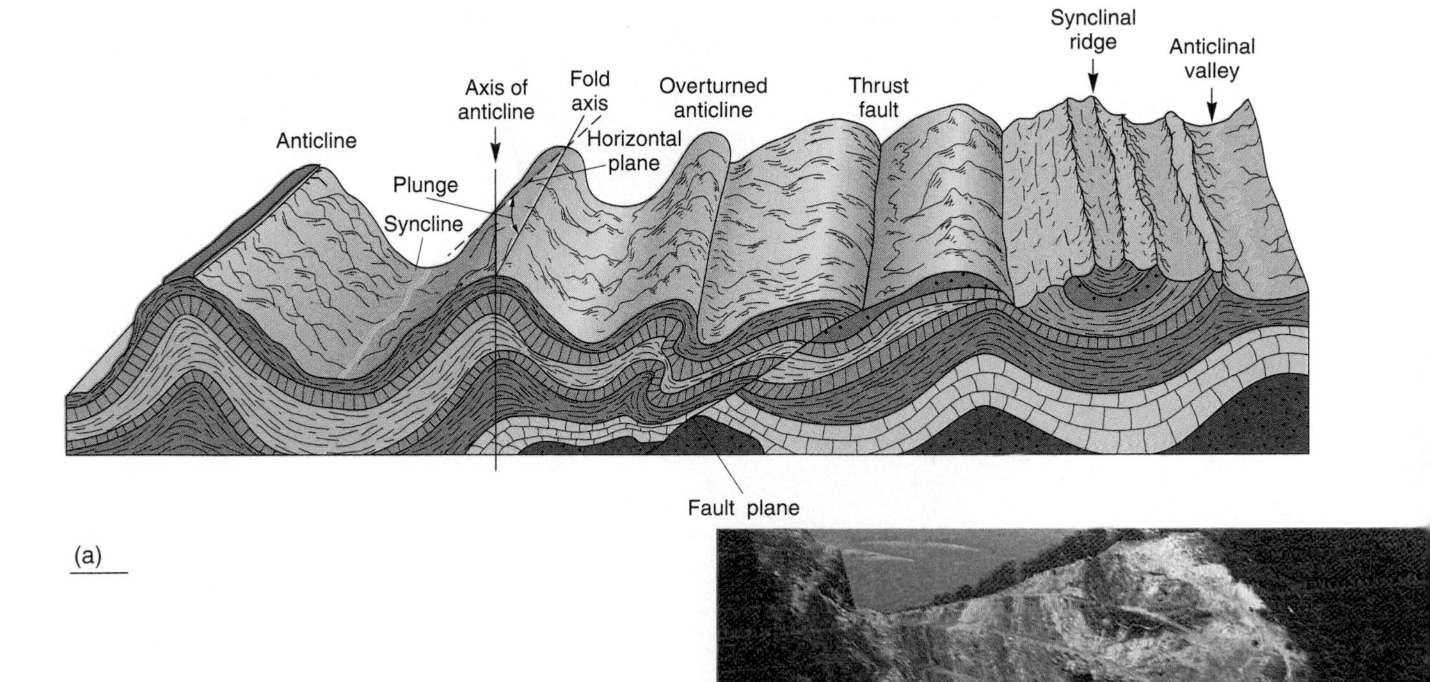

Figure 12-6
Folded landscape and the basic types of fold structures (a); a roadcut exposes a synclinal ridge near Hancock, Maryland (b). [Photo by John Thrasher.]

far enough that they actually overturn on their own strata, as you can see in the center of the figure ("overturned anticline"). Further stress eventually fractures the rock strata along distinct lines, and some overturned folds are thrust up, causing a considerable shortening of the original strata ("thrust fault"). The Canadian Rocky Mountains and the Appalachian Mountains illustrate well the complexity of the resulting folded landscape. An interstate highway roadcut dramatically exposes a synclinal ridge in western Maryland (Figure 12-6b).

Satellites let us view many of these structures from an orbital perspective, as in Figure 12-7. Northwest of the Strait of Hormuz, just north of the Persian Gulf, are the Zagros Mountains of Iran. Originally, this area was a dispersed terrane that had separated from the Eurasian plate. However, the collision produced by the northward push of the Arabian block is now shoving this terrane back into Eurasia and forming an active margin known as the Zagros crush zone, a zone of continued collision more than 400 km (250 mi) wide. In the satellite image, anticlines form the parallel ridges; active weathering and erosion processes expose underlying strata.

Earth's continental crust also is subjected to broad warping forces, which produce an up-and-down bending of strata that is too broad in extent to be considered folding. Such forces can be the result of mantle convection, isostatic adjustments,

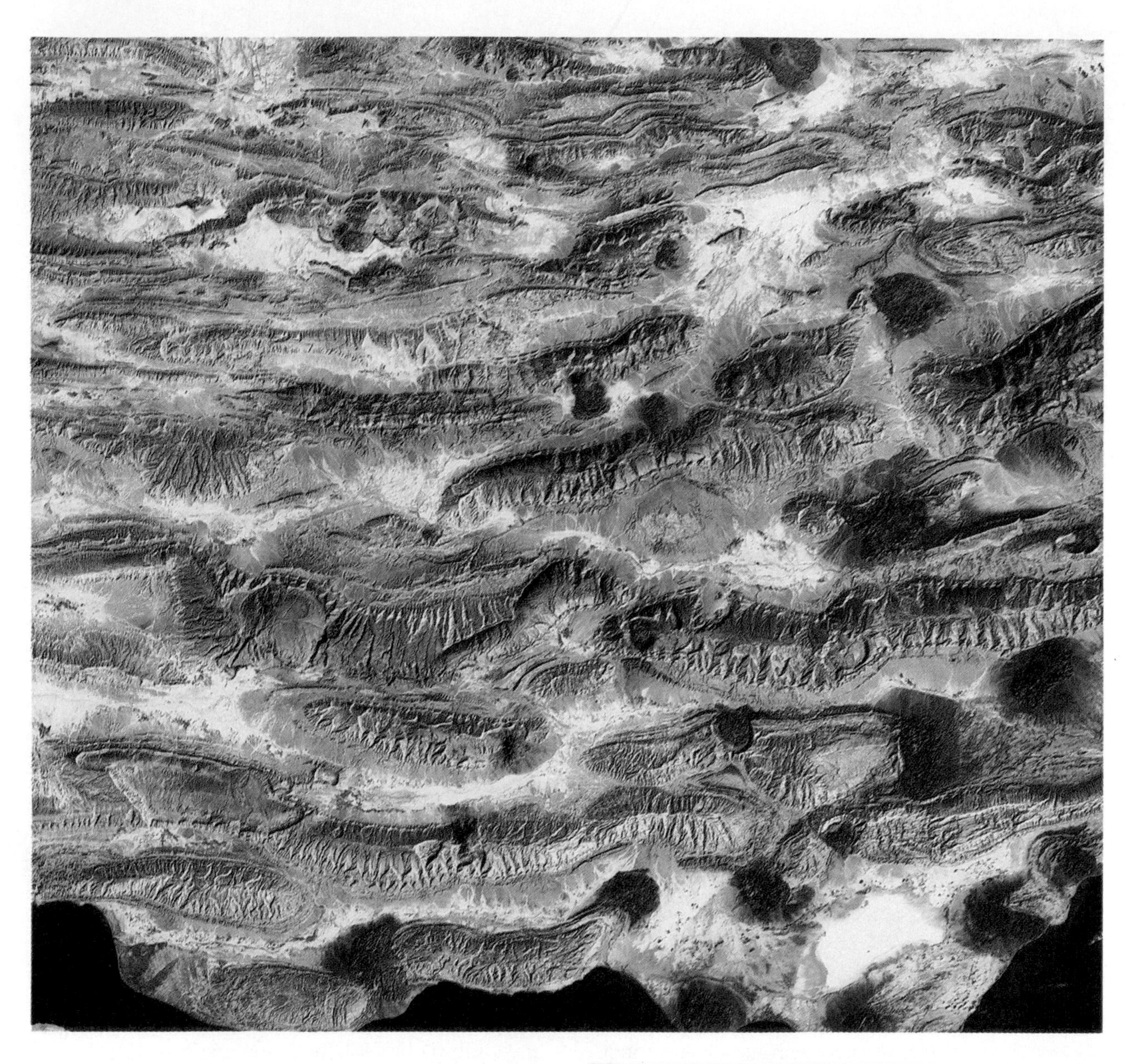

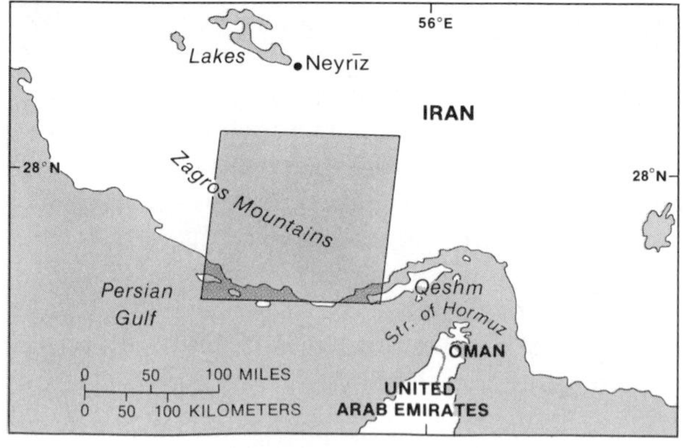

Figure 12-7
Southern Zagros Mountains in the Zagros crush zone between the Arabian and Eurasian plates. [NASA image.]

Figure 12-8
(a) An upwarped dome and a structural basin; (b) the Richat dome structure of Mauritania. [(a) After Joseph E. Van Riper, *Man's Physical World,* p. 436, copyright 1971 by McGraw-Hill; (b) NASA photo]

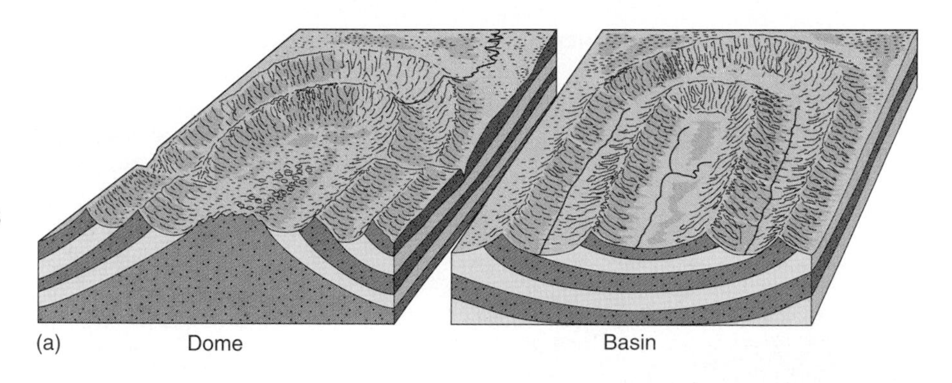

and/or swelling associated with hot spot activity. Warping features range from small, individual, foldlike structures called *basins* and *domes* (Figure 12-8a) up to regional features the size of the Ozark Mountain complex in Arkansas and Missouri, the Colorado Plateau in the West, or the Black Hills of South Dakota. The isostatic rebound of the Hudson Bay region represents a broad upwarping that will eventually lead to a draining of the bay. The Richat dome structure in the West African nation of Mauritania, shown in Figure 12-8b, is an example of this type of deformation.

Faulting. When rock strata are strained beyond their ability to remain a solid unit, they fracture, and one side is displaced relative to the other side in a process known as **faulting.** Thus, fault zones are areas of crustal movement. At the moment the

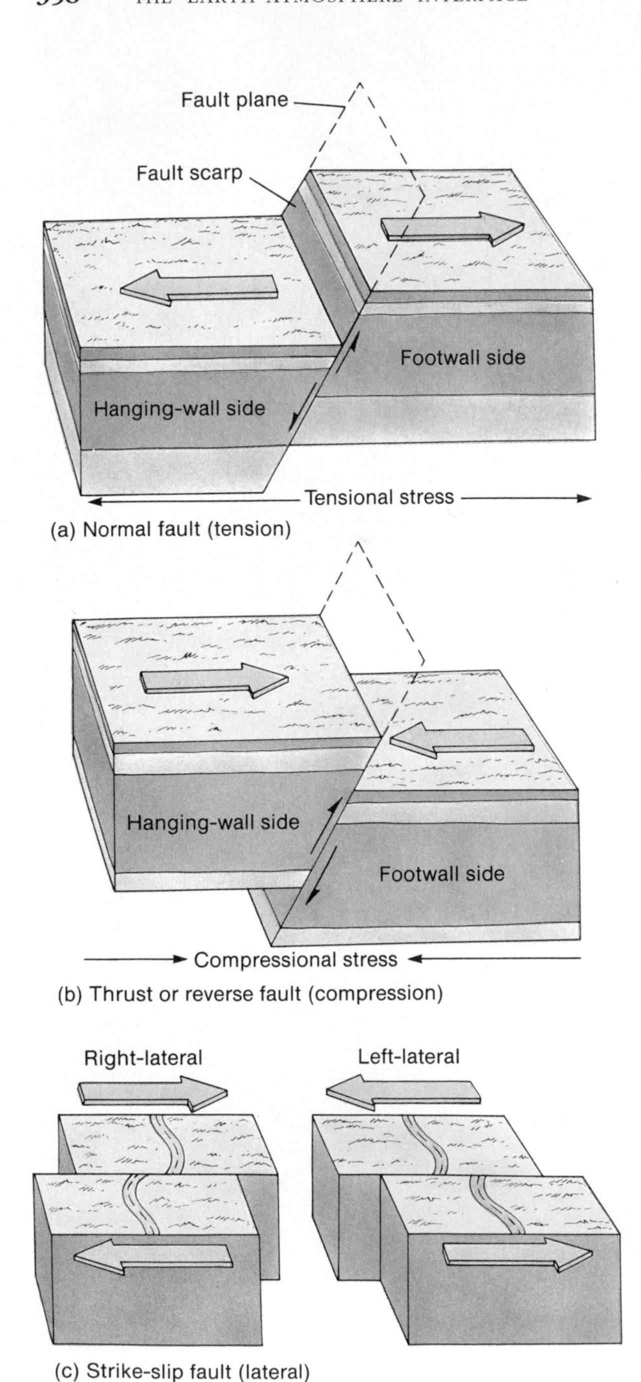

Figure 12-9
Basic types of faults: (a) normal fault (tension); (b) reverse fault (compression); and (c) strike-slip fault (transcurrent or horizontal). In England, when miners worked along a reverse fault (b), they would stand on the lower side (footwall) and hang their lanterns on the upper side (hanging wall), giving rise to these terms.

fault line shifts, a sharp release of energy occurs, called an **earthquake** or quake (Figure 12-9).

The surface along which the two sides of a fault move is called the *fault plane*. The tilt and orientation of the fault plane provides the basis for naming the three basic types of faults. A *normal fault*, or tension fault, moves vertically along an inclined fault plane (Figure 12-9a). The overlying block is called the hanging wall; it shifts downward relative to the footwall block. ("Hanging wall" and "footwall" are old coal-mining terms.) The exposed fault plane sometimes is evident along the base of faulted mountains, where individual ridges are truncated by the movements of the fault and appear as triangular facets at the ends of the ridges. A cliff formed by faulting is commonly called a *fault scarp*, or *escarpment*.

Compressional forces associated with converging plates produce hanging wall blocks that move upward relative to the footwall block. This is called a *reverse fault*, or compression fault (Figure 12-9b). On the surface it appears similar in form to a normal fault, although more collapse and landslides may occur from the hanging wall component. If the fault plane forms a low angle relative to the horizontal, the fault is termed a *thrust fault*, or overthrust fault, indicating that the overlying block has shifted far over the underlying block (Figure 12-6, "thrust fault"). In the Alps, three to five such overthrusts are in evidence from the compressional forces of the ongoing African plate collision with the Eurasian plate. Beneath the Los Angeles basin, such overthrust faults produce a high risk of earthquakes.

If movement along a fault line is horizontal, as produced by a transform fault, it is called a **strike-slip fault,** or transcurrent fault (Figure 12-9c). The movement is described as *right-lateral* or *left-lateral,* depending on the motion perceived when you observe movement on one side of the fault relative to the other side. Although strike-slip faults do not produce cliffs (scarps), they can create linear *rift valleys,* as is the case with the San Andreas fault system of California (Figure 12-10). The rift is clearly visible in the photograph where the edges of the North American and Pacific plates are grinding past one another as a result of transform faults associated with a former sea-floor spreading center. In its westward drift, the North American plate

Figure 12-10
The San Andreas fault in the eastern margin of the central Coast Ranges of California. This view is toward the north, so the Pacific plate is on the left (the west side). The Pacific plate is moving northward relative to the North American plate on the right (the east side). [Photo by Randall Marrett.]

rode over portions of this center, as shown in Figure 12-15. Consequently, the San Andreas system is described as a series of transform (associated with former spreading center), strike-slip (horizontal motion), right-lateral (one side moving to the right hand of the other) faults.

In the interior western United States, the Basin and Range Province experienced tensional forces caused by uplifting and thinning of the crust, which cracked the surface to form aligned pairs of normal faults and a distinctive landscape (please refer to Figures 15-19 and 15-20). The term **horst** is applied to upward-faulted blocks; **graben** refers to downward-faulted blocks. Examples of horst and graben landscapes include the Great Rift Valley of East Africa (associated with crustal spreading), which extends northward to the Red Sea, and the Rhine graben through which the Rhine River flows in Europe.

The Sierra Nevada of California and the Grand Tetons of Wyoming are examples of **tilted fault block** mountain ranges, in which a normal fault on one side of the range has produced a tilted landscape of dramatic relief (Figure 12-11). Slowly cooling magma intruded into those blocks and formed granitic cores of coarsely crystalline rock. After tremendous tectonic uplift and the removal of overlying material through weathering, erosion, and transport, those granitic masses are now exposed. In some areas, the overlying material originally covered these batholiths by more than 7500 m (25,000 ft).

Orogenesis

Orogenesis literally is the birth of mountains (*oros* comes from the Greek for "mountain"). An *orogeny* is a mountain-building episode that thickens continental crust. It can occur through large-scale deformation and uplift of the crust. It also may include the capture of migrating terranes and cementation of them to the continental margins, and the intrusion of granitic magmas to form plutons. These granite masses often are exposed by erosion following uplift. Uplift is the final act of the orogenic cycle. Earth's major chains of folded and faulted mountains, called *orogens,* correlate remarkably with the plate tectonics model.

No orogeny is a simple event; many involve previous developmental stages dating back millions of years, and the processes are ongoing today. Major orogens include the Rocky Mountains, produced

Figure 12-11
Example of a tilted fault block: the Teton Range in Wyoming. [Photo by Galen Rowell.]

during the Laramide orogeny (40–80 million years ago); the Sierra Nevada in the Sierra Nevadan orogeny (35 million years ago, with older batholithic intrusions dating back 130–160 million years); the Appalachians and the Ridge and Valley Province formed by the Alleghany orogeny (250–300 million years ago, preceded by at least two earlier orogenies); and the Alps of Europe in the Alpine orogeny (20–120 million years ago and continuing to the present, with many earlier episodes).

Examination of a world map reveals two large mountain chains. The relatively young mountains along the western margins of the North and South American plates stretch from the tip of Tierra del Fuego to the massive peaks of Alaska, forming the **Cordilleran system.** The mountains of southern Asia, China, and northern India continue in a belt through the upper Middle East to Europe and the European Alps, constituting the **Eurasian-Himalayan system** (see Figure 12-14).

Types of Orogenies. Figure 12-12 illustrates convergent plate-collision patterns associated with orogenesis. Shown in (a) is the *oceanic plate-continental plate collision* type of orogenesis. This is occurring along the Pacific coast of the Americas and has formed the Andes, the Sierra of Central America, the Rockies, and other western mountains. We see folded sedimentary formations and intrusions of magma forming granitic plutons at the heart of these mountains. Their buildup was augmented by accretion of displaced terranes, which were cemented during their collision with the continental mass. Also, note the associated volcanic activity inland from the subduction zone.

Shown in (b) is the *oceanic plate–oceanic plate collision,* where two portions of oceanic crust collide. Such collisions can produce either simple volcanic island arcs or more complex arcs like Japan that include deformation and metamorphism of rocks, and granitic intrusions. These processes have formed the chains of island arcs and volcanoes that continue from the southwestern Pacific to the western Pacific, the Philippines, the Kurils, and through portions of the Aleutians.

Both the *oceanic-continental* and *oceanic-oceanic* collision types are active around the Pacific Rim. Both are thermal in nature, because the diving plate melts and migrates back toward the surface as molten rock. This region around the Pacific is known as the **circum-Pacific belt** or, more popularly, the **ring of fire.**

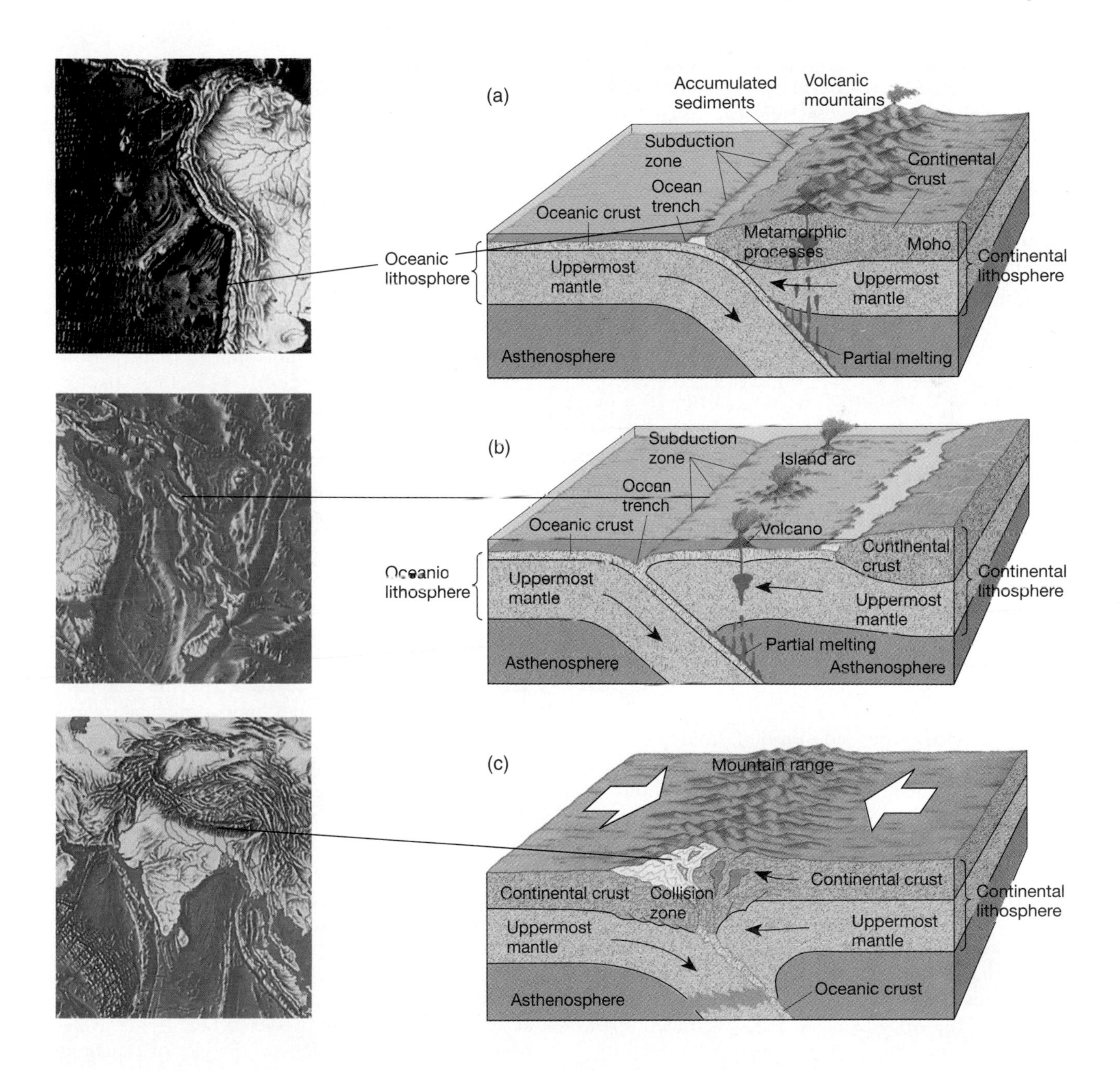

Figure 12-12
Types of plate convergence: (a) oceanic-continental (example: Nazca plate–South American plate collision and subduction); (b) oceanic-oceanic (example: New Hebrides Trench near Vanuatu, 16° S 168° W); and, (c) continental-continental (example: India plate and Eurasian landmass collision and resulting Himalayan Mountains). [Inset illustrations derived from *Floor of the Oceans* © 1980 by Marie Tharp. Reproduced by permission of Marie Tharp, 1 Washington Ave., South Nyack, NY 10960.]

Shown in (c) is the *continental plate–continental plate collision,* which occurs when two large continental masses collide. Here the orogenesis is quite mechanical; large masses of continental crust are subjected to intense folding, overthrusting, faulting, and uplifting. Deformation of shallow and deep marine sediments and basaltic oceanic crust is produced by crushing as the plates converge. As mentioned earlier, the collision of India with the Eurasian landmass produced the Himalayan Mountains. That collision is estimated to have shortened the overall continental crust by as much as 1000 km (620 mi), and to have produced sequences of thrust faults at depths of 40 km (25 mi). The disruption created by that collision has reached far under China, and frequent earthquakes there signal the continuation of this collision. The Himalayas feature the tallest above-sea-level mountains on Earth, including Mount Everest at 8848 m (29,028 ft) elevation.

As orogenic belts increase in elevation, weathering and erosion processes work to reduce the mountains. The mountain mass is in continual isostatic adjustment as it builds and wears away (see Figure 11-5).

Appalachians. The old, eroded, fold-and-thrust belt of the eastern United States contrasts with the younger mountains of the western portions of North America. As noted, the **Alleghany orogeny** followed at least two earlier orogenic cycles of uplift and the accretion of several captured terranes. (In Europe this is called the Hercynian orogeny.) The original material for the Appalachians and Ridge and Valley Province resulted from the collisions that produced Pangaea. In fact, the Atlas Mountains of northwest Africa were originally connected to the Appalachians, but have rafted apart, embedded in their own tectonic plates. Similarly, folded and faulted rock structures in the British Isles and Greenland demonstrate a past relationship with the Appalachians.

The linear folds of the Appalachian system are well displayed on the satellite image, topographic map, and digitized relief map in Figure 12-13. These dissected ridges are cut through by rivers, forming *water gaps* that greatly influenced migration and settlement patterns during the early days of the United States. The initial flow of people, goods, and ideas was guided by this topography.

World Structural Regions

A brief examination of the distribution of world structural regions helps summarize the information presented about the three rock-forming processes (igneous, sedimentary, metamorphic), plate tectonics, landform construction, and orogenesis. Figure 12-14a defines seven essential structural regions, highlights Earth's major mountain systems, and allows interesting comparison with the chapter-opening illustration of the ocean floor. The Cordilleran system is easily identified on the composite satellite image of the Western Hemisphere (Figure 12-14b).

EARTHQUAKES

Crustal plates do not move smoothly past one another. Instead, stress causes strain in the rocks along the plate boundaries until the sides release and lurch into new positions. In the Liaoning Province of northeastern China, ominous indications of possible tectonic activity began in 1970. Foreboding symptoms included land uplift and tilting, increased numbers of minor tremors, and changes in the region's magnetic field—all of this after almost 120 years of quiet.

These precursors of coming tectonic events continued for almost five years before Chinese scientists took the bold step of forecasting an earthquake. Finally, on February 4, 1975, at 2:00 P.M., some 3 million people were evacuated in what turned out to be a timely manner; the quake struck at 7:36 P.M., within the predicted time frame. Ninety percent of the buildings in the city of Haicheng were destroyed, but thousands of lives were saved, and success was proclaimed—an earthquake had been forecasted and preparatory action taken, for the first time in history.

Only 17 months later, at Tangshan in the northeastern province of Hebei (Hopei), an earthquake occurred on July 28, 1976, without warning. No preliminary activity occurred from which a forecast could be prepared. Consequently, this quake killed

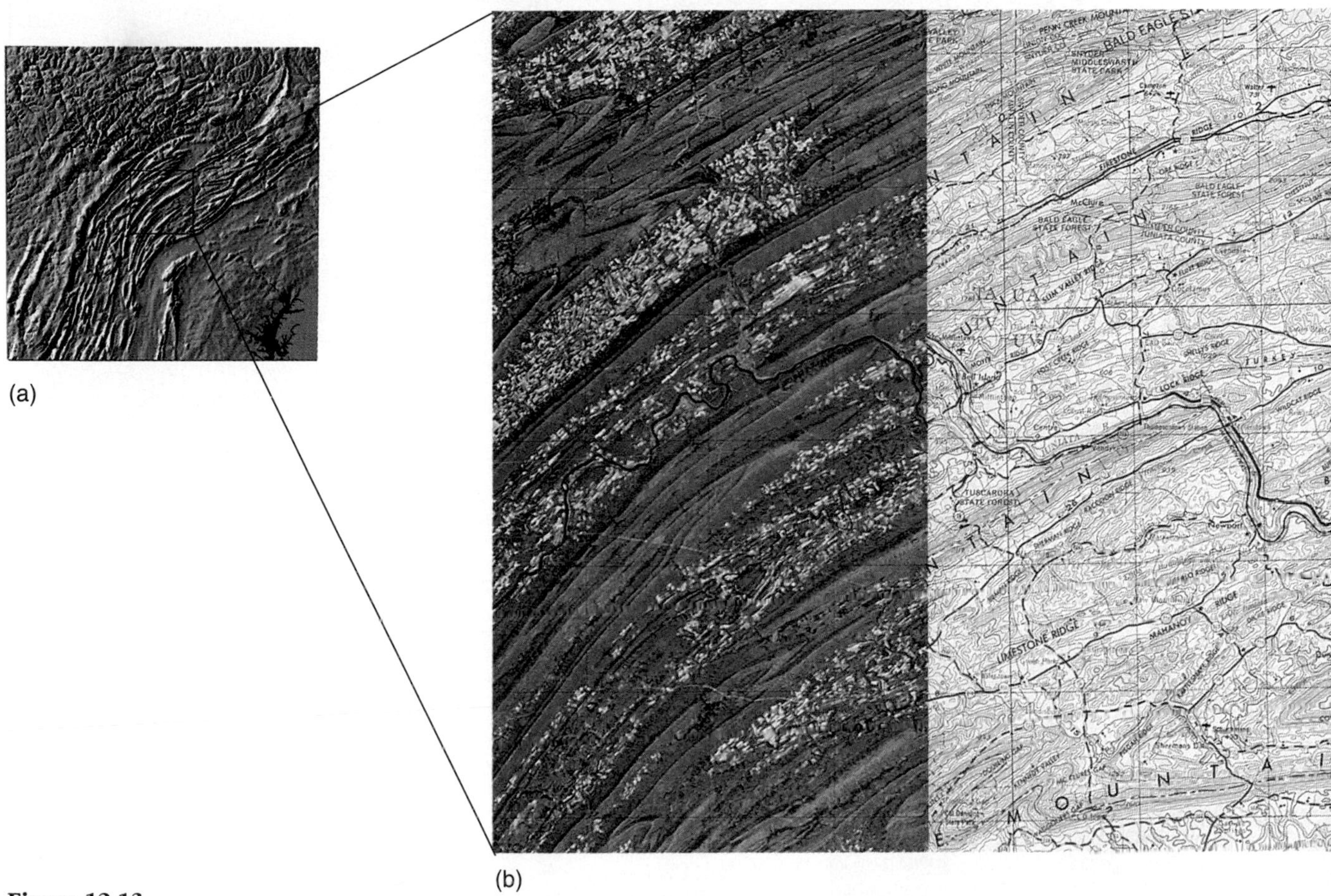

Figure 12-13
Appalachian Mountain region of central Pennsylvania shown on (a) a digital shaded-relief map originally prepared at a scale of 1:3,500,000 and (b) split *Landsat* image and U.S. Geological Survey topographic quadrangle map (original at 1:250,000 scale). The highly folded nature of this entire region is clearly visible on these small and intermediate scale illustrations. [(a) USGS digital terrain map I-2206, 1992; (b) U.S. Geological Survey and NASA.]

a quarter of a million people! It also destroyed 95% of the buildings and 80% of the industrial structures and severely damaged more than half the bridges and highways. The jolt was strong enough to throw people against the ceilings of their homes. An old, undetected fault had ruptured and offset 1.5 m (5 ft) along an 8 km (5 mi) stretch through Tangshan, devastating large areas just 145 km (90 mi) southeast of Beijing, China's capital city. How is it possible that these two tectonic events produced such different human consequences?

Earthquake Essentials

Earthquakes associated with faulting are referred to as tectonic earthquakes. Their vibrations are transmitted as waves of energy throughout Earth's interior and are detected with a **seismograph,** an instrument that records vibrations in the crust. Earthquakes are rated on two different scales, an intensity scale and a magnitude scale.

Intensity scales are useful in classifying and describing damage to terrain and structures following an earthquake. Earthquake intensity is rated on the

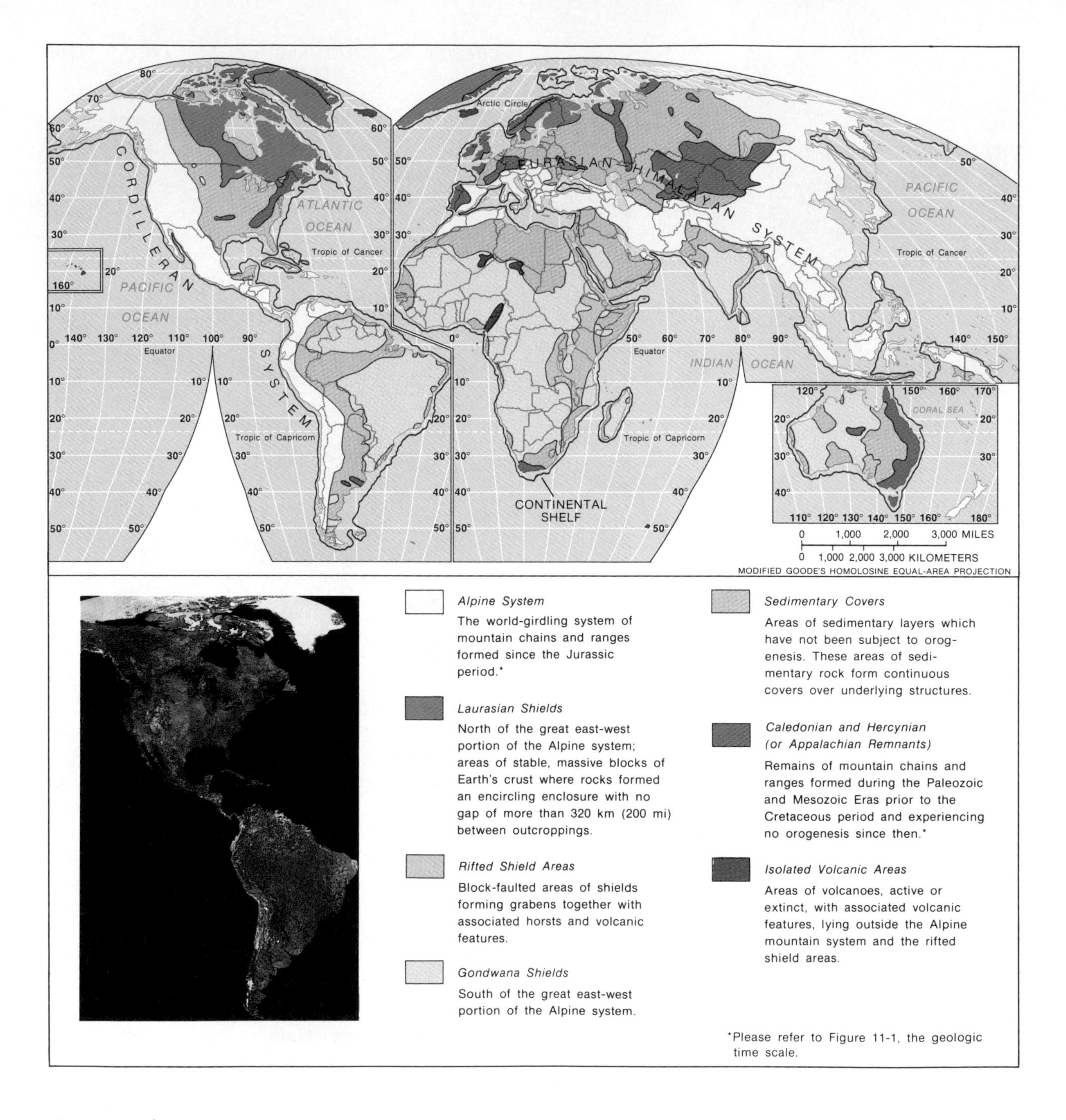

Figure 12-14
World structural regions and major mountain systems. Because each structural region in this figure includes related landforms adjacent to the central feature of the region, some of the regions appear larger than the structures themselves. Structural regions in the Western Hemisphere are visible on this composite Landsat image inset. [After Richard E. Murphy, "Landforms of the World," Map Supplement No. 9, *Annals of the Association of American Geographers* 58, no. 1 (March 1968). Adapted by permission. Inset image courtesy of EROS Data Center and the National Geographic Society.]

TABLE 12-1
Earthquake Characteristics and Frequency Expected Each Year

Characteristic Effects in Populated Areas	Approximate Intensity (modified Mercalli scale)	Approximate Magnitude (Richter scale)	Number per Year
Nearly total damage	XII	>8.0	1 every few years
Great damage	X–XI	7–7.9	18
Considerable-to-serious damage to buildings; railroad tracks bent	VIII–IX	6–6.9	120
Felt-by-all, with slight damage to buildings	V–VII	5–5.9	800
Felt-by-some to felt-by-many	III–IV	4–4.9	6200
Not felt, but recorded	I–II	2–3.9	500,000

SOURCE: Charles F. Richter, *Elementary Seismology,* Freeman, 1958, and others.

arbitrary *Mercalli scale,* a Roman numeral scale from I to XII representing "barely felt" to "catastrophic total destruction." It was designed in 1902 and modified in 1931 to be more applicable to conditions in North America (Table 12-1).

Earthquake *magnitude* is estimated according to a system designed by Charles Richter in 1935. First, the amplitude of seismic waves is recorded on a seismograph located at least 100 km (62 mi) from the epicenter of the quake. That measurement is then charted on the **Richter scale** (Table 12-1). The scale is open ended and logarithmic; that is, each whole number on the scale represents a 10-fold increase in the measured wave amplitude. Translated into energy, each whole number demonstrates a 31.5-fold increase in the amount of energy released. Thus, a 3.0 on the Richter scale represents 31.5 times more energy than a 2.0 and 992 times more energy than a 1.0. It is difficult to imagine the power released by the Tangshan quake, which was rated a 7.8 on the Richter scale.

Table 12-1 presents the characteristics and frequency of earthquakes that are expected annually worldwide. Table 12-2 lists a sampling of significant earthquakes. Note each earthquake's location, magnitude, and intensity rating (if available), and total loss of life (if known). Note that the table includes more earthquakes in the period following 1960. This is not because of an increase in overall frequency; rather, it reflects an effort to include recent events affecting increased population densities in vulnerable areas.

The subsurface area along a fault plane, where the motion of seismic waves is initiated, is called the *focus,* or hypocenter (see Figure 12-16). The area at the surface directly above this subsurface location is the *epicenter.* Shock waves produced by an earthquake radiate outward from both the focus and epicenter areas. An *aftershock* may occur after the main shock, sharing the same general area of the epicenter. A *foreshock* also is possible, preceding the main shock. (Before the June 1992 Landers earthquake in southern California, at least two dozen foreshocks occurred along that portion of the fault.) The greater the distance from an epicenter, the less severe the shock that is experienced. At a distance of 40 km (25 mi), only 1/10 of the full effect of an earthquake normally is felt. However, if the ground is unstable, distant effects can be magnified, as they were in Mexico in 1985 and in San Francisco in 1989.

Mexico City, currently the world's second-largest city, is positioned on the soft, moist sediments of an ancient lake bed. On September 19, 1985, during rush-hour traffic, two major earthquakes (8.1 and 7.6 on the Richter scale) struck 400 km (250 mi) southwest of Mexico City. The epicenter was on the sea floor off Mexico and Central America, and the old lake bed beneath Mexico City magnified the shock waves by more

TABLE 12-2
A Sampling of Significant Earthquakes

Year	Date	Location	Deaths	Intensity	Magnitude
1556	Jan. 23	Shensi, China	830,000	—	—
1737	Oct. 11	Calcutta, India	300,000	—	—
1811	Dec. 16	New Madrid, Missouri	—	—	—
1812	Jan. 23	New Madrid, Missouri	—	—	—
1812	Feb. 7	New Madrid, Missouri	Several	XI–XII	—
1857	Jan. 9	Fort Tejon, California	—	X–XI	—
1865	Oct. 8	San Francisco, California	—	IX	—
1870	Oct. 21	Montreal to Quebec, Canada	—	IX	—
1886	Aug. 31	Charleston, South Carolina	—	IX	6.7
1906	Apr. 18	San Francisco, California	3000	XI	8.25
1923	Sept. 1	Kwanto, Japan	143,000	XII	8.2
1939	Dec. 27	Erzincan, Turkey	40,000	XII	8.0
1960	May 21–30	Southern Chile	5700	—	8.5
1964	Mar. 28	Southern Alaska	131	X–XII	8.6
1970	May 31	Northern Peru	66,000	—	7.8
1971	Feb. 9	San Fernando, California	65	VII–IX	6.5
1972	Dec. 23	Managua, Nicaragua	5000	X–XII	6.2
1975	Feb. 4	Liaoning Province, China	Few	X	7.4
1976	Jul. 28	Tangshan, China	250,000	XI–XII	7.6
1978	Sept 16	Iran	25,000	X–XII	7.7
1985	Sept. 19	Mexico City, Mexico	7000	IX–XII	8.1
1988	Dec. 7	Armenia–Turkey border	30,000	XII	6.9
1989	Oct. 17	Loma Prieta (near Santa Cruz, California)	67	VII–IX	7.1
1991	Apr. 22	Puerto Limón, Costa Rica	52	IX	7.4
1991	Oct. 20	Uttar Pradesh, India	1700	IX–XI	6.1
1992	June 28	Landers and Big Bear, California	1	VII	7.5, 6.5

than 500%. As a result, 250 buildings collapsed, some 8000 structures experienced damage, and about 7000 people perished.

The Nature of Faulting. We earlier described specific types of faults and faulting motions. How a fault actually breaks is still under investigation, but the basic process is described by the **elastic-rebound theory.** Generally, two sides along a fault appear to be locked by friction, resisting any movement despite the powerful motions of adjoining pieces of crust. This stress continues to build strain along the fault surfaces, storing elastic energy like a wound-up spring. When movement finally does occur as the strain buildup exceeds the frictional lock, energy is released abruptly, returning both sides of the fault to a condition of less strain.

Think of the fault plane as a surface with irregularities that act as sticking points, preventing movement, similar to two pieces of wood held together by drops of glue rather than an even coating of glue. Research scientists at the USGS and the University of California have identified these small areas of high strain as **asperities.** They are the points that break and release the sides of the fault.

If the fracture along the fault line is isolated to a small asperity break, the quake will be small in magnitude. However, if the break involves the release of strain along several asperities, the quake will be greater in extent and will involve the shifting of massive amounts of crust. Clearly, as some asperities break (perhaps recorded as small foreshocks), the strain is increased on surrounding asperities. Thus, small earthquakes in an area may be precursors to a major quake.

The San Francisco Earthquakes. In 1906 an earthquake devastated San Francisco, a city of 400,000 people at the time, and three-quarters of a million today. Faulting was evident over 435 km (270 mi), and this prompted intensive research to discover the nature of faulting. (Realize that this occurred six years before Wegener's continental drift hypothesis was proposed.) The elastic-rebound theory developed as a result, and today quakes such as the one near San Francisco in 1989 can be interpreted.

The fault system involved was the San Andreas, which provides a good example of the evolution of a spreading center overridden by an advancing continental plate (Figure 12-15). As shown in the figure, the East Pacific Rise developed as a spreading center with associated transform faults (a), while the North American plate was progressing westward after the breakup of Pangaea. Forces then shifted the transform faults toward a northwest-southeast alignment along a weaving axis (b). Finally, the western margin of North America overrode those shifting transform faults (c). In relative terms, the motion along the fault is right-lateral, whereas in absolute terms the North American plate is still moving westward.

The earthquake that disrupted the 1989 World Series, mentioned at the outset of this chapter, involved a portion of the San Andreas fault approximately 16 km (9.9 mi) east of Santa Cruz and 95 km (59 mi) south of San Francisco (Figure 12-16a). The quake occurred at 5:04 P.M. Pacific Standard Time and registered 7.1 on the Richter scale. A fault had ruptured at a focus unusually deep for the San Andreas system—more than 18 km (11.5 mi) below the surface.

Unlike previous earthquakes—such as the one in 1906, when the plates shifted a maximum of 6.4 m (21 ft) relative to each other—there was no evidence of plate movement at the surface. Instead, the fault plane solution suggested in Figure 12-16a shows the two plates moving approximately 2 m (6 ft) past each other deep below the surface, with the Pacific plate thrusting 1.3 m (4.3 ft) upward. This is an unusual motion for the San Andreas fault, especially when compared to the apparent motion in 1906, and perhaps is a clue that this fault is more complex than previously thought. In only 15 seconds, more than 2 km (1.2 mi) of freeway overpass and a section of the San Francisco–Oakland Bay Bridge collapsed, $8 billion in damage was generated, 14,000 people were displaced from their homes, 4000 were injured, and 67 people were killed (Figure 12–16b).

The damage from this quake was extreme in the region of the Loma Prieta epicenter and in the San Francisco Bay Area, where houses and freeways are built on fill and bay mud. As in Mexico City, unstable soils magnified the effects of a distant quake. Scientists now think that seismic waves were reflected off the Moho crust-mantle boundary from the focal point of the quake. These reflected waves were at such an angle that both distant cities (San Francisco and Mexico City) were damaged.

One potentially positive discovery from this disaster was that Stanford University scientists reported unusually great changes in Earth's magnetic field—about 30 times more than normal—three hours before the main shock. These measurements were made 7 km (4.3 mi) from the epicenter, raising hopes that short-term warnings might be possible in the future.

The Southern California Earthquakes of 1992—and the Future. Since the mid-1980s the Coachella Valley area of California, about 193 km (120 mi) east of Los Angeles, experienced seven earthquakes greater than 6 on the Richter scale. A point near Joshua Tree, a small town north of the Coachella Valley, was hit by a 6.1 magnitude quake on April 22, 1992. Many small aftershocks rippled along the faultline between Coachella and Joshua Tree.

Two months later, on June 28, 1992, a 7.5-magnitude earthquake rocked the relatively unpopulated area near Landers, California, just north of the epicenters for the earlier quakes (see the inset map of southern California in Figure 12-15c). This event was the single largest quake in California in 30 years. However, because of the remote location, both injury and damage were low. Involved were four different faults, some segments unknown until this event, that ruptured along 69 km (43 mi) with right-lateral displacement of 6.1 m (20 ft)—comparable to the 1906 earthquake that devastated San Francisco.

Three hours passed; then a 6.5 magnitude earthquake hit near Big Bear City 60 km (37 mi) to the northwest of Landers. Aftershocks were almost continuous between these epicenters. For yet un-

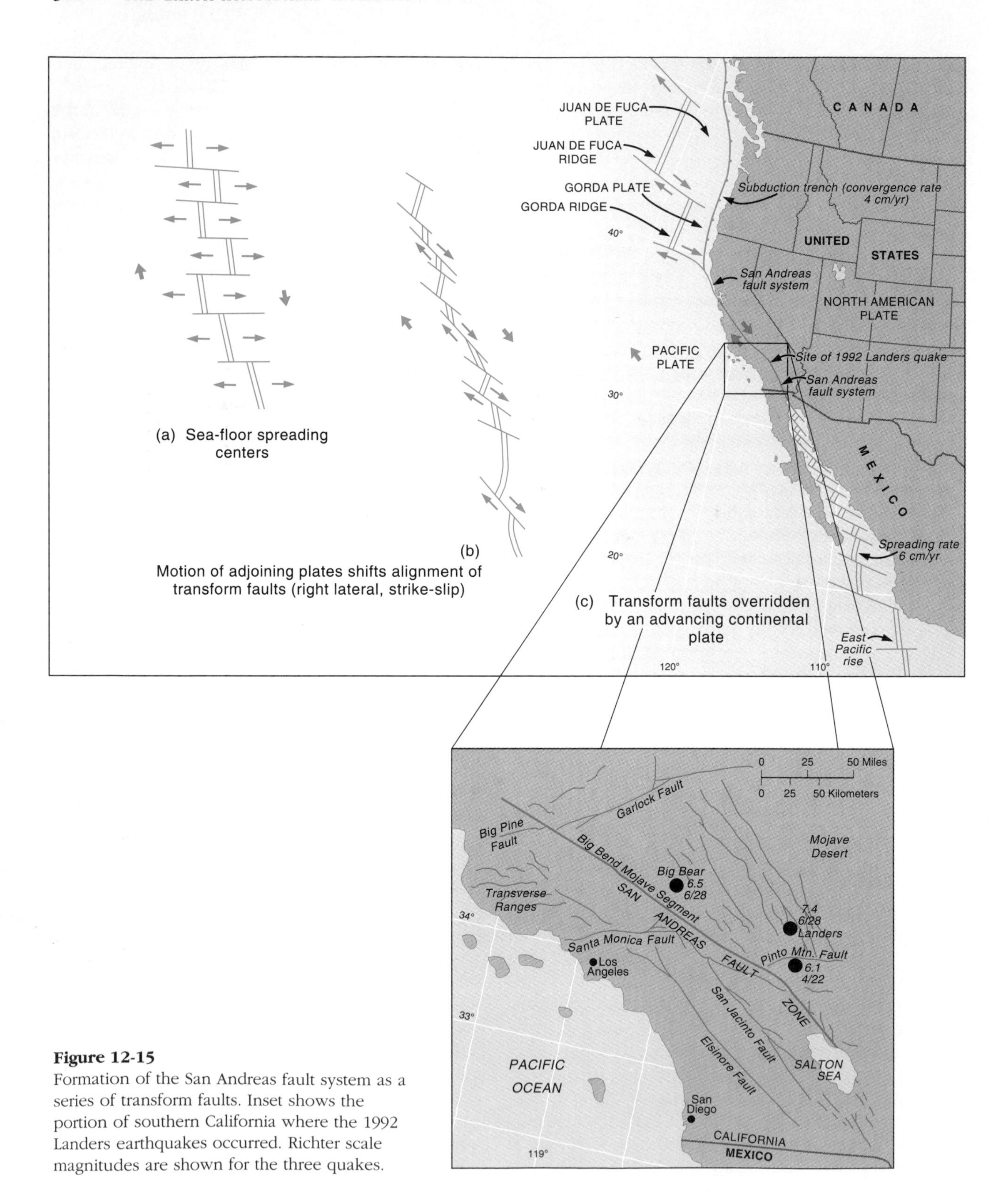

Figure 12-15
Formation of the San Andreas fault system as a series of transform faults. Inset shows the portion of southern California where the 1992 Landers earthquakes occurred. Richter scale magnitudes are shown for the three quakes.

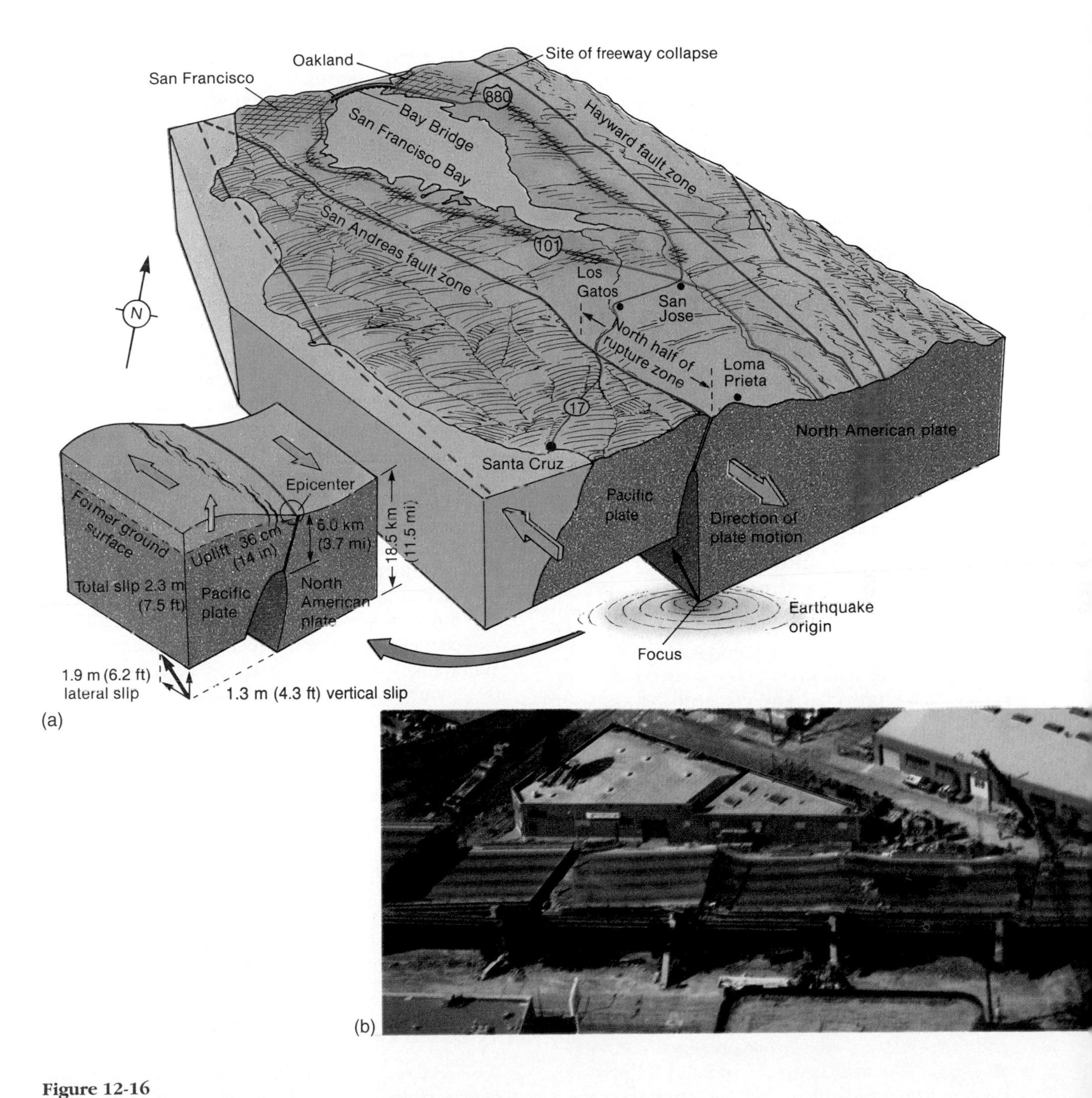

Figure 12-16
The 1989 Loma Prieta, California, earthquake: (a) fault-plane diagram of the lateral and thrust movement at depth for this earthquake; (b) Route 880 Cypress Freeway collapse. [(a) After "The Loma Prieta Earthquake of October 17, 1989," U.S. Geological Survey pamphlet by Peter J. Ward and Robert A. Page. Washington, DC: November 1989, p. 1. (b) Photo courtesy of California Department of Transportation.]

explained reasons, related earthquakes were recorded over the next few weeks in Mammoth Lakes 645 km (400 mi) to the north, Mount Shasta in northern California, southern Nevada and Utah, and 1810 km (1125 mi) distant in Yellowstone National Park, Wyoming!

USGS scientists now think that all the southern California earthquakes are related and that the earlier quakes were precursors to the June 28 events. Portions of the San Andreas' system now appear clamped between two activated faults, producing strain along a segment of plate boundary. The occurrence of these 1992 quakes has increased strain in the region and not relieved it. The region should anticipate a major earthquake (a 7.5 Richter rating or higher), with potentially devastating consequences. Scientists want to determine if these events are building up to the proverbial "big one" for southern California.

Paleoseismology and Earthquake Forecasting

Principal zones of earthquake activity occur near plate boundaries, as the shaded areas in Figure 11-19 demonstrate. The correlation is such that earthquakes have provided key diagnostic evidence in understanding plate tectonics, and in turn, that correlation offers a long-term forecast of earthquake sites. The challenge is to discover how to predict the *specific time and place* for a quake in the short term.

One approach is to examine the history of each plate boundary and determine the frequency of earthquakes in the past, a study called **paleoseismology.** Paleoseismologists then construct maps that provide an estimate of expected earthquake activity. Areas that are quiet and overdue for an earthquake are termed **seismic gaps;** such an area forms a gap in the earthquake occurrence record and is therefore a place that possesses accumulated strain. The area along the Aleutian Trench subduction zone had three such gaps until the great 1964 Alaskan earthquake (8.6 on the Richter scale) filled one of them.

The areas around San Francisco and northeast of Los Angeles represent other such gaps where the fault system appears to be locked by friction and is accumulating strain. The 1989 Loma Prieta earthquake was predicted in 1988 by the U.S. Geological Survey as having a 30% chance of occurring with a 6.5 magnitude, within 30 years. The actual quake filled a portion of the seismic gap in that region.

Except for isolated successes in China, Japan, and the United States, earthquake prediction and forecasting has proven to be a much more difficult task than scientists originally imagined. Prediction is thought to be a matter of assessing earthquake *precursors*. Scientists still are trying to assemble an accurate model of the microevents that precede an earthquake, including the detection of small asperity breaks. The complexity of the interactions along a fault continue to challenge science.

Earthquake Preparedness and Planning

Actual implementation of an action plan to reduce deaths, injuries, and property damage from earthquakes is very difficult. The political environment adds complexity; sadly, the impact of an accurate earthquake prediction would be viewed as a negative threat to a region's economy (Figure 12-17). If we examine the potential socioeconomic impact of earthquake prediction on an urban community, it is difficult to imagine a chamber of commerce, bank, real estate agent, tax assessor, or politician who would privately welcome such a prediction. Long-range planning is a complex subject.

A valid and applicable generalization seems to be that humans are unable or unwilling to perceive hazards in a familiar environment. Such an axiom of human behavior certainly helps explain why large populations continue to live and work in earthquake-prone settings in developed countries. Similar questions also can be raised about populations in areas vulnerable to other disasters.

VOLCANISM

We are reminded of Earth's internal energy by the recent violent eruptions of Mount Pinatubo and Mount Mayon (Philippines, 1991 and 1993 respectively), Mount Unzen (Japan, 1991), Mount Hudson (Chile, 1992), Mount Spurr and Mount Redoubt (Alaska, 1992), and Galeras volcano (Colombia, 1993). A **volcano** forms at the end of a central vent or pipe that rises from the asthenosphere

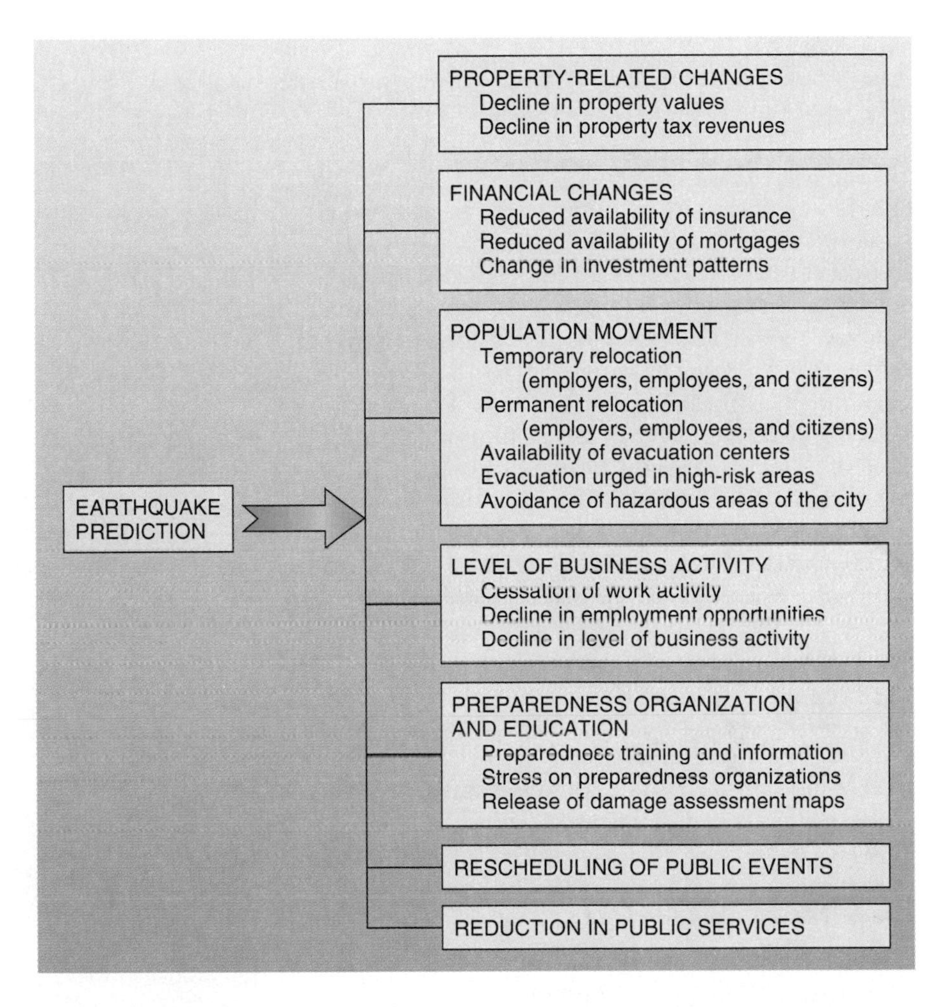

Figure 12-17
Socioeconomic impacts and adjustment to an earthquake prediction. [After J. E. Haas and D. S. Mileti, *Socioeconomic Impact of Earthquake Prediction on Government, Business, and Community,* Boulder: University of Colorado. Institute of Behavioral Sciences, 1976. Used by permission.]

through the crust into a volcanic mountain. It usually forms a **crater,** or circular surface depression at the summit. Magma rises and collects in a magma chamber deep below the volcano until conditions are right for an eruption.

This subsurface magma produces tremendous heat and in some areas it boils groundwater, as seen in the thermal features of Yellowstone National Park. The steam given off is a potential energy source if it is accessible. Such **geothermal energy** has provided heating and electricity for more than 90 years in Iceland, New Zealand, and Italy. North of San Francisco, some 1300 megawatts of electrical capacity are generated by steam from geothermal wells that is fed through turbines.

The magma that actually issues from the volcano is termed **lava** (molten rock). It can occur in many different textures and forms, which accounts for the varied behavior of volcanoes and the different landforms they build. In this section we will look at five related landforms and their origins: cinder cones, calderas, composite volcanoes, shield volcanoes, and plateau basalts.

Lava, gases, and **tephra** (pulverized rock and clastic materials ejected violently during an eruption) pass through the vent to the surface and build a volcanic landform. One such landform is a **cinder cone,** a small cone-shaped hill usually less than 450 m (1500 ft) high, with a truncated top formed from cinders that accumulate during moderately explosive eruptions. Cinder cones are made of tephra and scoria (a vesicular, cindery rock), as exemplified by Paricutín in southwestern Mexico. Another distinctive landform is a large basin-

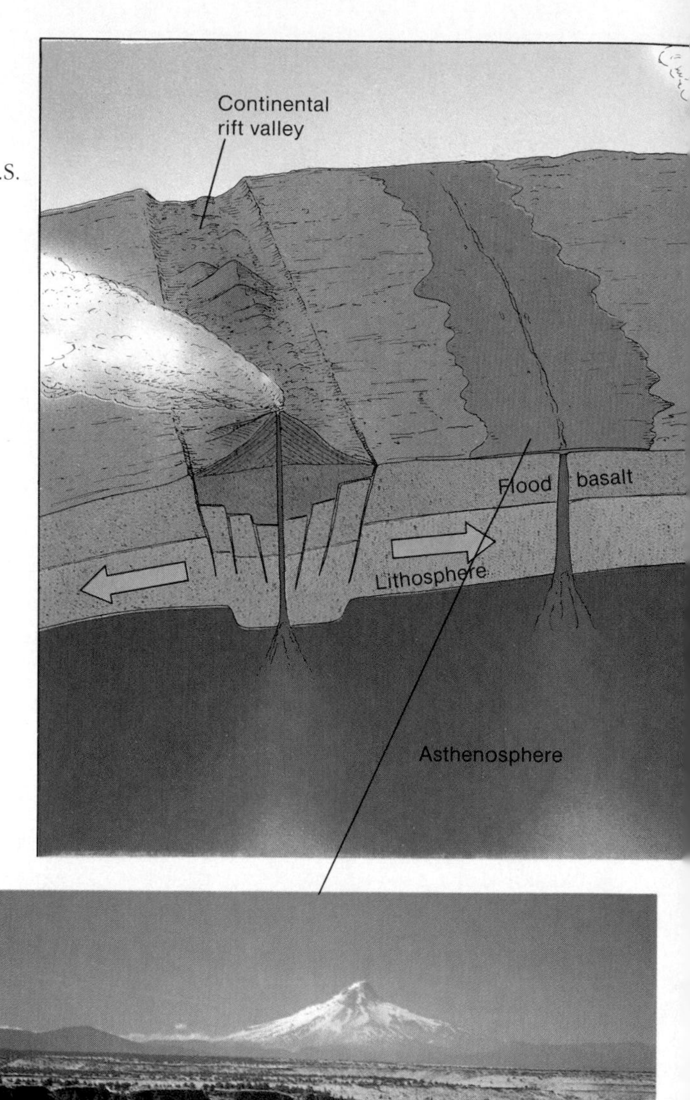

Figure 12-18
Principal mechanisms for each type of volcanic activity. [After U.S. Geological Survey, *The Dynamic Planet,* 1989.]

shaped depression called a **caldera** (Spanish for "kettle"). It forms when summit material on a volcanic mountain collapses inward after an eruption or other loss of magma, forming a caldera such as Crater Lake in southern Oregon (see Figure 12-25).

Over 1300 volcanoes exist on Earth, although fewer than 600 are active (have had at least one eruption in recorded history). North America has about 70 (mostly inactive) along the western margin of the continent, Mount Saint Helens in Washington State being a famous example. In an average year about 50 volcanoes erupt worldwide, varying from modest activity to major explosions. Eruptions in remote locations and at depths on the sea floor go largely unnoticed, but the occasional eruption of great magnitude near a population center makes headlines. More than 200,000 people have been killed from volcanic eruptions in the past 400 years.

Volcanoes also produce some benefits. These include materials that contribute to fertile soils, as in Hawaii; conditions that can provide geothermal energy; and even new real estate added to Iceland, Japan, Hawaii, and elsewhere.

Locations of Volcanic Activity

The location of volcanic mountains on Earth is a function of plate tectonics and hot-spot activity. Volcanic activity occurs in three areas: (1) along subduction boundaries at continental plate–oceanic plate convergence (Mount Saint Helens) or oceanic plate–oceanic plate convergence (volcanoes in the Philippines and Japan); (2) along sea-floor spreading centers on the ocean floor (Iceland) and areas of rifting on continental plates (the rift zone in east Africa); and (3) at hot spots, where individual plumes of magma rise through the crust (Hawaii). Figure 12-18 illustrates these three types of locations, which you can compare to the active volcano sites and plate boundaries shown in Figure 11-19. Figures 12-19 to 12-22 are included in Figure 12-18 to illustrate various aspects of volcanic activity.

Figure 12-19
Plateau basalts (in foreground) characteristic of the Columbia Plateau in Oregon. Mount Hood in the background is one of the Cascade Range volcanoes, south of Mount Saint Helens, formed from different magmas. [Photo by author.]

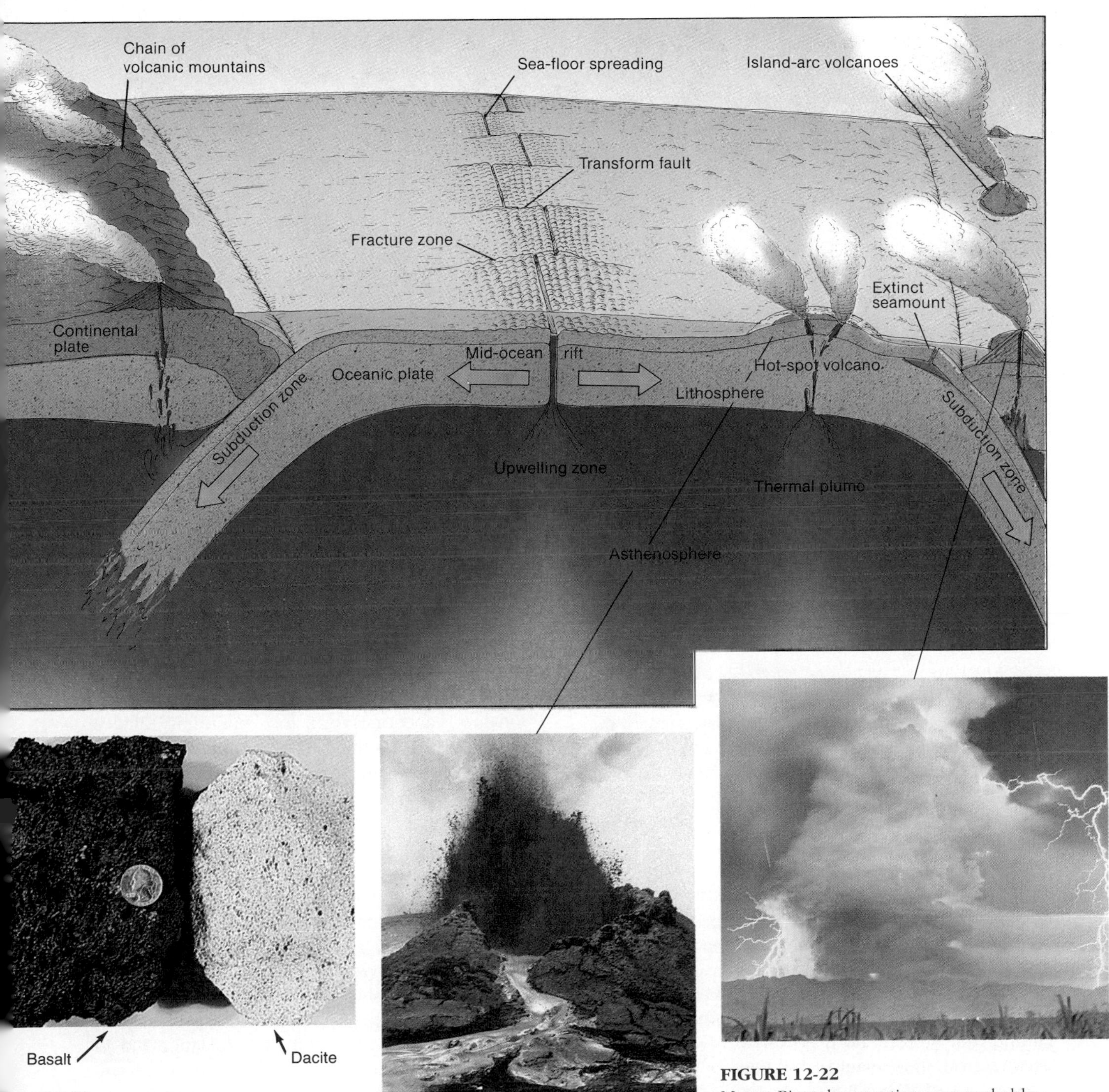

ure 12-20
alt from Hawaii and dacite
ı Mount Saint Helens.
otograph by J. D.
ggs/U.S. Geological Survey.]

Figure 12-21
Kilauea fountain eruption from East Riff spatter cone in Hawaii. [Photo by Kepa Maly/U.S.G.S.]

FIGURE 12-22
Mount Pinatubo eruption was probably the largest this century. Free electrons in the ash cloud generate lightning, adding drama to the explosions. [Photo by Reuters/Bettmann.]

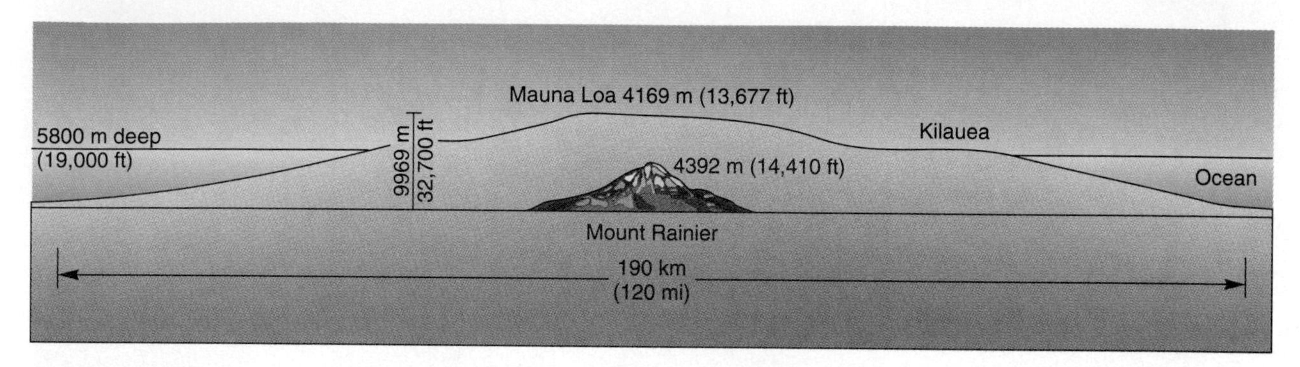

Figure 12-23
Comparison of Mauna Loa and Mount Rainier. [After U.S. Geological Survey, *Eruption of Hawaiian Volcanoes*. Washington, DC: Government Printing Office, 1986.]

Types of Volcanic Activity

The variety of forms among volcanoes makes them hard to classify; most fall in transition areas between one type and another. Even during a single eruption, a volcano may behave in several different ways. The primary factors in determining an eruption type are (1) the chemistry of its magma, which is related to the magma's source, and (2) its viscosity. Viscosity is magma's resistance to flow, or degree of fluidity; it ranges from low viscosity (very fluid) to high viscosity (thick and flowing slowly). We will consider two types of eruptions—effusive and explosive—and the characteristic landforms they build.

Effusive Eruptions. These are the relatively gentle eruptions that produce enormous volumes of lava annually on the sea floor and in places like Hawaii. Direct eruptions from the asthenosphere produce a low-viscosity magma which is very fluid and yields a dark, basaltic rock (less than 50% silica and rich in iron and magnesium). Gases readily escape from this magma because of its texture, causing an **effusive eruption** that pours out on the surface, with relatively small explosions and little tephra. However, dramatic fountains of basaltic lava sometimes shoot upward, powered by jets of rapidly expanding gases (Figure 12-21).

An effusive eruption may come from a single vent or from a flank eruption through side vents in surrounding slopes. If such vents form a linear opening, they are called *fissures;* these sometimes create a dramatic curtain of fire during eruptions. In Iceland, active fissures are spread throughout the plateau landscape. In Hawaii, rift zones capable of erupting tend to converge on the central crater, or vent. The interior of such a crater is often a sunken caldera, which may fill with low-viscosity magma during an eruption, forming a molten lake, which may then overflow lava downslope.

On the island of Hawaii the continuing Kilauea eruption represents the longest eruptive episode in Hawaiian history, active since 1823! During 1989–1990, lava flows from Kilauea actually consumed several visitor buildings in the Hawaii Volcanoes National Park and homes in Kalapana, a nearby town. Kilauea has produced more lava than any other vent on Earth.

A typical mountain landform built from effusive eruptions is gently sloped, gradually rising from the surrounding landscape to a summit crater, similar in outline to a shield of armor lying face up on the ground, and therefore called a **shield volcano.** The shield shape and size of Mauna Loa in Hawaii is distinctive when compared to Mount Rainier in Washington, which is a different type of volcano (we will explain shortly) and the largest in the Cascade Range (Figure 12-23). The height of the shield is the result of successive eruptions, flowing one on top of another. Mauna Loa is one of five shield volcanoes that make up the island of Hawaii, and it has taken at least 1 million years to accumulate its mass. Mauna Kea is slightly taller,

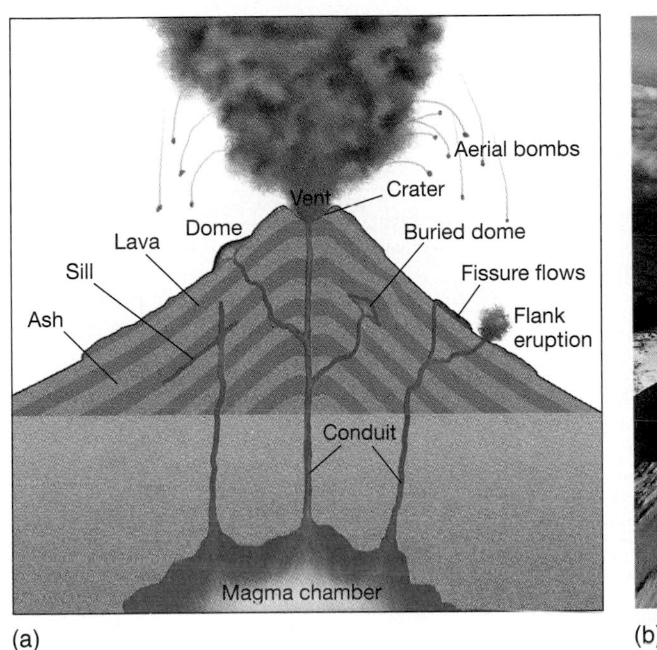

Figure 12-24
A typical composite volcano with its cone-shaped form (a). An eruption of Mount Shishaldin, Alaska, a composite volcano (b). [Photo courtesy of AeroMap U.S., Inc., Anchorage, Alaska.]

but Mauna Loa is the most massive single mountain on Earth.

In rifting areas, effusive eruptions send material out through elongated fissures, forming extensive sheets on the surface. The Columbia Plateau of the northwestern United States represents the eruption of **plateau basalts,** or flood basalts (Figure 12-19). These extensive sheet eruptions also episodically flow from the mid-ocean ridges, forming new sea floor in the process. Worldwide, the volume of material produced in effusive fissure eruptions along spreading centers is far greater than that of the continental eruptions. Iceland continues to form in this effusive manner, with upwelling basaltic flows creating new Icelandic real estate.

Explosive Eruptions. Volcanic activity along subduction zones produces the well-known explosive volcanoes. Magma produced by the melting of subducted oceanic plate and other materials is thicker (more viscous) than magma from effusive volcanoes; it is 50–75% silica and high in aluminum. Consequently, it tends to block the magma conduit inside the volcano, trapping and compressing gases, causing pressure to build and a possible **explosive eruption.** This magma forms a lighter dacitic rock at the surface (see *dacite* in Table 11-2), as illustrated in the comparison in Figure 12-20.

The term **composite volcano** is used to describe explosively formed mountains. (They are sometimes called *stratovolcanoes,* but because shield volcanoes also can exhibit a stratified structure, *composite* is the preferred term.) Composite volcanoes tend to have steep sides; they are more conical than shield volcanoes and therefore are also known as *composite cones.* If a single summit vent remains stationary, a remarkable symmetry may develop as the mountain grows in size, as demonstrated by Mount Orizaba in Mexico, Mount Shishaldin in Alaska, Mount Fuji in Japan, the pre-1980-eruption shape of Mount Saint Helens in Washington, and Mount Mayon in the Philippines (Figure 12-24).

TABLE 12-3
Composite Volcano Eruptions

Date	Name or Location	Number of Deaths	Amount extruded (mostly pyroclastics) in km^3 (mi^3)
Prehistoric	Yellowstone, Wyoming	Unknown	2400 (576)
4600 B.C.	Mount Mazama (Crater Lake, Oregon)	Unknown	50–70 (12–17)
1900 B.C.	Mount Saint Helens	Unknown	4 (0.95)
A.D. 79	Mount Vesuvius, Italy	20,000	3 (0.7)
1500	Mount Saint Helens	Unknown	1 (0.24)
1815	Tambora, Indonesia	66,000	80–100 (19–24)
1883	Krakatoa, indonesia	36,000	18 (4.3)
1902	Mont Pelée, Martinique	29,000	Unknown
1912	Mount Katmai, Alaska	Unknown	12 (2.9)
1943–1952	Paricutin, Mexico	0	1.3 (0.30)
1980	Mount Saint Helens	54	4 (0.95)
1985	Nevado del Ruiz, Colombia	23,000	1 (0.24)
1991	Mount Unzen, Japan	10	2 (0.5)
1991	Mount Pinatubo, Philippines	800	12 (3.0)
1992	Mount Spurr/Mount Shishaldin, Alaska	0	Current
1992	Mount Redoubt, Alaska	0	Current
1993	Galeras Volcano, Columbia	—	Current
1993	Mount Mayon, Philippines	—	Current

As the magma in a composite volcano forms plugs near the surface, those blocked-off passages build tremendous pressure, keeping the trapped gases compressed and liquified. When the blockage can no longer hold back this pressurized inferno, explosions equivalent to megatons of TNT blast the tops and sides off these mountains, producing much less lava but larger amounts of tephra, which includes volcanic ash (<2 mm in diameter), dust, cinders, lapilli (up to 32 mm in diameter), scoria (volcanic slag), pumice, and **pyroclastics** (explosively ejected rock pieces). Table 12-3 presents some notable composite volcano eruptions. Focus Study 12-1 details the much-publicized 1980 eruption of Mount Saint Helens.

Unlike the volcanoes in Hawaii Volcanoes National Park, where tourists gather at observation platforms to watch the relatively calm effusive eruptions, composite volcanoes do not invite close inspection and can explode with little warning:

- In A.D. 79, Mount Vesuvius buried the city of Pompeii, Italy, in three days of tephra and pyroclastic eruptions, even though the volcano had been dormant for centuries.
- In 1902 on the Caribbean island of Martinique, the beautiful port city of Saint-Pierre was devastated in a few minutes by an eruption of Mont Pelée. That eruption featured a lateral blast of incandescent ash and super-hot gases, known as a "glowing cloud" or **nuée ardente** in French. Despite months of precursory rumbling, ash, hot mudflows, and minor bursts from the mountain, politicians discouraged citizens from fleeing, and all but two of 30,000 people in the town perished.
- In 1883 in Indonesia, an entire group of islands centered on the volcano Krakatoa (between the big islands of Java and Sumatra) was obliterated in two days of explosions. The ash cloud reached into the mesosphere,

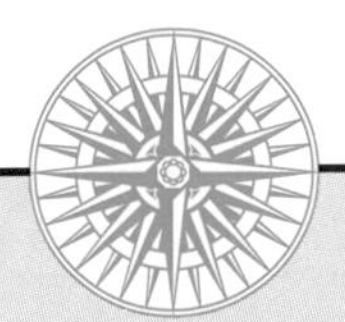

Focus Study 12-1 The 1980 Eruption of Mount Saint Helens

Probably the most studied and photographed composite volcano on Earth is Mount Saint Helens, located 70 km (45 mi) northeast of Portland, Oregon, and 130 km (80 mi) south of the Tacoma-Seattle area of Washington. Mount Saint Helens is the youngest and most active of the Cascade Range of volcanoes, which form a line from Mount Lassen in California to Mount Baker in Washington. The Cascade Range is the product of the Juan de Fuca sea-floor spreading center off the coast of northern California, Oregon, Washington, and British Columbia and the plate subduction that occurs offshore, as identified in Figure 12-15c.

Associated events began on March 20, 1980, with a sharp earthquake at 3:48 P.M. registering 4.1 on the Richter scale. The first eruptive outburst occurred one week later, beginning with a 4.5 Richter quake and continuing with a thick black plume of ash and the development of an initial crater about the size of a football field. April 1 marked the first harmonic tremor, or volcanic earthquake, from which it was inferred that magma was on the move within the mountain. *Harmonic tremors* are slow, steady vibrations, unlike the sharp releases of energy associated with earthquakes and faulting. A growing bulge on the north side of the mountain indicated the direction of the magma flow within the volcano. A bulge represents the greatest risk from a composite volcano, for it could mean a potential lateral burst across the landscape.

On Sunday, May 18, at 8:27 A.M., the area north of the mountain was rocked by a 5.0 Richter quake, the strongest to that date. The mountain, with its 245 m (800 ft) bulge, was shaken, but nothing else happened. Then a second quake (5.1) hit five minutes later, loosened the bulge, and launched the eruption. David Johnson, a volcanologist with the U.S. Geological Survey, was only 8 km (5 mi) from the mountain, servicing monitoring instruments, when he saw it begin. He radioed headquarters in Vancouver, Washington, saying, "Vancouver, Vancouver, this is it!" but he perished in the eruption that followed.

As the contents of the mountain exploded, a surge of hot gas (about 300°C or 570°F), steam-filled ash, pyroclastics, and a nuée ardente moved north-

FIGURE 1
Scorched earth and tree blowdown area, covering some 38,950 hectares (95,000 acres). [Photo by author.]

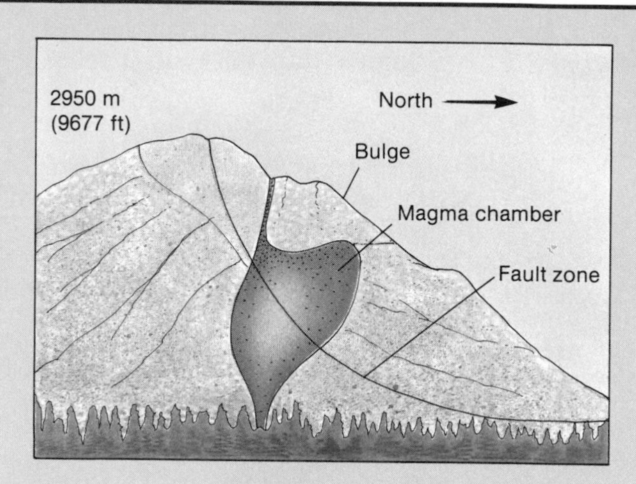
2950 m
(9677 ft)
North
Bulge
Magma chamber
Fault zone

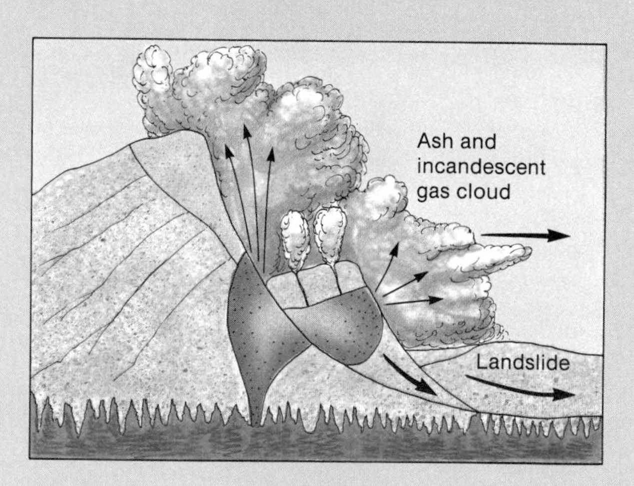
Ash and
incandescent
gas cloud
Landslide

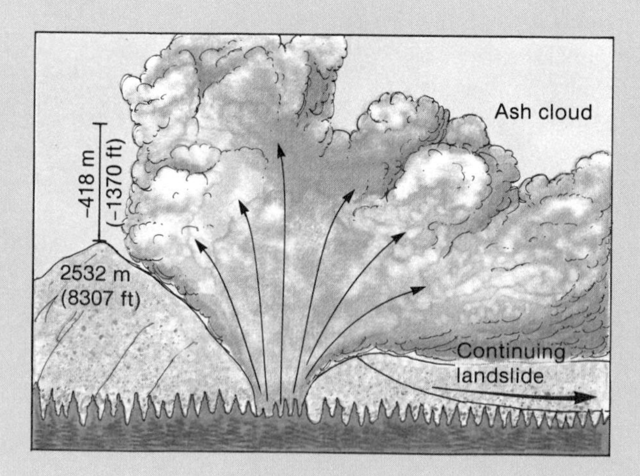
Ash cloud
−418 m
(−1370 ft)
2532 m
(8307 ft)
Continuing
landslide

FIGURE 2
The Mount Saint Helens eruption sequence and corresponding schematics. [Photo sequence by Keith Ronnholm.]

ward, hugging the ground and traveling at speeds up to 400 kmph (250 mph) for a distance of 28 km (17 mi) (Figure 1).

The slumping north face of the mountain produced the greatest landslide witnessed in recorded history; about 2.75 km^3 (0.67 mi^3) of rock, ice, and trapped air, all fluidized with steam, tumbled down at speeds approaching 250 kmph (155 mph). Landslide materials did not come to rest until they reached 21 km (13 mi) into the valley, blanketing the forest, covering a lake, and filling the rivers below. A series of photographs, taken at 10-second intervals from the east looking west, records this sequence (Figure 2). The eruption continued with intensity for nine hours, first clearing out old rock from the throat of the volcano and then blasting new material.

As destructive as such eruptions are, they also are constructive, for this is the way in which a volcano builds height. Before the eruption, Mount Saint Helens was 2950 m (9677 ft) tall; the eruption blew away 418 m (1370 ft). But today, Mount Saint Helens is growing by building a lava dome within the crater. The thick lava rapidly and repeatedly plugs and breaks in a series of dome eruptions that may continue for several decades. The dome already is over 300 m (1000 ft) high, so that a new mountain is being born from the eruption of the old. An ever resilient ecology is recovering as plants and animals alike reclaim the devastated area.

and the sounds of that August blast were heard in central Australia, the Philippines, and even 4800 km (3000 mi) away in the Indian Ocean! (The sound took some 4 hours to reach that listener, who noted the noise in his ship's log.) What was not blown into the sky fell back into the collapsed remains of a large caldera 275 m (900 ft) under the sea.

- Around 4500 B.C. in southern Oregon, a large caldera was left after Mount Mazama's violent eruption. However, that caldera was fed by snowmelt, rainwater, and subsurface water to form a lake 9.7 km (6 mi) in diameter and 610 m (2000 ft) deep—the beautiful Crater Lake (Figure 12-25). Later a cone formed a small island called Wizard Island in the middle of the lake.
- In June 1991, following 600 years of dormancy, Mount Pinatubo in the Philippines erupted. The summit of the 1460 m (4795 ft) volcano exploded, devastating many surrounding villages and permanently closing Clark Air Force Base, operated by the United States. Fortunately, scientists from the U.S. Geological Survey assisted local scientists in accurately predicting the eruption, allowing evacuation of the surrounding countryside and saving thousands of lives. Although volcanoes are regional events, their spatial implications are great, for their debris can spread worldwide. The single volcanic eruption of Mount Pinatubo was significant to the global environment and energy budget, as discussed in Chapters 3, 4, 5, and 10.

SUMMARY—Tectonics and Volcanism

Earth's surface is dramatically shaped by tectonic forces generated within the planet. Significant scientific and technological breakthroughs give scientists the ability to analyze tectonic and volcanic activity. The continents are formed as a result of upwelling material from below generated by plate

Figure 12-25
Crater Lake caldera and Wizard Island, Oregon, formerly Mount Mazama. [Photo by Ray Atkeson/American Landscapes.]

tectonics and enlarged through accretion of dispersed terranes. The crust is deformed by folding, broad warping, and faulting, which produce characteristic landforms. Earth's landscape features a great variety of relief and topography. Orogenesis, the process of mountain building, has produced Earth's major mountain belts—the Cordilleran system of the Americas and the Eurasian-Himalayan system that stretches from the Alps in Europe to the Himalayas in southern Asia.

Earthquakes generally occur along plate boundaries and cause disasters that affect the world's population. More is being learned all the time about the nature of faulting: stress and the buildup of strain, irregularities along fault plane surfaces, the way faults rupture, and the relationship among active faults. Earthquakes have claimed many lives and caused a great deal of damage through the centuries, so earthquake prediction and improved planning are active concerns of seismology, the study of earthquake waves and Earth's interior.

Volcanoes offer direct evidence of the makeup of the asthenosphere and uppermost mantle. Dramatic examples of volcanic activity have highlighted the first few years of the 1990s. Volcanoes are of two general types, based on the chemistry and the viscosity of the magma involved—effusive (like Kilauea in Hawaii) and explosive (like Mount Pinatubo in the Philippines). Volcanic activity has produced some destructive moments in history, but constantly creates new sea floor, land, and soils.

KEY TERMS

Alleghany orogeny
anticline
asperities
caldera
cinder cone
circum-Pacific belt
composite volcano
continental platforms
continental shield
Cordilleran system
crater
earthquake
effusive eruption
elastic-rebound theory
Eurasian-Himalayan system
explosive eruption
faulting
folding
geothermal energy
graben
horst
lava
nuée ardente
ocean basins
orogenesis
paleoseismology
plateau basalts
pyroclastics
relief
Richter scale
ring of fire
seismic gaps
seismograph
shield volcano
strike-slip fault
syncline
tephra
terranes
tilted fault block
topography
volcano
Wrangellia terranes

REVIEW QUESTIONS

1. How does the ocean floor map (chapter-opening illustration) bear the imprint of the principles of plate tectonics? Describe and explain.
2. Define "tectonic" and use the concept in two or three sentences.
3. What is a craton? Relate this structure to continental shields and platforms and describe these regions in North America.
4. What are the migrating terranes that came together to form continental masses? Give an example of such a terrane.
5. Diagram a folded landscape in cross section, and identify the features of the folding.
6. Define the four basic types of faults. How do these relate to earthquakes and seismic activity?
7. How did the Basin and Range Province evolve in the western United States? What other examples exist of this type of landscape?
8. Define orogenesis. What is meant by the birth of mountain chains?
9. Name some major orogenies.
10. Identify and detail Earth's two large mountain chains.
11. How are plate boundaries related to episodes of mountain building? Do different types of plate boundaries produce differing orogenic episodes?
12. Relate tectonic processes to the formation of the Appalachians and the Alleghany orogeny.

13. Describe the differences between the two earthquakes that occurred in China in 1975 and 1976.
14. Differentiate between the Mercalli and Richter scales. How are these used to describe an earthquake?
15. What is the relationship between an epicenter and the focus of an earthquake?
16. What local conditions in Mexico City and San Francisco severely magnified the energy felt from each quake?
17. How do the elastic-rebound theory and asperities help explain the nature of faulting?
18. Describe the San Andreas fault and its relationship to sea-floor spreading movements.
19. How does the seismic gap concept relate to expected earthquake occurrences? Have any gaps correlated with earthquake events in the recent past? Explain.
20. What do you see as the biggest barrier to effective earthquake prediction?
21. What is a volcano?
22. Where do you expect to find volcanic activity in the world? Why?
23. Compare effusive and explosive eruptions. Why are they different? What types of landforms are produced by each type?
24. Describe several recent volcanic eruptions.

SUGGESTED READINGS

Anderson, Don L. "The San Andreas Fault," *Scienitific American* 225 (November 1971): 52–66.

Bolt, Bruce A. *Earthquakes—A Primer.* San Francisco: W. H. Freeman, 1978.

Bullard, Fred M. *Volcanoes of the Earth.* Austin: University of Texas Press, 1976.

Decker, Robert W., Thomas L. Wright, and Peter H. Stauffer. *Volcanism in Hawaii.* 2 vols. U.S. Geological Survey Professional Paper 1350. Washington, DC: Government Printing Office, 1987.

Dvorak, J. J., Carl Johnson, and R. I. Tilling. "Dynamics of Kilauea Volcano," *Scientific American* 267, no. 2 (August 1992): 46–53.

Editors of Time-Life Books. *Volcano.* Planet Earth Series. Alexandria, VA: Time-Life Books, 1982.

Findley, Rowe. "Eruption of Mount St. Helens," *National Geographic* 159, no. 1 (January 1981): 3–65.

Francis, Peter, and Stephen Self. "Collapsing Volcanoes," *Scientific American* 256 (June 1987): 90–97.

Francis, Peter, and Stephen Self. "The Eruption of Krakatoa," *Scientific American* 249 (November 1983): 172–87.

Grove, Noel, and R. H. Ressmeyer. "Volcanoes—Crucibles of Creation," *National Geographic* 182, no. 6 (December 1992): 3–41.

Hill, R. I., I. H. Campbell, G. F. Davies, and R. W. Griffiths. "Mantle Plumes and Continental Tectonics," *Science* 256 (April 10, 1992): 186–93.

Hoblitt, Richard P. *Observations of the Eruptions of July 22 and August 7, 1980, at Mount St. Helens, Washington.* U.S. Geological Survey Professional Paper 1335. Washington, DC: Government Printing Office, 1986.

Keller, Edward A. *Environmental Geology,* 6th ed. Columbus, OH: Macmillan Publishing Company, 1992.

Macdonald, G. A., A. T. Abbott, and D. C. Cox. *Volcanoes in Hawaii.* Honolulu: University of Hawaii Press, 1983.

Maranto, Gina. "Inferno in Paradise—Double Eruption of Hawaii's Volcanoes," *Discover* 5, no. 6 (June 1984): 88–93.

McDowell, Bart. "When the Earth Moves," *National Geographic* 169, no. 5 (May 1986): 638–75.

Murphy, J. B., and R. D. Nance. "Mountain Belts and the Supercontinent Cycle," *Scientific American* 266, no. 4 (April 1992): 84–91.

Newell, Reginald E., and Adarsh Deepak, eds. *Mount St. Helens Eruptions of 1980*. Washington, DC: National Aeronautics and Space Administration, 1982.

Richter, Charles F. *Elementary Seismology*. San Francisco: W. H. Freeman, 1958.

Stein, R. S., and R. S. Yeats. "Hidden Earthquakes," *Scientific American* 260, no. 6 (June 1989): 48–57.

U.S. Geological Survey Staff. "The Loma Prieta, California, Earthquake: An Anticipated Event," *Science* 247 (January 19, 1990): 286–93.

Walker, Bryce. *Earthquake*. Planet Earth Series. Alexandria, VA: Time-Life Books, 1982.

Wallace, Robert E., ed. *The San Andreas Fault System, California*. U.S. Geological Survey Professional Paper 1515. Washington, DC: Government Printing Office, 1990.

Ward, Peter L., and Robert A. Page. *The Loma Prieta Earthquake of October 17, 1989*. Washington, DC. Government Printing Office, 1990.

Alabama Hills and Inyo Mountains of eastern California. [*Photo by author.*]

13

WEATHERING, KARST LANDSCAPES, AND MASS MOVEMENT

One of the benefits you receive from a physical geography course is an enhanced appreciation of the scenery. Whether you go on foot, or by car, train, or plane, travel is an opportunity to experience Earth's varied landscapes and witness the active processes that produce them. We now begin a five-chapter examination of various denudation processes that make up the exogenic system. This chapter examines weathering and mass movement of the lithosphere. The other four chapters look at river systems, wind-influenced landscapes, coastal processes and landforms, and glaciated regions.

We have discussed the internal processes of our planet and how the landforms of the crust are produced. However, as the landscape is formed, a variety of external processes simultaneously operate to wear it down. Perhaps you have noticed highways in mountainous and cold climates that appear rough and broken. Roads that experience freezing weather seem to pop up in chunks each winter and develop potholes. Or, maybe you have seen older marble structures such as tombstones, etched and dissolved by rainwater. Physical and chemical weathering processes similar to these are important to the overall reduction of the landscape and the release of essential minerals from bedrock.

Mass movement of surface materials rearranges landforms, providing often dramatic reminders of the power of nature. A news broadcast may bring you word of an avalanche in Colombia, mudflows in Los Angeles, a landslide in Turkey, or pyroclastic flows on the slopes of a volcano in Japan. The processes that produce such mass movement are continually operating in and on the landscape. The reduction and rearrangement of landforms and slopes begins this chapter.

LANDMASS DENUDATION

Geomorphology is a science that analyzes and describes landforms—their origin, evolution, form, and spatial distribution. It is an important aspect of physical geography. **Denudation** is a general term referring to all processes that cause reduction or rearrangement of landforms. The principal denudation processes affecting surface materials include *weathering, mass movement, erosion, transportation,* and *deposition,* as produced by the agents of moving water, air, waves, ice, and the pull of gravity.

Early Hypotheses

The idea that Earth is billions of years old and continually changes is modern. Centuries ago it was popularly held that Earth was less than 7000 years old and that when change occurred it was in response to cataclysmic events. The theory of uniformitarianism and the contrasting philosophy of catastrophism are discussed in Chapter 11. Acceptance of the vast span of geologic time led people to look anew at the process of change on Earth.

American geologist and ethnologist John Wesley Powell (1834–1902) was an early director of the U.S. Geological Survey, explorer of the Colorado River, and a pioneer in understanding the landscape (Figure 13-1). In 1875 he put forward the idea of **base level,** or a level below which a stream cannot erode its valley further. The hypothetical *ultimate base level* is sea level (which is the average level between high and low tides). You can imagine base level as a surface extending inland from sea level, inclined gently upward, under the continents. Ideally, this is the lowest practical level for all denudation processes.

Powell also recognized that not every landscape degraded down to sea level, for other base levels seemed to be in operation. A *local,* or temporary, *base level* may control a regional landscape and the lower limit of local streams. That local base level might be a river, a lake, a hard and resistant rock structure, or a human-made dam. In arid landscapes, with their intermittent precipitation, local control is provided by valleys, plains, or other low points (Figure 13-2).

Interactions between the structural elements of the land and denudation processes are complex. They represent a kind of struggle between the internal and external processes, among the resis-

FIGURE 13-1
John Wesley Powell developed his base level concept during extensive exploration of the West. He first ventured down the Colorado River in 1869 in heavy oak boats, drifting by Dead Horse Point and the bend in the river shown in the photo. [Photo by Scott T. Smith.]

tance of materials, weathering action on rock, and erosional processes.

Ideally, the endogenic processes build and create *initial* landscapes, whereas the exogenic processes work toward low relief, an ultimate condition of little change, and the stability of *sequential* landscapes. Various theories have been proposed to model these denudation processes and account for the appearance of the landscape at different developmental stages, not only for humid temperate climates, but also for coastal, arid desert, and equatorial landscapes.

William Morris Davis (1850–1934), a famed American geographer and geomorphologist, proposed the geographic cycle of landscape development. He later named it the *erosion cycle,* or the **geomorphic cycle model.** Davis theorized that a landscape goes through an initial uplift that is ac-

FIGURE 13-2
Ultimate and local base level concepts.

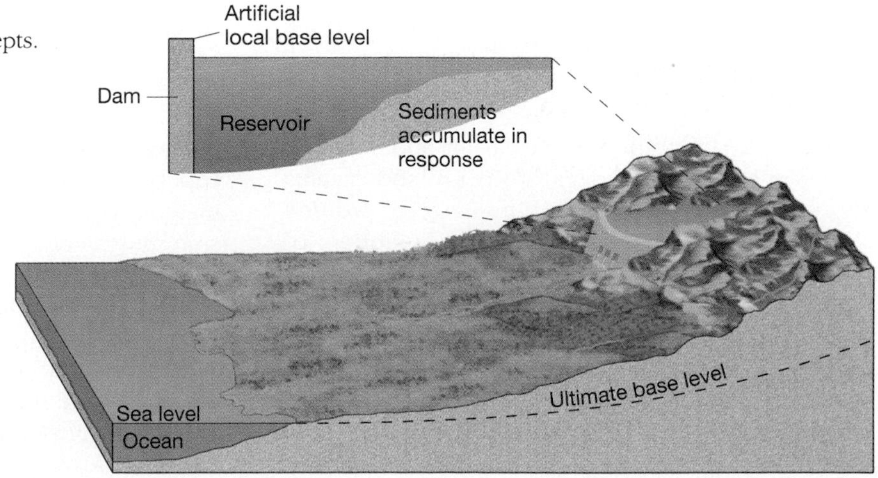

companied by erosion or removal of materials. The landscape then enters a prolonged period of stability (he later modified this idea). The raised elevation of the landscape is such that streams begin flowing more rapidly downhill, cutting both headward (upstream) and downward. Davis believed that slopes reduce progressively with a decreasing angle, ridges and divides becoming rounded and lowered over time. According to this cyclic model, the landscape eventually evolves into an old erosional surface (detailed further in Chapter 14).

In contrast to Davis, geomorphologist Walther Penck (1888–1923) believed that slopes do not gradually reduce in angle over time, but retreat at a constant angle, essentially parallel to former positions. He stressed the relationship between rate of uplift and parallel slope retreat.

Davis's theory, which helped launch the science of geomorphology and was innovative at the time, was too simple and did not account for the processes being observed as systems theory entered geomorphology. Academic support for his cyclic and evolutionary landform development model declined by 1960. Although not generally accepted today, his thinking about the evolution of landscapes is still influential and some of his terminology is still used.

Contemporary geomorphologists consider slope and landform stability a function of the resistance of rock materials to the attack of denudation processes. A balance among force, form, and process known as a *dynamic equilibrium* is operative on the landscape.

Dynamic Equilibrium Approach to Landforms

In reality, a landscape behaves as an open system, with highly variable inputs of energy and materials: uplift creates the *potential energy of position* above base level and therefore a disequilibrium of relief and energy. The Sun provides *radiant energy* that is converted into heat energy. The hydrologic cycle imparts *kinetic energy* through mechanical motion. *Chemical energy* is made available from the atmosphere and various reactions within the crust. In response to this input of energy and materials, landforms constantly adjust toward a condition of equilibrium. As physical factors fluctuate in an area, the surface constantly responds in search of an equilibrium condition, with every change producing compensating actions and reactions. The balancing act between tectonic uplift and the reduction rates of weathering and erosion, between the resistance of crust materials and the ceaseless attack of denudation processes, is summarized in the **dynamic equilibrium model.**

A dynamic equilibrium demonstrates a trend over time. According to current thinking, landscapes in a dynamic equilibrium show ongoing adaptations to the ever-changing conditions of rock structure, climate, local relief, and elevation. Various endogenic episodes, such as faulting, or exogenic episodes, such as a heavy rainfall, may provide new sets of relationships for the landscape. Following such destabilizing events a landform system arrives at a **geomorphic threshold**—the point at which there is enough energy to overcome available resistance. The system breaks through to a new set of equilibrium relationships. At this point the landform and its slopes enter a period of adjustment and realignment. The pattern over time is a sequence of (a) equilibrium stability, (b) a destabilizing event, (c) a period of adjustment, and (d) development of a new and different condition of equilibrium stability.

Slow, continuous-change events, such as soil development and erosion, tend to maintain an approximate equilibrium condition. Although less frequent, dramatic events such as a major landslide or dam collapse require longer recovery times before an equilibrium is re-established. (Figure 1-5 graphically illustrates the distinction between a steady-state equilibrium and a dynamic equilibrium).

Slopes. Material loosened by weathering is susceptible to erosion and transportation. However, if material is to move downslope, the forces of erosion must overcome the force of friction, the cohesion of particles to each other, and the inertial resistance to movement (Figure 13-3a). If the angle is such that the pull of gravity overcomes frictional forces, or if the impact of raindrops dislodges material, or if the force of water flowing downhill lifts and carries material, then erosion of particles and transport downslope can occur.

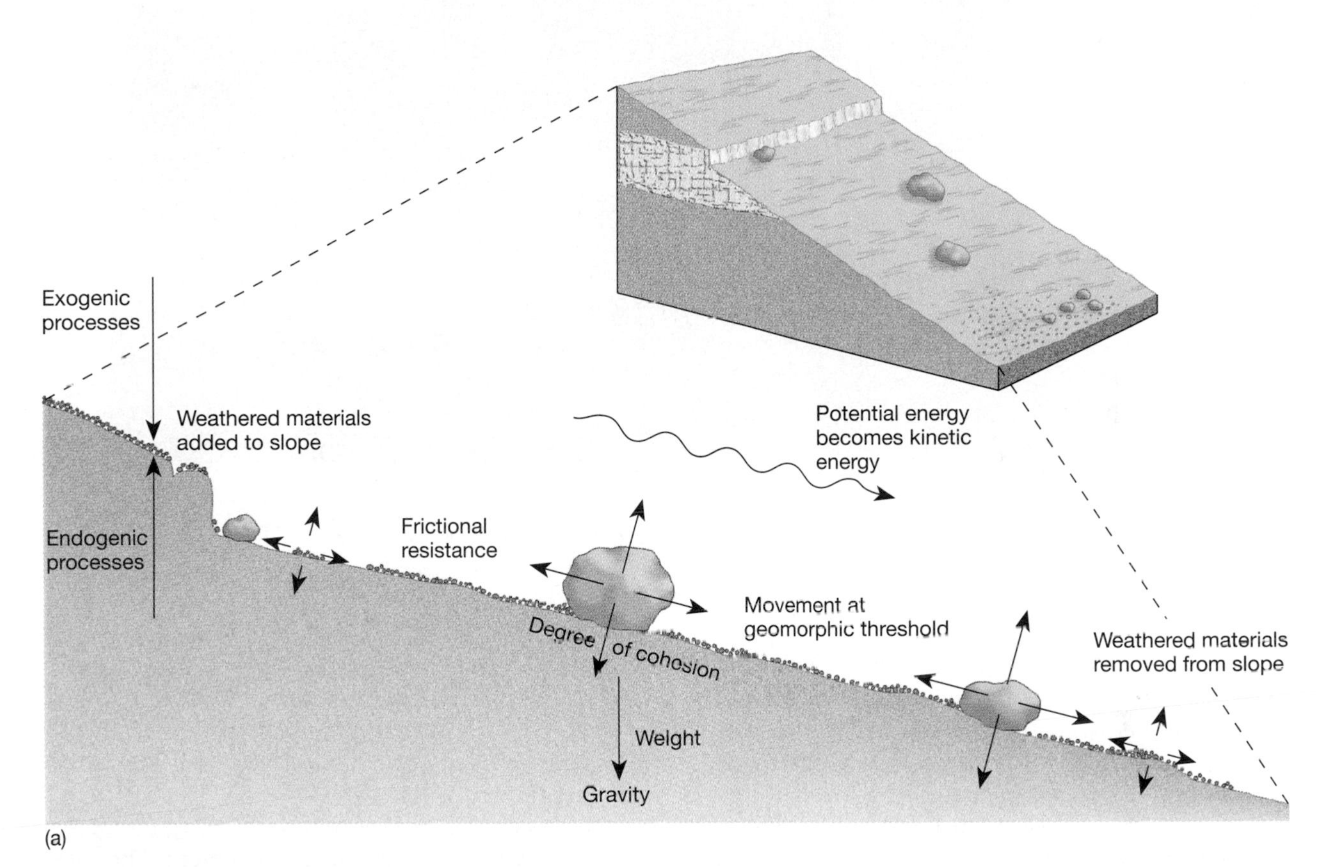

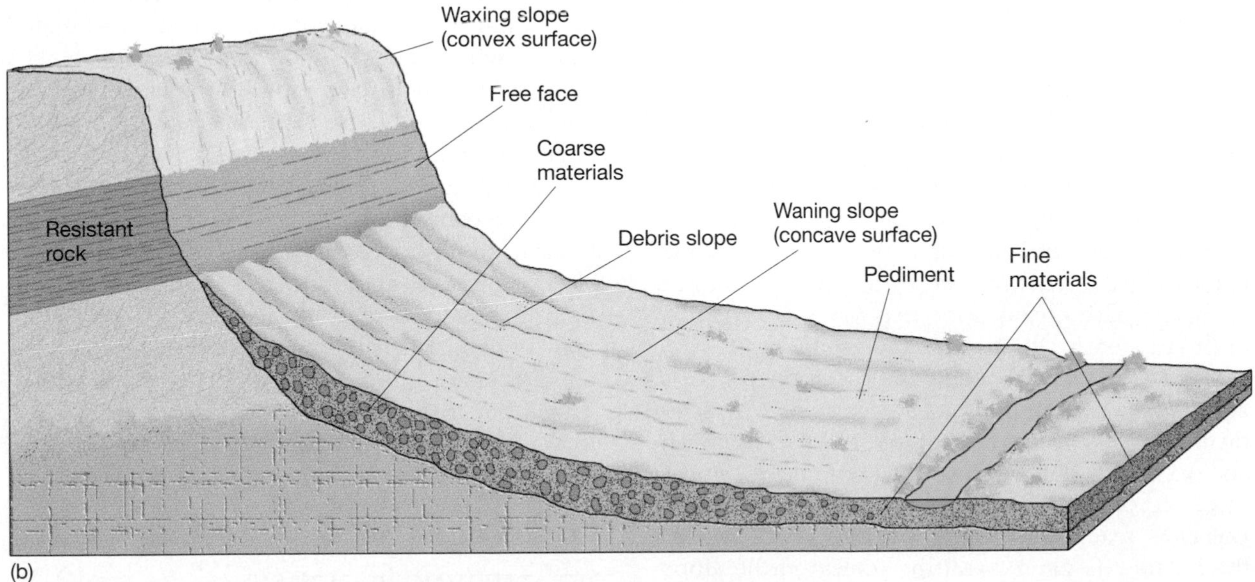

FIGURE 13-3
(a) Forces (noted by arrows) act on materials along an inclined slope; (b) the principal elements of slope form.

FIGURE 13-4
A slope in disequilibrium. Unstable, saturated soils gave way, leaving the roadbed destroyed and the guardrail suspended in air. [Photo by author.]

Slopes, or hillslopes, are curved, inclined surfaces that bound landforms. Figure 13-3b illustrates essential slope forms that may vary with conditions of rock structure and climate. Slopes generally feature an upper *waxing slope* near the top. This convex surface curves downward and grades into the *free face* below. The presence of a free face indicates an outcrop of resistant rock that forms a steep scarp or cliff.

Downslope from the free face is a *constant slope,* or *debris slope,* that receives rock fragments and materials from above. In humid climates constantly moving water carries away material as it arrives, whereas in arid climates, graded debris slopes persist and accumulate. A debris slope grades into a *waning slope,* a concave surface along the base of the slope that forms a *pediment,* or broad, gently sloping erosional surface.

Slopes are open systems and seek an *angle of equilibrium* among the forces described here. Conflicting forces work together on slopes to establish an optimum compromise incline that balances these forces. A geomorphic threshold (change point) is reached when any of the conditions in the balance is altered. All the forces on the slope then compensate by adjusting to a new dynamic equilibrium.

Slopes, then, are shaped by the relationship between rates of weathering and breakup of slope materials and the rates of mass movement and erosion of those materials. A slope is thought of as stable if its strength is greater than these denudation processes, and unstable if materials are weaker than these processes.

The recently disturbed hillslope in Figure 13-4 is in the midst of compensating adjustment. The disequilibrium was created by the failure of saturated slopes, which a day earlier were under water. (A dam break downstream rapidly emptied the reservoir and exposed these slopes, leading to their ultimate failure.)

Why hillslopes are shaped in certain ways, how slope elements evolve or age, and how hillslopes behave during rapid, moderate, or slow uplift are topics of active study and research. Modern geomorphology transcends the search for slope history and sequential development to emphasize individual processes and the role of materials resistance in soil and rock. Now, with the concepts of base level, landmass denudation, and slope development in mind, let's examine specific processes that operate to reduce landforms.

WEATHERING PROCESSES

Rocks at or near Earth's surface are exposed to both physical and chemical weathering processes. **Weathering** encompasses a group of processes by

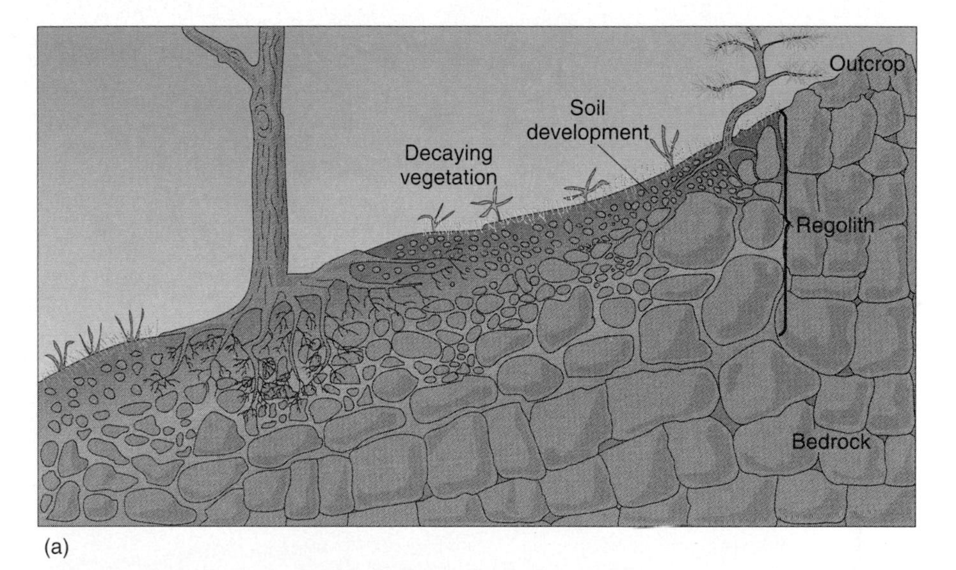

FIGURE 13-5
(a) A typical cross section of a hillside. (b) Sand dunes; their parent material is in the background. [Photo by author.]

which surface and subsurface rock disintegrates into mineral particles or dissolves into minerals in solution. Weathering does not transport the weathered materials; it simply generates these raw materials for transport by the agents of water, wind, waves, and ice—all influenced by gravity. In most areas, the upper surface of bedrock is partially weathered to broken-up rock called **regolith.** Loose surface material comes from further weathering of regolith and from transported and deposited regolith (Figure 13-5a). In some areas regolith may be missing or undeveloped, thus exposing an outcrop of unweathered bedrock.

Bedrock is the *parent rock* of weathered regolith and any soils that might develop from it (traceable through similarities in composition while the soils are relatively youthful). For example, the sand and soils in Figure 13-5b derive their color and character from the parent rock in the background. This unconsolidated fragmental material known as **sediment,** combines with weathered rock to form the **parent material** from which soil evolves.

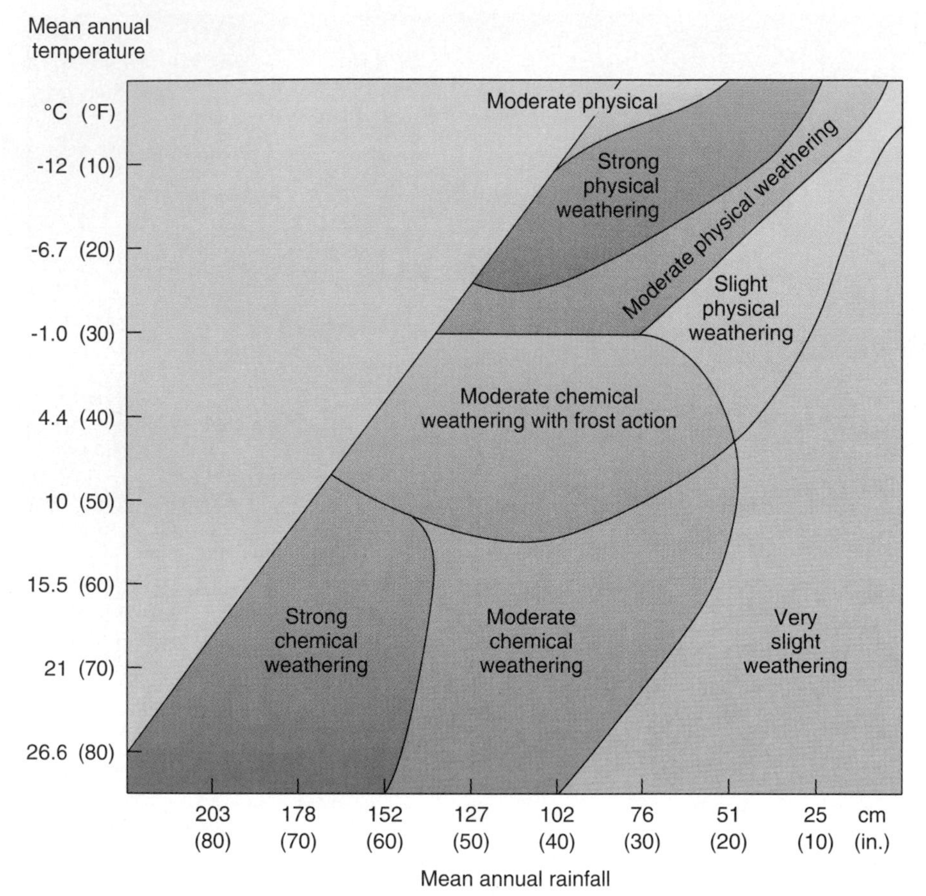

FIGURE 13-6
Relationship among temperature and rainfall and various types of weathering. (Note that values on both horizontal and vertical axes are in descending order.) [After L. C. Peltier, *Annals of the Association of American Geographers* 40, (1950): 214–36. Adapted by permission.]

Weathering is greatly influenced by the character of the bedrock: hard or soft, soluble or insoluble, broken or unbroken. Also important in determining weathering rates are climatic elements, including the amount of precipitation, overall temperature patterns, and any freeze-thaw cycles. Other significant factors are the presence and position of the water table and the topographic orientation of each exposed slope (which controls exposure to Sun, wind, and precipitation). Slope and exposure produce differences in temperature and moisture efficiency that influence weathering processes. These effects are especially noticeable in the middle and higher latitudes. Slopes facing away from the Sun's rays tend to be more moist and vegetated than those slopes that receive direct sunlight.

Figure 13-6 shows the relationship between annual precipitation and temperature and the physical (mechanical) or chemical weathering processes in an area. You can see that physical weathering dominates in drier, cooler climates, whereas chemical weathering dominates in wetter, warmer climates. Extreme dryness reduces most weathering to a minimum, as is experienced in desert climates *(BW low-latitude arid desert).* In the hot, wet tropical and equatorial rain-forest climates *(Af tropical rain forest),* most rocks weather rapidly, and the weathering extends deep below the surface.

Another form of weathering is the freeze-thaw action that alternately expands the water in rocks (freezing) and then contracts it (thawing), creating forces great enough to mechanically split the rocks. Freezing actions are important in humid microthermal climates *(Df humid continental)* and in subarctic and polar regimes *(subarctic Dfc* and *Dw* climates and *E polar* climate).

Vegetation is also a factor in weathering. Although vegetative cover can protect rock, it also

FIGURE 13-7
Cliff dwelling site in Canyon de Chelly, Arizona, used by the Anasazi people until about 900 years ago. The dark streaks on the rock are thin coatings of *desert varnish* composed of iron oxides with traces of manganese and silica. [Photo by author.]

can provide organic acids from the partial decay of organic matter and thus hasten chemical weathering. Plant roots can enter crevices and break up a rock, exerting enough pressure to drive rock segments apart, thereby exposing greater surface area to other weathering processes. (You may have observed how tree roots can heave the sections of a sidewalk or driveway sufficiently to raise and crack the cement.) The differing resistances of rock, coupled with these variations in the intensity of weathering, result in **differential weathering** as shown in Figure 11-10a.

Jointing in rock is important for weathering processes. **Joints** are fractures or separations in rock without displacement of the sides. The presence of these usually plane (flat) surfaces greatly increase the surface area of rock exposed to weathering.

Physical Weathering Processes

When rock is broken and disintegrated without any chemical alteration, the process is called **physical weathering** or mechanical weathering. By breaking up rock, physical weathering greatly increases the surface area on which chemical weathering may operate. In the complexity of nature, physical and chemical weathering usually operate together in a combination of processes. We separate them here for the convenience of discussion.

Crystallization. Especially in arid climates, where dry weather draws moisture to the surface of rock formations, some minerals develop crystals. As the water evaporates, salt crystals form from the dissolved mineral salts. As this process continues over time and the crystals grow and enlarge, they exert a force great enough to spread apart individual mineral grains and begin breaking up the rock. Such *crystallization,* or *salt-crystal growth,* is a form of physical weathering. In the Colorado Plateau of the Southwest, deep indentations continue to develop in sandstone cliffs, especially above impervious layers, where salty water is forced out of the rock strata. Crystallization then loosens the sand grains, and subsequent erosion and transportation by water and wind complete the sculpturing process. Over 1000 years ago, Native Americans built entire villages in these weathered niches at several locations, including Mesa Verde in Colorado and Arizona's Canyon de Chelly (pronounced "canyon duh shay," Figure 13-7).

Hydration. Another process involving water, but little chemical change, is **hydration.** It is a physical weathering process in which water is added to a mineral. This addition of water initiates swelling and stress within the rock, mechanically forcing grains apart as the constituents expand. Hydration can lead to granular disintegration and further susceptibility of the rock to chemical weathering, especially by oxidation and carbonation.

Frost Action. Water expands as much as 9% of its volume as it freezes (see Chapter 7, "Ice, the Solid Phase"). This expansion creates a powerful mechanical force called **frost action,** or freeze-thaw action, that can exceed the tensional strength of rock. Repeated cycles of water freezing and thawing break rock segments apart. The work of ice probably begins in small openings, gradually expanding until rocks are cleaved (split). Figure 13-8 shows this action on blocks of rock, causing *joint-block separation* along preexisting joints and fractures in the rock. This weathering action, called *frost-wedging,* pushes portions of the rock apart. Cracking and breaking can be in any shape, depending on the rock structure. The softer supporting rock underneath the slabs in the photo already has weathered physically—an example of differential weathering.

Several cultures have used this principle of frost action to quarry rock. Pioneers in the early American West drilled holes in rock, poured water in them, and then plugged them. During the cold winter months, expanding ice broke off large blocks along lines determined by the hole patterns. In spring the blocks were hauled to cities for construction. Frost action also produces unwanted fractures that damage pavement and split water pipes.

Because nature directs its own frost-action program, spring can be a risky time to venture into mountainous terrain. As ice melts with the rising temperatures, newly fractured rock pieces fall without warning and may even start rock slides. The falling rock pieces may shatter on impact—another form of physical weathering (Figure 13-9).

In arctic and subarctic climates, frost action dominates soil conditions and physical weathering. Chapter 17 covers these *periglacial regions.*

Pressure-Release Jointing. To review from Chapter 11, rising magma that is deeply buried and subjected to high pressures forms intrusive igneous rocks called plutons, which cool slowly and produce coarse-grained, crystalline granitic rocks. As the landscape is subjected to uplift, the weathering regolith that covers such a pluton can be eroded

FIGURE 13-8
Physical weathering along joints in rock produces discrete blocks. [Photo by author.]

FIGURE 13-9
Shattered rock debris, the product of a large rockfall in Yosemite National Park. [Photo by author.]

and transported away, eventually exposing the pluton as a mountainous batholith. As the tremendous weight of the overburden is removed from the granite, the pressure of deep burial is relieved. The granite responds with a slow but enormous heave, and in a process known as *pressure-release jointing,* layer after layer of rock peels off in curved slabs or plates, thinner at the top of the rock structure and thicker at the sides. As these slabs weather, they slip off in a process called **sheeting** (Figure 13-10a). This *exfoliation process* creates arch-shaped and dome-shaped features on the exposed landscape, forming an **exfoliation dome** (Figure 13-10b). Such domes probably are the largest single weathering features on Earth (in areal extent).

FIGURE 13-10
(a) Exfoliation processes loosen slabs of granite, freeing them for further weathering and downslope movement. (b) Exfoliated plates of rock are visible in layers, characteristic of dome formations in granites. View is from the east side of Half Dome in Yosemite National Park, California. [Photos by author.]

(a)

(b)

Chemical Weathering Processes

Chemical weathering refers to actual decomposition and decay of the constituent minerals in rock due to chemical alteration of those minerals, always in the presence of water. A familiar example of chemical weathering is the eating away of cathedral facades and the etching of tombstones caused by increasingly acid precipitation. Water is essential to chemical weathering and, as shown in Figure 13-6, the chemical breakdown becomes more intense as both temperature and precipitation increase. Although individual minerals vary in susceptibility, no rock-forming minerals are completely unresponsive to chemical weathering.

Spheroidal weathering of rock is a chemical weathering process, although physical processes are also involved. The sharp edges and corners of rocks are weathered by water that penetrates joints and fractures, creating a spheroidal appearance. The product resembles exfoliation but is not the result of pressure-release jointing. Figure 13-11 illustrates spheroidal weathering as chemical processes begin to dissolve cementing materials in joints in the granite. This in turn opens up spaces for frost action during winter.

A boulder can be attacked from all sides, shedding spherical shells of decayed rock like the layers of an onion. As the rock breaks down, more and more surface area is made susceptible. This type of weathering is visible in the photograph that opens this chapter; note the spheroidal rock formations.

Hydrolysis. When minerals chemically combine with water, the process is called **hydrolysis**. Hydrolysis is a decomposition process that breaks down silicate minerals in rocks. Compared to hydration, a physical process in which water is simply absorbed, the hydrolysis process involves active participation of water in chemical reactions to produce different compounds and minerals.

In Chapter 11, Table 11-2 (second line) shows the mineral ingredients in igneous rocks and their varying resistance to such weathering. On the ultramafic end of the table, olivine and peridotite are most susceptible to chemical weathering, with stability gradually increasing toward the felsic minerals. Among the feldspars, calcium feldspar has the least resistance, followed by sodium feldspar; potassium feldspar is the most resistant of the

FIGURE 13-11
Chemical weathering processes work the joints in granite, leading to spheroidal weathering and rounding of individual rock segments. [Photo by author.]

feldspar family. On the far left side of the table, quartz is very resistant to chemical weathering. You can see that, because of the nature of the constituent minerals, basalt weathers faster chemically than does granite.

When weaker constituent minerals in rock dissolve by hydrolysis, the interlocking crystal network in the rock fails, and *granular disintegration* takes place. Such disintegrating granite may appear etched, corroded, and softened. By-products of this chemical weathering of granite include clay, a soluble potassium bicarbonate salt, and silica. As clay is formed from the minerals in granite (especially feldspars), quartz particles are left behind. The resistant quartz may wash downstream, eventually becoming sand on some distant beach.

Oxidation. An example of chemical weathering occurs when the oxygen dissolved in water oxidizes (combines with) certain metallic elements to form oxides. This is a chemical weathering process known as **oxidation.** Perhaps most familiar is the "rusting" of iron in rocks or soil that produces a reddish-brown stain of iron oxide (Fe_2O_3). Such iron-bearing rocks then break down as their structure is disrupted because these oxides are physically weaker and more susceptible to further weathering than the original iron-bearing mineral.

Carbonation and Solution. The simplest form of chemical weathering occurs when a mineral dissolves into **solution**—for example, when sodium chloride (common table salt) dissolves in water. Water is called the *universal solvent* because it is capable of dissolving at least 57 of the natural elements and many of their compounds. It readily dissolves carbon dioxide, thereby yielding precipitation containing carbonic acid (H_2CO_3). This acid is strong enough to react with many minerals, especially limestone, in a process called **carbonation.** Such chemical weathering transforms minerals containing calcium, magnesium, potassium, and sodium into *carbonates.*

When rainwater attacks formations of limestone (which is calcium carbonate, $CaCO_3$), the constituent minerals are dissolved and the reaction forms a calcium bicarbonate solution. Walk through an old cemetery and you can observe the carbonation of marble, a metamorphic form of limestone. Weathered limestone and marble formations appear pitted and weathered wherever adequate water is available for carbonation. In this era of human-induced increases of acid deposition by precipitation, carbonation processes are greatly enhanced (see Focus Study 3-2, "Acid Deposition—A Blight on the Landscape").

Entire landscapes composed of limestone are dominated by the chemical weathering process of carbonation. These are the regions of karst topography.

KARST TOPOGRAPHY AND LANDSCAPES

Limestone is so abundant on Earth that many landscapes are composed of it (Figure 13-12). These areas are quite susceptible to chemical weathering. Such weathering creates a specific landscape of pitted, bumpy surface topography, poor surface drainage, and well-developed solution channels underground. Remarkable labyrinths of underworld caverns also may develop. These are the hallmarks of **karst topography,** originally named for the Krš Plateau in Yugoslavia where these processes were first studied. Approximately 15% of Earth's land area has some developed karst, with outstanding examples found in southern China, Japan, Puerto Rico, Cuba, the Yucatán of Mexico, Kentucky, Indiana, New Mexico, and Florida.

For a limestone landscape to develop into karst topography, several important prerequisites must exist. The limestone formation must contain 80% or more calcium carbonate for solution processes to proceed effectively. Complex patterns of joints in the otherwise impermeable limestone are needed for water to concentrate its solution activity and to form routes to subsurface drainage channels. Local surface relief also is important to conditions in the subsurface area between the ground surface and the water table, known as the *vadose zone,* or zone of aeration, above the water table. Finally, vegetation cover can determine the amount of organically derived acids that enhance solution processes. The role of climate in providing optimum environmental conditions for karst processes remains the subject of debate.

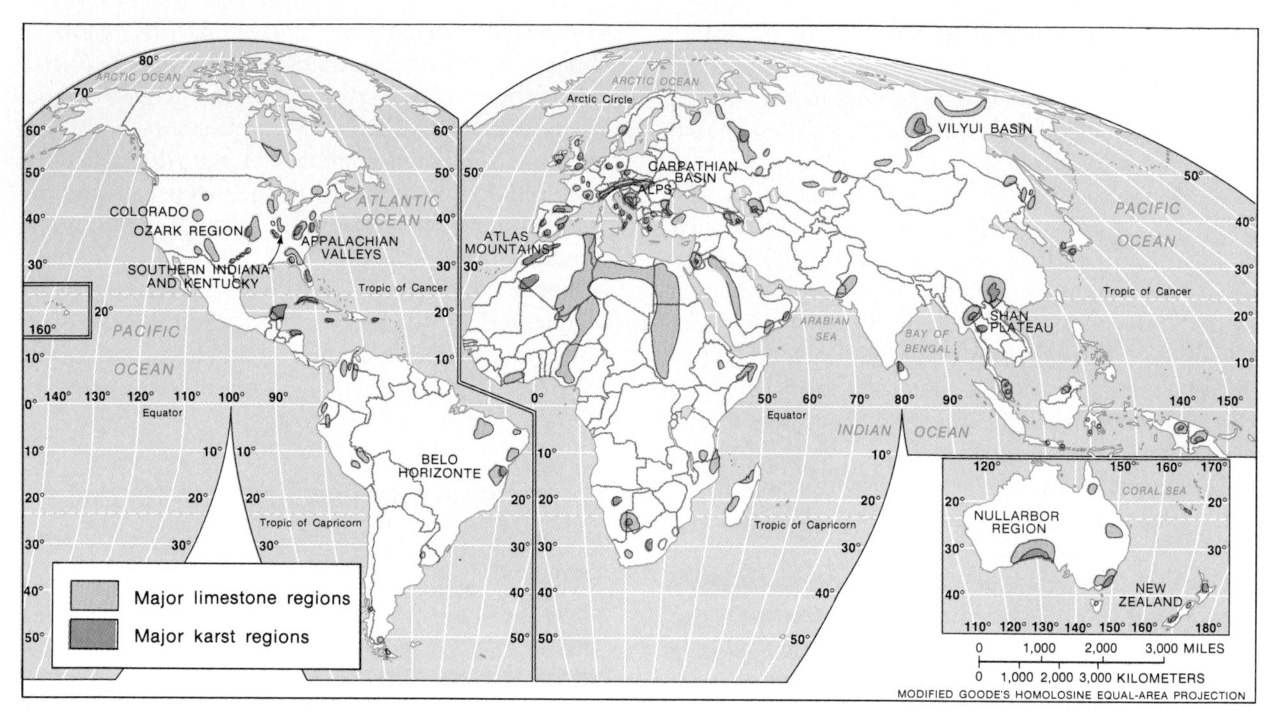

FIGURE 13-12
Distribution of karst landscapes worldwide. [After R. E. Snead, *Atlas of the World Physical Features*, p. 76. Copyright © 1972 by John Wiley & Sons. Adapted by permission of John Wiley & Sons, Inc.]

As with all weathering processes, time is a factor. Earlier in this century, karst landscapes were thought to progress through evolutionary stages of development, as if they were aging. Today they are thought to be locally unique, a result of specific conditions, and there is little evidence that different regions evolve sequentially along similar lines. Nonetheless, mature karst landscapes do display certain characteristic forms (Figure 13-13).

The weathering of limestone landscapes creates many **sinkholes,** which form in circular depressions. If these collapse through the roof of an underground cavern, the term *doline* is also used, and the landscape is referred to as *doline karst*. A gently rolling limestone plain might be pockmarked with sinkholes and dolines 2–100 m (7–330 ft) deep and 10–1000 m (33–3300 ft) wide. The area southwest of Orleans, Indiana, has over 1000 sinkholes and dolines in just 2.6 km^2 (1 mi^2). The Lost River in southern Indiana, a "disappearing stream," flows more than 12.9 km (8 mi) underground before it resurfaces; its dry bed can be seen on the lower left of Figure 13-13b.

Through solution and collapse, dolines may coalesce to form *uvalas,* or valley sinks, which are depressions up to several kilometers in diameter sometimes referred to as a *karst valley*. In Florida, several sinkholes have made the news because lowered water tables have caused their collapse into underground solution caves, taking with them homes, businesses, and even new cars from an auto dealership. One such sinkhole collapsed in a Florida suburban area in March 1993 (Figure 13-13c). A complex doline karst landscape in which dolines intersect is called a *cockpit karst*. The dolines can be symmetrically shaped in certain cir-

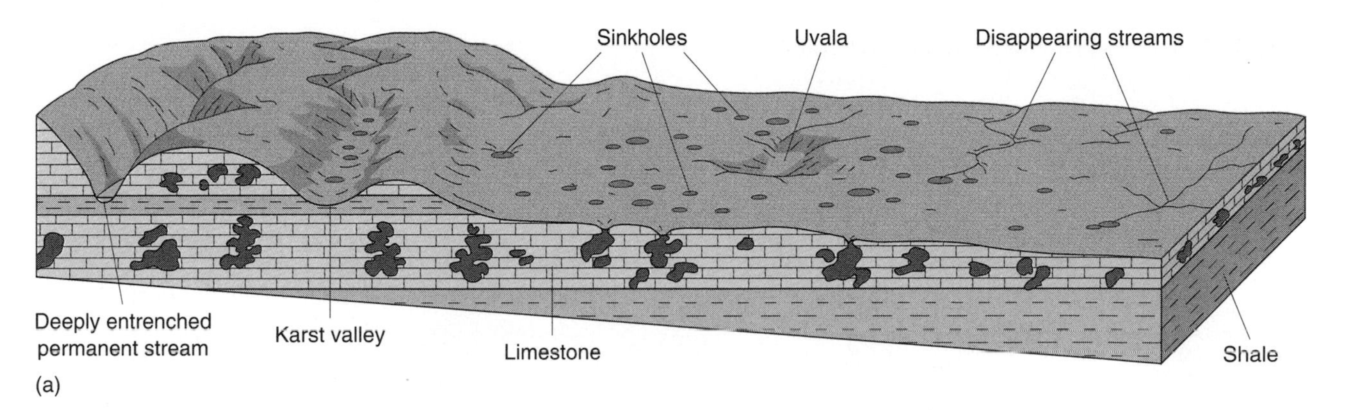

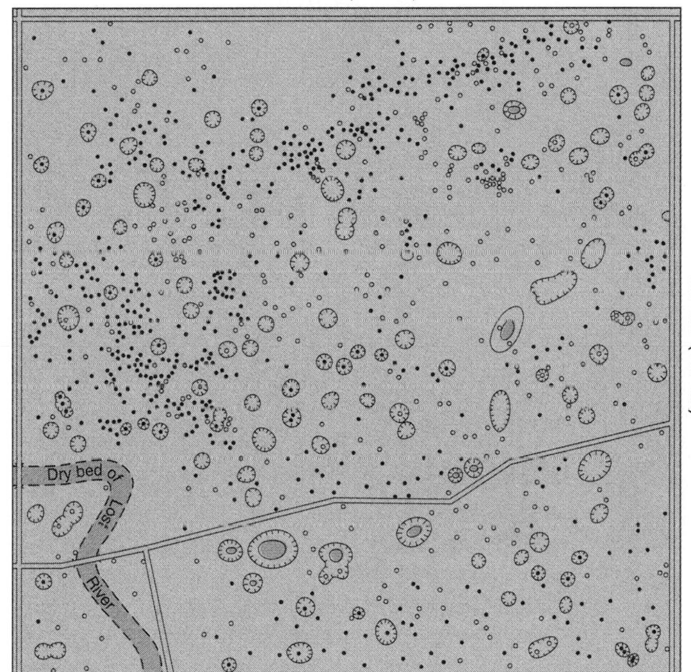

Dolines 30.5 m (100 ft) or more across and 3–9.1 m (10–30 ft) deep

Sinkhole ponds

Small sinkholes with visible openings

Small sinkholes without visible openings

(b)

(c)

FIGURE 13-13
(a) Idealized features of a portion of karst topography in southern Indiana; (b) map of 1022 sinkholes occurring in a 1 mi^2 area southwest of Orleans, Indiana. (c) The Orange City sinkhole, in east-central Florida, formed March 7, 1993, and enlarged to 40 m (130 ft) in diameter. [(a) and (b) adapted from William D. Thornbury, *Principles of Geomorphology*. (a) Illustration by William J. Wayne, p. 326; (b) after C. A. Malott, p. 321. Copyright © 1954 by John Wiley & Sons. (c) Photo by Thomas M. Scott, Florida Geological Survey.]

FIGURE 13-14
Cockpit karst topography near Arecibo, Puerto Rico, used for the dish antenna of a radio telescope. The Arecibo Observatory is part of the National Astronomy and Ionosphere Center, which is operated by Cornell University under contract with the National Science Foundation. [Photo courtesy of Cornell University.]

cumstances; one at Arecibo, Puerto Rico, is perfectly shaped for a radio telescope installation (Figure 13-14).

Another type of karst topography forms in the wet tropics, where deeply jointed, thick limestone beds are weathered into gorges and isolated resistant blocks. These resistant cones and towers are most remarkable in several areas of China, where an otherwise lower-level plain is interrupted by *tower karst* up to 200 m (660 ft) high (Figure 13-15).

Caves and Caverns

Most caves in the world are formed in limestone rock, because it is so easily dissolved by carbonation. The largest limestone caverns in the United States are Mammoth Cave in Kentucky, Carlsbad Caverns in New Mexico, and Lehman Cave in Nevada. Limestone formations in which the Carlsbad Caverns formed were themselves deposited 200 million years ago when the region was covered by shallow seas. Regional uplifts associated with the Rockies (the Laramide orogeny 40–80 million years ago) then brought the region above sea level, leading to active cave formation (Figure 13-16).

Caves generally form just beneath the water table, where a later lowering of the water level exposes them to further development. The *dripstones* that form under such conditions are produced as water containing dissolved minerals slowly drips from the cave ceiling. Calcium carbonate precipitates out of the evaporating solution, literally a molecular layer at a time, and accumulates at a point below on the cave floor. The depositional features called *stalactites* grow down from the ceiling and *stalagmites* build up from the

FIGURE 13-15
Tower karst of the Kwangsi Province, Yangzhou, China. [Photo by Wolfgang Kaehler.]

FIGURE 13-16
Carlsbad Caverns in New Mexico, where a series of underground caverns include rooms over 1200 m (4000 ft) long and 190 m (625 ft) wide. Although a national park since 1930, unexplored portions remain. [Photo by Scott T. Smith.]

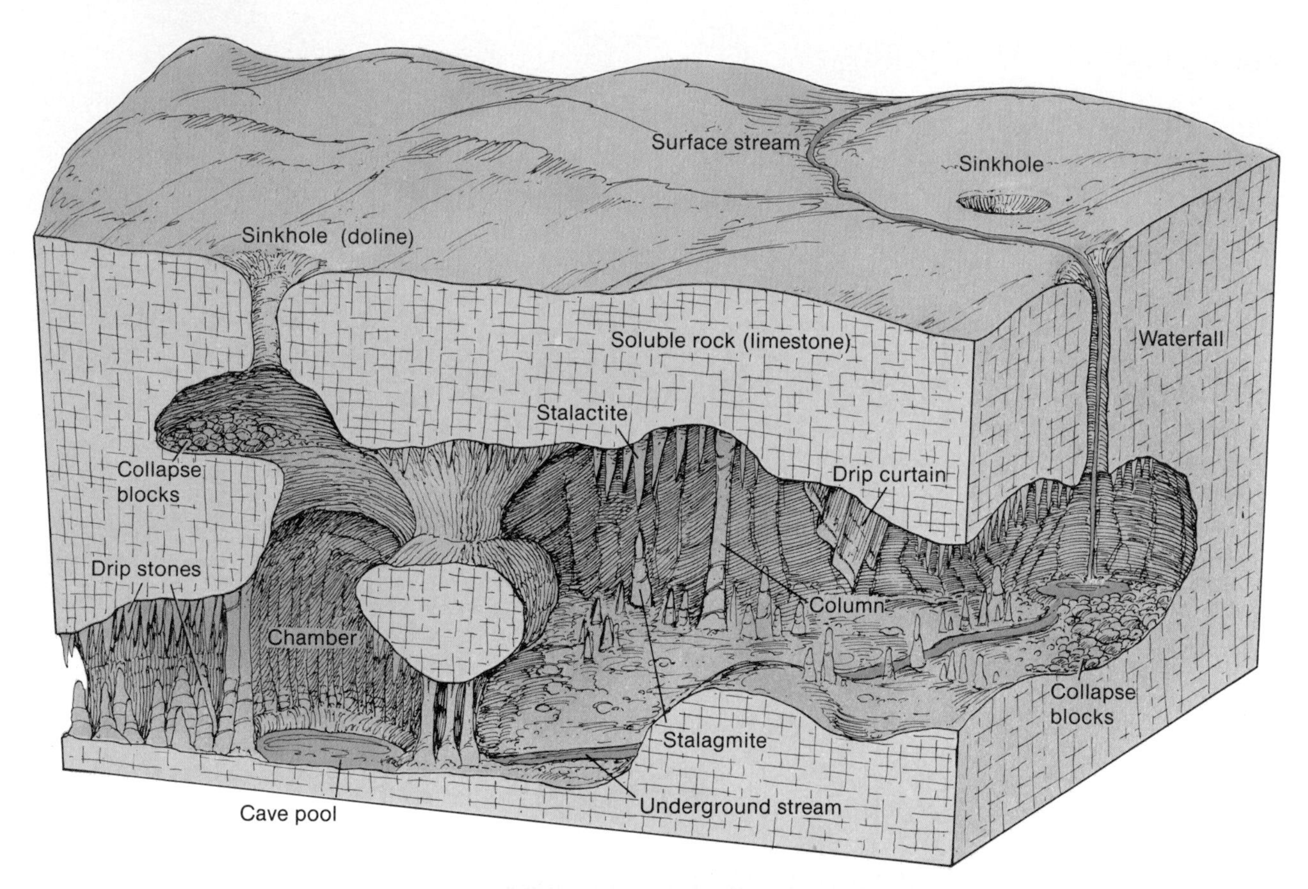

FIGURE 13-17
Formation of underground caves and related forms in limestone. [Inset photo of sinkhole in pasture by Thomas M. Scott, Florida Geological Survey.]

floor, sometimes growing until they connect and form a continuous *column* (Figure 13-17). A dramatic subterranean world is thus created.

The exploration and scientific study of caves is called *speleology*. Although investigations involve physical and biological scientists, amateur cavers, or "spelunkers," have made many important discoveries. Cave habitats are unique, nearly closed, self-contained ecosystems with simple food chains and great stability. The mystery, intrigue, and excitement of caves lies in the dark passageways, enormous chambers that narrow to tiny crawl spaces, strange formations, and underwater worlds only accessed by cave-diving.

MASS MOVEMENT PROCESSES

On November 13, 1985, at 11 P.M., after a year of earthquakes, 10 months of harmonic tremors, a growing bulge on its northeast flank, and 2 months of small summit eruptions, Nevado del Ruiz in central Colombia, South America, violently erupted in a lateral explosion. The volcano, northernmost of two dozen dormant peaks in the Cordilleran Central of Colombia, had erupted six times during the past 3000 years, killing 1000 people during its last eruption in 1845.

But on this night, the familiar tephra, lava, and the blast itself were not the problem. The hot eruption melted about 10% of the ice on the mountain's snowy peak, liquefying mud and volcanic ash, sending a hot mudflow downslope. Such a flow is called a *lahar,* an Indonesian word referring to flows of volcanic origin. This lahar moved rapidly down the Lagunilla River toward the villages below. The wall of mud was at least 40 m (130 ft) high as it approached Armero, a regional center with a population of 25,000. The city slept as the lahar buried its homes and citizens: 23,000 people were killed there and in other afflicted river valleys, thousands were injured, and 60,000 were left homeless. The debris flow from the volcano is today a permanent grave for its victims.

Although less devastating to human life, the mudflow generated by the eruption of Mount Saint Helens also was a lahar. Not all mass movements are this destructive, but such processes are important in the overall denudation of the landscape.

Mass Movement Mechanics

Physical and chemical weathering processes create an overall weakening of surface rock, which makes it more susceptible to the pull of gravity. The term **mass movement** applies to all unit movements of materials propelled and controlled by gravity, such as the lahar just described. These movements can range from dry to wet, slow to fast, small to large, and from free-falling to gradual or intermittent in motion. The term mass movement is sometimes used interchangeably with *mass wasting*. To combine the concepts, we can say that mass movement of material works to waste slopes and provide raw material for erosion, transportation, and deposition.

The Role of Slopes. All mass movements occur on slopes. If we try to pile dry sand on the beach, the grains will flow downward until an equilibrium is achieved. The steepness of the slope that results when loose sand comes to rest depends on the size and texture of the grains; this is called the **angle of repose.** This angle represents a balance of driving and resisting forces and commonly ranges between 33° and 37° (measured from a horizontal plane).

The driving force in mass movements is gravity, working in conjunction with the weight, size, and shape of the grains or surface material, the degree to which the slope is oversteepened, and the amount and form of moisture available—whether frozen or fluid. The greater the slope angle, the more susceptible the surface material is to mass movement. The resisting force is the shearing strength of slope material, that is, its cohesiveness and internal friction working against gravity and mass movement. To reduce shearing strength is to increase shearing stress, which eventually reaches the point at which gravity overcomes friction.

If the rock strata in a slope are underlain by clays, shales, and mudstones, all of which are highly susceptible to hydration (physical swelling in response to water added to a mineral), the rock strata will tend toward mass movement because less driving force energy is required. However, if the rock strata are such that material is held back from slipping, then more driving force energy may be required, such as that generated by an earthquake.

In the Madison River Canyon near West Yellowstone, Montana, on the Wyoming border, a blockade of dolomite (a magnesium-rich carbonate rock) had held back a deeply weathered and oversteepened slope (40–60° slope angle) for untold centuries (white area in Figure 13-18). Then, shortly after midnight on August 17, 1959, an earthquake measuring 7.5 on the Richter scale broke the dolomite structure along the foot of the slope. This released 32 million m^3 (1.13 billion ft^3) of mass, which moved downslope at 95 kmph (60 mph), creating gale force winds through the canyon. The material continued more than 120 m

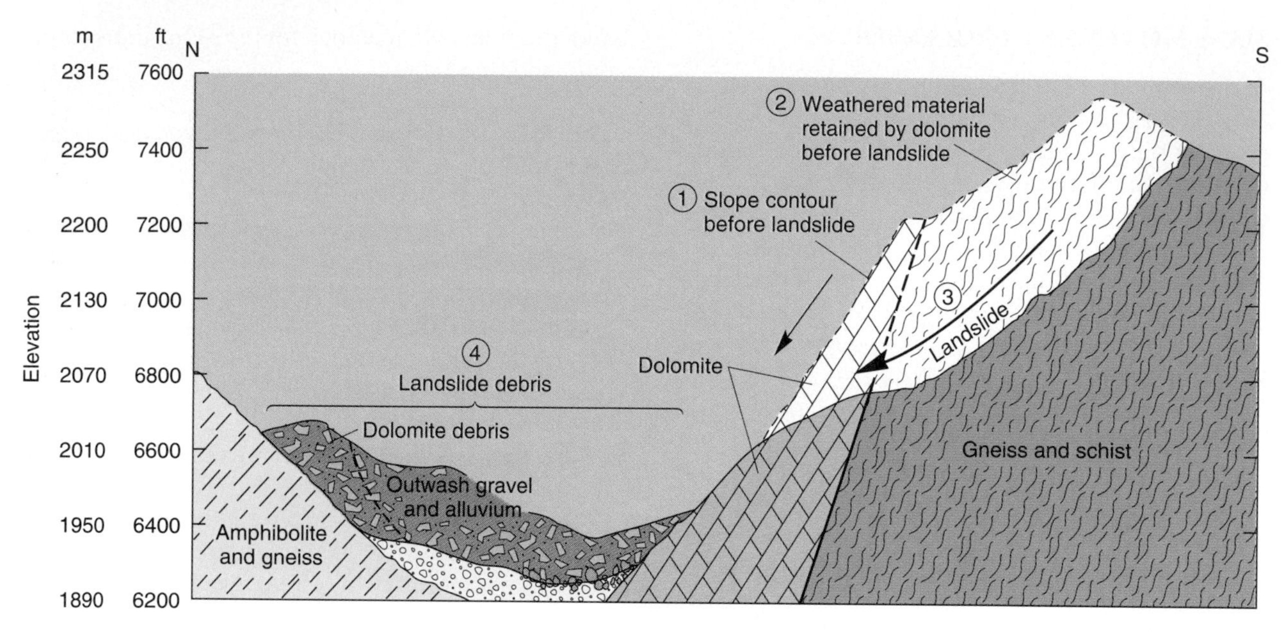

FIGURE 13-18
Cross section of the geologic structure of the Madison River Canyon at the location of the landslide triggered by the 1959 earthquake: (1) prequake slope contour, (2) weathered material that failed, (3) direction of landslide, and (4) landslide debris block the canyon and dam the Madison River. [After Jarvis B. Hadley, *Landslides and Related Phenomena Accompanying the Hebgen Lake Earthquake of 17 August 1959,* U.S. Geological Survey Professional Paper 435-K, p. 115.]

(390 ft) up the opposite canyon slope, entombing campers beneath about 80 m (260 ft) of rock and debris.

The mass of material also effectively dammed the Madison River and created a new lake, dubbed Quake Lake. The landslide debris dam established a new temporary base level for the canyon. A channel was quickly excavated by the U.S. Army Corps of Engineers to prevent a disaster below the dam, for if Quake Lake overflowed the landslide dam, the water would quickly erode a channel and thereby release the entire contents of the new lake onto farmland downstream. These events convey a dramatic example of the role of slopes and tectonic forces in creating mass land movements.

Classes of Mass Movements

For convenience, four basic classifications of mass movement are used: fall, slide, flow, and creep. Each involves the pull of gravity working on a mass until the critical shearing strength is reduced to the point that the mass falls, slides, flows, or creeps downward. The Madison River Canyon event was a type of slide, whereas the Nevado del Ruiz lahar mentioned earlier was a flow. Figure 13-19 summarizes the relationship between moisture content and rate of movement in producing the different classes of mass movement. Submarine mass movements, in contrast to these subaerial processes, are disorganized, turbulent flows of mud accumulating as deep-sea depositional fans along the continental slope.

Falls and Avalanches. A **rockfall** is simply a quantity of rock that falls through the air and hits a surface. During a rockfall individual pieces fall independently, and characteristically form a pile of irregular broken rocks called a *talus cone* at the base of a steep slope (Figure 13-20).

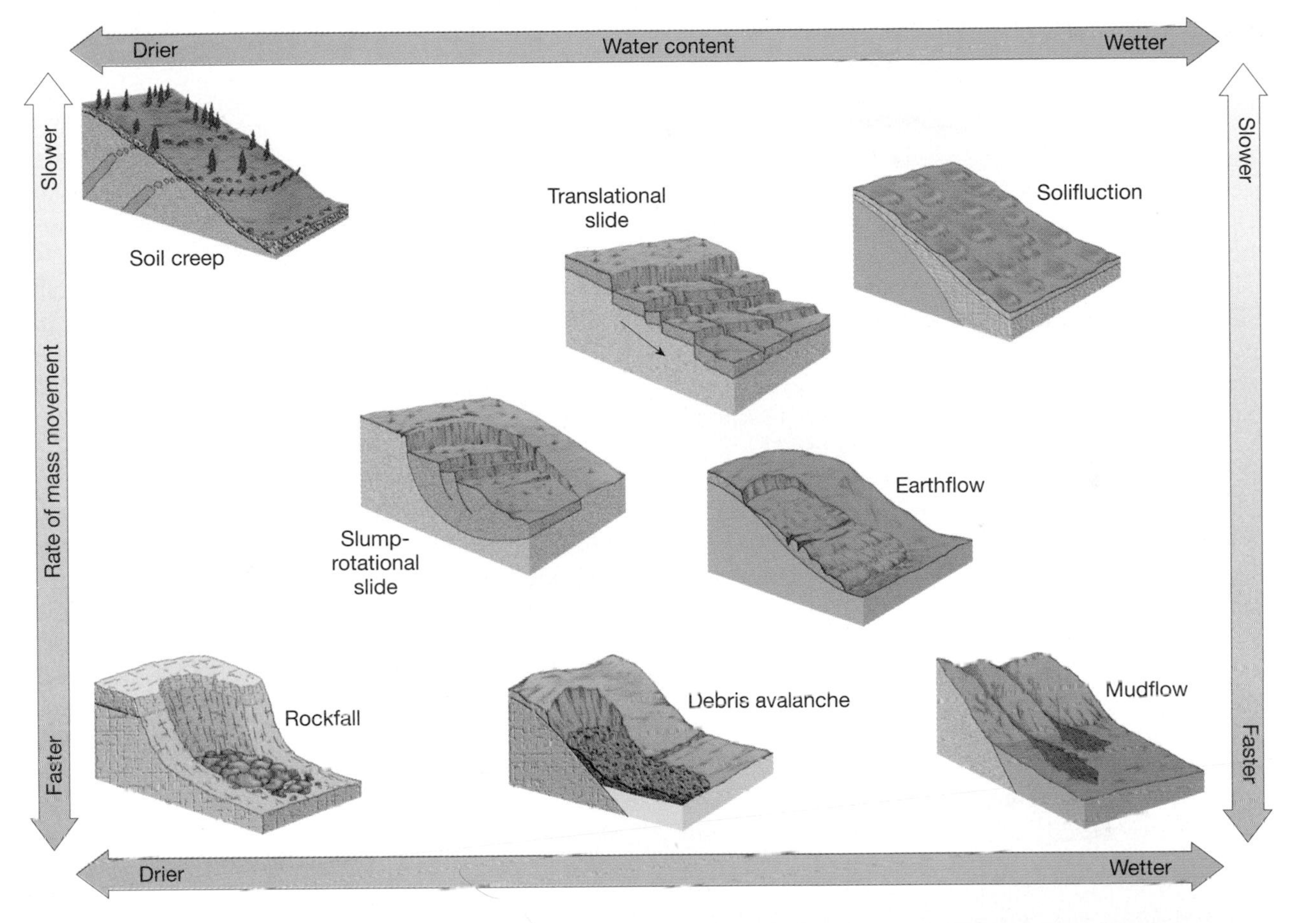

FIGURE 13-19
Principal types of mass movement and mass wasting events. Variations in water content and rates of movement produce a variety of forms.

FIGURE 13-20
Rockfall and talus cone at the base of a steep slope. [Photo by author.]

A **debris avalanche** is a mass of falling and tumbling rock, debris, and soil. It is differentiated from a debris slide or landslide by the tremendous velocity achieved by the onrushing materials. These speeds often result from ice and water that fluidize the debris. The extreme danger of a debris avalanche results from these tremendous speeds and lack of warning.

In 1962 and again in 1970, debris avalanches roared down the west face of Nevado Huascarán, the highest peak in the Peruvian Andes. The 1962 debris avalanche contained an estimated 13 million m^3 (460 million ft^3) of material, burying the city of Ranrahirca and eight other towns, killing 4000 people. The 1970 event was initiated by an earthquake. Upward of 100 million m^3 (3.53 billion ft^3) of debris buried the city of Yungay, where 18,000 people perished (Figure 13-21). This avalanche attained velocities of 300 kmph (185 mph), which is especially incredible when you consider the quantity of material involved and the fact that some boulders weighed thousands of tons. The avalanche covered a vertical drop of 4144 m (13,600 ft) and a horizontal distance of 16 km (10 mi) in only a few minutes.

Another dramatic example of a debris avalanche, one that led to no loss of life, occurred west of the Saint Elias Range, north of Yakutat Bay in Alaska. Several dozen rockfalls, large debris avalanches, and snow avalanches were triggered by a 7.1 Richter-scale earthquake in 1979. Approximately 4.7 km^2 of the Cascade Glacier was covered by the largest single avalanche caused by the earthquake (Figure 13-22). The photo reveals characteristic grooves, lobes, and large rocks associated with these fluid avalanches.

Slide. A sudden rapid movement of a cohesive mass of regolith and/or bedrock that is not saturated with moisture is a **landslide** —a large amount of material failing simultaneously. Surprise is what creates the danger, for the downward pull of gravity wins the struggle for equilibrium in an instant of time. Slides occur in one of two basic forms—translational or rotational (Figure 13-19). *Translational slides* involve movement along a planar (flat) surface roughly parallel to the angle of the slope. The Madison Canyon landslide described earlier was a translational slide. Flow and creep patterns also are considered translational in nature.

FIGURE 13-21
A 1970 debris avalanche falls 4100 m (2.5 mi) down the west face of Nevado Huascarán, burying the city of Yungay, Peru. The same area was devastated by a similar avalanche in 1962 and in pre-Columbian times. A great danger remains for the cities and towns in the valley from possible future mass movements. [Photo by George Plafker.]

FIGURE 13-22
A debris avalanche covers portions of the Cascade Glacier, west of the Saint Elias Range in Alaska. The one pictured here was the largest of several dozen triggered by a 1979 earthquake. [Photo by George Plafker.]

Rotational slides occur when surface material moves along a concave surface. Frequently, underlying clay presents an impervious surface to percolating water. As a result, water flows along the clay surface, undermining the overlying block. The simplest form of rotational slide is a rotational slump, in which a small block of land shifts downward. The upper surface of the slide appears to rotate backward and often remains intact. The surface may rotate as a single unit, or it may present a stepped appearance.

On March 27, 1964, Good Friday, the Pacific plate plunged a bit further beneath the North American plate near Anchorage, Alaska. More than 80 mass movement events were triggered by the resulting earthquake. One particular housing subdivision in Anchorage experienced a sequence of translational movements accompanied by rotational slumping. Figure 13-23 shows one view of the ground failure throughout the housing area.

Flow. When the moisture content of moving material is high, the suffix -flow is used, as in *earthflow* and *mudflow,* illustrated in Figure 13-19. Heavy rains frequently saturate barren mountain slopes and set them moving, as was the case east of Jackson Hole, Wyoming, in the spring of 1925. A slope above the Gros Ventre (pronounced "grow vaunt") River broke loose and slid downward as a unit. The slide occurred because sandstone formations rested on weak shale and siltstone, which became moistened and soft, offering little resistance to the overlying strata.

Because of melted snow and rain, the water content of the Gros Ventre landslide was high enough to classify it as an earthflow. About 37 million m^3 (1.3 billion ft^3) of wet soil and rock moved down one side of the canyon and surged 30 m (100 ft) up the other side. The earthflow dammed the river and formed a lake, as did the landslide across the Madison River in the Hebgen Lake area

FIGURE 13-23
Anchorage, Alaska, landslide, illustrates mass movement patterns throughout a housing subdivision. [Courtesy of U.S. Geological Survey.]

in 1959. However, in 1925 equipment was not available to excavate a channel, so the new lake filled. Two years later, the lake water broke through the temporary base-level dam, transporting a tremendous quantity of debris over the region downstream.

Creep. A persistent mass movement of surface soil is called **soil creep.** Individual soil particles are lifted and disturbed by the expansion of soil moisture as it freezes, by cycles of moistness and dryness, by diurnal temperature variations, or even by grazing livestock or digging animals. In the freeze-thaw cycle, particles are lifted at right angles to the slope by freezing soil moisture, as shown in Figure 13-24. However, when the ice melts, the particles are pulled straight downward by the force of gravity. As the process is repeated, the soil cover gradually creeps its way downslope. The overall wasting of a slope may cover a wide area and may cause fence posts, utility poles, and even trees to lean downslope. Various strategies are used to arrest the mass movement of slope material—grading the terrain, building terraces and retaining walls, planting ground cover—but the persistence of creep always wins.

Human-Induced Mass Movements

Every highway roadcut or surface-mining activity that exposes a slope can hasten mass wasting because the oversteepened surfaces are thrown into a new search for equilibrium. Imagine the disequilibrum in slope relationships created by the highway roadcut pictured in Figure 12-6b. Large open-pit strip mines—such as the Bingham Copper Mine west of Salt Lake City, the Berkeley Pit in Butte, Montana, and the extensive strip mining for coal in the East and West—are examples of human-induced mass movements, generally called **scarification.** At the Bingham Copper Mine, a mountain literally was removed. The disposal of *tailings* (mined ore of little value) and waste material is a significant problem with such large excavations

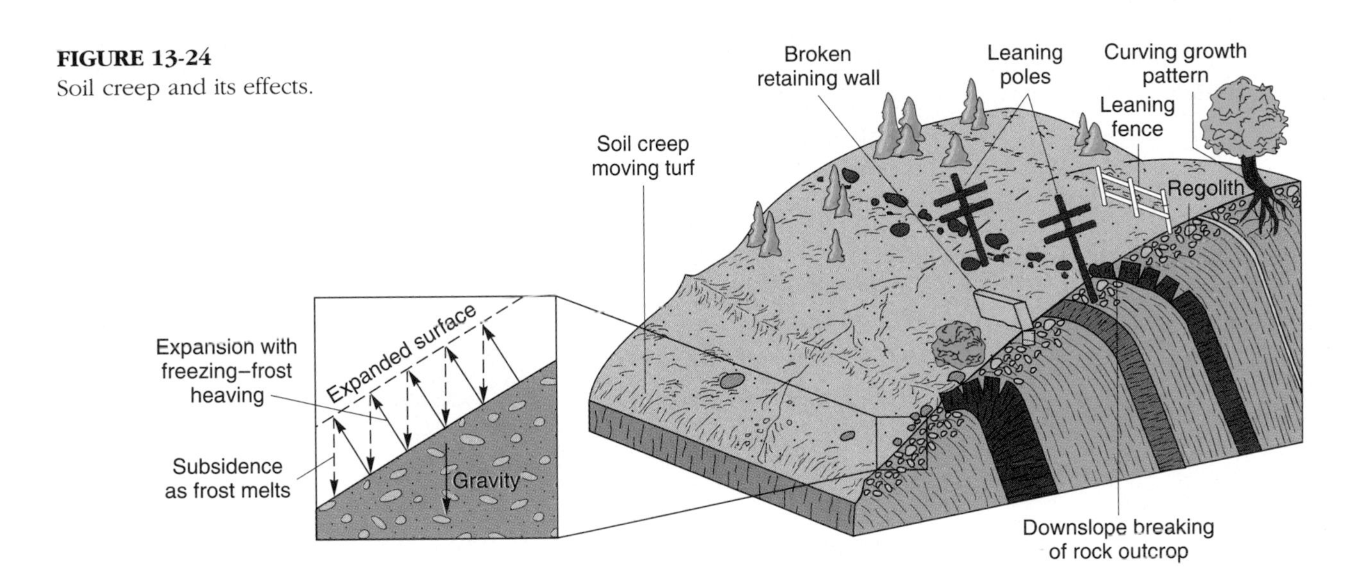

FIGURE 13-24
Soil creep and its effects.

because the tailing piles prove unstable and susceptible to further weathering, mass wasting, or wind dispersal. Wind dispersal is a particular problem with uranium tailings in the West due to their radioactivity.

Land subsidence and collapse in mined areas produce further mass movements of land. This is a major problem in portions of the Appalachians where buildings, utility poles, and streets, as well as drainage patterns, are affected.

SUMMARY—Weathering, Karst Landscapes, and Mass Movement

The exogenic system, powered by solar energy and gravity, tears down the landscape through processes of landmass denudation involving weathering, mass movement, erosion, transportation, and deposition. Agents of change include moving air, water, waves, and ice. Geomorphology is the science that analyzes and describes the origin, evolution, form, and spatial distribution of landforms.

Various models to characterize landmass denudation and slope reduction have been attempted. The principles of slope morphology are guided by the dynamic equilibrium model. Slopes are shaped by the relationship between rate of weathering and breakup of slope materials and the rate of mass movement and erosion of those materials. A slope is thought of as stable if its strength is greater than these denudation processes; it is unstable if materials are weaker than these processes.

Both physical and chemical weathering derive material from parent bedrock that is essential for soil formation. Physical weathering refers to the actual breakup of rock into smaller pieces, whereas chemical weathering is the chemical altering of minerals in rock. The physical action of water when it freezes (expands) and thaws (contracts) is a powerful agent in shaping the landscape. At the same time, the chemical action of carbon dioxide dissolved in rainwater attacks susceptible rock, such as limestone. Karst topography refers to distinctively pitted and weathered limestone landscapes. The formation of massive caverns is related to these processes.

Mass movement of Earth's surface produces some dramatic incidents, including rockfalls, landslides, mudflows, and soil creep, all of which are mass movements of soil and debris and are important aspects of overall landmass denudation processes. In addition, human mining activity has created massive scarification of landscapes.

The next four chapters look at the work of rivers, wind, waves, and ice as geomorphic agents that use erosion, transport, and deposition processes to form unique landscapes.

KEY TERMS

angle of repose
base level
bedrock
carbonation
chemical weathering
debris avalanche
denudation
differential weathering
dynamic equilibrium model
exfoliation dome
frost action
geomorphic cycle model
geomorphic threshold
geomorphology
hydration
hydrolysis
joints
karst topography
landslide
mass movement
oxidation
parent material
physical weathering
regolith
rockfall
scarification
sediment
sheeting
sinkholes
slopes
soil creep
solution
spheroidal weathering
weathering

REVIEW QUESTIONS

1. Define geomorphology and describe its relationship to physical geography.
2. How do the base level and the local base level control how far a stream can erode its channel?
3. Define landmass denudation. What processes are included in the concept?
4. How does the construction of a reservoir alter the relationship between a stream and base level?
5. Give a brief overview of Davis's geomorphic cycle model. What was his basic thinking in setting up this cyclic model? What was Davis's principal contribution to the models of landmass denudation?
6. Describe how a landscape operates in response to constantly adjusting conditions. Do you think a landscape ever reaches a stable, old-age condition? Explain.
7. When a hillslope system reaches a geomorphic threshold, what occurs?
8. Relative to slopes, what is meant by an "angle of equilibrium"? Can you apply this to the photograph in Figure 13-4?
9. Describe the processes of physical and chemical weathering on an open expanse of bedrock. How does regolith develop? How is sediment derived?
10. Describe the relationship between climatic conditions and rates of weathering activities.
11. What is the interplay between the resistance of rock structures and weathering variabilities?
12. Why is freezing water such an effective physical weathering agent?
13. What are the weathering processes that produce a granite dome? Describe the sequence of events that takes place.
14. What is hydrolysis? How does it affect rocks?
15. Describe the development of limestone topography. What is the name applied to such landscapes? From what was this name derived?

16. Differentiate among sinkholes, dolines, uvalas, and cockpit karst. Within which form is the radio telescope at Arecibo, Puerto Rico?
17. Name and describe the type of mudflow associated with volcanic eruptions.
18. Define the role of slopes in mass movements—angle of repose, driving force, resisting force, and so on.
19. What events occurred in the Madison River Canyon in 1959?
20. What are the classes of mass movement? Describe each briefly and differentiate among these classes.
21. Describe the difference between a landslide and what happened on the slopes of Nevado Huascarán.
22. What is scarification, and why is it considered a type of mass movement? Give an example of scarification.

SUGGESTED READINGS

Alden, William C. "Landslide and Flood at Gros Ventre, Wyoming," *Transactions of the American Institute of Mining Engineers* 76 (February 1928): 347–61.

Davis, William Morris. "The Geographical Cycle," *Journal of Geography* 14, no. 5 (November 1899): 481–504.

Fairbridge, Rhodes W., ed. *The Encyclopedia of Geomorphology*. Encyclopedia of Earth Sciences Series, vol. 3. New York: Reinhold Book Company, 1968.

Jackson, Donald Dale, and the editors of Time-Life Books. *Underground Worlds*. Planet Earth Series. Alexandria, VA: Time Life Books, 1982.

Kerr, Richard A. "Landslides from Volcanoes Seen as Common," *Science* 224 (20 April 1984): 275–76.

McDowell, Bart. "Eruption in Colombia," *National Geographic* 165, no. 5 (May 1986): 640–53.

Peterson, Cass. "Killing Mountains for Coal—Federal Strip Mining Laws Have Not Healed Kentucky's Scars," *Washington Post National Weekly Edition,* June 29, 1987: 6–7.

Plafker, George and G. E. Ericksen. *"Nevados Huascarán Avalanches, Peru."* In B. Voight, ed., *Rockslides and Avalanches*, vol. 1 of *Natural Phenomena: Developments in Geotechnical Engineering 14A,* pp. 277–315. Amsterdam: Elsevier Scientific Publishing Co., 1978. (See maps and photo pp. 278–83.)

Rice, R. J. *Fundamentals of Geomorphology,* 2nd ed. White Plains, NY: Longman, 1988.

Schumm, Stanley A. "Geomorphic Thresholds and Complex Response of Drainage Systems," *Proceedings of the 4th Annual Geomorphology Symposium,* pp. 299–310. Binghamton, NY, 1973.

Spencer, Edgar Winston. *The Dynamics of the Earth*. New York: Thomas Y. Crowell, 1972.

Thornbury, William D. *Principles of Geomorphology*. New York: John Wiley & Sons, 1965.

U.S. Geological Survey, National Park Service, Coast and Geodetic Survey, and Forest Service. *The Hebgen Lake, Montana, Earthquake of August 17, 1959*. U.S. Geological Survey Professional Paper 435. Washington, DC: Government Printing Office, 1964.

Missouri River at Judith Landing, Montana. [*Photo by Scott T. Smith.*]

14

River Systems and Landforms

Earth's rivers and waterways form vast arterial networks that both shape and drain the continents, transporting the by-products of weathering, mass movement, and erosion. To call them Earth's lifeblood is no exaggeration, inasmuch as rivers redistribute mineral nutrients important for soil formation and plant growth and generally serve society in many ways. Not only do rivers provide us with essential water supplies, but they also receive, dilute, and transport wastes, provide critical cooling water for industry, and form one of the world's most important transportation networks. Rivers have been of fundamental importance throughout human history. This chapter discusses the dynamics of river systems and their landforms.

At any one moment approximately 1250 km^3 (300 mi^3) of water flows through Earth's waterways. Even though this volume represents only 0.003% of freshwater and 0.0001% of all water, the work performed by this energetic flow makes it the dominant agent of landmass denudation. Figure 9-14 portrays North America's surface runoff, our basic water supply. Of the world's rivers, those with the greatest flow volume per unit of time, or *discharge,* are the Amazon and the Paraná of South America, the Zaire (Congo) of Africa, the Ganges of India, and the Chang Chiang (Yangtze) of Asia (see Table 9-3). In North America, the greatest discharges are from the Missouri-Ohio-Mississippi, Saint Lawrence, and Mackenzie river systems.

FLUVIAL PROCESSES AND LANDSCAPES

Stream-related processes are termed **fluvial** (from the Latin *fluvius,* meaning "river"). Geographers seek to describe recognizable patterns and the fluvial processes that created them. Fluvial systems, like all natural systems, have characteristic processes and produce predictable landforms. Yet, a stream system behaves with randomness, unpredictability, and disorder as described within chaos theory.

Insolation is the driving force of fluvial systems, operating through the hydrologic cycle and working under the influence of gravity. Individual streams vary greatly, depending on the climate in which they operate, the variety of surface composition and topography over which they flow, the nature of vegetation and plant cover, and the length of time they have been operating in a specific setting.

Denudation by wind, water, and ice that dislodges, dissolves, or removes surface material is called **erosion.** Thus, streams produce *fluvial erosion,* which supplies weathered and wasted sediments for **transport** to new locations, where they are laid down in a process known as **deposition.** A stream is a mixture of water and solids—carried in solution, suspension, and by mechanical transport. **Alluvium** is the general term for the clay, silt, and sand transported by running water.

The work of streams modifies the landscape in many ways. Landforms are produced by the erosive action of flowing water and the deposition of stream-transported materials. Rivers create floodplains during episodic flood events, which humans attempt to moderate and control. Let's begin our study by examining a basic fluvial unit—the drainage basin.

The Drainage Basin System

Streams are organized into areas or regions called **drainage basins.** A drainage basin is the spatial geomorphic unit occupied by a river system. A drainage basin is defined by ridges that form *drainage divides;* that is, the ridges are the dividing lines that control into which basin precipitation drains. Drainage divides define a **watershed,** the catchment area of the drainage basin. The United States and Canada are divided by several **continental divides;** these are extensive mountain and highland regions that separate drainage basins, sending flows either to the Pacific, or to the Gulf of Mexico and the Atlantic, or to Hudson Bay and the Arctic Ocean (Figure 14-1). The principal drainage basins in the United States and Canada are mapped in Figure 14-2. These form water-resource regions and provide a spatial framework for water studies and water planning.

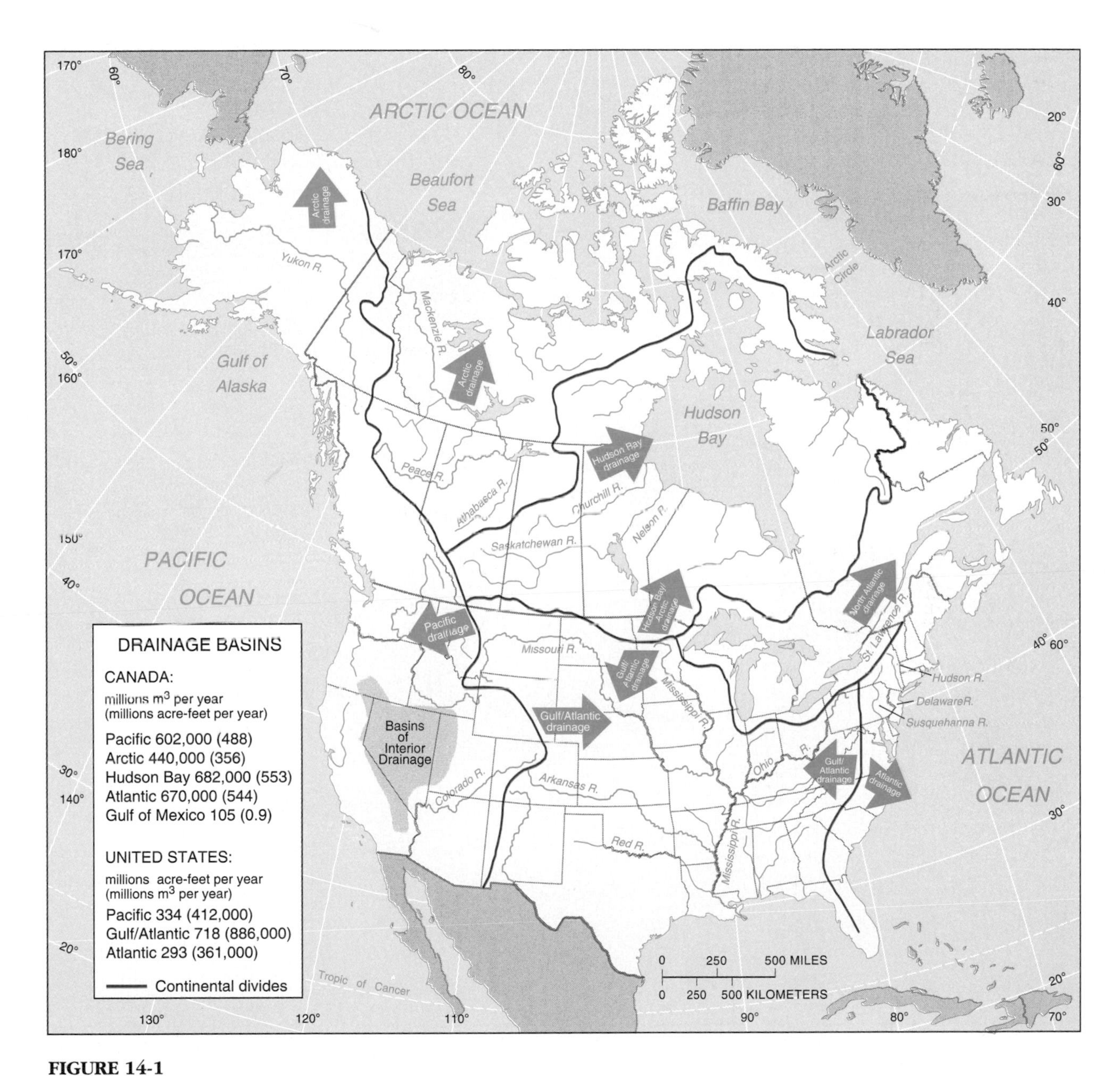

FIGURE 14-1
Continental divides (purple lines) separate the major drainage basins that empty into the Pacific, Atlantic, Gulf of Mexico, and to the north through Canada into Hudson Bay and the Arctic Ocean. [After U.S. Geological Survey and *The National Atlas of Canada,* Energy, Mines, and Resources Canada.]

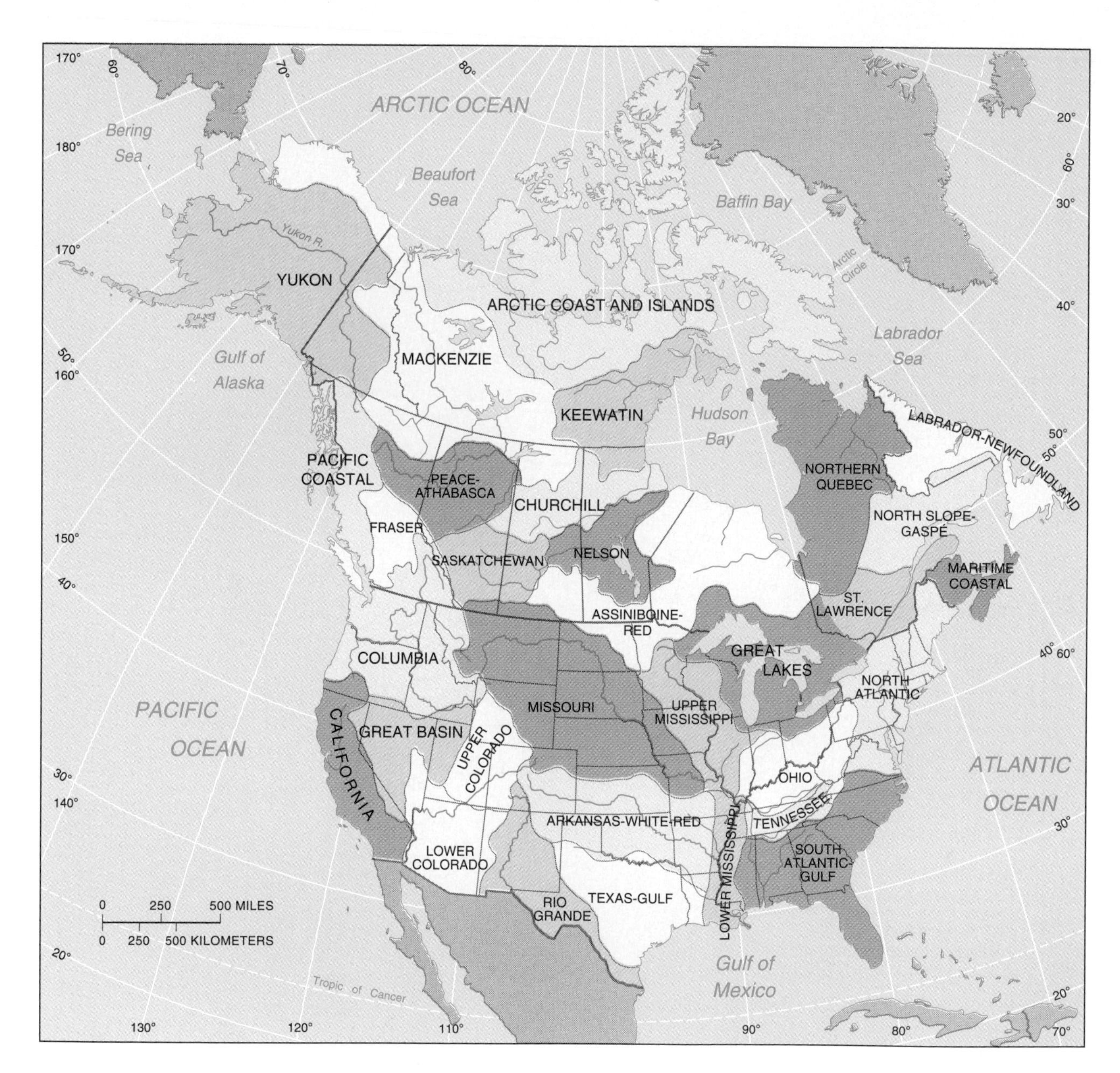

FIGURE 14-2
Major drainage basins in the United States and Canada. [Data from Environment Canada, *Currents of Change—Inquiry on Federal Water Policy—Final Report 1985*, and from U.S. Water Resources Council, *The Nation's Water Resources*.]

A major drainage basin system, such as the one created by the Mississippi-Missouri-Ohio river system, is made up of many smaller drainage basins, which in turn comprise even smaller basins, each divided by specific watersheds. Each drainage basin gathers and delivers its precipitation and sediment to a larger basin, concentrating the volume in the main stream. For example, rainfall in north-central Pennsylvania generates the streams that flow into the Allegheny River. The Allegheny River then joins with the Monongahela River at Pittsburgh to form the Ohio River. The Ohio flows southwest and connects with the Mississippi River at Cairo, Illinois, and eventually flows on past New Orleans to the Gulf of Mexico (Figure 14-1). Each contributing tributary adds its discharge and sediment load to the larger river. In our example, sediment weathered and eroded in north-central Pennsylvania is transported thousands of kilometers and accumulates as the Mississippi delta on the floor of the Gulf of Mexico.

Drainage basins are open systems whose inputs include precipitation, the minerals and rocks of the regional geology, and changes of energy with both the uplift and subsidence provided by tectonic activities. System outputs of water and sediment leave through the mouth of the river. Change that occurs in any portion of a drainage basin can affect the entire system as the stream adjusts to carry the appropriate load relative to discharge and velocity. If a region is uplifted or brought to a geomorphic threshold by other factors, the relationships within the drainage basin system are put into a disequilibrium, producing a transition period. A stream drainage system exhibits a constant struggle toward an equilibrium among interacting variables of discharge, transported load, channel characteristics, and channel slope.

An example of a specific drainage basin is the Delaware River basin, within the Atlantic Ocean drainage region (Figure 14-3). The Delaware River headwaters are in the Catskill Mountains of New York. This basin encompasses 33,060 km^2 (12,890 mi^2) and includes parts of five states in the stream's length, 595 km (370 mi) from headwaters to the mouth.

The entire basin lies within a humid, temperate climate and receives an average annual precipita-

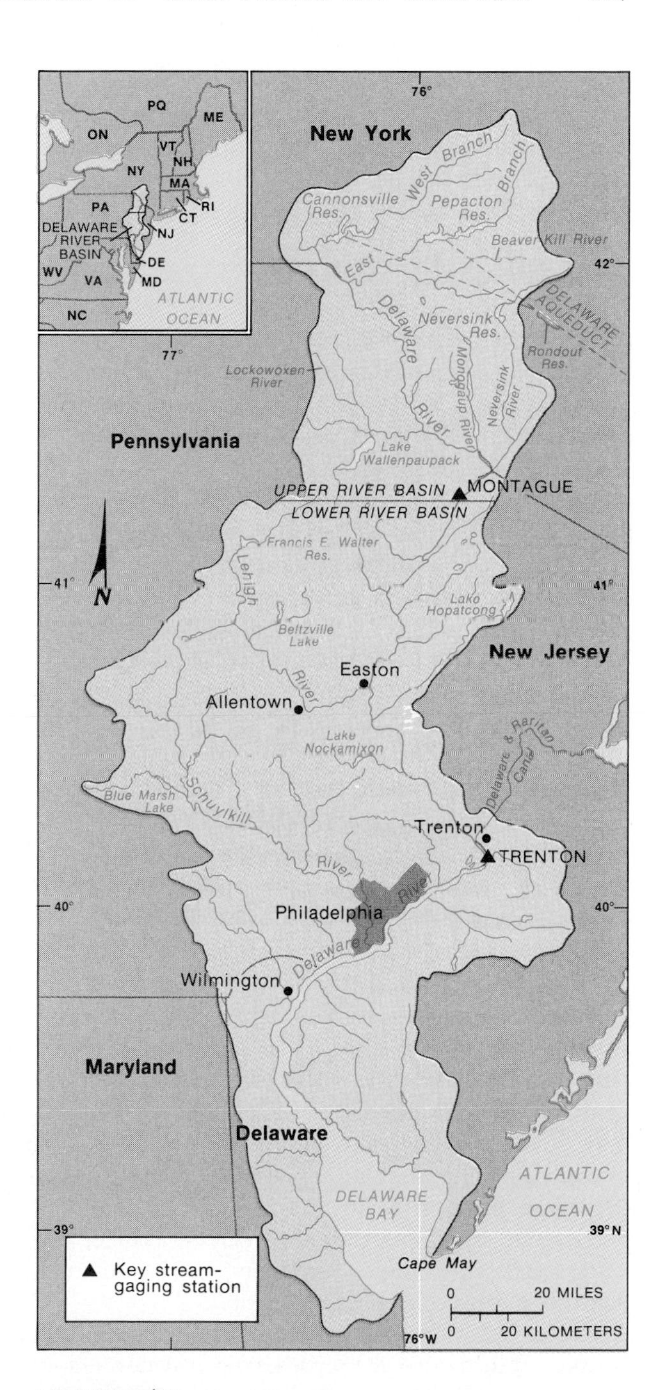

FIGURE 14-3
The Delaware River drainage basin. [After U.S. Geological Survey, 1986, "Hydrologic Events and Surface Water Resources," *National Water Summary 1985,* Water Supply Paper 2300, Washington, DC: Government Printing Office, p. 30.]

tion of 120 cm (47.2 in.). Topography varies from coastal plains of low relief to the mountains of the Appalachians in the north. The river provides water for an estimated 20 million people in the region, within and beyond the basin itself. Several major conduits export water from the Delaware River. Note on the map the Delaware Aqueduct to New York City (in the north) and the Delaware & Raritan Canal (near Trenton). Several reservoirs in the drainage basin enhance water availability and control of the drainage basin system. Delaware Bay, which eventually enters the Atlantic Ocean, marks the end of the river system.

The USGS launched a study of the Delaware River basin in 1988 to research the potential impact of future climate change, namely global warming, on water resources. The study includes changes in streamflow, irrigation demand, reduction in soil moisture storage, possible saltwater intrusion near the ocean, and problems associated with sea-level rise.

Drainage Density and Patterns. One measure of the overall efficiency of a drainage basin is expressed as the **drainage density,** which is determined by dividing the length of all the stream channels by the area of their drainage basin. The number and length of channels in a given area are an expression of the landscape's regional topographic texture and surface appearance. Initially, water moves downslope in a thin film as **sheet flow,** or overland flow. This surface runoff concentrates in *rills,* or small-scale indentations, which may develop further into *gullies* and *stream courses* with specific drainage patterns. The resultant **drainage pattern** is an arrangement of channels determined by slope, differing rock resistance, climatic and hydrologic variability, and structural controls imposed by the landscape.

The seven most common drainage patterns encountered in nature are represented in Figure 14-4. The most familiar pattern is *dendritic* (a); this treelike pattern is similar to that of many natural systems, such as capillaries in the human circulatory system or the vein patterns in leaves. Energy expended by this drainage system is efficient because the overall length of the branches is minimized. The *Landsat* image in Figure 14-5 is of the

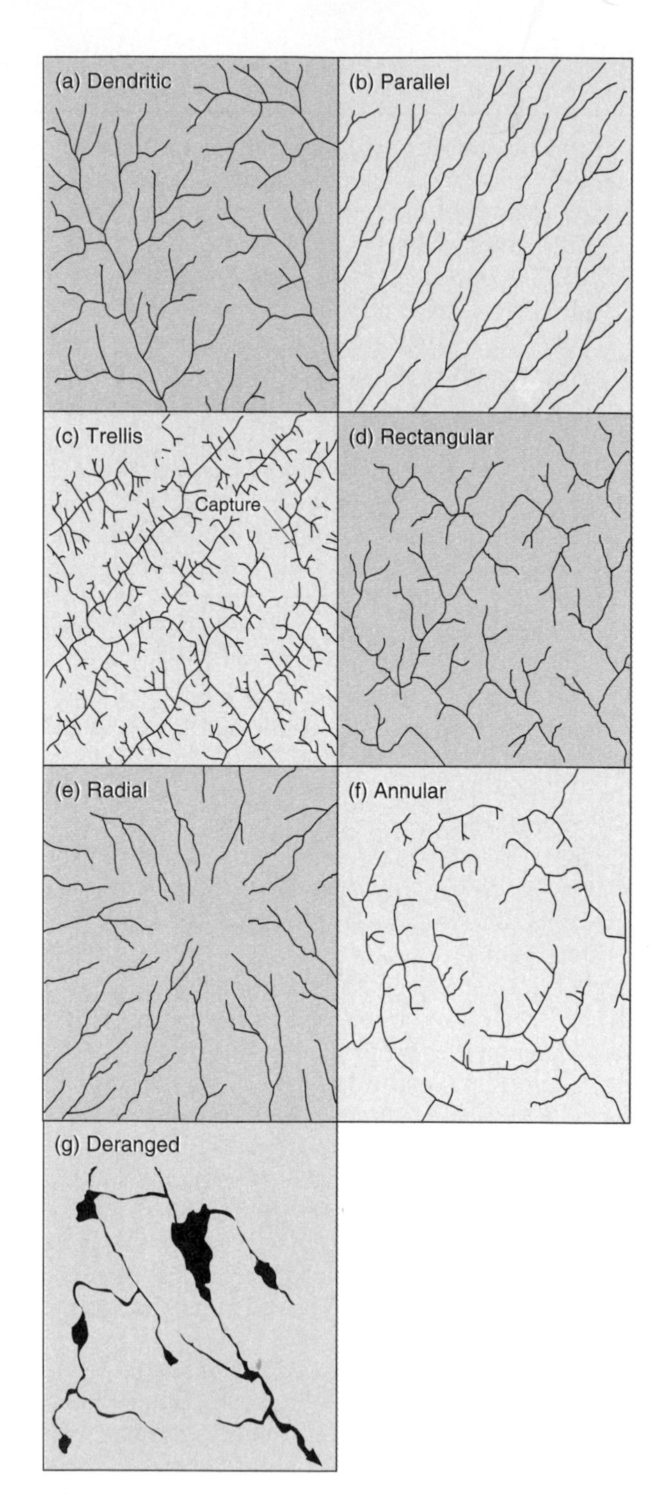

FIGURE 14-4
The seven most common drainage patterns. [After A. D. Howard, "Drainage Analysis in Geological Interpretation: A Summation," *Bulletin of American Association of Petroleum Geologists* 51, 1967, p. 2248. Adapted by permission.]

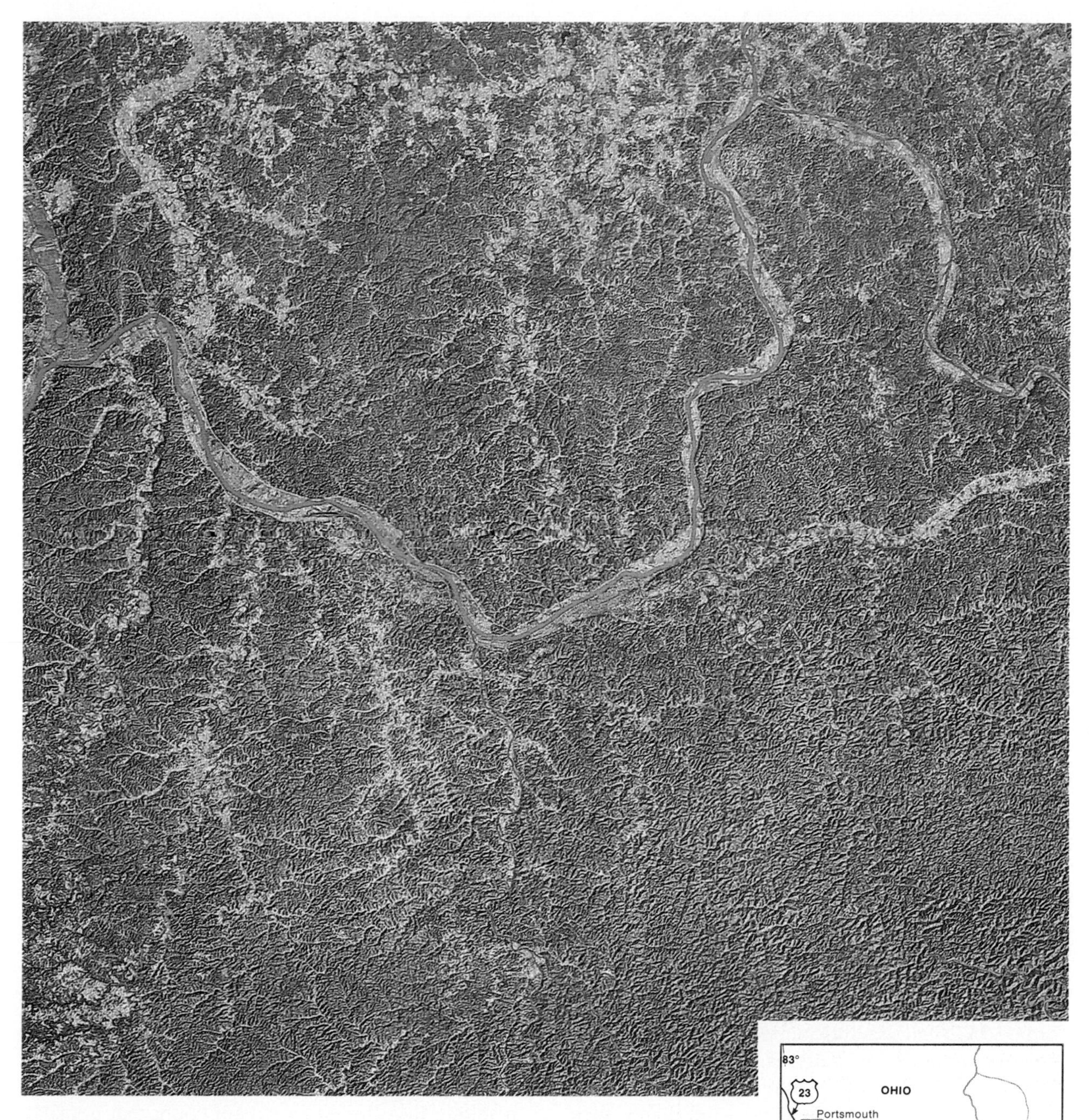

FIGURE 14-5
Drainage pattern of a highly dissected topography around the junction of the West Virginia, Ohio, and Kentucky borders, all of which are formed by rivers. [*Landsat* image from NASA.]

Ohio River area near the junction of the West Virginia, Ohio, and Kentucky borders. The high-density drainage pattern and intricate dissection of the land occur because of the region's generally level sandstone, siltstone, and shale strata, which are easily eroded, given the humid mesothermal climate. Fluvial action, along with other denudation processes, is responsible for this dissected topography.

The *trellis* drainage pattern (c) is characteristic of dipping or folded topography, which exists in nearly parallel mountains of the Ridge and Valley Province of the East, where drainage patterns are influenced by rock structures of variable resistance and folded strata. Figure 12-13 presents an orbital image of this region that shows this distinctive drainage pattern. The principal streams are directed by the parallel folded structures, whereas smaller streams are at work on nearby slopes, joining the main streams at right angles.

The sketch in Figure 14-4c suggests that the headward-eroding part of one stream could break through a drainage divide and *capture* the headwaters of another stream in the next valley, and indeed this does happen. The sharp bends in two of the streams in the illustration are called *elbows of capture* and are evidence that one stream has breached a drainage divide. This type of capture, or stream piracy, can occur in other drainage patterns.

The remaining drainage patterns in Figure 14-4 are caused by other specific structural conditions. *Parallel* drainage (b) is associated with steep slopes and some relief. A *rectangular* pattern is formed by a faulted and jointed landscape, directing stream courses in patterns of right-angle turns (d). A *radial* drainage pattern (e) results from streams flowing off a central peak or dome, such as occurs on a volcanic mountain. *Annular* patterns (f) are produced by structural domes, with concentric patterns of rock strata guiding stream courses. In areas having disrupted surface patterns, such as the glaciated shield regions of Canada and northern Europe, a deranged pattern (g) is in evidence, with no clear geometry in the drainage and no true stream valley pattern.

These seven drainage patterns are directed by the structure and relief of the land. Drainage patterns also occur that are in discordance with the landscape through which they flow. A drainage system may appear *superimposed* on older, buried structures uncovered by erosion. Streams in the Appalachians are considered superimposed in this manner. When a stream flows across younger, uplifted structures or retains its course through regions of recent mountain building the stream is described as *antecedent*. Examples of antecedence are the Columbia River through the Cascade mountains of Washington and the River Arun that cuts across the Himalayas.

Stream Orders. Another important way to understand the linear aspect of stream channels in a drainage basin is the concept of **stream orders,** which are convenient hierarchies of stream size and relationships in a given basin. These orders are a result of specific stream processes, meaning that a knowledge of stream orders is valuable to an understanding of the processes in operation. In 1945 the American hydrologist Robert E. Horton introduced this quantitative methodology, which was modified by the geomorphologist Arthur Strahler in the 1950s, and subsequently by others.

Think of a nesting of streams, from smallest to largest, as they join to form a main stream (Figure 14-6). A *first-order stream* has no tributaries and therefore is without smaller branches; it is the smallest stream in the system and has the smallest drainage basin. A *second-order stream* receives at least two first-order tributaries. A *third-order stream* is formed by the joining of two second-order streams; a *fourth-order stream* is formed by the joining of two third-order streams; and so forth. (However, note that the addition of a first-order stream to a second-order stream does not result in a third-order stream.) As you move upstream, the number of tributaries increases, resulting in a first-order stream count that is usually four or five times greater than the second-order count. Figure 14-6 illustrates one version of this hierarchical system.

Stream order designations proceed numerically until the main stream is reached; that order designation then applies all the way to the sea. The complexity of the Mississippi River system, the largest in the United States, makes it a 10th-order main stream. The Amazon River is identified as a 12th-order system. These higher designations generally result from increased length and drainage area.

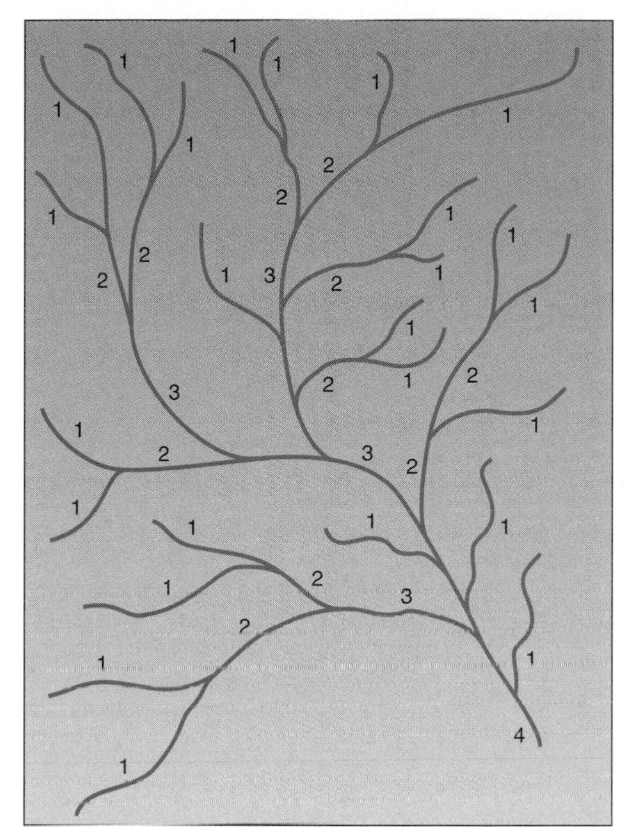

FIGURE 14-6
Stream orders of a characteristic fourth-order stream.

Streamflow Characteristics

A mass of water positioned above base level in a stream has potential energy. As the water flows downstream under the influence of gravity, this energy becomes kinetic energy. The rate of this potential energy conversion depends on the steepness of the stream channel.

Stream channels vary in width and depth. The streams that flow in them vary in velocity and in the sediment load they carry. All of these factors may increase with increasing discharge. Discharge is calculated by multiplying the velocity of the stream by its width and depth for a specific cross section of the channel as stated in the simple expression:

$$Q = wdv$$

where Q = discharge; w = channel width; d = channel depth; and v = stream velocity. As Q increases, some combination of channel width, depth, and stream velocity increases. Discharge is expressed either in cubic meters per second (m^3/s) or cubic feet per second (cfs).

Figure 14-7 illustrates the relationship of discharge to factors of width, depth, and velocity. It shows graphically that mean velocity actually increases with greater discharge downstream, despite the common misconception that streamflow becomes more sluggish. (The increased velocity downstream often is masked by reduced turbulence in the water.)

The cross-section profile of a stream varies over

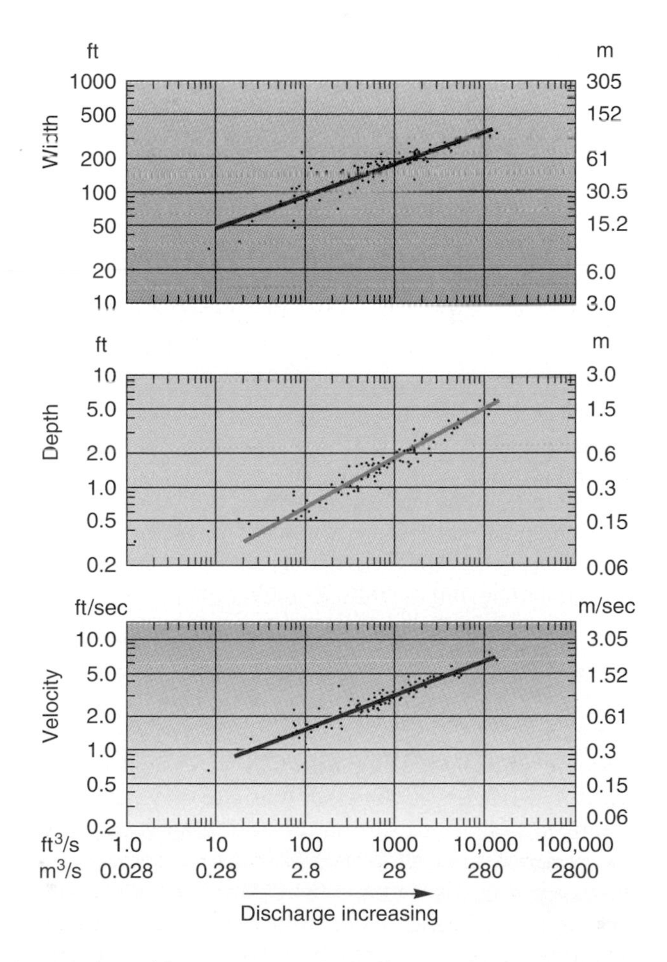

FIGURE 14-7
Relation of stream velocity, depth, and width to stream discharge at Powder River, Locate, Montana. Discharge is shown in cubic meters per second (m^3/s) and cubic feet per second (cfs). [After Luna Leopold and Thomas Maddock, Jr., 1953, "The Hydraulic Geometry of Stream Channels and Some Physiographic Implications," *U.S. Geological Survey Professional Paper 252,* Washington, DC: Government Printing Office, p. 7.]

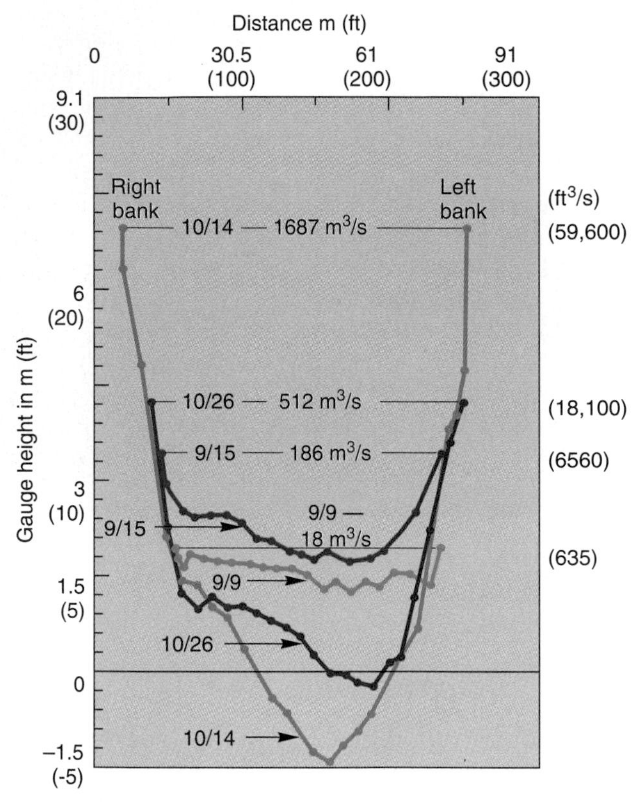

FIGURE 14-8
Stream channel cross sections showing the progress of a 1941 flood on the San Juan River near Bluff, Utah. [After Luna Leopold and Thomas Maddock, Jr., 1953, "The Hydraulic Geometry of Stream Channels and Some Physiographic Implications," *U.S. Geological Survey Professional Paper 252,* Washington, DC: Government Printing Office, p. 32.]

time, especially during heavy floods. Figure 14-8 shows changes in the San Juan River channel in Utah that occurred during a flood. The increase in discharge increases the velocity and therefore the carrying capacity of the river as the flood progresses. As a result, the river's ability to scour materials from its bed is enhanced. Such scouring represents a powerful clearing action, especially in the excavation of alluvium from the streambed by a stream in flood.

You can see in the figure that the San Juan River's channel was deepest on October 14, when floodwaters were highest. Then, as the discharge returned to normal, the kinetic energy of the river was reduced, and the bed again filled as sediments were redeposited. The process depicted in Figure 14-8 moved a depth of about 3 m (10 ft) of sediment from this cross section of the stream channel. Such channel adjustments occur as the stream system continuously works toward an equilibrium to balance discharge, velocity, and sediment load.

Stream Erosion. The erosional work of a stream carves and shapes the landscape through which it flows. Several types of erosional processes are operative. **Hydraulic action** is the work of *turbulence* in the water—the eddies of motion. Running water causes friction in the joints of the rocks in a stream channel. A hydraulic squeeze-and-release action works to loosen and lift rocks. As this debris moves along, it mechanically erodes the streambed further, through a process of **abrasion,** with rock particles grinding and carving the streambed.

The upstream tributaries in a drainage basin usually have small and irregular discharges, with most of the stream energy expended in turbulent eddies. As a result, hydraulic action in these upstream sections is at maximum, even though the coarse-textured load of such a stream is small. The downstream portions of a river, however, move much larger volumes of water past a given point and carry larger suspended loads of sediment. Thus, both volume and velocity are important determinants of the amount of energy expended in the erosion and transportation of sediment (Figure 14-9).

Figure 14-10 shows the relationships of velocity and erosion/deposition. It illustrates that sediment particles of various grain sizes are deposited at velocities below the lower line on the graph, whereas they are eroded at velocities above the upper line. In between is the "transition zone of transportation," where grains in motion remain in motion. Interestingly, finer clays have such a cohesiveness among particles that a stream can actually erode coarser sands, as well as noncohesive silts and clays, more easily than the fine clays.

Stream Transport. You may have watched a river or creek after a heavy rainfall, the water colored brown by the heavy sediment load being transported. The amount of material available to a

(a)

(b)

FIGURE 14-9
The relationship between volume and velocity in (a) a turbulent mountain stream, Alanje Rio, Costa Rica, and (b) the downstream portion of the Mississippi River near Natchez. [(a) Photo by Stephen J. Krasemann; (b) photo by author.]

stream is dependent upon topographic relief, the nature of rock and materials through which the stream flows, climate, vegetation, and the types of processes at work in a drainage basin. *Competence,* which is a stream's ability to move particles of specific size, is a function of stream velocity. The total possible load that a stream can transport is its *capacity.* Eroded materials are transported by four processes: solution, suspension, saltation, and traction (Figure 14-11).

Solution refers to the **dissolved load** of a stream, especially the chemical solution derived from minerals such as limestone or dolomite or from soluble salts. The main contributor of material in solution is chemical weathering. The undesirable salt content that hinders human use of some

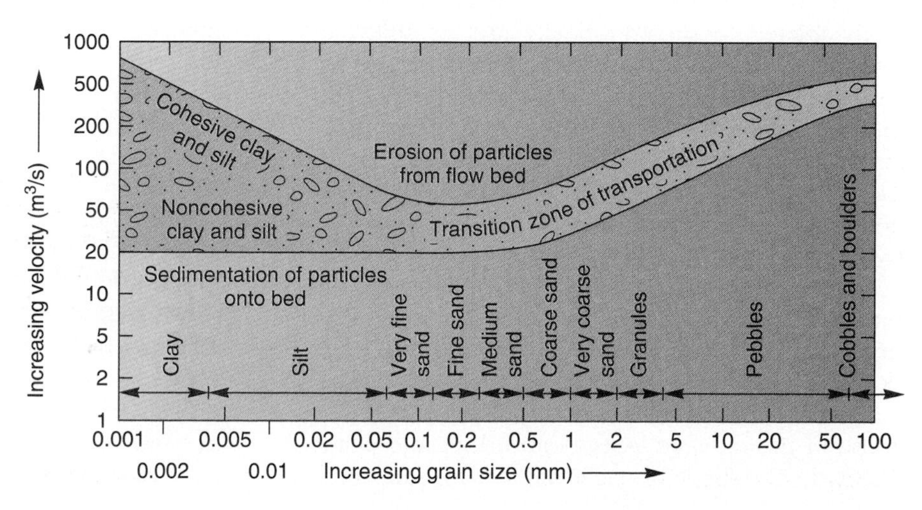

FIGURE 14-10
Relationship of velocity and sediment size to erosion, transportation, and deposition. [After Åke Sundborg (after F. Hjulström), "The River Klarälven: A Study of Fluvial Processes," *Geografiska Annaler 38,* no. 187 (1956). Adapted by permission.]

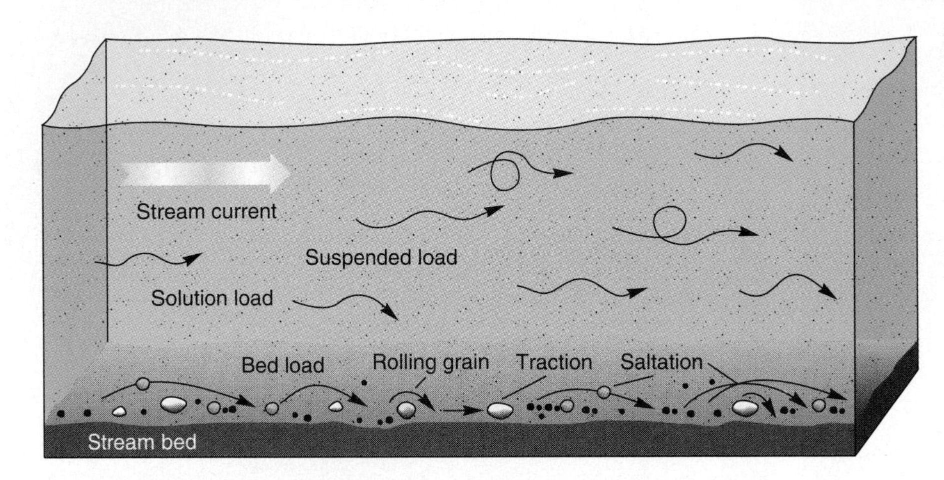

FIGURE 14-11
Fluvial transportation of eroded materials through saltation, traction, suspension, and solution.

rivers comes from dissolved rock formations and from springs in the stream channel.

The **suspended load** consists of fine-grained, clastic particles physically held aloft in the stream, with the finest particles not deposited until the stream velocity slows to near zero. Turbulence in the water, with random upward motions, is an important mechanical factor in holding a load of sediment in suspension.

The **bed load** refers to those coarser materials that are dragged along the bed of the stream by **traction** or are rolled and bounced along by **saltation** (from the Latin *saltim,* which means "by leaps or jumps"). At times the distinction between traction and saltation of bed load is difficult to determine. Particles transported by saltation are too large to remain in suspension, a determination directly related to stream velocity and its ability to retain particles in suspension. With increased kinetic energy, parts of the bed load are rafted up and become suspended load. This was demonstrated in the flood-induced channel deepening of the San Juan River shown in Figure 14-8. Saltation is also a process in the transportation of materials by wind (see Chapter 15).

The first Spanish explorers to visit the Grand Canyon reported in their journals that they were kept awake at night by the thundering sound of boulders tumbling along the Colorado River bed. Such sounds today are substantially lessened because of the reduced velocity and discharge of the Colorado resulting from the many dams and control facilities that now trap sediments and reduce bed load capacity.

Figure 14-12 shows the annual water discharge and suspended sediment load for the Colorado River at Yuma, Arizona, from 1905 to 1964. The completion of Hoover Dam in the 1930s dramatically reduced suspended sediment. The later construction of Glen Canyon Dam upstream from Hoover Dam further reduced stream *competence* and *capacity.* In fact, Lake Powell, which formed upstream behind the artificial base level of Glen Canyon Dam, is forecast to fill with sediment over the next 100 years. River flows through the Grand Canyon today frequently are blue-green rather than the muddy red of the past and can fluctuate up to 4.3 m (14 ft) at a time as hydroelectric production at Glen Canyon adjusts to daily electrical demands in distant cities.

Flow Characteristics and Channel Patterns. Flow patterns in a stream are generally of two types, laminar and turbulent. A streamlined flow of water is called a *laminar flow,* in which individual clay and other fine particles move along evenly in generally parallel flows. *Stream lines,* an imaginary set of lines connecting water molecules or fluid particles in a stream, remain distinct and in the direction of flow. Most natural streams have few stretches of laminar flow except for a thin layer near the bed of the channel.

In streams with flows of increased velocity—shallower streams, or in places where the channel

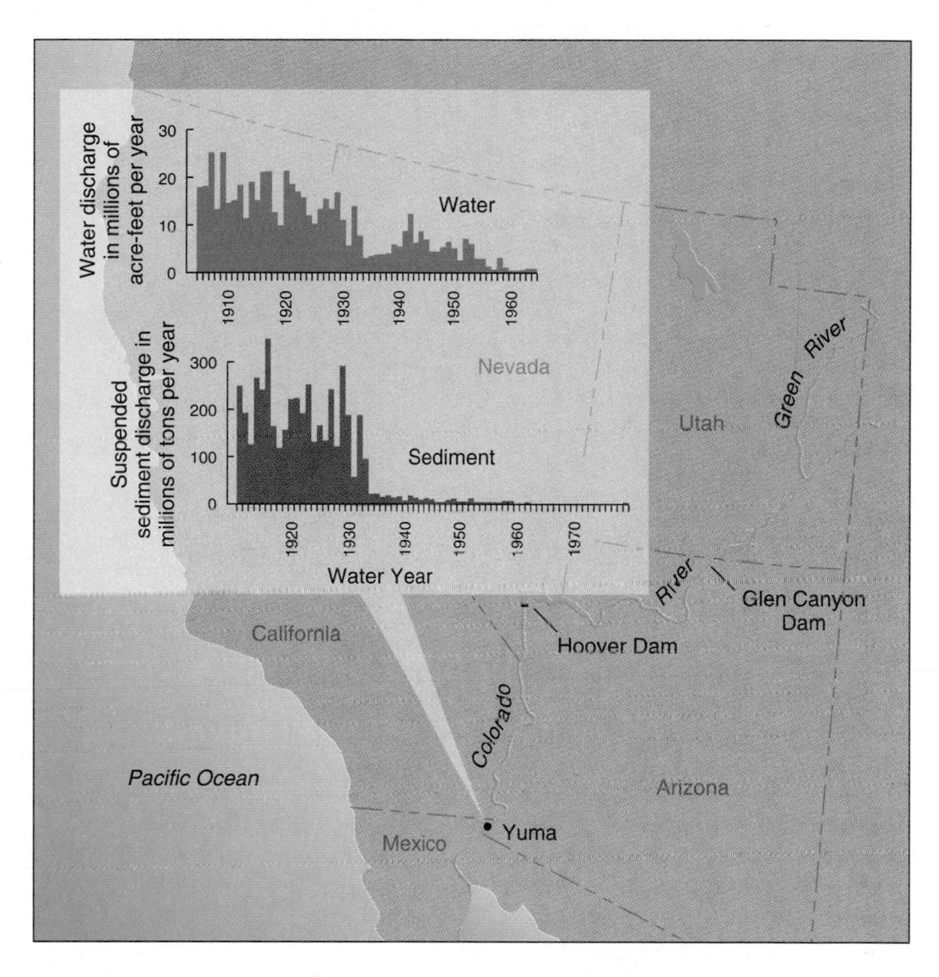

FIGURE 14-12
Water and sediment discharge at Yuma, Arizona, 1905–1964. [After U.S. Geological Survey, 1985, "Hydrologic Events, Selected Water-Quality Trends, and Ground Water Resources," *National Water Summary 1984,* Geological Survey Water Supply Paper 2275, Washington, DC: Government Printing Office, p. 55.]

is rough, as in a section of rapids—the flow becomes *turbulent.* Small eddies are caused by friction between streamflows and the channel sides and bed. Complex turbulent flows propel sand, pebbles, and even boulders, increasing the action of suspension, traction, and saltation. And as bed load material is rolled, lifted, and dragged along, channel abrasion is further enhanced. The *laminar velocity* of a stream is the greatest flow rate at which the flow remains laminar; above this rate flows become turbulent.

Viewed in cross section, the greatest velocities in a stream are near the surface at the center, corresponding with the deepest part of the stream channel (Figure 14-13). Velocities decrease closer to the sides and bottom of the channel because of the frictional drag on the water flow. In a curving stream, the maximum velocity line migrates from side to side along the channel, deflected by the curves.

Where slopes are gradual, stream channels assume a sinuous (snakelike) form weaving across the landscape. This action produces a **meandering stream,** from the Greek *maiandros,* after the ancient Maiandros River in Asia Minor (the present-day Menderes River in Turkey) that had a meandering channel pattern. The outer portion of each meandering curve is subject to the greatest scouring erosive action and can be the site of a steep bank called a **cut bank** (Figure 14-13). On the other hand, the inner portion of a meander receives sediment fill, forming a deposit called a **point bar.** As meanders develop, these scour-and-fill features gradually work their way downstream.

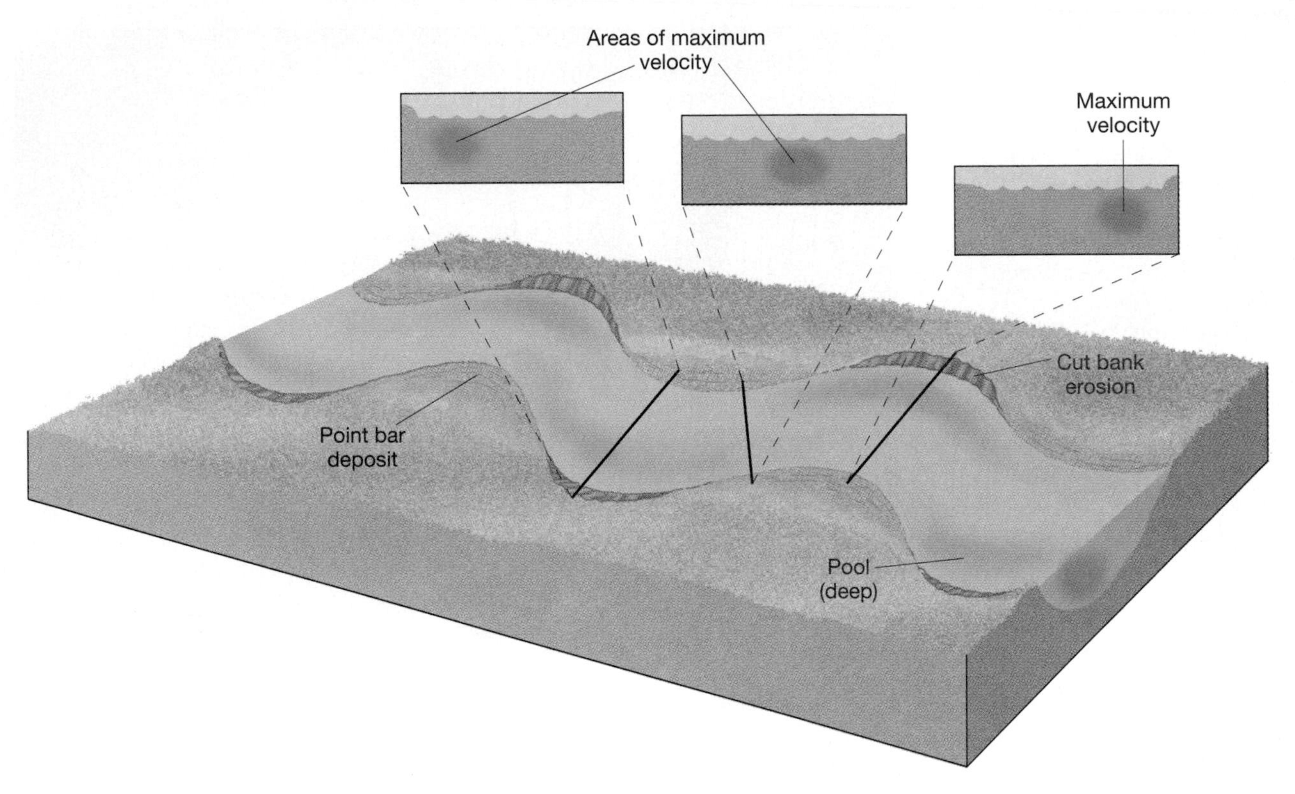

FIGURE 14-13
Aerial view and cross sections of a meandering stream, showing the location of maximum flow velocity, point bar deposits, and areas of cut bank erosion.

If the load in a stream exceeds its capacity, sediments accumulate as an **aggradation** in the stream channel (the opposite of degradation) as the channel builds up through deposition. With excess sediment, a stream becomes a maze of interconnected channels laced with sediments that form a **braided stream** pattern. Braiding often occurs when there is some reduction of discharge that affects a stream's transportation ability, such as under diurnal or seasonal conditions, or as a result of a mass-movement event such as a landslide upstream, or the presence of noncohesive banks of sand or gravel. Locally, braiding also may occur with the addition of a new sediment load, which frequently is associated with glacial meltwaters, such as in the Chitina River in Alaska, pictured in Figure 14-14.

Stream Gradient

Every stream has a degree of inclination or **gradient,** which is the rate of decline in elevation from its headwaters to its mouth, generally forming a concave-shaped slope (Figure 14-15). Characteristically, the *longitudinal profile* of a stream features a steeper slope upstream and a more gradual slope downstream. This constantly descending curve assumes a concave shape for complex reasons related to the stream's ability to do just enough work to accomplish the transport of the load it receives. The longitudinal profile of streams can be expressed mathematically.

Theoretically, a **graded stream** condition occurs when the load carried by the stream and the landscape through which it flows become mutually ad-

FIGURE 14-14
Braided stream pattern in Chitina River, Wrangell-Saint Elias National Park, Alaska. This stream reflects excessive sediment load associated with glacial meltwaters. [Photo by Tom Bean.]

justed, forming a state of dynamic equilibrium among erosion, transported load, deposition, and the stream's capacity.

Both high-gradient and low-gradient streams can achieve a graded condition. The differences in slope in each case are the result of variation in a stream's discharge and the nature of the transported load. Slope is the critical factor in a graded stream's maintenance of an equilibrium condition between water and solid materials. A stream's profile offers geographers specific diagnostic characteristics of slope, discharge, and load. J .H. Mackin, a geomorphologist, expressed this well:

> A graded stream is one in which, over a period of years, *slope* is delicately adjusted to provide, with available *discharge* and with prevailing *channel characteristics*, just the velocity required for transportation of the load supplied from the drainage basin.*

Attainment of a graded condition does *not* mean that the stream is at its lowest gradient, but rather that it represents a balance among erosion, transportation, and deposition over time along a specific portion of the stream.

One problem with applying the graded stream concept in an absolute sense, however, is that an individual stream can have both graded and ungraded portions and may have graded sections without having an overall graded slope. In fact, variations and interruptions in a graded *profile of*

* J. H. Mackin, "Concept of the Graded River," *Geological Society of America Bulletin*, 59 (1948): 463.

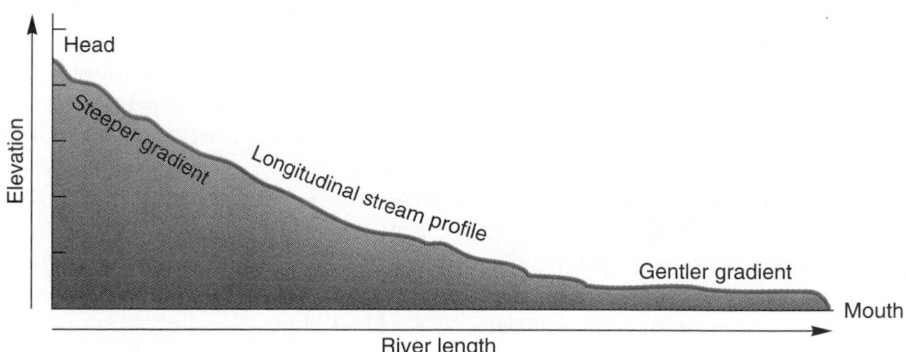

FIGURE 14-15
Idealized cross section of the longitudinal profile of a stream showing its gradient. Upstream segments have a steeper gradient; downstream the gradient is more gentle. The middle and lower portions in the illustration appear graded, or in dynamic equilibrium.

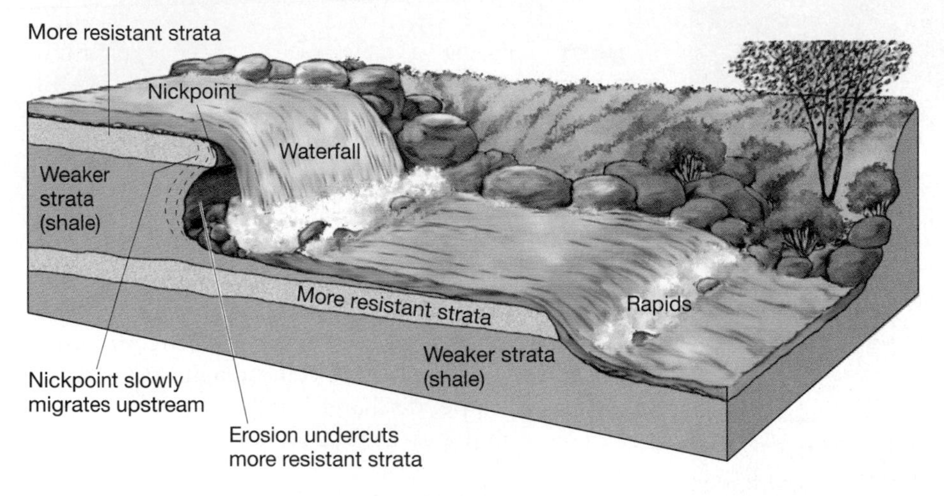

FIGURE 14-16
Longitudinal stream profile showing nickpoints produced by resistant rock strata. Potential energy is converted into kinetic energy and concentrated at the nickpoint, accelerating erosion, which eventually eliminates the feature.

equilibrium occur as a rule rather than the exception, making a universally acceptable definition difficult.

Nickpoints. When the longitudinal profile of a stream shows an abrupt change in gradient, such as at a waterfall or an area of rapids and cascades, the point of interruption is termed a **nickpoint** (also spelled knickpoint). At a nickpoint potential energy conversion to kinetic energy is concentrated and in turn works to eliminate the nickpoint. Figure 14-16 shows a stream with two such interruptions. Nickpoints can result from a stream flowing across a zone of hard, resistant rock, or from various tectonic uplift episodes, such as might occur along a fault line. Temporary blockage in a channel, caused by a landslide or a logjam, could also be considered a nickpoint; when the logjam breaks, the stream quickly readjusts its channel to its former grade.

One of the more interesting and beautiful features associated with a gradient break is a waterfall. At its edge, a stream becomes free-falling, moving at high velocity and causing increased abrasion on the channel below. This action generally undercuts the waterfall, and eventually the rock ledge at the lip of the fall collapses, causing the waterfall to shift a bit farther upstream (Figure 14-16). Thus, nickpoints migrate upstream. The height of the waterfall is gradually reduced as debris accumulates at its base.

At Niagara Falls on the Ontario-New York border, glaciers advanced and receded over the region, exposing resistant rock strata underlain by less resistant shales. As the less resistant material continues to weather away, the overlying rock strata collapse, allowing the falls to erode farther upstream toward Lake Erie. In fact, the falls have retreated more than 11 km (6.8 mi) from the steep face of the Niagara escarpment (long cliff) during the past 12,000 years. Flows over the American Falls at Niagara have even been reduced to a trickle from time to time by upstream flow controls so that the escarpment could be examined for possible reinforcement to save the waterfall (Figure 14-17). As this example demonstrates, a nickpoint should be thought of as a relatively temporary and mobile feature on the landscape.

The Davis Geomorphic Cycle. Whether or not streams follow some cyclic pattern to an ideal grade was an issue addressed earlier in this century by William Morris Davis, whose evolutionary concepts of denudation (Chapter 13) also included fluvial processes. Drawing from G. K. Gilbert, an important geomorphologist of the era, Davis incorporated the graded-stream concept into his model, identifying denudation stages of *youth, maturity,* and *old age.*

Following uplift and a related drop in base level, Davis thought that a stream would quickly downcut narrow V-shaped canyons in its youth, leaving

FIGURE 14-17
Niagara Falls, with the American Falls portion almost completely shut off by upstream controls; Horseshoe Falls in the background is still flowing. [Photo courtesy of the New York Power Authority.]

broad interfluvial uplands. Features such as waterfalls, rapids, and lakes are eliminated early as the stream evolves to a more graded profile.

In Davis's mature stage, the parent stream achieves a nearly graded profile and begins to widen its valley by cutting laterally into banks and the surrounding upland areas. A floodplain wider than the stream channel forms as the stream scours from side to side. Late maturity is achieved when the floodplain broadens and the low stream gradient produces a wide, meandering flow pattern and floodplain.

A most controversial aspect of Davis's model, and one largely rejected by later geomorphic thinking, is his final stage known as a **peneplain,** or an old erosional surface. It is doubtful that a peneplain ever has been formed, nor has significant evidence been produced to demonstrate the existence of one in the past, although many have been proposed and debated. The problem is that the tectonic processes do not remain stable or inactive for the long periods of time that his evolutionary model requires.

If renewed uplift or a drop in sea level occurs, the new base level imposed on the system stimulates renewed erosional activity, so that the region is *rejuvenated.* With rejuvenation, river meanders actively return to downcutting and become **entrenched meanders** in the landscape (Figure 14-18). Davis viewed the resulting rejuvenated landscape as one of multicyclic geomorphology, or a composite topography.

Today, the functional dynamic equilibrium model has replaced Davis's cyclic model. Land-

(a)

(b)

FIGURE 14-18
Rejuvenated (uplifted) Colorado Plateau landscape incised by entrenched meanders of the San Juan River near Mexican Hat, Utah. [Aerial photo by Betty Crowell; surface photo by Randall Christopherson.]

scapes simply do not provide enough clear evidence to indicate a particular landform history, as inferred from cyclic models. Instead, the dynamic equilibrium model emphasizes the effects of individual processes interacting in a slope system. Stream form and behavior result from complex interactions of slope, discharge, and load, all of which are variable within different climates and with different rock types.

However, as suggested by Schumm and Lichty, two modern geomorphologists, the validity of cyclic or functional landscape models may depend on the time frame. Over the long span of geologic time, cyclic models of evolutionary development might explain, for example, aspects of the disappearance of entire mountain ranges through denudation. But these generalizations would not be applicable to *steady time,* the adjustments ongoing in a portion of a drainage basin. In between these two time frames lies the realm of *graded time,* or the conditions of dynamic equilibrium.

Stream Deposition

Deposition is the next logical event after weathering, mass movement, erosion, and transportation. In deposition, a stream deposits alluvium, or unconsolidated sediments, thereby creating specific depositional landforms.

As discussed earlier, stream meanders tend to migrate downstream through the landscape. As a result, the landscape near a meandering river bears meander scars of residual deposits from the previous river channels. Former point-bar deposits leave low-lying ridges, creating a bar-and-swale relief (ridges and slight depressions). The photograph of the Itkillik River in Alaska, Figure 14-19a, shows both meanders and meander scars.

When a meandering stream erodes its outside bank as the curve migrates downstream, the neck of land created by the looping meander eventually erodes through and forms a *cutoff* (Figure 14-19b). When the former meander becomes isolated from

(a)

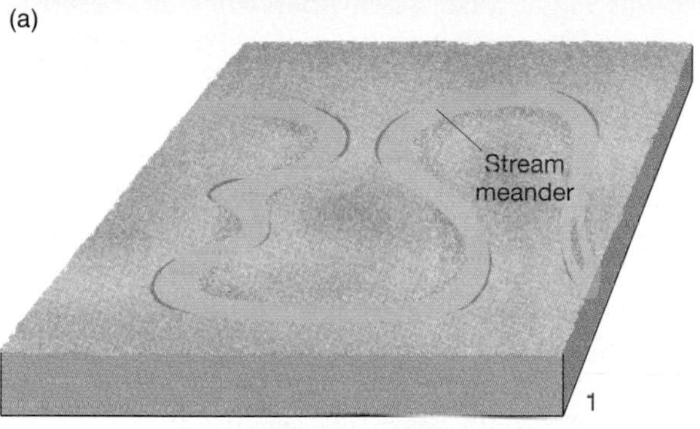

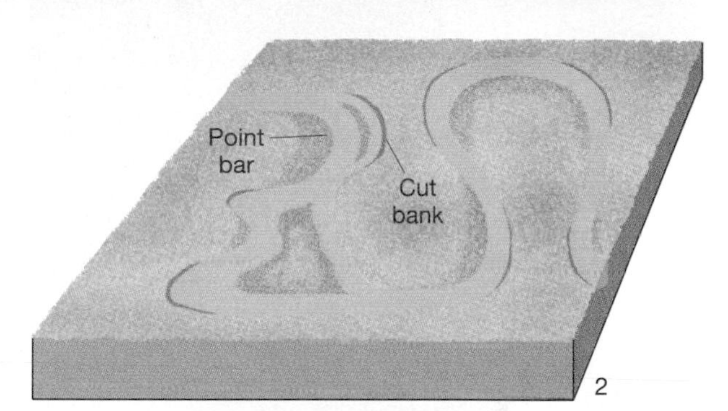

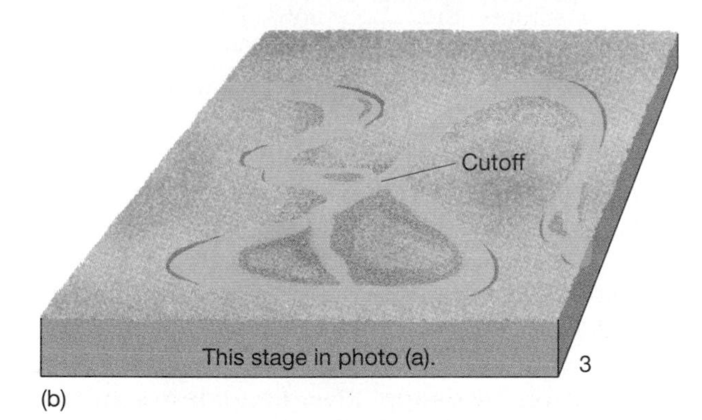

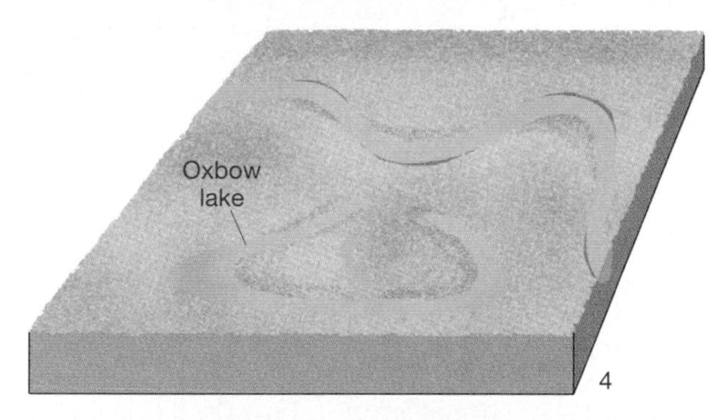

(b)

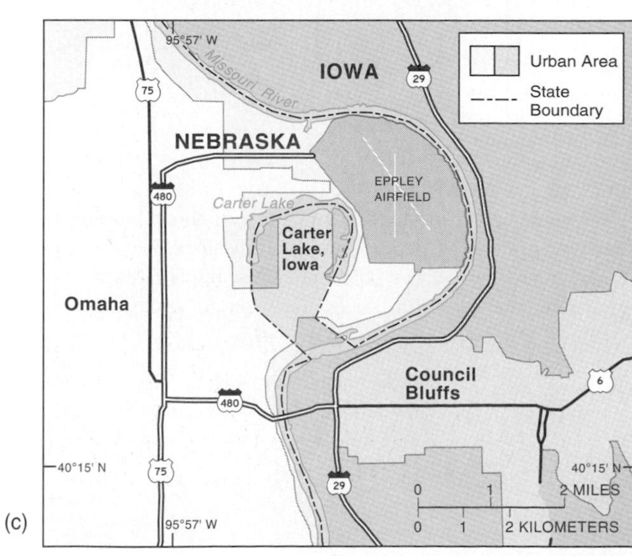

(c)

FIGURE 14-19
The evolution of meanders into an oxbow lake. (a) Itkillik River in Alaska; (b) development of a river meander and oxbow lake simplified in four stages. (c) Carter Lake, Iowa, sits within the curve of a former meander that was cut off by the Missouri River. The city and oxbow lake remain part of Iowa even though they are stranded within Nebraska. [(a) U.S. Geological Survey photo.]

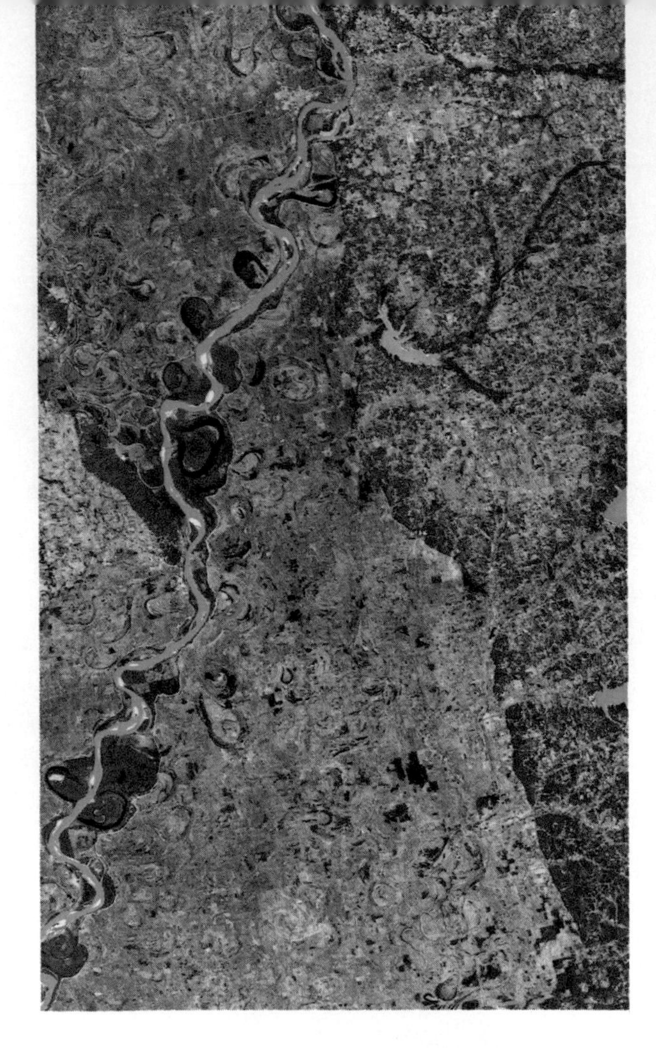

FIGURE 14-20
The Mississippi River forms a portion of the Mississippi-Arkansas border near Senatobia, Mississippi. Characteristic meander patterns and scars of former channels are visible in the *Landsat* image. [Image by GEOPIC, Earth Satellite Corporation.]

the rest of the river, the resulting **oxbow lake** may gradually fill in with silt or may again become part of the river when it floods. The Mississippi River is many miles shorter today than it was in the 1830s because of artificial cutoffs that were dredged across these necks of land to improve navigation and safety.

Streams often are used as natural political boundaries, but it is easy to see how disagreements might arise when boundaries are based on river channels that shift around. For example, the Ohio, Missouri, and Mississippi rivers can shift their positions and, therefore, the boundaries based upon them quite rapidly during times of flood. Carter Lake, Iowa, provides us with a case in point (Figure 14-19c). The Nebraska-Iowa border was originally placed mid-channel in the Missouri River. In 1877, the meander loop that curved around the town of Carter Lake was cut off by the river, leaving the town surrounded by Nebraska. The new oxbow lake was called Carter Lake and still constitutes the state line. Today, the state boundary follows the former channel meander, placing the town of Carter Lake, Iowa clearly within Nebraska!

Boundaries should always be fixed by surveys independent of river locations. Such surveys have been completed along the Rio Grande near El Paso, Texas, and along the Colorado River, between Arizona and California, permanently establishing political boundaries separate from changing river locations.

The lower Mississippi River in northwestern Mississippi, shown in the *Landsat* image in Figure 14-20, exhibits characteristic meandering scars: meander bends, oxbow lakes, natural levees, point bars, and cut banks. This portion of the river forms the Mississippi-Arkansas border.

Floodplains. The low-lying area near a stream channel that is subjected to recurrent flooding is a **floodplain.** It is formed when the river leaves its channel during times of high flow. Thus, when the river channel changes course or when floods occur, the floodplain is inundated. When the water

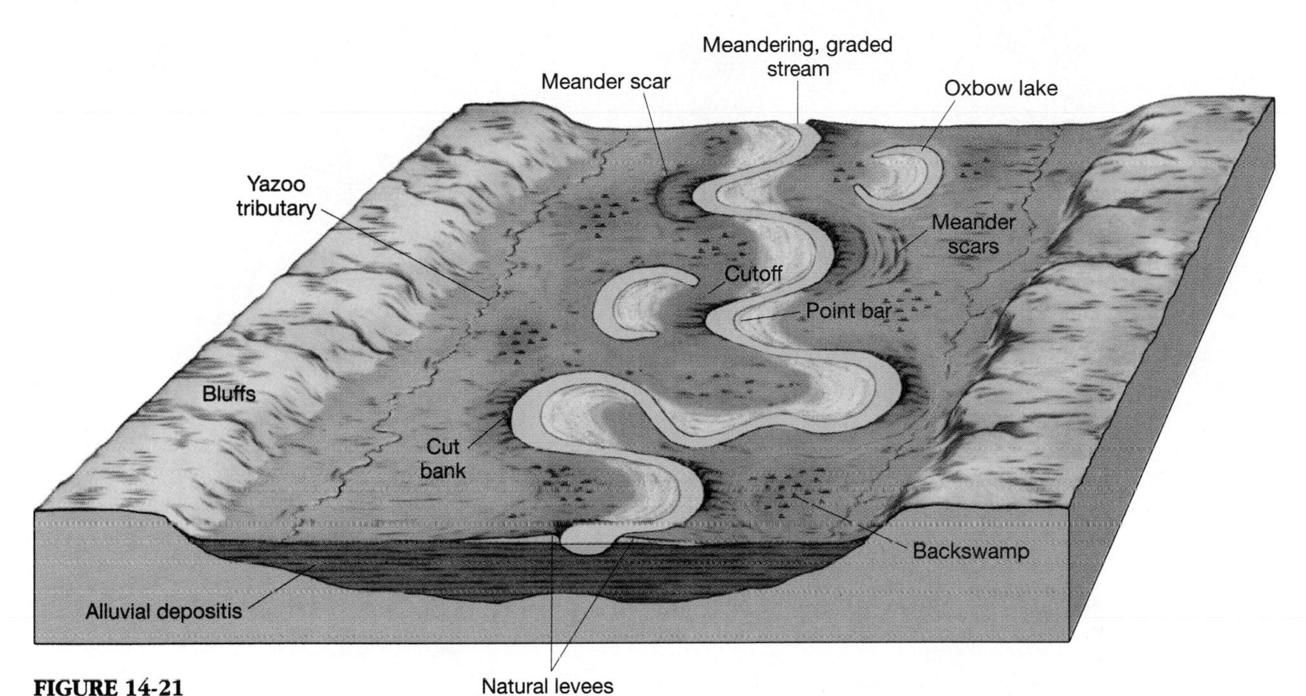

FIGURE 14-21
Typical floodplain landscape and related landscape features.

recedes, alluvial deposits generally mask the underlying rock. Figure 14-21 illustrates a characteristic floodplain, with the present river channel embedded in the plain's alluvial deposits.

On either bank of most streams, **natural levees** develop as by-products of flooding. When flood waters arrive, the river overflows its banks, loses velocity as it spreads out, and drops a portion of its sediment load to form the levees. Larger sand-sized particles drop out first, forming the principal component of the levees, with finer silts and clays deposited farther from the river. Successive floods increase the height of the levees and may even raise the overall elevation of the channel bed so that it is *perched* above the surrounding floodplain.

Notice on Figure 14-21 an area labeled **backswamp** and a stream called a **yazoo tributary.** The natural levees and elevated channel of the river prevent this tributary from joining the main channel, so it flows parallel to the river and through the backswamp area. (The terminology was derived from the Yazoo River in the southern part of the Mississippi floodplain.)

Society often chooses to build cities and conduct activities on floodplains despite the threat of flooding. People often are encouraged by government assurances of artificial protection from floods or of disaster assistance if there are floods. Government assistance may be provided in building artificial levees on top of natural levees. Artificial levees do increase the capacity in the channel, but they also lead to even greater floods when they are topped or when they fail. The catastrophic floods along the Mississippi River and its tributaries in 1993 illustrate the risk of floodplain settlement (Figure 14-22).

Perhaps the best use of some floodplains is for agriculture because inundation generally delivers nutrients to the land with each new alluvial de-

FIGURE 14-22
A farm 32 km (20 mi) south of Des Moines, Iowa, is inundated during the Midwest floods of 1993. [Photo by Les Stone/Sygma.]

posit. However, some floodplains that are covered with sand and gravel are less suitable for agriculture. Are there river floodplains in the vicinity where you live? If so, how would you assess their status relative to people's hazard perception and local planning and zoning?

Stream Terraces. Several factors may rejuvenate stream energy and stream-landscape relationships so that a stream again scours downward with increased erosion. The resulting entrenchment of the river into its own floodplain produces **alluvial terraces** on either side of the valley, which look like topographic steps above the river. Alluvial terraces generally appear paired at similar elevations on either side of the valley (Figure 14-23). If more than one set of paired terraces is present, the assumption is that the valley has gone through more than one episode of rejuvenation.

If the terraces on either side of the valley do not

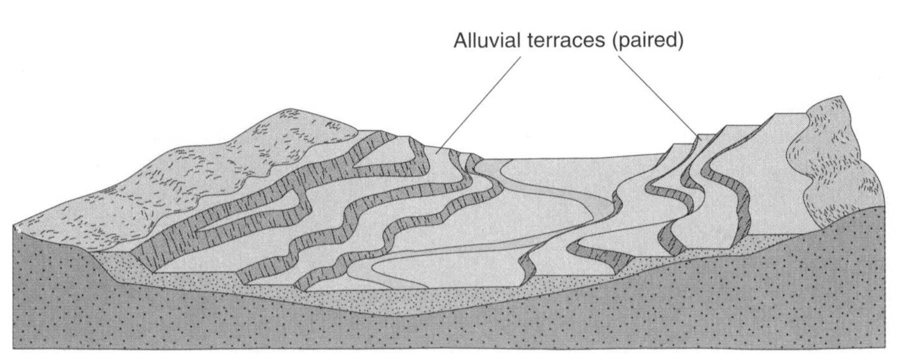

FIGURE 14-23
Alluvial terraces are formed as a stream cuts into alluvial fill. [After W. M. Davis, *Geographical Essays*. New York: Dover, 1964 (1909), p. 515.]

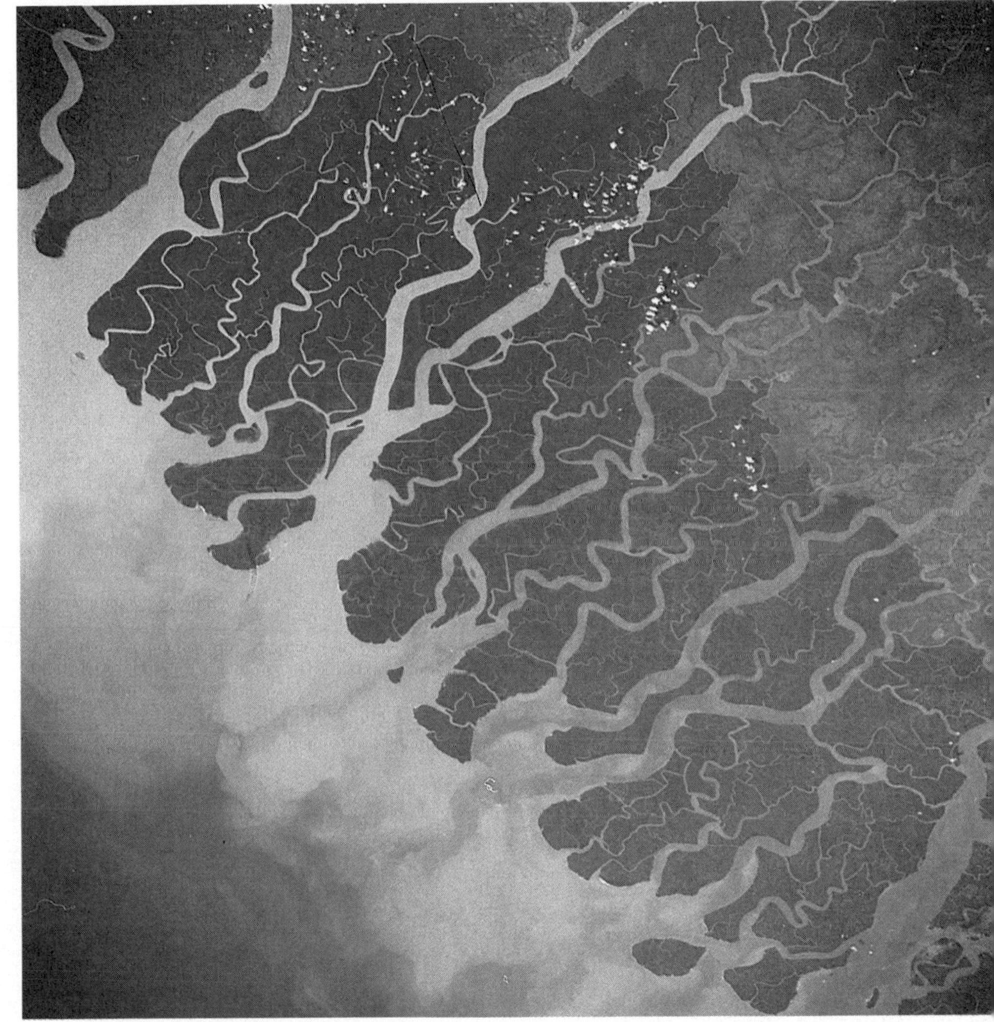

FIGURE 14-24
The complex distributary pattern in the "many mouths" of the Ganges River delta in Bangladesh and extreme eastern India is visible from orbit. [Space Shuttle photo courtesy of the National Aeronautics and Space Administration.]

match in elevation, then entrenchment actions must have been continuous as the river meandered from side to side, with each meander cutting a terrace slightly lower in elevation in the downstream direction. Thus, alluvial terraces represent an original depositional feature, a floodplain, which is subsequently eroded by a stream that has experienced changes in stream load and capacity.

River Deltas. The mouth of a river marks the point where the river reaches a base level. Its forward velocity rapidly decelerates as it enters a larger body of standing water, with the reduced velocity causing its transported load to be in excess of its capacity. Coarse sediments drop out first, with finer clays carried to the extreme end of the deposit. This depositional plain formed at the mouth of a river is called a **delta,** named after the Greek letter *delta:* Δ, the triangular shape of which was perceived by Herodotus in ancient times to be similar to the shape of the Nile River delta.

Each flood stage deposits a new layer of alluvium over the surface of the delta so that it grows outward. At the same time river channels divide into smaller channels known as *distributaries,* which appear as a reverse of the dendritic drainage pattern discussed earlier. The Ganges River delta features an intricate braided pattern of distributaries. Alluvium carried from deforested slopes upstream provides excess sediment that forms the many deltaic islands (Figure 14-24). The Nile River delta, described by Herodotus, forms an

FIGURE 14-25
The arcuate Nile River delta. Intensive agricultural activity is noted in false-color (red) in the delta and along the Nile River floodplain. [Image by GEOPIC, Earth Satellite Corporation.]

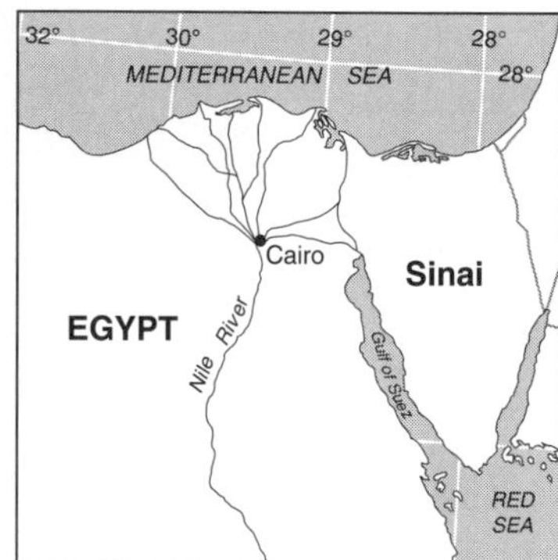

arcuate (arc-shaped) *delta* (Figure 14-25). Also arcuate in form are the Danube River delta in Romania as it enters the Black Sea and the Ganges and Indus river deltas. In another distinct form the Tiber River has an *estuarian delta,* or one that is in the process of filling an **estuary,** which is the seaward mouth of a river where the river's freshwater encounters seawater.

The Mississippi River has produced an elongate deltaic form called a *bird-foot delta,* a long channel with many distributaries and sediments carried beyond the tip of the delta into the Gulf of Mexico. Over the past 120 million years the Mississippi has deposited sediments downstream all the way from southern Illinois. During the past 5000 years, seven distinct deltaic complexes have formed along the Louisiana coast. The seventh and current subdelta has been building for at least the last 500 years. Each lobe reflects distinct course changes in the Mississippi River, probably where the river broke through its natural levees. In 1966, Kolb and Lopik, two engineering geologists, prepared a map of this recent deltaic history, which shows the relatively smaller size and difference in configuration of the present delta (Figure 14-26a).

The Mississippi River delta is therefore dynamic over time. *Landsat* images taken in 1973 and 1989 demonstrate the changes that occurred over a 16-year span (Figure 14-27). The main channel persists because of much effort and expense directed at maintaining the artificial levee system.

The tremendous weight of the sediments in the

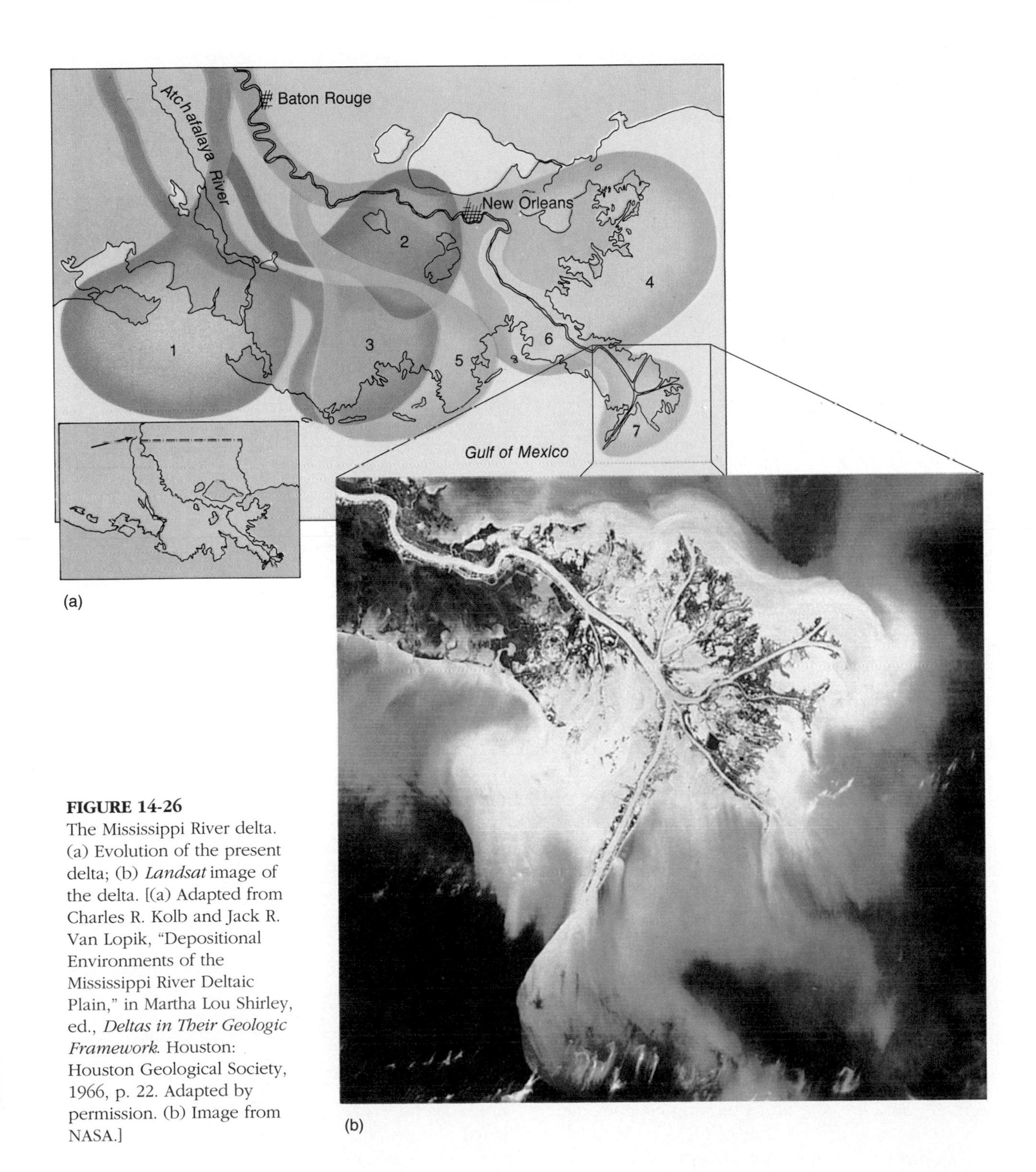

FIGURE 14-26
The Mississippi River delta. (a) Evolution of the present delta; (b) *Landsat* image of the delta. [(a) Adapted from Charles R. Kolb and Jack R. Van Lopik, "Depositional Environments of the Mississippi River Deltaic Plain," in Martha Lou Shirley, ed., *Deltas in Their Geologic Framework*. Houston: Houston Geological Society, 1966, p. 22. Adapted by permission. (b) Image from NASA.]

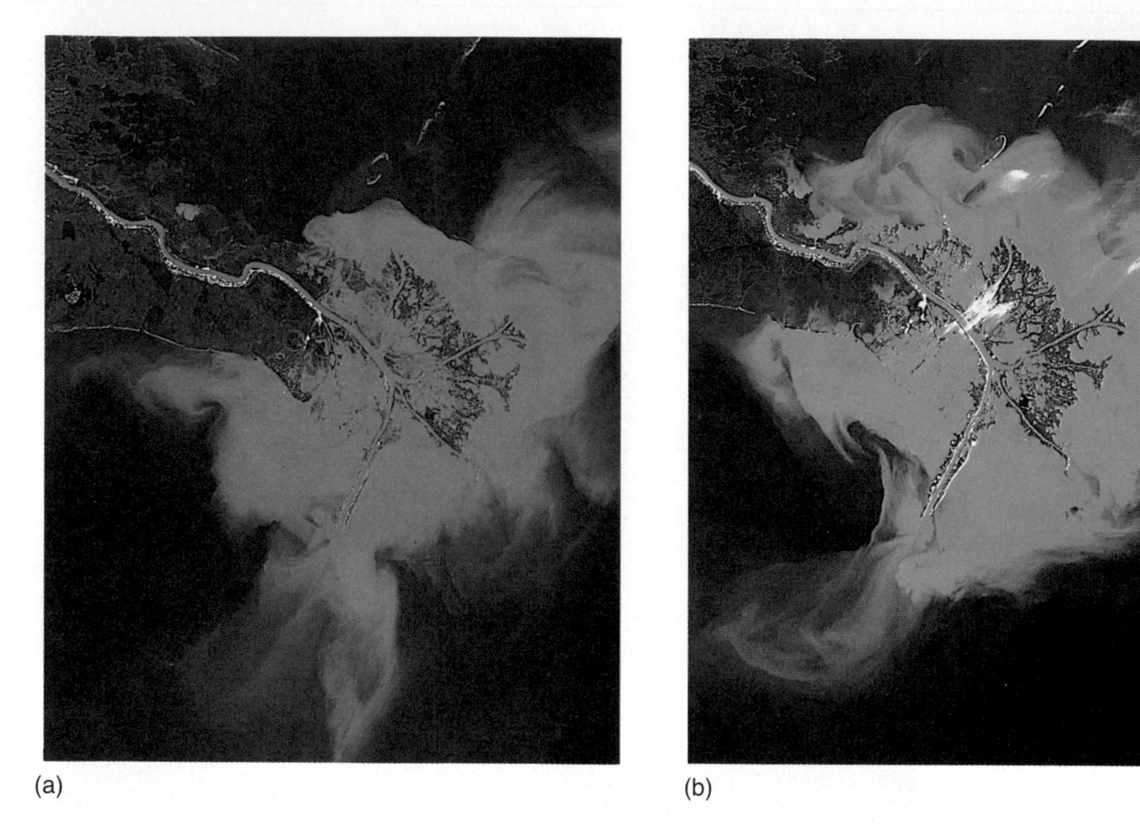

(a) (b)

FIGURE 14-27
The bird-foot delta of the Mississippi River exhibits change over time as shown in these two *Landsat* images from (a) 1973 and (b) 1989. The continuous supply of sediments, focused by controlling levees, extends ever farther into the Gulf of Mexico. [*Landsat* images from EROS Data Center.]

Mississippi River is creating isostatic adjustments in the crust. The entire region of the delta is subsiding, thereby placing ever-increasing stress on natural and artificial levees and other structures along the lower Mississippi. Severe problems are a certainty for existing and planned settlements unless further intervention or relocation efforts take place. Past protection and reclamation efforts by the U.S. Army Corps of Engineers apparently have only worsened the flood peril, as demonstrated by the 1993 floods in the Midwest.

An additional problem for the lower Mississippi Valley is the possibility, in a worst-case flood, that the river could break from its existing channel and seek a new route to the Gulf of Mexico. An obvious alternative is the Atchafalaya River, now blocked off from the Mississippi at 320 km (200 mi) from its mouth. It would be less than one-half the distance to the Gulf (see arrow in Figure 14-26a). The principal causes of such an event would be more than just water and include sediment deposition. Recent reports indicate that the occurrence of a major flood is only a matter of time and that people should prepare for the river to change channel.

The Amazon River, which exceeds 175,000 m^3/s (6.2 million cfs) discharge and carries sediments far into the Atlantic, lacks a true delta. Its mouth, 160 km (100 mi) wide, has formed a subaqueous (underwater) deltaic plain deposited on a sloping continental shelf. As a result, the mouth is braided into a broad maze of islands and channels (see Space Shuttle photo in Figure 9-15). Other rivers that lack significant sediment or whose discharge into the ocean faces strong erosive currents also lack deltaic formations. The Columbia River of the

U.S. Northwest lacks a delta because of offshore currents.

FLOODS AND RIVER MANAGEMENT

Throughout history, civilizations have settled floodplains and deltas, especially since the agricultural revolution that occurred some 10,000 years ago when the fertility of floodplain soils was discovered. Early villages generally were built away from the area of flooding, or on stream terraces, because the floodplain was the location of intense farming. However, as commerce grew, sites near rivers became important for transportation, as port and dock facilities and river bridges to related settlements were built. Also, because water is a basic industrial raw material used for cooling and for diluting and removing wastes, waterside industrial sites became desirable. However, all these human activities on vulnerable flood-prone lands require planning to reduce or avoid disaster.

Catastrophic floods continue to be a threat. In Bangladesh, intense monsoonal rains and tropical cyclones in 1988 and 1991 created devastating floods over the country's vast alluvial plain (130,000 km^2 or 50,000 mi^2). One of the most densely populated countries on Earth, Bangladesh was more than three-fourths covered by floodwaters. Excessive forest harvesting in the upstream portions of the Ganges-Brahmaputra river watersheds increased runoff and added to the severity of the flooding. Over time the increased load carried by the river was deposited in the Bay of Bengal, creating new islands (see Figure 14-24). These islands, barely above sea level, became sites of new settlements and farming villages. When the recent floodwaters finally did recede, the lack of freshwater—coupled with crop failures, disease, and pestilence—led to famine and the death of tens of thousands of people. About 30 million people were left homeless and many of the alluvial-formed islands were gone.

Floods and floodplains are rated statistically in terms of expected intervals between flooding. A *10-year flood* is one that is expected to occur once every 10 years (i.e., it has a 10% probability of occurring in any one year). Such a frequency would cause the floodplain to be labeled as one of moderate threat. A 50-year or 100-year flood is of greater and perhaps catastrophic consequence, but is also less likely to occur in a given year. These statistical estimates are probabilities that events of varying significance will occur randomly during any single year of the multiyear frame of reference. Of course, two decades might pass without a 10-year flood, or 10-year flood volumes could occur three years in a row. The record-breaking Mississippi River Valley floods in 1993 easily exceeded a 1000-year flood probability of occurrence.

Streamflow Measurement

A **flood** is a high-water level that overflows the natural (or artificial) banks along any portion of a stream. Understanding flood patterns for a drainage basin is as complex as understanding the weather, for floods and weather are equally variable, and both include a level of unpredictability. However, to develop the best possible management of flooding, the behavior of each large watershed and stream is measured and analyzed. Unfortunately, such data often are not available for small basins or for the changing landscapes of urban areas.

The key is to measure *streamflow*—the height and discharge of a stream (Figure 14-28). A **staff gauge,** a pole placed in a stream bank and marked with water heights, is used to measure stream level. With a fully measured cross section, stream level can be used to determine discharge. (Remember: discharge is equal to width times depth times velocity; accuracy improves when the entire margin of the wetted perimeter of the channel cross section is considered.) A **stilling well** is sited on the stream bank and a gauge is mounted in it to measure stream level. A movable current meter can be used to sample velocity at various locations. Approximately 11,000 stream gaging stations are used in the United States (an average of over 200 per state). Of these, 7000 have continuous stage and discharge recorders operated by the U.S. Geological Survey. Many of these stations automatically send telemetry data to satellites, from which information is retransmitted to regional centers. Environment Canada's Water Survey of Canada maintains more than 3000 gaging stations.

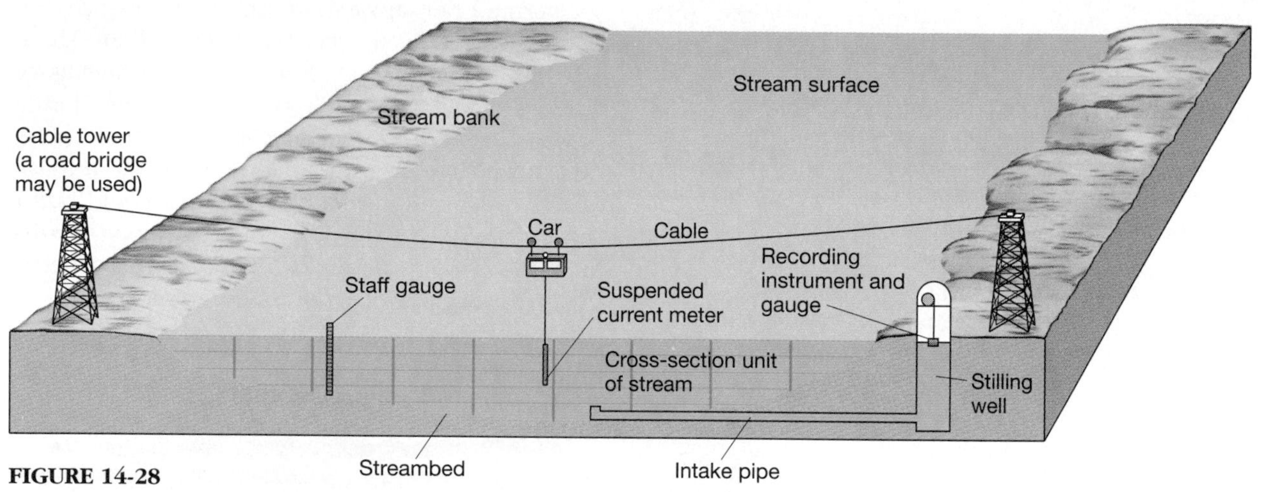

FIGURE 14-28
A typical streamflow measurement installation: staff gauge, stilling well with recording instrument, and suspended current meter.

FIGURE 14-29
Effect of urbanization on a typical stream hydrograph. Normal base flow is indicated with a dark blue line. The purple line indicates discharge after a storm, prior to urbanization. Following urbanization stream discharge dramatically increases, as shown by the light blue line.

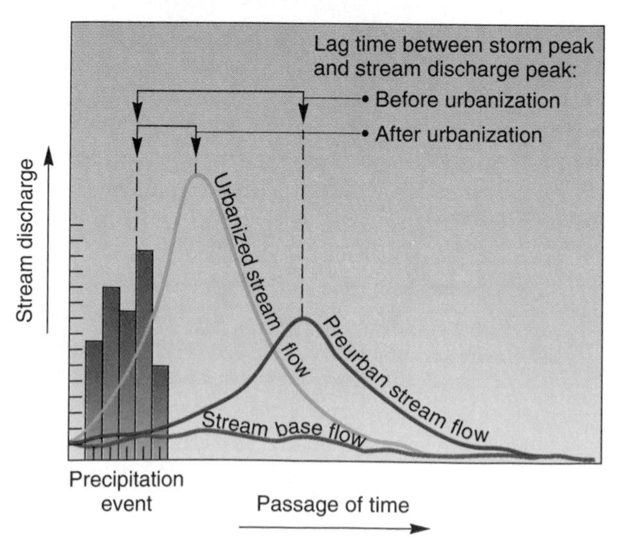

Hydrographs. If we study streamflow measurements, we can understand channel characteristics as conditions vary. A graph of stream discharge over a time period for a specific place is called a **hydrograph.** The hydrograph in Figure 14-29 shows the relationship between stream discharge and precipitation input. During dry periods, at low-water stages, the flow is described as *base flow* and is largely maintained by contributions from the local water table. When rainfall occurs in some portion of the watershed, the runoff collects and is concentrated in streams and tributaries. The amount, location, and duration of the rainfall episode determine the *peak flow*. Also important is the nature of the surface in a watershed; for example, a hydrograph for a specific portion of a stream changes after a forest fire or urbanization of the watershed.

Human activities have enormous impact on water flow in a basin. The effects of urbanization are quite dramatic, both increasing and hastening peak flow (Figure 14-29). In fact, urban areas produce runoff patterns quite similar to those of deserts. The sealed surfaces of the city drastically reduce infiltration and soil moisture recharge, behaving much like the hard, nearly barren surfaces of the desert.

A useful engineering tool is the **unit hydrograph,** which depicts a unit depth, such as a cm or in., of effective rainfall spread uniformly over a given drainage basin and received during a specific period (1 to 24 hours, depending on basin size). The unit hydrograph demonstrates how a basin behaves under various rainfall patterns and is used to design water-management facilities and to assess potential impact on cities downstream.

Additional Flood Considerations

Hurricane Camille, discussed in Chapter 8, is of interest here as well. In 1969 Camille's remnants combined with an existing low-pressure area and moved west-to-east over the entire James River basin in Virginia. According to unofficial measurements rainfall amounts reached 78.7 cm (31 in.) within the three-day storm period. The basin's narrow, steep valleys quickly concentrated runoff into a flood surge that peaked at Richmond, Virginia, as a 1000-year flood occurrence. That flood on the James River was the greatest since record keeping began in 1771.

> **Most of the residents of the mountain hollows, hamlets, and towns were asleep when the storm began. Little warning was possible. . . . Rapidly rising streams and landslides caused by the unprecedented rainfall destroyed homes as the occupants slept. . . the raging floods passed out of the small headwater streams and consolidated in the normally placid James River. . . .***

Ironically, the remnants of Hurricane Agnes in 1972—only three years later—produced severe floods in that same region which also had a magnitude reaching the 1000-year probability of occurrence. So much for feeling safe with probability estimates for potential disasters!

The benefit of any levee, bypass, or other project intended to prevent flood destruction is measured in avoided damage and is used to justify the cost of the protection facility. Thus, ever-increasing damage leads to the justification of ever-increasing flood control structures. All such strategies are subjected to cost-benefit analysis, but bias is a serious drawback because such an analysis usually is prepared by an agency or bureau with a vested interest in building more flood-control projects.

As suggested in an article titled "Settlement Control Beats Flood Control,"* there are other ways to protect populations than with enormous, expensive, sometimes environmentally disruptive projects. Strictly zoning the floodplain is one approach, but flat, easily developed floodplains near pleasant rivers might be perceived as desirable for housing, and thus weaken political resolve. A zoning strategy would set aside the floodplain for farming or passive recreation, such as a riverine park, golf course, or plant and wildlife sanctuary, or for other uses that are not hurt by natural floods. This study concludes that "urban and industrial losses would be largely obviated by setback levees and zoning and thus cancel the biggest share of the assessed benefits which justify big dams." Focus Study 14-1 takes a closer look at floodplain management.

*H. J. Thompson, "The James River Flood of August 1969," *Weatherwise* 22, no. 5 (October 1969): 183.

*Walter Kollmorgen, *Economic Geography* 29, no. 3 (July 1953): 215.

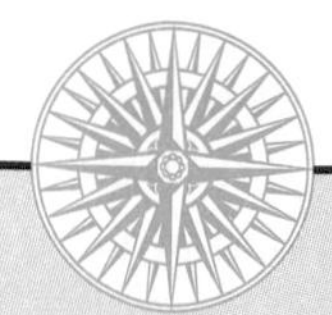

Focus Study 14-1 Floodplain Management Strategies

Detailed measurements of streamflows and floods have been rigorously kept in the United States only for about 100 years, in particular since the 1940s. At any selected location along any given stream, the *probable maximum flood* (PMF) is a hypothetical flood of such a magnitude that there is virtually no possibility it will be exceeded. Because floods are produced by the collection and concentration of rainfall, hydrologists speak of a corollary, the *probable maximum precipitation* (PMP) for a given drainage basin, which is an amount of rainfall so great that it will never be exceeded.

These probable parameters for a flood in a specific location along a stream are used by hydrologic engineers to establish a *design flood* against which to take protective measures. For urban areas near creeks, planning maps often include survey lines for a 50-year or a 100-year floodplain; such maps have been completed for most U.S. urban areas. The designation of a design flood usually is used to enforce planning restrictions and special insurance requirements. Unfortunately, the scenario all too often goes like this: (1) minimal zoning precautions are not carefully supervised; (2) a flooding disaster occurs; (3) the public is outraged at being caught off guard; (4) businesses and homeowners are surprisingly resistant to stricter laws and enforcement; and (5) eventually another flood refreshes the memory and promotes more knee-jerk planning. As strange as it seems, there is little indication that our risk perception improves as the risk increases.

Strategies

A strategy in some larger river systems is to develop artificial floodplains by constructing *bypass channels* to accept seasonal or occasional floods. When not flooded, the bypass channel can serve as farmland, often benefiting from the occasional soil-replenishing inundations. When the river reaches flood stage, large gates called *weirs* are opened, allowing the water to enter the bypass channel. This alternate route relieves the main channel of the burden of carrying the entire discharge.

Dams and reservoirs are common streamflow control methods within a watershed. For conservation purposes, a *dam* holds back seasonal peak flows for distribution during low-water periods. In this way streamflows are regulated to assure year-round water supplies. Dams also are constructed for flood control, to hold back excess flows for later release at more moderate discharge levels. Adding *hydroelectric power* production to these functions of conservation and flood control can define a modern multipurpose reclamation project.

The function of a *reservoir* is to provide flexible storage capacity within a watershed to regulate river flows, especially in a region with variable precipitation. Figure 1 shows one reservoir during drought conditions and during a time of wetter weather six years later.

Reservoir Considerations

Unfortunately, reservoir construction also involves negative consequences. The area upstream from a dam becomes permanently drowned. In mountainous regions, this may mean loss of white water rapids and recreational sections of a river. In agricultural areas the ironic end result may be that a hectare of farmland is inundated upstream in order to preserve a hectare of farmland downstream. Furthermore, dams built in warm and arid climates lose substantial water to evaporation, compared to the free-flowing streams they replace. Reservoirs in the southwestern United States can lose 3–4 meters (10–13 feet) of water a year. Also, sedimentation can reduce the effective capacity of a reservoir and can shorten a dam's life span, as mentioned earlier regarding Glen Canyon Dam.

Large multi-purpose projects invariably produce political conflict over territorial rights and questions of public trust versus private right to the environment. Vast scenes of environmental disruption for the sake of economic gain no longer appear popular with the public.

The James Bay Project in central and northern Québec is a case in point. Launched over 20 years ago by Hydro-Québec and only one-third complete at this time, the project might eventually include 215 dams, 25 power stations, and 20 river diversions. Many unexpected environmental problems have arisen because no environmental impact studies were completed at the outset. The early stages remain a huge experiment with fragile ecosystems. The public learned well into the planning phase of corporate interests seeking inexpen-

(a)

(b)

FIGURE 1
Comparative photographs of the New Hogan reservoir, central California, during (a) dry and (b) wet weather conditions. [Photos by author.]

sive public power supplies and that much of the power was for export to the United States, all at public expense. The second major phase of the James Bay Project, the Great Whale project, may never be completed because of the success of conservation programs begun by utilities in the northeastern United States and court challenges to assess impacts first before further construction. In addition, the state of New York in 1992 withdrew its offer to buy 1000 megawatts of power from the Hydro-Québec project.

One final concern is the need for great care in geologic assessment of the dam site, for dam failures occur more often than many realize. For example, in 1972 two dams failed, one near Rapid City, South Dakota, and another at Buffalo Creek, West Virginia, killing 237 and 118 people, respectively. The General Accounting Office estimates that about 1900 unsafe dams exist near urban areas.

The Teton Dam, near Rexburg, Idaho, collapsed June 5, 1976, releasing more than 303 billion liters (80 billion gallons) of water, destroying 41,000 hectares (100,000 acres) of farmland, killing 16,000 head of livestock, and causing more than $1 billion in property damage (Figure 2). Congressional testimony at the time disclosed: "The principal human cause of failure of the Teton Dam was very poor site selection." After engineering surveys disclosed specific geologic problems at the chosen site, construction of Teton Dam continued anyway. The dam survived less than one month after filling began!

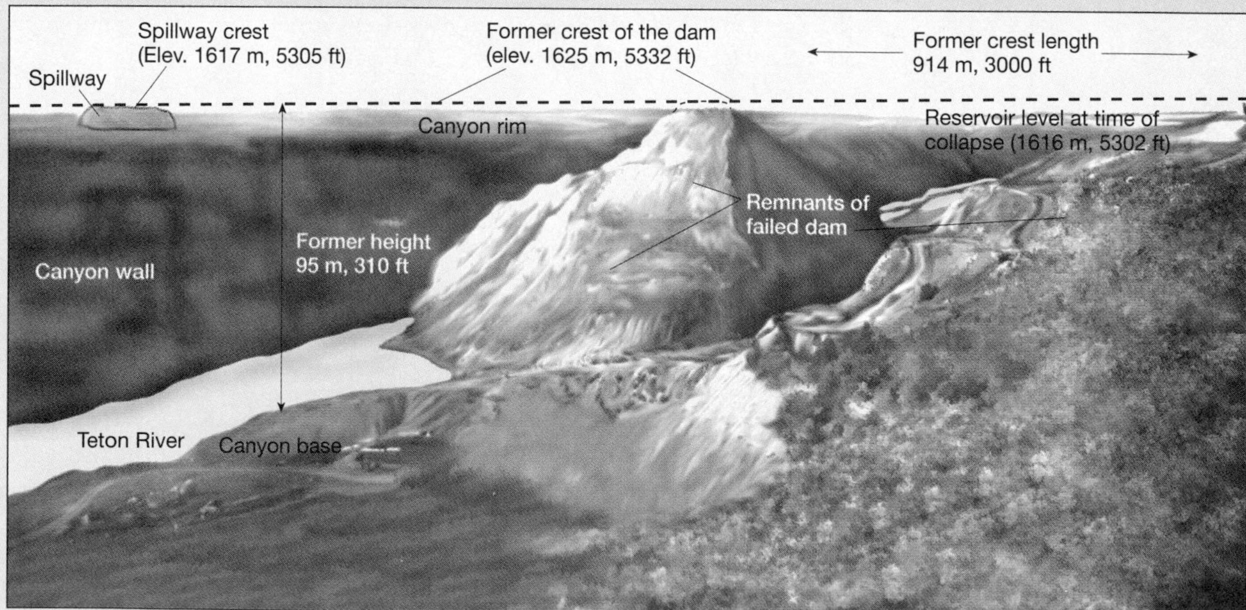

FIGURE 2
The failed remains of Teton Dam in Idaho. [Photo by author.]

SUMMARY—River Systems and Landforms

River systems, fluvial processes and landscapes, floodplains, and river control strategies are important topics as populations inhabit affected areas and as demands for limited water resources increase. The basic fluvial system is a drainage basin, an open system. Within it, stream tributaries are ordered to better understand the processes in operation in the basin.

Overland flow gathers and concentrates into the main stream channel in a manner consistent with topographic relief, the nature of rock and materials through which the stream flows, climate, vegetation, and the types of processes at work in a drainage basin. Streams establish a graded profile over distance, with interruptions triggering adjustments. Slope is the critical factor in a graded stream's maintenance of an equilibrium condition between water and solid materials. Stream form and operation result from complex interactions of slope, discharge, load, and channel characteristics, all variable within different climates and with different rock types. These functional considerations are embodied in the dynamic equilibrium approach to understanding the fluvial landscape.

Various landforms are associated with the action of flowing water: terraces, levees, deltas, and floodplains. Floodplains have been an important site of human activity throughout history. Rich soils, bathed in fresh nutrients by floodwaters, attract agricultural activity and urbanization. Floodplains are settled, despite our knowledge of historical devastation by floods, raising issues of human hazard perception. Collective efforts by government agencies undertake to reduce flood probabilities. Such management attempts include the construction of artificial levees, bypasses, straightened channels, diversions, dams, and reservoirs. Society is still learning how to live in a sustainable way with Earth's dynamic river systems.

KEY TERMS

abrasion
aggradation
alluvial terraces
alluvium
backswamp
bed load
braided stream
continental divides
cut bank
delta
deposition
dissolved load
drainage basins
drainage density
drainage pattern
entrenched meanders
erosion
estuary
flood
floodplain
fluvial
graded stream
gradient
hydraulic action
hydrograph
meandering stream
natural levees
nickpoint
oxbow lake
peneplain
point bar
saltation
sheet flow
staff gauge
stilling well
stream orders
suspended load
traction
transport
unit hydrograph
watershed
yazoo tributary

REVIEW QUESTIONS

1. What role is played by rivers in the hydrologic cycle?
2. Define the term *fluvial.* What is a fluvial process?
3. What is the basic spatial geomorphic unit of an individual river system? How is this unit delimited? Define the key terms used.
4. Follow the Allegheny River system to the Gulf of Mexico and analyze the pattern of tributaries, and describe the channel.
5. Describe a drainage basin. Define the various patterns that commonly appear in nature.
6. What is a stream order designation? How does the concept apply to major river systems such as the Mississippi or the Amazon?
7. What was the impact of flood discharge on the channel of the San Juan River near Bluff, Utah? Why did these changes take place?
8. How does stream discharge do its erosive work? What are the processes at work on the channel?
9. Differentiate between stream competence and stream capacity.
10. How does stream transport of sediments occur? What processes are at work?
11. Describe the flow characteristics of a meandering stream. What is the nature of the flow in the channel, the erosional and depositional features, and the typical landforms created?
12. Explain these statements: (a) all streams have a gradient but not all streams are graded, and (b) graded streams may have ungraded segments.
13. How is Niagara Falls an example of a nickpoint? Without human intervention, what do you think would eventually take place at Niagara Falls?
14. What did William Morris Davis intend to illustrate with his geomorphic cycle model? How did he construct the stages of landmass denudation?
15. What are the arguments against the formation of an old erosional surface such as a peneplain?
16. What is meant by the idea that the validity of cyclic or equilibrium models depends on the time frame being considered? Explain and discuss.
17. Describe the evolution of a floodplain. How are natural levees, oxbow lakes, backswamps, and yazoo tributaries formed?
18. What is a river delta? What are the various delta forms? Give some examples.
19. Has the Mississippi delta been stable for long? Explain.
20. Describe the Ganges River delta. What factors upstream explain its form and pattern? Assess the consequences of settlement on this delta.
21. What do you see as the major consideration regarding floodplain management? How would you describe the general attitude of society toward natural hazards and disasters?
22. Specifically, what is a flood? How are such flows measured and tracked?
23. Differentiate between a hydrograph from a natural terrain and one from an urbanized area.
24. What do you think the author of the article "Settlement Control Beats Flood Control" meant by the title? Explain your answer, using information presented in the chapter.

SUGGESTED READINGS

Burton, Ian, and Robert W. Kates. "The Perception of Natural Hazards in Resource Management," *Natural Resources Journal* 3, no. 3 (January 1964): 412–41.

Chorley, Richard J., Stanley A. Schumm, and David E. Sugden. *Geomorphology*. New York: Methuen, 1985.

Chow, Ven Te, ed. *Applied Hydrology—A Compendium of Water-Resources Technology*. New York: McGraw-Hill, 1964.

Clark, Champ, and the editors of Time-Life Books. *Flood*. Planet Earth Series. Alexandria, VA: Time-Life Books, 1982.

Davis, William Morris, *Geographical Essays*. Ed. by Douglas Wilson Johnson. New York: Dover Publications, 1964 (reprint of 1909 edition). (Note especially pp. 249–50, and 381–412.)

Deutsch, Morris, Donald R. Wiesnet, and Albert Rango, eds. "Satellite Hydrology," *Proceedings of the Fifth Annual William T. Pecora Memorial Symposium on Remote Sensing*. Minneapolis: American Water Resources Association, 1981.

Kazman, Raphael G. *Modern Hydrology*. New York: Harper & Row, 1965.

Leopold, Luna B. *Hydrology for Urban Planning—A Guidebook on the Hydrologic Effects of Urban Land Use*. Circular 554. Washington, DC: U.S. Geological Survey, 1968.

Leopold, Luna B., and Thomas Maddock, Jr. "The Hydraulic Geometry of Stream Channels and Some Physiographic Implications," *Geological Survey Professional Paper* 252. Washington, DC: Government Printing Office, 1953.

McCabe, G. J., Jr., and D. M. Wolock. "Effects of Climate Change and Climatic Variability on the Thornthwaite Moisture Index in the Delaware River Basin," *Climatic Change* 20 (1992): 143–153.

Pearse, P. H., F. Bertrand, and J. W. MacLaren. *Currents of Change—Final Report, Inquiry on Federal Water Policy*. Ottawa: Environment Canada, 1985.

Pringle, Laurence, and the editors of Time-Life Books. *Rivers and Lakes*. Planet Earth series. Alexandria, VA: Time-Life Books, 1985.

Rudloe, Jack, and Anne Rudloe. "Trouble in Bayou Country: Louisiana's Atchafalaya." *National Geographic,* September 1979, 377–97.

Schumm, Stanley A. *The Fluvial System*. New York: John Wiley & Sons, 1977

Shirley, Martha Lou, and James A. Ragsdale. *Deltas in Their Geologic Framework*, 2 vols. Houston: Houston Geological Society, 1966.

Wolock, D. M., G. J. McCabe, Jr., and others. "Sensitivity of Water Resources in the Delaware River Basin to Climate Change," *Proceedings of the USGS/Japan Workshop on the Effects of Global Climate Change on Hydrology and Water Resources at a Catchment Scale*. Washington, DC: USGS, 1992.

Bisti Wilderness, New Mexico. [*Photo by Jack W. Dykinga.*]

15

Eolian Processes and Arid Landscapes

Wind is an agent of erosion, transportation, and deposition. Its effectiveness has been the subject of much debate; in fact, wind at times was thought to produce major landforms. Presently, wind is regarded as a relatively minor exogenic agent, but it is significant enough to deserve our attention. Eolian processes modify and move material accumulations in deserts as well as along coastline beaches. The wind contributes to soil formation in distant places, bringing fine material from regions where glaciers deposited it. Elsewhere, fallow fields give up their soil resource to destructive wind erosion. In this chapter we examine the work of wind, associated processes, and resulting landforms.

We consider desert landscapes, where water remains the major erosional force, but where an overall lack of moisture and stabilizing vegetation allows wind processes to operate effectively to create extensive sand seas and dunes of infinite variety. The polar regions are deserts as well, with unique features related to their cold, dry environment—aspects covered in Chapter 17.

Arid landscapes display unique landforms and life forms: ". . . instead of finding chaos and disorder the observer never fails to be amazed at a simplicity of form, an exactitude of repetition and a geometric order. . . ."*

THE WORK OF WIND

The work of the wind—erosion, transportation, and deposition—is called **eolian** (also spelled *aeolian;* named for Aeolus, the ruler of the winds in Greek mythology). Much eolian research was accomplished by a British major, Ralph Bagnold, who was stationed in Egypt in 1925. Bagnold was an engineering officer who spent much of his time in the deserts west of the Nile, where he measured, sketched, and developed hypotheses about the wind and desert forms. His often-cited work, *The Physics of Blown Sand and Desert Dunes,* was published in 1941 following the completion of wind-tunnel simulations in London.

The actual ability of wind to move materials is small compared to that of other transporting agents such as water and ice, because air is so much less dense than these other media. Yet over time, wind accomplishes enormous work. Bagnold studied the ability of wind to transport sand over the surface of a dune. Figure 15-1 shows that a wind of 50 kmph (30 mph) can move approximately one-half ton of sand per day over a one-meter-wide section of dune. The graph also demonstrates how rapidly the amount of transported sand increases with wind speed.

Figure 15-1
Sand movement relative to wind velocity, as measured over a meter-wide (39 in.) strip of ground surface. [After Ralph A. Bagnold, 1941, *The Physics of Blown Sand and Desert Dunes.* London: Methuen & Co. Adapted by permission.]

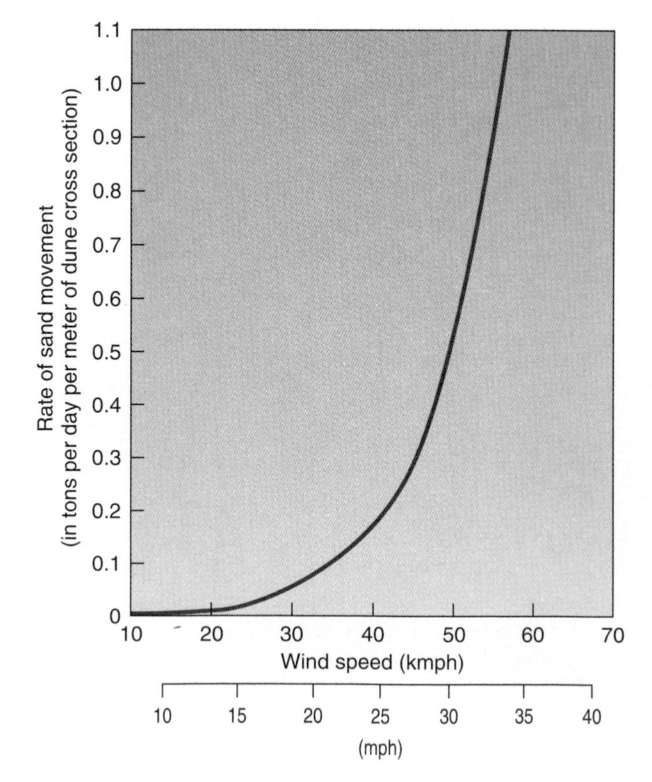

*From Ralph A. Bagnold, *The Physics of Blown Sand and Desert Dunes,* Methuen, 1941.

The highest-velocity winds are needed to move both the largest and the smallest particles, with intermediate-sized grains moved most easily. The reason is simple: small particles are difficult to move because they exhibit a mutual cohesiveness and usually present a smooth surface to the wind. Both wind velocity and turbulence increase with height.

Eolian Erosion

Two principal wind-erosion processes are **deflation,** the removal and lifting of individual loose particles, and **abrasion,** the grinding of rock surfaces with a "sandblasting" action by particles captured in the air. Deflation and abrasion produce a variety of distinctive landforms and landscapes.

Deflation. Deflation literally blows away unconsolidated or noncohesive sediment. After wind deflation and occasional sheetwash do their work on an arid landscape, a **desert pavement** is formed from the less mobile pebble and gravel concentration left behind. Desert pavement, which resembles a cobblestone street, protects underlying sediment from further deflation (Figure 15-2). Desert pavements are so common that many provincial names have been used for them—for example, *gibber plain* in Australia, *gobi* in China, and in Africa, *lag gravels* or *serir* (or *reg* desert if some fine particles remain). Water is an important factor in the formation of desert pavement, washing away finer materials and working to concentrate and cement remaining rock pieces.

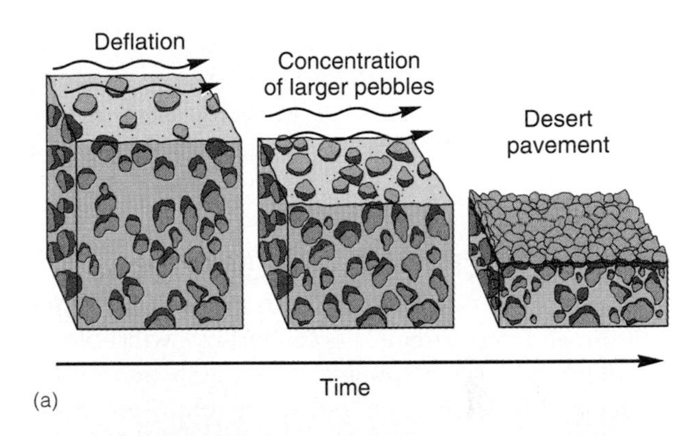

(a)

(b)

Figure 15-2
Desert pavement. (a) How desert pavement is formed from larger rocks and fragments left after deflation and sheetwash; (b) a typical desert pavement. [Photo by author.]

A serious environmental impact of the 1991 Persian Gulf War was the disruption of desert pavement. Thousands of square kilometers of stable desert pavement were shattered by bombardment with many thousands of tons of explosives and disrupted by the movement of heavy equipment. The resulting loosened sand and silt is now available for deflation, and thus threatens cities and farms with increased dust and sand accumulations.

Heavy recreational activity also produces much disruption in fragile desert landscapes, especially in the arid lands of the United States. Over 14 million off-the-road vehicles (ORV) are in use in the United States. Such vehicles crush plants and animals, disrupt desert pavement leading to greater deflation, and create ruts that easily concentrate sheetwash to form gullies. Possible planning measures to restrict use to specific areas, preserving the remaining desert, are at the center of public and political controversy.

Wherever wind encounters loose sediment, deflation also may form basins. Called **blowout depressions,** these range from small indentations of less than a meter up to areas hundreds of meters wide and many meters deep. Chemical weathering, although operating slowly in the desert, is important in the formation of a blowout, for it removes the cementing materials that give particles their cohesiveness. Large depressions occurring in the Sahara are at least partially formed by deflation. The enormous Qattara Depression just inland from the Mediterranean Sea in the Western Desert of Egypt, which covers 18,000 km^2 (6950 mi^2), is now about 130 m (427 ft) below sea level at its lowest point.

Abrasion. Sandblasting is commonly used to clean stone surfaces on buildings or to remove unwanted markings from streets. Abrasion by wind-blown particles is nature's version of sandblasting and is especially effective at polishing exposed rocks when the abrading particles are hard and angular. Variables that affect the rate of abrasion include the hardness of vulnerable surface rocks and the wind velocity and constancy. Abrasive action is restricted to the area immediately above the ground, usually no more than a meter or two in height because sand grains are only lifted a short distance. Rocks exposed to eolian abrasion appear pitted, grooved, or polished and usually are aerodynamically shaped in a specific direction, according to the flow of airborne particles. Rocks that bear such evidence of eolian erosion are called **ventifacts.** The chapter opening photograph shows wind-sculpted clay formations in the Bisti Wilderness of northwestern New Mexico.

On a larger scale, deflation and abrasion are capable of streamlining rock structures that are aligned parallel to the most effective wind direction, leaving behind distinctive, elongated ridges called **yardangs.** These can range from meters to kilometers in length and up to many meters in height. Abrasion is concentrated on the windward end of each yardang, with deflation operating on the leeward portions. The Sphinx in Egypt perhaps partially formed as a yardang, suggesting a head and body to the ancients. Some scientists think this shape led them to complete the bulk of the sculpture artificially with masonry.

Many other outstanding examples of yardangs occur worldwide: the Ica Valley of southern Peru has yardangs reaching 100 m (330 ft) in height and several kilometers in length, and yardangs in the Lūt Desert of Iran attain 150 m (490 ft) height. Some of these larger formations are detectable on satellite imagery. In fact, remote sensing by orbiting spacecraft has disclosed curious features on Mars that suggest yardangs.

Eolian Transportation

As mentioned in Chapter 6, atmospheric circulation is capable of transporting fine material such as volcanic debris worldwide within days. The distance that wind is capable of transporting particles varies greatly with their size. Wind exerts a drag or frictional pull on surface particles. Only the finest dust particles travel significant distances, and consequently the finer material suspended in a *dust storm* is lifted much higher than the coarser particles of a *sand storm,* which may be lifted only about 2 m (6.5 ft). People living in areas of frequent dust storms are faced with very fine particles infiltrating their homes and businesses through even the smallest cracks. (Figure 3-8 illustrates such a dust storm in Nevada and blowing alkali dust in the Andes.) People living in desert regions, where frequent sand storms occur, contend with the sandblasting of painted surfaces and etched windows.

Deflation and wind transport of soil produced a catastrophe in the American Great Plains in the 1930s—the Dust Bowl. Over a century of overgrazing and intensive agricultural development left soil susceptible to drought and eolian processes. The deflation of many centimeters of soil occurred in southern Nebraska, Kansas, Oklahoma, Texas, and eastern Colorado. The transported dust darkened the skies of midwestern cities and drifted over farmland.

The term *saltation* was used in Chapter 14 to describe movement of particles along streambeds by water. The term also is used in eolian processes to describe the wind transport of grains along the ground, grains usually larger than 0.2 mm (0.008 in.). About 80% of wind transport of particles is accomplished by this skipping and bouncing action (Figure 15-3). In comparison with fluvial transport, in which saltation is accomplished by hydraulic lift, eolian saltation is executed by aerodynamic lift, elastic bounce, and impact (compare Figures 15-3 and 14-11).

Saltating particles crash into other particles, knocking them both loose and forward. This causes **surface creep,** which slides and rolls particles too large for saltation and represents about 20% of the material being transported. Once in motion, particles continue to be transported by lower wind ve-

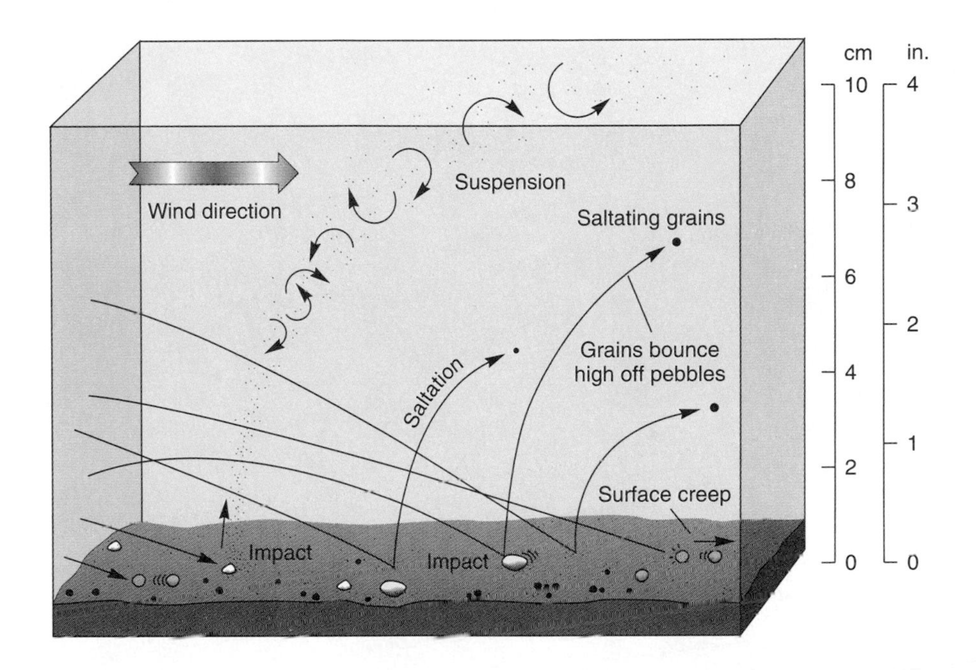

Figure 15-3
Eolian saltation and surface creep—forms of sediment transportation.

locities. In a desert or along a beach, you can hear the myriad saltating grains of sand produce a slight hissing sound, almost like steam escaping, as they bounce along and collide with surface particles.

Sand erosion and transport from a beach are slowed by conservation measures such as the introduction of stabilizing native plants, the use of fences, and the restriction of pedestrian traffic to walkways (Figure 15-4).

Through processes of weathering, erosion, and transportation, mineral grains are removed from parent rock and redistributed elsewhere. In Figure 13-5b, you can see the relationship between the composition and color of the sandstone in the

Figure 15-4
Preventing further erosion and transport of coastal dunes through stabilizing native plants, fences, and confining pedestrian traffic to walkways. [Photo by author.]

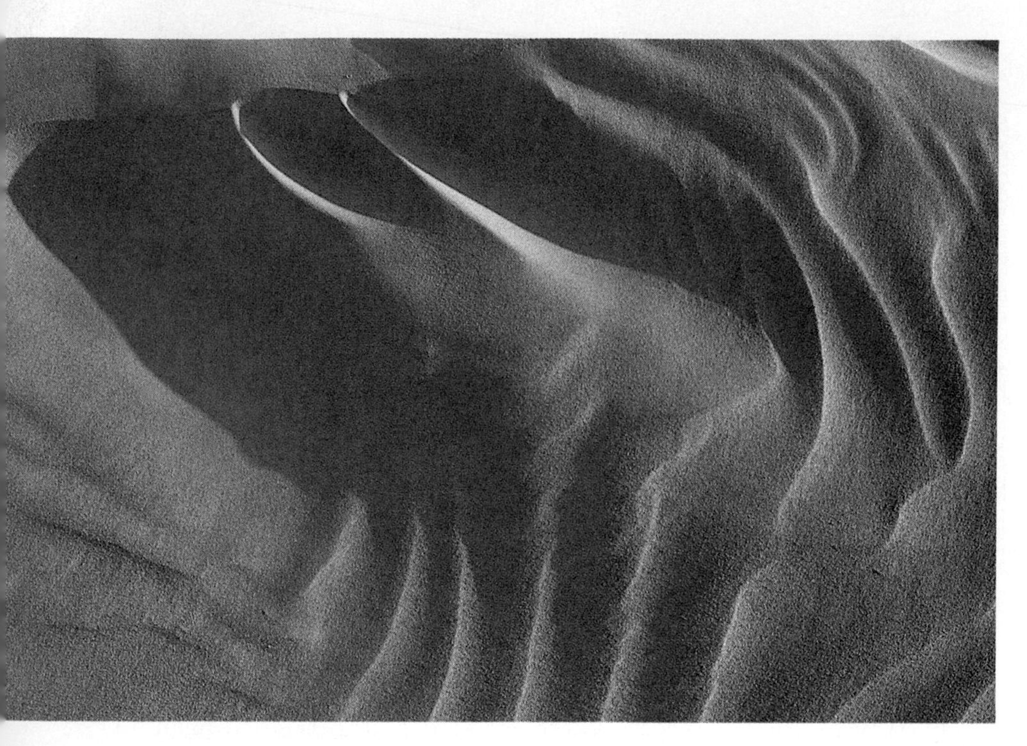

Figure 15-5
Sand ripple patterns later may become lithified into fixed patterns in rock. The area in the photo is approximately 1 m wide. [Photo by author.]

background and the derived sandy surface in the foreground. Wind action does not significantly participate in the weathering process that frees individual grains of sand from the parent rock, but it is active in relocating the weathered grains.

Eolian Depositional Landforms

The smallest features shaped by individual saltating grains are **ripples** (Figure 15-5). Ripples form in crests and troughs positioned transversely (at a right angle) to the direction of the wind. Their formation is influenced by the length of time particles are airborne. Eolian ripples are different from fluvial ripples because the impact of saltating grains is very slight in water.

A common assumption is that most deserts are covered by sand. Instead, desert pavements predominate across most subtropical arid landscapes; only about 10% of desert areas are covered with sand. Sand grains generally are deposited as transient ridges or hills called dunes. A **dune** is a wind-sculpted accumulation of sand. An extensive area of dunes, such as that found in North Africa, is characteristic of an **erg desert,** which means **sand sea.** The Grand Erg Oriental in the central Sahara exceeds 1200 m (4000 ft) in thickness and covers 192,000 km^2 (75,000 mi^2). This sand sea has been active for over 1.3 million years and has average dune heights of 120 m (400 ft). Similar sand seas are active in Saudi Arabia, such as the Ar Rub' al Khālī Erg (Figure 15-6).

Dune Movement and Form. Dune fields, whether in arid regions or along coastlines, tend to migrate in the direction of effective, sand-transporting winds. In this regard, stronger seasonal winds or winds from a passing storm may prove more effective than average prevailing winds. When saltating sand grains encounter small patches of sand their kinetic energy (motion) is dissipated and they accumulate. As height increases above 30 cm (12 in.) a **slipface** and characteristic dune features form.

Study the dune in Figure 15-7 and you can see that winds characteristically create a gently sloping *windward side* (stoss side), with one or more steeply sloped *slipfaces* on the *leeward side.* A

Figure 15-6
The Grand Ar Rub' al Khālī Erg that dominates southern Saudi Arabia. Effective southwesterly winds shape the pattern and direction of the transverse and barchanoid dunes. [SPOT image by CNES, Reston, Virginia. Used by permission.]

dune usually is asymmetrical in one or more directions. The angle of a slipface is the angle at which loose material is stable—its *angle of repose*. Thus, the constant flow of new material makes a slipface a type of *avalanche slope*. As sand moves over the crest of the dune to the brink, it builds up and avalanches as the slipface continually adjusts, seeking its angle of repose (usually 30–34°). In this way, a dune migrates downwind with the effective wind, as suggested by the successive dune profiles in Figure 15-7.

Dunes that move actively are called *freedunes* and reflect most dynamically the interaction between fluid atmospheric winds and moving sand. However, because the sand is moving close to the ground, it may encounter an obstruction such as stabilizing vegetation or a rock outcrop, resulting in a *tied dune,* or one that is fixed in place.

> **. . . I see hills and hollows of sand like rising and falling waves. Now at midmorning, they appear paper white. At dawn they were fog gray. This evening they will be eggshell brown.***

The ever-changing form of these eolian deposits is part of their beauty, but their many wind-shaped styles have made classification difficult. Although many terms exist, we can simplify dune forms into three classes—*crescentic, linear,* and *star* dunes (Figure 15-8).

*Janice E. Bowers, *Seasons of the Wind*. Flagstaff, AZ: Northland Press, 1985, p. 1.

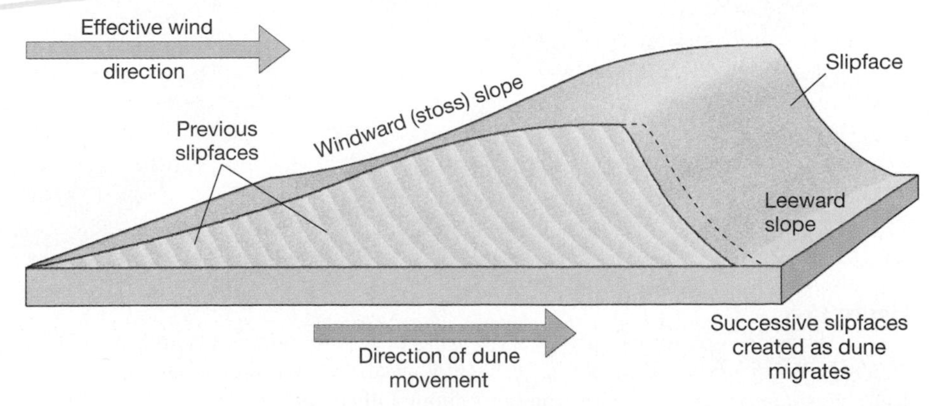

Figure 15-7
Dune cross section, showing the pattern of successive slipfaces as the dune migrates in the direction of the effective wind.

Class	Type	Description
Crescentic	Barchan	Crescent-shaped dune with horns pointed downwind. Winds are constant with little directional variability. Limited sand available. Only one slipface. Can be scattered over bare rock or desert pavement or commonly in dune fields.
	Transverse	Asymmetrical ridge, transverse to wind direction (right angle). Only one slipface. Results from relatively ineffective wind and abundant sand supply.
	Parabolic	Role of anchoring vegetation important. Open end faces upwind with U-shaped "blow-out" and arms anchored by vegetation. Multiple slipfaces, partially stabilized.
	Barchanoid Ridge	A wavy, asymmetrical dune ridge aligned transverse to effective winds. Formed from coalesced barchans; look like connected crescents in rows with open areas between them.

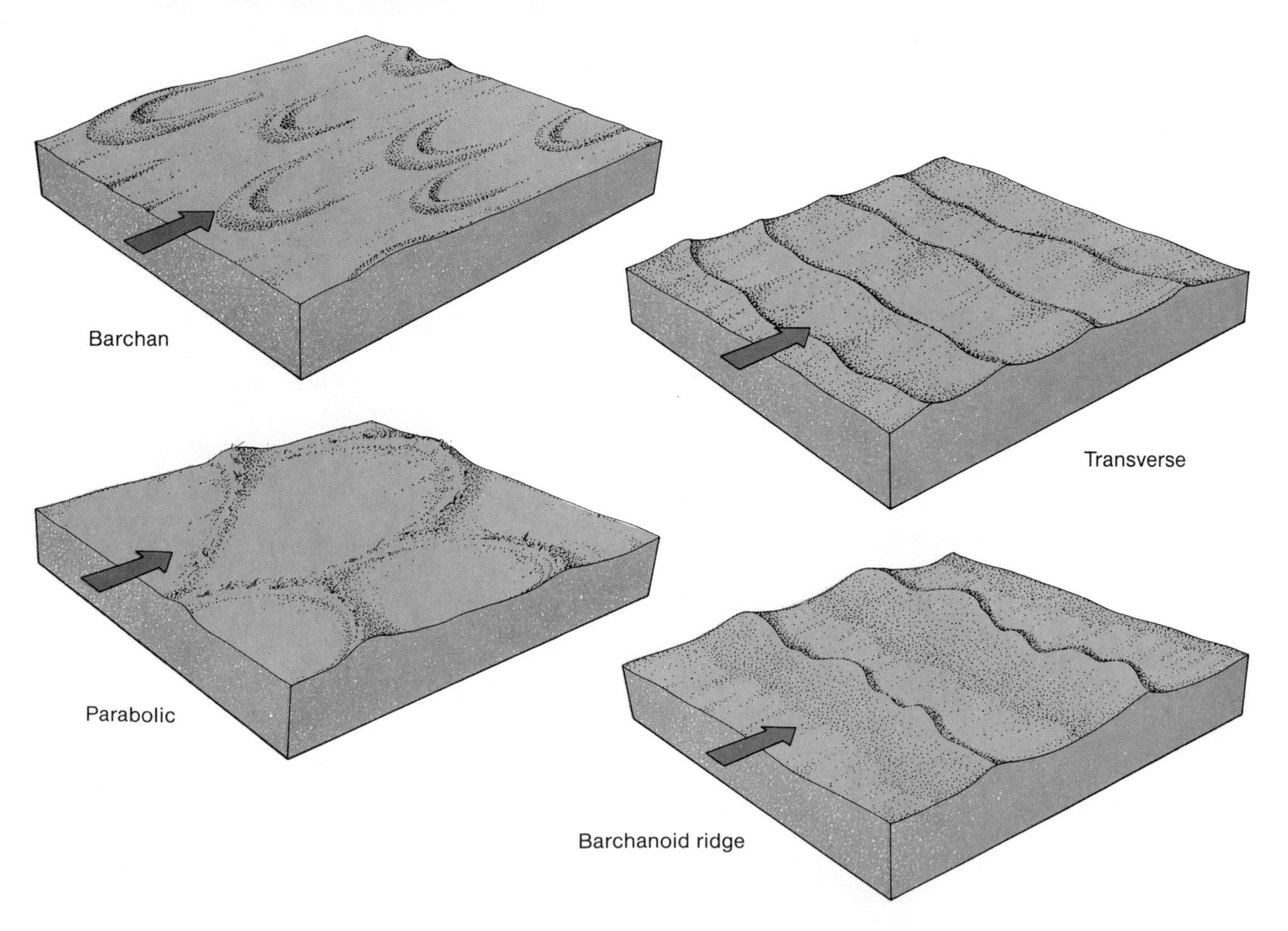

Figure 15-8
Major classes of dune forms. Arrows show wind direction. [Adapted from Edwin D. McKee, *A Study of Global Sand Seas,* U.S. Geological Survey Professional Paper 1052. Washington, DC: U.S. Government Printing Office, 1979.]

Class	Type	Description
Linear	Longitudinal	Long, slightly sinuous, ridge-shaped dune, aligned parallel with the wind direction; two slipfaces. Can be 100 m high and 100 km long. The "draas" at the extreme is up to 400 m high. Results from strong effective winds varying in one direction.
	Seif	After Arabic word for "sword"; a more sinuous crest and shorter than longitudinal dunes. Rounded towards upwind direction and pointed downwind.
Star Dune		The giant of dunes. Pyramidal or star-shaped with 3 or more sinuous radiating arms extending outward from a central peak. Slipfaces in multiple directions. Results from effective winds shifting in all directions. Tend to form isolated mounds in high effective winds and connected sinuous arms in low effective winds.
Other	Dome	Circular or elliptical mound with no slipface. Can be modified into barchanoid forms.
	Reversing	Asymmetrical ridge form intermediate between star dune and transverse dune. Wind variability can alter shape between forms.

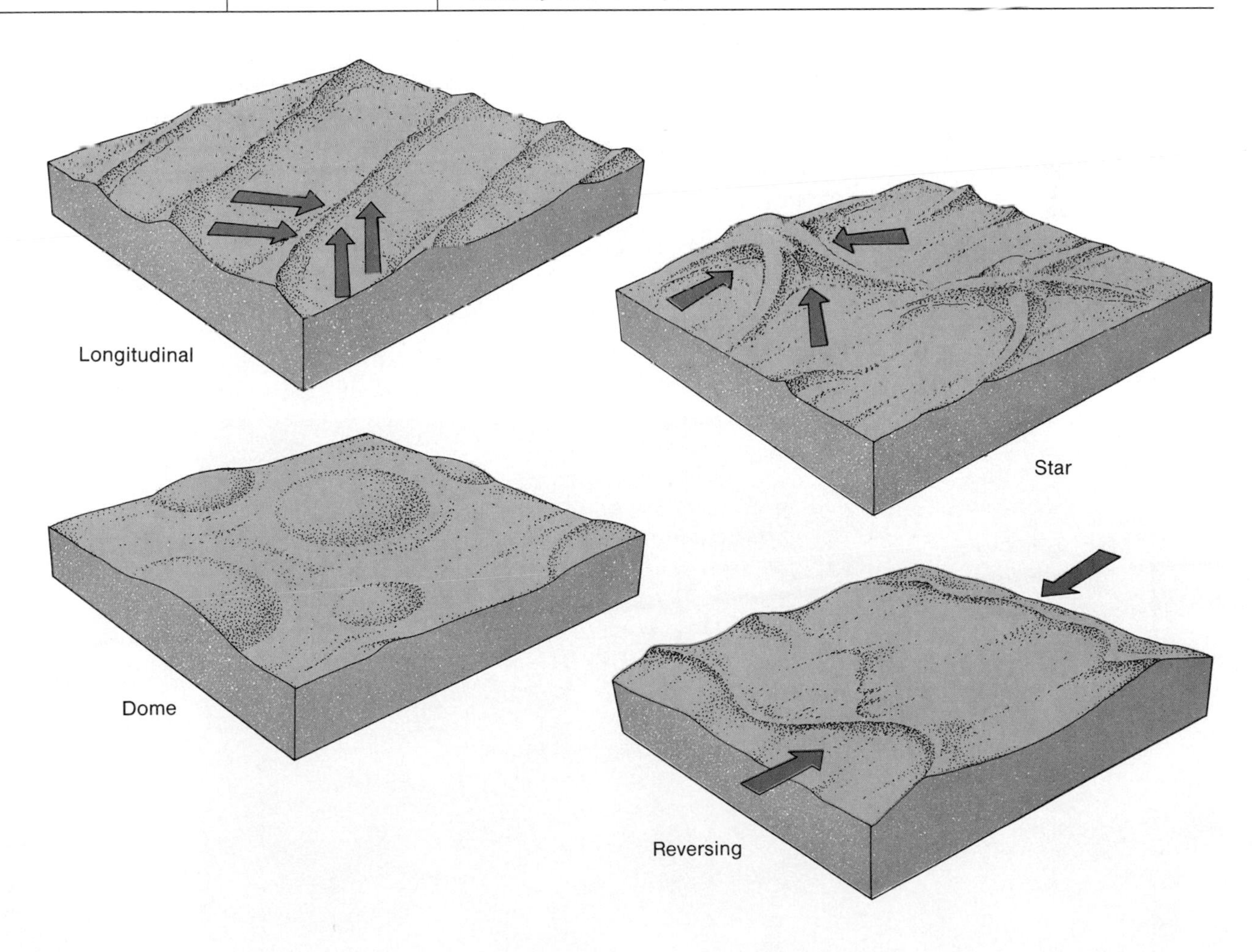

Crescentic dunes are crescent-shaped mounds or ridges of sand that form in response to a fairly unidirectional wind pattern. The crescentic group is the most common class, with related forms including *barchan* dunes (limited sand), *transverse* dunes (abundant sand), *parabolic* dunes (vegetation controlled), and *barchanoid* ridges (rows of coalesced barchans).

Linear dunes generally form sets of long parallel ridges separated by sheets of sand or bare ground, although they may occur as isolated ridges. Linear dunes characteristically are much longer than they are wide, with some exceeding 100 km (60 mi) in length. The effective winds which produce these dunes are principally bidirectional, so the slipface alternates from side to side. Related forms include *longitudinal dunes* (common name) and the *seif* (from the Arabic word for "sword"), which is a sharper, narrower version of a linear dune with a more sinuous crest.

Star dunes are the mountainous giants of the sandy desert. They form in response to complicated, changing wind patterns and have multiple slipfaces. They are pinwheel-shaped, with several radiating arms rising and joining to form a common central peak or crest. The best examples of star dunes are in the Sahara, where they approach 200 m (650 ft) in height (Figure 15-9).

Figure 15-10 correlates active sand regions with deserts (tropical, continental interior, and coastal). It is interesting to note the limited extent of desert area covered by active sand dunes—only about 10% of all continental land between 30° N and 30° S. Also noted on the map are dune fields in humid climates such as along coastal Oregon, the south shore of Lake Michigan, along the Gulf and Atlantic coastlines, in Europe, and elsewhere.

These same dune-forming principles and terms (e. g., dune, barchan, slipface) apply to snow-covered landscapes. *Snow dunes* form as wind deposits snow in drifts. In semiarid farming areas, capturing drifting snow with fences and tall stubble left in fields contributes significantly to soil moisture upon melting.

Figure 15-9
Star dune in the Namib Desert of Namibia in southwestern Africa. [Photo by Comstock.]

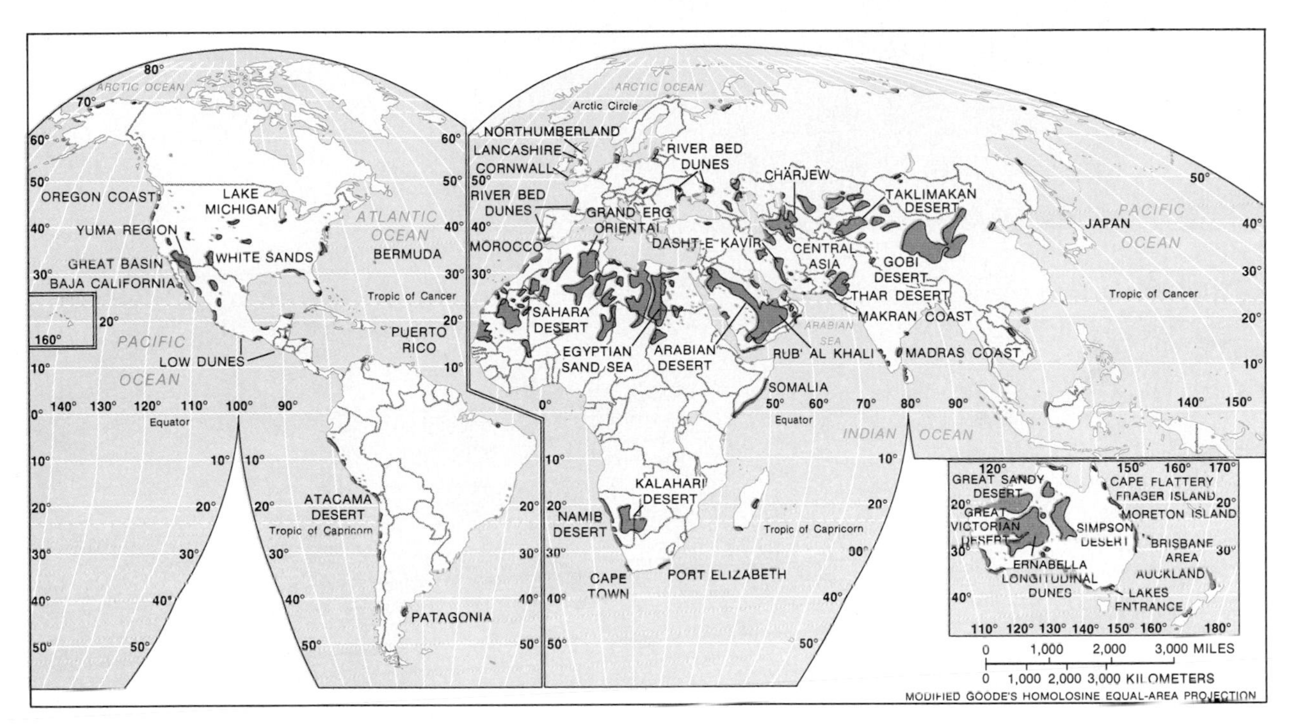

Figure 15-10
Worldwide distribution of active and stable sand regions. [After R. E. Snead, *Atlas of World Physical Features,* p. 134, copyright © 1972 by John Wiley & Sons. Adapted by permission of John Wiley & Sons, Inc.]

Loess Deposits

Pleistocene glaciers advanced and retreated in many parts of the world, leaving behind large glacial outwash deposits of fine-grained clays and silts (<0.06 mm or 0.0023 in.). These materials were blown great distances by the wind and redeposited in unstratified, homogeneous deposits, named **loess** (pronounced "luss") by peasants working along the Rhine River Valley in Germany where such deposits were first encountered. No specific landforms are created; instead, loess appears to cover existing landforms with a thick blanket of material that assumes the general topography of the preexisting landscape. Because of its own binding strength (coherent structure), loess weathers and erodes into steep bluffs, or vertical faces. At Xian (Shaanxi) China, a loess wall is excavated for dwelling space (Figure 15-11a). When a bank is cut into a loess deposit, it generally will stand vertically, although it can fail if saturated (Figure 15-11b).

Figure 15-12 shows the worldwide distribution of loess deposits. Significant accumulations of loess throughout the Mississippi and Missouri valleys form continuous deposits 15–30 m (50–100 ft) thick. Loess deposits also occur in eastern Washington State and Idaho. This silt explains the fertility of the soils in these regions, for loess deposits are well drained, deep, and characterized by excellent moisture retention. Loess deposits also cover much of Ukraine, central Europe, China, the Pampas-Patagonia regions of Argentina, and lowland New Zealand.

In Europe and North America, loess is thought to be derived mainly from glacial and periglacial sources. The vast deposits of loess in China, covering more than 300,000 km^2 (115,800 mi^2), are

(a)

(b)

Figure 15-11
Loess deposits. (a) Loess formation in Xian (Shaanxi), China, has strong enough structure to be excavated for dwelling rooms. (b) A loess bluff along the Arikaree River in extreme northwestern Cheyenne County, Kansas. [(a) Photo by Betty Crowell, (b) photo by Steve Mulligan.]

Figure 15-12
Worldwide loess deposits. Small dots represent small scattered loess formations. [After R. E. Snead, *Atlas of World Physical Features,* p. 138, copyright © 1972 by John Wiley & Sons. Adapted by permission of John Wiley & Sons, Inc.]

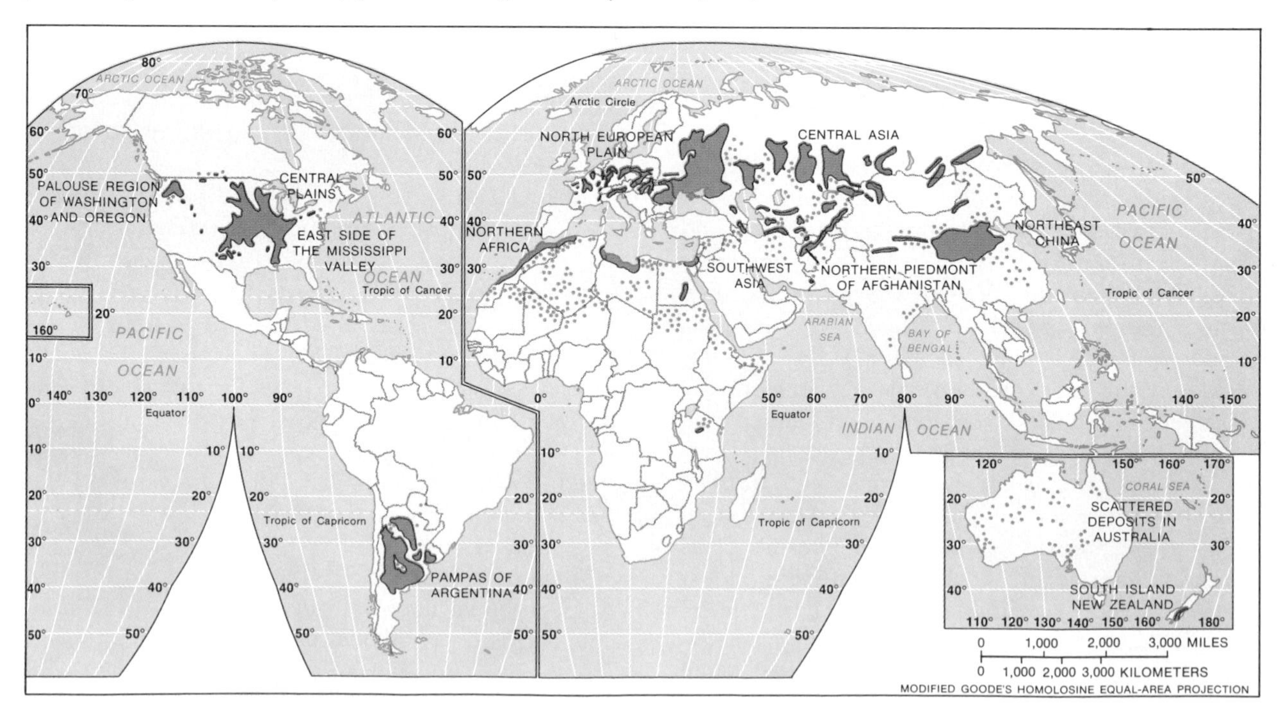

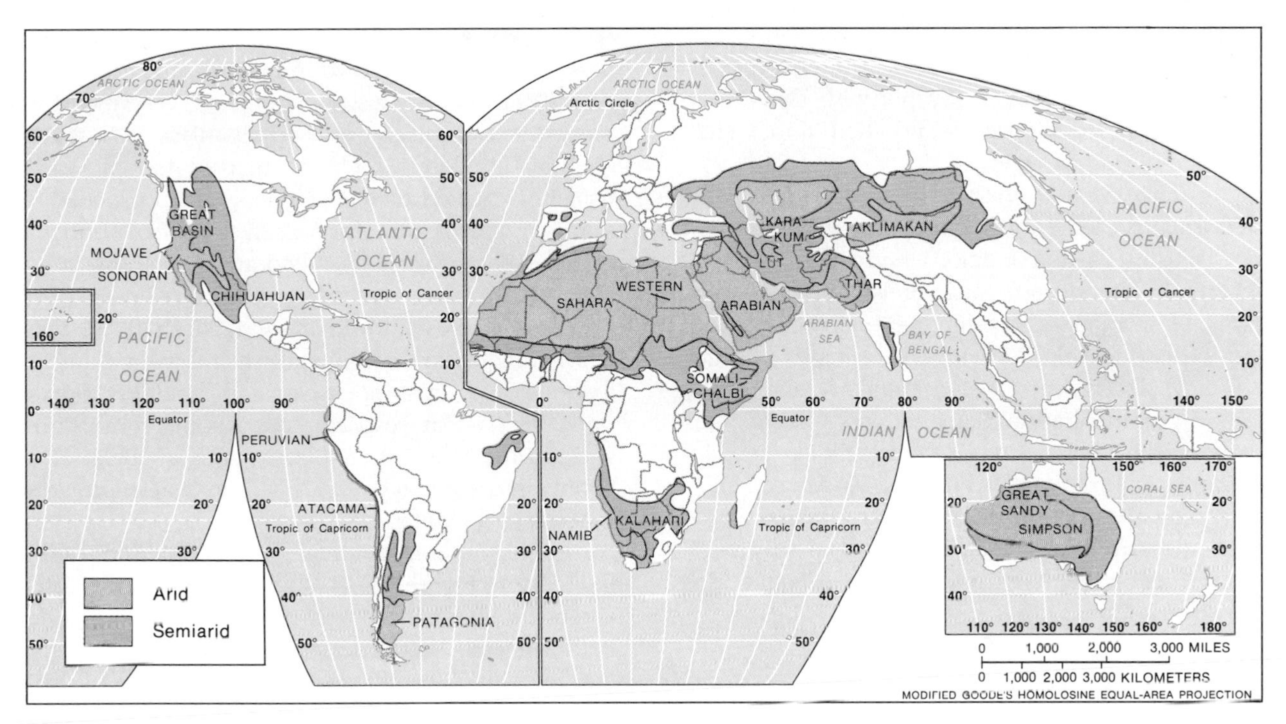

Figure 15-13
Worldwide distribution of arid lands (*BW arid desert climates*) and semiarid lands (*BS semiarid steppe climates*) based on the Köppen climatic classification system.

thought to be derived from desert rather than glacial sources. Accumulations in the Loess Plateau of China exceed 300 m (984 ft) thickness, forming some complex weathered badlands and some good agricultural land. These wind-blown deposits are interwoven with much of Chinese history and society. New research also has discovered that plumes of wind-blown dust from African deserts have moved across the Atlantic Ocean to enrich soils of the Amazon rain forest of South America, the southeastern United States, and the Caribbean islands.

When subjected to overgrazing or to dryness during episodes of drought, these fine sediments can be lifted by winds to form severe dust storms. This is what occurred in the Dust Bowl of the American Great Plains during the 1930s. Such episodes can have devastating consequences that disrupt economic activities, cause loss of topsoil, and even bury farmsteads.

OVERVIEW OF DESERT LANDSCAPES

Dry climates occupy about 26% of Earth's land surface and, if all semiarid climates are considered, perhaps as much as 35% of all land, constituting the largest single climatic region on Earth (see Figures 10-4 and 10-5 for the location of these *BW arid deserts* and *BS semiarid steppe* climate regions, and Figure 20-4 for the distribution of these desert environments).

The spatial distribution of these dry lands is related to subtropical high-pressure cells between 15° and 35° N and S (see Figures 6-9 and 6-11), to rain shadows on the lee side of mountain ranges (see Figure 8-8), and to areas at great distance from moisture-bearing air masses, such as central Asia. Figure 15-13 portrays this distribution according to the modified Köppen climate classification used in this text. These areas possess unique landscapes created by the interaction of intermittent precipitation events, weathering processes, and wind. Rugged, hard-edged desert landscapes of cliffs and scarps contrast sharply with the vegetation-covered, rounded and smoothed slopes characteristic of humid regions.

The daily surface energy balance for El Mirage, California, presented in Figure 4-12a, highlights the high sensible heat conditions and intense ground heating in the desert. Such areas receive a high input of insolation through generally clear skies and experience high radiative heat losses at night. A typical desert water balance shows high potential evapotranspiration demand, low precipitation supply, and prolonged summer deficits (see, for example, Figure 9-8e for Phoenix, Arizona). Fluvial processes in the desert generally are dominated by intermittent running water, with hard, poorly vegetated desert pavement yielding high runoff during rainstorms.

Desert Fluvial Processes

Precipitation events in a desert may be rare indeed, a year or two apart, but when they do occur, a dry streambed fills with a torrent called a **flash flood.** Such channels may fill in a few minutes and surge briefly during and after a storm. Depending on the region, such a dry streambed is known as a **wash,** an **arroyo** (Spanish), or a **wadi** (Arabic). A desert highway that crosses a wash usually is posted to warn drivers not to proceed if rain is in the vicinity, for a flash flood can suddenly and without warning sweep away anything in its path.

When washes fill with surging flash flood waters, a unique set of ecological relationships quickly develops. Crashing rocks and boulders break open seeds that respond to the timely moisture and germinate. Other plants and animals also spring into brief life cycles as the water irrigates their limited habitats.

At times of intense rainfall, remarkable scenes fill the desert. Figure 15-14 shows two photographs taken just one month apart in a large sand dune field in Death Valley, California. The rainfall event that intervened produced 2.57 cm (1.01 in.) of precipitation in one day, in a place that receives only 4.6 cm (1.83 in.) in an average year. The river in the photograph continued to run for hours and then collected in low spots on the hard, underly-

Figure 15-14
The Stovepipe Wells dune field of Death Valley. (a) The day following a 2.57 cm (1.01 in.) rainfall and (b) the same location one month later. [Photos by author.]

(a)

(b)

ing clay surfaces. The water was quickly consumed by the high evaporation demand so that, in just a month, these short-lived watercourses were dry and covered with accumulations of alluvial materials.

As runoff water evaporates, salt crusts may be left behind on the desert floor at low elevations. This intermittently wet-and-dry low area in a region of closed drainage is called a **playa,** site of an *ephemeral lake* when water is present. Accompanying our earlier discussion of evaporites, Figure 11-11 shows such a playa in Death Valley, covered with salt precipitate just one month after this record rainfall event.

Permanent lakes and continuously flowing rivers are uncommon features in the desert, although the Nile River and the Colorado River are notable exceptions. Both these rivers are *exotic streams* with their headwaters in a wetter region. The Colorado River and its problem of overuse is described in detail in Focus Study 15-1.

Alluvial Fans. In arid climates, with their intermittent water flow, a particularly noticeable fluvial landform is the **alluvial fan,** or alluvial cone, which occurs at the mouth of a canyon where it exits into a valley. The fan is produced by flowing water that loses velocity as it leaves the constricted channel of the canyon and therefore drops layer upon layer of sediment along the base of the mountain block. Water then flows over the surface of the fan and produces a braided drainage pattern, shifting from channel to channel with each moisture event (Figure 15-15). A continuous apron, or **bajada** (Spanish for "slope"), may form if individual alluvial fans coalesce into one sloping surface (see Figure 15-20). Fan formation of any sort is reduced in humid climates because perennial streams constantly carry away much of the alluvium.

An interesting aspect of an alluvial fan is the natural sorting of materials by size. Near the mouth of the canyon at the apex of the fan, coarser materials are deposited, grading slowly to pebbles and finer gravels with distance out from the mouth. Then sands and silts are deposited, with the finest clays and salts carried in suspension and solution all the way to the valley floor. Minerals dissolved in solution accumulate there as precipitates left by evaporation, forming evaporite deposits.

Well-developed alluvial fans also may be a major source of groundwater. Some cities—San Bernardino, California, for example—are built on alluvial fans and extract their municipal water supplies from them. However, because water resources in an alluvial fan are recharged from surface supplies and these fans are in arid regions, groundwater mining and overdraft beyond recharge rates are common. In other parts of the world, such water-bearing alluvial fans are known as *qanat* (Iran), *karex* (Pakistan), and *foggara* (western Sahara).

Desert Landscapes

Contrary to popular belief, deserts are not wastelands, for they abound in specially adapted plants and animals. Moreover, the limited vegetation, intermittent rainfall, intense insolation, and distant vistas produce starkly beautiful landscapes. And, all deserts are not the same: for example, North American deserts have more vegetation cover than do the generally barren Saharan expanses.

Mountainous deserts are found in interior Asia, from Iran to Pakistan, and in China and Mongolia. In South America, lying between the ocean and the Andes, is the rugged Atacama Desert. Deserts also occur worldwide as topographic plains, such as the Great Sandy and Simpson deserts of Australia, the Arabian Desert, the Kalahari Desert, and portions of the extensive Taklimakan Desert, which covers some 270,000 km^2 (105,000 mi^2) in the central Tarim Basin of China.

The shimmering heat waves and related mirage effects in the desert are products of light refraction through layers of air that have developed a temperature gradient near the hot ground. The desert's enchantment is captured in the book *Desert Solitaire:*

> **Around noon the heat waves begin flowing upward from the expanses of sand and bare rock. They shimmer like transparent, filmy veils between my sanctuary in the shade and all the sun-dazzled world beyond. Objects and forms viewed through this tremulous flow appear somewhat displaced or distorted. . . . the great Balanced Rock floats a few**

Figure 15-15
Alluvial fan in a desert landscape (right); topographic map of Cedar Creek alluvial fan (below). Topographic map is the Ennis Quadrangle, 15-minute series, scale 1:62,500, contour interval = 40 ft; latitude/longitude coordinates for mouth of canyon are 45°22' N, 111°35' W. [Photo by author; U.S. Geological Survey map.]

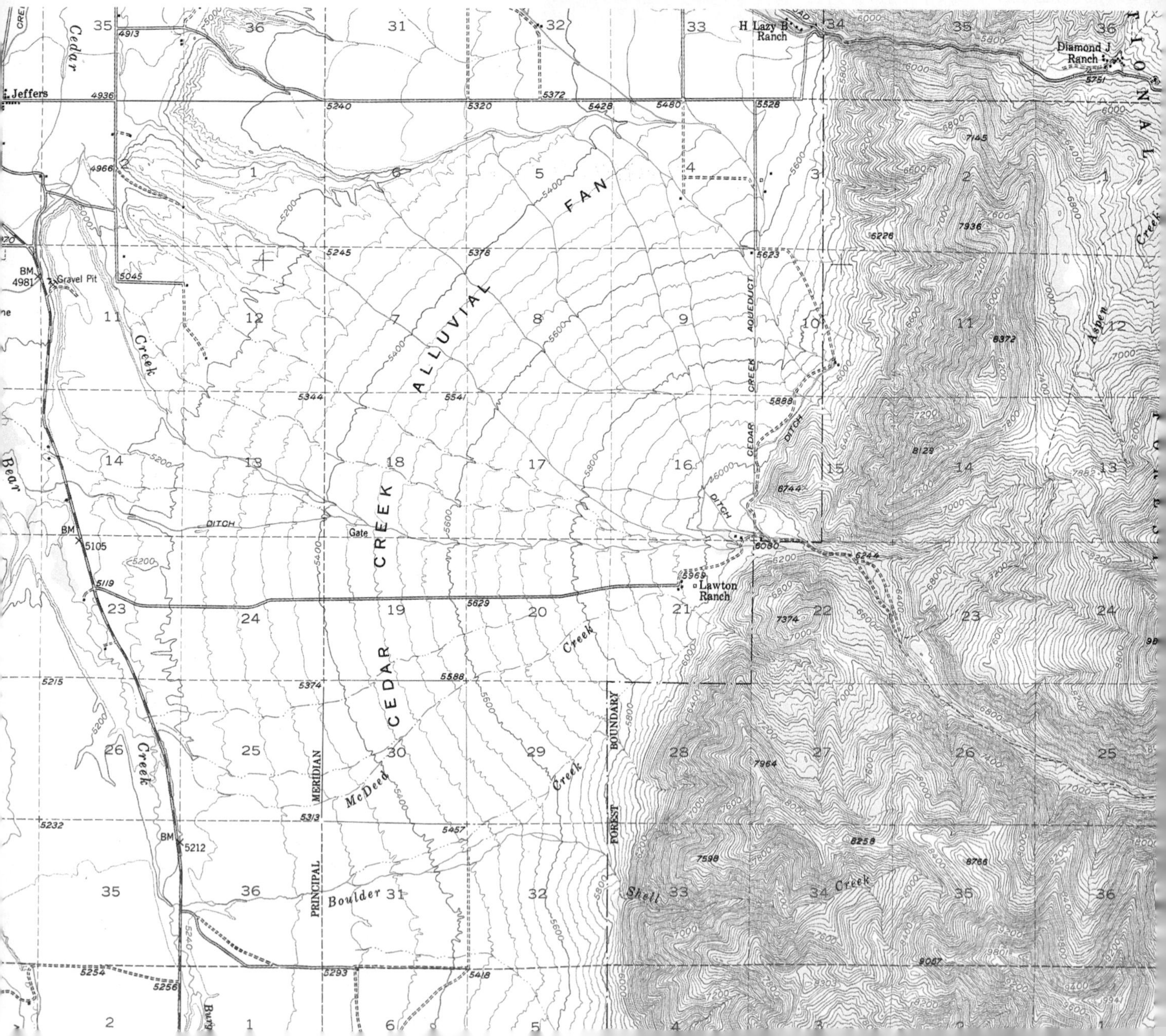

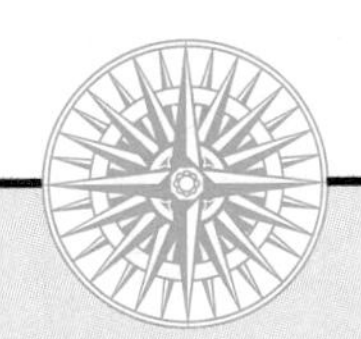

Focus Study 15-1 The Colorado River: A System Out of Balance

An exotic stream is one that headwaters in a humid region of water surpluses but then flows mostly through arid lands for the rest of its journey to the sea. Exotic streams have fewer incoming tributaries and a reduction in discharge downstream as compared to other rivers. The East African mountains and plateaus and the Colorado Rockies represent such source areas for the Nile River and the Colorado River, respectively. The Colorado rises on the high slopes of Mount Richthofen (3962 m or 13,000 ft) in Rocky Mountain National Park and flows almost 2317 km (1440 mi) to where a trickle of water disappears in the sand, kilometers short of its former mouth in the Gulf of California (Figure 1).

Orographic precipitation totaling 102 cm (40 in.) per year (mostly snow) falls in the Rockies, feeding the Colorado headwaters. But at Yuma, Arizona, near the river's end, annual precipitation is a scant 8.9 cm (3.5 in.), an extremely small amount when compared to the annual potential evapotranspiration demand in the Yuma region of 140 cm (55 in.).

From its source region, the Colorado River quickly leaves the humid Rockies and spills out into the arid desert of western Colorado and eastern Utah. At Grand Junction, Colorado, on the Utah border, annual precipitation is only 20 cm (8 in.). After carving its way through the intricate labyrinth of canyonlands in Utah, the river enters Lake Powell, 945 m (3100 ft) lower in elevation than the river's source area upstream in the Rockies some 982 km (610 mi) away. The Colorado then flows through the Grand Canyon chasm, formed by its own erosive power. The Canyon's mystical beauty is evident in photos on pages 208 and 309.

West of the Grand Canyon, the river turns south, tracing its final 644 km (400 mi) as the Arizona-California border. Along this stretch sits Hoover Dam, just east of Las Vegas; Davis Dam, built to control the releases from Hoover; Parker Dam for the water needs of Los Angeles; three more dams for irrigation water (Palo Verde, Imperial, and Laguna); and finally Morelos Dam at the Mexican border. Mexico owns the end of the river and whatever water is left. Overall, the drainage basin encompasses 641,025 km^2 (247,500 mi^2) of mountain, basin and range, plateau, canyon, and desert landscapes, in parts of seven states and two countries.

A Brief History

John Wesley Powell (1834–1902), the first to successfully navigate the Colorado River through the Grand Canyon, was named the first director of the U.S. Bureau of Ethnology and later director of the U.S. Geological Survey (1881–1892). Powell perceived that the challenge of the West was too great for individual efforts and believed that solutions to problems such as water availability could be met only through private cooperative efforts. His 1878 study (reprinted 1962), *Report of the Lands of the Arid Region of the United States,* is a conservation landmark.

Today, Powell probably would be skeptical of the intervention by government agencies in building large-scale reclamation projects. An anecdote in Wallace Stegner's *Beyond the Hundredth Meridian* relates that at an 1893 international irrigation conference held in Los Angeles, Powell observed development-minded delegates bragging that the entire West could be conquered and reclaimed from nature. Powell spoke against that sentiment: "I tell you, gentlemen, you are piling up a heritage of conflict and litigation over water rights, for there is not sufficient water to supply the land." He was booed from the hall. But history has shown Powell to be correct.

The Colorado River Compact was signed by six of the seven basin states in 1923. (The seventh, Arizona, signed in 1944, the same year as the Mexican Water Treaty.) With this compact the Colorado River basin was divided into an upper basin and a lower basin, arbitrarily separated for administrative purposes at Lees Ferry (Figure 1). Congress adopted the Boulder Canyon Act in 1928, authorizing Hoover Dam as the first major reclamation project on the river. Also authorized was the All-American Canal into the Imperial Valley, which required an additional dam.

Los Angeles then began its project to bring Colorado River water 390 km (240 mi) from still another dam and reservoir on the river to their city. Shortly after Hoover Dam was finished and downstream enterprises were thus offered flood protection, the other projects were quickly completed. There are now eight major dams on the river and many irrigation works. The latest effort to redistribute Colorado River water is the Central Arizona Project, which carries water to the Phoenix area.

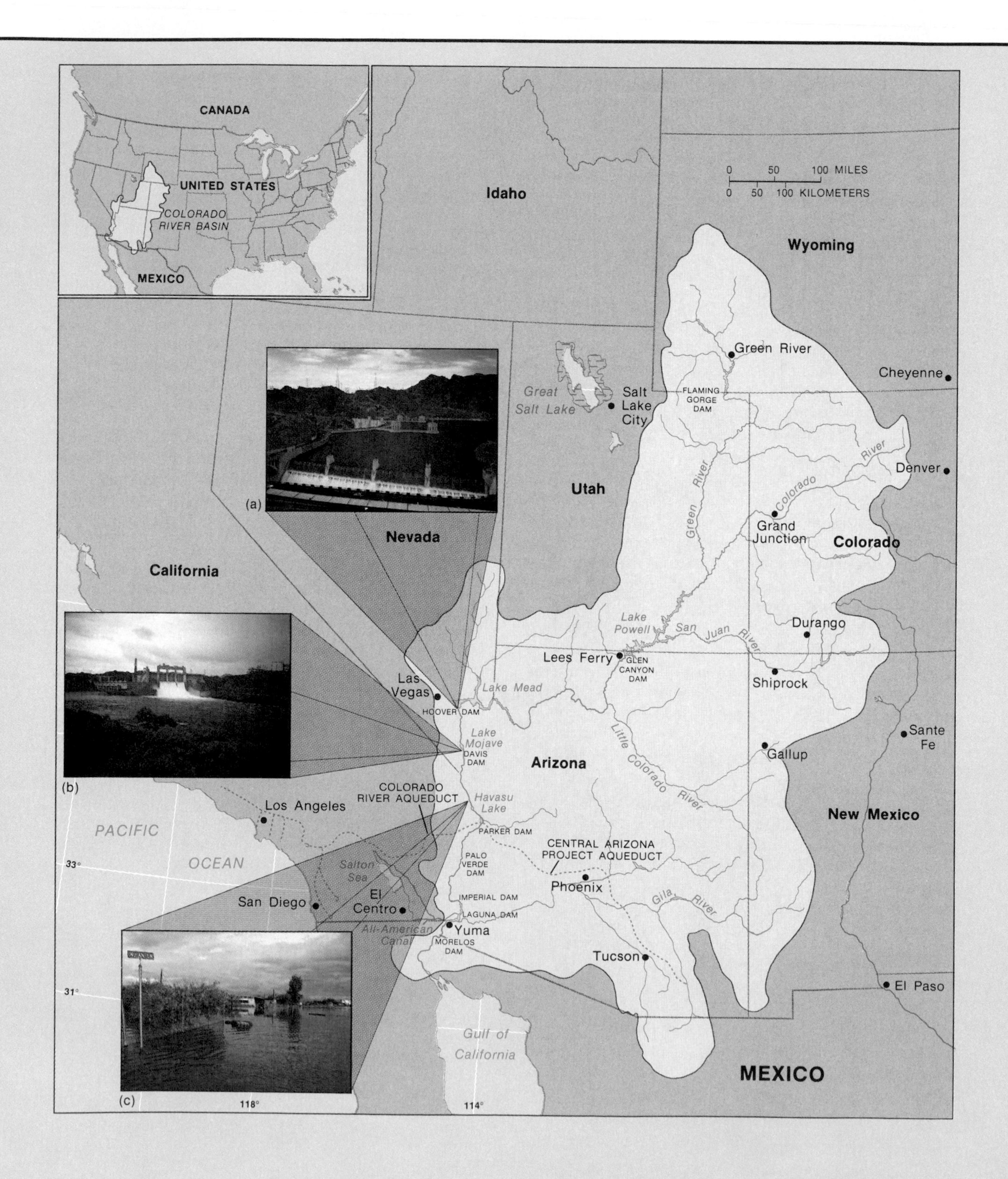
CANADA
UNITED STATES
COLORADO RIVER BASIN
MEXICO
0 50 100 MILES
0 50 100 KILOMETERS
Idaho
Wyoming
Green River
Cheyenne
Great Salt Lake
Salt Lake City
FLAMING GORGE DAM
Utah
Green River
Colorado River
Denver
Grand Junction
Colorado
Nevada
California
Lake Powell
San Juan River
Durango
Lees Ferry
GLEN CANYON DAM
Shiprock
Las Vegas
Lake Mead
HOOVER DAM
Lake Mojave
DAVIS DAM
Arizona
Little Colorado River
Gallup
Sante Fe
COLORADO RIVER AQUEDUCT
Havasu Lake
Los Angeles
PACIFIC OCEAN
PARKER DAM
New Mexico
PALO VERDE DAM
CENTRAL ARIZONA PROJECT AQUEDUCT
Salton Sea
Phoenix
33°
San Diego
El Centro
IMPERIAL DAM
Gila River
LAGUNA DAM
All-American Canal
Yuma
MORELOS DAM
Tucson
31°
El Paso
Gulf of California
MEXICO
118°
114°
(a)
(b)
(c)

TABLE 1
Estimated Colorado River Budget

Water Demand	Quantity (maf)*
In-basin consumptive uses (75% agricultural)**	11.5
Central Arizona Project (rising to 2.8 maf)	1.0
Mexican allotment (1944 Treaty)	1.5
Evaporation from reservoirs	1.5
Bank storage at Lake Powell	0.5
Phreatophytic losses (water-demanding plants)	0.5
Budgeted total demand	16.5
(1930–1980 average flow of the river	13.0)

*1 million acre-feet = 325,872 gallons; 1.24 million liters.
**7.0 maf in lower basin; includes 5.5 pumped out-of-basin, 5.1 of which goes to California.

Flow Characteristics

Exotic streamflows are highly variable, and the Colorado is no exception. For 1917, the discharge measured at Lees Ferry totaled 24 million acre-feet (maf), whereas in 1934 it reached only 5.03 maf. In 1977, the discharge dropped to 5.02 but it rose to an all-time high of 24.5 maf in 1984. In addition to this variability, approximately 70% of the year-to-year discharge occurs between April and July.

The average flows between 1906 and 1930 were almost 18 maf a year; averages dropped to 13 maf during the last 50 years. As a planning basis for the Colorado River Compact, the government used average river discharges from 1914 up to the treaty signing in 1923, an exceptionally high 18.8 maf. That amount was perceived as more than enough for the upper and lower basins each to receive 7.5 maf and, later, enough for Mexico to receive 1.5 maf in the 1944 Mexican Water Treaty.

We might question whether proper long-range planning should rely on the providence of high variability. Tree-ring analyses of past climates have disclosed that the only other time Colorado discharges were at the 1914–1923 level was between A.D. 1606 and 1625! However, government thinking about Colorado River flows seems never to have accepted this reality, for the dependable flows of the river have been consistently overestimated. This predicament is shown in an estimated budget for the river (Table 1): clearly the situation is out of balance, for there is not enough discharge to meet budgeted demands.

Water Losses at Glen Canyon Dam

Glen Canyon Dam was completed and began water impoundment in 1963, 27.4 km (17 mi) north of the basin division point at Lees Ferry. The deep, fluted, inner gorges of Glen, Navajo, Labyrinth, Cathedral, and myriad other canyons slowly were flooded by advancing water. The Bureau of Reclamation stated that the dam's purpose was to regulate flows between the basins, although many thought that Lake Mead, behind Hoover Dam, could have served this administrative-regulatory function without the risk of additional water losses.

The porous Navajo sandstone that forms the bulk of the container for the Lake Powell reservoir at Glen Canyon Dam absorbs an estimated 4% of the river's overall annual discharge as *bank storage,* water that is

Figure 1
The Colorado River basin showing division of the upper and lower basins near Lees Ferry in northern Arizona. Photos portray (a) Hoover Dam spillways in operation; (b) Davis Dam; and (c) flooded Topock Estates subdivision near Needles, California, in 1983. [Photos by author.]

not retrievable in any practical sense. The higher the lake level, the greater the loss into the sandstone. To date, approximately 10 maf have disappeared into this highly permeable rock. In addition, Lake Powell is an open body of water in an arid desert, where hot, dry winds accelerate evaporative losses. Another 4% of the river's overall discharge is lost annually in this manner from Lake Powell alone.

A third aspect of water loss involves the now-permanent sand bars and banks that have stabilized along the course of the regulated river. Water-demanding plants known as *phreatophytes* are now established and extracting an additional 4% of the flow of the river. Combined, these losses total 1.5 maf, or 12% of the long-term average flow—all attributable to Glen Canyon Dam and Lake Powell. Given the chronic deficits in the overall Colorado River budget, we can now seriously question the positive impact of this single project.

Flood Control in 1983

Intense precipitation and heavy snowpack in the Rockies, attributable to the 1982–1983 El Niño (see Focus Study 10-1), led to record-high discharge rates on the Colorado, testing the controllability of one of the most regulated rivers in the world. Federal reservoir managers were not prepared for the high discharge, having set aside Lake Mead's primary purpose (flood control) in favor of competing water and power interests. What ensued was a human-caused flood. Davis Dam, which regulates releases from Hoover Dam, was within 30 cm (1 ft) of overflow by June 22, 1983, a real problem for a structure made partially of earth fill.

The only time the spillways at Hoover Dam had ever operated was during a test more than 40 years before, when the reservoir capacity was artificially raised for the test; now they were opened to release the floodwaters. In addition, Glen Canyon Dam was over capacity and at risk; it actually was damaged by the volume of discharge tearing through its spillways. The decision to increase releases to save these and other upstream facilities doomed towns and homeowners along the river, especially in subdivisions near Needles, California.

We might wonder what John Wesley Powell would think if he were alive today to witness such errant attempts to control the mighty and variable Colorado. He foretold such a " . . . heritage of conflict and litigation."

inches above its pedestal, supported by a layer of superheated air. The buttes, pinnacles, and fins in the windows area bend and undulate beyond the middle ground like a painted backdrop stirred by a draft of air.*

The buttes, pinnacles, and mesas characteristic of arid and semiarid landscapes are resistant horizontal rock strata that have eroded differentially. Removal of the less-resistant sandstone strata produces unusual desert sculptures—arches, pedestals, and delicately balanced rocks (Figure 15-16). Specifically, the upper layers of sandstone along the top of an arch or butte are more resistant to weathering and protect the sandstone rock beneath. The removal of all surrounding rock through differential weathering leaves enormous buttes as residuals on the landscape. If you imagine a line intersecting the tops of the Mitten Buttes shown in Figure 15-17 you can gain some idea of the quantity of material that has been removed. These buttes exceed 300 m (980 ft) in height, similar to the Chrysler Building in New York City or First Canadian Place in Toronto.

Desert landscapes are places where stark erosional remnants stand above the surrounding terrain as knobs, hills, or "island mountains." Such a bare, exposed rock, called an **inselberg,** is exemplified by Uluru (Ayers) Rock in Australia, pictured in Figure 1-1.

In a desert area, weak surface material also may weather to a complex, rugged topography, usually of relatively low and varied relief. Such a landscape is called a **badland**, probably so-named be-

*Edward Abbey, *Desert Solitaire*. New York: McGraw-Hill Book Co., 1968, p. 154. Copyright © 1968 by Edward Abbey.

Figure 15-16
Balanced Rock in Arches National Park where writer-naturalist Edward Abbey worked as a ranger years before it became a park. [Photo by John Gerlach/DRK].

Figure 15-17
Mitten Buttes, Merrick Butte, and rainbow in Monument Valley, Navajo Tribal Park, along the Utah-Arizona border. [Photo by author.]

(a)

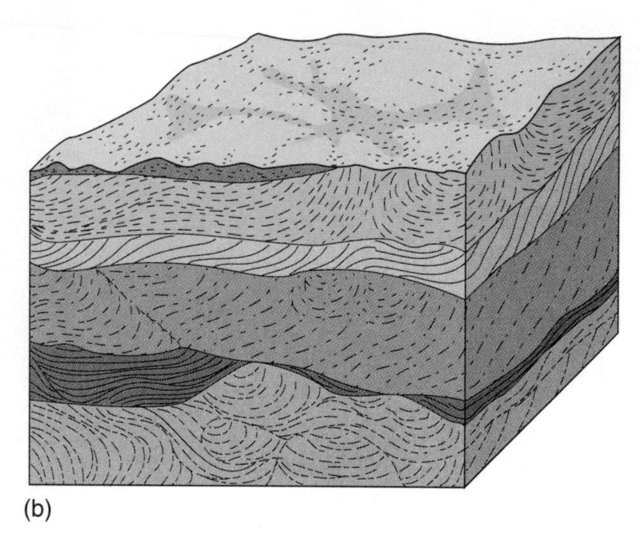
(b)

Figure 15-18
Cross bedding (a) in sandstone and (b) in a sketch to highlight the bedding pattern in the rock. [Photo by author.]

cause it offered little economic value and was difficult to traverse in nineteenth-century wagons (see this chapter's opening photo of badlands in New Mexico). The Badlands region of the Dakotas is of this form, as are portions of Death Valley soils shown in Figure 18-21.

In addition, sand dunes that existed in some ancient deserts have lithified, forming sandstone structures that bear the imprint of cross-stratification. When such a dune was accumulating, sand cascaded down its slipface, and distinct bedding planes (layers) were established that remained after the dune lithified (Figure 15-18). Ripple marks, animal tracks, and fossils also are found preserved in these sandstones, which originally were eolian-deposited sand dunes.

Basin and Range Province. A **province** is a large region that shares several geologic or physiographic traits, making appropriate the assignment of a regional name. The **Basin and Range Province** of the western United States consists of arid and semiarid basins and mountain ranges that lie in the rain shadow of mountains to the west. The province has a dry climate, few permanent streams, and interior drainage patterns—drainage basins that lack any outlet to the ocean (see Figure 14-1 map). The Basin and Range Province was a major barrier to early settlers in their migration westward.

As the North American plate moved westward, it overrode former oceanic crust and hot spots at such a rapid pace that slabs of subducted material literally were run over. The associated extension and spreading of the crust created a landscape fractured by many faults.

The present basin-and-range landscape features nearly parallel sequences of *horsts* (upward-faulted blocks, or ridges) and *grabens* (downward-faulted blocks, or valleys). They cover almost 800,000 km^2 (308,800 mi^2), although the pattern of faults is not that simple, as Figure 15-19 shows.

Basin-and-range relief is abrupt, and rock structures are angular and rugged. As the ranges erode, transported materials accumulate to great depths in the basins, gradually producing extensive desert plains. The basin's elevation averages 1220–1525 m (4000–5000 ft) above sea level, with mountain

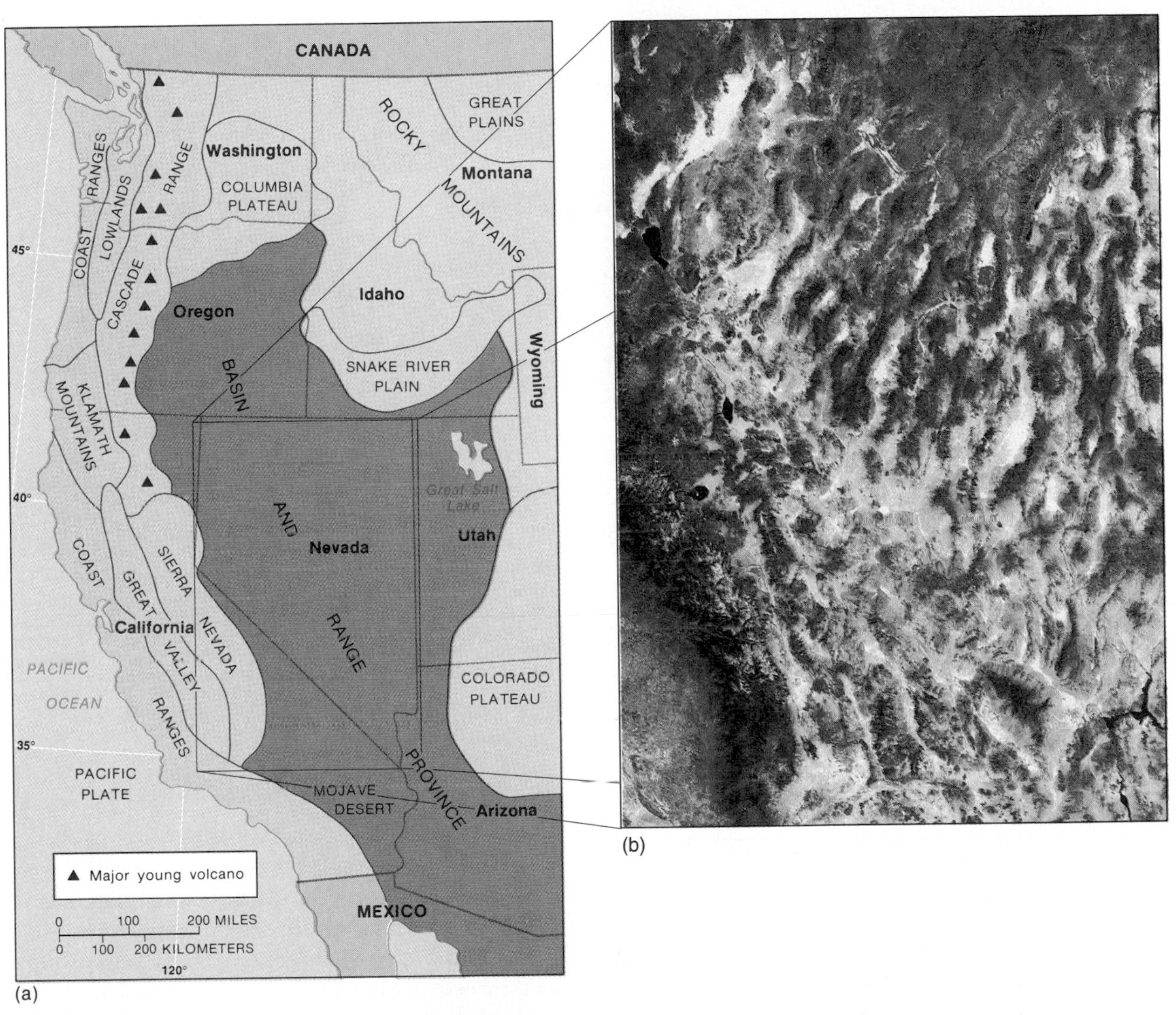

Figure 15-19
Basin and Range Province in the western United States. (a) Map and (b) *Landsat* image of the area. Recent scientific discoveries demonstrate that this province extends south through northern and central Mexico. [(b) Image from NASA.]

crests rising higher by some 915–1525 m (3000–5000 ft). Death Valley is an extreme, the lowest of these basins, with an elevation of −86 m (−282 ft). However, to the west of the valley, the Panamint Range rises to 3368 m (11,050 ft) at Telescope Peak—over 3 vertical kilometers (2 mi) of desert relief! John McPhee captured the feel of this desert province in his book *Basin and Range:*

> Supreme over all is silence. Discounting the cry of the occasional bird, the wailing of a pack of coyotes, silence—a great spatial silence—is pure in the Basin

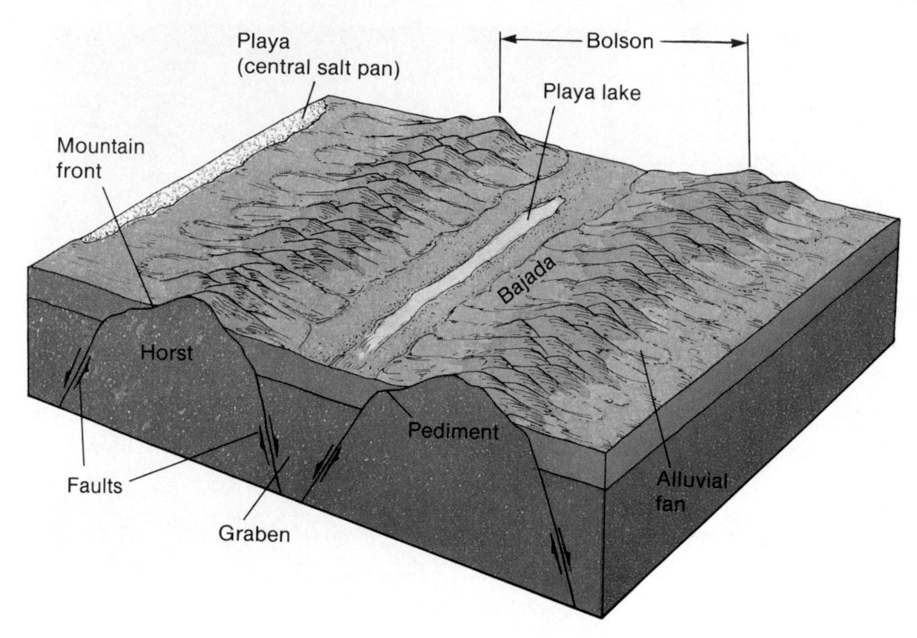

Figure 15-20
A bolson in the mountainous desert landscape of the Basin and Range Province. Parallel normal faults produce a series of horsts and grabens.

and Range. It is a soundless immensity with mountains in it. You stand . . . and look up at a high mountain front, and turn your head and look fifty miles down the valley, and there is utter silence.*

Figure 15-20 illustrates typical basin-and-range terrain. Note the **bolson,** a slope-and-basin area between the crests of two adjacent ridges in a dry region of interior drainage. The figure also identifies a playa (central salt pan), a bajada (coalesced alluvial fans), and a **pediment,** the area created as a mountain front retreats through weathering and erosional action. A pediment is an area of bedrock that is layered with a thin veneer, or coating, of alluvium. It is an erosional surface, as opposed to the depositional surface of the bajada. Often the distinction between the two is obscured by the continuous cover of alluvial deposits, but that distinction becomes significant when it is necessary to drill in search of groundwater.

* John McFee, *Basin and Range*. New York: Farrar, Straus, Giroux, 1981, p. 46.

SUMMARY—Eolian Processes and Arid Landscapes

Winds are produced by the movement of the atmosphere in response to pressure differences. Wind is an agent of erosion, transportation, and deposition. Fallow fields give up their soil resource to destructive wind erosion. Eolian processes modify and move sand accumulations along coastline beaches as well as in deserts. In arid and semiarid climates and along some coastlines where sand is available it accumulates in dunes, which appear in a variety of forms.

Eolian transported materials contribute to soil formation in distant places. Wind-blown loess deposits occur worldwide and provide a basis for soil development. These fine-grained clays and silts are moved by the wind many kilometers, where they are redeposited in unstratified, homogeneous deposits that cover the existing landscape.

Dry climates occupy about 26% of Earth's land surface and, if all semiarid climates are considered, perhaps as much as 35% of all land, constituting the largest single climatic region on Earth. Desert landscapes are regions of special plant and animal adaptations and unique rock structures that appear stark and angular, bared of vegetation and other coverings found in the more humid regions of the world. Wind and water operate together in dry regions. Although water events are infrequent, running water is still the major erosional agent in deserts. Rapid population growth and urbanization in the arid southwestern United States, as well as the Middle East, are producing much stress on available water resources. The Colorado River, an exotic stream, is budgeted beyond its average annual discharge, an issue of growing concern in the West. These fragile environments also are susceptible to disruption, as evidenced in the Persian Gulf region following the recent war.

KEY TERMS

abrasion
alluvial fan
arroyo
badland
bajada
Basin and Range Province
blowout depressions
bolson
deflation
desert pavement
dune
eolian
erg desert
flash flood
inselberg
loess
pediment
playa
province
ripples
sand sea
slipface
surface creep
ventifacts
wadi
wash
yardangs

REVIEW QUESTIONS

1. Who was Ralph Bagnold? What was his contribution to eolian studies?
2. Explain the term *eolian* and its application in this chapter. How would you characterize the ability of the wind to move material?
3. Describe the erosional processes associated with moving air.
4. Explain deflation and the evolutionary sequence that produces desert pavement, ventifacts, and yardangs.
5. Differentiate between a dust storm and a sand storm.
6. What is the difference between eolian saltation and fluvial saltation?
7. Explain the concept of surface creep.
8. What is the difference between an erg and a reg desert? Which type is a sand sea? Are all deserts covered by sand? Explain.
9. What are the three classes of dune forms? Describe the basic types of dunes within each class. What do you think is the major shaping force for sand dunes?

10. Which form of dune is the mountain giant of the desert? What are the characteristic wind patterns that produce such dunes?
11. How are loess materials generated? What form do they assume when deposited? Name a few examples of significant loess deposits.
12. Characterize desert energy and water balance regimes. What are the significant patterns of these arid landscapes in the world?
13. How would you describe the budget of the Colorado River? What was the basis for agreements regarding distribution of the river? Why has thinking about river flows been so optimistic?
14. Describe a desert bolson from crest to crest. Draw a simple sketch with the components of the landscape labeled.
15. Where in the Basin and Range Province do we find the greatest topographic relief?

SUGGESTED READINGS

Abbey, Edward, and the editors of Time-Life Books. *Cactus Country*. American Wilderness Series. New York: Time-Life Books, 1973.

Bagnold, Ralph A. *The Physics of Blown Sand and Desert Dunes*. London: Methuen & Co., 1941, 1954.

Cooke, R. U., and Andrew Warren. *Geomorphology in Deserts*. Los Angeles: University of California Press, 1973.

Dunbier, Roger. *The Sonoran Desert: Its Geography, Economy, and People*. Tucson: University of Arizona Press, 1968.

Hamblin, W. Kenneth. *The Earth's Dynamic Systems,* 6th ed. New York: Macmillan Publishing Company, 1992.

Hamilton, William J., III. "The Living Sands of the Namib," *National Geographic* 164, no. 3 (September 1983): 364–77.

Hundley, Norris. *Water and the West: The Colorado River Compact and the Politics of Water in the American West*. Berkeley: University of California Press, 1975.

Kahrl, William L. *Water and Power*. Berkeley: University of California Press, 1982.

Krutch, Joseph Wood. *The Voice of the Desert*. Wm. Sloane Associates, 1954.

Lancaster, Nicholas. "On Desert Sand Seas," *Episodes* 11, no. 1 (March 1988): 12–18.

Leopold, A. Starker, and the editors of Life. *The Desert*. Life Nature Library. New York: Time, Inc., 1962.

Mabbutt, J. A. *Desert Landforms*. Cambridge, MA: MIT Press, 1977.

McKee, Edwin D. *A Study of Global Sand Seas*. U.S. Geological Survey Professional Paper 1052. Washington, DC: Government Printing Office, 1979.

Page, Jake, and the editors of Time-Life Books. *Arid Lands*. Planet Earth Series. Alexandria, VA: Time-Life Books, 1984.

Péwé, T. L. *Desert Dust: Origin, Characteristics, and Effects on Man*. Geological Survey Special Paper 186. Washington, DC: Government Printing Office, 1981.

Powell, John Wesley. *The Exploration of the Colorado River and Its Canyons,* formerly titled *Canyons of the Colorado*. 1895. Reprint. New York: Dover Publications, 1961.

Stegner, Wallace. *Beyond the Hundredth Meridian*. New York: Houghton Mifflin, 1954.

United Nations Educational, Scientific, and Cultural Organization (UNESCO). Many publications and studies of interest by the Arid Zone Research Programme Paris.

Wallace, Robert, and the editors of Time-Life Books. *The Grand Canyon*. American Wilderness Series. New York: Time-Life Books, 1972.

Coastal cliffs, Nullarbor National Park, South Australia. [*Photo by M. P. Kahl.*]

16

COASTAL PROCESSES AND LANDFORMS

The interaction of vast oceanic and atmospheric masses is dramatic along a shoreline. At times, the ocean attacks the coast in a stormy rage of erosive power; at other times, the moist sea breeze, salty mist, and repetitive motion of the water are gentle and calming. Few have captured the confrontation between land and sea as well as Rachel Carson:

> The edge of the sea is a strange and beautiful place. All through the long history of Earth it has been an area of unrest where waves have broken heavily against the land, where the tides have pressed forward over the continents, receded, and then returned. For no two successive days is the shoreline precisely the same. Not only do the tides advance and retreat in their eternal rhythms, but the level of the sea itself is never at rest. It rises or falls as the glaciers melt or grow, as the floors of the deep ocean basins shift under its increasing load of sediments, or as the Earth's crust along the continental margins warps up or down in adjustment to strain and tension. Today a little more land may belong to the sea, tomorrow a little less. Always the edge of the sea remains an elusive and indefinable boundary.

Despite such variability, the realities of commerce and access to sea routes, fishing, and tourism prompt many people to settle near the ocean. A population survey by the United Nations estimates that about two-thirds of Earth's present population lives in or very near coastal regions. Therefore, an understanding of coastal processes and landforms is critical. And because the exogenic processes along coastlines oftentimes produce dramatic change, these processes are essential to consider in planning and development.

You will find in this chapter coastal processes organized in a system of specific inputs (components and driving forces), actions (movements and processes), and outputs (results and consequences). We conclude with a look at the considerable human impact on coastal environments.

*From "The Marginal World," in *The Edge of the Sea* by Rachel Carson. Copyright © 1955 by Rachel Carson, © renewed 1983 by Roger Christie. Boston: Houghton Mifflin, p. 11.

COASTAL SYSTEM COMPONENTS

Most of Earth's coastlines are relatively new and are the setting for continuous change. The land, ocean, atmosphere, Sun, and Moon interact to produce tides, currents, waves, erosional features, and depositional features along the continental margins. A dynamic equilibrium among the energy of waves, wind, and currents, the supply of materials, the slope of the coastal terrain, and the fluctuation of relative sea level produces coastline features of infinite variety.

Inputs to the Coastal System

Inputs to the coastal environment include many of the elements already discussed in this text. Solar energy remains the driving force of both the atmosphere and the hydrosphere. Prevailing winds and active weather patterns are produced by conversion of insolation to kinetic and mechanical energy; and in turn, ocean currents and waves are generated principally by atmospheric winds. Climatic regimes are influential in coastal geomorphic processes. The nature of coastal rock is important in determining rates of erosion and sediment production. All of these inputs occur within the ever-present influence of gravity's pull, not only from Earth but also from the Moon and the Sun. Gravity provides the potential energy of position for materials in motion and produces the tides.

The Coastal Environment

The coastal environment is called the **littoral zone.** (*Littoral* comes from the Latin word for shore.) The littoral zone spans both land and water. Landward, it extends to the highest water line that occurs on shore during a storm. Seaward, it extends to the point at which storm waves can no longer move sediments on the sea floor (usually at depths of approximately 60 m or 200 ft). The specific contact line between the sea and the land is the *shoreline,* and adjacent land is considered the *coast.* Figure 16-1 illustrates the littoral zone and includes specific components that are discussed later in the chapter.

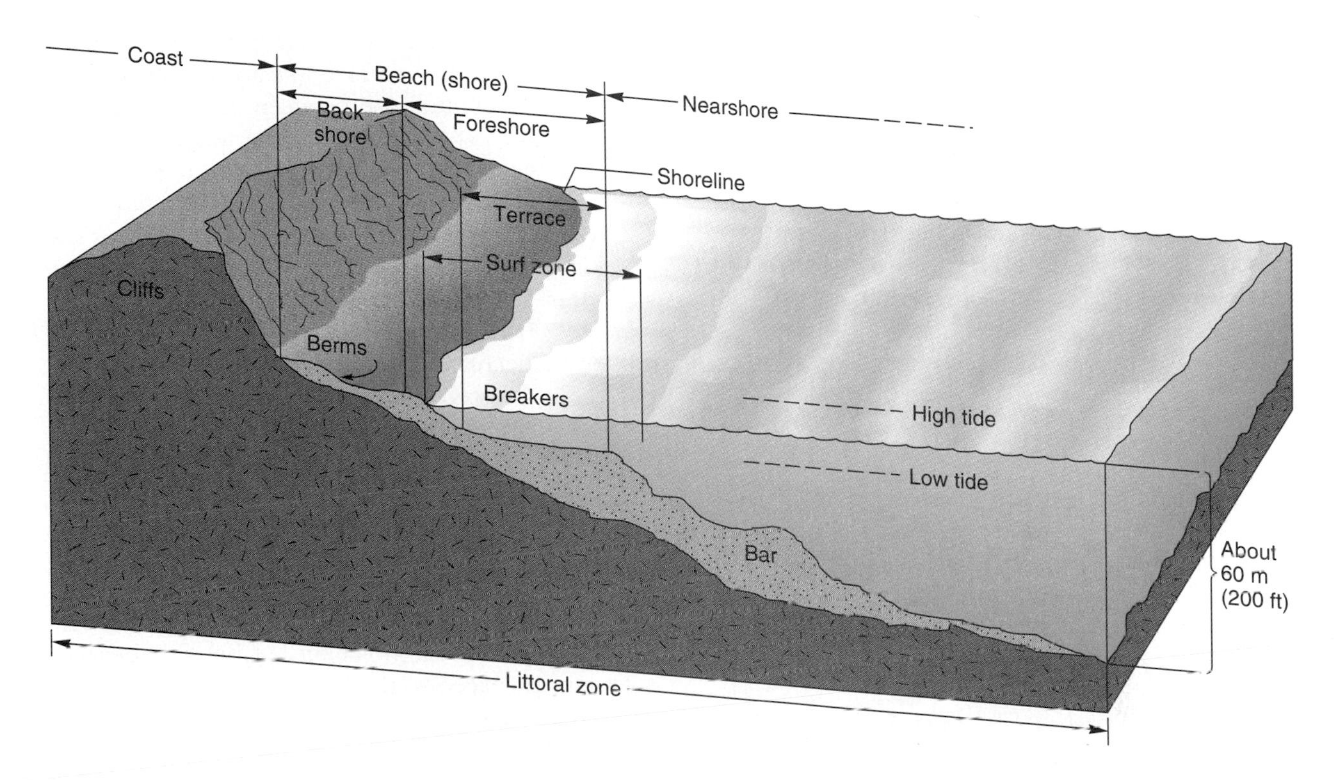

FIGURE 16-1
The littoral zone includes the coast, beach, and nearshore environments.

Because the level of the ocean varies, the littoral zone is redefined naturally from time to time. A rise in sea level is a coastal transgression that causes *submergence* of land, whereas a drop in sea level is a regression that produces coastline *emergence*. Submergent and emergent coastlines each have distinctive features which are discussed later in the chapter. Changes in the littoral zone also are initiated by uplift and subsidence of the coast itself.

Sea level is a relative term and reflects changes in both water quantity and land elevations. Because this is not a straightforward measurement, there exists no integrated international system to coordinate the determination of exact sea level over time. NOAA is establishing a cooperative Global Absolute Sea Level Monitoring System. The goal is to accurately monitor global changes in absolute sea level, discussed later in this chapter.

Mean sea level is a calculated value based on average tidal levels recorded hourly at a given site over a period of at least 19 years, which is one full lunar tidal cycle. Mean sea level varies spatially from place to place because of ocean currents and waves, tidal variations, air temperature and pressure differences, and ocean temperature variations. The mean sea level of the U.S. Gulf Coast is about 25 cm (10 in.) higher than that of the east coast of Florida, which has the lowest mean sea level in North America. Sea level rises as one moves northward up the east coast, measuring about 38 cm (15 in.) higher in Maine than in Florida. Along the U.S. west coast, mean sea level in San Diego is about 58 cm (23 in.) higher than it is in Florida, rising to about 86 cm (34 in.) higher in Oregon than in Florida. Overall, the Pacific coast of North America has a mean sea level that averages about 66 cm (26 in.) higher than the average sea level along the Atlantic coast.

At present, the overall mean sea level for the United States is calculated at approximately 40 locations along the coastal margins of the continent.

These sites are being upgraded with new equipment in the Next Generation Water Level Measurement System, specifically along the United States and Canadian Atlantic coasts, Bermuda, and the Hawaiian Islands. These measurements are augmented by new remote sensing technology including the *TOPEX/Poseidon* satellite launched in August 1992. Of great assistance are the *NAVSTAR* satellites that make up the new Global Positioning System (GPS), which allows correlation from a network of ground-based measurements.

COASTAL SYSTEM ACTIONS

The complex fluctuation of tides, the movement of solar energy-driven winds, waves, and ocean currents, and the occasional impact of storms all are powerful actions within the coastal system. These forces shape landforms from gentle beaches to steep cliffs, and sustain delicate ecosystems.

Tides

Earth's relationship to the Sun and the Moon and the reasons for the seasons are discussed in Chapter 2. These astronomical relationships produce the pattern of **tides,** the complex daily oscillations in sea level that are experienced to varying degrees around the world. Tides also are influenced by the size, depth, and topography of ocean basins, by latitude, and by shoreline configuration.

Tidal action is a distinctive energy agent for geomorphic change. As tides flood (rise) and ebb (fall), the migration of the shoreline landward and seaward causes significant changes that affect sediment erosion and transportation. Figure 16-2 illustrates the relationship among the Moon, the Sun, and Earth and the generation of variable tidal bulges on opposite sides of the planet.

Tides are produced by the gravitational pull exerted on Earth and its oceans by both the Sun and the Moon. Although the Sun's influence is only about half that of the Moon (46%) because of the Sun's greater distance from Earth, it still is a significant force. Figure 16-2a shows the Moon and the Sun in *conjunction* (lined up with Earth), a position in which their gravitational forces add together.

Earth and Moón share a common center of mass around which they revolve in stable orbits. Gravity is the force of attraction between the mass of these two bodies, whereas inertia—the tendency of objects to stay still if still or moving in the same direction if moving—keeps them separated. The gravitational effect on the side of Earth facing the Moon and Sun is greater than that experienced by the far side where inertial forces are slightly greater. This difference is because gravitational influences decrease with distance. It is this difference in the net force of gravitational attraction and inertia that generates the tides. Earth's solid and fluid surfaces actually experience some stretching as a result of this difference in gravitational pull. The fluid atmosphere and water are pulled outward on the side of Earth facing toward the Moon and Sun.

The corresponding tidal bulge on Earth's opposite side is primarily the result of the far-side water remaining in position (being left behind) because its inertia exceeds the gravitational pull of the Moon and Sun. In effect, from this inertial point of view, as the near-side water and Earth are drawn toward the Moon and Sun, the far-side water is left behind due to the slightly weaker gravitational pull. This produces the two opposing tidal bulges on opposite sides of Earth.

Tides appear to move in and out along the shoreline, but they do not actually do so. Instead, the surface rotates into and out of the relatively "fixed" tidal bulges as Earth changes its position in relation to the Moon and Sun. Every 24 hours and 50 minutes, any given point on Earth rotates through two bulges as a direct result of this rotational positioning. Thus, every day, most coastal locations experience two high (rising) tides known as **flood tides,** and two low (falling) tides known as **ebb tides.** The difference between consecutive high and low tides is considered the *tidal range.*

The combined gravitational effect is strongest in the conjunction alignment and results in the greatest tidal range between high and low tides, known as **spring tides.** (Spring means to "spring forth"; it has no relation to the season of the year.) Figure 16-2b shows the other alignment that gives rise to spring tides, when the Moon and Sun are at *opposition.* In this arrangement, the Moon and Sun cause separate tidal bulges, affecting the water

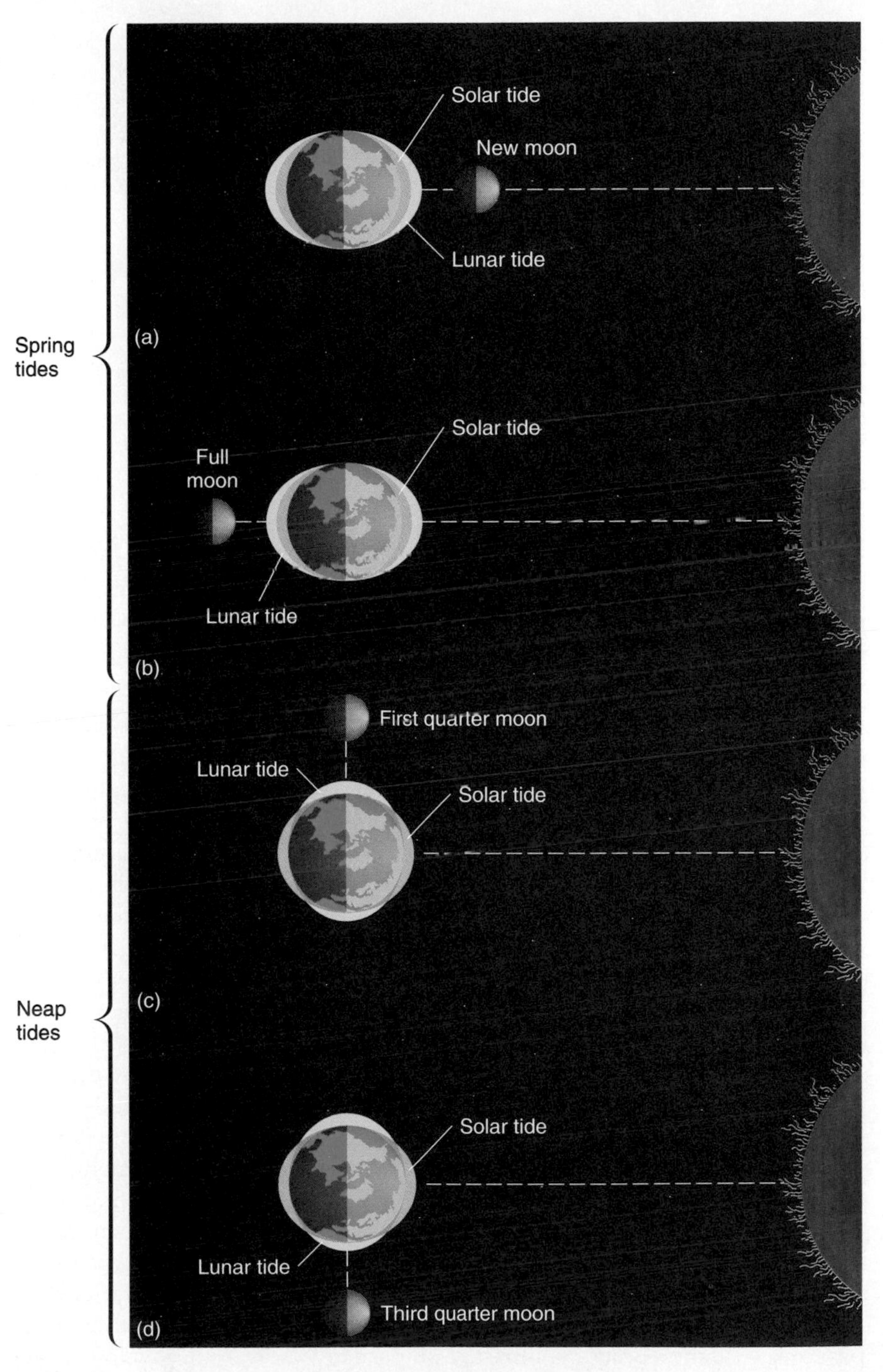

FIGURE 16-2
Sun, Moon, and Earth relationships combine to produce spring tides (a), (b) and neap tides (c), (d). (The tides depicted are not illustrated to scale.)

(a)

(c)

(b)

FIGURE 16-3
Tidal range is great in some bays and estuaries, such as in the Bay of Fundy at flood tide (a) and ebb tide (b). These ranges are ideal for turning turbines and generating electricity, as at the Annapolis Tidal Generating Station, in operation since 1984 (c). [(a) and (b) Photos by Imagery; (c) courtesy of Nova Scotia Power Incorporated.]

nearest each of them. In addition, each bulge is augmented by the left-behind water resulting from the pull of the body on the opposite side.

When the Moon and the Sun are neither in conjunction nor in opposition, but are more-or-less in the positions shown in (c) and (d), their gravitational influences are offset and counteract each other somewhat, producing a lesser tidal range known as **neap tide.**

Tides also are influenced by other factors, so there is a great variety of tidal ranges. For example, some locations may experience almost no difference between high and low tides, whereas the highest tides occur when open water is forced into partially enclosed gulfs or bays. The Bay of Fundy in Nova Scotia records the greatest tidal range on Earth, a difference of 16 m (52.5 ft) (Figure 16-3a and b). Tidal prediction is especially important to ships because the entrance to many ports is limited by shallow water and thus high tide is required for passage. At other times, tall-masted ships need a low tide to clear overhead bridges. Tides also exist in large lakes, but the tidal range is small, making tides difficult to distinguish from changes caused by wind. Lake Superior, for instance, has a tidal variation of only about 5 cm (2 in.).

A tidal current entering a bay or river is visible as a **tidal bore** and can cause turbulence as it moves within a narrowing, constricting estuary or bay, against an opposing tidal current, or against a river's discharge. Such a current entering the mouth of the Amazon River can raise the water level in the river channel a meter or more. Exceptional tidal bores occur in the Chian Tang Kiang River in China, measuring 2.5–3.5 m (8.2–11.5 ft) higher than normal river level.

Tidal Power. The fact that sea level changes daily with the tides suggests an opportunity: could these predictable flows be harnessed to produce electricity? The answer is yes, given the right conditions. The bay or estuary under consideration must have a narrow entrance suitable for the construction of a dam with gates and locks, and it must experience a tidal range of flood and ebb tides large enough to turn turbines, at least a 5 m (16 ft) range.

About 30 locations in the world are suited for tidal power generation, although at present only three of them are actually producing electricity, two outside North America—an experimental 1-megawatt station in Russia at Kislaya-Guba Bay, on the White Sea, since 1969, and a facility in the

Rance River estuary on the Brittany coast of France since 1967. The tides in the Rance estuary fluctuate up to 13 m (43 ft), and power production has been almost continuous there since 1967, providing an electrical generating capacity of a moderate 240 megawatts (about 20% of the installed electrical capacity generated at Hoover Dam).

According to studies completed by the Canadian government, the present cost of tidal power at ideal sites is economically competitive with that of fossil fuels, although certain environmental concerns must be addressed. Among several favorable sites on the Bay of Fundy, one plant is in operation. The Annapolis Tidal Generating Station was built in 1984 to test electrical production using the tides. Nova Scotia Power Incorporated operates the 20-megawatt plant (Figure 16-3c). Such sites as Observatory Bay near Prince Rupert, Sechelt Inlet near Vancouver, both in British Columbia, and Ungava Bay in Labrador are future possibilities. Further expansion depends on the growth in electrical demand, the price of fossil fuels, the resolution of problems with nuclear power, and consideration of environmental demands.

Waves

Wind friction on the surface of the ocean generates movement in the water. These undulations of ocean water called **waves** travel in *wave trains,* or groups of waves. At a smaller scale, a moving boat creates a wake, which consists of waves; at a larger scale, storms around the world generate large groups of wave trains. A stormy area at sea is called a *generating region* for these waves, which radiate outward from their formation center. As a result, the ocean is crisscrossed with intricate patterns of waves traveling in all directions. The wave forms seen along a coast may be the product of a storm center thousands of kilometers away.

Regular patterns of smooth, rounded waves are called **swells**—these are the mature undulations of the open ocean. As waves leave the generating region, wave energy continues to run in these swells, which can range from small ripples to very large flat-crested waves.

A deep-water wave leaving a generating region tends to extend its wavelength many meters (see inset in Figure 16-4 noting crest, trough, wavelength, and wave height). Tremendous energy occasionally accumulates to form unusually large waves. One moonlit night in 1933, the U.S. Navy tanker Ramapo, running downwind in the direction of wave travel for stability, was en route to San Diego and reported a wave higher than their mainmast at about 34 m (112 ft)!

Water within a wave in the open ocean is not really migrating but is transferring energy from molecule to molecule through the water in simple cyclic undulations, which form *waves of transition* (Figure 16-4). Individual water particles move forward only slightly, forming a vertically circular pattern. The diameter of the paths formed by the orbiting water particles decreases with depth. As a deep-ocean wave approaches the shoreline and enters shallower water (10–20 m or 30–65 ft), the drag of the bottom shortens the wavelength and slows the speed of the circular motion, flattening the orbital paths to elliptical shapes. As the peak of each wave rises, a point is reached when its height exceeds its vertical stability and the wave falls into a characteristic **breaker,** crashing onto the beach (Figure 16-4b).

In a breaker the orbital motion of transition gives way to form *waves of translation,* in which energy and water actually move forward toward shore as water cascades down from the wave crest. In conjunction with the energy of arriving waves, the slope of the shore determines wave style: plunging breakers indicate a steep bottom profile, whereas spilling breakers indicate a gentle, shallow bottom profile.

As various wave trains move along in the open sea, they interact, a phenomenon called *interference.* Waves sometimes align so that the wave crests and troughs from one wave train are in phase with those of another. When this in-phase wave interference occurs, the heights of the waves are increased. The resulting "killer waves" or "sleeper waves," can sweep in unannounced and overtake unsuspecting beachcombers. When wave trains are out of phase with each other, waves are dampened and their heights are reduced. When you observe the breakers along a beach, the changing beat of the surf actually is produced by the patterns of wave interference that occurred in far-distant areas of the ocean.

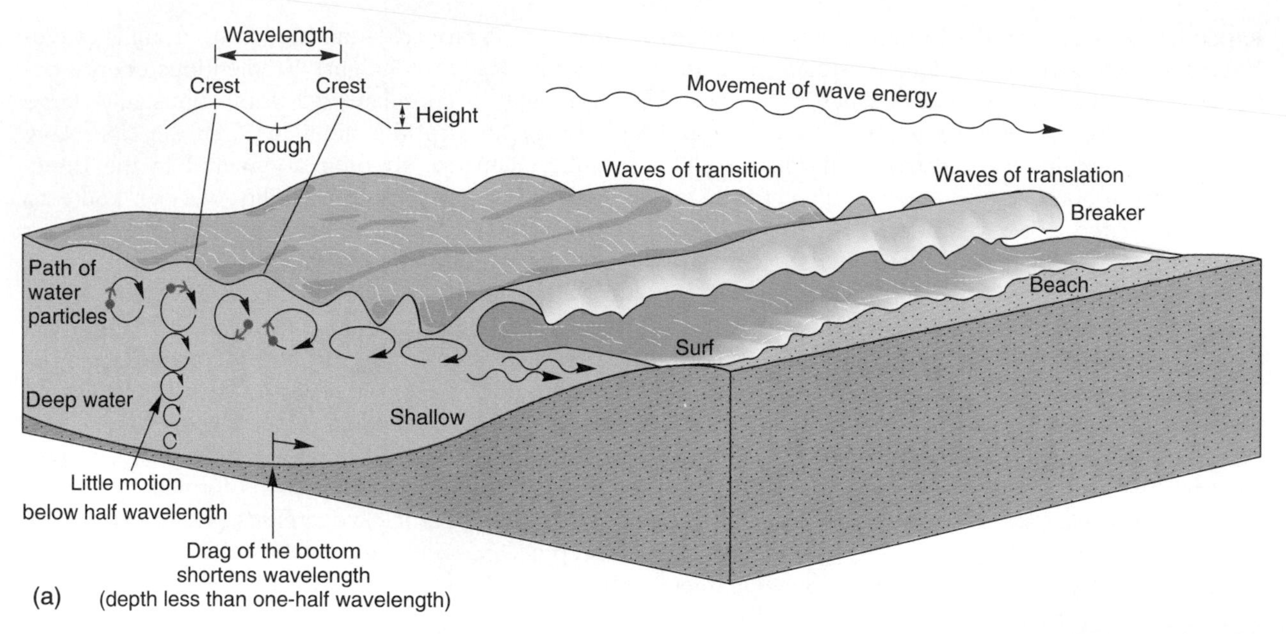

FIGURE 16-4
Wave structure and the orbiting tracks of water particles change from circular motions and swells in deep waves of transition to more elliptical orbits in shallow waves of translation (a). Cascades of waves and breakers work the shore (b). [Photo by Bobbé Christopherson.]

Wave Refraction. Generally, wave action results in coastal straightening. As waves approach an irregular coast, they bend around **headlands,** which are protruding landforms generally composed of more resistant rocks (Figure 16-5). The submarine topography refracts approaching waves. This produces patterns that focus energy around the undersea extension of headlands, and dissipate energy in coves and submerged coastal valleys. Thus, headlands represent a specific focus of wave attack along a coastline. Waves tend to disperse their energy in *coves* and *bays* on either side of headlands. This **wave refraction** (wave bending) along a coastline redistributes wave energy so that different sections of the coastline are subjected to variations in erosion potential.

As waves of translation approach a coast, they usually arrive at some angle other than parallel to it (Figure 16-6). As the waves enter shallow water, they are refracted and generate a current parallel to the coast, zigzagging as inflow and outflow in the prevalent direction of the incoming waves. This **longshore current,** or *littoral current,* depends on the effective wind and wave direction. Such a current is generated only in the surf zone and works in combination with wave action to

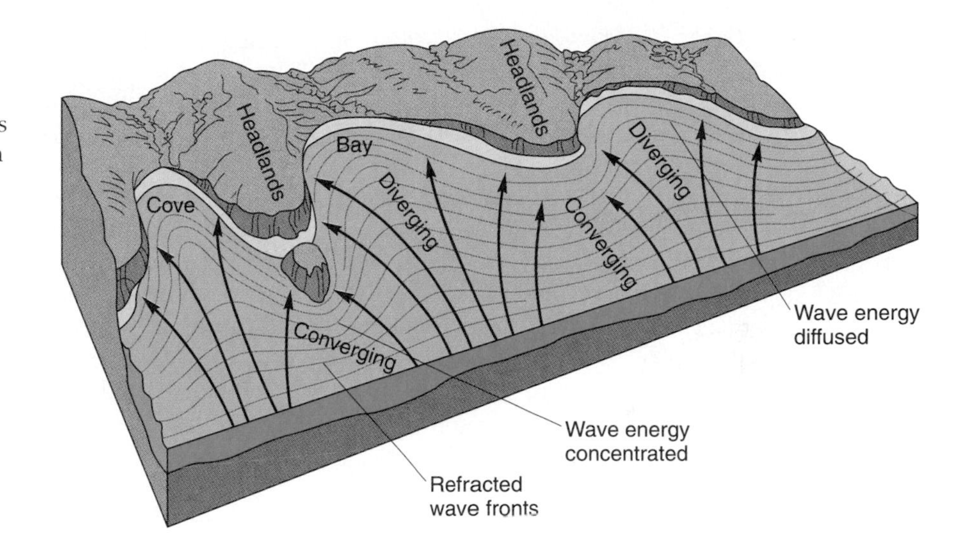

FIGURE 16-5
The process of coastal straightening brought about by wave refraction. Wave energy is concentrated as it converges on headlands and is diffused as it diverges in coves and bays.

transport large amounts of sand, gravel, sediment, and shells along the shore.

Particles on the beach also are moved along as **beach drift,** shifting back and forth between water and land in the effective wind and wave direction with each *swash* and *backwash* of surf. Individual sediment grains trace arched paths along the beach. You have perhaps stood on a beach and heard the sound of myriad grains and seawater, especially in the backwash of surf. These dislodged materials are available for transport and eventual deposition in coves and inlets and can represent a significant volume.

Tsunami, or Seismic Sea Wave. A particular type of wave that greatly influences coastlines is the **tsunami.** *Tsunami* is Japanese for "harbor wave," named for its devastating effect in harbors. Statistically, this sea wave occurs an average of every 3.5 years in the Pacific basin. Specifically for Hawaii,

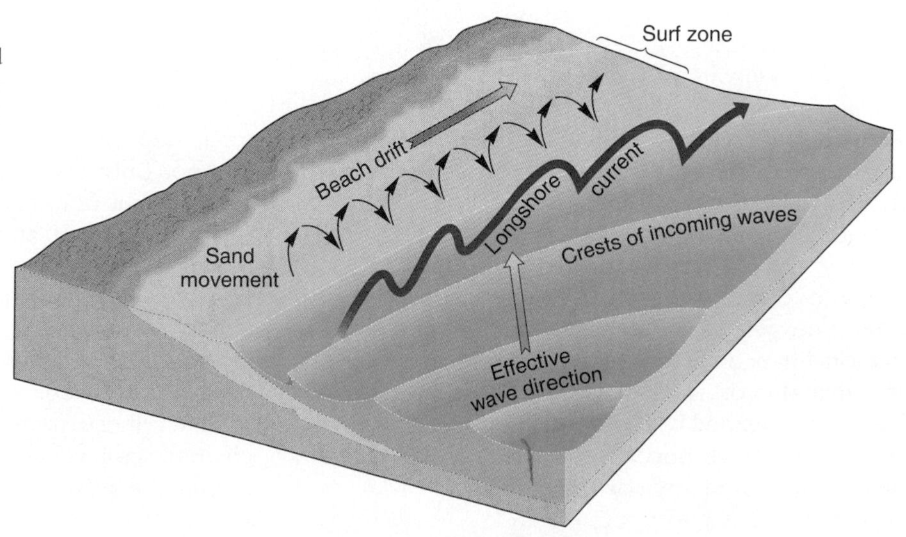

FIGURE 16-6
Longshore currents are produced as waves approach the surf zone and shallower water. Littoral and beach drift result as substantial volumes of material are moved along the shore.

the U.S. Army Corps of Engineers has reported 41 damaging occurrences during the past 142 years. Tsunamis, often incorrectly referred to as "tidal waves," are formed by sudden and sharp motions in the seafloor, caused by earthquakes, submarine landslides, or eruptions of undersea volcanoes. Thus they are called *seismic sea waves*. In September 1992 the citizens of Casares, Nicaragua, were surprised by a 12 m (39 ft) tsunami that took 270 lives. An earthquake along an offshore subduction zone triggered this killer wave.

A large undersea disturbance usually generates a solitary wave of great wavelength or sometimes a group of two or three long waves. These waves generally exceed 100 km (60 mi) in wavelength but only a meter or so in height. Because of their great wavelength, tsunamis are affected by the topography of the deep-ocean floor and are refracted by rises and ridges. They travel at great speeds in deep-ocean water—velocities of 600–800 kmph (375–500 mph) are not uncommon—but often pass unnoticed on the open sea because their great length makes the slow rise and fall of water hard to observe.

However, as a tsunami approaches a coast, the shallow water forces the wavelength to shorten. As a result, the wave height may increase up to 15 m (50 ft) or more. The impact of such a wave has the potential for coastal devastation, resulting in property damage and death. In the Alaskan earthquake of 1964, a tsunami was generated by undersea landslides and radiated across the Pacific Ocean. It reached Crescent City, California, in 4 hours 3 minutes; Hilo, Hawaii, in 5 hours 24 minutes; Hokkaido, Japan, in 6 hours 44 minutes; and La Punta, Peru, in 15 hours 34 minutes. Curious onlookers died in both California and Oregon as a result of the distant Alaskan earthquake.

Because tsunami travel such great distances so quickly and are undetectable in the open ocean, accurate forecasts are difficult and occurrences are often unexpected. A warning system now is in operation for nations surrounding the Pacific, where the majority of tsunami occur. Warnings always should be heeded, despite the many false alarms, for the causes lie beneath the ocean and are difficult to monitor in any consistent manner.

Sea Level Changes

Over the long term, sea level fluctuations expose a great range of coastal landforms to tidal and wave processes. Such changes are initiated either by tectonic forces that alter coastlines and basins, or by glacio-eustatic processes that determine the quantity of water stored as ice. Because of complex interactions, it is difficult to isolate individual factors. However, as average global temperatures cycle through cold or warm climatic spells, the quantity of ice locked up in Antarctica, Greenland, and mountain glaciers can increase or decrease accordingly.

If Antarctica and Greenland ever were completely ice-free (ice sheets completely melted), sea level would rise at least 65 m (215 ft). Indeed, sea level has risen throughout the most recent epoch, the Holocene (the last 10,000 years), and generally ever since the Pleistocene Ice Age, which began about 1.65 million years ago. At the peak of the last Pleistocene glaciation about 18,000 years B.P. (before the present), sea level was about 130 m (430 ft) lower than it is today. Just 100 years ago it was 38 cm (15 in.) lower along the coast of southern Florida. Venice, Italy, has experienced a rise of 25 cm (10 in.) since 1890. Elsewhere, increases of 10–20 cm (4–8 in.) are common. This rate of rise of 20–40 cm (8–16 in.) per 100 years is about six times faster than is observed from historical records or recent geologic time.

Chapters 5 and 10 present the climatic implications of an enhanced greenhouse-effect warming pattern on high-latitude and polar environments, which are critical areas because they contain enormous deposits of ice. With warming, sea level will rise not only because of melted ice, but also because of the thermal expansion of warmer water (water is densest at 4°C or 39°F, and expands at higher temperatures). Measurements by the Scripps Institute of Oceanography demonstrate a 0.8C° (1.44F°) warming since 1950 off southern California. Similar ocean warming trends are being reported worldwide.

Given these trends and the predicted climatic change, sea level could rise as much as 6 m (20 ft) during the next century. However, it is generally believed that the rise will be less than 2 m (6.5 ft). Nonetheless, even this amount represents a potentially devastating change for many coastal locations, because a rise of only 0.3 m (1 ft) would cause shorelines worldwide to move inland an average of 30 m (100 ft)!

A 1 m (3.2 ft) rise could inundate 15% of Egypt's arable land, 17% of Bangladesh, many island nations and communities, and 20,000 km^2 (7800 mi^2) of land along U. S. shores, at a staggering loss of $650 billion in North America alone. However, great uncertainty exists in these forecasts. The Intergovernmental Panel on Climate Change (IPCC, discussed in Chapters 5 and 10) places the estimated range of expected sea-level rise at between 30 and 110 cm (1–4 ft). Despite this uncertainty, planning should start now along coastlines worldwide because preventive strategies are cheaper than possible destruction. Insurance underwriters have begun the process by refusing coverage for shoreline properties vulnerable to rising sea level.

COASTAL SYSTEM OUTPUTS

Coastlines are active portions of continents, with energy and material being continuously delivered to a narrow environment; as a result, these areas are unstable and constantly changing. The action of tides, currents, wind, waves, and changing sea level produces a variety of erosional and depositional landforms.

Erosional Coastal Processes and Landforms

The active margins of the Pacific along the North and South American continents are characteristic coastlines affected by erosional landform processes. Erosional coastlines tend to be rugged, of high relief, and tectonically active, as expected from their association with the leading edge of drifting lithospheric plates (see plate tectonics discussion in Chapter 11). Figure 16-7 presents features commonly observed along an erosional coast. *Sea cliffs*

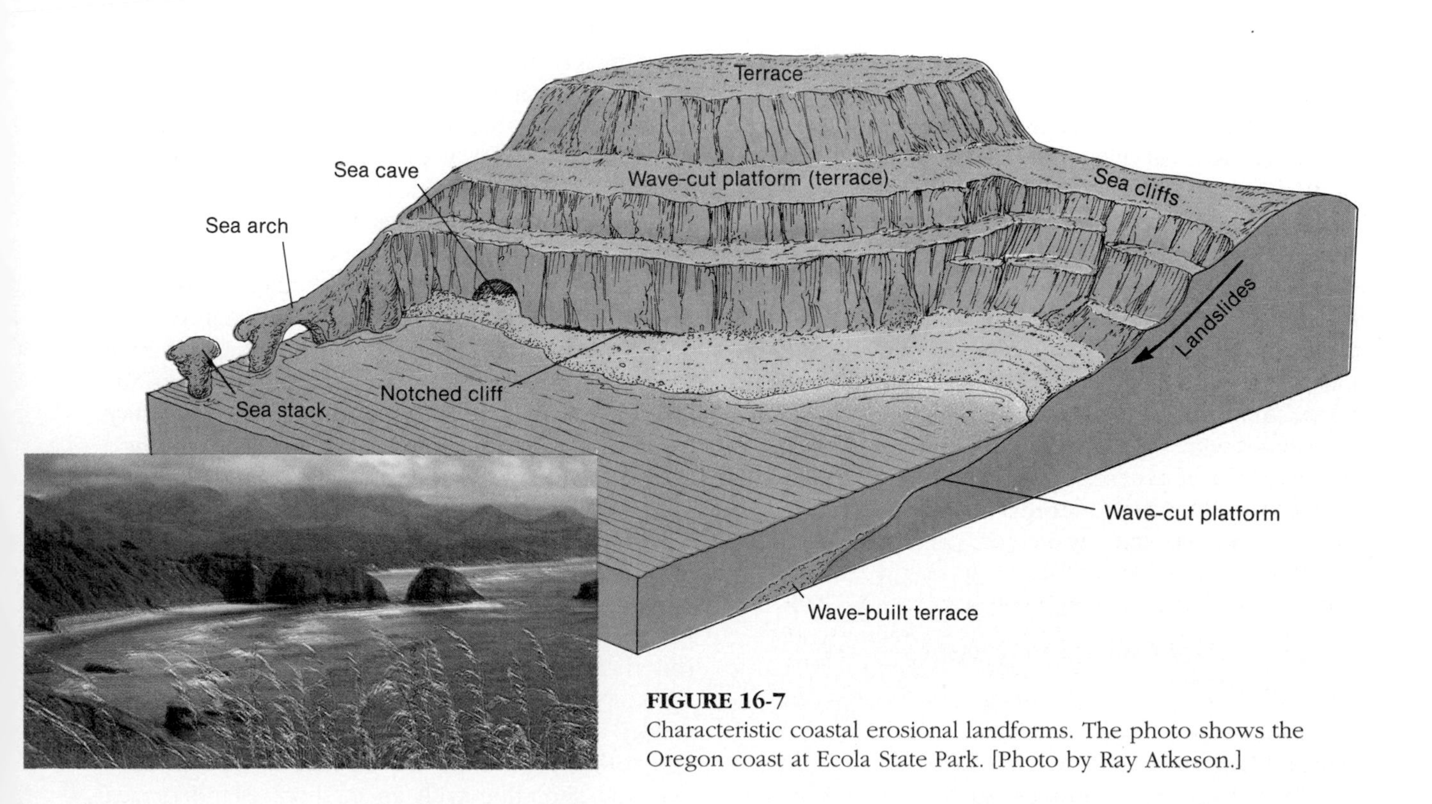

FIGURE 16-7
Characteristic coastal erosional landforms. The photo shows the Oregon coast at Ecola State Park. [Photo by Ray Atkeson.]

FIGURE 16-8
Marine terraces formed as wave-cut platforms along the Pacific coast. [Photo by author.]

are formed by the undercutting action of the sea. As indentations are produced at water level, such a cliff becomes notched, leading to subsequent collapse and retreat of the cliff. Other erosional forms evolve along cliff-dominated coastlines, including *sea caves, sea arches,* and *sea stacks*. As erosion continues, arches may collapse, leaving isolated stacks in the water. The coasts of southern England and Oregon are prime examples of such erosional landscapes (Figure 16-7).

Wave action can cut a horizontal bench in the tidal zone, extending from a sea cliff out into the sea. Such a structure is called a **wave-cut platform,** or *wave-cut terrace*. If the relationship between the land and sea level has changed over time, multiple platforms or terraces may arise like stairsteps back from the coast. These marine terraces are remarkable indicators of an emerging coastline, with some upper terraces more than 365 m (1200 ft) above sea level. A tectonically active region, such as the California coast, has many examples of multiple wave-cut platforms (Figure 16–8).

Depositional Coastal Processes and Landforms

Depositional coasts generally are located near onshore plains of gentle relief, where sediments are available from many sources. Such is the case with the Atlantic and Gulf coastal plains of the United States, which lie along the relatively passive, trailing edge of the North American lithospheric plate. These depositional coasts also are influenced by erosional processes and inundation, particularly during storm activity.

Characteristics of wave- and current-deposited landforms are illustrated in Figure 16-9. Such landforms may involve sediments of varying sizes. A **barrier spit** consists of material deposited in a long ridge extending out from a coast attached at one end; it partially crosses and blocks the mouth of a bay. Classic examples include Sandy Hook, New Jersey (south of New York City), and Cape Cod, Massachusetts. The photo in Figure 16-9 shows a typical barrier spit forming part way across the mouth of Prion Bay in Southwestern National Park in Tasmania. A spit becomes a **bay barrier,** sometimes referred to as a *baymouth bar,* if it completely cuts off the bay from the ocean and forms an inland **lagoon.** Tidal flats and salt marshes are characteristic low-relief features wherever tidal influence is greater than wave action. Spits and barriers are made up of materials that have been eroded and transported by *littoral drift* (beach and longshore drift combined). For much sediment to accumulate, offshore currents must be weak.

A **tombolo** occurs when sediment deposits connect the shoreline with an offshore island or sea

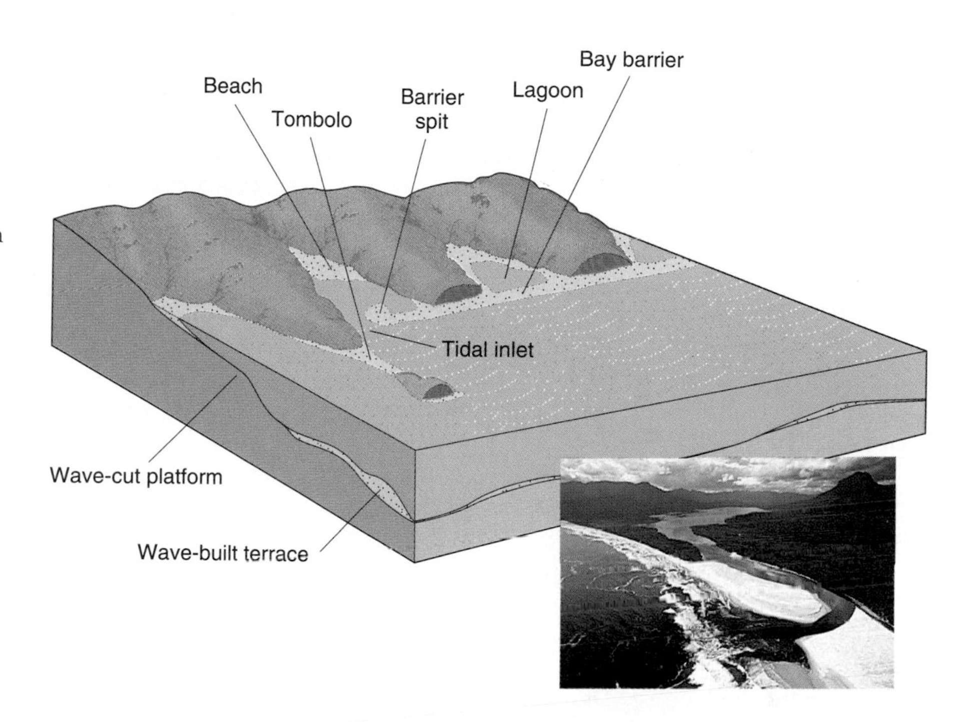

FIGURE 16-9
Characteristic coastal depositional landforms. The photo depicts a typical sand spit developing across the mouth of the New River, forming the New River Lagoon and Prion Bay at Southwestern National Park in Tasmania. [Photo by Reg Morrison/Auscape International.]

stack (Figure 16-10). It forms when sediments accumulate on a wave-built terrace that extends below the water.

Beaches. Of all the features associated with a depositional coastline, beaches probably are the most familiar. Beaches vary in type and permanence, especially along coastlines dominated by wave action. Technically, a **beach** is that place along a coast where sediment is in motion, deposited by waves and currents and attached to the margin of the continent along its entire length. Material from the land temporarily resides there while it is in active transit along the shore. You may have experienced a beach at some time, along a sea coast, a lake shore, or even a stream. Perhaps you have even built your own "landforms" in the sand, only to see them washed away by tides and waves.

The beach zone ranges, on average, from 5 m (16 ft) above high tide to 10 m (33 ft) below low tide, although specific definition varies greatly

FIGURE 16-10
A tombolo at Point Sur along the central California coast where a bar connects the shore with an island. [Photo by Bobbé Christopherson.]

along individual shorelines. Globally beaches are dominated by sands of quartz (SiO_2) because it is the most abundant mineral on Earth, resists weathering, and therefore remains after other minerals are removed. In volcanic areas, beaches are derived from wave-processed lava. These black-sand beaches are found in Hawaii and Iceland, for example. Many beaches, such as those in southern France and western Italy, are composed of pebbles and cobbles—a type of "shingle beach." Finally, some shores have no beaches at all; scrambling across boulders and rocks may be the only way to move along the coast. The coast of Maine and portions of the Atlantic provinces of Canada are classic examples, composed of resistant granite rock that is scenically rugged but with few beaches.

A beach acts to stabilize a shoreline by absorbing wave energy, as is evident by the amount of material that is in almost constant motion (see "sand movement" in Figure 16-6). Some beaches are continuous and stable. Others cycle seasonally: they accumulate during the summer only to be moved offshore by winter storm waves, forming a submerged bar, and again replaced the following summer. Protected areas along a coastline tend to accumulate sand, which can lead to large coastal sand dunes. Prevailing winds often drag such coastal dunes inland, sometimes burying trees and highways.

Locations of beach elements—the backshore, berm, and foreshore (which includes the surf zone) are shown in Figure 16-1. Beach profiles fall into two broad categories, according to prevailing conditions. During quiet, low-energy periods, both backshore and foreshore areas experience net accumulations of sediment and are well defined. However, a storm brings increased wave action and high-energy conditions to the beach, usually reducing its profile to a sloping foreshore only. Sediments from the backshore area do not disappear but may be moved by wave action to the nearshore (submerged) area, only to be returned by a later low-energy surf. Therefore, beaches are in constant flux, especially in this era of rising sea levels.

Changes in coastal sediment transport can thwart human activities—beaches are lost, harbors closed, and coastal highways and beach houses can be inundated with sediment. Consequently, various strategies are employed to interrupt longshore currents and beach drift. The goal is either to halt sand accumulation or to force accumulation in a desired way through construction of engineered structures—"hard" shoreline protection.

Figure 16-11 illustrates common approaches: *jetties* to block material from harbor entrances; *groins* to slow drift action along the coast, and a **breakwater** to create a zone of still water near the coastline. However, interrupting the coastal drift that is the natural replenishment for beaches may lead to unwanted changes in sediment distribution downcurrent. Careful planning and impact assessment should be part of any strategy for preserving or altering a beach.

Beach nourishment refers to the artificial placement of sand along a beach. Through such efforts, a beach that normally experiences a net loss of sediment will instead show a net gain. In contrast to hard structures this hauling of sand to replenish a beach is considered "soft" shoreline protection.

Enormous energy and material must be committed to counteract the relentless energy that nature invests along the coast. Years of human effort and expense can be erased by a single storm, as occurred when offshore islands along the Gulf Coast were completely eliminated by Hurricane Camille in 1969. In Florida, the city of Miami and surrounding Dade County have spent almost $70 million since the 1970s rebuilding their beaches, and needless to say, the effort is continuous. To maintain a beach 200 m (660 ft) wide, net sand loss per year is determined and a schedule of needed replenishment is set. For Miami Beach an eight-year replenishment cycle is maintained. During Hurricane Andrew in 1992 the replenished Miami Beach is estimated to have prevented millions of dollars in shoreline structural damage.

Unforeseen environmental impact may accompany the addition of new and foreign sand types to a beach. If the new sands do not match the existing varieties, untold disruption of coastal marine life is possible. The U.S. Army Corps of Engineers, which operates the Miami replenishment program, is running out of "borrowing areas" for sand that matches the natural sand of the beach. A proposal to haul a different type of sand from the Bahamas is being studied as to possible environmental consequences.

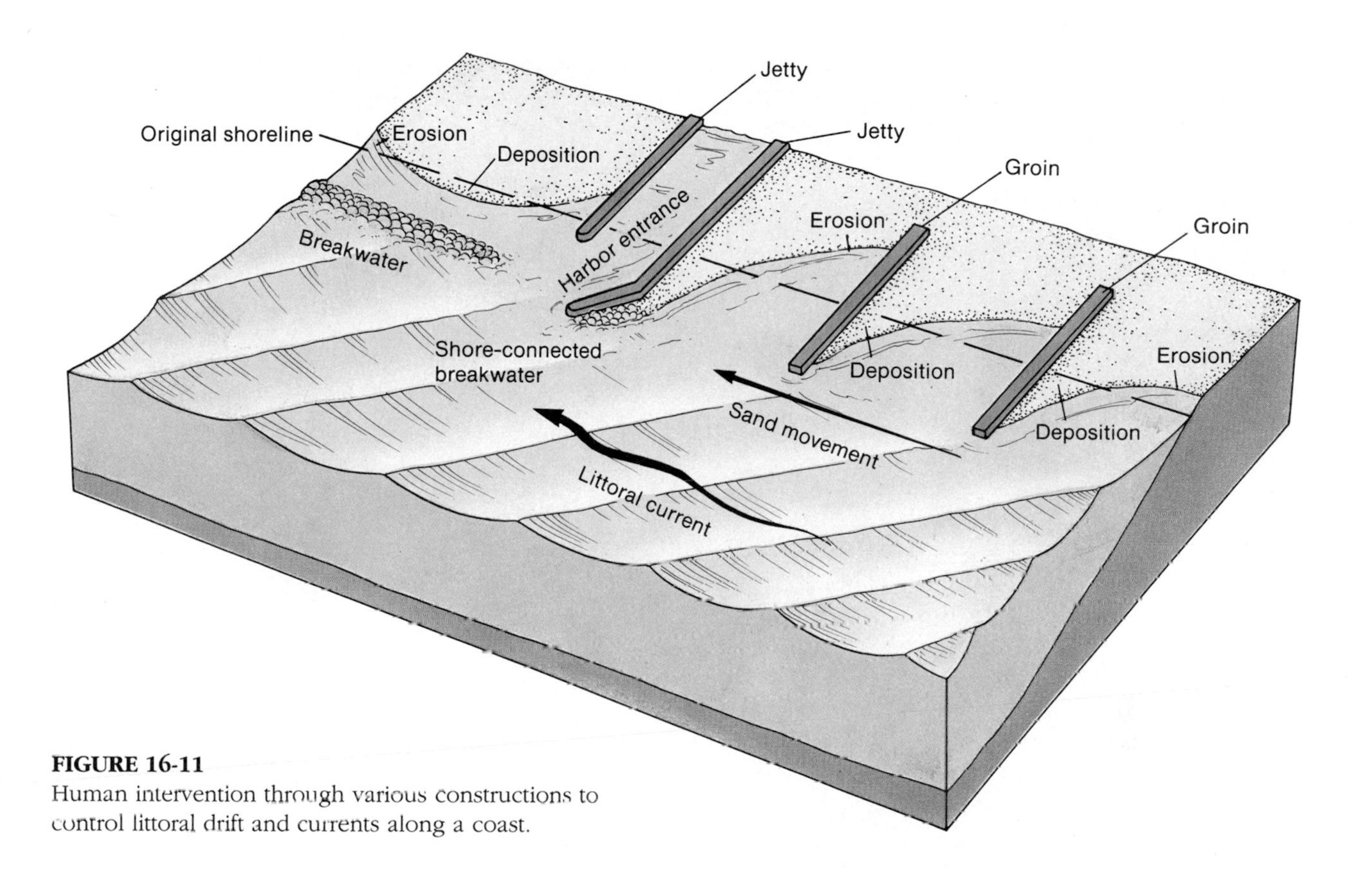

FIGURE 16-11
Human intervention through various constructions to control littoral drift and currents along a coast.

Barrier Forms. The image in Figure 16-12 illustrates the many features of barrier chains. These chains are long, narrow, depositional features, generally of sand, that form offshore roughly parallel to the coast. Common forms are **barrier beaches,** or the broader, more extensive landform, **barrier islands.** Tidal variation in the area usually is moderate to low, with adequate sediment supplies coming from nearby coastal plains. On the landward side of a barrier formation are tidal flats, marshes, swamps, lagoons, coastal dunes, and beaches. Barrier beaches appear to adjust to sea level and may naturally shift position from time to time in response to wave action and longshore currents. Breaks in a barrier form inlets, connecting a bay with the ocean.

Barrier beaches and islands are quite common worldwide, lying offshore of nearly 10% of Earth's coastlines. Examples are found offshore of Africa, India (east coast), Sri Lanka, Australia, the north slope of Alaska, and the shores of the Baltic and Mediterranean Seas. The most extensive chain of barrier islands is along the Atlantic and Gulf Coast states, extending some 5000 km (3100 mi) from Long Island to Texas and Mexico. One particular portion of this chain, shown in Figure 16-12, is the Outer Banks off North Carolina's coast. It includes Cape Hatteras, across Pamlico Sound from the mainland. The area presently is designated as one of three national seashore reserves supervised by the National Park Service. Figure 16-12 displays key depositional forms: spit, island, beach, lagoon, sound, and inlet.

Various ideas have been proposed to explain the formation of barrier islands. They may begin as an offshore bar or low ridge of submerged sediment near shore and then gradually migrate toward shore as sea level rises. Because many barrier islands seem to be migrating landward, they are an unwise choice for homesites or commercial building. Nonetheless, they are a common choice, even though they take the brunt of storm energy and actually act as protection for the mainland. The hazard represented by the settlement of barri-

FIGURE 16-12
Landsat image of barrier island chain along the North Carolina coast. Hurricane Emily swept past Cape Hatteras in August 1993 causing damage and beach erosion. [Image from NASA.]

er islands was made graphically clear when Hurricane Hugo assaulted South Carolina in 1989.

Hurricane Hugo made a destructive pass over Puerto Rico, the Virgin Islands, and the northeastern Caribbean and then turned northwestward. The storm attacked the Grand Strand barrier islands off the northern half of South Carolina's coastline, most affecting Charleston and the South Strand portion of the islands. The storm made landfall as the worst hurricane to strike there in 35 years.

Beachfront houses, barrier-island developments, and millions of tons of sand were swept away; up to 95% of the single-family homes in Garden City were destroyed. The southern portion of Pawleys Island was torn away, and one of every four homes was destroyed. Some homes that had escaped the last major storm, Hurricane Hazel in 1954, were lost to Hugo's peak winds of 209 kmph (130 mph). In comparable dollars, Hugo caused nearly $4 billion in damage, compared to Hazel's $1 billion. (See the section on Hurricane Andrew in Chapter 8.) The increased damage partially resulted from expanded construction during the intervening years between the two storms. Due to increased development and real estate appreciation, each future storm can be expected to cause ever-increasing capital losses.

Emergent and Submergent Coastlines

Coastlines can be classified in several ways. Most significant is the relationship between rising or falling sea level and whether the continental margins are uplifting or subsiding. This discussion simplifies coastlines to two general types—emergent and submergent. Please note that coastal landforms may exhibit traits of both emergence and submergence.

Emergent Coastlines. Emergent coastlines are indicated by sea cliffs and wave-cut platforms, such as terraces and benches, that appear raised above water level (Figure 16-7). Such coastlines develop either because sea level lowers or the land rises. Emergent coastlines are common during glaciations, when more of Earth's water is tied up as ice, thus lowering worldwide sea level. Note the exposed shorelines 18,000 years ago depicted in Figure 17-28. Emergent coasts also may be of tectonic origin, especially common around the Pacific Rim; former depositional features thus may appear high above sea level. Other emergent coastlines may be isostatically responding (rising) to the melting and retreat of great expanses of ice by uplifting submerged coastal valleys. Areas of northern Canada and Scandinavia are still emerging in this way at a centimeter (0.4 inch) or so a year, which is a rather rapid rate, geologically speaking.

Submergent Coastlines. Submergent coasts are drowned by the sea, either because of sea-level rise or land subsidence. Characteristically, submergent coastlines form *embayments,* or the inundation of lowlands near the coast by the sea, generally near the mouth of a river. Evidence relates such coastlines to interglacial periods and times of higher temperatures, with corresponding lower volumes of ice and rising sea levels. A Spanish term, *ria* (for "river"), is applied to this type of coast, which often is penetrated by a river entering the sea and therefore forming an *estuary,* where seawater and freshwater mix. Familiar estuaries include the flooded lower reaches of the Hudson River around New York City, and Delaware Bay and Chesapeake Bay to the south, which also contain tributary estuaries (Figure 16-13).

If sea level continues to rise over the next century, many estuaries will expand inland, and new ones could form, submerging even more coastline. Despite isostatic rising, the increase in sea level since the last ice age has left many glaciated coasts drowned with glacially carved valleys called *fjords.* Their great depth, to hundreds of meters, precludes any development of depositional features. Fjords exist in South Island of New Zealand, southern Chile, British Columbia, Alaska, Greenland, and Scandinavia.

Organic Processes: Coral Formations

Not all coastlines form due to purely physical processes. Some form as the result of organic processes, such as coral growth. A **coral** is a simple marine animal with a small, cylindrical, saclike body (polyp); it is related to other marine invertebrates,

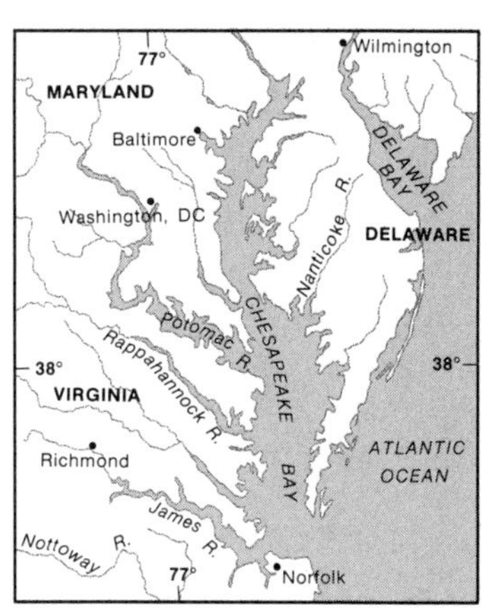

FIGURE 16-13
Landsat image of Chesapeake Bay, showing a *ria* coast of submergence. [Image from NASA.]

such as anemones and jellyfish. Corals secrete calcium carbonate ($CaCO_3$) from the lower half of their bodies, forming a hard external skeleton. Corals function in a *symbiotic* (mutually helpful) relationship with simple red-brown-to-green algae—that is, they live in close association with the algae, and both are codependent for survival. Corals cannot photosynthesize, but do ingest some of their own nourishment. Algae perform photosynthesis and convert solar energy to chemical energy in the system, providing the coral with about 60% of its nutrition and assisting the coral with calcification processes. In return, corals provide the algae with nutrients.

Although both solitary and colonial corals exist, it is the colonial forms that produce enormous structures, varying from treelike and branching forms to round and flat shapes. Through many generations, live corals near the ocean's surface build on the foundation of older corals below, which in turn may rest upon a volcanic seamount or some other submarine feature built up from the ocean floor.

Figure 16-14 shows the distribution of currently living coral formations. The distribution is specifically controlled by several factors. Corals thrive in warm tropical oceans, so the difference in ocean temperature between the western coasts and eastern coasts of continents is critical to their distribution. Western coastal waters tend to be cooler,

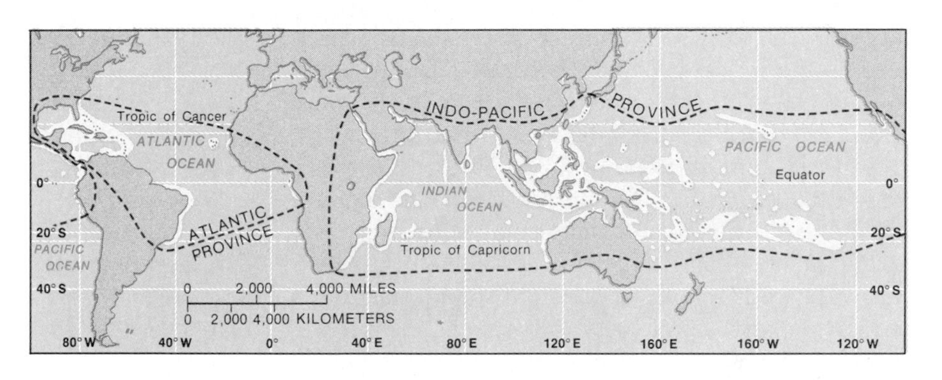

FIGURE 16-14
Worldwide distribution of living coral formations. Yellow areas also include most of the prolific reef growth and atoll formation. [After J. L. Davies, *Geographical Variation in Coastal Development*. Essex, England: Longman House, 1973. Adapted by permission.]

thereby discouraging coral activity, whereas eastern coastal currents are warmer and thus enhance coral growth. Living colonial corals range from about 30° N to 30° S and occupy a very specific ecological zone: 10–55 m (30–180 ft) depth, 27–40‰ salinity, and 18–29°C (64–85°F) water temperature. Corals require clear, sediment-free water and consequently do not locate near the mouths of sediment-charged freshwater tributaries.

Coral Reefs. An organically derived sedimentary formation of coral rock is called a *reef* and can assume one of several distinctive shapes. The principal shapes are fringing reefs, barrier reefs, and atolls. In 1842 Charles Darwin presented a hypothesis for an evolution of reef formation. He thought that as reefs developed around a volcanic island and the island itself gradually subsided, an equilibrium between the subsidence of the island and the upward growth of the corals is maintained. This generally accepted idea is portrayed in Figure 16-15 with specific examples of each stage.

Fringing reefs, sometimes called table reefs, are extensive platforms of coral rock that are attached

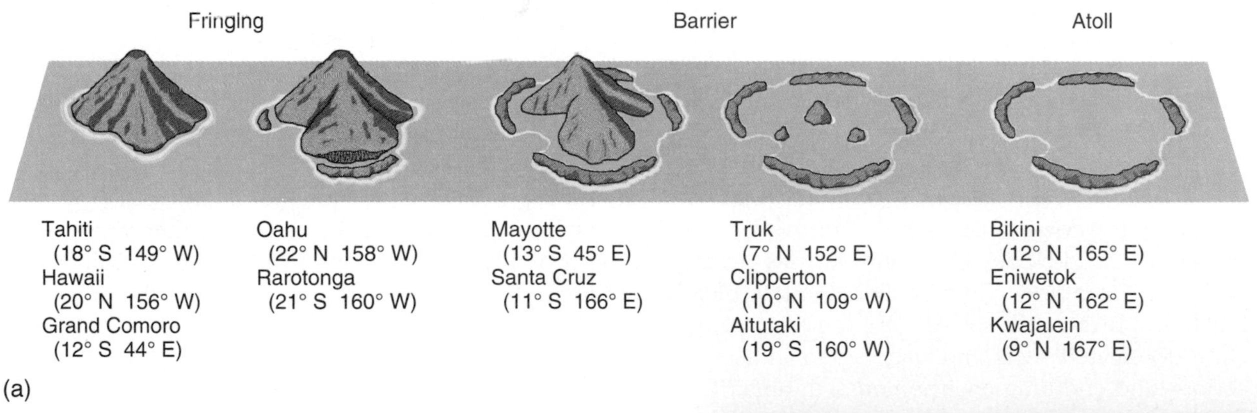

(b)

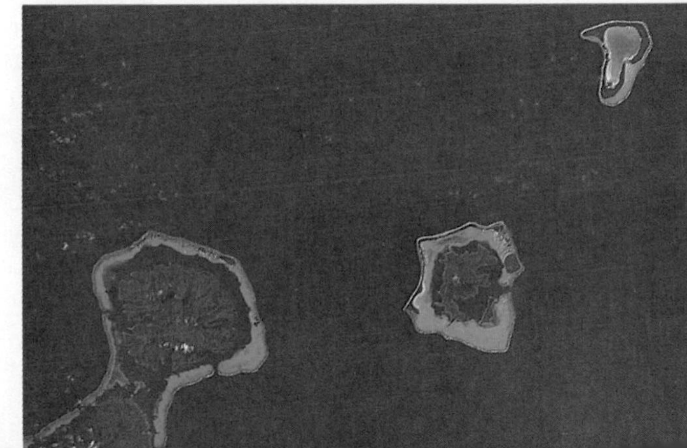

FIGURE 16-15
(a) Common coral formations as in a sequence of reef growth formed around a subsiding volcanic island: fringing reefs, barrier reefs, and atolls. (b) Orbital photograph of Bora Bora, Society Islands, 16°30′ S, 151°45′ W. [(a) After D. R. Stoddart, *The Geographical Magazine* LXIII (1971): 610. Adapted by permission. (b) Space Shuttle photo from NASA.]

to land. The largest of these is the Bahama platform in the western Atlantic, covering some 96,000 km^2 (37,000 mi^2). *Barrier reefs* align with the shoreline, separated from the shore by partially enclosed lagoons; they range from 300 to 1000 m wide (1000–3300 ft). The Great Barrier Reef along the shore of Queensland State in Australia exceeds 2025 km (1260 mi) length, is 16–145 km (10–90 mi) wide, and includes at least 200 coral-formed islands and keys (coral islets or barrier islands). An **atoll** frequently is circular, or ring-shaped, with a central lagoon; it often is built on an undersea volcanic island (Figure 16-15b).

Coral Bleaching. Scientists at the University of Puerto Rico and NOAA are tracking an unprecedented bleaching and dying-off of corals worldwide. The Caribbean, Australia, Japan, Indonesia, Kenya, Florida, Texas, and Hawaii are experiencing this phenomenon. The bleaching is due to a loss of colorful algae from within and upon the coral itself. Normally colorful corals have turned stark white as nutrient-supplying algae are expelled by the host coral. Exactly why the coral ejects its living partner is unknown. Possibilities include local pollution, disease, sedimentation, and changes in salinity.

Another possible cause is the 1–2C° (1.8–3.6F°) warming of sea-surface temperatures, as stimulated by greenhouse warming of the atmosphere. During the 1982–1983 ENSO (Chapter 10) areas of the Pacific Ocean were warmer than normal and widespread coral bleaching occurred. Coral bleaching worldwide is continuing as average ocean temperatures climb higher. Further bleaching through the decade will affect most of Earth's coral-dominated reefs and may be an indicator of serious and enduring environmental trauma. If present trends continue most of the living corals on Earth could perish by A.D. 2000.

Salt Marshes and Mangrove Swamps

Narrow, vegetated strips occupy many coastal areas, shorelines along estuaries, and other wet inland sites worldwide. Their great biological productivity stems from trapped organic matter and sediments that permit a coastal marsh to greatly outproduce a wheat field in raw vegetation per acre. These wetlands are productive ecosystems, are quite fragile ecologically, and are threatened by human tendencies to settle and develop them.

Wetlands are saturated with water enough of the time to support hydrophytic vegetation (plants that grow in water or very wet soil), form poorly drained soils, and sustain wetland processes. Other than coastlines, wetlands form as northern bogs (peatlands with high water tables), potholes in prairie lands, cypress swamps (standing or gently flowing water), river bottomlands and floodplains, and arctic and subarctic environments that experience permafrost during the year. In Canada and the United States, wetland distribution is generally along gradients of temperature (north-south) and precipitation (east-west). Only about half of the wetlands that existed in 1800 are left, because of development and destruction.

Coastal wetlands are of two general types—salt marshes and mangrove swamps. In the Northern Hemisphere, **salt marshes** tend to form north of the 30th parallel, whereas **mangrove swamps** form equatorward of that point. This is dictated by the occurrence of freezing conditions, which control the survival of mangrove seedlings. Roughly the same latitudinal limits apply in the Southern Hemisphere.

Salt marshes usually form in estuaries and behind barrier beaches and spits. An accumulation of mud produces a site for the growth of *halophytic* (salt-tolerant) plants. Plant growth then traps additional alluvial sediments and adds to the salt marsh area. Because salt marshes are in the intertidal zone (between the farthest reaches of high and low tides), sinuous, branching channels are produced as tidal waters flood into and ebb from the marsh (Figure 16-16).

Sediment accumulation on tropical coastlines provides the site for mangrove trees, shrubs, and other small trees. The prop roots of the mangrove are constantly finding new anchorages. They are visible above the water line but reach below the water surface, providing a habitat for a multitude of specialized lifeforms. Mangrove swamps often secure and fix enough material to form islands (Figure 16-17).

FIGURE 16-16
Salt marsh, a productive ecosystem commonly occurring poleward of 30° latitude in both hemispheres. [Photo by author.]

FIGURE 16-17
Mangroves tend to grow equatorward of 30° latitude. (a) Mangroves form along the East Alligator River (12° S latitude) in Kakadu National Park, Northern Territory, Australia. (b) Mangroves retain sediments and can form anchors for island formations such as Aldabra Island (9° S), Seychelles. [(a) Photo by Belinda Wright/ DRK Photo, (b) photo by Wolfgang Kaehler.]

(b)

Mangrove losses are estimated by the World Resources Institute and the U.N. Environment Programme at between 40% (e.g., Cameroon and Indonesia) to nearly 80% (e.g., Bangladesh and Philippines) since pre-agricultural times. Deliberate removal was a common practice by many governments because of a falsely conceived fear of disease or pestilence in these swamplands. The development of Marco Island, described later, is an example of mangrove loss due to urbanization.

HUMAN IMPACT ON COASTAL ENVIRONMENTS

The development of modern societies is closely linked to estuaries, wetlands, barrier beaches, and coastlines. Estuaries are important sites of human settlement because they provide natural harbors, a food source, and convenient sewage and waste disposal. Society depends upon daily tidal flushing of estuaries to dilute the pollution created by waste disposal. Thus, the estuarine and coastal environment is vulnerable to abuse and destruction if development is not carefully planned. In the United States and Canada about 50% of estuarine zones now have been altered or destroyed, and by the end of this decade most barrier islands will have been developed and occupied. A useful comparison is the fact that only 28 of 280 barrier islands in the United States were developed to any degree before World War II.

Barrier islands and coastal beaches do migrate over time. Houses that were more than a mile from the sea in the Hamptons along the southeastern shore of Long Island, New York, are now within only 30 m (100 ft) of it. The time remaining for these homes can be quickly shortened by a single hurricane or by a sequence of storms with the intensity of a series that occurred in 1962 or the severe nor'easter that struck the U.S. and Canadian Atlantic coast in December 1992 as discussed in Chapter 8 (see Figure 8-16). Nonetheless, despite our understanding of beach and barrier-island migration, the effect of storms, and warnings from government agencies, coastal planning and zoning proceeds. Society behaves as though beaches and barrier islands are stable, fixed features, or as though they can be engineered to be permanent. Experience has shown that severe erosion generally cannot be prevented, as we have seen with the cliffs of southern California, the shores of the Great Lakes, the New Jersey shore, the Gulf coast, and the devastation of coastal South Carolina by Hurricane Hugo. Shoreline planning is the topic of Focus Study 16-1.

Marco Island, Florida: An Example of Impact

An example of intense barrier-island development is the Marco Island area on the coast of southwestern Florida, about 24 km (14.9 mi) south of the city of Naples. Figure 16-18a shows Marco Island as it was prior to 1952: approximately 5300 acres of subtropical mangrove habitat and barrier island terrain, formed from river sediments and old shell and reef fragments. Then development began in 1962, despite tropical storms that occasionally assault the area, as Hurricane Donna had done in 1960. Development included artificial landfill for housing sites and general urbanization of the entire island. Scientists regarded the island and estuarine habitat as highly productive and unique in its mixture of mangrove species, endangered bird species, and a productive aquatic environment. Consequently, numerous challenges and court actions attempted to halt development. But construction activities continued, and today Marco Island is completely developed (Figure 16-18b).

In 1984 a Geographic Information System (GIS, see Chapter 2) was used to study Marco Island, incorporating existing maps, remote-sensing techniques, and a knowledge of the ecology of an estuarine mangrove community. In Figure 16-18, you can compare the 1952 and 1984 digitized inventory maps and a 1984 color-infrared aerial photograph, to see the impact of urban development on this island. Note that the only remaining natural community is restricted to a very limited portion along the extreme perimeter.

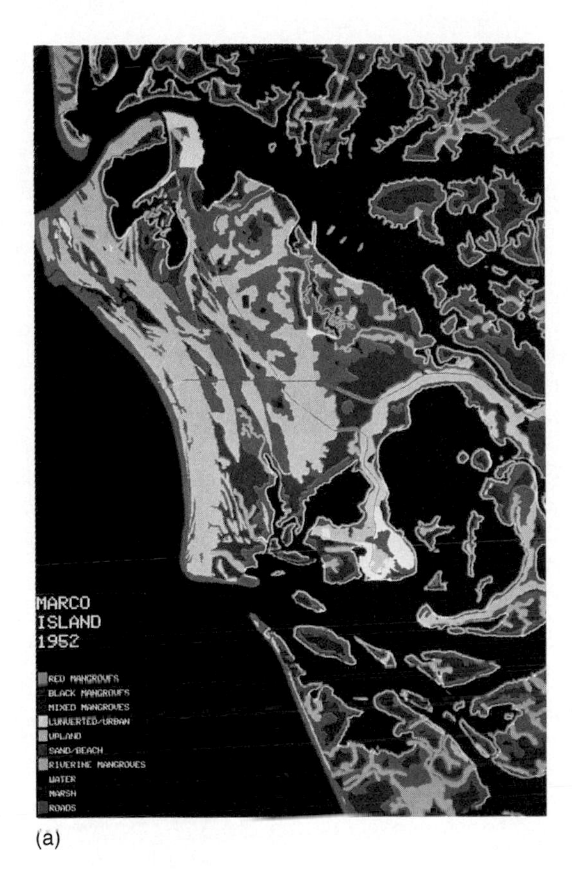

(a)

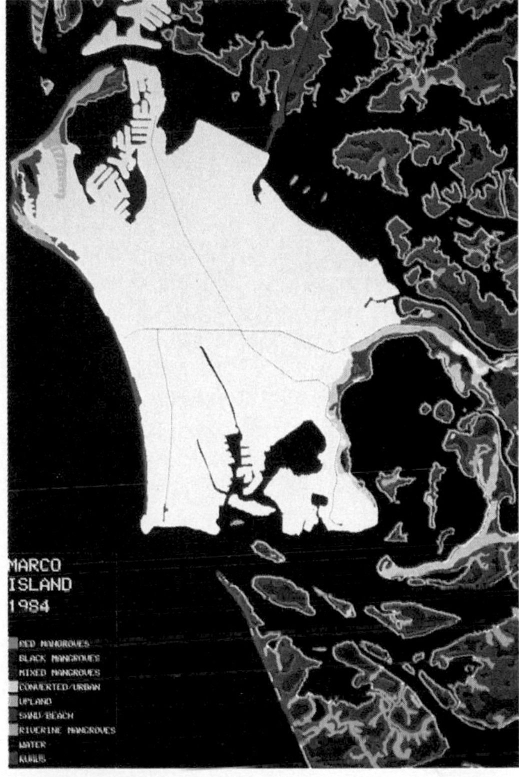

(b)

(c)

FIGURE 16-18
Digitized inventories of Marco Island, Florida, showing dramatic development from (a) 1952 to (b) 1984; (c) 1984 color-infrared photograph of the island. [From Samuel Patterson, *Mangrove Community Boundary Interpretation and Detection of Areal Changes on Marco Island, Florida,* Biological Report 86(10), for National Wetlands Research Center, U.S. Fish and Wildlife Service, August 1986, p. 23, 46, 49.]

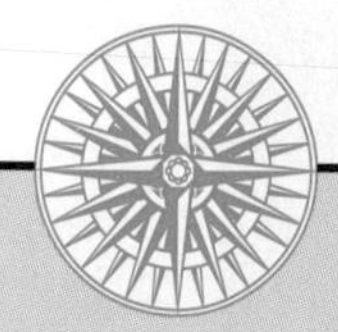

FOCUS STUDY 16-1 An Environmental Approach to Shoreline Planning

Coastlines are places of wonderful opportunity. They also are zones of specific constraints. Unfortunately, poor understanding of this resource and a lack of environmental analysis often go hand in hand. Landscape architect Ian McHarg, in *Design with Nature*, discusses the New Jersey shore and provides an excellent example of how proper understanding of a coastal environment could have avoided problems from major storms and from human development. Much of what is presented here is characteristic of similar coastal areas elsewhere. Figure 1 illustrates the New Jersey shore from ocean to back bay and describes each specific zone's tolerance to human use.

Beaches and Dunes—Where to Build?

Sand beaches are the primary natural defense against the ocean in this environment; they act as bulwarks against the pounding of a stormy sea. But they are susceptible to pollution and require environmental protection to control nearshore dumping of dangerous materials and inappropriate construction on the beach itself. The shoreline tolerates recreation, but cannot tolerate construction because of its shifting, changing nature during storms, daily tidal fluctuations, and the potential effects of a rising sea level. In recent years, high water levels in the Great Lakes and along the southern California coast certainly attest to the vulnerability of shorelines to erosion and inundation. Wave erosion attacks cliff formations and undermines, bit by bit, the foundations of houses and structures that were built too close to the edge (Figure 2).

The primary dune along a coast also is intolerant of heavy use, even the passage of people trekking to the beach; it is fragile, easily disturbed, and vulnerable to erosion. Delicate plants work to hold the sand in place, which eventually accumulates to increase dune height. Primary dunes are like human-made dikes in the Netherlands: they are the primary defense against the sea, so they should not be disturbed by development or heavy traffic. Carefully controlled access points to the beach should be enforced, for even foot traffic can

FIGURE 1
Coastal environment—a planning perspective from ocean to bay. [After *Design with Nature* by Ian McHarg. Copyright © 1969 by Ian L. McHarg. Adapted by permission of Doubleday, a division of Bantam Doubleday Dell Publishing Group, Inc.]

Ocean	Beach	Primary dune	Trough	Secondary dune	Backdune	Bayshore	Bay
Tolerant	Tolerant	Intolerant	Relatively tolerant	Intolerant	Tolerant	Intolerant	Tolerant
Intensive recreation Subject to pollution controls	Intensive recreation No building Intolerant of construction	No passage, breaching, or building	Limited recreation Limited structures	No passage, breaching, or building	Most suitable for development	No filling	Intensive recreation

FIGURE 2
Coastal erosion and a failed house foundation. [Photo by author.]

cause destruction. (See the beach stabilization example in Figure 15-4.)

The trough behind the primary dune is relatively tolerant, with limited recreation and building potential. The plants that fix themselves to the surface send roots down to fresh groundwater reserves. Thus, if construction should inhibit the surface recharge of that supply, the natural protective ground cover could fail and destabilize the environment; or subsequent saltwater intrusion might contaminate well water. Clearly, groundwater resources and the location of recharge aquifers must be considered in planning.

Behind the trough is the secondary dune, a second line of defense against the sea. It, too, is tolerant of some use yet is vulnerable to destruction through improper use. The backdune along the New Jersey shore is a more suitable location for development than any of the zones between it and the sea. Further inland are the bayshore and the bay, where there should be no dredging and filling and only limited dumping of treated wastes and toxics should be permitted. In reality, the opposite of such careful assessment and planning prevails.

A Scientific View and a Political Reality

Prior to an intensive government study completed in 1962, no analysis had been completed outside academic circles. Thus, what is common knowledge to geographers, botanists, biologists, and ecologists in the classroom and laboratory still has not filtered through to the general planning and political processes. As a result, on the New Jersey shore and along much of the Atlantic and Gulf coasts, improper development of the fragile coastal zone led to extensive destruction during the storms of March 1962 and the numerous hurricanes of this century. In 1962, damage was massive from the Carolinas to Massachusetts.

When the storms of 1962 hit, scientists were studying the area. They forecast a strong probability that the Cape May portion of the New Jersey shore could be 90 m (295 ft) further inland by the year 2010, less than two decades from now. There is a 15% chance that the shore could be even twice as far inland (180 m or 600 ft) by that time. Similar estimates persist along the entire east coast, making it clear that society must develop in a manner that synthesizes ecology and economics if these coastal environments are to be sustained.

South Carolina enacted their Beach Management Act of 1988 (modified 1990) to apply some of the principles McHarg described in his analysis of the New Jersey shore. Vulnerable coastal areas are protected from new construction or rebuilding. Now guided by law are structure size, replacement limits for damaged structures, and placement of structures on lots, although liberal interpretation is allowed. The act sets standards for different coastal forms: beach and dunes, eroding shorelines, or a hypothetical baseline along a coast that lacks dunes or has unstabilized inlets. In the first two years more than 70 lawsuits protested the act as an invalid seizure of private property without compensation. Implementation has been difficult, political pressure intense, and results mixed. Similar measures in other states have faced the same difficult path.

The key to protective environmental planning and zoning is the allocation of responsibility and cost in the event of a disaster. An ideal system places hazard tax on land based on assessed risk and restricts the government's responsibility to fund reconstruction or an individual's right to reconstruct on frequently damaged sites. Comprehensive mapping of erosion-hazard areas is one place for the federal government to begin to avoid ever-increasing costs to society from reoccurring disasters. This requires amending the National Flood Insurance Act. It should include not only coastlines but floodplains, earthquake and volcanic hazard areas, sites with toxic contamination, and important scenic and scientific areas.

Ian McHarg, ecologist and landscape architect, summarizes the situation well:

> **May it be that these simple ecological lessons will become known and incorporated into ordinance [law] so that people can continue to enjoy the special delights of life by the sea.***

*From *Design with Nature* by Ian McHarg, p. 17. Copyright © 1969 by Ian L. McHarg. Published by Doubleday, a division of Bantam Doubleday Dell Publishing Group, Inc.

Summary—Coastal Processes and Landforms

Rachel Carson called the edge of the sea "A strange and beautiful place." The coastal environment exists where the tide-driven, wave-driven sea confronts the land. Indeed, physical processes in operation along coastlines produce relatively rapid change, much faster than the rate of change inland. Physical geographers are interested in the spatial aspects of these processes, the landscapes created, and the implications of each unique coastline to planning and possible development.

Coastal processes operate as open systems with *inputs* of solar energy through the fluid agents of air and water and sea level, *actions* of tidal and sea-level variation and wave motion, and *outputs* of erosional, depositional, and organic landforms and physical features. These features include terraces, embayments, lagoons, cliffs, and beaches. Throughout geologic time, coastal margins have been subjected to numerous changes in sea level, with each change altering wave and tidal processes.

Reefs, islands, and coastlines are produced by organic formations built by corals, salt marshes, and mangroves. Presently, sea level is rising and sea-surface temperatures are increasing. The varied interaction of ocean and continent produces a fragile balance that can be both hazardous to humans and damaged by human occupation. The coastal zone, like the river floodplain, requires careful assessment of natural factors for sound use. Ideally, ecological planning can enhance long-term economic goals and our enjoyment of Earth's wonderful coastal environments.

KEY TERMS

atoll
barrier beaches
barrier islands
barrier spit
bay barrier
beach
beach drift
breaker
breakwater
coral
ebb tides
flood tides
headlands
lagoon
littoral zone
longshore current
mangrove swamps
mean sea level
neap tide
salt marshes
spring tides
swells
tidal bore
tides
tombolo
tsunami
wave-cut platform
wave refraction
waves
wetlands

REVIEW QUESTIONS

1. What are the key terms used to describe the coastal environment?
2. Define mean sea level. How is this value determined? Is it constant or variable around the world? Explain.
3. What interacting forces generate the pattern of tides?
4. What characteristic tides are expected during a new Moon or a full Moon? During the first-quarter and third-quarter phases of the Moon? What is meant by a flood tide? An ebb tide?
5. Is tidal power being used anywhere to generate electricity? Explain briefly how such a plant would utilize the tides to produce electricity. Are there any sites in North America?
6. What is a wave? How are waves generated, and how do they travel across the ocean? Discuss the process of wave formation and transmission.
7. Describe the refraction process that occurs when waves reach an irregular coastline.
8. Define the components of beach drift.
9. Explain how seismic sea waves attain such tremendous velocities. Why are they given a Japanese name?
10. What is meant by an erosional coast? What are the expected features of such a coast?
11. What are some of the depositional features encountered along a coastline?
12. How do people attempt to modify littoral drift? What are the positive and negative impacts of these actions?
13. Describe a beach—its form, composition, function, and evolution.
14. What success has Miami had with beach replenishment? Is it a practical strategy?
15. Based on the information in the text and any other sources at your disposal, do you think barrier islands and beaches should be used for development? If so, under what conditions? If not, why not?

16. After the Grand Strand off South Carolina was destroyed by Hurricane Hazel in 1954, settlements were rebuilt, only to be hit by Hurricane Hugo 35 years later, in 1989. Why do these recurring events happen to human populations? Compare the impact of the two storms.
17. Describe the western and eastern coasts of North America as emergent or submergent coastlines.
18. How are corals able to construct reefs and islands?
19. Evaluate a trend in corals that is troubling scientists.
20. Why are the coastal wetlands poleward of 30° N and S latitude different from those that are equatorward? Describe the differences.
21. Describe the condition of Marco Island, Florida, at the present time. Do you think a rational model was used to assess the environment prior to development?
22. What type of environmental analysis is needed for rational development and growth in a region like the New Jersey shore? Evaluate South Carolina's approach to coastal hazards and protection.

SUGGESTED READINGS

Bernstein, Joseph. "Tsunamis," *Scientific American* 190, no. 3 (August 1954): 60–64.

Blockson, Charles L. "Sea Change in the Sea Islands," *National Geographic* (December 1987): 735–63.

Brown, B. E., and J. C. Ogden. "Coral Bleaching," *Scientific American* 258, no. 1 (January 1993): 64–70.

Cameron, S. D. "The Living Beach," *Canadian Geographic* 113, no. 2 (March-April 1993): 66–79.

Committee on Restoration of Aquatic Ecosystems. *Restoration of Aquatic Ecosystems: Science, Technology, and Public Policy*. Washington, DC: National Research Council, November 1991.

Dolan, Robert, Bruce Hayden, and Harry Lins. "Barrier Islands," *American Scientist* 68, no. 1 (January-February 1980): 16–25.

Greely, A. W. "Hurricanes on the Coast of Texas," *National Geographic* (November 1900): 442–45.

Kaufman, W., and O. H. Pilkey, Jr. *The Beaches are Moving: The Drowning of America's Shoreline*. Durham, NC: Duke University Press, 1983.

McGee, W. J. "The Lessons of Galveston," *National Geographic* (October 1900): 377–83.

McHarg, Ian L. *Design with Nature*. Garden City, NY: Doubleday, 1969.

Mitchell, J. G. "Our Disappearing Wetlands," *National Geographic* 182, no. 4 (October 1992): 2–45.

Platt, R. H., T. Bentley, and H. C. Miller. "The Failings of U.S. Coastal Erosion Policy," *Environment 33,* no. 9 (July 1992): 7–10.

Rowntree, Rowan A. "Coastal Erosion: The Meaning of a Natural Hazard in the Cultural and Ecological Context," *Natural Hazards*. Edited by Gilbert F. White. New York: Oxford University Press, 1974.

Sackett, Russell, and the editors of Time-Life Books. *Edge of the Sea*. Planet Earth Series. Alexandria, VA: Time-Life Books, 1983.

Schwartz, Maurice L. *The Encyclopedia of Beaches and Coastal Environments*. Stroudsburg, PA: Hutchinson Ross, 1982.

Thurman, Harold V. *Introductory Oceanography,* 5th ed. Columbus, OH: Macmillan Publishing Company, 1988.

Wanless, H. R. "The Inundation of Our Coastlines," *Sea Frontiers* 35, no. 5 (September-October 1989): 264–71.

Warrick, R. A., E. M. Barrow, and T. M. L. Wigley, eds. *Climate and Sea Level Change—Observations, Projections, and Implications.* Cambridge, England: Cambridge University Press, 1993.

Whipple, A. B. C., and the editors of Time-Life Books. *Restless Oceans.* Alexandria, VA: Time-Life Books, 1983.

Ziegler, J. M. "Origin of the Sea Islands of the Southeastern United States," *Geographical Review* 49 (1959): 222–37.

Icefall and waterfall at terminus of Bow Glacier, Banff National Park, Alberta. [*Photo by Martin G. Miller.*]

17

Glacial and Periglacial Processes and Landforms

A large measure of the fresh water on Earth is frozen, with the bulk of that ice sitting restlessly in just two places—Greenland and Antarctica. The remaining ice covers various mountains and fills some alpine valleys. More than 29 million km^3 (7 million mi^3) of water is tied up as ice, or about 77% of all fresh water. These deposits of ice provide an extensive frozen record of Earth's climatic history over the past several million years and perhaps some clues to its climatic future.

This chapter focuses on Earth's extensive ice deposits—their formation, movement, and the ways in which they function to produce various erosional and depositional landforms. Glaciers are transient landforms themselves and leave in their wake a variety of landscape features. The fate of glaciers is intricately tied to changes in global temperature and ultimately concerns us all.

Approximately 20% of Earth's land area is subject to freezing conditions and frost action characteristic of periglacial regions. We examine in this chapter the cold, near-glacial world of permafrost and periglacial processes. Finally, we see reminders of the last ice age over many parts of the globe. A focus study details methods used to decipher past climates—the science of paleoclimatology.

RESERVOIRS OF ICE

A **glacier** is a large mass of perennial (year-round) ice, resting on land or floating shelflike in the sea adjacent to land. Glaciers are not frozen lakes or groundwater ice but form by the continual accumulation of snow that recrystallizes into an ice mass. They move under the pressure of their own great mass and the pull of gravity. Today, about 11% of Earth's land area is dominated by these slowly flowing streams of ice. During colder episodes in the past, as much as 30% of continental land was covered by glacial ice because below-freezing temperatures prevailed at lower latitudes, allowing snow to accumulate.

A **snowline** is the lowest elevation where snow remains year-round—specifically, the lowest line where winter snow accumulation persists throughout the summer. Glaciers form in such areas of permanent snow, both at high latitudes and high elevations, even forming on some high mountains along the equator, such as in the Andes Mountains of South America or on Mount Kilimanjaro in Tanzania. In equatorial mountains, the snowline is around 5000 m (16,400 ft); on midlatitude mountains, such as the European Alps, snowlines average 2700 m (8850 ft); and in southern Greenland snowlines are down to 600 m (1970 ft).

Types of Glaciers

Glaciers are as varied as the landscape itself. They fall within two general groups, based on their form, size, and flow characteristics: alpine glaciers and continental glaciers.

Alpine Glaciers. With few exceptions, a glacier in a mountain range is called an **alpine glacier,** or *mountain glacier*. The name comes from the Alps, where such glaciers abound. Alpine glaciers form in several subtypes. One prominent type is a **valley glacier,** an ice mass constricted within a valley that was formed by stream action. Such glaciers range in length from only 100 m (325 ft) to over 100 km (60 mi). The snowfield that continually feeds the glacier with new snow is at a higher elevation. In Figure 17-1a, at least a dozen valley glaciers are identifiable in the *Landsat* image of the Alaska Range and several are named. Figure 17-1b is a high-altitude photograph of the Eldridge and Ruth glaciers, filling valleys as they flow from source areas near Mount McKinley.

As a valley glacier flows slowly downhill, the mountains, canyons, and river valleys beneath its mass are profoundly altered by its passage. Some of the glacial debris created by its erosive excavation can be seen as dark streaks on the ice as the glacier transports the debris for deposition elsewhere.

Most alpine glaciers originate in a mountain snowfield that is confined in a bowl-shaped recess. This scooped-out erosional landform at the head

FIGURE 17-1
Glaciers in south-central Alaska. (a) Valley glaciers in the Alaska Range of Denali National Park; (b) oblique IR image of Eldridge and Ruth glaciers, with Mount McKinley at upper left, taken at 18,300 m (60,000 ft). [(a) *Landsat* image from NASA; (b) Alaska High Altitude Aerial Photography from EROS Data Center.]

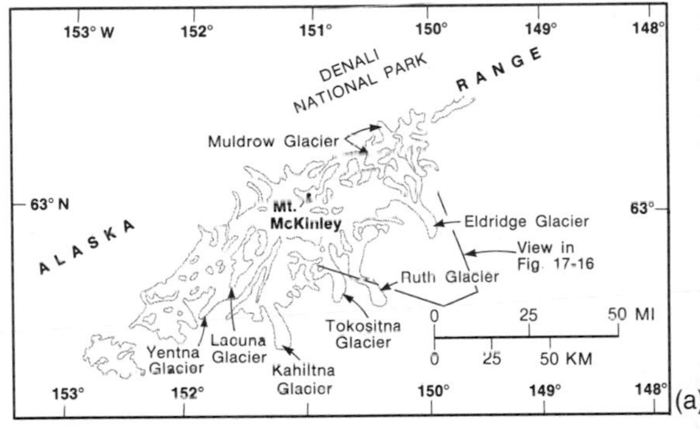

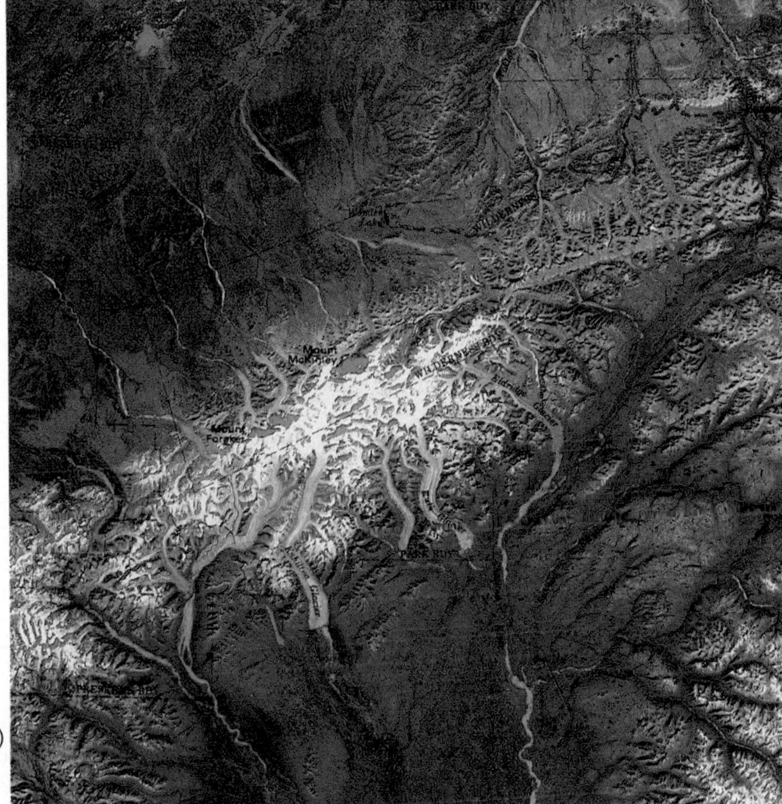

(a)

(b)

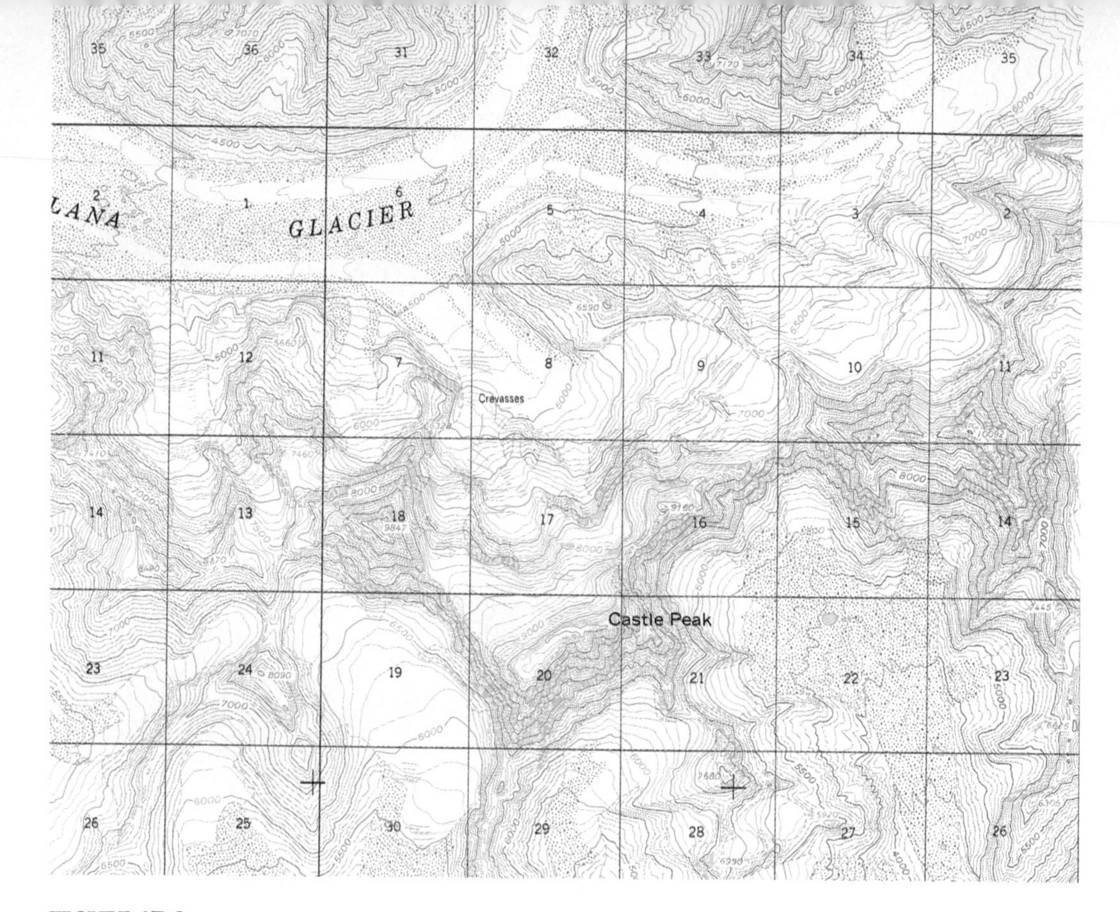

FIGURE 17-2
Numerous cirque glaciers depicted on a topographic map of southeastern Alaska (McCarthy C-7 Quadrangle, 1:63,360 scale, 100-ft contour interval). The blue areas represent active glaciers, and stripped brown areas on the ice are moraines. [Courtesy of U.S. Geological Survey.]

of a valley is called a **cirque.** A glacier that forms in a cirque is called a *cirque glacier*. Several cirque glaciers may feed a valley glacier, as shown on the topographic map in Figure 17-2 (note section 8 on the map). The Kuskulana Glacier, which flows off the map to the west (sections 1 and 2), is supplied by numerous cirque glaciers feeding several tributary valley glaciers.

Wherever several valley glaciers extend beyond their confining valleys and coalesce at the base of a mountain range, a *piedmont glacier* is formed and spreads freely over the nearby lowlands. Malaspina Glacier is an excellent example of a piedmont glacier (Figure 17-3). The debris deposits on the surface of the ice form beautiful streaked patterns as the glacier fans out over 5000 km^2 (1950 mi^2) of the coastal plain. This glacier is at the foot of the Mount Saint Elias Range in southern Alaska, ending in Yakutat Bay. A *tidal glacier*, such as the Columbia Glacier on Prince William Sound in Alaska, ends in the sea, *calving* (breaking off) to form floating ice called **icebergs.** Icebergs usually form wherever glaciers meet the ocean.

Continental Glaciers. On a larger scale than individual alpine glaciers, a continuous mass of ice is known as a **continental glacier** and in its most extensive form is called an **ice sheet.** Most glacial ice exists in the snow-covered ice sheets that blanket 80% of Greenland (1.8 million km^3, or 0.43 million mi^3) and 90% of Antarctica (13.9 million km^3, or 3.3 million mi^3). Antarctica alone has 91% of all the glacial ice on the planet.

These two ice sheets represent such an enormous mass that large portions of each landmass have been isostatically depressed more than 2000 m (6500 ft) below sea level. Each ice sheet is more than 3000 m (9800 ft) deep, burying all but the

FIGURE 17-3
Malaspina Glacier in southeastern Alaska. This piedmont glacier is nearly the size of Rhode Island and covers some 5000 km^2 (1950 mi^2); its thickest point exceeds 600 m (1970 ft). About 70% of Malaspina's ice comes from the Seward Glacier (far upper center), which is fed by ice fields in the Saint Elias Mountain Range. [High-altitude photo courtesy of AeroMap U.S., Inc., Anchorage, Alaska.]

highest peaks. An ice sheet is the most extensive form of continental glacier.

Two additional types of continuous ice cover associated with mountain locations are designated as *ice caps* and *ice fields*. Both ice caps and ice sheets completely bury the underlying landscape, although an **ice cap** is roughly circular and covers an area of less than 50,000 km^2 (19,300 mi^2). The actively volcanic island country of Iceland features several ice caps (Figure 17-4a).

The photograph in Figure 17-4b shows ridges and peaks visible above the buried terrain of the Andes in the southern Patagonian **ice field,** which extends in a characteristic elongated pattern in a mountainous region. The term *nunatak* refers to peaks that are visible above the glaciers. The Patagonian ice field is one of Earth's largest, stretching 360 km (224 mi) in length between 46° and 51° S latitude and up to 90 km (56 mi) in width. An ice field is not extensive enough to form the characteristic dome of an ice cap.

Continuous ice sheets, caps, and fields are drained by **outlet glaciers** that form around their periphery. An outlet glacier flows out from a continental glacier, taking the form of a valley glacier, sometimes moving down a rock-walled valley to the sea or to lowlands. Outlet glaciers flow radially from the edges of a dome-shaped ice mass and are visible in Greenland and Antarctica.

GLACIAL PROCESSES

A glacier is composed of dense ice that is formed from snow and water through a process of compaction, recrystallization, and growth. A glacier is an example of an open system, with *inputs* of snow and moisture and *outputs* of melting ice and evaporation that produce a mass balance, readily visible in the glacier's physical stature. A glacier's *budget* consists of net gains or losses of glacial ice, which determine whether the glacier expands or retreats. A glacier is a dynamic body, moving re-

(a)

FIGURE 17-4
Ice cap and ice field. (a) The Vatnajökull ice cap in southeastern Iceland (*jökull* means ice cap in Danish). (b) The southern Patagonian ice field of Argentina. [(a) *Landsat* image from NASA; (b) photo by Cosmonauts G. M. Greshko and Yu V. Romanenko, *Salyut 6*.]

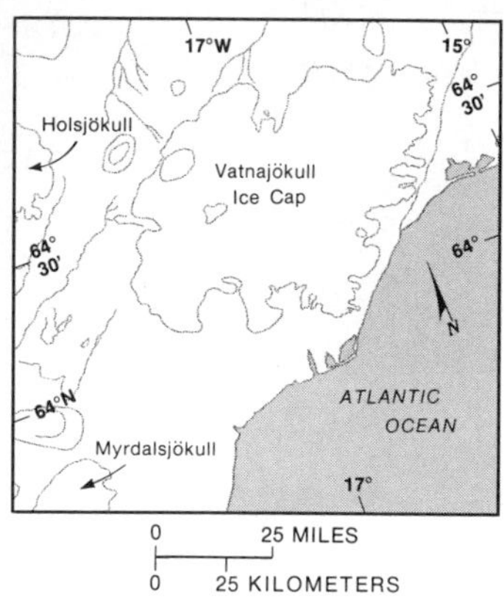

(b)

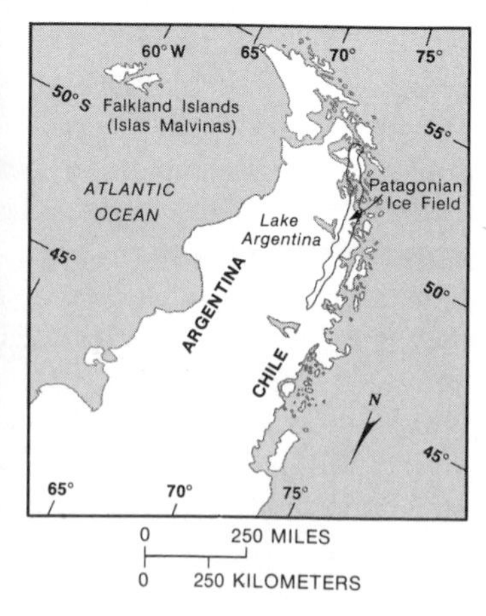

lentlessly downslope at rates which vary within its mass, greatly modifying the landscape through which it flows. Let's look at glacial ice formation, mass balance, movement, and erosion before we discuss specific landforms produced by these processes.

Formation of Glacial Ice

The essential input to a glacier is precipitation that accumulates in a *snowfield,* a glacier's formation site. Snowfields usually are at the highest elevation of an ice sheet, ice cap, or head of a valley glacier, usually in a cirque. Highland snow accumulation sometimes is the product of orographic processes, as discussed in Chapter 8. Avalanches from surrounding mountain slopes can add to the snowfield. As the snow accumulation deepens in sedimentary-like layers, the increasing thickness results in increased weight and pressure on underlying portions. Rain and summer snowmelt then contribute water, which stimulates further melting, and that meltwater seeps down into the snowfield and refreezes.

Snow that survives the summer and into the following winter begins a slow transformation into glacial ice. Air spaces among ice crystals are pressed out as snow packs to a greater density. The ice crystals recrystallize under pressure and go through a process of regrowth and consolidation. In a transition step to glacial ice, snow becomes **firn,** which has a compact, granular texture.

As this process continues, many years pass before denser glacial ice is produced. Formation of **glacial ice** is analogous to formation of metamorphic rock: sediments (snow and firn) are pressured and recrystallized into a dense metamorphic rock (glacial ice). In Antarctica, glacial ice formation may take 1000 years because of the dryness of the climate (minimal snow input), whereas in wet climates this time is reduced to just a few years because of the volume of new snow constantly being added to the system.

Glacial Mass Balance

A glacier is fed by snowfall and other moisture sources and is wasted by losses from its upper and lower surfaces and along its margins. A snowline called a **firn line** is visible across the surface of a glacier, indicating where the winter snows and ice accumulation survived the summer melting season. In a glacier's lower elevation, losses of mass occur because of surface melting, internal and basal melting, sublimation (recall from Chapter 7, "Heat Properties," that this is the direct evaporation of ice), wind removal by *deflation,* and the calving of ice blocks. The combined effect of these losses is called **ablation.**

A glacier's area of accumulation is, logically, at higher elevations where lower temperatures occur. The zone where accumulation gain balances ablation loss is the **equilibrium line** (Figure 17-5a). This area of a glacier generally coincides with the firn line. Glaciers achieve a positive net balance of mass—grow larger—during colder periods with adequate precipitation. In warmer times, the equilibrium line migrates to a higher elevation and the glacier retreats—grows smaller—due to its negative net balance. Internally, gravity continues to move a glacier forward even though its lower terminus is in retreat.

Highways of ice that flow from accumulation areas high in the mountains are marked by trails of transported debris called *moraines* (Figure 17-5b and c). A *lateral moraine* accumulates along the sides. A *medial moraine* forms down the middle when two glaciers merge and their lateral moraines combine.

Glacial Movement

We generally think of ice as those small, brittle cubes from the freezer, but glacial ice is quite different. In fact, glacial ice behaves in a plastic manner, for it distorts and flows in response to weight and pressure from above and the degree of slope below. Rates of flow range from almost nothing to a kilometer or two per year on a steep slope. The rate of accumulation of snow in the formation area is critical to the pace of glacial movement.

The movement of a glacier is not simply a process of ice slipping downhill. The greatest movement within a valley glacier occurs *internally,* below the rigid surface layer, which fractures as the underlying zone moves plastically forward

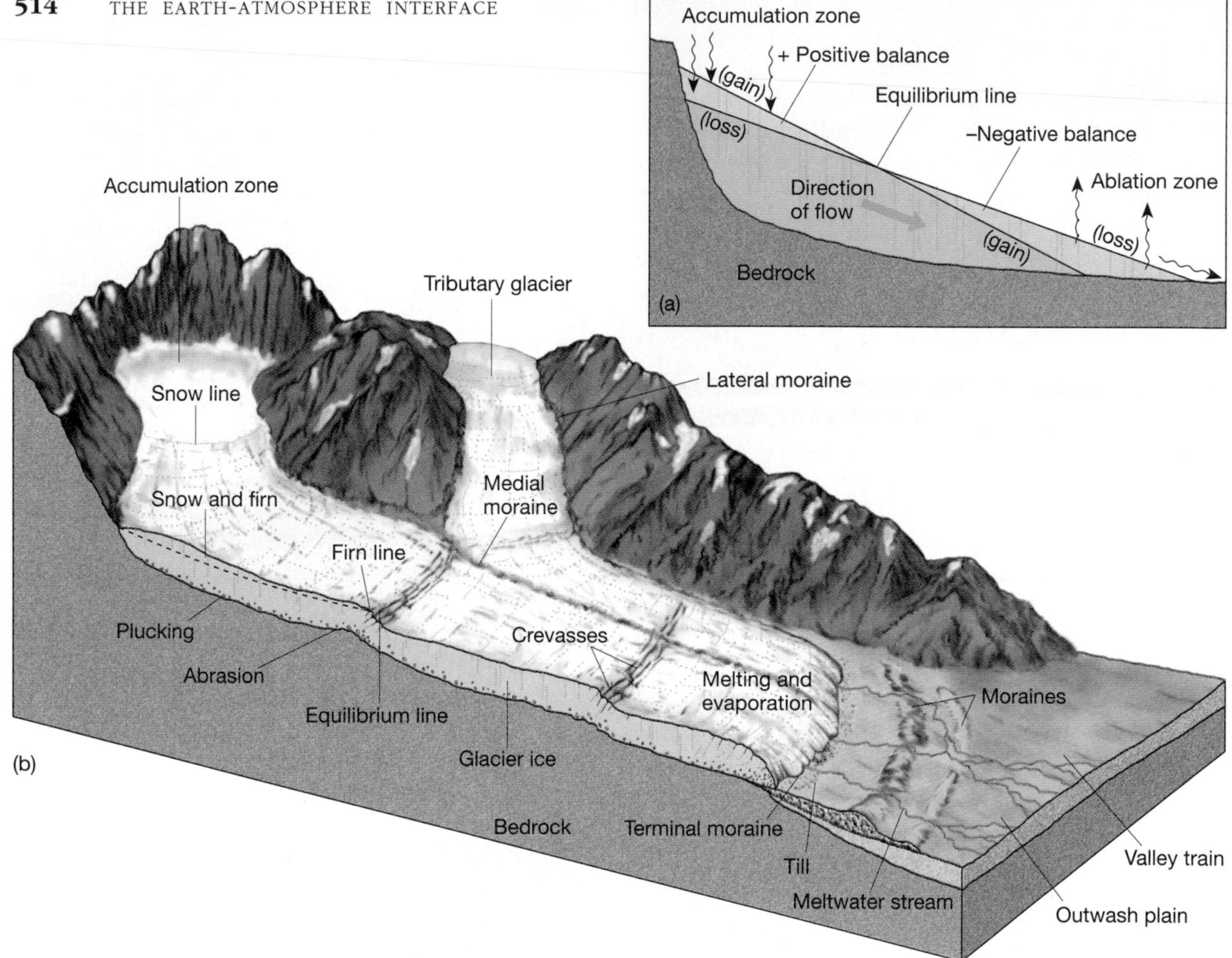

FIGURE 17-5
(a) Annual mass balance of a glacial system, showing the relationship between accumulation and ablation and the location of the equilibrium line. (b) Cross section of a typical retreating alpine glacier. (c) John Hopkins Glacier, Glacier Bay National Park, Alaska, demonstrates many of the features in the illustration. [(c) Photo by Frank S. Balthis.]

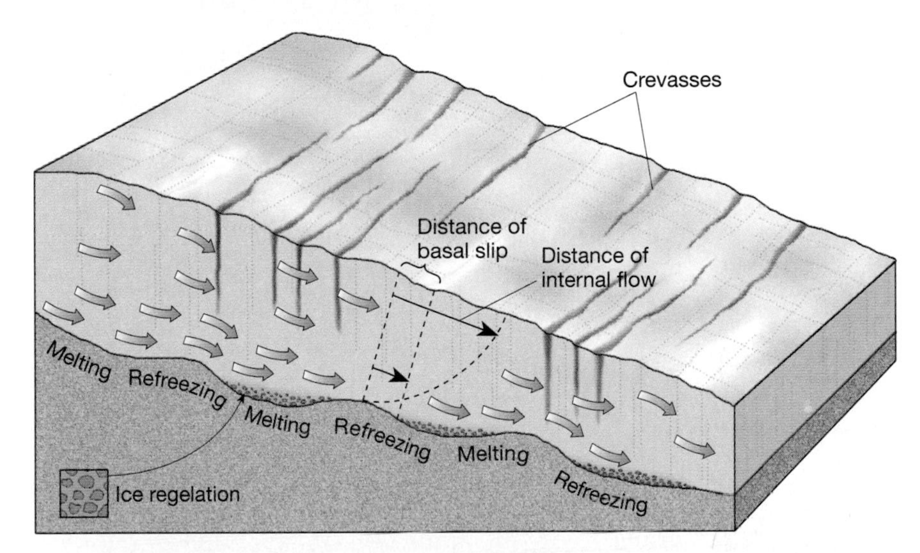

FIGURE 17-6
(a) Cross section of a glacier, showing the nature of its forward motion at the surface and along its basal layer. (b) Surface crevasses and cracks are evidence of a glacier's forward motion (near the Don Sheldon Amphitheater, Denali National Park, Alaska). [(b) Photo by Michael Collier.]

(Figure 17-6a). At the same time, the base creeps and slides along, varying its speed with temperature and the presence of any lubricating water beneath the ice. This *basal slip* usually is much less rapid than the internal plastic flow of the glacier.

In addition, unevenness in the landscape beneath the ice may produce variations in pressure, melting some of the basal ice at one moment, only to have it refreeze later; this process is called *ice regelation*—an important factor in glacial downslope movement. The basal ice layer, which can extend tens of meters above the base of the glacier, has a much greater debris content than occurs in the glacial ice above.

A flowing glacier can develop vertical cracks known as **crevasses** (Figure 17-6b). These result from friction with valley walls, or tension from extension as the glacier passes over convex slopes, or compression as the glacier passes over concave slopes. Traversing a glacier, whether an alpine glacier or an ice sheet, is dangerous because a thin veneer of snow sometimes masks the presence of

FIGURE 17-7
Yaetna Glacier in southeastern Alaska forms from two merging glaciers in the distance. The glacier is covered with morainal debris. [High-altitude photo courtesy of AeroMap U.S., Inc., Anchorage, Alaska.]

a crevasse. One type of crevasse occasionally associated with a cirque glacier is known as a *bergschrund,* or *headwall crevasse;* it forms when snow and firn compact and pull away from ice that remains frozen to rock at the head of the glacier.

Tributary valley glaciers merge to form a *compound* valley glacier. The flowing movement of a compound valley glacier is different from that of a river with tributaries. As the picture of the Yeatna Glacier in Alaska demonstrates (Figure 17-7), tributary glaciers flow into a compound glacier, but merge alongside one another by extending and thinning, rather than blending as do rivers. You can see this because each tributary maintains its own patterns of transported debris (dark streaks). This is visible on the John Hopkins Glacier as well (Figure 17-5c).

Glacier Surges. Although glaciers flow plastically and smoothly most of the time, some will lurch forward with little or no warning in a **glacial surge.** This is not quite as abrupt as it sounds; in glacial terms, a surge can be tens of meters per day. The Jakobshavn Glacier in Greenland, for example, is known to move between 7 and 12 km (4.3 and 7.5 mi) a year. Such surges appear more common today than previously thought.

In the spring of 1986 Hubbard Glacier and its tributary Valerie Glacier surged across the mouth of Russell Fjord in Alaska, cutting it off from contact with Yakutat Bay. This area in southeastern Alaska is fed by annual snowfall in the Saint Elias mountain range that averages more than 850 cm (335 in.) a year. Consequently, the surge event had been predicted for some time, but the rapidity of the surge, largely triggered by rapid flows from Valerie, was surprising. The fjord was dubbed Russell Lake for the time it remained isolated (Figure 17-8). The glacier's movement exceeded 34 m (112 ft) per day during the peak surge, an enormous increase over its normal rate of 15 cm (6 in.) per day.

The exact cause of such a glacial surge is still being studied. Some surge events result from a buildup of water pressure in the basal layers of the glacier. Sometimes that pressure is enough to actually float the glacier slightly during the surge. As a surge begins, icequakes are detectable, and ice faults are visible along the margins that separate the glacier from the surrounding stationary terrain.

Glacial Erosion. Glacial erosion is similar to a large excavation project, with the glacier hauling debris from one site to another for deposition. As rock structure fails along joint planes, the passing glacier mechanically plucks the material and carries it away. There is evidence that rock pieces ac-

FIGURE 17-8
The Hubbard Glacier surged across Russell Fjord in Alaska, effectively damming the fjord and temporarily creating Russell Lake. [*Landsat* IR image from EROS Data Center.]

tually freeze to the basal layers of the glacier in this **glacial plucking** process and, once embedded, allow the glacier to scour and sandpaper the landscape as it moves, a process called **abrasion.** This abrasion and gouging produces a smooth surface on exposed rock, which shines with *glacial polish* when the glacier retreats. Larger rocks in the glacier act much like chisels, working the underlying surface to produce glacial striations parallel to the flow direction (Figure 17-9).

GLACIAL LANDFORMS

The processes of glacial erosion and deposition produce unique landform features. You might expect all glaciers to create the same landforms, but this is not so. Alpine and continental glaciers each generate their own characteristic landscapes.

Erosional Landforms Created by Alpine Glaciation

Geomorphologist William Morris Davis characterized the stages of a valley glacier in a set of drawings published in 1906 and redrawn here in Figure 17-10. Illustration (a) shows a typical river valley with characteristic V-shape and stream-cut tributary valleys that exist before glaciation. Illustration (b) shows that same landscape during a later period of active glaciation. Glacial erosion and transport are actively removing much of the regolith (weathered

FIGURE 17-9
Glacial polish and striations, examples of glacial erosion. [Photo courtesy of U.S. Geological Survey.]

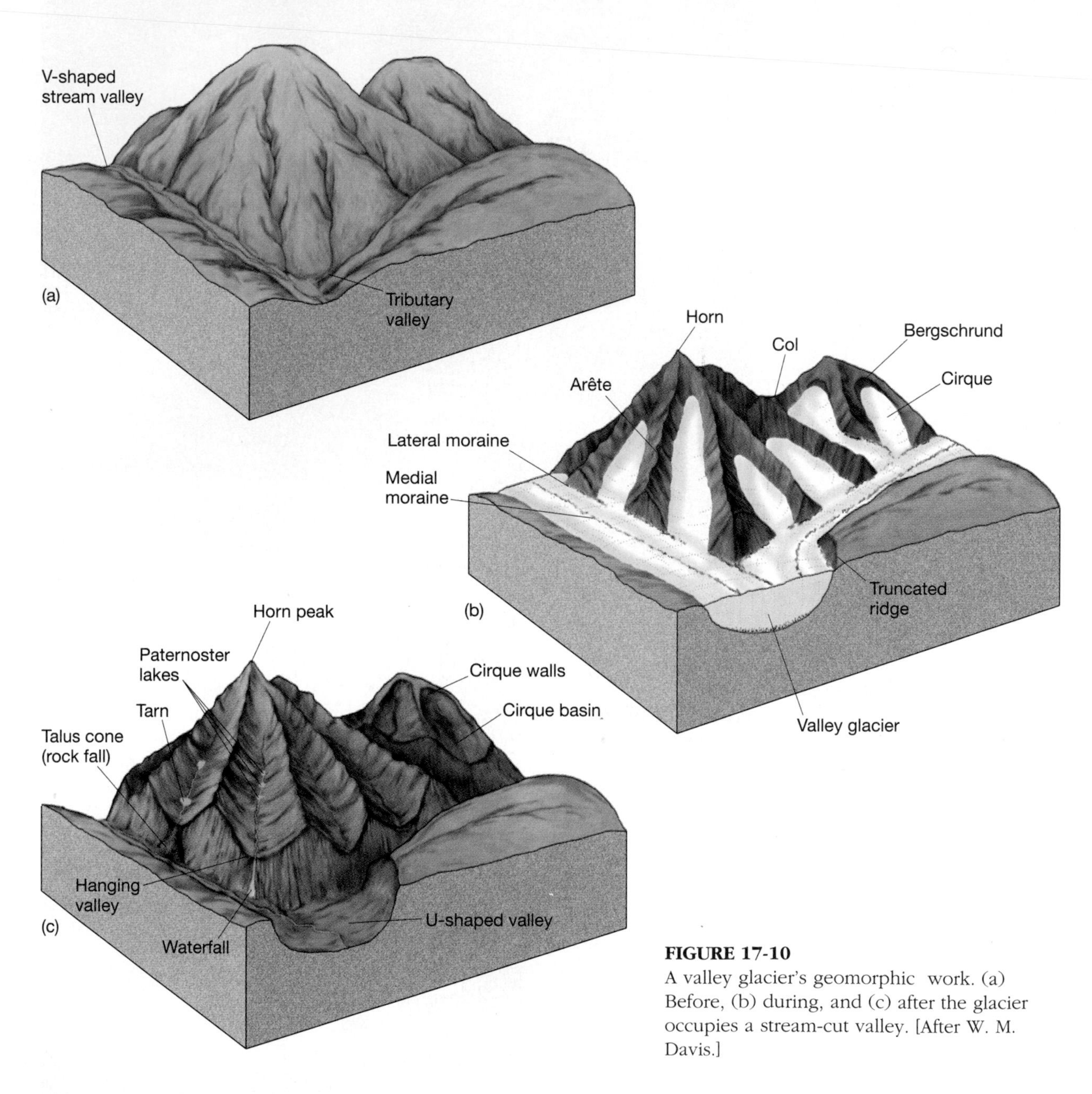

FIGURE 17-10
A valley glacier's geomorphic work. (a) Before, (b) during, and (c) after the glacier occupies a stream-cut valley. [After W. M. Davis.]

bedrock) and the soils that covered the stream valley landscape. Illustration (c) shows the same landscape at a later time when climates have warmed and ice has retreated. The glaciated valleys now are U-shaped, greatly changed from their previous stream-cut form. You can see the oversteepened sides and the straightened course of the valley. The physical weathering associated with the freeze-thaw cycle has loosened much rock along the steep cliffs, and the rock has fallen to form *talus cones* along the valley sides during the postglacial period.

The cirques where the valley glacier originated are clearly visible in (c). As the cirque walls have worn away, an **arête** has formed, a sharp ridge that divides two cirque basins (b). Arêtes ("knife-

FIGURE 17-11
Bridalveil Creek hanging valley and Bridalveil Falls, part of the scenic beauty of California's Yosemite National Park with its glaciated stream valleys. [Photo by author.]

edge" in French) become the sawtooth and serrated ridges in glaciated mountains. Two eroding cirques may reduce an arête to form a pass or saddlelike depression, called a **col.** A pyramidal peak called a **horn** results when several cirque glaciers gouge an individual mountain summit from all sides. The most famous example is the Matterhorn in the Swiss Alps, but many others occur worldwide.

A small mountain lake, especially one that collects in a cirque basin behind risers of rock material, is called a **tarn** (c). Small, circular, stair-stepped lakes in a series are called **paternoster lakes** because they look like a string of rosary (religious) beads forming in individual rock basins aligned down the course of a glaciated valley. Such lakes may have been formed by the differing resistance of rock to glacial processes or by the damming effect of glacial deposits.

Figure 17-10c also shows tributary valleys left stranded high above the glaciated valley floor, following the removal of previous slopes and stream courses by the glacier. These *hanging valleys* are the sites of spectacular waterfalls as streams plunge down the steep cliffs (Figure 17-11).

Where a glacial trough intersects the ocean, the glacier can erode even the landscape below sea level. As the glacier retreats, the trough floods and forms a deep **fjord** in which the sea extends inland, filling the lower reaches of the steep-sided valley (Figure 17-12). The fjord may be flooded further by a rising sea level or by changes in the

FIGURE 17-12
Tidewater terminus and fjord at Tracy Arm, Sawyer Glacier, Tongass National Forest, Alaska. [Photo by Tom Bean.]

elevation of the coastal region. All along the glaciated coast of Alaska, glaciers now are in retreat, thus opening many new fjords that previously were blocked by ice. Coastlines with notable fjords include those of Norway, Greenland, Chile, the South Island of New Zealand, Alaska, and British Columbia.

Depositional Landforms Created by Alpine Glaciation

You have just seen how glaciers excavate tremendous amounts of material and create fascinating landforms in the process. Glaciers also produce another set of landforms when they melt and deposit their eroded and transported debris cargo, which is carried toward the *terminus* (end) of the glacier (Figure 17-5b). Where the glacier melts, this debris accumulates to mark the former margins of the glacier—the end and sides. **Glacial drift** is the general term for all glacial deposits, both unsorted and sorted. Direct ice deposits appear unstratified and unsorted and are called **till.** Sediments deposited by glacial meltwater are sorted and are termed **stratified drift.**

As a glacier flows to lower elevation, a wide assortment of rock fragments are *entrained* (carried along) on the surface of the ice or embedded in the base of the glacier. As the glacier melts, this unsorted cargo of *ablation till* is lowered to the ground surface, sometimes covering the clay-rich *lodgement till* deposited along the base. The rock material is poorly sorted and is difficult to cultivate for farming, but the clays and finer particles can provide a basis for soil development.

Till is the term for this unsorted glacially deposited sediment. **Moraine** is the name for specific landforms produced by the deposition of these sediments. Several types of moraines are exhibited in Figure 17-13, here produced by the Hole-in-the-Wall Glacier. A **lateral moraine** forms along each side of a glacier. If two glaciers with lateral moraines join, a **medial moraine** may form as lateral moraines combine. Can you identify these elements on the active glacier in Figure 17-5c?

FIGURE 17-13
Medial, lateral, and ground morainal deposits of Hole-in-the-Wall Glacier, Wrangell-Saint Elias National Park, Alaska. Note the kettle pond in the foreground. [Photo by Tom Bean.]

(a)

(b)

FIGURE 17-14
Peyto Glacier in Alberta, Canada (a). Note valley train, braided stream, and glacial meltwater. (b) A glacial erratic in Maine. [(a) Photo by Wolfgang Kaehler; (b) photo by Chuck Breitsprecher.]

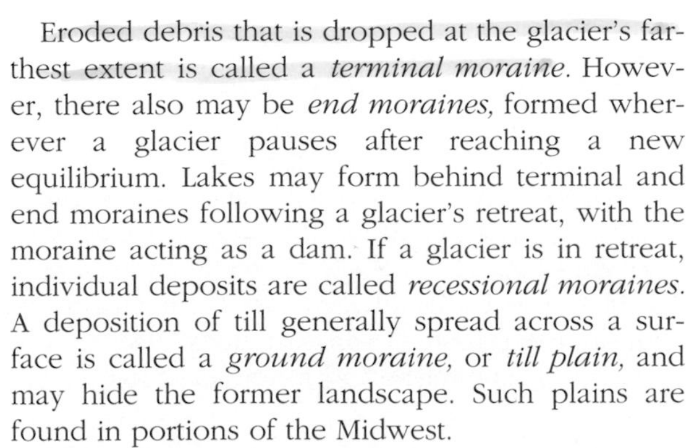

Eroded debris that is dropped at the glacier's farthest extent is called a *terminal moraine*. However, there also may be *end moraines,* formed wherever a glacier pauses after reaching a new equilibrium. Lakes may form behind terminal and end moraines following a glacier's retreat, with the moraine acting as a dam. If a glacier is in retreat, individual deposits are called *recessional moraines*. A deposition of till generally spread across a surface is called a *ground moraine,* or *till plain,* and may hide the former landscape. Such plains are found in portions of the Midwest.

All of these types of till are unsorted and unstratified debris. In contrast, sorted and stratified glacial drift, characteristic of stream-sorted material, can be deposited beyond a terminal moraine. This type of meltwater-deposited material downvalley from the terminus of a valley glacier is called a *valley train deposit.* Peyto Glacier in Alberta produces such a valley train that continues into Peyto Lake (Figure 17-14a). Distributary stream channels appear braided across its surface. The picture also shows the milky meltwater associated with glaciers, charged with finely ground rock ("rock flour"). Meltwater is produced by glaciers at all times, not just when they are retreating.

Retreating glaciers also leave behind large rocks (sometimes house-sized), boulders, cobbles, and smaller rocks that are different in composition and origin from the ground on which they are deposited. These **erratics** were an early clue to researchers that blankets of ice once had covered the land (Figure 17-14b). Louis Agassiz (1807–1873), a professor of natural history and the author of two books on glaciers, was an early proponent of a continental glaciation hypothesis as the explanation for many of Europe's erosional and depositional features. Today his grave is marked by one of his clues—appropriately, a polished and striated erratic boulder.

Erosional and Depositional Features of Continental Glaciation

The extent of continental glaciation in North America and Europe is portrayed in Figure 17-27. When these huge sheets of ice advanced and retreated, they produced many of the erosional and depositional features characteristic of alpine glaciation. However, because continental glaciers formed under different circumstances, the intricately carved alpine features, lateral moraines, and medial moraines all are lacking in continental glaciation. Table 17-1 compares the erosional and depositional features of alpine and continental glaciers.

Figure 17-15 illustrates some of the most common erosional and depositional features associated with the passage of a continental glacier. Many relatively flat plains of *unsorted* coarse till are formed behind terminal moraines. Characteristics of these **till plains** are low, rolling relief, and deranged drainage patterns (see Chapter 14, Figure 14-4g) Beyond the morainal deposits, glacio-fluvial **outwash plains** of *stratified drift* feature stream channels that are meltwater-fed, braided, and overloaded with debris deposited across the landscape.

Figure 17-15a shows a sinuously curving, narrow ridge of coarse sand and gravel called an **esker.** It forms along the channel of a meltwater stream that flows beneath a glacier, in an ice tunnel, or between ice walls beneath the glacier. As a glacier retreats, the steep-sided esker is left behind in a pattern roughly parallel to the path of the glacier. The ridge may not be continuous and in places may even appear branched following the course set by the subglacial watercourse. Commercially valuable deposits of sand and gravel are quarried from some eskers.

TABLE 17-1
Features of Valley Glaciation and Continental Glaciation Compared

Features	Alpine (Valley) Glacier	Continental Glaciation
	Erosional	
Striations, polish, etc.	Common	Common
Cirques	Common	Absent
Horns, arêtes, cols	Common	Absent
U-shaped valleys, truncated spurs, hanging valleys	Common	Rare
Fjords	Common	Absent
	Depositional	
Till	Common	Common
Terminal moraines	Common	Common
Recessional moraines	Common	Common
Ground moraines	Common	Common
Lateral moraines	Common	Absent
Medial moraines	Common, easily destroyed	Absent
Drumlins	Rare or absent	Locally common
Erratics	Common	Common
Stratified drift	Common	Common
Kettles	Common	Common
Eskers, crevasse fillings	Rare	Common
Kames	Common	Common
Kame terraces	Common	Present in hilly country

SOURCE: Adapted from L. Don Leet, Sheldon Judson, and Marvin Kauffman. *Physical Geology,* 5th ed., © 1978, p. 317. Reprinted by permission of Prentice-Hall, Inc., Englwood Cliffs, NJ.

Another feature of outwash plains is a **kame,** a small hill, knob, or mound of poorly sorted sand and gravel that is deposited directly by water, by ice in crevasses, or in ice-caused indentations in the surface (Figure 17-15c). Kames also can be found in deltaic forms and in terraces along valley walls.

Sometimes an isolated block of ice, perhaps more than a kilometer across, persists in a ground moraine, an outwash plain, or valley floor after a

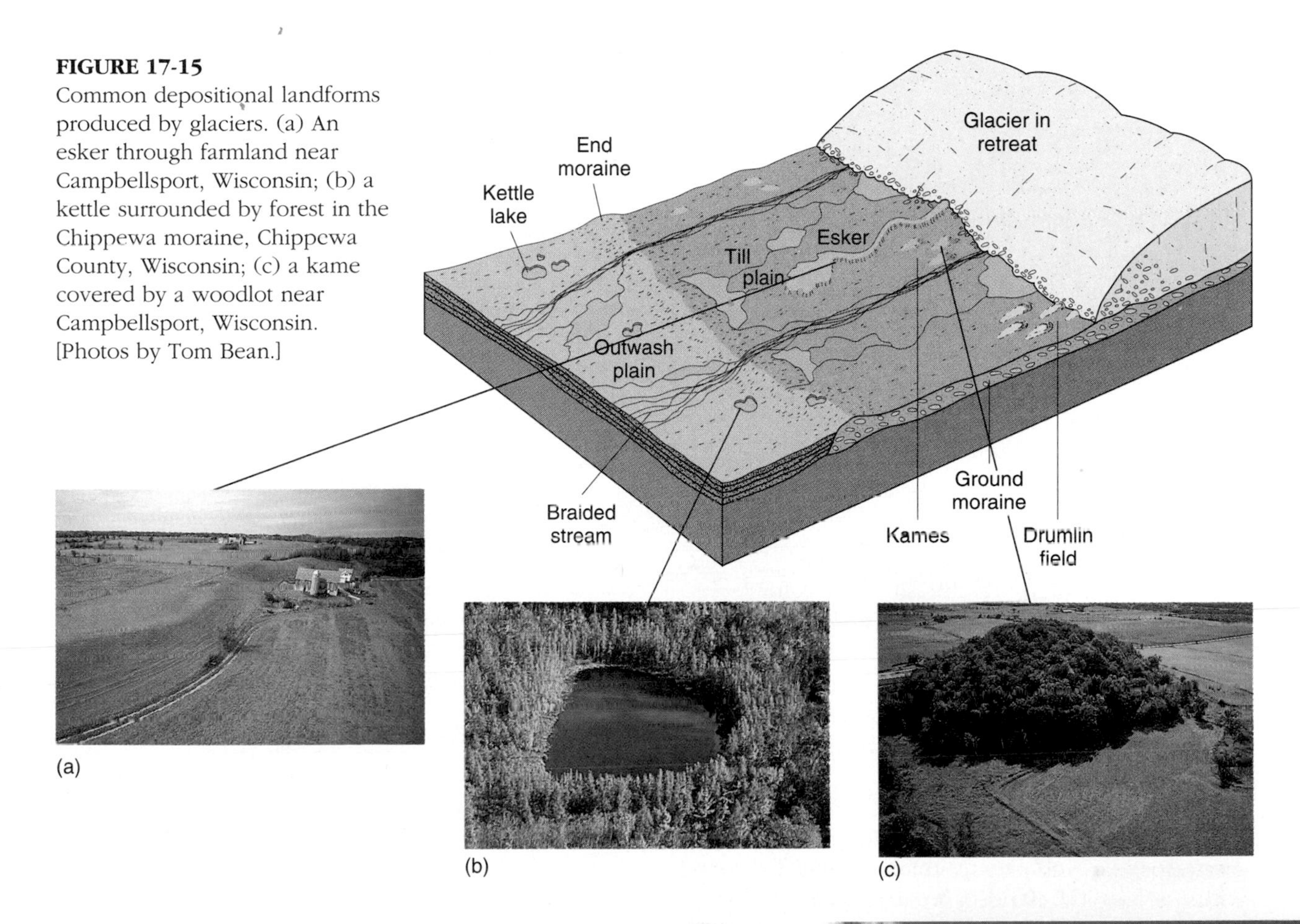

FIGURE 17-15
Common depositional landforms produced by glaciers. (a) An esker through farmland near Campbellsport, Wisconsin; (b) a kettle surrounded by forest in the Chippewa moraine, Chippewa County, Wisconsin; (c) a kame covered by a woodlot near Campbellsport, Wisconsin. [Photos by Tom Bean.]

glacier retreats. Perhaps 20 to 30 years are required for it to melt. In the interim, material continues to accumulate around the melting ice block. When the block finally melts, it leaves behind a steep-sided hole. Such a feature then frequently fills with water. This is called a **kettle** (Figure 17-15b). Thoreau's Walden Pond, mentioned in the quotation in Chapter 7, is such a kettle.

Glacial action also forms streamlined hills, one erosional, called a roche moutonnée, and the other depositional, called a drumlin. A **roche moutonnée** ("sheep rock" in French) is an asymmetrical hill of exposed bedrock. (Its gently sloping upstream side (*stoss* side) has been polished smooth by glacial action, whereas its downstream side (*lee* side) is abrupt and steep where rock pieces were plucked by the glacier (Figure 17-16).

A **drumlin** is deposited till that has been streamlined in the direction of continental ice

FIGURE 17-16
Roche moutonnée, as exemplified by Lambert Dome in the Tuolumne Meadows area of Yosemite National Park, California. [Photo by author.]

FIGURE 17-17
Topographic map south of Williamson, New York, featuring numerous drumlins (7.5-minute series quadrangle map, originally produced at a 1:24,000 scale, 10-ft contour interval). [Courtesy of U.S. Geological Survey.]

movement, blunt end upstream and tapered end downstream. Multiple drumlins (called *swarms*) occur across the landscape in portions of New York and Wisconsin, among other areas. Sometimes their shape is that of an elongated teaspoon bowl, lying face down. They attain lengths of 100–5000 m (330 ft to 3.1 mi) and heights up to 200 m (650 ft). Figure 17-17 shows a portion of a topographic map for the area south of Williamson, New York, which experienced continental glaciation. In studying the map, can you identify the numerous drumlins? In what direction do you think the continental glaciers moved across this region?

Along the margins of permanent ice, cold-climate processes dominate, with their resultant landforms. Let us now look at these cold-dominated landscapes.

PERIGLACIAL LANDSCAPES

The term **periglacial** was coined by scientist W. V. Lozinski in 1909 to describe cold-climate processes, landforms, and topographic features along the margins of glaciers, past and present. These periglacial regions occupy over 20% of Earth's land surface (Figure 17-18). The areas are either near permanent ice or at high elevation, and have ground that is seasonally snow free. Under these conditions, a unique set of periglacial processes operate, including permafrost, frost action, and ground ice.

Climatologically, these regions are in *Dfc, Dfd subarctic,* and *E polar* climates (especially *ET tundra* climate). Such climates occur either at high latitude (tundra and boreal forest environments) or high elevation in lower-latitude mountains (alpine environments). These periglacial regions are dominated by processes that relate to physical weathering, mass movement (Chapter 13), climate (Chapter 10), and soil (Chapter 18).

Permafrost

When soil or rock temperatures remain below 0°C (32°F) for at least two years, a condition of **permafrost** develops. An area that has permafrost but is not covered by glaciers is considered periglacial.

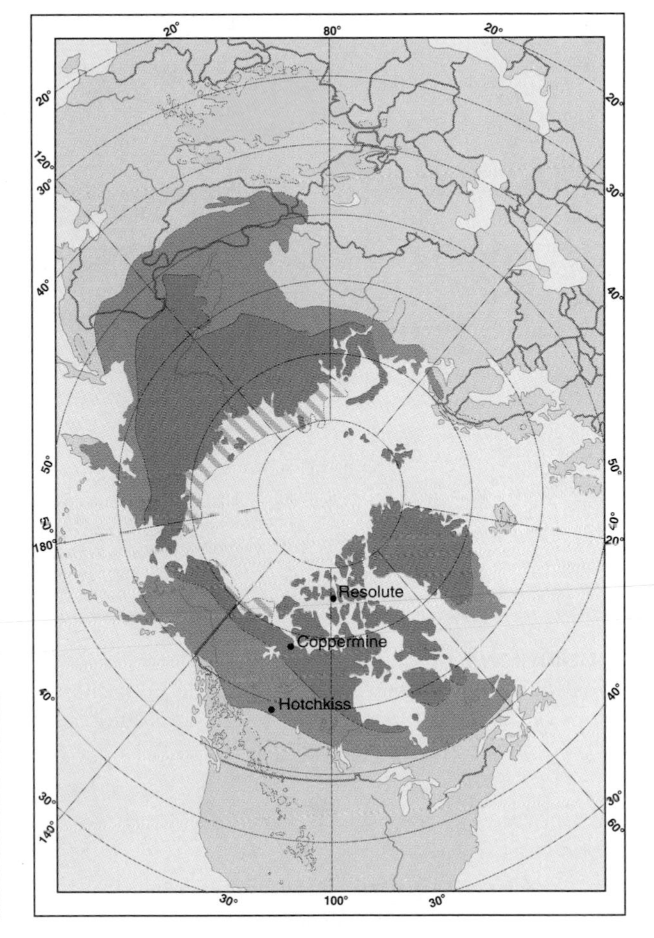

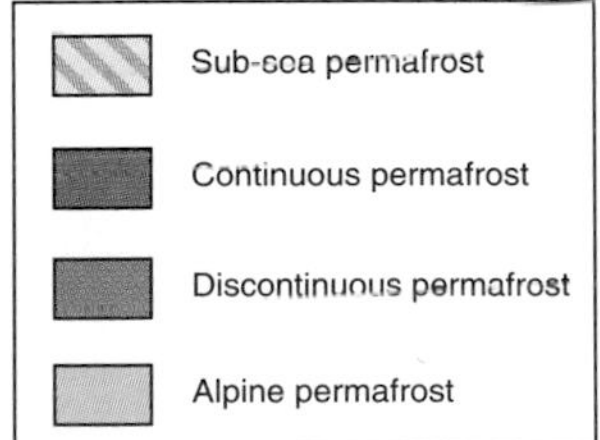

FIGURE 17-18
Distribution of permafrost in the Northern Hemisphere. Alpine permafrost is noted except for small alpine occurrences in Hawaii, Mexico, Europe, and Japan that are too small to show on the map. Sub-sea permafrost occurs in the ground beneath the Arctic Ocean along the margins of the continent noted. Noted in Figure 17-19, the towns of Resolute and Coppermine, Northwest Territories, and Hotchkiss, Alberta, are located on the map. [Adapted from Troy L. Péwé, "Alpine Permafrost in the Contiguous United States: A Review," *Arctic and Alpine Research* 15, no. 2 (May 1983): 146. © Regents of the University of Colorado. Used by permission.]

Note that this criterion is based solely on temperature and not on whether water is present. Other than high latitude and low temperatures, two other factors contribute to permafrost: the presence of fossil permafrost from previous ice-age conditions and the insulating effect of snow cover or vegetation that inhibits heat loss.

Permafrost regions are divided into two general categories, continuous and discontinuous, that merge along a general transition zone. *Continuous permafrost* describes the region of the most severe cold and is perennial, roughly poleward of the −7°C (19°F) mean annual temperature isotherm. Continuous permafrost affects all surfaces except those beneath deep lakes or rivers in the areas shown in Figure 17-18. Continuous permafrost may exceed 1000 m (over 3000 ft) in depth, averaging approximately 400 m (1300 ft).

Discontinuous permafrost occurs in unconnected patches that gradually coalesce poleward toward the continuous zone. Equatorward, along lower-latitude margins, permafrost becomes scattered or sporadic until it gradually disappears—roughly equatorward of the −1°C (30.2°F) mean annual temperature isotherm. In the discontinuous zone permafrost is absent on south-facing slopes, areas of warm soil, or areas insulated by snow. As much as 50% of Canada and 80% of Alaska is affected by permafrost of either type. In central Eurasia, the effects of continentality and elevation produce discontinuous permafrost that extends equatorward of the 50th parallel. Areas of discontinuous permafrost feature a mixture of *cryotic* (frozen) and noncryotic ground.

In addition to these two types, zones of high-altitude *alpine permafrost* extend to lower latitudes,

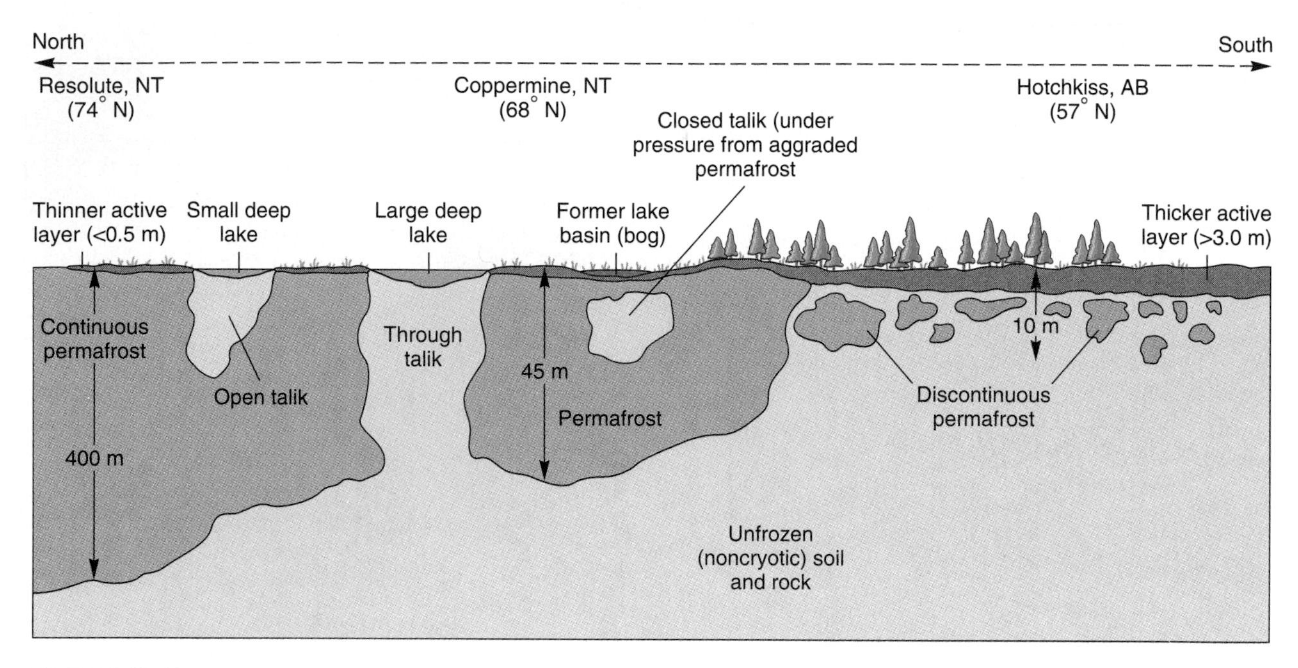

FIGURE 17-19
Cross section of a typical periglacial region in northern Canada showing typical forms of permafrost, active layer, talik, and ground ice. The three sites noted are located on the map in Figure 17-18.

as shown on the map. Microclimatic factors such as slope orientation and snow cover are important in the alpine environment. In the Mackenzie Mountains of Canada (62° N), continuous permafrost extends down to an elevation of 1200 m (4000 ft) and discontinuous permafrost occurs throughout. The Colorado Rockies (40° N) experience continuous permafrost down to an elevation of 3400 m (11,150 ft) and discontinuous permafrost to 1700 m (5600 ft).

We have looked at the spatial distribution of permafrost; let us now examine permafrost below the ground. Figure 17-19 is a stylized cross section from approximately 75° N to 55° N, and the three sites named are located on the map in Figure 17-18. The **active layer** is the zone of seasonally frozen ground that exists between the subsurface permafrost layer and the ground surface. The active layer is subjected to consistent daily and seasonal freeze-thaw cycles. This cyclic melting of the active layer affects as little as 10 cm (3.9 in.) depth in the north (Ellesmere Island, 78° N), up to 2 m (6.6 ft) in the southern margins (55° N) of the periglacial region, and 15 m, in the alpine permafrost of the Colorado Rockies (40° N).

The depth and thickness of the active layer and permafrost zone change slowly in response to climatic change. Higher temperatures produce degradation (reduction) in permafrost and a thickness increase of the active layer. Lower temperatures gradually cause aggradation (increase) in permafrost depth and a reduced thickness of the active layer. Although somewhat sluggish in response, this layer constitutes a dynamic open system keyed to energy gains and losses in the subsurface environment. Most permafrost exists in some state of disequilibrium with conditions in the environment and therefore is in active adjustment to inconstant climatic conditions.

A **talik** is an unfrozen portion of the ground

that may occur above, below, or within a body of discontinuous permafrost or beneath a body of water in the continuous region. Taliks are found beneath deep lakes and may extend to bedrock and noncryotic soil under large deep lakes (Figure 17-19). Taliks form connections between the active layer and groundwater, whereas in continuous permafrost groundwater is essentially cut off from water at the surface. In this way, permafrost disrupts aquifers and taliks leading to water supply problems.

Ground Ice and Frozen Ground Phenomena

In regions of permafrost, subsurface water that is frozen is termed **ground ice.** The moisture content of areas with ground ice may vary from nearly none in regions of drier permafrost to almost 100% in saturated soils. From the area of maximum energy loss, freezing progresses through the ground along a *freezing front,* or boundary between frozen and unfrozen soil. The presence of frozen water in the soil initiates geomorphic processes associated with *frost action* and the expansion of water volume as it freezes (Chapters 7 and 13).

Ground ice is found in lenses (horizontal bodies) and veins (channels extending in any direction). Ground ice may be exhibited as the common *pore ice* (subsurface water frozen in pore spaces), *segregated ice* (layers of buried ice that increase in mass by accreting water as the ground freezes, producing layers of relatively pure ice), *intrusive ice* (the freezing of water injected under pressure, as in a pingo, discussed shortly), and *wedge ice* (surface water entering a crack and freezing). Various aspects of frost action as a geomorphic agent are discussed in Chapter 13, ("Physical Weathering Processes," and in the section "Classes of Mass Movements" that describes soil creep).

Some forms of ground ice may be at the surface, as in a glacier covered by debris, or in episodes of *icing,* where a river or spring forms freezing layers of surface ice in the winter. Such icings can occur on slopes as well as level ground. Spring flooding along rivers can result from the presence of icings, a particular flood problem in rivers draining into the Arctic Ocean.

Frost Action Processes. The 9% expansion of water as it freezes produces strong mechanical forces that fracture rock and disrupt soil at and below the surface. Frost-action shatters rock, producing angular pieces that form a *block field,* or *felsenmeer,* accumulating as part of the arctic and alpine periglacial landscape, particularly on mountain summits and slopes.

If sufficient water undergoes the phase change to ice, the soil and rocks embedded in the water are subjected to *frost-heaving* (vertical movement) and *frost-thrusting* (horizontal motions). Boulders and slabs of rock generally are thrust to the surface. Layers of soil (soil horizons) may appear disrupted as if stirred or churned by frost action, a process termed *cryoturbation.* Frost action also produces a contraction in soil and rock, opening up cracks for ice wedges to form. Also, there is a tremendous increase in pressure in the soil as ice expands, particularly if there are multiple freezing fronts trapping unfrozen soil and water between them.

An **ice wedge** develops when water enters a crack in the permafrost and freezes. Thermal contraction in ice-rich soil forms a tapered crack—wider at the top, narrowing toward the bottom. Repeated seasonal freezing and melting progressively expands the wedge, which may widen from a few millimeters to 5–6 m and up to 30 m (100 ft) in depth (Figure 17-20). Widening may be small each year, but after many years the wedge can become significant. The annual thickness of ice added is marked by thin layers of sediment that form *foliations* in the wedge. In summer, when the active layer thaws, the wedge itself may not be visible. However, the presence of a wedge beneath is noticeable where the ice-expanded sediments form raised, upturned ridges.

Large areas of frozen ground (soil-covered ice) can develop a heaved-up, circular, ice-cored mound called a **pingo.** It rises above the flat landscape, occasionally exceeding 60 m (200 ft) height. Pingos rise as freezing water and ice expand, sometimes as a result of pressure developed by

(a)

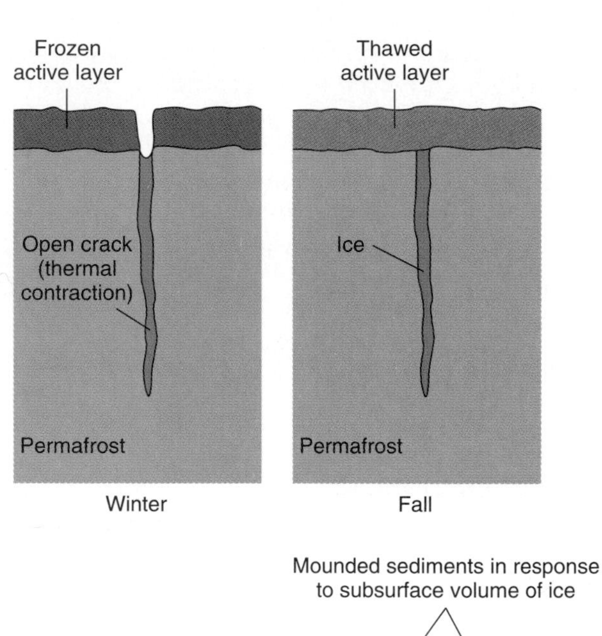

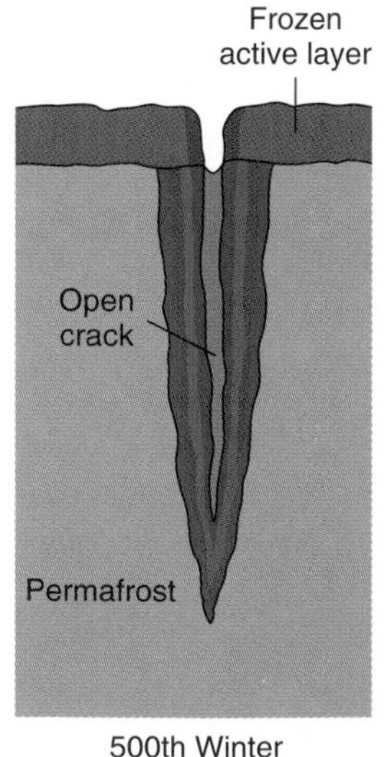

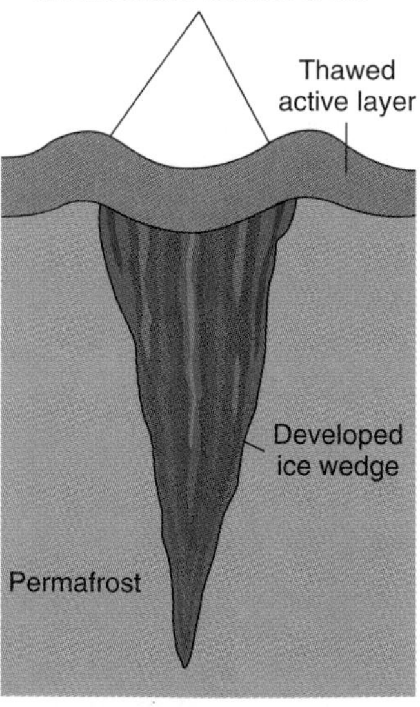

(b)

FIGURE 17-20
An ice wedge and ground ice in northern Canada (a). Sequential illustration of ice-wedge formation (b). [(a) Photo by Hugh M. French. (b) Adapted from A. H. Lachenbruch. "Mechanics of thermal contraction and ice-wedge polygons in permafrost," *Geological Society of America Bulletin Special Paper* 70 (1962).]

freezing artesian water injected into permafrost (Figure 17-21). A **palsa** (from the Swedish word for "elliptical") is a rounded or elliptical mound of peat that contains thin perennial ice lenses rather than an ice core as in a pingo. Palsas can be 2–30 m (6–100 ft) wide by 1–10 m (3–30 ft) high and usually are covered by soil or vegetation over a cracked surface.

The expansion and contraction of frost action results in the transport of stones and boulders. As the water-ice volume changes and the ice wedge deepens, coarser particles in the active layer of the soil are moved toward the surface. An area with a system of ground ice and frost action develops sorted and unsorted accumulations of rock at the surface that take the shape of polygons called **patterned ground.** Patterned ground may include stone polygons of arranged and sized rocks at the surface that coalesce into stone polygon nets (Figure 17-22). Various terms are in use to describe such ice-wedge and stone polygon forms: nets, circles, hummocks (vegetation covered), and stripes (elongated polygons formed on a hillslope). Ice

FIGURE 17-21
An ice-cored pingo resulting from hydraulic pressures that have pushed the mound upward above the landscape. Coastal erosion has exposed the ice core, near Tuktoyaktuk, Mackenzie Delta, Canada. [Photo by Hugh M. French.]

(a)

(b)

FIGURE 17-22
Patterned ground. (a) Aerial view of polygonal nets formed in Alaska, a fairly widespread frozen-ground phenomenon. (b) Surface appearance of patterned ground, near Richardson Mountains, Yukon Territory. [(a) Photo courtesy of U.S. Geological Survey; (b) photo by Joyce Lundberg.]

wedges sometimes are found beneath the perimeter of each cell; however, questions still exist as to how such patterned ground actually develops, or the degree to which such ground forms are the vestiges of past eras.

Hillslope Processes. Soil drainage is poor in areas of permafrost and ground ice. The active layer of soil and regolith is saturated with soil moisture during the thaw cycle and the whole layer commences to flow from higher to lower elevation if the landscape is slightly inclined. Such soil flows are in general called **solifluction;** in the presence of ground ice, the more specific term **gelifluction** is applied. In this ice-bound type of soil flow, movement up to 5 cm per year can occur on slopes as gentle as a degree or two. Gentle downslope movement of saturated surface material can occur in various climatic regimes outside of periglacial regions.

The cumulative effect of this landflow can be an overall flattening of a rolling landscape, with identifiable sagging surfaces and scalloped and lobed patterns in the downslope soil movements (Figure 17-23). Other types of periglacial mass movement

FIGURE 17-23
Gelifluction (solifluction) lobes on a hillside near the Yukon–Northwest Territories border. [Photo by Joyce Lundberg.]

include failure in the active layer, producing translational and rotational slides and rapid flows associated with melting ground ice. Periglacial mass movement processes are related to slope dynamics and processes discussed in Chapter 13.

Thermokarst Landscapes. As ground ice melts, various irregular features develop across the landscape, a condition called **thermokarst.** Note that the term has nothing to do with the solution processes and chemical weathering associated with limestone (karst); rather it refers to *thermal* subsidence and erosion processes caused by ground ice melting. Thermokarst topography is of hummocky, irregular relief marked by cave-ins, bogs, small depressions, and pits. Standing water and small lakes in the indentations resulting from poor drainage and ice-wedge melting are indicators of thermokarst processes. More thermokarst landforms are found in Siberia and Scandinavia than North America. In Canada and Alaska, rounded *thaw lakes,* or cave-in lakes, are thermokarst features.

Humans and Periglacial Landscapes

Human populations in areas that experience frozen ground phenomena encounter various difficulties. Because thawed ground in the active layer above the permafrost zone frequently shifts in periglacial environments, the maintenance of road beds and railroad tracks is a particular problem. In addition, any building placed directly on frozen ground will begin to melt itself into the defrosting soil. Thus, the melting of permafrost can create subsidence in structures and complete failure of building integrity (Figure 17-24).

Proper construction in periglacial regions dictates placing structures above the ground to allow air circulation beneath. This airflow allows the ground to cycle through its normal annual temperature pattern and not melt from conduction of building heat. Likewise, the Alaskan oil pipeline was constructed above the ground on racks to avoid melting the frozen ground, causing shifting that could rupture the line, disrupt oil delivery to the port, and cause severe environmental damage (Figure 17-25).

THE PLEISTOCENE ICE AGE EPOCH

The most recent episode of cold climatic conditions began about 1.65 million years ago, launching the Pleistocene epoch. At the height of the Pleistocene, ice sheets and glaciers covered 30% of Earth's land area, amounting to more than 45 million km^2 (17.4 million mi^2). Periglacial regions during the last ice age covered about twice their present areal extent. The Pleistocene is thought to have been one of the more prolonged cold periods in Earth's history. At least 18 expansions of ice occurred over Europe and North America, each obliterating and confusing the evidence from the one before.

The term **ice age** is applied to any such extended period of cold (not a single cold spell). It is a period of generally cold climate that includes one or more *glacials,* interrupted by brief warm spells known as *interglacials.* Each glacial and interglacial is given a name that is usually based on the location where evidence of the episode is prominent—for example, "Wisconsinan glacial."

FIGURE 17-24
Building failure due to the melting of permafrost south of Fairbanks, Alaska. [Adapted from U.S. Geological Survey. Photo courtesy of O. J. Ferrians, Jr., illustration based on U.S. Geological Survey pamphlet "Permafrost" by Louis L. Ray.]

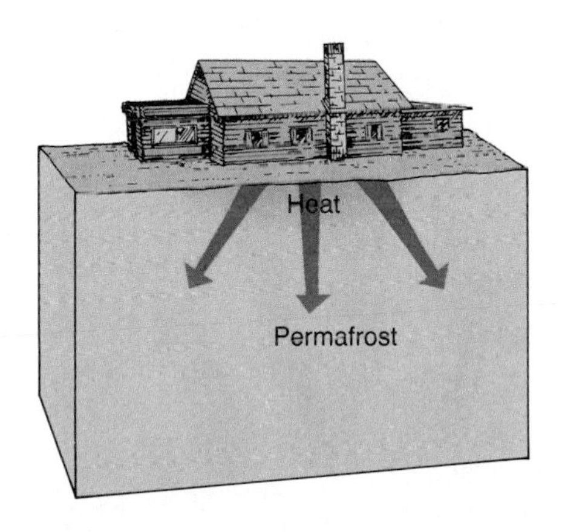

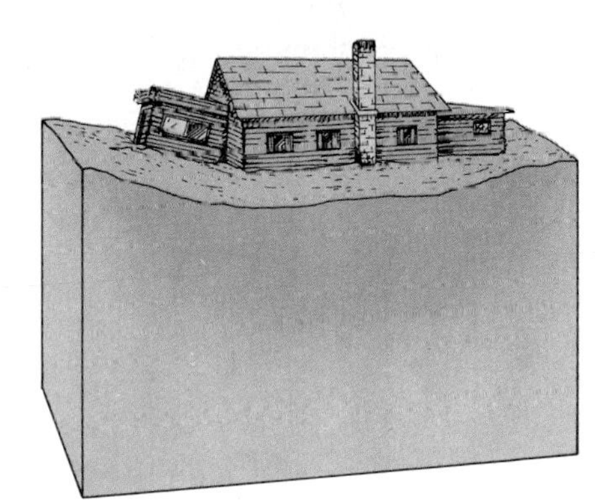

(a)

(b)

FIGURE 17-25
Proper construction in periglacial environments requires placement of buildings, water, and sewage lines above ground in "utilidors," Inuvik, Northwest Territories (a), and support of the Trans-Alaska oil pipeline on racks (b). [(a) Photo by Joyce Lundberg; (b) photo by Peter J. Williams.]

Traditionally, four major glacials and three interglacials were acknowledged for the Pleistocene epoch. In the United States the glacials were named the Nebraskan, Kansan, Illinoian, and Wisconsinan. In Europe, similar episodes coincided with those in North America, although they were given different names. Now, modern techniques have opened the way for a new chronology and understanding. Currently, glaciologists acknowledge the Illinoian glacial and Wisconsinan glacial and the Sangamon interglacial between them. These span the 300,000-year period prior to the Holocene (Figure 17-26). The Illinoian consisted of two glacials (that occurred during oxygen-isotope stages 6 and 8, shown in the figure), as did part of the Wisconsinan (stages 2 and 4), which is dated at 10,000 to 35,000 years ago. The oxygen-isotope stages on the chart are numbered back to stage 23 at approximately 900,000 years ago.

The continental ice sheets over Canada, the United States, Europe, and Asia are illustrated on the polar map projection in Figure 17-27. In North America, the Ohio and Missouri river systems mark the southern terminus of continuous ice. The edge of an ice sheet is not even; instead it expands and retreats in ice lobes. Such lobes were positioned where the Great Lakes are today and contributed to the formation of these lakes when the ice retreated from their gouged isostatically depressed basins.

Scientists worldwide participated in the ten-year *CLIMAP* project (*C*limate: *L*ong-Range *I*nvestigation, *M*apping, *a*nd *P*rediction). Using oxygen-isotope ratios from sea-floor core samples, they constructed sea-surface temperature maps for a time

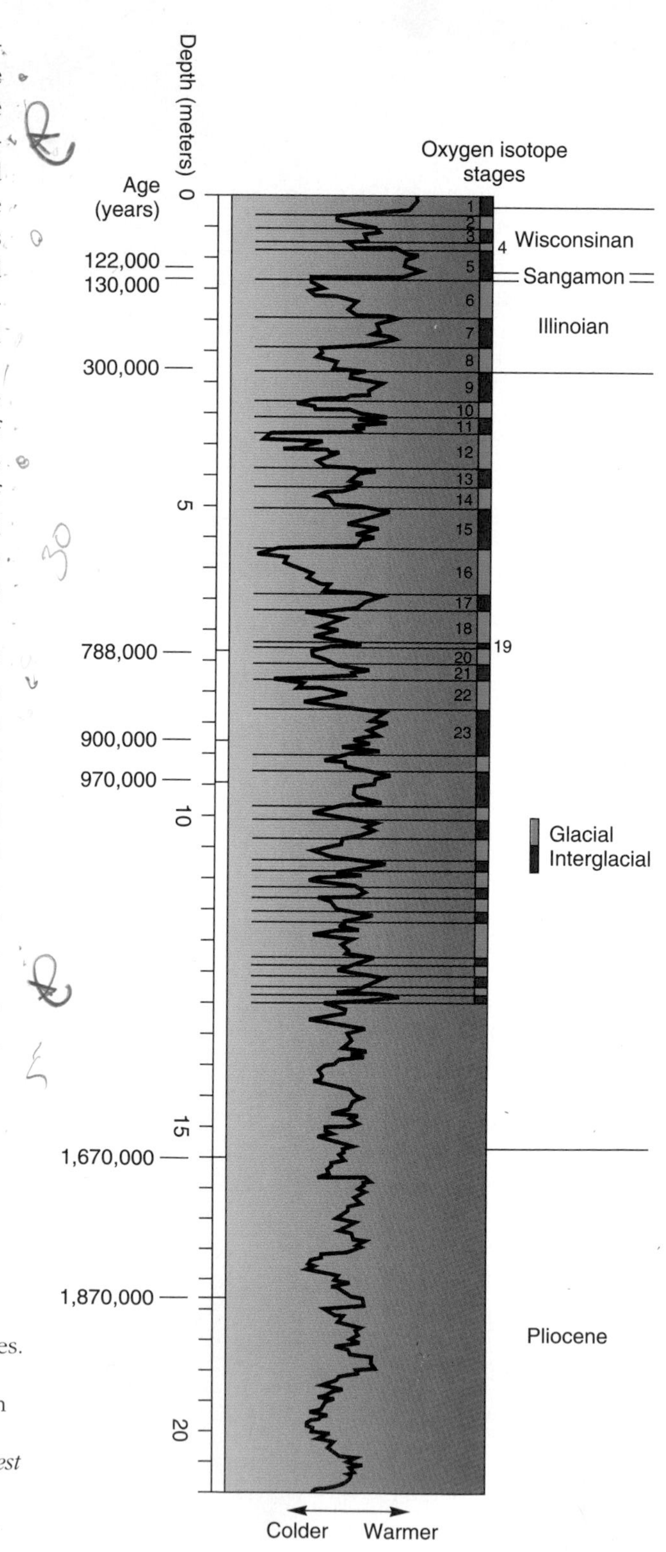

FIGURE 17-26
Pleistocene temperatures, determined by oxygen-isotope fluctuations in planktonic foraminifera from deep-sea cores. Twenty-three stages cover 900,000 years, with names assigned for the past 300,000 years. [After N. J. Shackelton and N. D. Opdyke, *Oxygen-Isotope and Paleomagnetic Stratigraphy of Pacific Core V28-239, Late Pliocene to Latest Pleistocene,* Geological Society of America Memoir 145. Copyright © 1976 by the Geological Society of America. Adapted by permission.]

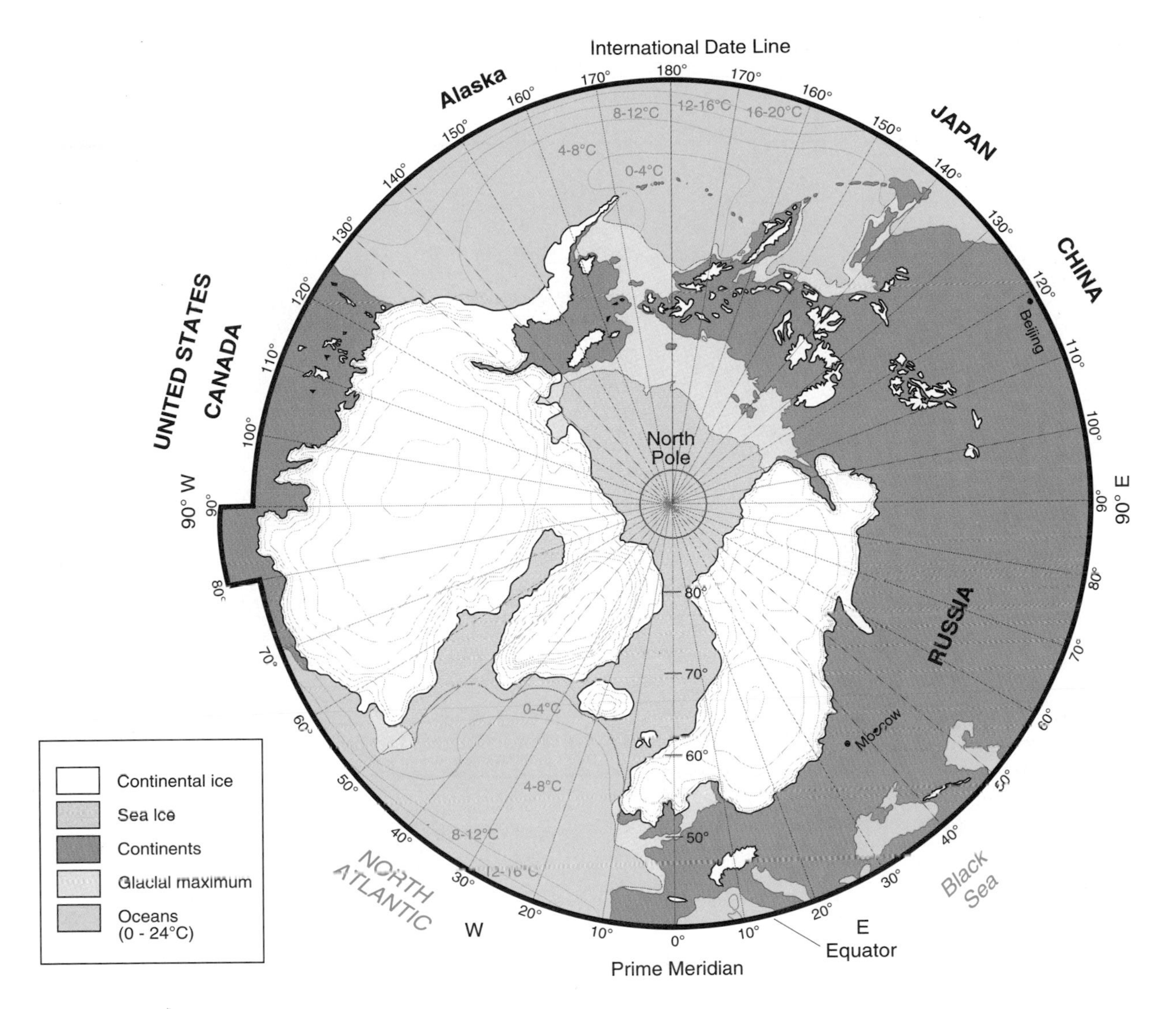

FIGURE 17-27
Extent of Pleistocene glaciation in the Northern Hemisphere 18,000 years ago viewed from a polar perspective. [From Andrew McIntyre, *CLIMAP Project*, Lamont-Dougherty Earth Observatory. Copyright © 1981 by the Geological Society of America. Reprinted by permission.]

18,000 years ago, when the last major extent of Pleistocene glacial ice prevailed. Sea levels at that time were approximately 100 m (330 ft) lower than today because so much of Earth's water was frozen and tied up in the glaciers (see tan areas along continental margins in Figure 17-28). In fact, sea ice extended southward into the North Atlantic and Pacific, and northward in the Southern Hemisphere about 50% farther than it does today (light brown ocean areas in Figure 17-28). Sea-surface temperatures during February were 1.4C° (2.5F°) lower than today's global average, and in August were 1.7C° (3.1F°) lower.

As these glaciers retreated, they exposed a drastically altered landscape: the rocky soils of New England, the polished and scarred surfaces of Canada's Atlantic Provinces, the sharp crests of the Sawtooth Range and Tetons of Idaho and Wyoming, the scenery of the Canadian Rockies and the Sierra Nevada, the Great Lakes of the United States and Canada, the fjords of New Zealand, Norway, and Chile, the Matterhorn of Switzerland,

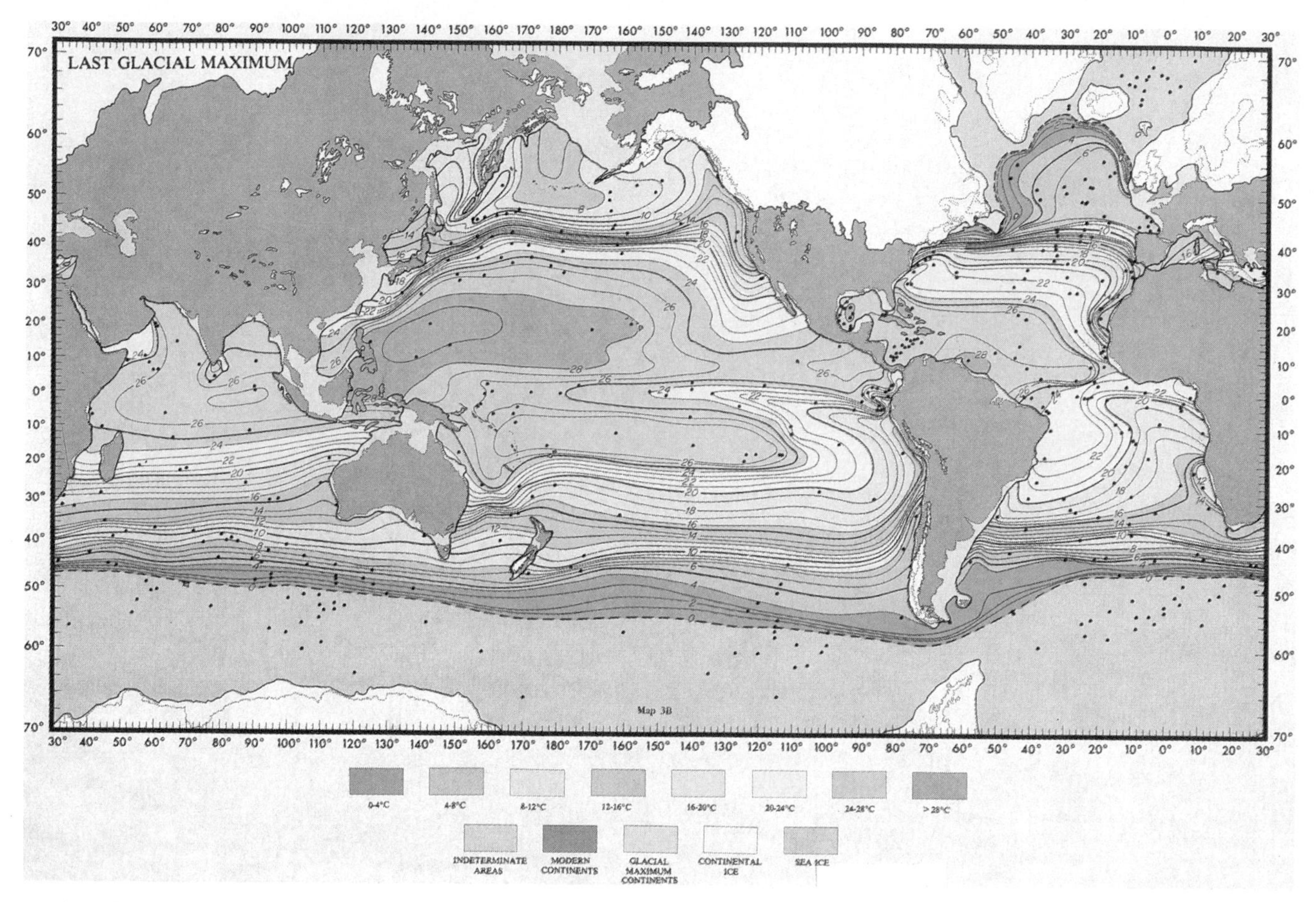

FIGURE 17-28
Ocean temperatures (August) and the location of glaciers 18,000 years ago. White areas reflect the maximum extent of the most recent glaciation. The lowered level of the ocean during this maximum uncovered continental shelf areas (the cream-colored areas). [From Andrew McIntyre, *CLIMAP Project,* Lamont-Dougherty Earth Observatory. Copyright © 1981 by the Geological Society of America. Reprinted by permission.]

and much more. Study of these glaciated landscapes is important, for we can better understand paleoclimatology (past climates) and discover the mechanisms that produce ice ages and climatic change. Focus Study 17-1 elaborates on this.

Pluvial Lakes (Paleolakes)

The term **pluvial** refers to wetter periods of greater effective moisture availability or moisture efficiency, principally during the Pleistocene epoch. During these periods, lake levels increased in arid and seasonally arid regions. The drier periods between these significantly moist climatic times are called *interpluvials.* Interpluvials are marked with **lacustrine deposits,** lake sediments that form terraces along former shorelines, with deltas from tributary streams, or with wave-cut cliffs and beaches.

Earlier researchers attempted to correlate pluvial and glacial ages, given their coincidence during the Pleistocene. However, few sites actually

Focus Study 17-1 Deciphering Past Climates: Paleoclimatology

The Vikings were favored by a medieval warming episode as they sailed out into the less frozen North Atlantic to settle Iceland and Greenland between A.D. 900 and 1200, and inadvertently ventured onto the North American continent as well. However, between the years 1250 and 1740, a "little ice age" took place, burying many key mountain passes of Europe under deep ice, and lowering snowlines about 200 m (650 ft). Evidence that Earth's climate has fluctuated in and out of warm and cold ages now is being traced back through ice cores, layered deposits of silts and clays, changing paleomagnetism, the geologic sequences locked in rocks, and radioactive dating methods. And one interesting fact is emerging from these studies: humans (*Homo erectus* and *Homo sapiens* of the last 1.9 million years) have never experienced Earth's normal climate, that is, the climate most characteristic of Earth's 4.6-billion-year span.

Apparently, Earth's climates were quite moderate until the last 1.2 billion years, when cyclic temperature patterns with cycles of 200–300 million years became more pronounced. The most recent cold episode was the Pleistocene epoch, which began in earnest 1.65 million years ago and through which we may still be progressing. The Holocene epoch began 10,000 years ago and may represent either an end to the Pleistocene, or merely an interglacial (Figure 1). Several episodes of less severe cold have occurred during the Holocene.

The mechanisms that bring on an ice age are the subject of much research and debate. Because past occurrences of low temperatures appear to have followed a pattern, researchers have looked for causes that also are cyclic in nature. They have identified a complicated mix of interacting variables that appear to influence long-term climatic trends.

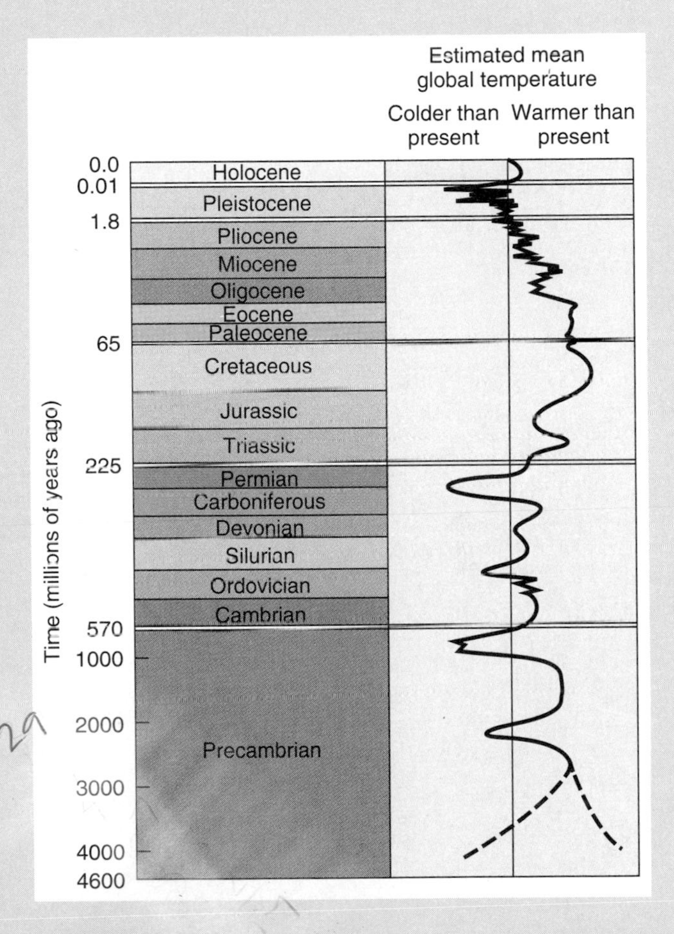

FIGURE 1
A generalization of Earth's surface temperatures over geologic time. [After L. A. Frakes, *Climates Throughout Geologic Time*. Amsterdam: Elsevier, 1979. Adapted by permission.]

Galactic and Earth-Sun Relationships

As our Solar System revolves around the distant center of the Milky Way, it crosses the plane of the galaxy approximately every 32 million years. At that time Earth's plane of the ecliptic aligns parallel to the galaxy's plane, and we pass through periods of increased interstellar dust and gas, which may have some climatic effect.

Other possible astronomical factors were proposed by Milutin Milankovitch (1879–1954), a Yugoslavian astronomer who compiled extensive formulations about Earth-Sun orbital relationships. Milankovitch wondered whether the development of an ice age was related to seasonal astronomical factors—Earth's revolution, rotation, and tilt—extended over a longer time span. Earth's elliptical orbit about the Sun is not constant. The shape of the ellipse varies by more than 17.7 million km (11 million mi) during a 100,000-year cycle, stretching out to an extreme ellipse (Figure 2a). In addition, Earth's axis "wobbles" through a 26,000-year

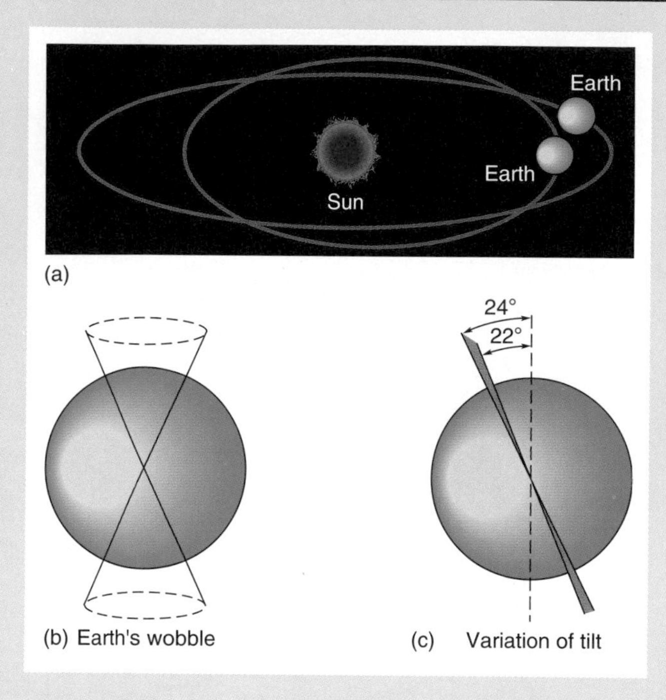

FIGURE 2
Astronomical factors that possibly affect broad climatic cycles. (a) Earth's elliptical orbit varies widely during a 100,000-year cycle, stretching out to an extreme ellipse. (b) Earth's 26,000-year axial wobble. (c) Variation in Earth's axial tilt every 40,000 years.

cycle, in a movement much like that of a spinning top winding down. Earth's slowly wobbling axis inscribes an inverted cone in space, a movement called precession (Figure 2b).

Finally, Earth's present axial tilt of 23.5° varies from 22° to 24° during a 40,000-year period (Figure 2c). Milankovitch calculated, without the aid of today's computers, that the interaction of these Earth-Sun relationships creates a 96,000-year climatic cycle. His model assumes that changes in astronomical relationships affect the amounts of insolation received.

Milankovitch died in 1954, his ideas still not accepted by a skeptical scientific community. Now, in the era of modern computers, remote sensing satellites, and worldwide efforts to decipher past climates, Milankovitch's valuable work has stimulated much research in an attempt to explain climatic cycles.

Solar Variability

If the Sun significantly varies its output over the years, as some other stars do, that change would seem a convenient and plausible cause of ice-age timing. However, the lack of evidence that the Sun behaves in a significantly variable manner over long cycles argues against this hypothesis. Nonetheless, inquiry about the Sun's variability actively continues. We do know that fluctuations in solar wind have some effect on Earth's weather. Analyzing the effects of the solar maximum of 1990 and early 1991 may further our understanding.

Geophysical Factors

Major glaciations also can be associated with the migration of landmasses to higher latitudes. Chapters 11 and 12 explain that the shape and orientation of landmasses and ocean basins have changed greatly during Earth's history. Continental plates have drifted from equatorial locations to polar regions and vice versa, thus producing cold and warm episodes on the land. Gondwana (the southern half of Pangaea) experienced extensive glaciation that left its mark on the rocks of parts of Africa, South America, India, Antarctica, and Australia. Landforms in the Sahara, for example, bear the markings of even earlier glacial activity, partly explained by the fact that portions of Africa were centered near the South Pole during the Ordovician period, 465 million years ago (see Figure 11-16a).

Geographical-Geological Factors

Geographical-geological factors include the terrestrial causes of atmospheric variables, increases in relief associated with mountain building, sea level changes, and alterations of oceanic circulation patterns. One such terrestrial source, a volcanic eruption, might produce lower summer temperatures for a year or two, leading to a buildup of long-term snow cover at high latitudes. These high-albedo surfaces then would reflect more insolation away from Earth, to further enhance cooling. The climatic effects of the volcanic eruptions in Japan (Mount Unzen) and the Philippines (Mount Pinatubo) in the summer of 1991 are being closely studied for the climatic effects they caused.

Furthermore, episodes of mountain building over the past billion years have forced mountain summits above the snowline, where snow remains after the summer melt. Mountain chains can affect downwind weather patterns and jet stream circulation, which in turn guides weather systems. In addition, oceanic circulation patterns have changed. For example, the Isthmus of Panama formed about 3 million years ago and effectively separated the circulation of the Atlantic and Pacific oceans. Such changes in circulation and ocean basin configuration affect air mass formation and temperature patterns.

Also linked to a terrestrial source is the fluctuation of atmospheric greenhouse gases, triggering higher or lower temperatures. An ice core taken at Vostok, the Russian research station near the geographic center of Antarctica, contained trapped air samples from approximately 160,000 years ago and showed carbon dioxide levels varying from more than 290 ppm to a low of nearly 180 ppm (Figure 3). Higher levels of CO_2 generally are thought to correlate with each interglacial. During the 1990s, CO_2 reached 360 ppm, higher than at any time in the past 160,000 years, principally due to anthropogenic sources. A more extensive ice core, drilled in Greenland (GISP 2), over 3000 m (9000 ft) deep and spanning 250,000 years, was completed in 1993 and promised to unlock more climatic clues.

Thus, Earth's climate is unfolding as a multicyclic system controlled by an interacting set of cooling and warming processes, all founded on galactic alignments, Earth-Sun relationships, geophysical, and geographical-geological factors. Within this complex of change, scientists are attempting to decipher past climatic rhythms and present and future climatic trends.

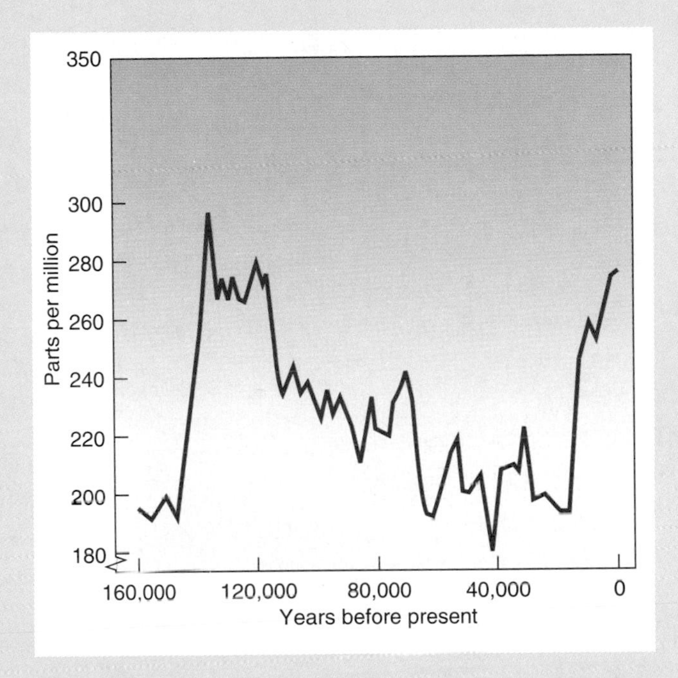

FIGURE 3
Previous CO_2 levels revealed in air bubbles trapped in Antarctic ice removed from the Vostok ice core. [Adapted by permission from *Nature* 329, p. 408. Copyright © 1988 Macmillan Magazines Ltd.]

demonstrate such a simple relationship. For example, in the western United States, the estimated ice melt amounts to only a small portion of the water volume in pluvial lakes of the region, and these lakes tend to predate glacial times and correlate instead with wetter periods, or periods thought to have had lower evaporation rates. The term *paleolake* is increasingly used in the scientific literature to describe these lakes and to separate their occurrence from specific glacial stages.

Evidence of pluvial lakes exists in North and South America, Africa, Asia, and Australia. The Caspian Sea in Kazakhstan and southern Russia (about 40° N 50° E) has visible pluvial shorelines about 80 m (265 ft) above today's lake level, which is 30 m (100 ft) below mean world sea level. In North America, the two largest late Pleistocene paleolakes were in the Basin and Range Province of the West. Figure 17-29 portrays these two lakes—Lake Bonneville and Lake Lahontan—and other pluvial lakes at their highest levels. They attained their greatest extent 15,000 to 25,000 years ago. Since then, climatic shifts have been too small to be formally identified as another pluvial cycle. Pluvial lakes also occurred in Mexico, with one even larger than Lake Bonneville.

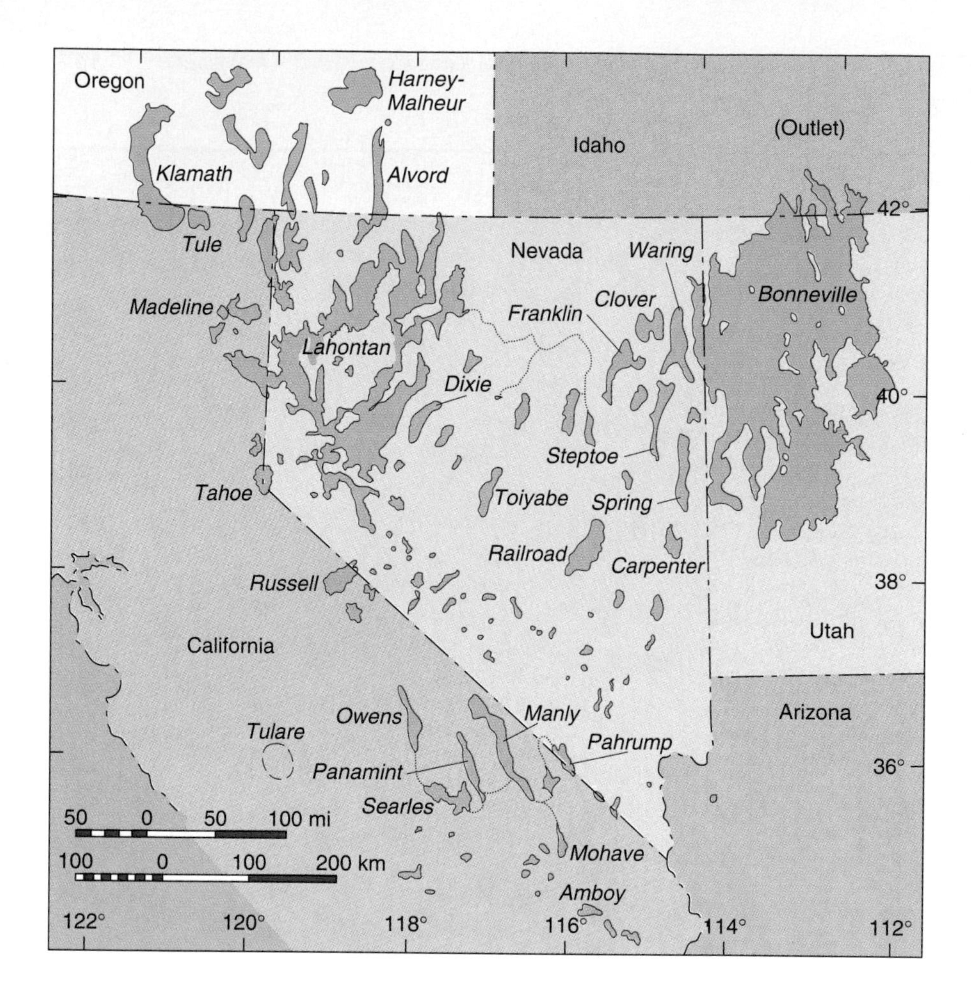

FIGURE 17-29
Pluvial lakes (paleolakes) of the western United States at their highest level 15,000–25,000 years ago. Lake Lahontan and Lake Bonneville were the largest. [After R. F. Flint, *Glacial and Pleistocene Geology*. Copyright © 1957 by John Wiley & Sons. Adapted by permission of John Wiley & Sons, Inc.]

The Great Salt Lake, near Salt Lake City, Utah, and the Bonneville Salt Flats in western Utah are significant remnants of Lake Bonneville, which at its greatest extent covered more than 50,000 km^2 (19,500 mi^2) and reached depths of 300 m (1000 ft), spilling over into the Snake River drainage to the north.

The present Great Salt Lake experienced a dynamic trend change during the 1980s. The lake level began rising in 1981, following periods of increased rainfall and snowpack input from the adjoining Wasatch Mountains. Definite shifts in storm tracks which favored increased precipitation were a feature of the 1982–1986 period. Evaporation is the lake's only output in the closed Salt Lake basin. By the mid-1980s, the lake achieved its highest level in historic times, flooding an interstate highway, transcontinental railroad tracks, resorts, and thousands of acres of grazing land. Fortunately, the years following the highest lake level had high evaporation rates and precipitation dropped to 50% below average. These ancient pluvial lakes associated with past glacial times remain today as climatic reminders of environmental change.

ARCTIC AND ANTARCTIC REGIONS

Climatologists use environmental criteria to define the Arctic and the Antarctic regions (Figure 17-30). The Arctic area is defined by the 10°C (50°F) isotherm for July because it coincides with the visible treeline, which is the boundary between the northern forests and tundra climates. The Arctic Ocean is covered by floating sea ice (frozen seawater) and glacier ice (frozen freshwater) in icebergs that formed at the edge of the surrounding land. This ice pack thins in the summer months and sometimes breaks up.

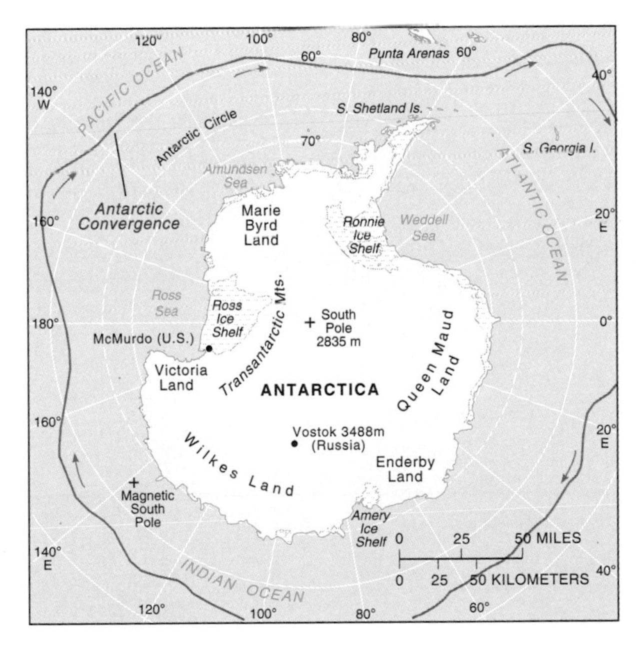

FIGURE 17-30
The Arctic and Antarctic regions.

The environmental criterion that defines the Antarctic region is the *antarctic convergence,* a narrow zone that extends around the continent as a boundary between colder antarctic water and warmer subantarctic water. This boundary follows roughly the 10°C (50°F) February isotherm and is located near 60° S latitude. The Antarctic region that is covered with sea ice represents an area greater than North America, Greenland, and Western Europe combined!

The Antarctic Ice Sheet

Antarctica is much colder overall than the Arctic, and is underlain by a continent-sized landmass instead of a sea. In simplest terms, Antarctica can be thought of as a continent covered by a single enormous glacier, although it contains distinct regions such as the East Antarctic and West Antarctic ice sheets, which respond differently to slight climatic variations.

The edges of the ice sheet that enter bays along the coast form extensive **ice shelves,** with sharp ice cliffs rising up to 30 m (100 ft) above the sea. Large tabular islands of ice are formed when sections of the shelves break off and move out to sea (Figure 17-31). These ice islands can be very large; one in the late 1980s exceeded the size of Rhode Island.

FIGURE 17-31
Icebergs break off from the Filchner Ice Shelf into the Weddell Sea, Antarctica (78° S 40° W). [SPOT Image courtesy of CNES. All rights reserved. Used by permission.]

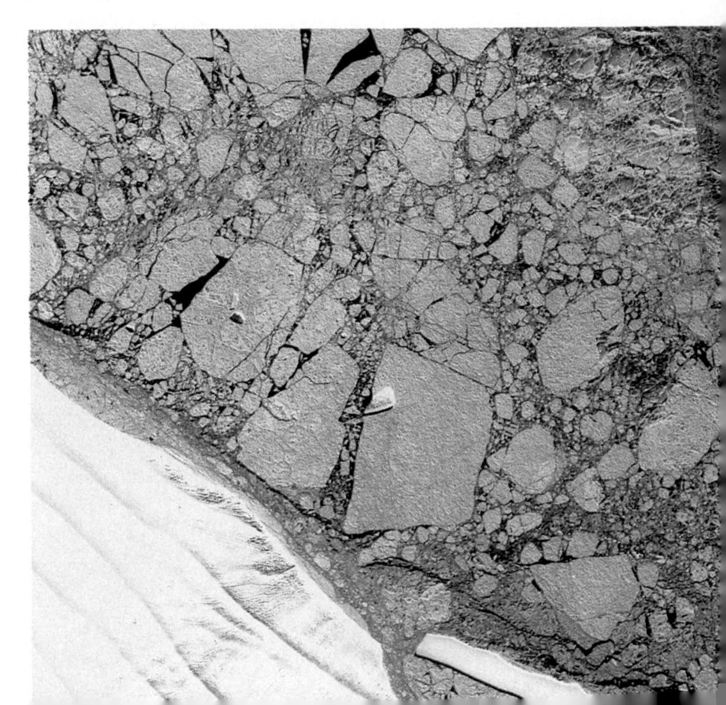

Individual outlet glaciers operate around the periphery of the ice sheet in Antarctica. The Byrd Glacier moves through the Transantarctic Mountains, draining an area of more than 1,000,000 km^2 (386,000 mi^2) as it flows 180 km (112 mi) and drops 5000 m (3.1 mi) in elevation. This glacier exceeds 22 km (13.7 mi) in width at its narrowest point and, where its thickness is 1000 m (3300 ft), it maintains a flow rate of approximately 840 m (2750 ft) per year, as fast as some surging glaciers.

SUMMARY— Glacial and Periglacial Processes and Landforms

More than 77% of the freshwater on Earth is frozen. Ice presently covers about 11% of Earth's surface, and periglacial features occupy another 20% of ice-free, cold-dominated landscapes. Glaciers form by the accumulation and recrystallization of snow into dense glacial ice that flows plastically. They move under the pressure of their own mass and the pull of gravity. Large ice sheets cover Greenland and Antarctica, and less-extensive ice caps and ice fields dominate mountain landscapes. Alpine glaciers occupy valleys at high elevations and high latitudes.

A glacier is an open system with inputs and outputs that can be analyzed through observation of the expansion and wasting of the glacier itself. A glacier is fed by snowfall and is wasted by losses from its upper and lower surfaces and along its margins. Snow that survives the summer and into the following winter undergoes a slow transformation into glacial ice. Formation of glacial ice is analogous to the formation of metamorphic rock: sediments (snow and firn) are pressured and recrystallized into a dense metamorphic rock (glacial ice).

Extensive valley glaciers have reshaped mountains worldwide, carving V-shaped stream valleys into U-shaped glaciated valleys. The passage of ice over a landscape excavates and creates profound changes, producing many distinctive erosional and depositional landforms.

Modern active glaciers and the present physical landscape as molded by past glaciers provide evidence for the study of past climates—paleoclimatology. During the Pleistocene epoch, extensive continental ice covered portions of the United States, Canada, Europe, and Russia. Glaciers worldwide are being watched more closely than ever because of their direct link to changing trends in world climates and the potential increase in mean sea level that can be brought on by the melting of even small amounts of stored ice.

KEY TERMS

ablation
abrasion
active layer
alpine glacier
arête
cirque
col
continental glacier
crevasses
drumlin
equilibrium line
erratics
esker
firn
firn line
fjord
gelifluction
glacial drift
glacial ice
glacial plucking
glacial surge
glacier
ground ice
horn
ice age
ice cap

ice field
ice sheet
ice shelves
ice wedge
icebergs
kame
kettle
lacustrine deposits
lateral moraine
medial moraine
moraine
outlet glaciers
outwash plains
palsa
paternoster lakes
patterned ground
periglacial
permafrost
pingo
pluvial
roche moutonnée
snowline
solifluction
stratified drift
talik
tarn
thermokarst
till
till plains
valley glacier

REVIEW QUESTIONS

1. Describe the location of most of the freshwater on Earth today.
2. What is a glacier? What is implied about existing climate patterns in a glacial region?
3. Differentiate between an alpine glacier and a continental glacier.
4. Name the three types of continental glaciers. What is the basis for dividing continental glaciers into types? Which type is Antarctica?
5. How is an iceberg generated?
6. Trace the evolution of glacial ice from fresh fallen snow.
7. What is meant by glacial mass balance? What are the basic inputs and outputs underlying that balance?
8. What is meant by a glacial surge? What do scientists think produces surging episodes?
9. How does a glacier accomplish erosion?
10. Describe the evolution of a V-shaped stream valley to a U-shaped glaciated valley. What kinds of features are visible after a glacier has retreated?
11. Differentiate between two forms of glacial drift—till and outwash.
12. What is a morainal deposit? What specific forms of moraines are created by alpine and continental glaciers?
13. What are some common depositional features encountered in a till plain?
14. Compare a roche moutonnée and a drumlin.
15. In terms of climatic types, describe the areas on Earth where periglacial landscapes occur. Include both higher latitude and higher altitude climate types.
16. Define two types of permafrost and differentiate their occurrence on Earth. What are the characteristics of each?
17. Describe the active zone in permafrost regions and relate the degree of development to specific latitudes.
18. What is a talik? Where might you expect to find taliks and to what depth do they occur?

19. Under what circumstances would you find a talik in the zone of continuous permafrost?
20. What is the difference between permafrost and ground ice?
21. Describe the role of frost action in the formation of various landform types in the periglacial region.
22. Relate some of the specific problems humans encounter in developing periglacial landscapes.
23. What is paleoclimatology? Describe Earth's past climatic patterns. Are we experiencing a normal climate pattern in this era or have scientists noticed any significant trends?
24. Define an ice age. When was the last one on Earth? How is the modern age classified? Explain "glacial" and "interglacial" in your answer.
25. Give an overview of what science has learned about the various causes of ice ages by listing and explaining at least four possible climate change factors.
26. What did the *CLIMAP* project determine and plot on the map shown in Figure 17-28?
27. What criteria define the Arctic and Antarctic regions? Is there any coincidence in these criteria and the distribution of Northern Hemisphere forests on the continents?

SUGGESTED READINGS

"The Antarctic," *Scientific American* 207, no. 3 (September 1962): Entire issue.

Bailey, Ronald H., and the editors of Time-Life Books. *Glacier.* Planet Earth Series. Alexandria, VA: Time-Life Books, 1982.

Budyko, M. I. *The Earth's Climate: Past and Future.* International Geophysics Series, vol. 29. New York: Academic Press, 1982.

Chorlton, Windsor, and the editors of Time-Life Books. *Ice Ages.* Planet Earth Series. Alexandria, VA: Time-Life Books, 1983.

Central Intelligence Agency. *Atlas of Polar Regions.* Washington, DC: CIA—National Foreign Assessment Center, 1978.

CLIMAP Project Members. "Seasonal Reconstructions of the Earth's Surface at the Last Glacial Maximum." *CLIMAP Project,* compiled by Andrew McIntyre. Geological Society of America Map and Chart Series MC-36. Palisades, NY: 1981.

CLIMAP Project Members. "The Surface of the Ice-Age Earth," *Science* 191, No. 4232 (March 19, 1976): 1131–37.

Embleton, C., and C. A. M. King. *Glacial Geomorphology.* London: Edward Arnold, 1975.

Estes, John E., ed. *Manual of Remote Sensing.* Vol. 2. *Interpretation and Applications.* Falls Church, VA: American Society of Photogrammetry, 1983.

Flint, R. F. *Glacial and Quaternary Geology.* New York: John Wiley & Sons, 1971.

Kamb, Barclay, C. F. Raymond, W. D. Harrison, Herman Englehardt, K. A. Echelmeyer, N. Humphrey, M. M. Brugman, and T. Pfeffer. "Glacier Surge Mechanism: 1982–1983 Surge of Variegated Glacier, Alaska," *Science* 227, No. 4686 (February 1, 1985): 469–479.

Lamb, H. H. *Climate: Present, Past and Future.* Vol. 2. *Climate History and the Future.* London: Methuen, 1977.

Ley, Wiley, and the editors of Time-Life Books. *The Poles.* Life Nature Library. New York: Time, 1962.

Oliver, John E., and Rhodes W. Fairbridge, eds. *Encyclopedia of Earth Sciences.* Vol. 11. *The Encyclopedia of Climatology.* New York: Van Nostrand Reinhold, 1987.

Péwé, Troy L. "Alpine Permafrost in the Contiguous United States: A Review," *Arctic and Alpine Research* 15, no. 2 (May 1983): 145–56.

Radok, Uwe. "The Antarctic Ice," *Scientific American* 253, no. 2 (August 1985): 98–105.

Sharp, Robert P. *Living Ice—Understanding Glaciers and Glaciation.* Cambridge: Cambridge University Press, 1988.

Short, Nicholas M., and Robert W. Blair, Jr., eds. *Geomorphology from Space—A Global Overview of Regional Landforms.* Washington, DC: National Aeronautics and Space Administration, 1986.

Stanley, Steven M. *Earth and Life Through Time,* 2nd ed. New York: W. H. Freeman, 1989.

Swithinbank, Charles. *Antarctica.* U.S. Geological Survey Professional Paper 1386 B. Washington, DC: Government Printing Office, 1988.

Williams, Peter J., and Michael W. Smith. *The Frozen Earth—Fundamentals of Geocryology.* Cambridge: Cambridge University Press, 1989.

PART 4

SOILS, ECOSYSTEMS, AND BIOMES

Hoh Rain Forest, Olympic National Park. [*Photo by Jack W. Dykinga.*]

Earth is the home of the only known biosphere in the Solar System—a unique, complex, and interactive system of abiotic and biotic components working together to sustain a tremendous diversity of life. Energy enters the system through conversion of solar energy by photosynthesis in the leaves of plants. Soil is the essential link among the lithosphere, plants, and the rest of Earth's physical systems. Thus, soil helps sustain life. Life is organized into a feeding (trophic) hierarchy from producers to consumers, ending with decomposers. Taken together, the soils, plants, animals, and all abiotic components produce functioning aquatic and terrestrial ecosystems, known as biomes. Specific questions facing society today relate to preservation of the diversity of life in the biosphere and the survival of the biosphere itself—important applied topics considered in Part 4.

Wheat farming, contoured fields, Palouse County, Washington. [*Photo by J. Irwin/Allstock.*]

18

THE GEOGRAPHY OF SOILS

Earth's landscape generally is covered with soil. **Soil** is a dynamic natural material composed of fine particles in which plants grow, and which contains both mineral and organic matter. If you have ever planted a garden, tended a house plant, or been concerned about famine and soil loss, this chapter will interest you. A knowledge of soil is at the heart of agriculture and food production.

Physical and chemical weathering in the upper lithosphere provides the raw mineral ingredients for soil formation. These are called *parent materials,* and their composition, texture, and chemical nature are important. Soils subsequently acquire physical properties that can distinguish them from their source materials as they develop through time, and in response to local relief and topography. Air, water, nutrients, and solid materials all cycle through soil and provide a critical sustaining structure for plants, animals, and human life.

Soil fertility is the ability of soil to support plant productivity. Soil has this ability when it contains organic substances and clay minerals that absorb water and certain elemental ions needed by plants. Much effort and many dollars are expended to create fertile soil conditions, yet the future of Earth's most fertile soils is threatened because soil erosion is increasing worldwide. Some 35% of farmlands are losing soil faster than it can form—a loss exceeding 22.75 billion metric tons (25 billion tons) per year. Production increases from artificial fertilizers and new crop designs partially mask this effect, but such compensations are nearing the end of their usefulness. Soil depletion and loss are at record levels from Iowa to China, Peru to Ethiopia, the Middle East to the Americas. The impact on society probably will be significant.

Soil science is interdisciplinary, involving principles of physics, chemistry, biology, mineralogy, hydrology, taxonomy, climatology, and cartography. Physical geographers are interested in determining the spatial patterns formed by soil types and the physical factors that interact to produce them. As an integrative science, physical geography is well suited for this task.

An important reality is that soil science must deal with a substance whose characteristics vary from kilometer to kilometer, and even centimeter to centimeter. Thus, people interested in soil need information specific to their area. In many locales, an *agricultural extension service* can provide this information and perform a detailed soil analysis. Soil surveys and local soil maps are available for most counties in the United States and the Canadian provinces. (Your local phone book may list the U.S. Department of Agriculture, or Agriculture Canada's Canada Soil Survey, or you may contact the appropriate department at a local college or university.)

SOIL CHARACTERISTICS

Simple theoretical models are the basis for soil classifications. The following model represents general assumptions about soil structure, soil properties, and soil formation processes.

Soil Profiles

Just as a book cannot be evaluated by its cover, soils should not be studied at the surface level only, for this can produce a misleading picture. Instead, a soil profile selected for study should extend from the surface to the lowest extent of plant roots, or to the point where regolith or bedrock is encountered. Such a profile, known as a **pedon,** is imagined as a hexagonal column encompassing from 1 to 10 m^2 in surface area (Figure 18-1). At the sides of the pedon, the various layers of the soil profile are visible in cross section.

A pedon is the basic sampling unit in soil surveys. Many pedons together in one area make up a polypedon, which has distinctive characteristics differentiating it from surrounding polypedons. These **polypedons** are the essential soil individuals, constituting an identifiable series of soils in an area. A polypedon has a minimum dimension of about 1 m^2 and no specified maximum size. It is the soil unit used in preparing local soil maps.

Soil Horizons

Each layer exposed in a pedon is a **soil horizon.** A horizon is roughly parallel to the pedon's sur-

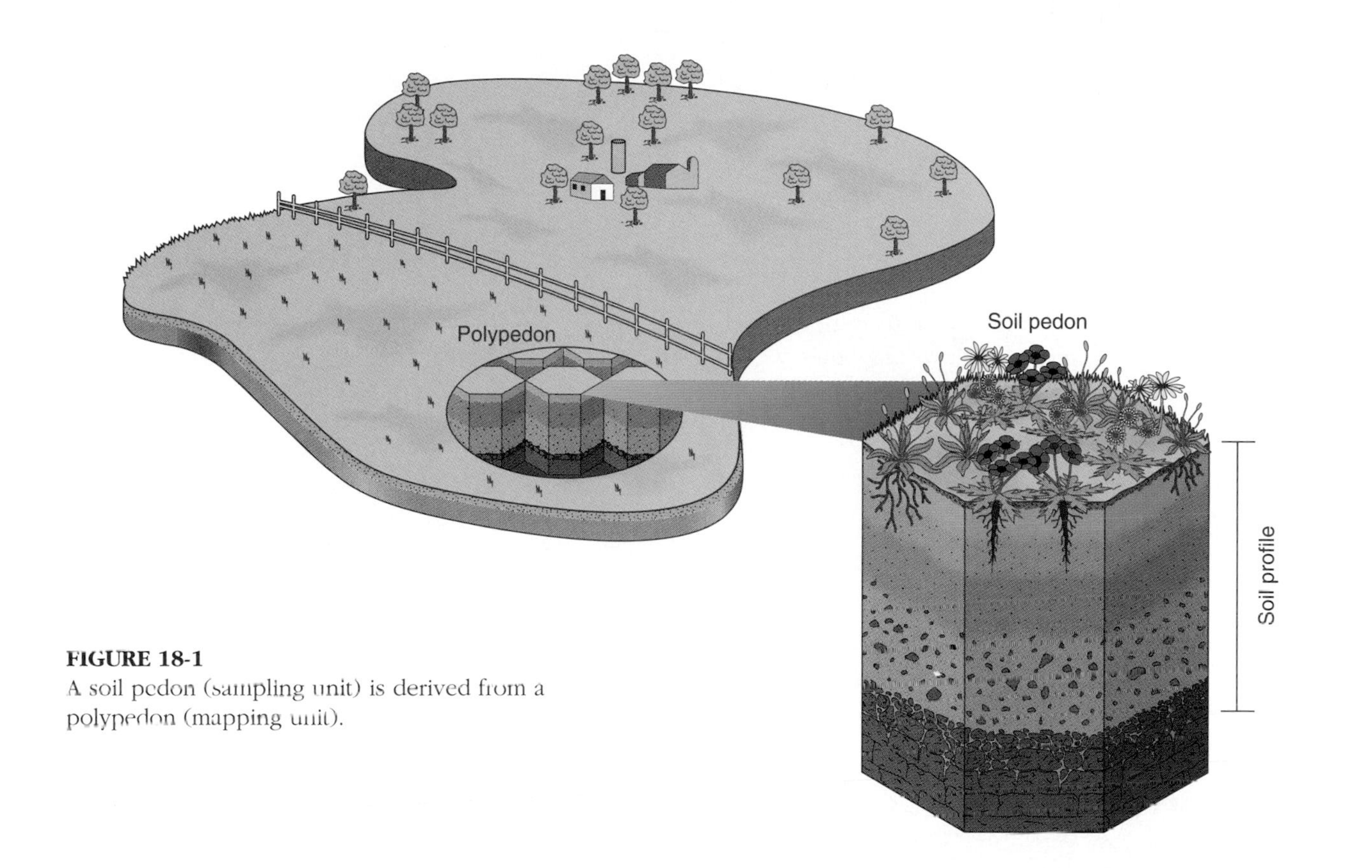

FIGURE 18-1
A soil pedon (sampling unit) is derived from a polypedon (mapping unit).

face and has characteristics distinctly different from horizons directly above or below. The boundary between horizons usually is visible in the field, using the properties of color, texture, structure, consistence (meaning soil consistency), porosity, the presence or absence of certain minerals, moisture, and chemical processes. Soil horizons are the building blocks of soil classification. However, because they can vary in an almost endless number of ways, a simple model is useful. Figure 18-2 presents an ideal pedon and soil profile, with letters assigned to each soil horizon for identification.

At the top of the soil profile is the O horizon, named for its organic composition, derived from plant and animal litter that was deposited on the surface and transformed into humus. **Humus** is a mixture of decomposed organic materials in the soil and is usually dark in color. Consumer and decomposer microorganisms are busy at work on this organic debris, performing a portion of the humification (humus-making) process. The O horizon is 20–30% or more organic matter, important because of its water-absorbing ability and nutrients.

At the bottom of the soil profile is the R horizon, representing either unconsolidated material or con-

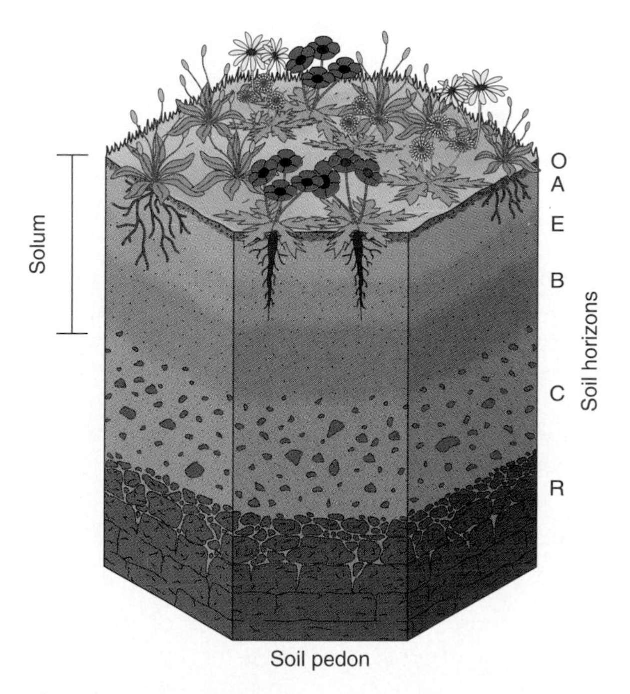

FIGURE 18-2
Typical O, A, E, B, C, and R soil horizons within a developed soil pedon.

solidated bedrock of granite, sandstone, limestone, or other rock. When bedrock weathers to form regolith, it may or may not contribute to overlying soil horizons.

The A, E, B, and C horizons mark differing mineral strata between O and R. These middle layers are composed of sand, silt, clay, and other weathered by-products. (In Chapter 11, Table 11-3 presents a description of grain sizes of these weathered soil particles.)

In the A horizon, the presence of humus and clay particles is particularly important, for they provide essential chemical links between soil nutrients and plants. The A horizon usually is darker and richer in organic content than lower horizons. The lower portion of this horizon grades into the E horizon, which is a bit more pale and is made up of coarse sand, silt, and resistant minerals. Clays and oxides of aluminum and iron are leached (removed) from the E horizon and migrate to lower horizons with water as it percolates through the soil. This process of rinsing through upper horizons and removing finer particles and minerals is termed **eluviation;** thus the designation E for this horizon. The greater the precipitation in an area, the higher the rate of eluviation that occurs in the E horizon.

In contrast to the A and E horizons, B horizons demonstrate an accumulation of clays, aluminum, iron, and possibly humus. These horizons are dominated by **illuviation,** a depositional process. B horizons may exhibit reddish or yellowish hues because of the illuviated presence of mineral and organic oxides. Some materials occurring in this soil zone also may have formed in place from weathering processes rather than arriving there by translocation, or migration. In the humid tropics, these layers often develop to some depth. The combination of the A and E horizons with their eluviation removals and the B horizon with its illuviation accumulations is designated the **solum,** considered the true definable soil of the pedon. The A, E, and B horizons are most representative of active soil processes.

The C horizon is weathered bedrock or weathered parent material, excluding the bedrock itself. This zone is identified as *regolith* (although this term sometimes is used to include the solum as well). The C horizon is not much affected by soil operations in the solum and lies outside the biological influences experienced in the shallower horizons. Plant roots and soil microorganisms are rare in the C horizon; it lacks clay concentrations and generally is made up of carbonates, gypsum, or soluble salts, or of iron and silica, which form cemented soil structures. In dry climates, calcium carbonates commonly form the cementing material of these hardened layers.

Soil scientists using the U.S. system employ letter suffixes to further define special conditions within each soil horizon—for example, Ap or Bt (Table 18-1). Soil horizon and horizon suffix designations used in the Canadian System of Soil Classification are presented later in this chapter. Before looking further at soil horizons, let us examine soil properties.

Soil Properties

To help identify some of these soil properties, it is helpful if you can observe an actual soil profile. A good place to see a soil profile is at an active construction site or excavation, perhaps on your campus, or at a roadcut.

Soil Color. Color is important, for it sometimes reflects composition and chemical makeup. If you look at exposed soil in a garden or along a highway roadcut, color may be the most obvious trait. Among the many possible hues are the red and yellow soils of the southeastern United States, which are high in ferric (iron) oxides; the black-prairie, richly organic soils of portions of the U.S. grain-growing regions and Ukraine; or the white-to-pale soils attributable to the presence of aluminum oxides and silicates. Color can be deceptive too, for soils of high humus content are often dark, yet clays of warm-temperate and tropical regions with less than 3% organic content are some of the world's blackest soils.

To eliminate as much subjectivity as possible and to improve accuracy, soil scientists describe a soil's color by comparing it to a *Munsell Color Chart* (developed by artist and teacher Albert Munsell in 1913). These charts display 175 different colors arranged by *hue* (the dominant spectral color, like red), *value* (degree of darkness or lightness), and *chroma* (purity and strength of the

TABLE 18-1
Soil Horizon Suffixes

Symbol	Description
b	A buried soil horizon; soils that resulted from some previous set of soil-forming processes and that now are buried beneath another solum
ca	Accumulation of carbonates of alkaline elements, commonly calcium; usually lime ($CaCO_3$) or dolomite ($MgCO_3$) in concentrations greater than that of the parent material; may appear in A, B, or C horizons
f	Frozen soil; applied to layers in the soil that are permanently frozen, as in permafrost
g	Strong gleying; condition usually found in stagnant, poorly aerated water; usually dull gray or blue-gray because of the removal of oxygen from iron compounds; may appear mottled or spotted
h	Accumulations of humus; distinct coating of sand and silt particles with illuviated humus or organic material in the B horizon; appears as a dark coating
ir	Accumulation of iron; distinct coating or cementing of silts and sands with illuviated iron; appears reddish in color
m	Strong cementation; applied to hardened horizons that are more than 90% continuous, such as caliche or desert hardpan; not applicable to the firmness associated with fragipans
p	Plowing or other disturbance; used with the A horizon to designate disturbance by cultivation or pasturage; Ap used even when plowing may have gone into what was once B horizon
sa	Accumulation of soluble salts; similar to ca, in that salts are present in an amount greater than that of parent materials
t	Illuviated clay, accumulations of silicate clays as masses, pore fillings, or coatings on sand and silt within B horizons; transported there from horizons above
x	Fragipan formation; firmness or brittleness found in A2, B, and C horizon

SOURCE: *Soil Taxonomy*, Agricultural Handbook No. 436, U.S. Department of Agriculture, 1975.

color saturation, which increases with decreasing grayness). Each color is identified by a name and a Munsell notation, and checked at various depths within a pedon.

Soil Texture. Soil texture, perhaps a soil's most permanent attribute, refers to the size and organization of particles in the soil. Individual mineral particles are called *soil separates;* those smaller than 2 mm (0.08 in.) in diameter, such as very coarse sand, are considered part of the soil, whereas larger particles are identified as pebbles, gravel, or cobbles.

Figure 18-3 shows a soil triangle, with sand, silt, and clay concentrations distributed along each side. Each corner of the triangle represents a soil consisting solely of the particle size noted (rarely are true soils composed of a single separate). The figure includes the common designation **loam**, which is a balanced mixture of sand, silt, and clay beneficial to plant growth. A sandy loam with clay content below 30% usually is considered ideal by farmers because of its water-holding characteristics and cultivation ease.

For example, one soil type found in Indiana is called the Miami silt loam. Samples from it are plotted on the soil texture triangle (points 1, 2, and 3). A sample taken near the surface in the A horizon is recorded at point 1; in the B horizon at point 2; and in the C horizon at point 3. Textural analyses of these samples are summarized in Table 18-2. Note that silt dominates the surface, clay the B horizon, and sand the C horizon.

The U.S. Department of Agriculture *Soil Survey Manual* presents guidelines for estimating soil texture by feel, a relatively accurate method when used by an experienced individual. However, laboratory methods using graduated sieves and separation by sedimentation in water allow more precise measurements. Soil texture is important in determining water retention and water transmission traits.

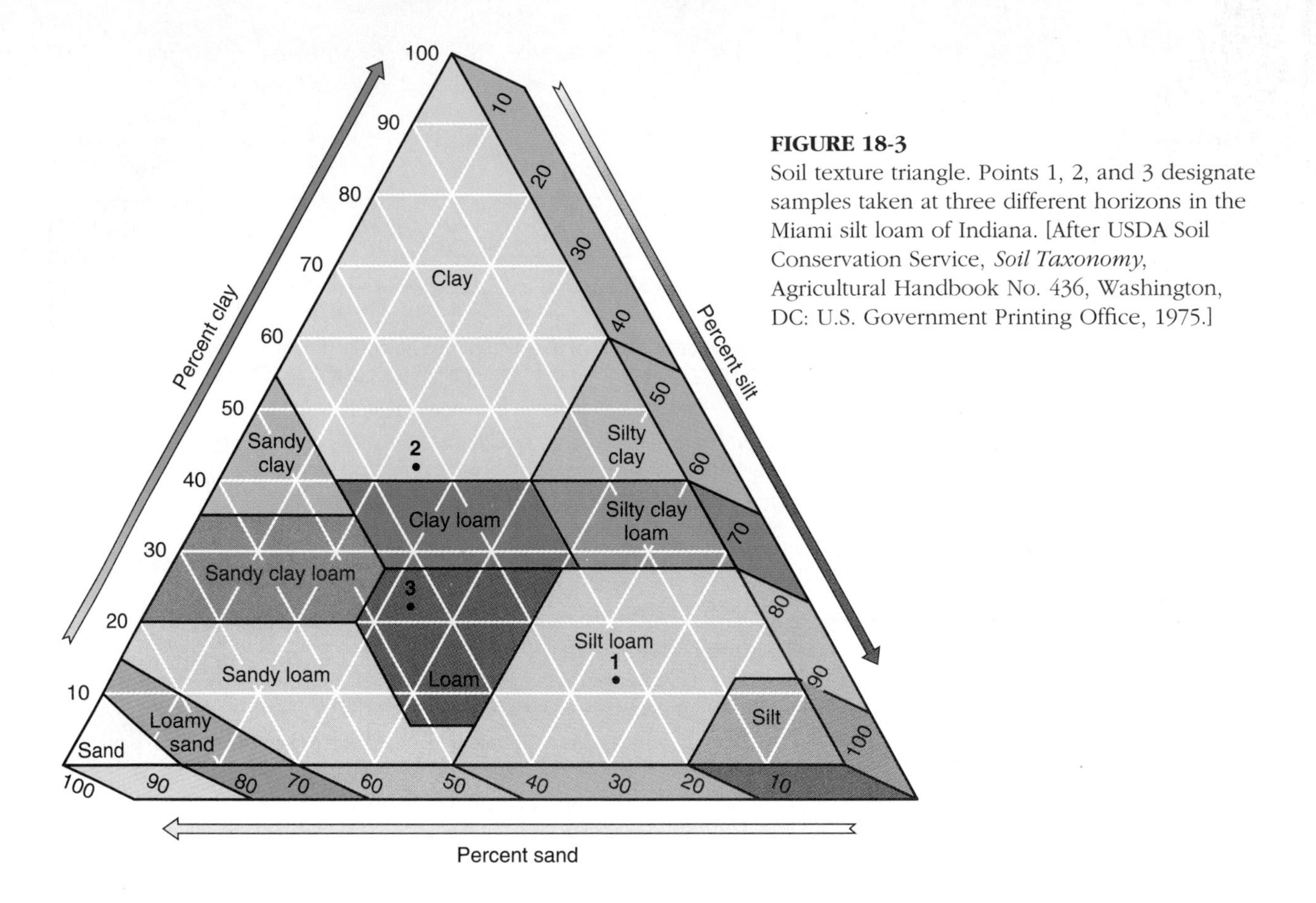

FIGURE 18-3
Soil texture triangle. Points 1, 2, and 3 designate samples taken at three different horizons in the Miami silt loam of Indiana. [After USDA Soil Conservation Service, *Soil Taxonomy,* Agricultural Handbook No. 436, Washington, DC: U.S. Government Printing Office, 1975.]

Soil Structure. Soil structure refers to the *arrangement* of soil particles. Structure is important because it can partially modify the effects of soil texture. The term *ped* describes an individual unit of soil particles; it is a tiny natural lump or cluster of particles held together. The shape of these peds determines which of the structural types the soil exhibits (Figure 18-4). Peds separate from each other along zones of weakness, creating voids that are important for moisture storage and drainage. Spheroidal peds (peds that are rounded) have more pore space and greater permeability. They are therefore more productive for plant growth than are coarse, blocky, prismatic, or platy peds, despite comparable fertility. Terms used to describe soil structure include fine, medium, or coarse, with structural grades of adhesion within aggregates ranging from weak, to moderate, to strong.

Soil Consistence. *Soil consistence,* which refers to cohesion in soil, is a product of texture and structure. (In soil science, the term *consistence* is used to describe the consistency of soil particles.) Consistence reflects a soil's resistance to mechanical stress and manipulation under varying moisture

TABLE 18-2
Textural Analysis of Miami Silt Loam

Sample Points	% Sand	% Silt	% Clay
1 = A horizon	21.5	63.4	15.0
2 = B horizon	31.1	25.0	43.4
3 = C horizon	42.4	34.0	23.5

SOURCE: Joseph E. Van Riper, *Man's Physical World,* p. 570. Copyright © 1971 by McGraw-Hill. Adapted by permission.

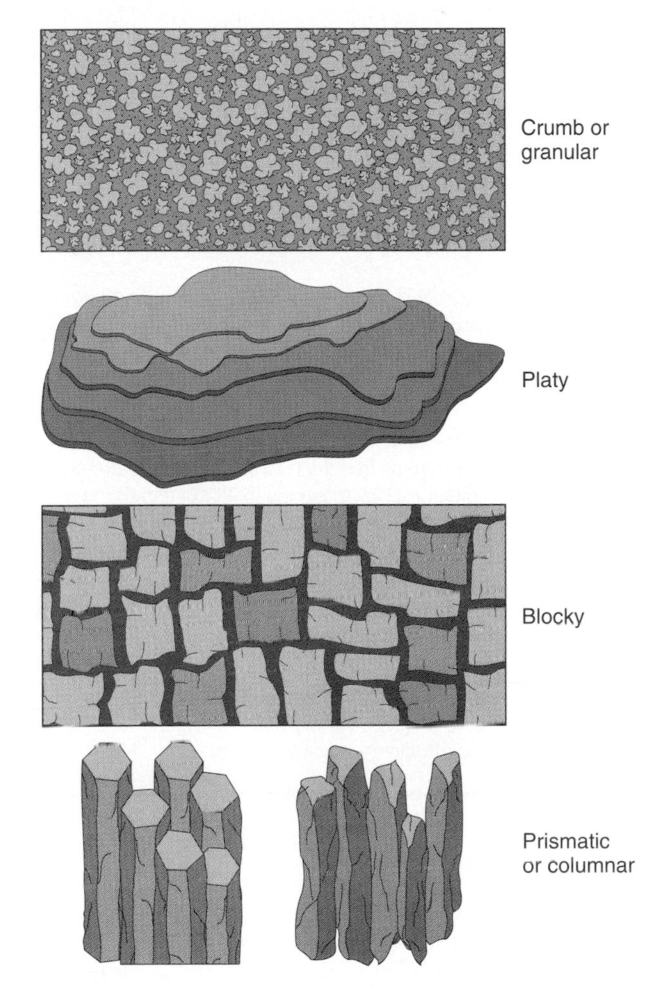

FIGURE 18-4
Types of soil structure.

conditions. Wet soils are variably sticky when held between the thumb and forefinger, ranging from a little adherence to either finger, to sticking to both fingers, to stretching when the fingers are moved apart. Plasticity, the quality of being molded, is roughly measured by rolling a piece of soil between your fingers and thumb to see whether it rolls into a thin strand.

Moist soil is soil that is filled to about half of field capacity, and its consistence grades from loose (noncoherent), to *friable* (easily pulverized), to firm (not crushable between thumb and forefinger). Finally, a dry soil is typically brittle and rigid, with consistence ranging from loose, to soft, to hard, to extremely hard.

The cementation of soil particles that occurs in various horizons is a function of consistence and usually is described as continuous or discontinuous. Soils are noted as weakly cemented, strongly cemented, or *indurated* (hardened). Calcium carbonate, silica, and oxides or salts of iron and aluminum all can serve as cementing agents.

Soil Porosity. Soil porosity, permeability, and moisture storage are discussed in Chapter 9. Pores, or voids, in the soil horizon control the flow of water, its intake and drainage, and air ventilation. In addition to pore size, other factors are important: *continuity* (whether the pores are interconnected), shape (whether pores are spherical, irregular, or tubular), *orientation* (whether tubular pore spaces are vertical, horizontal, or random), and *pore location* (whether within or between soil peds). Porosity is increased and improved by the biotic actions of plant roots, animal activity such as the tunneling of gophers or worms, and human intervention through soil manipulation (plowing, adding humus or sand, or planting special crops). When you work in the garden or with house plants, much of your soil preparation work is done to improve soil porosity.

Soil Moisture Regimes. You will find it helpful to review Figures 9-5 and 9-6 to assist your understanding of this section. Recall that plants operate most efficiently when the soil body is at field capacity, and that field capacity is determined by knowing the soil type. The effective rooting depth of a plant determines the amount of soil moisture to which the plant's roots are exposed. If soil moisture is drawn down below field capacity, soil moisture utilization rates become increasingly inefficient until the wilting point is reached, beyond which plants cannot extract the water bonded to each soil particle.

Soil moisture regimes and their associated climate types shape the biotic and abiotic properties of the soil more than any other factor. Based on Thornthwaite's water-balance principles (see "The Water Balance Concept" in Chapter 9), the U.S. Soil Conservation Service recognizes five soil moisture regimes (Table 18-3).

TABLE 18-3
Principal Soil Moisture Regimes

Name	Description
Aquic (L. *aqua,* "water")	The groundwater table lies at or near the surface, so the soil is almost constantly wet, as in bogs, marshes, and swamps, a reducing environment with virtually no dissolved oxygen present. Commonly, groundwater levels fluctuate seasonally; a small borehole will produce standing, stagnant water.
Aridic (torric) (L. *aridis*, "dry," and L. *torridus*, "hot and dry")	Soils in this regime are dry more than half the time, with soil temperatures at a depth of 50 cm (19.7 in.) above 5°C (41°F). In some or all areas, soils are never moist for as long as 90 consecutive days. This regime occurs mainly in arid climates, although where surface structure inhibits infiltration and recharge or where soils are thin and therefore dry, this regime occurs in semiarid regions as well.
Udic (L. *udus,* "humid")	Soils have little or no moisture deficiency throughout the year, specifically during the growing season. The water balance exhibits soil moisture surpluses that flush through the soils during one season of the year. If the water balance exhibits a moisture surplus in all months of the year, the regime is called *perudic,* with adequate soil moisture always available to plants.
Ustic (L. *ustus,* "burnt," implying dryness)	This regime is intermediate between aridic and udic regimes; it includes the semiarid and tropical wet-dry climates. Moisture is available but is limited, with a prolonged deficit period following the period of soil moisture utilization during the early portion of the growing season. Temperature is important in deteriming the moisture efficiency of available water in this borderline moisture regime.
Xeric (L. *xeros,* "dry")	This regime applies to those few areas that experience a Mediterranean climate, dry and warm summer, rainy and cool winter. There are at least 45 consecutive dry days during the four months following the summer solstice. Winter rains effectively leach the soils.

SOURCE: *Soil Taxonomy,* Agricultural Handbook No. 436, U.S. Department of Agriculture, 1975.

Soil Chemistry

The atmospheric component of soil is mostly nitrogen, oxygen, and carbon dioxide. Nitrogen concentrations in the soil are about the same as in the atmosphere. Oxygen is less and carbon dioxide is greater in concentration because of ongoing respiration processes.

Water present in soil pores is the *soil solution.* It is the medium in the soil for chemical reactions. This solution is critical to plants as their source of nutrients and is the foundation of *soil fertility.* Carbon dioxide and various organic constituents combine with water to produce carbonic and organic acids, respectively. These acids are then active in soil processes, as are dissolved alkalies and salts.

To understand how the soil solution behaves, let's go through a quick chemistry review. An *ion* is an atom, or group of atoms, that carries an electrical charge. An ion has either a positive charge or a negative charge. For example, when NaCl (sodium chloride) is dissolved in solution, it separates into two ions: Na^+, a *cation* (positively charged ion), and Cl^- an *anion* (negatively charged ion). Some ions in soil carry single charges, whereas others carry double or even triple charges (e.g., sulfate, SO_4^{2-}; aluminum, Al^{3+}).

Soil colloids are important for retention of ions in soil. These tiny particles of clay and organic material carry a negative electrical charge and consequently are attracted to any positively charged ions in the soil (Figure 18-5). The positive ions, many metallic, are critical to plant growth. If it were not for the soil colloids, the positive ions would be leached away by the soil solution and thus would be unavailable to plant roots.

Clay colloids and organic colloids have different levels of chemical activity. Individual clay colloids are thin and platelike, with parallel surfaces that are negatively charged. They are more chemically active than silt and sand particles but less active than organic colloids. Colloids can exchange cations between their surfaces and the soil solution, an ability called **cation-exchange capacity**

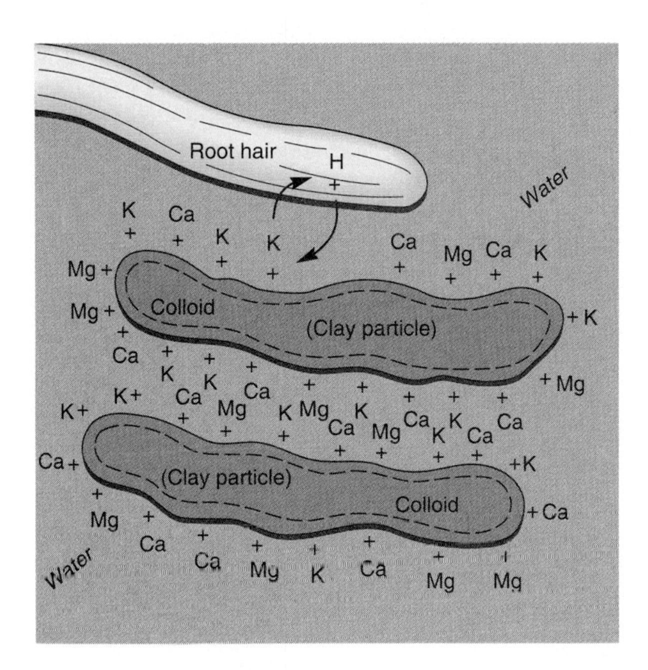

FIGURE 18-5
Typical soil colloids retaining mineral ions by adsorption.

(CEC). A high CEC means that the soil colloids can store or exchange more cations from the soil solution, an indication of good soil fertility (unless there is a complicating factor, such as the soil being too acid). Cations attach to the surfaces of the colloids by **adsorption.** The metallic cations are adsorbed by the soil colloids.

Soil Acidity and Alkalinity. A soil solution may contain significant hydrogen ions (H^+), cations that indirectly stimulate acid formation. These ions will displace critical base cations (such as calcium, magnesium, potassium, and sodium). The result is a soil rich in hydrogen ions: an *acid soil.* On the other hand, a soil high in base cations is a *basic* or *alkaline* soil. Such acidity and alkalinity is expressed on the pH scale (Figure 18-6).

Pure water is nearly *neutral,* with a pH of 7.0. Readings below 7.0 represent increasing acidity. Readings above 7.0 represent increasing alkalinity. Acidity usually is regarded as strong at 5.0 or lower on the pH scale, whereas 10.0 or above is considered strongly alkaline.

Today, the major contribution of acidity to soil is through acid deposition by precipitation. Acid rain actually has been measured below pH 3.0—an incredibly low value for natural precipitation. Because most crops are sensitive to specific pH levels, acid soils below pH 6.0 require treatment to raise the pH. This is done by the addition of bases in the form of minerals that are rich in base cations, usually lime (calcium carbonate). Increased acidity in the soil solution accelerates the chemical weathering of mineral nutrients and increases their depletion rates, as in soils in the region of the northern forests.

Soil Formation Factors and Management

There are three soil-forming factors: dynamic (climatic and biologic), passive (parent material, topography and relief, and time), and the human component. These three factors work together as a system to form soils. The role of parent material in providing weathered minerals to form soils was

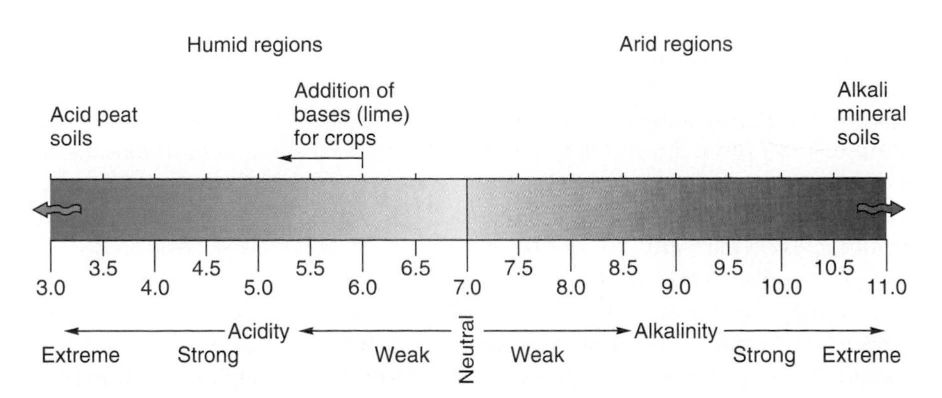

FIGURE 18-6
Soil pH scale, measuring acidity (lower pH) and alkalinity (higher pH).

discussed earlier in this chapter. The roles played by climate and humans are considered in the soil order discussions that follow.

Climate types correlate closely with soil types worldwide. The moisture, evaporation, and temperature regimes of climates determine the chemical reactions, organic activity, and eluviation rates of soils. Not only is the present climate important, but many soils exhibit the imprint of past climates, sometimes over thousands of years. Most notable is the effect of glaciations, which among other contributions produced the loess soil materials that have been transported thousands of kilometers by wind to their present locations, as described and mapped in Chapter 15.

The organic content of soil is determined by vegetation growing in that soil and by animal and bacterial activity. The chemical makeup of the vegetation contributes to the acidity or alkalinity of the soil solution. For example, broadleaf trees tend to increase alkalinity, whereas needleleaf trees tend to produce higher acidity. Thus, when civilization moves into new areas and alters the natural vegetation by logging or plowing, the affected soils are likewise altered. In many cases they are permanently changed.

Landforms also affect soil formation, mainly through slope and orientation. Slopes that are too steep cannot have full soil development because gravity and erosional processes remove materials. Slopes that are nearly level inhibit soil drainage. As for orientation, in the Northern Hemisphere a southern slope exposure is warmer (slope faces the southern Sun), which affects water balance relationships. North-facing slopes are colder, causing slower snowmelt and lower evaporation rates, which result in more moisture for plants than on south-facing slopes.

Human intervention has a significant impact on slope utilization. Millennia ago, farmers in every culture on Earth learned to plow slopes "on the contour"—to plow around a slope at the same elevation, not vertically up and down the slope (see the chapter opening photo). The reasons behind contour plowing are to reduce *sheet erosion* (which removes topsoil) and simultaneously to improve water retention in the soil. However, the demands of today's larger farm equipment have forced farmers to alter their practice of contouring because the equipment cannot make the tight turns that such proper plowing requires. Other effective soil-holding practices also are being eliminated to accommodate the larger equipment; methods in decline include terracing (similar to contour plowing), planting tree rows as wind breaks, and planting shelter belts.

All of the identified factors in soil development (climate, biological activity, parent material, landforms and topography, and human activity) require *time* to operate. A few centimeters thickness of prime farmland soil may require *500 years* for maturation. Yet these same soils are being lost at a few centimeters *per year* to sheetwash and gullying when the soil-holding vegetation is removed. Over the same period, exposed soils may be completely leached of needed cations, thereby losing their fertility.

The U.S. General Accounting Office recently estimated that from 3 to 5 million acres of prime farmland are lost each year in the United States through mismanagement or conversion to nonagricultural uses. A 1984 report to the Canadian Environmental Advisory Council estimated that the organic content of cultivated prairie soils has declined by as much as 37–48% compared with noncultivated native soils. In Ontario and Quebec, losses of organic content increased to as much as 50%, and losses are even higher in the Atlantic Provinces, which were naturally low in organic content at the outset of cultivation.

About half of all cropland in the United States and Canada is experiencing excessive rates of soil erosion. This assessment is mirrored worldwide, where approximately 1.2 billion hectares (3 billion acres) of soil are suffering from human-induced erosion and loss to nonagricultural uses, depletion, and contamination (Figure 18-7).

Human influence is ever-present, and our spatial impact on the precious soil resource must be recognized, with protective measures taken. There is a need for cooperative international action.

SOIL CLASSIFICATION

Classification of soils is complicated by the variety of interactions that create thousands of distinct soils—well over 15,000 soil types in the United States and Canada alone. Not surprisingly, a num-

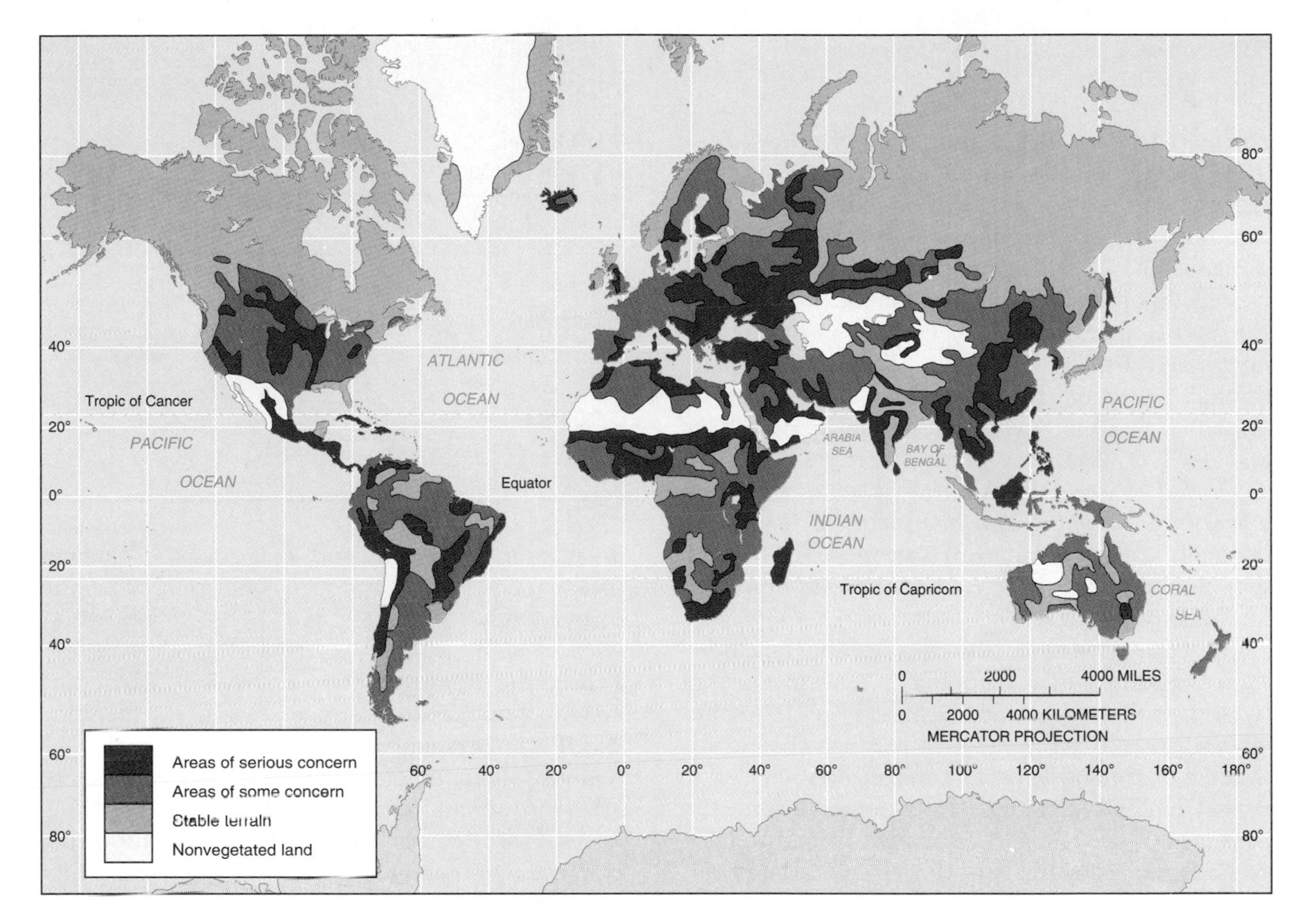

FIGURE 18-7
Approximately 1.2 billion hectares (3.0 billion acres) of Earth's soils suffer some degree of degradation through erosion caused by human misuse and abuse. [*A Global Assessment of Soil Degradation*, adapted from United Nations Environment Programme, International Soil Reference and Information Centre, "Map of Status of Human-induced Soil Degradation," Sheet 2, Nairobi, Kenya, 1990.]

ber of different classification systems are in use worldwide. The United States, Canada, the United Kingdom, Germany, Australia, Russia, and the United Nations Food and Agricultural Organization (FAO) each has its own soil classification system. In particular, the National Soil Survey Committee of Canada developed a system suited to its great expanses of boreal forest, tundra, and cooler climatic regimes, detailed later.

A Brief History of Soil Classification

In 1883 the Russian soil scientist V. V. Dukuchaev published a monograph that organized soils into rough groups based on observable properties, most of which resulted from climatic and biological soil-forming processes. The Russians contributed greatly to modern soil classification because they were first to consider soil an independent natural body with a definite individual soil genesis that could be organized in an orderly global pattern. By contrast, American and European scientists were still considering soil as a geologic product, or simply a mixture of chemical compounds. The Russians determined that broad interrelationships exist among the physical environment, vegetation, and soils.

The first formal U.S. system is credited to Dr. Curtis F. Marbut, who derived his approach from Dukuchaev. Published in the 1935 *Atlas of Ameri-*

can Agriculture and the *Yearbook of Agriculture—1938,* Marbut's classification of great soil groups recognized the importance of soils as products of dynamic natural processes. However, as the database of soil information grew, the inadequacies of this system became apparent. In 1951 the U.S. Soil Conservation Service began researching a new soil classification system. That process went through seven exhaustive reviews, culminating in what is called the Seventh Approximation in 1960 and 1964. The current version was published in 1975: *Soil Taxonomy—A Basic System of Soil Classification for Making and Interpreting Soil Surveys,* generally called simply **Soil Taxonomy.** Much of the information in this chapter is derived from that keystone publication.

Canadian efforts at soil classification began in 1914 with the partial mapping of soils in Ontario by A. J. Galbraith. Efforts to develop a taxonomic system spread across the country, anchored by academic departments at universities in each province. Regional differences emerged, hampered by a lack of specific soil details. By 1936 only 1.7% of Canadian soil, totaling 15 million hectares (37 million acres), was surveyed.

Canadian scientists needed a taxonomic system based on observable and measurable properties in soils specific to Canada. This meant a departure from efforts in the United States, as formalized by Marbut in 1938. Canada's first taxonomic soil classification system was introduced in 1955, splitting away from the United States during the Fourth Approximation stage of development. Classification work progressed through the Canada Soil Survey Committee after 1970 and was replaced by the Expert Committee on Soil Survey in 1978, all under Agriculture Canada.

U.S. Soil Taxonomy. The U.S. Soil Taxonomy system is based on observable soil properties actually seen in the field. Thus, it is open to addition, change, and modification as the sampling database grows. The classification system divides soils into six categories, creating a hierarchical sorting system (Table 18-4). Each *soil series* (the smallest, most detailed category) ideally includes only one polypedon, but may include portions continuous with adjoining polypedons in the field. An important aspect of the system is that it recognizes the interaction of humans and soils and the changes that humans have introduced, both purposely and inadvertently.

TABLE 18-4
U.S. Soil Taxonomy

Category	Number Included
Soil orders	10
Suborders	47
Great groups	230
Subgroups	1200
Families	6000
Soil series	13,000

Pedogenic Regimes. Before the Soil Taxonomy system, **pedogenic regimes** were used, featuring specific soil-forming processes keyed to climatic regions. Although each pedogenic process may be active in several soil orders and in different climates, we discuss them within the soil order where they commonly occur. The principal pedogenic regimes are:

- *laterization:* a leaching process in humid and warm climates, discussed with Oxisols.
- *salinization:* a process that concentrates salts in soils in areas with excessive POTET rates, discussed with Aridisols.
- *calcification:* a process that produces an illuviated accumulation of calcium carbonates, discussed with Mollisols and Aridisols.
- *podzolization:* a process of soil acidification associated with forest soils in cool climates, discussed with Spodosols.

Such climate-based regimes are convenient for relating climate and soil processes. However, *the Soil Taxonomy system recognizes that there is great uncertainty in basing soil classification on such inferences.*

Diagnostic Soil Horizons

To identify specific soil types within the Soil Taxonomy classification, the U.S. Soil Conservation Service describes "diagnostic horizons" in a pedon.

TABLE 18-5
Major Features of Diagnostic Horizons

Name	Derivation	Characteristics
		Surface Horizons (Epipedons)
Mollic (A)	*mollis,* "soft"	Thick, dark-colored, humus-rich; base cation dominant; strong strucuture
Anthropic (A)	*anthropos,* "having to do with humans"	Mollic-like, modified by long and continued human use and addition of water to soil
Umbric (A)	*umbra,* "shade," hence "dark"	Identical to mollic, except low base cation content; dark coloration not related to humus
Histic (O)	*histos,* "tissue"	Thin layer of peat or muck in wet places; saturated at least 30 consecutive days a year
Plaggen (A)	*plaggen,* "sod"	Human-made >50 cm thick, produced by long, continuing manuring (W. Europe)
Ochric (A)	*ochros,* "pale"	Light color, low humus content, hard and massive when dry, too marginal to fit other epipedons
		Subsurface Horizons
Argillic	argilla, "white clay"	Illuviated clay accumulation
Agric	*ager,* "field"	Illuvial layer formed below cultivation
Natric	*natrium,* "sodium"	Argillic that is high in sodium
Spodic	*spodos,* "wood ashes"	Accumulation of organic matter, Fe, and Al
Placic	*plax,* "flatstone"	Thin, dark-reddish, iron-cemented pan
Cambic	*cambiare,* "to exchange"	Altered or changed by physical movement or chemistry
Oxic	*oxide,* "oxide"	Highly weathered mix of Fe, Al oxides; >30 cm thick
Duripan	*durus,* "hard"	Silica-cemented hardpan; not softened by water
Fragipan	*fragilis,* "brittle"	Weakly cemented brittle pan; loam texture
Albic	*albus,* "white"	Light, pale layer lacking clay, Fe, and Al
Calcic	calcium	Accumulation of $CaCO_3$ or $CaCO_3 \cdot MgCO_3$
Gypsic	gypsum	Accumulation of gypsum (hydrous calcium sulfate)
Salic	salts	Accumulation of >2% soluble salts
Plinthite	*plinthos,* "brick"	Iron-rich, humus-poor; becomes ironstone under repeated wet/dry cycles

SOURCE: *Soil Taxonomy,* Agricultural Handbook No. 436, U.S. Department of Agriculture, 1975.

A diagnostic horizon often reflects a distinctive physical property (color, texture, structure, consistence, porosity, moisture), or a dominant soil process.

In the solum, two diagnostic horizons may be identified: the epipedon and the subsurface. The **epipedon** (literally, over the soil) is the diagnostic horizon that forms at the surface and may extend through the A horizon, even including all or part of an illuviated B horizon. The epipedon is visibly darkened by organic matter and sometimes is leached of minerals. There are six recognized epipedons (Table 18-5). Not included in the epipedon are alluvial deposits, eolian deposits, or areas disturbed by cultivation. They are excluded because soil-forming processes have not had sufficient time to erase these transient characteristics.

The second type of diagnostic horizon is the **subsurface diagnostic horizon.** It originates below the surface at varying depths and may include part of the A and/or B horizons. Many subsurface diagnostic horizons have been identified.

Table 18-5 summarizes both surface and subsurface diagnostic horizons and their general characteristics, as described in the Soil Taxonomy. The presence or absence of either diagnostic horizon usually distinguishes a soil for classification.

The Eleven Soil Orders of Soil Taxonomy

At the heart of the Soil Taxonomy are eleven general soil orders, which are listed in Table 18-6 along with other helpful information. Their worldwide distribution is shown in Figure 18-8. You will

TABLE 18-6
Soil Taxonomy Soil Orders

Order	Pronunciation	Derivation of Term	Marbut Equivalent (Canadian)	General Location and Climate	World Land Area (%)	U.S. Land Area (%)	Description
Oxisols	*ox*	Fr. *oxide,* "oxide" Gr. *oxide,* "acid or sharp"	Latosols lateritic soils	Tropical soils, hot humid areas	9.2	0.02	Maximum weathering and eluviation, oxic horizon, continuous plinthite layer
Aridisols	*arid*	L. *aridos,* "dry"	Reddish desert, gray desert, sierozems	Desert soils, hot dry areas	19.2	11.5	Limited alteration of parent material, low climate activity, ochric epipedon, light color, subsurface illuviation of carbonates
Mollisols	*mollify*	L. *mollis,* "soft"	Chestnut, chernozem (Chernozemic)	Grassland soils; subhumid, semiarid lands	9.0	24.6	Noticeably dark with organic material, base saturation high, mollic epipedon; friable surface with well-structured horizons
Alfisols	*alf*alfa	Invented syllable	Gray-brown podzolic, degraded chernozem (Luvisol)	Moderately weathered forest soils, humid temperate forests	14.7	13.4	B horizon high in clays, moderate to high degree of base saturation, orgillic or natric horizons, no pronounced color change with depth
Ultisols	*ulti*mate	L. *ultimus,* "last"	Red-yellow podzolic, reddish-yellow lateritic	Highly weathered forest soils, subtropical forests	8.5	12.9	Similar to Alfisols, B horizon high in clays, generally low amount of base saturation, strong weathering in orgillic horizons, redder than Alfisols,

Spodosols	*odd*	Gr. *spodos* or L. *spodus,* "wood ash"	Podzols, brown podzolic (Podzol)	Northern conifer forest soils, cool humid forests	5.4	5.1	Illuvial B horizon of Fe/Al clays, humus; without structure, partially cemented; spodic horizon; highly leached, strongly acid; coarse texture of low bases
Entisols	rec*ent*	Invented syllable	Azonal soils, tundra	Recent soils, profile undeveloped, all climates	12.5	7.9	Limited development; inherited properties from parent material; ochric epipedon; few specific properties; young soils lacking horizons
Inceptisols	*incept*ion	L. *inceptum,* "beginning"	Ando, subarctic brown forest, lithosols, some humic gleys (Brunisol, Cryosol with permafrost, Gleysol wet)	Weakly developed soils, humid regions	15.8	18.2	Intermediate development; embryonic soils, cambic horizon, but few diagnostic features; further weathering possible
Vertisols	in*vert*	L. *verto,* "to turn"	Grumusols (1949) tropical black clays	Expandable clay soils; subtropics, tropics; sufficient dry period	2.1	1.0	Forms large cracks on drying, self-mixing action, contains >30% in swelling clays, ochric epipedon
Histosols	*histo*logy	Gr. *histos,* "tissue"	Peat, muck, bog (Organic)	Organic soils, wet places	0.8	0.5	Peat or bog, >20% organic matter, muck with clay >40 cm thick, surface organic layers, no diagnostic horizons
Andisols	*and, and*esite	L. *Ando,* "volcanic ash"	—	Areas affected by frequent volcanic activity (Formerly within Inceptisols and Entisols)	<1.0	<1.0	Volcanic parent materials, particularly ash and volcanic glass; weathering and mineral transformation important; high CEC and organic content, generally fertile.

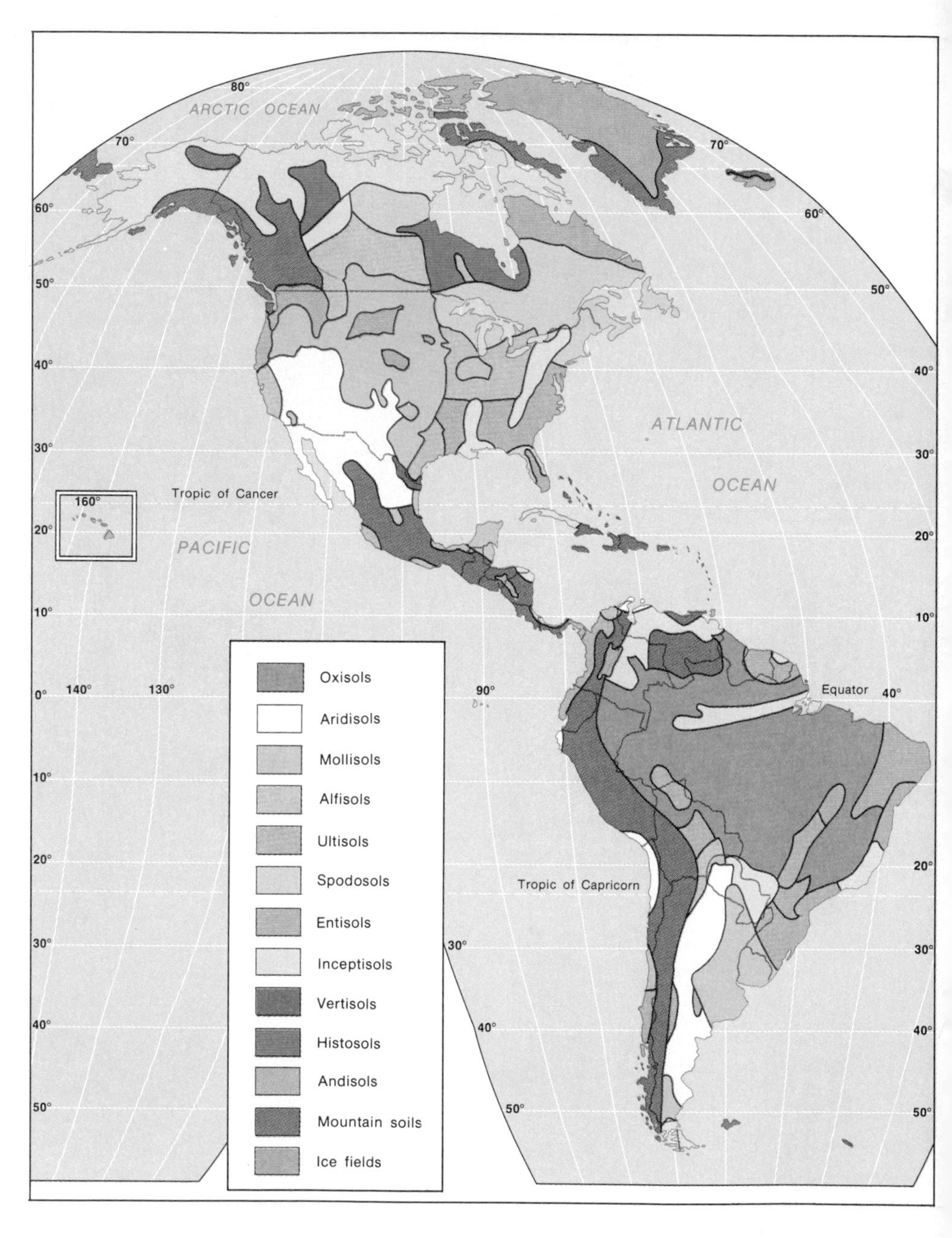

FIGURE 18-8
Worldwide distribution of the 11 soil orders. Volcanic ash related soils are finely distributed and not completely represented on this map. [Adapted from U.S. Soil Conservation Service.]

find this table and map useful in the following descriptions. Because the taxonomy evaluates each soil order on its own characteristics, there is no priority to the classification. You will find the 11 orders loosely arranged here by latitude, beginning along the equator as is done in Chapter 10, "Global Climate Systems," and Chapter 20, "Terrestrial Biomes."

Oxisols. The remarkable moisture and temperature intensity and uniform daylength of equatorial latitudes greatly affect soils. These generally old landscapes, exposed for many thousands to hundreds of thousands of years, are deeply developed and have greatly altered minerals (except in certain newer volcanic soils in Indonesia—Andisols). The soil structure at these latitudes usually lacks dis-

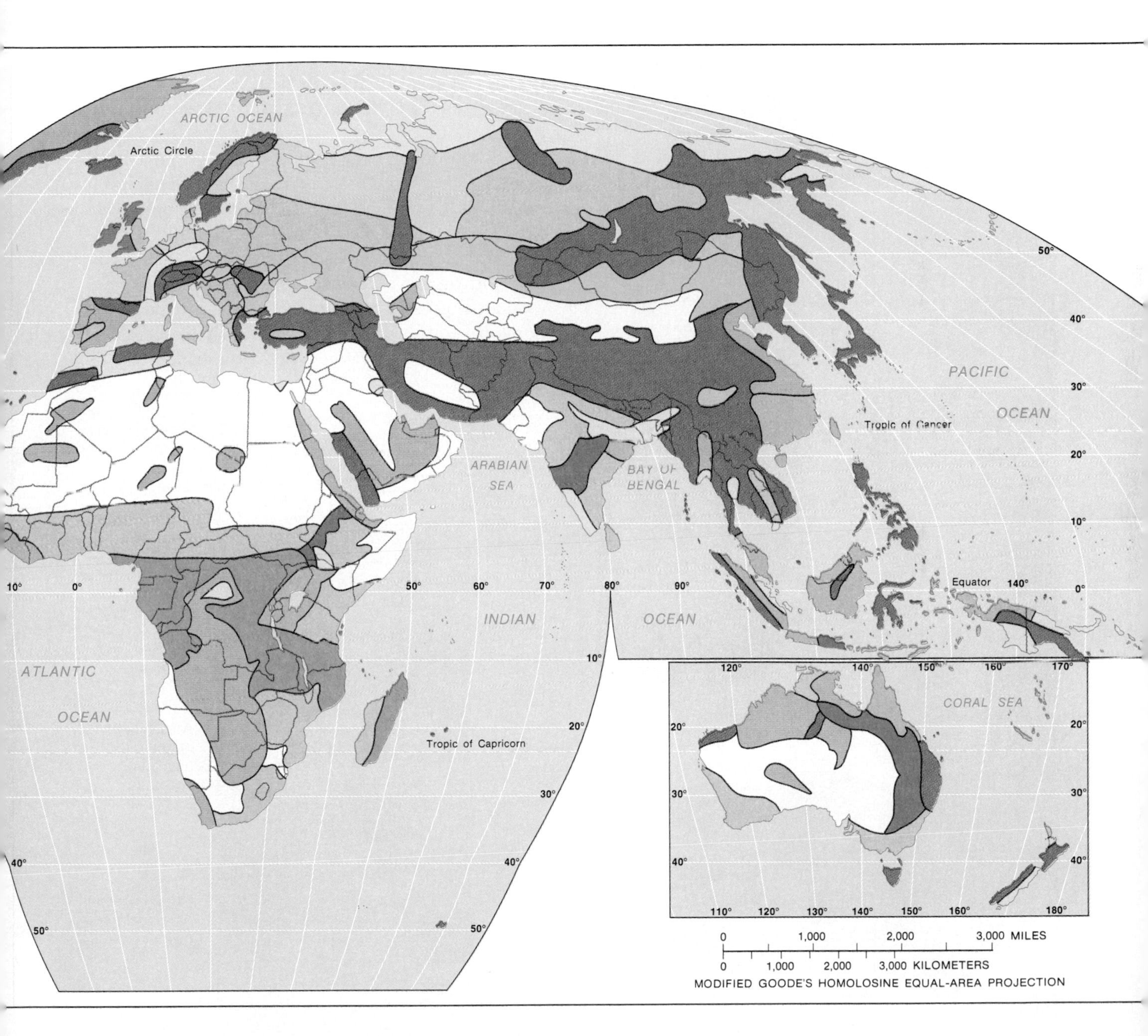

tinct horizons where these soils are well drained (Figure 18-9). These soils are called **Oxisols** (tropical soils) and feature five suborders. Related vegetation is the luxuriant and diverse tropical and equatorial rain forest.

The large amount of precipitation leaches soluble minerals and soil constituents (such as silica) from the A horizon. Typical soils are reddish and yellowish from the iron and aluminum oxides left behind, with a weathered clay-like texture, sometimes in a granular soil structure that is easily broken apart. The high degree of eluviation removes base cations and colloidal material to illuviated horizons. Thus, Oxisols are low in CEC (cation-exchange capacity) and fertility, except in regions augmented by alluvial or volcanic materials. Fig-

(a)

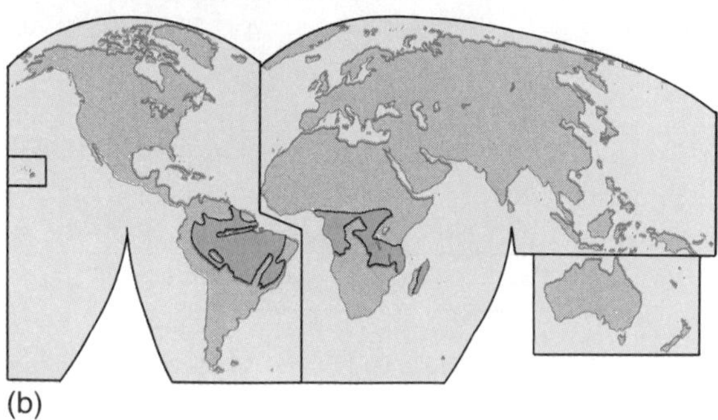

(b)

FIGURE 18-9
Oxisol. (a) Deeply weathered Oxisol profile in central Puerto Rico and (b) general map of worldwide distribution. [Photo from Marbut Collection, Soil Science Society of America, Inc.]

ure 18-10 illustrates **laterization,** the leaching process described here, which operates in well-drained soils in warm, humid tropical and subtropical climates.

The subsurface diagnostic horizon in an Oxisol is highly weathered, containing iron and aluminum oxides, at least 30 cm (12 in.) thick and within 2 m (6.5 ft) of the surface (Figure 18-9). If these horizons are exposed and subjected to repeated wetting and drying sequences, an *ironstone hardpan* (or hardened soil layer in the lower A or in the B horizon) soil structure called a **plinthite** results (from the Greek *plinthos,* meaning "brick"). This

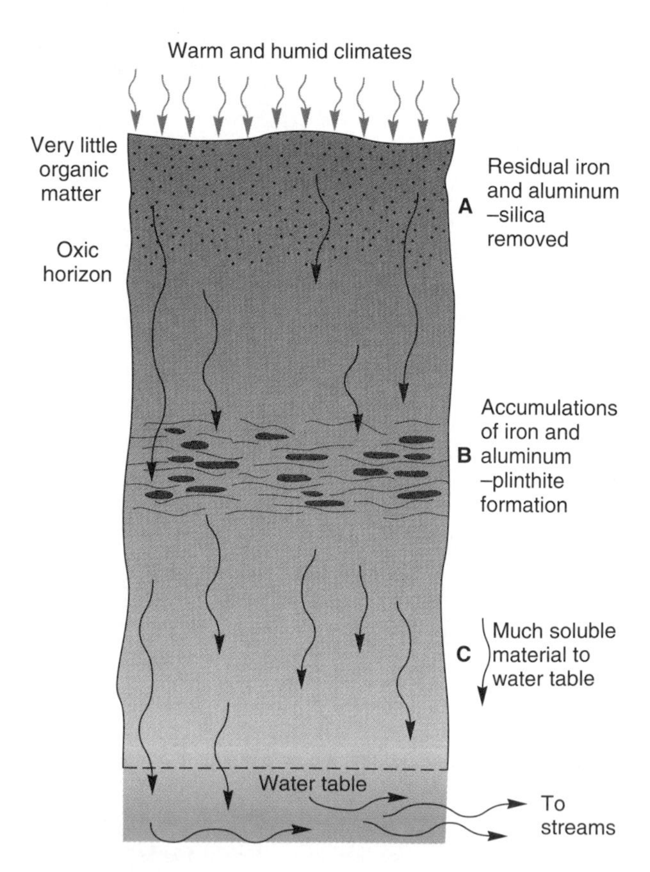

FIGURE 18-10
Laterization process characteristic of tropical and subtropical climate regimes.

FIGURE 18-11
Plinthite quarrying in India. [Photo by Henry D. Foth.]

form of soil, also called a *laterite,* can be quarried in blocks and used as a building material (Figure 18-11).

Simple agricultural activities can be conducted in these soils with care. Early slash-and-burn shifting cultivation practices were adapted to these soil conditions and formed a unique style of crop rotation. The scenario went like this: people in the tropics cut down (slashed) and burned the rain forest in small tracts and then cultivated the land with stick and hoe. After several years the soil lost fertility by leaching, and the people moved on to the next tract to repeat the process. After many years of movement from tract to tract, the group returned to the first patch to begin the cycle again. This practice protected the limited fertility of the soils somewhat, allowing periods of recovery to follow active production.

However, this orderly native pattern of land rotation was halted by the invasion of foreign plantation interests, development by local governments, vastly increased population pressures, and conversion of vast new tracts to pasturage. Permanent tracts of cleared land put tremendous pressure on the remaining tracts of forest and brought disastrous consequences. When Oxisols are disturbed in such a regime of high temperature and moisture, soil loss can exceed a thousand tons per square kilometer per year, not to mention the greatly increased rate of extinction of plant and animal species that accompanies such destruction. The regions dominated by the Oxisols and rain forests are rightfully the focus of much worldwide environmental attention at this time.

Aridisols. The largest single soil order occurs in the world's deserts. **Aridisols** (desert soils) occupy approximately 19% of Earth's land surface and some 12% of the land in the United States (Figure 18-8). A pale, light soil color is characteristic of the ochric epipedon, and a large variety of subsurface diagnostic horizons can be identified (Figure 18-12).

Not surprisingly for desert soils, the water balance in Aridisol regions has marked periods of soil moisture deficit and generally inadequate soil moisture for plant growth. High annual values of potential evapotranspiration and low amounts of precipitation produce very shallow development of soil horizons. Usually there is no period greate

(a)

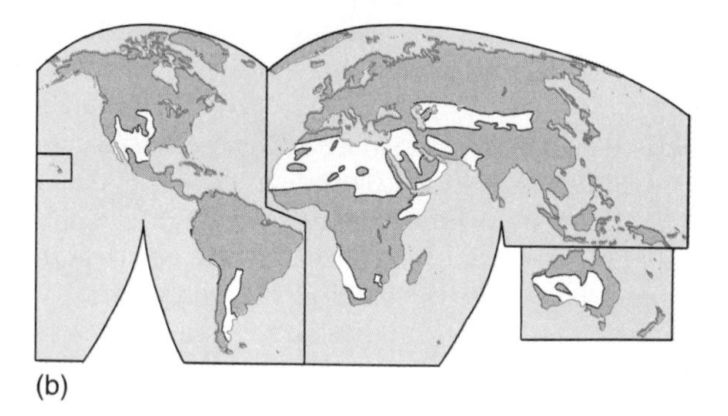

(b)

FIGURE 18-12
Aridisol. (a) Profile from central Arizona and (b) general map of worldwide distribution. [Photo from Marbut Collection, Soil Science Society of America, Inc.]

than three months when the soils have adequate moisture. Aridisols also lack organic matter of any consequence. Although the soils are not leached often because of the low precipitation, they are leached easily when exposed to excessive water, for they lack a significant colloidal structure.

A soil process that occurs in Aridisols and nearby soil orders is salinization. **Salinization** results from excessive POTET rates in the deserts and semiarid regions of the world. Salts dissolved in soil water are brought to surface horizons and deposited there as surface water evaporates. These deposits appear as subsurface *salic horizons,* which will damage and kill plants when the horizons occur near the root zone. The salt accumulation associated with a desert playa is an example of extreme salinization (see Chapter 15).

Obviously, salinization complicates farming in Aridisols. The introduction of irrigation water may either waterlog poorly drained soils or lead to salinization. Nonetheless, vegetation does grow where soils are better drained and lower in salt content. In the Nile and Indus river valleys, for example, Aridisols are intensively farmed with a careful balance of these environmental factors, although thousands of acres of once-productive land not so carefully treated now sit idle and salt-encrusted.

If large capital investments are made in water, drainage, and fertilizers, Aridisols possess much agricultural potential. However, certain agricultural practices may present problems of a different sort. In the California valleys of Imperial, Coachella, and San Joaquin, and in the Wellton-Mohawk district of Arizona, soils were actually dug up, underlain with drainage tiles, and then replaced (Figure 18-13). The installed tiles and pipes drain away excess irrigation water, which has been purposely applied to leach salts away from plant roots. The problem now is to find a place to dump the salt- and chemical-laden drainage water. (See Focus Study 18-1.)

Such practices have fueled political and environmental struggles all over the West. From the 1950s through the early 1970s, Colorado River water was made so salty by drainage water that it killed many irrigated crops across the border in Mexico. In California, the Kesterson Wildlife Refuge, protected by

FIGURE 18-13
Agriculture in arid lands of the Coachella Valley of southeastern California. Providing drainage for excessive water application in such areas often is necessary to prevent salinization of the rooting zone. [Photo by author.]

FIGURE 18-14
Mollisol. (a) Profile from central Iowa and (b) general map of worldwide distribution. [Photo from Marbut Collection, Soil Science Society of America, Inc.]

(a)

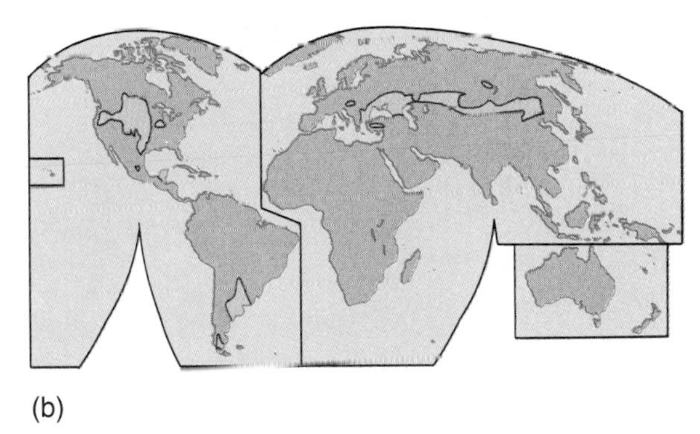
(b)

international agreements, was reduced to a toxic waste dump in the early 1980s. Focus Study 18-1 elaborates on the Kesterson tragedy. Ironically, in the Wellton-Mohawk district, if the government had taken out of production the irrigated land that receives subsidized federal water, it actually would have been cheaper than desalinization of the drainage water that eventually contaminated the Colorado River.

Mollisols. Mollisols (grassland soils) are some of Earth's most significant agricultural soils. There are seven recognized suborders, not all of which bear the same degree of fertility. The dominant diagnostic horizon is called the *mollic epipedon,* which is a dark, organic surface layer some 25 cm (10 in.) thick (Figure 18-14). As the Latin name implies, Mollisols are soft, even when dry, with granular or

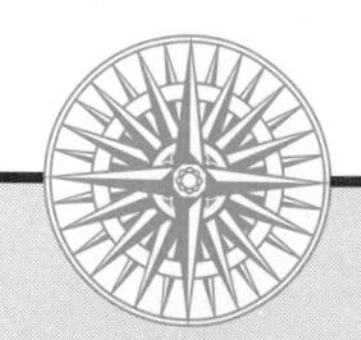

FOCUS STUDY 18-1 Selenium Concentrations in Western Soils

Irrigated agriculture has increased greatly since 1800, when only 8 million hectares (19.8 million acres) were irrigated worldwide. Today approximately 200 million hectares (494 million acres) are irrigated, and that figure will increase greatly by the turn of the century. Representing about 15% of Earth's agricultural land, irrigated lands account for nearly 35% of the harvest. Two related problems common in irrigated lands are salinization and waterlogging, especially in arid lands that are poorly drained. However, in many areas production has decreased and even ended because of the salt buildup in the soils. A few examples include areas along the Tigris and Euphrates rivers, the Indus River valley, sections of South America and Africa, and the western United States.

About 95% of the irrigated acreage in the United States is west of the 98th meridian. This region is increasingly troubled with salinization and waterlogging problems. But at least nine sites in the West, particularly California's western San Joaquin valley, are experiencing related contamination of a more serious nature—increasing selenium concentrations. The soils in these areas were derived from former marine sediments that formed shales in the adjoining Coast Ranges. As parent materials weathered, selenium-rich alluvium washed into the semiarid valley, forming the soils that needed only irrigation water to become productive. After 1960, such large-scale irrigation efforts intensified, resulting in subtle initial increases in selenium concentrations in soil and water. In trace amounts, selenium is a dietary requirement for animals and humans, but in higher amounts it is toxic to both. Toxic effects were reported during the 1980s in some domestic animals grazing on grasses grown in selenium-rich soils in the Great Plains.

Drainage of agricultural wastewater poses a particular problem in semiarid and arid lands where river discharge is inadequate to dilute and remove field runoff. One solution to prevent salt accumulations and waterlogging is to place field drains beneath the soil at depths of 3–4 meters (10–13 feet). Such drains collect gravitational water from fields that have been purposely overwatered to keep salts away from the effective rooting depth of the crops.

But drainage water must go somewhere, and for the San Joaquin Valley this problem triggered a 15-year controversy. Potential drain outlets to the ocean, San Francisco Bay, and an area in the central portion of the valley north of the irrigated lands all failed to pass environmental impact assessments under Environmental Policy Act requirements. Nonetheless, by the late 1970s, about 130 km (80 mi) of the San Luis drain was finished, even though no formal plan nor adequate funding had been completed for its outlet! In the absence of such a plan, large-scale irrigation continued, supplying the field drains with selenium-laden runoff that made its way to the Kesterson National Wildlife Refuge in the northern San Joaquin Valley east of San Francisco, where the drain terminated.

It only took three years before the wildlife refuge was officially declared a contaminated toxic waste site. Aquatic life forms (e.g., marsh plants, plankton, and insects) had taken in the selenium, which thus made its way into the diets of higher life forms in the refuge. According to U.S. Fish and Wildlife Service scientists, the toxicity moved through the food chain and genetically damaged and killed wildlife. For example, birth defects and death were widely reported in all varieties of birds that nested at Kesterson; approximately 90% of the exposed birds perished or were injured. Because this wildlife refuge was a major migration flyway and stopover point for birds from throughout the hemisphere, this destruction of the refuge also violated several multinational protection treaties.

Such damage to wildlife presents a real warning to human populations—remember where we are in the food chain. The field drains were sealed and removed in 1986, following court action that forced the federal government to uphold existing laws. But irrigation water immediately began backing up in the fields, producing both waterlogging and selenium contamination. Since 1985 more than 0.6 million hectares (1.5 million acres) of irrigated Aridisols and Alfisols have gone out of production in California, marking the end of several decades of irrigated farming in climatically marginal lands. Severe cutbacks in irrigated acreage no doubt will continue, underscoring the need to preserve prime farmlands elsewhere.

Frustrated agricultural interests have asked the federal government to finish the drain, either to San Francisco Bay or to the ocean. However, neither option appears capable of passing an environmental impact analysis. Another strategy is to allow irrigated lands to start pumping again into the former wildlife refuge because it was now a declared toxic dump site. Government agencies are still wrestling with these options in the 1990s. There are nine such threatened sites in the West, and Kesterson was simply the first to fail.

crumbly peds, loosely arranged when dry. These humus-rich organic soils are high in base cations (calcium, magnesium, and potassium) and have a high CEC. In terms of water balance, we can think of mollic soils as representing an intermediate stage between humid and arid soil moisture budgets.

Soils of the steppes and prairies of the world belong to this soil group: the North American Great Plains, the Pampas of Argentina, and the region from Manchuria in China through to Europe. Agriculture ranges from large-scale commercial grain farming to grazing along the drier portions of the soil order. With fertilization or soil-building cultural practices, high crop yields are common. The "fertile triangle" of Ukraine, Russia, and western portions of the Commonwealth of Independent States stretches from the Caspian Sea westward in a widening triangle toward Central Europe. The soils here are known as the *chernozem* in other classification systems (Figure 18-15). The remainder of the Russian landscape presents varied soils of lower fertility and productivity.

In North America, the Great Plains straddle the 98th meridian, which is coincident with the 51 cm (20 in.) isohyet of annual precipitation—wetter to the east and drier to the west. The Mollisols here mark the historic division between the short- and tall-grass prairies. The relationship among Mollisols, Aridisols (to the west), and Alfisols (to the east) is shown in Figure 18-16. The illustration also presents some of the important graduated changes

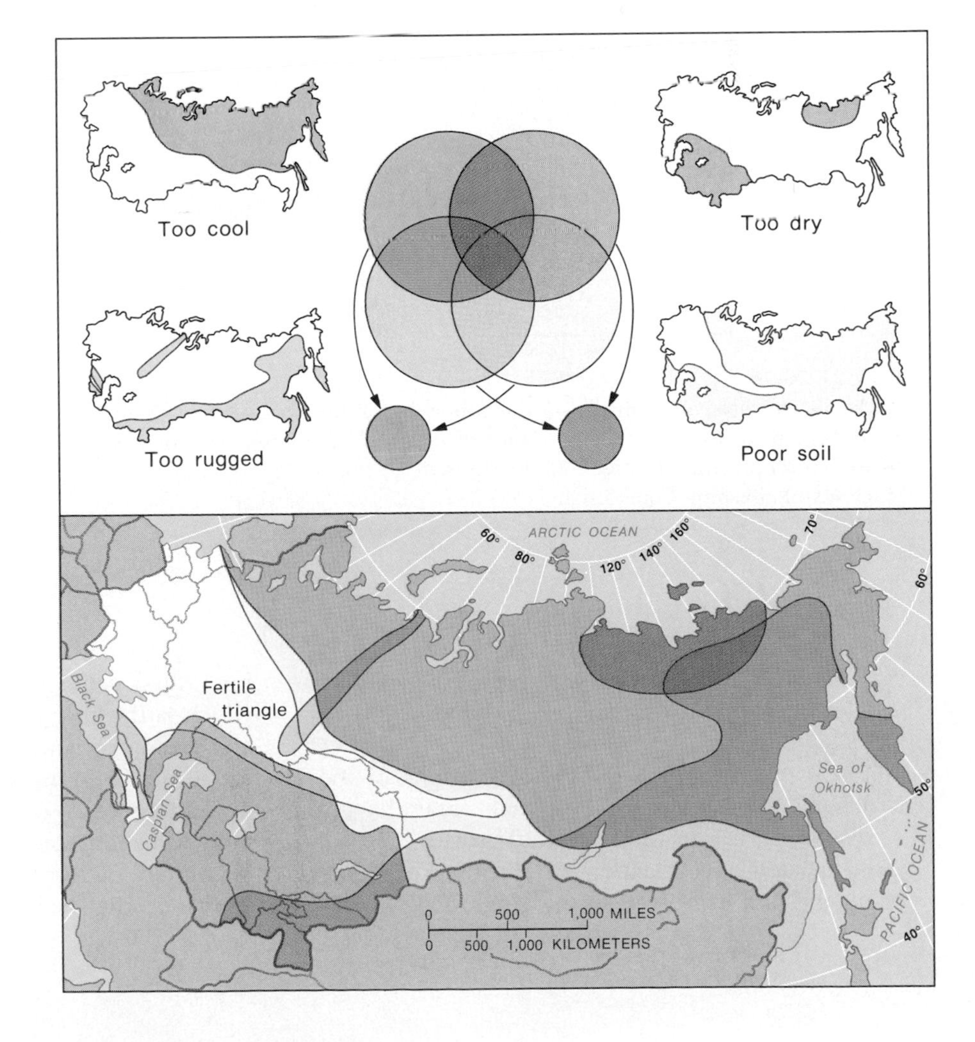

FIGURE 18-15
The fertile triangle of western portions of the Commonwealth of Independent States (former Soviet Union). The white area is the optimum; areas with any color or combination are limited for the reason shown. [After P. W. English and J. A. Miller, *World Regional Geography: A Question of Place,* p. 183. Copyright © 1989 by John Wiley & Sons. Adapted by permission of John Wiley & Sons, Inc.]

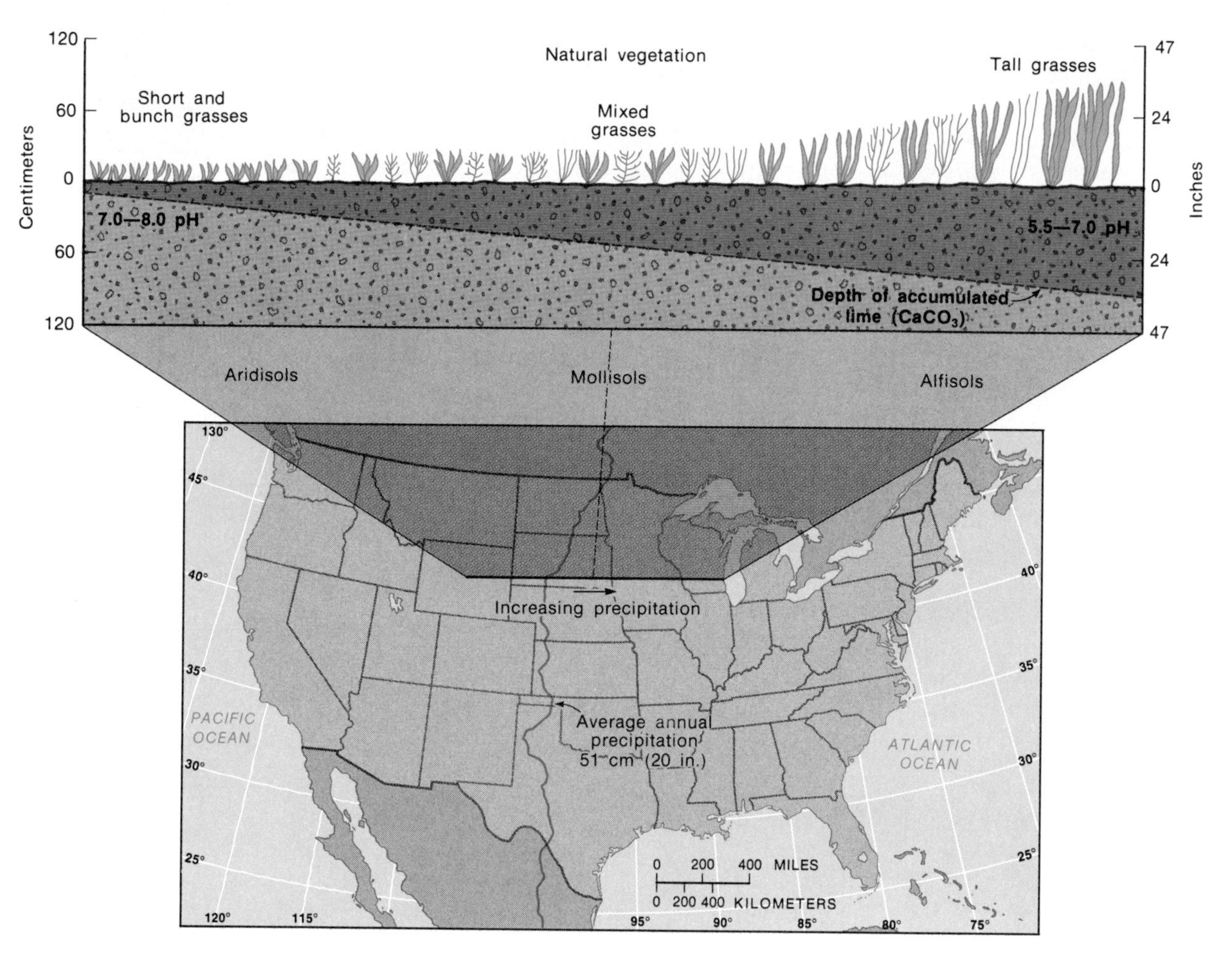

FIGURE 18-16
Aridisols, Mollisols, and Alfisols—a soil continuum in the north-central United States. [Adapted with permssion of Macmillan Publishing Company from *The Nature and Properties of Soils,* 10th ed., by Nyle C. Brady. Copyright © 1990 by Macmillan Publishing Company.]

that denote these different soil regions, including the level of pH concentration and the depth of available lime.

A soil process characteristic of portions of the Mollisols, and adjoining marginal areas having Aridisols, is calcification. **Calcification** involves the illuviated accumulation of calcium carbonate or magnesium carbonate in the B and C horizons, forming a subsurface diagnostic calcic horizon, which is thickest along the boundary between dry and humid climates (Figure 18-17). When cemented or hardened, these deposits are called **caliche,** or *kunkur;* they occur in widespread soil formations in central and western Australia, the Kalahari region of interior Southern Africa, and the High Plains of the west-central United States, among

POTET equal to or greater than PRECIP

Dark color, high in bases

O Dense sod cover of

A interlaced grasses and roots

E

Calcic horizon; possible formation of caliche

B Accumulation of excess calcium carbonate

C

FIGURE 18-17
Calcification process in Aridisol/Mollisol soils in climatic regimes with potential evapotranspiration equal to or greater than precipitation.

other places. (Soils dominated by the processes of calcification and salinization were formerly known as *pedocals.)*

(a)

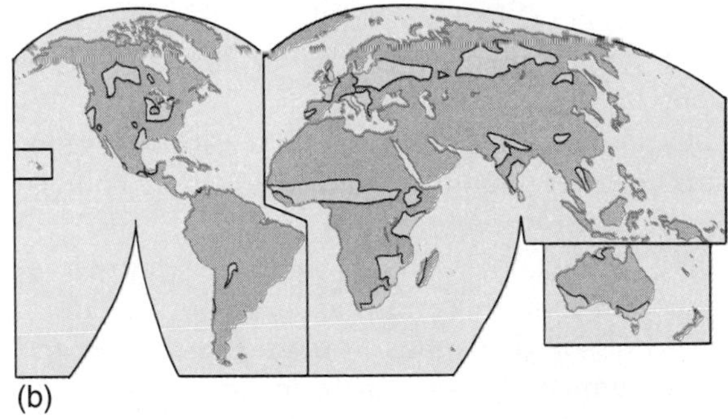

(b)

FIGURE 18-18
Alfisol. (a) Profile from central California and (b) general map of worldwide distribution. [Photo from Marbut Collection, Soil Science Society of America, Inc.]

Alfisols. Alfisols (moderately weathered forest soils) are the most wide-ranging of the soil orders, extending in five suborders from near the equator to high latitudes. Representative Alfisol areas include Boromo and Burkina Faso (interior western Africa); Fort Nelson, British Columbia; the states near the Great Lakes, and the valleys of central California. Most Alfisols have a grayish brown-to-reddish ochric epipedon and are considered moist versions of the Mollisol soil group. Moderate eluviation is present, as well as a subsurface *argillic horizon* of illuviated clays because of a pattern of increased precipitation (Figure 18-18).

Alfisols have moderate-to-high reserves of base cations and are fertile. However, productivity depends on specific patterns of moisture and temperature. Alfisols usually are supplemented by a moderate application of lime and fertilizers in areas of active agriculture. Some of the best farmland in the United States stretches from Illinois, Wisconsin, and Minnesota through Indiana, Michigan, and Ohio to Pennsylvania and New York. This land produces grains, hay, and dairy products. The soil here is an Alfisol subgroup called *Udalfs,* which are characteristic of humid continental/hot summer climates.

The *Xeralfs,* another Alfisol subgroup, are associated with the moist winter/dry summer pattern of the Mediterranean climate. These naturally productive soils are farmed intensively for subtropical fruits, nuts, and special crops that can grow only in a few locales worldwide (e.g., grapes, citrus, artichokes, almonds, figs).

Ultisols. Farther south in the United States are the **Ultisols** (highly weathered forest soils) and their five suborders. An Alfisol might degenerate into an Ultisol, given time and exposure to increased weathering under moist conditions. These soils tend to be reddish because of residual iron and aluminum oxides in the A horizon (Figure 18-19).

The increased precipitation in Ultisol regions means greater mineral alteration, more eluvial leaching, and therefore a lower level of base cations, leading to infertility. Fertility is further reduced by certain cultural practices and the effects of growing soil-damaging crops such as cotton and tobacco, which deplete nitrogen and expose soil to erosion. However, these soils respond well if subjected to good management—for example, crop rotation that restores fixed nitrogen and cultivation practices that prevent sheetwash and soil erosion. Much needs to be done relative to sustainable management of these soils.

Spodosols. The **Spodosols** (northern coniferous forest soils) and their four suborders occur generally to the north and east of the Alfisols. They are in cold and forested moist regimes (*Dfb humid continental mild summer* climates) in northern North America and Eurasia, Denmark, the Netherlands, and southern England. Because there are no

(a)

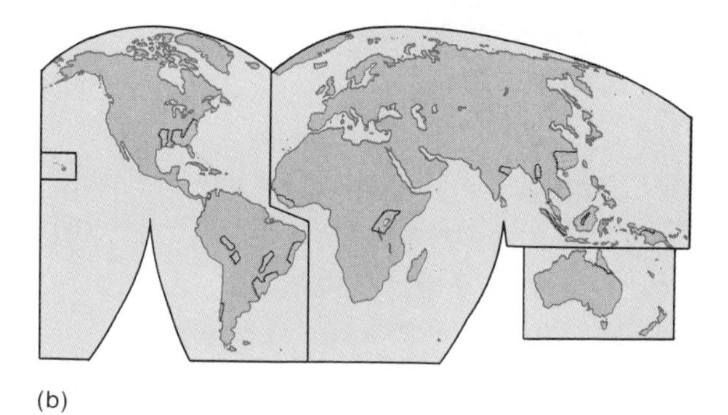

(b)

FIGURE 18-19
Ultisol. (a) Profile from the upper coastal plain of central North Carolina and (b) general map of worldwide distribution. [Photo from Marbut Collecton, Soil Science Society of America, Inc.]

(a)

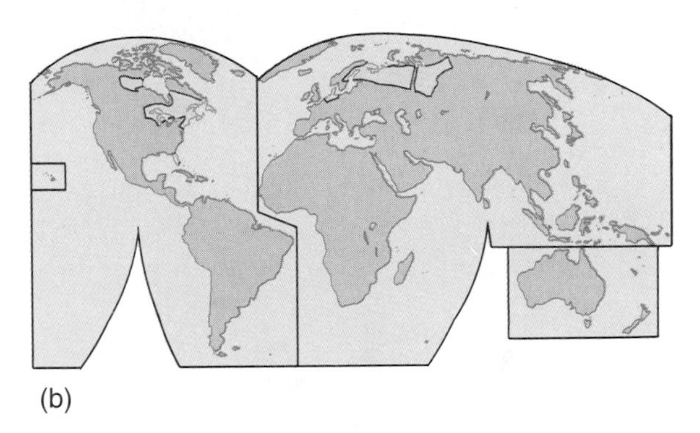

(b)

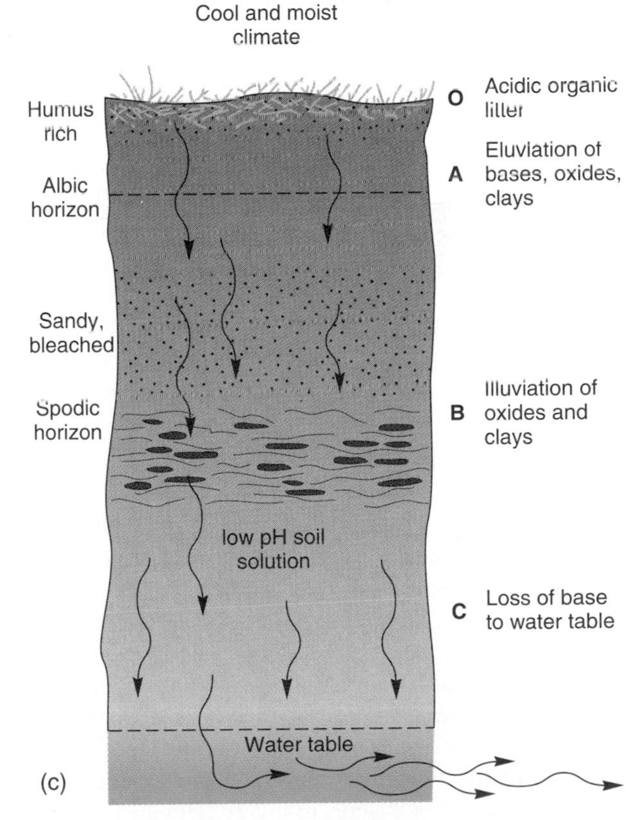

(c)

FIGURE 18-20
Spodosol. (a) Profile from northern New York and (b) general map of worldwide distribution; (c) podzolization process, typical in cool and moist climatic regimes. [Photo from Marbut Collection, Soil Science Society of America, Inc.]

comparable climates in the Southern Hemisphere, this soil type is not identified there. Spodosols form from sandy parent materials, shaded under evergreen forests of spruce, fir, and pine. Spodosols with more moderate properties form under mixed or deciduous forests (Figure 18-20a and b).

Spodosols lack humus and clay in the A horizons. An eluviated *albic horizon,* sandy and leached of clays and irons, lies in the A horizon instead and overlies a *spodic horizon* of illuviated organic matter and iron and aluminum oxides. The surface horizon receives organic litter from base-poor, acid-rich trees, which contribute to acid accumulations in the soil. The low pH (acidic) soil solution effectively removes clays, iron, and aluminum, which are passed to the

FIGURE 18-21
A characteristic Entisol forming from a shale parent material in the desert near Zabriskie Point, Death Valley. [Photo by author].

upper diagnostic horizon. An ashen-gray color is common in these subarctic forest soils and is characteristic of a formation process called **podzolization** (Figure 18-20c).

If agriculture is to be attempted, the low base-cation content of Spodosols requires the addition of nitrogen, phosphate, and potash, and perhaps crop rotation as well. A soil *amendment* such as limestone can significantly increase crop production in these acidic soils. For example, the yields of several crops (corn, oats, wheat, and hay) grown in specific Spodosols in New York State were increased up to a third with the application of 1.8 metric tons (2 tons) of limestone per 0.4 hectare (1.0 acre) per 6-year rotation.

Entisols. The **Entisols** (recent, undeveloped soils) lack vertical development of horizons. The five suborders of this soil group are based on differences in parent materials and climatic conditions, although the presence of Entisols is not climate dependent, for they occur in many climates worldwide. Entisols are true soils that have not had sufficient time to generate the usual horizons.

Entisols generally are poor agricultural soils, although those formed from river silt deposits are quite fertile. The conditions that have inhibited complete development also have prevented adequate fertility—too much or too little water, poor structure, and insufficient accumulation of weathered nutrients. Active slopes, alluvium-filled floodplains, poorly drained tundra, tidal mud flats, dune sands and erg (sandy) deserts, and plains of glacial outwash all are characteristic regions that have these soils. Figure 18-21 shows an Entisol in a desert climate where shales formed the parent material.

Inceptisols. Inceptisols (weakly developed soils) and their six suborders are inherently infertile. They are weakly developed young soils, although they are more developed than the Entisols. Inceptisols include a wide variety of different soils, all holding in common a lack of maturity with evidence of weathering just beginning. Inceptisols are associated with moist soil regimes and are regarded as eluvial because they demonstrate a loss of soil constituents throughout their profile but retain some weatherable minerals. This soil group has no distinct illuvial horizons and therefore shows no evidence of the translocation (movement) of clays, organics, or other minerals.

Inceptisols include the soils of most of the arctic tundra; glacially derived till and outwash materials from New York down through the Appalachians; and alluvium on the Mekong and Ganges floodplains.

Andisols. Andisols (volcanic parent materials) and seven suborders occur in areas of volcanic activity. These soils formerly were classified as Inceptisols and Entisols, but in 1990 they were placed in this new order. Andisols are derived from volcanic ash and glass. Previous soil horizons

FIGURE 18-22
Fertile Andisols planted with sugar cane, Kauai, Hawaii. Hawaii is the largest producer of sugar cane in the United States. [Photo by Wolfgang Kaehler.]

frequently are found buried by ejecta from repeated eruptions. Volcanic soils are unique in their mineral content and in their recharge by eruptions. For examples, see Figure 9-13 (ash-covered fertile Andisols in the Palouse region of Washington and Oregon), and Figure 19-21 (recovery and succession of pioneer species in the new soil north of Mount Saint Helens).

Weathering and mineral transformations are important in this soil order. Volcanic glass weathers readily into allophane (a noncrystalline aluminum silicate clay mineral that acts as a colloid) and oxides of aluminum and iron. Andisols feature a high CEC and high water-holding ability and develop moderate fertility, although phosphorus availability is an occasional problem. The fertile fields of Hawaii produce sugar cane and pineapple as important cash crops in soils of this order (Figure 18-22). Andisol distribution is small in area extent; however, such soils are locally important around the "ring of fire" surrounding the Pacific Rim.

Vertisols. Vertisols (expandable clay soils) are heavy clay soils. They contain more than 30% *swelling clays* (which swell significantly when they absorb water), such as *montmorillonite*. They are located in regions experiencing highly variable soil moisture balances through the seasons. These soils occur in subhumid-to-semiarid moisture conditions and under moderate-to-high temperature patterns. Vertisols frequently form under savanna and grassland vegetation in tropical and subtropical climates and are sometimes associated with a distinct dry season following a wet season. Individual Vertisol units are limited in extent, although their overall occurrence is widespread.

Vertisol clays are black when wet (but not because of organics, rather because of specific mineral content) and range to brown and dark gray. These deep clays swell when moistened and shrink when dried. In the process, vertical cracks form, which widen and deepen as the soil dries, producing cracks 2–3 cm (0.8–1.2 in.) wide and up to 40 cm (16 in.) deep. Loose material falls into these cracks, only to disappear when the soil again expands and the cracks close. After many such cycles, soil contents tend to invert or mix vertically, bringing lower horizons to the surface. (Figure 18-23).

Despite the fact that clay soils are plastic and heavy when wet, with little available soil moisture for plants, Vertisols are high in bases and nutrients and thus are some of the better farming soils wherever they occur. For example, they occur in a narrow zone along the coastal plain of Texas and in a section along the Deccan region of India. Vertisols often are planted with grain sorghums, corn, and cotton.

Histosols. Histosols (organic soils), including four suborders, are formed from accumulations of thick organic matter. In the midlatitudes, when conditions are right, beds of former lakes may turn into Histosols, with water gradually replaced by organic material to form a bog and layers of peat. (Lake succession and bog/marsh formation are discussed in Chapter 19.) Histosols also form in small, poorly drained depressions. Here, Histosols have produced conditions ideal for significant deposits of *sphagnum peat* to form. This material can be cut, baled, and sold as a soil amendment. In other locales, dried peat has served for centuries as a low-grade fuel. The area southwest of Hudson Bay in Canada is typical of Histosol formation.

THE CANADIAN SYSTEM OF SOIL CLASSIFICATION (CSSC)

The **Canadian System of Soil Classification (CSSC)** provides taxa for all soils presently recognized in Canada and is adapted to Canada's expanses of forest, tundra, prairie, frozen ground, and colder climates. As in the U. S. Soil Taxonomy system, the CSSC classifications are based on observable and measurable properties found in real soils rather than idealized soils that may result from the interactions of genetic processes. The system is flexible in that its framework can accept new findings and information in step with progressive developments in the soil sciences. The system is arranged in a nested, hierarchical pattern to allow generalization at several levels of detail.

Categories of Classification in the CSSC

Categorical levels are at the heart of a taxonomic system. These categories are based on soil profile properties organized at five levels of generalization. Each level is referred to as a *category of classification*. The levels in the CSSC are briefly

(a)

(b)

FIGURE 18-23
Vertisol. (a) Profile in the Lajas Valley of Puerto Rico and (b) general map of worldwide distribution. [Photo from Marbut Collecton, Soil Science Society of America, Inc.]

TABLE 18-7
Three Mineral Horizons and Mineral Horizon Suffixes Used in the CSSC

Symbol	Mineral Horizon Description
A	Forms at or near the surface, experiences *eluviation,* or leaching, of finer particles or minerals. Several subdivisions are identified, with the surface usually darker and richer in organic content than lower horizons ***(Ah)***; or, a paler, lighter zone below which reflects removal of organic matter with clays and oxides of aluminum and iron leached (removed) to lower horizons ***(Ae).***
B	Experiences *illuviation,* a depositional process, as demonstrated by accumulations of clays ***(Bt),*** sesquioxides of aluminum or iron, and possibly an enrichment of organic debris ***(Bh),*** and the development of soil structure. Coloration is important in denoting whether hydrolysis, reduction, or oxidation processes are operational for the assignment of a descriptive suffix.
C	Exhibits little effect from pedogenic processes operating in the ***A*** and ***B*** horizons, except the process of gleysation (associated with poor drainage and the reduction of iron, denoted ***(Cg)***, and the accumulation of calcium and magnesium carbonates ***(Cca)*** and more soluble salts ***(Cs)*** and ***(Csa).***

Symbol	Horizon Suffix Description
b	A buried soil horizon.
c	Irreversible cementation of a pedogenic horizon, e.g., cemented by $CaCO_3$.
ca	Lime accumulation of at least 10 cm thickness that exceeds in concentration that of the unenriched parent material by at least 5%.
cc	Irreversible cemented concretions, typically in pellet form.
e	Used with ***A*** mineral horizons ***(Ae)*** to denote eluviation of clay, Fe, Al, or organic matter.
f	Enriched principally with illuvial iron and aluminum combined with organic matter, reddish in upper portions and yellowish at depth, determined through specific criteria. Used with ***B*** horizons alone.
g	Gray to blue colors, or prominent mottling, or both, produced by intense chemical reduction. Various applications to ***A, B,*** and ***C*** horizons.
h	Enriched with organic matter: accumulation in place or biological mixing ***(Ah)*** or subsurface enrichment through illuviation ***(Bh).***
j	A modifier suffix for ***e, f, g, n,*** and ***t*** to denote limited change or failure to meet specified criteria denoted by that letter.
k	Presence of carbonates as indicated by visible effervescence with dilute hydrocholoric acid (HCl).
m	Used with ***B*** horizons slightly altered by hydrolysis, oxidation, or solution, or all three to denote a change in color or structure.
n	Accumulation of exchangeable calcium (Ca) in ratio to exchangeable sodium (Na) that is 10 or less, with the following characteristics: prismatic or columnar structure, dark coatings on ped surfaces, and hard consistence when dry. Used with ***B*** horizons alone.
p	***A*** or ***O*** horizons disturbed by cultivation, logging, and habitation. May be used when plowing intrudes on previous ***B*** horizons.
s	Presence of salts, including gypsum, visible as crystals or veins, or surface crusts of salt crystals, and by lowered crop yields. Usually with ***C*** but may appear with any horizon and lowercase suffixes.
sa	A secondary enrichment of salts more soluble than Ca or Mg carbonates, exceeding unenriched parent material, in a horizon at least 10 cm thick.
t	Illuvial enrichment of the ***B*** horizon with silicate clay that must exceed in overlying ***Ae*** horizon by 3 to 20% depending on the clay content of the ***Ae*** horizon.
u	Markedly disrupted by physical or faunal processes other than cryoturbation.
x	Fragipan formation—a loamy subsurface horizon of high bulk density and very low organic content. When dry it has a hard consistence and seems to be cemented.
y	Affected by cryoturbation (frost action) with disrupted and broken horizons and incorporation of materials from other horizons. Application to ***A, B,*** and ***C*** horizons and in combination with other suffixes.
z	A frozen layer.

described as adapted from *The Canadian Soil Classification System,* 2nd ed. Publication 1646. Ottawa: Supply and Services Canada, 1987, p. 16.

- *Order:* each of nine soil orders has pedon properties that reflect the soil environment and effects of active soil-forming processes.
- *Great Group:* subdivisions of each order reflect differences in the dominant processes or other major contributing processes. As an example, in Luvic Gleysols (great group name followed by order) the dominant process is gleying—reduction of iron and other minerals—resulting from poor drainage under either grass or forest cover with Aeg and Btg horizons (see Table 18-7).
- *Subgroup:* differentiated by content and arrangement of horizons that indicate the relation of the soil to a great group or order, or the subtle transition toward soils of another order.
- *Family:* subdivisions of subgroups; parent material characteristics such as texture and mineralogy, soil climatic factors, and soil reactions are important..
- *Series:* subdivisions of the family are differentiated by the detailed features of the pedon—the essential soil sampling unit. Pedon horizons fall within a narrow range of color, texture, structure, consistence, porosity, moisture, chemical reaction, thickness, and composition.

Soil Horizons

Soil horizons are named and standardized as diagnostic in the classification process. Several mineral and organic horizons and layers are used in the CSSC. Three *mineral horizons* are recognized by capital letter designation, followed by lowercase suffixes for further description. Principal soil-mineral horizons and suffixes are presented in Table 18-7.

Four *organic horizons* are identified in the Canadian classification system. O is further defined through subhorizon designations. Note that for organic soils, such layers are identified as *tiers*. These organic horizons are detailed in Table 18-8.

TABLE 18-8
Four Organic Horizons Used in the CSSC

Symbol	Description
O	Organic mateials, mainly mosses, rushes, and woody materials
L	Mainly discernible leaves, twigs, and woody materials
F	Partially decomposed, somewhat recognizable ***L*** materials
H	Indiscernible organic materials
O is further defined through subhorizon designations:	
Of	Readily identifiable fibric materials
Om	Mesic materials of intermediate decomposition
Oh	Humic material at an advanced stage of decomposition—low fiber, high bulk density

The Nine Soil Orders of the CSSC

The nine orders of the CSSC, and related great groups, are summarized in Table 18-9 with a general description of properties, an estimated percentage of land area for the soil order, a fertility assessment, and any applicable Soil Taxonomy equivalent. These nine soil orders appear in our key terms list and in the glossary.

Figure 18-24 is a generalized map of the distribution of principal soil orders in relation to physiographic regions in Canada. Please consult the *National Atlas of Canada,* 5th edition, for a detailed map of Canadian soils. For further study note the sources available from agencies in Ottawa that are listed as suggested readings at the end of this chapter.

SUMMARY—The Geography of Soils

Soil is the portion of the land surface in which plants can grow. It is a dynamic natural body composed of fine materials and contains both mineral and organic matter. Soils range in depth from a few millimeters to many meters, in some locales they may be completely absent. The basic soil individual is the polypedon, made up of numerous

TABLE 18-9
Nine Orders of the Canadian System of Soil Classification

Order Great Group	Characteristics*	Fertility
Chernozemic (Russia, chernozem) Brown (more moist) Dark Brown Black Dark Gray (less moist) (38 subgroups)	Well to imperfectly drained soils of the steppe-grassland-forest transition. Southern Alberta, Saskatchewan, Manitoba, Okanagan Valley, B.C., Palouse Prairie, B.C. Accumulation of organic matter in surface horizons. Most frozen during some winter months with soil-moisture deficits in the summer. A diagnostic **Ah** is typical (although **Ahe, Ap** are present) at least 10 cm thick or 15 cm if disturbed by cultivation. Mean annual temperature >0°C and usually less than 5.5°C. (5.1%, 470,000 km^2; Soil Taxon. = Mollisols.)	High; wheat growing
Solonetzic (Russian, *solonetz)* Solonetz Solodized Solonetz Solod (27 subgroups)	Solonetz denotes saline or alkaine soils. Well to imperfectly drained mineral soils developed under grasses in semiarid to subhumid climates. Limited areas of central and north-central Alberta. Noted for a **B** horizon that is very hard when dry but swells to a sticky, low-permeability mass when wet. A saline **C** horizon reflects nature of parent materials. (0.8%, 73,700 km^2; Soil Taxon. = Natric horizon of Mollisols and Alfisols.)	Variable (medium) about 50% cultivated, remainder in pasture
Luvisolic Gray Brown Luvisol Gray Luvisol (18 subgroups)	Eluviation-illuviation processes produce a light-colored **Ae** horizon and a diagnostic **Bt** horizon. Soils of mixed deciduous-coniferous forests. Major occurrence is the St. Lawrence lowland. Luvisols do *not* have a solonetzic **B** horizon, evidence of Gleysolic order and gloying, or organics less than in the Organic order. Permafrost within 1 m of surface and 2 m if soils are cryoturbated. (10.3%, 950,000 km^2; Soil Taxon. = Boralfs, Udalfs-suborders of Alfisols.)	High
Podzolic (Russian, *podzol)* Humic Ferro-Humic Podzol Humo-Ferric Podzol (25 subgroups)	Soils of coniferous forests and sometimes heath, leaching of overlying horizons occurs in moist, cool-to-cold climates. Iron, aluminum and organic matter from **L, F,** and **H** horizons are redeposited in podzolic **B** horizon. A diagnostic **Bh, Bhf,** or **Bf** is present depending on great group. Dominant in western British Columbia, Ontario, and Quebec. (22.6%, 2,083,000 km^2; Soil Taxon. = Spodosols, some Inceptisols.)	Low to medium depending on acidity
Brunisolic (French, "brown") Melanic Brunisol Eutric Brunisol Sombric Brunisol Dystric Brunisol (18 subgroups)	Sufficiently developed to distinguish from Regosolic order. Soils under forest cover with brownish-colored **Bm** horizons although various colors are possible. Also, can be with mixed forest, shrubs, and grass. Diagnostic **Bm Bfj,** thin **Bf,** or **Btj** horizons differentiate from soils of other orders. Well to imperfectly drained. Lack the podzolic **B** horizon of podzols although surrounded by them in St. Lawrence lowland. (8.8%, 811,000 km^2; Soil Taxon. = Inceptisols, some Psamments (Aquents in Entisols.)	Medium (variable)
Regosolic (Greek, *rhegos)* Regosol Humic Regosol (8 subgroups)	Weakly developed limited soils, the result of any number of factors: young materials, fresh alluvial deposits, material instability, mass-wasted slopes, or dry, cold climatic conditions. Lack solonetzic, illuvial, or podzolic **B** horizons. Lack permafrost within 1 m of surface, or 2 m if cryoturbated. May have **L, F, H.** or **O** horizons, or an **Ah** horizon if less than 10 cm thick. Buried horizons possible. Note: dominant in Northwest Territories and northern Yukon, now designated as Cryosols under CSSC. (1.3%, 120,000 km^2; Soil Taxon. = Entisols.)	Low (variable)
Gleysolic (Russian, *glei*) Luvic Gleysol Humic Gleysol Gleysol (13 subgroups)	Defined on the basis of color and mottling that results from chronic reducing conditions inherent in poorly drained mineral soils under wet conditions. High water table and long periods of water saturation. Rather than continuous they appear spotty within other soil orders, and occasionally may dominate an area. A diagnostic **Bg** horizon is present. (1.9%, 175,000 km^2; Soil Taxon. = Various aquic suborders, a reducing moisture regime.)	High to medium
Organic Fibrisol Mesisol Humisol Folisol (31 subgroups)	Peat, bog, and muck soils, largely composed of organic material. Most water-saturated for prolonged periods. Are widespread in association with poorly to very poorly drained depressions, although Folisols are found under upland forest environments. Exceed 17% organic carbon and 30% organic matter overall. (4.2%, 387,000 km ^{2}; Soil Taxon. = Histolsols.)	High to medium given drainage, available nutrients
Crysolic (Greek, kyros) Turbic Cryosol Static Cryosol Organic Cryosol (15 subgroups)	Dominate the northern third of Canada, with permafrost close to the surface of mineral and organic soil deposits. Generally found north of the tree line, or in fine-textured soils in subarctic forest, or in some organic soils in boreal forests. **Ah** horizon lacking or thin. Cryoturbation common, often denoted by patterned ground circles, polygons, and stripes. Subgroups based on degree of cryoturbation and the nature of mineral or organic soil material. (45%, 4,150,000 km^2; Soil Taxon. = Cryoquepts, Inceptisols; and pergelic temperature regime in several suborders.)	Not applicable

*Estimated percent and square kilometers of Canada's land area and Soil Taxonomy equivalent are given in parentheses.

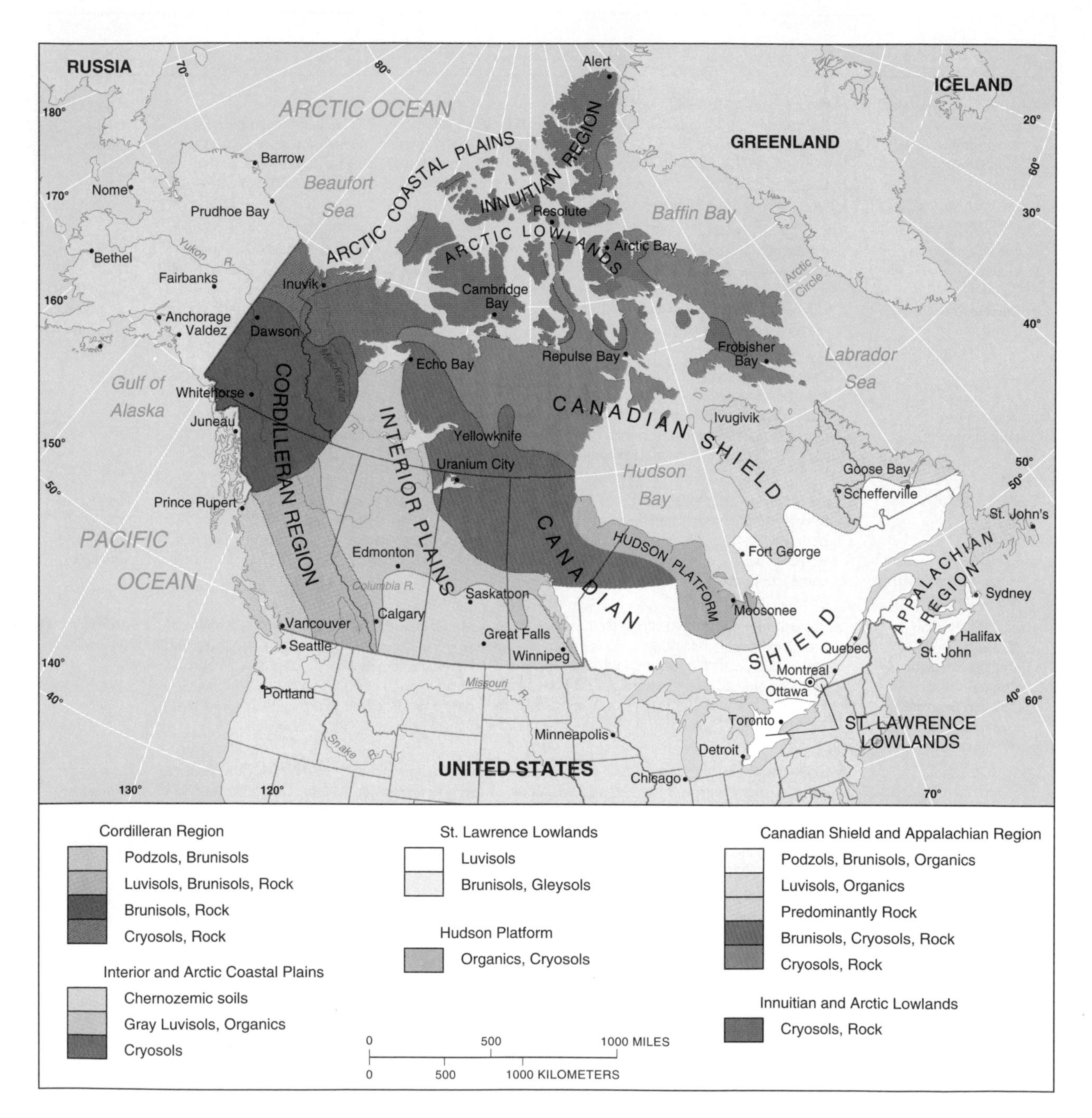

FIGURE 18-24
Principal soil regions of the Canada System of Soil Classification (CSSC) as related to Canada's major physiographic regions. [After maps prepared by the Land Resource Research Institute, Geological Survey of Canada, and the Canada Soil Survey Committee.]

sampling units called pedons. Soils are structured in vertical horizons that permit analysis and classification. The key properties for analysis are color, texture, structure, consistence, porosity, moisture regimes, and chemistry.

Environmental factors that affect soil formation include parent materials, climate, vegetation, topography, and time. Human influence is having great impact on Earth's prime soils. A new survey completed by the United Nations Environment Programme identified significant areas of concern relative to soil erosion and loss caused by human misuse and abuse.

The past century is marked by repeated efforts to develop soil classifications not only in the United States and Canada but worldwide. Presently, the Soil Taxonomy classification system is used in the United States, built around an analysis of various diagnostic horizons and eleven soil orders. In the Canadian System of Soil Classification nine principal orders are identified and mapped.

The fertility of soils (ability to sustain plant productivity) is critical to society. Essential soils for agriculture and their fertility are threatened by mismanagement, destruction, and conversion to other uses. Much soil loss is preventable through the application of known technologies, improved agricultural practices, and sustainable government policies.

KEY TERMS

- adsorption
- calcification
- caliche
- Canadian System of Soil Classification
    - Brunisolic
    - Chernozemic
    - Cryosolic
    - Gleysolic
    - Luvisolic
    - Organic
    - Podzolic
    - Regosolic
    - Solonetzic
- cation-exchange capacity (CEC)
- eluviation
- epipedon
- humus
- illuviation
- laterization
- loam
- pedogenic regimes
- pedon
- plinthite
- podzolization
- polypedon
- salinization
- soil
- soil colloids
- soil fertility
- soil horizon
- Soil Taxonomy
    - Alfisols
    - Andisols
    - Aridisols
    - Entisols
    - Histosols
    - Inceptisols
    - Mollisols
    - Oxisols
    - Spodosols
    - Ultisols
    - Vertisols
- solum
- subsurface diagnostic horizon

REVIEW QUESTIONS

1. Soils provide the foundation for animal and plant life and therefore are critical to Earth's ecosystems. Why is this true?
2. Explain some of the details supporting concern for the loss of our most fertile soils.
3. Define polypedon and pedon, the basic units of soil.
4. Characterize the principal aspects of each soil horizon. Where does the main accumulation of organic material occur? The formation of humus? What is the eluviated layer? The illuviated layer? The solum, or true soil?
5. How is soil color analyzed and compared?
6. Define a soil separate. What are the various sizes of particles in soil? What is loam? Why is loam regarded so highly by agriculturalists?
7. What is a possible method for determining soil consistence?
8. Summarize the five soil moisture regimes common in mature soils.
9. What are soil colloids? How do they relate to cations and anions in the soil? Explain the cation-exchange capacity.
10. Briefly describe the contribution of the following factors and their effect on soil formation: parent material, climate, vegetation, landforms, time, and humans.
11. Summarize the brief history of soil classification described in this chapter. What led soil scientists to develop the new Soil Taxonomy classification system?
12. What is the basis of the Soil Taxonomy system? How many orders, suborders, great groups, subgroups, families, and soil series are there?
13. Define an epipedon and a subsurface diagnostic horizon. Give a simple example of each.
14. Locate each soil order on the map as you give a general description of it.
15. How was slash-and-burn cultivation, as practiced in the past, a form of crop and soil rotation and a conserving of soil properties?
16. Describe the salinization process in arid and semiarid soils. What associated soil horizons develop?
17. Which of the soil orders are associated with Earth's most productive agricultural areas?
18. What is the significance of the 51 cm (20 in.) isohyet in the Midwest relative to soils, pH, and lime content?
19. Describe the podzolization process associated with northern coniferous forest soils. What characteristics are associated with the surface horizons? What strategies might enhance these soils? Name associated soil orders in the Soil Taxonomy and the CSSC.
20. What overlapping factors produce the soils of the fertile triangle in the western portion of the Commonwealth of Independent States?
21. What former Inceptisols now form a new soil order? Describe these soils: location, nature, and formation processes. Why do you think they were separated into their own order?
22. Why did Canada develop its own system of soil classification? Describe a brief history of the events that led up to the adoption of the CSSC.
23. Which soil order is associated with the formation of a bog? Explain its use as a low-grade fuel.
24. Why has a selenium contamination problem arisen in western soils? Explain the impact of cultural practices, and tell why you think this is or is not a serious problem.

SUGGESTED READINGS

Bentley, C. F., ed. *Photographs and Conditions of Some Canadian Soils.* University of Alberta Extension Series Publication B791. Ottawa K1P 5H4: Canadian Society of Soil Science, 1979.

Botkin, Daniel B., and Edward A. Keller. *Environmental Studies—Earth as a Living Planet.* Columbus:, OH: Macmillan Publishing Company, 1987.

Brady, Nyle C. The *Nature and Properties of Soils,* 10th ed. New York: Macmillan Publishing Company, 1990.

Clayton, J. S., et al.*Soils of Canada,* Vol. 1 *Soil Report* and Vol. 2 *Soil Inventory.* A Cooperative Project of the Canada Soil Survey Committee and the Soil Research Institute, Research Branch, Agriculture Canada. Ottawa K1A OS9: Supply and Services Canada, 1977. Boxed set with two large color wall maps titled "Soils of Canada" and "Soil Climates of Canada" including maps of soil temperature and soil moisture.

Expert Committee on Soil Science, Agriculture Canada Research Branch. *The Canadian System of Soil Classification,* 2nd ed. Publication 1646. Ottawa K1A OS9: Supply and Services Canada, 1987. Replacing Agriculture Canada Publications 1455 (1974) and 1646 (1978).

Expert Committee on Soil Survey, Agriculture Canada Research Branch. *The Canadian Soil Information System (CanSIS) Manual for Describing Soils in the Field.* Publication 1459. Ottawa K1A OS9: Supply and Services Canada, 1979.

Fairbridge, Rhodes W., and Charles W. Finkl, Jr. *The Encyclopedia of Soil Science.* Part 1, Physics, Chemistry, Biology, Fertility, and Technology. *Encyclopedia of Earth Sciences,* vol. 12. Stroudsburg, PA: Dowdon, Hutchinson and Ross, 1979.

Finkl, Charles W., Jr., ed. "Soil Classification," *Benchmark Papers in Soil Science,* vol. 1. Stroudsburg, PA: Hutchinson Ross, 1982.

Foth, Henry D. *Fundamentals of Soil Science.* 8th ed. New York: John Wiley & Sons, 1990.

Gersmehl, Philip J. "Soil Taxonomy and Mapping," *Annals of the Association of American Geographers* 67 (September 1977): 419–28.

Gibbons, Boyd. "Do We Treat Our Soil like Dirt?" *National Geographic.* September 1984: 350–89.

Harris, Tom. "Selenium: The Poisoning of America," Trouble in Paradise Series, *Sacramento Bee* (December 6–10, 1988).

Research Branch, Canada Department of Agriculture. *Glossary of Terms in Soil Science.* Publication 1459. Ottawa K1A OC7: Information Canada, 1976.

Singer, Michael J., and Donald N. Munns. *Soils—An Introduction,* 2nd ed. New York: Macmillan Publishing Company, 1987.

Soil Survey Staff and the Agronomy Department of Cornell University. *Keys to Soil Taxonomy.* Soil Management Support Services Technical Monograph 6. Ithaca, NY: Cornell University, 1987.

Soil Survey Staff. *Soil Taxonomy—A Basic System of Soil Classification for Making and Interpreting Soil Surveys.* Agricultural Handbook No. 436. Washington, DC: Government Printing Office, 1975.

Steila, Donald, and Thomas E. Pond. *The Geography of Soils—Formation, Distribution, and Management.* Savage, MD: Rowman & Littlefield, 1989.

Flowering bromeliads in El Yunque Rain Forest, Puerto Rico. [*Photo by Tom Bean.*]

19

ECOSYSTEM ESSENTIALS

Diversity is an impressive feature of the living Earth. The diversity of organisms is a response to the interaction of the atmosphere, hydrosphere, and lithosphere, which produces diverse conditions within which the biosphere exists.

This sphere of life and organic activity extends from the ocean floor to about 8 km (5 mi) in the atmosphere. The biosphere includes myriad ecosystems from simple to complex, each operating within general spatial boundaries. An **ecosystem** is a self-regulating association of living plants and animals and their nonliving physical environment. In an ecosystem a change in one component causes changes in others, as systems adjust to new operating conditions.

Earth itself is the largest ecosystem within the natural boundary of the atmosphere. Natural ecosystems are open systems for both energy and matter, with almost all ecosystem boundaries functioning as transition zones rather than as sharp demarcations. Smaller ecosystems—for example, forests, seas, mountain tops, deserts, beaches, islands, lakes, ponds—make up the larger whole.

Ecology is the study of the relationships between organisms and their environment and among the various ecosystems in the biosphere. The word *ecology,* developed by German naturalist Ernst Haeckel in 1869, is derived from the Greek *oikos* (household, or place to live) and *logos* (study of). **Biogeography,** essentially a spatial ecology, is the study of the distribution of plants and animals, the diverse spatial patterns they create, and the physical and biological processes, past and present, that produce this distribution across Earth. To better understand an ecosystem, biogeographers must look across the ages and reassemble Earth's tectonic plates to recreate previous environmental relationships.

The degree to which modern society understands ecosystems will help determine our success as a species and the long-term survival of a habitable Earth. Gilbert White, a well-known geographer, researcher, and activist, and Mostafa Tolba, director of the United Nations Environment Programme, joined in this assessment:

> The time is ripe to step up and expand current efforts to understand the great interlocking systems of air, water, and minerals nourishing the Earth. . . . Moreover, without vigorous action toward that goal, nations will be seriously handicapped in trying to cope with proven and suspected threats to ecosystems and to human health and welfare resulting from alterations in the cycles of carbon, nitrogen, phosphorus, sulfur, and related materials. . . . Society depends upon this life-support system of planet Earth.*

ECOSYSTEM COMPONENTS AND CYCLES

An ecosystem is a complex of many variables, all functioning independently yet in concert, with complicated flows of energy and matter (Figure 19-1).

> Life devours itself: everything that eats is itself eaten; everything that can be eaten is eaten; every chemical that is made by life can be broken down by life; all the sunlight that can be used is used. . . . The web of life has so many threads that a few can be broken without making it all unravel, and if this were not so, life could not have survived the normal accidents of weather and time, but still the snapping of each thread makes the whole web shudder, and weakens it. . . . You can never do just one thing: the effects of what you do in the world will always spread out like ripples in a pond.**

An ecosystem is composed of both biotic and abiotic components. Nearly all depend upon an input of solar energy; the few limited ecosystems that exist in dark caves or on the ocean floor depend upon chemical reactions. Ecosystems are divided into subsystems, with the biotic portion composed of producers, consumers, and decomposers. The abiotic flows in an ecosystem include gaseous and sedimentary nutrient cycles. Figure 19-2 illustrates the essential elements of an ecosystem.

---

*Gilbert F. White and Mostafa K. Tolba, "Global Life Support Systems," *United Nations Environment Programme Information,* no. 47, Nairobi, Kenya: United Nations, 1979, p. 1.

**Friends of the Earth and Amory Lovins, The United Nations Stockholm Conference: *Only One Earth*. London: Earth Island Limited, 1972, p. 20.

FIGURE 19-1
The intricate web of a spider mirrors the web of life; "the snapping of each thread makes the whole web shudder." [Photo by author.]

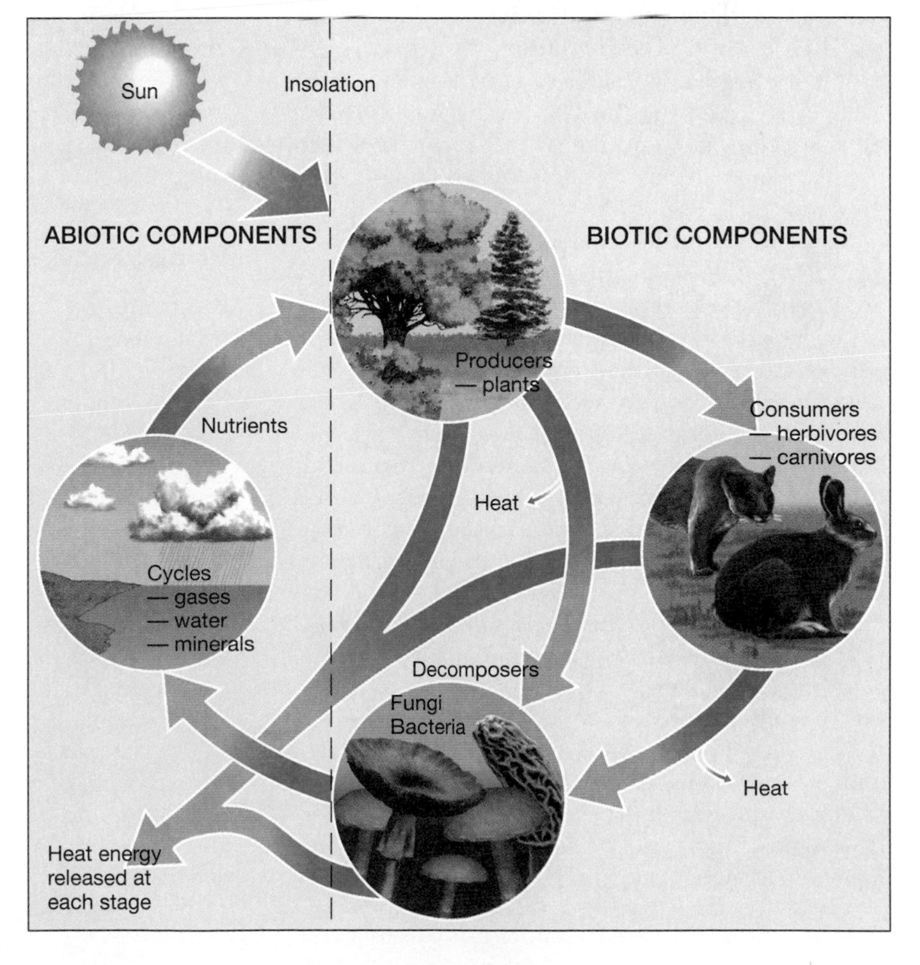

FIGURE 19-2
Abiotic and biotic components of ecosystems. Solar energy is the input that drives the biosphere. Heat is the energy output from this life system.

Communities

A convenient biotic subdivision within an ecosystem is a **community,** which is formed by relationships among populations of living animals and plants in an area. An ecosystem is the interaction of many communities with the abiotic physical components of its environment. For example, in a forest ecosystem a specific community may exist on the forest floor, whereas another community functions in the canopy of leaves high above. Similarly, within a lake ecosystem, the plants and animals that flourish in the bottom sediments form one community, whereas those near the surface form another. A community is identified in several ways—by its physical appearance, the number of species and the abundance of each, the complex patterns of their interdependence, and the trophic (feeding) structure of the community.

Within a community, two concepts are important: habitat and niche. **Habitat** is the specific physical location of an organism, the type of environment in which it resides or is biologically suited to live. In terms of physical and natural factors, most species have specific habitat parameters with definite limits and a specific regimen of sustaining nutrients.

Niche refers to the function, or occupation, of a life form within a given community. It is the way an organism obtains and sustains the physical, chemical, and biological factors it needs to survive. Similar habitats produce comparable niches. In a stable community, no niche is left unfilled. The principle of *competitive exclusion* states that no two species can occupy the same niche successfully in a stable community. Thus, closely related species are separated at some distance from one another. In other words, each species operates to reduce competition.

An individual species must satisfy several aspects in its niche. Among these are a habitat niche, a trophic (food) niche, and a reproductive niche. For example, the red-wing blackbird *(Agelaius phoeniceus)* occurs throughout the United States and most of Canada in habitats of meadow, pastureland, and marsh. These birds nest in blackberry tangles and thick vegetation in freshwater marshes, sloughs, and fields. Red-wing blackbirds feed on weed seeds and cultivated seed crops throughout the year, adding insects to their diet during the nesting season. They disperse seeds of many plants during their travels.

Some species have *symbiotic relationships,* an arrangement that mutually benefits and sustains each organism. For example, lichen (pronounced "liken") is made up of algae and fungi. The algae is the producer and food source, and the fungus provides structure and support. Their mutually beneficial relationship (mutualism) allows the two to occupy a niche in which neither could survive alone. Lichen developed from an earlier parasitic relationship in which the fungi broke into algae cells directly. Today the two organisms have evolved into a supportive harmony and symbiotic relationship (Figure 19-3).

By contrast, parasitic relationships eventually may kill the host, thus destroying the parasite's own niche and habitat. An example is mistletoe (*Phoradendron*), which lives on and may kill various kinds of trees. Some scientists are questioning whether our human society and the physical systems of Earth constitute a global-scale symbiotic relationship (sustainable) or a parasitic one (nonsustainable).

Plants: The Essential Biotic Component

Plants are the critical biotic link between life and solar energy. Ultimately the fate of the biosphere rests on the success of plants and their ability to capture sunlight.

Beginnings. The first plants were simple single- or multiple-celled structures known as *cyanobacteria* (formerly known as blue-green algae). They began the harvest of sunlight and release of free oxygen about 3.3 billion years ago. Long, slow development ensued as these early aquatic forerunners of plants migrated from the subsurface to the water surface, increasingly protected by the evolving functional layers of the atmosphere.

The last billion years of Earth's history have seen an explosion of biotic diversity. The move to rocky coastlines and then inland required complicated and integrated multicellular organisms possessing evolved structures for survival on land. Specialized

FIGURE 19-3
Lichen, an example of a symbiotic relationship. [Photo by Bobbé Christopherson.]

tissues developed for food production, body support, and anchorage. Land plants and animals became common about 430 million years ago, according to fossilized remains. The equisetum plant (horsetail, or scouring rush) is one example of an ancient (Carboniferous period) plant of primitive structure that still survives today (Figure 19-4).

As plants evolved, **vascular plants** developed conductive tissues and true roots for internal transport of fluid and nutrients. At present there are almost 250,000 species of them. This great diversity and complexity compounds the difficulty of classification for convenient study.

Leaf Activity. Photochemical reactions take place within plant leaves. Flows of carbon dioxide, water, light, and oxygen enter and exit the surface of each leaf (see Figure 1-4).

The upper and lower surfaces of a leaf are covered with a layer of epidermal cells that protect the inner workings of the leaf and reduce water loss. The largest concentration of light-responsive, photosynthetic cells rests below the upper layers of the leaf. These are *chloroplast* bodies, and within each resides a green, light-sensitive pigment called **chlorophyll.** Within this pigment, light stimulates photochemistry.

FIGURE 19-4
Equisetum, a plant of ancient origin—over 300 million years ago. [Photo by author.]

Gases flow into and out of a leaf through small pores called **stomata** (singular: stoma), which usually are more numerous on the lower side of the leaf. A stoma is surrounded by small guard cells that open and close the pore, depending on the plant's needs at the moment. Veins in the leaf bring in water and nutrient supplies and carry off the sugars produced by photosynthesis. The veins in each leaf connect to the stems and branches of the plant and thus to the main circulation system.

Water that moves through a plant exits the leaves through the stoma and evaporates from leaf surfaces, thereby assisting heat regulation within the plant. As water evaporates from the leaves, a pressure deficit is created that allows atmospheric pressure to push water up through the plant all the way from the roots, like a soda straw. We can only imagine the complex operation of a 100 m (328 ft) tree!

Photosynthesis and Respiration. Photosynthesis unites carbon dioxide and oxygen (derived from water in the plant) under the influence of certain wavelengths of visible light. The process releases oxygen and produces energy-rich organic material. The name is descriptive: *photo-* refers to sunlight, and *-synthesis* describes the bringing together of materials to form compounds and produce reactions within plant leaves.

Only about one-quarter of the light energy arriving at the surface of a leaf is useful to the light-sensitive chlorophyll. Chlorophyll absorbs only the orange-red and violet-blue wavelengths for photochemical operations, and reflects predominantly green hues (and some yellow). This is why trees and other vegetation look green. Understandably, competition for light is a dominant factor in the formation of plant communities. This competition is expressed in their height, orientation, and structure.

Photosynthesis essentially follows this equation:

$$\underset{\text{(Carbon dioxide)}}{CO_2} + \underset{\text{(Water)}}{H_2O} \xrightarrow[\text{energy}]{\text{light}} \underset{\text{(Carbohydrate)}}{x(CH_2O)} + \underset{\text{(Oxygen)}}{O_2}$$

From the equation, you can see that photosynthesis removes carbon (in the form of CO_2) from Earth's atmosphere. The quantity is prodigious: approximately 91 billion metric tons (100 billion tons) per year. Carbohydrates, the organic result of the photosynthetic process, are combinations of carbon, hydrogen, and oxygen. They can form simple sugars, such as glucose ($C_6H_{12}O_6$). Glucose, in turn, is used by plants to build starches, which are more complex carbohydrates and the principal food storage substance in plants. *Primary productivity* refers to the rate at which energy is stored in such organic substances.

Plants not only store energy; they must consume some of this energy by converting carbohydrates through respiration to derive energy for their other operations. Thus, **respiration** is essentially a reverse of the photosynthetic process:

$$\underset{\text{(Carbohydrate)}}{x(CH_2O)} + \underset{\text{(Oxygen)}}{O_2} \rightarrow \underset{\text{(Carbon dioxide)}}{CO_2} + \underset{\text{(Water)}}{H_2O} + \underset{\text{(Heat)}}{\text{energy}}$$

In respiration, plants oxidize stored energy, releasing carbon dioxide, water, and energy as heat. The overall growth of a plant depends on a surplus of carbohydrates beyond what is lost through plant respiration. Figure 19-5 presents a simple schematic of this process, which produces plant growth.

The *compensation point* is the break-even point between the production and consumption of organic material. Each leaf operates above the compensation point, with unproductive leaves being eliminated by the plant—something each of us has no doubt experienced with a house plant that received inadequate water or light. The difference between photosynthetic production and respiration loss is called *net photosynthesis*. The amount varies, depending on controlling environmental factors such as light, water, temperature, soil fertility, and the plant's site, elevation, and competition from other plants and animals.

Plant productivity increases as light availability increases—up to a point. When the light level is too high, *light saturation* occurs and most plants actually reduce their output in response. Some plants adapt better to shade, whereas others flourish in full sunlight. Crops such as rice, wheat, and sugar cane do well with high light intensity. Figure 19-6 portrays the general disposition of energy received by green plants, how it is utilized, and the final disposition of net primary production.

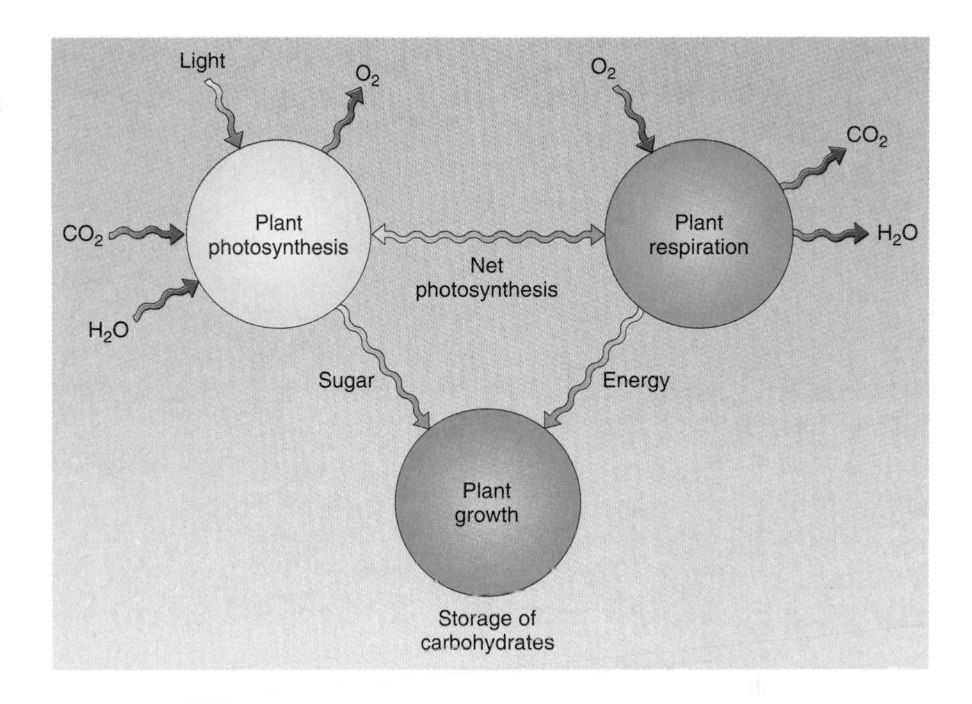

FIGURE 19-5
The relationship of photosynthesis and respiration determines plant growth.

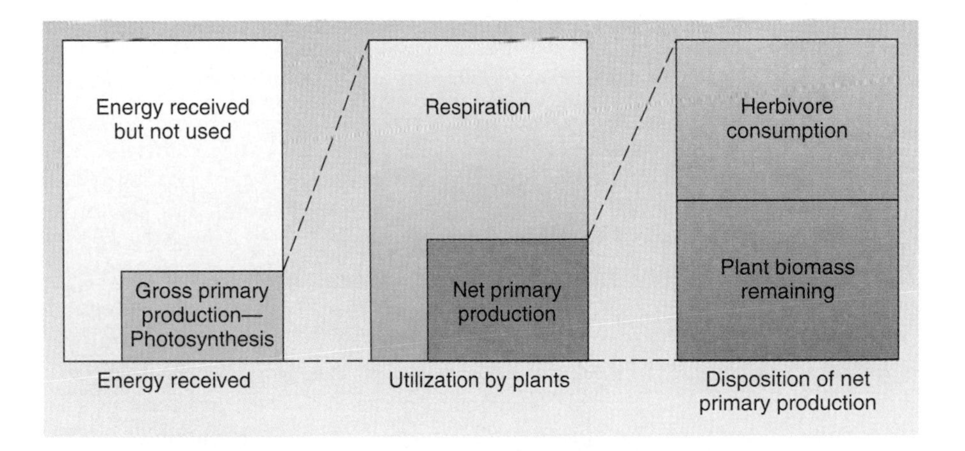

FIGURE 19-6
Energy receipt, utilization, and disposition by green plants.

Net Primary Productivity. The net photosynthesis for an entire community is its **net primary productivity.** This is the amount of useful chemical energy (biomass) that the community generates for the ecosystem. **Biomass** is the net dry weight of organic material; it is biomass that feeds the food chain.

Net primary productivity is mapped in terms of *fixed carbon per square meter per year*. Study Figure 19-7 and you can see that on land, net primary production tends to be highest in the tropics at sea level and decreases toward higher latitudes. But precipitation also affects productivity, as evidenced on the map by the correlations of abundant precipitation with high productivity and reduced precipitation with low productivity, the latter observable in subtropical deserts. Even though deserts are high in solar radiation, other controlling factors are important, namely water availability and soil conditions.

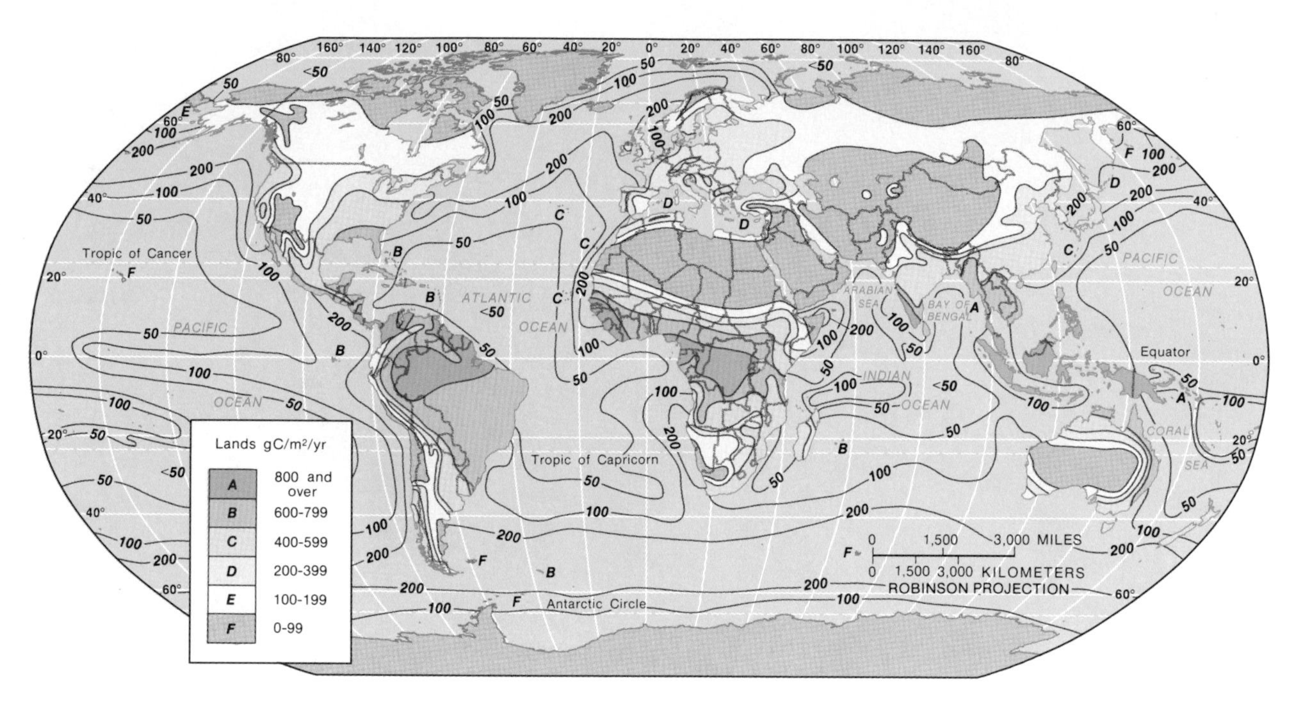

FIGURE 19-7
Worldwide net primary productivity in grams of carbon per square meter per year (approximate values). [After D. E. Reichle, *Analysis of Temperate Forest Ecosystems*. Heidelberg, Germany: Springer-Verlag, 1970. Adapted by permission.]

In the oceans, productivity is limited by differing nutrient levels. Regions with nutrient-rich upwelling currents generally are the most productive. The map shows that the tropical ocean and areas of subtropical high pressure are quite low in productivity.

In temperate and high latitudes, the rate at which carbon dioxide is fixed by vegetation varies seasonally, increasing in spring and summer as plants flourish with increasing solar input and, in some areas, more available (nonfrozen) water, and decreasing in late fall and winter. Rates in the tropics are high throughout the year, and turnover in the photosynthesis-respiration cycle is faster, exceeding by many times the rates experienced in a desert environment or in the far northern limits of the tundra. A lush acre of sugar cane in the tropics might fix 18 metric tons (20 tons) of carbon in a year, whereas desert plants in an equivalent area might achieve only 1% of this amount.

Table 19-1 lists various ecosystems, their net primary productivity per year, and an estimate of net total biomass worldwide. Net primary productivity is estimated at 170 billion metric tons of dry organic matter per year. Please compare the various ecosystems, especially cultivated land, with most of the natural communities. Net productivity is generally regarded as the most important aspect of of any type of community, and the distribution of productivity over Earth's surface is an important subject of biogeography.

Abiotic Ecosystem Components

Critical in each ecosystem is the flow of energy and the cycling of matter in the form of nutrients and water. Nonliving abiotic components set the stage for ecosystem operations.

Light, Temperature, Water, and Climate. The pattern of solar energy receipt is crucial in both terrestrial and aquatic ecosystems. Solar energy en-

TABLE 19-1
Net Primary Production and Plant Biomass on Earth

Ecosystem	Area (10^6 km^2)*	Net Primary Productivity per Unit Area (g/m^2/yr)**		World Net Primary Production (10^9/t/yr)***
		Normal Range	Mean	
Tropical rain forest	17.0	1000–3500	2200	37.4
Tropical seasonal forest	7.5	1000–2500	1600	12.0
Temperate evergreen forest	5.0	600–2500	1300	6.5
Temperate deciduous forest	7.0	600–2500	1200	8.4
Boreal forest	12.0	400–2000	800	9.6
Woodland and shrubland	8.5	250–1200	700	6.0
Savanna	15.0	200–2000	900	13.5
Temperate grassland	9.0	200–1500	600	5.4
Tundra and alpine region	8.0	10–400	140	1.1
Desert and semidesert scrub	18.0	10–250	90	1.6
Extreme desert, rock, sand, ice	24.0	0–10	3	0.07
Cultivated land	*14.0*	*100–3500*	*650*	*9.1*
Swamp and marsh	2.0	800–3500	2000	4.0
Lake and stream	2.0	100–1500	250	0.5
Total continental	**149**		**773**	**115.17**
Open ocean	332.0	2–400	125	41.5
Upwelling zones	0.4	400–1000	500	0.2
Continental shelf	26.6	200–600	360	9.6
Algal beds and reefs	0.6	500–4000	2500	1.6
Estuaries	1.4	200–3500	1500	2.1
Total marine	**361.0**		**152**	**55.0**
Grand total	**510.0**		**333**	**170.17**

*1 km^2 = 0.39 mi^2
**1 g per m^2 = 8.9 lb per acre
***1 metric ton = 1.1023 tons
SOURCE: Robert H. Whittaker, *Communities and Ecosystems*. Heidelberg: Springer-Verlag, 1975, p. 224. Reprinted by permission.

ters an ecosystem by way of photosynthesis, with heat dissipated from the system at many points. Of the total energy intercepted at Earth's surface and available for work, only about 1.0% is actually fixed by photosynthesis as chemical energy (energy stored as carbohydrates in plants).

The duration of Sun exposure is the **photoperiod.** Along the equator, days are almost always 12 hours in length; however, with distance from the equator, seasonal effects become pronounced, as discussed in Chapter 2. Plants have adapted their flowering and seed germination to seasonal changes in insolation. Some plant seeds germinate only when daylength reaches a certain number of hours. A plant that responds in the opposite manner is the poinsettia, *Euphorbia pulcherrima,* which requires at least two months of 14-hour nights to start flowering.

Other components also are important to ecosystem processes. Air and soil temperatures determine the rates at which chemical reactions proceed (see Chapter 18). Significant temperature factors are seasonal variation, duration and pattern of minimum and maximum values, and the average (Chapter 5).

Operations of the hydrologic cycle and water availability depend on precipitation/evaporation rates and their seasonal distribution (see Chapters

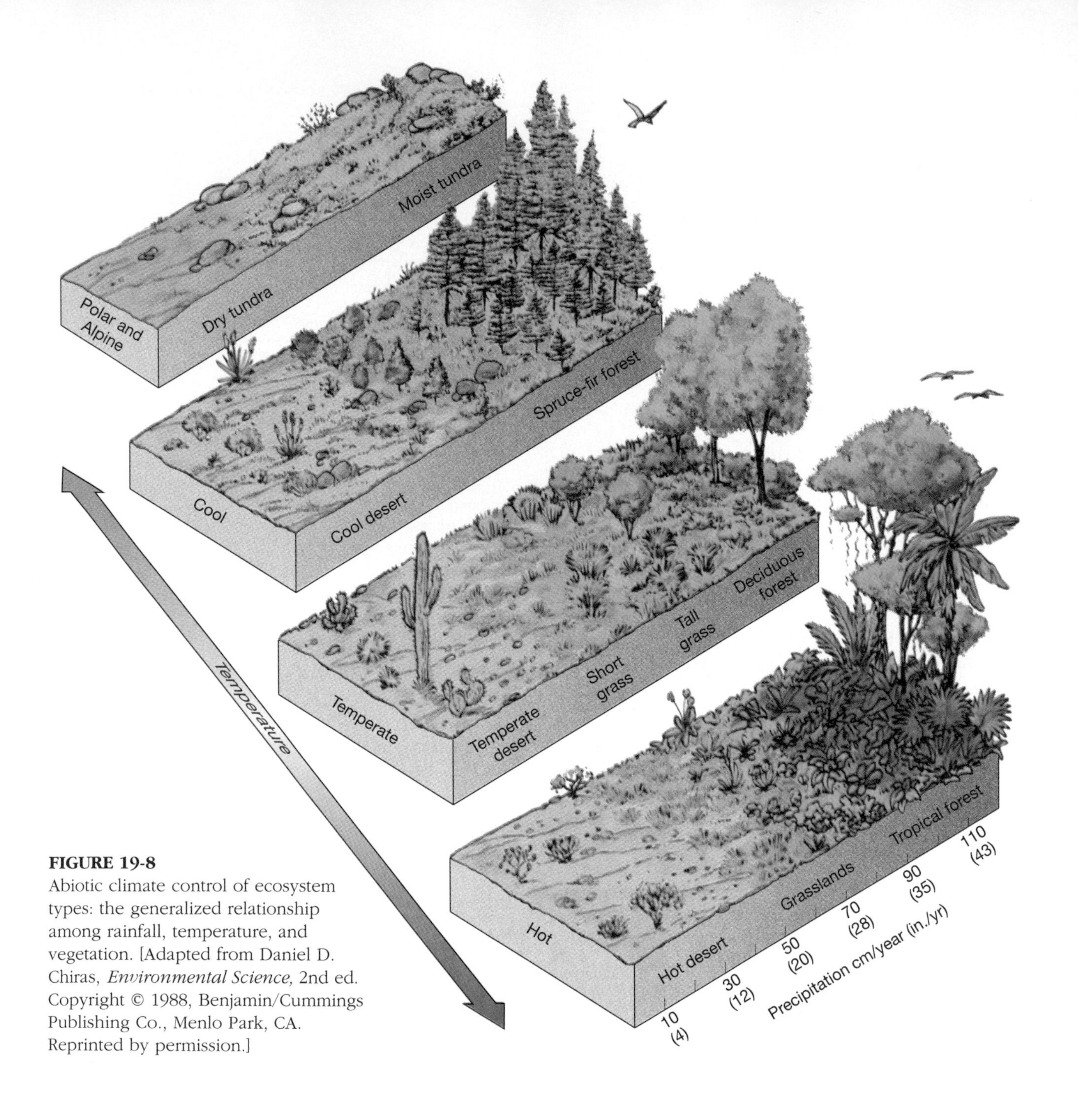

FIGURE 19-8
Abiotic climate control of ecosystem types: the generalized relationship among rainfall, temperature, and vegetation. [Adapted from Daniel D. Chiras, *Environmental Science,* 2nd ed. Copyright © 1988, Benjamin/Cummings Publishing Co., Menlo Park, CA. Reprinted by permission.]

7, 8, and 9). Water quality is important—its mineral content, salinity, and levels of pollution and toxicity. Also, daily weather patterns over time create regional climates (see Chapter 10), which in turn affect the pattern of vegetation and ultimately influence soil development. All of these factors work together to establish the parameters for ecosystems that may develop in a given location.

Figure 19-8 illustrates the general relationship among temperature, precipitation, and vegetation. In general terms, can you identify the characteristic vegetation type and related temperature and moisture relationship that fits the area of your town or school?

Alexander von Humboldt (1769–1859), an explorer, geographer, and scientist, deduced that

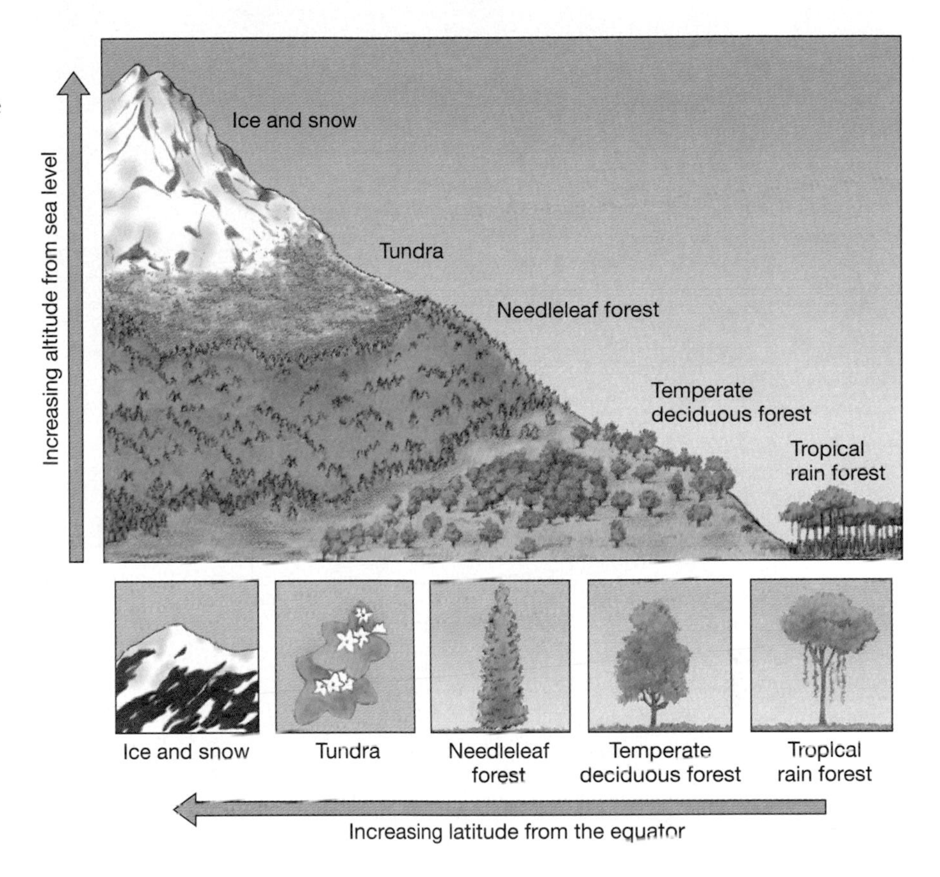

FIGURE 19-9
Progression of plant community life zones with increasing altitude or latitude.

plants and animals recur in related groupings wherever similar conditions occur in the abiotic environment. He described the similarities and dissimilarities of vegetation and the associated uneven distribution of various organisms. After several years of study in the Andes Mountains of Peru, he described a distinct relationship between altitude and plant communities, his *life zone* concept. As he climbed the mountains, he noticed that the experience was similar to that of traveling away from the equator toward higher latitudes (Figure 19-9).

This zonation of plants with altitude is noticeable on any trip from lower valleys to higher elevations. Each **life zone** possesses its own temperature, precipitation, and insolation relationships and therefore its own biotic communities. The Grand Canyon in Arizona provides a good example. The inner gorge at the bottom of the canyon (600 m or 2000 ft in elevation) has life forms characteristic of the lower Sonoran Desert of northern Mexico. However, the north rim of the canyon (2100 m or 7000 ft in elevation) is dominated by ecosystems similar to those of southern Canadian forests. On the summits of the nearby San Francisco Mountains (3600 m or 12,000 ft in elevation), the vegetation is similar to that of the arctic tundra of northern Canada.

Beyond these general conditions, each ecosystem further produces its own *microclimate,* specific to individual sites. For example, in forests the insolation reaching the ground is reduced. A pine forest cuts light by 20–40%, whereas a birch-beech forest reduces it by as much as 50–75%. Forests also are about 5% more humid than nonforested landscapes, have warmer winters and cooler summers, and experience reduced winds. Slope and exposure are important, too, for they translate into differences in temperature and moisture efficiency, especially in middle and higher latitudes. With all other factors equal, slopes facing away from the Sun's rays tend to be more moist and vegetated.

FIGURE 19-10
Fern and forest trail, indicative of a microecosystem. The microclimate at the forest floor is drastically changed when the forest in cleared. [Photo by author.]

Such highly localized *microecosystems* are evident along a mountain trail, where changes in exposure and moisture can be easily seen (Figure 19-10).

Gaseous and Sedimentary Cycles. The most abundant natural elements in living matter are hydrogen (H), oxygen (O), and carbon (C). Together, these elements make up more than 99% of Earth's biomass; in fact, all life (organic molecules) contains hydrogen and carbon. In addition, nitrogen (N), calcium (Ca), potassium (K), magnesium (Mg), sulfur (S), and phosphorus (P) are important *nutrients,* elements necessary for the growth and development of a living organism.

Several important cycles of chemical elements operate. Oxygen, carbon, and nitrogen each have gaseous cycles, part of which are in the atmosphere. Other elements have sedimentary cycles which principally involve the mineral and solid phases (major ones include phosphorus, calcium, and sulfur). Some elements combine gaseous and sedimentary cycles. These processes are called **biogeochemical cycles,** because they involve chemical reactions in both life and Earth systems. The chemical elements themselves recycle over and over again in life processes.

Oxygen and carbon cycles we consider together because they are so closely intertwined through photosynthesis and respiration (Figure 19-11). The atmosphere is the principal reserve of available oxygen. Larger reserves of oxygen than in the atmosphere exist in Earth's crust, but they are unavailable, being tied up in combination with other elements. Unoxidized reserves of fossil fuels and sediments also contain oxygen.

The greatest pool of carbon is in the ocean—about 39,000 billion tons, or about 93% of the total. All of this carbon is bound chemically in carbon dioxide, calcium carbonate, and other compounds. The ocean absorbs carbon dioxide through the functioning (photosynthesis) of small phytoplankton organisms. The atmosphere, which serves as the integrating link in the cycle, contains only about 700 billion tons of carbon at any moment, far less than fossil fuels and oil shales (12,000 billion tons) or living and dead organic matter (2275 billion tons). Carbon dioxide in the atmosphere is produced by the respiration of plants and animals, volcanic activity, and fossil fuel combustion by industry and transportation. In organic matter, carbon is stored in carbohydrate molecules; in fossil fuels it exists in hydrocarbon molecules. Carbon also is stored in certain carbonate minerals, such as limestone.

The carbon dioxide dumped into the atmosphere by human activity constitutes a vast geochemical experiment, using the real-time atmosphere as a laboratory. Since 1970, we have added to the atmospheric pool an amount of carbon equivalent to more than 25% of the total amount

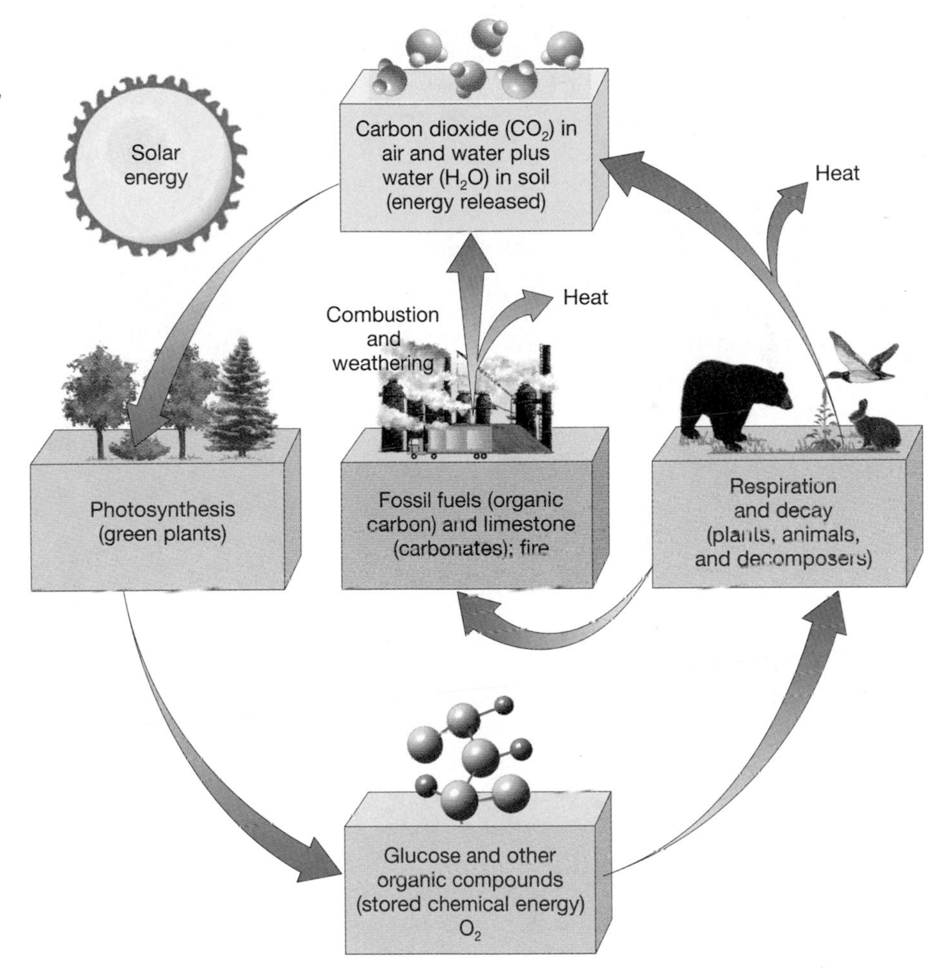

FIGURE 19-11
The carbon and oxygen cycles, simplified.

added since 1880. Annual emissions by society have now reached 6 billion tons and represent our greatest single atmospheric waste product. Furthermore, about 50% of the carbon dioxide emitted since the beginning of the industrial revolution and not absorbed by oceans and organisms is still in the atmosphere, enhancing Earth's natural greenhouse effect. This is an important topic of discussion in Chapters 3, 5, and 10.

The *nitrogen cycle* involves the major constituent of the atmosphere, 78.084% of each breath we take. Nitrogen also is important in the makeup of organic molecules, especially proteins, and therefore is essential to living processes. However, this vast atmospheric reservoir is inaccessible *directly* to most organisms. The key link to life is provided by nitrogen-fixing bacteria, which live principally in the soil and are associated with the roots of certain plants—for example, the legumes such as clover, alfalfa, soybeans, peas, beans, and peanuts. Bacteria colonies reside in nodules on the legume roots and chemically combine the nitrogen from the air in the form of nitrates (NO_3) and ammonia (NH_3). Plants use these fixed forms of nitrogen to produce their own organic matter. Anyone or anything feeding on the plants thus ingests the nitrogen. Finally, the nitrogen in the organic wastes of these consuming organisms is freed by denitrifying bacteria, which recycle it back to the atmosphere.

To improve agricultural yields, inorganic fertilizers are used. These fertilizers are chemically produced through artificial nitrogen fixation at factories. The annual production of inorganic fertilizers far exceeds the ability of natural denitrification systems—present manufacturing of inorganic fertilizers is now doubling every 8 years. The surplus of usable nitrogen accumulates in Earth's ecosystems. Some is present as excess nutrients, washed from soil into waterways and eventually the ocean. This excess nitrogen load begins a water pollution process that involves an unwanted growth of algae and phytoplankton, increased biological oxygen demand, loss of dissolved oxygen reserves, and eventual disruption of the aquatic ecosystem. In addition, excess nitrogen compounds in air pollution are a component in acid deposition, further altering the nitrogen cycle in soils and waterways (see Chapter 3).

Limiting Factors. The term **limiting factor** identifies the one physical or chemical abiotic component that most inhibits biotic operations, through its lack or excess. A few examples include low temperatures at high elevations, the lack of water in a desert, the excess water in a bog, the amount of iron in ocean surface environments, the phosphorus content of soils in the eastern United States, or the general lack of active chlorophyll above 6100 m (20,000 ft). In most ecosystems, precipitation is the limiting factor, although variation in temperatures and soil characteristics certainly affect vegetation patterns.

Each organism possesses a range of tolerance for each variable in its environment. This is illustrated vividly in Figure 19-12, which shows the geographic range for two tree species and two bird species. The coast redwood *(Sequoia sempervirens)* is limited to a narrow section of the Coast Ranges in California, covering barely 9500 km^2 (3667 mi^2) and concentrated in areas that receive summer advection fog. The red maple *(Acer rubrum),* on the other hand, thrives over a large area under varying conditions of moisture and temperature, thus demonstrating a broader tolerance to environmental variations.

The mallard duck *(Anas platyrhynchos)* and the snail kite *(Rostrhamus sociabilis)* also demonstrate a variation in tolerance and range. The mallard, a *generalist,* feeds from widely diverse sources, is easily domesticated, and is found throughout most of North America in at least one season of the year. By contrast, the snail kite is a *specialist* that feeds only on one specific type of snail. This single food source, then, is its limiting factor. Note its small habitat area near Lake Okeechobee in Florida.

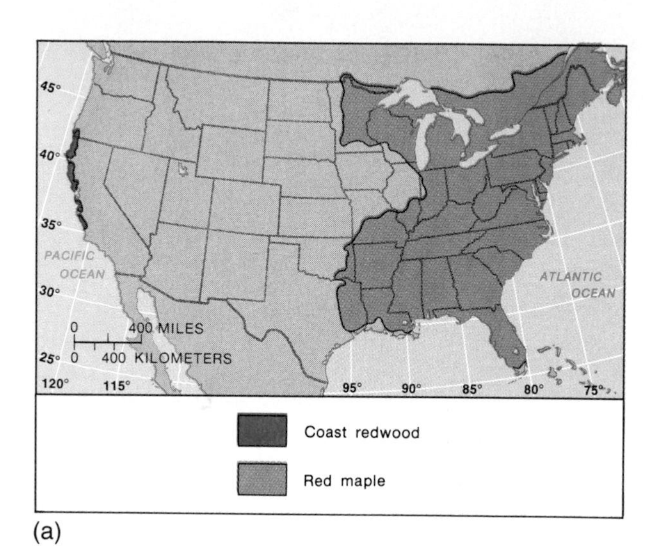

(a)

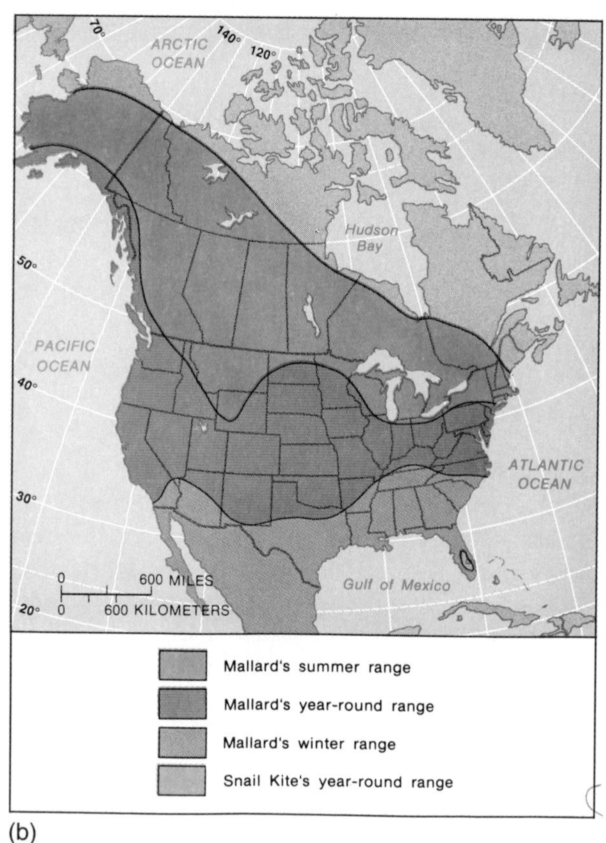

(b)

FIGURE 19-12
Biotic distribution. (a) Coast redwood and red maple; (b) mallard ducks and snail kites.

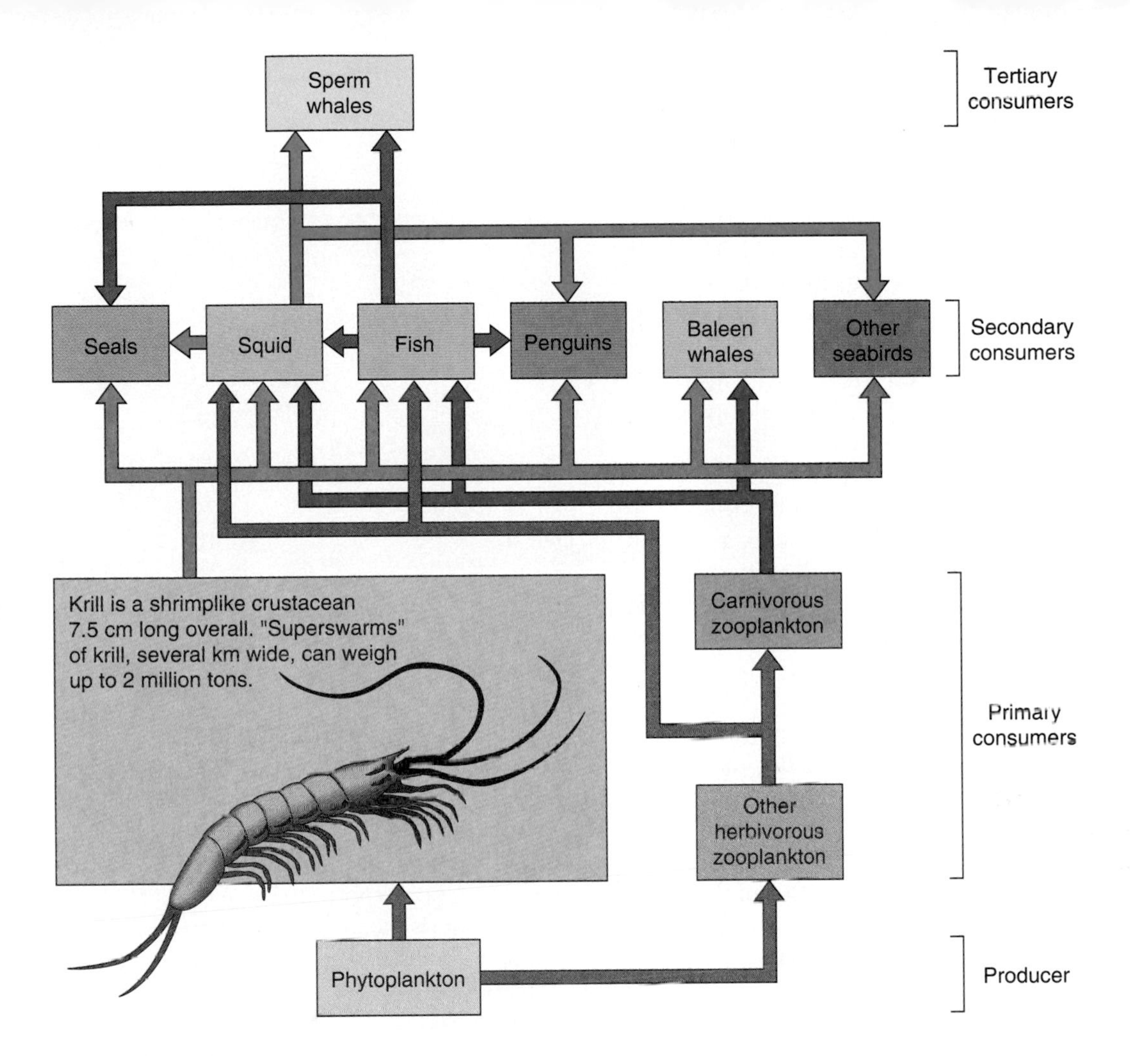

FIGURE 19-13
The food web of the krill in Antarctic waters, from phytoplankton *producers* (bottom) through various consumers. [After *State of the Ark* by Lee Durrell. Copyright © 1986 by Gaia Books Ltd. Adapted by permission of Doubleday, a division of Bantam Doubleday Dell Publishing Group, Inc.]

Biotic Ecosystem Operations

The abiotic components of energy, atmosphere, water, weather, climate, and minerals support the biotic components of each ecosystem and their constituent soils, vegetation, and life forms.

Producers, Consumers, and Decomposers. Organisms that are capable of using carbon dioxide as their sole source of carbon are called **autotrophs,** or **producers.** They chemically fix carbon through photosynthesis. Organisms that depend on autotrophs for their carbon are called **heterotrophs,** or **consumers.** Autotrophs are the essential producers in an ecosystem—capturing light and generating heat energy and converting it to chemical energy, incorporating carbon, forming new plant tissue and biomass, and freeing oxygen, all as a part of photosynthesis.

From these producers, which manufacture their own food, energy flows through the system along a circuit called the **food chain,** reaching consumers and eventually decomposers. Ecosystems generally are structured in a **food web,** a complex network of interconnected food chains. In a food web, consumers participate in several different food chains. Organisms that share the same basic foods are said to be at the same *trophic* (feeding, nutrition) *level.*

As an example, look at the food web based on krill (Figure 19-13). Krill is a shrimplike crustacean about 7 cm (2.8 in.) long that is a major food source for an interrelated group of organisms, including whales, fish, seabirds, seals, and squid in the Antarctic region. All of these organisms participate in numerous other food chains as well, some consuming and some being consumed.

Phytoplankton begin this chain by harvesting solar energy in photosynthesis. The krill, along with other organisms, feed on the phytoplankton. Krill in turn are fed upon by the next trophic level. Because they are a protein-rich, plentiful food, krill are increasingly sought by factory ships, such as those from Japan and Russia. The annual harvest currently surpasses half a million tons. The impact on the food web of further increases in the harvest is uncertain, for little is known about krill—their reproduction rate, life span, and availability. The possible effects on krill of increased ultraviolet radiation resulting from the seasonal hole in the ozone layer above Antarctica is under investigation. All of these interrelationships need to be clarified before further harvesting continues.

Primary consumers feed on producers. Because producers are always plants, the primary consumer is called a **herbivore,** or plant eater. A **carnivore** is a *secondary consumer* and primarily eats meat. A *tertiary consumer* eats secondary consumers and is referred to as the "top carnivore" in the food chain, like the sperm whale in the krill web. A consumer that feeds on both producers (plants) and consumers (meat) is called an **omnivore**—a role occupied by humans, among others.

Any assessment of world food resources depends on the level of consumer being targeted. Using humans as an example, many can be fed from wheat harvested from an acre of land, which produces about 810 kg (1800 pounds) of grain. However, if herbivores eat that grain, only 82 kg (180 pounds) of biomass is produced which can feed far fewer people (Figure 19-14). In terms of energy, only about 10% of the calories in plant matter survive from the primary to the secondary trophic level. When humans consume the meat, there is a further loss of biomass and added inefficiency. And more energy is lost to the environment at each progressive step in the food chain. You can see that an omnivorous diet is quite expensive in terms of biomass and energy.

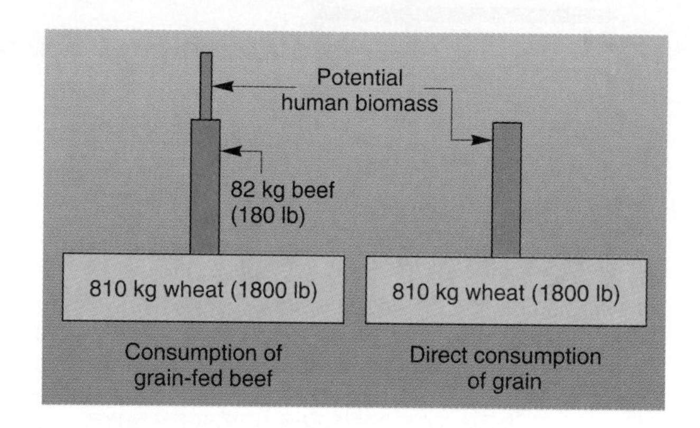

FIGURE 19-14
Biomass pyramids illustrating the difference between direct and indirect consumption of grain.

Food chain concepts are becoming politicized as world food issues grow more critical. Today, approximately half of the cultivated acreage in the United States and Canada is planted for animal consumption. This includes more than 80% of the annual corn and nonexported soybean harvest. In addition, some lands cleared of rain forests in Central and South America have been converted to pasture to produce beef for export to some restaurants, stores, and fast-food outlets in developed countries. Thus, life-style decisions and dietary patterns in the United States and Canada are perpetuating certain food chains, not to mention the destruction of valuable resources, both here and overseas.

Decomposers are the final links in the chain; they are the microorganisms—bacteria, fungi, insects, worms, and others—that digest and recycle the organic debris and waste in the environment. Waste products, dead plants and animals, and other organic remains are their principal food source. Material is released by the decomposers and enters the food chain—and the cycle continues.

Clearly, portions of some food chains are exceptionally simple, such as eating grains directly. Others are more complex than the Antarctic krill food chain. And familiarity is no indication of simplicity: the home gardener's tomatoes may be eaten by a tomato hornworm, which is then plucked off by a passing robin, which is later eaten by a hawk—and so it goes, in seemingly endless cycles.

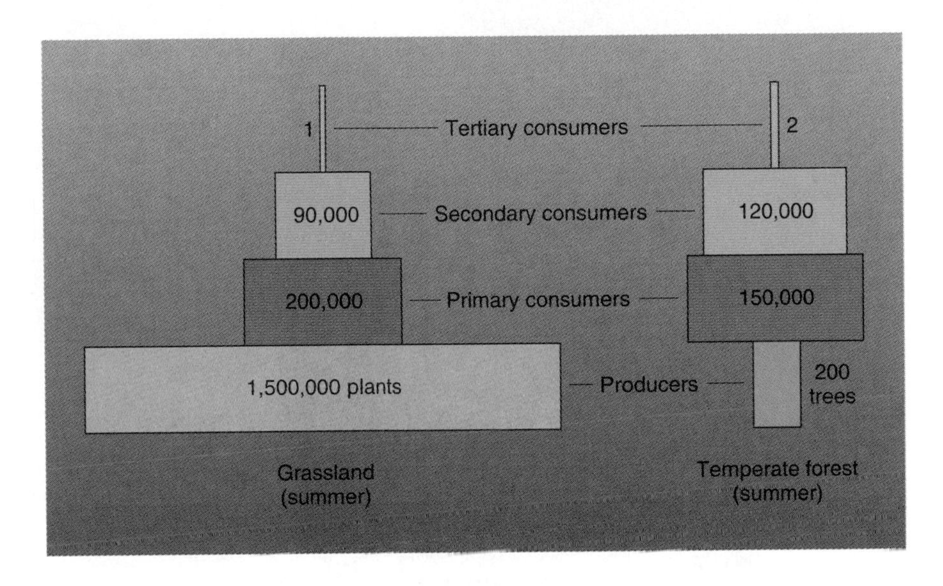

FIGURE 19-15
Ecological pyramids for 0.1 hectare (0.25 acre) of grassland and forest in the summer. Numbers of consumers are shown. [Adaptation of Figure 3-15a from *Fundamentals of Ecology* by Eugene P. Odum, copyright © 1971 by Saunders College Publishing, a division of Holt, Rinehart and Winston, Inc. Adapted by permission of the publisher.]

Ecological Relationships. A study of food chains and webs is a study of who eats what, and where they do the eating. Figure 19-15 demonstrates the summertime distribution of populations in two ecosystems, grassland and temperate forest. The stepped population pyramid is characteristic of summer conditions in such ecosystems. You can see the decreasing number of organisms at each successive higher trophic level. The base of the temperate forest pyramid is narrow, however, because most of the producers are large, highly productive trees and shrubs, which are outnumbered by the consumers they can support in the chain.

A basic approach to ecological study is to analyze a community's metabolism. Metabolism is the way in which a community uses energy and produces food for continued operation (the sum of all its chemical processes). In 1957 H. T. Odum completed one of the best-known studies of this type for Silver Springs, Florida. Figure 19-16a, an adaptation of his work, shows that the energy source for the community is insolation. The amount of light absorbed by the plants for photosynthesis is 24% of the total amount arriving at the study site. Of the absorbed amount, only 5% is transformed into the community's gross photosynthetic production—and this is only 1.2% of the total insolation input. The net plant production (gross production minus respiration) is 42.4% of the gross production, or 8833 kilocalories per square meter per year. That amount then moves through the food chain of herbivores and carnivores, with the top carnivores receiving only 0.24% of the net plant production of the system.

You can see that at each trophic level some of the biomass flows to the decomposers. The largest portion of the biomass is exported from the system in the form of community respiration. In addition, a relatively small portion leaves the system as organic particles by downstream export from the system. The biomass pyramid shown in Figure 19-16b portrays the same community from a standing crop perspective, which is an important structural way of examining the existing biomass at each trophic level.

Concentration in Food Chains. When a typical ecosystem of producers and consumers has certain chemical pesticides applied to the producer population, the food web functions to concentrate some of these chemicals. Many chemicals are degraded or diluted in air and water and thus are rendered relatively harmless. However, other chemicals are long-lived, stable, and soluble in the fatty tissues of consumers. These become increasingly concentrated at each higher trophic level. Figure 19-17 illustrates this *biological amplification,* or *magnification.*

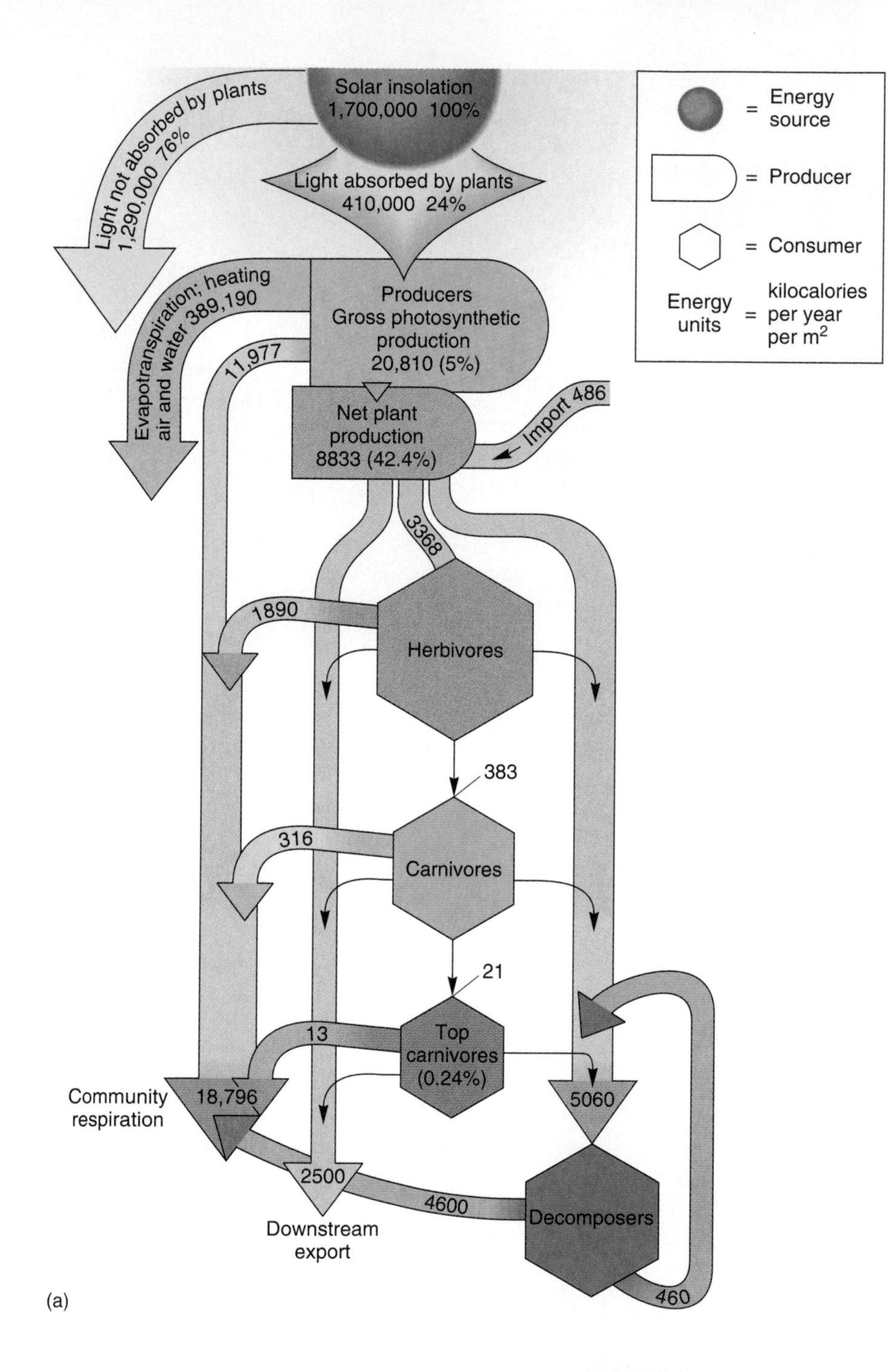

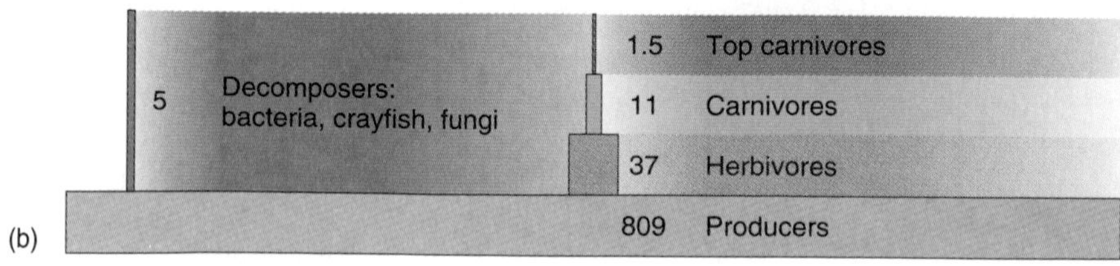

FIGURE 19-16
Analysis of community metabolism (a) and standing crop distribution in grams per square meter (b) for Silver Springs, Florida. [Adaptation of Figures 6-1a and 6-1b from *Fundamentals of Ecology* by Eugene P. Odum, copyright © 1971 by Saunders College Publishing, a division of Holt, Rinehart and Winston, Inc. Adapted by permission of the publisher.]

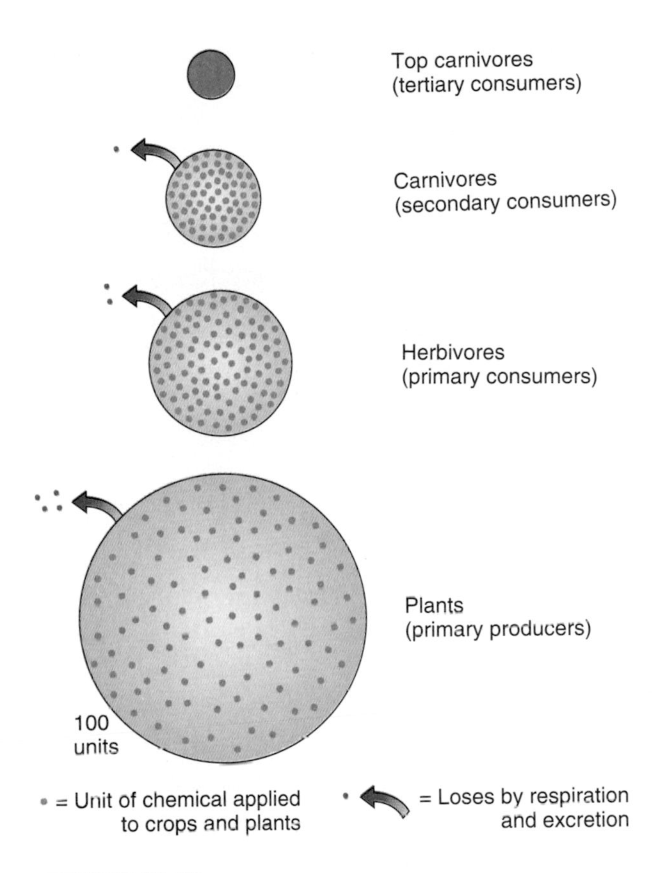

FIGURE 19-17
Chemical concentration in food chains. Chemical residues are passed along a simple food chain to the top carnivores.

DDT is one such chemical. Although banned in some countries, it still is widely used in others. In addition, some other organic and synthetic chemicals, radioactive debris, and heavy metals such as lead and mercury become concentrated in the food chain, with only small amounts being lost to the environment through respiration and excretion. Thus, a food chain can efficiently poison the organism at the top. Many species are threatened in this manner, and, of course, humans are at the top of the food chain and can receive concentrated chemicals passed along in what is consumed.

STABILITY AND SUCCESSION

A rough equilibrium is achieved in an ecosystem by the constant interplay of increasing growth toward the potential in a community and decreasing growth caused by resisting factors that force limits in a community. Both biotic and abiotic factors affect growth (Figure 19-18). Far from being static, Earth's ecosystems are dynamic and ever-changing. Over time, communities of plants and animals have adapted to such variation, evolved, and in turn shaped their environments.

Ecosystem Stability

In a given ecosystem, a community moves toward maximum biomass and relative stability because

FIGURE 19-18
Ecosystem and population balance in a community.

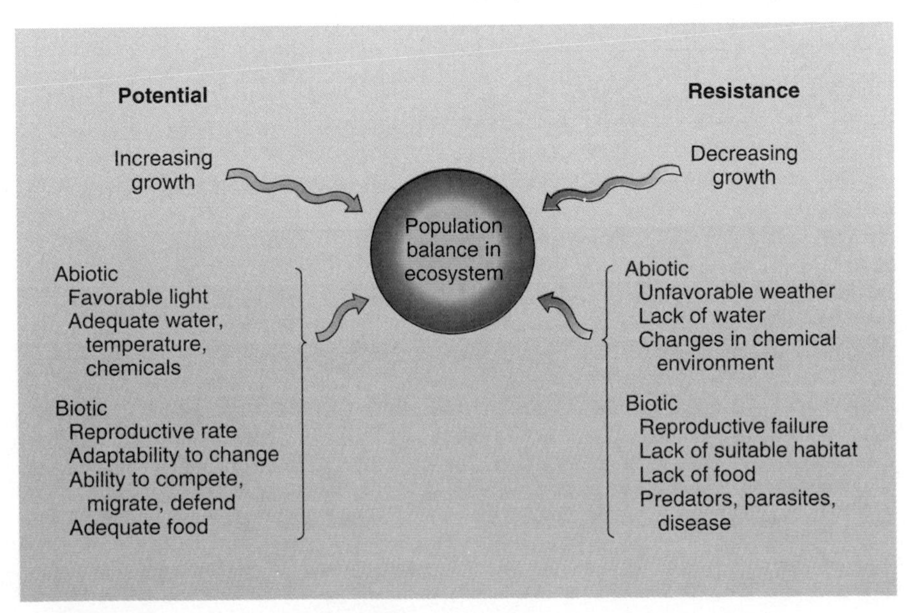

(a)

(b)

FIGURE 19-19

Natural and human disruption of stable communities. (a) Natural disruption of the forest by the Mount Saint Helens volcanic eruption in 1980 and by logging in myriad clear-cut tracts of national forest. About 10% of the old-growth forests remain in the northwest, as identified through orbital images and GIS analysis. (b) A logged pine forest—an example of clear-cut timber harvesting disrupting a stable community. [(a) *Landsat* image of a portion of the Gifford Pinchot National Forest and the Mount Saint Helens National Volcanic Monument, April 29, 1992, centered at approximately 46.5° N 122° W, courtesy of Compton J. Tucker, NASA Goddard Space Flight Center, Greenbelt, Maryland. (b) Photo by author.]

this has optimum survival value for the community. However, the tendency for birth and death rates to balance and the composition of species to remain stable, *inertial stability,* does not necessarily foster the ability to recover from change, *resilience.* Examples of stable communities include a redwood forest, a pine forest at a high elevation, and a tropical rain forest near the equator. Yet, cleared tracts recover slowly (if ever) and therefore have poor resilience. Figure 19-19 shows how clear-cut tracts of former forest have altered drastically the microclimatic conditions, making regrowth of similar species difficult. In contrast, a midlatitude grassland is low in stability; yet when burned, its resilience is high because the community recovers rapidly.

Another aspect related to stability is **diversity.** The more diverse the species population (both in number of species and quantity of members in each species), the more risks are spread over the entire community, because several food sources exist at each trophic level. In other words, greater diversity in an ecosystem results in greater stability. An artificially produced monoculture community, such as a field of wheat, is singularly vulnerable to failure or attack. Humans simplify communities by eliminating diversity and in this way we place more ecosystems at risk of unwanted change and perhaps failure. In some regions, a simple return to multiple crops brings more stability to the ecosystem.

Agricultural Ecosystems. A modern *agricultural ecosystem* not only is vulnerable to failure due to its lack of ecological diversity, but also creates an enormous demand for energy, chemical pesticides and herbicides, artificial fertilizer, and irrigation water. The practice of harvesting and removing biomass from the land interrupts the cycling of materials into the soil. This net loss of nutrients must be artificially replenished.

Energy from fossil fuels is used to manufacture chemicals, to operate mechanized farm equipment, and to run pumps for wells and irrigation. (Focus Study 9-1 discusses the problem of fuel costs for pumping water from the Ogallala Aquifer.) High energy cost and fuel availability problems can lead to crop failures. Unfortunately the promise of increased yields through a *green revolution* in agriculture is dampened by these investment requirements.

Climate Change. Certainly the distribution of plant species is affected by changes in climate. Plant communities have survived wide climate swings in the past. Consider the beginning of the Tertiary period, 75 million years ago. Warm, humid conditions and tropical forests dominated northward to southern Canada, pines were in the Arctic, and deserts were few. Then, between 15 and 50 million years ago, deserts began developing in the southwestern United States, and mountain-building processes were creating higher elevations, with leeward slopes and rainshadow aridity affecting plants.

The movement of Earth's tectonic plates created climate changes that played an important role in the evolution and distribution of plants and animals (see Figure 11-16a–e). Europe and North America were joined in Pangaea and positioned near the equator where vast swamps formed (site of coal deposits today). The southern mass of Gondwana was extensively glaciated as it drifted at high latitudes in the Southern Hemisphere. This glaciation left matching glacial scars and specific distributions of plants and animals across South America, Africa, India, and Australia. The diverse and majestic dinosaurs dispersed with the drifting continents.

The key question is: as temperature patterns change, how fast can plants either adapt to the new conditions, or migrate to remain within their specific habitats?

Adaptation by species is important in evolution. Through mutation and natural selection, species adapted or failed to adapt to changing environmental conditions over millions of years. The current pace of global change is occurring fast, at the rate of decades instead of millions of years. Thus, we see a die-out along disadvantageous margins with a succession to different species. The displaced species may colonize new regions made more advantageous by climate change—a migration through succession to maintain a suitable habitat. Rapid environmental change can lead to outright extinction of plants and animals unable to adapt or disperse.

As an example, a transition area between prairie grassland and northern forest presently exists in central Minnesota, but with increasing temperatures this transition zone is expected to move northward 400–600 km (250–375 mi) over the next 40 years, an adjustment of more than 100 km (60 mi) per decade. Agricultural lands producing wheat, corn, soybeans, and other commodities also will shift. Society will have to adapt to new cropping patterns.

A study completed by biologist Margaret Davis

on North American forests suggests that trees will have to respond quickly if temperatures increase. Shifts in the climate inhabited by certain species could attain 100–400 kilometers (60–250 mi) during the next 100 years. Some species, such as the sugar maple, may migrate northward, disappearing from the United States except for Maine, and moving into eastern Ontario and Quebec. Davis prepared a map showing the possible impact of increasing temperatures on the distribution of beech trees (Figure 19-20).

> **Changes in the geographical distributions of plants and animal species in response to future greenhouse warming threaten to reduce biotic diversity. . . . The risk posed by CO_2-induced warming depends on the distances that regions of suitable climate are displaced northward [in the Northern Hemisphere] and on the rate of displacement. . . . If the change occurs too rapidly for colonization of newly available regions, population sizes may fall to critical levels, and extinction will occur.***

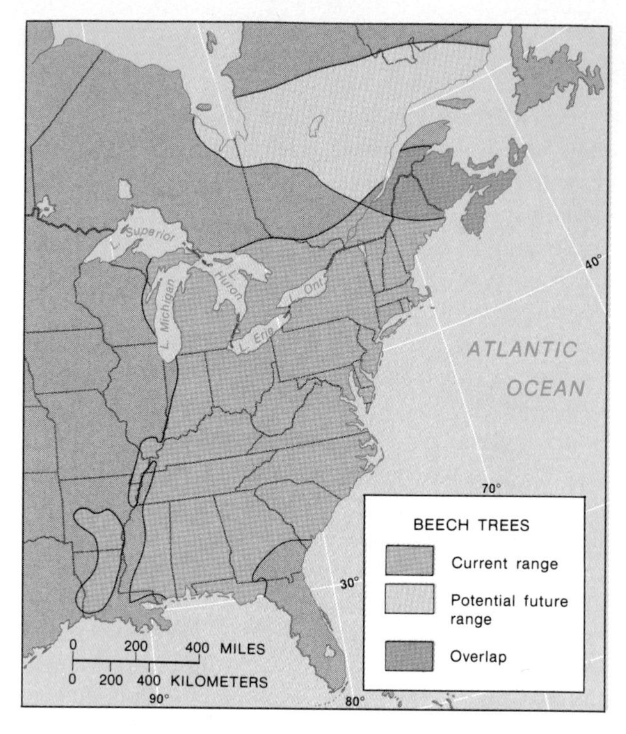

FIGURE 19-20
Present and predicted distribution of beech trees in North America, given estimated climate changes resulting from a doubling of CO_2 using the Goddard Fluid Dynamics Laboratory (GFDL) general circulation model. [After *Science* 243 (10 February 1989): 735. "How Fast Can Trees Migrate?" by Leslie Roberts, 1989. Copyright 1989 by the AAAS.]

Ecological Succession

Ecological succession occurs when older communities of plants and animals (usually simpler) are replaced by newer communities (usually more complex). Each successive community of species modifies the physical environment in a manner suitable for the establishment of a later community of species. Changes apparently move toward a more stable and mature condition, to be optimum for a specific environment. This end product in an area is called the *ecological climax,* with plants and animals forming a *climax community*—a stable, self-sustaining, and symbiotically functioning community with balanced birth, growth, and death—which itself can be disturbed.

A generalized ecological succession provides a convenient model with which to characterize community development. However, given the complexity of natural ecosystems, real succession involves much more than a series of predictable stages ending with a specific "monoclimax" community. Instead, there may be several final stages, or a "polyclimax" condition. Climax communities are properly thought of as being in *dynamic equilibrium,* and at times may even be out of phase with the immediate physical environment due to a lag time in adjustment.

Succession often requires an initiating disturbance, such as strong winds or a storm, or a practice such as prolonged overgrazing. Early communities may actually inhibit the growth of other species, causing a temporary stability, but when existing organisms are disturbed or removed, new communities can emerge. At such times of transition, the interrelationships among species produce elements of chance, and species having an adaptive edge will succeed in the competitive struggle for light, water, nutrients, space, time, reproduction, and survival. Thus, the succession of plant

*M. B. Davis and C. Zabinski, "Changes in the Geographical Range Resulting from Greenhouse Warming: Effects on Biodiversity in Forests," in R. L. Peters and T. E. Lovejoy, eds., *Global Warming and Biological Diversity*. New Haven: Yale University Press, 1992, p. 297.

and animal communities is an involved process with many interactive variables.

As Earth cycled through glacial and interglacial ages (Chapter 17), the ecological succession in all ecosystems was affected repeatedly. Imagine the milder climate in the midlatitudes at the beginning of an interglacial, with lush herb and shrub vegetation slowly giving way to pioneer trees of birch, aspen, and pine. Winds and animals dispersed seeds and changing communities readily spread in response to changing conditions. During the warm interglacial, trees increased shade and the organic content of soil increased. Deciduous trees such as oak, elm, and ash spread freely on well-drained soils; willow, cottonwood, and alder grew on poorly drained land.

As climates cooled with the approach of the next glacial, soils became more acidic and podzolization processes (see Chapter 18) dominated, helped by spruce and fir trees. Ecosystems slowly returned to the vegetation community that existed at the beginning of the interglacial. The increased cold produced open regions of disturbed ecosystems. Highly acidic soils were disrupted by freezing and the advance of ice.

The Holocene (last 10,000 years) is apparently atypical because of the arrival of humans and the development of human societies and agriculture. Civilization is accelerating an artificial succession, similar in some ways to the closing centuries of an interglacial, although at a greatly accelerated pace. The challenge for biogeographers is to understand such long-term successional change during a time of short-term rapid climate change and global warming.

Terrestrial Succession. An area of bare rock and soil without any vestige of a former community can be a site for primary succession. The initial community is called a **pioneer community.** It may be found in sites such as a new surface created by mass movements of land, in areas exposed by a retreating glacier, on devastated lands such as those north of Mount Saint Helens in Washington State (Figure 19-21), on cooled lava flows, or on lands disturbed by surface mining, clear-cut logging, and land development. Mosses and lichens often begin primary succession on bare rock (see Figure 19-3). However, succession from a previously functioning community is more common. An area whose natural community has been destroyed or disturbed, but still has the underlying soil intact, may experience *secondary succession* (Figure 19-22).

In terrestrial ecosystems, regrowth begins with pioneer species and further soil development (Figure 19-21). As succession progresses, a different set of plants and animals with different niche requirements may adapt. Further niche expansion follows as the community matures. For example, when a previously farmed field is permitted to lie fallow (unused), it initially fills with annual weeds, then perennial weeds and grasses, followed by shrubs and perhaps trees.

Figure 19-22 illustrates such a sequence for the southeastern United States. Secondary succession begins on an abandoned farm of formerly plowed fields (left side of illustration). Crabgrass and ragweed quickly take hold and do well in direct sunlight. These slowly give way to grasses and shrubs that invade and stabilize the soil, adding nutrients and organic matter. Pines eventually dominate the land from year 25 through the first century. The shade created by the pine forest produces conditions in which seed germination becomes more difficult. Shade-tolerant, slow-growing oak and hickory hardwoods readily take root beneath the pines. As these hardwoods grow they eventually shade the pine forest, which slowly dies back in reduced light conditions. A fairly stable climax forest of oak and hickory is in place between 150 and 200 years (right side of illustration). But despite the convenient categorization shown in the figure, you should think of succession as continuous, with succeeding communities overlapping in time and space.

One challenge in a study of plant and animal communities is to devise criteria that allow researchers to distinguish between a *successional community* and a *climax community*. The relative stability of the community appears to be the key factor. Unfortunately, however, that stability must be assessed over a period of time that greatly exceeds human life spans and may be interrupted from time to time by phenomena ranging from storms to volcanoes to widespread disease in one or more species. Focus Study 19-1 addresses fire, which is one such natural interruption.

FIGURE 19-21
Succession and recovery of pioneer species in the devastated area north of Mount Saint Helens, three years after the 1980 eruption. [Photo by author.]

FIGURE 19-22
Typical secondary succession of principal plants in the southeastern United States. [Adaptation of Figure 9-4 from *Fundamentals of Ecology* by Eugene P. Odum, copyright © 1971 by Saunders College Publishing, a division of Holt, Rinehart and Winston, Inc. Adapted by permission of the publisher.]

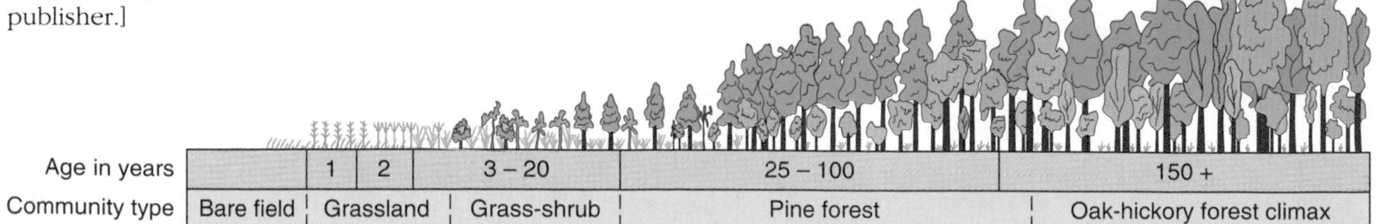

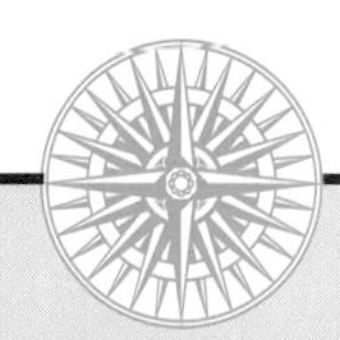

Focus Study 19-1 Fire Ecology and Succession

It is estimated that one-fourth of Earth's land area experiences fire each year. Over the past 50 years, *fire ecology* has been the subject of much scientific research and experimentation. Today, fire is recognized as a natural component of most ecosystems and not the enemy of nature it once was popularly considered to be. In fact, in many forests, undergrowth is purposely burned in controlled "cool fires" to remove fuel that could enable a catastrophic and destructive "hot fire." In contrast, when fire suppression and prevention strategies are rigidly followed, they can lead to abundant undergrowth accumulation, which allows the potential total destruction of a forest by a major fire.

Fire ecology imitates nature by recognizing fire as a dynamic ingredient in community succession. The U.S. Department of Agriculture's Forest Service first recognized the principle of fire ecology in the early 1940s and formally implemented the practice in 1972. Their challenge became one of controlling fires to secure reproduction and to prevent accumulation of forest litter and brush. Controlled ground fires now are widely regarded as wise forest management practice and are used across the country (Figure 1).

The geographer Carl Sauer (1889–1975) wrote in 1955 about the use of fire throughout the history of civilization. He stated that people often used fire deliberately to increase the food supply on their land:

> **Mature woody growth provides less food for humans and ground animals than do fire-disturbed sites, with protein-rich young growth and stimulated seed production. . . game yields are usually greatest where the vegetation is kept in an immediate state of ecological succession.***

Sauer contended that the origin and preservation of grasslands, which occur in various climates and over many soil types, are mainly the result of human-set fires. Their origin, however, remains controversial.

Modern society's demand for fire prevention to protect property goes back to European forestry of the 1800s. Fire prevention became an article of faith for forest managers in North America. However, in studies

---

*Carl O. Sauer, "Agency of Man on Earth," in William L. Thomas, *Man's Role in Changing the Face of the Earth*. Chicago: University of Chicago Press, 1956, p. 54.

FIGURE 1
Fire ecology practices of the U.S. Forest Service. Controlled burning has been used for several decades. [Photo by author.]

FIGURE 2
Yellowstone National Park in northwestern Wyoming: (a) fire of 1988 and (b) map of burned vegetation. [Courtesy of the National Park Service.]

(a)

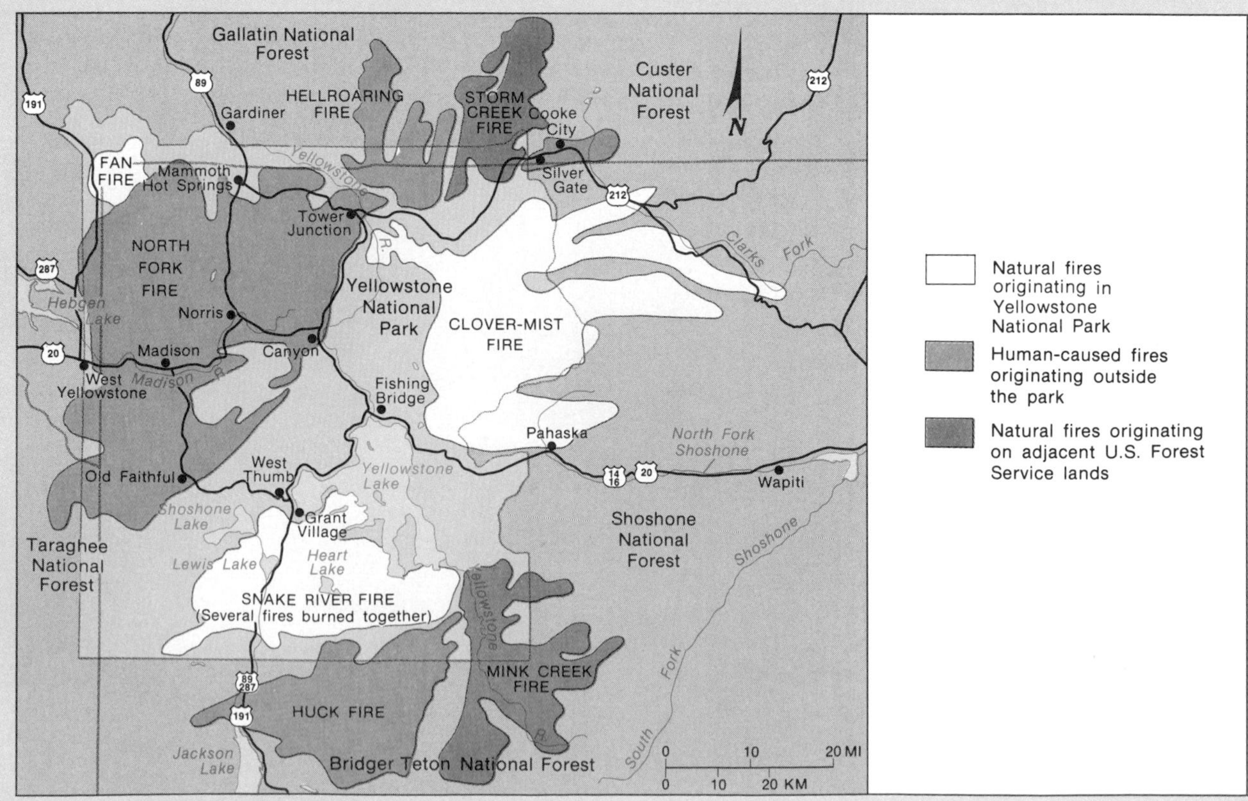

(b)

of longleaf pine forest that stretches in a wide band from the Atlantic coastal plain to Texas, fire was discovered to be an integral part of their regrowth following lumbering. In fact, in some pine species, such as the knobcone pine, seed dispersal does not occur unless assisted by a forest fire! Heat from the fire opens the cones, releasing seeds so they can fall to the ground for germination.

Nonetheless, after some 72,000 forest fires in the western United States in 1988, especially those that charred portions of the highly visible Yellowstone National Park, an outcry was heard from forestry and recreational interests. The demand was for the Forest Service and ecologists to admit they were wrong and to abandon fire ecology. Critics called fire ecology practices the government's "let burn" policy. In truth, about 20% of the acreage in Yellowstone Park actually burned (180,000 out of 900,000 hectares, or 440,000 of 2,200,000 acres). Only about half of that area experienced the worst fire damage (Figure 2). This expanse was far less than originally reported in the media.

In its final report on the fire, a government interagency task force concluded that, "an attempt to exclude fire from these lands leads to major unnatural changes in vegetation . . . as well as creating fuel accumulation that can lead to uncontrollable, sometimes very damaging wildfire." Thus, participating federal land managers and others reaffirmed their stand that fire ecology is a fundamentally sound concept.

Aquatic Succession. Lakes and ponds exhibit another form of ecological succession. A lake experiences successional stages as it fills with nutrients and sediment and as aquatic plants take root and grow, capturing more sediment and adding organic debris to the system (Figure 19-23). This gradual enrichment through various stages in water bodies is known as **eutrophication.** For example, in moist climates a floating mat of vegetation grows outward from the shore as the lake fills, to form a bog. Cattails and other marsh plants become established, and partially decomposed organic material accumulates in the basin, with additional vegetation bordering the remaining lake surface. A meadow may form as the peat bog solidifies; willow trees follow, and perhaps cottonwood trees; and eventually the lake may evolve into a forest community.

The progressive stages in lake succession are named *oligotrophic* (low nutrients), *mesotrophic* (medium nutrients), and *eutrophic* (high nutrients). Each stage is marked by an increase in primary productivity and resultant decreases in water transparency so that photosynthesis becomes concentrated near the surface. Energy flow shifts from production to respiration in the eutrophic stage, with oxygen demand exceeding oxygen availability.

The nutrient levels vary in different areas of a lake: oligotrophic conditions occur in deep water, and eutrophic conditions along the shore, in shallow bays, or in locations where sewage or other nutrient inputs occur. Even large water bodies may have eutrophic areas along the shore. As society dumps sewage and pollution in waterways, the nutrient load is enhanced beyond the cleansing ability of natural biological processes, thus producing *cultural eutrophication* and hastening the succession in aquatic systems.

SUMMARY—Ecosystem Essentials

Earth's biosphere is unique in the Solar System; its ecosystems are the heart of life. Ecosystems are self-regulating associations of living plants and animals and their nonliving physical environment. Earth's ecosystems are woven in a complex, interdependent web, with each life form mutually tied to the success of the others. The diversity of life on Earth is impressive.

Biogeography, essentially a spatial ecology, is the study of the distribution of plants and animals and the diverse spatial patterns they create. Plants and animals have developed complex community relationships centered in specific habitats and nich-

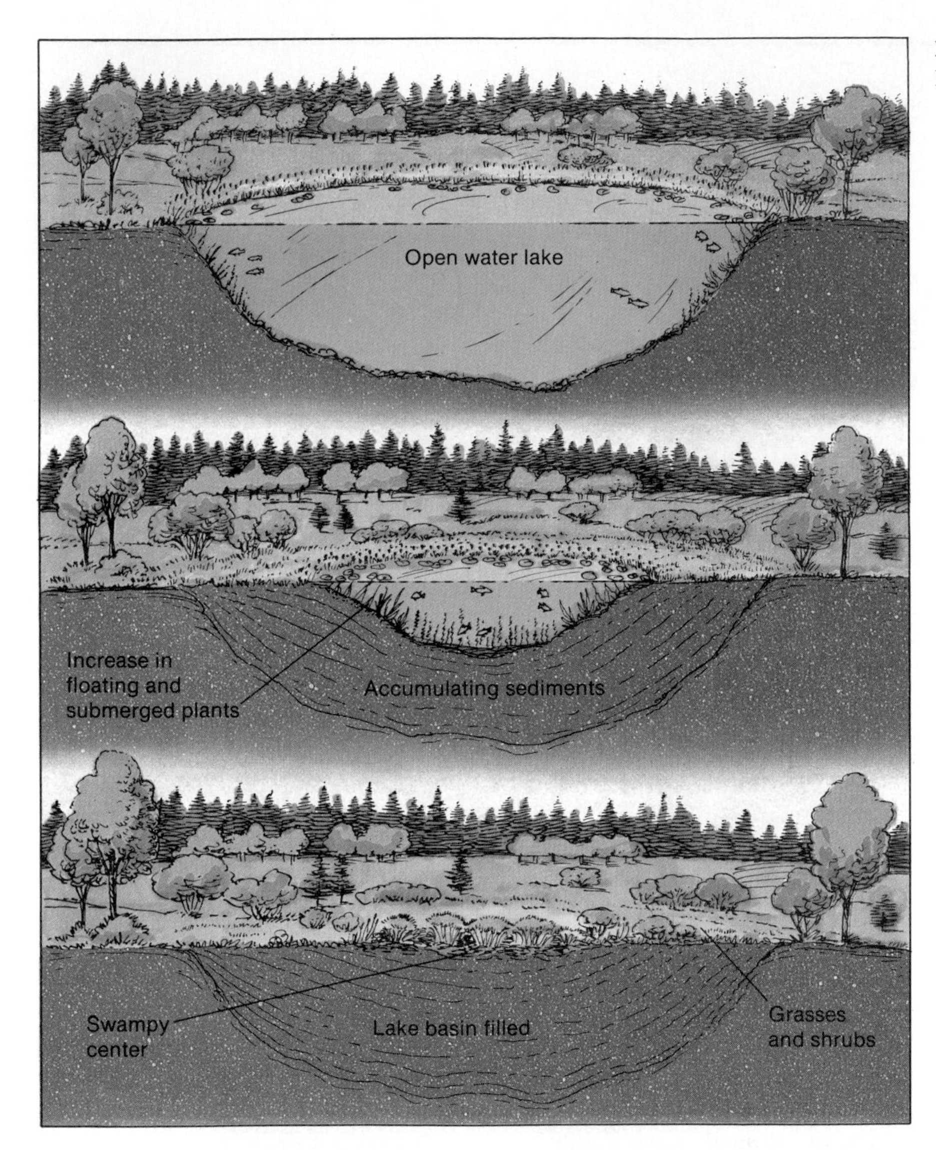

Figure 19-23
Lake bog succession.

es. Plants perform photosynthesis, providing energy and oxygen to drive biological processes. Net primary productivity may be the most important aspect of a community. Light, temperature, water, and the cycling of gases and nutrients constitute the supporting abiotic components of ecosystems. Biomass and population pyramids characterize the flow of energy and the numbers of producers, consumers, and decomposers operating in an ecosystem.

Ecosystems appear to move toward stability and balance, with individual terrestrial and aquatic ecosystems capable of progressing through a succession of stages. Biogeographers trace communities across the ages, considering plate tectonics and ancient dispersals of plants and animals. Past glacial and interglacial climatic episodes have created a long-term succession that is a challenge for researchers to decipher. Human activity intercedes at almost every level of natural ecosystem operations, producing artificial conditions, ecosystems that lack diversity, and intermediate stages of interrupted succession.

KEY TERMS

autotrophs
biogeochemical cycles
biogeography
biomass
carnivore
chlorophyll
community
consumers
decomposers
diversity
ecological succession
ecology
ecosystem
eutrophication
food chain
food web
habitat
herbivore
heterotrophs
life zone
limiting factor
net primary productivity
niche
omnivore
photoperiod
photosynthesis
pioneer community
producers
respiration
stomata
vascular plants

REVIEW QUESTIONS

1. What is the relationship between the biosphere and an ecosystem? Define *ecosystem* and give some examples.
2. What does biogeography include? Describe its relationship to ecology.
3. Briefly summarize what an ecosystem is and what its operation implies about the complexity of life.
4. Define a community within an ecosystem.
5. What do the concepts of habitat and niche involve? Relate them to some specific plant and animal communities.
6. Describe symbiotic and parasitic relationships in nature. Draw an analogy between these relationships and human societies on our planet. Explain.
7. What are the principal abiotic components in terrestrial ecosystems?
8. Describe what Humboldt found that led him to propose the life-zone concept. What are life zones? Explain the interaction among altitude, latitude, and the types of communities that develop.
9. What are biogeochemical cycles? Describe several of the essential cycles.
10. What role is played in an ecosystem by autotrophs and heterotrophs?
11. What is a limiting factor? How does it function to control the spatial distribution of plant and animal species? Explain the concept of tolerance in this connection.
12. Define a vascular plant. How many species are there on Earth?
13. How do plants function to link the Sun's energy to living organisms? What is formed within the light-responsive cells of plants?
14. What is net photosynthesis? How does the concept relate to photosynthesis, respiration, or net primary productivity?
15. What are the broad spatial patterns worldwide of net primary productivity?

16. Describe the relationship among producers, consumers, and decomposers in an ecosystem. What is the trophic nature of an ecosystem? What is the place of humans in a trophic system?
17. What are biomass and population pyramids? Describe how these models help explain the nature of food chains and communities of plants and animals.
18. Follow the flow of energy and biomass through the Silver Springs, Florida, ecosystem. Describe the pathways in Figure 19-16a.
19. What is meant by ecosystem stability?
20. How does ecological succession proceed? What are the relationships that exist between existing communities and new, invading communities? Explain the principle of competition.
21. Summarize the process of succession in a body of water. What is meant by cultural eutrophication?
22. Discuss the concept of fire ecology in the context of the Yellowstone National Park fires of 1988. What were the findings of the government task force?

SUGGESTED READINGS

Audesirk, Gerald, and Teresa Audesirk. *Biology—Life on Earth,* 3rd ed. New York: Macmillan Publishing Company, 1993.

Berner, Robert A., and Antonio C. Lasaga. "Modeling the Geochemical Carbon Cycle," *Scientific American* 260 (March 1989): 74–84.

Bolin, Bert. "The Carbon Cycle," *Scientific American* 223 (September 1970): 124–32.

Bormann, F. Herbert, and Gene E. Likens. "The Nutrient Cycles of an Ecosystem," *Scientific American* (October 1970): 92–101.

Christensen, Norman L. *Ecological Consequences of the 1988 Fires in the Greater Yellowstone Area—Final Report.* Denver: Greater Yellowstone Coordinating Committee, 1988.

Clapham, W. B., Jr. *Natural Ecosystems,* 2nd ed. New York: Macmillan Publishing Company, 1983.

Delwiche, C. C. "The Nitrogen Cycle," *Scientific American* 223 (September 1970): 136–46.

Ehrlich, Paul R, and Anne H. Ehrlich, and John P. Holdren. *Ecoscience—Population, Resources, Environment.* San Francisco: W. H. Freeman, 1977.

Farb, Peter, and the editors of Life. *Ecology.* Life Nature Library. New York: Time, 1963.

Findley, Rowe. "Will We Save Our Own [Forests]?" *National Geographic* 78, no. 3 (September 1990): 106–36.

Jeffery, David. "Yellowstone—The Great Fires of 1988," *National Geographic* 175, no. 2 (February 1989): 255–73.

Miller, G. Tyler, Jr. *Living in the Environment,* 6th ed. Belmont, CA: Wadsworth, 1990.

Mills, Susan M., and the Greater Yellowstone Coordinating Committee, eds. *The Greater Yellowstone Postfire Assessment.* Yellowstone National Park: National Park Service, 1989.

Odum, Eugene P. *Ecology and Our Endangered Life-Support Systems.* Sunderland, MD: Sinauer Associates, 1989.

Peters, R. L, and T. E. Lovejoy, eds. *Global Warming and Biological Diversity*. New Haven: Yale University Press, 1992.

Smith, R. C., B. B. Prézelin, K. S. Baker, et al. "Ozone Depletion: Ultraviolet Radiation and Phytoplankton Biology in Antarctic Waters," *Science* 255 (February 12, 1992): 952–59.

Taylor, J. A, ed. *Biogeography—Recent Advances and Future Directions*. Totowa, NJ: Barnes and Noble Books, 1984.

Whittaker, Robert H. *Communities and Ecosystems,* 2nd ed. New York: Macmillan Publishing Company, 1975.

Williams, Ted. "Incineration of Yellowstone," *Audubon* January 1989: 38–85.

Fog at sunrise. Daniel Boone National Forest, Red River Gorge, Kentucky. [*Photo by Scott T. Smith.*]

20

Terrestrial Biomes

Comparison of two quotations, written little more than a century apart in 1874 and 1992, suggests the danger of ignoring the interrelationships among plants and their environment:

> **We have now felled forest enough everywhere, in many districts far too much. Let us now restore this one element of material life to its normal proportions, and devise means of maintaining the permanence of its relations to the fields, the meadows, and the pastures, to the rain and the dews of heaven, to the springs and rivulets with which it waters the Earth.***

> **Forests worldwide have been and are being threatened by uncontrolled degradation and conversion to other types of land uses, influenced by increasing human needs, agricultural expansion, and environmentally harmful mismanagement. . . .The impacts of loss and degradation of forests are in the form of soil erosion, loss of biological diversity, damage to wildlife habitats and degradation of watershed areas, deterioration of the quality of life and reduction of the options for development. . . .The present situation calls for urgent and consistent action for conserving and sustaining forest resources.****

Both these selections offer a warning; unfortunately, the first one was ignored. In 1874, Marsh perceived the need for understanding and managing the environment at a time when Earth was viewed as offering few limits to human activity. In contrast, scientists today measure ecosystems carefully and build elaborate computer models to simulate the evolving real-time human-environment experiment. This at a time when we know that Earth's natural systems pose limits to human activity.

Climate change is discussed in numerous sections of this text. The effect of these changes on ecosystems is likely to be significant. Two Harvard researchers assessed the impact of these alterations of the atmosphere:

> **Based on more than a decade of research, it is obvious that the CO_2-rich atmosphere of our future will have direct and dramatic effects on the composition and operation of ecosystems. According to the best scientific evidence, we see no reason to be sanguine about the response of these habitats to our changing environment.***

Humans are Earth's major biotic agent, influencing all ecosystems. What will the quotation about Earth's plant communities be like in another century? In this chapter we explore Earth's major terrestrial ecosystems, their appearance and structure, location, and the present status of related plants, animals, and environment.

BIOGEOGRAPHICAL REALMS

A **biogeographical realm** of plants and animals is a geographic region where a group of species has evolved. From such a center, species migrate worldwide according to their niche requirements, reproductive success, competition, and climatic and topographic barriers. Recognition that such distinct regions of flora and fauna exist was an early beginning of biogeography as a discipline.

The map in Figure 20-1 illustrates the botanical (plant) and zoological (animal) regions forming major biogeographical realms. Each realm contains many distinct ecosystems that distinguish it from other realms.

The Australian realm is unique, giving rise to 450 species of *Eucalyptus* among its plants and 125 species of marsupials among its animals. Australia's uniqueness is the result of its early isolation from the other continents. During critical evolutionary times, Australia drifted away from Pangaea (see Chapter 11) and never again was reconnected by a land bridge, even when sea level was lowered during repeated glacial ages. New Zealand's isola-

*George Perkins Marsh, *The Earth as Modified by Human Action*. New York: Scribner's, 1874, pp. 385–86.

**United Nations Conference on Environment and Development, "Topic 11. Combatting Deforestation," *Agenda 21—Adoption of Agreements on Environment and Development* Drafts, Rio de Janeiro: United Nations, A/CONF.151/4 (Part II), Section II. Conservation and Management of Resources for Development, paragraph 11.12 and 11.13, May 1, 1992, p. 31.

*Fakhri A. Bazzaz and Eric D. Fajer, "Plant Life in a CO_2-Rich World," *Scientific American* 256, no. 1 (January 1992): p. 74.

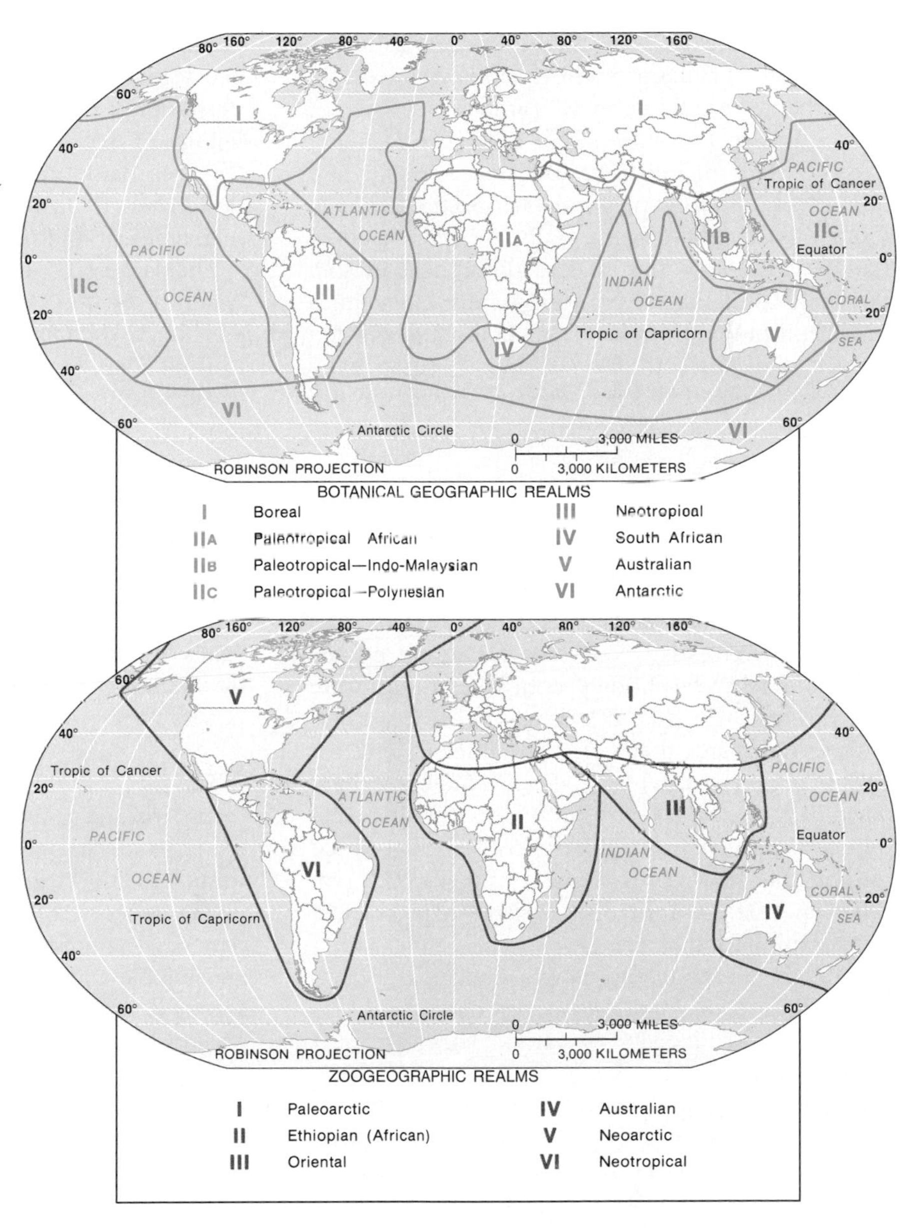

FIGURE 20-1
Biogeographical realms of plants (after R. D. Good, 1947) and animals (after L. F. deBeaufort, 1951). [Adaptation of Figure 14-1 from *Fundamentals of Ecology* by Eugene P. Odum, copyright © 1971 by Saunders College Publishing, a division of Holt, Rinehart and Winston, Inc. Adapted by permission of the publisher.]

tion from Australia also undoubtedly explains why no native marsupials exist in New Zealand.

Alfred Wallace (1823–1913), the first scholar of zoogeography, used the stark contrast in animal species among islands in present-day Indonesia to delimit a boundary between the Oriental and Australian realms, thinking that deep water in the straits had prevented species crossover. However, his boundary line actually is the site of a wide transition zone, where one region grades into the other. Wallace himself recognized the existence of such transition zones. Boundaries between natural systems are, more accurately, "zones of shared traits." Thus, despite distinctions among individual realms, it is best to think of their borders as transition zones of mixed identity and composition, rather than as rigidly defined boundaries.

A boundary transition zone between adjoining ecosystems is an **ecotone.** Because ecotones are defined by different physical factors, they vary in width. Climatic ecotones usually are more gradual than physical ecotones, where differences in soil, moisture availability, or topography sometimes form abrupt boundaries. An ecotone between prairies and northern forests may occupy many kilometers of land. The ecotone is an area of tension as similar species of plants and animals compete for the resource base.

Aquatic Ecosystems

The major ecosystems that make up biogeographical realms are conveniently divided for study into aquatic and terrestrial ecosystems. Oceans, estuaries, and freshwater bodies compose the **aquatic ecosystems.** The *photic layer* is the active upper zone of the water that receives sufficient light for *phytoplankton* to survive. These are the one-celled, chlorophyll-based primary producers that float suspended in the water. Primary consumers include innumerable herbivores, some as small as 0.2 mm (0.008 in.). These form the basis of a productive and complex food chain.

Kenneth Sherman and Lewis Alexander, scientists with the U.S. Marine Fisheries Service Laboratory and the University of Rhode Island, respectively, conceived the **large marine ecosystem (LME)** to foster better management and to improve the sustainability of both biotic and abiotic aquatic resources. An LME is an oceanic region having unique organisms, floor topography, currents, areas of nutrient-rich upwelling circulations, or areas of significant predation, including human (Figure 20-2). Such areas also include nearly enclosed bodies of water, such as the Baltic and Mediterranean seas, which are so threatened by pollution.

Thirty LMEs, each encompassing more than 200,000 km^2 (77,200 mi^2), have been identified (21 are shown in Figure 20-2). The goal of such ecosystem designations is to encourage resource planners and managers to consider complete ecosystems, not just a targeted species. Poor understanding of aquatic ecosystems has led to the failure of some species, such as herring in the Georges Bank area of the Atlantic. Also, LMEs are designated to insure long-term data collection for management and further research.

The largest of eleven protected areas in North America is the Monterey Bay National Marine Sanctuary, established in 1992 within the California Current LME. Stretching along 645 km (400 mi) of coast and covering 13,500 km^2 (5300 mi^2), the sanctuary is home to 27 species of marine mammals, 94 species of shorebirds, 345 species of fish, and the largest sampling of invertebrates in any one place in the Pacific. This represents one of the most species-rich and diverse marine communities on Earth. The Florida Keys Marine Sanctuary is the second largest such protected area in the United States. All these protected areas are administered by the Sanctuaries and Reserves Division of NOAA.

Terrestrial Ecosystems

Plants are the most visible part of the biotic landscape, a key aspect of Earth's **terrestrial ecosystems.** In their growth, form, and distribution, plants reflect Earth's physical systems: its energy patterns; atmospheric composition; temperature and winds; air masses; water quantity, quality, and seasonal timing; soils; regional climates; geomorphic processes; and ecosystem dynamics.

A brief review of the ways that living organisms relate to the environment and each other is helpful at this point (see Chapter 19). A **community** is formed by interacting populations of plants and animals in an area. An **ecosystem** involves the interplay between a community of plants and animals and its abiotic physical environment. Each plant and animal occupies an area in which it is biologically suited to live—its **habitat;** and within that habitat it performs a basic operational function—its **niche.**

A large, stable terrestrial ecosystem is known as a biome. A **biome** is characterized by specific plant and animal communities and their interrelationship with the physical environment. Each

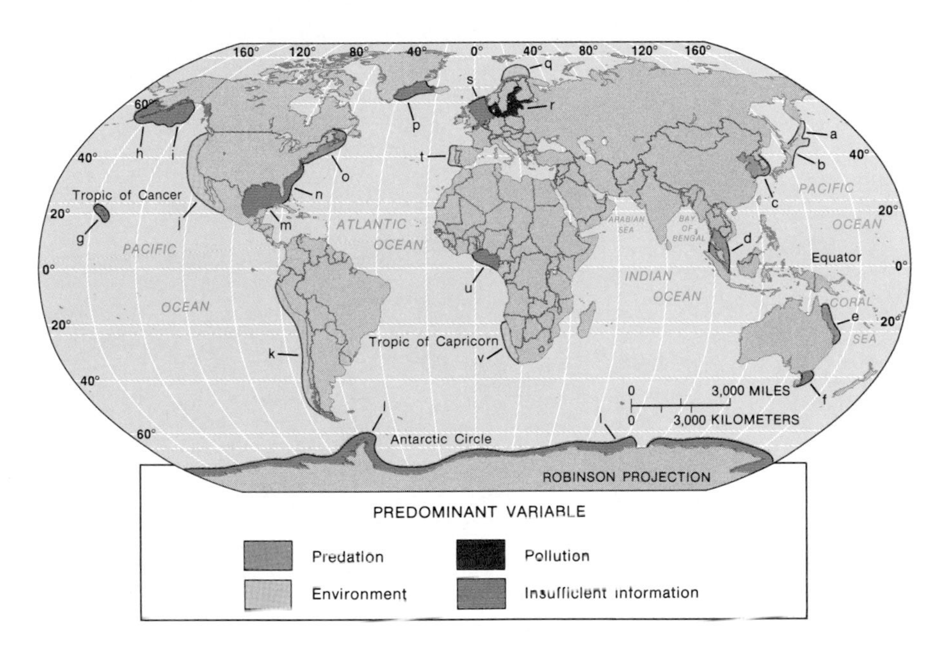

FIGURE 20-2
Large marine ecosystems (LMEs): *a.* Oyashio Current; *b.* Kuroshio Current; *c.* Yellow Sea; *d.* Gulf of Thailand; *e.* Great Barrier Reef; *f.* Tasman Sea; *g.* Insular Pacific; *h.* East Bering Sea; *i.* Gulf of Alaska; *j.* California Current (including the Monterey Bay National Marine Sanctuary); *k.* Peru-Chile Current; *l.* Antarctic; *m.* Gulf of Mexico; *n.* Southeast Continental Shelf; *o.* Northeast Continental Shelf; *p.* East Greenland Sea; *q.* Barents Sea; *r.* Baltic Sea; *s.* North Sea; *t.* Iberian Coastal; *u.* Gulf of Guinea; *v.* Benguela Current. [After Kenneth Sherman, *World Resources 1988–89.* New York: World Resources Institute, 1988, p. 147. Adapted by permission.]

biome is usually named for its *dominant vegetation.* We can generalize Earth's wide-ranging plant species into six broad biomes: *forest, savanna, grassland, shrubland, desert, and tundra.* Because plant distributions are responsive to environmental conditions and reflect variation in climatic and other abiotic factors, the biome-related world climate map in Figure 10-5 is a helpful reference for this chapter.

We further define these general biomes into more specific vegetation units called **formation classes.** These units refer to the structure and appearance of dominant plants in a terrestrial ecosystem, for example, equatorial rain forest, northern needleleaf forest, Mediterranean shrubland, arctic tundra. Each formation includes numerous plant communities, and each community includes innumerable plant habitats. Within those habitats, Earth's diversity is expressed in 250,000 plant species. Despite this intricate complexity we can generalize Earth's numerous formation classes into 10 global terrestrial biome regions as portrayed in Figure 20-4 and detailed in Table 20-1.

More specific systems are used for the structural classification of plants. Such *life-form* designations are based on the outward physical properties of individual plants or the general form and structure of a vegetation cover. These physical life forms, portrayed in Figure 20-3, include *trees* (large woody main trunk, perennial, usually exceeding 3

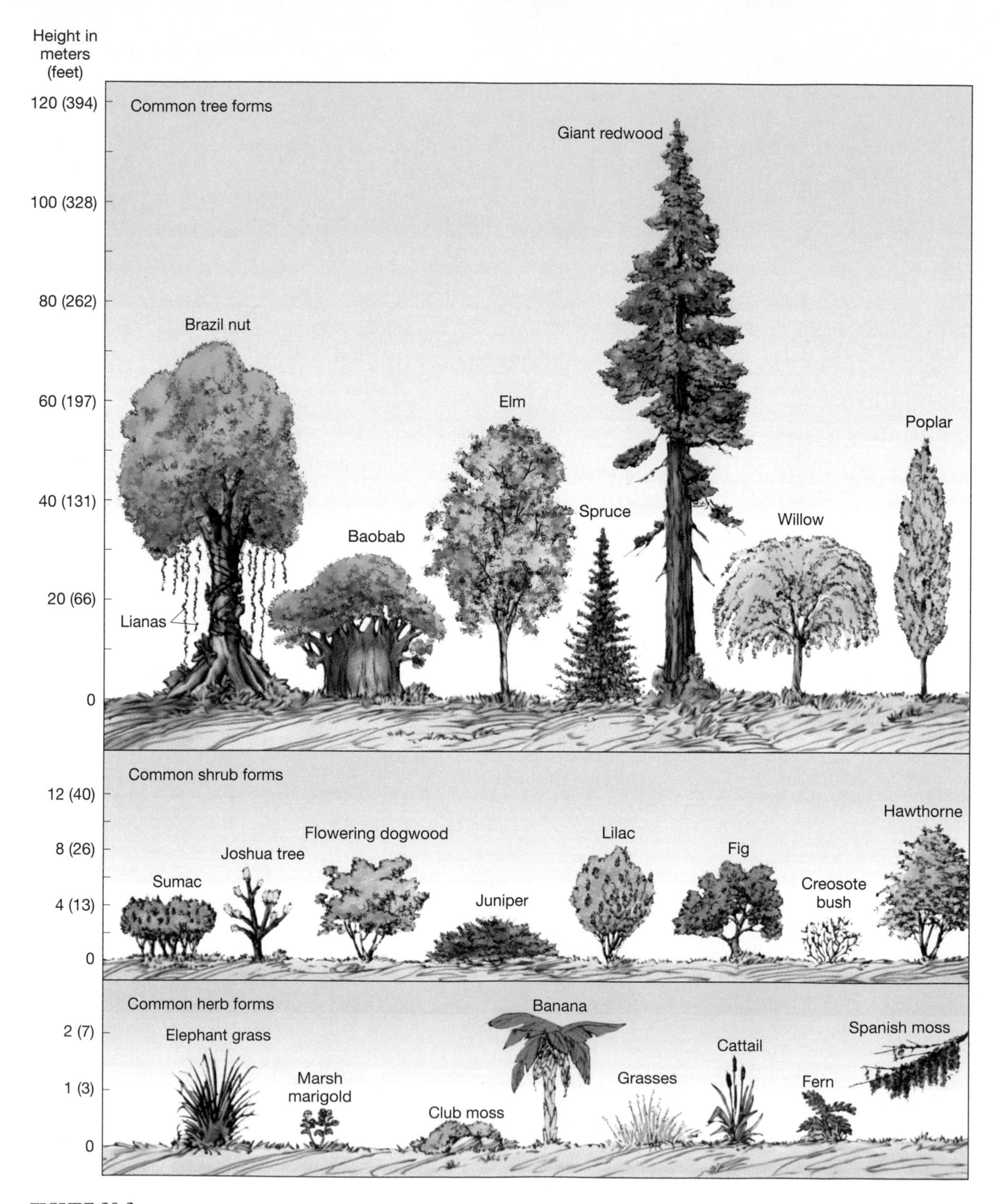

FIGURE 20-3
Examples of plant life forms. [After William M. Marsh and J. Dozier, *Landscape: An Introduction to Physical Geography,* copyright © 1981, p. 273. Adapted by permission of John Wiley & Sons, Inc.]

m or 10 ft); *lianas* (woody climbers and vines); *shrubs* (smaller woody plants; branching stems at the ground); *herbs* (small plants without woody stems above ground); *bryophytes* (mosses, liverworts); *epiphytes* (plants growing above the ground on other plants, using them for support); and *thallophytes,* which lack true leaves, stems, or roots (bacteria, fungi, algae, lichens).

EARTH'S MAJOR TERRESTRIAL BIOMES

Earth's diverse plant and animal communities and their interrelationship with the physical environment are simplified into 10 major terrestrial biomes. These regions generally are based on vegetation *formation classes.* Few natural communities of plants and animals remain; most biomes have been greatly altered by human intervention. Thus, the "natural vegetation" identified on many biome maps reflects ideal climax vegetation potential, given the physical factors prevalent in a region. Even though human practices have greatly altered these ideal forms, it is valuable to study the natural biomes to better understand the natural environment and to assess the extent of human-caused alteration.

In addition, knowing the ideal in a region guides us to a closer approximation of natural vegetation in the plants we introduce. In the United States and Canada, we humans tend to perpetuate a type of transition community, somewhere between a grassland and a forest. We plant trees and lawns and then must invest energy, water, and capital to sustain such artificial modifications of nature.

The global distribution of Earth's major terrestrial biomes is portrayed in Figure 20-4. Table 20-1 describes each biome on the map and summarizes other pertinent information from throughout this text—a compilation from many chapters, for Earth's biomes are a synthesis of the environment and biosphere. The sections that follow expand on the table and describe each biome in greater detail.

(*Note:* "The Living Earth" composite image appears inside the front cover of this text. A computer artist and scientists at the Jet Propulsion Laboratory worked with hundreds of thousands of satellite images to produce a cloudless view of Earth in the true colors of a local summer day. Compare the map in Figure 20-4 with this remarkable composite image and see what correlations you can make.)

Equatorial and Tropical Rain Forest

Earth is girdled with a lush biome—the **equatorial and tropical rain forest.** In a climate of consistent daylength (12 hours), high insolation, and average annual temperatures around 25°C (77°F), plant and animal populations have responded with the most diverse body of life on the planet. The Amazon region is the largest tract of equatorial and tropical rain forest, also called the *selva.* Rain forests also cover equatorial regions of Africa, Indonesia, the margins of Madagascar and Southeast Asia, the Pacific coast of Ecuador and Colombia, and the east coast of Central America, with small discontinuous patches elsewhere. The cloud forests of western Venezuela are such tracts of rain forest at high elevation, perpetuated by high humidity and cloud cover. Undisturbed tracts of rain forest are rare.

In his 1960 book *The Forest and the Sea,* Marston Bates compared rain forests and oceans and found remarkable similarities in these very dissimilar environments. Biomass in a rain forest is concentrated high up in the canopy, that dense mass of overhead leaves, just as life in the sea is concentrated in the photic layer near the ocean's surface. The sea and the forest both have a vertical distribution of life, dependent on a competitive struggle for sunlight. The treetops of the rain forest are generally analogous to the upper (pelagic) region of the open ocean, where photosynthesis fixes energy for a dependent food chain. The floor of the rain forest and the floor of the deep ocean also are roughly parallel in that both are dark or deeply shaded, and a place of fewer life forms. This general analogy by Bates gives an idea of the structure of life in the rain forest.

Rain forests feature ecological niches distributed by height rather than horizontally, because of the competition for light. The canopy has a rich variety of plants and animals. Lianas (vines) branch from tree to tree, binding them together with cords that can reach 20 cm (8 in.) in diameter.

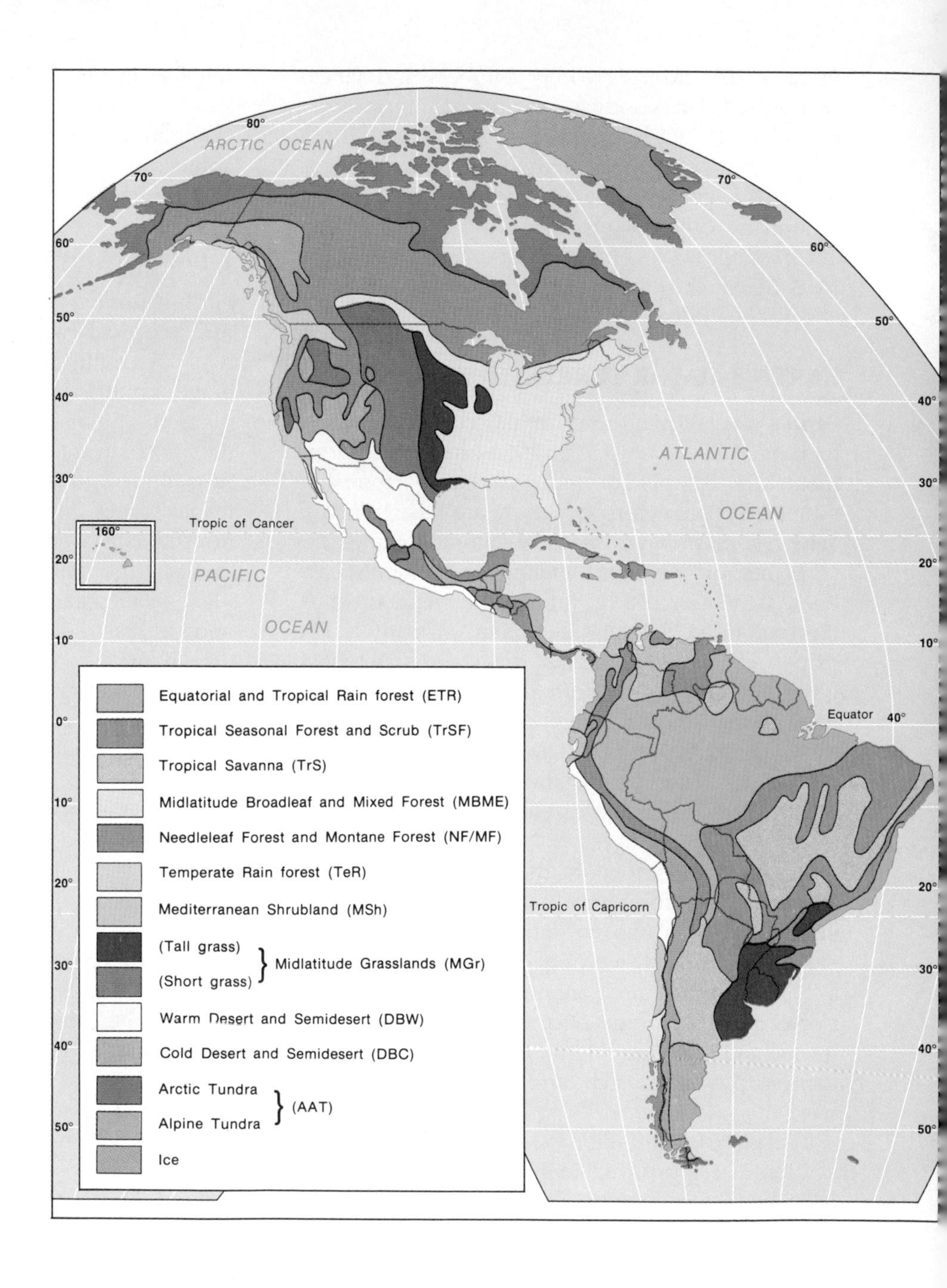

FIGURE 20-4
The 10 major global terrestrial biomes.

Epiphytes flourish there, too: such plants as orchids, bromeliads, and ferns that live completely above ground, supported physically but not nutritionally by the structures of other plants. Windless conditions on the forest floor make pollination difficult—insects, other animals, and self-pollination predominate.

The rain forest canopy usually is considered to have three levels. These include a high level that averages 50–60 m (165–200 ft) above ground, a

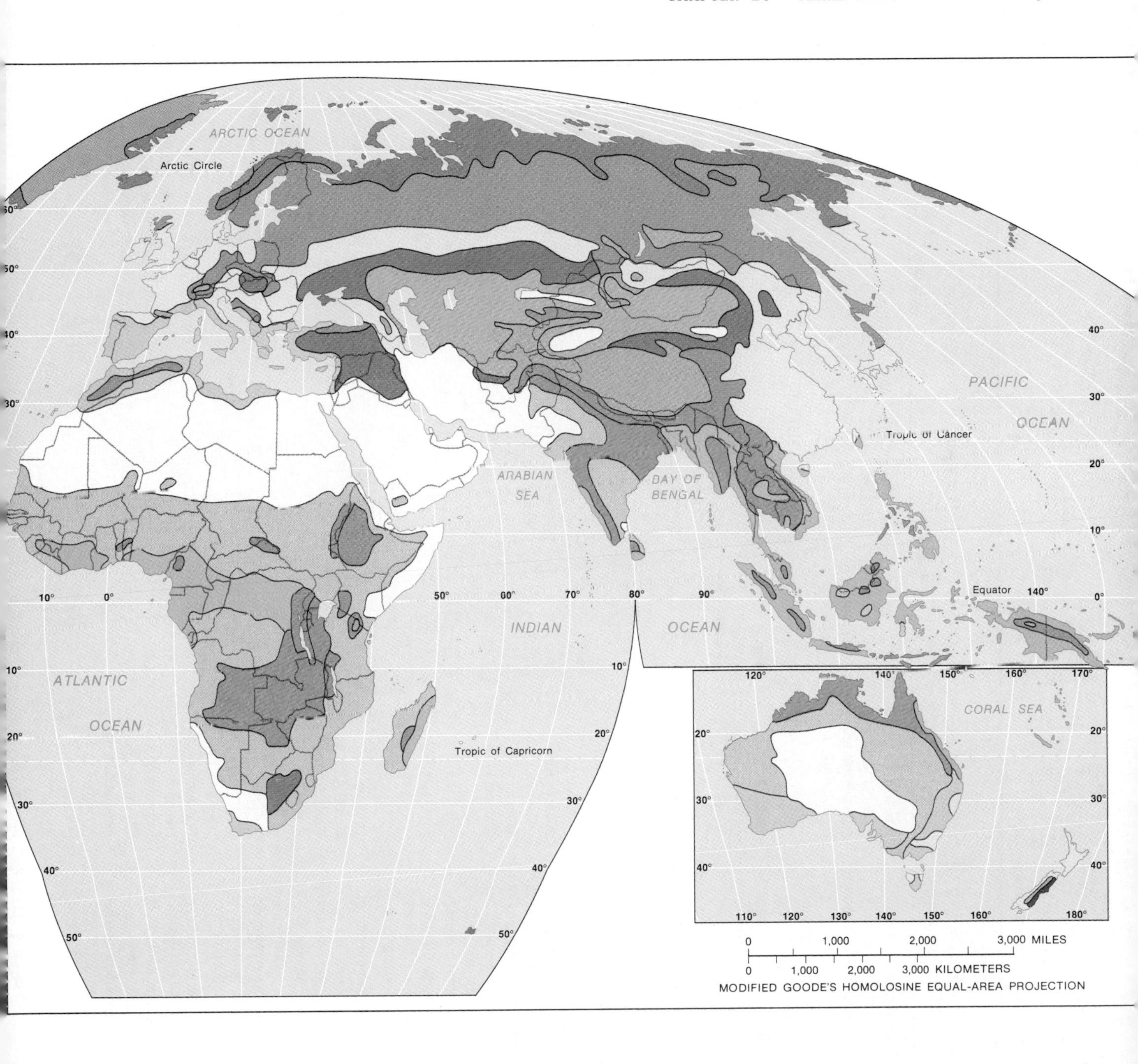

middle level from 20 to 40 m (65–130 ft), and a lower level from 5 to 15 m (15–50 ft)—see Figure 20-5. Of the three, the middle canopy is the most continuous, with its broad leaves blocking much of the light and creating a darkened forest floor.

The upper level is not continuous but features tall trees whose high crowns rise above the middle canopy. The lower level is composed of small seedlings, ferns, bamboo, and the like, leaving the litter-strewn ground level in deep shade and fairly

TABLE 20-1
Major Terrestrial Biomes and Their Characteristics

Biomes and Ecosystems (map symbol)	Vegetation Characteristics	Soil Orders	Köppen Climate Designation	Annual Precipitation Range	Temperature Patterns	Water Balance
Equatorial and Tropical Rain Forest (ETR) Evergreen broadleaf forest Selva	Leaf canopy thick and continuous; broadleaf evergreen trees; vines (lianas), epiphytes, tree ferns, palms	Oxisols Ultisols (on well-drained uplands)	Af Am (limited dry season)	180–400 cm (>6 cm/mo)	Always warm (21–30°C; avg. 25°C)	Surpluses all year
Tropical Seasonal Forest and Scrub (TrSF) Tropical monsoon forest Tropical deciduous forest Scrub woodland and thorn forest	Transitional between rain forest and grasslands; broadleaf, some deciduous trees; open parkland to dense undergrowth; acacias and other thorn trees in open growth	Oxisols Ultisols Vertisols (in India) Some Alfisols	Am Aw Borders BS	130–200 cm (<40 rainy days during 4 driest months)	Variable, always warm (>18°C)	Seasonal surpluses and deficits
Tropical Savanna (TrS) Tropical grassland Thorn tree scrub Thorn woodland	Transitional between seasonal forests, rain forests, and semiarid tropical steppes and desert; trees with flattened crowns, clumped grasses, and bush thickets; fire association	Alfisols (dry: Ustalfs) Ultisols Oxisols	Aw BS	90–150 cm, seasonal	No cold weather limitations	Tends toward deficits, therefore fire and drought susceptible
Midlatitude Broadleaf and Mixed Forest (MBME) Temperate broadleaf Midlatitude deciduous Temperate needleleaf	Mixed broadleaf and needleleaf trees; deciduous broadleaf, losing leaves in winter; southern and eastern evergreen pines demonstrate fire association	Ultisols Some Alfisols Podzols (red and yellow)	Cfa Cwa Dfa	75–150 cm	Temperate, with cold season	Seasonal pattern with summer maximum PRECIP and POTET; no irrigation needed
Northern Needleleaf Forest and Montane Forest (NF/MF) Taiga Boreal forest Other montane forests and highlands	Needleleaf conifers, mostly evergreen pine, spruce, fir; Russian larch, a deciduous needleleaf	Spodosols Histosols Inceptisols Gleysols Alfisols (Boralfs: cold) Podzols	Subarctic Dfb Dfc Dfd	35–100 cm	Short summer, cold winter	Low POTET, moderate PRECIP, moist soils, some waterlogged and frozen in winter; no deficits

Temperate Rain Forest (TeR) West coast forest Coast redwoods (U.S.)	Narrow margin of lush evergreen and deciduous trees on windward slopes; redwoods, tallest trees on Earth	Spodosols Inceptisols (mountainous environs) Podzols	Cfb Cfc	150–500 cm	Mild summer and mild winter for latitude	Large surpluses and runoff
Mediterranean Shrubland (MSh) Sclerophyllous shrubs Australian Eucalyptus forest	Short shrubs, drought adapted, trending to grassy woodlands; chaparral	Alfisols (Xeralfs) Mollisols (Xerolis) Luvisols	Csa Csb	25–65 cm	Hot, dry summers, cool winters	Summer deficits, winter surpluses
Midlatitude Grasslands (MGr) Temperate grassland Sclerophyllous shrub	Tall grass prairies and short grass steppes, highly modified by human activity; major areas of commercial grain farming; plains, pampas, and veld	Mollisols Aridisols Chernozemic	Cfa Dfa	25–75 cm	Temperate continental regimes	Soil moisture utilization and recharge balanced; irrigation and dry farming in drier areas
Warm Desert and Semidesert (DBW) Subtropical desert and scrubland	Bare ground graduating into xerophytic plants including succulents, cacti, and dry shrubs	Aridisols Entisols (sand dunes)	BWh BWk	<2 cm	Average annual temperature, around 18°C, highest temperatures on Earth	Chronic deficits, irregular precipitation events, PRECIP <1/2 POTET
Cold Desert and Semidesert (DBC) Midlatitude desert, scrubland, and steppe	Cold desert vegetation includes short grass, and dry shrubs	Aridisols Entisols	BSh BSk	2–25 cm	Average annual temperature around 18°C	PRECIP >1/2 POTET
Arctic and Alpine Tundra (AAT)	Treeless; dwarf shrubs, stunted sedges, mosses, lichens, and short grasses; alpine, grass meadows	Inceptisols Histosols Entisols (permafrost) Organic Cryosols	ET Dwd	15–80 cm	Warmest month <10°C, only 2 or 3 months above freezing	Not applicable most of the year, poor drainage in summer
Ice			EF			

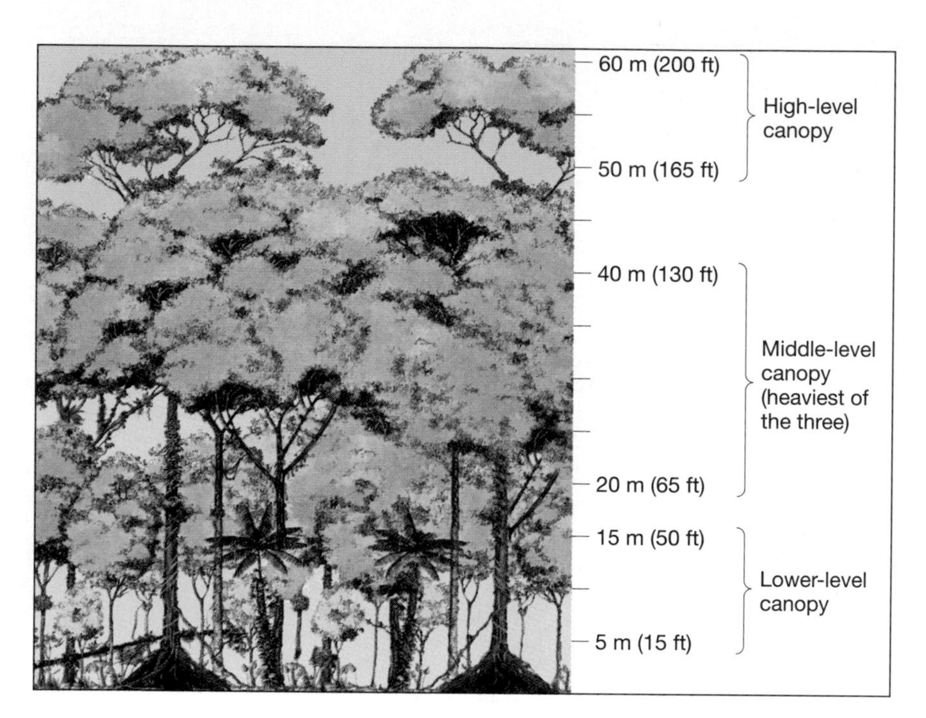

FIGURE 20-5
The three levels of a rain forest canopy.

open. A look at aerial photographs of a rain forest, or views along river banks covered by dense vegetation, aided by the false Hollywood-movie imagery of the jungle, makes it difficult to imagine the shadowy environment of the actual rain forest floor, which receives only about 1% of the sunlight arriving at the canopy (Figure 20-6a). The constant moisture, rotting fruit and moldy odors, strings of thin roots and vines dropping down from above, windless air, and echoing sounds of life in the trees, together create a unique environment.

The smooth, slender trunks of rain forest trees are covered with thin bark and buttressed by large wall-like flanks that grow out from the trees to brace the trunks (Figure 20-6b). These buttresses form angular open enclosures, a ready habitat for various animals. There are usually no branches for at least the lower two-thirds of the tree trunks.

The wood of many rain forest trees is extremely hard, heavy, and dense—In fact, some species will not even float. (Exceptions are balsa and a few others, which are very light.) Varieties of trees include mahogany, ebony, and rosewood. Logging is difficult because individual species are widely scattered; a species may occur only once or twice per square kilometer. Selective cutting is required for species-specific logging, whereas pulpwood production takes everything. Conversion of the forest to pasture is usually accomplished with destruction by fire.

Rain forests represent approximately one-half of Earth's remaining forests, occupying about 7% of the total land area worldwide. This biome is stable in its natural state, resulting from the long-term residence of these continental plates near equatorial latitudes and their escape from glaciation. The soils, principally Oxisols, are essentially infertile, yet they support rich vegetation. The trees have adapted to these soil conditions with root systems able to capture nutrients from litter decay at the soil surface. High precipitation and temperatures work to produce deeply weathered and leached soils characteristic of the laterization process, with a clayey texture sometimes breaking up into a granular structure, and lacking in nutrients and colloidal material. With much investment in fertilizers, pesticides, and machinery, these soils can be productive.

(a)

(b)

FIGURE 20-6
The rain forest. (a) Equatorial rain forest is thick along the Amazon River where light breaks through to the surface, producing a rich gallery of vegetation. (b) The rain forest floor in Corcovado, Costa Rica, with typical buttressed trees and lianas. [(a) Photo by Wolfgang Kaehler; (b) photo by Frank S. Balthis.]

The animal and insect life of the rain forest is diverse, ranging from small decomposers (bacteria) working the surface to many animals living almost completely in the upper stories of the trees. These tree dwellers are referred to as *arboreal,* from the Latin for tree, and include sloths, monkeys, lemurs, parrots, and snakes. Beautiful birds of many colors, tree frogs, lizards, bats, and a rich insect life that includes over 500 species of butterflies are found in rain forests. Surface animals include pigs (bushpigs and the giant forest hog in Africa, wild boar, bearded pig in Asia, and peccary in South America), species of small antelopes (bovids), and mammal predators (tiger in Asia, jaguar in South America, and leopard in Africa and Asia).

The present human assault on Earth's rain forests has put this diverse fauna and the varied flora at risk. It also jeopardizes an important recycling system for atmospheric carbon dioxide, and a potential source of valuable pharmaceuticals and many types of foods.

Deforestation of the Tropics. More than one-half of Earth's original rain forest is gone, cleared for pasture and cattle grazing, timber, fuel, farming, and wood. An area slightly less than the size of Wisconsin or Florida is lost a year (168,900 km^2, 65,200 mi^2) and about a third more is disrupted by selective cutting of canopy trees. When astronauts in orbit look down on the rain forests at night, they see thousands of human-set fires. During the day the lower atmosphere in these regions appears choked with the smoke from these fires (Figure 20-7a and b). These fires are used to clear land for agriculture, which is intended to feed the domestic population as well as to produce cash exports of beef, rubber, coffee, and other commodities. Edible fruits are not abundant in an undisturbed rain forest, but cultivated clearings produce bananas (plantains), mangos and jack fruit, guava, and starch-rich roots such as manioc and yams.

However, poor soil fertility means that the lands are quickly exhausted under intensive use and are

FIGURE 20-7
The Shuttle *Discovery* captured these photographs October 1, 1988. (a) Thousands of fires produce smoke that hides the Amazon rain forest from the astronauts' view. (b) A widespread smoke plume rises from a single rain forest fire burning an area greater than that of the 1988 Yellowstone fire. (c) A *Landsat 5* satellite image of a portion of the State of Rondônia in western Brazil recorded in 1987. Highway BR364, the main artery in the region, passes through the towns of Jaru and Ji-Paraná near the center of the image at about 62° E 11° S. The branching pattern of feeder roads and cleared forest appear in a light blue-grey color. [(a) and (b) Space Shuttle photos from NASA; (c) courtesy of GEOPIC, Earth Satellite Corporation.]

(a)

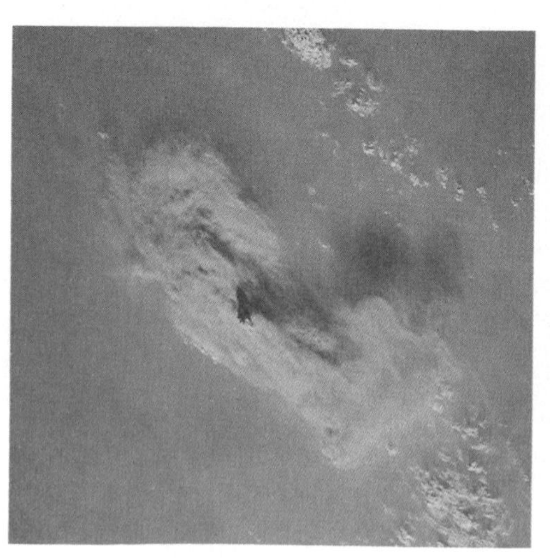

(b)

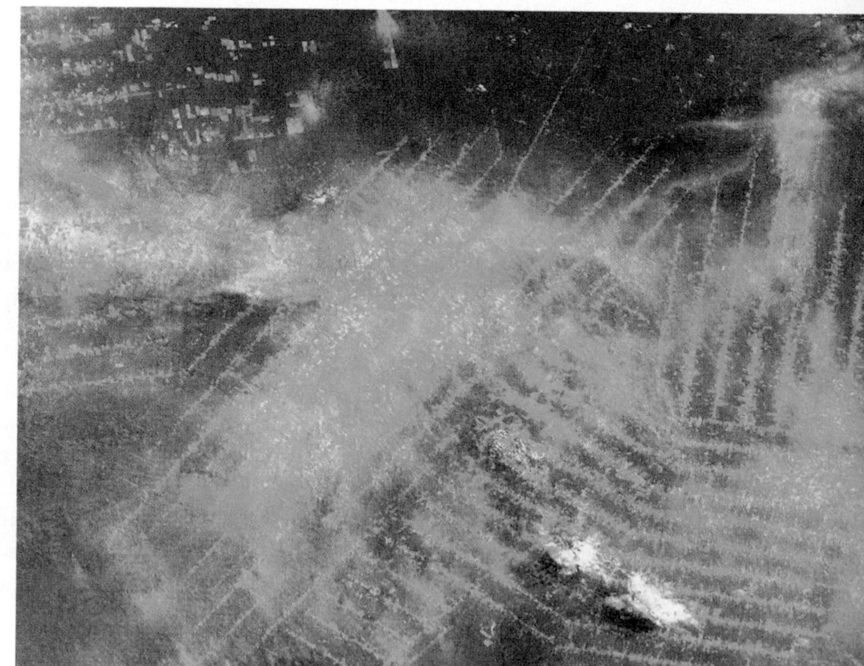

(c)

then generally abandoned in favor of newly burned and cleared lands, unless the lands are artificially maintained. Unfortunately, the dominant trees require from 100 to 250 years to establish themselves after major disturbances. Following clearing, the massive growth of low bushes intertwined with vines and ferns means that a new forest may be slow to follow.

Figure 20-7c is a satellite image of a portion of western Brazil recorded in 1987, the peak year for losses in the Brazilian rain forest. This image gives a sense of the level of rain forest destruction in progress. Scientists with Goddard Space Flight Center and the Brazilian government recently completed a digitized database for the country to facilitate a remote sensing-based GIS analysis of these rain forest losses.

The United Nations Food and Agricultural Organization (FAO) estimates that every year approximately 16.9 million hectares (41.7 million acres) are destroyed, and more than 5 million hectares (12.3 million acres) are selectively logged. This total—averaged for the period 1980–1991 in 76 countries that contain 97% of all rain forest—represents a 0.9% loss of equatorial and tropical rain forest worldwide each year (up from the previous average of 0.6%). If this destruction continues unabated, these forests will be completely removed by about A.D. 2050! By continent, forest losses are estimated at more than 50% in Africa, over 40% in Asia, and 40% in Central and South America. Brazil, Colombia, Mexico, and Indonesia lead the list of lesser-developed countries that are removing their forests at record rates, although these statistics are disputed by the countries in question.

Another threat to the rain forest biome and indigenous peoples emerged in 1991: exploration for and development of oil reserves. U.S. oil corporations are going ahead in Yasuni National Park, Ecuador (near the equator at 77° W) with road building and drilling. One estimate of the ultimate petroleum reserve there is 1.5 billion barrels, or enough to satisfy about three months of the U. S. demand. Similar projects are being considered in Peru.

To slow this continuing catastrophe of deforestation, the Tropical Forestry Action Plan was initiated in 1985 by the Food and Agriculture Organization (FAO), the UN Development Programme, the World Bank, and the World Resources Institute. That plan instituted a more accurate worldwide survey of the rate of deforestation, using orbiting satellites for remote measurement. Workshops also have been conducted to stimulate action. But we have barely begun to reverse the trend. Focus Study 20-1 looks more closely at efforts to curb increasing rates of species extinction, much of which is directly attributable to the loss of rain forests.

Tropical Seasonal Forest and Scrub

A varied biome on the margins of the rain forest is the **tropical seasonal forest and scrub,** which occupies regions of less and more erratic rainfall. The shifting intertropical convergence zone (ITCZ) brings precipitation with the seasonally shifting high Sun and dryness with the low Sun, producing a seasonal pattern that features areas of moisture deficits, some leaf loss, and dry season flowering. The term *semideciduous* is appropriate. Areas of this biome have fewer than 40 rainy days during their four consecutive driest months, yet heavy monsoon downpours characterize their summers (see Chapter 6, especially Figure 6-20). The Köppen climates *Am tropical monsoon* and *Aw tropical savanna* apply to these transitional communities between rain forests and tropical grasslands.

Portraying such a varied biome is difficult. In many areas the natural biome is disturbed by human clearing, so that the savanna grassland adjoins the rain forest directly. The biome does include a gradation from wetter to drier areas: monsoonal forests, open woodlands and scrub woodland, thorn forests, and finally drought-resistant scrub species (Figure 20-8). The monsoonal forests average 15 m (50 ft) high with no continuous canopy of leaves, graduating into open orchardlike parkland with grassy openings or into areas choked by dense undergrowth. In more open tracts, a common tree is the acacia, with its flat-topped appearance and usually thorny stems. Most of the trees in this biome have the look of the fabric portion of an upside-down umbrella, as do trees in the tropical savanna.

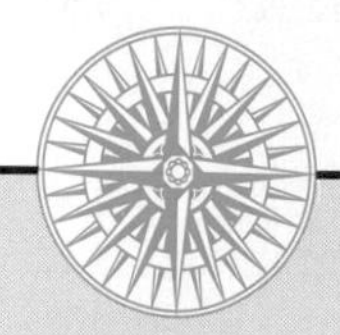

Focus Study 20-1 Biodiversity and Biosphere Reserves

When the first European settlers landed in the Hawaiian Islands in the late 1700s, 43 species of birds were counted. Today, 15 of those species are extinct, and 19 more are threatened or endangered, with only one-fifth of the original species relatively healthy.

As more is learned about Earth's ecosystems and their related communities of plants and animals, more is known of their value to civilization. They are a major source of new foods, new chemicals, new medicines and drugs, and specialty woods, and of course they are indicators of a healthy, functioning biosphere.

International efforts are underway to study and preserve specific segments of the biosphere: Rain Forest Action Network, Natural World Heritage Sites, Wetlands of International Importance, the Nature Conservancy, and Biosphere Reserves are just a few of the active organizations. The motivation to set aside natural sanctuaries is directly related to concern over the increase in the rate of species extinctions. We are facing a loss of genetic diversity that may be unparalleled in Earth's history, even compared to the major extinctions in the geologic record.

Species Threatened

Table 1 summarizes the known and estimated species on Earth. Scientists have classified only 1.4 million species of an estimated 10–30 million overall; the latter figure represents an increase in what scientists once thought to be the diversity of life on Earth. And those yet-to-be-discovered species represent a potential future resource for society. However, in 1982, the *Global 2000 Report,* prepared by the Council of Environmental Quality under a commission originally established by President Carter, stated that between 15 and 20% of all plant and animal species on Earth would be extinct by the year 2000. That percentage equals 500,000 to 2 million species, many of which have not yet even been identified. This rate of species extinction averages one every 30 minutes! The effects of pollution, loss of wild habitats, excessive grazing, poaching, and collecting are at the root of such devastation. Approximately 60% of the extinctions are attributable to the clearing and loss of rain forests alone.

Both black rhinos and white rhinos in Africa exemplify species in jeopardy. Rhinos once grazed over

TABLE 1
Known and Estimated Species on Earth

Form of Life	Known Species	Estimated Total Species
Insects and other invertebrates	989,761	30 million (insect species), extrapolated from surveys in forest canopy in Panama; most believed unique to tropical forests
Vascular plants	248,400	At least 10–15% of all plants believed to be undiscovered
Fungi and algae	73,900	Not available
Microorganisms	36,600	Not available
Fishes	19,056	21,000 based on assumption that 10% of fish remain undiscovered (the Amazon and Orinoco rivers alone may account for 2000 additional species)
Birds	9040	Probably 2% more (98% of total known)
Reptiles and amphibians	8962	Probably less than 5% more (95% of total known)
Mammals	4000	Probably less than 5% more (95% of total known)
Miscellaneous chordates	1273	Not available
TOTAL	1,390,992	10 million, considered a conservative count; with insect estimates perhaps over 30 million

SOURCE: Reprinted from *State of the World 1988,* Worldwatch Institute Report on Progress Toward a Sustainable Society, Project Director: Lester R. Brown, by permission of W. W. Norton & Company, Inc. Copyright © 1988 by Worldwatch Institute.

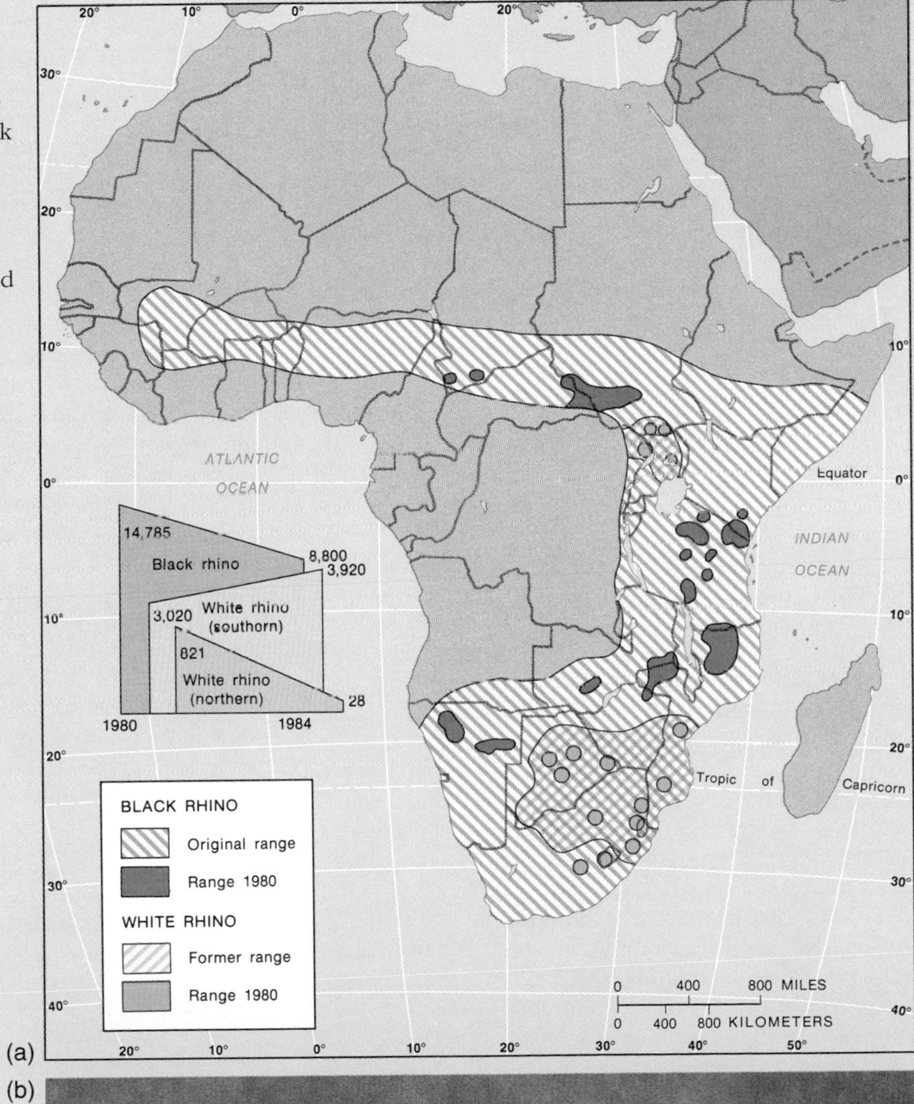

FIGURE 1
The rhinoceros in Africa.
(a) Failing population and receding distribution of both the black and white rhinos. (b) Black rhino and young in Tanzania, escorted by oxpecker birds.
[(a) Illustration after *State of the Ark* by Lee Durrell. Copyright © 1986 by Gaia Books Ltd. Adapted by permission of Doubleday, a division of Bantam Doubleday Dell Publishing Group, Inc.
(b) Photo by Stephen J. Krasemann/DRK Photo.]

much of the savanna grasslands and woodlands. Today, they survive only in protected districts and are threatened even there (Figure 1). There are less than two dozen white rhinos in existence now in sub-Saharan ranges; the number of black rhinos is approximately 8000, but that is less than half the number in 1980. These large land mammals are nearing extinction and will survive only as a dwindling zoo population. The limited genetic pool that remains complicates further reproduction.

Biosphere Reserves

To slow this type of extinction, formal natural reserves are a possible strategy. Setting up such a *biosphere reserve* involves principles of *island biogeography*. Island communities are special places for study because of their spatial isolation and the relatively small number of species present. They resemble natural experiments because the impact of individual factors, such as civilization, can be more easily assessed on islands than over larger continental areas. According to the equilibrium theory of island biogeography, developed by biogeographers R. H. MacArthur and E. O. Wilson, the number of species should increase with the size of the island, decrease with increasing distance from the nearest continent, and remain about the same over time, even though composition may vary. Clearly, human intervention throws natural balances out of equilibrium.

Studies of islands also can assist in the study of mainland ecosystems, for in many ways a park or biosphere reserve, surrounded by modified areas and artificial boundaries, is like an island. Indeed, a biosphere reserve is conceived as an ecological island in the midst of change. The intent is to establish a core in which genetic material is protected from outside disturbances, surrounded by a buffer zone that is, in turn, surrounded by a transition zone and experimental research areas.

It is predicted that new, undisturbed reserves will not be possible after A.D. 2010, because pristine areas will be gone. Coordinated by the Man and the Biosphere (MAB) Programme of UNESCO, nearly 300 such biosphere reserves covering some 12 million hectares (30 million acres) are now operated voluntarily in 76 countries. Not all protected areas are ideal bioregional entities. Some are simply imposed on existing park space and some remain in the planning stage although they are officially designated.

Some of the best examples have been in operation since the late 1970s and range from the struggling Everglades National Park in Florida, to the Changbai Reserve in China, to the Tai Forest on the Ivory Coast. Added to these efforts, the work of the Nature Conservancy in acquiring land for preservation is a valuable part of the reserve process.

The ultimate goal, about half achieved, is to establish at least one reserve in each of the 194 distinctive *biogeographical communities* presently identified. The United Nations list of national parks and protected areas is compiled by UNESCO and the World Conservation Monitoring Centre. Presently 6930 areas covering 657 million hectares (1.62 billion acres) are designated in some protective form, representing about 4.8% of national land area on the planet. Even though these are not all set aside to the degree of biosphere reserves, they do demonstrate progress toward preservation of our planet's plant and animal heritage.

> The preservation of species diversity is a problem that must today be confronted by one species, *Homo sapiens*. . . . the diversity of species is worth preserving because it represents a wealth of knowledge that cannot be replaced. Moreover, today's extinctions are unlike those in previous eras, in which long periods of recovery could follow extinctions. The present situation is an inexorably irreversible one in which human overpopulation will destroy most species unless we plan for protection immediately. Accepting that the goal is worthwhile requires that more energy be devoted to planning and priorities and less to emotionalism and indignation [on all sides].*

---

* Daniel E. Koshland, Jr., "Preserving Biodiversity," editorial, *Science* 253, no. 5021 (August 16, 1991): 717.

FIGURE 20-8
Open thorn forest and savanna in Kenya. [Photo by Gael Summer-Hebdon.]

Local names are given to these communities: the *caatinga* of the Bahia State of northeast Brazil, the *chaco* area of Paraguay and northern Argentina, the *brigalow scrub* of Australia, and the *dornveld* of southern Africa. The map in Figure 20-4 shows areas of this biome in Africa, extending from eastern Angola through Zambia to Tanzania; in southeast Asia and portions of India, from interior Myanmar (Burma) through northeastern Thailand; and in parts of Indonesia. The trees throughout most of this biome are not good for lumber but may be valuable for fine cabinetry, especially teak wood. In addition, some of the plants with dry season adaptations produce usable waxes and gums, such as carnauba and palm-hard waxes. Animal life includes the koalas and cockatoos of Australia and the elephants, rodents, and ground-dwelling birds elsewhere.

Tropical Savanna

Large expanses of grassland interrupted by trees and shrubs aptly describes the **tropical savanna,** a transitional biome between the rain forests and tropical seasonal forests, and semiarid tropical steppes and deserts. The savanna biome also includes treeless tracts of grasslands, and in very dry savannas, grasses grow discontinuously in clumps, with bare ground between them. The trees of the savanna woodlands are characteristically flat topped (Figure 20-9), although one relatively rare tree associated with the tropical savanna, the baobab tree, is not.

Savannas covered more than 40% of Earth's land surface before human intervention but were especially modified by human-caused fire. Fires occur annually throughout the biome. The timing of these fires is important. Early in the dry season they are beneficial and increase tree cover; if late in the season they are very hot and kill trees and seeds. Savanna trees are adapted to resist the "cooler" fires. Elephant grasses averaging 5 m (16 ft) high and forests are assumed once to have penetrated much farther into the dry regions, for they are known to survive there when protected. Savanna grasslands are much richer in humus than the

FIGURE 20-9
Savanna landscape of the Serengeti Plains, with wildebeest, zebras, and thorn forest. [Photo by Stephen Cunha.]

wetter tropics and are better drained, thereby providing a better base for agriculture and grazing. Sorghums, wheat, and groundnuts (peanuts) are common commodities.

Tropical savannas receive their precipitation during less than six months of the year, when they are influenced by the ITCZ. The rest of the year they are under the drier influence of shifting subtropical high-pressure cells. Savanna shrubs and trees are frequently *xerophytic,* or drought resistant, with various adaptations to protect them from the dryness: small, thick leaves, rough bark, or waxy leaf surfaces.

Africa has the largest region of this biome, including the famous Serengeti Plains and the Sahel region. Sections of Australia, India, and South America also are part of the savanna biome. Some of the local names for these lands include the *Llanos* in Venezuela, stretching along the coast and inland east of Lake Maricaibo and the Andes; the *Campo Cerrado* of Brazil and Guiana; and the *Pantanal* of southwestern Brazil.

Particularly in Africa, savannas are the home of large land mammals that graze on savanna grasses or feed upon the grazers themselves: lion, cheetah, zebra, giraffe, buffalo, gazelle, wildebeest, antelope, rhinoceros, and elephant (Figure 20-9). Birds include the ostrich, martial eagle (largest of all eagles), and secretary bird. Many venomous species of snakes occur, as does the crocodile.

However, in our lifetime we may see the reduction of these animal herds to zoo stock only, because of poaching and habitat losses. The loss of both the black and white rhino is discussed in Focus Study 20-1. Establishment of large tracts of savanna as biosphere reserves is critical for the preservation of natural aspects of this biome and its associated fauna.

Midlatitude Broadleaf and Mixed Forest

Moist continental climates support a mixed broadleaf and needleleaf forest in areas of warm-to-hot summers and cool-to-cold winters. This **midlatitude broadleaf and mixed forest** includes several distinct communities. Relatively lush evergreen broadleaf forests occur in narrow margins along the Gulf of Mexico, southern Japan, eastern and northeastern China, southeastern Australia, New Zealand, and central Chile. Northward of this broadleaf margin in the Northern Hemisphere are mixed deciduous and evergreen needleleaf stands, specifically associated with sandy soils and burning. When areas are given fire protection, broadleaf trees quickly take over.

Pines of several varieties—longleaf, shortleaf, pitch, and loblolly—predominate in the southeastern and Atlantic coastal plains of the United States. Farther north into New England and westward in a narrow belt to the Great Lakes, white and red pines and eastern hemlock are the principal evergreens, mixed with deciduous varieties of oak, beech, hickory, maple, elm, chestnut, and many others.

These mixed stands contain valuable timber for logging, but their distribution has been greatly altered by human activity. Native stands of white pine in Michigan and Minnesota were removed before 1910; only later reforestation sustains their presence today. In northern China these forests have almost disappeared as a result of centuries of occupation. The forest species that once flourished in China are similar to species in eastern North America: oak, ash, lime, walnut, elm, maples, and birch once flourished there.

This biome is quite consistent in appearance from continent to continent and at one time represented the principal vegetation of the *Cfa, Cwa humid subtropical hot summer climates, Cfb marine west coast,* and *Cwb cool summer, winter drought* climatic regions of North America, Europe, and Asia.

However, the level of deforestation in historic times is well illustrated by the two maps in Figure 20-10. The first shows the forest cover estimated for Europe in A.D. 900. The second map is the dis-

FIGURE 20-10
A comparison of Europe's forests in (a) A.D. 900 and (b) A.D. 1900. [From H. C. Darby, "The Clearing of the Woodland in Europe," in *Man's Role in Changing the Face of the Earth,* edited by William L. Thomas, Jr. Copyright © 1956 by the University of Chicago Press. All rights reserved. Reprinted by permission.]

(a)

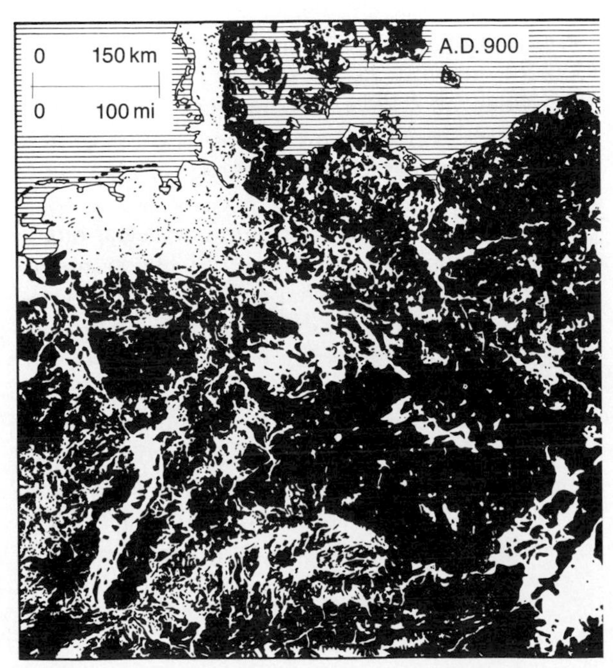

(b)

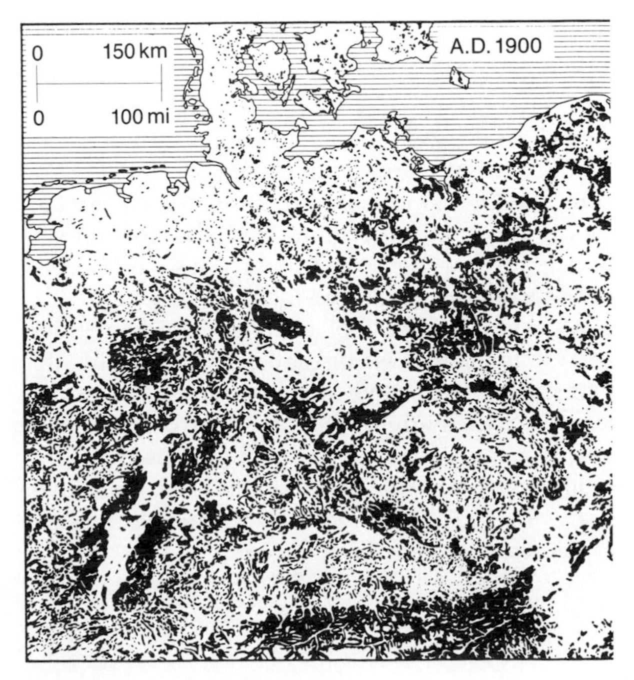

FIGURE 20-11
Boreal forest of Canada. [Photo by author.]

tribution of the remaining forest, A.D. 1900. The comparison dramatically illustrates human modification of natural vegetation. This deforestation was principally for agricultural purposes, as well as construction materials and fuel.

A wide assortment of mammals, birds, reptiles, and amphibians is distributed throughout this biome. Representative animals (some migratory) include fox, squirrel, deer, opossum, bear, rodents, and a great variety of birds. To the north of this biome, poorer soils and colder climates favor stands of coniferous trees and a gradual transition to the northern needleleaf forests.

Northern Needleleaf Forest and Montane Forest

Stretching from the east coast of Canada and the Atlantic provinces westward to Alaska and continuing from Siberia across the entire extent of Russia to the European Plain is the **northern needleleaf forest,** also called the **boreal forest.** A more open form of boreal forest, transitional to arctic and subarctic regions, is termed the **taiga.** The Southern Hemisphere, lacking *D humid microthermal* climates except in mountainous locales, has no biome designated as such. However, **montane forests** of needleleaf trees exist worldwide at high elevation (Figure 20-11).

Boreal forests of pine, spruce, fir, and larch occupy most of the subarctic climates on Earth that are dominated by trees. Although these forests are similar in formation, individual species vary between North America and Eurasia. The larch (Larix) is interesting because it loses its needles in the winter months, perhaps as a defense against the extreme cold of its native Siberia (see the Verkhoyansk climographs in Figures 10-17 and 5-9); larches also occur in North America. The Sierra Nevada, Rocky Mountains, Alps, and Himalayas have similar forest communities occurring at lower latitudes. Douglas and white fir are associated with the western mountains in the United States and Canada. Economically, these forests are important for lumbering, with saw timber occurring in the southern margins of the biome and pulpwood throughout the middle and northern portions. Present timber practices and whether or not these yields are sustainable are issues of increasing controversy.

Certain regions of the northern needleleaf biome experience the permafrost and gelifluction (solifluction) discussed in Chapter 17. When coupled with rocky and poorly developed soils, these conditions generally limit the existence of trees to those with shallow rooting systems. The summer thaw of surface layers results in muskeg (moss-covered) bogs and Histolic (organic) soils of poor drainage and stability. Soils of the taiga are typically Spodosols (Podzolic), characteristically acidic and leached of humus and clays.

FIGURE 20-12
Temperate Hoh Rain Forest in Olympic National Park, Washington, in the northwestern United States. Club moss, big leaf maples, wood sorrel, and sword ferns are pictured. [Photo by Jack W. Dykinga.]

Representative fauna in this biome include wolf, moose (the largest deer), bear, lynx, wolverine, marten, small rodents, and migratory birds during the brief summer season. Birds include hawks and eagles, goshawks, several species of grouse, and owls. About 50 species of insects particularly adapted to the presence of coniferous trees inhabit the biome.

Temperate Rain Forest

The **temperate rain forest** biome is recognized by its lush forests at middle and high latitudes, occurring only along narrow margins of the Pacific Northwest in North America (Figure 20-12). Some similar types exist in southern China, small portions of southern Japan, New Zealand, and a few areas of Chile. This biome contrasts with the diversity of the equatorial and tropical rain forest in that only a few species make up the bulk of the trees. The rain forest of the Olympic Peninsula in Washington State is a mixture of broadleaf and needleleaf trees, huge ferns, and thick undergrowth. Precipitation approaching 400 cm (160 in.) per year on the western slopes, moderate air temperatures, summer fog, and an overall maritime influence produce this moist, lush vegetation community. Animals include bear, badger, deer, wild pig, wolf, bobcat, and fox. The trees are home to numerous birds and mammals.

The tallest trees in the world occur in this biome—the coastal redwoods *(Sequoia sempervirens)*. Their distribution is shown on the map in Figure 19-12a. These trees can exceed 1500 years of age and typically range in height from 60 to 90 m (200–300 ft), with some exceeding 100 m (330 ft). Virgin stands of other representative trees—Douglas fir, spruce, cedar, and hemlock—have been reduced to a few remaining valleys in Oregon and Washington, less than 10% of the original

FIGURE 20-13
Chaparral vegetation associated with the Mediterranean dry summer climate. [Photo by author.]

forest. A study released in 1993 by the U.S. Forest Service focused attention on the failing ecology of these forest ecosystems and suggested that timber management plans should include ecosystem preservation as a priority. The government held a forestry summit in Portland, Oregon in 1993, attempting to bring the disparate sides of the logging-versus-environment conflict together for compromise and for better economic and ecological planning.

Mediterranean Shrubland

The **Mediterranean shrubland,** also referred to as a temperate shrubland, occupies those regions poleward of the shifting subtropical high-pressure cells. As those cells shift poleward with the high Sun, they cut off available storm systems and moisture. Their stable high-pressure presence produces the characteristic dry summer climate—Köppen's *Csa, Csb Mediterranean dry summer*—and establishes conditions conducive to numerous fires. Plant ecologists think that this biome is well adapted to frequent fires, for many of its characteristically deep-rooted plants have the ability to resprout from their roots after a fire. Earlier stands of evergreen woodlands (holm or evergreen oak) no longer are dominant.

The dominant shrub formations that occupy these regions are short, stunted, and tough in their ability to withstand hot-summer drought. The vegetation is called *sclerophyllous* (from *sclero* or "hard" and *phyllos* for "leaf"); it averages a meter or two in height and has deep, well-developed roots, leathery leaves, and uneven low branches.

Typically, the vegetation varies between woody shrubs covering more than 50% of the ground and grassy woodlands with 25–60% coverage. In California, the Spanish word *chaparro* for "scrubby evergreen" gives us the word **chaparral** for this vegetation type (Figure 20-13). This scrubland includes species such as manzanita, toyon, red bud, ceanothus, mountain mahogany, blue and live oaks, and the dreaded poison oak. A counterpart to chaparral in the Mediterranean region, called *maquis,* includes live and cork oak trees, as well as pine and olive trees. In Chile, such a region is called *mattoral;* in southwest Australia, *mallee scrub.* Of course, in Australia the bulk of

FIGURE 20-14
Midlatitude grassland under cultivation. [Photo by author.]

the eucalyptus species is sclerophyllous in form and structure in whichever climate it occurs.

As described in Chapter 10, Mediterranean climates are important in commercial agriculture for specific varieties of subtropical fruits, vegetables, and nuts, with many food types produced only in this biome (e.g., artichokes, olives, almonds). Larger animals, such as different types of deer, are grazers and browsers, with coyote, wolf, and bobcat as predators. Many rodents, other small animals, and a variety of birds also proliferate.

Midlatitude Grasslands

Of all the natural biomes, the **midlatitude grasslands** are most modified by human activity. Here are the world's "breadbaskets"—regions of grain and livestock production (Figure 20-14). In these regions, the only naturally occurring trees were deciduous broadleafs along streams and other limited sites. These regions are called grasslands because of the original predominance of grasslike plants:

> **In this study of vegetation, attention has been devoted to grass because grass is the dominant feature of the Plains and is at the same time an index to their history. Grass is the visible feature which distinguishes the Plains from the desert. Grass grows, has its natural habitat, in the transition area between timber and desert . . . . The history of the Plains is the history of the grasslands.***

In North America, tall grass prairies once rose to heights of 2 m (6.5 ft) and extended westward to about the 98th meridian, with shortgrass prairies farther west. The 98th meridian is roughly the location of the 50 cm (20 in.) isohyet, with wetter conditions to the east and drier to the west (see Figure 18-16).

The deep sod of those grasslands posed problems for the first settlers, as did the climate. The self-scouring steel plow, introduced in 1825 by John Deere, allowed the interlaced grass sod to be broken apart, freeing the soils for agriculture. Other inventions were critical to opening this region and solving its unique spatial problems: barbed wire (the fencing material for a treeless prairie), well-drilling techniques developed by Pennsylvania oil drillers but used for water wells, windmills for pumping, and railroads to conquer the distances.

Few patches of the original prairies (tall grasslands) or steppes (short grasslands) remain within this biome. For prairies alone, the reduction of natural vegetation went from 100 million hectares

*From *The Great Plains* by Walter Prescott Webb, p. 32. © Copyright 1959 by Walter Prescott Webb, published by Ginn and Company, Needham Heights, MA. Used by permission of Silver, Burdett & Ginn, Inc.

(250 million acres) down to a few areas of several hundred hectares each. A "prairie national park" was considered as a preservationist effort along the Kansas-Oklahoma border, but plans were put on hold. The map in Figure 20-4 shows the natural location of these former prairie and steppe grasslands.

Outside North America, the Pampas of Argentina and Uruguay and the grasslands of Ukraine are characteristic midlatitude grassland biomes. Figure 18-15 portrays the interacting variables of the physical environment that produce the grasslands and Mollisols of the Russian and Ukrainian fertile triangle.

In each region of the world where these grasslands have occurred, human development of them was critical to territorial expansion. This is the home of large grazing animals including deer, antelope, pronghorn, and bison, although the almost complete annihilation of the latter is part of American history. Grasshoppers and other insects feed on the grasses and crops as well, and gophers, prairie dogs, ground squirrels, turkey vultures, grouse, and prairie chickens are on the land; predators include coyote and badger, and birds of prey such as hawks, eagles, and kites.

Desert Biomes

Earth's **desert biomes** cover more than one-third of its land area. In Chapter 15 we examined desert landscapes and in Chapter 10, desert climates. On a planet with such a rich biosphere, the deserts stand out as unique regions where fascinating adaptations for survival have developed. Much as a group of humans in the desert might behave with short supplies, plant communities also compete for water and site advantage. Some desert plants, called *ephemerals,* wait years for a rainfall event, at which time their seeds quickly germinate, develop, flower, and produce new seeds, which then rest again until the next rainfall event. The seeds of some xerophytic species open only when fractured by the tumbling, churning action of flash floods cascading down a desert arroyo, and of course such an event produces the moisture that a germinating seed needs.

Perennial desert plants employ other adaptive features to cope with the desert, such as long, deep tap roots (e.g., the mesquite), succulence (i.e., thick, fleshy, water-holding tissue such as that of cacti), spreading root systems to maximize water availability, waxy coatings and fine hairs on leaves to retard water loss, leafless conditions during dry periods (e.g., palo verde and ocotillo), reflective surfaces to reduce leaf temperatures, and tissue that tastes bad to discourage herbivores.

The creosote bush *(Lorrea divaricata)* sends out a wide pattern of roots and contaminates the surrounding soil with toxins that prevent the germination of other creosote seeds, possible competitors for water. When a creosote bush dies and is removed, surrounding plants or germinating seeds work to occupy the abandoned site, but must rely on infrequent rains to remove the toxins.

The faunas of both warm and cold deserts are limited by the extreme conditions and include few resident large animals, exceptions being the desert bighorn sheep (in nearby mountains) and the camel, which can lose up to 30% of its body weight in water without suffering (for humans a 10–12% loss is dangerous). Some representative desert animals are the ring-tail cat, kangaroo rat, lizards, scorpions, and snakes. Most of these animals are quite secretive and become active only at night, when temperatures are lower. In addition, various birds have adapted to desert plants and other available food sources—for example, roadrunners, thrashers, ravens, wrens, hawks, grouse, and nighthawks.

Desertification. We are witnessing an unwanted expansion of the desert biome. This is due principally to agricultural practices (overgrazing and inappropriate agricultural activities), improper soil-moisture management, erosion and salinization, deforestation, and the ongoing climatic change. A process known as **desertification** is now a worldwide phenomenon along the margins of semiarid and arid lands.

As a result, the southward expansion of the Sahara through portions of the Sahel region has left many African peoples on land that no longer experiences the rainfall patterns of just two decades earlier. Regions at risk of desertification stretch from Asia and central Australia to portions of

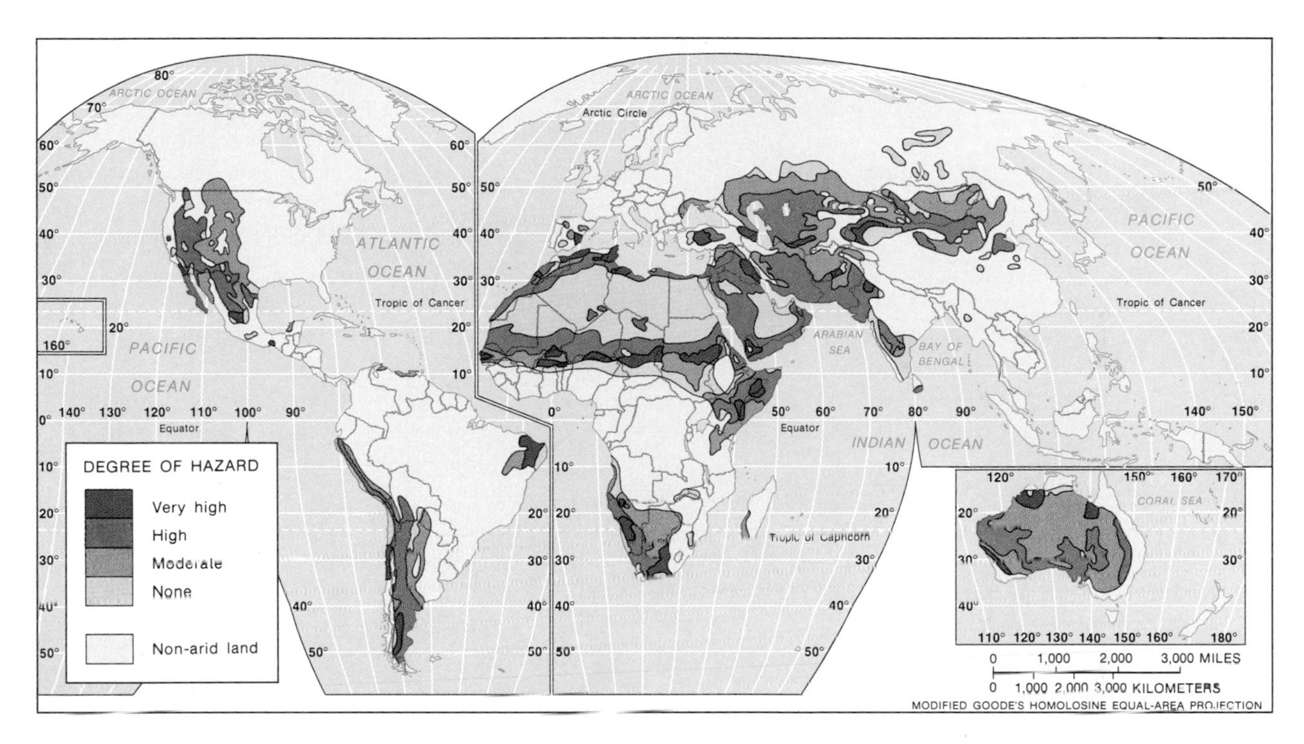

FIGURE 20-15
Worldwide desertification estimates by the United Nations. [Data from U.N. Food and Agriculture Organization (FAO), World Meteorological Organization (WMO), United Nations Educational, Scientific, and Cultural Organization (UNESCO), Nairobi, Kenya, 1977; adapted by permission of Merrill, an imprint of Macmillan Publishing Company, from *The Cultural Landscape,* by James M. Rubenstein, pp. 468–69. Copyright © 1989 by Merrill Publishing.]

North and South America. The United Nations estimates that desertified lands have covered some 800 million hectares (2 billion acres) since 1930; many millions of additional hectares are added each year. The process is increasing at an alarming rate. Of course, an immediate need is to improve the database for a more accurate accounting of the problem.

Figure 20-15 is drawn from a map prepared for the 1977 U.N. Conference on Desertification held in Nairobi, Kenya, and a conference paper titled "Plan of Action to Combat Desertification." Desertification areas are ranked: moderate hazard with an average 10–25% drop in agricultural productivity; high hazard with a 25–50% drop; and very high, representing more than a 50% decrease. Desertification includes consideration of the soil degradation portrayed on the map in Figure 18-7. Because human activities and economies, especially unwise grazing practices, appear to be the major cause of desertification, actions to slow the process are readily available with leadership.

Earth's deserts are subdivided into desert and semidesert associations, to distinguish those with expanses of bare ground from those covered by xerophytic plants of various types. The two broad associations are further separated into warm deserts, principally tropical and subtropical, and cold deserts, principally midlatitude.

Warm Desert and Semidesert. Earth's warm desert and semidesert biomes are associated with

FIGURE 20-16
Cold desert scene in the interior Great Basin of the western United States. [Photo by author.]

the presence of dry air and low precipitation from subtropical high-pressure cells. These areas are very dry, as evidenced by the Atacama Desert of northern Chile, where only a minute amount of rain has ever been recorded—a 30-year annual average of only 0.05 cm (0.02 in.)! Like the Atacama, the true deserts of the Köppen *BW desert* classification are under the influence of the descending, drying, and stable air of high-pressure systems from 8 to 12 months of the year. Remember that these dry regions are defined by lower amounts of precipitation that fail to satisfy higher amounts of potential evapotranspiration. *BW deserts* receive precipitation that is less than one-half of potential evapotranspiration. Semiarid *BS steppes* receive precipitation that is more than one-half of annual potential evapotranspiration.

Vegetation ranges from almost none in the arid deserts to numerous xerophytic shrubs, succulents, and thorn tree forms. A few of the subtropical deserts—such as those in Chile, Western Sahara, and Namibia—are right on the sea coast and are influenced by cool offshore ocean currents. As a result, these true deserts experience summer fog that mists the plant and animal populations with needed moisture.

The equatorward margin of the subtropical high-pressure cell is a region of transition to savanna, thorn tree, scrub woodland, and tropical seasonal forest. Poleward of the warm deserts, the subtropical cells shift to produce the Mediterranean dry summer regime along west coasts and may grade into cool deserts elsewhere.

Cold Desert and Semidesert. The cold desert and semidesert biomes tend to occur at higher latitudes where seasonal shifting of the subtropical high is of some influence less than six months of the year. Specifically, interior locations are dry because of their distance from moisture sources or their location in rain shadow areas on the lee side of mountain ranges such as the Sierra Nevada of the western United States, the Himalayas, and the Andes. The combination of interior location and rain shadow positioning produces the cold deserts of the Great Basin of western North America (Figure 20-16).

Winter snows, although generally light, are possible in the cold deserts. Summers are hot, with highs from 30° to 40°C (86°–104°F). Nighttime lows, even in the summer, can cool 10°–20C° (18°–36F°) from the daytime high. The dryness,

FIGURE 20-17
Tundra on the Kamchatka Peninsula, Russia. The Uzon Caldera is in the center-background mountain range. [Photo by Wolfgang Kaehler.]

generally clear skies, and sparse vegetation lead to high radiative heat loss and cool evenings. Many areas of these cold deserts that are covered by sagebrush and scrub vegetation were actually dry shortgrass regions in the past, before extensive grazing forever altered their appearance. The look of the deserts in the upper Great Basin is the result of more than a century of such cultural practices.

Arctic and Alpine Tundra

The **arctic tundra** is found in the extreme northern area of North America and Russia, bordering on the Arctic Ocean and generally north of the 10°C (50°F) isotherm for the warmest month. Daylength varies greatly throughout the year, seasonally changing from almost continuous day to continuous night. The region, except a few portions of Alaska and Siberia, was covered by ice during all of the Pleistocene glaciations. There is no comparable tundra biome on the continent of Antarctica, except on an extreme portion of Palmer Land across Drake Passage from the tip of South America. Winters in this biome, an *ET tundra* climate classification, are long and cold; cool summers are brief.

Winter is governed by intensely cold continental polar air masses and stable high-pressure anticyclonic conditions. A growing season of sorts lasts only 60–80 days, and even then frosts can occur at any time. Vegetation is fragile in this flat, treeless world; soils are poorly developed periglacial surfaces, which are underlain by permafrost. In the summer months only the surface horizons thaw, thus producing a mucky surface of poor drainage. Roots can penetrate only to the depth of thawed ground, usually about a meter. The surface is shaped by freeze-thaw cycles that create the frozen ground phenomena and gelifluction (solifluction) processes discussed in Chapter 17.

Tundra vegetation is characterized by low, ground-level herbaceous plants and some woody species (Figure 20-17). Representative are sedges, mosses, arctic meadow grass, snow lichen, and dwarf willow. Owing to the short growing season, some perennials form flower buds one summer and open them for pollination the next. Animals of the tundra include musk-ox, caribou, wolf, lemmings, and other small rodents (important food for the larger carnivores), fox, weasel, rabbit, reindeer, snowy owls, ptarmigans, polar bears, and of course mosquitoes. The tundra is an important breeding ground for geese, swans, and other waterfowl.

Alpine tundra is similar in many ways to arctic tundra, although alpine tundra can occur at lower latitudes because it is associated with high elevation. This biome usually is described as above the timberline, that elevation above which trees cannot grow. Alpine timberlines increase in elevation with decreasing latitude in both hemispheres. Alpine tundra communities occur in the Andes near the equator, the White Mountains of California, the Rockies, the Alps, and Mount Kilimanjaro of equatorial Africa, as well as mountains through the Middle East to Asia. Alpine tundra can experience permafrost in the rocky soils and periglacial soil flow. Alpine meadows feature grasses and stunted shrubs, such as willows and heaths. Because alpine locations are frequently windy sites, many plants appear sculpted by the wind.

Because the tundra biome is of such low productivity, it is fragile. Disturbances such as tire tracks, hydroelectric projects, and mineral exploitation leave marks that persist for hundreds of years. As development continues, the region will face even greater challenges from oil spills, contamination, and disruption.

Plans continue for the Arctic National Wildlife Refuge (ANWR) on Alaska's North Slope above the Arctic Circle, bordering on the Beaufort Sea, and adjoining the Yukon Territory. (See photo facing the title page of this text.) The refuge area sustains almost 200,000 caribou, polar and grizzly bears, musk-ox, and wolves. The ANWR remains the only portion of Alaska's Arctic coast that is not open to oil and gas exploration. But controversy over this refuge continues as political and corporate pressures mount to conduct oil exploration there. The resulting disruption would be devastating and the cost of development would price the oil at well over $30 per barrel, far above current market prices. The estimated oil reserve in the ANWR could be offset in only two years by a 2-mile-per-gallon increase in automobile efficiency. Consequently, economic activities should be weighed against the nature of the ecosystem itself, its limitations and its uniqueness.

SUMMARY—Terrestrial Biomes

Earth is the only planet in the Solar System with a biosphere. An impressive feature of the living Earth is its diversity, a diversity that biogeographers categorize into discrete spatial biomes for analysis and study. The interplay among supporting physical factors within Earth's ecosystem determines the distribution of plant and animal communities. Plants assume adaptive associations and forms as each plant competes for individual success in its habitat and niche.

Biomes are Earth's major terrestrial ecosystems, each named for its dominant plant community. The 10 major biomes are generalized from numerous formation classes that describe vegetation units on the basis of structure and appearance. Specifically, plant life forms describe the outward physical properties of individual plants or the general form and structure of a vegetation cover. Ideally a biome represents a climax community of natural vegetation. In reality, few undisturbed biomes exist in the world, for most are modified by human activity. Many of Earth's plant and animal communities are experiencing an accelerated rate of change that could produce dramatic alterations within our lifetime.

The equatorial and tropical rain forest biome is undergoing rapid deforestation. Because the rain forest is Earth's most diverse biome and important to the climate system, such losses are creating great concern among citizens, scientists, and nations. Efforts are underway worldwide to set aside and protect remaining representative sites within most of Earth's principal biomes.

KEY TERMS

aquatic ecosystems
arctic tundra
biogeographical realms
biome
boreal forest
chaparral
community
desert biomes
desertification
ecosystem
ecotone
equatorial and tropical rain forest
formation classes
habitat
large marine ecosystem (LME)
Mediterranean shrubland
midlatitude broadleaf and mixed forest
midlatitude grasslands
montane forests
niche
northern needleleaf forest
taiga
temperate rain forest
terrestrial ecosystems
tropical savanna
tropical seasonal forest and scrub

REVIEW QUESTIONS

1. Reread the two opening quotations in this chapter. What clues do you have to the path ahead for Earth's forests? Is our future direction controllable? Explain.
2. What is a biogeographical realm? How is the world subdivided according to plant and animal types?
3. What is a large marine ecosystem? What is the purpose of such a classification?
4. Distinguish between formation classes and life-form designations as a basis for spatial classification.
5. Define *biome*. What is the basis of the designation?
6. Use the integrative chart in Table 20-1 and the world map in Figure 20-4 to correlate the description of each biome with its specific spatial distribution.
7. Describe the equatorial and tropical rain forests. Why is the rain forest floor somewhat clear of plant growth? Why are logging activities for specific species so difficult there?
8. What are the issues surrounding deforestation of the rain forest? What is the impact of these losses on the rest of the biosphere? What new threat to the rain forest has emerged?
9. What do *caatinga, chaco, brigalow,* and *dornveld* refer to? Explain.
10. Describe the role of fire or fire ecology in the tropical savanna biome and the midlatitude broadleaf and mixed forest biome.
11. Why does the northern needleleaf forest biome not exist in the Southern Hemisphere? Where is this biome located in the Northern Hemisphere, and what is its relationship to climate type?
12. In which biome do we find Earth's tallest trees? Which biome is dominated by small, stunted plants, lichens, and mosses?
13. What type of vegetation predominates in the Mediterranean dry summer climates? Describe the adaptation necessary for these plants to survive.
14. What is the significance of the 98th meridian in terms of North American grasslands? What types of inventions were necessary for humans to cope with the grasslands?

15. Describe some of the unique adaptations found in a desert biome.
16. What is desertification? Explain its impact.
17. What physical weathering processes are specifically related to the tundra biome? What types of plants and animals are found there?
18. What is the relationship between island biogeography and biosphere reserves? Describe a biosphere reserve.
19. Compare the map in Figure 20-4 with the composite satellite image inside the front cover of this text. What similarities can you identify between the local summertime portrait of Earth's biosphere and the biomes identified on the map? Take the biomes as listed in the legend for the map and work through them one at a time.

SUGGESTED READINGS

Attenborough, David, Philip Whitfield, Peter D. Moore, and Barry Cox. *The Atlas of the Living World.* Boston: Houghton Mifflin, 1989.

Barbour, Michael G., and William D. Billings, eds. *North American Terrestrial Vegetation.* Cambridge: Cambridge University Press, 1988.

Colinvaux, Paul A. "The Past and Future Amazon," *Scientific American* 260, no. 5 (May 1989): 102–108.

Dansereau, Pierre. *Biogeography: An Ecological Perspective.* New York: Ronald Press, 1957.

Editors of Time-Life Books. *Grasslands and Tundra.* Alexandria, VA: Time-Life Books, 1985.

Ehrlich, P. R., and E. O. Wilson. "Biodiversity Studies: Science and Policy," *Science* 253 (August 16, 1991): 758–62. The entire "Research News" section of this issue is related to biodiversity and species extinction.

Ellis, William S. "Rondônia: Brazil's Imperiled Rain Forest," *National Geographic* 174, no. 6 (December 1988): 772–79.

Food and Agricultural Organization and United Nations. "Second Interim Report on the State of the Tropical Forests," *Forest Resources Assessment 1990 Project.* Tenth World Forestry Conference, Paris. New York: United Nations, September 1991, rev. October 15, 1991.

Page, Jake, and the editors of Time-Life Books. *Forest.* Alexandria, VA: Time-Life Books, 1983.

Perry, Donald R. "The Canopy of the Tropical Rain Forest," *Scientific American* 251, no. 5 (November 1984): 138–47.

Postal, Sandra, and Lori Heise. *Reforesting the Earth.* Worldwatch Paper 83. Washington, DC: Worldwatch Institute, 1988.

Sherman, Kenneth, Lewis M. Alexander, and Barry D. Golds, eds. *Large Marine Ecosystems—Patterns, Processes, and Yields.* Washington, DC: American Association for the Advancement of Science, No. 90-30S, 1990.

Skole, David, and Compton Tucker. "Tropical Deforestation and Habitat Fragmentation in the Amazon: Satellite Data from 1978 to 1988," *Science* 260 (25 June 1993): 1905-10.

Tucker, Compton J., John R. G. Townshend, and Thomas E. Goff. "African Land-Cover Classification Using Satellite Data," *Science* 227 (25 January 1985): 369–76.

White, Peter T. "Tropical Rain Forests: Nature's Dwindling Treasures," *National Geographic* 163, no. 1 (January 1983): 2–47.

World Resources Institute, The World Conservation Union, United Nations Environment Programme (UNEP), FAO and UNESCO. *Global Biodiversity Strategy—Guidelines for Action to Save, Study, and Use Earth's Biotic Wealth Sustainably and Equitably*. Washington, DC: World Resources Institute, 1992.

Paddy rice farming and rice terraces in Bali, Indonesia. [*Photo by Wolfgang Kaehler.*]

21

The Human Denominator

During my space flight, I came to appreciate my profound connection to the home planet and the process of life evolving in our special corner of the Universe, and I grasped that I was part of a vast and mysterious dance whose outcome will be determined largely by human values and actions.*

Earth is observed and studied from many profound vantage points, as this astronaut experienced on the 1969 Apollo IX mission. Our vantage point in this book is physical geography. We have examined Earth through its energy, atmosphere, water, weather, climate, endogenic and exogenic systems, soils, ecosystems, and biomes—all of which leads to an examination of the planet's most abundant large animal, *Homo sapiens.*

Because human influence is so pervasive, we can consider the totality of our impact the *human denominator.* Just as the denominator in a fraction tells how many parts a whole is divided into, so the growing human population and the increasing demand for resources suggest how much the whole Earth system must adjust. Yet, Earth's resource base remains relatively fixed.

In his 1980 book and PBS television series, astronomer Carl Sagan asked:

What account would we give of our stewardship of the planet Earth? We have heard the rationales offered by the nuclear superpowers. We know who speaks for the nations. But who speaks for the human species? Who speaks for Earth?**

We might attempt an answer: perhaps those who have studied Earth and know the operations of the global ecosystem should speak for Earth. Or perhaps those who have the greatest stake in the continued functioning of that ecosystem—the children—should address the importance of a workable future. Some might say that these questions of technology, politics, and future thinking belong outside of academia, that our job here is merely to learn how Earth's processes work. However, biologist-ecologist Marston Bates addressed this line of thought in 1960:

Then we came to humans and their place in this system of life. We could have left humans out, playing the ecological game of "let's pretend humans don't exist." But this seems as unfair as the corresponding game of the economists, "let's pretend nature doesn't exist." The economy of nature and the ecology of humans are inseparable and attempts to separate them are more than misleading, they are dangerous. Human destiny is tied to nature's destiny and the arrogance of the engineering mind does not change this. Humans may be a very peculiar animal, but they are still a part of the system of nature.*

Interestingly, most of the environmental problems being studied by scientists already have recognized solutions or strategies for positive resolution, if only this knowledge could be applied in a constructive way. Unfortunately, many of these solutions are not being acted upon by society. We do not seem to assess the true long-term costs of present economic thinking. A role for physical geography is to explain Earth systems through spatial analysis and through a synthesis of information from many disciplines—hopefully to assist an informed citizenry toward more considered judgments.

There is an international environmental awareness and concern in people that is in turn driving new action by government. The George H. Gallup International Institute completed an extensive 22-nation, 22,000-person survey that exposed some fallacies in the conventional wisdom. Survey re-

*Rusty Schweickart, "Our Backs Against the Bomb, Our Eyes on the Stars," *Discovery,* July 1987: 62. Reprinted by permission.

**Carl Sagan, *Cosmos.* New York: Random House, 1980, p. 329. Reprinted by permission.

*Marston Bates, *The Forest and the Sea.* New York: Random House, 1960, p. 247. Reprinted by permission.

FIGURE 21-1
Cormorant contaminated with oil from the *Exxon Valdez* tanker accident in Prince William Sound, Alaska. [Photo by Gary Braasch.]

sults established that ". . . people in both poor and rich nations give priority to environmental protection over economic growth," ". . . accept responsibility for contributing to problems, and believe that citizen efforts can contribute significantly to a healthier planet," and that ". . . environmental problems are at the very top of the list of problems in virtually all nations." Majorities in 20 nations are willing to "endorse environmental protection at the risk of slowing down economic growth."*

AN OILY DUCK

At first glance the chain of events that exposes wildlife to oil contamination seems to stem from a technological problem. An oil tanker splits open at sea and releases its petroleum cargo, which is moved by ocean currents toward shore, where it coats coastal waters, beaches, and animals (Figure 21-1). In response, concerned citizens mobilize and try to save as much of the spoiled environment as possible, but the real problem goes far beyond the physical facts of the spill.

On March 24, 1989, in Prince William Sound off the southern coast of Alaska in clear weather and calm seas, a single-hulled supertanker operated by Exxon Corporation struck a reef that was outside the normal shipping lane and spilled 42 million liters (11 million gallons) of oil. It took only 12 hours for the *Exxon Valdez* to spill its contents, yet a reasonable cleanup will take years and billions of dollars.

Because contingency emergency plans were not in place, and promised equipment was unavailable, response by the oil industry took over 12 hours to activate, about the same time that it took the ship to empty. This initial response by Alyeska Pipeline Service Company and Exxon proved inadequate to the task. Badly needed equipment was unavailable in dry docks, in warehouses, and buried in snow. In the years following the spill, evidence of this initial inaction and confusion was documented. Total damage now is estimated at between $15 and $20 billion, although Exxon settled with the government for $1.125 billion.

*R. E. Dunlap, G. H. Gallup, Jr., and A. M. Gallup, *The Health of the Planet Survey—A Preliminary Report on Attitudes Toward the Environment and Economic Growth Measured by Surveys of Citizens in 22 Nations to Date.* A George H. Gallup Memorial Survey. Princeton, NJ: George H. Gallup International Institute, July 1992.

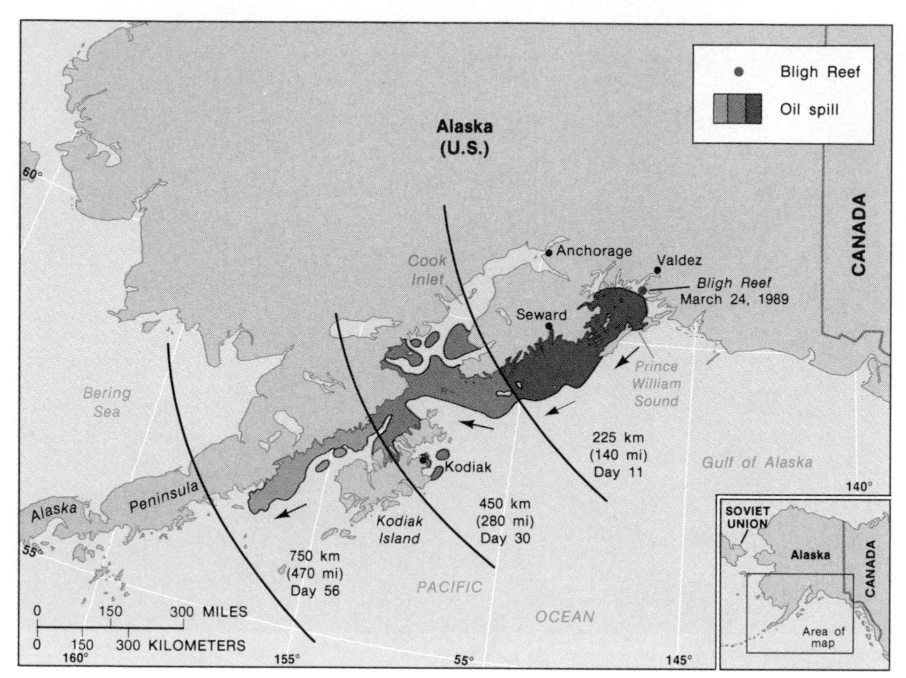

FIGURE 21-2
Track of spreading oil for the first 56 days of the 1989 Alaskan oil spill.

Eventually, almost 2415 km (1500 mi) of sensitive coastline were ruined. For perspective, had this spill occurred farther south, enough oil spilled to blacken every beach and bay along the Pacific Coast from southern Oregon to the Mexican border (Figure 21-2).

The death toll for animals was massive: at least 3000 sea otters killed (or about 20% of the resident otters), 300,000 birds, and uncounted fish, shellfish, plants, and aquatic microorganisms. Sublethal effects, namely mutations, now are appearing in fish. This latter side effect of the spill is serious because salmon fishing is the main economy in Prince William Sound, not oil. Conflicting scientific reports emerged in 1993 as to long-term damage in the region—scientific studies commissioned by Exxon disagreed with scientific studies prepared outside the industry.

The Larger Picture

The immediate effect of the oil spill on wildlife was contamination and death, but the issues involved are bigger than damaged ecosystems. Let's ask some fundamental questions: Why was the oil tanker there in the first place? Why is petroleum being imported into the continental United States from Alaska and abroad in such quantity? Is the demand for petroleum products based on real need, or does it reflect the business strategies of transnational oil corporations? The U.S. demand for oil is higher per capita than the demand of any other country; is this demand inflated by waste and inefficiency?

A great many factors influence our demand for oil. Well over half of our imported oil goes for transportation. Despite this enormous and expensive consumption, the 1980s saw a major rollback of auto efficiency standards, a reduction in gasoline prices, large reductions in funding for rapid transit development, and the continuing slow demise of America's passenger rail system. Thus, waste, low prices, and a lack of alternatives all combined to spur demand for petroleum. In addition, land-use patterns continue to foster diffuse sprawl, thereby placing stress upon existing transportation systems.

The demand for fossil fuels in the 1980s also was affected by the slowing of domestic conservation programs, the elimination of research for energy alternatives such as solar and wind, and even delays of a law requiring small appliances to be

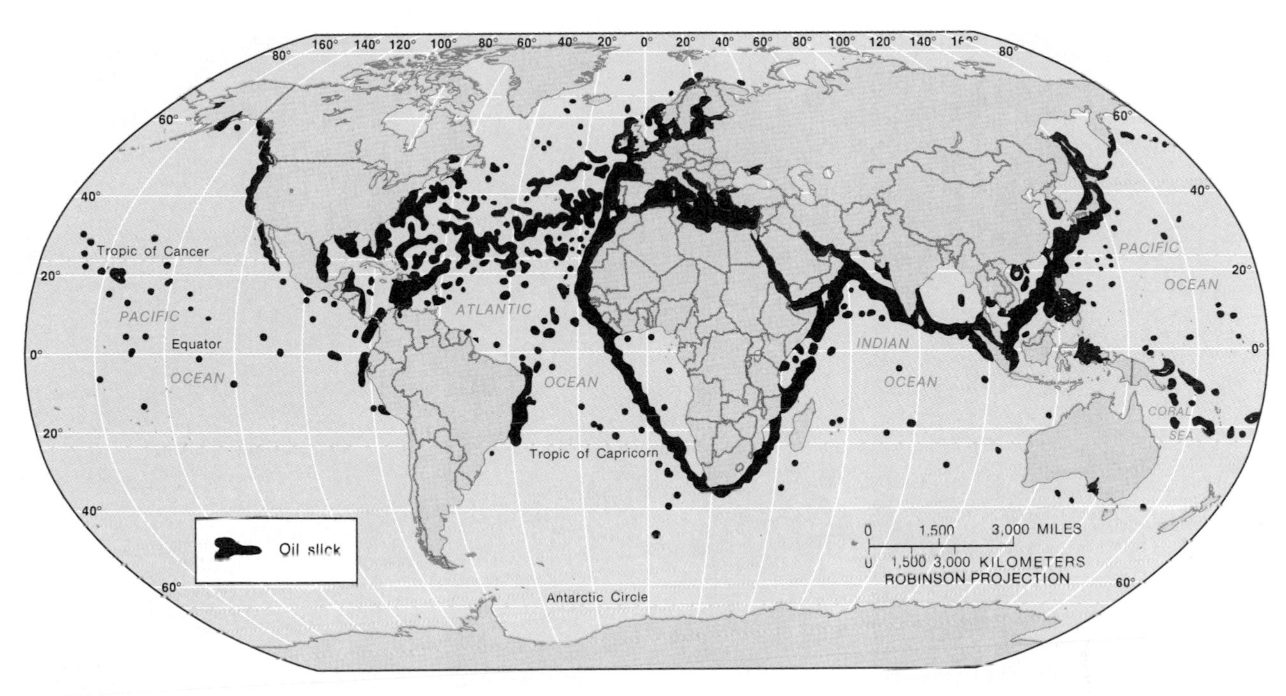

FIGURE 21-3
Location of visible oil slicks worldwide. [Data from Organization for Economic Cooperation and Development, *The State of the Environment*. Paris: Organization for Economic Cooperation and Development, 1985.]

more energy efficient. Conservation plans again were politically blocked in the Department of Energy in 1990 and 1991. All of these manipulations increased fossil fuel demand (5.4 billion barrels a year for oil alone) at a time (1989) when the U.S. Geological Survey estimated that the United States had only 16 years of actual and expected domestic petroleum reserves remaining, or 32 years if 50% of U.S. consumption is imported. Increased imports make the United States more vulnerable to foreign conflicts and war.

Thus, the immediate problem of cleaning oil off a duck in Prince William Sound becomes a national and international concern with far-reaching significance. And while we search for answers, oil slicks continue their contamination. In just one year following the *Exxon Valdez* disaster, nearly 76 million liters (20 million gallons) of oil were spilled in 10,000 accidents worldwide (Figure 21-3).

In addition, an even greater spill than that of the *Exxon Valdez* occurred off the coast of Morocco (in northwestern Africa) in the same year. Into those warm tropical waters that vessel spilled 140.6 million liters (37 million gallons) of oil. A few examples of recent spills include the *Aegean Sea,* off Spain in 1992, 83.6 million liters (22 million gallons); the *Maersk Navigator,* in Indonesian waters in 1993, 30 million liters (7.8 million gallons); and the *Braer,* off the Shetland Islands in 1993, 91 million liters (24 million gallons).

In the Persian Gulf War of 1991 we saw over 1.1 billion liters (300 million gallons) flow into the Persian Gulf and additional millions of liters pour onto the land and burn into the atmosphere in purposely set fires (Figure 21-4). Public opinion holds the second largest corporation in America responsible for what happened in Prince William Sound and a dictator in the Middle East as responsible for the tragedy in the Persian Gulf. Yet, there is apparent hypocrisy in our outrage over these events, since we continue to stand at the pump filling our automobiles with gas, thus creating the

FIGURE 21-4
Progress of Kuwait oil well fires and smoke over thousands of square kilometers on April 28, May 30, and July 1, 1991. [Satellite images courtesy of EOSAT, Earth Observation Satellite Company. Used by permission.]

demand for imports. Physical geography is capable of analyzing all the spatial aspects of these events and ironies.

GAIA HYPOTHESIS

Just as the abiotic spheres affect the biosphere, so do living processes affect abiotic functions. All of these interactive effects in concert influence Earth's overall ecosystem. In essence, the planetary ecosystem sets the physical limits for life, which in turn evolves and helps to shape the planet. Thus, Earth can be viewed as one vast, self-regulating organism—a global symbiosis, or mutualism. This controversial concept is called the *Gaia hypothesis*. It was proposed in 1979 by James Lovelock, a British astronomer and inventor, and elaborated by American biologist Lynn Margulis.

Gaia is the ultimate synergistic relationship, in which the whole greatly exceeds the sum of the individual interacting components. The hypothesis contends that life processes control and shape inorganic physical and chemical processes, with the biosphere so interactive that a very small mass can affect a very large mass. Thus, Lovelock and Margulis think that the material environment and the evolution of species are tightly joined; as the species evolve through natural selection, they in turn affect their environment. The present composition of the atmosphere is given as proof of this coevolution of living and nonliving systems.

From the perspective of physical geography, this hypothesis permits a view of all of Earth and the spatial interrelationships among systems. Such a perspective is necessary for analyzing specific environmental issues. For instance, in deciding the fate of the Arctic National Wildlife Refuge (ANWR), we must weigh the supply, demand, and importation of oil, the public trust aspect of such wilderness, our will for conservation and efficiency, and

FIGURE 21-5
In Alaska's Arctic National Wildlife Refuge, Alaska, Mount Chamberlin, the second highest peak of the Brooks Range, overlooks tundra in the foreground. Is this region destined for petroleum exploration and development or for continued preservation? [Photo by Scott T. Smith.]

our view of our place *in* nature or *outside* of nature (Figure 21-5). All these variables interact synergistically, producing both wanted and unwanted spatial results.

One disturbing aspect of this unity, however, is that any biotic threats to the operation of an ecosystem tend to move toward extinction *themselves,* thus preserving the system overall. Earth systems operation and feedback tend to eliminate offensive members. The degree to which humans represent a planetary threat, then, becomes a topic of great concern, for Earth (Gaia) will prevail, regardless of the outcome of the human experiment.

> **The maladies of Gaia do not last long in terms of her life span. Anything that makes the world uncomfortable to live in tends to induce the evolution of those species that can achieve a new and more comfortable environment. It follows that, if the world is made unfit by what we do, there is the probability of a change in regime to one that will be better for life but not necessarily better for us.***

The debate is vigorous regarding the true applicability of this hypothesis to nature. Regardless, it remains philosophically intriguing in its portrayal of the relationship between humans and Earth.

*James Lovelock, *The Ages of Gaia—A Biography of Our Living Earth*. New York: W. W. Norton, 1988, p. 178. Used with permission.

THE NEED FOR INTERNATIONAL COOPERATION

We already have seen dramatic examples of international environmental problem solving: the Limited Test Ban Treaty of 1962, which bans atmospheric testing of nuclear weapons, and the 1987 and 1990 treaties proposing bans on ozone-destroying chlorofluorocarbons are examples. The International Geosphere-Biosphere Program (IGBP) represents another integrative effort; its goal is to improve understanding of the entire natural system and to discern how the many and varied subsystems interact. All major concerns need cooperative attention. For example, the escalating greenhouse effect that is discussed throughout this text and influences so many global systems certainly deserves international research and mutual action.

The largest gathering of nations and nongovernmental organizations ever held was about Earth—the United Nations Conference on Environment and Development (UNCED), the "Earth Summit of 1992." This important event and resulting agreements are surveyed in Focus Study 21-1. (See Appendix B for the numerous groups, organizations, and international action programs related to geography and environmental concerns.)

An important corollary to international efforts is the linkage of academic disciplines. A positive step

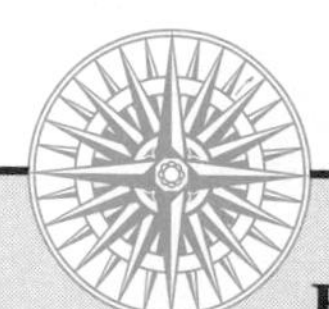

Focus Study 21-1 Earth Summit 1992

Ten thousand delegates from over 160 countries, the leaders of 100 nations, 9000 journalists, and the world's attention, all converged at Rio Centro, 40 minutes outside Rio de Janeiro, June 1–12, 1992 (Figure 1). An additional 1000 nongovernmental organizations (NGOs) with over 50,000 attendees assembled at Flamingo Park for a parallel, unofficial gathering dubbed the "Global Forum." The setting in Rio was ironic, for many of the problems on the table at the two conferences were evident on the streets of the Rio metropolitan region, where air pollution, water pollution, toxics, noise, disparity of income and wealth, and a struggle for health and education occur in abundance. Maurice F. Strong, a Canadian and Secretary-General of the UNCED, summarized in his conference address:

> The people of our planet, especially our youth and the generations which follow them, will hold us accountable for what we do or fail to do at the Earth Summit in Rio. Earth is the only home we have, its fate is literally in our hands. . . .The most important ground we must arrive at in Rio is the understanding that we are all in this together.

Setting the Stage

The U.N. General Assembly in 1987 achieved a landmark in global planning by agreeing to hold the Summit. The idea for some type of global meeting on the environment was put forward at the 1972 U.N. Conference on the Human Environment held in Stockholm, also chaired by Maurice Strong. *Our Common Future** set the tone for the 1992 Earth Summit:

> The Earth is one but the world is not. We all depend on one biosphere for sustaining our lives. Yet each community, each country, strives for survival and prosperity with little regard for its impact on others. Some consume the Earth's resources at a rate that would leave little for future generations. Others, many more in number, consume far too little and live with the prospect of hunger, squalor, disease, and early death.

To the detriment of the final agreements that emerged from the UNCED Earth Summit, the United States administration blocked specific plans for cutting emissions of greenhouse gases and timetables for sustainable forestry goals. It was left to the Japanese and the nations of the European Community at the Earth Summit to push for timetables to reduce damaging emissions, to exercise leadership, and to take advantage of the growing economic opportunity of marketing environmental technology. With all this in mind, let's overview the five key agreements that emerged from Rio.

The Five Earth Summit Agreements

1. Climate Change Framework—This legally binding agreement is a first-ever attempt to evaluate and address global warming on an international scale. As of August 1992, 154 nations, including the United States and Canada, signed the Convention on Climate Change.

Initially the goal was to set specific timetables and limits for cutting emissions of greenhouse gases. The United States objected to specific targets for controlling CO_2 emissions throughout the pre-summit sessions, citing economic uncertainties and unknown costs. The European Community, Canada, Japan, and the majority of attending nations favored stabilizing emissions at 1990 levels by the year 2000. The Office of Technology Assessment and National Academies of Science and Engineering, in separate assessments, concluded that the United States could hold to 1990 levels by the year 2015 at little or no additional cost. Richard Kerr summarized:

> The philosophy that many scientists contacted by *Science* are now espousing amounts to buying some insurance—in the form of no-cost or low-cost reductions in greenhouse gas emissions—against the possibility that the higher predictions of global warming turn out to be right. *(Science* 256 (May 22, 1992) 1140.

2. Biological Diversity—This legally binding agreement is a first international attempt to protect Earth's biodiversity. It provides more equitable rights among nations in biotechnology and the genetic wealth of tropical ecosystems in particular.

Out of 161 signatories, the United States, Vietnam, Singapore, and Kiribati refused to sign the original treaty. This was a perplexing stand for the U.S. administration to take. Biodiversity is one of the more divisive issues separating developed countries in the Northern Hemisphere from the predominantly developing na-

---

*World Commission on Environment and Development, *Our Common Future*. Oxford and New York: Oxford University Press, 1987, p. 27.

(a)

(b)

FIGURE 1
Leaders and delegates at the 1992 UNCED Earth Summit in Rio de Janeiro. (a) A portion of the delegates that attended from many nations. (b) Inset photo of Maurice Strong, Secretary-General of the UNCED (second from left), a representative of indigenous peoples, and other delegates. [(a) Photo by Reuters/Bettmann; (b) photo by Ricardo Funavi/Imagens Da Terra/Impact Visuals.]

tions of the equatorial and tropical regions and Southern Hemisphere. The developing countries want to be compensated for drugs and medicines that are derived from plants and animals harvested from the indigenous genetic wealth in the tropics. This return of some profit from transnational corporations and rich-nation enterprises would in turn produce a potential incentive for further protection of these critical biomes.

3. Management, Conservation, and Sustainable Development of All Types of Forests—This non-binding agreement guides world forestry practices toward a more sustainable future of forest yields and diversity.

The conflict between industrialized nations and developing nations is a classic confrontation of a "north" versus "south" dichotomy. How can the rich nations continue to clear-cut their forests, yet turn to the developing countries, such as Brazil, and virtually ask them to place their lands in a national park or biosphere reserve? How can the industrialized nations continue to produce excessive CO_2, far beyond reasonable per capita limits, yet turn to the developing countries and ask them to cease destroying a principle sink for CO_2? The developing countries of the "south" insist on an egality (political and economic equality) of forest practice. Along with sustainable timber practices, countries of the "north" need to begin government-sponsored paper recycling and packaging-reform programs. Such recycling now is part of the forestry debate.

4. The "Earth Charter"—This is a non-binding statement of environmental and economic principles. These 27 principles establish an ethical basis for a sustainable human-Earth relationship. An important emphasis in this charter is the need to include *environmental costs* in economic assessments. Impoverishment of air, soil, water, and ecosystems sometimes is mistaken for progress. The notion that the environment is not an inexhaustible mine of resources to be tapped indefinitely is gaining many adherents. In terms of *natural capital*—air, water, timber, fisheries, petroleum—Earth is indeed a finite physical system.

5. Agenda 21 (Sustainable Development)—This 800 pages of non-binding action program, is a guide for the nations of the world during the remainder of this century and the imminent 21st century. The idea of "sustainable development," as opposed to a business proposal of "sustainable growth," is examined in Agenda 21.

Agenda 21 covers many key topics: *energy conservation and efficiency* to reduce consumption and related pollution; *climate change; stratospheric ozone depletion; transboundary air pollution; ocean and water resource protection; soil losses and increasing desertification; deforestation; regulations for safely handling radioactive waste and disposal; hazardous chemical exports for disposal in developing countries;* and *disparities of wealth* and *the plague of poverty.*

Agenda 21 also addresses the difficult question of financing sustainable development. Developing countries are asking the developed nations to spend 0.7% of their gross domestic product—approximately $125 billion per year—to assist them in implementing the Earth Charter and Agenda 21.

The Future

From the Earth Summit emerged a new organization—the *U.N. Commission on Sustainable Development*—to oversee the promises made in the five documents and agreements. Most of the participating countries completed State of the Environment Reports (SERs) and gathered environmental statistics for publication. These reports are an invaluable resource that will direct further research efforts in many countries.

Considering these environmental problems and possible world actions, these are challenging times for humanity as we ponder our relationship to the home planet. Over the long term we no longer can sustain human activity through old paradigms. The truth is that society knows many of the economic-ecological solutions to problems. The study of physical and human geography is central to this assessment.

Asking whether the Earth Summit succeeded or failed is the wrong question. The occurrence of this largest-ever official gathering of Earthlings and the five years of preparatory effort and study that set the agenda are in themselves significant accomplishments. The challenge is one of education; the lesson is one of compromise and some sacrifice. Members of society should work to move the solutions for environmental and developmental problems off the bench and into play. The Earth Summit process continues as a good beginning.

in that direction is an *Earth systems science* approach, as illustrated in this text. Exciting progress toward an integrated understanding of Earth's formation and the operation of its physical systems is happening right now, driven by insights drawn from our remote-sensing capabilities. Never before has society been able to monitor Earth's physical geography so thoroughly.

Modern warfare has had a definite impact on the environment: scars from World Wars I and II still are visible on the landscapes of Europe and Asia, and more than 33 million bomb craters and massive defoliation provide environmental evidence of the more recent struggle in Vietnam. In fact, during the Vietnam War more than 2.3 times the equivalent tonnage of all munitions used in World War II were detonated. In terms of more modern weapons, just one of the planned MX missiles carries 3 megatons of explosive power, the equivalent of all the firepower of all the combatants in World War II! The 1991 Persian Gulf War subjected the desert biome to 88,500 tons of bombs. We must consider the environmental consequences of such weapons to air, water, land, and biotic systems, especially in light of the conventional war devastation in the Middle East, where the ecosystems were damaged extensively—with unknown long-term consequences.

In 1975, the National Academy of Sciences published its study on the potential impact of modern nuclear war on stratospheric ozone. This was the beginning of public awareness that modern technological warfare is not an environmentally sound activity. This NAS study states that a nuclear war could lead to the reduction of 40–70% of the stratospheric ozone, a catastrophic change for food chains, plants, animals, and humans.

In addition, from 1982 to the present, scientific publications have described the nuclear winter hypothesis. This hypothesis indicates that a "nuclear winter" might follow the detonation of only modest numbers of nuclear weapons. As you have learned, the atmosphere is very dynamic. The soot and ash from the many urban fires following a nuclear exchange would enter the atmosphere and raise Earth's albedo, absorbing insolation in the stratosphere and upper troposphere, reradiating it to space, and thus cooling Earth's surface to below freezing even during midlatitude summer months (as we saw to a much lesser degree following the eruption of Mount Pinatubo in 1991). As the nuclear winter hypothesis developed, others suggested a milder although still significant "nuclear autumn" as a consequence.

Refinement of the nuclear winter hypothesis continues, but the warnings are clear. One of the original nuclear winter study teams recently affirmed that "severe environmental anomalies—possibly leading to more human casualties globally than the direct effects of nuclear war—would be not just a remote possibility, but a likely outcome."* These potential consequences demand that the methods of conflict resolution chosen by nations must acknowledge the limiting factors imposed by Earth's ecosystems. Modern technological conflicts no longer can be thought of only in a regional context, for Earth's systems effectively spread the consequences around the globe. The real need for international cooperation should consider our symbiotic relationships with each other and with Earth's resilient, yet fragile, life-support systems.

*R. P. Turco, O. B. Toon, T. P. Ackerman, J. B. Pollack, and C. Sagan, "Climate and Smoke: An Appraisal of Nuclear Winter," *Science* 247, no. 4939 January 22, 1990): 174.

WHO SPEAKS FOR EARTH?

Public awareness and education is a growing positive force on Earth. Geography awareness is improving each year. The National Geographic Society is conducting an annual National Geography Bee to promote geography education and global awareness to millions of 6th to 8th graders and their parents and teachers. There are presently 60 geographic alliances in 47 states coordinating geographic education among K–12, community college, college, and university teachers and students.

The global concern about environmental impacts prompted *Time* magazine in 1989 to deviate from its 60-year tradition of naming a prominent citizen as its person of the year, instead naming Earth the "Planet of the Year." The magazine featured 33 pages of Earth's physical and human geography, covering the endangered status of many ecosystems and cultures (Figure 21-6). Importantly,

FIGURE 21-6
Time magazine cover for January 2, 1989, naming Earth "Planet of the Year." [Copyright © 1988 The Time Inc. Magazine Company. Reprinted by permission.]

Time also offered positive policy strategies for consideration. We seem on the brink of a new age in Earth awareness.

Carl Sagan asked, "Who speaks for Earth?" He answered with this perspective:

> **We have begun to contemplate our origins: starstuff pondering the stars; organized assemblages of ten billion billion billion atoms considering the evolution of atoms; tracing the long journey by which, here at least, consciousness arose. Our loyalties are to the species and the planet. We speak for Earth. Our obligation to survive is owed not just to ourselves but also to that Cosmos, ancient and vast, from which we spring.***

May we all perceive our spatial importance within Earth's ecosystems and do our part to maintain a life-supporting Earth far into the next century.

---

*Carl Sagan, *Cosmos*. New York: Random House, 1980, p. 345. Reprinted by permission.

REVIEW QUESTIONS

1. What part do you think technology, politics, and future thinking should play in an academic physical geography course?
2. According to the discussion in the chapter, what worldwide-scale factors led to the accident of the *Exxon Valdez?* Describe the complexity of that event.
3. What is meant by the Gaia hypothesis? Describe any concepts from this text that pertain to this hypothesis.
4. Relate the content of the various chapters in this text to the integrative *Earth systems science* concept. Which chapters in this text help you better understand the global spread of pollution?
5. Relative to possible oil and gas exploration in the Arctic National Wildlife Refuge (ANWR), assume a geographical (spatial) perspective and analyze the pros and cons of such development. In your analysis, examine both supply-side and demand-side issues, as well as environmental and strategic factors.
6. Explain the potential spatial impact of nuclear warfare on the environment. What is the nuclear winter hypothesis, and what are its potential implications for the environment?
7. This chapter states that we already know many of the solutions to the problems we face. Why do you think we are not implementing those solutions?
8. Who speaks for Earth?

SUGGESTED READINGS

Association of American Geographers. "Statement of the Association of American Geographers on Nuclear War." Adopted by AAG, May 6, 1986, Minneapolis.

Barney, Gerald O. *The Global 2000 Report to the President*. New York: Penguin Books, 1982.

Blackburn, Anne M. *Pieces of the Global Puzzle—International Approaches to Environmental Concerns*. Golden, CO: Fulcrum, 1986.

Brown, Lester R. *State of the World—1993*. Worldwatch Institute Series. New York: W. W. Norton, 1993. (Published every year.)

Canby, Thomas Y. "The Persian Gulf After the Storm," *National Geographic* 180, no. 2 (August 1991): 2–35.

Carnegie Commission on Science, Technology, and Government. *International Environmental Research and Assessment—Proposals for Better Organization and Decision Making*. New York: Carnegie Corporation, July 1992.

Hodgson, Bryan. "Alaska's Big Spill—Can the Wilderness Heal?" *National Geographic* 177, no. 1 (January 1990): 5–43.

Kates, Robert W., and Ian Burton, eds. *Geography, Resources, and Environment*. Vol. 1. *The Selected Writings of Gilbert F. White*. Vol. 2. *Themes from the Work of Gilbert F. White*. Chicago: University of Chicago Press, 1986.

Keeble, J. "A Parable of Oil and Water," and Sankovitch, N. "Lax Regulation of Oil Vessels and Processing Facilities Continues." *The Amicus Journal* 15, no. 1 (Spring 1993): 35–43.

Lee, Douglas B. "Tragedy in Alaskan Waters," *National Geographic* 176, no. 2 (August 1989): 260–63.

Lovelock, James. *The Ages of Gaia—A Biography of Our Living Earth.* New York: W. W. Norton, 1988.

Lovelock, James *Gaia—A New Look at Life on Earth.* London: Oxford Press, 1979.

National Academy of Sciences and National Research Council. *Long-Term Worldwide Effects of Multiple Nuclear-Weapons Detonations.* A report for the Department of Transportation's Climatic Impact Assessment Program. Washington, DC: National Academy of Sciences, 1975.

Parker, J. and C. Hope. "The State of the Environment— A Survey of Reports from Around the World," *ENVIRONMENT* 34, no. 1, (January/February 1992).

Renner, M. G. "Saving the Earth, Creating Jobs," *World Watch* (January-February 1992): 10–17.

Repetto, Robert. "Accounting for Environmental Assets," *Scientific American* 266, no. 6 (June 1992): 94–100

Schell, Jonathan. *The Fate of the Earth.* New York: Alfred A. Knopf, 1982.

Turco, R. P., et al. "Climate and Smoke: An Appraisal of Nuclear Winter," *Science* 247, no. 4939 (January 12, 1990): 166–74.

Turco, R. P., et al. "The Climatic Effects of Nuclear War." *Scientific American* 251, no. 2 (August 1984): 33–43.

Turco, R. P., et al. "Nuclear Winter: Global Consequences of Multiple Nuclear Explosions." *Science* 222, no. 4630 (December 23 , 1983): 128392.

United Nations. *Drafts—Agenda 21, Rio Declaration, Forest Principles.* New York: United Nations, E.92.I.16, 92-1-100482-9, June 1992.

United Nations. *The Global Partnership for Environment and Development—A Guide to Agenda 21.* Geneva: United Nations Conference on Environment and Development, April 1992.

Westing, Arthur H., and E. W. Pfeiffer. "The Cratering of Indochina," *Scientific American* 226, no. 5 (May 1972): 21–29.

APPENDIX A

CLIMATE AND WATER BALANCE DATA

Tropical Rain Forest (Af)

Belau (Palau), Caroline Islands (Af): pop. 13,900, lat. 7° 30′ N, long. 134° 30′ E, elev. 32 m (104 ft)													
	Jan	Feb	Mar	Apr	May	Jun	Jul	Aug	Sep	Oct	Nov	Dec	Annual
Temperature °C	26.9	26.7	27.2	27.8	27.5	27.5	26.9	27.5	27.2	27.2	27.2	26.9	27.2
(°F)	(80.5)	(80.0)	(81.0)	(82.0)	(81.5)	(81.5)	(80.5)	(80.5)	(81.0)	(81.0)	(81.0)	(80.5)	(81.0)
PRECIP cm	38.8	23.9	17.3	19.3	39.4	31.5	50.5	35.6	39.9	37.6	30.0	32.3	395.5
(in.)	(15.3)	(9.4)	(6.8)	(7.6)	(15.5)	(12.4)	(19.9)	(14.0)	(15.7)	(14.8)	(11.8)	(12.7)	(155.7)
POTET cm	14.0	12.4	14.5	14.7	15.2	15.0	14.7	14.7	14.2	14.5	13.7	14.0	172.0
(in.)	(5.5)	(4.9)	(5.7)	(5.8)	(6.0)	(5.9)	(5.8)	(5.8)	(5.6)	(5.7)	(5.4)	(5.5)	(67.7)

Salvador (Bahia), Brazil (Af): pop. 1,507,000, lat. 12° 59′ S, long. 38° 31′ W, elev. 9 m (30 ft)													
	Jan	Feb	Mar	Apr	May	Jun	Jul	Aug	Sep	Oct	Nov	Dec	Annual
Temperature °C	26.0	26.3	26.3	25.8	24.8	23.8	23.0	22.9	23.6	24.5	25.1	25.6	24.8
(°F)	(78.8)	(79.3)	(79.3)	(78.4)	(76.6)	(74.8)	(73.4)	(73.2)	(74.5)	(76.1)	(77.2)	(78.1)	(76.6)
PRECIP cm	7.4	7.9	16.3	29.0	29.7	19.6	20.6	11.2	8.4	9.4	14.2	9.9	183.6
(in.)	(2.9)	(3.1)	(6.4)	(11.4)	(11.7)	(7.7)	(8.1)	(4.4)	(3.3)	(3.7)	(5.6)	(3.9)	(72.3)
POTET cm	13.7	12.4	13.5	11.9	10.7	9.1	8.6	8.6	9.4	11.2	11.9	13.2	134.4
(in.)	(5.4)	(4.9)	(5.3)	(4.7)	(4.2)	(3.6)	(3.4)	(3.4)	(3.7)	(4.4)	(4.7)	(5.2)	(52.9)

Tropical Monsoon (Am)

Rangoon, Myanmar (Burma) (Am): pop. 2,459,000, lat. 16° 47′ N, long. 96° 10′ E, elev. 23 m (76 ft)													
	Jan	Feb	Mar	Apr	May	Jun	Jul	Aug	Sep	Oct	Nov	Dec	Annual
Temperature °C	24.3	25.2	27.2	29.8	29.5	27.8	27.6	27.1	27.6	28.3	27.8	25.0	27.3
(°F)	(75.7)	(77.4)	(81.0)	(85.6)	(85.1)	(82.0)	(81.7)	(80.8)	(81.7)	(82.9)	(82.0)	(77.0)	(81.1)
PRECIP cm	0.8	0.5	0.5	1.8	25.9	52.3	49.3	57.4	39.9	20.8	3.3	0.3	252.7
(in.)	(0.3)	(0.2)	(0.2)	(0.7)	(10.2)	(20.6)	(19.4)	(22.6)	(15.7)	(8.2)	(1.3)	(0.1)	(99.5)
POTET cm	9.4	9.9	14.5	16.8	17.8	16.0	16.3	15.2	14.7	15.2	13.7	10.2	169.4
(in.)	(3.7)	(3.9)	(5.7)	(6.6)	(7.0)	(6.3)	(6.4)	(6.0)	(5.8)	(6.0)	(5.4)	(4.0)	(66.7)

Mayagüez, Puerto Rico (Am): pop. 101,000, lat. 18° 12′ N, long. 67° 9′ W, elev. 24 m (79 ft)													
	Jan	Feb	Mar	Apr	May	Jun	Jul	Aug	Sep	Oct	Nov	Dec	Annual
Temperature °C	23.7	23.4	23.6	24.4	25.3	25.9	26.0	26.2	26.4	26.2	25.3	24.5	25.1
(°F)	(74.7)	(74.1)	(74.5)	(75.9)	(77.5)	(78.6)	(78.8)	(79.2)	(79.5)	(79.2)	(77.5)	(76.1)	(77.2)
PRECIP cm	4.7	6.1	8.1	14.3	19.0	23.6	29.2	27.9	26.6	24.3	15.7	6.3	205.8
(in.)	(1.9)	(2.4)	(3.2)	(5.6)	(7.5)	(9.3)	(11.5)	(11.0)	(10.5)	(9.6)	(6.2)	(2.5)	(81.0)
POTET cm	8.9	8.0	9.4	10.7	12.9	13.6	14.2	14.3	13.6	13.0	10.8	9.9	139.3
(in.)	(3.5)	(3.1)	(3.7)	(4.2)	(5.1)	(5.4)	(5.6)	(5.6)	(5.4)	(5.1)	(4.3)	(3.9)	(54.8)

Tropical Savanna (Aw)

Calcutta, Inda (Aw): pop. 3,305,000, lat. 22° 32′ N, long. 88° 20′ E, elev. 6 m (20 ft)													
	Jan	Feb	Mar	Apr	May	Jun	Jul	Aug	Sep	Oct	Nov	Dec	Annual
Temperature °C	20.2	23.0	27.9	30.1	31.1	30.4	29.1	29.1	29.2	27.9	24.0	20.6	26.8
(°F)	(68.4)	(73.4)	(82.2)	(86.2)	(88.0)	(86.7)	(84.4)	(84.4)	(84.6)	(82.2)	(75.2)	(69.1)	(80.2)
PRECIP cm	1.3	2.3	2.8	4.3	12.2	25.9	30.2	30.5	29.0	16.0	3.6	0.3	159.0
(in.)	(0.5)	(0.9)	(1.1)	(1.7)	(4.8)	(10.2)	(11.9)	(12.0)	(11.4)	(6.3)	(1.4)	(0.1)	(62.6)
POTET cm	5.3	7.4	15.2	17.3	19.3	18.8	18.3	17.5	16.3	14.7	8.9	5.6	163.3
(in.)	(2.1)	(2.9)	(6.0)	(6.8)	(7.6)	(7.4)	(7.2)	(6.9)	(6.4)	(5.8)	(3.5)	(2.2)	(64.3)

Managua, Nicaragua (Aw): pop. 682,000, lat. 12° 09' N, long. 86° 17' W; elev. 56 m (184 ft)													
	Jan	Feb	Mar	Apr	May	Jun	Jul	Aug	Sep	Oct	Nov	Dec	Annual
Temperature °C	25.7	26.7	27.3	28.2	28.3	27.0	26.7	27.1	27.0	26.2	26.4	26.3	26.8
(°F)	(78.3)	(80.0)	(81.1)	(82.8)	(82.9)	(80.6)	(80.1)	(80.8)	(80.6)	(79.2)	(79.5)	(79.3)	(80.4)
PRECIP cm	0.3	0.3	0.5	0.3	15.0	23.4	12.7	11.2	21.3	30.5	4.1	1.0	120.4
(in.)	(0.1)	(0.1)	(0.2)	(0.1)	(5.9)	(9.2)	(5.0)	(4.4)	(8.4)	(12.0)	(1.6)	(0.4)	(47.4)
POTET cm	11.7	12.2	14.5	15.5	16.3	14.7	15.0	15.0	14.0	12.7	12.2	12.5	166.4
(in.)	(4.6)	(4.8)	(5.7)	(6.1)	(6.4)	(5.8)	(5.9)	(5.9)	(5.5)	(5.0)	(4.8)	(4.9)	(65.5)

Miami, Florida (Aw): pop. 385,000, lat. 25° 46' N, long. 80° 11' W, elev. 2 m (7 ft)													
	Jan	Feb	Mar	Apr	May	Jun	Jul	Aug	Sep	Oct	Nov	Dec	Annual
Temperature °C	19.4	20.0	21.1	23.3	25.6	27.2	27.8	27.8	27.2	25.6	22.2	20.0	23.9
(°F)	(67.0)	(68.0)	(70.0)	(74.0)	(78.0)	(81.0)	(82.0)	(82.0)	(81.0)	(78.0)	(72.0)	(68.0)	(75.0)
PRECIP cm	5.1	4.8	5.8	9.9	16.3	18.8	17.3	17.8	24.1	20.8	7.1	4.3	151.9
(in.)	(2.0)	(1.9)	(2.3)	(3.9)	(6.4)	(7.4)	(6.8)	(7.0)	(9.5)	(8.2)	(2.8)	(1.7)	(59.8)
POTET cm	4.7	5.0	7.4	10.3	13.6	16.2	17.2	16.8	14.7	11.6	7.6	5.7	130.8
(in.)	(1.9)	(2.0)	(2.9)	(4.1)	(5.4)	(6.4)	(6.8)	(6.6)	(5.8)	(4.6)	(3.0)	(2.2)	(51.5)

Humid Subtropical (Cfa)

New Orleans, Louisiana (Cfa): pop. 557,000, lat. 29° 57′ N, long. 90° 04′ W, elev. 3 m (9 ft)

	Jan	Feb	Mar	Apr	May	Jun	Jul	Aug	Sep	Oct	Nov	Dec	Annual
Temperature°C	13.3	14.4	17.2	21.1	24.4	27.8	28.3	28.3	26.7	22.8	16.7	13.9	21.1
(°F)	(56.0)	(58.0)	(63.0)	(70.0)	(76.0)	(82.0)	(83.0)	(83.0)	(80.0)	(73.0)	(62.0)	(57.0)	(70.0)
PRECIP cm	12.2	10.7	16.8	13.7	13.7	14.2	18.0	16.3	14.7	9.4	10.2	11.7	161.3
(in.)	(4.8)	(4.2)	(6.6)	(5.4)	(5.4)	(5.6)	(7.1)	(6.4)	(5.8)	(3.7)	(4.0)	(4.6)	(63.5)
POTET cm	2.2	2.6	4.9	8.4	12.7	16.8	18.0	17.1	13.9	8.8	4.0	2.4	111.8
(in.)	(0.9)	(1.0)	(1.9)	(3.3)	(5.0)	(6.6)	(7.1)	(6.7)	(5.5)	(3.5)	(1.6)	(0.9)	(44.0)

Milan, Italy (Cfa): pop. 1,561,000, lat. 45° 48′ N, long. 9° 12′ E, elev. 116 m (381 ft)

	Jan	Feb	Mar	Apr	May	Jun	Jul	Aug	Sep	Oct	Nov	Dec	Annual
Temperature°C	1.3	2.9	7.9	12.2	17.0	20.7	23.0	22.0	18.4	12.4	6.6	3.0	12.3
(°F)	(34.3)	(37.2)	(46.2)	(54.0)	(62.6)	(69.3)	(73.4)	(71.6)	(65.1)	(54.3)	(43.9)	(37.4)	(54.1)
PRECIP cm	4.8	5.6	7.6	7.1	7.9	10.9	5.8	8.4	6.4	12.7	10.9	8.9	97.0
(in.)	(1.9)	(2.2)	(3.0)	(2.8)	(3.1)	(4.3)	(2.3)	(3.3)	(2.5)	(5.0)	(4.3)	(3.5)	(38.2)
POTET cm	0.3	0.5	2.8	5.3	9.7	12.7	15.0	13.0	8.9	4.6	1.8	0.8	73.7
(in.)	(0.1)	(0.2)	(1.1)	(2.1)	(3.8)	(5.0)	(5.9)	(5.1)	(3.5)	(1.8)	(0.7)	(0.3)	(29.0)

Kingsport, Tennessee (Cfa): pop. 32,000, lat. 36° 30′ N, long. 82° 30′ W, elev. 391 m (1284 ft)

	Jan	Feb	Mar	Apr	May	Jun	Jul	Aug	Sep	Oct	Nov	Dec	Annual
Temperature°C	4.3	4.8	8.2	14.1	18.6	23.0	24.6	23.9	21.3	15.0	8.5	4.5	14.2
(°F)	(39.7)	(40.6)	(46.8)	(57.4)	(65.5)	(73.4)	(76.3)	(75.0)	(70.3)	(59.0)	(47.3)	(40.1)	(57.6)
PRECIP cm	9.7	9.9	9.7	8.4	10.4	9.7	13.2	11.2	6.6	6.6	6.6	9.9	111.9
(in.)	(3.8)	(3.9)	(3.8)	(3.3)	(4.1)	(3.8)	(5.2)	(4.4)	(2.6)	(2.6)	(2.6)	(3.9)	(44.1)
POTET cm	0.7	0.8	2.4	5.7	9.7	13.2	15.0	13.3	9.9	5.5	1.2	0.7	78.1
(in.)	(0.3)	(0.3)	(0.9)	(2.2)	(3.8)	(5.2)	(5.9)	(5.2)	(3.9)	(2.2)	(0.5)	(0.3)	(30.7)

Marine West Coast (Cfb)

Valdivia, Chile (Cfb): pop. 103,000, lat. 39° 48′ S, long. 73° 14′ W, elev. 13 m (43 ft)

	Jan	Feb	Mar	Apr	May	Jun	Jul	Aug	Sep	Oct	Nov	Dec	Annual
Temperature°C	16.5	16.2	14.6	11.6	9.7	7.9	7.6	8.0	9.1	11.6	14.0	16.2	11.9
(°F)	(61.7)	(61.2)	(58.3)	(52.9)	(49.5)	(46.2)	(45.7)	(46.4)	(48.4)	(52.9)	(57.2)	(61.2)	(53.4)
PRECIP cm	10.2	5.1	9.7	17.8	43.9	37.1	45.2	31.8	22.6	9.9	8.1	7.1	248.9
(in.)	(4.0)	(2.0)	(3.8)	(7.0)	(17.3)	(14.6)	(17.8)	(12.5)	(8.9)	(3.9)	(3.2)	(2.8)	(98.0)
POTET cm	6.4	6.4	6.9	5.6	5.1	3.8	3.8	3.8	3.8	4.8	5.3	6.1	60.2
(in.)	(2.5)	(2.5)	(2.7)	(2.2)	(2.0)	(1.5)	(1.5)	(1.5)	(1.5)	(1.9)	(2.1)	(2.4)	(23.7)

Edinburgh, Scotland (Cfb): pop. 445,000, lat. 55° 57′ N, long. 3° 13′ W, elev. 134 m (439 ft)

	Jan	Feb	Mar	Apr	May	Jun	Jul	Aug	Sep	Oct	Nov	Dec	Annual
Temperature°C	3.0	3.0	5.0	7.6	10.1	12.7	14.7	14.3	12.5	9.7	6.5	4.8	8.7
(°F)	(37.4)	(37.4)	(41.0)	(45.7)	(50.2)	(54.9)	(58.5)	(57.7)	(54.5)	(49.5)	(43.7)	(40.6)	(47.7)
PRECIP cm	4.8	3.6	3.3	3.3	4.8	4.6	8.9	9.1	4.8	5.1	6.1	7.4	65.8
(in.)	(1.9)	(1.4)	(1.3)	(1.3)	(1.9)	(1.8)	(3.5)	(3.6)	(1.9)	(2.0)	(2.4)	(2.9)	(25.9)
POTET cm	1.3	1.3	2.8	4.8	7.6	10.2	11.7	10.2	7.4	4.8	2.8	1.8	64.8
(in.)	(0.5)	(0.5)	(1.1)	(1.9)	(3.0)	(4.0)	(4.6)	(4.0)	(2.9)	(1.9)	(1.1)	(0.7)	(25.5)

Mediterranean Dry, Warm Summer (Csa)

Sacramento, California (Csa): pop. 330,000, lat. 38° 35′ N, long. 121° 21′ W, elev. 11 m (36 ft)													
	Jan	Feb	Mar	Apr	May	Jun	Jul	Aug	Sep	Oct	Nov	Dec	Annual
Temperature°C	8.4	11.2	12.9	15.6	19.1	22.3	24.8	24.2	22.7	18.5	12.6	8.6	16.8
(°F)	(47.1)	(52.2)	(55.3)	(60.1)	(66.3)	(72.2)	(76.6)	(75.6)	(72.9)	(65.3)	(54.7)	(47.5)	(62.2)
PRECIP cm	10.7	7.4	5.6	3.6	1.0	0.3	0.3	0.3	0.8	2.3	5.8	7.6	45.5
(in.)	(4.2)	(2.9)	(2.2)	(1.4)	(0.4)	(0.1)	(0.1)	(0.1)	(0.3)	(0.9)	(2.3)	(3.0)	(17.9)
POTET cm	1.5	2.3	4.0	6.0	8.5	11.5	14.0	12.7	9.9	6.6	3.0	1.5	81.5
(in.)	(0.6)	(0.9)	(1.6)	(2.4)	(3.3)	(4.5)	(5.5)	(5.0)	(3.9)	(2.6)	(1.2)	(0.6)	(32.1)

Athens, Greece (Csa): pop. 3,000,000, lat. 37° 58′ N, long. 23° 43′ E, elev. 107 m (351 ft)													
	Jan	Feb	Mar	Apr	May	Jun	Jul	Aug	Sep	Oct	Nov	Dec	Annual
Temperature°C	9.3	9.9	11.3	15.3	20.0	24.6	27.6	27.4	23.5	19.0	14.7	11.0	17.8
(°F)	(48.7)	(49.8)	(52.3)	(59.5)	(68.0)	(76.3)	(81.7)	(81.3)	(74.3)	(66.2)	(58.5)	(51.8)	(64.0)
PRECIP cm	6.1	3.6	3.8	2.3	2.3	1.3	0.5	0.8	1.5	5.1	5.6	7.1	39.9
(in.)	(2.4)	(1.4)	(1.5)	(0.9)	(0.9)	(0.5)	(0.2)	(0.3)	(0.6)	(2.0)	(2.2)	(2.8)	(15.7)
POTET cm	1.5	1.8	2.8	5.1	9.7	14.5	18.3	16.8	11.2	6.9	3.8	2.0	93.0
(in.)	(0.6)	(0.7)	(1.1)	(2.0)	(3.8)	(5.7)	(7.2)	(6.6)	(4.4)	(2.7)	(1.5)	(0.8)	(36.6)

Mediterranean Dry, Cool Summer (Csb)

Portland, Oregon (Csb): pop. 366,000, lat. 45° 31′ N, long. 122° 40′ W, elev. 9 m (30 ft)													
	Jan	Feb	Mar	Apr	May	Jun	Jul	Aug	Sep	Oct	Nov	Dec	Annual
Temperature°C	4.4	6.7	8.9	12.2	15.0	17.8	20.0	20.0	17.8	13.3	8.3	5.6	12.8
(°F)	(40.0)	(44.0)	(48.0)	(54.0)	(59.0)	(64.0)	(68.0)	(68.0)	(64.0)	(56.0)	(47.0)	(42.0)	(55.0)
PRECIP cm	13.7	12.4	10.7	6.1	4.8	4.1	1.0	1.5	4.6	8.9	15.2	18.0	101.3
(in.)	(5.4)	(4.9)	(4.2)	(2.4)	(1.9)	(1.6)	(0.4)	(0.6)	(1.8)	(3.5)	(6.0)	(7.1)	(39.9)
POTET cm	0.9	1.7	3.4	5.4	8.1	10.1	12.3	11.3	8.1	5.4	2.6	1.3	70.6
(in.)	(0.4)	(0.7)	(1.3)	(2.1)	(3.2)	(4.0)	(4.8)	(4.4)	(3.2)	(2.1)	(1.0)	(0.5)	(27.8)

Capetown, Union of South Africa (Csb): pop. 1,108,000, lat. 33° 55′ S, long. 18° 22′ E, elev. 53 m (173 ft)													
	Jan	Feb	Mar	Apr	May	Jun	Jul	Aug	Sep	Oct	Nov	Dec	Annual
Temperature°C	20.3	20.0	18.8	16.1	14.0	12.6	11.6	12.3	13.7	15.0	17.6	19.3	15.9
(°F)	(68.5)	(68.0)	(65.8)	(61.0)	(57.2)	(54.7)	(52.9)	(54.1)	(56.7)	(59.0)	(63.7)	(66.7)	(60.7)
PRECIP cm	1.0	1.5	1.3	5.3	8.9	8.4	8.4	7.4	4.6	3.0	1.8	1.0	52.6
(in.)	(0.4)	(0.6)	(0.5)	(2.1)	(3.5)	(3.3)	(3.3)	(2.9)	(1.8)	(1.2)	(0.7)	(0.4)	(20.7)
POTET cm	7.6	7.1	7.9	6.4	5.6	4.8	4.3	4.3	4.8	5.1	6.1	6.9	70.6
(in.)	(3.0)	(2.8)	(3.1)	(2.5)	(2.2)	(1.9)	(1.7)	(1.7)	(1.9)	(2.0)	(2.4)	(2.7)	(27.8)

Humid Continental Hot Summer (Dfa)

Belgrade, Yugoslavia (Dfa): pop. 1,204,000, lat. 44° 50′ N, long. 20° 30′ E, elev. 139 m (456 ft)													
	Jan	Feb	Mar	Apr	May	Jun	Jul	Aug	Sep	Oct	Nov	Dec	Annual
Temperature°C	−0.2	1.6	6.2	12.2	17.1	20.5	22.6	22.0	18.3	12.5	6.8	2.5	11.8
(°F)	(31.6)	(34.9)	(43.2)	(54.0)	(62.8)	(68.9)	(72.7)	(71.6)	(64.9)	(54.5)	(44.2)	(36.5)	(53.2)
PRECIP cm	4.8	4.6	4.6	5.3	7.4	9.7	6.1	5.6	5.1	5.6	6.1	5.6	70.4
(in.)	(1.9)	(1.8)	(1.8)	(2.1)	(2.9)	(3.8)	(2.4)	(2.2)	(2.0)	(2.2)	(2.4)	(2.2)	(27.7)
POTET cm	0	0.3	2.0	5.3	9.7	12.7	14.7	13.2	8.9	4.8	1.8	0.5	72.9
(in.)	(0)	(0.1)	(0.8)	(2.1)	(3.8)	(5.0)	(5.8)	(5.2)	(3.5)	(1.9)	(0.7)	(0.2)	(28.7)

Chicago, Illinois (Dfa): pop. 3,005,000, lat. 41° 47′ N, long. 87° 45′ W, elev.186 m (610 ft)													
	Jan	Feb	Mar	Apr	May	Jun	Jul	Aug	Sep	Oct	Nov	Dec	Annual
Temperature°C	−3.9	−2.8	2.3	8.6	14.4	20.0	23.2	22.4	18.7	12.4	4.6	−1.4	9.9
(°F)	(25.0)	(27.0)	(36.1)	(47.5)	(57.9)	(68.0)	(73.8)	(72.3)	(65.7)	(54.3)	(40.3)	(29.5)	(49.8)
PRECIP cm	4.9	4.8	6.8	7.4	9.0	9.3	8.4	8.0	7.6	6.7	5.9	5.0	83.8
(in.)	(1.9)	(1.9)	(2.7)	(2.9)	(3.5)	(3.7)	(3.3)	(3.1)	(3.0)	(2.6)	(2.3)	(2.0)	(33.0)
POTET cm	0	0	0.6	3.7	7.9	12.2	14.6	13.2	9.0	5.1	1.2	0	67.5
(in.)	(0)	(0)	(0.2)	(1.5)	(3.1)	(4.8)	(5.7)	(5.2)	(3.5)	(2.0)	(0.5)	(0)	(26.6)

Humid Continental Mild Summer (Dfb)

Montreal, Quebec, Canada (Dfb): pop. 2,818,000, lat. 45° 30′ N, long. 73° 35′ W, elev. 57 m (187 ft)													
	Jan	Feb	Mar	Apr	May	Jun	Jul	Aug	Sep	Oct	Nov	Dec	Annual
Temperature°C	−10.0	−9.4	−3.3	5.6	13.3	18.3	21.1	19.4	15.0	8.3	0.6	−6.7	6.0
(°F)	(14.0)	(15.1)	(26.1)	(42.1)	(55.9)	(64.9)	(70.0)	(66.9)	(59.0)	(46.9)	(33.1)	(19.9)	(42.8)
PRECIP cm	9.6	7.7	8.8	6.6	8.0	8.7	9.5	8.8	9.3	8.7	9.0	9.1	103.8
(in.)	(3.8)	(3.0)	(3.5)	(2.6)	(3.1)	(3.4)	(3.7)	(3.5)	(3.7)	(3.4)	(3.5)	(3.6)	(40.8)
POTET cm	3	0	0	2.7	8.1	11.9	13.9	12.1	7.7	3.7	0.2	0	60.3
(in.)	(0)	(0)	(0)	(1.1)	(3.2)	(4.7)	(5.5)	(4.8)	(3.0)	(1.5)	(0.1)	(0)	(23.7)

Fargo, North Dakota (Dfb): pop. 62,600, lat. 46° 52' N, long. 96° 47' W, elev. 273 m (895 ft)													
	Jan	Feb	Mar	Apr	May	Jun	Jul	Aug	Sep	Oct	Nov	Dec	Annual
Temperature°C	−13.9	−11.7	−3.9	5.5	12.8	18.3	21.7	20.6	15.0	7.8	−2.2	−10.6	5.0
(°F)	(7.0)	(11.0)	(25.0)	(42.0)	(55.0)	(65.0)	(71.0)	(69.0)	(59.0)	(46.0)	(28.0)	(13.0)	(41.0)
PRECIP cm	1.5	1.8	2.3	4.8	5.6	7.6	5.8	6.9	4.3	3.3	2.3	1.5	47.5
(in.)	(0.6)	(0.7)	(0.9)	(1.9)	(2.2)	(3.0)	(2.3)	(2.7)	(1.7)	(1.3)	(0.9)	(0.6)	(18.7)
POTET cm	0	0	0	3.1	7.8	11.9	14.0	12.4	7.6	3.3	0	0	60.1
(in.)	(0)	(0)	(0)	(1.2)	(3.1)	(4.7)	(5.5)	(4.9)	(3.0)	(1.3)	(0)	(0)	(23.7)

Stockholm, Sweden (Dfb): pop. 1,387,000, lat. 59° 20' N, long. 18° 03' E, elev. 52 m (170 ft)													
	Jan	Feb	Mar	Apr	May	Jun	Jul	Aug	Sep	Oct	Nov	Dec	Annual
Temperature°C	−2.9	−3.1	−0.7	4.4	10.1	14.9	17.8	16.6	12.2	7.1	2.8	0.1	6.6
(°F)	(26.8)	(26.4)	(30.7)	(39.9)	(50.2)	(58.8)	(64.0)	(61.9)	(54.0)	(44.8)	(37.0)	(32.2)	(43.9)
PRECIP cm	4.3	3.0	2.5	3.0	3.3	4.6	6.1	7.6	6.1	4.8	5.3	4.8	55.6
(in.)	(1.7)	(1.2)	(1.0)	(1.2)	(1.3)	(1.8)	(2.4)	(3.0)	(2.4)	(1.9)	(2.1)	(1.9)	(21.9)
POTET cm	0	0	2.5	3.0	7.9	11.7	14.0	11.7	7.4	3.8	1.3	0.3	59.7
(in.)	(0)	(0)	(1.0)	(1.2)	(3.1)	(4.6)	(5.5)	(4.6)	(2.9)	(1.5)	(0.5)	(0.1)	(23.5)

Subarctic (Dfc, Dwc, Dwd)

Churchill, Manitoba, Canada (Dfc): pop. 1300, lat. 58° 47′ N, long. 94° 11′ W, elev. 35 m (115 ft)													
	Jan	Feb	Mar	Apr	May	Jun	Jul	Aug	Sep	Oct	Nov	Dec	Annual
Temperature°C	−28.3	−26.7	−21.1	−10.0	−1.1	6.1	12.2	11.1	6.1	−2.8	−14.4	−23.9	−8.9
(°F)	(−19.0)	(−16.0)	(−6.0)	(14.0)	(30.0)	(43.0)	(54.0)	(52.0)	(43.0)	(27.0)	(6.0)	(−11.0)	(16.0)
PRECIP cm	1.3	1.5	2.3	2.3	2.3	4.8	5.6	6.9	5.8	3.6	2.5	1.8	40.6
(in.)	(0.5)	(0.6)	(0.9)	(0.9)	(0.9)	(1.9)	(2.2)	(2.7)	(2.3)	(1.4)	(1.0)	(0.7)	(16.0)
POTET cm	0	0	0	0	0	7.1	11.2	9.6	5.1	0	0	0	33.0
(in.)	(0)	(0)	(0)	(0)	(0)	(2.8)	(4.4)	(3.8)	(2.0)	(0)	(0)	(0)	(13.0)

Tobol'sk, Siberia, Russia (Dwc): pop. 72,000, lat. 58° 09′ N, long. 68° 11′ E, elev. 44 m (144 ft)													
	Jan	Feb	Mar	Apr	May	Jun	Jul	Aug	Sep	Oct	Nov	Dec	Annual
Temperature°C	−18.5	−17.3	−10.8	1.3	9.6	16.0	18.1	15.7	9.4	1.6	−9.4	−16.8	−0.1
(°F)	(−1.3)	(0.9)	(12.6)	(34.3)	(49.3)	(60.8)	(64.6)	(60.3)	(48.9)	(34.9)	(15.1)	(1.8)	(31.9)
PRECIP cm	1.8	1.5	2.0	2.5	4.1	6.1	7.4	6.4	5.3	3.8	3.6	2.5	47.0
(in.)	(0.7)	(0.6)	(0.8)	(1.0)	(1.6)	(2.4)	(2.9)	(2.5)	(2.1)	(1.5)	(1.4)	(1.0)	(18.5)
POTET cm	0	0	0	1.3	8.1	13.5	15.2	11.7	6.1	1.0	0	0	56.9
(in.)	(0)	(0)	(0)	(0.5)	(3.2)	(5.3)	(6.0)	(4.6)	(2.4)	(0.4)	(0)	(0)	(22.4)

Oymyakon, Siberia, Russia (Dwd): pop. 3,000, lat. 63° 28′ N, long. 142° 49′ E, elev. 726 m (2382 ft)													
	Jan	Feb	Mar	Apr	May	Jun	Jul	Aug	Sep	Oct	Nov	Dec	Annual
Temperature°C	−47.2	−42.9	−34.2	−15.4	1.4	11.6	14.8	10.9	1.6	−16.2	−35.0	−44.0	−16.3
(°F)	(−53.0)	(−45.2)	(−29.6)	(4.3)	(34.5)	(52.9)	(58.6)	(51.6)	(34.9)	(2.8)	(−31.0)	(−47.2)	(2.7)
PRECIP cm	0.8	0.5	0.5	0.3	1.0	3.3	4.1	3.8	2.0	1.3	1.0	0.8	19.3
(in.)	(0.3)	(0.2)	(0.2)	(0.1)	(0.4)	(1.3)	(1.6)	(1.5)	(0.8)	(0.5)	(0.4)	(0.3)	(7.6)
POTET cm	0	0	0	0	3.0	13.5	15.7	10.9	2.3	0	0	0	45.5
(in.)	(0)	(0)	(0)	(0)	(1.2)	(5.3)	(6.2)	(4.3)	(0.9)	(0)	(0)	(0)	(17.9)

Polar Tundra (ET) and Polar Marine (EM)

Barrow, Alaska (ET): pop. 2200, lat. 71° 18′ N, long. 156° 47′ W, elev. 9 m (31 ft)													
	Jan	Feb	Mar	Apr	May	Jun	Jul	Aug	Sep	Oct	Nov	Dec	Annual
Temperature°C	−26.7	−27.8	−26.1	−17.8	−7.8	0.6	3.9	3.3	−1.1	−8.3	−18.3	−23.9	−12.2
(°F)	(−16.0)	(−18.0)	(−15.0)	(0)	(18.0)	(33.0)	(39.0)	(38.0)	(30.0)	(17.0)	(−1.0)	(−11.0)	(10.0)
PRECIP cm	0.5	0.5	0.3	0.3	0.3	1.0	2.0	2.3	1.5	1.3	0.5	0.5	10.9
(in.)	(0.2)	(0.2)	(0.1)	(0.1)	(0.1)	(0.4)	(0.8)	(0.9)	(0.6)	(0.5)	(0.2)	(0.2)	(4.3)
POTET cm	0	0	0	0	0	4.8	8.4	2.8	0	0	0	0	16.0
(in.)	(0)	(0)	(0)	(0)	(0)	(1.9)	(3.3)	(1.1)	(0)	(0)	(0)	(0)	(6.3)

Macquarie Island, Australia (EM): pop. <100, lat. 54° 30′ S, long. 158° 56′ E, elev. 6 m (20 ft)													
	Jan	Feb	Mar	Apr	May	Jun	Jul	Aug	Sep	Oct	Nov	Dec	Annual
Temperature°C	6.4	6.1	5.6	4.4	4.2	2.8	2.8	2.8	3.1	3.3	4.4	5.6	4.3
(°F)	(43.5)	(43.0)	(42.0)	(40.0)	(39.5)	(37.0)	(37.0)	(37.0)	(37.5)	(38.0)	(40.0)	(42.0)	(39.8)
PRECIP cm	10.2	8.9	10.4	9.7	8.4	7.4	8.1	8.1	9.7	4.2	7.1	9.9	106.2
(in.)	(4.0)	(3.5)	(4.1)	(3.8)	(3.3)	(2.9)	(3.2)	(3.2)	(3.8)	(3.3)	(2.8)	(3.9)	(41.8)
POTET cm	8.4	6.6	5.8	4.1	3.3	2.0	2.3	2.8	3.6	4.6	6.1	7.9	56.6
(in.)	(3.3)	(2.6)	(2.3)	(1.6)	(1.3)	(0.8)	(0.9)	(1.1)	(1.4)	(1.8)	(2.4)	(3.1)	(22.3)

Polar Ice Cap (EF)

Amundsen Base, Antarctica (EF): pop. transitory, lat. 90° 00′ S, long. 0° 00′, elev. 2800 m (9186 ft)													
	Jan	Feb	Mar	Apr	May	Jun	Jul	Aug	Sep	Oct	Nov	Dec	Annual
Temperature °C	−28.2	−38.5	−55.2	−57.7	−56.2	−59.0	−58.5	−60.4	−59.3	−51.4	−38.7	−28.1	−49.3
(°F)	(−18.8)	(−37.3)	(−67.4)	(−71.9)	(−69.2)	(−74.2)	(−73.3)	(−76.7)	(−74.8)	(−60.5)	(−37.7)	(−18.6)	(−56.7)
PRECIP cm	0.1	0.2	0	0	0	0	0	0	0	0	0	0.1	0.4
(in.)	(0)	(0.1)	(0)	(0)	(0)	(0)	(0)	(0)	(0)	(0)	(0)	(0)	(0.2)
POTET cm	← near zero →												
(in.)													
McMurdo, Anarctica (EF): pop. transitory, lat. 77° 50′ S, long. 166° 36′ W, elev. 45 m (148 ft)													
	Jan	Feb	Mar	Apr	May	Jun	Jul	Aug	Sep	Oct	Nov	Dec	Annual
Temperature °C	−3.6	−8.7	−18.5	−22.5	−23.1	−24.6	−27.2	−29.0	−23.4	−20.4	−9.6	−3.7	17.9
(°F)	(25.5)	(16.3)	(− 1.3)	(− 8.5)	(− 9.6)	(−12.3)	(−17.0)	(−20.2)	(−10.1)	(− 4.7)	(−14.7)	(25.3))	(−0.2)
PRECIP cm	1.1	0.4	0.6	0.6	1.3	0.5	0.5	1.1	(1.2)	0.8	0.6	0.7	9.4
(in.)	(0.4)	(0.2)	(0.2)	(0.2)	(0.5)	(0.2)	(0.2)	(0.4)	(0.5)	(0.3)	(0.2)	(0.3)	(3.7)
POTET cm	← near zero →												
(in.)													

Hot Low-Latitude Desert (BWh)

Alice Springs, Northern Territory, Australia (BWh). pop. 24,000, lat. 23° 42′ S, long. 133° 53′ E, elev. 579 m (1901 ft)													
	Jan	Feb	Mar	Apr	May	Jun	Jul	Aug	Sep	Oct	Nov	Dec	Annual
Temperature °C	28.6	27.8	24.7	19.7	15.3	12.2	11.7	14.4	18.3	22.8	25.8	27.8	20.7
(°F)	(83.5)	(82.0)	(76.5)	(67.5)	(59.5)	(54.0)	(53.0)	(58.0)	(65.0)	(73.0)	(78.5)	(82.0)	(69.4)
PRECIP cm	4.3	3.3	2.8	1.0	1.5	1.3	0.8	0.8	0.8	1.8	3.0	3.8	25.1
(in.)	(1.7)	(1.3)	(1.1)	(0.4)	(0.6)	(0.5)	(0.3)	(0.3)	(0.3)	(0.7)	(1.2)	(1.5)	(9.9)
POTET cm	17.8	14.7	11.9	6.6	3.6	2.0	1.8	3.3	5.6	10.2	14.0	17.0	108.5
(in.)	(7.0)	(5.8)	(4.7)	(2.6)	(1.4)	(0.8)	(0.7)	(1.3)	(2.2)	(4.0)	(5.5)	(6.7)	(42.7)
Yuma, Arizona (BWh): pop. 42,000, lat. 32° 40′ N, long. 114° 36′ W, elev. 61 m (200 ft)													
	Jan	Feb	Mar	Apr	May	Jun	Jul	Aug	Sep	Oct	Nov	Dec	Annual
Temperature °C	12.7	14.8	17.9	21.5	25.1	29.6	33.2	32.8	29.7	23.3	17.1	13.2	22.6
(°F)	(54.9)	(58.6)	(64.2)	(70.7)	(77.2)	(85.3)	(91.8)	(91.0)	(85.5)	(73.9)	(62.8)	(55.8)	(72.7)
PRECIP cm	1.1	1.1	0.8	0.3	0.1	0	0.5	1.4	1.0	0.8	0.6	1.2	8.9
(in.)	(0.4)	(0.4)	(0.3)	(0.1)	(0)	(0)	(0.2)	(0.6)	(0.4)	(0.3)	(0.2)	(0.5)	(3.5)
POTET cm	1.3	2.3	4.6	8.2	13.6	18.9	21.1	20.0	16.4	9.0	3.4	1.5	110.3
(in.)	(0.5)	(0.9)	(1.8)	(3.2)	(5.4)	(7.4)	(8.3)	(7.9)	(6.5)	(3.5)	(1.3)	(0.6)	(43.4)

Cold Midlatitude Desert (BWk)

San Luis Potosí, Mexico (BWk): pop. 2,000,000, lat. 22° 09′ N, long. 100° 58′ W, elev. 1877 m (6157 ft)													
	Jan	Feb	Mar	Apr	May	Jun	Jul	Aug	Sep	Oct	Nov	Dec	Annual
Temperature°C	12.9	15.0	17.0	20.4	21.5	20.9	19.5	19.6	18.4	17.2	14.8	13.9	17.6
(°F)	(55.2)	(59.0)	(62.6)	(68.7)	(70.7)	(69.6)	(67.1)	(67.3)	(65.1)	(63.0)	(58.6)	(57.0)	(63.7)
PRECIP cm	1.0	0.6	1.0	0.8	3.3	7.4	5.7	4.3	7.5	2.0	1.0	1.2	35.8
(in.)	(0.4)	(0.2)	(0.4)	(0.3)	(1.3)	(2.9)	(2.2)	(1.7)	(3.0)	(0.8)	(0.4)	(0.5)	(14.1)
POTET cm	3.6	4.5	6.4	8.8	10.4	9.8	9.0	8.9	7.0	6.2	4.4	3.9	82.9
(in.)	(1.4)	(1.8)	(2.5)	(3.5)	(4.1)	(3.9)	(3.5)	(3.5)	(2.8)	(2.4)	(1.7)	(1.5)	(32.6)

Reno, Nevada (BWk): pop. 115,000, lat. 39° 30′ N, long. 119° 47′ W, elev.1342 m (4404 ft)													
	Jan	Feb	Mar	Apr	May	Jun	Jul	Aug	Sep	Oct	Nov	Dec	Annual
Temperature°C	−0.6	2.2	5.0	8.9	12.8	16.7	21.1	19.4	15.6	10.6	4.4	0.6	10.0
(°F)	(31.0)	(36.0)	(41.0)	(48.0)	(55.0)	(62.0)	(70.0)	(67.0)	(60.0)	(51.0)	(40.0)	(33.0)	(50.0)
PRECIP cm	2.5	2.5	1.8	1.3	1.3	1.0	0.5	0.5	0.5	1.5	1.5	2.3	17.8
(in.)	(1.0)	(1.0)	(0.7)	(0.5)	(0.5)	(0.4)	(0.2)	(0.2)	(0.2)	(0.6)	(0.6)	(0.9)	(7.0)
POTET cm	0	0.7	2.2	4.0	7.1	10.1	13.0	11.7	7.8	4.3	1.7	0.2	62.8
(in.)	(0)	(0.3)	(0.9)	(1.6)	(2.8)	(4.0)	(5.1)	(4.6)	(3.1)	(1.7)	(0.7)	(0.1)	(24.7)

Hot Low-Latitude Steppe (BSh)

Monterrey, Mexico (BSh): pop. 1,702,000, lat. 25° 40′ N, long. 100° 19′ W, elev. 538 m (1765 ft)													
	Jan	Feb	Mar	Apr	May	Jun	Jul	Aug	Sep	Oct	Nov	Dec	Annual
Temperature°C	16.4	17.1	20.3	23.7	26.1	28.1	28.4	28.3	25.9	22.4	17.7	15.6	22.6
(°F)	(61.5)	(63.7)	(68.5)	(74.7)	(79.0)	(82.6)	(83.1)	(82.9)	(78.6)	(72.3)	(63.9)	(60.1)	(72.7)
PRECIP cm	0.8	2.3	1.3	1.8	1.8	5.1	4.0	7.6	11.9	9.4	2.3	0.8	49.0
(in.)	(0.3)	(0.9)	(0.5)	(0.7)	(0.7)	(2.0)	(1.6)	(3.0)	(4.7)	(3.7)	(0.9)	(0.3)	(19.3)
POTET cm	3.6	4.1	6.9	10.7	14.7	17.0	17.8	16.8	12.7	8.6	4.3	3.3	120.4
(in.)	(1.4)	(1.6)	(2.7)	(4.2)	(5.8)	(6.7)	(7.0)	(6.6)	(5.0)	(3.4)	(1.7)	(1.3)	(47.4)

Walgett, New South Wales, Austrailia (BSh): pop. 2000, lat. 30° 01' S, long. 148° 07' E, elev. 133 m (436 ft)													
	Jan	Feb	Mar	Apr	May	Jun	Jul	Aug	Sep	Oct	Nov	Dec	Annual
Temperature°C	28.1	27.2	24.4	19.7	15.0	11.7	10.8	12.5	19.2	20.6	24.4	27.2	20.0
(°F)	(82.5)	(81.0)	(76.0)	(67.5)	(59.0)	(53.0)	(51.5)	(54.5)	(66.5)	(69.0)	(76.0)	(81.0)	(68.0)
PRECIP cm	5.3	4.8	4.1	3.0	3.8	4.1	3.3	2.8	2.5	3.0	3.8	4.3	45.0
(in.)	(2.1)	(1.9)	(1.6)	(1.2)	(1.5)	(1.6)	(1.3)	(1.1)	(1.0)	(1.2)	(1.5)	(1.7)	(17.7)
POTET cm	17.8	14.5	11.9	6.6	3.6	1.8	1.8	2.5	6.3	8.4	12.7	17.3	104.6
(in.)	(7.0)	(5.7)	(4.7)	(2.6)	(1.4)	(0.7)	(0.7)	(1.0)	(2.5)	(3.3)	(5.0)	(6.8)	(41.2)

N'Djamena (Fort Lamy), Chad (BSh): pop. 512,000, lat. 12° 08' N, long. 15° 02' E, elev.300 m (984 ft)													
	Jan	Feb	Mar	Apr	May	Jun	Jul	Aug	Sep	Oct	Nov	Dec	Annual
Temperature°C	23.5	25.9	30.1	32.7	32.3	30.5	27.5	26.2	27.1	28.6	27.1	24.1	27.9
(°F)	(74.3)	(78.6)	(86.2)	(90.9)	(90.1)	(86.9)	(81.5)	(79.2)	(80.8)	(83.5)	(80.8)	(75.4)	(82.2)
PRECIP cm	0	0	0	0.5	3.5	6.6	15.5	25.7	10.4	2.3	0	0	64.5
(in.)	(0)	(0)	(0)	(0.2)	(1.4)	(2.6)	(6.1)	(10.1)	(4.1)	(0.9)	(0)	(0)	(25.4)
POTET cm	8.6	10.9	17.0	18.3	19.1	17.8	15.7	13.5	14.2	15.5	13.5	9.4	173.0
(in.)	(3.4)	(4.3)	(6.7)	(7.2)	(7.5)	(7.0)	(6.2)	(5.3)	(5.6)	(6.1)	(5.3)	(3.7)	(68.1)

Cold Midlatitude Steppe (BSk)

Tehran, Iran (BSk): pop. 5,433,000, lat. 35° 40′ N, long. 51° 26′ E, elev. 1206 m (3957 ft)													
	Jan	Feb	Mar	Apr	May	Jun	Jul	Aug	Sep	Oct	Nov	Dec	Annual
Temperature°C	3.5	5.2	10.2	15.4	21.2	26.1	29.5	28.4	24.6	18.3	10.6	4.9	16.5
(°F)	(38.3)	(41.4)	(50.4)	(59.7)	(70.2)	(79.0)	(85.1)	(83.1)	(76.3)	(64.9)	(51.1)	(40.8)	(61.7)
PRECIP cm	3.8	2.3	3.6	3.0	1.3	0.3	0	0	0	0.5	2.8	2.8	20.3
(in.)	(1.5)	(0.9)	(1.4)	(1.2)	(0.5)	(0.1)	(0)	(0)	(0)	(0.2)	(1.1)	(1.1)	(8.0)
POTET cm	0.3	0.5	2.3	5.3	10.7	16.0	19.8	17.5	12.2	6.6	2.0	0.5	93.7
(in.)	(0.1)	(0.2)	(0.9)	(2.1)	(4.2)	(6.3)	(7.8)	(6.9)	(4.8)	(2.6)	(0.8)	(0.2)	(36.9)

Lanzhou, China (BSk): pop. 2,000,000, lat. 36° 03′ N, long. 103° 51′ E, elev.1506 m (4941 ft)													
	Jan	Feb	Mar	Apr	May	Jun	Jul	Aug	Sep	Oct	Nov	Dec	Annual
Temperature°C	−6.3	−1.4	5.5	12.0	17.3	20.1	22.8	21.4	16.3	10.3	1.6	−5.0	9.0
(°F)	(20.7)	(29.5)	(41.9)	(53.6)	(63.1)	(60.6)	(73.0)	(70.6)	(61.4)	(50.5)	(34.9)	(23.0)	(49.3)
PRECIP cm	0.3	0.3	0.8	1.3	1.8	3.8	6.6	0.1	5.6	1.5	0.3	0	30.7
(in.)	(0.1)	(0.1)	(0.3)	(0.5)	(0.7)	(1.5)	(2.6)	(3.6)	(2.2)	(0.6)	(0.1)	(0)	(12.1)
POTET cm	0	0	2.0	5.6	9.7	12.4	14.2	12.4	7.9	4.1	0.5	0	67.6
(in.)	(0)	(0)	(0.8)	(2.2)	(3.8)	(4.9)	(5.6)	(4.9)	(3.1)	(1.6)	(0.2)	(0)	(26.6)

Kemmerer, Wyoming (BSk): pop. 3,300, lat. 41° 48′ N, long. 110° 32′ W, elev. 2120 m (6954 ft)													
	Jan	Feb	Mar	Apr	May	Jun	Jul	Aug	Sep	Oct	Nov	Dec	Annual
Temperature°C	−8.3	−6.6	−2.8	3.9	9.2	13.3	17.2	16.1	11.5	5.7	−1.9	−5.5	4.3
(°F)	(17.0)	(20.1)	(27.0)	(39.0)	(48.6)	(55.9)	(63.0)	(60.9)	(52.7)	(42.3)	(28.5)	(22.0)	(39.8)
PRECIP cm	1.8	1.7	1.9	2.1	2.6	2.8	1.9	2.1	1.6	2.1	1.7	1.7	24.0
(in.)	(0.7)	(0.7)	(0.7)	(0.8)	(1.0)	(1.1)	(0.7)	(0.8)	(0.6)	(0.8)	(0.7)	(0.7)	(9.5)
POTET cm	0	0	0	2.7	6.4	9.1	11.9	10.4	6.6	3.4	0	0	50.5
(in.)	(0)	(0)	(0)	(1.1)	(2.5)	(3.6)	(4.7)	(4.1)	(2.6)	(1.3)	(0)	(0)	(19.9)

Highland (H)

La Paz, Bolivia (H): pop. 788,000, lat. 16° 30′ S, long. 68° 10′ W, elev. 4103 m (13,461 ft)													
	Jan	Feb	Mar	Apr	May	Jun	Jul	Aug	Sep	Oct	Nov	Dec	Annual
Temperature°C	9.0	9.0	9.0	9.0	8.0	7.0	8.0	8.0	9.0	10.0	10.0	10.0	9.0
(°F)	(48.2)	(48.2)	(48.2)	(48.2)	(46.4)	(44.6)	(46.4)	(46.4)	(48.2)	(50.0)	(50.0)	(50.0)	(48.2)
PRECIP cm	13.9	10.8	5.6	2.2	0.9	0.5	0.4	1.7	3.4	3.7	4.8	7.6	55.5
(in.)	(5.5)	(4.2)	(2.2)	(0.9)	(0.4)	(0.2)	(0.2)	(0.7)	(1.3)	(1.5)	(1.9)	(3.0)	(21.9)

Lhasa, (Tibet) China (H): pop. 382,000, lat 29° 40′ N, long. 91° 07′ E, elev. 3685 m (12,090 ft)													
	Jan	Feb	Mar	Apr	May	Jun	Jul	Aug	Sep	Oct	Nov	Dec	Annual
Temperature°C	−2.0	1.0	5.0	8.0	12.0	17.0	16.0	16.0	14.0	9.0	4.0	0	8.0
(°F)	(28.4)	(33.8)	(41.0)	(46.4)	(53.6)	(62.6)	(60.8)	(60.8)	(57.2)	(48.2)	(39.6)	(32.0)	(46.4)
PRECIP cm	0	1.3	0.8	0.5	2.5	6.3	12.2	8.9	6.6	1.3	0.2	0	40.6
(in.)	(0)	(0.5)	(0.3)	(0.2)	(1.0)	(2.5)	(4.8)	(3.5)	(2.6)	(0.5)	(0.1)	(0)	(16.0)

Appendix B
Information Sources, Organizations, and Agencies

This is a sampling of information sources for further geographic inquiry. Consult your campus library and instructor for related sources and research pathways.

General Sources

Academic Press Dictionary of Science and Technology edited by Christopher Morris. Published by Academic Press, San Diego, CA, 1992.

Canada and the World. An atlas resource by Geoffrey J. Matthews and Robert Morrow published by Prentice Hall Canada, Scarborough, Ontario, Canada, 1985.

Canada Yearbook. Published annually by minister of Supply and Services. Available from Publication Sales and Services, Statistics Canada, Ottawa, Canada K1A 0T6. Also available from Statistics Canada: *Human Activity and the Environment—1986 and State of the Environment, Report to Canada* (May 1986).

Climate Alert. Published by Climate Institute, 316 Pennsylvania Avenue SE, Suite 403, Washington, DC 20003. A clearinghouse of information on world climate.

The Climates of Canada. Compiled by Canada's chief climatologist David Phillips. Available from Minister of Supply and Services Canada, Ottawa, Canada K1A 0S9, 1990.

The Complete Guide to Environmental Careers. The CEIP Fund. Island Press, Covela, CA 95428.

Congressional Directory. Published after every biennial congressional election. Available from U.S. Government Printing Office, Washington, DC 20402.

The Encyclopedia of Environmental Studies by William Ashworth. Published by Facts of File, New York, NY, 1991.

Environment. Heldref Publications, 4000 Albemarle St. NW, Washington, DC 20016.

1992 Environmental Almanac (Information Please). Compiled by the World Resources Institute and published by Houghton Mifflin Company, New York, NY, 1992.

Geographical Bibliography for American Libraries. Chauncy D. Harris, ed. A joint project of the Association of American Geographers and the National Geographic Society, Washington, 1985. AAG, 1710 16th Street, NW, Washington, DC 20009.

Instant Information. 10,000 addresses, both public and private. Compiled by Joel Makower and Alan Green, published by Prentice Hall Press, New York, 1987.

The Island Press Bibliography of Environmental Literature. Compiled by Joseph Miller, Sarah Friedman, and others at the Yale School of Forestry and Environmental Studies. Published by Island Press, Washington, DC, 1993.

Source Book on the Environment—A Guide to the Literature. Kenneth A. Hammond, George Macinko, and Wilma B. Fairchild, University of Chicago Press, 1978.

State of the Environment—A View Toward the Nineties. A report from the Conservation Foundation, 1987, sponsored by the Charles Stewart Mott Foundation. Conservation Foundation, 1250 42nd Street NW, Washington, DC 20076.

State of the World 1993—Report on Progress Toward A Sustainable Society. Published annually by the Worldwatch Institute (Lester R. Brown, project director),

1776 Massachusetts Avenue NW, Washington, DC 20036.

Statistical Abstract of the United States. Published annually by the U.S. Bureau of the Census, Department of Commerce, Washington, DC. Also available from U.S. Government Printing Office, Washington, DC 20402.

U.S. Government Manual. Published by the Office of the Federal Register with annual updates. Available through the Superintendent of Documents, U.S. Government Printing Office, Washington, DC 20402.

U.S. Government Printing Office. North Capitol and H Streets NW, Washington, DC 20401. Superintendent of Documents provides monthly catalog of all U.S. government publications. Many local bookstores and offices.

The World Almanac. Published annually by World Almanac and Book of Facts, a Scripps Howard company, New York.

World Resources 1992–93. Published annually by the World Resources Institute, in collaboration with the United Nations Environment Programme, 1735 New York Avenue NW, Washington, DC. 20006.

Geography Organizations

Alaska Geographical Society, P. O. Box–EEE, Anchorage, AK 99509. One of many state and provincial geographical societies operating in the United States and Canada.

American Geographic Society, 156 Fifth Avenue, Suite 600, New York, NY 10010–7002. Publishes *Focus* quarterly.

Association of American Geographers, 1710 16th Street NW, Washington, DC 20009. Principal academic and professional geography organization—offers student memberships and career information. Publishes *Annals* and *The Professional Geographer.*

Canadian Association of Geographers, McGill University, 805 Sherbrooke Street West, Montréal, Québec H3A 2K6. Publishes *The Canadian Geographer* and *The Operational Geographer–The Canadian Journal for Practising Geographers.*

Geographic alliances, 60 state chapters. Information is available from college and university geography departments; generally coordinated by the National Geographic Society.

Institute of British Geographers, 1 Kingsington Gore, London, SW7 2AR, England. Publishes *Transactions of the Institute of British Geographers* and *Area.*

National Council for Geographic Education, Indiana University of Pennsylvania, Indiana, PA 15705. Publishes the *Journal of Geography.*

National Geographic Society, 17th and M Streets NW, Washington, DC 20036. Publishes maps, atlases, *National Geographic* Magazine, and *Geography Education Update* through their Geographic Education Program.

Royal Canadian Geographical Society, 39 McArthur Ave., Vanier, Ontario, Canada K1L 8L7. Publishes *Canadian Geographic.*

Royal Geographical Society, 1 Kingsington Gore, London SW7 2AR, England.

Organizations of Interest

American Association for the Advancement of Science, 1333 H Street NW, Washington, DC 20005. Publishes *Science.*

American Chemical Society, 1155 16th Street NW, Washington, DC 20036.

American Forestry Association, 1319 118th Street NW, Washington, DC 20036.

American Geophysical Union, 2000 Florida Avenue NW, Washington, DC 20009.

American Meteorological Society, 45 Beacon Street, Boston, MA 02108.

British Association for the Advancement of Science, Fortress House, 23 Savile Row, London WIX 1AB.

Canadian Coalition on Acid Rain, 112 St. Clair Avenue West, Toronto, Ontario, Canada M4K 2Y3.

Center for Marine Conservation, Inc., 1725 DeSales Street NW, Suite 500, Washington, DC 20036.

Center for Science in the Public Interest, 1501 16th Street NW, Washington, DC 20036

Climate Institute, 316 Pennsylvania Avenue, SE, Suite 403, Washington, DC 20003.

Conservation Foundation, 1250 24th Street NW, Washington, DC 20037.

Conservation International (tropical ecosystems), 1015 18th Street NW, Suite 1000, Washington, DC 20036.

Cousteau Society, 930 West 21st Street, Norfolk, VA 23517.

Environmental Action Foundation, 1525 New Hampshire Avenue NW, Washington DC 20036.

Environmental Defense Fund, 257 Park Avenue South, New York, NY 10010.

Friends of the Earth, Inc., 530 Seventh Avenue SE, Washington, DC 20003.

Greenpeace Canada, 427 Bloor Street West, Toronto, Ontario, Canada M5S 1X7.

Greenpeace, U.S.A., Inc., 1436 U Street NW, Washington, DC 20009.

League of Conservation Voters, 1150 Connecticut Avenue NW, Suite 201, Washington, DC 20036.

National Parks and Conservation Association, 1015 31st Street NW, Washington, DC 20007. (Natural history associations are organized in conjunction with most national parks.)

National Recycling Coalition, 1101 30th Street NW, Washington, DC 20007.

National Weather Association, 4400 Stamp Road, Suite 404, Temple Hills, MD 20748.

National Wildlife Federation, 1400 16th Street NW, Washington, DC 20036. Publishes an annual updated list of all agencies and organizations dealing with natural resources—the *Conservation Directory*.

Natural Resources Defense Council, 122 East 42nd Street, New York, NY 10168. Publishes *The Amicus Journal*.

The Nature Conservancy, 1800 North Lynn Street, Arlington, VA 22209.

Rainforest Action Network, 301 Broadway, Suite A, San Francisco, CA 94133.

Rocky Mountain Institute, 1739 Snowmass Creek Road, Snowmass, CO 81654. Research in energy-related issues and conservation.

Sierra Club, 730 Polk Street, San Francisco, CA 94109.

Soil Conservation Society of America, 7515 Northeast Ankeny Road, Ankeny, IA 50021.

Union of Concerned Scientists, 26 Church Street, Cambridge, MA 02238.

United Nations Environment Programme, Office of Public Information, New York Liaison Office, Room DC 2–0803, United Nations, New York, NY 10017; or, UNEP, P.O. Box 30552, Nairobi, Kenya.

The Wilderness Society, 1400 I Street NW, Washington, DC 20005.

Worldwatch Institute, 1776 Massachusetts Avenue NW, Washington, DC 20036.

World Wildlife Fund, 1250 24th Street NW, Washington, DC 20037.

World Wildlife Fund Canada, 60 St. Clair Avenue, East, Suite 201, Toronto, Ontario, Canada M4T 1N5.

Government Agencies

Council of Environmental Quality, 722 Jackson Place NW, Washington, DC 20006.

Department of Agriculture, 14th Street and Independence Avenue SW, Washington, DC 20250.
- Forest Service
- Soil Conservation Service

Department of Commerce, 14th Street between Constitution Avenue and E Street NW, Washington, DC 20230.
- National Environmental Satellite, Data, and Information Service
- National Geophysical Data Center
- National Oceanic and Atmospheric Administration
  - National Hurricane Center
  - National Climate Program Office
- National Oceanographic Data Center
- National Severe Storms Forecast Center
- National Weather Service

Department of Energy, 1000 Independence Avenue SW, Washington, DC 20585.

Department of the Interior, 1800 C Street NW, Washington, DC 20240.
- Bureau of Land Management
- Bureau of Mines
- Bureau of Reclamation
- Fish and Wildlife Service
- National Park Service

Energy, Mines, and Resources Canada, 580 Booth Street Ottawa, Ontario, Canada K1A 0E4.

Environment Canada, Ottawa, Ontario K1A OH3.

Environmental Protection Agency, 401 M Street SW, Washington, DC 20460

Geological Survey of Canada, Headquarters: 601 Booth Street, Ottawa, Ontario K1A OE8. Many divisions, centres, and branches performing specialized functions. Provincial survey offices nationwide.

International Centre for Ocean Development, 5670 Spring Garden Road, 9th Floor, Halifax, Nova Scotia, Canada B3J 1H6.

National Academy of Sciences, Washington, DC 20550.

Office of Coastal Zone Management, 1825 Connecticut Avenue, Suite 700, Washington, DC 20235.

Solar Energy Research Institute, 6536 Cole Boulevard, Golden, CO 80401.

U.S. Geological Survey, 12201 Sunrise Valley Drive, Reston, VA 22092. Many regional offices, bookstores, and mapping centers.

APPENDIX C

Weather Map, Saturday, April 2, 1988

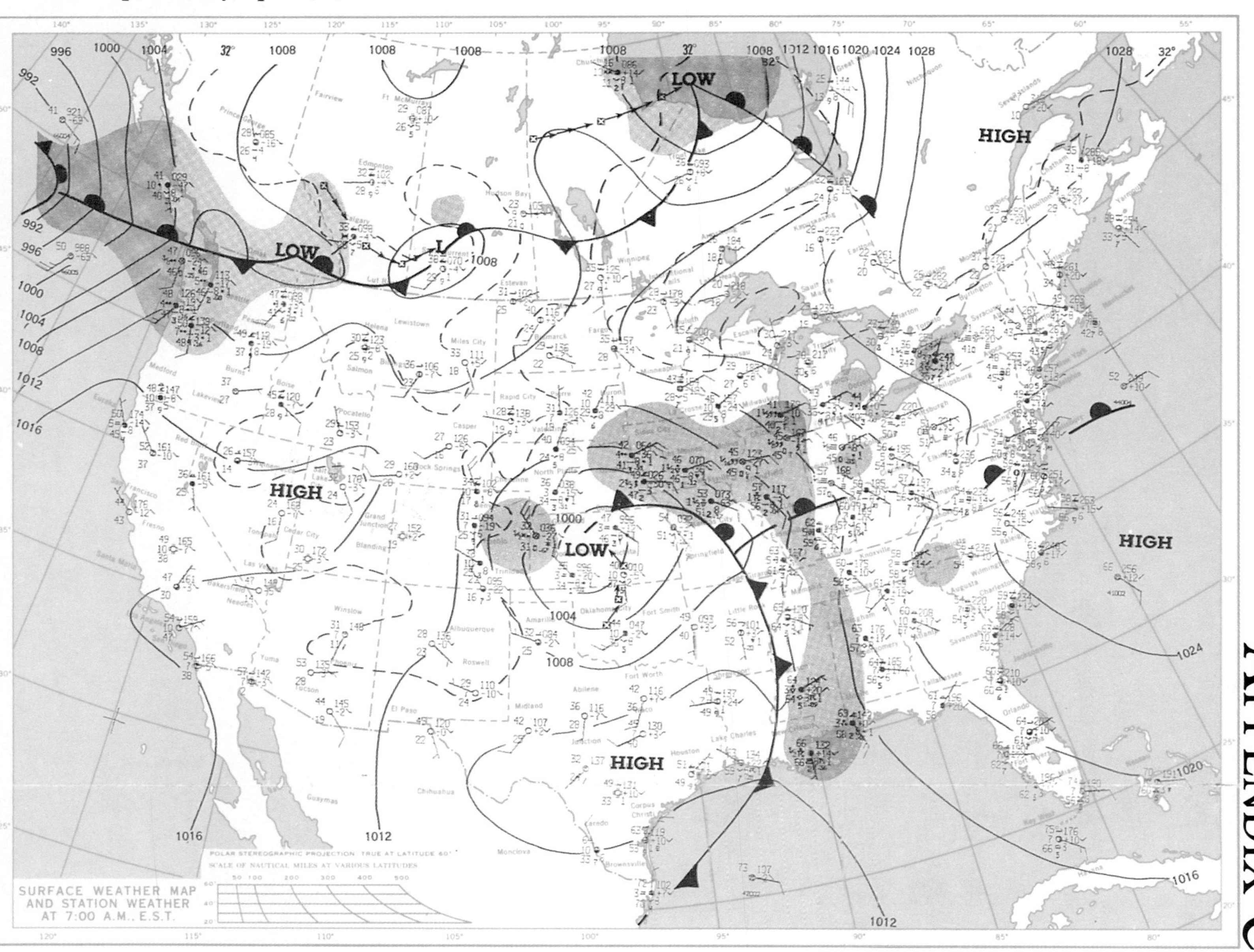

GLOSSARY

The chapter in which each term appears **boldfaced** is designated in parentheses, followed by a specific definition relevant to the key term's usage in the chapter.

Abiotic (1) Nonliving.

Ablation (17) Loss of glacial ice through melting, sublimation, wind removal by deflation, or the calving off of blocks of ice.

Abrasion (14, 15, 17) Mechanical wearing and erosion of bedrock accomplished by the rolling and grinding of particles and rocks carried in a stream, moved by wind in a "sandblasting" action, or imbedded in glacial ice.

Absolute humidity (7) Actual amount of water vapor in the air, expressed as the weight of water vapor (grams) per unit volume of air (cubic meter)—an expression of weight per volume. Saturation absolute humidity refers to the maximum possible vapor content in grams per cubic meter.

Absorption (4) Assimilation and conversion of radiation from one form to another in a medium. In the process the temperature of the absorbing surface is raised, thereby affecting the rate and quality of radiation from that surface.

Active layer (17) A zone of seasonally frozen ground that exists between the subsurface permafrost layer and the ground surface. The active layer is subjected to consistent daily and seasonal freeze-thaw cycles.

Actual evapotranspiration (9) Actual amount of evaporation and transpiration that occurs; derived in the water-balance equation by subtracting the deficit from potential evapotranspiration (ACTET).

Adiabatic (8) Pertaining to the heating and cooling of a descending or ascending parcel of air through compression and expansion, without any exchange of heat between the parcel and the surrounding environment.

Adsorption (18) The process whereby cations become attached to soil colloids, or the adhesion of gas molecules and ions to solid surfaces with which they come into contact.

Advection (4) Horizontal movement of air or water from one place to another.

Advection fog (7) Active condensation formed when warm, moist air moves laterally over cooler water or land surfaces, causing the lower layers of the overlying air to be chilled to the dew-point temperature.

Aggradation (14) The general building up of a land surface because of the deposition of material; opposite of degradation. When the sediment load of a stream exceeds the stream's capacity, the stream channel is filled through this process.

Air mass (8) A distinctive, homogeneous body of air in terms of temperature and humidity, that takes on the moisture and temperature characteristics of its source region.

Air pressure (3) Pressure produced by the motion, size, and number of gas molecules and exerted on surfaces in contact with the air. Normal sea level pressure, as measured by the height of a column of mercury (Hg), is expressed as 1013.2 millibars, 760 mm of Hg, or 29.92 inches of Hg. Air pressure can be measured with mercury or aneroid barometers.

Albedo (4) The reflective quality of a surface, expressed as the relationship of incoming to reflected insolation and stated as a percentage; a function of surface color, angle of incidence, and surface texture.

Aleutian low (6) See subpolar low-pressure cell.

Alfisols (18) Moderately weathered forest soils that are moist versions of Mollisols, with productivity dependent on specific patterns of moisture and temperature; rich in organics; most wide-ranging of the 10 soil orders in the Soil Taxonomy classification.

Alleghany orogeny (12) See orogenesis.

Alluvial fan (15) Fan-shaped fluvial landform at the mouth of a canyon, particularly noticeable in arid landscapes where streams are intermittent.

Alluvial terraces (14) Level areas that appear as topographic steps above the river, created by a stream as it scours with renewed downcutting into its floodplain; composed of unconsolidated alluvium.

Alluvium (14) General descriptive term for clay, silt, and sand transported by running water and deposited in sorted or semisorted sediment on a floodplain, delta, or in a stream bed.

Alpine glacier (17) A glacier confined in a mountain valley or walled basin, consisting of three subtypes: valley glacier (within a valley), piedmont glacier (coalesced at the base of a mountain, spreading freely over nearby lowlands), and outlet glacier (flowing outward from a continental glacier).

Altitude (2) The angular distance between the horizon (a horizontal) and the Sun (or any point).

Analemma (2) A convenient device to track the passage of the Sun, the Sun's declination, and the positive and negative equation of time throughout the year.

Andisols (18) A soil order in the Soil Taxonomy derived from volcanic parent materials in areas of volcanic activity. A new order formed in 1990 of soils previously considered under Inceptisols and Entisols. Andisols are derived from volcanic ash and glass.

Anemometer (6) A device used to measure wind velocity.

Aneroid barometer (3) A device to measure air pressure using a partially emptied, sealed cell (see air pressure).

Angle of repose (13) The steepness of a slope that results when loose particles come to rest; an angle of balance between driving and resisting forces, ranging between 33° and 37° from a horizontal plane.

Antarctic geographic zone (1) The region poleward of the Antarctic Circle (66.5° S latitude), including Antarctica, which is covered by Earth's greatest ice sheet.

Antarctic high (6) A consistent high-pressure region centered over Antarctica; source region for an intense polar air mass that is dry and associated with the lowest temperatures on Earth.

Anthropogenic atmosphere (3) Earth's next atmosphere, so named because humans appear to be the principal causative agent.

Anticline (12) Upfolded strata in which layers slope away from the axis of the fold, or central ridge.

Anticyclone (6) A dynamically or thermally caused area of high atmospheric pressure with descending and diverging air flows.

Aphelion (2) The most distant point in Earth's elliptical orbit about the Sun; reached on July 4 at a distance of 152,083,000 km (94.5 million mi); variable over a 100,000-year cycle.

Apparent temperature (5) The temperature subjectively perceived by each individual, also known as sensible temperature.

Aquatic ecosystems (20) An association of plants and animals and their nonliving environment in a marine setting.

Aquiclude (9) A body of rock that does not conduct water in usable amounts; an impermeable layer.

Aquifer (9) Rock strata permeable to groundwater flow.

Aquifer recharge area (9) The surface area where water enters an aquifer to recharge the water-bearing strata in the groundwater system.

Arctic geographic zone (1) The region poleward of the Arctic Circle (66.5° N latitude), dominated by the Arctic Ocean, which is covered by floating sea ice that thins in the summer and by glacial ice in the form of icebergs.

Arctic tundra (20) A biome in the northernmost portions of North America, Europe, and Russia, featuring low ground-level herbaceous plants as well as some woody plants.

Arête (17) A sharp ridge that divides two cirque basins. Means fish bone in French Arêtes form sawtooth and serrated ridges in glaciated mountains.

Aridisols (18) Largest single soil order in the Soil Taxonomy classification and typical of dry climates; low in organic matter and dominated by calcification and salinization.

Arroyo (15) A Spanish term for an intermittently dry streambed or wash.

Artesian water (9) Pressurized groundwater that rises in a well or a rock structure above the local water table;

may flow out onto the ground (see piezometric surface).

Asperities (12) Sticking or friction points of high strain, which are the points that give way and release energy between two sides of a fault. Earthquake magnitude is related to the spatial extent to which asperities fail.

Asthenosphere (11) Region of the upper mantle known as the plastic layer; the least rigid portion of Earth's interior; shatters if struck yet flows under extreme heat and pressure.

Atmosphere (1) The thin veil of gases surrounding Earth that forms a protective boundary between outer space and the biosphere; generally considered to be below 480 km (300 mi).

Atoll (16) A circular, or ring-shaped, island enclosing a central lagoon; created by coral colonies.

Auroras (2) A spectacular glowing light display in the ionosphere, stimulated by the interaction of the solar wind with oxygen and nitrogen gases; aurora borealis (northern) and aurora australis (southern).

Autotrophs (19) Organisms capable of using carbon dioxide directly as their sole source of carbon; known as producers. They fix carbon through photosynthesis.

Autumnal (September) equinox (2) The time around September 22–23 when the Sun's declination crosses the equatorial parallel; all places on Earth experience days and nights of equal length. The Sun rises at the South Pole and sets at the North Pole.

Available water (9) The portion of capillary water that is accessible to plant roots; usable water held in soil moisture storage.

Axial parallelism (2) Earth's axis is parallel to itself throughout the year; the North Pole points to near Polaris.

Axis (2) An imaginary line, extending through Earth from the geographic North Pole to the geographic South Pole, around which Earth rotates.

Azores high (6) A subtropical high-pressure cell that forms in the Northern Hemisphere in the eastern Atlantic (see Bermuda high); associated with warm, clear water and large quantities of sargassum, or gulf weed, characteristic of the Sargasso Sea.

Backswamp (14) A low-lying, swampy area of a floodplain; adjacent to a river, with the river's natural levees on one side and the sides of the valley on the other (see yazoo tributary).

Badland (15) Rugged topography, usually of relatively low, varied relief and barren of vegetation; associated with arid and semiarid regions and rocks with low resistance to weathering.

Bajada (15) A continuous apron of coalesced alluvial fans, formed along the base of mountains in arid climates; presents a gently rolling surface from fan to fan.

Barrier beaches (16) Narrow, long depositional features, generally of sand, which form offshore and are roughly parallel to the coast; may appear as barrier islands and long chains of barrier beaches.

Barrier island (16) Generally, a broadened barrier beach.

Barrier spit (16) A depositional form that develops when transported sand in a barrier beach or island is deposited in long ridges that are attached at one end to the mainland and partially cross the mouth of a bay.

Basalt (11) A common extrusive igneous rock; its mafic composition is fine-grained, comprising the bulk of the ocean floor crust, lava flows, and volcanic forms; gabbro in its intrusive form.

Base level (13) A hypothetical level below which a stream cannot erode its valley, and thus the lowest operative level for denudation processes; in an absolute sense represented by sea level extending back under the landscape.

Basin and Range Province (15) A region of dry climates, few permanent streams, and interior drainage patterns in the western United States; formed by a sequence of horsts and grabens.

Batholith (11) The largest plutonic form exposed at the surface; an irregular intrusive mass (>100 km^2; >40 mi^2) that invades crustal rocks, cooling slowly so that crystals develop.

Bay barrier (16) An extensive sand spit that encloses a bay, cutting it off completely from the ocean and forming a lagoon; produced by littoral drift and wave action; sometimes referred to as a *baymouth bar.*

Beach (16) The portion of the coastline where an accumulation of sediment is in motion.

Beach drift (16) Material, sand, gravel, and shells, that are moved by the longshore current in the effective direction of the waves.

Beaufort wind scale (6) A descriptive scale for the visual estimation of wind speeds; originally conceived in 1806 by Admiral Beaufort of the British Navy.

Bed load (14) Coarse materials that are dragged along the bed of a stream by traction or by the rolling and bouncing motion of saltation; involves particles too large to remain in suspension.

Bedrock (13) The rocks of Earth's crust that are below the soil and are basically unweathered; such solid crust sometimes exposed as an outcrop.

Bermuda high (6) A subtropical high-pressure cell that forms in the western North Atlantic (see Azores high).

Biogeochemical cycles (19) The various circuits of flowing elements and materials (carbon, oxygen, nitrogen, phosphorus, water) that combine Earth's biotic and abiotic systems; the cycling of materials is continuous and renewed through the biosphere and the life processes.

Biogeographical realms (20) Eight regions of the biosphere, each representative of evolutionary core areas of related flora (plants) and fauna (animals); a broad geographical classification scheme.

Biogeography (19) The study of the distribution of plants and animals and related ecosystems; the geographical relationships with related environments over time.

Biomass (19) The total mass of living organisms on Earth or per unit area of a landscape; also, the weight of the living organisms in an ecosystem.

Biome (20) A large terrestrial ecosystem characterized by specific plant communities and formations; usually named after the predominant vegetation in the region.

Biosphere (1) That area where the atmosphere, lithosphere, and hydrosphere function together to form the context within which life exists; an intricate web that connects all organisms with their physical environment.

Biotic (1) Living.

Blowout depressions (15) Eolian erosion whereby deflation forms basins in areas of loose sediment. Size may range up to hundreds of meters.

Bolson (15) The slope and basin area between the crests of two adjacent ridges in a dry region.

Boreal forest (20) See northern needleleaf forest.

Brackish (7) Seawater with a salinity of less than 35‰; for example, the Baltic Sea.

Braided stream (14) A stream that becomes a maze of interconnected channels laced with excess sediments. Braiding often occurs with a reduction of discharge that affects a stream's transportation ability or a mass-movement event such as a landslide upstream.

Breaker (16) The point where a wave's height exceeds its vertical stability and the wave breaks as it approaches the shore.

Breakwater (16) A structure built to block the movement of longshore currents and beach drift; used to protect harbor entrances and beach deposits.

Brine (7) Seawater with a salinity of more than 35‰; for example, the Persian Gulf.

Brunisolic (18) A soil order in the Canadian System of Soil Classification formed under forest cover with brownish-colored horizons, although various colors are possible. Also, can be with mixed forest, shrubs, and grass; well to imperfectly drained.

Calcification (18) The illuviated accumulation of calcium carbonate or magnesium carbonate in the B and C soil horizons.

Caldera (12) An interior sunken portion of a composite volcanic crater; usually steep-sided and circular, sometimes containing a lake; also found in conjunction with shield volcanoes.

Caliche (18) A cemented or hardened subsurface diagnostic calcic soil horizon; found in the southwestern United States, usually in arid and semiarid climates.

Canadian System of Soil Classification (18) A taxa for all soils presently recognized in Canada and adapted to Canada's particular forest, tundra, prairie, frozen ground, and colder climates. The CSSC is organized at five levels of generalization (order, great group, subgroup, family, and series) with each level referred to as a *category of classification*. The system is described in *The Canadian Soil Classification System,* 2nd edition, published by Agriculture Canada, 1987.

Capillary water (9) Soil moisture, most of which is accessible to plant roots; held in the soil by surface tension and cohesive forces between water and soil (see also available water, field capacity, and wilting point).

Carbonation (13) A process of chemical weathering by a weak carbonic acid (water and carbon dioxide) that reacts with many minerals, especially limestone, containing calcium, magnesium, potassium, and sodium transforming them into *carbonates*.

Carbon dioxide (3) A natural by-product of life processes and complete combustion; the principal radiatively active gas in the greenhouse effect; CO_2.

Carbon monoxide (3) An odorless, colorless, tasteless combination of carbon and oxygen produced by the incomplete combustion of fossil fuels or other carbon-containing substance; CO.

Carnivore (19) A secondary consumer that principally eats meat for sustenance. The top carnivore in a food chain is considered a tertiary consumer.

Cartography (1) The making of maps and charts; a specialized science and art that blends aspects of geography, engineering, mathematics, graphics, computer science, and artistic specialties.

Catastrophism (11) A philosophy that attempts to fit the vastness of Earth's age and the complexity of the rock record into a very shortened time span through a belief in short-lived and catastrophic worldwide events.

Cation-exchange capacity (CEC) (18) The ability of soil colloids to exchange cations between their surfaces and the soil solution; a measured potential.

Chaparral (20) Dominant shrub formations of Mediterranean dry summer climates; characterized by sclerophyllous scrub and short, stunted, and tough forests; derived from the Spanish *chapparo*; specific to California.

Chemical weathering (13) Decomposition and decay of the constituent minerals in rock through chemical alteration of those minerals. Water is essential, with rates keyed to temperature and precipitation values. Processes include hydrolysis, oxidation, carbonation, and solution.

Chemosynthesis (2) Production of organic compounds by living tissue without involving sunlight; the action of purple-sulfur bacteria, for example.

Chernozemic (18) A soil order in the Canadian System of Soil Classification formed in well to imperfectly drained soils of the steppe grassland-forest transition. Accumulation of organic matter in surface horizons. Most frozen during some winter months with soil-moisture deficits in the summer. Mean annual temperature >0°C and usually less than 5.5°C.

Chinook winds (8) North American term for a warm, dry, downslope air flow; characteristic of the rain shadow region on the leeward side of mountains; known as föhn, or foehn, winds in Europe.

Chlorofluorocarbon compounds (CFCs) (3) Large manufactured molecules (polymers) containing chlorine, fluorine, and carbon; inert and possessing remarkable heat properties; also known as hologens. After slow transport to the stratospheric ozone layer CFCs react with ultraviolet radiation freeing chlorine atoms that act as a catalyst to produce reactions that destroy ozone.

Chlorophyll (19) A light-sensitive pigment that resides within the chloroplast bodies of plants in leaf cells; the basis of photosynthesis.

Cinder cone (12) A landform of tephra and scoria, usually small and cone-shaped and generally not more than 450 m (1500 ft) in height; with a truncated top.

Circle of illumination (2) The division between lightness and darkness on Earth; a day-night great circle.

Circum-Pacific belt (12) A tectonically and volcanically active region encircling the Pacific Ocean; also known as the "ring of fire."

Cirque (17) A scooped-out, amphitheater-shaped basin at the head of an alpine glacier valley; an erosional landform.

Cirrus (7) Wispy filaments of ice-crystal clouds that occur above 6000 m (20,000 ft); appear in a variety of forms, from feathery hairlike fibers to veils of fused sheets of ice crystals, and sometimes to a dappled mackerel sky.

Classification (10) The process of ordering or grouping data or phenomena in related classes; results in a regular distribution of information; a taxonomy.

Climate (10) The consistent, long-term, behavior of weather over time, including its variability; in contrast to weather, which is the condition of the atmosphere at any given place and time.

Climatic regions (10) Areas of similar climate, which contain characteristic regional weather and air mass patterns.

Climatology (10) A scientific study of climate and climatic patterns and the consistent behavior of weather and weather variability and extremes over time in one place or region; and the effects climate and climate change has on human society and culture.

Climographs (10) Graphs that plot daily, monthly, or annual temperature and precipitation values for a selected station; may also include additional weather information.

Closed system (1) A system that is shut off from the surrounding environment so that it is entirely self-contained in terms of energy and materials; Earth is a closed material system (see open system).

Col (17) Formed by two headward eroding cirques that reduce an arête (ridge crest) to form a high pass or saddlelike narrow depression.

Cold front (8) The leading edge of a cold air mass; identified on a weather map by a line marked with a series of triangular spikes, pointing in the direction of frontal movement.

Collision-coalescence process (7) Principal process of raindrop formation predominant in clouds that form at temperatures above freezing. Larger moisture droplets sweep through the cloud, combining with smaller droplets, gradually coalescing and growing in size.

Community (19, 20) A convenient biotic subdivision within an ecosystem; formed by interacting populations of animals and plants in an area.

Composite volcano (12) A volcano formed by a sequence of explosive volcanic eruptions; steep-sided, conical in shape; sometimes referred to as a stratovolcano, although composite is the preferred term.

Condensation nuclei (7) Necessary microscopic particles on which water vapor condenses to form moisture droplets; can be sea salts, dust, soot, or ash.

Conduction (4) The slow molecule-to-molecule transfer of heat through a medium, from warmer to cooler areas.

Cone of depression (9) The depressed shape of the water table around a well after active pumping from an aquifer. The water table adjacent to the well is drawn down during the process of water removal.

Confined aquifer (9) An aquifer that is bounded above and below by impermeable layers of rock or sediment.

Consumers (19) Organisms in an ecosystem that depend on autotrophs (organisms capable of using carbon dioxide as their sole source of carbon) for their source of nutrients; also called *heterotrophs*.

Consumptive uses (9) Water losses that cause water to be removed from a water budget at one point and to remain unavailable further downstream.

Continental divides (14) A ridge or elevated area that determines the drainage patterns of drainage basins; specifically, that ridge in North America that separates drainage to the Pacific in the west from drainage to the Atlantic and Gulf in the east and to Hudson Bay and the Arctic Ocean in the north.

Continental drift (11) A proposal by Wegener in 1912 stating that Earth's landmasses have migrated over the past 200 million years from a supercontinent he called Pangaea to the present configuration; a widely accepted concept today (see plate tectonics).

Continental glacier (17) A continuous mass of unconfined ice, covering at least 50,000 km^2 (19,500 mi^2); most extensive as ice sheets covering Greenland and Antarctica.

Continentality (5) A qualitative designation applied to stations that lack the temperature-moderating effects of the sea and that exhibit a greater range between minimum and maximum temperatures, both daily and annually.

Continental platforms (12) The broadest category of landforms, including those masses of crust that reside above or near sea level and the adjoining undersea continental shelves along the coastline.

Continental shield (12) Generally old, low-elevation heartland regions of continental crust; various cratons (granitic cores) and ancient mountains exposed at the surface.

Contour lines (1) Isolines on a topographic map that connect all points at the same elevation relative to a reference elevation called the vertical datum. The spaces between contour lines portray slope and are known as contour intervals.

Convection (4) Vertical transfer of heat from one place to another through the actual physical movement of air; involves a strong vertical motion.

Coordinated Universal Time (UTC) (1) The official reference time in all countries formerly known as Greenwich Mean Time outside the United Kingdom; now measured by 5 primary standard atomic clocks whose time calculations are collected in Paris, France, by the Bureau International de l'Heure. UTC assumes a prime meridian (0° longitude) passing through Greenwich, London, England.

Coral (16) A simple cylindrical marine animal with a saclike body that secretes calcium carbonate to form a hard external skeleton; lives symbiotically with nutrient-producing algae. Presently going through a bleaching and dying cycle worldwide, apparently due to increasing water temperatures in the ocean.

Cordilleran system (12) One of two large mountain systems on Earth; refers to the relatively young mountains along the western margins of North and South America, from the tip of Tierra del Fuego to the massive peaks of Alaska.

Core (11) The deepest inner portion of Earth, representing one-third of its entire mass; differentiated into two zones—a solid iron inner core surrounded by a dense, molten, fluid metallic-iron outer core.

Coriolis force (6) The apparent deflection of moving objects on Earth from a straight path, in relationship to the differential speed of rotation at varying latitudes. Deflection is to the right in the Northern Hemisphere and to the left in the Southern Hemisphere; it produces a maximum effect at the poles and zero effect along the equator.

Crater (12) A circular surface depression formed by volcanism (accumulation, collapse, or explosion); usually located at a volcanic vent or pipe; can be at the summit or on the flank of a volcano.

Crevasses (17) Vertical cracks that develop in a glacier as a result of friction between valley walls, or tension forces of extension on convex slopes, or compression forces on concave slopes.

Crust (11) Earth's outer shell of crystalline surface rock, ranging from 5 to 60 km (3 to 38 mi) in thickness from oceanic crust to mountain ranges. Average density of continental crust is 2.7 g per cm^3, whereas oceanic crust is 3.0 g per cm^3.

Cryosolic (18) A soil order in the Canadian System of Soil Classification formed with permafrost close and composed of mineral and organic soil deposits. Generally found north of the tree line, or in fine textured soils in subarctic forest, or in some organic soils in boreal forests; dominate the northern third of Canada. Cryoturbation common often denoted by patterned ground circles, polygons, and stripes.

Cumulonimbus (7) A towering, precipitation-producing cumulus cloud that is vertically developed across altitudes associated with other clouds; frequently associated with lightning and thunder and thus designated a thunderhead.

Cumulus (7) Bright and puffy cumuliform clouds up to 2000 m in altitude (6500 ft).

Cut bank (14) A steep bank formed along the outer portion of a meandering stream; produced by lateral erosive action of a stream.

Cyclogenesis (8) An atmospheric process that describes the birth of a midlatitude wave cyclone; usually along the polar front; strengthening and development of a wave cyclone also associated with the eastern slope of the Rockies, other north-south mountain barriers, and along the North American and Asian east coasts.

Cyclone (6) A dynamically or thermally caused low-pressure area of converging and ascending air flows (see wave cyclone and tropical cyclone).

Daylength (2) Duration of exposure to insolation, varying during the year depending on latitude; an important aspect of seasonality.

Daylight saving time (1) Time is set ahead one hour in the spring and set back one hour in the fall in the Northern Hemisphere. Time is set ahead on the first Sunday in April and set back on the last Sunday in October—only Hawaii, Arizona, portions of Indiana, and Saskatchewan exempt themselves. In Europe the last Sundays in March and September generally are used.

Debris avalanche (13) A mass of falling and tumbling rock, debris, and soil; can be dangerous because of the tremendous velocities achieved by the onrushing materials; differentiated from a debris slide by velocity.

Declination (2) The latitude that receives direct overhead (perpendicular) insolation on a particular day; migrates annually through the 47° of latitude between the tropics.

Decomposers (19) Microorganisms that digest and recycle the organic debris and waste in the environment: includes bacteria, fungi, insects, and worms.

Deep cold zone (7) The deepest part of the ocean, where temperatures gradually lower to near freezing; represents about 80% of the oceanic mass.

Deficit (9) In a water balance, the amount of unmet, or unsatisfied, potential evapotranspiration (DEFIC).

Deflation (15) A process of wind erosion that removes and lifts individual particles, literally blowing away unconsolidated, dry, or noncohesive sediments.

Delta (14) A depositional plain formed where a river enters a lake or an ocean; named after the triangular shape of the Greek letter delta.

Denudation (13) A general term that refers to all processes that cause degradation of the landscape: weathering, mass movement, erosion, and transportation; at the heart of exogenic processes.

Deposition (14) The process whereby weathered, wasted, and transported sediments are laid down. Specific depositional landforms result from the actions of air, water, and ice agents.

Desert biomes (20) Arid landscapes of uniquely adapted dry-climate plants and animals.

Desert pavement (15) An arid landscape surface formed when wind deflation and sheetwash remove smaller particles, leaving residual pebbles and gravels to concentrate at the surface; resembles a cobblestone street.

Desertification (20) The expansion of deserts worldwide, related principally to poor agricultural practices (i.e., overgrazing and inappropriate agricultural practices), improper soil-moisture management, erosion and salinization, deforestation, and the ongoing climatic change; an unwanted semipermanent invasion into neighboring biomes.

Dew-point temperature (7) The temperature at which a given mass of air becomes saturated, holding all the water it can hold. Any further cooling or addition of water vapor results in active condensation.

Differential weathering (13) The effect of different resistances in rock, coupled with variations in the intensity of physical and chemical weathering, produces a landscape exhibiting uneven landforms.

Diffuse radiation (4) The downward component of scattered incoming insolation from clouds and the atmosphere; casts a shadowless light on the ground.

Discharge (9) The measured volume of flow in a river that passes by a given cross section of a stream in a given unit of time; usually denoted in cubic meters per second or cubic feet per second.

Dissolved load (14) Materials carried in chemical solution in a stream derived from minerals such as limestone and dolomite, or from soluble salts.

Diversity (19) A principle of ecology: the more diverse the species population (both in number of species and quantity of members in each species), the more risks are spread over the entire community, which results in greater overall stability in an ecosystem, as compared to a monoculture of limited diversity.

Doldrums (6) A region of equatorial calm associated with the intertropical convergence zone; so named for the difficulty encountered in the era of sailing ships.

Downwelling current (6) An area of the sea where a convergence or accumulation of water thrusts excess water downward; occurs, for example, at the western end of the equatorial current or along the margins of Antarctica.

Drainage basin (14) The basic spatial geomorphic unit of a river system; distinguished from a neighboring basin by ridges and highlands that form divides delimiting the catchment area of the drainage basin, or its watershed.

Drainage density (14) A measure of the overall operational efficiency of a drainage basin; determined by the ratio of combined channel lengths to the unit area.

Drainage pattern (14) A geometric arrangement of streams in a region; determined by slope, differing rock resistance to weathering and erosion, climatic and hydrologic variability, and structural controls of the landscape.

Draw down (9) See cone of depression.

Drumlin (17) A depositional landform related to glaciation that is composed of till and is streamlined in the direction of continental ice movement; blunt end upstream and tapered end downstream with a rounded summit.

Dry adiabatic rate (DAR) (8) The rate at which a parcel of air that is less than saturated cools (if ascending) or heats (if descending); a rate of 10C° per 1000 m (5.5F° per 1000 ft) (see also adiabatic).

Dune (15) A depositional feature of sand grains deposited in transient mounds, ridges, and hills; characteristic of the erg deserts of North Africa. Extensive areas of sand dunes are called sand seas.

Dust dome (4) A dome of airborne pollution associated with every major city; may be blown by winds into elongated plumes downwind from the city.

Dynamic equilibrium model (13) The balancing act between tectonic uplift and reduction rates of erosion, between the resistance of crust materials and the work of denudation processes. Landscapes evidence ongoing adaptation to rock structure, climate, local relief, and elevation.

Earthquake (12) A sharp release of energy that produces shaking in Earth's crust at the moment of rupture along a fault or in association with volcanic activity. Movement of the crust produces tectonic earthquakes, whereas the movement of magma produces volcanic earthquakes. Earthquake magnitude is estimated by the Richter scale; intensity is described by the Mercalli scale.

Ebb tides (16) Tides that are falling during the daily tidal cycle.

Ecological succession (19) The process whereby different and usually more complex assemblages of plants and animals replace older and usually simpler communities. Changes apparently move toward a more stable and mature condition.

Ecology (19) The science that studies the interrelationships among organisms and their environment and among various ecosystems.

Ecosphere (1) Another name for the biosphere.

Ecosystem (19, 20) A self-regulating association of living plants, animals, and their nonliving physical and chemical environment.

Ecotone (20) A boundary transition zone between adjoining ecosystems that may vary in width and represent areas of tension as similar species of plants and animals compete for the resources. Climatic ecotones usually are more gradual then physical ecotones, where changes in soil, moisture availability, or topography sometimes form abrupt boundaries.

Effusive eruption (12) An eruption characterized by low viscosity, basaltic magma, with characteristic low-gas content readily escaping. Lava pours forth onto the surface with relatively small explosions and little tephra; tends to form shield volcanoes.

Elastic-rebound theory (12) A concept describing the faulting process, in which the two sides of a fault appear locked despite the motion of adjoining pieces of crust, but with accumulating strain they rupture suddenly, snapping to new positions relative to each other.

Electromagnetic spectrum (2) All the radiant energy produced by the Sun placed in an ordered range and divided according to wavelengths; travels to Earth at the speed of light in form of waves.

Eluviation (18) The downward removal of finer particles and minerals from the upper horizons of soil.

Empirical classification (10) A classification based on weather statistics or other data; used to determine general climate categories.

Endogenic system (11) The system internal to Earth, driven by radioactive heat derived from sources within the planet. In response, the surface is fractured, mountain building occurs, and earthquakes and volcanoes are activated.

Entisols (18) A soil order in the Soil Taxonomy classification that specifically lacks vertical development of horizons; usually young or undeveloped; found in active slopes, alluvial-filled floodplains, poorly drained tundra.

Entrenched meanders (14) Excavated into the landscape, such as incised river meanders; thought to be evidence of stream rejuvenation.

Environmental lapse rate (3) The actual lapse rate in the lower atmosphere at any particular time under local weather conditions; may deviate above or below the average normal lapse rate of 6.4C° per 1000 m (3.5F° per 1000 ft).

Eolian (15) Caused by wind; refers to the erosion, transportation, and deposition of materials; spelled aeolian in some countries. Such processes tend to operate more efficiently in arid and semiarid climates.

Epipedon (18) The diagnostic soil horizon that forms at the surface; not to be confused with the A horizon; may include all or part of the illuviated B horizon.

Equal area (1) A trait of a map projection; indicates the equivalence of all areas on the surface of the map, although shape is distorted.

Equatorial geographic zone (1) Generally referring to that area within the tropics that straddles the equator, stretching roughly 10° of latitude north and south.

Equatorial and tropical rain forest (20) A lush biome of tall broadleaf evergreen trees and diverse plants and animals. The dense canopy of leaves is usually arranged in three levels.

Equatorial countercurrent (6) A strong countercurrent, generated by the western intensification pile-up, that travels east along the full extent of the Pacific, Atlantic, and Indian oceans; usually flows alongside or just beneath the westward-flowing surface current.

Equatorial low-pressure trough (6) A thermally caused low-pressure area that almost girdles Earth, with air converging and ascending all along its extent; also called the intertropical convergence zone (ITCZ).

Equilibrium (1) The status of a balanced energy and material system; maintains the system's general structure and character; a balance between form and process.

Equilibrium line (17) The area of a glacier where accumulation (gain) and ablation (loss) are balanced.

Erg desert (15) Sandy deserts, or areas where sand is so extensive that it constitutes a sand sea.

Erosion (14) Denudation by wind, water, and ice, which dislodges, dissolves, or removes surface material.

Erratics (17) Large rocks and boulders left behind by retreating glaciers and strewn across the glaciated surface; usually different in composition and origin from the ground on which they were deposited.

Esker (17) A sinuously curving, narrow deposit of coarse gravel that forms along a meltwater stream channel, developing in a tunnel beneath the glacier. A retreating glacier leaves such ridges behind in a pattern that usually parallels the path of the glacier; may appear branching.

Estuary (14) The point at which the mouth of a river enters the sea and freshwater and seawater are mixed; a place where tides ebb and flow.

Eurasian-Himalayan system (12) One of two major mountain chains on Earth, stretching from southern Asia, China, and northern India and continuing in a belt through the upper Middle East to Europe and the European Alps; includes the world's highest peak, Mount Everest in the Himalayas, at an elevation of 8848 m (29,028 ft).

Eustasy (7) Refers to worldwide changes in sea level that are not related to movements of land but rather to a rise and fall in the volume of water in the oceans, principally through changes in ice volume.

Eutrophication (19) A natural process in which lakes receive nutrients and sediment and become enriched; the gradual filling and natural aging of water bodies.

Evaporation (9) The movement of free water molecules away from a wet surface into air that is less than saturated; the phase change of water to water vapor; vaporization below the boiling point of water.

Evaporation fog (7) A fog formed when cold air flows over the warm surface of a lake, ocean, or other body of water; forms as the water molecules evaporate from the water surface into the cold, overlying air; also known as a steam fog or sea smoke.

Evaporation pan (9) A standardized pan from which evaporation occurs, with water automatically replaced and measured; an evaporimeter.

Evaporites (11) Chemical sediments formed from inorganic sources when water evaporates and leaves behind a residue of salts previously in solution.

Evapotranspiration (9) The merging of evaporation and transpiration water loss into one term (see potential and actual evapotranspiration).

Evolutionary atmosphere (2) Earth's atmosphere between 3.3 and 4.0 billion years ago; thought by scientists to have been composed principally of water vapor and lesser amounts of carbon dioxide and nitrogen; an anaerobic environment.

Exfoliation dome (13) A dome-shaped feature of weathering, produced by the response of granite to the overburden removal process, which relieves pressure from the rock. Layers of rock sluff off in slabs or shells in a sheeting process.

Exogenic system (11) The external surface system, powered by insolation, that energizes air, water, and ice and sets them in motion, under the influence of gravity. Includes all processes of landmass denudation.

Exosphere (3) An extremely rarefied outer atmospheric halo beyond the thermopause, which occurs at an altitude of 480 km (300 mi); probably composed of hydrogen and helium atoms, with some oxygen atoms and nitrogen molecules present near the thermopause.

Exotic stream (9) A river that rises in a humid region and flows downstream through an arid region, with discharge decreasing toward the mouth; for example, the Nile River and the Colorado River.

Explosive eruption (12) A violent and unpredictable eruption, the result of magma that is thicker, stickier, higher in gas content and silica, and more viscous than that of an effusive eruption; tends to form blockages within a volcano; produces composite volcanic landforms.

Faulting (12) The process whereby displacement and fracturing occurs between two portions of Earth's crust; usually associated with earthquake activity.

Feedback loops (1) When a portion of the system output cycles back as an information input, causing changes that guide further system operations (see negative and positive feedback).

Field capacity (9) Water held in the soil by hydrogen bonding against the pull of gravity, remaining after water drains from the larger pore spaces, or storage capacity; the available water for plants. Field capacity is specific to each soil type and is an amount that can be determined by soil surveys.

Firn (17) Snow of a granular texture that is transitional in the slow transformation from snow to glacial ice; snow that has persisted through a summer season in the zone of accumulation.

Firn line (17) The snow line that is visible on the surface of a glacier, where winter snows survive the summer ablation season; analogous to a snowline on land.

Fjord (17) A drowned glaciated valley, or glacial trough, along a coast which is filled by the sea.

Flash flood (15) A sudden and short-lived torrent of water that exceeds the capacity of a stream channel; associated with desert and semiarid washes.

Flood (14) A high water level that overflows the natural (or artificial) banks along any portion of a stream.

Floodplain (14) A low-lying area near a stream channel, subject to recurrent flooding; alluvial deposits generally mask underlying rock.

Flood tides (16) Rising tides during the daily tidal cycle.

Fluvial (14) Stream-related processes; from the Latin *fluvius* for "river" or "running water."

Fog (7) A cloud, generally stratiform, in contact with the ground, with visibility usually restricted to less than 1 km (3300 ft).

Folding (12) A process that bends and deforms beds of various rocks subjected to compressional forces.

Foliated (11) Planes in rock are aligned, or appear as wavy striations, following metamorphism; describes the mineral structure in metamorphic rocks.

Food chain (19) The circuit along which energy flows from producers, who manufacture their own food, to consumers; a one-directional flow of chemical energy, ending with decomposers.

Food web (19) A complex network of interconnected food chains, with consumers generally having several different food chains available to them.

Formation class (20) That portion of a biome that concerns the plant communities only, subdivided by size, shape, and structure of the dominant vegetation present.

Friction force (6) The effect of drag by the wind as it moves across a surface; may be operative through 500 m (1640 ft) of altitude. Surface friction slows the velocity of the wind and therefore reduces the effectiveness of the Coriolis force.

Front (8) The leading edge of an advancing air mass; a line of contrasting weather conditions.

Frost action (13) A powerful mechanical force produced as water expands as much as 9% of its volume as it freezes and can exceed the tensional strength of rock. Repeated cycles of water freezing and thawing breaks rock segments apart.

Funnel cloud (8) The visible swirl extending from the bottom side of a cloud, which may or may not develop into a tornado. A tornado is a funnel cloud that has extended all the way to the ground.

Fusion (2) The process of forcibly joining, under extreme temperature and pressure, positively charged hydrogen and helium nuclei; occurs naturally in thermonuclear reactions within stars, such as our Sun.

Gelifluction (17) Refers to soil flow processes in periglacial environments; progressive, lateral movement. A type of solifluction under periglacial conditions of permafrost and frozen ground.

General circulation model (GCM) (10) Complex climate models that use mathematical models originally established for forecasting weather. GCMs produce generalizations of reality and produce predictive forecasts of future weather and climate conditions. Four complex GCMs (three-dimensional models) are in use in the United States and four in other countries

Genetic classification (10) A type of classification that uses causative factors to determine climatic regions; for example, an analysis of the effect of interacting air masses on climate.

Geodesy (1) The science that determines Earth's shape and size through surveys, mathematical means, and remote sensing.

Geographic information system (GIS) (2) A computer-based data processing tool or methodology used for gathering, manipulating, and analyzing geographic information to produce a holistic, interactive analysis.

Geography (1) The science that studies the interdependence among geographic areas, natural systems, processes, society, and cultural activities over space—a spatial science. The five themes of geographic education include: location, place, movement, regions, and human-Earth relationships.

Geoid (1) A word that describes Earth's shape, literally, the shape of Earth is Earth-shaped. A theoretical surface at sea level that extends through the continents; deviates from a perfect sphere.

Geologic cycle (11) A general term characterizing the vast cycling at the Earth-atmosphere interface.

Geologic time scale (11) A listing of eras, periods, and epochs that span Earth's history; reflects the relative relationship of various layers of rock strata and the absolute dates as determined by scientific methods such as radioactive isotopic dating.

Geomorphic cycle model (13) A conceptual model proposed by William Morris Davis to characterize landscape development as a cyclic pattern from initial uplift to final erosional surface, or peneplain. He named the evolutionary phases youth, maturity, and old age.

Geomorphic threshold (13) The threshold up to which landforms change before lurching to a new set of relationships, with rapid realignments of landscape materials and slopes.

Geomorphology (13) The science that analyzes and describes the origin, evolution, form, classification, and spatial distribution of landforms.

Geostrophic winds (6) Winds moving between pressure areas along paths that are parallel to the isobars. In the upper troposphere the pressure gradient force equals the Coriolis force so that the amount of deflection is proportional to air movement.

Geothermal energy (12) The energy potential of boiling steam produced by subsurface magma in near contact with groundwater. Active examples of geothermal energy development include Iceland, New Zealand, Italy, and northern California.

Glacial drift (17) The general term for all glacial deposits, both unsorted and sorted.

Glacial ice (17) A hardened form, very dense in comparison to normal snow or firn; under pressure within a glacier and capable of downhill movement.

Glacial plucking (17) A process of rock pieces actually freezing to the basal layers of a glacier; once embedded, they become the tools the glacier uses to scour and sandpaper the landscape through which it flows.

Glacial surge (17) The rapid, lurching, unexpected movement of a glacier. Causes are still debatable, but it appears that some surge is due to water pressure in the basal layers of the glacier.

Glaciation (7) The formation of ice crystals in a rising cumulonimbus cloud as it enters a region of the atmosphere that is below freezing; not to be confused with the geomorphic term for a glacial phase.

Glacier (17) A large mass of perennial ice resting on land or floating shelflike in the sea adjacent to the land; formed from the accumulation and recrystallization of snow, which then flows slowly under the pressure of its own weight and the pull of gravity.

Glacio-eustatic (7) Changes in sea level in response to the amount of water stored on Earth as ice; the more water that is bound up in glaciers and ice sheets, the lower the sea level.

Gleysolic (18) A soil order in the Canadian System of Soil Classification formed under chronic reducing conditions inherent in poorly drained mineral soils and wet conditions, high water table, and long periods of water saturation. Rather than continuous they appear discontinuous within other soil orders, defined on the basis of color and mottling.

Graben (12) Pairs or groups of faults that produce downward faulted blocks; characteristic of the basins of the interior western United States (see horst and Basin and Range Province).

Graded stream (14) A condition in a stream of mutual adjustment between the load carried by a stream and the related landscape through which the stream flows, forming a state of dynamic equilibrium among erosion, transported load, deposition, and the stream's capacity. In reality, an individual stream may have both graded and ungraded portions in its course.

Gradient (14) The drop in elevation from a stream's headwaters to its mouth, ideally forming a concave slope. Characteristically the upper portion has a greater slope than the downstream channel.

Granite (11) A coarse-grained (slow-cooling) intrusive igneous rock of 25% quartz and more than 50% potassium and sodium feldspars; characteristic of the continental crust.

Gravitational water (9) That portion of surplus water that percolates downward from the capillary zone, pulled by gravity from the soil surface area to the groundwater zone.

Gravity (2) The natural force exerted by the mass of an object; produced in an amount proportional to an object's mass.

Great circle (1) Any circle of circumference drawn on a globe with its center coinciding with the center of the globe. An infinite number of great circles can be drawn, but only one parallel is a great circle—the equator.

Greenhouse effect (4) The process whereby radiatively active gases absorb and delay the loss of heat to space, thus keeping the lower troposphere moderately warmed through the radiation and reradiation of infrared wavelengths. The approximate similarity between this process and that of a greenhouse explains the name.

Greenwich Mean Time (GMT) (1) Former world standard time, now known as Coordinated Universal Time (UTC) (see Coordinated Universal Time).

Ground ice (17) Subsurface water that is frozen in regions of permafrost. The moisture content of areas with ground ice may vary from nearly absent in regions of drier permafrost to almost 100% in saturated soils.

Groundwater mining (9) Pumping an aquifer beyond its flow and recharge capacities; an overutilization of a groundwater resource.

Gulf Stream (5) A strong northward-moving warm current off the east coast of North America, which carries its water far into the North Atlantic.

Gutenberg discontinuity (11) The transition zone between the outer core and the lower mantle.

Gyres (6) The dominant circular ocean currents beneath subtropical high-pressure cells in both hemispheres; offset to the western margin of each ocean basin.

Habitat (19, 20) That physical location in which an organism is biologically suited to live. Most species have specific habitat parameters, or limits.

Hadley cell (6) The vertical convection cell in each hemisphere that is generated by the low-pressure system of converging and ascending air along the equator, which then subsides and diverges in the subtropical regions.

Hail (8) A type of precipitation formed when a raindrop is repeatedly circulated above and below the freezing level in a cloud, with each cycle adding more ice to the hailstone until it becomes too heavy to stay aloft.

Hair hygrometer (7) An instrument for the measurement of relative humidity; based on the principle that human hair will change as much as 4% in length between 0 and 100% relative humidity.

Headlands (16) Protruding landforms that are extensions of the coast; generally composed of more resistant rocks. Incoming wave energy focuses on and bends around these headlands.

Herbivore (19) The primary consumer in a food chain, which eats plant material formed by a producer that has synthesized organic molecules.

Heterosphere (3) A zone of the atmosphere above the mesopause, 80 km (50 mi) in altitude; composed of rarified layers of oxygen atoms and nitrogen molecules; includes the ionosphere, which functions to absorb short wavelengths of radiation.

Heterotrophs (19) Organisms that depend on autotrophs for their source of carbon since they do not fix (capture) their own; also known as consumers.

Histosols (18) A soil order in the Soil Taxonomy classification that is formed from thick accumulations of organic matter, such as beds of former lakes, bogs, and layers of peat.

Homosphere (3) A zone of the atmosphere from the surface up to 80 km (50 mi), composed of an even mixture of gases including nitrogen, oxygen, argon, carbon dioxide, and trace gases.

Horn (17) A pyramidal, sharp-pointed peak that results when several cirque glaciers gouge an individual mountain summit from all sides.

Horse latitudes (6) The calms of Cancer and the calms of Capricorn, zones of windless, hot calms beneath the subtropical high-pressure cells in both hemispheres; windless areas of great difficulty for sailing ships throughout history.

Horst (12) Upward-faulted blocks produced by pairs or groups of faults; characterized by the mountain ranges of the interior of the western United States (see graben and Basin and Range Province).

Hot spots (11) Individual points of upwelling material originating in the asthenosphere, not necessarily associated with spreading centers, although Iceland is astride a mid-ocean ridge. They tend to remain fixed relative to migrating plates; some 50 to 100 are identified worldwide; exemplified by Yellowstone National Park and Hawaii.

Human-Earth relationships (1) One of the oldest themes of geography (the human-land tradition); includes settlement patterns, resource utilization and exploitation, hazard perception and planning, and the impact of environmental modification and artificial landscape creation.

Humidity (7) Water vapor content of the air. The capacity of the air to hold water vapor is mostly a function of water vapor and air temperature.

Humus (18) A mixture of organic debris in the soil, worked by consumers and decomposers in the humification process; characteristically formed from plant and animal litter laid down at the surface.

Hurricane (8) A tropical cyclone that is fully organized and intensified in inward-spiraling rainbands; ranges from 160 to 960 km (100 to 600 mi) in diameter, with wind speeds in excess of 119 kmph (65 knots, or 74 mph); a label used specifically in the Atlantic and eastern Pacific.

Hydration (13) A process involving water, although not involving any chemical change. This is actually a physical weathering process; water is added to a mineral, which initiates swelling and stress within the rock, mechanically forcing grains apart as the constituents expand.

Hydraulic action (14) The erosive work accomplished by the turbulence of water; causes a squeezing and releasing action in joints in bedrock; capable of prying and lifting rocks.

Hydrograph (14) A graph of stream discharge (in cms or cfs) over a period of time (minutes, hours, days, years) at a specific place on a stream. The relationship between stream discharge and precipitation input is illustrated on the graph.

Hydrologic cycle (7) A simplified model of the flow of water and water vapor from place to place as energy activates system operations. Water flows through the atmosphere, across the land where it is also stored as ice, and within groundwater.

Hydrolysis (13) When minerals chemically combine with water, a decomposition process that causes silicate minerals in rocks to break down and become altered.

Hydrosphere (1) An abiotic open system that includes all of Earth's waters.

Hydrothermal activity (11) Igneous activity that involves heated or superheated water, such as occurs with geothermal activity. Deposition can occur through chemical reactions between the minerals and the oxygen in heated groundwater.

Hygroscopic nuclei (7) Condensation nuclei that are particularly attracted to moisture, such as certain salts derived from ocean spray.

Hygroscopic water (9) That portion of soil moisture that is so tightly bound to each soil particle that it is unavailable to plant roots; the water, along with some bound capillary water, that is left in the soil after the wilting point is reached.

Ice age (17) A cold episode, with accompanying alpine and continental ice accumulations, that has repeated roughly every 200 to 300 million years since the late Precambrian era (1.25 billion years ago); includes the most recent episode of cold during the Pleistocene Ice Age, which began 1.65 million years ago.

Ice cap (17) A dome-shaped form of a glacier, less extensive than an ice sheet (<50,000 km^2), although it buries mountain peaks and the local landscape. Several representative ice caps are located in Iceland.

Ice-crystal process (7) The feeding of ice crystals on supercooled cloud droplets in the process of raindrop formation. Since the saturation vapor pressure near an ice surface is lower than that near a water surface, supercooled water droplets evaporate near ice crystals, which then absorb the vapor.

Ice field (17) An extensive form of land ice, with mountain ridges and peaks visible above the ice.

Icelandic low (6) See subpolar low-pressure cell.

Ice sheet (17) An enormous continuous continental glacier. The bulk of glacial ice on Earth covers Antarctica and Greenland in two ice sheets.

Ice shelves (17) The most advanced margins of an ice sheet entering a bay along a coast. The farthest edge is marked by a sharp cliff up to 30 m (98 ft) in height above the sea; tabular islands are formed when sections of the shelf break off and move out to sea.

Ice wedge (17) Formed when water enters a thermal contraction crack in permafrost and freezes. The tapered crack is wider at the top, narrowing toward the bottom. Repeated seasonal freezing and melting progressively expand the wedge.

Icebergs (17) Floating ice created by calving ice or large pieces breaking off and floating adrift; a hazard to shipping because the bulk of the ice is submerged and can be irregular in form.

Igneous rocks (11) Rocks that solidify and crystallize from a hot molten state.

Illuviation (18) A depositional soil process as differentiated from eluviation; usually in the B horizon, where accumulations of clays, aluminum, iron, and some humus occur.

Inceptisols (18) An order in the Soil Taxonomy classification; weakly developed soils that are inherently infertile; usually young soils that are weakly developed, although they are more developed than Entisols.

Industrial smog (3) Air pollution associated with coal-burning industries; sulfur oxides, particulates, carbon dioxide, and exotics may comprise it.

Infiltration (7) Water access to subsurface regions of soil moisture storage through penetration of the soil surface.

Inselberg (15) Stark erosional remnants that stand above the surrounding terrain as knobs, hills, or "island mountains" as exemplified by Uluru (Ayers) Rock in Australia.

Insolation (2) Solar radiation that is intercepted by Earth.

Interception (7) Delays the fall of precipitation toward Earth's surface; caused by vegetation or other ground cover.

Internal drainage (9) Outflow of rivers through evaporation or subsurface gravitational flow in regions where rivers do not flow into the ocean. Portions of Africa, Asia, Australia, and the western United States have such drainage.

International Date Line (1) The 180° meridian; an important corollary to the prime meridian on the opposite side of the planet; established by the treaty of 1884 to mark the place where each day officially begins.

Intertropical convergence zone (ITCZ) (6) See equatorial low-pressure trough.

Ionosphere (3) A layer in the atmosphere above 80 km (50 mi) where gamma, X-ray, and some ultraviolet radiation is absorbed and converted into infrared, or heat energy, and where the solar wind stimulates the auroras.

Isobaric surface (6) An elevated surface on which all points have the same pressure, based on a datum plane of equal elevation above sea level, usually the 500 mb level. Along this constant-pressure surface isobars mark the paths of upper air winds.

Isobars (6) An isoline connecting all points of equal pressure.

Isostasy (11) A state of equilibrium formed by the interplay between portions of the lithosphere and the asthenosphere; the crust depresses with weight and recovers with the melting of the ice or removal of the load, in an uplift known as isostatic rebound.

Isotherm (5) An isoline connecting all points of equal temperature.

Jet stream (6) The most prominent movement in upper-level westerly wind flows; irregular, concentrated sinuous bands of geostrophic wind, traveling at 300 kmph (190 mph) that occur in both hemispheres.

Joints (13) Fractures or separations in rock without displacement of the sides of usually plane (flat) surfaces that greatly increase the surface area of rock exposed to a weathering processes.

Kame (17) A small hill of poorly sorted sand and gravel that accumulates in crevasses or in ice-caused indentations in the surface; may also be deposited by water in ice-contact deposits.

Karst topography (13) Distinctive topography formed in a region of limestone with poorly developed surface drainage. Weathering creates solution features that appear pitted and bumpy; originally named after the Krs Plateau of Yugoslavia.

Katabatic winds (6) Air drainage from elevated regions, flowing as gravity winds. Layers of air at the surface cool, become denser, and flow downslope, known worldwide by many local names.

Kettle (17) Forms when an isolated block of ice persists in a ground moraine, an outwash plain, or valley floor after a glacier retreats; as the block finally melts, it leaves behind a steep-sided hole that may frequently fill with water.

Kinetic energy (3) The energy of motion in a body; derived from the vibration of the body's own movement and stated as temperature.

Lacustrine deposit (17) Deposit associated with lake level fluctuations; for example, benches or terraces marking former shorelines.

Lagoon (16) A portion of coastal seawater that is virtually cut off from the ocean by a bay barrier or barrier beach; also, the water surrounded and enclosed by an atoll.

Landfall (8) The location along a coast where a storm moves onshore.

Landslide (13) A sudden rapid downslope movement of a cohesive mass of regolith and/or bedrock in a variety of mass-movement forms under the influence of gravity.

Land-water heating differences (5) The differences in the way land and water heat, as a result of contrasts in transmission, evaporation, mixing, and specific heat capacities. Land surfaces heat and cool faster than water and are characterized as having aspects of continentality, whereas water is regarded as producing a marine influence.

Large marine ecosystem (LME) (20) A proposed designation of regions of the oceanic ecosystem that have unique organisms, floor topography, currents, areas of nutrient-rich upwelling, or significant predation, either human or natural.

Latent heat (7) Heat energy that is absorbed in the phase change of water and is stored in one of the three states—ice, water, or water vapor; includes the latent heat of melting, freezing, vaporization, evaporation, and condensation.

Latent heat of condensation (7) The heat energy released in a phase change from water vapor to liquid; under normal sea-level pressure, 540 calories is released from 1 gram of water vapor changing phase to water; 585 calories is released from 1 gram of water vapor that condenses at 20°C (68°F).

Latent heat of vaporization (7) The heat energy required to change phase from liquid to water vapor at boiling; under normal sea-level pressure, 540 calories must be added to 1 gram of boiling water to achieve a phase change to water vapor.

Lateral moriane (17) Debris transported by a glacier that accumulates along the sides of the glacier.

Laterization (18) A pedogenic process operating in well-drained soils that are found in warm and humid regions; typical of Oxisols. High precipitation values leach soluble minerals and soil constituents, leaving behind iron and aluminum sesquioxides and usually reddish or yellowish colors.

Latitude (1) The angular distance measured north or south of the equator from a point at the center of Earth. A line connecting all points of the same latitudinal angle is called a parallel.

Lava (11, 12) Magma that issues from volcanic activity onto the surface; the extrusive rock that results when magma solidifies.

Leeward slope (8) The slope opposite the windward slope; usually drier in the rain shadow.

Life zone (19) An altitudinal zonation of plants and animals that form distinctive communities. Each life zone possesses its own temperature and precipitation relationships.

Lightning (8) Flashes of light, caused by tens of millions of volts of electrical charge igniting the air to temperatures of 15,000°C to 30,000°C.

Limestone (11) The most common chemical sedimentary rock (nonclastic); lithified calcium carbonate ($CaCO_3$), which is very susceptible to chemical weathering.

Limiting factor (19) The physical or chemical factor that most inhibits (either through lack or excess) biotic processes.

Lithification (11) The compaction, cementation, and hardening of sediments into sedimentary rock.

Lithosphere (1) Earth's crust and that portion of the upper mantle directly below the crust that extends down to 70 km (45 mi). Some use this term to refer to the entire Earth.

Littoral zone (16) A specific coastal environment; that region between the high water line during a storm and a depth at which storm waves are unable to move sea-floor sediments.

Longshore current (16) A current that forms parallel to a beach as waves arrive at an angle to the shore; generated in the surf zone by wave action, transporting large amounts of sand and sediment (also known as littoral current).

Living atmosphere (2) Earth's atmosphere between 0.6 and 3.3 billion years ago; thought by scientists to be principally composed of carbon dioxide, water vapor, and nitrogen gas (3.0 billion years ago); beginnings of photosynthesis.

Loam (18) A mixture of sand, silt, and clay in almost equal proportions, with no one texture dominant.

Location (1) A basic theme of geography dealing with the absolute and relative position of people, places, and things on Earth's surface; tells "where" something is.

Loess (15) Large quantities of fine-grained clays and silts left as glacial outwash deposits; subsequently blown by the wind great distances and redeposited as a generally unstratified, homogeneous blanket of material covering existing landscapes; in China loess is derived from desert lands.

Longitude (1) The angular distance measured east or west of a prime meridian from a point at the center of Earth. A line connecting all points of the same longitude is called a meridian.

Luvisolic (18) A soil order in the Canadian System of Soil Classification formed under mixed deciduous-coniferous forests by eluviation-illuviation processes that produce a light-colored Ae horizon and a diagnostic Bt horizon. Major area of occurrence is the St. Lawrence lowland.

Lysimeter (9) A device for measuring potential and actual evapotranspiration; isolates a portion of a field so that the moisture moving through the plot is measured.

Magma (11) Molten rock from beneath the surface of Earth; fluid, gaseous, under tremendous pressure, and either intruded into country rock or extruded onto the surface as lava.

Magnetic reversal (11) An important aspect of Earth's magnetic field. With an uneven regularity the field fades to zero, then phases back to full strength, but with the magnetic poles reversed. Reversals have occurred nine times during the past 4 million years.

Magnetosphere (2) Earth's magnetic force field, which is generated by dynamolike motions within the planet's outer core; works to deflect the solar wind.

Mangrove swamps (16) Wetland ecosystems between 30° N or S and the equator tend to form a distinctive community of mangrove plants, owing to the lack of freezing temperatures so that mangrove seedlings can survive.

Mantle (11) An area within the planet representing about 80% of Earth's total volume, with densities increasing with depth and averaging 4.5 g per cm^3; occurs above the core and below the crust; is rich in iron and magnesium oxides and silicates.

Map projection (1) The reduction of a spherical globe onto a flat surface in some orderly and systematic realignment of the latitude and longitude grid.

Marine (5) Descriptive of stations that are dominated by the moderating effects of the ocean and that exhibit a smaller minimum and maximum temperature range than continental stations (see land-water heating differences).

Mass movement (13) All unit movements of materials propelled and controlled by gravity; can range from dry to wet, slow to fast, small to large, and free-falling to gradual or intermittent; sometimes used interchangeably with mass wasting, although the latter term should be applied principally to gravitational movement of nonunified material.

Meandering stream (14) The sinuous, curving pattern common to graded streams, with the outer portion of each curve subjected to the greatest erosive action and the inner portion receiving sediment deposits. Former meanders remain on the landscape as meander scars.

Mean sea level (16) A calculated value based on the average of tidal levels recorded hourly at a given site over a long period of time, which must be at least a full lunar tidal cycle.

Medial moraine (17) Debris transported by a glacier that accumulates down the middle of the glacier when two glaciers merge and their lateral moraines combine.

Mediterranean shrubland (20) A major biome dominated by shrub formations; occurs in Mediterranean dry summer climates and is characterized by sclerophyllous scrub and short, stunted, tough forests.

Mercury barometer (3) A device that measures air pressure with a column of mercury in a tube that is inserted in a vessel of mercury. The surrounding air exerts pressure on the mercury in the vessel (see air pressure).

Meridian (1) See longitude.

Mesocyclone (8) A large rotating circulation initiated within a parent cumulonimbus cloud at the mid-troposphere level; generally produces heavy rain, large hail, blustery winds, and lightning; may lead to tornado activity.

Mesosphere (3) The upper region of the homosphere from 50 to 80 km (30 to 50 mi) above the ground; designated by temperature criteria and very low pressures, ranging from 0.1 to 0.001 mb.

Metamorphic rock (11) Preexisting rock, both igneous and sedimentary, that goes through profound physical and chemical changes under increased pressure and temperature. Constituent mineral structures may exhibit foliated or nonfoliated textures.

Meteorology (8) The scientific study of the atmosphere that includes a study of the atmosphere's physical characteristics and motions, related chemical, physical, and geological processes, the complex linkages of atmospheric systems, and weather forecasting.

Methane (5) A radiatively active gas that participates in the greenhouse effect; derived from the organic processes of burning, digesting, and rotting in the presence of oxygen; CH_4.

Microclimatology (4) The study of climates at or near Earth's surface.

Midlatitude geographic zone (1) A generalized geographic zone, extending roughly from 35° to 55° N and S latitudes, although the limits of this zone are somewhat arbitrary, given the distribution of land and water and differences in climate; home of a majority of the world's population.

Midlatitude broadleaf and mixed forest (20) A biome in moist continental climates in areas of warm-to-hot summers and cool-to-cold winters; includes several distinct communities. Relatively lush stands of broadleaf forests trend northward into needleleaf evergreen stands.

Midlatitude grasslands (20) The major biome most modified by human activity; so named because of the predominance of grasslike plants, although deciduous broadleafs appear along streams and other limited sites; location of the world's breadbaskets of grain and livestock production.

Mid-ocean ridges (11) Submarine mountain ranges that extend more than 65,000 km (40,000 mi) worldwide and average more than 1000 km (620 mi) in width; centered along sea-floor spreading centers as the direct result of upwelling areas of heat in the upper mantle.

Milky Way Galaxy (2) A flattened, disk-shaped mass estimated to contain up to 400 billion stars; about 100,000 light years in diameter and includes our Solar System.

Mineral (11) An element or combination of elements that form an inorganic natural compound; described by a specific formula and qualities of a specific nature.

Mixing zone (7) The surface layer of the ocean, representing only 2% of the oceanic volume; Sun-warmed and wind-driven; mixes rapidly compared to lower depths.

Model (1) A simplified version of a system, representing an idealized part of the real world.

Modern atmosphere (2) The fourth distinct atmosphere on Earth, having evolved from the living atmosphere about 0.6 billion years ago; lower layers composed of nitrogen, oxygen, argon, carbon dioxide, and trace gases; an additive mixture of gases.

Mohorovičić discontinuity (11) The boundary between the crust and the rest of the lithospheric upper mantle; named for the Yugoslavian seismologist Mohorovičić; a zone of sharp material and density contrasts.

Moist adiabatic rate (MAR) (8) The rate at which a parcel of saturated air cools in ascent; a rate of 6C° per 1000 m (3.3F° per 1000 ft). This rate may vary, with moisture content and temperature, from 4C° to 10C° per 1000 m (2F° to 6F° per 1000 ft) (see adiabatic).

Moisture droplets (7) Initial composition of clouds. Each droplet measures approximately 0.002 cm (0.0008 in.) in diameter and is invisible to the human eye.

Mollisols (18) A soil order in the Soil Taxonomy classification that has a mollic epipedon and a humus-rich organic content high in alkalinity; some of the world's most significant agricultural soils.

Monsoon (6) From the Arabic word *mausim,* meaning "season"; refers to an annual cycle of dryness and wetness, with seasonally shifting winds produced by changing atmospheric pressure systems; affects India, Southeast Asia, Indonesia, northern Australia, and portions of Africa.

Montane forests (20) Needleleaf forests associated with mountain elevations (see northern needleleaf forest).

Moraine (17) Marginal glacial deposits of unsorted and unstratified material, producing a variety of depositional landforms.

Movement (1) A major theme in geography involving migration, communication, and the interaction of people and processes across space.

Natural levees (14) Long, low ridges that occur on either side of a river in a developed floodplain; sedimentary (coarse gravels and sand) depositional by-products of river-flooding episodes.

Neap tides (16) Unusually low tidal ranges produced during the first and third quarters of the Moon, with an offsetting pull from the Sun; as opposed to spring tides.

Nebula (2) A cloud of dust and gas in space.

Negative feedback (1) A feedback loop that tends to slow or dampen response in a system; promotes self-regulation in a system; far more common than positive feedback in living systems.

Net primary productivity (19) The net photosynthesis (photosynthesis minus respiration) for a given community; considers all growth and all reduction factors that affect the amount of useful chemical energy (biomass) fixed in an ecosystem.

Net radiation (NET R) (4) The net all-wave radiation available; the final outcome of the radiation balance process between incoming and outgoing shortwave and longwave energy.

Niche (19, 20) The basic function, or occupation, of a lifeform within a given community; the way an organism obtains its food, air, and water.

Nickpoint (knickpoint) (14) The point at which the longitudinal profile of a stream is abruptly broken by a change in gradient; for example, a waterfall, rapids, or cascade.

Nimbostratus (7) A rain-producing, dark, grayish, stratiform cloud characterized by gentle drizzles.

Nitrogen dioxide (3) A reddish-brown choking gas produced in high-temperature combustion engines; can be damaging to human respiratory tracts and to plants; participates in photochemical reactions and acid deposition.

Noctilucent clouds (3) Rare and unusual shining bands of ice, which may be seen glowing at high latitudes long after sunset; formed within the mesosphere, where cosmic and meteoric dust acts as nuclei for the formation of ice crystals.

Normal lapse rate (3) The average rate of temperature decrease with increasing altitude in the lower atmosphere; an average value of 6.4C° per km, or 1000 m (3.5F° per 1000 ft); a rate that exists principally during daytime conditions.

Northern needleleaf forest (20) Forests of pine, spruce, fir, and larch, stretching from the east coast of Canada westward to Alaska and continuing from Siberia westward across the entire extent of Russia to the European Plain; called the taiga (a Russian word) or the boreal forest; principally in the D climates. Includes montane forests.

Nuclear winter hypothesis (5) An increase in Earth's albedo and upper atmospheric heating resulting in surface cooling; associated with the detonation of a relatively small number of nuclear warheads within the biosphere. Now encompasses a whole range of ecological, biological, and climatic impacts.

Nuée ardente (12) Incandescent ash and super-hot gases that often accompany explosive eruptions of composite volcanoes; also known as a glowing cloud.

Occluded front (8) In a cyclonic circulation, the overrunning of a surface warm front by a cold front and the subsequent lifting of the warm air wedge off the ground. Initial precipitation is moderate to heavy.

Ocean basins (12) The physical containers for Earth's oceans.

Oceanic trenches (11) The deepest single features of Earth's crust; associated with subduction zones. The deepest is the Mariana Trench near Guam, which descends to 11,033 m (36,198 ft).

Omnivore (19) A consumer that feeds on both producers (plants) and consumers (meat)—a role occupied by humans, among other animals.

Open system (1) A system with inputs and outputs crossing back and forth between the system and the surrounding environment. Earth is an open system in terms of energy.

Organic (18) A soil order in the Canadian System of Soil Classification formed as peat, bog, and muck soils, largely composed of organic material and water saturated for prolonged periods; usually in poorly to very poorly drained depressions.

Orogenesis (12) The process of mountain building that occurs when large-scale compression leads to deformation and uplift of the crust; literally the birth of mountains.

Orographic lifting (8) The uplift of migrating air masses in response to the physical presence of a mountain, which acts as a topographic barrier. The lifted air cools adiabatically as it moves upslope; may form clouds and produce increased precipitation.

Outgassing (7) The release of trapped gases from rocks, forced out through cracks, fissures, and volcanoes from within Earth; the terrestrial source of Earth's water.

Outlet glacier (17) A type of alpine glacier that flows outward from a continental glacier and appears as a valley glacier in form.

Outwash plain (17) Glaciofluvial deposits of stratified drift from meltwater-fed, braided, and overloaded streams with debris beyond a glacier's morainal deposits.

Overland flow (9) Surplus water that flows across a surface toward stream channels. Together with precipitation and subsurface flows, it constitutes the total runoff from an area.

Oxbow lake (14) A lake that was formerly part of the channel of a meandering stream; isolated when a stream eroded its outer bank forming a cutoff through the neck of a looping meander.

Oxidation (13) A chemical weathering process whereby oxygen dissolved in water oxidizes (combines with) certain metallic elements to form oxides; most familiar as the "rusting" of iron in a rock or soil that produces a reddish-brown stain of iron oxide (Fe_2O_3).

Oxisols (18) In the Soil Taxonomy classification a soil order of tropical soils that are old, deeply developed, and lacking in horizons wherever well-drained; heavily weathered, low in cation exchange capacity, and low in fertility.

Ozone layer (3) See ozonosphere.

Ozonosphere (3) A layer of ozone (O_3) occupying the full extent of the stratosphere (20 to 50 km or 12 to 30 mi) above the surface; the region of the atmosphere where ultraviolet wavelengths are principally absorbed and converted into heat.

Pacific high (6) Also known as the Hawaiian high; a high-pressure cell that dominates the Pacific in July, retreating southward in the Northern Hemisphere in January.

Paleoclimatology (10) The science that studies the climates of past ages.

Paleoseismology (12) The science that studies the frequency or lack of frequency along specific faults, location, and magnitude of past earthquakes.

Palsa (17) A rounded or elliptical mound of peat that contains thin perennial ice lenses rather than an ice core as in a pingo. Palsas can be 2 to 30 m wide by 1 to 10 m high and usually are covered by soil or vegetation over a cracked surface.

PAN (3) See peroxyacetyl nitrates.

Pangaea (11) The supercontinent formed by the collision of all continental masses approximately 225 million years ago; named by Wegener in 1912 in his continental drift theory.

Parallel (1) See latitude.

Parent material (13) The unconsolidated material, from both organic and mineral sources, that is the basis of soil development.

Paternoster lakes (17) Small, circular, stair-stepped lakes in a series, named because they look like a string of rosary (religious) beads, forming in individual rock basins aligned down the course of a glaciated valley.

Patterned ground (17) Areas in the periglacial environment where freezing and thawing of the ground create polygonal forms of arranged rocks at the surface that may coalesce into stone polygon nets.

Pediment (15) The area created when a mountain front retreats through erosional action in an arid landscape; an inclined bedrock plain layered with a thin veneer, or coating, of alluvium; different from the depositional surface of a bajada.

Pedogenic regimes (18) Specific soil-forming processes keyed to climatic regimes: laterization, calcification, salinization, and podzolization, among others.

Pedon (18) A soil profile extending from the surface to the lowest extent of plant roots or to the depth where regolith or bedrock is encountered; imagined as a hexagonal column encompassing from 1 to 10 square meters in surface area forming the basic soil sampling unit.

Peneplain (14) A final old erosional surface in Davis's geomorphic cycle; a very controversial concept and one that is unproven.

Percolation (7) The process by which water permeates through to the subsurface environment; vertical water movement through soil or porous rock.

Periglacial (17) Cold-climate processes, landforms, and topographic features along the margins of glaciers, past and present, that occupy over 20% of Earth's land surface. The areas are either near permanent ice or are at high elevation, and have ground that is seasonally snow free. Under these conditions, a unique set of periglacial processes operate, including permafrost, frost action, and ground ice.

Perihelion (2) That point in Earth's elliptical orbit about the Sun where Earth is closest to the Sun; occurs on January 3 at 147,255,000 km (91,500,000 mi); variable over a 100,000-year cycle.

Permafrost (17) Forms when soil or rock temperatures remain below 0°C (32°F) for at least two years in areas considered periglacial. Note that this criterion is based solely on temperature and not on whether water is present.

Permeability (9) The ability of water to flow through soil or rock; a function of the texture and structure of the medium.

Peroxyacetyl nitrates (PAN) (3) A whole family of pollutants formed from photochemical reactions involving nitric oxide (NO) and hydrocarbons (HC). They produce no known human health effects but are particularly damaging to plants.

Phase change (7) The change in phase, or state, between ice, water, and water vapor; involves the absorption or release of latent heat.

Photochemical smog (3) Air pollution produced by the interaction of ultraviolet light, nitrogen dioxide, and hydrocarbons; produces ozone and PAN through a series of complex photochemical reactions. Automobiles are the major source of the contributive gases.

Photogrammetry (2) An aspect of remote sensing that uses aerial photographs and orbital imagery to improve the accuracy of surface maps, usually at less expense and with greater ease than is possible with on-site surveys.

Photoperiod (19) The duration of daylight experienced at a given location.

Photosynthesis (19) The joining of carbon dioxide and oxygen in plants, under the influence of certain wavelengths of visible light; releases oxygen and produces energy-rich organic material (sugars and starches).

Physical geography (1) A science that studies the spatial aspects of the physical elements and processes that make up the environment: energy, air, water, weather, climate, landforms, soils, animals, plants, and Earth.

Physical weathering (13) The breaking and disintegrating of rock without any chemical alteration; sometimes referred to as mechanical or fragmentation weathering.

Piezometric surface (9) A pressure level in an aquifer, creating an imaginary line above the surface in some instances and defined by the level to which water rises in wells; caused by the fact that the water in a confined aquifer is under the pressure of its own weight.

Pingo (17) Large areas of frozen ground (soil-covered ice) that develop a heaved up, circular, ice core mound, rising above a periglacial landscape as freezing water and ice expand; sometimes as a result of pressure developed by freezing artesian water injected into permafrost; occasionally exceed 60 m height (200 ft).

Pioneer community (19) The initial community in an area; usually found on new surfaces or those that have been stripped of life—for example, surfaces created by mass movements of land, or land exposed by a retreating glacier, or cooled lava flows, or land disturbed by human activities.

Place (1) A major theme in geography focused on the tangible and intangible characteristics that make each location unique.

Plane of the ecliptic (2) A plane intersecting all the points of Earth's orbit.

Planetesimal hypothesis (2) The idea that early protoplanets formed from the condensing masses of a nebular cloud of dust, gas, and icy comets; a formation process now being observed in other parts of the galaxy.

Planimetric map (1) A basic map showing the horizontal position of boundaries, land-use activities, and political, economic, and social outlines.

Plate tectonics (11) The conceptual model that encompasses continental drift, sea-floor spreading, and related aspects of crustal movement; widely accepted as the foundation of crustal tectonic processes.

Plateau basalts (12) An accumulation of horizontal flows formed when lava spreads out from elongated fissures onto the surface in extensive sheets; associated with effusive eruptions; also known as flood basalts—for example, the Columbia Plateau.

Playa (15) An area of salt crust left behind on a desert floor usually in the middle of a bolson or valley; intermittently wet and dry, as runoff evaporates in a desert climate.

Plinthite (18) An ironstone hardpan structure formed in Oxisol subsurfaces or surface horizons that are exposed to repeated wetting and drying sequences; quarried in some areas for building materials.

Pluton (11) A mass of intrusive igneous rock that has cooled slowly in the crust; forms in any size or shape. The largest partially exposed pluton is a batholith.

Pluvial (17) Referring to a period in the past, principally during the Pleistocene, when there was greater moisture availability or efficiency; resulted in increased lake levels in arid and seasonally arid regions.

Podzolic (18) A soil order in the Canadian System of Soil Classification formed under coniferous forests and heath where leaching of overlying horizons occurs in moist cool to cold climates. Iron, aluminum, and organic matter from L, F, and H horizons are redeposited in a podzolic B horizon. Dominant in western British Columbia, Ontario, and Quebec.

Podzolization (18) A pedogenic process in cool, moist climates; forms a highly leached soil with strong surface acidity because of humus from acid-rich trees.

Point bar (14) The inner portion of a meander which receives sediment fill.

Polar easterlies (6) Variable weak, cold, and dry winds moving away from the polar region; an result of anticyclonic circulation.

Polar front (6) A significant zone of contrast between cold and warm air masses; roughly situated between 50° and 60° latitude.

Polar high-pressure cells (6) A weak, anticyclonic, thermally produced pressure system positioned roughly above each pole; is the region of the lowest temperatures on Earth.

Polar jet stream (6) A strong wind current located at the tropopause along the polar front; occurs between 7600 and 10,700 m (24,900 and 35,100 ft) in altitude and meanders between 30° and 70° N latitude (see jet stream).

Polypedon (18) The identifiable soil in an area, with distinctive characteristics differentiating it from surrounding polypedons forming the basic mapping unit; composed of many pedons.

Porosity (9) The total volume of available pore space in soil; expressed as a ratio and percentage; a result of the texture and structure of the soil.

Positive feedback (1) Feedback that amplifies or encourages responses in the system.

Potential evapotranspiration (9) The amount of moisture that would evaporate and transpire if adequate moisture were available; the amount lost under optimum moisture conditions—that is, the moisture demand (POTET).

Precipitation (9) Rain, snow, sleet, and hail—the moisture supply (PRECIP).

Pressure gradient force (6) Causes air to move from areas of higher barometric pressure to areas of lower barometric pressure due to pressure differences.

Prime meridian (1) An arbitrary meridian designated as 0° longitude; the point from which longitude is measured east or west; agreed on by the nations of the world in an 1884 treaty.

Primordial atmosphere (2) Earth's first atmosphere, derived from the formation nebula. Lighter gases escaped early, with water vapor, hydrogen, cyanide, ammonia, methane, and others predominant.

Process (1) A set of actions and changes that occur in some special order; central to modern geographic synthesis.

Producers (19) Organisms that are capable of using carbon dioxide as their sole source of carbon, which they chemically fix through photosynthesis to provide their own nourishment; also called *autotrophs.*

Province (15) A large region that shares several homogeneous traits, making it appropriate to assign a regional name.

Pyroclastics (12) Explosively ejected rock pieces and broken fragments launched aerially by a volcanic eruption.

Radiation fog (7) Formed by radiative cooling of the surface, especially on clear nights in areas of moist ground; occurs when the air layer directly above the surface is chilled to the dew-point temperature, thereby producing saturated conditions.

Radiatively active gases (5) Gases in the atmosphere such as carbon dioxide, methane, and water vapor, that absorb and radiate infrared wavelengths.

Rain gauge (9) A standardized device that catches and measures rainfall.

Rain shadow (8) The area on the leeward slopes of a mountain range; in the shadow of the mountains, where precipitation receipt is greatly reduced compared to windward slopes.

Reflection (4) The portion of arriving energy that returns directly back into space without being converted into heat or performing any work (see albedo).

Refraction (4) The bending effect that occurs when insolation enters the atmosphere or another medium; the same process by which a crystal, or prism, disperses the component colors of the light passing through it.

Region (1) A geographic theme that focuses on areas that display unity and internal homogeneity of traits; includes the study of how a region forms, evolves, and interrelates with other regions.

Regolith (13) Partially weathered rock overlying bedrock, whether residual or transported.

Regosolic (18) A soil order in the Canadian System of Soil Classification formed as the result of any number of factors: young materials, fresh alluvial deposits, material instability, mass-wasted slopes, dry, cold climatic conditions. Weakly developed limited soils that lack solonetzic, illuvial, or podzolic B horizons; buried horizons possible. *Note:* Dominant in Northwest Territories and northern Yukon, although now designated as Cryosols under CSSC.

Relative humidity (7) A term that reflects the ratio of water vapor actually in the air (content) compared to the maximum water vapor the air is able to be hold (capacity) by at that temperature; expressed as a percentage.

Relief (12) Elevation differences in a local landscape as an expression of the unevenness, height, and slope variation.

Remote sensing (2) Information acquired from a distance, without physical contact with the subject; for example, photography or orbital imagery.

Respiration (19) The process other than photosynthesis by which plants derive energy for their operations; essentially, the reverse of the photosynthetic process; releases carbon dioxide, water, and heat into the environment.

Revolution (2) The annual orbital movement of Earth about the Sun; determines year length and the length of seasons.

Rhumb line (1) A line of constant compass direction, or constant bearing, which crosses all meridians at the same angle. A portion of a great circle.

Richter scale (12) An open-ended, logarithmic scale that estimates earthquake magnitude; designed by Charles Richter in 1935.

Ring of fire (12) See circum-Pacific belt.

Ripples (15) The smallest features that form in response to the distance traveled (eolian or water agents) by individual saltating grains.

Roche moutonnée (17) An asymmetrical hill of exposed bedrock; displays a gently sloping upstream side that has been smoothed and polished by a glacier and an abrupt, steep downstream side.

Rock cycle (11) A model representing the interrelationships among the three rock-forming processes: igneous, sedimentary, and metamorphic.

Rockfall (13) Free-falling movement of debris from a cliff or steep slope, generally falling straight down or bounding downslope.

Rossby waves (6) Undulating horizontal motions in the upper-air westerly circulation at middle and high latitudes.

Rotation (2) The turning of Earth on its axis; averages 24 hours in duration; determines day-night relationships.

Salinity (7) The concentration of natural elements and compounds dissolved in solution, as solutes; measured by weight in parts per thousand (‰) in seawater.

Salinization (18) A pedogenic regime that results from high potential evapotranspiration rates in deserts and semiarid regions. Soil water is drawn to surface horizons, and dissolved salts are deposited as the water evaporates.

Saltation (14) The transport of sand grains (usually larger than 0.2 mm, or 0.008 in.) by stream or wind, bouncing the grains along the ground in asymmetrical paths.

Salt marshes (16) Wetland ecosystems characteristic of latitudes poleward of the 30th parallel.

Sand sea (15) An extensive area of sand and dunes; characteristic of Earth's erg deserts.

Saturated (7) Air that is holding all the water vapor that it can hold at a given temperature.

Scale (1) The ratio of the distance on a map to that in the real world; expressed as a representative fraction, graphic scale, or written scale.

Scarification (13) A term that is generally applied to human-induced mass movements of Earth materials, such as large-scale open-pit mining and strip mining.

Scattering (4) Deflection and redirection of insolation by atmospheric gases, dust, ice, and water vapor. The amount of scattering is a function of the wavelengths involved and the scattering media; the shorter the wavelength, the greater the scattering.

Scientific method (2) An approach that uses applied common sense in an organized and objective manner; based on observation, generalization, formulation of a hypothesis, and ultimately the development of a theory.

Sea-floor spreading (11) As proposed by Hess and Dietz, the mechanism driving the movement of the continents; associated with upwelling flows of magma along the worldwide system of mid-ocean ridges.

Secondary air mass (8) An air mass that becomes modified in terms of temperature and moisture characteristics with distance and time away from its source region.

Sediment (13) Fine-grained mineral matter that is transported by air, water, or ice.

Sedimentary rocks (11) One of three basic rock types; formed from the compaction, cementation, and hardening of sediments derived from former rocks; principally quartz, feldspar, and clay minerals.

Seismic gaps (12) Areas along fault zones that form a gap in earthquake occurrence and may therefore possess the greatest accumulated strain (see paleoseismology).

Seismic waves (11) The shock waves sent through the planet by an earthquake or underground nuclear test. Transmission varies according to temperature and the density of various layers within the planet.

Seismograph (12) A device that measures seismic waves of energy transmitted throughout Earth's interior.

Sensible heat (3) Heat that can be measured with a thermometer; a measure of the concentration of kinetic energy from molecular motion.

Sheet flow (14) Water that moves downslope in a thin film as overland flow, not concentrated in channels larger than rills.

Sheeting (13) A form of weathering associated with fracturing or fragmentation of rock by pressure release; often related to exfoliation processes.

Shield volcano (12) A symmetrical mountain landform built from effusive eruptions; gently sloped, gradually rising from the surrounding landscape to a summit crater; typical of the Hawaiian Islands.

Sinkholes (13) Nearly circular depressions created by the weathering of limestone karst landscapes; also known as dolines; may collapse through the roof of an underground space.

Sling psychrometer (7) A device for the measurement of relative humidity using two thermometers—a dry bulb and a wet bulb—mounted side by side.

Slipface (15) Formed as dune height increases above 30 cm (12 in.) on the leeward side at an angle at which loose material is stable—its angle of repose—usually 30° to 34°.

Slopes (13) Curved, inclined surfaces that bound landforms.

Small circles (1) Circles on a globe's surface that do not share Earth's center; for example, all parallels other than the equator.

Snowline (17) A temporary line marking the elevation where winter snowfall persists throughout the summer; seasonally, the lowest elevation covered by snow during the summer.

Soil (18) A dynamic natural body made up of fine materials covering Earth's surface in which plants grow; composed of both mineral and organic matter.

Soil colloids (18) Tiny clay and organic particles in soil; provide chemically active sites for mineral ion adsorption.

Soil creep (13) A persistent mass movement of surface soil where individual soil particles are lifted and disturbed by the expansion of soil moisture as it freezes, by cycles of moistness and dryness, by diurnal temperature variations, or even by grazing livestock or digging animals.

Soil fertility (18) The ability of soil to support plant productivity when it contains organic substances and clay minerals that absorb water and certain elemental ions needed by plants.

Soil horizon (18) The various layers exposed in a pedon; roughly parallel to the surface and identified as O, A, E, B, and C.

Soil moisture storage (9) The retention of moisture within soil; represents a savings account that can accept deposits (soil moisture recharge) or experiences withdrawals (soil moisture utilization) as conditions change.

Soil Taxonomy (18) A soil classification system based on observable soil properties actually seen in the field; a final version published in 1975 by the U.S. Soil Conservation Service.

Solar constant (2) The amount of insolation intercepted by Earth on a plane surface perpendicular to the Sun's rays when Earth is at its average distance from the Sun; a value of 1.968 calories per cm^2 per minute; averaged over the entire globe at the thermopause; or, 1370 watts per m^2.

Solar wind (2) Clouds of ionized (charged) gases emitted by the Sun and traveling in all directions from the Sun's surface. Effects on Earth include auroras, disturbance of radio signals, and possible influences on weather.

Solifluction (17) Gentle downslope movement of a saturated surface material (soil and regolith), in various climatic regimes, where temperatures are above freezing. Sometimes used in a generic sense to include gelifluction in periglacial environments.

Solonetzic (18) A soil order in the Canadian System of Soil Classification formed in well to imperfectly drained mineral soils developed under grasses in semiarid to subhumid climates and denoting saline or alkaline soils. Limited areas of central and north-central Alberta.

Solum (18) A true soil profile in the pedon; ideally, a combination of A and B horizons.

Solution (13) The dissolved load of a stream; especially the chemical solution derived from limestone, dolomite, or salt deposits.

Spatial analysis (1) The examination of phenomena in the context of spatial interactions, patterns, and variations over area and/or space; the key integrative approach of geography.

Specific heat (5) The increase of temperature in a material when energy is absorbed. Because water requires far more heat to raise its temperature than does a comparable volume of land, water is said to have a higher specific heat.

Specific humidity (7) The mass of water vapor (in grams) per unit mass of air (in kilograms) at any specified temperature. The maximum mass of water vapor that a kilogram of air can hold at any specified temperature is termed its maximum specific humidity.

Specific yield (9) The groundwater that is available for utilization at the surface; expressed as a percentage of the total volume of a saturated aquifer.

Speed of light (2) Specifically, 299,792 km per second (186,282 mi per second), or more than 9.4 trillion km per year (5.9 trillion mi per year)—a distance known as a light-year.

Spheroidal weathering (13) A chemical weathering process in which the sharp edges and corners of boulders and rocks are weathered in thin plates that creates a rounded, spheroidal form.

Spodosols (18) A soil order in the Soil Taxonomy classification that occurs in northern coniferous forests; best developed in cold, moist, forested climates of Dfb, Dfc, Dwc, Dwd; lacks humus and clay in the A horizons, with high acidity associated with podzolization processes.

Spring tides (16) The highest tidal ranges, which occur when the Moon and the Sun are in conjunction (at new Moon) or in opposition (at full Moon) stages.

Squall line (8) A zone slightly ahead of a fast-advancing cold front, where wind patterns are rapidly changing and blustery and precipitation is strong.

Stability (8) The condition of a parcel of air relative to whether it remains where it is or changes its initial position. The parcel is stable if it resists displacement upwards, unstable if it continues to rise; relates to adiabatic and lapse rate processes.

Staff gauge (14) A simple device to measure stream height; uses a pole stuck vertically into a stream, with various elevations marked on its face.

Steady-state equilibrium (1) The condition that occurs in a system when rates of input and output are equal and the amounts of energy and stored matter are nearly constant around a stable average.

Stilling well (14) A device positioned on a stream bank and used to measure stream height.

Stomata (19) Small openings on the undersides of leaves, through which water and gases pass.

Storm surges (8) Large quantities of seawater pushed inland by the strong winds associated with a tropical cyclone.

Stratified drift (17) Sediments deposited by glacial meltwater that appear sorted; a specific form of *glacial drift*.

Stratigraphy (11) An analysis of the sequence, spacing, and spatial distribution of rock strata; offer clues as to the ages and origins of rocks and structures.

Stratocumulus (7) A lumpy, grayish, low-level cloud, patchy with sky visible, sometimes present at the end of the day.

Stratosphere (3) That portion of the homosphere that ranges from 20 to 50 km (12.5 to 30 mi) above Earth's surface, with temperatures ranging from −57°C (−70°F) at the tropopause to 0°C (32°F) at the stratopause. The functional ozonosphere is within the stratosphere.

Stratus (7) A stratiform cloud generally below 2000 m (6500 ft).

Stream orders (14) A way of expressing the linear aspects of channels in a drainage basin; a convenient hierarchy of stream size.

Strike-slip fault (12) Horizontal movement along a faultline, that is, movement in the same direction as the fault; also known as a transcurrent fault. Such movement is described as right-lateral or left-lateral, depending on the relative motion observed.

Subantarctic geographic zone (1) Related to the geographic region between 55° and 66.5° S latitude (Antarctic Circle).

Subarctic geographic zone (1) Related to the geographic region between 55° and 66.5° N latitude (Arctic Circle).

Subduction zone (11) An area where two plates of crust collide and the denser oceanic crust dives beneath the less dense continental plate, forming deep oceanic trenches and seismically active regions.

Sublimation (7) A process in which ice evaporates directly to water vapor or water vapor freezes to ice (deposition).

Subpolar low-pressure cell (6) A region of low pressure centered approximately at 60° latitude in the North Atlantic near Iceland and in the North Pacific near the Aleutians, as well as in the Southern Hemisphere. Air flow is cyclonic, counterclockwise circulation that weakens in summer and strengthens in winter; source area of cyclonic activity affecting Europe and North America.

Subsolar point (2) The only point receiving perpendicular insolation at a given moment—the Sun directly overhead.

Subsurface diagnostic horizon (18) A soil horizon that originates below the epipedon at varying depths; may be part of the A and B horizons; important in soil description as part of the Soil Taxonomy.

Subtropical geographic zone (1) Pertaining to a geographic region that ranges from 23.5° to 35° latitude in both hemispheres; dominated by Earth's arid and semiarid lands.

Subtropical high-pressure cells (6) Dynamic high-pressure areas covering roughly the region from 20° to 35° N and S latitudes; responsible for the hot, dry areas of Earth's arid and semiarid deserts.

Subtropical jet stream (6) A wind current that occurs in the subtropical latitudes, near the boundary between the tropics and the midlatitudes.

Sulfur dioxide (3) A colorless gas detected by its pungent odor; produced by the combustion of fossil fuels that contain sulfur as an impurity; can react in the atmosphere to form sulfuric acid, a component of acid deposition.

Summer (June) solstice (2) The time when the Sun's declination is at the Tropic of Cancer, at 23.5° N latitude; June 20–21 each year.

Sunrise (2) That moment when the disk of the Sun first appears above the horizon.

Sunset (2) That moment when the disk of the Sun totally disappears.

Sunspots (2) Magnetic disturbances on the surface of the Sun; occurring in an average 11-year cycle; related flares, prominences, and outbreaks produce surges in solar wind.

Supercooled droplets (7) Found in clouds at temperatures ranging from −15° to −35°C. Such droplets form ice crystals in colder clouds and provide an essential mechanism of raindrop formation.

Surface creep (15) A form of eolian transport that involves particles too large for saltation; a process whereby individual grains are impacted by moving grains and slide and roll.

Surplus (9) The amount of moisture that exceeds potential evapotranspiration; moisture oversupply when soil moisture storage is at field capacity.

Suspended load (14) Fine particles held in suspension in a stream. The finest particles are not deposited until the stream velocity nears zero.

Swells (16) Regular patterns of smooth, rounded waves in open water; can range from small ripples to very large waves.

Syncline (12) A trough in folded strata, with beds that slope toward the axis of the downfold.

System (1) Any ordered, interrelated set of materials or items existing separate from the environment, or within a boundary; energy transformations and energy and matter storage and retrieval occur within a system.

Taiga (20) See northern needleleaf forest.

Talik (17) An unfrozen portion of the ground that may occur above, below, or within a body of discontinuous permafrost or beneath a body of water in the continuous region, such as a deep lake; may extend to bedrock and noncryotic soil under large deep lakes.

Tarn (17) A small mountain lake, especially one that collects in a cirque basin behind risers of rock material.

Tectonic processes (11) Driven by internal energy from within Earth; refers to large-scale movement and deformation of the crust.

Temperate rain forest (20) A major biome of lush forests at middle and high latitudes; occurs along narrow margins of the Pacific Northwest in North America, among other locations; a mixture of broadleaf and needleleaf trees and thick undergrowth, including the tallest trees in the world.

Temperature (5) A measure of sensible heat energy present in the atmosphere and other media, indicates the average kinetic energy of individual molecules within the atmosphere.

Temperature inversion (3) A reversal of the normal decrease of temperature with increasing altitude; can occur anywhere from ground level up to several thousand meters.

Tephra (12) Pulverized rock and clastic materials ejected violently during a volcanic eruption—all pyroclastics from a volcano.

Terranes (12) Migrating crustal pieces, dragged about by processes of mantle convection and plate tectonics. Displaced terranes are distinct in their history, composition, and structure from the continents that accept them.

Terrestrial ecosystem (20) A self-regulating association characterized by specific plant formations; usually named for the predominant vegetation and known as a biome when large and stable.

Thermal equator (5) A line on an isothermal map that connects all points of highest mean temperature.

Thermocline transition zone (7) A region in the depths of an ocean or a lake where a strong vertical temperature gradient occurs; lacks the motion of the surface.

Thermokarst (17) Topography of hummocky, irregular relief marked by cave-ins, bogs, small depressions, and pits formed as ground ice melts. Not related to solution processes and chemical weathering associated with limestone (karst); rather it refers to thermal subsidence and erosion processes caused by ground ice melting.

Thermopause (2) A zone approximately 480 km (300 mi) in altitude that serves conceptually as the top of the atmosphere; an altitude used for the determination of the solar constant.

Thermosphere (3) A region of the heterosphere extending from 80 to 480 km (50 to 300 mi) in altitude; contains the functional ionosphere layer.

Thunder (8) The violent expansion of suddenly heated air, created by lightning discharges, sending out shock waves as an audible sonic bang.

Tidal bore (16) An incoming tidal current that is constricted as it enters a river channel forming a measurable wave.

Tides (16) Patterns of daily oscillations in sea level produced by astronomical relationships among the Sun, the Moon, and Earth; experienced in varying degrees around the world.

Till (17) Direct ice deposits that appear unstratified and unsorted; a specific form of *glacial drift*.

Till plains (17) Large, relatively flat plains composed of unsorted glacial deposits behind a terminal or end moraine. Low-rolling relief and unclear drainage patterns are characteristic.

Tilted fault block (12) A tilted landscape produced by a normal fault on one side of a range; for example, the Sierra Nevada and the Grand Tetons.

Tombolo (16) A landform created when coastal sand deposits connect the shoreline with an offshore island outcrop or sea stack.

Topographic map (1) A map that portrays physical relief through the use of elevation contour lines that connect all points at the same elevation above or below a vertical datum, such as mean sea level.

Topography (12) The undulations and configurations that give Earth's surface its texture; the heights and depths of local relief including both natural and human-made features.

Tornado (8) An intense, destructive cyclonic rotation, developed in response to extremely low pressure; associated with mesocyclone formation.

Total runoff (9) Surplus water that flows across a surface toward stream channels; formed by overland flow, combined with precipitation and subsurface flows into those channels.

Traction (14) A type of sediment transport that drags coarser materials along the bed of a stream.

Trade winds (6) Northeast and southeast winds that converge in the equatorial low pressure trough, forming the intertropical convergence zone.

Transform fault (11) An elongate zone along which faulting occurs between mid-ocean ridge crests; produces a relative horizontal motion with no new crust formed or destroyed; strike-slip motion, either left or right lateral.

Transmission (4) The passage of shortwave and longwave energy through the atmosphere or water.

Transparency (5) The quality of light able to shine through a medium. Light penetrates the oceans to an average depth of 60 m (200 ft).

Transpiration (9) The movement, as water vapor, out through the pores in leaves of water drawn by plant roots from the soil moisture storage.

Transport (14) The actual movement of weathered and eroded materials by air, water, and ice.

Tropical geographic zone (1) The geographic region ranging between 10° and 23.5° N or S latitude.

Tropical cyclone (8) A cyclonic circulation originating in the tropics, with winds between 30 and 64 knots (39 to 73 mph); characterized by closed isobars, circular organization, and heavy rains (see hurricane and typhoon).

Tropical savanna (20) A major biome containing large expanses of grassland interrupted by trees and shrubs; a transitional area between the humid rain forests and tropical seasonal forests and the drier, semiarid tropical steppes and deserts.

Tropical seasonal forest and scrub (20) A variable biome on the margins of the rain forests, occupying regions of lesser and more erratic rainfall; the site of transitional communities between the rain forests and tropical grasslands.

Tropic of Cancer (2) The northernmost point of the Sun's declination during the year; 23.5° N latitude.

Tropic of Capricorn (2) The southernmost point of the Sun's declination during the year; 23.5° S latitude.

Troposphere (3) The home of the biosphere; the lowest layer of the homosphere, containing approximately 90% of the total mass of the atmosphere; extends up to the tropopause, marked by a temperature of −57°C (−70°F); occurring at an altitude of 18 km (11 mi) at the equator, 13 km (8 mi) in the middle latitudes, and at lower altitudes near the poles.

True shape (1) A map property showing the correct configuration of coastlines; a useful trait of conformality for navigational and aeronautical maps, although areal relationships are distorted.

Tsunami (16) A seismic sea wave, traveling at high speeds across the ocean, formed by sudden and sharp motions in the seafloor, such as sea-floor earthquakes, submarine landslides, or eruptions from undersea volcanoes.

Typhoon (8) A tropical cyclone in excess of 65 knots (74 mph) that occurs in the western Pacific; same as a hurricane except for location.

Ultisols (18) A soil order in the Soil Taxonomy classification, featuring highly weathered forest soils, principally in the Cfa climatic classification. Increased weathering and exposure can degenerate an Alfisol into the reddish color and texture of these more humid to tropical soils. Fertility is quickly exhausted when Ultisols are cultivated.

Unconfined aquifer (9) The zone of saturation in water-bearing rock strata, with no impermeable overburden and recharge generally accomplished by water percolating down from above.

Uniformitarianism (11) An assumption that physical processes active in the environment today are operating at the same pace and intensity that has characterized them throughout geologic time; proposed by Hutton and Lyell.

Unit hydrograph (14) A model hydrograph characterizing a drainage basin that receives a unit depth of effective rainfall over a specific unit of time, uniformly spread over the drainage basin.

Upslope fog (7) Forms when moist air is forced to higher elevations along a hill or mountain and is thus cooled; for example, along the Appalachians and the eastern slopes of the Rockies.

Upwelling current (6) Ocean currents that cause surface waters to be swept away from the coast by surface divergence or offshore winds. Cool, deep waters, which are generally nutrient rich, rise to replace the vacating water.

Urban heat islands (4) Urban microclimates, which are warmer on the average than areas in the surrounding countryside because of various surface characteristics; other climatic and weather anomalies are also related to urbanization.

Valley fog (7) The settling of cooler, more dense air in low-lying areas, produces saturated conditions and fog.

Valley glacier (17) A type of alpine, or mountain, glacier within the confines of a valley; can range from 100 m (328 ft) to 100 km (62 mi) in length.

Vapor pressure (7) That portion of total air pressure that results from water vapor molecules; expressed in millibars (mb). At a given temperature the maximum capacity of the air is termed its saturation vapor pressure.

Vascular plants (19) Those plants with internal fluid and material flows through their tissues; almost 250,000 species on Earth.

Ventifacts (15) The individual pieces and pebbles etched and smoothed by eolian erosion—abrasion by wind-blown particles.

Vernal (March) equinox (2) The time when the Sun's declination crosses the equatorial parallel and all places on Earth experience days and nights of equal length; around March 20–21 each year. The Sun rises at the North Pole and sets at the South Pole.

Vertisols (18) A soil order in the Soil Taxonomy classification that features expandable clay soils; composed of more than 30% swelling clays such as montmorillonite; occurs in regions that experience highly variable soil moisture balances through the seasons.

Volcano (12) A landform at the end of a conduit or pipe which rises from below the crust and vents to the surface. Magma rises and collects in a magma chamber deep below, resulting in eruptions that are effusive or explosive forming the mountain landform.

Wadi (15) An Arabic term for an intermittently dry streambed or wash.

Warm front (8) The leading edge of an advancing warm air mass, which is unable to push cooler, passive air out of the way; tends to push the cooler, underlying air into a wedge shape; noted on weather maps with a series of rounded knobs heading in the direction of the frontal movement.

Wash (15) An intermittently dry streambed that fills with torrents of water after rare precipitation events in arid lands.

Water detention (9) The retaining of excess surplus water in surface puddles and indentations.

Watershed (14) The catchment area of a drainage basin; delimited by divides.

Waterspout (8) An elongated, funnel-shaped circulation formed when a tornado takes place over water.

Water table (9) The upper limit of groundwater; that contact point between zones of saturation and aeration in an unconfined aquifer.

Wave-cut platform (16) A flat, or gently sloping, table-like bedrock surface that develops in the tidal zone, where wave action cuts a bench that extends from the cliff base out into the sea.

Wave cyclone (8) An organized area of low pressure, with converging and ascending air flow producing an interaction of air masses; migrates along storm tracks. Such lows or depressions form the dominant weather pattern in the middle and higher latitudes of both hemispheres.

Wavelength (2) Measurement of waveform; the actual distance between the crests of successive waves. The number of waves passing a fixed point in one second is called the frequency of the wavelength.

Wave refraction (16) A process that concentrates energy on headlands and disperses it in coves and bays; a coastal straightening process.

Waves (16) Undulations of ocean water produced by the conversion of solar energy to wave energy; produced in a generating region or a stormy area of the sea.

Weather (8) The short-term condition of the atmosphere, as compared to climate, which reflects long-term atmospheric conditions and extremes. Temperature, air pressure, relative humidity, wind speed and direction, daylength, and Sun angle are important measurable elements that contribute to the weather.

Weathering (13) The related processes by which surface and subsurface rock disintegrate, or dissolve, or are otherwise broken down. Rocks at or near Earth's surface are exposed to physical, organic, and chemical weathering processes.

West Antarctic ice sheet (10) A vast grounded ice mass held back by the Ross ice shelf in Antarctica.

Westerlies (6) The predominant wind flow pattern from the subtropics to high latitudes in both hemispheres.

Western intensification (6) The piling up of ocean water along the western margin of each ocean basin, to a height of about 15 cm (6 in.); produced by the trade winds that drive the oceans westward in a concentrated channel.

Wetlands (16) Unique, narrow, vegetated strips occupying many coastal areas and estuaries worldwide; highly productive ecosystems with an ability to trap organic matter, nutrients, and sediment.

Wilting point (9) That point in the soil moisture balance when only hygroscopic water and some bound capillary water remains. Plants wilt and eventually die after prolonged stress from a lack of available water.

Wind (6) The horizontal movement of air relative to Earth's surface; produced essentially by air pressure differences from place to place; also influenced by the Coriolis force and surface friction.

Wind chill factor (5) An indication of the enhanced rate at which body heat is lost to the air. As wind speed increases, heat loss from the skin also increases.

Wind vane (6) A device to determine wind direction.

Windward slope (8) The intercepting slope of a mountain range, relative to the prevailing winds (see orographic lifting).

Winter (December) solstice (2) That time when the Sun's declination is at the Tropic of Capricorn, at 23.5° S latitude, December 21–22 each year.

Withdrawal (9) The removal of water by society from the supply, after which it is used for various purposes and then is returned to the water supply.

Wrangellia terranes (12) One of many terranes that became cemented together to form present day North America and the Wrangell Mountains; arriving from a far locale approximately 10,000 km (6200 mi) away; a former volcanic island arc and associated marine sediments.

Yardangs (15) Streamlined rock structures, formed by deflation and abrasion; appear elongated and aligned with the most effective wind direction.

Yazoo tributary (14) A small tributary stream draining alongside a floodplain; blocked from joining the main river by its natural levees and elevated stream channel (see backswamp).

Zone of aeration (9) A groundwater zone above the water table, which may or may not hold water in pore spaces.

Zone of saturation (9) A groundwater zone below the water table, in which all pore spaces are filled with water.

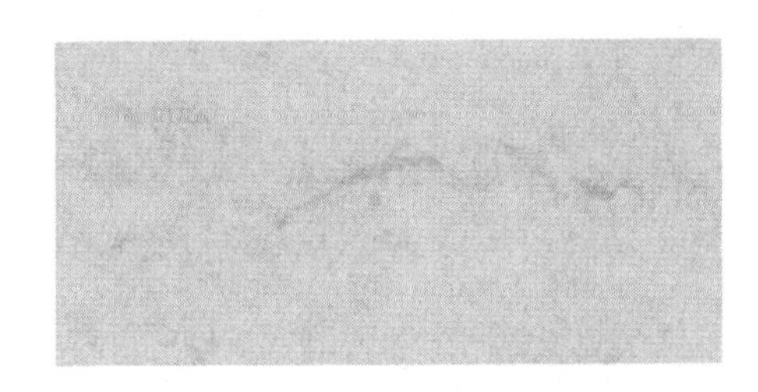

INDEX

B

C

F

J

K

L

Q

R

S

V

W

X

Y

Z